1st International Conference on Electromechanical Engineering and its Applications (ICEMEA 2020)

IOP Conference Series: Materials Science and Engineering
Volume 765

Baghdad, Iraq
25 - 26 February 2020

ISBN: 978-1-7138-3502-8
ISSN: 1757-8981

Printed from e-media with permission by:

Curran Associates, Inc.
57 Morehouse Lane
Red Hook, NY 12571

Some format issues inherent in the e-media version may also appear in this print version.

No changes have been made to the content of these proceedings. There may be changes to pagination and minor adjustments for aesthetics.

Printed with permission by Curran Associates, Inc. (2021)

For permission requests, please contact the Institute of Physics at the address below.

Institute of Physics
Dirac House, Temple Back
Bristol BS1 6BE UK

Phone: 44 1 17 929 7481
Fax: 44 1 17 920 0979

techtracking@iop.org

Additional copies of this publication are available from:

Curran Associates, Inc.
57 Morehouse Lane
Red Hook, NY 12571 USA
Phone: 845-758-0400
Fax: 845-758-2633
Email: curran@proceedings.com
Web: www.proceedings.com

TABLE OF CONTENTS

ELECTROMECHANICAL SYSTEM, AND MECHATRONICS ENGINEERING

ENERGY AND RENEWABLE ENERGY

NAVIGATION AND GUIDANCE COMPUTER AND COMMUNICATION ENGINEERING

IOP Conf. Series: Materials Science and Engineering **765** (2020) 011001 doi:10.1088/1757-899X/765/1/011001

Preface

International Conference on electromechanical engineering and its applications [ICEMEA-2020] was held on February 25-26th, 2020, Baghdad, Iraq. This conference aims to provide an international forum for researchers, experts, students, and engineers all over the world to exchange sophisticated ideas and discuss recent developments in the fields of electromechanical engineering and its applications.

The ICEMEA-2020 conference received **127** submissions from various countries and regions over the world. All submissions went through fair review process, where each paper received peer- review per 3-5 reviewers. Based on the review reports, the technical committee of conference selected **70** high quality papers for presentation at the ICEMEA-2020. These papers cover the topics of the conference. There are six keynote speakers from different regions of the world and oral presentations. The keynote speakers are recognized in their research fields, who have presented invaluable achievements in their specialization. The proceedings would be published by IOP Conference Series: Materials Science and Engineering (MSE).

The success of the ICEMEA-2020 conference was based on the fund of the Electrometrical Engineering Department and the University of Technology, as well as the hard work of the conference committees.

We would like to thank all the members of Technical Program Committee and organizers for their enormous contributions to the conferences. Our great gratitude also goes to many volunteers and staffs for their hard work that given to the ICEMEA-2020. Particular thanks are for authors, speaker and participants for their great contributions to success of the ICEMEA-2020.

IOP Conf. Series: Materials Science and Engineering **765** (2020) 011002 doi:10.1088/1757-899X/765/1/011002

Peer review statement

All papers published in this volume of *IOP Conference Series: Materials Science and Engineering* have been peer reviewed through processes administered by the proceedings Editors. Reviews were conducted by expert referees to the professional and scientific standards expected of a proceedings journal published by IOP Publishing.

Electron-Ion Trap for Combined Density and Composition Sensor of the Upper Atmosphere

[1]V.E. Dreyzin., [2] Al Kadhimi Ali Noori Mohammed, [3]Ahmed Raisan Hussein
[1]*Dr. Techn. Sciences, Professor. of SOUTHWEST State University. Department of Electronics, radio engineering and communication system. SOUTHWEST State University. (E-mail: drejzin-ve@yandex.ru). +79103148445, ,305016, Kursk, street Pavlovskogo, 1, ap.93. Kursk /Russia.*
[2]*post-graduate student, SOUTHWEST State University»*, Department *of Electronics, radio engineering and communication system. SOUTHWEST State University.*
(E-mail:alinoorimm@yahoo.com). +79051540006, 305040, 50 Let Oktyabrya Street, 94. Kursk/Russia.
[(2)] *(E-mail: alinoorimm@yahoo.com). +9647801620595 Ministry of Oil, Baghdad Oil Training Institute (BOTI), ALwazirya behind the Ministry of work. Baghdad /Iraq.*
[3]*Dr. Head of electrical engineering Departement/ college of engineering /Misan university. (E-mail: Alhusseinahmed70@uomisan.edu.iq). +9647708861738. Misan/Iraq.*

Abstract. The urgency and problems of instrumental studies of the density and composition of the upper atmosphere (ionosphere) are shown, which consist in the absence of instrumental means of direct measurements of the composition electron-ion trap combined density sensor of the atmosphere at altitudes from 150 to 500 km. In this article describes the possible design and study of the electron-ion trap combined density sensor and the composition of the upper atmosphere and simulation of the processes occurring in it. The simulation of the electric field between the electrodes of the trap and the motion of charged particles in it is carried out. The calculation of the maximum speed and energy of the particles, which the trap holds all charged particles, even in the case of the most unfavourable direction of their speed – along the gap between the electrodes. It is shown that this critical energy does not depend on the particle mass, and the probability of passage of charged particles with energy above the critical energy through the trap is estimated.

Keywords: earth atmosphere, upper layers, air density and composition, measurements, sensor, ionization, vacuum gauge converter.

Introduction:

Measuring the density and composition electron-ion trap combined density sensor of the earth's atmosphere is extremely important for understanding the complex processes occurring in the upper atmosphere under the influence of solar and galactic radiation, for predicting the further development of the global ecosystem of the Earth, its interaction with solar radiation and near space, for studying the anthropogenic impact on it, as well as for the further development of near space, which is becoming increasingly involved in human activities.

However, direct measurements of the density and composition of the earth's atmosphere were studied only to heights of 30-70 km, i.e. within the troposphere and stratosphere (using equipment installed on high-altitude aircraft, stratospheric balloons and geophysical rockets) .[1]

Strain gauges were successfully used at altitudes up to 20 km to measure atmospheric pressure (air density), at altitudes up to 50-70 km – thermoelectric vacuum gauges. To measure the composition of the air at these altitudes – spectrometric instruments, but at high altitudes-up to 500-1000 km, where the pressure becomes less than 1 Pa. [2]

The structure of the ionosphere undergoes daily, seasonal, and solar cycle changes. Significant ionospheric electric currents and wave processes occur in it. The registration of these changes, the construction of local altitude profiles of the concentration of neutral and charged particles (electrons, protons and ions of various types), their dependences on the latitude and longitude of the area, time of day, season, solar activity and other factors are extremely important for building local and global models the state of the upper atmosphere and a deeper understanding of the processes occurring in them. That is why direct instrumental studies of the density and composition of the atmosphere at these altitudes are so important, carried out using scientific equipment placed on manned and unmanned satellites. But the satellite's orbits are located at altitudes of more than 350-400 km. [3]

And altitudes from 150 to 500 km, where the most important processes of interaction of the atmosphere with solar and galactic radiation occur, were the least studied. Therefore, it is at these sites that equipment placed on micro-and nanosatellites launched from the international space station (ISS) during astronauts ' spacewalks or from manned or unmanned spacecraft or geophysical rockets, and orbiting the Earth in gradually decreasing orbits for several months, can play an important role in research. However, it is not possible to use existing industrial ionization vacuum sensors for these purposes due to a number of reasons: significant weight and size characteristics and power consumed from the power source, low mechanical strength and insufficient vibration resistance of all ionization primary converters with a hot cathode. But the most important problem is undoubtedly the influence of charged particles on the readings of these sensors in the ionosphere.[4]

The main problem preventing the use of industrial vacuum ionization transducers to measure the density of air in the ionosphere is the high concentration of charged particles in it: ions, free electrons and protons. Given that the coefficient of ionization of neutral particles in ion-electron Converter cold cathode is very small, getting in the active zone of ionization of the Converter from ambient air a substantial fraction of the ions, the concentration of which in the ionosphere reaches a few percent of the number of neutral particles, can distort the readings on the vacuum gauge dozens of times. To avoid this, an electron-ion trap is installed before entering the active zone of the Converter. In this case, the ionization sensor readings will be correct, because the ion current of the sensor will be determined only by the concentration of neutral atoms and molecules, and gamma radiation, from which the trap does not protect and the intensity of which at these heights is significantly higher than at the earth's surface, will only facilitate the ignition of the discharge in the working area of the sensor, without affecting its readings. [5]. At the same time, the trap of charged particles serves not only to increase the reliability of the vacuum transducer readings, but also allows to determine the concentration of negatively charged particles (mainly electrons) and positively charged particles: ions and protons in parallel with the determination of the concentration of neutral particles. The same high voltage source as the vacuum transducer must be used to power the trap.

In this series of articles, we consider the possibility of creating a combined miniature sensor density and composition of the upper atmosphere, allowing the ionosphere to separately measure the concentration of neutral and charged particles (electrons, ions and protons) and suitable for installation on nanosatellites. The sensor consists of an inversely magnetron vacuum transducer, which allows to measure the concentration of neutral particles and electron-ion trap, which does not pass charged particles into the active zone of the vacuum transducer, which preserves the reliability of its readings and at the same time allows to determine the concentration of charged particles. Mathematical modeling of its operation in the ionosphere is carried out and its main characteristics are determined on the basis of experimental study of its model sample in a vacuum chamber.

An exemplary design of a combined Converter-density and composition of the air environment

This Converter is an upgraded vacuum-gauge Converter, equipped with an electrostatic trap for charged particles. The trap should intercept all charged particles contained in the ionosphere, not passing them into the active zone of the vacuum transducer, but completely pass their neutral atoms and molecules.

IOP Conf. Series: Materials Science and Engineering **765** (2020) 012001 doi:10.1088/1757-899X/765/1/012001

Schematically, the design of such a trap of charged particles attached to the vacuum transducer is shown in 'figure 1'.

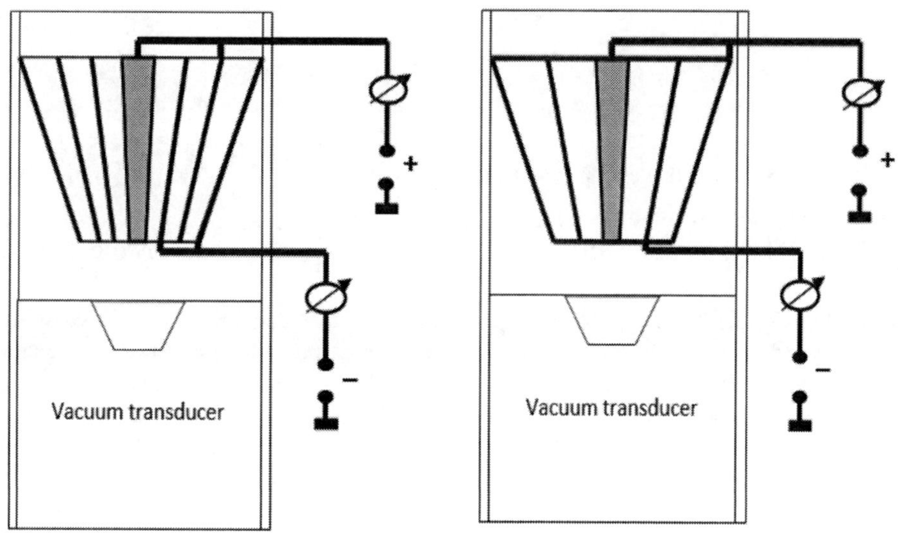

Figure 1. Variants of the charged particle trap design: four-electrode (left) and three-electrode (right).

The trap itself consists of several electrodes in the form of nested hollow truncated cones connected to a high-voltage DC power source. Positive electrodes (anodes) collect negatively charged particles (electrons and negative ions), and negative electrodes (cathodes) – positively charged particles (positive ions and protons), the Corresponding electric currents are measured by additional measuring channels, which makes it possible to determine the concentration of positively and negatively charged particles. The upper end of the trap is open to the outside environment. Charged particles entering the interelectrode space of the trap are deflected by a strong electric field to the electrodes (negative to the anodes, positive to the cathodes).

Accordingly, only neutral particles fall into the working volume of the vacuum transducer. This ensures the correctness of its readings proportional to the concentration of neutral atoms and molecules in the air. Thus, it is possible not to be afraid of difficulty of ignition of the discharge because of lack of the "seed" electrons necessary for its ignition. They will be much more than in the operation of such converters in ground conditions, since at these heights the gamma background is repeatedly increased, and when interacting with the substance of gamma rays of the structural elements of the Converter (mainly the magnet and the pole overlays) from their inner wall layers will fly into the working volume of the Converter electrons knocked out of atoms due to the photoelectric effect, Compton effect and the effect of the birth of electron-positron pairs. This will facilitate the extension of the transducer measurement limit to the lower side. Thus, this combined Converter will correctly measure not only the concentration of neutral atoms and molecules in the ionized air, but also the concentration of negatively charged particles (electrons and negative ions) and positively charged particles (positive ions and protons).

Macroscopic mathematical model of charged particle trap.

The calculation of the charged particle trap is possible both with the use of an analytical approach and with the use of standard electric field simulation programs. Since high accuracy of this model is not required, we will use an analytical approach. Its errors will be determined by the edge curvatures of

IOP Conf. Series: Materials Science and Engineering **765** (2020) 012001 doi:10.1088/1757-899X/765/1/012001

the electric field not taken into account. But, since the gaps between the electrodes are much smaller than the height of the electrodes, these errors will be small.
To calculate the electric field in the interelectrode space, we will use the known formula of the electric field of a coaxial cylindrical capacitor: [6]

$$E(h) = \frac{U}{r \ln \frac{R_l(h)}{R_u(h)}} \quad (1)$$

where U is the voltage on the capacitor plates;

- r – current radius;
- $R1$ and $R2$ are the radii of the inner and outer electrodes, respectively.
- E= electric field.

In the proposed design of the trap, the gaps between adjacent electrodes are reduced with a decrease in height, which provides an increase in the electric field as the particles penetrate deeper into the trap.

To calculate the field in such a design, it is possible to replace conical surfaces with cylindrical ones with stepwise decreasing diameters, as shown in 'figure 2' (left). In this case, the expression (1) is valid for each step of such surfaces. From it is easy to go to the continuous function $E(h)$, if you set the linear functions $R_l(h)$ and $R_u(h)$, where h – the current height of the cones: [7]

$$E(h) = \frac{U}{r \ln \frac{R_l(h)}{R_u(h)}} \quad (2)$$

The height of the trap will be divided into 10 sections, and the calculations will be carried out for the radii corresponding to the middle of these sections. The formula for finding the radii of any of these cones at an arbitrary height h is easy to obtain from the geometric construction shown in 'figure 2' (right). Based on the geometric construction shown in 'figure 2' (right), current radii of the cones of the trap is calculated as: 'figure 2' (right), current trap cone radii should be calculated as: [7]

$$R(h) = R_l + \frac{R_u - R_l}{H} \cdot h \quad (3)$$

where: $R(h)$ is the radius of the cone at an arbitrary height h;

- R_u is the radius of the cone at the upper end;
- R_l is the radius of the cone at the lower end;
- R_{av} is the radius of the cone at the average end;
- H *is* the total height of the truncated cone.

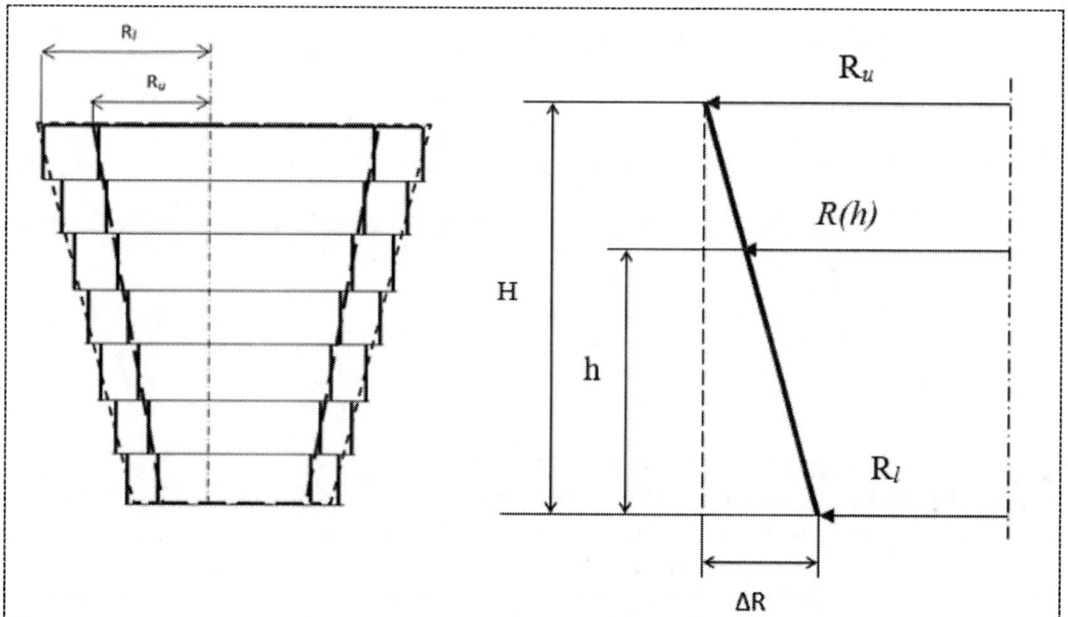

Figure 2. The approximation of conical surfaces of stepped cylindrical surfaces (left) and the determination of the radii of the conical electrode at an arbitrary height h (on the right).

For a four-electrode trap, the geometric dimensions of the electrodes are shown in table 1 (the thickness of the electrodes is assumed to be 0.1 mm).
where: D1, Du is the refer to dimensions of the electrodes

Table 1. Geometric dimensions of the charged particle trap

№ cones		1	2	3	4
upper	D_l, mm	19	15	11	7
	D_u, mm	18,8	14,8	10,8	-
lower	D_l, mm	10	7	4	1
	D_u, mm	9,8	6,8	3,8	-

Take the total height H for all cones equal to 40 mm and divide it by height into 10 sections, and the current radii will be found for the middle of these sections. Then for each section it is possible to approximate conical surfaces cylindrical and for calculation of an electric field of each section to apply expression (2). The value of the electric voltage is the same as for the vacuum transducer and equal to 1000 V. Below (table. 2 and 'figure 3') here are the results of the calculation of the electric field for one of the pairs of electrodes (These measurements are calculated in units of mm).
where: E_l is the electric field lower;

- E_u is the electric field upper;
- E_{av} is electric field average.

For the first cone (inner surface) R_u = 9.4 mm, R_l = 4.9 mm H for all cones 40 mm, **h** is shown in the tables. Consequently, for the height h = 38 mm we get for the first cone (inner surface)

$$R_2(38) = 4{,}9 + \frac{9{,}4-4{,}9}{40} \cdot 38 = 9{,}175$$

For the minimum height h=2 mm, we obtain:

$$R_2(2) = 4{,}9 + \frac{9{,}4-4{,}9}{40} \cdot 2 = 5{,}125$$

Table 2. Calculation of the electric field for the first pair of electrodes

h, mm	$R_1(h)$, mm	$R_2(h)$, mm	$R_{av}(h)$, mm	E_l,B/mm	E_u, B/mm	E_{av},B/mm
38	7.3	9.175	8.237	476	598	537
34	6.9	8.725	7.812	490	619	554
30	6.5	8.275	7.387	501	638	569
26	6.1	7.825	6.962	513	658	585
22	5.7	7.375	6.535	525	680	602
18	5.3	6.925	6.112	541	707	624
14	4.9	6.475	5.687	555	734	644
10	4.5	6.025	.262	568	761	664
6	4.1	5.575	4.837	584	794	689
2	3.7	5.125	4.412	599	829	714

For charged particles entering the space between the electrodes, the case is critical when they fall along the gap between the electrodes. In all other cases, the particles collide with the electrodes and are either neutralized (if the electrode has the opposite charge) or repelled from it (if they have the same charges) and fly to the opposite electrode and there are neutralized. Therefore, below is the calculation of the limit energies of charged particles for the critical case when their initial velocity is directed along the gap Between the electrodes.

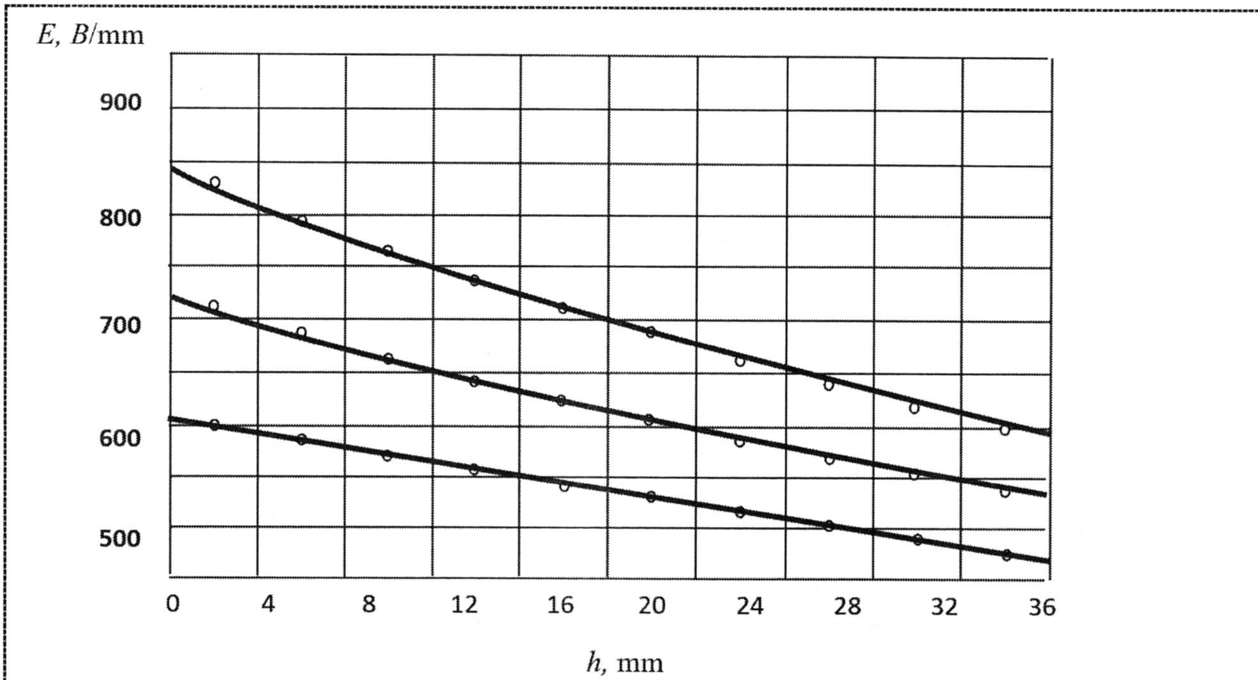

Figure 3. Dependences $E=f(h)$ for the first pair of electrodes of a four-electrode trap

The calculation of the maximum energy of charged particles trapped

When moving in a transverse electric field, a force acts on the charged particle, deflecting it in the transverse direction. Even in a uniform electric field, the charged particle moves uniformly accelerated, and in this case the field increases as it approaches the internal electrode and as the particle sinks into the trap. However, given that high accuracy is not required, we assume that the field is uniform and equal to the corresponding mid-gap between the electrodes at half the height of the trap. Given these conditions, the particle will move along the r axis uniformly accelerated [8].

$$r = \frac{1}{2} mqEt^2 \tag{4}$$

Assuming that during this time the component of the particle velocity along the z axis remains unchanged and equal to *V*, we find the current height of the trap *(H – h)* =*VT*, at which the particle reaches the opposite electrode.

$$H - h_T = V\sqrt{\frac{2md}{qE}}\,. \tag{5}$$

Passing from the velocity to the energy of the particle $E_p = \frac{mV^2}{2}$, where $V = \sqrt{\frac{2E_p}{m}}$

Receive:

$$(H - h_T) = \sqrt{\frac{4E_p d_{av}}{qE}} \tag{6}$$

Finding out E_p of the particles obtained

$$E_p = \frac{(H-h_T)^2 qE}{4d}\,. \tag{7}$$

At $h_T = 0$, we obtain the maximum kinetic energy of the particles captured by this trap in the most unfavourable case, when their velocity is directed along the gap between the electrodes of the trap:

$$E_{p\,max} = \frac{H^2 qE}{4d_{av}} \tag{8}$$

As you can see, it no longer appears the mass of the particle, but only its charge. Therefore, for both electrons and single-charged ions, the limit energy at which they will be trapped will be the same. This is an important conclusion because it allows us to determine the boundary energy of any charged particles. Substituting in (10) numerical values $H = 40$ мм, $q = e = 1.6{\cdot}10^{-19}$, $d_{av} = 1{,}65$ мм, $E = 620$ V/mm (for the first pair of electrodes), we obtain

$$E_{p\,max} = \frac{40^2{\cdot}1{,}6{\cdot}10^{-19}{\cdot}620}{4{\cdot}1{,}65} = 2{,}4\cdot 10^{-14} = 150\ \text{Kev}$$

where: *e* is the electric charge = $1.6{\cdot}10^{-19}$

Of course, this expression is approximate, since the electric field strength will increase as the particle approaches the oppositely charged electrode and moves deeper into the trap, since *E* depends on both *r* and *h*. This means that the real value of the critical energy of the particles captured by the trap will be higher than this estimate.

But the most important thing is that here we consider the most unfavourable case of the direction of the particle velocity relative to the position of the trap. In fact, the direction of particle velocity relative to the trap can be considered chaotic, since the position of nanosatellites, which is supposed to place the projected sensor, relative to the earth's surface is not stabilized, and the particles themselves move quite chaotically, although the earth's magnetic field affects the charged particles within the ionosphere, causing them to move in cyclotron orbits along the magnetic field lines. But in relation to the velocity vector of the spacecraft itself, the directions of their velocity can be considered equally probable. This means that the probability of the above case of the direction of the initial velocity of the particle along the gap between the electrodes can be estimated as the ratio of the solid angle θ, in which in the absence of the field the particle would not collide with the trap electrodes, to the solid angle 2π. Given the trap geometry, this angle θ can be estimated to be approximately 1^0 for a four-electrode trap and 1.5^0 for a three-electrode trap. Therefore, even if we limit the solid angle at which the particle can fall into the trap 120 degrees (instead of 180^0), the probability of coincidence of the particle velocity with the direction of the gap between adjacent electrodes can be estimated as 0.8 % for a four-electrode trap and 1.25 % for a three-electrode trap. This means that with a probability of about 99 %, this trap will capture charged particles of any energy, and only for the remaining 1 % of the particles, the energy of the captured particles will be limited to 0.15 MeV. Therefore, it is safe to say that this trap will delay and neutralize almost all charged particles. Given that the concentration of charged particles in the ionosphere is about 50...1000 times lower than that of neutral particles, the error in measuring the concentration of neutral particles due to the neutralization of charged particles and their entry into the active zone of the vacuum transducer will not exceed one percent.

Conclusion

Summarizing the development of the combined sensor of the density and composition of the upper atmosphere (ionosphere), the following conclusions can be drawn:

1. Within the ionosphere, all characteristics of the atmosphere, including density, composition, temperature, degree of ionization, radius of cyclotron rotation in the earth's magnetic field, etc. are highly variable and depend not only on the height, but also on the time of day, time of year, latitude, solar activity and many other factors, the influence of most of which are poorly studied due to the complexity of direct measurements at these heights and the lack of adequate measuring instruments suitable for installation on micro-and nanosatellites. This determines the relevance of the development of such tools.
2. The main factors hindering the use of existing industrial devices (ionization vacuum meters and mass spectrometers) for these purposes are a high degree of ionization and a strong rarefaction of the upper atmosphere. Especially difficult for direct measurements are the heights from 150 to 500 km, where the most complex and poorly studied processes of interaction of the earth's atmosphere with solar and galactic radiation, affecting both the weather and long-range radio communication and, ultimately, the entire ecosystem of the earth, and it is at these heights that the most effective scientific equipment can be placed EA micro-and nano-satellites.
3. The experimental studies of the prototype of this transducer in a vacuum chamber made it possible to determine the real ionization coefficient of neutral particles in this transducer and confirmed the correctness of the developed mathematical model of the processes occurring in it.

Acknowledgements

We would like to thank the University of the SOUTHWEST State University for supporting us with our article success, as well as the teaching staff in the Department of Electronics, radio engineering and communication system.

References

[1] Krivolutsky A.A., Klyuchnikova A.V., Zakharov G.R., Vyushkova T.Yu., Kuminov A.A. Dynamical response of the middle atmosphere to solar proton Event of July 2000: Three-dimensional model simulations, Advances in Space Research, 37, pp. 1602-1613, 2006.

[2] Andreeva E., Franke J., Yeh K., Kunitsyn V. Some features of the equatorial anomaly revealed by ionospheric tomography. Geophys. Res. Lett., v. 27(16), pp. 2465-2468, 2000.

[3] Bilitza D. International Reference Ionosphere 2000. Radio Sci., v. 36(2), pp. 261-275, 2001.

[4] V.E. Dreyzin, Al Kadhimi Ali Noori Mohammed. Prospects for studying the density and composition of the upper atmosphere using small spacecraft. Innovative information and communication technologies. Proceedings of the XV International scientific and practical conference. 2018. P. 289-293.

[5] V.E. Dreyzin., Bondar O., Ovsyannikov, Yu. a., etc. Eliminating the influence of variations ambient temperature in thermoelectric power vacuum gauge, Sensors and Systems. №5, 2009 p. 48-51.

[6] Yeh K., Franke S., Andreeva E., Kunitsyn V. An investigation of motions of the equatorial anomaly. Geophys. Res. Lett., v. 28, pp. 4517–4520, 2001.

[7] Tereshchenko E., Khudukon B., Gurevich A., Zybin K., Frolov V., Myasnikov E., Muravicva N., Carlson H. Radio tomography and scintillation studies of ionospheric electron density modification caused by a powerful HF-wave and magnetic zenith effect at mid-latitudes. Phys. Lett., A325, pp. 381-388. 2004.

[8] Ahmadov R. and Kunitsyn V. Simulation of generation and propagation of acoustic gravity waves in the atmosphere during a rocket flight. Int. J. Geomagn. Aeron, v. 5(2), G12002, doi:10.1029/2004G1000064, 2004.

IOP Conf. Series: Materials Science and Engineering **765** (2020) 012002 doi:10.1088/1757-899X/765/1/012002

Stability system design of flying body elevator control

Muhammad A. R Yass

Hussian Abdulaziz Abrahem

Electromechanical Engineering Department, University of Technology, Baghdad, Iraq

corresponding author's e-mail : 50251@uotechnology.edu.iq

Abstract. This paper, suggest the solution of a dynamic non -homogenous longitudinal pitch-mode flying body equation to analysis motion stability by determining the transfer function of a pitch mode control system from feedback control system diagram .It used a flying actuator design system to estimate optimum system stability and control using the time response method and root locus diagram for different gain values (K) and Gyro sensitivity values (GR) to obtain the best stability, peak value and time response. The results shows that the best stability and control behavior are achieved when K=1.4 and GR= 1.2.

Keywords: control system, stability, Aerodynamic, response

1. Introduction

Any unstable flight needs a control system depending on flying body shape and design characteristics.

The mathematical model to estimate the flying body transfer function derived from flying body design properties presents the dynamic motion generated from a flying body equation of motion.

The set of six equations consist of very complicated and nonlinear differential equation which can be linearized into longitudinal and lateral equations.

A control system design with feedback controller and actuator was designed for pitch angle for a lower time setting ,lower low peak over value limited by elevator and actuator slewing.

1.1. Mathematical Analysis

The longitudinal equations of motion [1].

$$\frac{mv}{sq}i-\frac{d}{2U}cm_{\alpha}\alpha+\frac{Iy}{sqd}\theta-\frac{d}{2U}cmq\theta^{\circ}=CFxa\text{..}(1)$$

$$-(C2u\overline{U})+\left[\frac{mu}{sq}-\frac{d}{2u}C_2x)\alpha-C_{2\alpha}\alpha\right]$$

$$+ \left[(\frac{mu}{sq} - \frac{d}{2u} C_2 q)\theta^{\circ} - \ddot{C}_w (\sin(H))\theta \right] = CF_{2a} \text{..} (2)$$

$$(-C_{mo}{}' V + (\frac{-d}{2U} C_{m\alpha} \alpha - C_{m\alpha} \alpha) + (\frac{Iy}{Sqd} \ddot{\theta} - \frac{d}{2U} C_{mq} {}^{\circ} e) = cma \text{..........................} (3)$$

$$(13.78s + 0.088)'_u(s) - 0.392'_\alpha(s) + 0.7\theta(s) = 0 \text{ ... ...} (4)$$
$$1.48'_u(s) - (13.78s + 4.46)'_\alpha(s) - 13.78s\theta(s) = 0 \text{ ....} (5)$$
$$0 + (0.0552s + 0.619)'_\alpha(s) + (0.514s^2 + 0.192s)\theta(s) = 0 \text{ ... ...} (6)$$

The only nonzero solution of these simultaneous equations requires that the determinate of the coefficient be zero. Thus

$$\begin{vmatrix} 13.78s + 0.088 & -0.392 & 0.74 \\ 1.48 & 13.78s + 4.46 & -13.78s \\ 0 & 0.0552s + 0.619 & 0.514s^2 + 0.192s \end{vmatrix} = 0$$

Expanding this determinant, the following quartic equation is obtained.

$$9754s^4 + 79s^3 + 128.9s^2 + 0.998s + 0.677 = 0 \text{ ... .} (7)$$

Dividing through by **97.5**, the equation reduces to

$$s^4 + 0.811s^3 + 1.32s^2 + 0.012s + 0.00695 = 0 \text{ ... ...} (8)$$

The transfer function for 6, input to $'\alpha$output in determinant form is

$$\frac{'_\alpha(S)}{\delta_e} = \frac{\begin{vmatrix} 13.78s + 0.088 & -0.392 & 0.74 \\ 1.48 & -0.246 & -13.78s \\ 0 & -0.710 & 0.514s^2 + 0.192s \end{vmatrix}}{\nabla}$$

$$\frac{'_\alpha(S)}{\delta_e} = \frac{-0.01785(s^3 + 77.8s^2 + 0.496s + 0.446)}{s^2 + 0.00466s + 0.0053)(s^2 + 0.806s + 1.311)} \text{ ... ...} (9)$$

Factoring the numerator

$$\frac{'_\alpha(S)}{\delta_e} = \frac{-0.01785(s + 77.79)(s^2 + 0.0063s + 0.0057)}{s^2 + 0.00466s + 0.0053)(s^2 + 0.806s + 1.311)} \text{ ... ...} (10)$$

Going to the alternate form, Eq. [7] becomes

$$\frac{'_\alpha(S)}{\delta_e} = \frac{-1.14\left(\frac{s}{77.79} + 1\right)\left[\left(\frac{s}{0.0755}\right)^2 + \frac{2(0.041)}{0.0755}s + 1\right]}{\left[\left(\frac{s}{0.073}\right)^2 + \frac{2(0.032)}{0.073}s + 1\right]\left[\left(\frac{s}{1.14}\right)^2 + \frac{2(0.352)}{1.145}s + 1\right]} \text{ ... ...} (11)$$

For input to θ output

$$\frac{\theta(S)}{\delta_e} = \frac{\begin{vmatrix} 13.78s + 0.088 & -0.392 & 0 \\ 1.48 & 13.78s + 4.46 & -0.246 \\ 0 & 0.0552s + 0.619 & -0.710 \end{vmatrix}}{\nabla}$$

Expanding and factoring

$$\frac{\theta(S)}{\delta_e} = \frac{-1.31(s + 0.016) + (s + 0.3)}{s^2 + 0.00466s + 0.0053)(s^2 + 0.806s + 1.311)} \quad \ldots\ldots\ldots\ldots\ldots\ldots (12)$$

Or the alternate form

$$\frac{\theta(S)}{\delta_e(s)} = \frac{-0.95\left(\frac{s}{0.016} + 1\right)\left(\frac{s}{0.3} + 1\right)}{\left[\left(\frac{s}{0.073}\right)^2 + \frac{2(0.032)}{0.073}s + 1\right]\left[\left(\frac{s}{1.145}\right)^2 + \frac{2(0.352)}{1.145}s + 1\right]} \quad \ldots\ldots\ldots (13)$$

Substituting for q and simplifying, Eq. [10] becomes

$$\zeta = -\frac{1}{4}\left(C_{mq} + C_{m\alpha} + \frac{2I_y}{Cm^2}C_{z\alpha}\right)\left(\frac{cm^2}{I_y\left(\frac{C_{mq}C_{z\alpha}}{2} - \frac{2mC_{mq}}{\rho Sc}\right)}\right)^{1/2} \quad \ldots\ldots\ldots (14)$$

$$\frac{'_\alpha(S)}{\delta_e} = \frac{\begin{vmatrix} -.246 & -13.78s \\ -0.710 & 0.514s^2 + 0.192s \end{vmatrix}}{\begin{vmatrix} 13.78s + 4.45 & -13.78s \\ 0.0552s + 0.619 & 0.514s^2 + 0.192s \end{vmatrix}} \quad \ldots\ldots\ldots (15)$$

Expanding

$$\frac{'_\alpha(S)}{\delta_e(s)} = \frac{-0.01782(s + 77.8)}{s^2 + 0.805s + 1.3325} \quad \ldots\ldots\ldots (16)$$

Going to the alternate form

$$\frac{'_\alpha(S)}{\delta_e(s)} = \frac{-1.05\left(\frac{s}{77.8} + 1\right)}{\left(\frac{s}{1.15}\right)^2 + \frac{2(0.35)}{1.15}s + 1} \quad \ldots\ldots\ldots (17)$$

A comparison of Eq. [10] and [12] shows excellent agreement substantiating the original assumption. The$\theta(s)/\delta e(s)$ transferfunction will now be evaluated.

$$\frac{\theta(S)}{\delta_e(s)} = \frac{\begin{vmatrix} 13.78s + 4.46 & -0.246 \\ 0.0552s + 0.619 & -0.710 \end{vmatrix}}{\begin{vmatrix} 13.78s + 4.45 & -13.78s \\ 0.0552s + 0.619 & 0.514s^2 + 0.192s \end{vmatrix}} \quad (18)$$

Expanding,

$$\frac{\theta(S)}{\delta_e(s)} = \frac{-1.39(s + 0.306)}{s(s^2 + 0.805s + 1.3325)} \quad (19)$$

Going to the alternate form

$$\frac{\theta(S)}{\delta_e(s)} = \frac{-0.321\left(\frac{s}{0.306} + 1\right)}{s\left[\left(\frac{s}{1.15}\right)^2 + \frac{2(0.35)}{1.15}s + 1\right]} \quad (20)$$

The magnitude plot of Eq. [16] is shown in Figure 1-12. A comparison of Figures 1, 11 and 1-12 shows very good agreement in the vicinity of the natural frequency of the short-period mode. It should be noted that Eq. [19] has a pole at s = 0, that is, a root of the denominator.

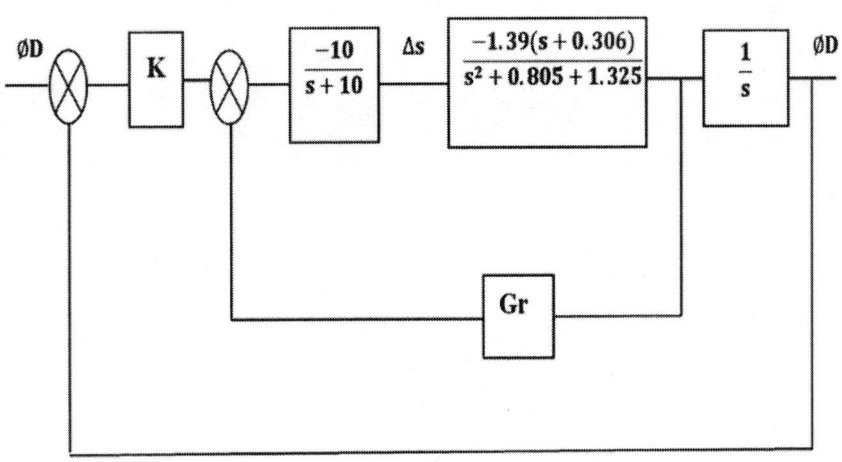

Figure (1):- Feedback control system Design

$$\frac{\theta(S)}{\delta_e(s)} = \frac{13.9k(s + 0.306)}{s^5 + 10.805s^4 + (9.375 + 13.9GR)s^3 + (13.255 + 4.253Gr + 13.9k)s^2 + 4.254ks}$$

$$F(x) = \frac{19.6x+6}{x^5+10.805x^4+25.16x^3+37.196x^2+6x}$$

$$F(x) = \frac{1}{x} + \frac{-0.24071+0.3223i}{x+1.215-1.62075i} + \frac{-0.24071-0.3223i}{x+1.215+1.62075i} + \frac{-0.0458}{x+8.1967} + \frac{-0.47276}{x+0.178408}$$

$$Q=1-de^{-zt}-ce^{-pt}+2e^{-gt}(-a\cosh t+b\sinh t)$$

IOP Conf. Series: Materials Science and Engineering **765** (2020) 012002 doi:10.1088/1757-899X/765/1/012002

a=0.24071 g=1.1215 z=8.12 c=0.4727
b=0.3223 h=1.6207 p=0.178 d=0.0458

2-Results and Dissection:

2.1-Longitudinal Response at Different K&GR:

Table (1): Constant Gyro Value(Equal=1),Different Gain Value(1-1.8)

GR	K	Characteristics Equations	Real Root	Imaginary Root	Steady Time	Max Peak Over Shoot	stability
1	1	$s^5+10.805s^4+23.725s^3+31.4s^2+4.25s$	0.15	1.06±1.52	13	1.25	stable
1	1.1	$s^5+10.805s^4+23.725+32.8s^2+4.68s$	0.156	1.04±1.57	13	1.3	stable
1	1.2	$s^5+10.805s^4+23.725+34.s^22+5.1s$	0.16	1.01±1.62	12.5	1.4	stable
1	1.3	$s^5+10.805s^4+23.725+35.58s^2+5.53s$	0.17	1±1.67	15	1.475	stable
1	1.4	$s^5+10.805s^4+23.725+36.96s^2+5.95s$	0.18	0.904±1.69	10.3	1.48	stable
1	1.6	$s^5+10.805s^4+23.725+38.96s^2+$	0.185	0.909±1.801	10	1.56	stable
1	1.7	$s^5+10.805s^4+23.725+39.74s^2+6.23$	0.189	0.92±1.832	11	1.6	stable
1	1.8	$s^5+10.805s^4+23.725+41.52s^2+7.6$	0.193	0.928±1.86	10	1.65	stable

Table (2): Constant Gyro Value(Equal=1.3),Different Gain Value(1-1.5)

GR	K	Characteristics Equations	Real Root	Imaginary Root	Steady Time	Max Peak Over Shoot	stability
1.3	1	$s^5+10.805s^4+27.445+32.6s^2+4.2s$	0.14	1.3±1.42	16	1.25	stable
1.3	1.1	$s^5+10.805s^4+27.445+35.4s^2+4.62s$	0.16	1.3±1.43	10	2	unstable
1.3	1.2	$s^5+10.805s^4+27.445+36.7s^2+5.1s$	1.161	1.3±1.48	9	1.35	stable
1.3	1.3	$s^5+10.805s^4+27.445+38.4s^2+5.5s$	0.17	1.36±1.49	9.8	0.75	unstable
1.3	1.4	$s^5+10.805s^4+27.445+39.6s^2+5.95s$	0.175	1.54±1.57	11	0.512	unstable
1.3	1.5	$s^5+10.805s^4+27.445+41s^2+6.3s$	0.18	1.32±1.63	9	0.8	unstable

Table (3): Constant Gyro Value(Equal=1.1),Different Gain Value(1-1.7)

GR	K	Characteristics Equations	Real Root	Complex Root	Steady Time	Max Peak Over Shoot	stability
1.1	1	$s^5+10.805s^4+24.66+31.83s^2+4.25s$	0.157	1.18±1.48	14	1.25	stable
1.1	1.1	$s^5+10.805s^4+24.66+32.6s^2+4.6s$	0.159	1.178±1.499	14	1.3	stable
1.1	1.2	$s^5+10.805s^4+24.66+36.5s^2+5.1s$	0.161	1.1±1.51	13.9	1.4	stable
1.1	1.3	$s^5+10.805s^4+24.66+37.6s^25+5.5s$	0.1637	1.098±1.598	15	1.43	stable
1.1	1.4	$0.1705s^5+10.805s^4+24.66+38.2s^2+5.95s$	0.169	1.078±1.63	11	1.42	stable
1.1	1.5	$s^5+10.805s^4+24.66+40.1s^2+6.3s$	0.172	1.0678±1.78	12	1.51	stable
1.1	1.6	$s^5+10.805s^4+24.66+41.07s^2+6.8s$	0.186	1.063±1.803	12	1.56	stable
1.1	1.7	$s^5+10.805s^4+24.66+42.93s^2+7.23s$	0.189	1.059±1.81	15	1.59	stable

Table (4): Constant Gyro Value(Equal=1.2),Different Gain Value(1-1.5)

GR	K	Characteristics Equations	Real Root	Complex Root	Steady Time	Max Peak Over Shoot	stability
1.2	1	$s^5+10.805s^4+26.005+31.8s^2+4.25s$	0.14	1.3±1.36	15	1.23	stable
1.2	1.1	$s^5+10.805s^4+26.005+34.6s^2+5.1s$	1.15	1.27±1.43	15	1.32	stable
1.2	1.2	$s^5+10.805s^4+26.005+36.7s^2+5.5s$	1.16	1.26±1.5	15	1.37	stable
1.2	1.3	$s^5+10.805s^4+26.005+37.4s^2+5.95s$	1.172	1.24±1.55	15	1.4	stable
1.2	1.4	$s^5+10.805s^4+26.005+38.9s^2+6.3s$	1.174	1.2±1.59	10	1.42	stable
1.2	1.5	$s^5+10.805s^4+26.005+40.2s^2+6.8s$	1.176	1.23±1.59	11	1.52	stable

IOP Conf. Series: Materials Science and Engineering **765** (2020) 012002 doi:10.1088/1757-899X/765/1/012002

Table (5): Constant Gyro Value(Equal=1.4),Different Gain Value(1-1.8)

GR	K	Characteristics Equations	Real Root	Complex Root	Steady Time	Max Peak Over Shoot	stability
1.4	1	$s^5+10.805s^4+28.83+33.1s^2+4.2s$	0.146	1.59±1.2	9	1.21	stable
1.4	1.1	$s^5+10.805s^4+28.83+33.4s^2+4.6s$	0.15	1.52±1.22	9	1.22	stable
1.4	1.2	$s^5+10.805s^4+28.83+35.7s^2+5.1s$	0.16	1.52±1.21	10	1.298	stable
1.4	1.3	$s^5+10.805s^4+28.83+37.2s^2+5.5s$	0.161	1.5±1.35	9.8	1.32	stable
1.4	1.5	$s^5+10.805s^4+28.83+38.6v+5.95s$	0.17	1.48±1.43	11	1.4	stable
1.4	1.6	$s^5+10.805s^4+28.83+40s^2+6.3s$	0.18	1.46±1.5	11	1.4	stable
1.4	1.7	$s^5+10.805s^4+28.83+41.4s^2+6.8s$	0.18	1.44±1.56	9	1.478	unstable
1.4	1.8	$s^5+10.805s^4+28.83+42.8s^2+7.23s$	0.193	1.42±1.62	9	1.52	stable

Table (6): Constant Gyro Value(Equal=1.5),Different Gain Value(1-1.5)

GR	K	Characteristics Equations	Real Root	Complex Root	Steady Time	Max Peak Over Shoot	stability
1.5	1	$s^5+10.805s^4+30.225+33.53s^2+4.25s$	0.143	1.402±1.52	15	1.22	stable
1.5	1.1	$s^5+10.805s^4+30.225+34.9s^2+4.67s$	0.15	1.41±1.53	15	1.288	stable
1.5	1.2	$s^5+10.805s^4+30.225+36.3v+5.1s$	0.162	1.42±1.55	14	1.3	stable
1.5	1.3	$s^5+10.805s^4+30.225+37.7s^2+5.5s$	0.164	1.43±1.61	13.5	1.33	stable
1.5	1.4	$s^5+10.805s^4+30.225+39s^2+5.95s$	0.17	1.439±1.72	14	1.83	stable
1.5	1.5	$s^5+10.805s^4+30.225+40.59s^2+6.38s$	0.178	1.51±1.799	13	1.41	stable

IOP Conf. Series: Materials Science and Engineering **765** (2020) 012002 doi:10.1088/1757-899X/765/1/012002

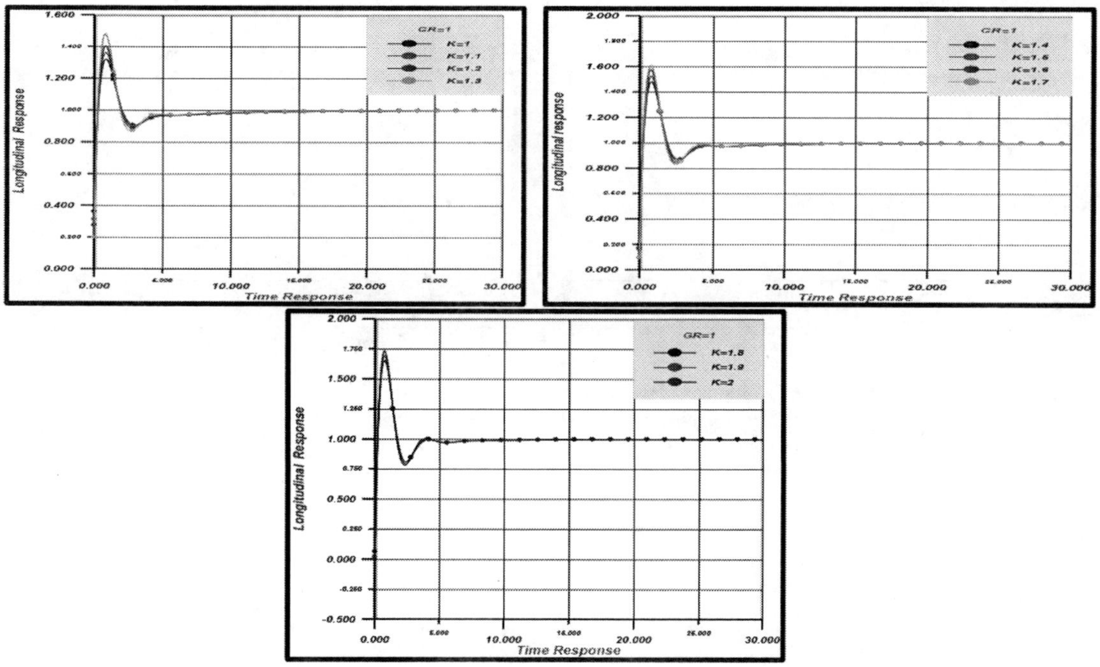

Figure (2) Longitudinal Response verses Time response for GR=1and Different Gain

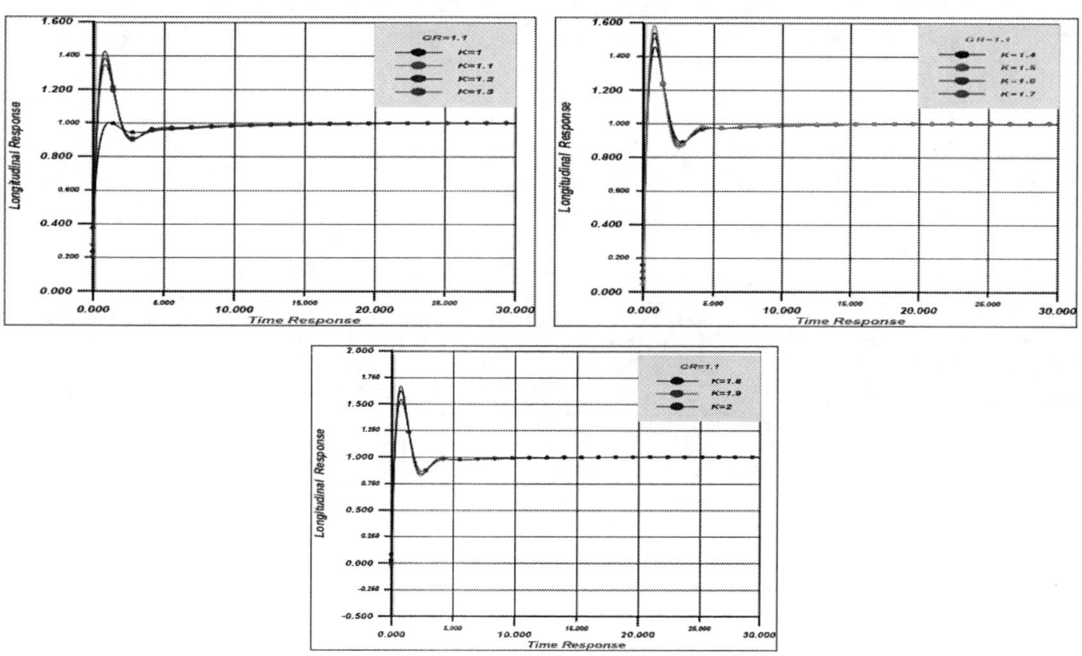

Figure (3) Longitudinal Response verses Time response for GR=1.1 and Different Gain Value

IOP Conf. Series: Materials Science and Engineering **765** (2020) 012002 doi:10.1088/1757-899X/765/1/012002

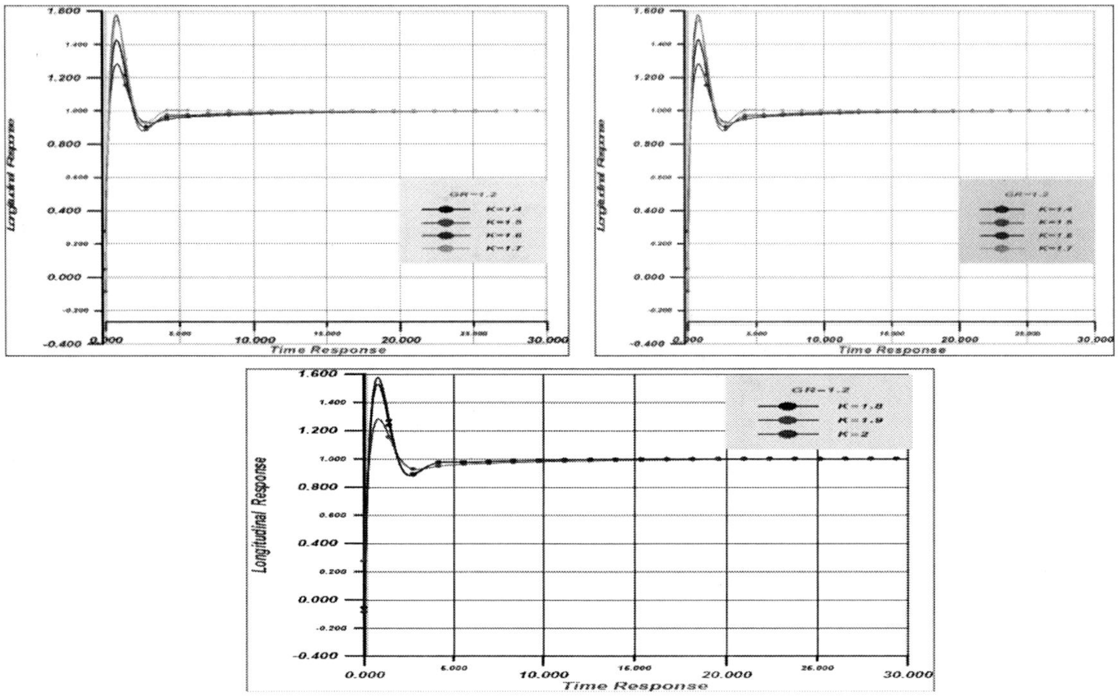

Figure (4) Longitudinal Response verses Time response for GR=1.3 and Different Gain Value

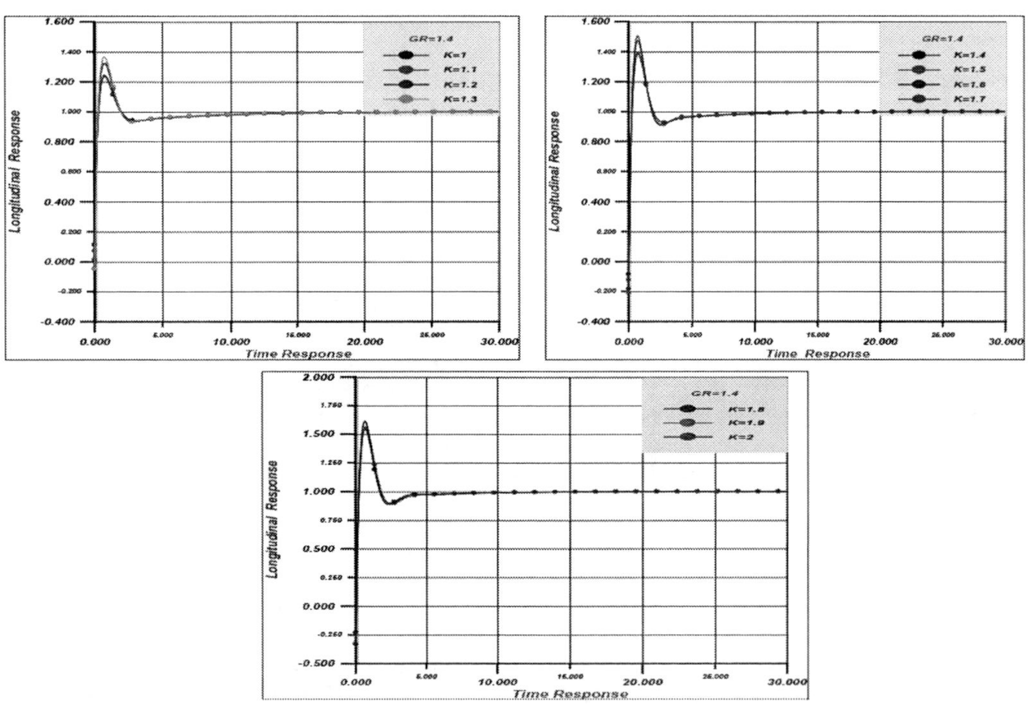

Figure (5) Longitudinal Response verses Time response for GR=1.4 and Different Gain Value

IOP Conf. Series: Materials Science and Engineering **765** (2020) 012002 doi:10.1088/1757-899X/765/1/012002

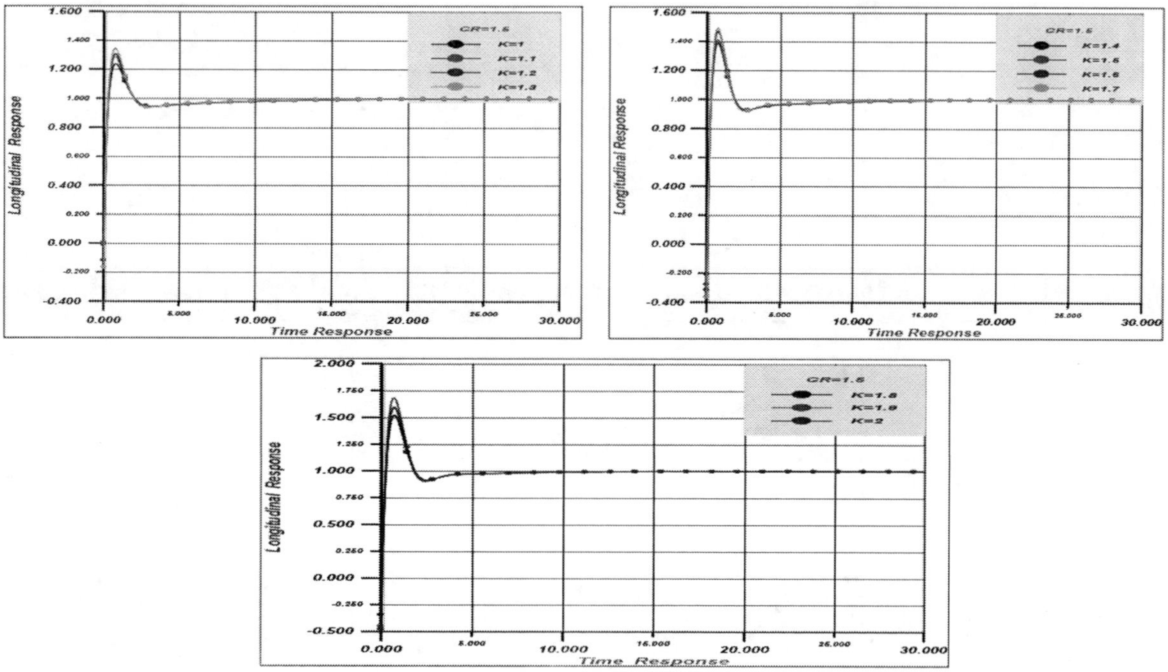

Figure (6) Longitudinal Response verses Time response for GR=1.5 and Different Gain Value

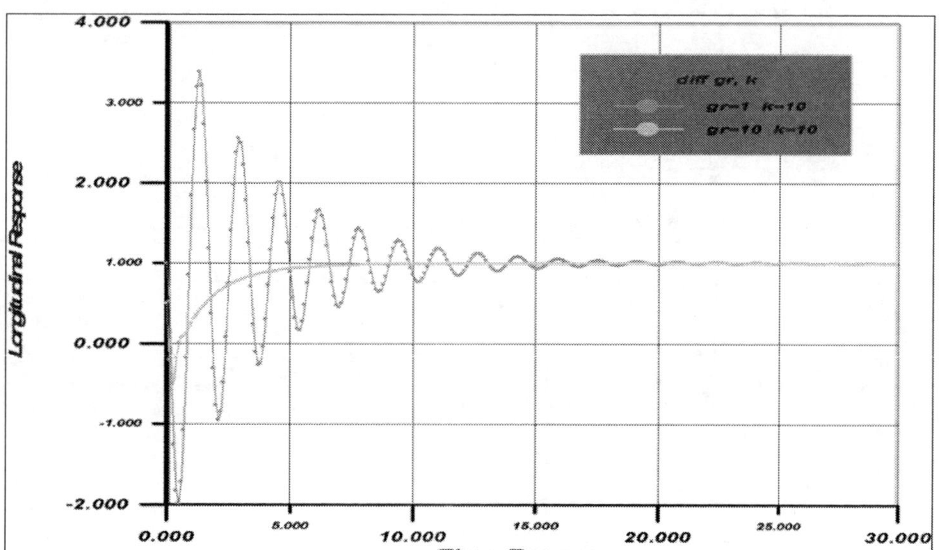

Figure (7) Longitudinal Response verses Time response for Different GR and Different Gain Value

IOP Conf. Series: Materials Science and Engineering **765** (2020) 012002 doi:10.1088/1757-899X/765/1/012002

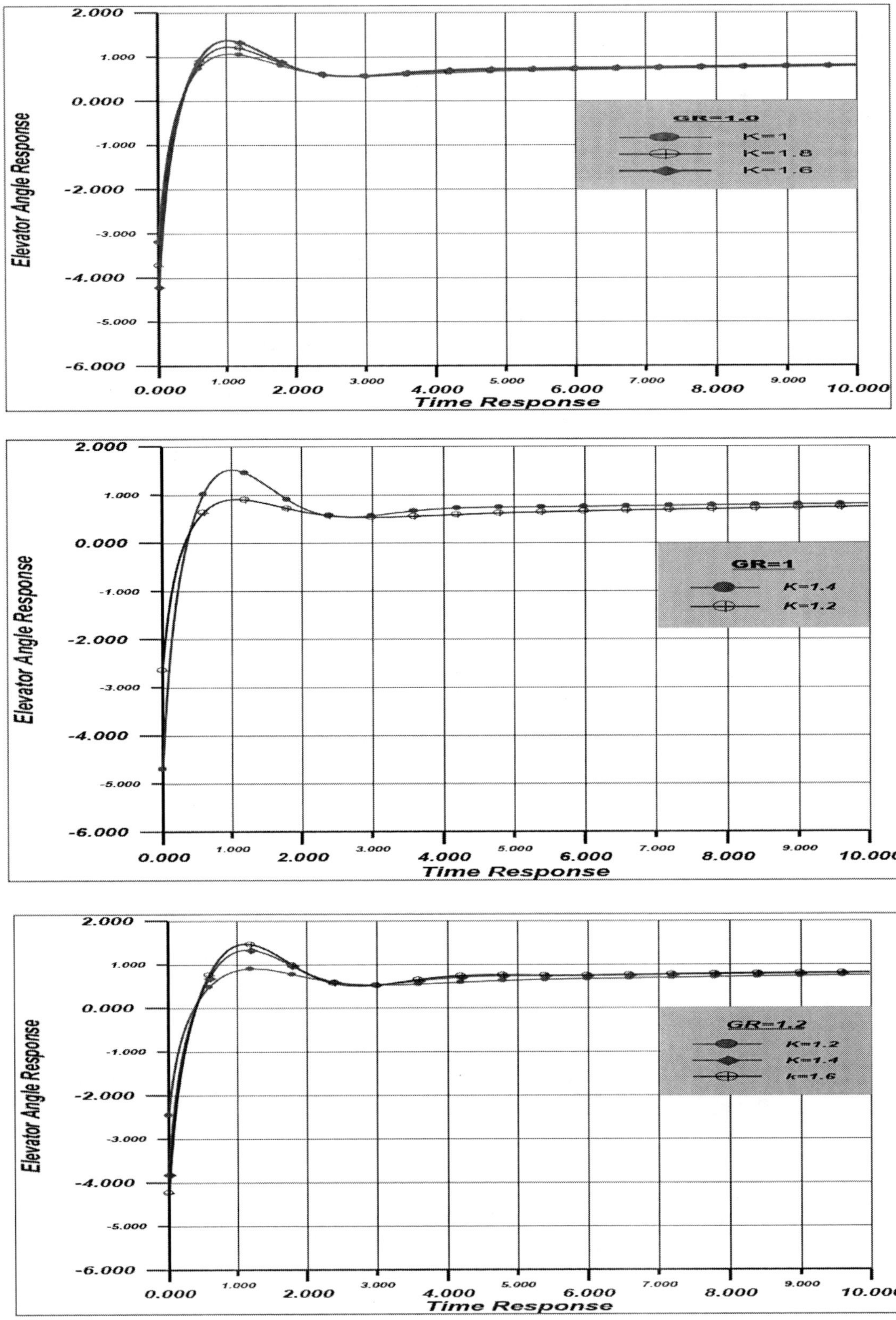

Figure (8) Elevator Angle Response for GR=1 and different "k"

2.2 Discussion

The stability flash point of the control system is the compensation between the time setting, and peak overshoot, which comes from the arrangement between the gyro sensitivity (GR) and gain (K) values. From table(1) and figure(1) for fixed value of (GR)r and different value of the real part of the complex root decreases and the imaginary part increases meaning the complex root shifts to the right of root locus diagram , so the stability decreases and construability increases .
Again from table (2—4)and Figure(2-4) the gyro sensitivity increase by(GR) 0.1 for different value of gain(k) gives that as the (GR) increases for a fixed value of (k) ,the real part of complex root increases and the imaginary part decreases meaning the complex root shifts to the left of the root locus diagram , stability increases , controllability decreases , peak value does not increases and all systems remain stable. From table(4-6) and figures(4-6) as the gyro sensitivity(GR) increases for different value of gain(k) and according to the calculation GR=1.2 and K=1.4 .The value cannot go beyond these number because they are limited by elevator deflection angle design equal to 5 rad. Therefore, the value of Gr=1.2,K=1.4 is the optimum value for the control system design which also satisfies the actuator slewing time(actuator deflection)which was equal to 45 rad/sec and design value equal to 90 rad/sec.

3. Conclusions

1. The characteristic equation of any control system depends on flying body seep and shape.
2. Variation of (GR) and (K) in the system depend on the characteristic system equation .
3. Maximum requirement actuator slewing rate depend on the value of gain (K)
4. An increase in the value of gain (K) decreases stability and increases controllability.
5. An increase in the value of gyro sensitivity (GR) increase stability and decrease controllability.
6. The value of Gr and K elevator deflection and actuator deflection.

List of symbols

Symbols	*Definition*	*Units*
b	*Wing Span*	ft
$\bar{c}$	*Wing Chord Line*	ft
S	*Wing Area*	ft^2
G	*Ground Force*	$slug.ft^2/\sec^2$
m	*mass*	$slug$
q	*Dynamic Pressure*	$slug/ft.\sec^2$
u	*Air Sped*	ft/sec
u_o	*Initial Air Sped*	ft/sec
p_o	*Roll Rate*	deg/sec
I_{xx}	$X-$ *Axis moment of Inertia*	$slug.ft^2$
I_{yy}	$Y-$ *Axis moment of Inertia*	$slug.ft^2$
I_{zz}	$Z-$ *Axis moment of Inertia*	$slug.ft^2$
ρ	*Air Density*	$slug/ft^3$
c_{m_q}	*Damping in pitch*	$1/$ *radian*
c_{m_α}	*Static longitudinal stability*	$1/$ *rad*
c_{n_r}	*Damping in yaw*	$1/$ *rad*
c_{n_β}	*Directional Stability*	$1/$ *rad*

α	$angle\ of\ attack$	deg
β	$angle\ of\ attack$	deg
r	$angle\ of\ attack$	deg
$\dot{\alpha}$	$Rate\ of angle\ of\ attack$	deg/sec
$\dot{\beta}$	$Rate\ of angle\ of\ attack$	deg/sec
$\dot{r}$	$Rate\ of angle\ of\ attack$	deg/sec
$\dot{q}$	$Rate\ of\ Dynamic\ Pressure$	$slug/ft.\mathrm{sec}^3$

References

[1]Etkin "Dynamics of Flight Stability and Control "John Willey, 2002.

http:\\www.Amazon Online Reader Dynamics of Flight Stability and Control.mht
[2]P.garnell and D.J. East "Guided weapon Control System" Royal military college of science, England, 2001.

www.amazon.com/Guided-Weapon-Control-Systems-Garnell/dp/0080254683 - 204k -
[3]John H. Block lock "Automatic Control of Aircraft and Missiles" John Willey and sonic, 2006.

http:\\www.Amazon_com Automatic Control of Aircraft and Missiles John H_ Blake lock Books.mht
[4]Alan Pope "Aerodynamic Of Supersonic Flight Optimum Aeronautical Publications"2002.

http:\\www .Amazon_com Used and New AERODYNAMICS OF SUPERSONIC FLIGHT an Introduction.mht
[5]J.J D'Azzo and C.H.Houpis "Feedback Control System Analysis and Synthesis" Mc Graw. Hill, .2006.

http:\\www.Customer Image Gallery for Feedback Control System Analysis and Synthesis.mht 7. John J.D'Azzo C.H. Houpis "linear Control System Analysis and Design"
[6]Gustav Doetsch. "Guide to the Applications of the Laplace and B-Transformation ", 2008.

http:\\www.Amazon_com Gustav Doetsch Books.mht
[7]J.J D'Azzo. C.H.Houpis "Linear Control System, Analysis and Design ", 2006.
[8]Flight Dynamic NOTES C. I. T U.k, 1986.

Algorithm For Constructing Modified Relative Phase Permeability For The Average Model Of Three-Phase Filtration In A Multilayered Layer

Yakupov Z.Ya[1].,, Hussain Abdulaziz Abrahem[2] , Bogomolov V.A. [1], Plokhotnikov S.P[1]

[1] Kazan National Research Technical University A.N. Tupolev - KAI, Department of Special Mathematics, Kazan, Russia ZYaYakupov@kai.ru , bogomolov@kfti.knc.ru , plokhotnikov@kstu.ru

[2]University of Technology – Iraq, Electromechanical Engineering Department, 50008@uotechnology.edu.iq

Abstract. The paper presents models of relative phase permeabilities for laboratory relative phase permeabilities. A technique has been developed for modifying the relative phase permeabilities taking into account the jet flow. Using a computational experiment, the applicability of the proposed averaged models was investigated. The models described in the article are applicable in numerical calculations for multilayered reservoirs in order to reduce the dimension of the problems under consideration.

Keywords: filtration, phase permeability, modified permeability, jet flow pattern, computational experiment.

1. Introduction

Oil fields have, as a given, a pronounced layered heterogeneity. This greatly complicates multivariate hydrodynamic three-dimensional calculations. The mathematical problem is much more complicated if we take into account the three-phase (gas, oil, water) flow. In this case, the calculations require detailed geological information about the formation and significant time calculations on computer. Therefore, in practice, models averaged over the thickness of the formation are often used, allowing reducing the dimension of the initial problem in certain areas of the multilayer field and thereby significantly simplifying numerical calculations [1-3]. It should be noted that despite the widespread use of foreign and domestic software systems created by Landmark Halliburton Int, Schlumberger, Roxar Software Solution, Tigers, CMG, LAURA, etc., the question of reducing the dimension of the problem remains relevant, especially when combining layers into bundles in some areas of deposits, as well as when solving multivariate problems of optimal development of real layers.

IOP Conf. Series: Materials Science and Engineering **765** (2020) 012003 doi:10.1088/1757-899X/765/1/012003

2. The purpose of the work

The purpose of the work is to construct mathematical relative phase permeabilities from laboratory (experimental) relative phase permeabilities, to create a modification technique for the constructed relative phase permeabilities taking into account the jet flow, using a computational experiment , investigate the applicability for calculating averaged models, including those with modified relative phase permeabilities, previously proposed for two-phase filtration in [1-3].

2.1 Algorithm for the experimental determination of relative phase permeabilities in the oil-water system.

To determine the experimental relative phase permeabilities by the method of unsteady displacement (unsteady-state displacement) perform the following operations [4] (in the oil-water system):

1. A core is made from a core.
2. An initial oil saturation was created in this sample. Steps for that are as follows:

Saturate the sample with a model of produced water, drive off the moving water using a centrifuge. Then the sample is saturated with kerosene. Subsequently, oil is pumped in the unit, replacing kerosene. Carry out the displacement of oil by water. Water is pumped through this oil-saturated sample. During the experiment, the amount of water injected, the amount of oil displaced, and the pressure drop across the sample are recorded.

After data processing, relative phase permeabilities are obtained. Similarly, core research is carried out for the oil-gas system. To simulate three-phase filtration, empirical formulas for carbonate rocks were obtained from the relative phase permeabilities obtained in the laboratory. Fig. 1 shows graphs of laboratory relative phase permeabilities for various cores and comparison of the graphs obtained by the formulas for relative phase permeabilities [3] in the form of functions:

$$k_{rw}(S_w) = k_{rw}^0\left[(S_w - S_{wc})/(1 - S_{wc} - S_{or} - S_{gc})\right]^{\alpha}$$
$$k_{ro}(S) = k_{ro}\left[(S_o - S_{or})/((1 - S_{wc} - S_{or} - S_{gc}))\right]^{\beta}$$
$$k_{rg}(S_g) = k_{rg}^0\left[(S_g - S_{gc})/((1 - S_{wc} - S_{or} - S_{gc})\right]^{\alpha}$$
$$k_{ro}(S_g) = k_{ro}\left[(S_o - S_{or})/((1 - S_{wc} - S_{or} - S_{gc})\right]^{\beta} \tag{1}$$

Where ,

k_{ro}^0 – maximum relative phase permeability of oil;

k_{rw}^0 – maximum relative phase permeability of water;

– maximum relative phase permeability of gas;

Sor – residual oil saturation;

Swc – saturation of bound water;

Sw, So – water and oil saturation,

Swc≤Sw≤1-Sor, ,

IOP Conf. Series: Materials Science and Engineering **765** (2020) 012003 doi:10.1088/1757-899X/765/1/012003

To obtain empirical formulas, it is necessary to select the following coefficients
 , , , Swc, Sgc, α, β, so that the graphs are as close to the laboratory as possible.
As a result, we obtained for the first sample

 , , , Swc=0,286, Sgc=0; αw=2.3, βw=1.3, αg=1.3, βg=3,

And for the second sample

 , , , Swc=0,247, Sgc=, αw=2.5, βw=2.8, αg=1.3, βg=3.

Basically, the obtained coefficients are close, except for the degrees. Degrees vary from 1 to 3. In Fig. 1 and Fig. 2 it is clearly seen that the obtained theoretical graphs are close to the laboratory graphs. From the mathematical two-phase relative phase permeabilities, three-phase relative phase permeabilities were obtained according to the Stone II model [4].

3. Computational experiment results

As an example, a ten-layer square-shaped layer was considered, which was developed by different water flooding systems. The problem was solved numerically for a given pressure differential between injection and producing wells, the external boundary was impenetrable. In the calculations, we used grids of blocks shown in Fig. 4: 11x11x10 (x, y, z) for the reference three-dimensional case; and 11x11x1 for the averaged two-dimensional case (rescaled). In the calculations, a completely implicit scheme was used.

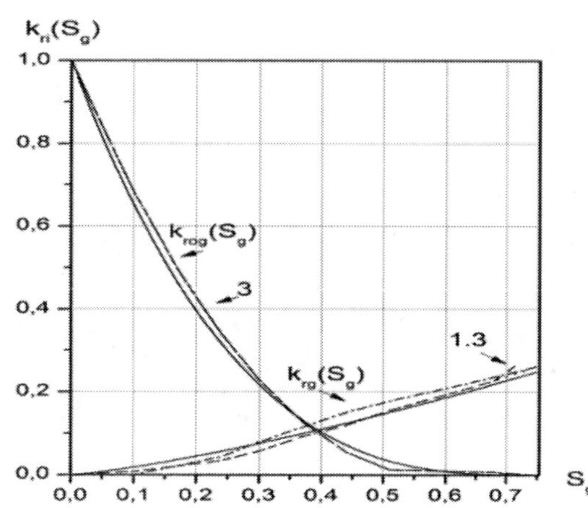

Fig. 1: Laboratory relative phase permeabilities RPP for carbonate rocks.

IOP Conf. Series: Materials Science and Engineering **765** (2020) 012003 doi:10.1088/1757-899X/765/1/012003

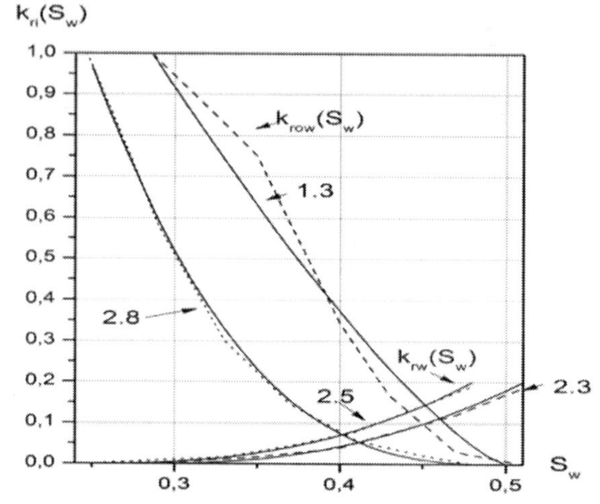

Fig. 2 : Mathematical relative phase permeabilities RPP for carbonate rocks.

Figure 3 shows graphs of modified RPP (oil-water) versus water saturation Sw for a uniform law with linear RPP . The graphs of the modified RPP for the oil-gas system were obtained in a similar form. The calculation formulas have the form [2,3]:

$$k_{rw}^{m} = k_{rw} \cdot [1 + \nu \cdot \sqrt{3} \cdot (1 - S_p)],$$

$$k_{ro}^{m} = k_{ro} \cdot [1 - \nu \cdot \sqrt{3} \cdot S_p]. \qquad (2)$$

Here the mobility of water S_p

The oil-water system is calculated

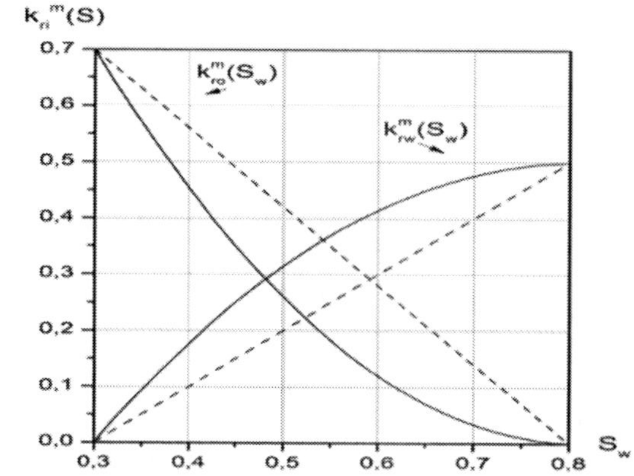

Fig. 3. Function graphs of modified RPP (oil-water system) from water saturation Sw at linear RPP (uniform law). The values specified in the model of physical parameters [2]:

IOP Conf. Series: Materials Science and Engineering **765** (2020) 012003 doi:10.1088/1757-899X/765/1/012003

128 – initial reservoir pressure, atm;
22 – formation temperature, C;
100 – temperature of injected water, C;
55 – bottom hole pressure at the producing well, atm .;
170 – bottom hole pressure at the injection well, atm .;
=0.5 – maximum relative phase permeability of water;
=0.7 – maximum relative phase permeability of oil;
=0.8 – maximum relative phase permeability of gas;
Sor=0.2 – residual oil saturation;
Swc=0.2 – saturation of bound water;
Sgc=0.1 – pinch gas saturation;
Sw,So,Sg – water, oil and gas saturation,

Swc $\leq$Sw$\leq$ 1- Sor.- Sgc.

Tasks were solved with physical parameters [2]: 1230 - reservoir depth, m; 1000 - distance between producing wells for a 5-point element, m; 707 - between production and injection wells for a 5-point element, m; 500 - distance between producing wells for a 9-point element, m; 500 and 707 - between production and injection wells for a 9-point element, (in meters).
The numerical calculations of the process within the framework of the Musket - Moris three-phase filtration model for five-point and nine-point water flooding systems were carried out for four different mathematical models.

3.1.Two-dimensional models :

1. C - averaged model with mathematical relative phase permeabilities and average absolute permeability over the thickness of the reservoir.
2. B is an averaged model described in detail in [2,3], with modified relative phase permeabilities for the β-probability distribution law for parameters of the form $\gamma = 1$, $\eta = 2$. This is a well-known uniform law; it was applied at the highest value of the coefficient of variation of vertical heterogeneity equal to 0.56. In this case, the absolute permeability of the layers was set by table No. 1.

3.2.Three-dimensional models :

1. A8 – the reference three-dimensional model with ten isolated layers of the same thickness (no vertical flows);
2. A7 - the same as the previous model, but with uninsulated layers.

Table 1. The distribution of permeability in layers for the A7-model for a uniform law, m darsy

K_1	K_2	K_3	K_4	K_5	K_6	K_7	K_8	K_9	K_{10}
4500	5500	3500	6500	2500	7500	1500	8500	500	9500

The calculations were performed for 8 different versions of the standards described in detail in [3], but Fig. 4 shows graphs of only two boundary models. All other graphs of standards are between the graphs of options A8, A7.

IOP Conf. Series: Materials Science and Engineering **765** (2020) 012003 doi:10.1088/1757-899X/765/1/012003

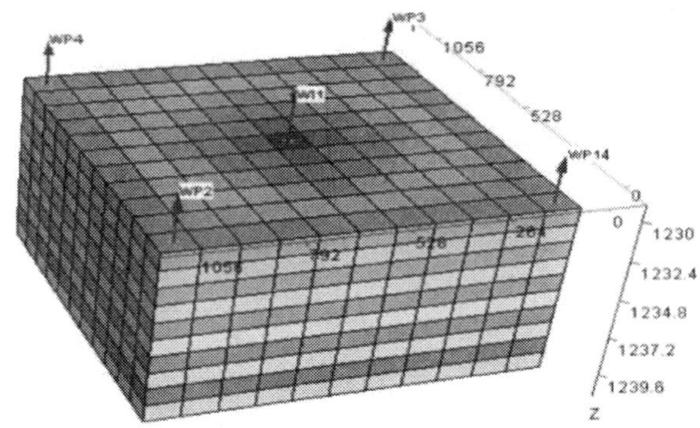

Fig. 4 Mesh of blocks 11x11x10 for five-point water flooding system (three-dimensional case)

It is clearly seen in Fig. 5 that the averaged solution B with modified relative phase permeabilities has a smaller error with respect to the reference solution A8, and solution C with mathematical relative phase permeabilities has a smaller error with respect to the reference solution A7. Similar results were obtained for the nine-point water flooding system, and for other development indicators, both for gas and oil. Solutions B and C limit all three-dimensional reference solutions Ai for which calculations have been performed.

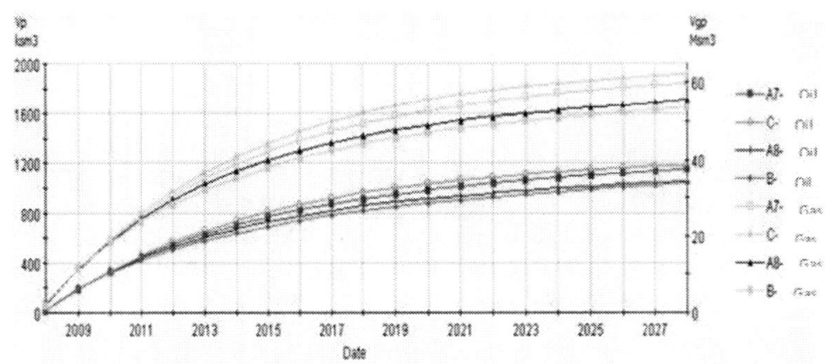

Fig. 5. Graphs of the functions of the dependence of the total volume of extracted oil and gas Vp on the time date for a five-point water flooding system, and accordingly for solutions - A7, A8, C, B

4. Conclusions.

A comparative analysis of the graphs of the development indicator functions for various three-phase models revealed the correct construction of averaged modified relative permeabilities in three-phase models. Numerical calculations on three-phase models showed good results, - the graphs of model B from the lower limit for the extracted volumes of oil and gas, and the graphs of model C from the upper limit for all 8 standards considered in the work. This indicates the validity of applying this averaging technique for a three-phase filtration model. Similar positive results in the numerical calculations of two-phase filtration took place in [2,3].

References:

[1] Hearn C.L. Simulation of Stratified Water-Flooding by Pseudo Relative Permeability Curves / C.L. Hearn //J. of Pet. Tech.- 1971.-July.- P. 805-813.

[2] Bogomolov V.A. Mathematical Simulation of Three-Phase filtration in Stratified Beds with Account for the scheme of jets / V.A. Bogomolov, S.P. Plokhotnikov, O.R. Bulgakova, D.S. Plokhotnikov // Journal of Engineering Physics and Thermophysics. – Springer. – 2011. Vol. 84. – No. 5. – pp.975-979.

[3] Plokhotnikov S. P. Mathematical Averaging of Coefficients of System of Elliptic and Parabolic Equations in Continuum Mechanics / S. P. Plokhotnikov, V. A. Bogomolov, R.Kh.Nizaev, O.I. Bogomolova, P.V.Malov // Lobachevskii Journal of Mathematics. – Springer. – 2019. - Vol.40, No 5, -pp. 553 – 561.

[4] Guidelines for the Creation of Permanent Geological and Technological Models of Oil and Gas and Oil Fields (Part 2. Filtration models) // - M.: VNIIOENG. - 2003. - 228s.

IOP Conf. Series: Materials Science and Engineering **765** (2020) 012004 doi:10.1088/1757-899X/765/1/012004

Drag Reduction and Flow Enhancement in Iraqi Crude Oil Pipelines using PMMA polymer and CNTs

Raheek I. Ibrahim[1]*, Manal K. Odah[1], Dhoha A. Shafeeq[1], Ali Dawood Salman[2]
[1]Electromechanical Eng. Dep. / University of Technology/ Baghdad- Iraq
[2]Institute of Environmental Engineering, University of Pannonia, Veszprém, Hungary

*Corresponding author email: 80058@uotechnology.edu.iq

Abstract. Crude oil flowing in pipelines suffers from a problem fluid flow pressure drop and high energy consumption for fluid pumping. The flow can be enhanced using either viscosity reduction or drag reduction techniques. Drag reduction (DR) can be considered as the most effective and applicable method. Such method contributes in reducing the frictional energy losses during the flow by addition of little amounts of drag reducing agents (DRAs).The present work focuses on comparing between the effect of PMMA only and PMMA with CNT on drag reduction In order to achieve this object, an experimental rig has been designed and implemented. It should be mentioned that the ring consists of: a crude oil Perspex pipe, oil pump, pressure sensors. Two additive materials including PMMA and CNT with different concentrations have been used to reduce the drag inside the oil pipe. polymethyl metha acrylate polymer (PMMA) with different concentrations 1000, 2000, 3000, 4000 mg/ L. It has been shown 3000ppm is the best where the percentage of drag reduction reached to 65%. While, A mixture of CNT and PMMA with (1000, 2000, 3000, 4000, 5000) mg/L are used and the results showed that 50% of DR is achieved at 2000 mg/L.

1. Inrtoduction

The first paragraph Pipeline can be considered as the safest tools to transport crude oil for long distance. Thereby such tools are usually utilized for crude oil exporting. When fluids are transported through the pipeline, fluid pressure will be decreed, due to the friction between the wall of oil pipe and fluids. Here, fluids must be transported under sufficient pressure to achieve the desired productivity, and then more pressure must be directed. However, pipeline design limitations lead to reduce the amount of pressure that can be used. It should be stated that the problems associated with pressure reduction are compounded when the fluids are transported over long distances. Such decreases results in increasing the volume of the equipment and costs[1,2].

The additives are polymers or nano materials. Polymer works as a DRAs by dampening present eddies in order to greatly reduce turbulence, as several scientists concluded [3-5]. It was also discovered that the efficiency of polymer was influenced by several variables such as: concentration impact, tube and geometry size, molecular polymer weight, chain flexibility and flow rate [3].Addition of little amount of long-chain polymers to a liquid flow will cause massive decreases within the frictional resistance at the wall [6]. Moreover, it reduces the hydrodynamic drag level when pumping oil or oil products, which provides the possibility of easy pipe operation, i.e. increasing the volume flow rate when piping through the overloaded pipeline section ; accelerating to compensate for some unavoidable time losses ; minimizing the loading and unloading of oil barges ; increasing the pipeline flow capacity in offshore oil development [4, 5].Ptasinski, P. K, et al, (2001) [7] studied hydrolyzed polyacrylamide (PAMH) in water, The main part of the setup consists of a smooth straight pipe of Plexiglas with a length of 34 m and an inner diameter of 40.37 mm. It was shown that the highest DR of 70% obtained with 435ppm.Liberatore, Matthew W, et al, (2004) [8] used polyacrylamide (PAM) in water with different concentration 200, 500 and 1000 ppm and 3m test section of rectangular pipe , 58% Of DR occurred at 1000ppm concentration.

Kim, Nam-Jin, et al, (2010) [9] applied polyethylene oxides (PEO) on water, inner diameter and length of pipe line were (17.1mm and 2000mm), respectively. The maximum drag reduction rate was 50% at 20ppm concentration.Nicholas B. Wyatt,et al, (2011) [10] used polyelectrolyte xanthan in water with three separate pipes of varying diameters and materials (1/4 iN. stainless steel, 1/2 iN. copper, and 1 iN. copper) , the result showed that maximum DR 35% obtained at 200ppm. Zadrazil, I, et al, (2012) [11] used PEO with diverse concentration 5, 10, 25, 50, 75, 125 and 250ppm ,a stainless steel horizontal pipe test section of length 8m and inner diameter of 25.3m , 68% of DR achieved at 250ppm. AneesA. Khadom,et al, (2014) [12] Investigated the impact of polyacrylamide (PAM) as a drag-reducing polymer on Iraqi crude oil flows in pipe lines. Five concentration of PAM (10,20,30,40,50) ppm and pipes made of carbon steel with length of 3 m and diverse inner diameters (0.0508, 0.0254 and 0.0191 m) were used. The maximum reduction in drag was 40.64% with 50 ppm of PAM polymer flowing in 0.0508 m I.D straight pipes. ŞERİFE, Z,et al, (2014) [13] .Utilized low molecular weight sodium carboxy methyl cellulose (CMC) in water with four concentrations (200 to 500 ppm) The flow system consists of a test section that is 6 m longwith 46 mm inner diameter of PVC pipe. Themaximum drag reduction obtained was 22% using 500 ppm CMC solution.Tian, Meng, et al, (2015) [14] applied Hydroxypropyl xanthan gum (HXG) whichprepared from xanthan gum (XG) in water, the test section was a smooth tube made of stainless steel (2.0 m length and 20.5 mm inner diameter).The maximum drag reduction for 1 g/L for XG and HXG were 68.1% and 72.8% respectively.L. Marylin Pumisacho,et al, (2016) [15] used hydrolyzed polyacrylamide (PAMH) in water with test pipe section of 34-mm inner diameter and 6-m length made of Plexiglas , at 20ppm the DR decreased by 35%. Siti S,et al, (2016)[16]. Studied the synthesizing of biopolymer, Carboxymethylcellulose (CMC) on water, and studied its effectiveness in drag reduction at different concentration and flow rate. At1000 ppm DR was 27.27%. Hassanean,M. H, et al,(2016) [1]. Studied the impact of polyalpha olefin on crude oil stream lines situated in the western Egyptian desert (field Fagour) owned by Khalda Petroleum Company with a pipeline length of 16 km from Fagour to Bouchis with a nominal 6 inch pipeline diameter and 12 km from Bouchis to Kalabsha with a nominal 8 inch pipeline diameter. Following the field application of 60 ppm of the DR%, the pressure drop was decreased by 36%.Danfu Caoa,et al, (2018) [17] . discussed the influence of Poly-α-olefin as DR on crude oil , a field experiment of the tubular dispersing device was carried out in a short-range product oil pipeline in China , the maximum DR was 56%.

Nano scale materials are defined as a set of materials in which at least one dimensions is less than approximately 100 nanometers. Nano materials can be nano scale in one dimension (eg. surface films), two dimensions (eg strands or fibrs), or three dimensions (eg. particles). Engineered nanomaterials are resources designed at the level of the molecular (nanometer) to take advantage of their small size and new properties that are generally not found in their conventional counterparts [18]. Nano materials additive can be used as drag reducers and can set the contact surface to make the fluid flow smoother and more effective [19].CNT with flexible polymer such as PMMA or xanthan gum (XG) can be used as a DRA[20].So after sonication it can be used as a good drag reduction agent especially when it would bedissolved in an organic liquid. Pouranfard A R., et al, (2014) [6]applied nanosilica on water, the test section of the experimental set-up consisted of a smooth unique pipe (5 layers composite pipe of poly ethyleneand aluminum). Four pipes with length of 9 m and the diameters 1.27, 1.905, 2.54, 3.175 cm, respectively parallel horizontal galvanized rough pipes were tested. At 1 % wt and 1.27 ID, the maximum DR was 24 %.Abdulbari, Hayder A,et al, (2015) [21] studied the effect of three nano powder additives (bismuth oxide ,iron oxide, silica andtitanium oxide) on DR, Straight micro channels of 50, 100, and 200μm with a length of 58.5 mm were employed in this study. The optimum DR% were calculated on the basis of the experimental results. The optimum percentages were as follows: bismuth oxides: ~65% DR, 200 ppm and 100μm microchannel size; iron oxides: 57% DR, 300 ppm, and 50μm microchannel size; titanium oxides: 57% DR, 200 ppm, and 50μm microchannel size; and silica: 55% DR, 200 ppm, and 50μm microchannel size.

Akindoyo, E. O,et al, (2016) [20]studied the effect of mixture Carbon Nanotubes (CNT) and Xanthan Gum (XG) on water using rotating disk apparatus. The CNT diameter was 20–40nm and the length was 10-30μm, the rotating disk was of 148mm diameter and 3mm thickness.The maximum drag reduction of approximately 50% was observed at 0.6 concentrations.

Yanuar, et al, (2017) [19] employed Al2O3 nanoparticles on water in a ratio of 100 ppm, 200 ppm and 300 ppm as well as 30 minutes, 60 minutes and 120 minutes mixing time. A circular pipe was used as a spiral pipe comparison and both were horizontally mounted. The ratio of the spiral pipe was (P / Di=10.8) and the inner circular pipe diameter was 3 mm. Nanoparticles mixing time and composition ratio in basic fluid affect the drag reduction results. Nanofluid flows through the test pipe showed high drag reduction at a mixture concentration of 300 ppm and 120 minutes of mixing time in the spiral pipe as 38 %.

2. Experimental set-up

Drag reduction was measured in a closed flow loop.The experiment is performed in a re-circulatory flow facility which is shown in Figure 1. This system consists of Control unit, oil pump, Pressure gauge, Pressure Sensor, PVC pipe with a diameter of 0.5 inch and length of 2m, , oil tank.

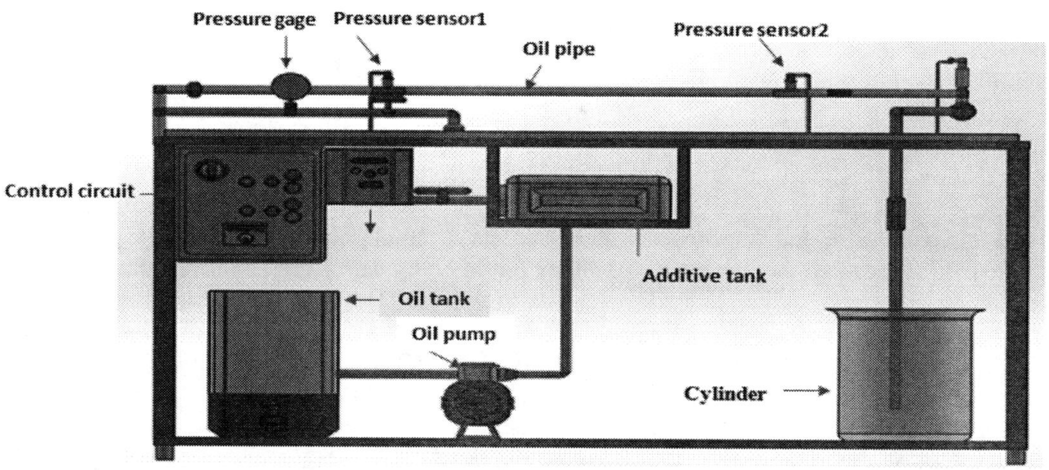

Figure 1. The experimental system.

IOP Conf. Series: Materials Science and Engineering **765** (2020) 012004 doi:10.1088/1757-899X/765/1/012004

3. Experimental work:

Firstly poly methylmethacrylate (PMMA) with density of 0.032g/cm3 and five concentration (1000,2000,3000,4000,5000) ppmis mixed with crude oil by magnetic stirrer for 20 minutes and adding to the oil tank to measure each of pressure drop, DR%, viscosity, density, friction factor, Reynolds number and power consumption.
Secondly carbon nano tube (CNT) with 90% purity, (5-10nm) inner diameter, (10-30nm) out diameter and 0.06g/cm3 bulk density, is mixed with PMMA in the rate of (25% of CNT with 75% of PMMA) and mixed by ultrasonic for 1 hour then mixed using magnetic stirrer for 30 minutes and adding to oil tank to measure each of pressure drop, viscosity, density, friction factor, Reynolds number and power consumption.
Table (1) shows the device specifications:

Table 1. Device specifications

device	Type	Specifications
Oil pump	WCB75	220/380V,750W,50L/min,Max Head 30m,0.3Mpa (43 psi)
Pressure Sensor	PA-21Y/81554.33	Rang 0-10 bar,Out /GND 4-20mA
Accurate electronicbalance	FRAGILE	230V50HZ,0.001g-500g
Magneticstirrer	78-1HOT PLATE &STIRRER	250V, 6A
Viscometer	Cannon-Fenske-routine	Size 200, Constant (cst/s)=0.1, Kinematic Vis (cst) from 20 to100
Pycnometer	Duran	Volume 25.81cm3
Ultrasonic	KQ3200E	220v, 50 HZ, 150 watt

4. Experimental Calculations:

Percentage of drag reduction

Pressure drop reading through the pressure sensor before and after additive .Percentage of drag reduction can be calculated by:[22]

$$DR = \frac{\Delta p_b - \Delta p_a}{\Delta p_b} \quad (1)$$

Where *pa* is a pressure drop before additive and pbis pressure drop after additive.

Power consumption, Reynolds number and friction factor

Flow rate Q (m3/s), Re Reynolds number density **ρ**, viscosity **μ**,u velocity, D diameter of pipe, f friction factor , L length of pipe [23,24]

IOP Conf. Series: Materials Science and Engineering **765** (2020) 012004 doi:10.1088/1757-899X/765/1/012004

$$P_E = Q.\Delta p \tag{2}$$

$$Re = \frac{\rho u D}{\mu} \tag{3}$$

$$f = \frac{\Delta P.D}{2\,L\,\rho\,.u2} \tag{4}$$

5. Results and discussion

5.1 Effect of PMMA

Figures 2 to7 show the effect of PMMA with five concentration (1000, 2000, 3000, 4000, 5000) ppm on friction factor, pressure drop, drag reduction, viscosity, Reynolds number and power consumption.

5.1.1 Effect of PMMA on friction factor:

Figure 2 shows the effect of PMMA on friction factor in which the polymer acts as lubricating oil, where the slippage between the crude oil molecules increases with each other and between the oil molecules with the pipe wall. Thereby, reducing friction it was noticed that when adding 3000ppm of PMMA the friction factor decreases from 0.0052 where there was no addition to be 0.0016. After 3000ppm the friction gradually increased until it reached 0.0036 at 5000ppm where increasing the polymer led to increase the viscosity of fluid, which increased the friction factor.

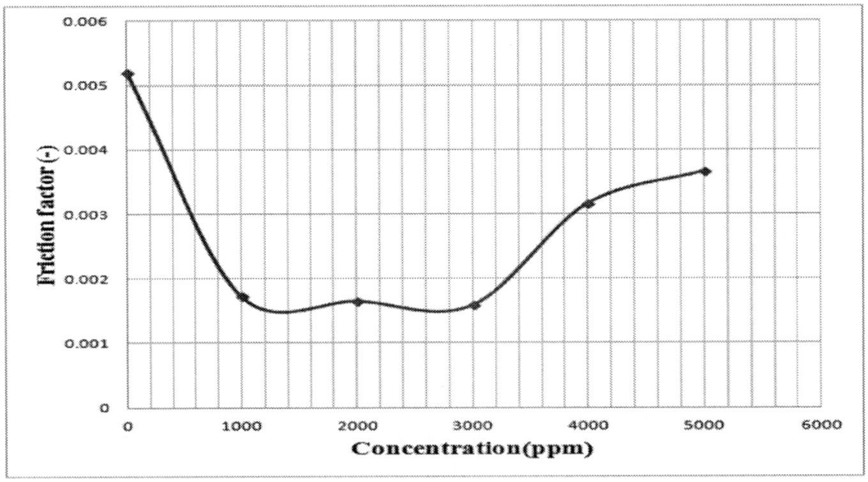

Figure 2. Effect of PMMA on friction factor

5.1.2 Effect of PMMA on pressure drop:

Figure 3 represents the effect of PMMA on pressure drop, where the pressure drop decreases from 46000Pa to 16000Pa at 3000ppm. The pressure drop is proportional to the friction factor, where the low friction factor leads to a drop in the pressure drop. This is because lowering the friction between the crude oil molecules with each other and the pipe wall saves the energy of the liquid particles and keeps the stream energy in the form of pressure. After 3000ppm, the pressure drop gradually increased to 3600at concentration of 5000ppm.

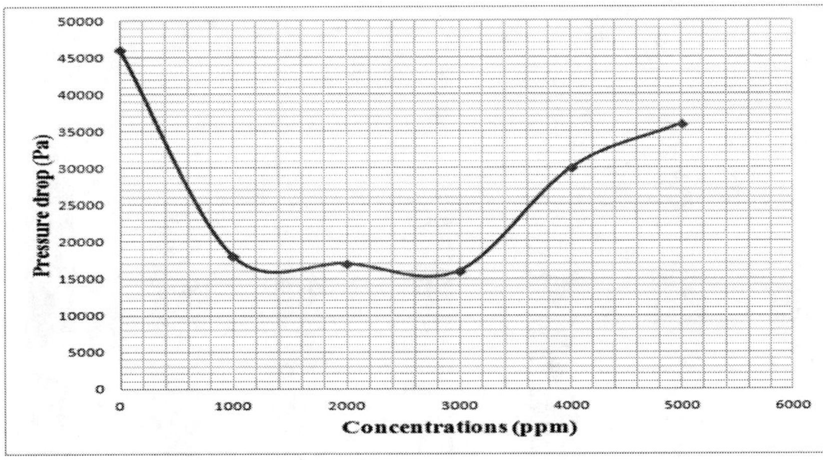

Figure 3. Impact of PMMA on pressure drop

5.1.3 Impact of PMMA on DR:

Figure 4 shows that adding 3000ppm of PMMA leads to increase the drag reduction rate by 65%. It should be noted that pressure drop reduction results in drag reduction. After 3000ppm the DR% decreased to 22% at concentration of 5000ppm.

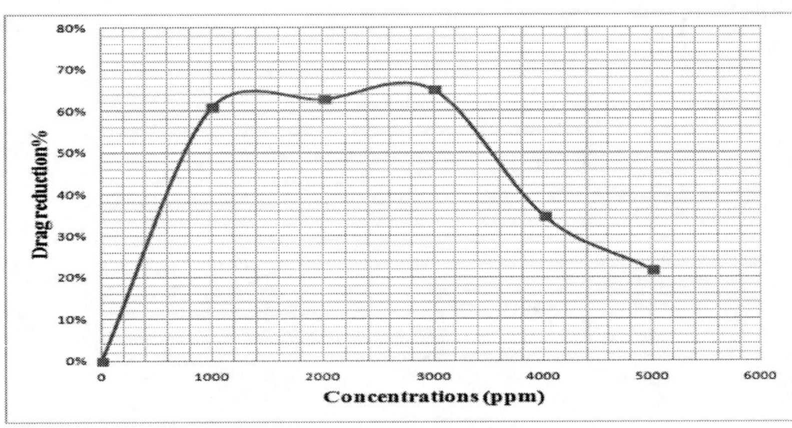

Figure 4. Impact of PMMA on DR

5.1.4 Effect of PMMA on viscosity:

Figure 5 shows the effect of PMMA on viscosity, where the viscosity of crude oil decreases from 21.5cst at no addition to 15.37cst at 3000ppm. This is due to the lubrication characteristics of the polymer which leads to increase sliding liquid particles with each other leading to reduce the viscosity. After 3000ppm the viscosity increased to 18.16cst.

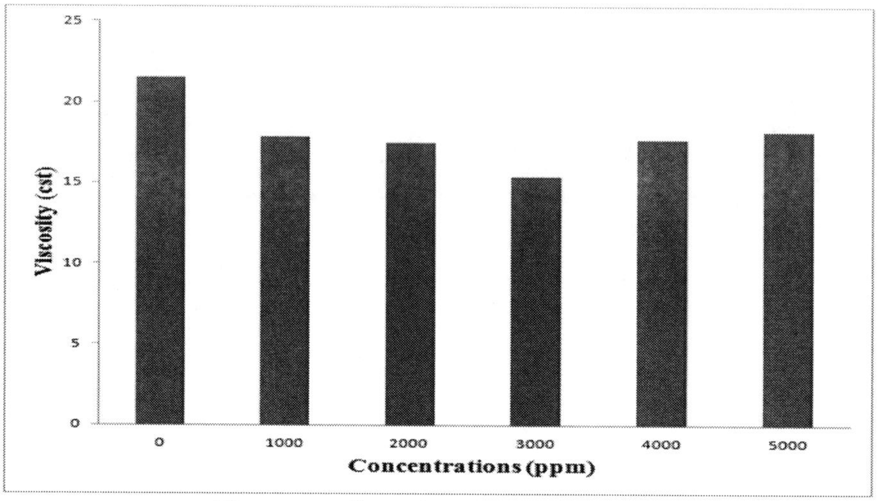

Figure 5. Effect of PMMA on viscosity

5.1.5 Effect of PMMA on Reynolds number:

Figure 6 shows that the Reynolds number values increase from 3264.12 without addition to 5218.22 at 3000ppm concentration of polymer addition. Thereby, an enhancement in flow rate of crude oil inside the pipe when using the polymer can be achieved. After 3000ppm the Reynolds number gradually decreased to 4140.5 at 5000ppm.

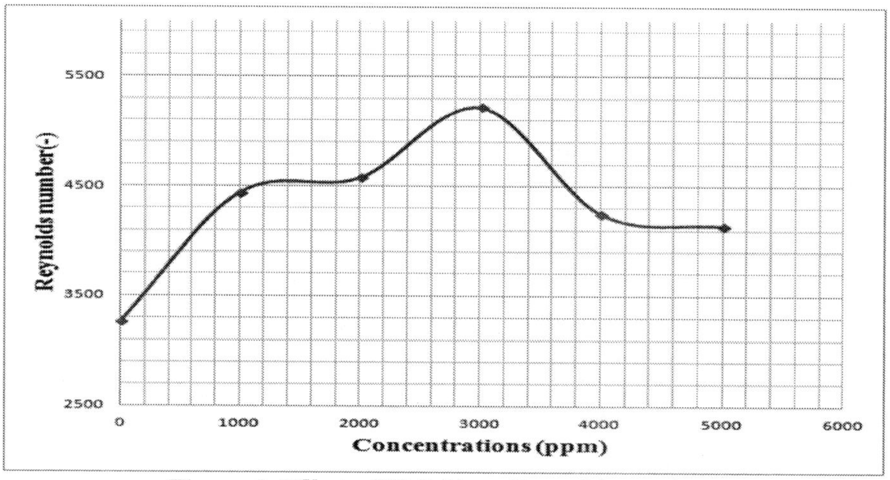

Figure 6. Effect of PMMA on Reynolds number

5.1.6 Effect of PMMA on power consumption:

Figure 7 shows that power consumption decreased from 32 watt at no addition to 11watt at 3000ppm. As the value of power consumption is proportional to pressure drop it decreased by 65.21% when using 3000ppm of PMMA. After 3000ppm the power consumption gradually increased to 25watt at 5000ppm.

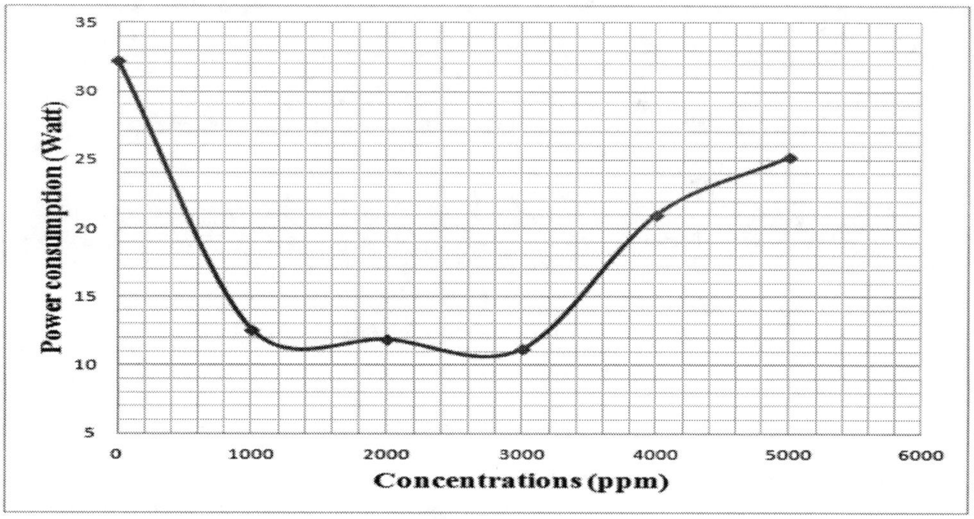

Figure 7. Effect of PMMA on pumping power consumption

5.2 Effect of PMMA with CNT

Figures 8 to13 show the effect of PMMA with Carbon nano tube using six concentrations as (1000, 2000, 3000, 4000, 5000, 6000) ppm on each of friction factor, pressure drop, drag reduction, viscosity, Reynolds number and power consumption.

5.2.1 Effect of PMMA with CNT on friction factor:

Figure 8 illustrates the effect of PMMA with CNT on friction factor. Clearly, the friction factor decreases from 0.0.0052 without addition to 0.0022 at 2000ppm. This is because of the nano particles encapsulate the inner surface of the pipe wall and acts as a buffer between the inner surface of the tube and the crude oil molecules, which makes the flow in the center of the tube, and consequently reduces the friction factor values. After 2000ppm the friction factor gradually increases to 0.00305 at 5000 and 6000 ppm. This behavior may be due to aggregations of nano particles within the pipe, which gives the reverse outcomes.

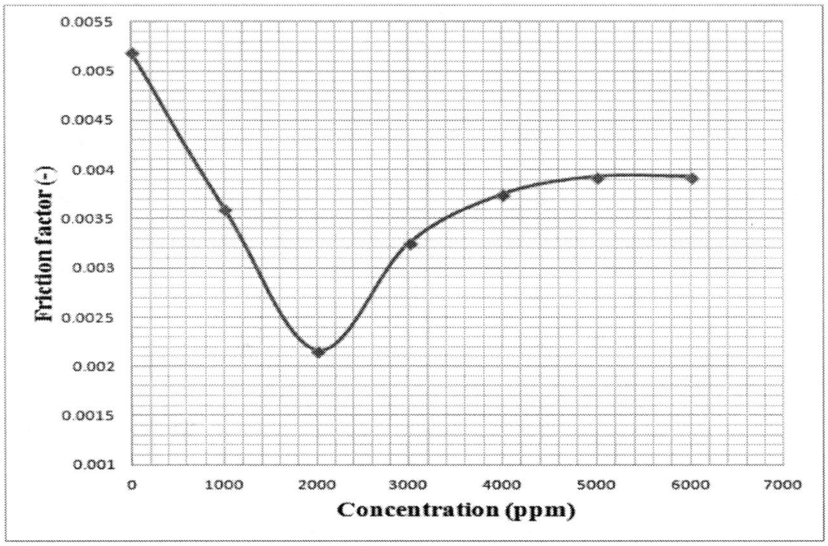

Figure 8. Effect of PMMA with CNT on friction factor

5.2.2 Effect of PMMA with CNT on pressure drop:

Figure 9 illustrates that pressure drop decreased from 46000Pa without addition to be 23000Pa at 2000ppm concentration of PMMA with CNT. The pressure drop is decreasing because the friction reduction inside the pipe, therefore the molecules of the fluid saving in their flow energy. After 2000ppm the friction factor gradually increased to 37000 at 5000 and 6000 ppm.

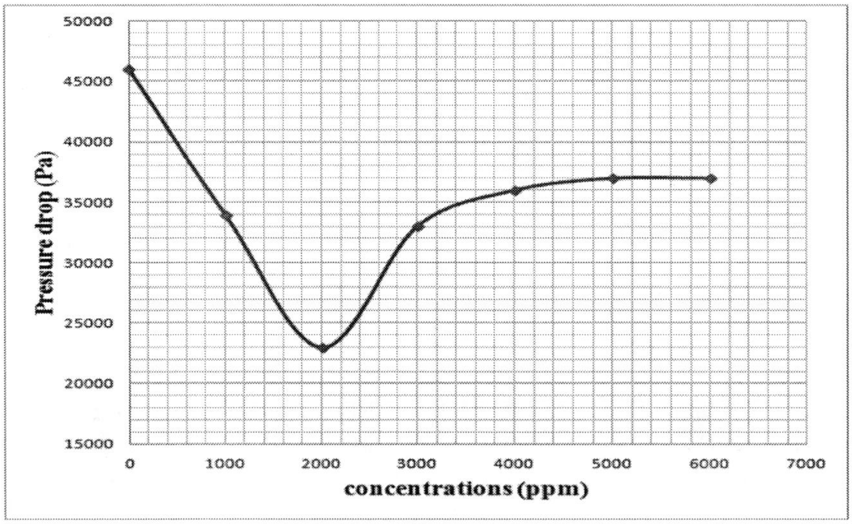

Figure 9. Effect of PMMA with CNT on pressure drop

IOP Conf. Series: Materials Science and Engineering **765** (2020) 012004 doi:10.1088/1757-899X/765/1/012004

5.2.3 Effect of PMMA with CNT on drag reduction:

Figure 10 shows the variation of the drag reduction with different concentrations of PMMA with CNT. It can be observed that 50% of DR is obtained when adding 2000ppm of PMMA with CNT to the crude oil. This DR% is considered as a good result in comparison with PMMA and fiber (28.26%). After 2000ppm the DR% gradually decreased and reached to 19.5% at 5000 and 6000ppm.

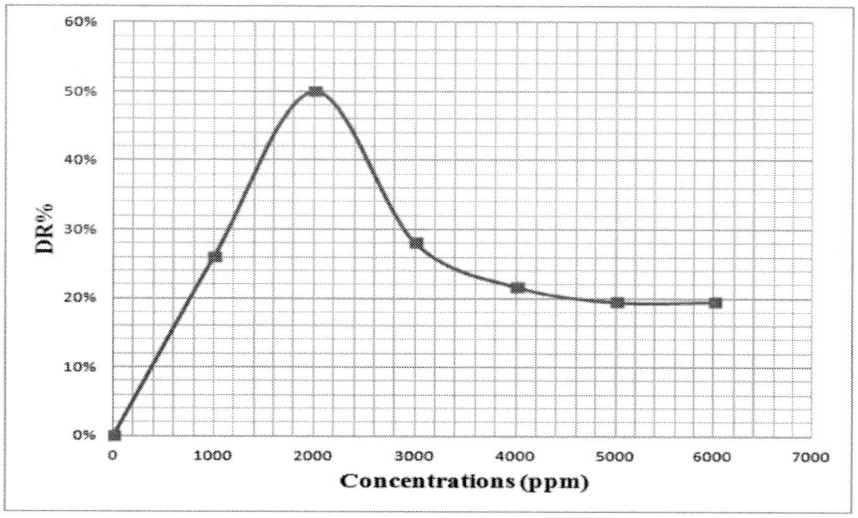

Figure 10. Effect of PMMA with CNT on drag reduction

5.2.4 Effect of PMMA with CNT on viscosity:

Figure 11 shows that the viscosity is decreasing from 21.5cst at zero addition to 14cst at 2000ppm of PMMA with CNT. This is because that CNT are formed in longitudinal chains that allow fluid layers to slide over each other and thus reduce the viscosity of the fluid. After 2000ppm the viscosity gradually increased to17cst.

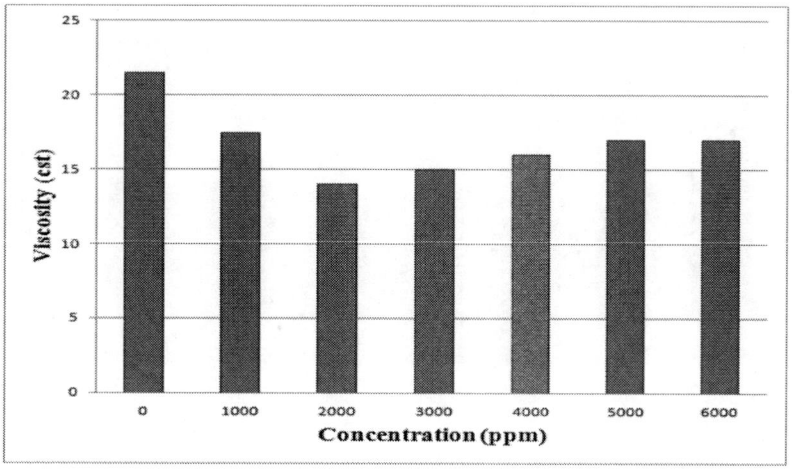

Figure 11. Effect of PMMA with CNT on viscosity

5.2.5 Effect of PMMA with CNT on Reynolds number:

Figure 12 shows that the Reynolds number increases from 3246.12 at zero addition to be 5585.64 at 2000 ppm. The values of Reynolds number reaches the turbulent region indicates a clear flow enhancement inside the crude oil pipe. After 2000ppm Reynolds number gradually decreased to 4305.07.

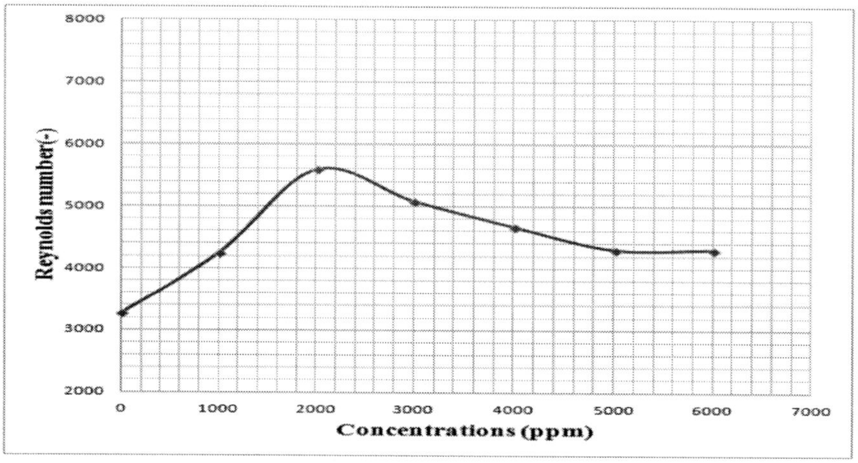

Figure 12. Effect of PMMA with CNT on Reynolds number

5.2.6 Effect of PMMA with CNT on power consumption:

Figure 13 shows that the power consumption decrease from 32 watt at zero addition to 17.94 watt at 2000ppm. The percentage of power consumption decreasing was 42%which indicates a good saving in power when additives material such as polymer and nano are used. After 2000ppm the power consumption gradually increased to 27 watt.

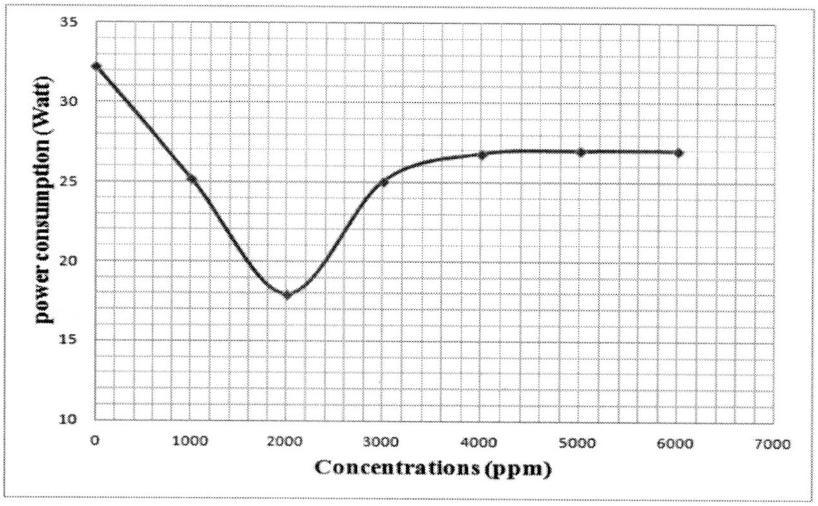

Figure 13. Effect of PMMA with CNT on power consumption

IOP Conf. Series: Materials Science and Engineering **765** (2020) 012004 doi:10.1088/1757-899X/765/1/012004

6. Conclusions

The addition of 3000ppm of PMMA to the crude oil leads to reduce the friction factor and the pressure drop to 0.0016 and 16000Pa, respectively. On the other hand, the DR% increases by 65% results in decreasing the power consumption to 8 watt. Also, the addition of 2000ppm of PMMA with CNT decreases the friction factor to be 0.0022, the pressure drop will be decreased to 23000Pa, while the DR% rate increases by 50% so the power consumption decrease to be 17.94watt.

References:

[1] Hassaneana H, Awada M, Marwanb H,Bhrana A and Kaoud M. Studying the rheological properties and the influence of drag reduction on a waxy crude oil in pipeline flow. 2016, Egyptian Journal of Petroleum 25.1: 39-44.

[2] Costalonga, Michell L, Bruno V, and Edson J. Drag reducing flows by polymer solutions in annular spaces. 2018, Journal of Fluids Engineering 140.5 , 051101.

[3] Mucharam, Leksono, Silvya R, and Rizki R. Drag reducer selection for oil pipeline based laboratory experiment. 2017 Modern Applied Science 12.1 , 112.

[4] Georgi V, Konstantin B, Olga V, and Pavel M. Laboratory evaluation of the drag reduction additives effectiveness.2015 Procedia Chemistry 15 , 371-377.

[5] Ivchenko, P. V., I. E. Nifant'ev, and A. V. Tavtorkin. Polyolefin drag reducing agents.2016, Petroleum Chemistry 56.9 , 775-787.

[6] Pouranfard, A. R., D. Mowla, and F. Esmaeilzadeh. An experimental study of drag reduction by nanofluids through horizontal pipe turbulent flow of a Newtonian liquid.2014 Journal of Industrial and Engineering Chemistry 20.2 ,

[7] Meng T, Bo F,Leiping J,Yongjun L,Xiaohui Q , Hao J and Kejing L. Rheological and drag reduction properties of hydroxypropyl xanthan gum solutions.2015, Chinese Journal of Chemical Engineering 23.9 , 1440-1446.

[8] Pioter P, Nieuwstadt F and Martien A. Experiments in turbulent pipe flow with polymer additives at maximum drag reduction.2001, Flow Turbulence and Combustion 66.2 , 159-182.

[9] Matthew W, Seunghyun B, Anthony J and Thomas J . Turbulent drag reduction of polyacrylamide solutions: effect of degradation on molecular weight distribution.2004, Journal of non-newtonian fluid mechanics123.2-3 , 175-183.

[10] Nam J, Sin K, Sang H, Kuan C Wongee C. Measurement of drag reduction in polymer added turbulent flow. 2009, International Communications in Heat and Mass Transfer 36.10 , 1014-1019.

[11] Wyatt, Nicholas B., Casey M. Gunther, and Matthew W. Liberatore. Drag reduction effectiveness of dilute and entangled xanthan in turbulent pipe flow.2011, Journal of Non-Newtonian Fluid Mechanics 166.1-2 , 25-31.

[12] Zadrazil I, Bismarck and G,Hewitt C. Shear layers in the turbulent pipe flow of drag reducing polymer solutions.2012, Chemical Engineering Science 72 , 142-154.

[13] Khadom, Anees A., and Ali A. Abdul-Hadi. Performance of polyacrylamide as drag reduction polymer of crude petroleum flow.2014, Ain Shams Engineering Journal 5.3, 861-865.

[14] Vural, Şerife Zeybek, Göknur Bayram, and Yusuf Uludağ. Experimental investigation of drag reduction effects of polymer additives on turbulent pipe flow using ultrasound Doppler velocimetry.2014, Turkish Journal of Chemistry 38.1 , 142-151.

[15] Pumisacho, L. Marylin, and Luis Fernando A. Azevedo. Turbulence Measurements in Pipe Flow With Drag Reducing Polymer Additives.2016, 14th International Conference on Nanochannels, Microchannels, and Minichannels. American Society of Mechanical Engineers.

[16] Salehudin, Siti Sarah, and Syahrir Ridha. Coconut Residue as Biopolymer Drag Reducer Agent in Water Injection System.2016. International Journal of Applied Engineering Research 11.13 , 8037.

IOP Conf. Series: Materials Science and Engineering **765** (2020) 012004 doi:10.1088/1757-899X/765/1/012004

[17] Danfu C, Chuanxian L, Hao L , and Fei Y . Effect of dispersing time on the prediction equation of drag reduction rate and its application in the short distance oil pipeline. 2018, Petroleum Science and Technology Journal, Vol.36, No. 16,pp.1312-1318.
[18] Alagarasi A. Chapter-Introduction To Nanomaterials. 2011.
[19] Yanuar, Sealtial M, Kurniawan T, Okky A and Rifqi H. Drag reduction of alumina nanofluid in spiral pipe with turbulent flow conditions.2017, AIP Conference Proceedings. Vol. 1826. No. 1. AIP Publishing.
[20] Akindoyo, E. O., and H. A. Abdulbari. Investigating the Drag Reduction Performance of Rigid Polymer–Carbon Nanotubes Complexes.2016, Journal of Applied Fluid Mechanics 9.3 , 1041-1049.
[21] Abdulbari, Hayder A., and F. L. W. Ming. Drag reduction properties of nanofluids in microchannels.2014, The Journal of Engineering Research , 12.2 , 60-67.
[22] Khadom, Anees A and Ali A. Performance of polyacrylamide as drag reduction polymer of crude petroleum flow.2014, Ain Shams Engineering Journal 5.3 , 861-865.
[23] Al-Kayiem, Hussain H., and Mohamed E. Ribbed double pipe heat exchanger: analytical analysis.2011, Journal of Engineering Science and Technology 6.1 , 39-49.
[24] Holland F, and Bragg R. Fluid flow for chemical and process engineers. Elsevier, 1995.

IOP Conf. Series: Materials Science and Engineering **765** (2020) 012005 doi:10.1088/1757-899X/765/1/012005

A mathematical model of the non-stationary filtration process for oil-supercritical CO2 system in a homogeneous porous medium under various modes of oil displacement

Radaev A.V. [1], Ali Kamil Sebur[2], Wisam Essmat Abdul-lateef[2], Hussain Abdulaziz Abrahem[3], Plokhotnikov S.P.[1], Salimyanov I.T.[1], Vaseneva N.A.[1], Sabirzyanov A.N.[1]

[1]kazan State Power Engineering University, Center for energy-saving equipment and technological innovation, Kazan, Russia, radaev_neftianik@mail.ru, belova-elena-n@mail.ru, inisal@yandex.ru, v-nadezhda@list.ru, sabirz@kstu.ru
[2]University of Technology – Department of Electromechanical Engineering, Iraq, 50278@uotechnology.edu.iq, 50110@uotechnology.edu.iq, 50008@uotechnology.edu.iq
E-mail: tejaratemrooz@gmail.com*

Abstract. The phase diagram of the system "transformer oil-supercritical CO2" is built. The modes of miscible and microbubble displacement of oil are determined. Oil was displaced by carbon dioxide in a supercritical state on isotherms 313 and 333 K in the pressure range up to 12 MPa at the experimental bench created by the authors. New data have been obtained on the coefficient of oil displacement by supercritical CO2 from a model of a homogeneous oil reservoir in a miscible and micro-bubble mode of supercritical CO2 filtration. It was revealed that the miscible regime of oil displacement by supercritical CO2 has the greatest technological efficiency. For the region of a mixed filtration regime, a mathematical model was created and solved for the process of filtering the unsteady flow of "supercritical CO2-oil" in a homogeneous terrigenous porous medium in the pressure range up to 12 MPa, temperatures up to 333 K, the calculated curve agrees with the experimental data on the oil displacement coefficient. The convergence and stability of the difference scheme of the computational process is ensured. It was found that in the transition region of supercritical CO2 filtration on the boundary curve in the phase diagram and in the region of microbubble filtration of supercritical CO2, the process of oil displacement is more intense, which is associated with the appearance of a supercritical sublayer on the pore surface, which leads to slipping of the fluid inside the pore. For the transition-filtering region on the boundary curve in the phase diagram and in the region of the microbubble displacement of oil by supercritical carbon dioxide, a method for increasing technological efficiency is proposed based on the injection of supercritical CO2 and water rims to prevent the transition of microbubble carbon dioxide to the free phase.

Keywords: Non-stationary multicomponent filtration, oil displacement coefficient, residual oil saturation coefficient, micro-bubble supercritical carbon dioxide filtration mode

IOP Conf. Series: Materials Science and Engineering **765** (2020) 012005 doi:10.1088/1757-899X/765/1/012005

1.Introduction

Since substances in a supercritical state are characterized by intermediate properties between a liquid and a gas—high density and low kinematic viscosity—supercritical fluids are currently one of the promising oil displacing agents. The high technological efficiency of using supercritical fluid systems for displacing viscous oils is due to their high dissolving and penetrating ability. Increasing the solubility of carbon dioxide transfers the oil-gas / supercritical fluid system into a single-phase state with the disappearance of the phase boundary and, as a result, to the regime of miscible oil displacement. For low viscosity oils, this means that in the process of its displacement, the existence of a mode of miscible displacement is possible as the most optimal from the point of view of material costs. Achieving a miscible displacement mode means that a single-phase region forms at the displacement front, and the phase boundary disappears, i.e., the transition from the region of supercritical (SC) CO2 to the region of the liquid phase occurs monotonously. The study of a miscible displacement regime by various gases in our country has been conducted [1,2]. In all the cited works, the results of experimental studies of oil displacement by carbon dioxide at various state parameters are presented along with triple diagrams of the composition of the oil-carbon dioxide system, which makes the identification of the mechanism of oil displacement possible. Currently, to describe the process of non-stationary filtration of a multicomponent flow in a miscible mode, models that describe compressible / incompressible fluid filtration, carbonated fluid filtration [3–5] are used, which form the scientific basis of secondary methods for increasing oil recovery (flow rejection, steam and thermal exposure, and so on.). The latter are characterized by the presence of threshold restrictions on the viscosity of reservoir oil, the permeability of a porous medium, and so on.To describe tertiary oil recovery enhancement methods, it is necessary to develop mathematical models based on experimental and calculated data on phase permeabilities, viscosity of reservoir systems, and so on. In addition, for the correct choice of a method for calculating the dynamic viscosity of gas-saturated oil and its phase permeability, data on phase equilibria in the oil-supercritical fluid system are absolutely necessary, which make it possible to unambiguously determine in which region of the phase diagram is the system currently located. The construction of such models is aimed at improving the methods for developing hard-to-recover oil deposits.

2.Mathematical model

A mathematical model of the process of unsteady isothermal filtration of an oil-supercritical CO2 stream is as follows.

We introduce the dimensionless quantity, which we define as:

$$x = \ln(\frac{r}{r_w})$$

The notation used in the development of this mathematical model is shown in Fig.1

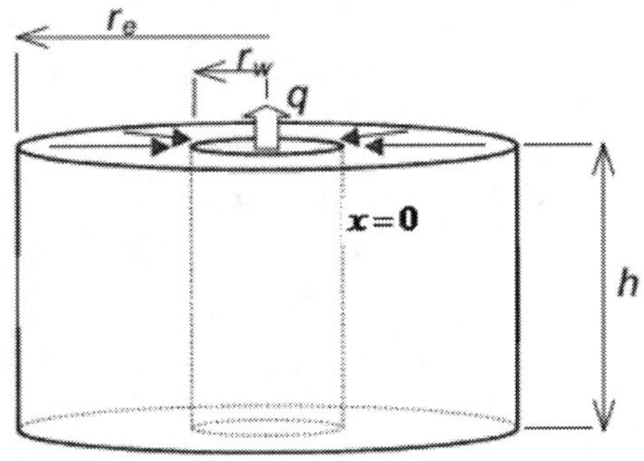

Fig. 1. Scheme of fluid motion in a porous medium.

This paper describes the process of non-stationary filtration of two phases – oil and the phase "oil-supercritical CO2". The physical description of the process is as follows: one-dimensional isothermal, non-inertial filtration. There are no mass transfer processes between the porous medium and its components; the phase movement obeys Darcy's law, the porosity of the rock skeleton is constant, the capillary pressure at the interface is absent, i.e.

The continuity equations for the oil phase:

$$\frac{1}{r_w^2} e^{-2x} \frac{\partial}{\partial x}\left(k \frac{k_{ro}}{B_o \mu_o} \frac{\partial p_o}{\partial x}\right) = \frac{\partial}{\partial t}\left(\frac{\phi S_o}{B_o}\right) \tag{1}$$

The continuity equations for the oil-supercritical CO2 phase:

$$\frac{1}{r_w^2} e^{-2x} \frac{\partial}{\partial x}\left(k\left(\frac{k_{rg}}{B_g \cdot \mu_g} \cdot \frac{\partial p_g}{\partial x} + \frac{R_{s0}}{B_0 \cdot \mu_0} \cdot \frac{\partial p_0}{\partial x}\right)\right) = \frac{\partial}{\partial t}\left[\phi\left(\frac{S_g}{B_g} + \frac{R_{s0} \cdot S_0}{B_0}\right)\right] \tag{2}$$

ϕ : porosity of the reservoir model;

B_g , B_o : volumetric coefficient of carbon dioxide and oil

k , k_{ro} , k_{rg} : the coefficient of absolute permeability of the porous medium and phase permeability of the oil phase and the oil – CK CO2 phase, respectively.

R_{so} : solubility of CK CO2 in oil, determined by the equation in [6].

S_g : gas saturation of a porous medium

μ_0 , μ_g : dynamic viscosity of gas-saturated oil and carbon dioxide;
The conditions of uniqueness are determined by the initial and boundary conditions and close the system of equations. The initial conditions are specified by the distribution of saturation at the initial moment of time:

IOP Conf. Series: Materials Science and Engineering **765** (2020) 012005 doi:10.1088/1757-899X/765/1/012005

$$S_0(r,t)\big|_{t=0} = S_{0вх}$$

P_{0IN} , S_{0IN} : accordingly, pressure and oil saturation at the entrance to the reservoir model.

The boundary conditions are specified as follows (displacement at a given flow rate).
At the entrance:

$$Q_t = Q_0 + Q_g = 2\pi hk \cdot \frac{k_{r0}}{\mu_0} \frac{\partial p_0}{\partial x}\bigg|_{x=\ln(\frac{r}{r_w})} \tag{3}$$

The mathematical model is closed by the expression for the saturations of the porous medium during two-component filtration.

$$S_g = 1 - S_0 \tag{4}$$

$$k = k^*_{ro}\left(\frac{S_0 - S_{0r}}{1 - S_{0r} - S_{gc}}\right)^{n_0} \tag{5}$$

$$k_{rg} = k^*_{rg}\left(\frac{1 - S_0 - S_{gc}}{1 - S_{0r} - S_{gc}}\right)^{n_g} \tag{6}$$

Where:

n0=1,654; ng=1,96; $k^*_{r0} = 0{,}92$ $k^*_{rg} = 0{,}272$

where the letter r denotes the residual saturation; the letter c is the critical saturation, 0; and g is oil and gas, respectively.

3. Results:

The above mathematical model describes the mode of miscible filtration (there is no capillary pressure at the phase boundary "transformer supercritical CO2 oil") shown in the phase diagram at point 4. The phase diagram of the transformer-supercritical CO2 oil system is shown in Fig. 2. It is a graphical representation of the states of a thermodynamic system in the space of the main state parameters: temperature T, pressure p.

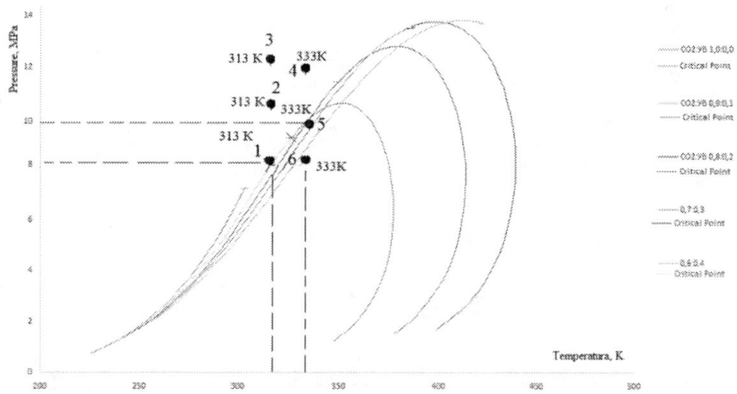

Fig. 2. Phase diagram of the system transformer oil - supercritical CO2.

The testing of the mathematical models oil displacement by supercritical CO2 was carried out as a result of numerical modeling in the licensed software product DELPHI. Comparisons with the results of experimental studies of the authors [7] were also conducted. The experimental conditions are shown table 1.

Table 1. Experimental conditions.

Experiment parameter	The designations	3	Values
Carbon dioxide injection pressure	MPa	121	12
Тем, Experiment temperature	K	313	333
Reservoir model length	m	0,7	0,7

The model was solved numerically by the IMPES method (implicitly by pressure, obviously by saturation). The results of numerical simulation are shown in Fig. 3

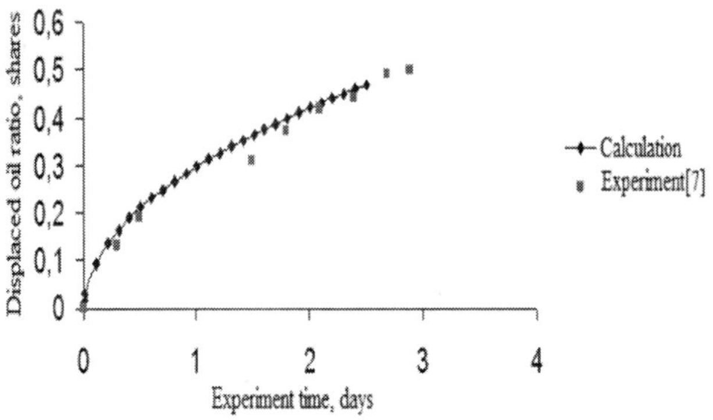

Fig. 3. Numerical simulation of transformer oil displacement by supercritical carbon dioxide

From the analysis of Fig. 3, it follows that the mathematical model developed by the authors of the process of non-stationary filtration of the oil-supercritical fluid flow describes the process of oil displacement by supercritical CO2 well; the maximum deviation does not exceed 9%. This is probably due to the fact that the model takes into account the processes occurring in the oil reservoir in a miscible displacement mode, primarily the dissolution of carbon dioxide in the reservoir oil. The convergence and stability of the difference scheme of the computational process used in this work is ensured by the analysis and carried out in [8]. The stability of the implicit calculation scheme by the IMPES method is ensured by the fulfillment of the condition:

$$\frac{\Delta t}{\Delta x} << 1 \qquad (7)$$

Where:

$\Delta t , \Delta x$ accordingly, a step in time and space.

4.Conclusions:

1. A model of the process of unsteady filtration of a multicomponent oil-water-supercritical fluid flow in a homogeneous and inhomogeneous porous medium in the mode of a mixing, transitional and microbubble flow regime with a viscous supercritical sublayer describing the phase transitions of the first kind of supercritical CO2 in temperature and pressure ranges developed up to 333 K14 MPa respectively.

2. To solve the model numerically, a methodology was developed for determining the dynamic viscosity and phase permeability of the oil-supercritical CO2 system in a wide temperature range using phase diagrams of the oil-supercritical CO2 system.

3. The model is numerically solved by the IMPES method in the licensed DELPHI software product in the pressure range up to 16 MPa and temperatures up to 333 K.

IOP Conf. Series: Materials Science and Engineering **765** (2020) 012005 doi:10.1088/1757-899X/765/1/012005

4. The mathematical model developed by the authors for the process of non-stationary filtration of the oil-supercritical fluid flows correctly and qualitatively and quantitatively describes the process of oil displacement by supercritical CO2 in the mode of miscible displacement; the maximum deviation does not exceed 5.1%.

5. References:

[1] Jousse T.P. The role of compressed gases as solvents.- M: Nedra, 1981.- 165 p.

[2] Stepanova G. S. Gas and water-gas methods of influence on oil reservoirs. M .: "Gazoil press", 2006.- 202 p.

[3] Aziz X, Settari E. Mathematical modeling of reservoir systems.- M: Nedra, 1982.-407 p .;

[4] Barenblatt G.I., Entov V.M., Ryzhik V.M. Theory of non-stationary filtration of liquid and gas. - M.: Nedra, 1972.- 288 p .;

[5] Guidelines for the creation of permanent geological and technological models of oil and gas and oil fields (Part 1. Geological models) .- M: VNIIOENG, 2003.- 164 p.

[6] Brusilovsky A.I. Phase transformations in the development of oil and gas fields. - M: Grail, 2002.- 575 s;

[7] Radaev A.V. Investigation of the optimal volumes of injection of supercritical carbon dioxide into a homogeneous porous medium at various degrees of water cut / N.R. Batrakov, I.D. Zakiev, R.R. Galimzyanov, A.A. Mukhamadiev, A.N. Sabirzyanov // Bulletin of Kazan Technological University, 2013.-T. 16.- No. 24.- S. 45-47.

[8] Gabriela B. Savioli, M. Susana Bidner Simulation of the oil and gas flow toward a well — A stability analysis // G.B. Savioli, M.S. Bidner / Journal of Petroleum Science and Engineering 48 (2005) 53–69.

Improvement of mechanical properties of Oil well cement by incorporate Nano-$CaCO_3$ prepared from eggshell waste

Ali Dawood Salman[1], Juzsakova Tatjána[1], Mohammad A. Al-Mayyahi [2], Raheek I. Ibrahim*[3], Thamer Adnan Abdullah[1], Eman H. Khader[4]
[1]Institute of Environmental Engineering, University of Pannonia, Veszprém, Hungary
[2] Department of Chemical and Petroleum Refining Engineering /College of Oil and Gas Engineering Basra University for Oil and Gas
[3]Electromechanical Engineering Department/ University of Technology/ Baghdad- Iraq
[4]Chemical Engineering Department, University of Technology, Iraq

*Raheek I. Ibrahim 80058@uotechnology.edu.iq

Abstract

The objectives of this work were to prepare, characterize the $CaCO_3$ nanoparticles as admixture derived from eggshell waste to improve the mechanical properties and performance of oil well cement (OWC) and to evaluate the interactions regarding the cement-admixture compatibility. The results suggest to use the combination of ultrafine grinding (ball milling) and heating treatment. The evaluation of the effectiveness of the addition of eggshell nanoparticles (ESNP) on the mechanical properties (compressive strength) of OWC was also an important objective. The high purity of eggshell powder was milled for 5–30 h to prepare ESNP with an average particle size of 40 nm after 25 h. The physico-chemical properties of the ESNP were characterized by different techniques including particle size analyzer (PSA), X-ray diffraction (XRD), X-ray fluorescence (XRF), scanning electron microscopy (SEM) and the Brunauer-Emmett-Teller (BET) methods. The ESPN was added in three different percentages to the OWC (2%, 6% and 10% as partial replacement of cement) and a w/c of 0.44. It was also revealed that the best percentage of ESNP addition not only resulted in much denser microstructure in OWC but also changed the formation of hydration products through presence of $CaCO_3$ nanoparticles, which increased the rate of reaction of tricalcium aluminate (C3A) to form a carboaluminate complex. In this way the rate of hydration increased, the total hydration products increased and consequently the strength as well. Hence the addition of ESNP contributed to the improvement of early-age compressive strength, microstructure and durability properties of cement. According to the results, it was found that the optimal addition of ESNP to OWC is 6%.

Keywords: Eggshell nanoparticles, Oil well cement, Ball milling, Mechanical properties,

1. Introduction:

The eggshell (ES) is one of the most common biomaterials in nature. It is a very interesting material for potential waste treatment, because it is a by-product in food industry and after the production of eggs it is considered waste **[1]**. Chicken eggshell (ES) is an agricultural byproduct that has been listed worldwide as one of the environmental issues. The chicken eggshell wastes were applied as raw materials for the preparation of nano-calcite ($CaCO_3$) **[2]**. In recent years, limited studies have been conducted on the additions of nano-$CaCO_3$ as partial replacement of cement in concrete on the hydration and compressive strength modifying. Currently, the commercial additives were imported and the price is kept increasing year by year **[3]**. Therefore, the researchers were continuously looking for potential additives such as nanoparticle to improve the cement properties Although the use of $CaCO_3$ was first considered as filler to partially replace the cement, studies have shown some advantages of using $CaCO_3$ in terms of strength gain, reaction rate accelerating effect and economic benefits as compared to cement and other supplementary cementitious materials **[4,5]**. Chemically, the presence of $CaCO_3$ nanoparticles increase the rate of reaction of tricalcium aluminate (C3A) to form carboaluminate complex thereby increase the total hydration products, delay the formation of micro-cracks, the total hydration products and consequently the strength **[6]**. In addition, it also reacts with tricalcium silicate (C3S) and accelerates setting and early strength development as shown in **Fig.1** **[7]**. Physically, **$CaCO_3$** nanoparticles filling the voids and pores in cement with the small size nanoparticles thus enhancing the strength and minimizing the porosity of the cement. In terms of durability properties, it was revealed that replacement of cement with limestone powder ($CaCO_3$) had significant effect on the resistance of sulphate attack and water absorption which is related to the filler effect, heterogeneous nucleation and the dilution effect of limestone powder (Ramezanianpour 2010) **[8,9]**. Elsewhere, Sato and Beaudoin (2011) carried out an investigation on the incorporation of micro- and nano-$CaCO_3$ with high volume of supplementary cementitious materials. In that experiment, cement was replaced with 50% fly ash and 50% slag and was incorporated with10 and 20% of micro- and nano-$CaCO_3$ by weight of the binders **[10]**. It was found that the replacement of cement with nano-$CaCO_3$ accelerated the early hydration of cement and enhanced the early development of modulus of elasticity as the amount of nano-$CaCO_3$ was increased. The presence of nano-$CaCO_3$ particles has been suggested to have a significant effect on the hydration kinetics of C3A and C3S which may cause acceleration of setting and early strength development **[11]**. Another study, Sato and Diallo (2010) reported the seeding effect of nano-$CaCO_3$ where the rapid growth of CSH was obtained on the surface of the C3S particles. This view is supported by Kawashima *et al.* (2013) who provided a basis for understanding the mechanical properties of high-volume fly ash when incorporated with nano-$CaCO_3$. It was shown that the addition of 5% nano-$CaCO_3$ with 30% fly ash-cement paste samples tested at 1, 3 and 7 days showed progressive development in compressive strength compared to control fly ash- cement paste **[12,13]**. Oil well cement (OWC), a kind of special cement-based materials, have relatively high compressive strengths but sensitive to tensile stresses inevitably. In the petroleum industry, OWC is typically utilized to fill the annular space between the pipe and rock formation, displaces the drilling fluids, support the casing and protect it against corrosion and impact loading, restricts the movement of fluids between formations, and isolates the productive and none productive zones **[14,15]**. The cement must withstand higher pressures and also resist more corrosive fluid attacks encountered in down hole applications **[16]**. Therefore, to improve the strength while maintaining higher ductility is a challenging task for possible application of OWC. According to the American Petroleum Institute Specification for materials and testing for well cements (API Specification 10A,

2002) nine special classes of cements were established (Class A–Class J) [**17**]. Also OWCs are classified into three grades based upon their C_3A contents [**18**]. At present, a common approach to reduce the brittleness is blending with various reinforcing additives such as fibers, nano-sized particles of different types of cement admixtures which enhance OWC slurry properties, to achieve successful placement and rapid compressive strength development for adequate zonal isolation during the lifetime of the well and enhance mechanical properties by decreasing the micro-porosity of nano-sized particles and reducing the crack propagation [**19,20**]. Therefore, the aims of this study was to prepare and evaluate the effects of ESNP as a novel reinforcing additive on mechanical properties development and microstructure of OWC.

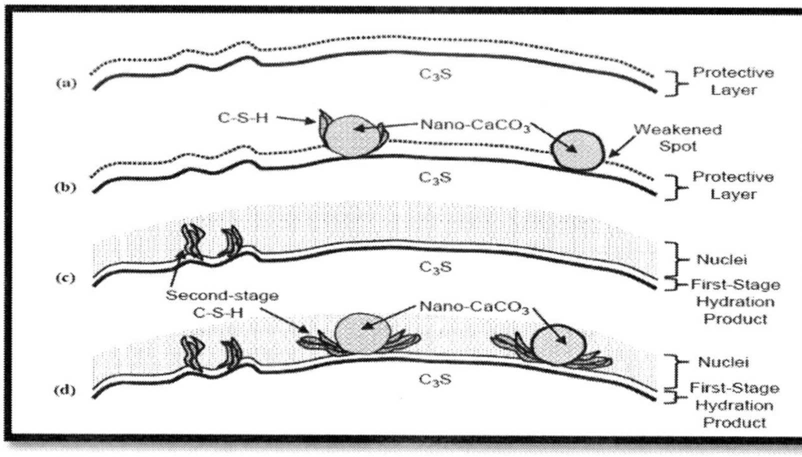

Fig.1 Schematic description of accelerating effect of addition of nano-$CaCO_3$ on hydration of C3S.

2. Materials and methods

2.1. Preparation of eggshells nanoparticles (ESNP)

In this study raw eggshell was collected from a kitchen home and immediately, were rinsed with clean water to remove the residue egg contains that attached on the egg shells. After that, it was dried in an oven at 70 °C for 30 min. to remove water content. The clean eggshells were pretreated by home grinder to obtain eggshell micro-particles (ESMP). The powder obtained were sieved to 300 μm sieve using a Sieve Shaker, micro -particles were then placed into the ball mill simultaneously with the grinding balls for a period of 5 hr., then powders was sent to mesh for the size lower than 300 μm in order to remove severe clusters that originated during milling process. The fraction was dried again in an oven at 70 °C for 30 min as shown in **Fig.2.** which represents the practical steps of preparation ball-milled eggshells nanoparticles. After 5 hour-milling; the ball-milled eggshells were tested using particle size analyzer (Malvern Instruments Master size 2000) to detect the particle size distribution. The meshing, drying and (PSA) are frequently done in every 5 h of milling while attaining the total time of 30 h. The mesh sizes employed were 600 μm, 300 μm, 140 μm, and 100 μm. The resulting powder was milled in a ball-milling machine (GW, Korea) for 5, 10, 15, 20, 25 and 30 h at a speed of 170 rpm. The crystalline phases of the ball-milled powders before and after the mechanical treatment were analyzed by using X-ray diffraction (XRD-6000, Shimadzu). The microstructure and morphology of the powders were studied using scanning electron microscope (SEM; JSM-6700F, JEOL, Japan). The composition of the prepared powders was analyzed by X-ray fluorescence spectrometry (Shimadzu XRF-1800). The chemical compositions and physical properties of eggshell nanoparticles of ESNP are summarized in **Table 1.**

Fig. 2. Practical steps of preparation ball-milled eggshells nanoparticles.

2.2. Mixing and preparing cement slurry of (OWC)

Cement slurries used in this study were prepared using oil well cement API Class G with a specific gravity of 3.14. Deionized distilled water was used for the mixing with a *w/c* = 0.44, and its temperature was maintained at 23±1°C using an isothermal container. The cement slurries were prepared using a high-shear blender type mixer running at a speed of 4000 r/min ±250 r/min. with bottom driven blades according to the following procedure. First, the weighted amount of cement and the solid admixture (ESNP) in three concentrations (2% ,6% and 10%) were dry mixed in a bowl by hand using a spatula for about 30 sec. The mixing water was subsequently poured into the blender. The mixing started at a slow speed for 20 sec. Manual mixing was conducted for 20 sec and a rubber spatula was used to remove material sticking to the wall of the mixing container to ensure homogeneity. Finally, mixing resumed for another 35 sec at high speed. This mixing procedure was strictly followed for all cement slurries. All mixing was conducted at a controlled ambient room temperature of 23±1°C. All the abbreviations are shown in **Table 1**. The chemical compositions and physical properties of Oil well cement API Class G, and eggshell nanoparticles are given in **Table2** A and B.

Table 1. Abbreviations

Symbols	Meaning
ESNP	Eggshell nanoparticles
ESNP2(40)	Eggshell nanoparticles according to the particle size analyzer for this group, higher v % particle size than 40nm with 2% nanoparticles.
OWC	Oil Well Cement
CM	Control mixture
C–S–H	Calcium silicate hydrate
w/c	water-to-cementitious materials ratio

Table2. The chemical compositions and physical properties of OWC and ESNP.

A	
Chemical Component (%)	
Oxides	**Wt.%**
CaO	64.20
SiO_2	19.40
Fe_2O_3	5.50
Al_2O_3	4.50
SO_3	2.80
MgO	2.00
Na_2O	0.10
K_2O	0.60
C2S	15

C_3S	61.33
C3A	1.72
C3AC$_4$AF +2 C_3A	21.8
Loss on ignition	0.49
Density (g/cm^3)	3.15

B	
Chemical Component (%)	
Eggshells nanoparticles (ESNP)	
CaO	97.8
MgO	0.5
P_2O_5	0.35
SO_3	0.19
Na_2O	0.14
SrO	0.02
Fe_2O_3	0.02
Average particle size	40nm
Specific surface area	45m^2/g
Color	whit

2.3. Specimen preparation, curing, and testing

The specification gives the standard procedure for sample condition prior testing. The slurries prepared according to API RP (10B-2) standard were placed in the moulds sized (2in x 2in x 2in) in a layer equal to half of mold depth and puddle for 25 times per specimen with pudding rod. The prepared slurries were placed in all specimen compartments before commenced the pudding operation. After pudding the layer, the remaining slurries was stirred using pudding rod and strike off the excess slurries on the mould top using a straight edge. The dry cover plate was placed on the mould top before the sample was ready for curing process. **Table 3.** Shows mixing proportion of prepared OWC samples. The specifications of equipment's used during this work have been given in **Table 4.**

Table 3. Mixing proportion of cement samples.

Mixes	**Mix proportion (Wt.%)**		**w/c**
	OWC	**ESNP**	
NM0	100	0	0.44
NM3	97	3	0.44
NM6	94	6	0.44
NM10	90	10	0.44

Table 4. Specifications of equipment's used.

Equipment	**Specification**
Ball mill	(GW, Korea) 250 rpm
Mechanical Shaker, Heidolph	UNIMAX 1010/PROMAX 1020, Germany
Ultrasonic	750 Watt Ultrasonic Processors – VCX Series
Electrical furnace	Hanna 211, Romania Electrical furnace (20–500) ^{o}C
X-ray fluorescence spectrometry	Shimadzu 1800
Mixer for preparation cement slurry	High speed mixer – 4000 rpm
Particle Size Analyzers (PSA)	Malvern Mastersizer 2000
Brunauer, Emmett and Teller (BET) method by surface area analyzer	Q Surf 1600, USA
Atomic Force Microscope (AFM)	Angstrom, Scanning Probe Microscope, Advanced Inc., AA 3000A, USA

IOP Conf. Series: Materials Science and Engineering **765** (2020) 012006 doi:10.1088/1757-899X/765/1/012006

Scanning electron microscope (SEM)	(SEM; JSM-6700F, JEOL, Japan)
X-ray diffraction XRD-6000	Shimadzu scan range is (10.0–60.0) deg.

3.Results and discussion

3.1 Characterization of ball-milled powders ESNP

As presented in **Table1**, the chemical composition of the ESNP from the X-ray fluorescence (XRF) shows that calcium oxide (CaO) was the most abundant component (97.8 %). The high amount of (CaO) is associated to the presence of calcium carbonate, which is the main component of eggshell. Thus, the eggshell waste sample can be considered from a chemical viewpoint a pure relatively natural carbonate-based material, as well as its composition is very similar to that of calcitic calcareous. Scanning electron microscope images of eggshell micro-particles (ESMP) and eggshell nanoparticles (ESNP) are presented in **Fig.3**.The SEM images shown in **Fig. 3A** illustrate morphology, texture, size and distribution of the ball-milled eggshell (ESMP) after 5 h of milling. It can be seen that some of ESMP particulates which have fused together resulted from constricting action of the ball-milling process were irregular in shape and their surface was rough. Furthermore, the diameter varied greatly, ranging from 0.7 μm to 10 μm. Conversely, the SEM images shown in **Fig. 3B** demonstrate ball-milled eggshell nanoparticles after 25 h of milling, different structures of the particles including oval and irregular shaped particles as well as agglomerates are due to constricting action of the ball-milling process with sizes ranging from 40 nm to 110 nm. As it can be visually perceived from the results the ESNP is not homogeneous and several primary particles seem to cluster or fuse at their faces. This tendency for agglomeration in the particles is induced by the Van-der-Waals forces acting between the individual particles. The purity and crystallinity of ESMP and ESNP were analyzed using X-ray diffraction (XRD) as illustrated in **Fig. 4A** and **B** which demonstrate the structural changes and degradation of the ball-milled eggshell after mechanical treatment for 25 hr., compared with mechanical treatment for 5 hr. XRD profile in **Fig. 4A** shows that, the pattern of calcite ($CaCO_3$), and demonstrates the presence of a large amount of highly crystalline calcite. The (012) and (104) basal reflections, initially are very strong, a decrease in intensity after bash is shown in **Fig. 4B**. The reflections (110), (113) and (202) decrease too, but are different from others after mechanical treatment for 25 h. This confirms the structural changes in the treated calcite (ESNP) compared to the ESMP. The distortion of structure is reflected in the line vacillation, in the decrease of peak intensity (area) and in the shifting of reflections. The losses of the crystallinity in the basal diffractions are due to the amorphization phenomenon. **The eggshell** powder obtained from mechanical treatment were verified by particle size analyzer instruments to evaluate particle size and size distribution during milling process. According to PSA and AFM results the eggshell powder has successfully been reduced the particle size to 40 nm within 25 hr. The size distribution of particles was in six main different size categories as shown in **Table 5, Fig.5** and **Fig.6.** Based on the milling pattern, it can be concluded that the mineral is brittle and can be characterized by reduction in particle size at first and then by agglomeration (clumping) on increasing the milling time. Agglomeration is the result of Van der Waals forces between particles at the surface.

IOP Conf. Series: Materials Science and Engineering **765** (2020) 012006 doi:10.1088/1757-899X/765/1/012006

Table 5 The average particle size of milled eggshell with milling time.

Groups	Milling Time (h)	Particle size group	Average particle size (nm)
ESNP1	5	(0.7-20) µm	500
ESNP2	10	(200-350) nm	170
ESNP3	15	(130-170) nm	90
ESNP4	20	(90-120) nm	60
ESNP5	25	(40-110) nm	40
ESNP6	30	(50-150) nm	70

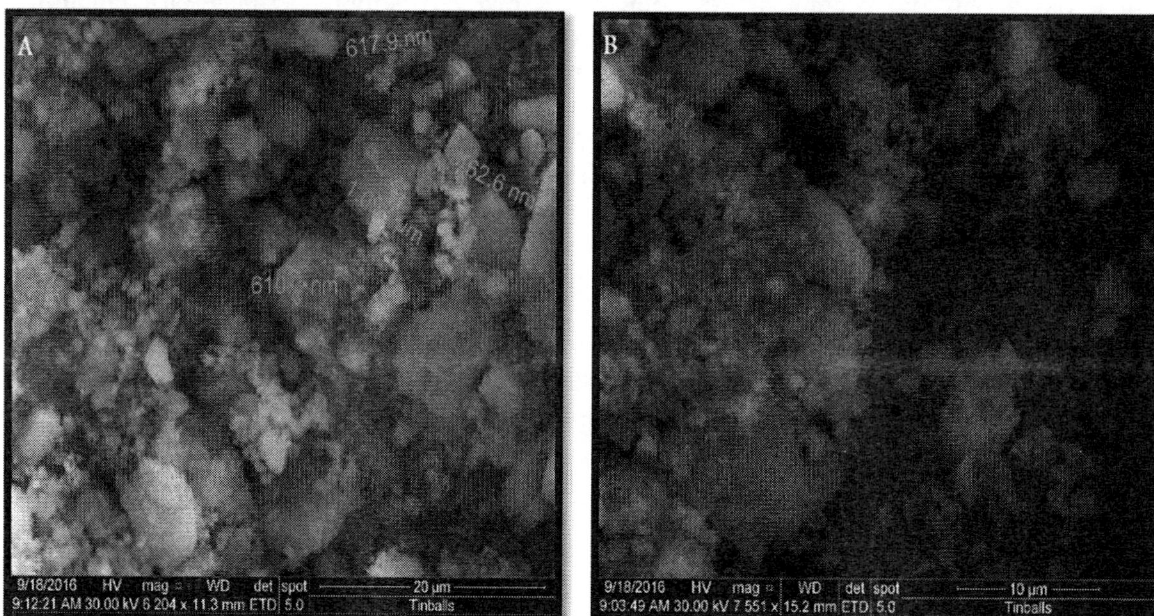

Fig. 3. SEM images of (A) eggshell microparticles (B) eggshell nanoparticles.

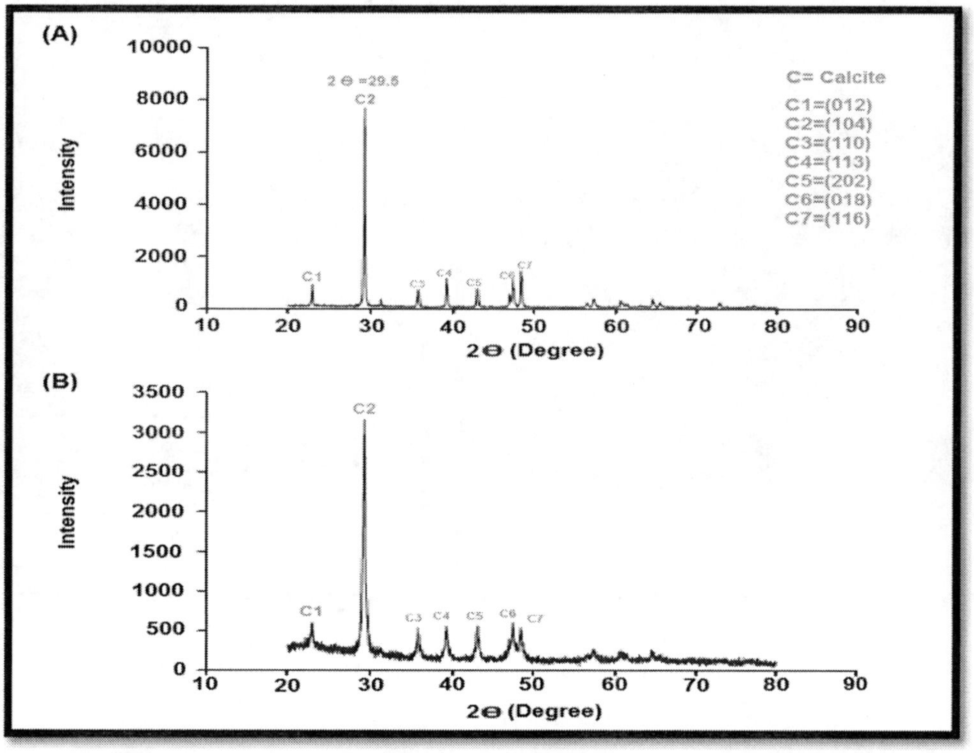

Fig.4. X-ray diffraction patterns of (A) eggshells micro-powder and (B) eggshells nano-powder.

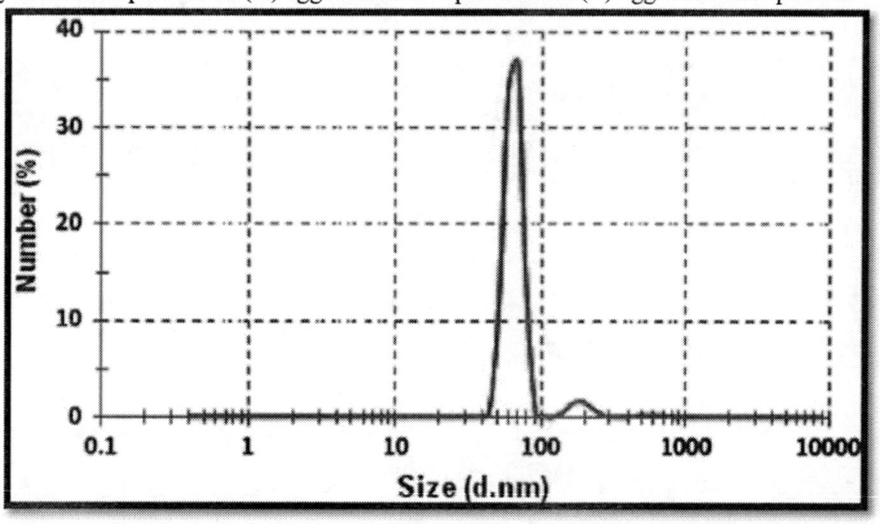

Fig. 5. Particle size analyzer of prepared eggshell nanoparticles after 25h of milling.

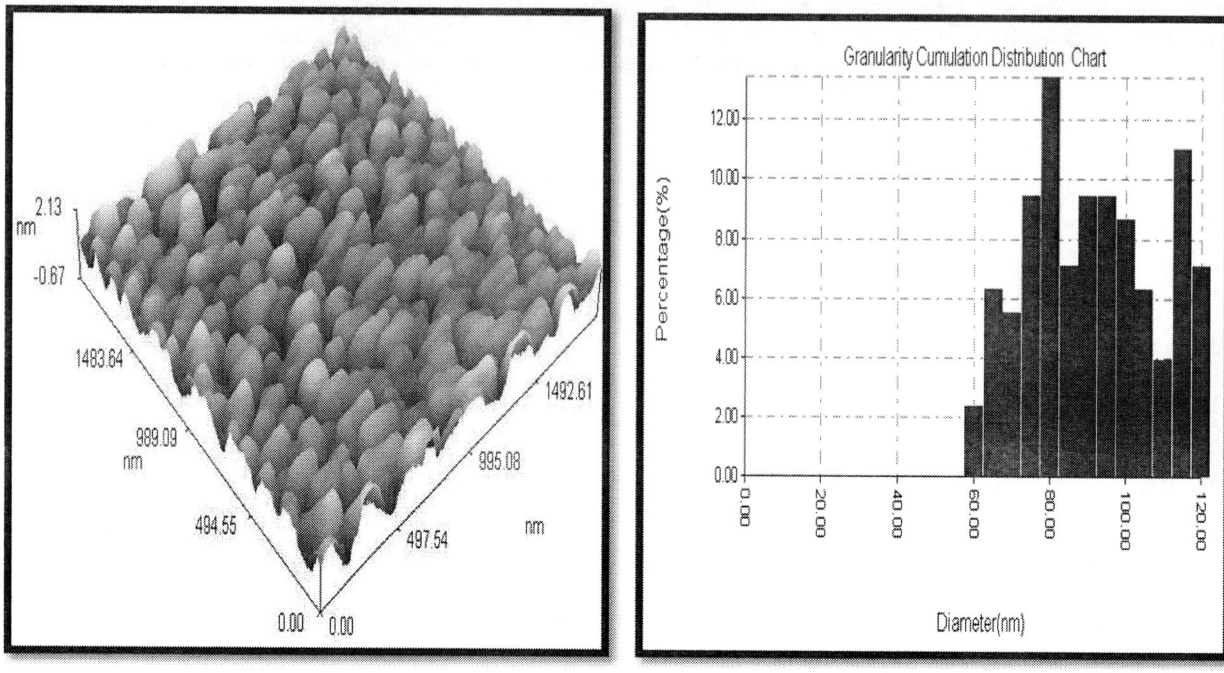

Fig. 6. AFM of ball-milled eggshell (ESNP) after mechanical treatment for 20 h.

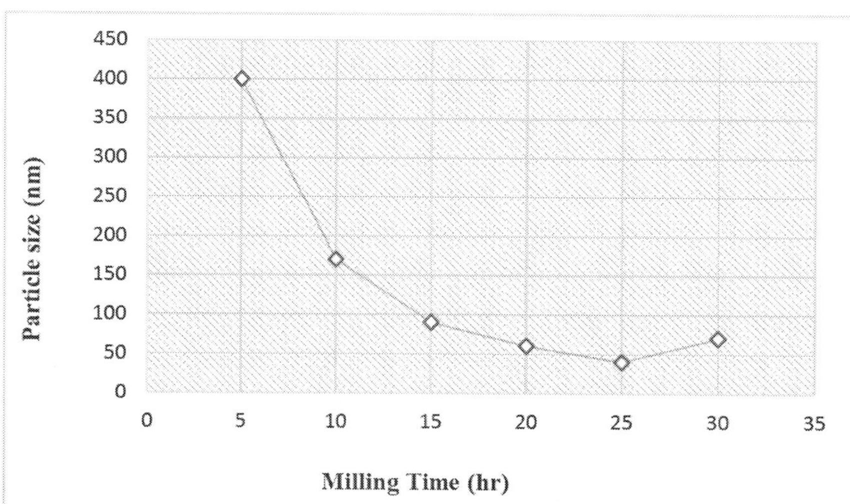

Fig.7 Average particle size of eggshell samples at different milling times according to particle size analyzer.

3.2. Characterization and analysis of OWC

3.2.1. SEM observations and microstructural analysis of control mixture

Fig. 8A shows the SEM micrographs of the OWC control mixture (CM) at the age of 8 h. at 60 °C and demonstrates the porous structure which is full of large size pores and the presence of $Ca(OH)_2$ or (CH) that is over-shadowed. Also, it can be seen from the same figures that the CH crystals are connected to the C–S–H gel which indicates that the hydration process is not completed and also explains the low records of compressive strength for the control mixtures. Also, the same photos show that the concentration of the CH is higher than the C–S–H gel concentration and that the CH hydrate needles cover a large area.

3.2.2 SEM observations and microstructure analysis of modified OWC
The SEM images of the OWC mixture after incorporating 6% of ESNP are shown in **Fig.8B** that was prepared with nominal particle size group of 40-110 nm. The images show that the microstructure of the mixture after incorporating ESNP at 8h and 60 ^{0}C is dense and more organized with a small number of $Ca(OH)_2$ crystals and small sized pores while in the control mixture the C–S–H gel existed in the form of clusters lapped and joined together by many CH needles hydrates. It can also be noticed from the reference figure that the CH needles are visible and there is a compact structure with the absence of the unhydrated crystals and voids. The structure is more uniform and homogeneous than that of the CM sample which explains the superior compressive strength results. This could be due to the high activity of many particles that promote the pozzolanic reaction to produce more C–S–H gel in order to reach high compressive strength at an early age which is confirmed by the strength results. The $CaCO_3$ nanoparticles consumes calcium hydroxide crystals, fills pores to increase the strength, reduces the size of the crystals at the interface zone and transmutes the calcium hydroxide feeble crystals to the C–S–H crystals, and improves the interfacial zone and cement paste structures.

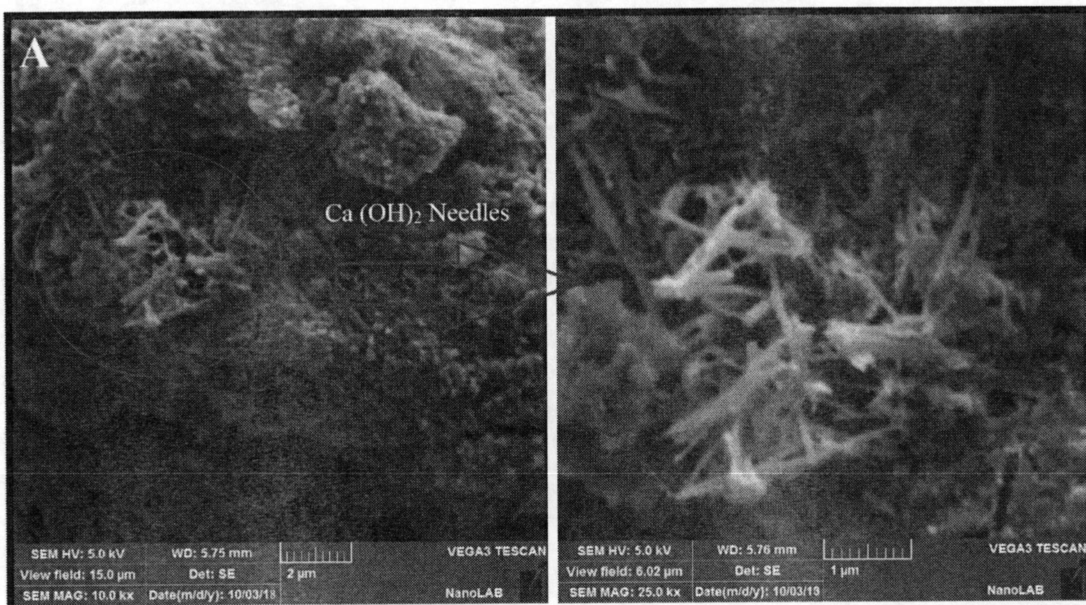

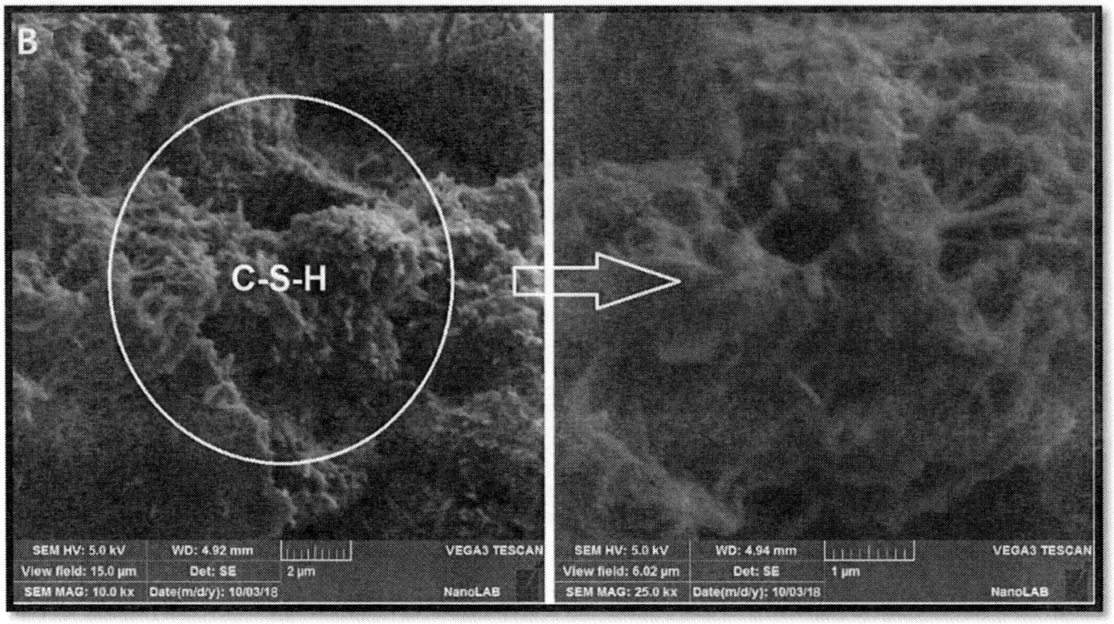

Fig. 8. SEM images of OWC (A), before incorporating NP, and (B) after incorporating NP at 8hr. at 60 ^{0}C.

3.2.3 X-ray diffraction of OWC

XRD analyses were conducted to investigate the activity and potential of incorporating ESNPs on selective OWC specimens after 8h of hydration. The samples were cured for 8h of hydration and 60 ^{0}C before being subjected to XRD technique. It is obvious from XRD profiles that the CH peak is almost decreased with the addition of NP, while the same is significantly present in the control mixture. It is therefore inferred from **Fig. 9A** and **B** that, the NPs reacts with CH produced during the hydration process. The pozzolanic reactivity of NP at early stage of hydration is significantly high and improves the microstructure of cementitious system, thereby enhancing the mechanical properties of the cementitious materials. For comparison, the peaks of CH at 2Θ =18 and 47 have been selected as shown in **Fig. 9A**. A sharp peak in CH is observed in the control mixture representing the pure hydration product (CH), which is released from the hydration of cement. Evidently, the intensity of the CH peak is decreased due to adding NPs as a cement replacement, which reflects the consumption of CH by pozzolanic reaction as shown in **Fig. 9B**. In agreement with the SEM results, which explain the absence of the unhydrated crystals (needles) and voids, in addition, more uniform and homogeneous mixtures have been obtained than from the CM sample.

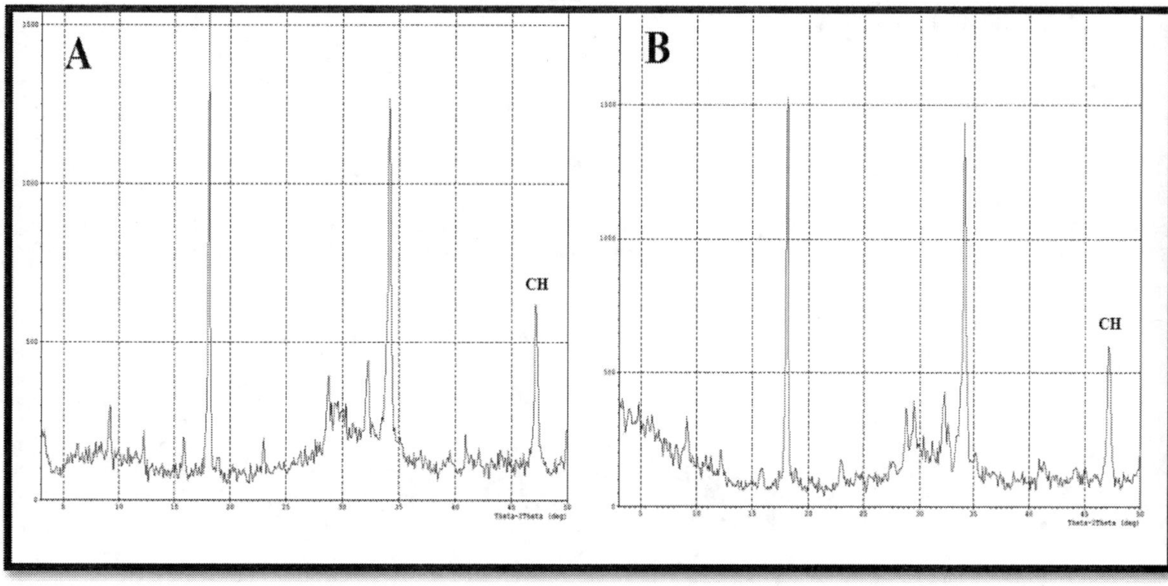

Fig. 9. XRD profile of OWC, (A) CH peaks before adding NP and (B) after adding NP.

3.3 Compressive strength development

All specimens were prepared with minimum particle size group (40-110 nm) due to impact of the particle size on strength development. The minimum particle size of NPs increases the packing efficiency decreases leading to a drop in the compressive strength.

3.3.1 Effect of ESNP dosage on compressive strength

Compressive strength development of the OWC incorporating different dosages of ESNP is given in **Table 6**. In comparison with that of CM, the modified OWC with NPs had exhibited higher compressive strengths. However, the strength of the samples increased with the increase in NPs concentration to best value up to 60^{0}C. By examining the strength of the OWC prepared in different percentages of NPs for the same particle size, it can be seen from **Fig. 10**, that mixtures having 6 % ESNP content displays higher compressive strength than their counterparts prepared with low NPs percentage. For example, the mixture with NPs of nominal particle size 40 nm exhibited a compressive strength of 3.61 MPa at 38 ^{0}C and 6 % NPs, while it recorded 2.7 MPa at 2% NPs at the same age and temperature. Conversely, a different observation was detected for mixtures which incorporated NPs where the compressive strength result at 6% ESPN content and temperature of 60 ^{0}C was 12.46 MPa, while slumped to 12.11 MPa at 10% of NPs. This means that on increasing the percentage of NPs is useful in increasing strength to a certain limit after which any increase in the NPs content leads to a decrease in the compressive strength.

Table 6. Compressive strength with different percentage of ESNPs

Test	Without Additives	With Additives		
	CM (0%)	2%	6%	10%
Compressive strength in (MPa) at	2.50	2.70	3.61	3.53

8 h at 38 ^{0}C and atmospheric pressure				
Compressive strength in (MPa)at 8 h at 60 ^{0}C and atmospheric pressure	10.40	11.12	12.46	12.11

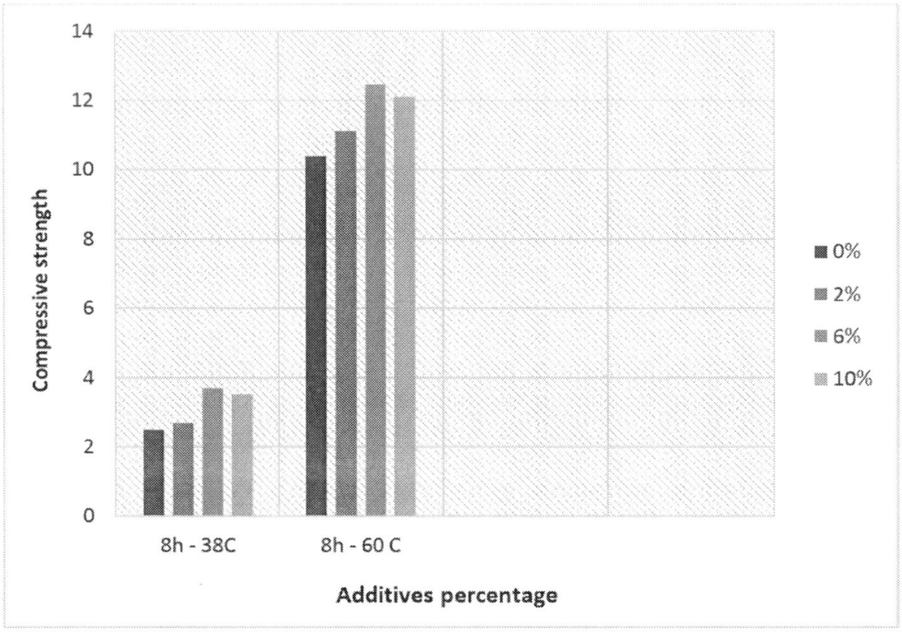

Fig. 10. Effect of ESNPs percentage on compressive strength.

3.4 Effect of ESNP on rapid chloride permeability (RCP) of OWC

All the OWC mixtures result with NPs were less permeable to chloride penetration than the CM, which indicates that adding NPs to OWC mixtures decreases their permeability. For instance, it can be depicted from **Table 6** that the CM recorded charges of 5307 coulombs which dropped to 3411 coulombs for the mixture ESNP2(40) and further decreased to 590 coulombs with the ESNP6(40) mixture. Also, it shows that increasing the percentage of NPs in the mixture, decreases its permeability which matches the results of compressive strength test. For example, the mixture ESNP2(40)produced with 2 % NPs reached high permeability charges 3411 coulombs which makes it moderately permeable while the mixture ESNP4(40) produced with 6 % NPs of the same nominal particle size displayed less charges of 600 coulombs which is classified as very low permeable. The test results are shown in **Fig.11** and listed in **Table 7.** In conclusion, the RCP decreases with adding $CaCO_3$ nanoparticles. It can also be concluded that the permeability of OWC mixtures decreases with the increase in the NPs content added to the mixture due to the effect of packing efficiency of the voids and cavities.

Table 7. Rapid chloride permeability test results.

Mixture I.D.	Cell No.	Charges passed (Columbus)	Permeability class
CM	1	5307	high
ESNP2(40)	2	3411	moderate
ESNP6(40)	3	600	very low
ESNP10(40)	4	590	very low

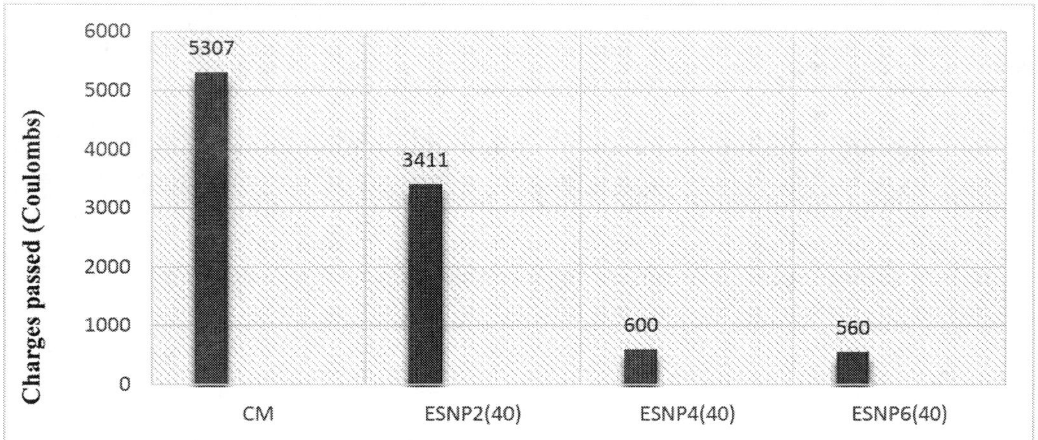

Fig. 11. Chloride ion permeability resistance of different concretes measured.

4. Conclusions

From this work, the following main conclusions are drawn.

- $CaCO_3$ nanoparticles was successfully obtained by ball milling method for 25 h with average particle size of 40 nm.
- The characterization of $CaCO_3$ nanoparticles shows spherical particles and those have irregular shapes as well as agglomerates fuse together and the range of external diameters is in the 40-110 nm range
- Incorporating $CaCO_3$ nanoparticles into OWC mixtures enhances their mechanical properties and performance by yielding higher compressive strengths than the control mixture exhibiting relatively high early strengths and resisting the acidity.
- Within the particle size range 40-110nm and dosage range studied the compressive strength generally increased with the addition of $CaCO_3$ nanoparticles. The addition of NPs in 6% at 40 nm particle size gives the highest the compressive strength for the OWC mixtures.

- The results supported the notion that the $CaCO_3$ nanoparticles act not only as filler, but also as an activator to support the hydration process. The presence of $CaCO_3$ nanoparticles increases the reaction rate of tricalcium aluminate (C3A) to form carboaluminate complex thereby increases the total hydration products and delays the formation of micro-cracks, and the hydration products and consequently the compressive strength. In addition, it also reacts with tricalcium silicate (C3S) and accelerates setting and early strength development.

References

[1] M.C.M. Nasvi, P.G. Ranjith, J. Sanjayan, A numerical study of triaxial mechanical behaviour of geopolymer at different curing temperatures: an application for geological sequestration wells, J. Nat. Gas Sci. Eng. 26, 1148–1160, 2015.

[2] J. Won, D. Lee, K. Na, I.-M. Lee, H. Choi, Physical properties of G-class cement for geothermal well cementing in South Korea, Renewable Energy 80, 123–131, 2015.

[3] C. Ma, L. Chen, B. Chen, Analysis of strength development in soft clay stabilized with cement-based stabilizer, Constr. Build. Mater. 71, 354–362, 2014.

[4] C. Ma, B. Chen, Properties of a foamed concrete with soil as filler, Constr. Build. Mater. 76, 61–69, 2015.

[5] W. Khaliq, H.A. Khan, High temperature material properties of calcium aluminate cement concrete, Constr. Build. Mater. 94, 475–487, 2015.

[6] A. Brandl, J. Cutler, A. Seholm, M. Sansil, G. Braun, A. Brandl, et al., Cementing solutions for corrosive well environments, Spe Drill. Complet. 26 (2), 208–219, 2011.

[7] Najat J. Saleh, Raheek I. Ibrahim, Ali D. Salman. Characterization of nano-silica prepared from local silica sand and its application in cement mortar using optimization technique. Advanced Powder Technology 26, 1123–1133, 2015.

[8] M.A. Caldarone, K.A. Gruber, R.G. Burg, Highreactivity metakaolin: a new generation mineral admixture, Concr. Int. 16, 1994.

[9] M.H. Zhang, V.M. Malhotra, Characteristics of a thermally activated aluminosilicate pozzolanic material and its use in concrete, Cem. Concr. Res. 25 (8), 1713–1725, 1995.

[10] F. Wang, S. Wang, Y. Meng, L. Zhang, Q. Wu, J. Hao, Mechanisms and roles of fly ash compositions on the adsorption and oxidation of mercury in flue gas from coal combustion, Fuel 163, 232–239, 2016.

[11] S.V. Vassilev, C.G. Vassileva, V.S. Vassilev, Advantages and disadvantages of composition and properties of biomass in comparison with coal: an overview, Fuel 158, 330–350, 2015.

[12] Najat J. Saleh, Raheek I. Ibrahim, Ali D. Salman. Optimization Process for Using Prepared Nano silica in Concrete. The 2nd International Conference of Buildings, Construction and Environmental Engineering, BCEE2-2015.

[13] M.R. Shatat, Hydration behavior and mechanical properties of blended cement containing various amounts of rice husk ash in presence of metakaolin, Arab. J. Chem., 2013.

[14] M.S. Morsy, Y.A. Al-Salloum, H. Abbas, S.H. Alsayed, Behavior of blended cement mortars containing nano-metakaolin at elevated temperatures, Constr. Build. Mater. 35, 900–905, 2012.

[15] Wu, Z.; Shi, C.; Khayat, K.H. Multi-scale investigation of microstructure, fiber pullout behavior, and mechanical properties of ultra-high-performance concrete with nano-$CaCO_3$ particles. Cem. Concr. Compos. 2018, 86, 255–265.

[16] Ge, Z.; Wang, K.; Sun, R.; Huang, D.; Hu, Y. Properties of self-consolidating concrete containing nano-CaCO3. J. Sustain. Cement-Based Mater. 2014, 3, 191–200.

[17] Yang, H.; Che, Y.; Leng, F. High volume fly ash mortar containing nano-calcium carbonate as a sustainable cementitious material: microstructure and strength development. Sci. Rep. 2018, 8, 16439.

[18] Ge, Z.; Wang, K.; Sun, R.; Huang, D.; Hu, Y. Properties of self-consolidating concrete containing nano-CaCO3 . J. Sustain. Cement-Based Mater. 2014, 3, 191–200.

[19] American Petroleum Institute. Recomended practice for Testing Well Cements, API Spec.10 B 1st edition, American Petroleum Institute, Washington DC (2010).

[20] Kim D H, Park C G. Strength, permeability, and durability of hybrid fiber-reinforced concrete containing styrene butadiene latex. Journal of Applied Polymer Science, 2013, 129(3): 1499–1505).

IOP Conf. Series: Materials Science and Engineering **765** (2020) 012007 doi:10.1088/1757-899X/765/1/012007

DESIGN AND IMPLEMENTATION OF AN AUTOMATIC CONTROL SYSTEM TO AVOID FOULING IN PIPES WITH BLOW DOWN HEAT RECOVERY IN STEAM BOILERS

Raheek I. Ibrahim[1], AbdulrahimT. Humod[2], and Najat A. Essa[3]
[1]Electromechanical Eng. Dep. / University of Technology/ Baghdad-Iraq.
[2] Electrical Eng. Dap. / University of Technology/ Baghdad-Iraq.
[3] Ministry of oil / Midland Oil Company (Md.O.C)

Abstract .The aim of the present work is to solve the problem of fouling in boiler pipes and improve energy efficiency of steam boilers in the south refineries company, Al-Basrah province, Iraq. An automatic control system with a novel electronic circuit has been proposed to achieve this aim. This system consists of an on-line conductivity sensor and a conductivity monitor connected to an additional resistance to measure the concentrations of dissolved solids in boiler water. A square wave feedback has been employed to determine the levels of solid concentrations and controls the surface blow down process through electronic circuit. This circuit consists of a D.C power supply, an amplifier, a rectifier, comparators and a driver sub-circuits for solenoid valve. The driver sub-circuit consists of two transistors and three relays. The solenoid valve is used with throttle valve to control the flow rate of boiler blow down water. Two flow meters are used to measure blow down and feed water flow rates. The design and implementation of a heat recovery unit is also included. It consists of a heat exchanger (shell and coiled tube) type used to recover heat energy from blow down water. To measure the inlet and outlet temperatures of feed water and blow down water, four (k-type) thermocouples were used. A pump was used to pump feed water from the storage tank in to the boiler. To automatically control water level inside the boiler two electrodes (water level sensor) were used integrated with the float less level switch circuit. The experimental results proved the effectiveness of the innovative control system in improving boiler efficiency and in fouling control through enhancing the blow down process and reducing indirect losses. The blow down quantity reduced from 2244.897 kg/h to 1111.111 kg/h. Which provided 27210.864 kg/day of blow down water. Hence, saved 158450201.2 ID/ year or 83.16% of the energy lost with blow-down water can be recovered using heat-recovery unit with an energy saving of 103411.8 MJ/day. Which will save a mass of fuel equals to 482.46 ton/ year. The automatic control system proved to be a good solution for saving energy and reducing harmful emissions to the environment and it contributes to the maintenance of sewage pipes from damage caused by the heat of discharged water.

1. INTRODUCTION

The most prominent targets of modern technology are to improve the efficiency of energy technology in industrial processes. They seek opportunities to reduce production costs without negatively affecting product. Petroleum refineries are more focused on improving utility consumption and reducing stack emissions. Private industries that generate their own steam need to optimize their energy generation and use by taking all the measures to reduce energy loss [1]. Since steam boilers are the main fuel consumers in the petroleum refineries where the petroleum industry uses about 40% of its energy to generate steam [2]. So optimization of a boiler system is one of the best solutions to save energy and reduce pollutions [3].

Blow down is the main energy loss at a place where steam is generated. Hence the amount of water to be blown down is to be optimized to reduce the heat energy waste [4].When steam is generated it leaves practically all the dissolved solids. If more solids are with feed water, they will be concentrated and may eventually reach a level where their solubility is exceeded hence, they deposit from the solution. Above a certain level of concentration, these solids encourage foaming and cause carryover of water into the steam. The deposits also lead to scale formation inside the boiler, resulting in localized overheating and finally causing damage to boiler tubes, steam traps, or process equipment [5, 6].

In order to reduce the level of suspended solids and total dissolved solids in the boiler water within the acceptable levels, a blow down process must be achieved. Blow down is water intentionally wasted from a boiler while the boiler is operating to avoid concentration of impurities [7].There are two sources of blow down water: bottom blow down which is the removal of the sludge accumulated at the bottom of a fire tube boiler, or in the mud drum of a water tube boiler. While surface blow down is often done as a continuous removal of dissolved impurities from the surface of water in a steam boiler [8].

Insufficient blow down may lead to the formation of deposits. While, excessive blow down will waste energy and chemicals added. Therefore, it is necessary to control the level of minerals concentration in boilers. There are two methods to control the boiler blow down, these are: Manual blow down control method: which requires checking of boiler water many times a day or according to a set schedule to adjust suspended solids, TDS, pH, Silica and Phosphates concentration including sludge formed in the boiler water and adjust blow down accordingly [9, 10] with manual boiler blow down control, operators are delayed in knowing when to conduct blow down or for how long. They cannot immediately respond to the changes in feed water conditions or variations in steam demand the other method is automatic blow down control: In which boiler water conductivity is continuously monitored and the blow down rate is accordingly adjusted. A probe measures the conductivity and provides feedback to the controller of the blow down valve. An automatic blow down control system can keep the blow down rate uniformly close to the maximum allowable dissolved solids level, while minimizing blow down and reducing energy losses [11].

Several studies focused on boiler blow down techniques. Lakshmi and Pillai (2015)[12]compared between manual and automatic boiler blow down methods to determine the rate of blow down in each case. Their results showed that there was about 1.28% difference between the two methods. This has led to an increase in boiler efficiency by using automatic blow down method.

Godwin and Johnson (2003)[13]invented a mechanical method for automatically controlling boiler blow down system by using bimodal feed water pumps. In this study the control process was done by removing boiler water at a predetermined rate equal to a selected percentage of the feed water. Gupta, et al. (2011)[14] suggested an automatic blow down system consisting of continuous monitoring of conductivity. Zainal and Nor(2012) [15]conducted an experimental study on two boilers for the optimization of boiler blow down frequency and its effect on the chemical values of the water steam cycle. Their results showed that boiler blow down must be carried out if one of the chemical parameters in the steam cycle and water was trending upwards and attained the limit value. Panda, et al. (2013) [16]proposed a design for the control system using neural network technology for optimization of blow down of a power plant boiler. The design proved to be efficient for calculation and optimization of the blow down since the blow down water contains a large quantity of energy. As well as, any boiler with continuous blow down exceeding 5% of the steam rate is a good candidate for the introduction of blow down waste heat recovery. Also, heat can be recovered to preheat boiler makeup water [17].Bahadori and B.vuthaluru (2010) [18] developed an accurate and simple method for estimation of heat recovery from blow down systems during steam generation. Madhav D., et al. (2013) [19] presented a design for heat recovery system consisting of a flash vessel and a heat exchanger to minimize heat losses. They used mathematical modeling to calculate energy recovered by this system. Sherin and prince (2013) [20] carried out an investigation on mathematical modeling of boiler and heat exchanger to generalize the direct impact of boiler blow down condensate, waste water heat recovery and fuel consumption. The results found that boiler efficiency has significantly improved up to 2% with every 6% rise in temperature of feed water due to blow down and a 1% reduction in fuel consumption. Also, Arunkumar S., et al.(2014) [21] designed a heat recovery system consisting of a plate type heat exchanger to recover heat from blow down and supply it to the boiler feed water. The results showed that 6.61kg/day total energy was saved in the process. Vandani, et al. 2015 [22], performed energy and exergy analysis for heat recovery of boiler blow down on a steam power plant to estimate the performance effect of heat recovery system. Their results showed, that the net generated power was increased by 0.72% with blow down recovery mechanism, and efficiency was raised by (31.68 – 31.91) %. Their results also showed that energy and exergy efficiencies were increased by 0.23 and 0.22, respectively. Exergy efficiency reached (30.66) %, which showed that, there was (1.86) percent increase with respect to a situation where a flash tank has been used. The south refineries company/ Al-Basrah province/ south of Iraq suffers from serious fouling problems in boiler pipes because of high salinity water in such locations.

The aim of the present work is to solve the problem of fouling in boiler pipes and improve energy efficiency in the south refineries company. This has been achieved through enhancing the boiler blow down process by designing and implementing an automatic control system with a novel electronic circuit to determine the concentration of dissolved solids in the boiler water at different levels. As well as, a heat exchanger of shell and coiled tube type is designed and manufactured to recover heat from surface blow down water. The proposed improving method utilizes an on-line conductivity sensor and a conductivity monitor that is connected to an additional resistance to measure the concentrations of dissolved solids in boiler water, then generation of a feedback signal to determine the level of solid concentrations and hence control the surface blow down process.

2. MATERIALS AND METHODS

2.1 Materials:

The materials include two types of water: blow down water and feed water. These waters were used to determine the maximum and minimum limits of concentrations of salt allowed in boiler water. The samples of these materials were taken from the boilers blow down and feed water from south refineries company- Al-Basrah/ Iraq. The properties of these samples are given in Table (1).

Table 1: Properties of blow down water and feed water supplied from South Refineries Company.

Property	**Blow down**	**Feed water**
pH	11.0	8.7
E.C (μs/cm)	302	5.5
B-ALK	60	-
M-ALK	80	-
T-H (ppm)	0.4	0.16
TDS (ppm)	211	7
Ca-H (ppm)	0.29	0.1
Mg-M (ppm)	0.11	0.06
Cl^- (ppm)	1.6	0.4
SiO_2 (ppm)	0.810	0.06
P_2O_5(ppm)	1.6	0.3

2.2 Experimental system:

Figure 1 is a schematic diagram of the experimental rig. The main parts used in the experimental system are: boiler, feed water tank, feed water pump, electrical control panel, heat recovery unit, and automatic control system which consists of: conductivity monitor, throttle valve and automatic control circuit. The measuring devices are conductivity monitor, flow meter, temperature controller, thermocouples, an oscilloscope and a digital multimeter.

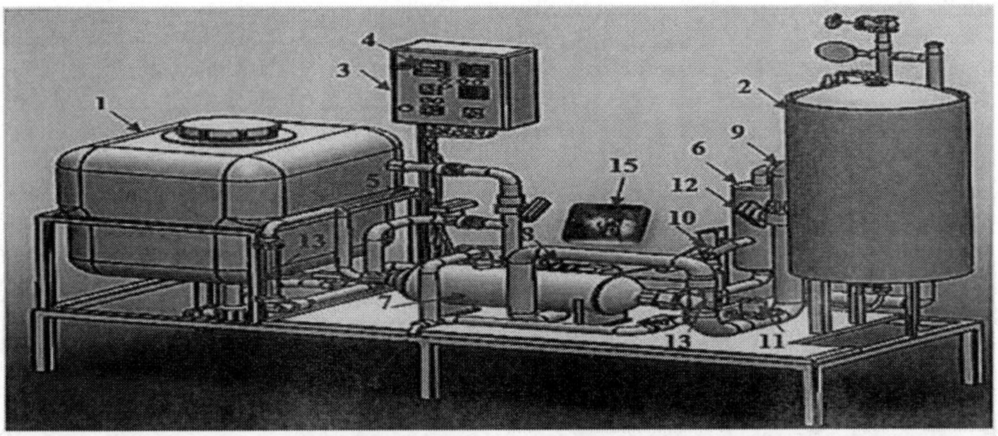

1-Feed water tank, 2- Boiler, 3- Electrical control Panel, 4- Online Conductivity monitor, 5- Flow regulating valve, 6- Sample cooling water, 7- Heat exchanger, 8- Feed water pump, 9- Surface blow down pipe, 10- Conductivity sensor probe, 11- Solenoid valve, 12-Throttling valve, 13- Water flow meter, 14- Blow down to drain, 15- Automatic control circuit.

Figure 1: The experimental rig.

The main parts of the automatic control system are:

I- Conductivity monitor: Intelligent microprocessor conductivity measurement instrument as shown in Figure 3 was used to measure the conductivity and TDS on-line. It has been used to monitor conductivity automatically and generate a square wave through additional resistance. Conductivity sensor of stainless material with (10) m cable was used to sense the concentrations of salts in the boiler water and provide feedback to the connectivity monitor; it was installed inside a cell flow.

II- Solenoid valve (drive actuator): An electromechanical controlled valve type (2w-250-25); was used to automatically control the flow rate of boiler blow down water.

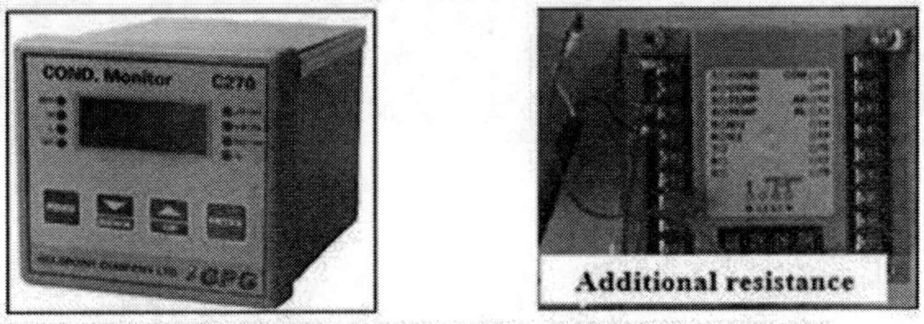

Figure 2: Conductivity monitor with additional resistance.

III- Electronic circuit: The control circuit is shown in Figure 3. It consists of: DC power supply circuit, amplifier circuit, rectifier circuit, comparator circuit and a driver circuit.

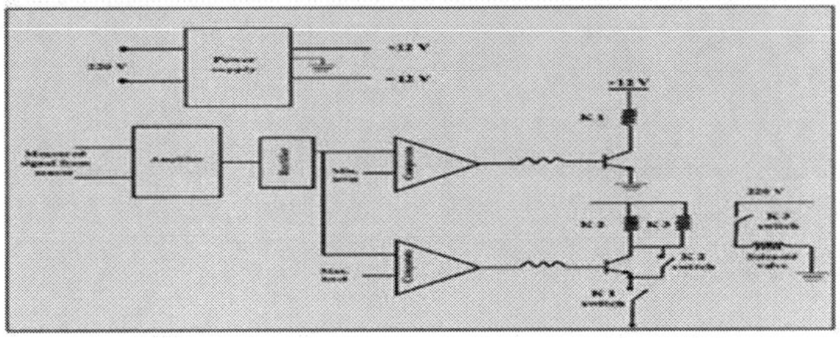

Figure 3: Block diagram of automatic control circuit.

Working principles of the automatic control circuit: The control circuit is shown in Figure4. The first step is to amplify the low signal square wave voltage from the conductivity sensor generated from conductivity monitor. The second step is to convert the amplified signal to direct voltage signal (DC voltage). The third step is to compare the DC voltage with a two set point values by comparator device. When the concentration of dissolved solids signal value reaches the lower set point value (minimum level), the comparator device for minimum level generates a signal (high) to activate relay (K1). At this condition the switch of K1 closes the solenoid valve which remains inactivated and blow down is in off state. When the concentration of dissolved solids reaches the maximum level, the comparator device of maximum level generates a signal (high) to activate relay (K2). When relays (K1 and K2) are in activated state, this will cause relay (K3)to be in activated state, and the switch of (K3) will be in closed condition or solenoid activated, and then blow down water will discharge outside the boiler. During the blow down process, feeding water is supplied to the boiler to compensate for the blow down water. The concentrations of dissolved solids begin to drop from the maximum level, relays (K1) and (K2) remain activated and the solenoid valve in ON state. When the concentrations are reduced to reach a value below the minimum level, relays (K1) and (K2) are not in activated state. This will cause relay K3 to be in an inactivated state, switch of K3 will be in open condition and solenoid valve in OFF state. The output voltage signals from comparators circuit (two binary digital) operate the blow down valve as in truth values shown in Table 2.

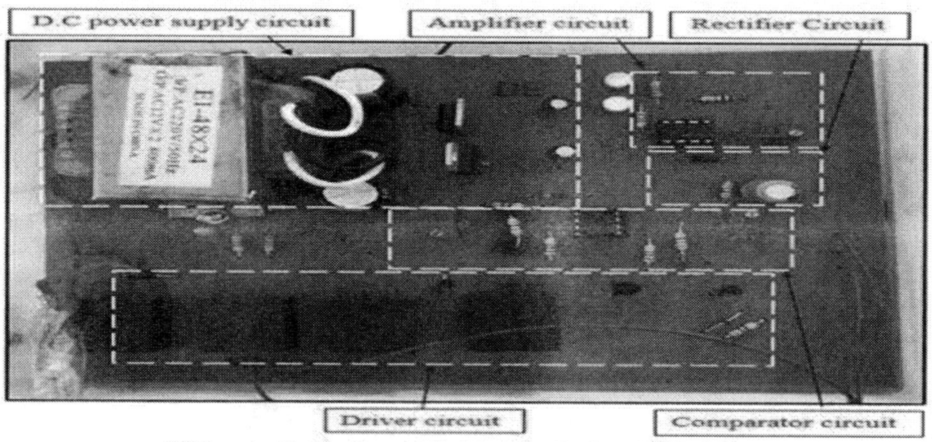

Figure 4: Automatic control circuit design.

Table 2: Truth table for operating of the blow down valve.

Initial state of solenoid valve	Maximum level	Minimum level	Solenoid valve
OFF	0	0	0
	0	1	0
	1	1	1
ON	1	1	1
	0	1	1
	0	0	0

IOP Conf. Series: Materials Science and Engineering **765** (2020) 012007 doi:10.1088/1757-899X/765/1/012007

The main parts of the heat recovery unit are: shell and coiled tube type heat exchanger as shown in Figure 5. It was used to heat the make-up water that enters at ambient temperature, and cools the blow down water before it is drained to sewers. Table 3 shows the data adopted for the design of the heat recovery unit.

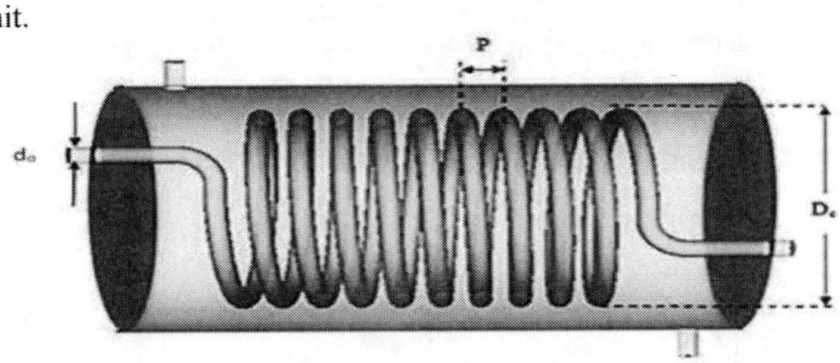

Figure 5: shell and coiled tube heat exchanger.

Table 3: Specifications of shell and coiled tube heat exchanger.

parameter	specification
Shell material	316L stainless steel
Inner and outer diameter of shell	13.6 cm , 14 cm
Length of shell	43 cm
Tube material	316L stainless steel
Number of coiled tubes	Two
Inner and outer diameter of coiled tube	7 mm , 10 mm
Length of coiled tube	30 cm
Number of turns	15
Outer and inner diameter of the coil	11 cm, 8 cm
Distance between two coils	1.5 cm
Pitch for each coil	20 mm

In the present work the blow down water (hot fluid) flows in the shell side and make-up water (cold fluid) is supplied through coiled tube side. The shell of the heat exchanger was isolated by aluminum foil thermal insulation rubber foam roll, with a thickness of 10 mm to reduce heat dissipation .Two flow meters are installed upstream of the heat exchanger to measure the flow rate of the hot and cold stream. Two PVC ball valves were used to control the flow rate of cold and hot water inside the flow meters. To measure the inlet and outlet temperatures of cold and hot water, four k- type thermocouples were inserted in the small holes drilled in the inlet and outlet tubes of heat exchanger. They are sealed to prevent any leakage. The thermocouples were joined to a digital temperature indicator installed in the electrical control board.

To operate and assess the pilot plant designed for automatic control of the surface blow down process in the steam boiler, various concentrations of samples as listed in Table 4 were used. These concentrations were the maximum levels of dissolved concentrations allowed in the boiler water. A number of experiments were carried out to recover the latent heat in the waste water discharged during the surface blow down process. The blow down water was supplied to the heat exchanger at atmospheric pressure. The experiments were conducted by passing feed water (cold fluid) through coils tube and the blow down (hot fluid) in the shell side; this was done for counter flow operation. Measuring the heat gained by feed water $QF.W$ and heat lost by blow down water $QB.D$ has been done by keeping the flow rates of feed water constant, while, blow down flow was changed from 0.06 m3/h to 0.14 m3/h with 0.02 m3/h interval. Next, the flow rates of blow down water was kept constant and the flow rates of feed water was varied from 0.1 m3/h to 0.5 m3/h with 0.1m3/h interval. All the data contained in this work are based on average values of five conducted experiments.

Table 4: conductivity and voltage values for measuring the TDS in water samples.

Sample No.	Mixing ratio (mL)	E.C (μs)	TDS (ppm)	Voltage of reference point at the max. level (volt)
1	100 B.D +150 F.W	86.5	60	2.7
2	150 B.D +150 F.W	130	91	4.1
3	175 B.D +75 F.W	146	102	4.6
4	200 B.D +50 F.W	163	114	5.1
5	225 B.D +25 F.W	175	122	5.5

3. RESULTS AND DISCUSSIONS

3.1 Automatic control circuit results:A test was carried out on each stage of the electronic control circuit. The amplifier stage, the rectifier stage, comparator stage and the driver stage were tested using an oscilloscope. Figures (6 a, and b) show the signal of an on line conductivity sensor for the additional resistance for minimum and maximum allowable levels of TDS in the boiler water. The average values of minimum and maximum voltages are ± 30 mV, and ± 0.6 V, respectively.
Figures (7 a, and b) show the output of the amplifier circuit which represent the amplified signal of the minimum and maximum levels of TDS, respectively with a gain of 10. The square voltage of minimum level was amplified to ± 0.3V and the voltage of maximum level was amplified to ± 6 V. Figures (8 a, and b) show the signals output of the rectifier circuit which represent the DC voltage signal of the minimum and maximum levels of TDS, respectively. The voltage of minimum level is 0.3 V and the voltage of maximum level is 6 V. Figures (9 a, b and c) show the signal output from comparators circuit compared with the voltage of a set point at minimum level of 0.3 V and a maximum level of 6 V. These signals are used to control the switches of relays K1, K2 and K3.If the measured voltage of the boiler water is lower than the voltage of minimum and maximum setting point, then the output signal of comparators are as depicted in Figure 9 (a), where relays K1, K2 and K3 are normally opened and solenoid valve is in (OFF state).
If the voltage measured of the boiler water was higher than the voltage of minimum set point and lower than the voltage of maximum set point. The output signal of comparators are depicted in Figure 9 (b), where switch K1 is closed, relays K2 and K3 are normally opened and solenoid valve isin(OFF state).If the voltage measured of the boiler water is higher than the voltage of minimum and maximum setting points, then the output signal of comparators is as depicted in Figure 9(c), where switches K1, K2 and K3 are activated and solenoid valve is at (ON state) during blow down process. Figures (10 a, and b) show the signal output from comparators circuit compared with the voltage of set point at minimum level of 0.3 V and maximum level of 6 V during feed water supply to the boiler. When the concentration is between maximum level and minimum level, the output of the comparators is as shown in Figure 10 (a). The relays K1, K2 and K3 are in activate state; solenoid valve is at (ON state). When the concentration of the boiler water becomes below the minimum level, the output of the comparators become as shown in Figure 10 (b). The relays K1, K2 and K3 are normally opened, solenoid valve is at (OFF state).

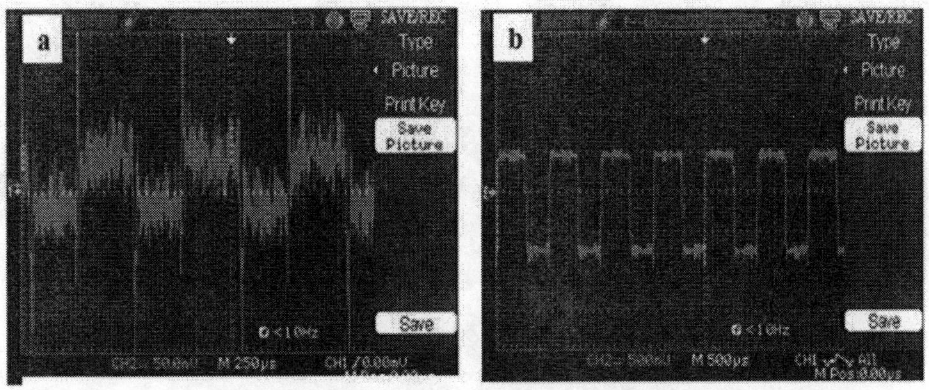

Figure 6: Measured voltages for (a) minimum level equal to ± 30 mV, (b) for maximum level equal to 0.6 V.

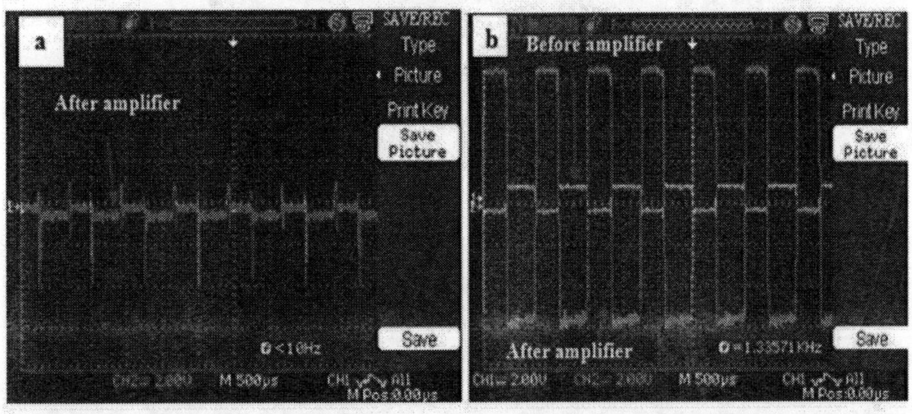

Figure 7: output voltages of the amplifier circuit (a) for minimum level equal to ± 0.3 V, (b) for maximum level equal to ± 6 V.

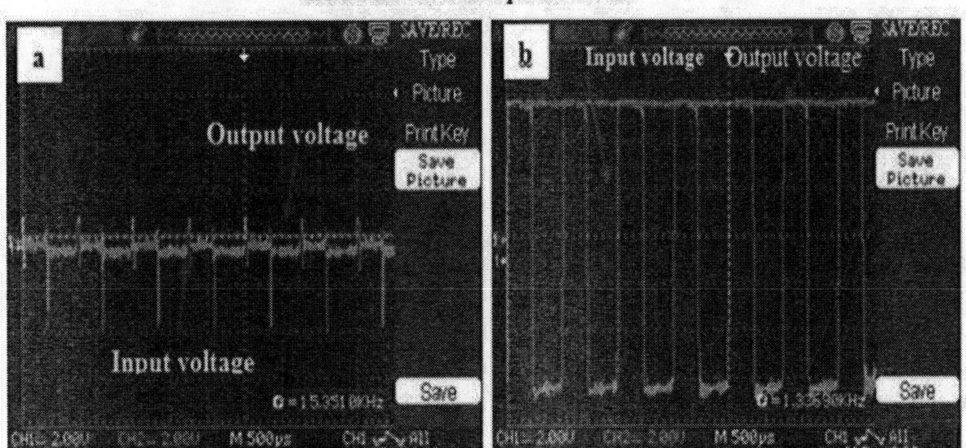

Figure 8: output voltage of the rectifier circuit (a) for minimum level equal to 0.3 V, (b) for maximum level equal to 6 V.

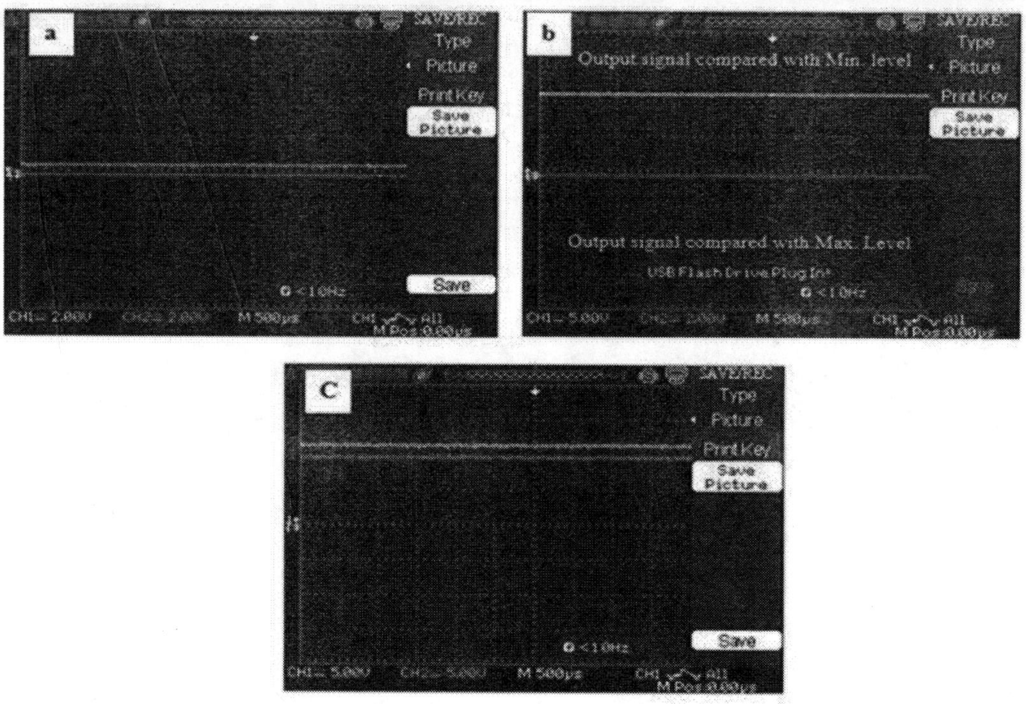

Figure 9: output of the comparators circuit: (a) for the measured voltage lower than minimum and maximum voltage of setting point, (b) for the measured voltage higher than minimum and lower than maximum voltage of setting point, (c) for the measured voltage higher than minimum and maximum voltage of setting point (during blow down process).

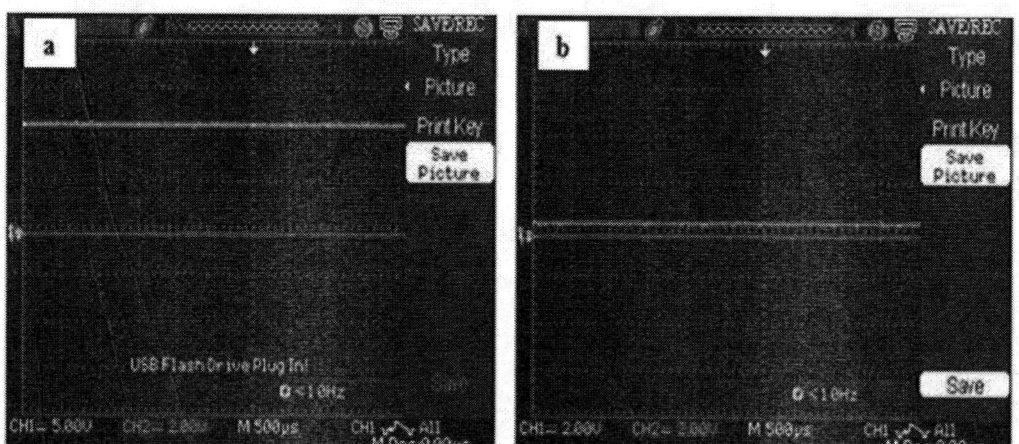

Figure 10: output of the comparator circuit: (a) during feed water supply to the boiler, (b) when finishing the blow down and feed water process.

3.2 Comparison between current treatment method and the proposed method using the innovative circuit:

Figures 11 and 12 show a comparison between the present methods of treatment for boiler water used in the South Refineries Company- AL Basra (manual blow down method) and the proposed method in the present study (automatic blow down method) to adjust the concentration of total dissolved solids (TDS) in boiler water.

Figure 11 shows the reading of the TDS concentration of boiler blow down water recorded along

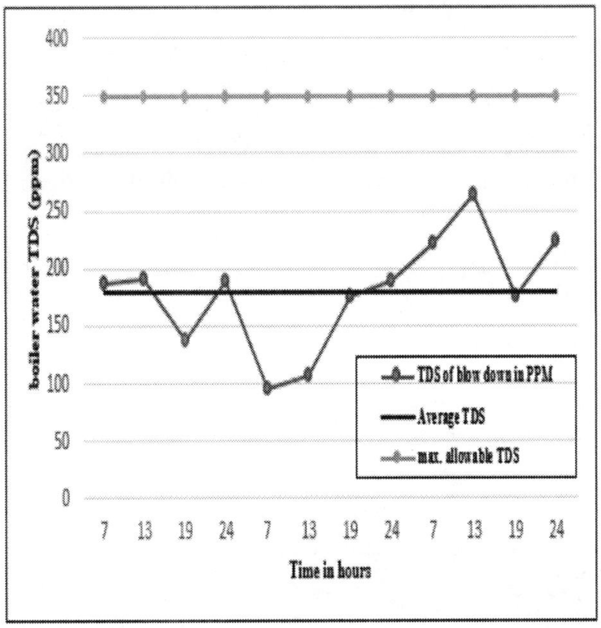

Fig. 11: Boiler water TDS using manual blow down control

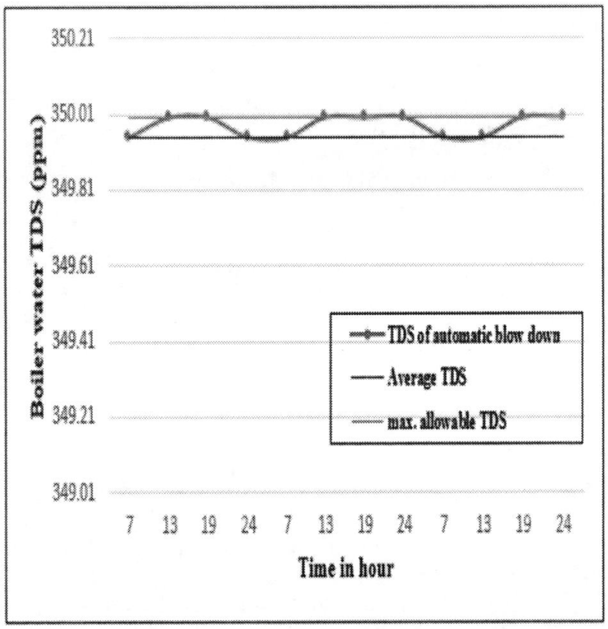

Fig. 12: Boiler water TDS using automatic blow down control

three days (20/3/2017 to 22/3/2017) for the manual blow down method of 30TPH water tube boiler in the South Refineries Company- AL Basrah. It is observed that there is a variation in the values of blow down water TDS. This difference in concentrations of blow down water is because the treatment done manually. Also the figure shows that the actual TDS blow down rate at which the manual blow down process occurred, is significantly below the maximum allowable TDS. Low level of blow down water TDS than the maximum allowable level of TDS leads to higher losses of the boiler water.

Figure 12 shows the level of concentration of the blow down water TDS by using the automatic control circuit to maintain the TDS in the boiler water at an allowable level. It is observed that the boiler water TDS can be adjusted to be close to the maximum allowable level. This will reduce the losses of boiler water resulting from the blow down process.

3.3 Effect of automatic control circuit on blow down rate:

The use of an electronic circuit is to control the concentrations of TDS in the boiler water to be close to the maximum allowable dissolved solids level. It reduced the rate of blow down water during surface blow down process by regulating water blow down volume in relation to amount of total dissolved solids. According to the operating conditions in the South Refineries Company- Iraq, listed in Table 5, the boiler water analysis was carried out for five water tube boilers operating in the production unit of thermal energy to estimate the value of savings in boiler water blow down when installing the automatic blow down control circuit.

Table 5: Blow down energy analysis supplied from South Refineries Company.

No.	Variables	quantity
1	Amount of steam generated by boiler	110000 kg/h
2	Maximum allowable TDS in boiler water as design	700 ppm
3	Feed water TDS	7 ppm
4	Average TDS maintained in boiler at manual blow down	350 ppm

3.4 Calculation of blow down rate and quantity during the manual and automatic blow down processes:

Using the data available in table (5) for a boiler, the blow down rate at manual process can be calculated as:

$$B.D\,\% = \frac{feed\ water\ TDS(ppm)}{\max permissible\ TDS\ in\ boiler\ water(ppm) - feed\ water\ TDS(ppm)}\,\% \quad \ldots\ldots\ldots\ldots\ldots (1)$$

Blow down rate at manual process $= \frac{7}{350-7} \times 100\% = 2.04\%$

$$B.D\ quantity = \frac{feed\ water\ TDS(ppm) \times amount\ of\ steam\ generated\ by\ boiler(\frac{kg}{h})}{\max permissible\ TDS\ in\ boiler\ water(ppm) - feed\ water\ TDS(ppm)} \quad \ldots\ldots\ldots\ldots (2)$$

$B.D$ quantity $= \frac{7 \times 110000}{350-7} = 2244.897\ kg/h$

And the blow down rate at automatic control process can be calculated as:

Blow down rate at automatic process $= \frac{7}{700-7} \times 100\% = 1.0\%$

Blow down quantity $= \frac{7 \times 110000}{700-7} = 1111.111\ kg/h$

Excess amount of B.D rate = 2.04 – 1.0 = 1.04 %

Excess amount of B.D quantity = 2244.897 – 1111.111 = 1133.786 kg/h

From the comparison between the results of blow down rate and quantity of blow down with manual blow down system and automatic blow down system, it is observed that the blow down rate is reduced by 1.04 percent by raising the average boiler water total dissolved solids. Also the blow down quantity is reduced by 1111.111 kg/h by utilization of the automatic control circuit.

3.5 Effect of Blow down flow rate on the heat energy rate:
The effect of volumetric flow rate on the heat energy lost by the blow down water in the shell side and the heat gained by the feed water in coiled tube side are illustrated in Figures13 to 16.
Figure 13 shows the effect of different coiled tube flow rates of feed water on heat energy lost $Q_{B.D}$ while keeping the shell side flow rate constant. It is observed that heat energy lost from blow down water increases with increasing volumetric flow rate of feed water because as the volume of feed water increases there is more heat being transferred from the blow down water. Figure 14 shows the effect of variation in the blow down water volumetric flow rate from 0.06 m^3/h to 0.14 m^3/h with an interval of 0.06 on heat energy gained by the feed water $Q_{F.W}$. It is observed that the heat energy

gained by feed water increases gradually with increasing volumetric flow rate of blow down water, because when the flow rate of blow down increases, there is a large amount of heat transferred to the feed water. Figures (15and 16) show the relationship between the heat energy lost$Q_{B.D}$ and heat energy gained $Q_{F.W}$ with the volume flow rate of blow down water and feed water. It is observed that the heat energy lost $Q_{B.D}$ and heat energy gained $Q_{F.W}$ increased when the flow rate of blow down water and flow rate of feed water were increased, respectively. This is because the heat energy rate is a function of mass flow rate, when the mass flow rate increases; it increases the heat energy rate.

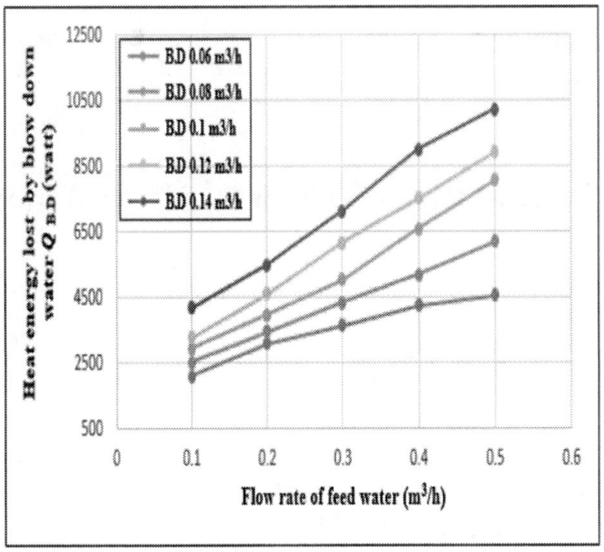

Figure 13: The effect of feed water flow rates on heat energy lost from blow down water

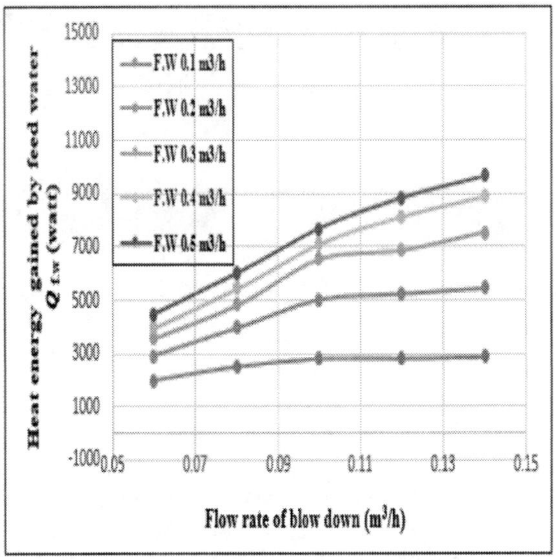

Figure 14: The effect of blow down water flow rates on heat energy gained by feed water

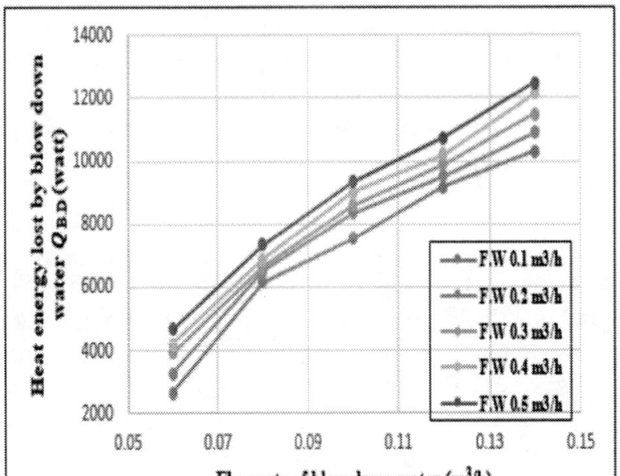

Figure 15: The effect of blow down water flow rates on heat energy lost

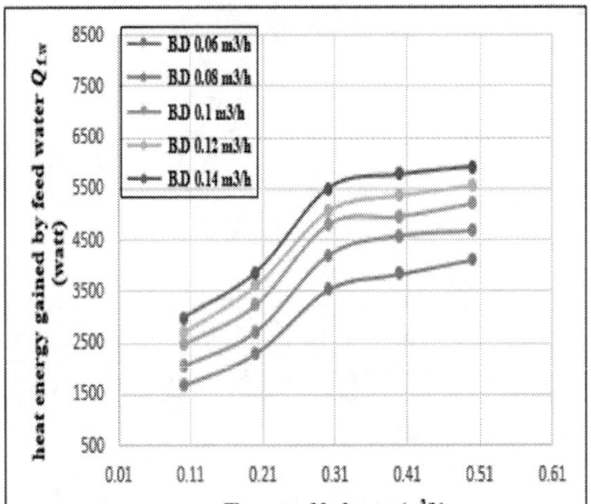

Figure 16: The effect of feed water flow rates on heat energy gained

3.6 Optimum conditions for heat energy transfer:

As a result of experiments conducted for heat exchange by changing the flow rate of blow down water and feed water, the optimum conditions for heat transfer are obtained by calculating the heat energy gained and lost by working fluids as shown in Table 6. The optimum conditions were volumetric flow rate of blow down water = 0.14 m^3/h and volumetric flow rate of feed water = 0.1 m^3/h. Where, at these values of flow rates maximum heat exchange has been obtained.

Table 6: Optimum operating conditions for heat exchanged.

Fluid	T_{in} ℃	T_{out} ℃	T_{mean}	ρ kg/m^3	$\dot{V}$ m^3/h	Cp kJ/kg.k	Δ T ℃	$\dot{m}$ kg/s
B.D water	90	57	73.5	975.7	0.14	4.190.4	33	0.038
Feed water	18	47	32.5	994.8	0.1	4.179.3	29	0.027

4. CONCLUSIONS

From the obtained results it can be concluded that the proposed design of automatic control system to control the surface blow down process proves to be an efficient system to regulate the volume of water discharging form the boiler in correspondence with the concentration of dissolved solids in boiler water. It provides the possibility of working the boiler close to the maximum allowable levels of dissolved solids concentration and this leads to minimize energy losses. The technology of the on-line conductivity monitoring device provided precision in measurement and reduced the complexities associated with the testing process. The innovative design of the electronic control circuit achieved high accuracy and flexibility in controlling the concentrations of dissolved solids in boiler water at desired levels. When installing the novel automatic control system, 1133.786 kg/h of blow down water was saved. This reduces the indirect losses and improves the efficiency of the boiler system. Also, installation of the heat recovery unit contributed in recovering83.15% of energy wasted with discharge blow down water from flash tank at 90℃. The automatic control system proved to be a good solution to save energy. There is also no need for blow down cooling before discharging to the sewer system.

ACKNOWLEDGMENT

The authors would like to express grateful thanks to the Iraqi Ministry of Oil/ South Refineries Company/ Al-Basrah for the financial support and practical information in accordance with graduate research (grant No. 1/ 2016).

REFERENCES

[1] Ernst W., Mariëlle Corsten, and C. Galitsky, "Energy efficiency improvement and cost saving opportunities for petroleum refineries", An Energy Star® Guide for energy and plant managers, U.S.A. (2015).

[2] Sachu Constantine and Kerry Phillips, "Steam system efficiency improvements in refineries in Fushun, China", Project design document, (2003).

[3] Ankitkumar C. Patel Prof. Imran Molvi An and Mistry, "Performance analysis of blow down heat recovery system", International Journal for Scientific Research & Development, Vol. 3,pp. 2720-2725,(2015).
[4] Sunudas T, M G Prince, "Optimization of boiler blow down and blow down heat recovery in textile sector", International Journal of Engineering Research and Applications, Vol. 3, pp.35-38,(2013).
[5] Thamer J. Mohammed, Usama B. Mahmoud, "Treatment and re-using of boiler blow down in thermal electric power plants", Diyala Journal of Engineering Sciences, First Engineering Scientific Conference, pp. 374-390, (2010).
[6] Giovanni Petrecca, "Energy Conversion and Management: Principles and Applications", DOI 10.1007/978-3-319-06560-1_1, Springer International Publishing Switzerland2014.
[7] Amit Kumar Jain, "An Approach towards Efficient Operation of Boilers", International Journalof Scientific and Engineering Research, Vol. 3, pp. 1-11, (2012).
[8] "Selection of Blow down Systems", Section 9 - blow down systems - EG 9.2, Engineering guidelines by industrial. Steam, 1998.
[9] Paul Eichamer, "Industrial Fired Boilers for General Refinery and Petrochemical Service", American petroleum Institute, (2011).
[10] Kiran G. Gayakwad, Prof.V.H.Patil, Dr.C.R.Patil, "Experimental Study on Heat Recovery from Continuous Blow down Water And Reduced Feed Water Consumption By Utilizing It In CFBC Boiler", International Journal of Engineering Research and Applications, Vol. 6, pp.15-19, (2016).
[11] N.C. Division of Pollution Prevention and Environmental Assistance, "Boiler Chemicals 101", fact sheet, (2004).
[12] Keerthi R Lekshmi, Vandana S Pillai, "Boiler Blow down Analysis in an Industrial Boiler", IOSR Journal of Engineering, Vol. 5, PP. 22-28, (2015).
[13] Douglas A. Godwin, David A. Johnson, "Boiler Water Blow Down Control System", Patent N0.2 US 6,655,322 B1, United States Patent, (2003).
[14] Rahul Dev Gupta, Sudhir Ghai, Ajai Jain, "Energy Efficiency Improvement Strategies for Industrial Boilers: A Case Study", Journal of Engineering and Technology, Vol. 1, pp. 52-56, (2011).
[15] Zainal Zakaria, Nor Ismail Hashim, "Corrosion-erosion on waste heat recovery boiler system via blow down optimization", Journal of Petroleum and Gas Engineering, Vol. 3(3), pp. 41-50, (2012).
[16] Sudodh Panda, Bikash Swain, Sandeep Mishra, "Blow down losses control in thermal power plants using neural network", Int. J. of Advancements in Research & Technology, volume 2, pp. 10-13, (2013).
[17] "Recover Heat from Boiler Blow down Water", Southern California Gas Company (A Sempra Energy Utility), (2012).
[18] Alireza Bahadori, Hari B. Vuthaluru, "A method for Estimation of Recoverable Heat from Blow down Systems during Steam Generation", International Journal of energy, Vol. 35, pp. 3501-3507, (2010).
[19] D.Madhav, L.Ramesh, M.Naveen, "Heat Recovery through Boiler Blow down Tank", International Journal of Engineering Trends and Technology (IJETT), Vol. 4I,(2013).
[20] Sunudas T, M G Prince, "Optimization of Boiler Blow down and Blow down Heat Recovery in Textile Sector", International Journal of Engineering Research and Applications, Vol. 3, pp.35-38, (2013).
[21] S. Arunkumar, R. Prakash, N. Jeeva, M. Muthu, B. Nivas, "Boiler Blow down Heat Recovery", IOSR Journal of Mechanical and Civil Engineering (Iosr-jmce), Vol. 11, pp. 83-85, (2014).
[22] Amin Mohammadi Khoshkar Vandani, Mokhtar Bidi, Fatemeh Ahmadi, "Exergy analysis and evolutionary optimization of boiler blow down heat recovery in steam power plants", Journal of energy conversion and management, Vol. 106, pp. 1-9, (2015).

IOP Conf. Series: Materials Science and Engineering **765** (2020) 012008 doi:10.1088/1757-899X/765/1/012008

Investigation of residual oil saturation in the process of filtration of viscous multicomponent flow with a wide range of pressures and temperatures

Radaev A.V. [1], Wisam Essmat Abdul-lateef [2], Ali Kamil Sebur [2], Hussain Abdulaziz Abrahem[2], Mukhamadiev A.A. [1], Sabirzyanov A.N. [1], Karibullina F.R. [1], Vaseneva N.A[1].

[1] Kazan state power engineering university, Center for energy-saving equipment and technological innovation, Kazan, Russia, radaev_neftianik@mail.ru
[2]University of Technology – Department of Electromechanical Engineering, Iraq, 50110@uotechnology.edu.iq,50278@uotechnology.edu.iq
50008@uotechnology.edu.iq

Abstract. An experimental setup was created to study the process of oil displacement by supercritical CO2 and fringes of supercritical CO2 and water from the model of homogeneous and inhomogeneous formation of different degrees of water saturation. The process of multicomponent filtration of oil-water-supercritical CO2 systems at 333 K isotherm in the pressure range of up to 14 MPa was studied at the experimental plant. The new data, the oil displacement efficiency of supercritical CO2 from a homogeneous porous medium with water content of 65% and rims supercritical CO2, and water from homogeneous porous medium with a water cut of 25%. The results showed that the use of supercritical CO2 in the filtration processes of multicomponent reservoir systems allows further displacement of oil from the watered porous medium.

Keywords: Supercritical fluid, multicomponent filtration, watered porous medium, fringing of supercritical CO2 and water, oil displacement coefficient, residual water, and oil saturation.

1. Introduction

The main share of oil currently, usually (more than 50%) and sometimes (about 85%), belongs to the category of hard to recover [1]. These include high-viscosity oils, oils confined to high-water reservoirs, and low-permeability reservoirs. The development of this kind of oil using secondary methods to increase oil recovery is difficult due to the presence of threshold restrictions.To overcome the threshold limits of reservoir oil viscosity, permeability, and degree of water-flooding, it is proposed that oil reservoirs use supercritical fluids extraction (SFE) SC-CO2. Its use meets the new licensing requirements coefficient of oil displacement, which should be more than 25%, and disposes large industrial carbon emissions.

IOP Conf. Series: Materials Science and Engineering **765** (2020) 012008 doi:10.1088/1757-899X/765/1/012008

2-Theoretical equations used

The thermo-physical properties of "oil-water-SC CO2" systems (heat capacity and thermal conductivity) are determined in accordance with the methods given in [10–16]. In accordance with [16], we assume that the thermos-physical properties of such systems obey the Neumann–Kopp additively rule. The heat capacity of the system is determined by the following equation [16]:

$$C_p^{\Sigma} = \sum_{i=1}^{n} C_{pi}\left(\frac{m_1}{m_1 + m_n}\right) \tag{1}$$

$C_1 ... C_n$ is heat capacity of dry rock, oil, respectively, и CO2;

$m_1 ... m_n$ is oil, water, and CO2, respectively, determined at a known porosity value of the rock.
Since the heat capacity of the rock at pressures up to 100 MPa practically does not depend on the pressure, we consider it constant.
The heat capacity of oil is determined by the equation [10]. For paraffin hydrocarbons,

$$C_p(P,T) = 1{,}268 + 1{,}3358 - \frac{\pi}{10}(1{,}256 \cdot \tau - 0{,}087) \tag{2}$$

$\tau = \frac{T}{T_{кр}}$ $\pi = \frac{P}{P_{кр}}$ - Critical temperature and pressure

For aromatic hydrocarbons,

$$C(P,T) = \frac{A}{\rho(P,T)} \tag{3}$$

Where:

$$A = 1{,}7 - B \cdot n_c \tag{4}$$

n_c is the number of carbon atoms;

B is the constant value, $B = 2{,}2 \cdot 10^{-3}$ for aromatic hydrocarbons;
$\rho(P,T)$ is the oil density at appropriate pressure and temperature.
The density of oil at high state parameters is determined as follows [11]:

$$\rho_н(P,T) = \rho_н + \Delta\rho_н \tag{5}$$

Where:

$\rho_н = \rho_T \cdot e^{Г \cdot (-1{,}84 \cdot 10^{-3})}$

$\rho_T = \rho_{293} - (1{,}83 - 0{,}0013 \cdot \rho_{293}) \cdot (T - 293)$ is the correction of oil density value for temperature;

$\rho_н^{293} = \dfrac{\rho_н + 1{,}83 \cdot (T - 293)}{1 + 0{,}00132 \cdot (T - 293)}$

$\Delta\rho_н = [1{,}87 \cdot [1 + 1{,}543 \cdot (P_{нас} - 0{,}1)] - 1{,}54 \cdot 10^{-3} \cdot \rho_н^{293}] \cdot (P - P_{нас})$ is the correction of oil density value for pressure.The heat capacity of carbon dioxide is determined in accordance with [12].

The thermal conductivity of the "rock-oil-SC CO2" system is also determined by the Neumann–Kopp rule in accordance with [13]. Therefore, the equation takes the following form:

$$\lambda_p^{\Sigma} = \sum_{i=1}^{n} \lambda_{pi} \left(\frac{m_1}{m_1 + m_n}\right) \quad (6)$$

The thermal conductivity of Sandstone is determined by the equation [14]:

$$\lambda_{эф}(P,T) = C(P) \cdot T^{n(P)} \quad (7)$$

Where

$C(P)$ and $T^{n(P)}$ is the values determined from the table in accordance with pressure and temperature.

The thermal conductivity of oils is determined by the equation [15]:

$$\lambda_{P,T} = [0{,}148 + 7 \cdot 10^{-4} \cdot (P - P_0)] \cdot \frac{0{,}689}{\sqrt{\tau}} \quad (8)$$

The thermal conductivity of carbon dioxide at high state parameters is determined in accordance with [12]. The thermal conductivity and heat capacity of water is determined by known methods [17,18].

3-Experimental system description

The existing experimental stands used for the study of the processes of oil displacement by various displacing agents does not fully meet the requirements [2] for studies of oil displacement by supercritical fluid systems of the moist layer, rims of supercritical CO2 and water, as well as studies of the process of displacement of oil from heterogeneous reservoirs [2], and existing experimental setup [3,4] has been upgraded. The scheme of the modernized experimental installation is shown in Fig. 1.

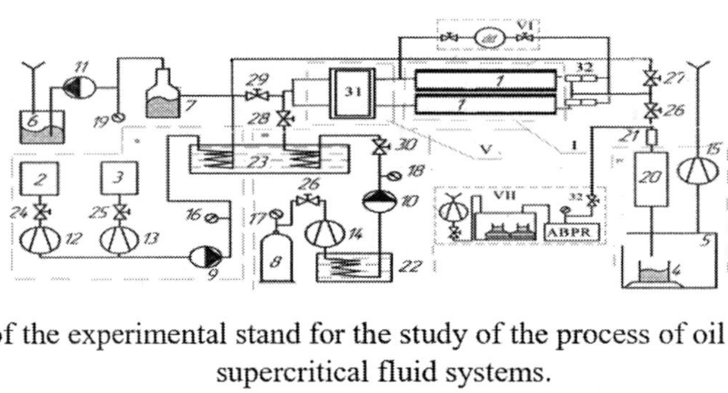

Fig.1. Scheme of the experimental stand for the study of the process of oil displacement using supercritical fluid systems.

1: reservoir model; 2: feed tank with water; 3: feed tank with oil; 4: collection of oil and emulsions; 5: receiving capacity of gases; 6: receiving capacity of oil; 7: Bunsen flask; 8: gas cylinder; 9: high pressure pump for pumping water or oil; 10: high pressure pump for pumping liquid carbon dioxide; 11: vacuum pump; 12: water flowmeter; 13: oil flowmeter; 14, 15: CO2-flowmeters 16–19: pressure gauges; 20: separator; 21: filter; 22: cryotherapy thermostat; 23: thermostat heating; 24–30, 32-shut-off valves, 31-mixer (high-pressure vessel); 32-back pressure regulator (BPR). Systems and units: I-reservoir model, II-water injection unit, III-CO2 injection unit, IV-oil and water saturation system, V-supercritical carbon dioxide and water fringing unit, VI-pressure gradient measurement unit, VII-backpressure and sampling unit.

The experimental stand consists of the following systems and units: reservoir models; reservoir model watering systems under reservoir conditions; supercritical fluid systems injection systems; sampling and analysis systems. The watering process is carried out by means of system II, consisting of a system of high-pressure pumps, feed vessels with oil and water and a thermostatic system, which provides the watering of the reservoir model at reservoir temperatures and pressures with a temperature measurement error of 0.5 0C, pressure up to 0.5 ATM.

The installation allows to carry out studies of the process of displacement of oil, CO2 and water from models of homogeneous and heterogeneous formation with a degree of water cut of more than 90% by supercritical fluid systems in the range of pressures up to 20 MPa and temperatures up to 100 0C.

4-Results and discussions

The main conclusions from the results of the work on the displacement of oil by supercritical CO2 from the model of an un-watered oil reservoir are the following: First, the increase in the viscosity of the model oil in the experiments – transformer oil from 9 to 18 MPa * s – did not lead to a decrease in displaced oil ratio (DOR) within the error of the experiment. Second, the main mechanism of increasing oil losses in experiments with transformer oil is the dissolution of supercritical CO2 in the oil, the approximation of the filtration regime to the mixing one and the increase in the mobility of gas-saturated transformer oil. Third, the simultaneous increase in temperature and pressure acts in opposite directions, which leads to immutability within the error of the DOR experiment: the decrease in solubility with increasing temperature is compensated by its increasing pressure. In addition, an important conclusion from the results is that the final DOR in these experiments did not exceed 20%. The reason for this phenomenon is micro-breaks of carbon dioxide during the entire time of the experiment, violating the mixing regime of the flow of "oil-supercritical carbon dioxide" systems, determined by a decrease in the temperature of the carbon dioxide coming out of the model reservoir.

Results of the studies on the watered reservoir are shown in Fig. 2 and 3. Oil displacement was carried out from the 65% water cut reservoir model.

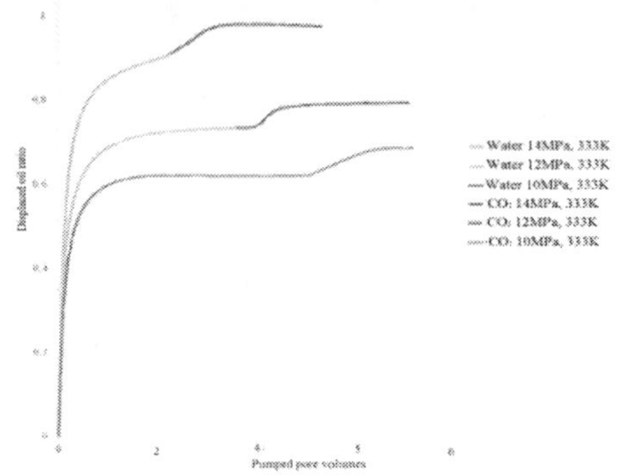

Fig. 2. The results of the oil displacement process studies from the model with 65% water cut by supercritical CO2 in the pressure range of 8–14 MPa, at isotherms 333 k.

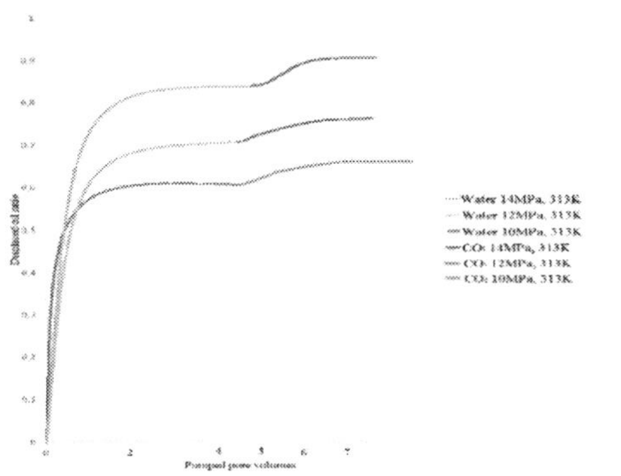

Fig.3. The results of the oil displacement process studies from the model with 65% water cut by supercritical CO2 in the pressure range of 8–14 MPa, at isotherms 313 k.

At the same time, the values of the oil displacement coefficients significantly exceeded those obtained in [5-8], which corresponds with the modern ideas about the high efficiency of supercritical fluid systems when injected into highly watered oil formations [9], the oil displacement coefficient at pressures of 10 and 12 MPa did not differ within the experimental error. However, as in the experiments on non-watered formations, the required amount of supercritical carbon dioxide at higher pressure was less: at a pressure of 10 MPa, 1.0 pumped pore volume was injected. Supercritical CO2, at a pressure of 12 MPa, about 0.8 pumped pore volume to achieve the same DOR within the error of the experiment (as shown in Fig. 4)

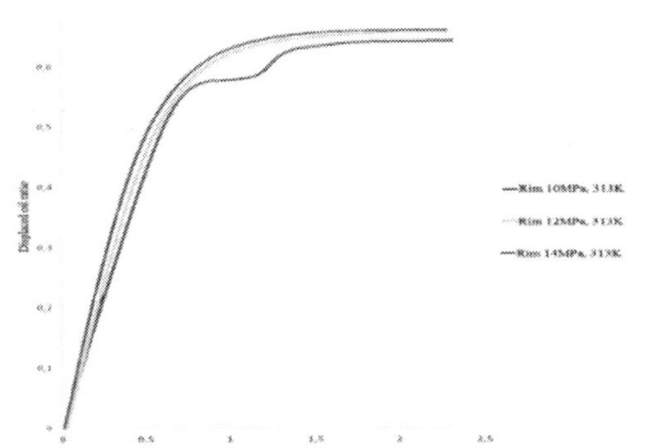

Fig.4. Value of residual oil saturation in experiments on non-watered formations within the error of the experiment.

IOP Conf. Series: Materials Science and Engineering **765** (2020) 012008 doi:10.1088/1757-899X/765/1/012008

The nature of the experimental curves is also the same, although the current values of DOR at a pressure of 12 MPa are formed at lower costs of the displacing agent, which ultimately leads to a total smaller volume of carbon dioxide expended. The mechanism of this kind of displacement processes according to modern ideas is that when supercritical CO2 is dissolved in oil, its dynamic viscosity decreases; when dissolved in water, the viscosity of water increases. Due to these processes, the viscosity of the displacing and displacing agent converges, leading to the alignment of the displacement front.It was found that throughout the studied pressure range, the increase in the oil displacement coefficient was the same within the error of the experiment, 9–10%; however, the required volume of injection of SFE SC-CO2 is different and decreases with increasing pressure. At a pressure of 10 MPa, more than 3 pumped pore volume of SFE, SC-CO2 was required to be pumped into the reservoir model; at 14 MPa, this volume decreased to about 1.5–2 pumped pore volume.The experiments also revealed a significant decrease in residual oil and water saturation during injection of supercritical CO2 and fringes of supercritical CO2 and water. The values of residual oil saturation when oil is displaced by supercritical CO2 and fringes of supercritical CO2 and water are shown in Fig.5 and 6, respectively.

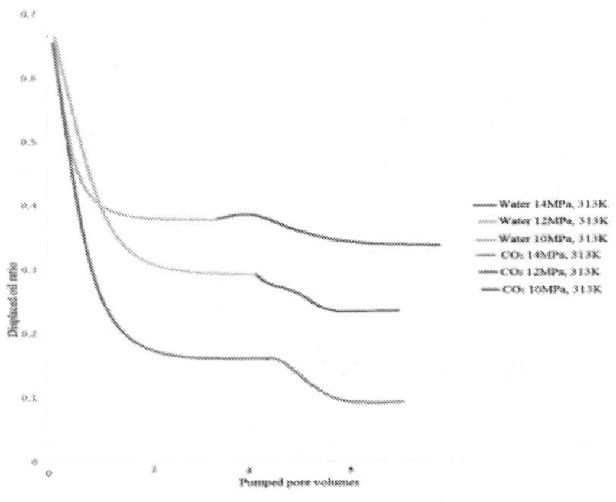

Fig.5. The value of residual oil saturation of a homogeneous porous medium during injection of supercritical CO2 into a watered formation

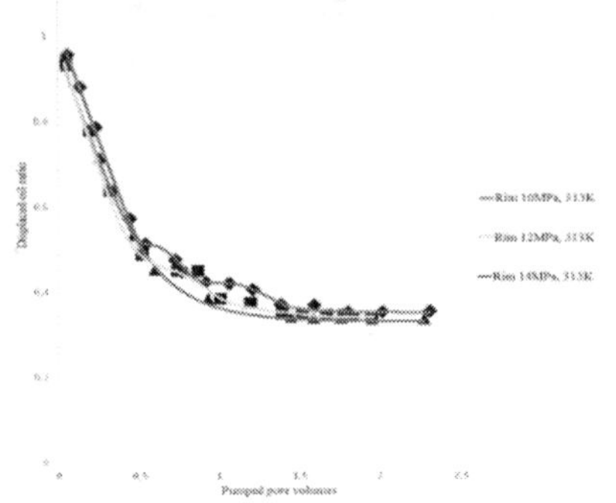

Fig 6. Value of residual oil saturation of homogeneous porous medium during injection of supercritical CO2 fringes and water.

5-Conclusions

The water saturation decreased very significantly, not less than 10 percent or more, because each measurement of the volume of displaced oil was marked by an exit from the reservoir of water during the experiment. It was found that the pressure change did not lead to a significant increase or decrease in the volume of displaced water within the error of the experiment, i.e., at a pressure of 14 MPa, the final volume of displaced water was approximately the same as at 8, 10, and 12 MPa, although its current value changed.

Three points are presented as follows:

1. The use of supercritical CO2 allows to significantly reduce the value of residual oil saturation of a homogeneous watered porous medium in a wide range of changes in temperatures and pressures of carbon dioxide injection.

2. Supercritical CO2 is the optimal agent for oil displacement from highly watered porous media and supercritical CO2 and water fringes for low-watered media.

3. An increase in the CO2 injection pressure at supercritical state parameters leads to an increase in the value of the displaced oil at all studied isotherms.

6-References:

[1] Muslimov R.Kh. Fundamental problems of the Russian oil industry and ways of transition to innovative development / R.Kh. Muslimov // Horizontal wells and hydraulic fracturing in improving the efficiency of oil field development: Materials of a scientific and practical conference. - Kazan: Slovo Publishing House, 2017.-320s.

[2] Gazizov A.A. Enhanced oil recovery in heterogeneous formations at a late stage of development, 2002.- 640 s.

[3] Radaev A.V. An experimental setup for studying the process of water-gas exposure during the displacement of viscous oils / I.A. Kondratiev, N.R. Batrakov, A.V. Radaev, R.R. Galimzyanov, A.A.

Mukhamadiev, A.N. Sabirzyanov // Bulletin of Kazan Technological University, 2013.- T.6.-№ 6.- P. 199-200;

[4] Effect of Thermobaric Conditions in a Uniform Bed on the Displacement of Low Viscosity Oil by Supercritical Carbon Dioxide / A. V. Radaev *, N. R. Batrakov, A. A. Muhamadiev, and A. N. Sabirzyanov Russian Journal of Physical Chemistry B, 2009, Vol. 3, No. 8, pp. 1134–1139

[5] Radaev A.V. The influence of thermobaric conditions in a homogeneous reservoir on the displacement of low-viscosity oil by supercritical carbon dioxide / A.V. Radaev, N.R. Batrakov, A.A. Mukhamadiev, A.N. Sabirzyanov // Supercritical fluids: theory and practice. - M: Nauka, 2009.-T.4.- No. 3.- S. 7-15;

[6] Radaev A.V. Investigation of the process of oil displacement by supercritical CO2 on a terrigenous reservoir model / N.R. Batrakov, I.D. Zakiev, R.R. Galimzyanov, A.V. Radaev, A.B. Kuzmin, A.A. Mukhamadiev, A.N. Sabirzyanov // Scientific session of KSTU, Kazan, 2014;

[7] Radaev A.V. Investigation of the displacement of hard-to-recover oils using supercritical CO2 / I.D., Zakiev A.V. Radaev, N.R. Batrakov // IX International Youth Scientific Conference "Tinchurin Readings." - Kazan, April 23-25, 2014;

[8] Radaev A.V. Investigation of the process of displacing viscous oils from carbonate reservoirs using supercritical CO2 // IX International Youth Scientific Conference “Tinchurinsky Readings.” Kazan, April 23-25, 2014;

[9] Ballint V. The use of carbon dioxide in oil production. M: Nedra, 1977.- p. 240.

[10] Mustafayev R.A. Thermophysical properties of hydrocarbons at high state parameters / R.A. Mustafaev. - M: Energoatomizdat. - 1991, 312 p .;

[11] Brot R.A. Determination of rheophysical parameters of gas-saturated oils / R.A. Brot S.E. Kutukov // Oil and Gas Business, 2005.-s. 2-12;

[12] Altunin V.V. Thermophysical properties of carbon dioxide / V.V. Altunin.- M .: Publishing house of standards, 1975.- 546 p .;

Thermal Analysis and Temperature Distribution of Two-Layer and Spoke-Type Ferrite Interior Permanent Magnet Machine

Hayder Abdulrahem
Electronic and electrical department
University of Thi-Qar
Nasiriyah, Iraq
Email: hayder.alrubyee@gmail.com

Patrick Chi-Kwong Luk
School of Water, Energy and Environment
(SWEE)
Cranfield University
Bedfordshire, U. K
Email: p.c.k.luk@cranfield.ac.uk

John Economou
Centre for Defence Engineering
Cranfield University
Shrivenham, Swindon SN6 8LA, U.K
Email: j.t.economou@cranfield.ac.uk

Abstract- The main requirements for a rotating electric machine are the torque/power density and the energy efficiency. Producing an efficient electric machine required not only studying and analysing the electromagnetic properties but also deeper thermal analysis. It has been found that attention to the thermal design can be improve the total machine performance. In this paper a thermal model of the two-layer and spoke-type ferrite interior permanent magnet (IPM) machines is developed. FEA models are built to run the thermal studies and temperature distribution of the machine under different operating conditions. A prototype machine is built and tested for the experimental validations of the FEA. A prototype machine is tested under various working conditions. The testing results validate analytical model and the FEA methods.

1. Introduction

The performances of a rotating electric machine were governed in the past by its electromagnetic design with less attention to thermal analysis. Motor designers have only superficially dealt with the thermal design aspects, maybe by specifying a limiting value of current density. Nowadays, an impressive effort is done to improve the efficiency even with a small fraction without compromising the torque density. With increasing availability of high-performance computing and the acute interest in efficiency improvement it is now necessary to analyze the thermal circuit to the same extent as the electromagnetic design. Both designs are in fact interrelated [1,2]. Not only are the losses depending on the temperatures and vice-versa, but more complex issues arise at the design stage However, the number of published technical papers relating to the thermal analysis of electric motors is still many orders of magnitude fewer than those associated with electromagnetic analysis. The published papers to date highlight a number of thermal design issues that are more difficult to analysis than others. Electric-motor thermal analysis can be divided into three basic types: analytical lumped-circuit, numerical methods and FEA calculations. The analytical approach has the advantage of being very fast to calculate; however, the developer of the network model must invest effort in defining a circuit that accurately models [2-5] the main heat-transfer paths. The analysis consists of the calculation of conduction, convection, and radiation resistances for different parts of the motor construction. The formulations for such resistances are really quite simple. The conduction resistance is equal to the path length divided by the product of the path area and the materials' thermal conductivity. The convection and radiation resistances are equal to one divided by the product of the surface area and the heat-transfer coefficient. An accurate FEA solution requires the knowledge of the thermal parameters and a superficial knowledge of the geometrical and material properties used in a machine construction is often not sufficient to give an accurate prediction of the thermal performances. The main role of FEA is in the accurate calculation of conduction heat transfer in complex geometric shapes, such as heat transfer through strands of copper in a slot. For this problem, FEA analysis can be used to calculate the equivalent thermal conductivity that can then be used in the network analysis [6]. A comprehensive understanding and modelling of the loss production mechanism are required to accurately predict the main heating source in the machine.

In this paper a thermal FEA model is proposed for the novel two-layer spoke-type ferrite IPM machines based on the equivalent thermal circuit., to predict the temperature distribution in IPM motors. Then, a model is presented It considers that: (1) the flux waveform in the yoke and tooth are trapezoidal with regard to time; (2) the eddy current loss is proportional to the square of time change rate of the magnitude of flux density. and finally, a prototype machine is tested to validate the developed analytical model of the two-layer spoke-type ferrite machine The specification of the machine is given below together with the experimental results. The experimental results show an acceptable prediction of the total temperature distribution using FEA model.

2. Machine Specification

The structure of one layer spoke type machine shown in. the **Error! Reference source not found.** general sizing parameters are listed in table I[9]

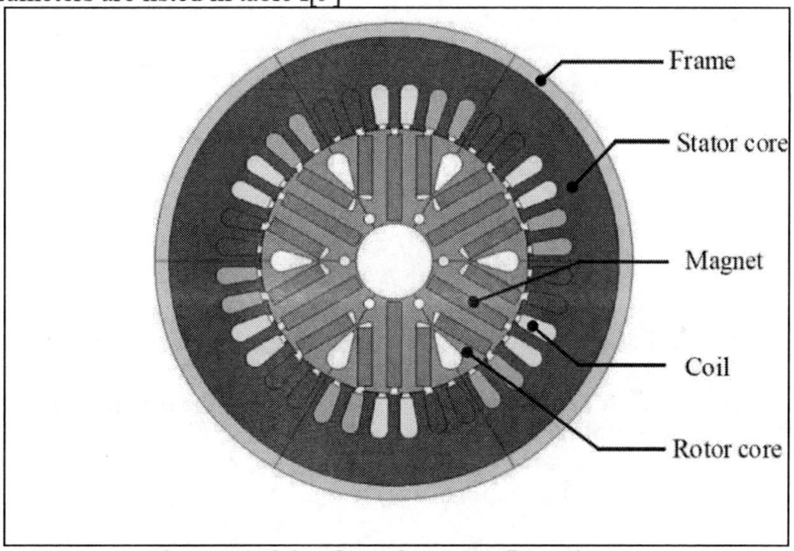

Fig. 1. Models of two-layer configurations

TABLE I GENERAL SIZING PARAMETERS

Parameters	Value	Unit
Stator Outer Diameter (Dso)	160	mm
Rotor Outer Diameter (Dro)	94.5	mm
Air gap Length (g)	0.25	mm
Stack Length (l)	90	mm
Base Rotating Speed	1500	rpm
Number of Poles (p)	6	
Number of Slots (q)	36	
Lamination Material	50CS470	
PM Material	Y30	

3. FEA Thermal Model

The cross-section of the machine is given in Fig1**Error! Reference source not found.** the winding is 3-phase with 6 coils per phase giving a single layer winding with a coil pitch of 6 slots. The rated current was 6.2A and the speed fixed at 1200rpm. The frame size was 160mm with an axial core length of 90mm To carry out an accurate thermal analysis, it is necessary at first to find the amount and location of heat generation. It would be essential to calculate the losses in copper and that caused by the magnetic field using the finite element method, and from there to carry out a thermal analysis using the resulting loss distribution. Table II shows the loss in each part of the motor at rated load and speed. Copper loss in the coils and iron loss in the core are the dominant heat sources, so this analysis mainly evaluates the effects of this heat. The loss in the coil is relatively large, about 2/3~3/4 of the total loss, so it is expected to be the main source of heat.

TABLE II LOSS IN EACH PART AT RATED LOAD

Heat generation parts	Iron loss in the rotor core		Iron loss in the stator core	Iron loss in the coils
Losses (W)	9.49		38.28	117.8

According to the structure of the designed ferrite IPM machine, a thermal circuit model is built and shown in Fig 2. The square blocks in the circuit represent the motor's core parts that generate and transmit heat, and they are meshed in FEA simulation. The circle blocks represent the surroundings and ambiences, and they absorb the heat from inner motor.

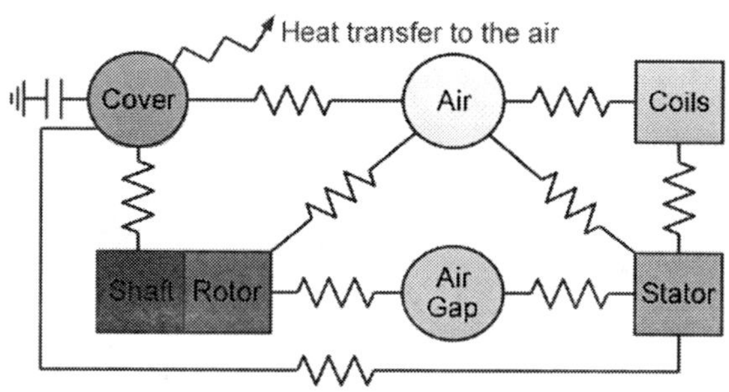

Fig .2. Thermal circuit model of the designed motor

The resistance signs are illustrated as thermal resistance among the square and circles blocks, such as thermal resistance between coil and stator core, stator core and cover, cover and air, and so on. The capacitor sign attached to cover is heat capacitor, and the cover's heat capacity is related to its volume, density and specific heat. All the heat will eventually transfer into the ambient air.

4. FEA Results

4.1. Temperature Distribution at Steady State

The simulation of thermal circuit model is carried out with rated motor load. The motor operates continuously until a thermal balanced state is reached. The final balanced state is also called steady state, and the temperature distribution of it is shown in Fig**Error! Reference source not found.** 3.

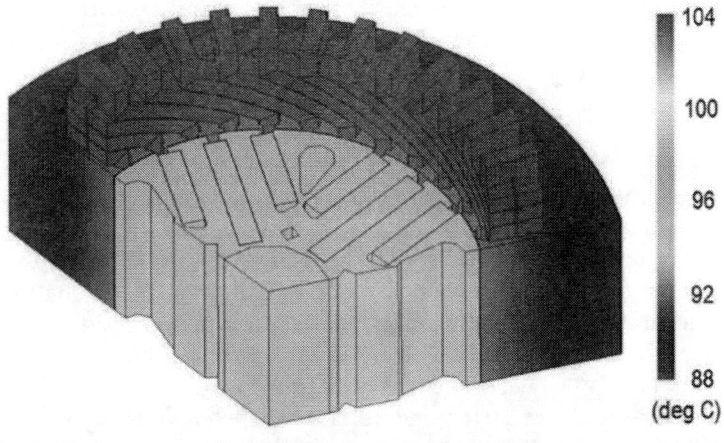

Fig. 3. Temperature distribution at a steady state

According to the temperature distribution, the motor is generally divided into three thermal regions: the copper coils, the rotor and the stator. The highest temperature region is the copper coils, as copper loss is the dominant source of heat. The highest temperature is about 104°C. Most of the heat from the coils is transferred to the cover through the stator core, and the rest goes through the air inside of motor to the cover and finally dissipates to the ambience. The moderate temperature region is the rotor. Since ferrite PM is non-conductive, no loss is generated in magnets. The heat accumulated in the rotor is mainly from its iron loss. As the majority of the rotor's surface area is encompassed by air, which has high heat transfer resistance, it is hard for the rotor to dispel the generated heat. The accrued heat increases the temperature of the rotor and the magnets imbedded. The rotor has higher temperature than the stator core. Most of the heat generated in the rotor flows through the air around it and then to the stator and cover, and the rest is transferred to cover through the bearings and shaft.

To further understand the process of temperature changing, the temperature of some typical points in the motor are measured and compared. The measuring points are shown in Fig 4,

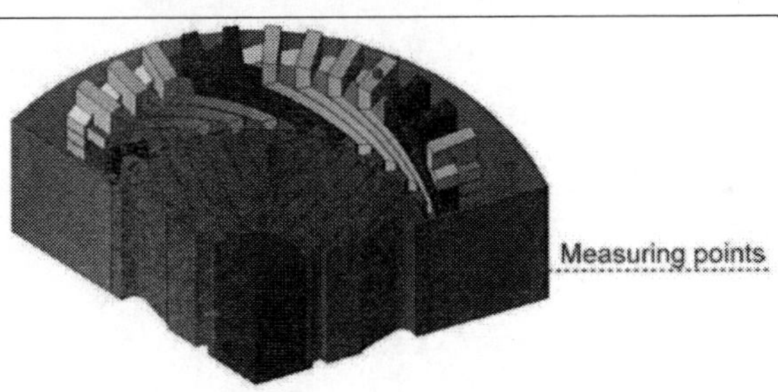

Fig. 4. Points of measuring temperature for each part

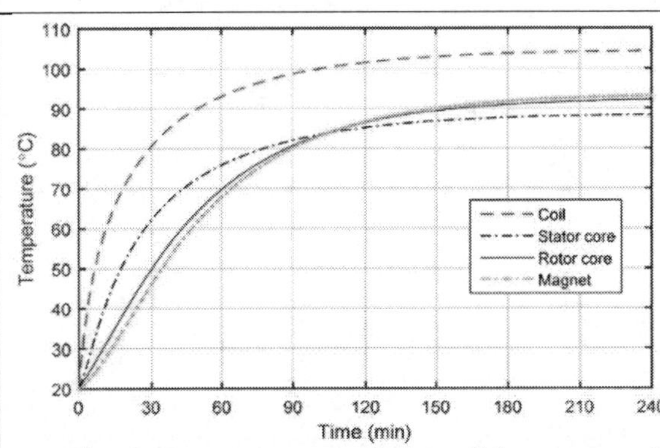

Fig. 5. Temperature variation of the measured points

which include points on rotor core, coil, magnet, and stator core. The initial temperature is 20°C. The temperature measurement lasts for four hours and it is recorded in Fig5.

The temperature of the four points increase fast in the first two hours and gradually become stable in the last two hours. The highest temperature appears in the coil. As the main source of heat comes from copper loss, the winding temperature is growing faster at the beginning and more than 10°C higher than other measuring points at the end. In the first 100 minutes, the temperature of stator core is higher than rotor core and magnet, but it is exceeded later. The losses in the stator are much higher than rotor, and the stator temperature rises much faster than rotor at first. However, the rotor has worse heat dissipation conditions as it is surrounded by air and the heat generated in rotor is difficult to transfer to the outside; while the stator core is directly (interference) fitted to motor housing that transfer the heat from stator efficiently. The accumulated heat in rotor makes it hotter than stator in relatively long running time. The results are consistent with the temperature distribution of steady state in Fig 5.

4.2. Continuous Operating Time at Over-Load Conditions

In some applications, motors are required to work in overload conditions for a short while. The over load ability is an essential characteristic to evaluate a motor and useful for motor selection. Larger output power requires larger input current. Although increasing the current produces a higher torque, it also boosts the heat generation in terms of copper losses and core losses. Owing to the large input current, the copper losses generated in the windings increase exponentially. Along with coil temperature surging up, the resistivity of the copper rises, which further deteriorates the copper losses. The iron losses in stator and rotor cores are also increased due to the large armature reaction of the designed IPM machine. Accordingly, the temperature of stator and rotor will increase dramatically with current, and the properties of the ferrite PMs will be worsening by high temperature. Therefore, it is vital to find the longest consecutive operating time under over load condition before causing any irreversible damage. The rated current is 6.2 A (RMS value), designated as 1C, and the corresponding average torque is 13.32 Nm, designated as 1pu. When the input current is increased from 1C to 3C, the output torque grows from 1pu to 3.1pu, which are summarized in Table III.

Table III Current and average torque

Input Current	1C	2C	3C
Average Torque (pu)	1	2.1	3.1

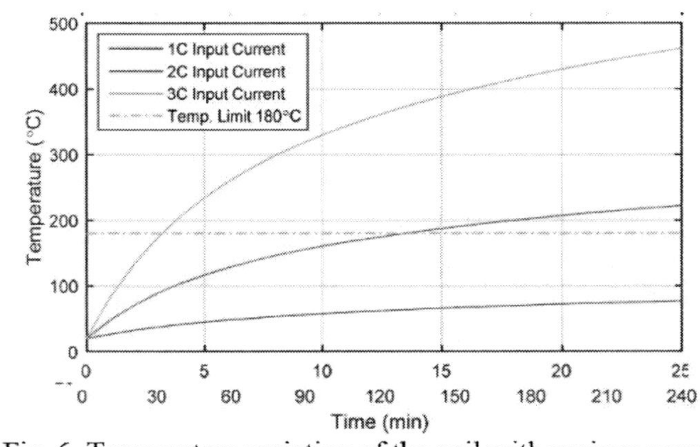

Fig. 6. Temperature variation of the coil with various currents

The enameled copper wire used in the designed motor is in thermal Class H (180°C), and the temperature limit is indicated in green dash-dot line in Fig 6.

The shape of the temperature variation lines with various currents is similar but slop of them is different. The larger the input current, the steeper the temperature changing line is. In rated load condition (input current is 1C) the temperature of the coil is always under the temperature limit; in double load condition (input current is 2C) it takes about 13 minutes to reach the temperature limit; and in triple load condition (input current is 3C) the coil's temperature exceeds the temperature limit in about 3 minutes, which is much shorter than double load condition. The magnet material used in the designed motor is ferrite Y30, the maximum working temperature of which is 180°C that is indicated in green dash-dot line Fig7.

IOP Conf. Series: Materials Science and Engineering **765** (2020) 012009 doi:10.1088/1757-899X/765/1/012009

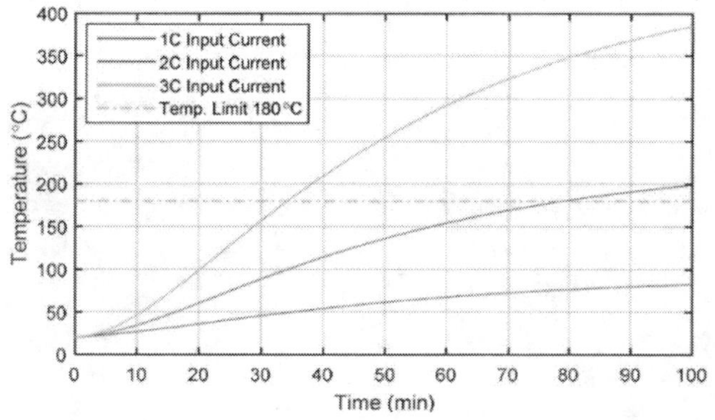

Fig. 7. Temperature variation of the magnet with various currents

Similar to the temperature variation of the coil, the temperature trend of the magnet with various input currents have similar shapes but different slops. In the rated load condition (input current is 1C) the magnet's temperature stays under 70°C; in double load condition (input current is 2C) it does not reach the temperature in the 78 minutes running time; and in triple load condition (input current is 3C) the magnet's temperature exceeds the temperature limit in no more than 34 minutes.

5. Experimental Validation

The stator, rotor, shaft and PMs of the two-layer ferrite IPM prototype machine are demonstrated in Fig 8, and the dimensional specification is listed in table I. The setup of the testing rig for loaded experiments is depicted in Fig 9 The IPM prototype machine is powered by TI HVDMC Kit. The torque sensor is connected directly to the motor shaft to measure the output torque. The programmable dynamometer is connected to the other side of the torque sensor to provide required loading.

(a) FEA model (b) Stator (c) Rotor and shaft (d) Rotor

Fig. 8. Models Of Two-Layer 2kW proposed machine

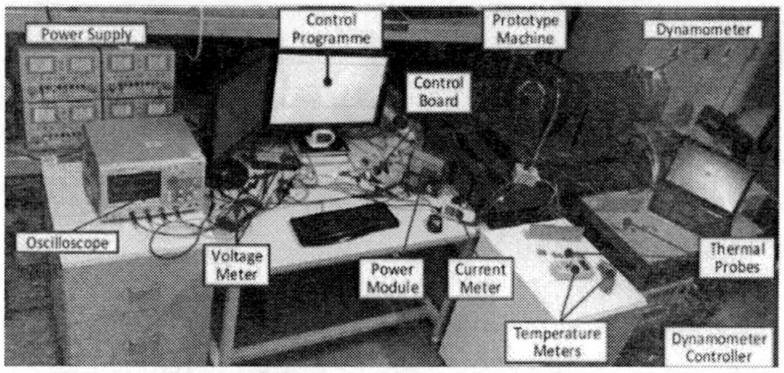

Fig. 9. Setup of the testing rig. Of IPM

IOP Conf. Series: Materials Science and Engineering **765** (2020) 012009 doi:10.1088/1757-899X/765/1/012009

The prototype machine is tested with different load conditions (0.5, 1, 2, and 3) times rated power. Fig 10 and Fig 11 show how the thermal sensors are installed in different positions of the tested motor to generate data for this work.

Fig. 10. Thermal props installation

Fig. 11. Motor with the thermal props installation

The Motor is thermally monitored by K-type wired thermal meter and sensors. Sixteen sensors are used on the end winding (one for each phase), other sensors are insert inside a stator slot. The last sensors group is in a hole positioned in the stator core. This measurement set up allows measurement of the winding and iron core temperatures during the tests. The housing temperature can be measured by Infrared thermal meter.

As shown in Fig 12, 16 thermal probes are divided in two groups, and each group has 8 probes and numbered from 1 to 8. Measure point 2, 3, 4, and 5 are fixed on the surface of copper windings, where point 2 is around the corner between winding and stator teeth. Measure point 1, 6, and 7 are fixed on the top surface of stator core, where point 6 is at the corner between winding and stator core. And measure point 8 is mounted on the inner surface of housing frame.

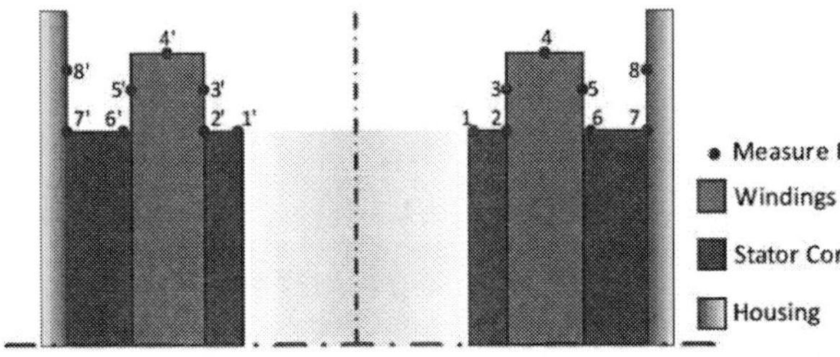

Fig 12 Cross-section of stator end and thermal measuring points

When all the thermal probes have been properly adhered inside of the motor housing, the prototype motor is reassembled and installed on the testing rig. As shown in Fig 9 the setup of the thermal testing consists of two groups of thermal probes, a wired thermal meter and a infrared thermal meter. The wires of the first and second group of thermal probes are bundled in red and blue heat-shrink tube respectively.

6. Experimental Results.

Several results were run over time however here we will concentrate on four sets of results. Since the relative measuring positions of the two groups of thermal probes are same, the temperature results will be the average of the two groups. Before the starting of the motor, the temperatures of all the parts of the machine are same, which is the ambient temperature 27.2 °C. To compare with simulation results conveniently, the beginning temperature is adjusted to 20 °C while the temperature increases are kept unchanged.

The thermal test of the designed ferrite IPM motor takes 240 minutes and the measuring results are displayed in below figures.

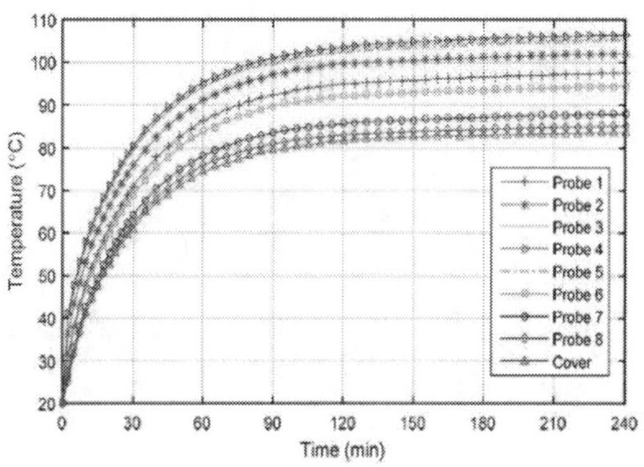

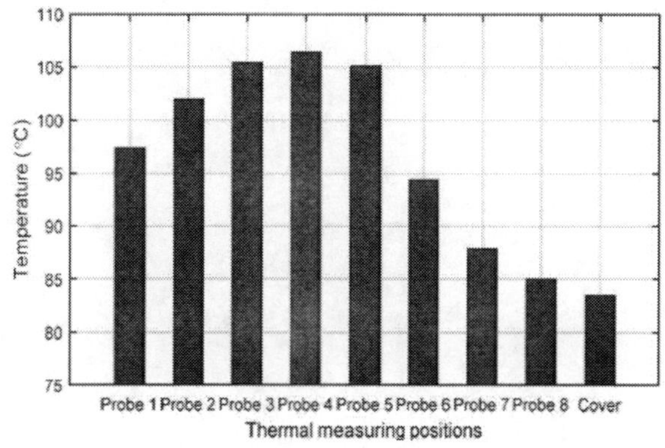

Fig 13 Measured temperature of the ferrite IPM motor Fig 14 Temperature distribution of the motor

As shown in Fig 13 the temperatures of the motor grow fast in the first hour and gradually slowdown in the second hour. After two hours of running, the motor is getting into thermal balanced state, and there is no much variation of its temperature in the last two hours. Although the temperature growing rates are different for these various measuring positions, there is no cross lines in the temperature changing curves. That means all the measuring points keep the same thermal gradients all the time, and the temperature distribution of the motor at steady state is depicted in Fig 14. The temperature sequence of the measurement is: Probe 4 > Probe 3 > Probe 5 > Probe 2 > Probe 1 > Probe 6 > Probe 7 > Probe 8 > Cover. Because the copper loss is the primary heat source, the temperature of the winding increases rapidly and is always higher than any other part of the motor. Since all the generated heat is transferred from the inside to the outside and the area near the housing is cooler, the temperature of point 5 is slightly lower than that of point 3. The point 2, at the corner of copper coil and stator tooth, is heated by the winding but is easier than point 3, 4 and 5 to dissipate its heat as it is adhered to stator core.

7. comparison with experimental

The thermal testing results of the designed ferrite IPM motor are compared with its simulation results that are obtained in previously. The comparison of copper coil temperature and stator core temperature are depicted in, Fig 14 and Fig 15 respectively

Table IV Comparison of steady state temperature

	Stator	Copper Coil
Analysis (°C)	98.5	115
Simulation (°C)	88	106
Experimental(°C)	90	108.5

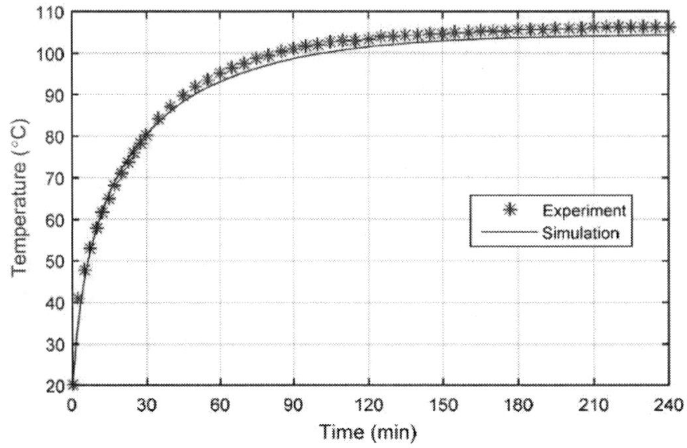

Fig. 15. Temperature comparison of copper coil

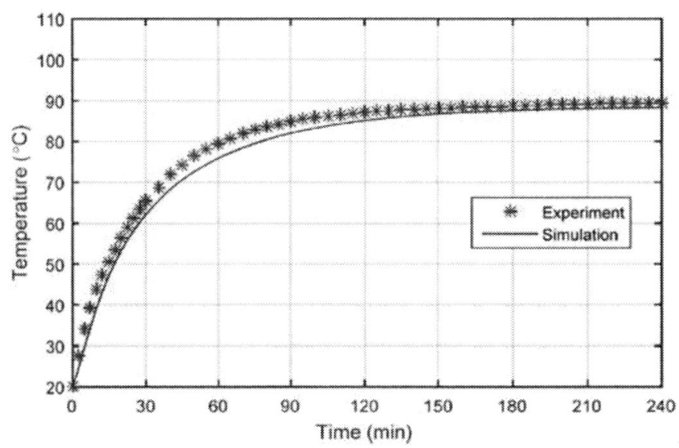

Fig. 16. Temperature comparison of stator core

The analytical results show about 7% higher temperature than the FEA one and 10 % higher than experimental test. There are two main possible explanations for this difference: firstly, it is difficult to calculate the accurate coefficient of heat transfer for the thermal equivalent circuit, secondly, the radiation effect has been ignored in analytical model. However, the results of the proposed analytical model still show a good agreement and Therefore, it can be an effective method to predict the temperature of the two-layer spoke-type ferrite machine in the design process

In the first half hour, the temperature results of experiment have good agreement with that of simulation, as shown in Fig 15 but after that the experimental temperature is about 2°C higher than the simulated temperature. This might be because the actual thermal testing is conducted in a closed room and the ambient temperature is increasing during the experiment, which is caused by the heat that is generated

from the running motor. But the ambient temperature in the simulation is set to be constant, no increase during the operating process.
Since there is no direct measurement that is carried out at the same position where is sampled in the thermal simulation. According to the sampling position in Fig 4, the stator core temperature at the corresponding position is estimated from the temperatures of its nearby measuring point 6 and 7 and compared with simulated results in Fig 16. The actual testing stator core temperature grows a litter faster than the simulated temperature, and all get into balanced state after around 150 minutes. The testing temperature is larger, and the largest difference is about 4°C happening around the 60th minute. And in the steady state the temperature is about 1.5°C. The difference of the stator core temperature may be caused from the inconsistency of lamination material.

8. Conclusion

This paper proposed a thermal equivalent circuit of a Two-Layer Spoke-Type Ferrite Interior Permanent Magnet Machine to estimate the temperature. A FEA model has been developed (using a combination of an electromagnetic analysis is and a thermal model) to Comair the analytical results. The results show that the proposed thermal equivalent circuit method can be a effective method to estimate the temperature distribution in the motor in the design process. From the FEA simulation and experiment test, it's seen that the high temperature region is the winding, as copper loss is the dominant source of heat. The moderate temperature region is the rotor, in which the heat is accumulated from its iron loss. Because the heat generated in the stator core are easier to be dissipated into ambient air, the temperature of which is relatively low. Moreover, the consecutive operating time of the ferrite IPM machine is determined by the thermal limitation of enamelled copper wire. The FEA results show that the motor can continually work for 29 minutes in double load condition, but it can only last for 5 minutes under triple load. Finally, a prototype machine is tested to validate the developed analytical model and FEA model of the two-layer spoke-type ferrite machine

REFERENCS

[1] J. J. Nelson, G. Venkataramanan, and A. M. El-Refaie, "Fast thermal profiling of power semiconductor devices using Fourier techniques," IEEE Trans. Ind. Electron., vol. 53, no. 2, pp. 521–529, Apr. 2006.

[2] Z. Gao, T. G. Habetler, R. G. Harley, and R. S. Colby, "A sensorless rotor temperature estimator for induction machines based on a current harmonic spectral estimation scheme," IEEE Trans. Ind. Electron., vol. 55, no. 1,pp. 407–416, Jan. 2008.

[3] P. Mellor, D. Roberts, and D. Turner, "Lumped parameter thermal model for electrical machines of TEFC design," Proc. Inst. Elect. Eng., vol. 138, no. 5, pp. 205–218, Sep. 1991.

[4] W. Fei, B. Xia and J. Shen, "Design of spoke-type ferrite interior permanent magnet synchronous machines considering the number of rotor poles," 2017 20th International Conference on Electrical Machines and Systems (ICEMS), Sydney, NSW, 2017, pp. 1-6. doi: 10.1109/ICEMS.2017.8056387

[5] A. Boglietti, A. Cavagnino, D. Staton, M. Shanel, M. Mueller, and C. Mejuto, "Evolution and modern approaches for thermal analysis of electrical machines," IEEE Trans. Ind. Electron., vol. 56, no. 3, pp. 871–882, 2009.

[6] C. Kral, T. G. Habetler, R. G. Harley, F. Pirker, and G. Pascoli, "Rotor temperature estimation of squirrel cage induction motors by means of a combined scheme of parameter estimation and a thermal equivalent model," in Proc. IEMDC, 2003, pp. 1049–1057.

[7] B. H. Lee, K. S. Kim, J. W. Jung, J. P. Hong, and Y. K. Kim, "Temperature estimation of IPMSM using thermal equivalent circuit," IEEE Trans. Magn., vol. 48, no. 11, pp. 2949–2952, 2012.

[8] A. Boglietti, A. Cavagnino, M. Lazzari, and M. Pastorelli, "A simplified thermal model for variable-speed self-cooled industrial induction motor," IEEE Trans. Ind. Appl., vol. 39, no. 4, pp. 945–952, Jul./Aug. 2003.
[9] B. Xia and P. C. Luk, "Analytical Model of Open-Circuit Characteristics of Two-Layer Spoke-Type Ferrite Interior Permanent Magnet Machines", IEEE International Electric Machines & Drives, Miami, USA, May 21-24,2017

Inductance Calculation and Torque Separation of a Unique Multi-Layer Spoke-type Interior Permanent Magnet Motor using the Frozen Permeability Method

Hayder Abdulrahem
University of Thi-Qar
Nasiryah, Iraq
Email:
hayder.alrubyee@gmail.com

Patrick Chi-Kwong Luk
School of Water, Energy and Environment (SWEE)
Cranfield University
Bedfordshire, U. K
Email: p.c.k.luk@cranfield.ac.uk

Mohammed K. Al-Saadi
Electromechanical Engineering Dept
University of Technology
Baghdad, Iraq
Email:
mohammed16377@gmail.com

Abstract—Frozen permeability approach (FP) is employed to separate the average torque components (reluctance and permanent magnet torque) and for the inductance calculation for a Unique Ferrite Multi-Layer Spoke-type Interior Permanent Magnet Motor. Finite element analyses (FEA) models are built to provide many results help to offer more insight analysis. Three factors are taking in account to analyse their effects: permanent magnetism, armature currents of stator and angular rotor position. Also, different saturation levels are analysed, to reveal the influence of magnetic saturation and crosscoupling. These detailed analyses are providing valuable insights for permanent machine design and optimization.

Keywords— interior permanent magnet; Frozen permeability method; torque separation; nonlinear magnet, inductance calculations.

1. Introduction

The energy efficiency of a rotating electric machine and the torque/power density are an essential requirement. Recently, significant effort has been delivered to enhance the efficiency even with a small fraction without losing the torque density. Due to dramatically development in computational power the need for efficiency improvement, the Permanent magnet synchronous machines (PMSMs) are gaining the most increasingly attractive in various applications to be promising alternative due their high-power density, higher efficiency and exceptional control ability [1-3].

A comprehensive understanding and modelling of the torque component and flux distribution including saturation and crosscoupling effect production mechanism are required to accurately predict the performance of such an electric machine. In case of on-load PM or armature field, its essential to calculate the average torque separation, the flux density or flux linkage components. Existing research papers have been reported three methods have been reported in describing the average torque separation methods [*1-3*]: 1) the constant PM flux linkage model; 2) the partial-coupling model [*1*], in which the PM flux linkage restricted to axis current changing only; and 3) the frozen permeability (FP) FE method. Since the PM and armature fields influence each other and change with the magnetic saturation and crosscoupling, both the constant PM flux linkage model and the partial-coupling model may result in significant inaccuracy. Hence, the FP method is often employed [*4*] in order to accurately separate the PM and armature fields and hence the torque components accounting for the magnetic saturation and crosscoupling.

In this paper, inductance flux and the average torque components separated are calculated by the analytical model and compared and when the FP method is employed. Different values of average torque components have been calculated using these two methods. The average torque separation can be implemented properly with the FP method, while the analytical model cannot. The analysis has been done using inset of PM prototype machine, which is described in Section 2-1. The FP method is

IOP Conf. Series: Materials Science and Engineering **765** (2020) 012010 doi:10.1088/1757-899X/765/1/012010

illustrated, and the procedure of the on-load field components separation are introduced and analysed in Section 2-2. The field separation is examined and compared in normal FE in Section 3-1 Analyses for the total torque calculation and for the average torque separation in Section 4. Finally, the key point are highlighted in Section 6.

2. Background

2.1. Proposed Machine

PMSMs are one of must promising machine due to its high torque density and efficiency comparing to the IMs, However, As the future supply of rare-earth materials is likely to be limited because of the strategical nature, the prices of rare-earth materials and highly unstable in the global market. Ferrite magnets are representing the most alternative potential for non-rare-earth PMSMs to replace the IMs. For ferrite motor, since the eddy current loss in PM poles is eliminated and eddy loss of armature winding is negligible, overall eddy loss is extremely low. Moreover, the coreless in the rotor is reduced due to the non-conductive ferrite pole layers. In this work a 2kW two-layer spoke-type ferrite prototype machine is used to evaluate the impact of both crosscoupling and magnetic saturation, as well as the PM variation and reluctance torque components. A commercial version of induction machine with a power of 1.5kW and 36 slots with 4 poles is utilized in our work to provide housing and stator of our prototype machine. Starting from the analytical models with the help of optimization algorithm, results in a design of 6 poles rotor, while the stator windings are re-winded using single layer distributed windings based on short-pitch winding [6]. Fig. 1 (a)&(b) illustrates the cross-section of the machine model as well as the 3-D model of the prototype machine respectively. The specifications and dimensions are detailed in table I. Two magnets (PM2) are added to form a V-shape in between two first layer PMs to reduce the use of PM material and improve the reluctance torque. The current is three phases ideal sinusoidal and the current phase with a maximum amplitude=8.8A and the advance angle (phi) is varied from 0 to 90 The q axis current iq is positive and the d axis id current is negative.

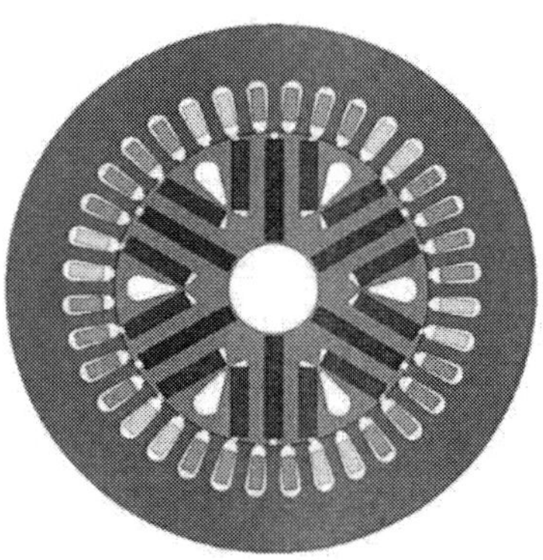

(a) FEA mode

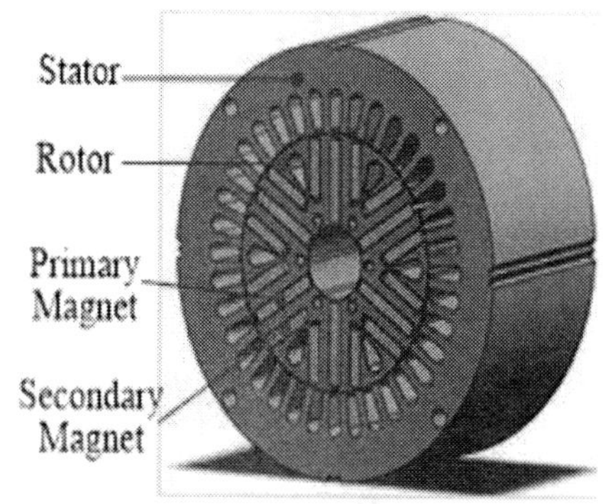

1(b) 3D model

Fig. 1. The proposed machine configuration

TABLE I GENERAL SIZING PARAMETERS

Parameters	*Value*	*Unit*
Stator Outer Diameter (Dso)	160	mm
Rotor Outer Diameter (Dro)	94.5	mm
Air gap Length (g)	0.25	mm
Stack Length (l)	90	mm
Base Rotating Speed	1500	rpm
Number of Poles (p)	6	

TABLE I GENERAL SIZING PARAMETERS

Parameters	*Value*	*Unit*
Number of Slots (q)	36	
Lamination Material	50CS470	
Torque	16	N.m
Rated speed	1500	rpm
PM Material	Y30	

2.2. Frozen Permeability Principle and Procedure

The principle of The FP method used to investigate the issues related with magnetic saturation and crosscoupling is illustrated in Fig 2 and will described in [1]

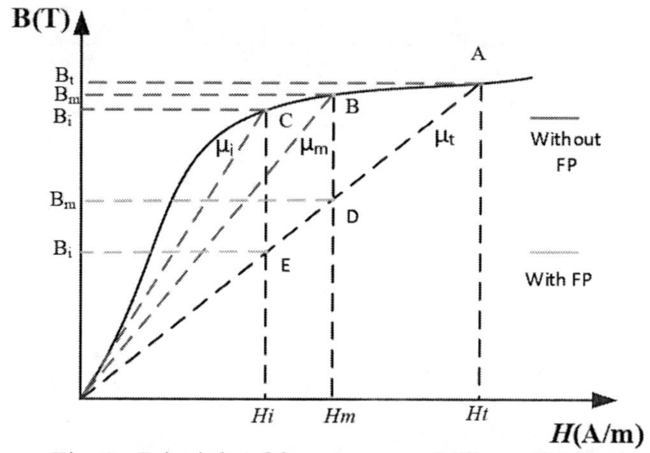

Fig. 2. Principle of frozen permeability method

Based on excluding FP method, and using only the PM excitation (H_{PM}), i.e., open circuit, this will make the resultant flux density is B_{PM} (point B). Also, the resultant flux density is Bi (point C) when the armature excitation (Hi) only. When the PM machine is working on load (point A), it is excited by both PM and current, i.e., $H_{all}=H_{PM}+H_i$. However, the resultant flux density Ball is lower than $B_{PM}+B_i$. It is not possible to decompose the on-load PM and armature field components. When using the FP method, firstly, μ_{all} and the on-load permeability is obtained and stored based on the on-load model (point A). Then, points D and E which are modeled by two linear analyses, can be solved. This solution is based on the frozen μ_{all} with one of two components, either the PM or armature excitation only to obtain B (FP, PM) or B (FP, i) respectively. The difficulty of nonlinear problem has been overcome by approximate it into linear one using freezing the permeability as µall, B_{all}=B (FP, PM) +B(FP,.i). This approximation helps to decompose the on-load PM and armature field components. Furthermore, we are including the influence of magnetic saturation and crosscoupling because μ_{all} varies according to the operation point. Fig. 3. Summarizes the procedure of the FEA computation which is based on the principle of the FP method.

IOP Conf. Series: Materials Science and Engineering **765** (2020) 012010 doi:10.1088/1757-899X/765/1/012010

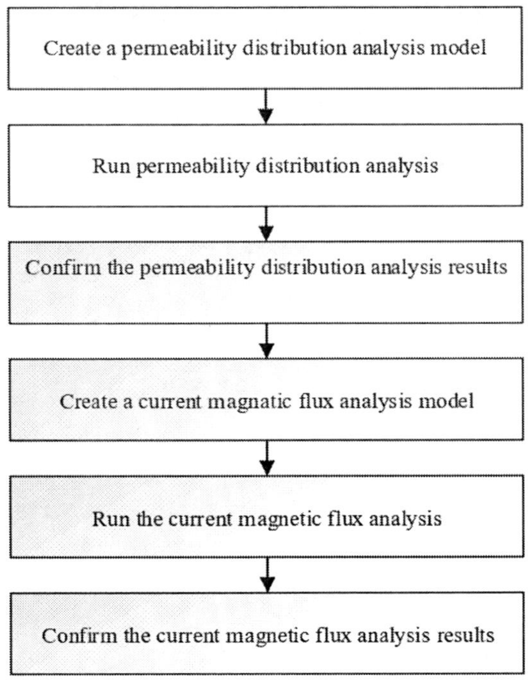

Fig.3. The procedure to obtaining the results

3. Analytical Calculations

3.1. Magnetizing inductances calculations

The saliency and reluctance torque are calculated after the d- and q-axis inductance have been calculated. In this work, starting with building analytical models. This model is used to evaluate the novel two-layer spoke-type ferrite IPM machines. The model has used to predict the armature reactions for prototype machine. The prediction is implemented based on the equivalent magnetic circuit theory in the d-q frame. the d-axis and q-axis magnetizing inductance can be obtained by:

$$\begin{cases} L_{md} = \frac{\Psi_{md}}{i_d} = \frac{4N_aR_{ro}l_{ef}}{p\cdot i_d}\sum_{v\neq m\cdot k}\frac{k_{wv}}{v}B_{gdv} = \sum_{v\neq m\cdot k} L_{mdv} \\ L_{mq} = \frac{\Psi_{mq}}{i_q} = \frac{4N_aR_{ro}l_{ef}}{p\cdot i_q}\sum_{v\neq m\cdot k}\frac{k_{wv}}{v}B_{gqv} = \sum_{v=6k\pm 1} L_{mqv} \end{cases} \quad (1)$$

To reveal the influence of flux distribution with different harmonic orders, their contributions to the q-axis magnetizing inductance are listed in Table II.

Table II MAGNETIZING INDUCTANCE CALCULATIONS

Harmonic order	**magnetizing inductance calculations**		
	Analytical Lmd	***Analytical Lmq***	***Analytical Uf***
Total	0.01381484	0.090871165	0.277662023

The analytical inductance values are considering to be constant under different operating conditions. Much higher reluctance torque can be expected because of high Lq/Ld salient ratio, while the one-layer machine will be majorly of torque components dependent on PM torque

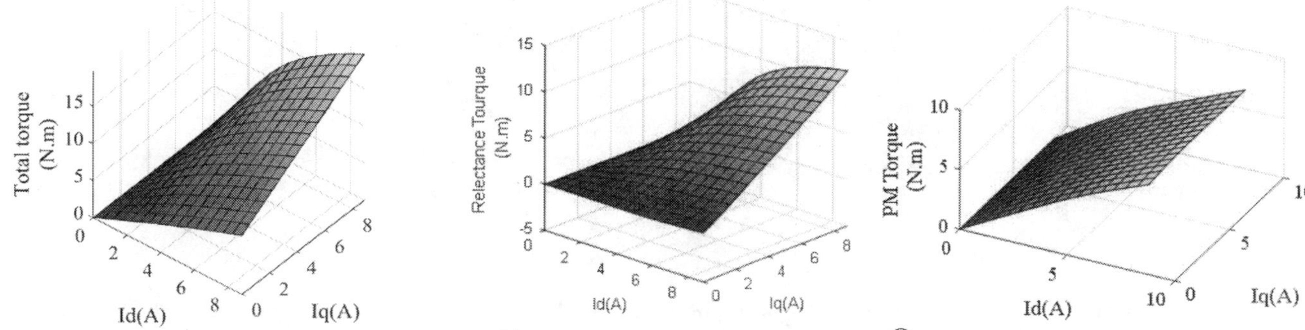

(a) (b) ©

Fig. 4. Torque profiles against id and iq by FEA (a) total torque (b)reluctance torque (c) PM torque.

3.2. Torque Production

The Permanent magnet motor has two components of electromagnetic torques: 1. The first torque component is PM. This component is resulting from the interactions between the field comes from armature windings and PM, 2. the second torque component is reluctance torque. This component is arising by the winding inductance changing with the rotor position due to the rotor saliency. The total torque production for a PMSM is calculated by [8]

$$\begin{cases} T_{pm} = \dfrac{mp}{4}\Psi_f i_q \\ T_r = \dfrac{mp}{4}(L_d - L_q)i_d i_q \end{cases} \quad (2)$$

Ferrite PM machine cannot achieve high torque density because of its low magnetic strength as compared to PM torque density the rare-earth counterpart. To improve the torque density, a larger difference between the d- and q-axis inductances is needed as indicated in the second term inside the brackets of Equation (1). Fig. 4. illustrates the changes of average torque components with the current. At high currents values (high id and high iq), the overall electromagnetic torque performances of the Machin are expected to be different from real values due to heavily saturated working condition. The results reveal that the increase of armature current, the reluctance torque will grow more quickly than PM torque. Consequently, the contribution of the two torque components could be quite different under different loading conditions.

4. Simulation Results

4.1. Field Distributions

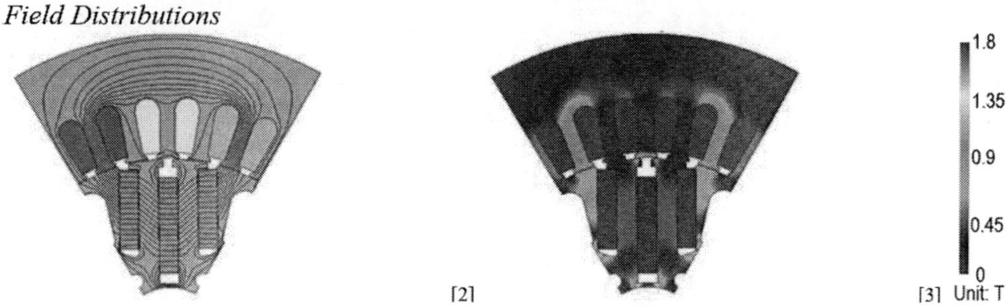

[1] [2] [3]

[4] (a). [5] (b) [6]

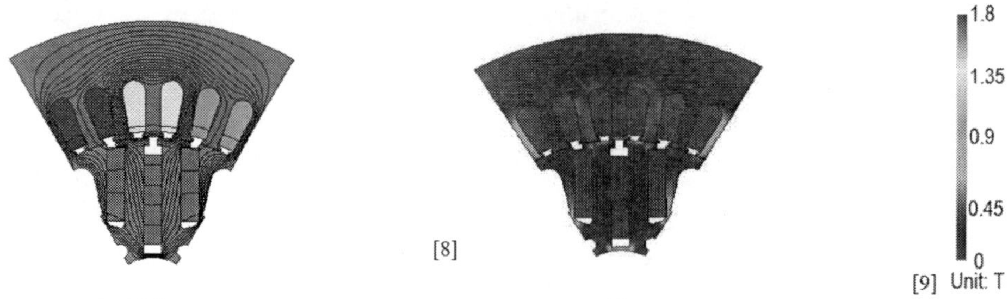

[7] [8] [9]

[10](c) [11](d) [12]

Fig. 5. field distributions when id=4.4A &iq=0A, 0. (a) and (b) whole field (current and PM) and (c) and (d) armature field current (FP)

Under on-load condition PM and armature field components can be calculated using the FP method. Fig 5 shows a contour plot of the magnetic flux density distribution and magnetic flux lines when the peak phase current is id=4.4 A, iq=0 A, and rotor position is 0 (electrical), results shows that the permeability distributions are not symmetrical in both d and q exes and the field distribution is asymmetrical accordingly. In Fig. 4(a) and (b) the influence of armature field can be seen. Therefore, the on-load PM field Fig. 4 (c) is asymmetric with the d-axis and has q-axis flux linkage, although the PM excitation

4.2. Magnetizing inductances calculations

Practically, the saturation and crosscoupling effects led to

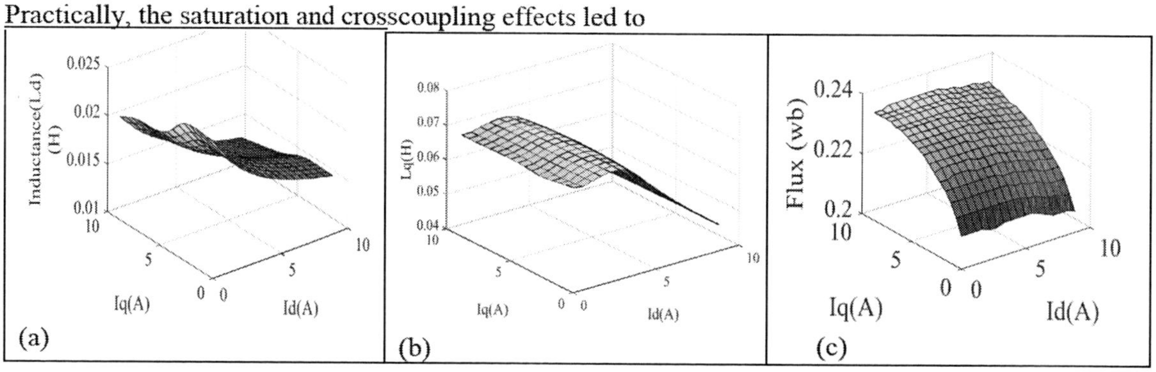

Fig. 6 Flux linkage and inductance of FEA (a)Lmd (b)Lmq (c)Uf

a change in these values. Fig6 demonstrates the variation of average flux linkages and inductances with the id and iq obtained by using FP method.

Fig 6-a shows that the Lmd affected more by id values than iq. For example, when iq was 1.65A, the Lmd has change from 0.023 to 0.018 when id varied from 0 to 8.8A.

5. Torqu Profile Predictions

5.1. Torque Production

The variation of average torque components with the

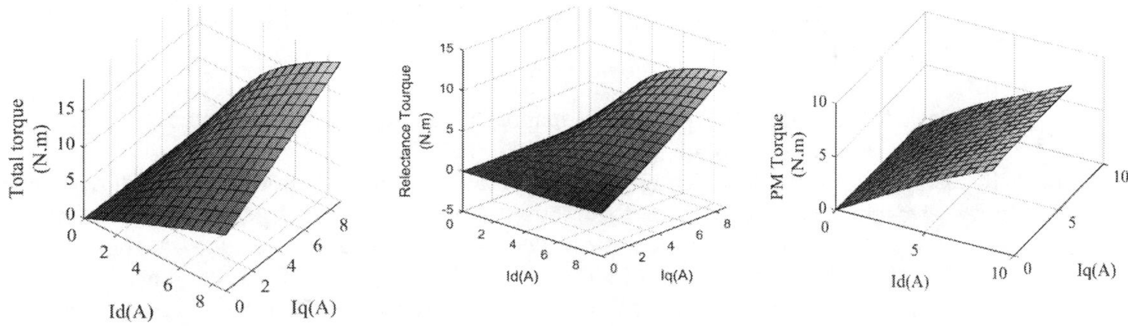

(a) (b) (c)

Fig. 7. Torque profiles against id and iq by FEA (a) total torque (b)reluctance torque (c) PM torque..
current resultant from FEA is shown in Fig. 7. T_{pm} and T_r are

influenced by the magnetic saturation. The (PM) torque component increase and become higher by increasing id. While it is less sensitive to the change in iq values. For example, when iq=1.1
Under zero id current, results show that The PM torque is almost identical for both models (analytical and FEA). This result can be representing the accuracy PM flux prediction. However, there is a slight difference between two model's reluctance torque because q-axis flux linkage and inductance are slightly lower than the FEA simulation. This difference will result in lower torque for the analytical method with the increase of id and machine loading. As a consequence, the error will increase because of saturation effect. The error will be n=more noticeable under heavy load conditions. The error may go up to 14% of the maximum torque output under severe saturation condition as depict in Fig. 5. However, with higher current angle, larger demagnetizing injected current the saturated is alleviated. Again, the predicted torque by analytical models coincides back with FEA. The trend of torque profile and the optimal current angle can be well predicted as compared with analytical, in spite of the error percentage increase with saturation. This good prediction can be very helpful for the preliminary design

6. Conclusion

In this paper, using the frozen permeability method, a model for the average torque separation and inductance calculating for novel two-layer spoke-type ferrite IPM machines are built. Firstly, FEA models are built and several simulations run under different load conditions to investigate the influences of the saturation and crosscoupling on the machine performances are deeply investigated. The developed procedure can offer overview design guidance for this developed type. Magnitude and phase of the current generate torque in an IPM motor. To enhance torque develop3ment large saliency ratio and flux focusing structure is designed which can provide similar portions of reluctance and PM torque Finally, various loading conditions is applied to the two-layer spoke-type ferrite IPM machine to validate the design by comparing test results with the derived analytical methods and torque performance.

References

[1] S. Javadi and M. Mirsalim, "Design and Analysis of 42-V Coreless Axial-Flux Permanent-Magnet Generators for Automotive Applications," IEEE Trans.Magn., vol. 46, no. 4, pp. 1015–1023, 2010.

[2] E. Sulaiman, T. Kosaka, and N. Matsui, "High Power Density Design of 6-Slot-8-Pole Hybrid Excitation Flux Switching Machine for Hybrid Electric Vehicles," IEEE Trans.Magn., vol. 47, no. 10. pp. 4453–4456, 2011.

[3] Z. Q. Zhu and D. Howe, "Electrical Machines and Drives for Electric, Hybrid, and Fuel Cell Vehicles," in Proc. the IEEE, vol. 95, no. 4. pp. 746–765, 2007.

[4] A. M. El-Refaie, "Fractional-slot concentrated-windings synchronous permanent magnet machines: Opportunities and challenges," IEEE Trans. Ind. Electron., vol. 57, no. 1, pp. 107–121, Jan. 2010.

[5] Z. Q. Zhu, "Fractional slot permanent magnet brushless machines and drives for electrical and hybrid propulsion systems," Int. J. Comput.Math. Electr. Electron. Eng., vol. 30, no. 1, pp. 9–31, 2011.
[6] D. G. Dorrell, M. Hsieh, and A. M. Knight, "Alternative rotor designs for high performance brushless permanent magnet machines for hybrid electric vehicles," IEEE Trans. Magn., vol. 48, no. 2, pp. 835–838, Feb. 2012.
[7] Bing Xia, Weizhong Fei and P. Luk, "Analysis and design of V-spoke ferrite interior permanent magnet machine for traction applications," 6th International Conference on Power Electronics Systems and Applications (PESA), Hong Kong, 2015, pp. 1-6
[8] Xia B., Luk PC. Analytical Model of Open-Circuit Characteristics of Two-Layer Spoke-Type Ferrite Interior Permanent Magnet Machines,IEEE International Electric Machines & Drives,Miami,USA,May 21-24,2017. IEEE International Electric Machines & Drives,Miami,USA,May 21-24,2017..
[9] Fei, W., D. Wu, B. Xia, and P.C.K. Luk. "Design of a multi-layer interior ferrite permanent magnet synchronous machine for traction applications", 7th IET International Conference on Power Electronics Machines and Drives (PEMD 2014), 2014.
[10] A.Wang, Y. Jia, and W. L. Soong, "Comparison of five topologies for an interior permanent-magnet machine for a hybrid electric vehicle," IEEE Trans. Magn., vol. 47, no. 10, pp. 3606–3609, Oct. 2011.

IOP Conf. Series: Materials Science and Engineering **765** (2020) 012011 doi:10.1088/1757-899X/765/1/012011

Theoretical and Experimental Analysis of Photovoltaic Module under Clouds Effects

Ali. H. Numan, Zahraa S. Dawood, Hashim A. Hussein,
Electromechanical Engineering Department, University of Technology, Baghdad, Iraq
E-mail: ahnuman2001@gmail.com , hashim171967@gmail.com

Corresponding author: zahraasalman777@gmail.com

Highlights

- Output power losses of PV module during irradiance transitions were studied.
- The maximum output power loss of PV module was around 62.26%.
- Experimental result were compared with theoretical results.

Abstract

Photovoltaic (PV) module is one of the most useful, sustainable and non-harmful product in the field of renewable energy therefore, it increased rapidly. Photovoltaic performance affected inversely by Partial shading which occurs due to clouds, birds, buildings , trees and dust deposition etc,. This paper investigates the performance of monocrystalline photovoltaic module under cloud effect theoretically and experimentally. The experiments are conducted in Baghdad during November, December and January in 2018and 2019. PV parameters are simulated by using MATLAB to determine this effect. The measured and simulated data are compared to verify the effectiveness and robustness of the proposed approach. The results show that the cloud has a significant impact on the performance of photovoltaic module. The clouds decrease the amount of direct radiation reaching to the surface of PV module. Therefore, output power of PV module in cloudy days is less than the power output in sunny days. The power output in the sunny days was 69.7W, and by comparing this value with measured values in (November, December and January), it is found that the measured values are reduced by 15.5, 59.65 and 62.26% respectively.

Keywords: PV module, Cloud effect, solar radiation, efficiency.

1. Introduction

As the world is confronting the issue of global warming, energy deficit and deterioration of environment and energy sources, there is a requirement for an alternative energy resource for generation power instead of using fossil fuels. An exciting alternative to diminish these negative effects is an expansion of the renewable energy utilizations, energy from the sun is the best option for electricity generation amonge renewable energy[3] because of it is clean **[4]**, available and inexhaustible [5]. Solar energy can be employed in various applications; such as, thermal field or electricity generation through solar Photovoltaic (PV) cells [6]. The great advantages such as freedom from environmental pollution, absence of moving or rotating parts and low maintenance cost represent the important reasons to increase attention about the Photovoltaic (PV) system over the last decade**[7]**, **[8]**. A PV cell is basically a semiconductor diode with a large barrier layer when expose to light portion of the energy in the light photons arriving at the cell is convert directly to DC electrical power **[9]**. The electrical energy produced by PV panels can be affected by relative humidity high, angle of irradiance **[10]**, temperature, excessive radiation, shading and dust **[11]**. In Iraq, the fluctuation of oil prices and their decline, air pollution due to emissions of greenhouse gases from diesel and gasoline, the difficulty of exploring the oil and gas sites on the one hand, and the aging of the electrical grid and lack of equipment lead to a small production of electrical energy. The severe shortage of energy forced the Iraqi citizen to rely on the private generators, which are drivers the diesel and gasoline. The emission results from excessive use of fossil fuels reflected on the Iraqi environment in addition to pollution from generators and motor vehicles **[15-19]**. Therefore, one of the rest solutions to reduce the risk of pollution of the environment is to find environmentally friendly sources. Solar energy represents the most important alternative source of energy in Iraq. This energy is free, has high power to be used by photovoltaic and in CPC station, available during the year **[17]**, **[18]**. There are many application of solar cell such as generating electricity in remote area and it can operate communication towers in this region. In addition, it can be used for outdoor street lighting, reducing the load on electricity grid in cities by converting part of parking lights, health clinic energies to the use of PV modules **[19]**. **P. Chrobak1 and et al (2016) [20]:** Studied the impact of cloud on the electrical power produced by 9 photovoltaic panels for the months of October, November, December, and January in year 2015 and 2016, respectively. The results showed that the average amount of the electrical power were 3.601, 2.652, 1.006 and 1.370 kWh, while the average amount of the solar radiation were 80.73, 52.14, 23.27 and 30.12 W/m^2 for the four months mentioned above. **M. H. Ali1 and A. I. Gaya (2016) [21]:** Investigated the effect of cloud on the performance of the PV panel at three different locations. The results showed that the reduction in the performance of the PV panel was 0.96% at central Mosque, 1.68% at Nana Hall and 3.77% at PG Common Room. **A. Bonkaney and et al (2017) [22]:** Studied the effect of dust and cloud on the performance of PV module in Niamy for the months from May to August (2015). The cloud prevents the direct component of solar radiation reaching the module's surface resulting in minimum output power. Therefore, the power producing in cloudy day is lower comparing to sunny days. Kari Lappalainen and Seppo Valkealahti [23]: Studied the mismatch losses of PV arrays with various layouts and electrical configurations during around 27,000 irradiance transitions identified in measured irradiance data. The overall effect of the mismatch losses caused by moving clouds on the energy production of PV plants was also studied. The study was conducted using a mathematical model of irradiance transitions and an experimentally verified MATLAB/ Simulink model of a PV module. The relative mismatch losses during the identified irradiance transitions ranged from 1.4% to 4.0% depending on the electrical configuration and layout of the PV array. The overall effect of the mismatch losses caused by moving clouds on the total electricity production of PV arrays was about 0.5% for the PV array with strings of 28 PV modules and substantially smaller for arrays with shorter strings.

This paper presents an experimental study that evaluates the effects of clouds on the performance of monocrystalline solar module during November, December and January (2018-2019) in Baghdad. Moreover, Matlab/Simulink is used for investigating this effect and compared theoretical results with the experimental results.

2. Modeling of photovoltaic module

PV module is the basic unit of power PV generation system. PV module has non-linear characteristics which depend on solar radiation and cell temperature. In this investigation, PV module with 36 series connected solar cell is chosen. Fig. (2). Shows the photovoltaic module model. The environmental data; such as, irradiance (G)and temperature (T) in addition to the electrical characteristics parameter of PV module such as, open circuit voltage (Voc), short circuit current (Isc), etc. It is having 23.42V, 5.192A and 1.5 as open circuit voltage (Voc) , short circuit current and ideality factor respectively, while series resistance value (Rs) is 0 ohm. Different cases are carried out using Matlab/ Simulink to determine I-V and P-V characteristics. To develop the model of PV module, Solar cell block from SimElectronics block set is employed. The parameter of solar cell is defined in (1) and (2) [24].

$$I = I_{ph} - I_{rs}\left[\exp\left(\frac{V+I\,R_s}{aV_T}\right) - 1\right] \qquad (1)$$

Where:

$$V_T = \frac{N_s\, k\, T_c}{q} \qquad (2)$$

I, I_{Ph} and I_{rs} are the output, photo-generated and the diode saturation currents respectively, V is the output voltage, R_S is the series resistance, N_S is the number of cells, V_T is the junction thermal voltage, A is the ideality factor, k is the Boltzman constant (1.3806503 ×10-23 J/K), T is the cell temperature and q is the electron charge (1.6021765 ×10-19 C). Figure 1. Shows photovoltaic module model with three bypass diodes configuration which used to investigate the cloud effect on the PV module.

The same module of PV module employed to study different cases of random irradiance level as illustrated in Table (2). It can be notice that the module divided to three groups connected with three bypass diodes, each group consist of 12 cells and resaved random irradiation levels.

Table (1). The level of radiation on PV module.

IOP Conf. Series: Materials Science and Engineering **765** (2020) 012011 doi:10.1088/1757-899X/765/1/012011

Case study	Solar Cell Irradiance (W/m^2)		
	Unshaded group A (12 cells)	Shaded group B (12 cells)	Shaded group C (12 cells)
Case 1	1000	1000	1000
Case 2	1000	900	800
Case 3	1000	800	600
Case 4	700	600	200
Case 5	800	950	300
Case 6	300	500	700
Case 7	300	200	400
Case 8	200	200	200

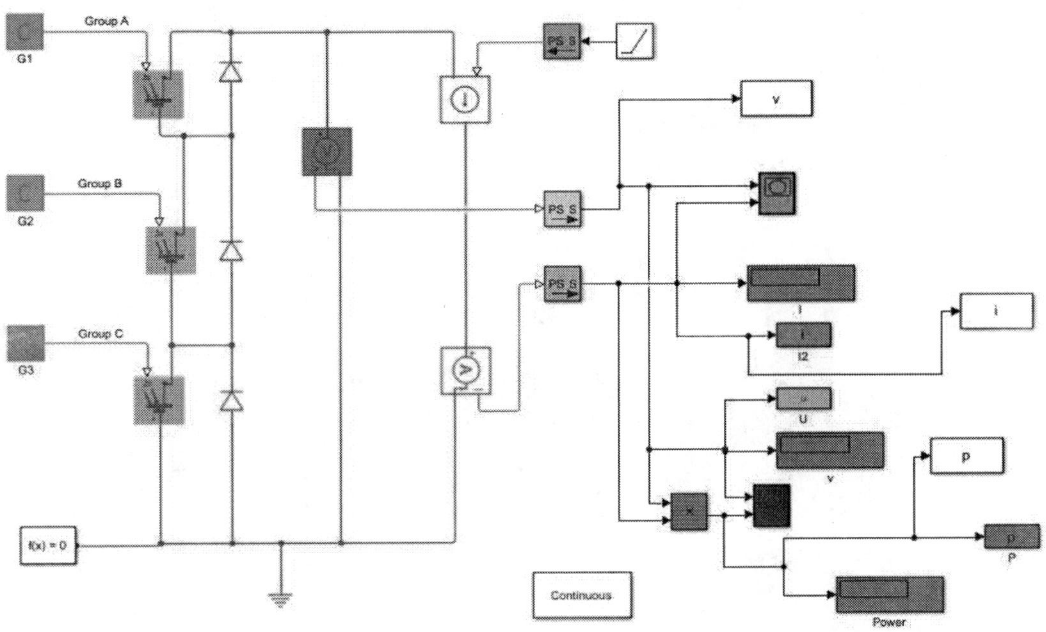

Fig. 1. Photovoltaic module model (36 cells with three bypass diodes)

3. Experimental set up

In order to obtain the I-V and P-V characteristics of PV panel under clouds effect, this requires current and voltage measurement. Solar radiation and ambient temperature affected on these characteristics therefore, it is necessary to measure these parameters. At environmental conditions, the change in atmospheric condition may lead to rapid change in the solar radiation therefore; the data must be recorded quickly in short time as possible.

The experimental set up as shown in figure 2. It was installed in Baghdad city (33.33° N latitude and 44.39° E longitude) for collecting the required data and consisted of Monocrystalline PV panel with 36 cell connected in series, each 12 cell connected to one bypass diode. The panel placed on a mobile metallic holder and installed to face the south with an inclination of 31.2° **Table 1** explain the electrical characteristics of the PV panel under standard test condition. The required data are radiation intensity, ambient and module temperatures, wind speed and relative humidity therefore, two digital multi- meter to measure current and voltage. Thermocouple type k was used to obtain the panels temperature using a digital data recorder. Relative humidity and ambient temperatures were measured by using UNI-T UT332 digital Thermo-hygrometer devise. The measurement of wind speed has been obtained using wind gage type Kaindl Windmaster 2.

Solar radiation was measured by solar meter pyranometer type TES 1333R and rheostat (100W, 10Ω) used as load to measure the maximum current and maximum voltage. Experimental has been carried in November, December and January (2018-2019).

Table.2. Electrical Characteristic of Monocrystalline solar panel

Maximum power (P max)	88 W
Open circuit voltage (VOC)	23.42 V
Short circuit current (ISC)	5.192 A
Voltage at maximum power (Vmp)	18. 33 V
Current at maximum power (Imp)	4.801 A
Voltage temperature coefficient (Kv)	-2.10 mV/cell/°C
Current temperature coefficient (Ki)	15 micro A/cm2/°C
No. of cells connected in series (NS)	36

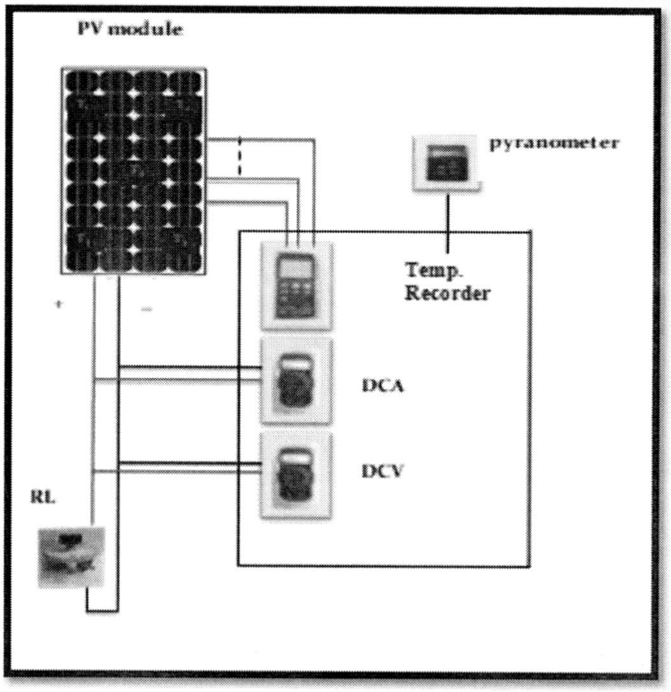

Fig. 2. The schematic view of the experimental setup.

4. Results and discussion

4.1 Theoretical results

When the irradiance on the module was non uniform irradiance as illustrated in Table (1), multi steps in the I-V characteristics and multiple peaks in the P-V characteristics curves are observed. Figure (3) shows the simulation results of the cases study in Table (3). The same figure appeared that the short circuit current negatively affected with decreasing the irradiance levels which lead to decreasing the maximum current and consequently, the maximum power decreasing. While the open circuit voltage less affected in compared with short circuit current under these conditions. The maximum power values for these cases were 88, 76, 56.9, 33, 44.3, 29, 16.1 and 15W for first, second, third, fourth, fifth, sixth, seventh, eighth case respectively.

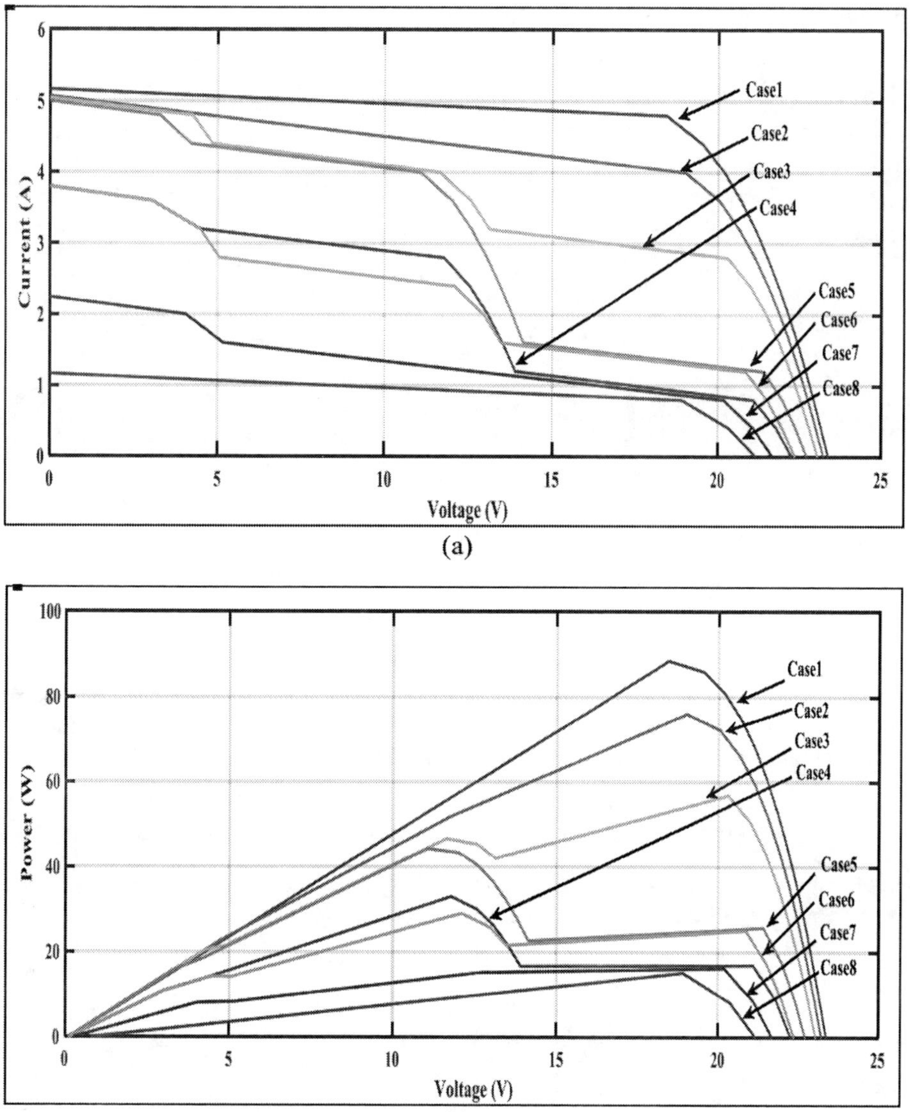

(b)

Figure (3): Photovoltaic module characteristics under non-uniform irradiation levels (a)I-V;(b)P-V.

4.2 Experimental results

Corresponding to the conditions at which the experiments were conducted, the I-V and P-V curves of 36 cell (3 group each one with 12 cells) under environmental condition with variable cell temperature and irradiance are analyzed. However, in Iraq, the clouds are abundant in the autumn and winter seasons (usually in October, November, December, January and February). The clouds are of little thickness at the beginning of autumn while, they are thicker in the cloudy winter days, especially in the months of December and January. Fig.4. Shows the I-V and P-V curves of monocrystalline solar module on separate plot during different days of years (2018-2019). From the figure, it can be notice that the current is affected by cloud significantly. It drop from 4.39A at 2nd June to 3.83, 1.93and 1.82A in 2nd November, December and January respectively. From the other hand, it can be noticed that power at2nd June was 69.7W; while, the power at

IOP Conf. Series: Materials Science and Engineering **765** (2020) 012011 doi:10.1088/1757-899X/765/1/012011

2nd November, December and January were 58.9, 28.26 and 25.74 respectively. This means that the power is decreasing by 15.5, 59.65 and 62.26% respectively. So, the amount of energy produced by photovoltaic solar module negatively affected with thickness of the cloud layer.

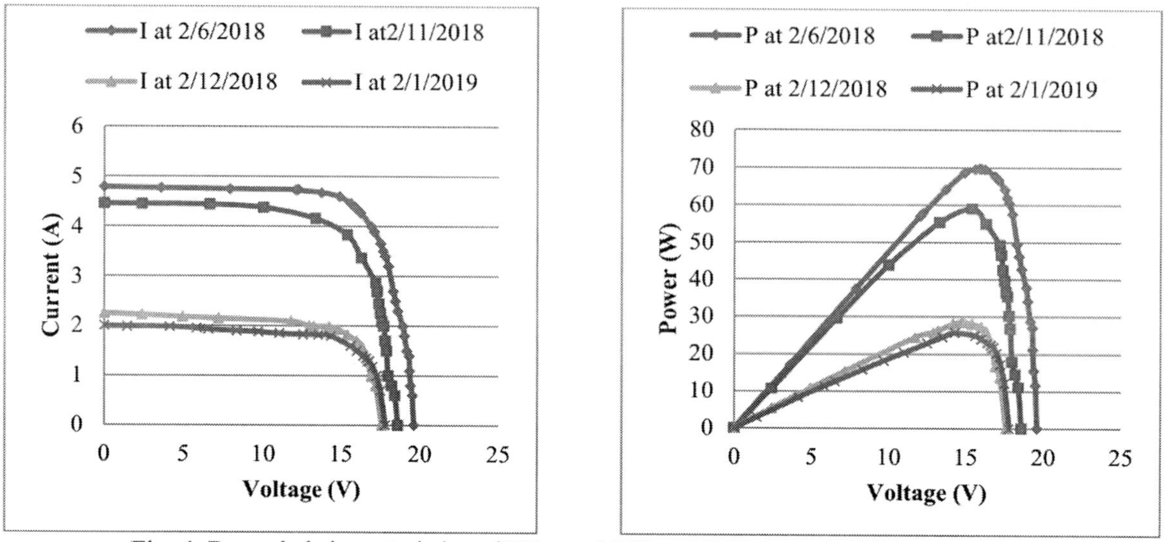

Fig. 4. Recorded characteristics of PV panel I-V and P-V through different months

It is worth noting that the clouds are randomly appeared during the day. For the purpose of studying their effect, the characteristics of the solar module were carried out on different months, days and times as shown in figs 5, 6 and 7. It can be noticed that the short circuit current decreases in cloudy day which cause reduction in the output power and consequently decrease the performance of the PV module.

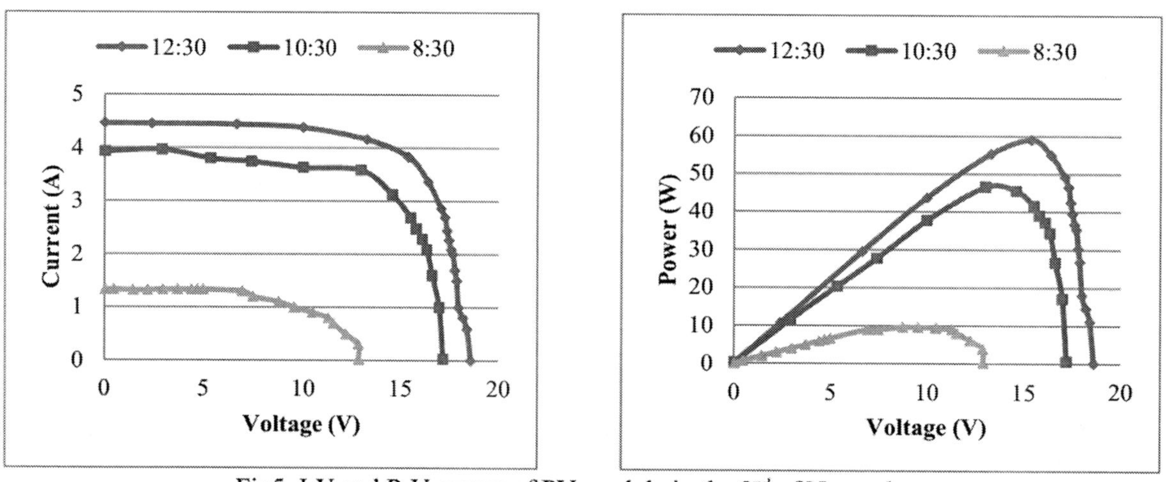

Fig5. I-V and P-V curves of PV module in the 2nd of November

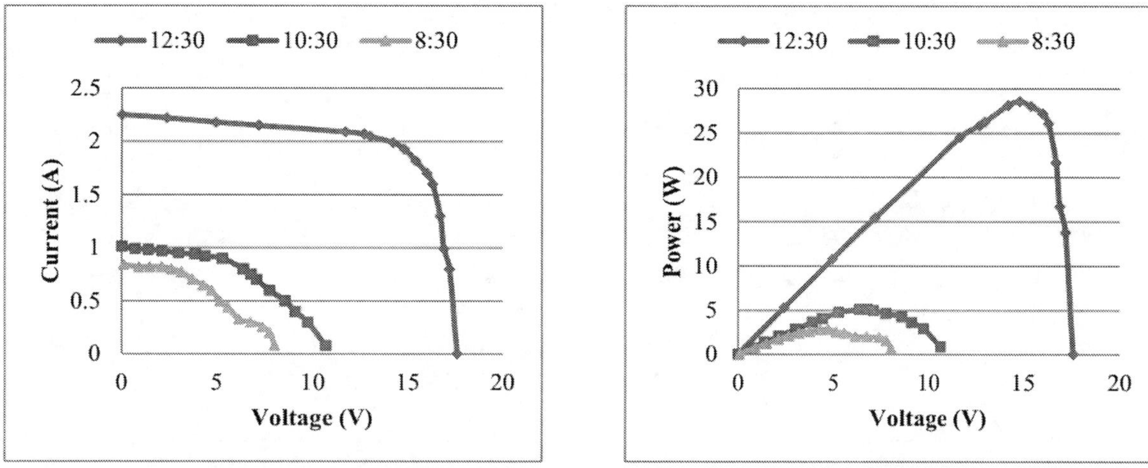

Fig. 6. I-V and P-V curves of PV module in the 2nd of December

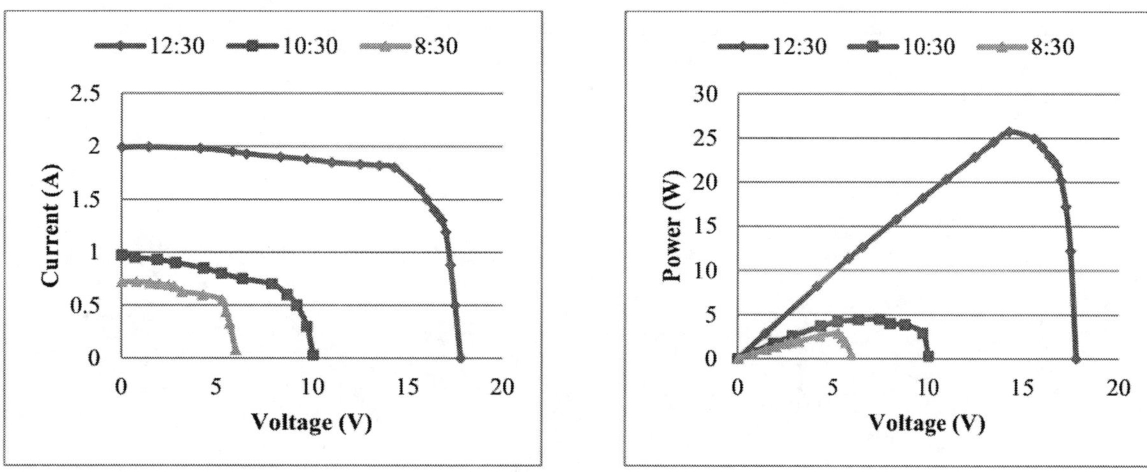

Fig.7. I-V and P-V curves of PV module in the 2nd of January

The above figures show that the currents value in November is higher than its value in December and January. The corresponding power also decreases in month of December and January due to increases thickness of clouds.

Table 3. Values obtained from experiment for the three different days.

Time	2nd November			2nd December			2nd January		
	P (W)	G (W/m²)	η (%)	P (W)	G (W/m²)	η (%)	P (W)	G (W/m²)	η (%)
8:30 AM	9.64	133	10.81	2.84	135	3.14	2.9	133	3.25
10:30 AM	46.57	620	11.21	5.12	150	5.1	5.53	160	5.16
12:30 AM	58.9	765	11.5	28.26	410	10.28	25.74	368	10.4

Figs. 8, 9 and 10. Shows both the simulated (red lines) and measured (blue lines) IV curves are plotted in the same figure for each experimental day at 12:30 am. The corresponding power also plotted in the separate figures. In general, the graphs appeared good agreement between the simulated and measured results.

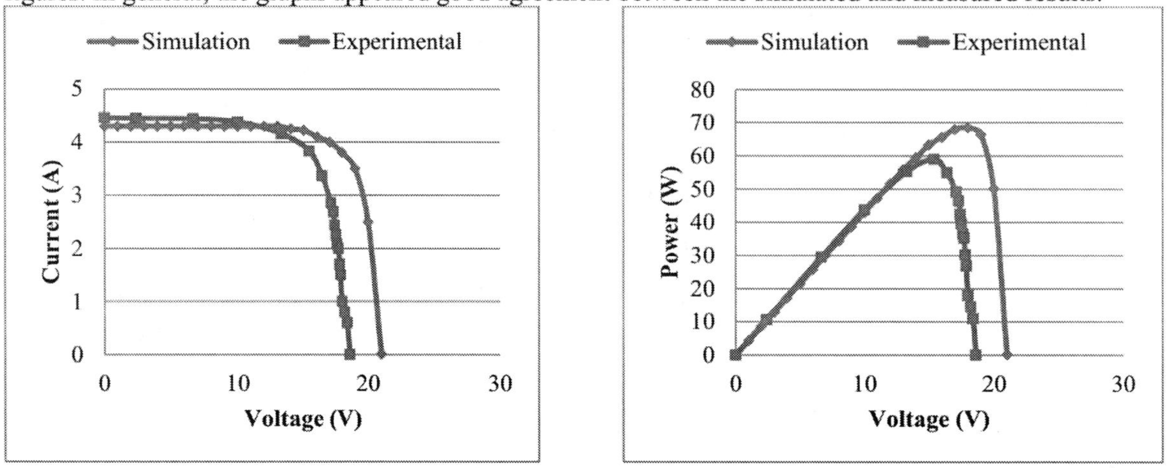

Fig.8. Electrical Characteristics obtained from simulation and experimental study on the day of 2nd November 2018 at 12:30, solar radiation= $765W/m^2$.

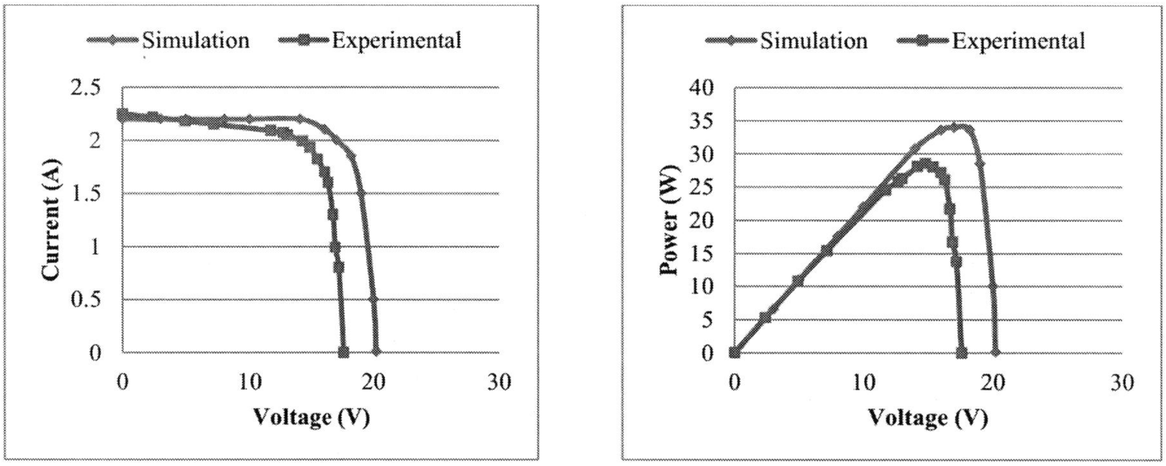

Fig. 9. Electrical Characteristics obtained from simulation and experimental study on the day of 2nd December 2018 at 12:30, Solar radiation = $410W/m^2$.

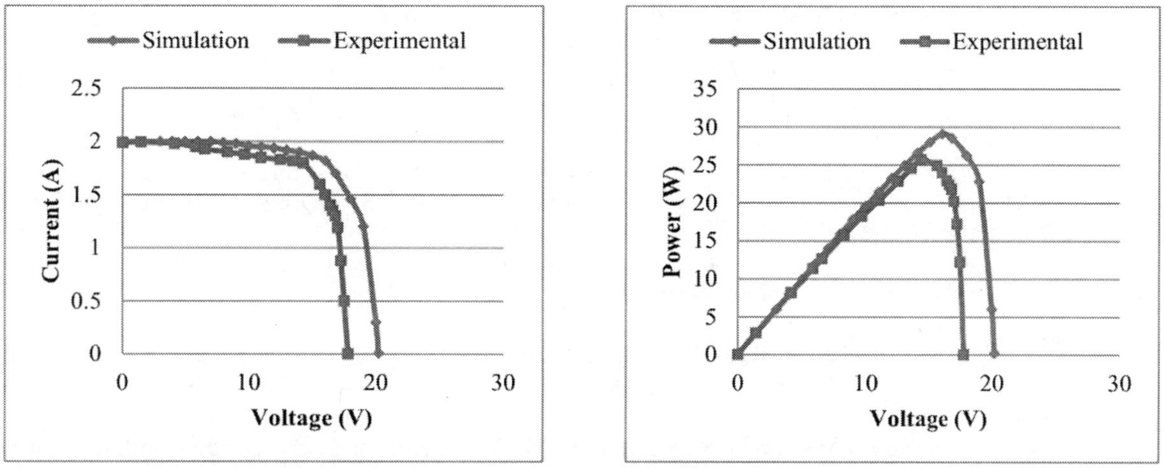

Fig.10. Electrical Characteristics obtained from simulation and experimental study on the day of 2nd January 2019 at 12:30, Solar radiation= 368W/m^2.

Conclusion

The electrical characteristics of monocrystalline solar module are evaluated under different cloudy days in Baghdad, Iraq. The I-V and corresponding P-V curves are plotted in separate figures for different days and times of years 2018 and 2019. At 12:30, the power output dropped by 15.5, 59.65 and 62.26% for 2nd November, December and January in comparing with its value at 2nd June. Besides, the simulations of parameters of the PV module are conducted using MATLAB to determine this impact, random irradiance levels have been applied. Furthermore the results reveal that there is a considerable agreement between the experimental and simulated results. The results show that the reduction in irradiance reaching to the surface of solar module under cloudy days leads to decrease the output power and efficiency significantly.

References

[1] Aurora, Domkundwar, A Course in Power Plant Engineering, Dhanpat Rai & Co. (P) LTD, Third Edition 2007, pp 31.1.

[2] Fundamentals of Photovoltaic Materials, National Solar Power Institute,Inc, Olivia, Mah NSPRI,2008.

[3] Dharmendra thakur, Amit arnav, Abhishek datta, E.V.V Ramanamurthy, "A Review on Immersion System to increase the efficiency of Solar Panels," *International Journal of Advanced Research,* vol. 4, pp. 312-325, 2016.

[4] Michael D. Oisamoje, Esther Eguono Oisamoje, "Exploring the Economic and Environmental Benefits of Solar Energy Generation in Developing Countries: The Nigerian Perspective," *Journal of Energy Technologies and Policy*, vol. 3, pp. 23-30, 2013.

[5] Shahzada Adnan, Azmat Hayat Khan, Sajjad Haider, and Rashed Mahmood, "Solar energy potential in Pakistan," *Journal of Renewable and Sustainable Energy*, vol. 4, pp. -, 2012.

[6] Ahmad Fudholi, Mariyam Fazleena Musthafa, Abrar Ridwan, Rado Yendra, Ari Pani Desvina, Rahmadeni, Tri Suyono and Kamaruzzaman Sopian "Energy and exergy analysis of air based photovoltaic thermal (PVT) collector: a review," *International Journal of Electrical and Computer Engineering (IJECE),* Vol. 9, No. 1, , pp. 109~117, 2019.

[7] M. Z. S. EL-Dein, M. Kazerani, and M. M. A. Salama, "Optimal photovoltaic array reconfiguration to reduce partial shading losses," *IEEE Trans. Sustain. Energy*, Vol. 4, No. 1, 2013.

[8] K. Sundareswaran, S. Peddapati and S. Palani, "MPPT of PV systems under partial shaded conditions through a colony of flashing fireflies", *IEEE Trans. on Energy Convers.*Vol. 29, No.2, 2014.

[9] Tanvir Ahmad, Sharmin Sobhan, Md. Faysal Nayan, "Comparative Analysis between Single Diode and Double Diode Model of PV Cell: Concentrate Different Parameters Effect on Its Efficiency," *Journal of Power and Energy Engineering*, vol. 4, pp. 31-46, 2016.

[10] H. A. Hussien, A. H. Numan and A. R. Abdulmunem, "Improving of the photovoltaic / thermal system performance using water cooling technique," *in IOP Conference Series: MaterialsScience and Engineering*, 2015.

[11] Fadliondi, Haris Isyanto, Budiyanto., "Bypass Diodes for Improving Solar Panel Performance, " *International Journal of Electrical and Computer Engineering (IJECE),* Vol. 8, No. 5, pp. 2703-2708. 2018.

[12] M. T. Chaichan, K. I. Abaas, B. A. Mohammed, "Experimental study of the effect of fuel type on the emitted emissions from SIE at idle period," *Al-Khwarizmi Engineering Journal*, vol. 13, No.1, pp. 1-12, 2017.

[13] M. T. Chaichan, H. A. Kazem, T. A. Abid, "The environmental impact of transportation in Baghdad, Iraq," Environment, Development and Sustainability, 2016. DOI: 10.1007/s10668-016-9900-x.

[14] K. A. Al-Aasadi, A. A. Alwaeli and H. A. Kazem, "Assessment of air pollution caused by oil investments in Basra Province-Iraq," *Journal of Novel Applied Sciences,* vol. 4, No. 1, pp. 82-86, 2015.

[15] M. T. Chaichan & H. A. Kazem, "Status and future prospects of renewable energy in Iraq," *Renewable and Sustainable Energy Reviews,* vol. 16, No. 1, pp. 6007–6012, 2012.

[16] A. A. Al-Waeely, S. D. Salman, W. K. Abdol-Reza, M. T. Chaichan, H. A. Kazem and H. S. Al-Jibori, "Evaluation of the spatial distribution of shared electrical generators and their environmental effects at Al-Sader City-Baghdad-Iraq," *International Journal of Engineering & Technology IJET-IJENS*, vol. 14, No. 2, pp. 16-23, 2014.

[17] H. M. S. Al-Maamary, H. A. Kazem, M. T. Chaichan, "Renewable energy and GCC States energy challenges in the 21st century: A review," *International Journal of Computation and Applied Sciences IJOCAAS,* vol.2, No. 1, pp. 11-18, 2017.

[18] Z. Abdmouleh, R. M. Alammari, A. Gastli, "Recommendations on renewable energy policies for the GCC countries," *Renewable and Sustainable Energy Reviews,* vol. 50, pp. 1181-1191, 2015. DOI: 10.1016/j.rser.2015.05.057

[19] Ali A. K. Al-Waeli, Kadhem A. Al-Asadi, Mariyam M. Fazleena, "The Impact of Iraq Climate Condition on the Use of Solar Energy Applications in Iraq: a Review," *International Journal of Science and Engineering Investigations,* vol. 6, No. 68, pp. 2251-8843, 2017.

[20] Pavel Chrobak, Jan Skovajsa1 and Martin Zalesak, "Effect of cloudiness on the production of electricity by photovoltaic panels", *MATEC Web of Conferences*, Vol. 7, NO. 02010, 2016.

[21] M. H. Ali1 and Abdulsalam Ibrahim Gaya, "Determination of Cloud Effect on the Performance of Photovoltaic Module," *IOSR Journal of Applied Physics (IOSR-JAP),*Vol. 8, No. 4, pp. 03-07, 2016.

[22] Abdoulatif Bonkaney, Saidou Madougou and Rabani," Adamou Impacts of Cloud Cover and Dust on the Performance of Photovoltaic Module in Niamey", *Hindawi Journal of Renewable Energy* Volume 2017.

[23] Kari Lappalainen and Seppo Valkealahti, " Photovoltaic mismatch losses caused by moving clouds" Solar Energy, Vol. 158, pp. 455–461, 2017.

[24] J. C. Teo , R. H. G. Tan, O. V. H. Mok , V. K. Ramachandaramurthy and C. Tan, "Impact of Partial Shading on the P-V Characteristics and the Maximum Power of a Photovoltaic String", MDPI, energies, Vol.11, No.7, 2018.

IOP Conf. Series: Materials Science and Engineering **765** (2020) 012012 doi:10.1088/1757-899X/765/1/012012

Impact of the Weather on the Combined Economic and Emission Optimization Problem of MG

Mohammed K. Al-Saadi Hashmia Sh. Dakheel Zainb B. Abdullah

Electromechanical Engineering Department, University of Technology, Baghdad, Iraq

E-mail: 50055@uotechnology.edu.iq, hashimia.shr@yahoo.com, btb_n@yahoo.com

Corresponding author: 50055@uotechnology.edu.iq

Abstract: The high integration of renewable energy resources and utilizing of storage devices with possible exchange power with the utility grid make the optimization problem of microgrid (MG) significantly difficult comparing with the conventional systems. This paper quantifies the impact of weather conditions on the combined economic and emission optimal operation of the MG. The optimization problem is formulated as mixed integer quadratic program (MIQP), where the unit commitment (UC) is taken into consideration in the optimization problem. In addition, realistic constraints are taken into account to make the problem close to the real scenario. The proposed optimization approach is applied to the low voltage MG, which includes distributed generators (DGs), renewable energy sources (wind, solar) and storage battery. The results reveal that the abundant of wind and solar energies leads to reduce the operation and emission cost.

1. Introduction

Decreasing of fossil fuel, increasing of load demands, environmental pollution, and global warming lead to draw more attention to use renewable and clean energy sources in the generation of electricity. The MG is an enabling technology to integrate these energy sources with distribution system. The MG is a small system comprises of renewable energy sources which are uncontrollable, such as wind and solar energy resources. The MG also has DGs such as diesel generators, micro turbine and fuel cell which are controllable sources. Besides, it involves storage devices. In addition, it has different types of loads which can be managed to make sure the system work safely and economically. The MG operates either autonomously in an isolated mode without any connection with the utility grid or operates in a connected mode and can exchange power with main grid. In both operation modes, the MG has to maintain power supply to its load. The penetration of the renewable energy sources, DGs, and storage devices with distribution networks creates many challenges in different aspects especially for optimal management of power flow in the MGs. Accordingly, new intelligent methods have been proposed to solve the optimal scheduling of the MG.

The optimization problem includes both continuous decision variables such as battery (charging / discharging) and discrete variables, such as UC. The UC, charging and discharging operations of battery, and trading energy with the main grid make the optimization problem significantly complex and a nonlinear. The outcome of the optimization of the MG result in enormous economic benefits for humans and environment. This urges researchers to develop approaches and algorithms to formulate and solve the optimization problem of MGs efficiently. So, many sophisticated algorithms have been presented to solve these complicated problems.

Many researchers have reported the optimization problem in MG in their studies. Authors in [1] proposed an optimal management approach to optimize the operation of low voltage MG with integration of renewable energy resources. Authors in [2] presented a method combining the UC and scheduling of generation into a single optimization approach. This method applied on Texas (ERCOT) power system which includes mix of conventional generation and wind generation. Reference [3] addressed economic dispatch problem with unit commitment in MGs. Lambda Iteration algorithm has been modified to use in this context and a genetic algorithm with some novel specific operator has been used to tackle the UC problem with different constraints and scenarios. In reference [4] a novel energy management system for MG with UC based on rolling horizon strategy was suggested. The optimization problem formulated as (MIP) and for each decision step the optimization problem was solved. The Benders decomposition technique has been used to solve the Multi- objective optimization approach with UC of a MG includes 33-bus. References [6, 7] proposed a model predictive control (MPC) approach to formulate optimization problem in MG with UC. The proposed method applied on connected MG consists of distributed generators, storage device and controlled. The aim of paper is to minimize the operating cost and satisfying the constraints.

In this paper, the impacts of weather conditions on the optimal operation of the MG are analyzed and the optimization problem is solved for the system for four different scenarios. Different renewable and low emission generation technologies with storage battery are considered in this study. Constraints are inserted in the formulation of problem to make the prosed system close to real one.

2. System modelling

It is necessary to model the distributed generators, storage battery, and constraints to formulate the optimization problem in the MG.

2.1. Operation cost of the distributed generators

The fuel cost of DGs is formulated as a function of output real power at interval t and it is modelled by quadratic function as follows [8, 9]:

$$C_{DG_i}(P_{DG_i}) = \sum_{i=1}^{N}\left(a_i + b_i . P_{DG_i}(t) + c_i . P_{DG_i}^2(t)\right) \quad (1)$$

where a_i (€/h), b_i (€/kWh), and c_i (€/kW^{2}h) are the fuel cost function coefficient, N is the number of distributed generators.

2.2. Cost of operating and maintenance of the distributed generators

This cost is assumed to be proportional with the produced power and it is calculated by the following equation [10]:

$$C_{OM,i}(P_i) = \sum_{i=1}^{N} K_{OMi} . P_{DGi}(t) \quad (2)$$

where K_{OMi} is the coefficient of operating and maintenance cost of ith distributed generators.

IOP Conf. Series: Materials Science and Engineering **765** (2020) 012012 doi:10.1088/1757-899X/765/1/012012

2.3. Model of storage battery

Storage battery has been modelled in several different ways. Criteria that have been used for choosing the model of battery depend on the details and the field of the study. Accordingly, the following equation is used in the field of optimal operation of MG [7]:

$$E_{bat}(t) = \left(E_{bat}(t-1) - \left(\frac{P_{bat-dis}(t)}{\eta_{dis}}\right).\Delta t + P_{bat-ch}(t).\eta_{ch}.\Delta t\right) \quad (3)$$

where $E_{bat}(t)$, $E_{bat}(t-1)$ are charge of battery at current and previous state. $P_{bat\text{-}ch}(t)$, $P_{bat\text{-}dis}(t)$ are the battery charging and discharging power respectively. η_{ch}, η_{dis} are charging and discharging efficiencies, Δt is the sampling time.

The battery is deteriorated due to charge and discharge operations. This degradation is converted to monetary concept using the following equation:

$$C_{bat} = C_{Degradation} \cdot E_{bat}(t) \quad (4)$$

where $C_{Degradation}$ is the degradation cost of battery (€/kWh), $E_{bat}(t)$ is the charging or discharging energies of the battery (-) for charging (+) discharging.

2.4. Interaction with the utility grid

MG can exchange power with the utility grid when it works in connecting mode. The trading power cost with the upstream grid is calculated using the following equation [8]:

$$C_g(P,t) = C_g(t).P_g(t) \quad (5)$$

where $C_g(t)$is the trading power price with the main grid, P_g (t) is the trading power with the main grid (+) when buying power and (–) when selling power to the utility grid.

2.5. Model of emission of the greenhouse gases

The emission of carbon dioxides CO_2, Sulphur dioxides SO_2, nitrogen oxides NO_x , and particle matter are caused by burning the fossil fuel which leads to environmental pollution. This emission depends on the power generated by the unit. The cost of emission at interval t for each greenhouse gases is calculated by the following equations [11]:

The emission cost of the CO_2 from i^{th} distributed generators

$$E_{CO2,i}(t) = \sum_{i=1}^{N} E_{CO2,i}.C_{CO2,i}.P_{DGi}(t) \quad 6)$$

where C_{CO2} (€/kW) is a cost of CO_2 emission, E_{CO2} (kg/kWh) is the emission rate of CO_2 from i^{th} generators.

The emission cost of the NO_x from i^{th} generators

$$E_{NO_X,i}(t) = \sum_{i=1}^{N} E_{NO_X,i}.C_{NO_X,i}.P_{DGi}(t) \quad (7)$$

where C_{NO_X} (€/kW) is the emission cost of NO_X, E_{NO_X} is the emission rate of NO_X from i^{th} generators.

The emission cost of SO_2 from i^{th} generators

$$E_{SO2,i}(t) = \sum_{i=1}^{N} E_{SO2,i}.C_{SO2,i}.P_{DGi}(t) \tag{8}$$

where C_{SO2} (€/kW) is a cost of SO2 emission, E_{SO2} is the emission rate of SO_2 from i^{th} generators.

The emission cost of particle matters from i^{th} generators

$$E_{PM,i}(t) = \sum_{i=1}^{N} E_{PM}.C_{PM,i}.P_{DGi}(t) \tag{9}$$

where C_{PM} (€/kW) is a cost of emission of particulate matter, E (kg/kWh) is the emission rate of particulate matter from i^{th} generators.

3. Objective function

The UC in the optimization problem is a mixed integer and its aim is to determine which generators are turned on or off, otherwise all generators are working through the entire scheduling time. The objective function of operation of the MG with is formulated with considering the UC as follows:

$$F = min \sum_{t=1}^{T} [\sum_{i=1}^{N} U_{DG,i}[(C_i(P_{DGi}(t) + OM_i(P_{DGi}(t)) + SU_{DG,i}(t) + SD_{DG,i}(t) + E_{CO2,i}(t) + E_{NO_{X,i}}(t) + E_{SO2}(t) + E_{PM,i}(t)] + C_{bat}(P_{bat}(t)) + C_{grid}(t)] \tag{10}$$

where SU_{DGi} and SD_{DGi} are startup and shut down cost of the DGs and they are calculated as follows [5, 12]:

$$SU_{DGi}(t) = Sc_i \,.\, (U_{DGi}(t) - U_{DGi}(t).U_{DGi}(t-1)) \tag{11}$$

$$SU_{DGi}(t) \geq 0 \tag{12}$$

$$SD_{DGi}(t) = Sd_i \,.\, (U_{DGi}(t-1) - U_{DGi}(t).U_{DGi}(t-1)) \tag{13}$$

$$SD_{DGi}(t) \geq 0 \tag{14}$$

where Sc_i and Sd_i are the startup and shut down cost of the ith generators (€).

4. Modelling of constraints

The following constraints are taken into consideration when formulation of the optimization problem:

4.1. Generating capacity constraint

$$U_{DGi}(t)\,.\,P_{DG_i-min} \leq P_{DG_i}(t) \leq U_{DGi}(t)\,.\,P_{DG_i-max} \tag{15}$$

4.2. Minimum up/ minimum down constraints

Each generator has a minimum up/down time constraint. The generator has to operate for a specific period of time after it switches on (minimum up) before it switches off. It also has to be off for a period (minimum down) before it switches again. These constraints can be formulated as following [6, 13]:

$$U_{DG_i}(t) - U_{DG_i}(t-1) \leq U_{DG_i}(\tau) \text{ (Off-on switch)} \tag{16}$$

IOP Conf. Series: Materials Science and Engineering **765** (2020) 012012 doi:10.1088/1757-899X/765/1/012012

$$U_{DG_i}(t-1) - U_{DG_i}(t) \leq 1 - U_{DGi}(\tau) \text{ (On-off switch)} \tag{17}$$

where τ =t+ 1,….., min (t + $T^{up}{}_i$ _ 1, T) in case of the minimum up time and τ = t+ 1,…., min (t + $T^{down}{}_i$ - 1, T) for the minimum down constraint.

4.3. Spinning reserve constrain

$$\sum_{i=1}^{N} U_{DG,i} . P_{DG,i}^{max}(t) = P_D(t) + R(t) \tag{18}$$

4.4. Power balance constrain

The summation of output power of each generation source, purchased and sold energy from the utility grid, and the charging/ discharging of batteries should satisfy the total load demand at each sampling time.

$$\sum_{t=1}^{T}[\left(\sum_{i=1}^{N} P_{DG_i}(t)\right) + P_w(t) + P_{PV}(t) + P_{bat}(t) + P_{grid}(t) = P_D(t)] \tag{19}$$

where P_{bat} (t) is (+) for discharging and (–) for charging, P_{grid} (t) is (+) for purchasing and (–) for selling.

4.5. Ramp rate limits constraint

It is inconvenient to increase the generated power of any i^{th} generators in a certain interval more than a specific value up–ramp limit (UR_i) compare to the previous generated level and neither deceases less than certain value down-ramp limit (DR_i) compare to previous generated power. This constraint should be met when solving the optimal problem of the MG and it is formulated as follows [14]:

$$-DR_i.\Delta T \leq P_{DGi}(t+1) - P_{DGi}(t) \leq UR_i.\Delta T \tag{20}$$

4.6. Exchange power with the main grid constraints

The limitation of power exchanges with the upstream grid should be satisfied for each time interval. The exchange power with utility grid at each period either purchasing power from the utility, selling power to the utility or neither of them. Therefore, two binary variables $u_{g\text{-}pur}(t)$, $u_{g\text{-}sell}(t) \in [0,1]$ are assigned to represent these operations and to prevent purchasing and selling power in the same time. This operation is formulated as follows:

$$u_{g-pur}(t).P_{g-pur,min} \leq P_{g-pur}(t) \leq u_{g-pur}(t).P_{g-pur,max} \tag{21}$$

$$u_{g-sell}(t).P_{g-sell,min} \leq P_{g-sell}(t) \leq u_{g-sell}(t).P_{g-sell,max} \tag{22}$$

$$u_{g-pur}(t) + u_{g-sell}(t) \leq 1 \tag{23}$$

4.7. Constraints of the Storage Battery

The storage devices have many constraints should be met and they are formulated as in the following.

A. Battery state charge constraint:

$$E_{bat-min} \leq E_{bat}(t) \leq E_{bat-max} \tag{24}$$

B. Charging and Discharging Power Constraints:

IOP Conf. Series: Materials Science and Engineering **765** (2020) 012012 doi:10.1088/1757-899X/765/1/012012

The exchange power with storage battery at each sampling period: charging, discharging or no exchange power. Two binary variable $u_{bat\text{-}ch}(t)$, $u_{bat\text{-}dis} \in [0,1]$ are assigned to formulate the status of battery operation and to prevent charging and discharging power simultaneously.

$$u_{bat-ch}(t).P_{bat-ch,min} \leq P_{bat-ch}(t) \leq u_{bat-ch}(t).P_{bat-ch,max} \quad (25)$$

$$u_{bat-dis}(t).P_{bat-dis,min} \leq P_{bat-dis}(t) \leq u_{bat-dis}(t).P_{bat-dis,max} \quad (26)$$

$$u_{bat-ch}(t) + u_{bat-dis}(t) \leq 1 \quad (27)$$

4.8. Emission constraints

The constraints of emission of greenhouse gas from DGs are formulated as follows:

$$\sum_{i=1}^{N} \beta_{CO2,i}.(P_{DGi}(t)) \leq L_{CO2} \quad (28)$$

$$\sum_{i=1}^{N} \beta_{NO_X,i}.(P_{DGi}(t)) \leq L_{NO_X} \quad (29)$$

$$\sum_{i=1}^{N} \beta_{SO2,i}.(P_{DGi}(t)) \leq L_{SO2} \quad (30)$$

$$\sum_{i=1}^{N} \beta_{PM,i}.(P_{DGi}(t)) \leq L_{PM} \quad (31)$$

where L_{co2} , L_{NOX} , L_{So2} , L_{PM} are the limitations of CO_2 , NO_X , SO_2 , PM in area of the MG.

5. Case study

Fig.1 shows the MG that is considered in this paper. It is a modified grid of the MG which has been presented in [15]. It consists of nineteen bus bars and three feeders: residential feeder supply residential loads, the industrial feeder supplies the workshops and Commercial feeder supplies commercial loads. The Load for each feeder and the total load are shown in Fig.2. This grid includes two diesel generators (DG), two micro turbines (MT), a fuel cell (FC), a wind turbine (WT), photovoltaic panels (PV) and a storage battery. ILOG CPLEX version 12.6 is used to solve (MIQP) optimization problem [16, 17]. In this paper, four scenarios are considered for weather as shown in Table 1. In all cases, the storage device should be fully charged at the beginning and end of control horizon.

Table 1 Scenarios of weather effect on optimal operation of MG

LC	Low wind and cloudy day
LS	Low wind sunny day
HS	Windy and sunny
HC	Windy but cloudy day

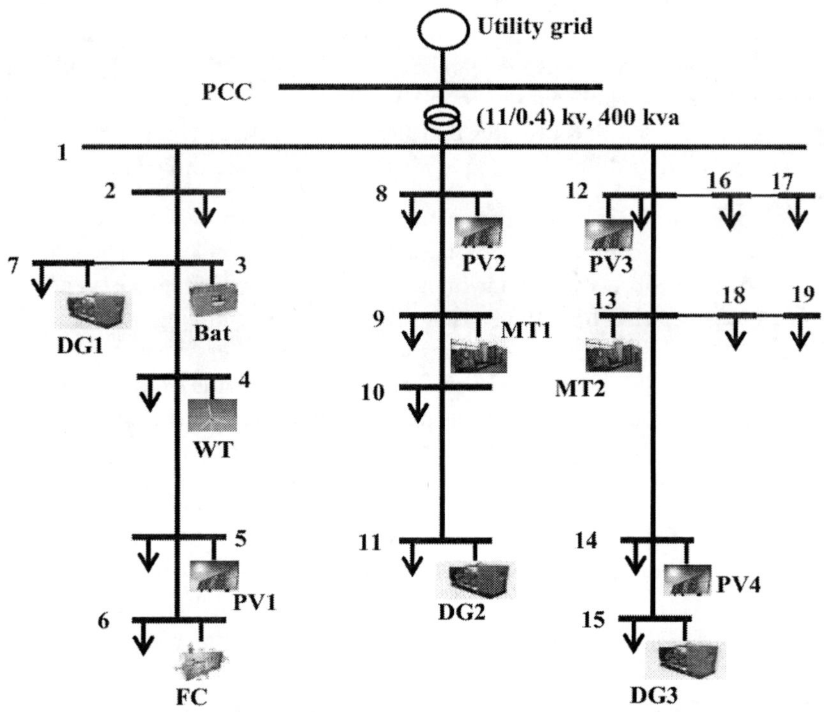

Fig. 1. MG under study

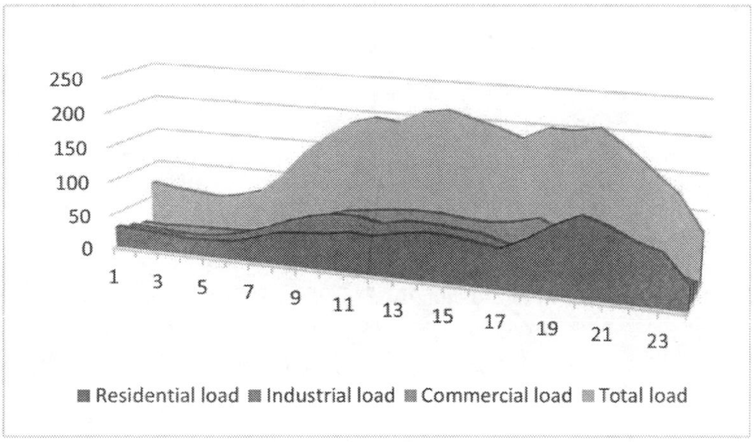

Fig. 2. Typical load curves for the feeders of the MG

IOP Conf. Series: Materials Science and Engineering **765** (2020) 012012 doi:10.1088/1757-899X/765/1/012012

6. Results and discussion

Figs. 3 to 6 show the optimal scheduling of DGs and interaction power with storage battery and the main grid. It can be noticed from Fig. 3 that the MG purchases the lowest power from the utility grid at hour 10 because the price has the highest value. It also can be seen from Fig. 5 that the MG sells energy to the main grid at hour 10 because the renewable energy abundant and the price at this hour has the highest value. In addition, it can be observed from these figures that the storage battery charges at hours 23 to 24 to fully charge to satisfy the constraint of fully charging at the end of the scheduling day.

Figs. 7 and 8 show the operating and emission cost of the four scenarios. It can be seen from these figures that the highest value of the operation cost is at hour 13 when renewables energy is low at low wind and cloudy weather. It also can be observed that the highest value of the emission cost is at hour 10 for all cases because at this hour the generation from the DGs is the highest value among other hours because the price has the highest value and the renewable energy has low value. Further, it can be seen that the UC makes micro turbines that they have the lowest fuel and emission cost provide higher energy than fuel cell and diesel generators. Furthermore, the UC also makes micro turbines work more operating hours than fuel cell and diesel generators. This leads to reduce the operating and emission cost.

Table 2 summarizes the obtained results from applying aforementioned analysis on four scenarios. It can be seen from the data in this table the lowest operating cost obtains in case of windy and sunny day because the energy from renewable sources is regarded as free operating cost.

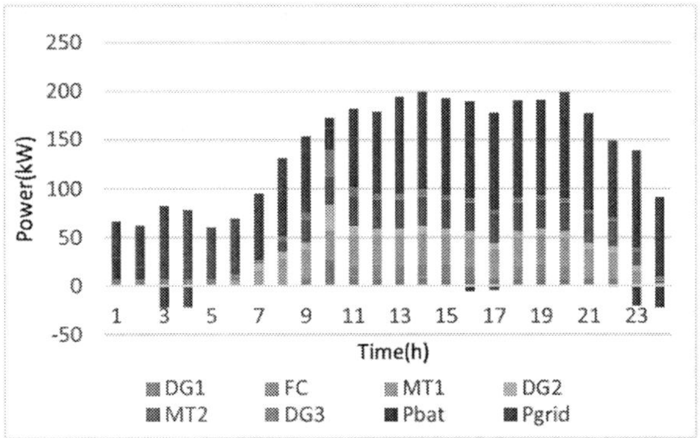

Fig. 3. Generation schedule for low wind and cloudy day

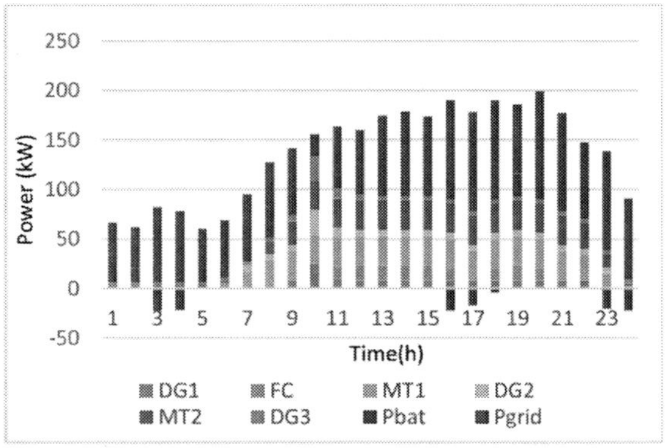

Fig. 4. Generation schedule for low wind and sunny day

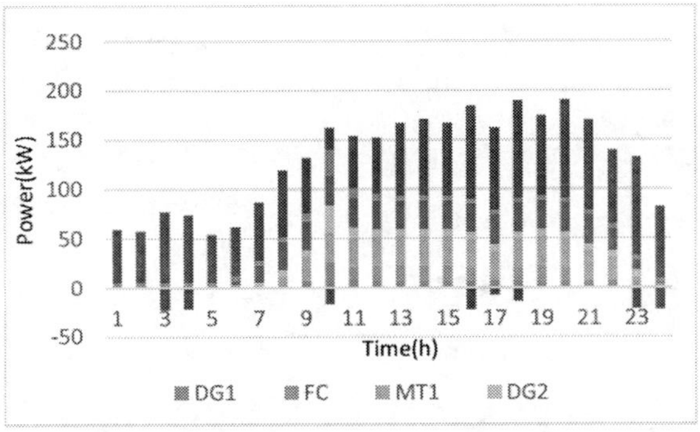

Fig. 5. Generation schedule for high wind and sunny day

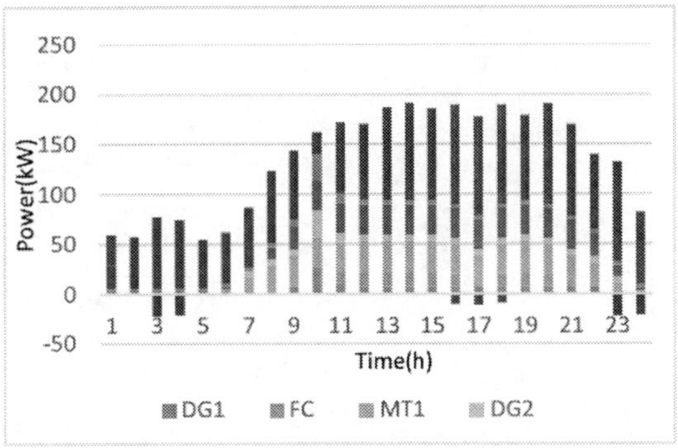

Fig. 6. Generation schedule for high wind and cloudy day

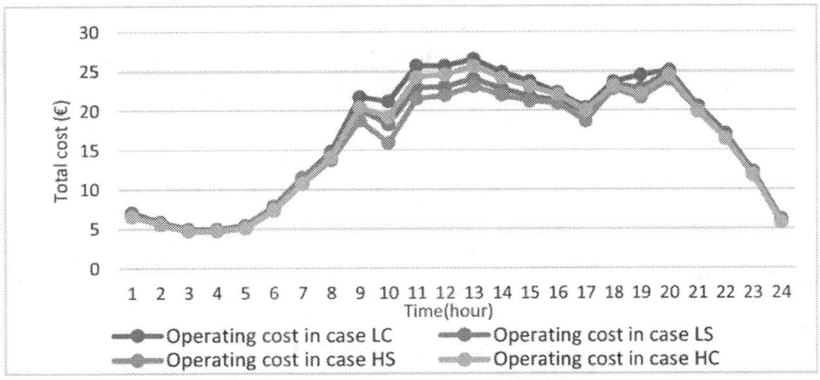

Fig. 7. Operating cost of four scenarios

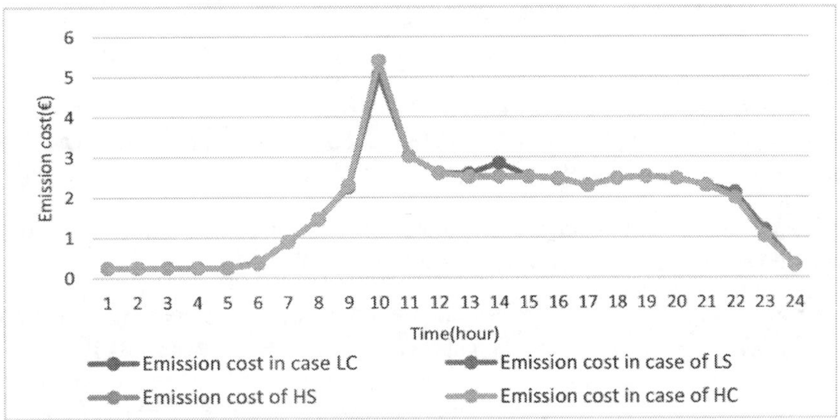

Fig. 8. Emission cost of four scenarios

Table 2 Result of the four scenarios

	Fuel cost (€/24h)	Emission cost (€/24h	Total cost (€/24h)
LC	404.02	43.41	447.43
LS	383.33	42.70	426.03
HS	364.87	42.59	407.46
HC	386.06	42.66	428.72

7. Conclusions

The impacts of weather conditions on the combined economic and emission optimization of the MG with consideration of the UC and constraints are analyzed. Comparison between different scenarios is carried out to quantify the effect of weather on the optimal operation of MG. The whole optimization problem is formulated as (MIQP). It can be concluded from the results that the implementing of the UC in the simulation leads to reduce the total operating and emission cost. In addition, it can be concluded that the weather has significant effect on the operating and emission cost and the optimal scheduling of DGs and exchanging power with the main grid and storage device. Further, the lowest operating and emission cost occurs when the renewable energies are abundant.

8. References

[1] M. W. Khan, et al, "Modelling and Optimal Management of Distributed Microgrid Using Multi-Agent System", Sustainable city and society, vol. 41, pp. 154-169, 2018.

[2] B. Palmintier, and M. Webster. "Impact of unit commitment constraints on generation expansion planning with renewables", power and energy society general meeting, 24-29 July 2011, San Diego, IEEE, p 1–7.

[3] J. P. Fossati, "Unit Commitment and Economic Dispatch in Micro Grids," Memoria de Trabajos. Difus. Cient. y Tec., vol. 10, pp. 83–96, 2012.

[4] R. Palma, et al, "A MG energy management system based on the rolling horizon strategy", IEEE Transactions on smart grid, vol. 4, no. 2, June 2013.

[5] A. Zakariazadeh, S. Jadid, and P. Siano, "Multi-objective scheduling of electric vehicles in smart distribution system," Energy Convers. Manag., vol. 79, pp. 43–53, Mar. 2014

[6] A. Parisio, E. Rikos, George, and L. Glielmo, "Use of model predictive control for experimental MG optimization", Applied Energy, vol. 115, p 37-46, 2014.

[7] A. Parisio, E. Rikos and L.Glielmo, "A Model Predictive Control Approach to MG Operation Optimization", IEEE Transactions on Control System Technology, vol.22, no.5, September 2014.

IOP Conf. Series: Materials Science and Engineering **765** (2020) 012012 doi:10.1088/1757-899X/765/1/012012

[8] W.Sua, J. Wangb, K. Zhangb, and A. Q. Huang, "Model predictive control-based power dispatch for distribution system considering plug-in electric vehicle uncertainty", Electric Power Systems Research, Vol. 106, p 29– 35, 2014.

[9 Mohammed K. Al-Saadi1, Patrick C. K. Luk, "Security-Constrained Two-Stage Stochastic Unified Active and Reactive Power Management System of the Microgrids", International Conference on Life System Modeling and Simulation, LSMS 2017 and International Conference on Intelligent Computing for Sustainable Energy and Environment, ICSEE 2017 Nanjing, China, September 22–24, 2017, Proceedings, Part III.

[10] H. Ren, *A*. Xiang, W.Teng and R. Cen, "Economic optimization with environmental cost for a MG", power and energy society general meeting, 22-26 July 2012, San Diego, IEEE, p 1–6.

[11] G. Liao "Solve environmental economic dispatch of Smart MG containing distributed generation system – Using chaotic quantum genetic algorithm", Electrical Power and Energy Systems, vol. 43, p 779–787, 2012.

[12] J. R. Santos, A.T. L, A. g. Expósito, j.L.Ramos "Finding Improved Local Minima of Power System Optimization Problems by Interior-Point Methods", IEEE Transactions on Power System, vol. 18, no. 1, February 2003.

[13] M. Carrión, and J. M. Arroyo "A Computationally Efficient Mixed-Integer Linear Formulation for the Thermal Unit Commitment Problem", IEEE Transactions on power system, Vol. 21, No. 3, August 2006.

[14] X. S. Han, H. B. Gooi and D.S. Kirschen, "Dynamic Economic Dispatch: Feasible and Optimal Solutions, IEEE transactions on power systems", vol. 16, no. 1, February 2001.

[15] S. Papathanassiou, N. Hatziargyriou and K. Strunz, "A Benchmark Low Voltage MG Network," Proceedings of the CIGRE Symposium: Power Systems with Dispersed Generation, Athens, 2005, pp. 1-8.

[16] IBM ILOG CPLEX Optimization Studio CPLEX User's Manual 12.4, USA, 2011.

[17] IBM ILOG CPLEX OPL Language V6.3, Reference manual, International Business Machines Corporation, USA, 2009.

IOP Conf. Series: Materials Science and Engineering **765** (2020) 012013 doi:10.1088/1757-899X/765/1/012013

Mitigation of Harmonic Distortion in a Power System Feeding Nonlinear Loads Using Shunt Passive Single Tuned Filters

Adel Ridha Othman

Electromechanical Engineering Department, University of Technology, Baghdad, Iraq
50063@uotechnology.edu.iq

Abstract. In this paper a distribution system of 2 MVA, 11 KV standalone generator is feeding a group of six branches of pulse width modulation (PWM) drive nonlinear loads, such a distribution power system is used in oil fields drilling rigs, each branch is fed through a step down transformer of 11/0.4 KV and of capacity proportional to its branch load rating. For this group of nonlinear loads, the total current harmonic distortion (ITHD) and total voltage harmonic distortion (VTHD) are calculated at the point of common coupling (PCC#1) which show that the system failed to meet IEEE Std 519 harmonic limits. A shunt passive single tuned filters are designed to mitigate (ITHD) and (VTHD). The addition of these filters to this power system were effective in making it compliance with IEEE Std 519.

1. Introduction

In an ideal power system the electrical power must be supplied with a sinusoidal voltage and current wave shape but due to the increased use of power electronics which are considered as nonlinear loads such as adjustable speed drive (ASD) and also the magnetization current drawn by transformers, there will be a current wave shape distortion and when this distorted current passes through the impedance that is between the nonlinear load and the system source, harmonic voltages are produced by impedance in the system for each harmonic. These voltages sum and when added to the nominal voltage produce voltage distortion. To overcome the harmonic distortion problem which cause many bad effect on power system, harmonic mitigation is important for both utilities and customers so shunt passive filters have been used to mitigate harmonic distortion [1].

2. Harmonic Sources

The sources of harmonics in power system are a lot and they have a common specification which is the nonlinear voltage – current relationship. The nonlinear load permit the current to flow through it discontinuously and doesn't correspond the voltage waveform. The nonlinear loads includes transformers and solid state devices for motor drives. These solid state harmonic producing devices is specified by the pulse number (p) and equation (1) calculates the characteristic harmonics (h) of the solid state device [2].

$$h = (n * p) \pm 1 \qquad n\ is\ an\ integer = 1, 2, 3, ... \qquad (1)$$

3. Voltage Total Harmonic Distortion (VTHD)

As given in equation (2), it is the ratio of the root-sum-square of the harmonics of the voltage to the root-mean-square of the fundamental voltage [3].

$$\text{VTHD} = \left(\sum_{h=2}^{h_{max}} (V_h)^2\right)^{\frac{1}{2}} . \ (V_1)^{-1} \quad \times 100\% \tag{2}$$

4. Current Total Harmonic Distortion (ITHD)

As given in equation (3), it is the ratio of the root-sum-square of the harmonics of the current to the root-mean-square of the fundamental current [3].

$$\text{ITHD} = \left(\sum_{h=2}^{h_{\max}} (I_h)^2\right)^{\frac{1}{2}} . \ (I_1)^{-1} \quad \times 100\% \tag{3}$$

4.1. Current Total Demand Distortion

As given in equation (4), it is the ratio of the root-sum-square of the harmonics of the current to the maximum demand load current

$$\text{ITDD} = \left(\sum_{h=2}^{h_{\max}} (I_h)^2 . \right)^{\frac{1}{2}} . \ (I_L)^{-1} \quad \times 100\% \tag{4}$$

5. Harmonics and Power Factor (PF)

PF is a measure of how effectively a load consumes electricity to do work. The higher the PF, the more work produced for a given voltage and current. PF is measured as the ratio between real power in kilowatts and apparent power in kilovoltamperes. For linear loads, which are defined as resistive, inductive, or capacitive, the apparent power in kilovoltamperes (S) is the phasor sum of the reactive power in kilovoltamperes reactive (Q) and the real power in kilowatts (P). The PF is P/S = Cos Φ, where Φ is the angle between S and P. This angle is the same as the displacement angle between the voltage and the current for linear loads and is therefore often referred to as displacement power factor (DPF). For nonlinear loads, the power phasor relationship becomes three dimensional with distortion reactive power H combining with both Q and P to produce the apparent power S which is the power system must deliver as given in equation (5).

$$S = (P^2 + Q^2 + H^2)^{1/2} \tag{5}$$

PF remains the ratio of kilowatts to kilovoltamperes, but the kilovoltamperes now has a harmonic component. True power factor (TPF) becomes the combination of DPF and distortion PF. DPF is still equal to Cos Φ, with Φ being the angle between the fundamental current and voltage.
DPF can be either leading or lagging. Distortion PF is then TPF (kW/kVA) divided by DPF. Distortion PF is neither leading nor lagging. For typical nonlinear loads, the DPF will be near unity. TPF, however, is normally very low because of the distortion component. [4].

6. Passive Filters

A shun filter is said to be tuned to the frequency that makes its inductive and capacitive reactances to be equal. The quality of the filter (Q) specify the sharpness of filter tuning so it may be either of high or low Q type. If (Q) is high then the filter is sharply tuned to one of the harmonic frequencies and it is called single tuned filter STF [5].

6.1. Single Tuned Filter (STF)

The single tuned filter (STF) is an R-L-C series filter and it is either a low pass or a band pass. The inductive and capacitive reactances of the STF at the tuned frequency are calculated from equations (6), equation (7) and equation (8 [1][5].

IOP Conf. Series: Materials Science and Engineering **765** (2020) 012013 doi:10.1088/1757-899X/765/1/012013

$$Z = j\omega_n L + \frac{1}{j\omega_n C} = R \tag{6}$$

$$\omega_n = (LC)^{-\frac{1}{2}} \tag{7}$$

If X_0 is the reactance of the capacitor or the reactor of the filter at its tuned frequency (ω_n)

$$X_0 = \omega_n L = \frac{1}{\omega_n C} = (L/C)^{-\frac{1}{2}} \tag{8}$$

The quality factor Q of the filter reactor is given in equation (9)

$$Q = \frac{X_0}{R} = \left(\frac{L}{C}\right)^{\frac{1}{2}} .(R)^{-1} \tag{9}$$

Equation (10) gives the detuning of the filter from the resonance frequency which is represented by a factor δ

$$\delta = \frac{\omega - \omega_n}{\omega_n} \tag{10}$$

δ depends on (i) variation in supply frequency; (ii) variations of L and C caused by aging and temperature and (iii) tolerances and finite step of tuning steps.

The variation of L and C values by 2% causes detuning as a change in frequency by 1%, so δ can be calculated from equation (11)

$$\delta = \frac{\Delta f}{f_n} + \frac{1}{2}\left(\frac{\Delta L}{L} + \frac{\Delta C}{C}\right) \tag{11}$$

7. IEEE Std 519 Harmonic Limits

IEEE Std 519 was first introduced in 1981 to provide direction on dealing with harmonics introduced by static power converters and other nonlinear loads so that power quality problems could be averted. It is being applied by consulting engineers and enforced by Utilities more frequently in recent years as the use of Variable Frequency Drives (VFD) and other non-linear loads has grown [3]. Referring to IEEE Std 519 tables and for the voltage of the simulated system which is 11 KV so the voltage total harmonic distortion (VTHD) limit is 5% [6].To define current distortion limits, IEEE Std 519 uses a short circuit ratio to establish a customer's size and potential influence on the voltage distortion of the system. The short circuit ratio (I_{SC}/I_L) is the ratio of short circuit current (I_{SC}) at the point of common coupling (PCC) with the utility to the customer's maximum load or demand current (I_L). Lower ratios or higher impedance systems have lower current distortion limits to keep voltage distortion at reasonable levels [3].

IOP Conf. Series: Materials Science and Engineering **765** (2020) 012013 doi:10.1088/1757-899X/765/1/012013

8. Simulation of the Power System without Filters

The one line diagram of the simulated power system without filters is shown in figure 1. The simulation results at PCC#1 of the voltage and current waveforms are shown in figure 2 and Figure 3 respectively. Figure 4 and Figure 5 show the harmonic spectrum of the voltage and current respectively at PCC#1.

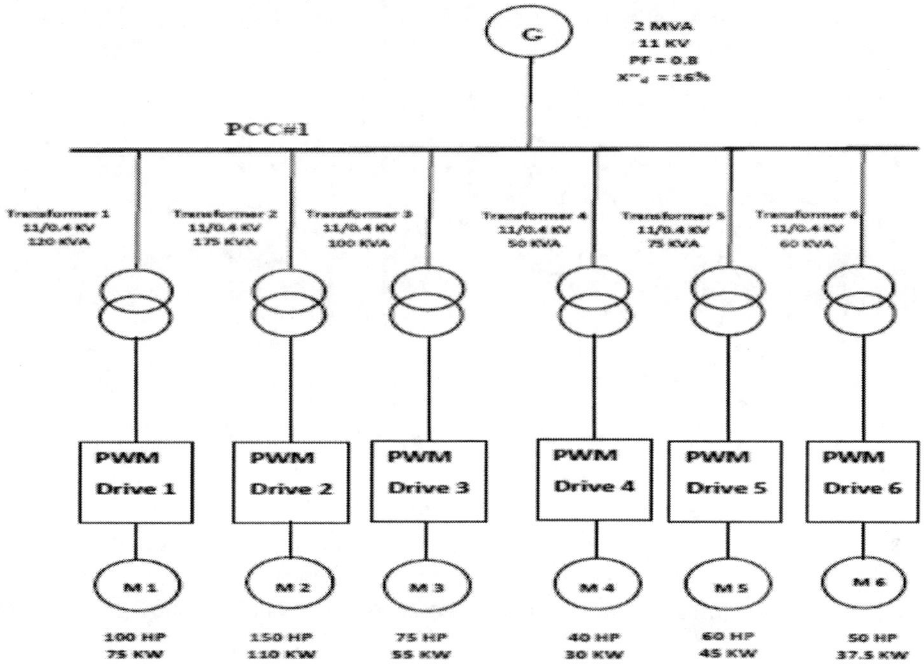

Figure 1. The Simulated Power System without Filters.

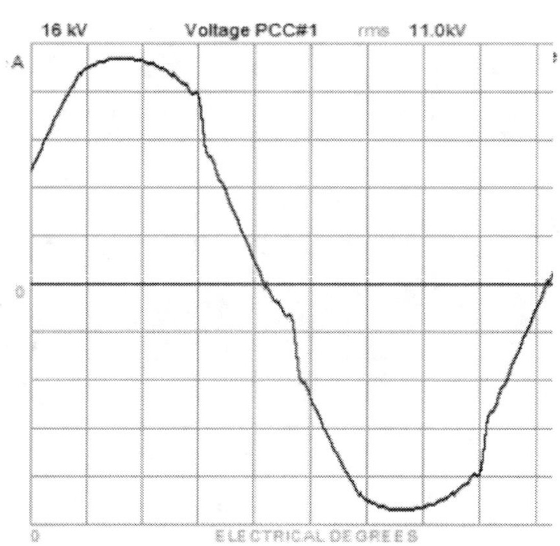

Figure 2. Voltage Waveform at PCC#1.

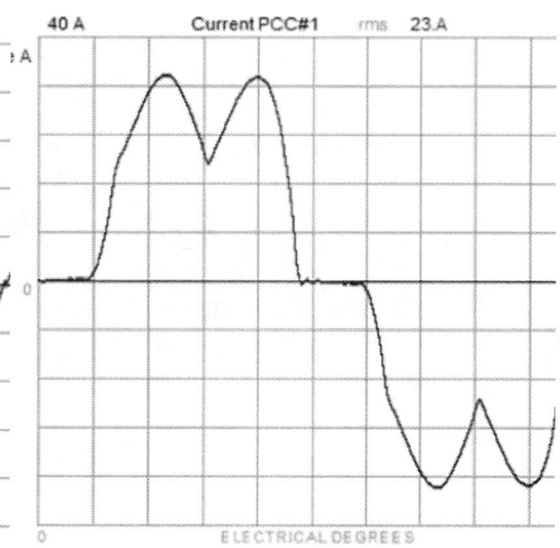

Figure 3. Current Waveform at PCC#1.

IOP Conf. Series: Materials Science and Engineering **765** (2020) 012013 doi:10.1088/1757-899X/765/1/012013

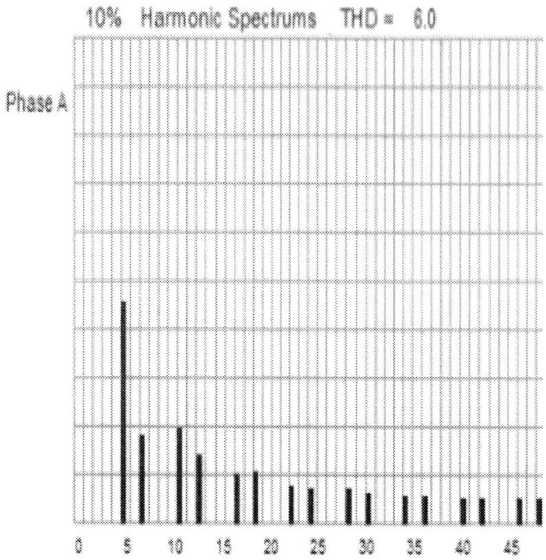

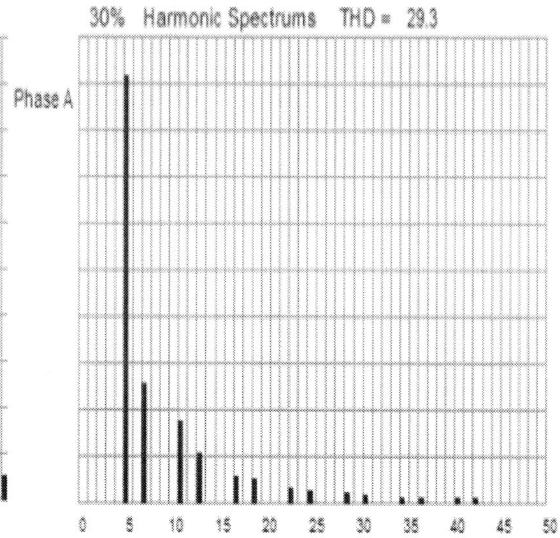

Figure 4. Harmonic Spectrum of Voltage. **Figure 5.** Harmonic Spectrum of Current.

Table 1 shows the calculated parameters at PCC#1 of the simulated power system. Table 2 shows the voltage individual harmonic components, the absolute value and in percentage of its fundamental component values at PCC#1. Table 3 shows the current individual harmonic components, the absolute and in percentage of its fundamental component values at PCC#1. Table 4 shows both of the total harmonic distortion and individual harmonic distortion values of both of the voltage and current at PCC#1, these values are compared to IEEE Std 519 limit values of VTHD and ITHD, it is shown that the VTHD and the 5'th harmonic component of the voltage are failed to compliance with IEEE Std 519. Also the ITHD, the 5'th and 11'th harmonic components of the current are failed to compliance with IEEE Std 519.

Table 1. Calculated Parameters at PCC#1

Total Current I_L (A)	22.8
Short Circuit Current I_{SC} (KA)	0.7
Short Circuit to Load Current Ratio, I_{SC} / I_L	30.0
Displacement Power Factor (DPF)	0.968 lag.
True Power Factor (TPF)	0.927
Apparent Power S (KVA)	436.3
Reactive Power (Q) (KVAR)	163.4
Active Power (P) (KW)	404.6

IOP Conf. Series: Materials Science and Engineering **765** (2020) 012013 doi:10.1088/1757-899X/765/1/012013

Table 2. Voltage Individual Harmonic Distortion Components

Harmonic #	V_{LL} rms	V_{LL} % Fund.	Harmonic #	V_{LL} rms	V_{LL} % Fund
1	10980.03	100.00	30	0.04	0.00
2	0.06	0.00	31	69.41	0.63
3	0.14	0.00	32	0.05	0.00
4	0.02	0.00	33	0.25	0.00
5	503.75	4.59	34	0.06	0.00
6	0.05	0.00	35	63.29	0.58
7	198.66	1.81	36	0.09	0.00
8	0.03	0.00	37	64.02	0.58
9	0.13	0.00	38	0.06	0.00
10	0.05	0.00	39	0.21	0.00
11	214.81	1.96	40	0.03	0.00
12	0.04	0.00	41	57.36	0.52
13	158.68	1.45	42	0.07	0.00
14	0.01	0.00	43	57.05	0.52
15	0.11	0.00	44	0.07	0.00
16	0.03	0.00	45	0.17	0.00
17	110.85	1.01	46	0.08	0.00
18	0.04	0.00	47	57.72	0.53
19	115.72	1.05	48	0.12	0.00
20	0.05	0.00	49	55.25	0.50
21	0.19	0.00	50	0.09	0.00
22	0.02	0.00			
23	83.10	0.76			
24	0.05	0.00			
25	78.05	0.71			
26	0.02	0.00			
27	0.17	0.00			
28	0.02	0.00			
29	78.79	0.72			

IOP Conf. Series: Materials Science and Engineering **765** (2020) 012013 doi:10.1088/1757-899X/765/1/012013

Table 3. Current Individual Harmonic Distortion Components.

Harmonic #	I_{LL} rms	I_{LL} % Fund.	Phase (degr.)	Harmonic #	I_{LL} rms	I_{LL} % Fund.	Phase (degr.)
1	21.83	100.00	-16.39	30	0.00	0.00	
2	0.00	0.00		31	0.12	0.56	16.87
3	0.01	0.02		32	0.00	0.00	
4	0.00	0.00		33	0.00	0.00	
5	5.99	27.45	85.24	34	0.00	0.00	
6	0.00	0.00		35	0.10	0.45	61.28
7	1.68	7.72	135.39	36	0.00	0.00	
8	0.00	0.00		37	0.09	0.42	77.03
9	0.00	0.00		38	0.00	0.00	
10	0.00	0.00		39	0.00	0.00	
11	1.15	5.27	- 177.74	40	0.00	0.00	
12	0.00	0.00		41	0.07	0.33	111.57
13	0.72	3.28	- 166.72	42	0.00	0.00	
14	0.00	0.00		43	0.07	0.31	132.97
15	0.00	0.00		44	0.00	0.00	
16	0.00	0.00		45	0.00	0.00	
17	0.38	1.73	- 110.16	46	0.00	0.00	
18	0.00	0.00		47	0.06	0.28	170.55
19	0.35	1.60	- 94.35	48	0.00	0.00	
20	0.00	0.00		49	0.06	0.26	-170.91
21	0.00	0.00		50	0.00	0.00	
22	0.00	0.00					
23	0.21	0.94	- 62.25				
24	0.00	0.00					
25	0.18	0.80	- 35.35				
26	0.00	0.00					
27	0.00	0.00					
28	0.00	0.00					
29	0.15	0.69	- 0.95				

Table 4. Compliance with IEE Std 519

The Calculated Parameter		Harmonic No.	Calculated Value [%]	IEEE Std 519 Limit [%]	
Voltage Harmonic Distortion (VTHD)		------	6.01	5.0	FAIL
Maximum individual Voltage Harmonic		5	4.6	3.0	FAIL
Current Total Harmonic Distortion (ITHD)		------	29.33	8.0	FAIL
Maximum Individual Current Harmonic	< 11	5	27.45	7.0	FAIL
Maximum Individual Current Harmonic	11 to 16	11	5.27	3.5	FAIL
Maximum Individual Current Harmonic	17 to 22	17	1.73	2.5	PASS
Maximum Individual Current Harmonic	23 to 34	23	0.94	1.0	PASS
Maximum Individual Current Harmonic	> 35	35	0.45	0.5	PASS

9. Single Tuned Filter (STF) Design

The single tuned filters that have to be connected to the PCC#1 consist of two resonant arms for the 5'th and 11'th harmonics components.

9.1, 5'th harmonic STF Design

From table 1 the

TPF = 0.927 (True Power Factor)

Active power = 404.6 KW

So the actual phase angle Φ_{TPF} = arc TPF = arc 0.927 = 22° then tan Φ_{TPF} = 0.4

If the required PF is to be equal 1 then $\Phi_{Required}$ = zero deg. and tan $\Phi_{Required}$ = zero deg.

The required compensation reactive power = load (KW) [tan Φ_{TPF} - tan $\Phi_{Required}$]

= 404.6 [0.4 – zero] = 161.84 KVAR

$$X_C = \frac{V_{LL}^2}{VAR} = \frac{11^2 \times 10^6}{161.84 \times 10^3} = 747\ \Omega$$

$$X_C = \frac{1}{2\pi fC}$$

$$\text{So } C = \frac{1}{2\pi f X_C} = \frac{1}{2\pi \times 50 \times 747} = 4.263\ \mu f$$

If the capacitance is to be divided between the two filters branches as two- third of the capacitance value in 5'th harmonic branch and one –third of the capacitance value in the 11'th harmonic branch so

$C_5 = 2.842\ \mu f$

And

$C_{11} = 1.421\ \mu f$

Now

$$X_{C5} = \frac{10^6}{2\pi \times 250 \times 2.842} = 224.117\ \Omega$$

$X_{L5} = X_{C5}$ at resonance frequency

So

$X_{L5} = 224{,}117\ \Omega$

$$L_5 = \frac{224.117}{2\pi \times 250} = 0.142\ \mathrm{H}$$

Now if the

The capacitor temperature coefficient ($\frac{\Delta C}{C} = 0.05\ \%$ per degree Celsius)

The inductor temperature coefficient ($\frac{\Delta L}{L} = 0.01\ \%$ per degree Celsius)

The frequency tolerance ($\frac{\Delta f}{f_n} = \pm 1\%$)

So δ can be calculated from equation (11)

$$\delta = \frac{\Delta f}{f_n} + \frac{1}{2}\left(\frac{\Delta L}{L} + \frac{\Delta C}{C}\right)$$

$$\delta = \left(\frac{1}{100}\right)[\,1 + 0.5\,(0.05 \times 20 + 0.01 \times 20)] = 0.016$$

IOP Conf. Series: Materials Science and Engineering **765** (2020) 012013 doi:10.1088/1757-899X/765/1/012013

The optimum quality factor (Q) which results in the lowest harmonic voltage is then obtained from the following equation [5].

$$Q = \frac{1 + \cos \Phi_{sy}}{2\delta \sin \Phi_{sy}}$$

Where Φ_{sy} is the phase angle of the system impedance which is less than 75° at any frequency. So

$$\mathrm{Q} = \frac{1 + \cos 75}{2\,(0.016) \sin 75} = 40.738$$

$$\mathrm{R_5} = \frac{\mathrm{X_{L5}}}{\mathrm{Q}} = \frac{224.117}{40.738} = 5.5\ \Omega$$

9.2. 11'th harmonic STF Design

$$\mathrm{C_{11}} = 1.421\ \mu f$$

$$\mathrm{X_{C11}} = \frac{10^6}{2\pi \times 550 \times 1.42\,1} = 203.743\ \Omega = \mathrm{X_{L11}}$$

$$\mathrm{L_{11}} = \frac{203.743}{2\pi \times 550} = 0.0589\ \mathrm{H}$$

$$\mathrm{R_{11}} = \frac{\mathrm{X_{L11}}}{\mathrm{Q}} = \frac{203.743}{40.738} = 5\ \Omega$$

10. Simulation of the Power System with Filters

A 5'th harmonic shunt passive STF and an 11'th harmonic shunt passive STF are designed to mitigate the VTHD and ITHD of this system to be compliance with IEEE Std 519. The power system is simulated with the two branches of the STF (5'th and 11'th) are connected to PCC#1. Table 5 shows the filters parameters C, L and R.

Table 5. Filters Parameters

The Filter	C (μf)	L (H)	R (Ω)
5'th Harmonic STF	2.842	0.142	5.5
11'th Harmonic STF	1.421	0.0589	5

The simulation results at PCC#1 of the voltage and current waveforms is shown in figures 6 and Figure 7 respectively. The harmonic spectrum of the voltage and current are shown in Figure 8 and Figure 9 respectively.

IOP Conf. Series: Materials Science and Engineering **765** (2020) 012013 doi:10.1088/1757-899X/765/1/012013

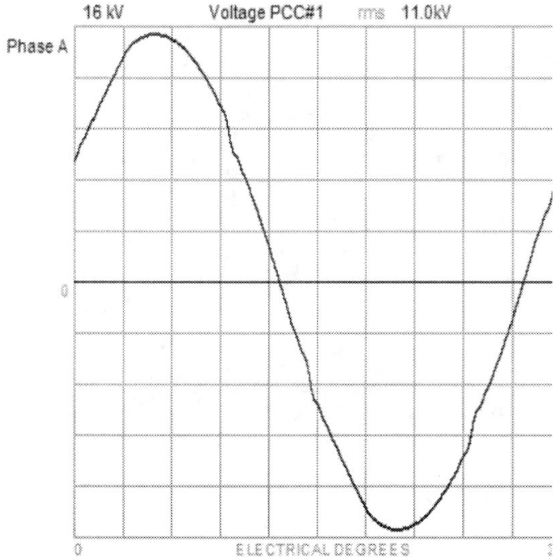

Figure 6. Voltage Waveform at PCC#1

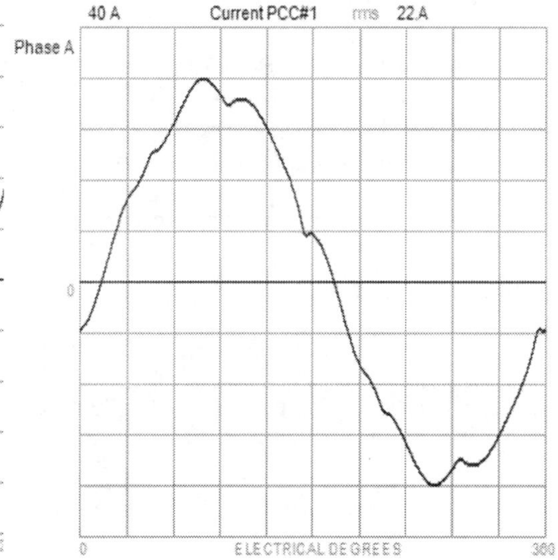

Figure 7. Current Waveform at PCC#1

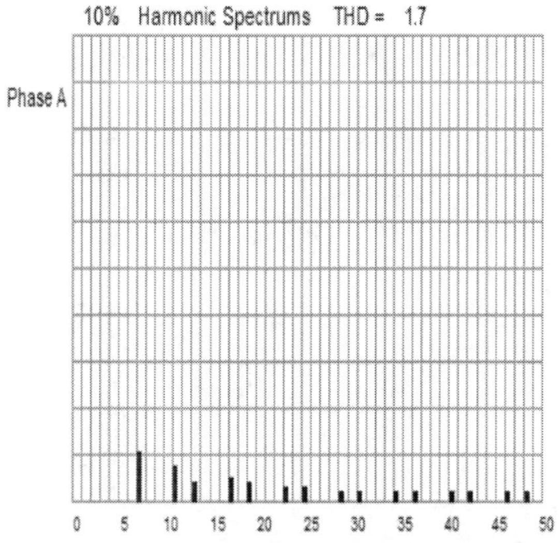

Figure 8. Harmonic Spectrum of Voltage

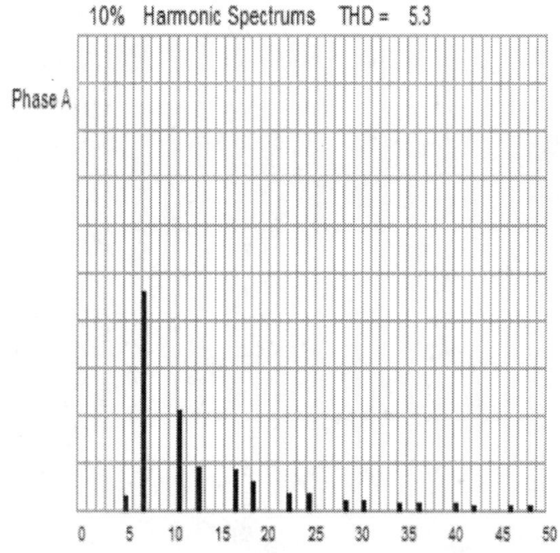

Figure 9. Harmonic Spectrum of Current

Table 6 shows the calculated parameters at PCC#1 of the simulated power system. Table 7 shows the voltage individual harmonic components, the absolute and in percentage of its fundamental component values at PCC#1. Table 8 shows the current individual harmonic components, the absolute and in percentage of its fundamental component values at PCC#1. Table 9 shows both of the total harmonic distortion and individual harmonic distortion values of both of the voltage and current at PCC#1, these values are compared to IEEE Std 519 limit values of VTHD and ITHD, it is shown that the power system is compliance with IEEE Std 519 for both of the voltage and current waveforms.

Table 6. Calculated Parameters at PCC#1.

Total Current I_L (A)	21.7
Short Circuit Current I_{SC} (KA)	0.7
Short Circuit to Load Current Ratio, I_{SC} / I_L	30.2
Displacement Power Factor (DPF)	0.978 lag.
True Power Factor (TPF)	0.976
Apparent Power S (KVA)	415.0
Reactive Power (Q) (KVAR)	90.3
Active Power (P) (KW)	405.0

Table 7. Voltage Individual Harmonic Distortion Components.

Harmonic #	V_{LL} rms	V_{LL} % Fund.	Harmonic #	V_{LL} rms	V_{LL} % Fund
1	10997.81	100.00	30	0.10	0.00
2	0.37	0.00	31	27.23	0.25
3	0.26	0.00	32	0.02	0.00
4	0.26	0.00	33	0.17	0.00
5	6.33	0.06	34	0.29	0.00
6	0.33	0.00	35	26.91	0.24
7	117.79	1.07	36	0.34	0.00
8	0.22	0.00	37	22.72	0.21
9	0.10	0.00	38	0.17	0.00
10	0.07	0.00	39	0.30	0.00
11	86.85	0.79	40	0.14	0.00
12	0.23	0.00	41	25.75	0.23
13	44.40	0.40	42	0.21	0.00
14	0.11	0.00	43	24.08	0.22
15	0.12	0.00	44	0.41	0.00
16	0.13	0.00	45	0.50	0.00
17	56.11	0.51	46	0.17	0.00
18	0.18	0.00	47	23.10	0.21
19	46.47	0.42	48	0.24	0.00
20	0.10	0.00	49	24.84	0.23
21	0.15	0.00	50	0.30	0.00
22	0.30	0.00			
23	33.63	0.31			
24	0.13	0.00			
25	37.85	0.34			
26	0.26	0.00			
27	0.36	0.00			
28	0.11	0.00			
29	26.18	0.24			

Table 8. Current Individual Harmonic Distortion Components

Harmonic #	I_{LL} rms	I_{LL} % Fund.	Phase (degr.)	Harmonic #	I_{LL} rms	I_{LL} % Fund	Phase (degr.)
1	21.70	100.00	-13.99	26	0.00	0.00	
2	0.00	0.02		27	0.00	0.00	
3	0.01	0.03		28	0.00	0.00	
4	0.00	0.01		29	0.05	0.23	-98.15
5	0.07	0.33	-5.36	30	0.00	0.00	
6	0.00	0.01		31	0.05	0.22	-77.53
7	1.00	4.61	-176.51	32	0.00	0.00	
8	0.00	0.01		33	0.00	0.00	
9	0.00	0.00		34	0.00	0.00	
10	0.00	0.00		35	0.04	0.19	-65.47
11	0.47	2.15	144.85	36	0.00	0.00	
12	0.00	0.00		37	0.03	0.15	-49.83
13	0.20	0.93	160.07	38	0.00	0.00	
14	0.00	0.00		39	0.00	0.00	
15	0.00	0.00		40	0.00	0.00	
16	0.00	0.00		41	0.03	0.15	-23.42
17	0.19	0.88	-160.74	42	0.00	0.00	
18	0.00	0.00		43	0.03	0.13	-17.33
19	0.14	0.65	-163.48	44	0.00	0.00	
20	0.00	0.00		45	0.00	0.00	
21	0.00	0.00		46	0.00	0.00	
22	0.00	0.00		47	0.02	0.11	14.49
23	0.08	0.38	-121.43	48	0.00	0.00	
24	0.00	0.00		49	0.03	0.12	20.90
25	0.09	0.39	-115.78	50	0.00	0.00	

Table 9. Compliance with IEE Std 519

The Calculated Parameter	Harmonic No.	Calculated Value [%]	IEEE Std 519 Limit [%]	
Voltage Harmonic Distortion (VTHD)	------	1.7	5.0	PASS
Maximum individual Voltage Harmonic	7	1.1	3.0	PASS
Current Total Demand Distortion (ITDD)	------	5.3	8.0	PASS
Maximum Individual Current Harmonic < 11	7	4.6	7.0	PASS
Maximum Individual Current Harmonic 11-16	11	2.1	3.5	PASS
Maximum Individual Current Harmonic 17- 22	17	0.9	2.5	PASS
Maximum Individual Current Harmonic 23-34	25	0.4	1.0	PASS
Maximum Individual Current Harmonic > 35	35	0.2	0.5	PASS

11. Conclusion

The 5'th harmonic shunt passive STF and the 11'th harmonic shunt passive STF have mitigated the VTHD and ITHD of this system to be compliance with IEEE Std 519. So the single tuned filters which have been used were effective in mitigating the distortion of both of the current and the voltage at the point of common coupling.

References

[1] Seema P. Diwan, Dr. H. P. Inamdar, and Dr. A. P. Vaidya, "Simulation Studies of Shunt Harmonic Filters: Six Pulse Rectifier Load – Power Factor Improvement and Harmonic Control" *Proc. of Int. Conf. on Advances in Electrical & Electronics 2010, ACEE.*

[2] Qual-Tech Engineers, Inc. www.QualTechEng.com.

[3] Tony Hoevenaars, P. Eng., Kurt LeDoux, P. Eng. And Matt Colosino, "Interpreting IEEE Std 519 and Meeting its Harmonic Limits in VFD Application", *copyright Material IEEE, Paper No. PCIC-2003-15.*

[4] Marek Fabris, Anthony H. Hoevnaars, and John L. Greenwald, " Oil Field Retrofit Of ESPs to Meet Harmonic Compliance" *IEEE Transactions on Industry Applications,* ***Vol. 52****, No. 1, January/Febreuary 2016.*

[5] Arillaga J. and. Watson N. R , "*Power System Harmonics*" Hoboken, NJ, USA; Wiley, 2003.

[6] IEEE Recommended Practice and Requirements for Harmonic Control in Electric Power Systems. IEEE Power and Energy Society, Approved 27 March 2014, IEEE-SA Standards Board

Portable field hospital fed from solar system depends on interleaved boost converter (IBC)

Carlo Y. Makdisi [(a)] , Marah F. Mariam [(b)] , Jaafar T. Maarouf [(c)]
(a): Electrical Power Engineering Department, Mechanical and Electrical Engineering Faculty, Tishreen University, Lattakia, Syria. makdisiecarlo@yahoo.com
(b): Electrical Power Engineering Department, Mechanical and Electrical Engineering, Tishreen University, Lattakia, Syria. marah.blackrose@gmail.com
(c):Electrical Power Engineering Department, Mechanical and Electrical Engineering, Tishreen University, Lattakia, Syria. Jaafar.maarouf@gmail.com

Abstract. In this paper we will suggest equipment for a portable field hospital that is electrically fed by solar panels. The solar panel is connected to the removable (Portable) field hospital via an interface, which is an Interleaved Boost Converter (IBC). The analytical study is carried out for the traditional Boost Converter(BC) and the Interleaved Boost Converter (IBC) using Matlab/Simulink To ensure the electric supply of the portable field hospital with suitable voltage values. The paper will explain what might happen if we doubled or tripled the main elements of the conventional boost converter. The expected results will reflect the implementation importance of the interleaved converter (IBC) in comparison with conventional converter. We will also compare the different types of interleaved converter (IBC), so that we get the best interior connection of the IBC, which achieves the suitable distribution of the total current between the coils, as well as the best size – weight specifications. The selected IBC will contain less ripple in the output voltage waveform , where we provide the potable field hospital feed without disturbances or breakdowns , which cause the IBC to be out of serves or fault occurring to any part of it.

Keywords: BC chopper, Interleaved Boost Converter (IBC), Portable Field Hospital (PFH), Hybrid Uninterruptible Power Supplies (HUPS), Intensive Care Unit (ICU).

1. Introduction:

The operating method of the conventional DC chopper and its internal construction was explained in [1],[2], the applications of the interleaved boost converters in PV systems and switched mode power supply are illustrated in [3]. In reference [5] the comparison of interleaved converter for battery charging of PHEV in production plants are studied, and from reference [6] how the magnetic coupling between two inductances in the converter, then we started with the design and simulation of 2-ph interleaved boost converters via MATLAB/Simulink in references [4],[8],[9] and from these references we concluded that the higher number of phases leads to better performance and output parameters, whereas the effect of magnetic coupling in three phase converters is established in reference [10], however the floating output in these converters with les THD% values which doesn't exceed 5% is configured in reference [11], and from reference [12] we understood the equations configuring these converters and its operation methods.

In this paper, portable hospital is suggested which is fed from interleaved boost converter connected to solar system. The aim of the suggested system is to provide the devices and equipment of the portable field hospital by AC and DC required electrical power to prevent the interruption of electrical feeding. The essence of this paper focuses on suitable IBC, therefore the different types of IBC are discussed in this paper.

2. Conventional boost converter:

This study describes interleaved boost converters, within multiple cases. first to understand the conventional converter we must understand the DC choppers, DC choppers are semiconductor switching devices which convert the fixed DC voltage into controlled DC voltage, it is used to control the power of traction and propulsion like in electrical vehicles and subway metro, transportations in mines and hybrid uninterruptible power supplies (HUPS).

This type of choppers are allowed to obtain an output voltage higher than the input voltage value, $\mathbf{V_o} > \mathbf{V_s}$. We always have in this type of choppers induction coil (L) at the input of the chopper circuit, as shown in figure (1).

2.1. Operation principle:

Within the switch-on period $\mathbf{0} < \boldsymbol{t} < \boldsymbol{t_1}$, the current of the induction coil increases to the maximum value at the end of this period, and during this period the coil will store electromagnetic power.
At the moment $\mathbf{t} = \mathbf{t_1} = \mathbf{t_{on}}$, the switch (MOSFET) is turned off for a period of time $\mathbf{t_2} = \mathbf{t_{off}}$, the stored energy in the coil (L) will be discharged , and the current starts to decrease passing through the diode D1 and the Load. [4],[12].

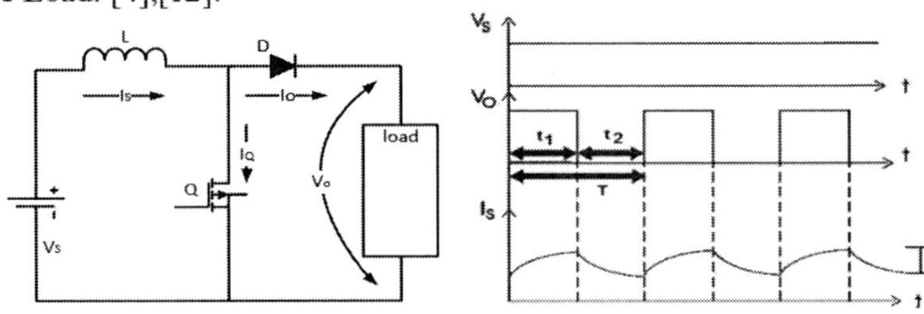

(a) (b)

Fig (1): a) the conventional chopper circuit diagram.
b) the output waves of the conventional chopper.

During the switch-on period the voltage value on the coil is expressed by the following equation:

$$U_L = \mathrm{L}\,\frac{\mathrm{di}}{\mathrm{dt}} \ldots\ldots\ldots(1)$$

And the current varies between two values $\mathrm{I_2}\&\mathrm{I_1}$:

$$\Delta\mathrm{I} = \mathrm{I_2} - \mathrm{I_1} = \frac{V_s}{\mathrm{L}} t_1 \ldots\ldots\ldots(2)$$

Through the switch-on period of SW, the load circuit is shorted , and $V_s = U_L$.
During the period $\mathrm{t_2}$,the instant load voltage is given as following:

$$\mathrm{V}_o = \mathrm{V}_s + \frac{\mathrm{L}.\Delta\mathrm{I}}{\mathrm{t_2}} = \mathrm{V}_s + \mathrm{L}.\frac{V_s}{\mathrm{L}.\mathrm{t_2}}\mathrm{t_1} = \mathrm{V}_s\left(1 + \frac{\mathrm{t_1}}{\mathrm{t_2}}\right) \ldots\ldots\ldots(3)$$

$$\mathrm{V}_o = \mathrm{V}_S.\frac{1}{1-k} \ldots\ldots\ldots(4)$$

Where:

$$\mathrm{k} = \frac{\mathrm{t_1}}{\mathrm{t1_+t_2}} = \frac{\mathrm{t_1}}{\mathrm{T}} \Rightarrow \mathrm{t_1} = \mathrm{k.T}, \qquad \mathrm{t_2} = \mathrm{T} - \mathrm{t_1} = \mathrm{T} - \mathrm{k.T} = \mathrm{T}(1-\mathrm{k})$$

When a capacitive filter is connected to the load, we can presume that the output voltage is constant and $\mathrm{V}_o = V_a$.
From the equation (4) we can observe that the load voltage is configured by the duty cycle (k), where the voltage increases with the increase of (k=D [duty cycle]), however when $k = 0$ the voltage will be minimum and $\mathrm{V}_o = V_S$.
From equation (4), the output voltage is changing by varying of (k=D) value.

2.2. Simulation of conventional interleaved boost converters using Matlab/Simulink:

Table (1) parameters for BC

Input voltage	$12\ [V]$
L (Inductive)	$76.8 \times 10^{-6}\ [H]$
Switching frequency	$\frac{1}{100000}\left[\frac{1}{Hz}\right]$
Duty Cycle	82%

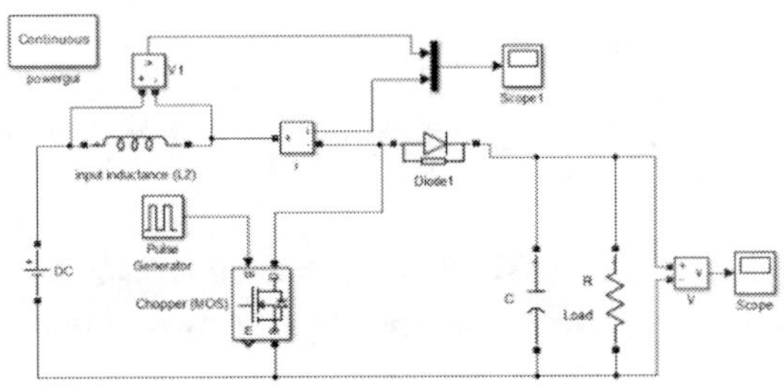

Fig (2) the simulation diagram of BC

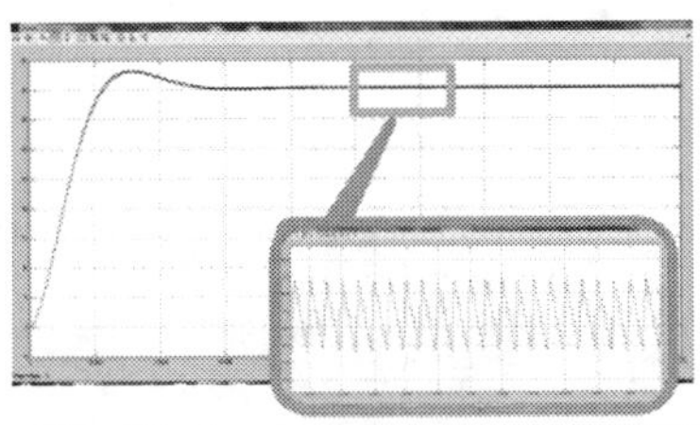

Fig (3) the output voltage of BC

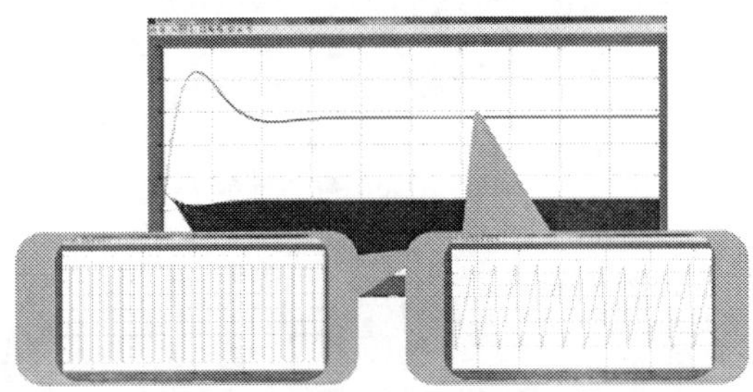

Fig (4) : - pink color the inductor (coil) current
- yellow color the inductor (coil) voltage

3. Double Boost Converters:

In this case we doubled the number of coils, diodes and MOSFETs(chopper), however we decreased the values of the coils to half, to observe the advantages of this change by simulating the circuit using MATLAB\ Simulink.

Table (2) parameters for 2-ph IBC.

Input voltage	$12\ [V]$
L (Inductive)	$38.4 \times 10^{-6}\ [H]$
Switching frequency	$\frac{1}{100000}\left[\frac{1}{Hz}\right]$
Duty Cycle	82%

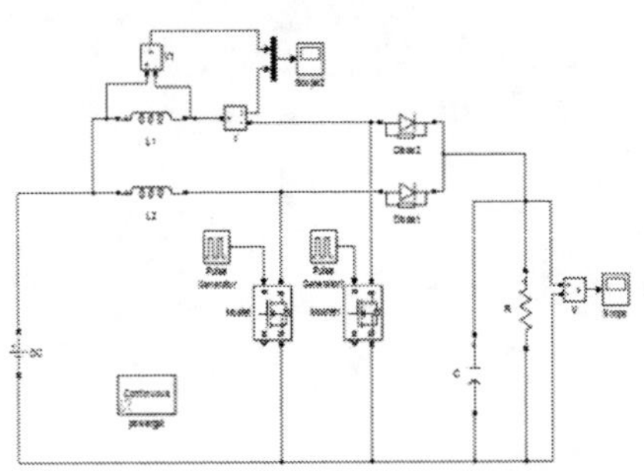

Fig (5) the simulation diagram of 2-ph IBC

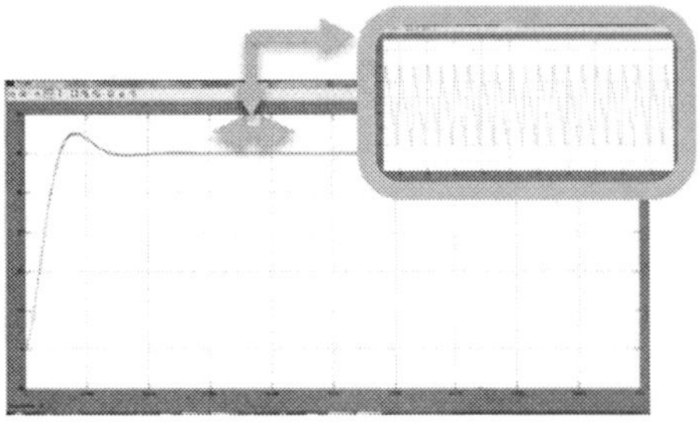

Fig (6) the output voltage of 2-ph IBC

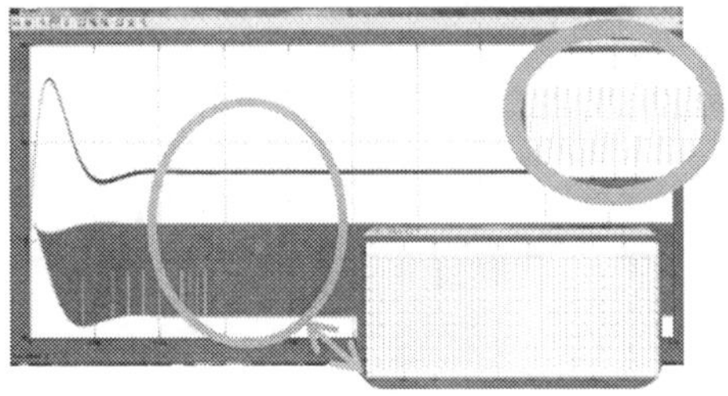

Fig (7) : - pink color the inductor (coil) current
- yellow color the inductor (coil) voltage

As we can observe the change was obvious, the output voltage value increased almost up to 25% more than the conventional boost converter.

We must keep in mind that the switching frequency has to be high so the voltage ripple can be as low as possible, in our case we chose switching frequency of 100 kHz, as the switching chopper is a MOSFET, it can hold up to 1 MHz, but unfortunately the MOSFET with these mentioned parameters is commercially unavailable.

3.1. Voltage ripple comparison :

In case of double boost interleaved converter if we changed the switching frequency of the MOSFET we would decrease the voltage ripple:[5],[6],[7]

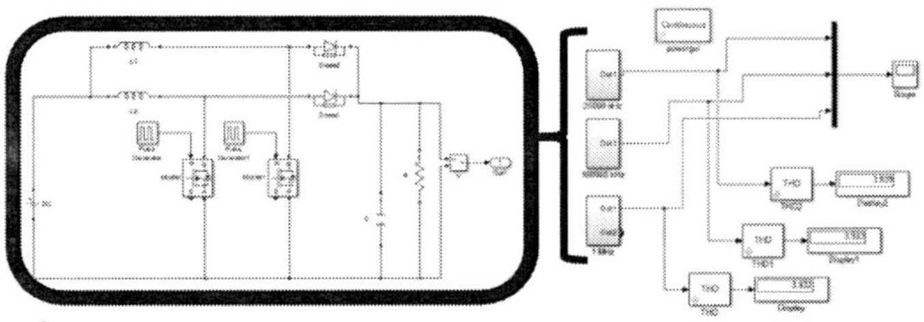

Fig (8) the simulation of the comparison

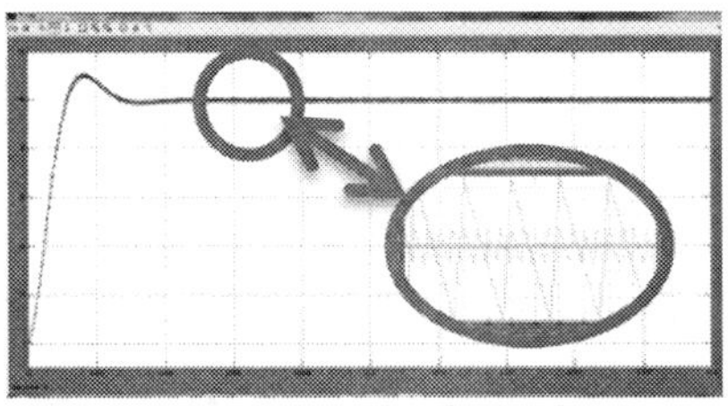

Fig (9) reducing ripple in the output voltage versus to higher switching frequency

Table (3) comparison parameters

Switching Frequency	25 $[kHz]$	100 $[kHz]$	1 $[MHz]$
V_{in}	12	12	12
V_{out}	50.46	50.1	49.99
THD %	3.939	3.933	3.932

3.2. The effect of the duty cycle:

If the switching frequency is maintained at a constant value of 100 kHz and the duty cycle D is varied of 67, 75 and 82 % we get how the average value of the voltage increases with the increasing of D in limited range up to 82% :[2],[9]

IOP Conf. Series: Materials Science and Engineering **765** (2020) 012014 doi:10.1088/1757-899X/765/1/012014

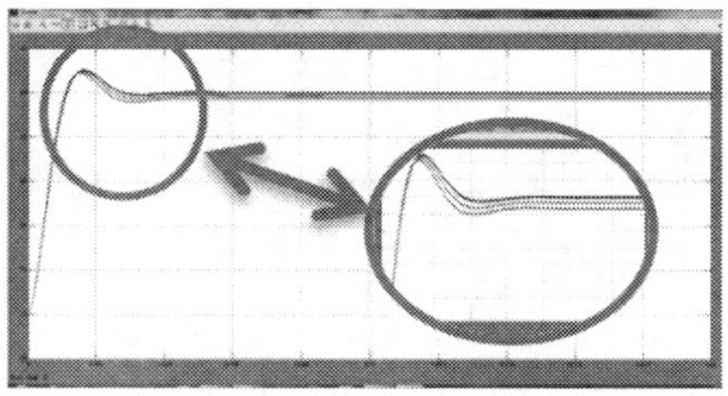

Fig (10) the output voltage versus to duty cycle values.

a. Black line: 82%
b. Pink line: 75%
c. Blue line: 67%

Table (4) comparison parameters

Duty cycle %	82	75	67
V_{in}	12	12	12
V_{out}	50.1	49.44	48.76
THD %	3.933	4.076	4.224

4. 3-ph IBC:

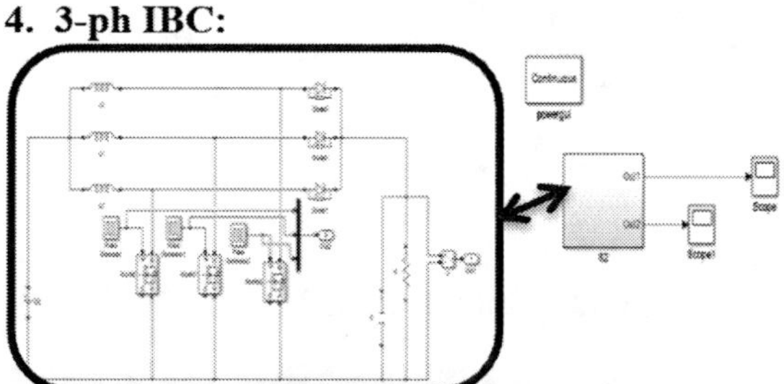

Fig (11) the simulation diagram for 3-ph IBC.

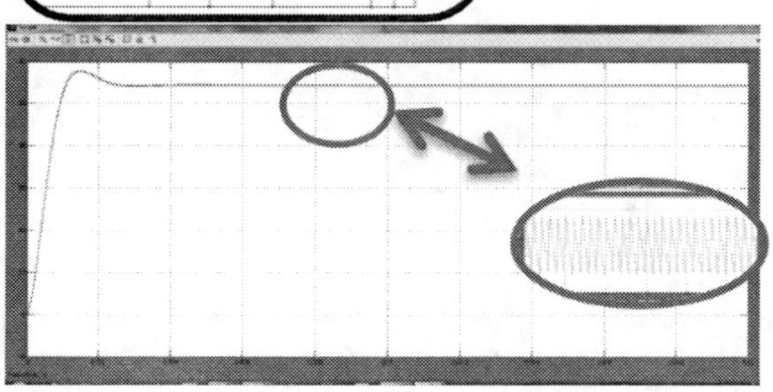

Fig (12) the output voltage of 3-ph IBC

So if we compared between the conventional boost converter and the double converter and the 3ph converter, with the same switching frequency of 100 kHz, and the same duty cycle of 82% [1], we would notice the increase of the average voltage value:

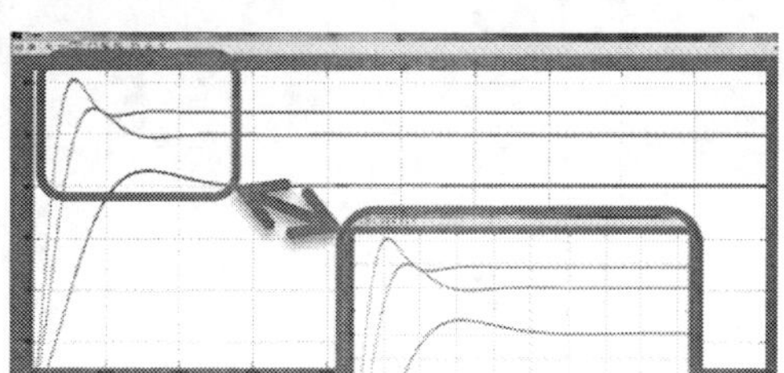

Fig (13) comparison between BC and 2-ph & 3-ph IBC

5. Study case:

IOP Conf. Series: Materials Science and Engineering **765** (2020) 012014 doi:10.1088/1757-899X/765/1/012014

After the analysing of the IBC operation of different topologies, we have to apply the previous study on the portable field hospital supplied via solar panel.

The box diagram of our study case is illustrated below:

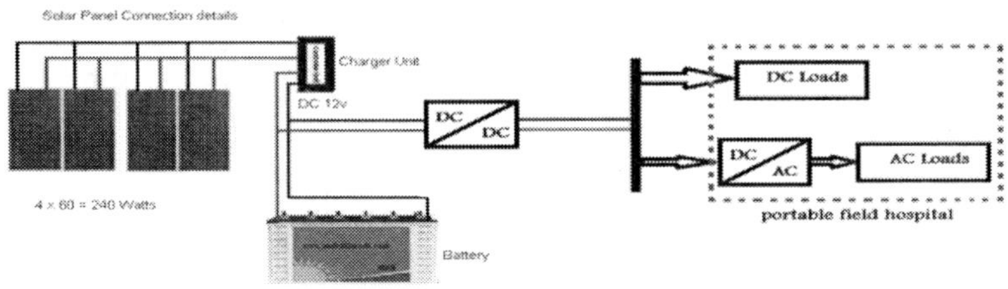

Fig (14) flow chart of the study case.

The general overview of the portable field hospital contains all devices and equipments is shown below, which is designed by DIALux

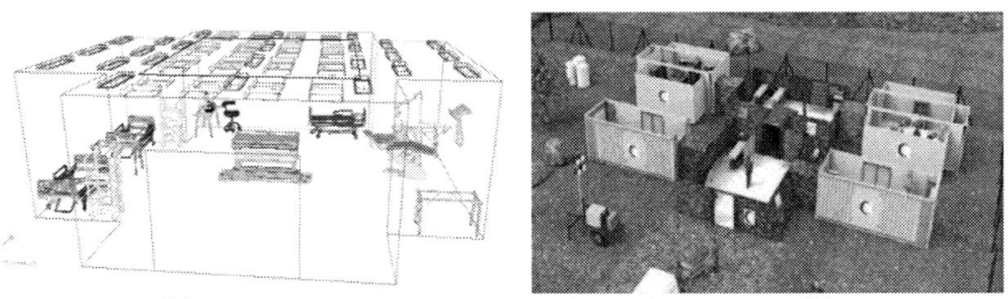

(a) (b)

fig (15) shows: (a) the general overview of portable field hospital
(b) the designed practical structure of the portable field hospital.

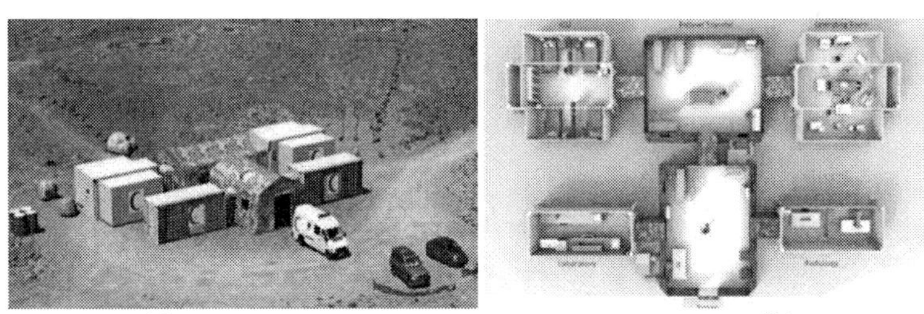

(a) (b)

Fig (16): (a) actual portable field hospital in real location.
(b) plan view of the portable field hospital.

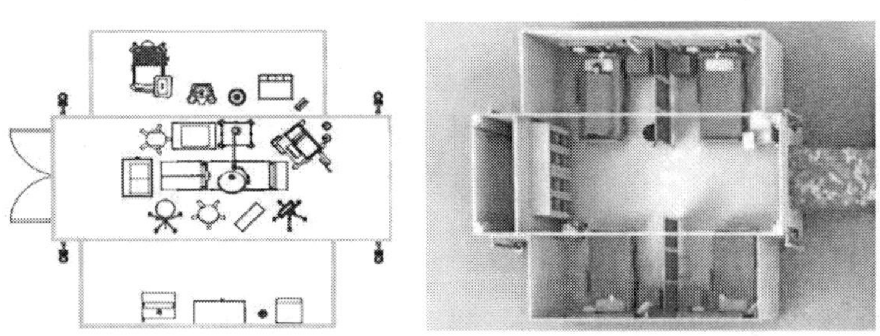

Fig (17): (a) plan view of the operating room.
(b) plan view of the intensive care unit (ICU).

IOP Conf. Series: Materials Science and Engineering **765** (2020) 012014 doi:10.1088/1757-899X/765/1/012014

The portable field hospital contains various loads (AC & DC), therefore we have to provide these loads by a suitable feeding. DC loads can be resourced directly from the output of IBC, while the AC loads must be supplied via inverter, and the output voltage of the inverter will be applied to the various AC loads, such as: operating room; ICU; and radiology room, as shown in figure (18).

The presence of inverter requires to make a simulation of this inverter to insure the waveform required by the mentioned previously AC loads.[3],[10],[11].

5.1. The simulation process of the portable field hospital using MATLAB \ Simulink:

The simulation for the diagram shown in figure (14) using MATLAB \ Simulink:

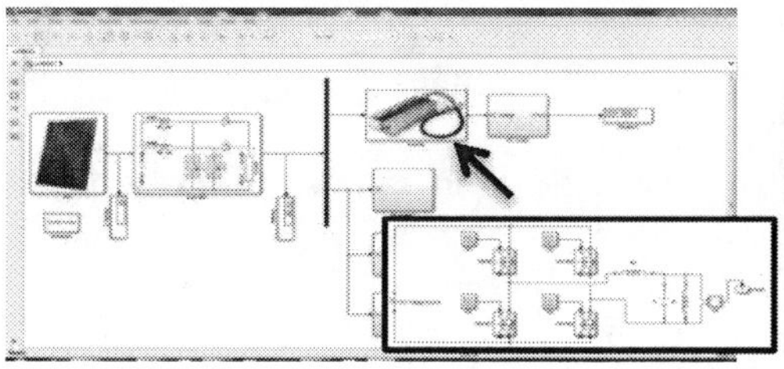

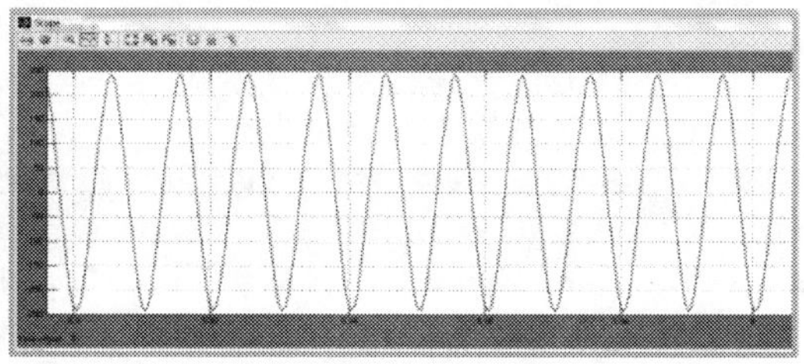

Fig (18) simulation of the portable field hospital. Fig (19) output voltage waveform of the inverter.

From figure (20) the suggested system consists of many components:

- Solar panels.
- Charger unit.
- Set of batteries.
- IBC (boost converter).
- Inverter.

6. *Result discussion:*

- The study had been carried out by using conventional boost converter and various topologies of IBC.
- From dedicated previous figures (2) and (5) the comparison shows that from double IBC , we can obtain higher voltage value, and less distortion (THD% = 3.933%).
- The total current in IBC is divided equally between the coils, therefore the power dissipation will be less than conventional BC.
- Figure (8) and (9) shows the reflection of increasing the duty cycle (D%) and varying the switching frequency on the total harmonic distortion (THD), since the output voltage will contain less harmonics with the increasing of D% and switching frequency.
- The simulation of the portable field hospital had been done by using MATLAB \ Simulink to provide to all loads to avoid the interruption of the electrical supply from any part of this hospital.

7. Recommendations and conclusions:

- The higher number of phases of IBC will lead to enhancing the output voltage of boost converter. But we have to consider the increasing of volume – weight of the converter cascade.
- There is no doubt, that increasing the number of phases of the boost converter in these systems will certainly improve the specifications and values of output voltage suitable for feeding different loads. In order to feed some of the proposed AC field facilities, it is necessary to introduce an appropriate inverter to convert the DC voltage into an appropriate AC voltage for these loads.
- The increasing of the IBC links (number of phases) leads to complexity of control circuit of the used switches related to their high number, therefore the future suggestion is to reduce the number of these switches and simplify the control circuit of the used switches.

IOP Conf. Series: Materials Science and Engineering **765** (2020) 012014 doi:10.1088/1757-899X/765/1/012014

- MPPT Algorithm can be discussed to obtain maximum efficiency from solar system in the future, might be a good recommendation to meet the reliable expectations.

8. References

[1] Steady-state Characterization of Multi-phase, Interleaved Dc-Dc converters for Photovoltaic Applications. Sairaj V. Dhople, Ali Davoudi and Patrick L. Chapman – 2009.
[2] Power Electronics Handbook 3rd ed. M. Rashid (B-H, 2011).
[3] High Voltage Gain Interleaved DC Boost Converter Application for Photovoltaic Generation System. Weerachat Khadmun, Wanchai Subsingha – 2012.
[4] Simple And Efficient Implementation Of Two-Phase Interleaved Boost Converter For Renewable Energy Source. Mounica Ganta, Pallam reddy Nirupa, Thimmadi Akshitha, Dr.R.Seyezhai – 2012.
[5] Comparison of Interleaved Boost Converter Topologies with Voltage Multiplier for Battery Charging Of PHEV. R. Seyezhai, V. Aarthi – 2014.
[6] Analysis of non-isolated two phase interleaved high voltage gain boost converter for PV application. A. Ramesh Babu, T.A.Raghavendiran – 2014.
[7] Basic Design and Review of Two Phase and Three Phase Interleaved Boost Converter for Renewable Energy Systems. Chitra.P, Seyezhai.R – 2014.
[8] Simulation of Incremental Conductance MPPT based Two phase Interleaved Boost Converter using MATLAB/Simulink. Sheik Mohammed, D. Devaraj – 2015.
[9] Modeling and simulation of choppers switching via MATLAB/SIMULINK. Ehsan Hosseini 2015.
[10] Review of Coupled Two and Three Phase Interleaved Boost Converter (IBC) and Investigation of Four Phase IBC for Renewable Application. P.Abishri, Umashankar S, Sudha Ramasamy – 2016.
[11] A floating-output interleaved boost DC–DC converter with high step-up gain. Ardavan Kianpour, Ghazanfar Shahgholian – 2017.
[12] Power electronics /1/ Carlo J. Makdisie – Tishreen university publication – 2017A reference
This reference has two entries but the second one is not numbered (it uses the 'Reference (no number)' style.

Evaluation the Influences of Temperature, Dust and Shading on Photovoltaic System Performance

H F Khazaal[1] H M Kadhim[2] M S Farhan[3] I Sh Baqer[4] S R Salih[5]

[1]Electrical Eng. Dept., College of Engineering, Wasit University, Iraq. E-mail: hf1964@yahoo.com
[2]Electrical Eng. Dept., College of Engineering, Wasit University, Iraq. E-mail: humammonam@gmail.com
[3]Electrical Eng. Dept., College of Engineering, Wasit University, Iraq. E-mail: mansur545@yahoo.com
[4]Electrical Eng. Dept., College of Engineering, Wasit University, Iraq. E-mail: ismaoll2000@yahoo.com
[5]Raban Al-Safina Company for Engineering and Technology, Iraq. E-mail: sameh@rabanalsafina.com

Abstract The proliferation of the use of Photovoltaic PV systems, as one of the sources of electricity that comes from renewable energy resources, attracting the researchers to investigate on the factors that have an effect on the performance of it. This paper presents an analytical study for evaluating the performance of PV system that generates 2.5 kWp through the deployment of this system to supply the electricity for the Engineering Consultant office, of the College of Engineering at Wasit University, Iraq (32.497 N, 45.842 E). The evaluation is conducted by monitoring the pre-defined system which consists of 10 PV modules as an array of two rows. The system is supervised for six months from 1st of Feb. until 31st of July, to assess the impact of temperature, shading, and dust on its' output power and performance. The PV array is omitted to 8° south-west, and 40° in elevation. The results shows that the PV array is operated at efficiency of 12.12% and the maximum efficiency reached 13.01 during April. The average daily power of it, for the months from Feb. until July are: 1.347 kW, 1.421kW, 1.436, 1.215 kw, 1.045kW and 1.042 kW, respectively. The minimum average output power is recorded in summer season (June and July), because of the high ambient temperature in these months. The maximum reduction of power due to shading at 4 pm is 52.77% in February followed by March 47.7 % and April 22.93 %, while the maximum reduction in efficiency due to shading were 51.82 %, 44.71% ,23.01% in February, March and April , respectively. The reduction in power due to the dust accumulated after 54 days exposure to environments conditions was 7.88%.

Keywords: **Photovoltaic system, solar cells, solar energy, renewable energy.**

1. Introduction

The use of photovoltaic systems, as a good alternative for supplying electricity from renewable energy resources instead of the generated electricity from fossil fuel is proliferated, that the PV systems have a role in reduction the global warming and pollution [1]. Also these systems can be deployed for supplying electricity to remote and rural populated villages [2,3]. Because of all the preceded reasons, these systems need to continues supervision, evaluation and assessment to its performance by means of output power and efficiency, through the investigation on the factors that have a direct effect on its' work. Many researches were conducted to assess the effect of different climate conditions on the performance of PV.

2. The effects of temperature and irradiance

Solar cells are influenced by environments conditions. There are two important factors affected the solar cell performance, which are: ambient temperature and solar radiation. When the temperature of the ambient of the cell rises above 25°C, the power of the cell will drops due to decrease of output voltage with a slight increase in current, but when the value of solar radiation increases the output current increases significantly while the output voltages increase very slightly. Therefore, the increase and decrease in power and efficiency are directly related to these two factors. The nature of the drop and rise is understood by calculating the temperature coefficients of voltage, current and power. Power and voltage coefficients are negative because the voltage is dramatically reduced by heat and consequently lowering the output power. While the temperature coefficient of the current is positive because as we mentioned, there is a very slight increase of the current with high temperature.

3. System's description and components

Figure 1 presents the solar array under test. Which represents an off -grid 2.5 kW_p PV system consists 10 poly-crystalline modules, 250W_p (Euronet) each. The array is tilted at 40° and oriented toward south-west 8°azmiuth at Al-Kut city, Iraq (32.497 N, 45.842 E). Each two modules in the array are connected in series to form a string, then the resultant strings are five pairs of panels connected in parallel. The experiment was done at Wasit University, College of Engineering. The inverter used in this work is (MUST PV18-5048 MPK, 5KVA, 4KW) including charge control. The battery bank consists 4 batteries (FARBER SMF-VRLA, 12V, 100AH) are connected in series as shown in figure 2. The specifications of modules and inverter including charge control specifications are tabulated in tables 1 and 2.

Figure 1: Off- grid PV array 2.5 kW_p

IOP Conf. Series: Materials Science and Engineering **765** (2020) 012015 doi:10.1088/1757-899X/765/1/012015

Figure 2: Inverter (right) and battery bank (left)

Table 1: Modules specifications presented by manufacture

P_{max}	$250W_p$
Tolerance	±3%
V_{mp}	30.8V
I_{mp}	8.12A
V_{oc}	37.6V
I_{sc}	8.94A
Dimensions	1600mm*992mm*35mm
NO. of cells in series	60
Test condition	1000 W/m^2,AM 1.5 , 25°C

Table 2: Inverter and charge control specifications

Model Name	PV18-5048 MPK
Rated power	5KVA/4KW
DC Input	48VDC,93A

AC output	230VAC,50 Hz,22A,1φ ,power factor: 0.8
AC charge mode	AC input:230 VAC,50Hz,35A,1φ DC Output:54 VDC,60A(max)
Solar charge mode	Rated current : 60A System voltage : 48VDC MPPT Voltage range:64~130VDC Max. Solar voltage(V_{OC}):145VDC I_{SC} PV :50A Max. Charge current :120 A

4. The methodology of the work

The measurements was taken from 08:00 am to 04:00 pm daily, for the period from Feb. 1st to July 31st. The measurements were including Irradiance, Ambient temperature, module temperature, and the output parameters (PV power and efficiency).

5. Efficiency of solar cell

The solar module efficiency (η) is the ratio between the output produced power "P" and the solar power "P_{solar}" available on the panel surface "A_c" **[4]**

$$\eta = \frac{P}{P_{solar}} = \frac{V.I}{A_c.G_t} \quad (1)$$

Where V and I are voltage and current, respectively which represent the output power, A_c is the surface area of module (m^2), and G_t is the total irradiance falling on module surface (W/m^2)

6. Results and discussion

The ambient temperature during the period of the test was found to be in the range from 17°C in Feb. to 47°C in July as shown in figure 3. The heat generated in the module depends on the amount of solar radiation incidents on the module, conversion efficiency of module and the properties of the panel material (absorption and reflection). In this study the average difference between module temperature and ambient temperature ranges between 15°C and 16 °C approximately. During June and July, at the afternoon when the ambient temperature ranges from 44 °C to 47 °C, and when the wind is absence ; the module temperature reached 70 °C.

Figure 4 presents average PV efficiency at each month. The efficiency was gradually increased from minimum value 11.68 % in February (due to the cloudy and rainy weather which is leading to reduce the irradiance), reaching the highest value 13.01% during April., then the efficiency decay gradually from May, and showed the lowest value 11.36% in July.

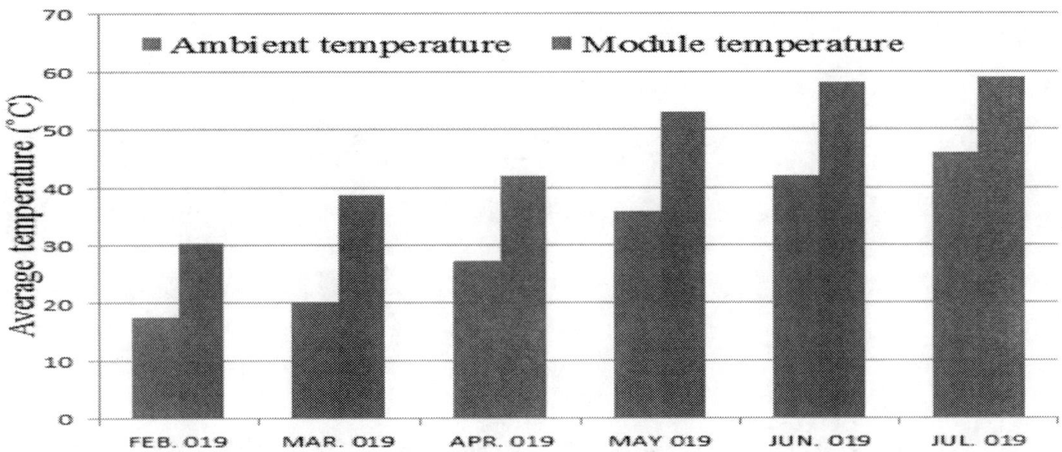

Figure 3: Average ambient and module temperature

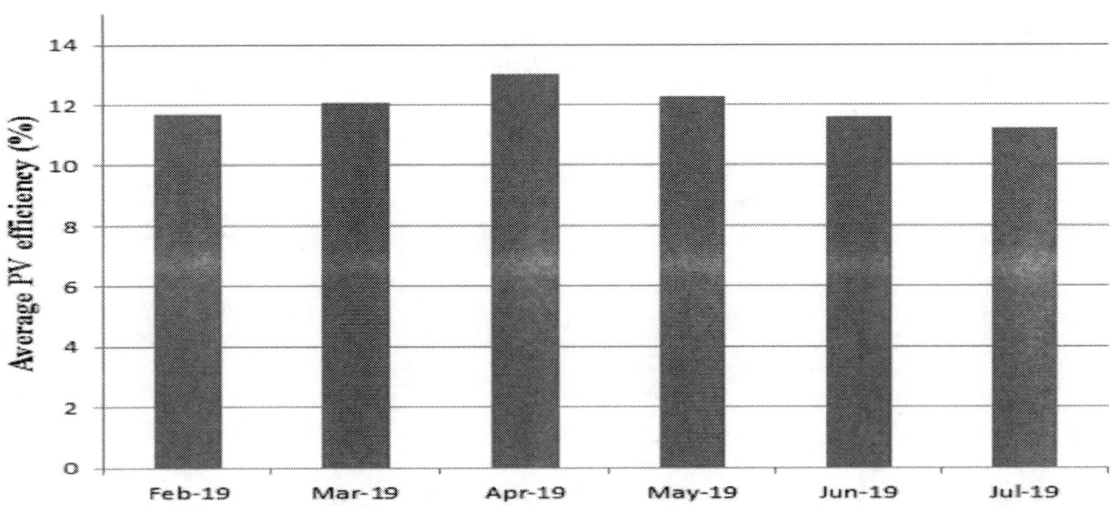

Figure 4: Average PV efficiency at each month

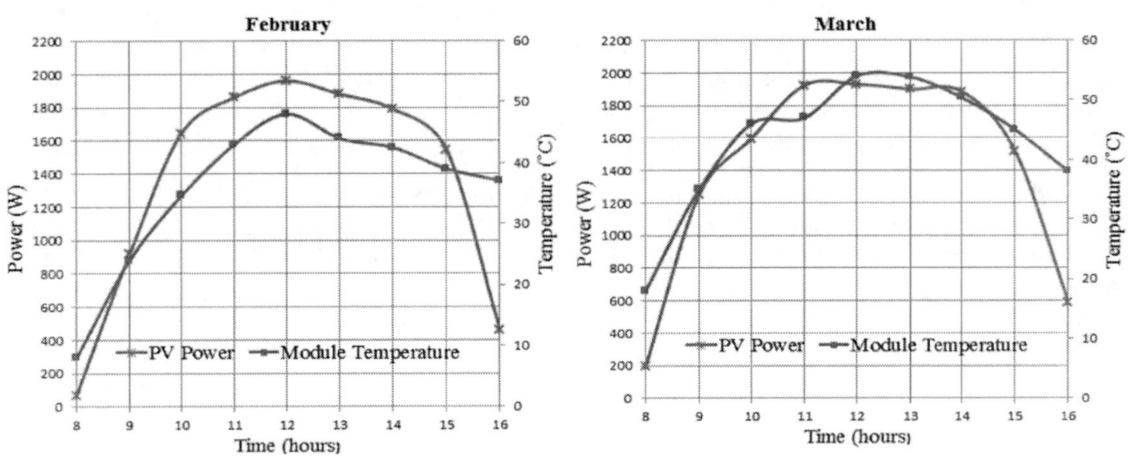

IOP Conf. Series: Materials Science and Engineering **765** (2020) 012015 doi:10.1088/1757-899X/765/1/012015

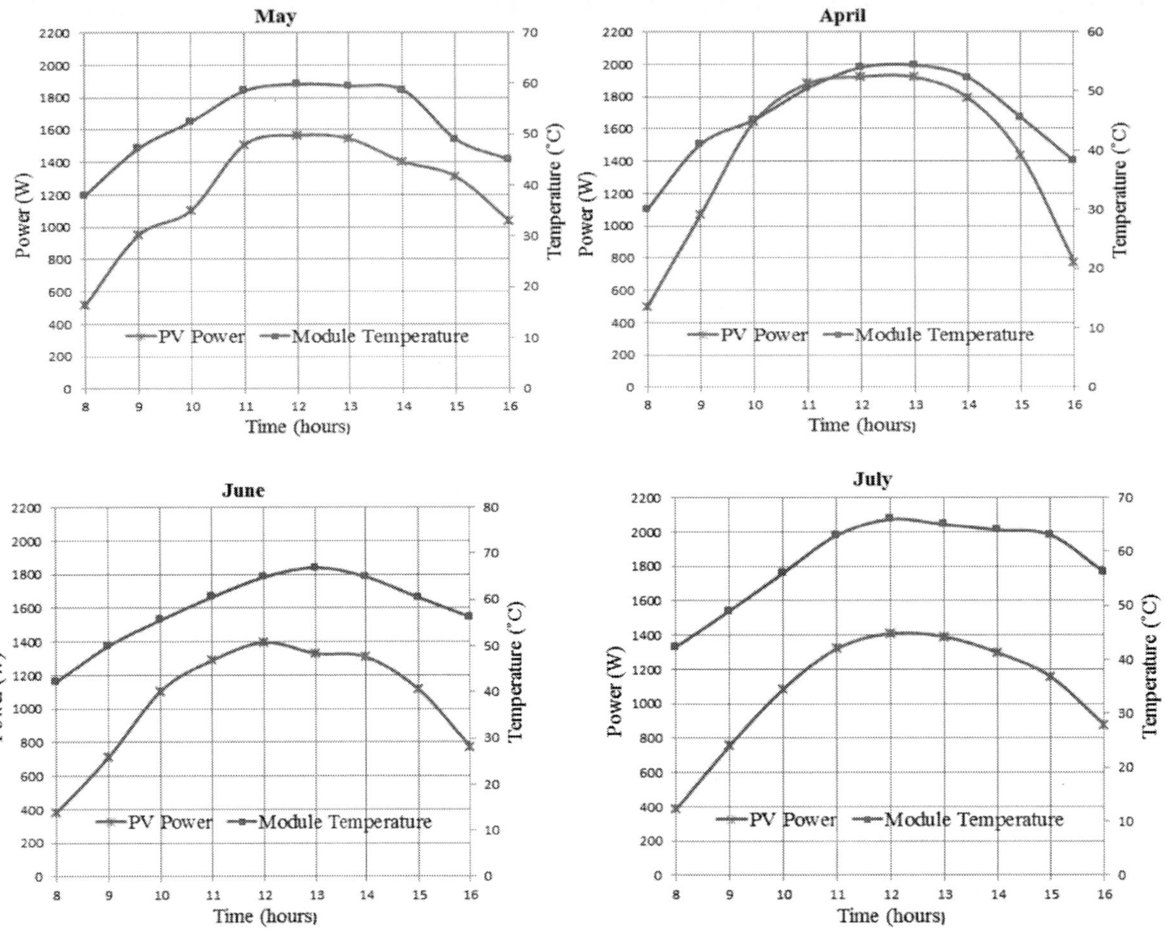

Figure 5: Average daily of PV power

Figure 5, shows that the minimum output power through the six months, is recorded until an 08:00 am, that the solar irradiance is low, due to partially shading that caused by the building of College of Engineering from the right until about 08:00 am. In February through March, all of the modules become completely sunny at about 08:30 am. While in the period from April until July the panels becomes completely sunny at 08:00 am. After 08:00 am, the power is gradually rise due to the continuous increase in solar radiation up to 12:00 midday and reach its peak at 01:00 pm where the radiation is at its peak at this time. And then the power is gradually decreases due to the decrease in solar radiation. The average maximum module temperature lies between 12:00 midday and 01:00 pm (this is for the period of study), because at this time the sun is perpendicular on the surface of the earth and we get at the same time the highest incident solar radiation.

The average of daily PV power for six months from (February to July) was 1.347 kW, 1.421kW, 1.436 kW, 1.215 kW, 1.045kW and 1.042 kW respectively. The minimum average output power was in summer season (June and July), because the average module temperature in these months is about 58°C. The module temperature sometimes reaches 70°C, during the days without winds. The maximum average power were recorded at April, because the average module temperature is about 45°C, and this is lowers than in other months (May, June and July) and higher incident irradiance than (February and March). Figure 5 shows that the large drop in power is at 04:00 pm, because of the shading that happens at this time in February, March and April. as the shading pattern on the modules is shown in figure 6.

Efficiency does not take a linear pattern of ups' and downs' such as power, because it depends on two important factors; namely radiation and power. The more radiation, leads to increase in power, and this is causes a decrease in efficiency according to equation No.1. Since the ratio of the output power to the radiation is not constant (because the radiation leads to increase the temperature of the module, and when the temperature reaches a high value, the power begins to decline despite the high radiation) so the curve of efficiency will be different from power curve.

It is noted that the efficiency of the module which is shown in figure 7, is at the lowest value when the temperature at its' peak from 12:00 midday to 01:00 pm except in the months of February, March and April, where the lowest efficiency is at 04:00 pm which happens due to the shading caused by the building of consultation office on the modules from left. In most months during the experiment, the highest efficiency is confined in two periods: the first period is in the morning from 08: am to 11:00 am, while the second period, is from 02:00 pm to 04:00 pm, because the temperature during these two periods is low compared to the period from 11:00 am to 01:00 pm in which the efficiency at the lowest value.

Figure 6: Shading on modules at 4 pm (left) and 07:30 am (right)

from figures 5 and 7, it can be seen that the highest drop in power and efficiency is at 04:00 pm in (February, March and April) due to the decrease in irradiance in shaded areas which is lead to a greatly reduced short circuit current. The difference in the level of radiation between the shaded and un-shaded areas is large, and the rate of decay in radiation intensity is approximately 23%. The highest reduction in power and efficiency due to the shading are in February, where the proportion of shading is almost one-third of the PV array area, and then the rate of shading decrease gradually during March and April.

Table 3: Reduction in power and efficiency due to the shading

Month	power	efficiency
February	52.77 %	51.82 %
March	47.7%	44.71 %
April	22.93	23.01%

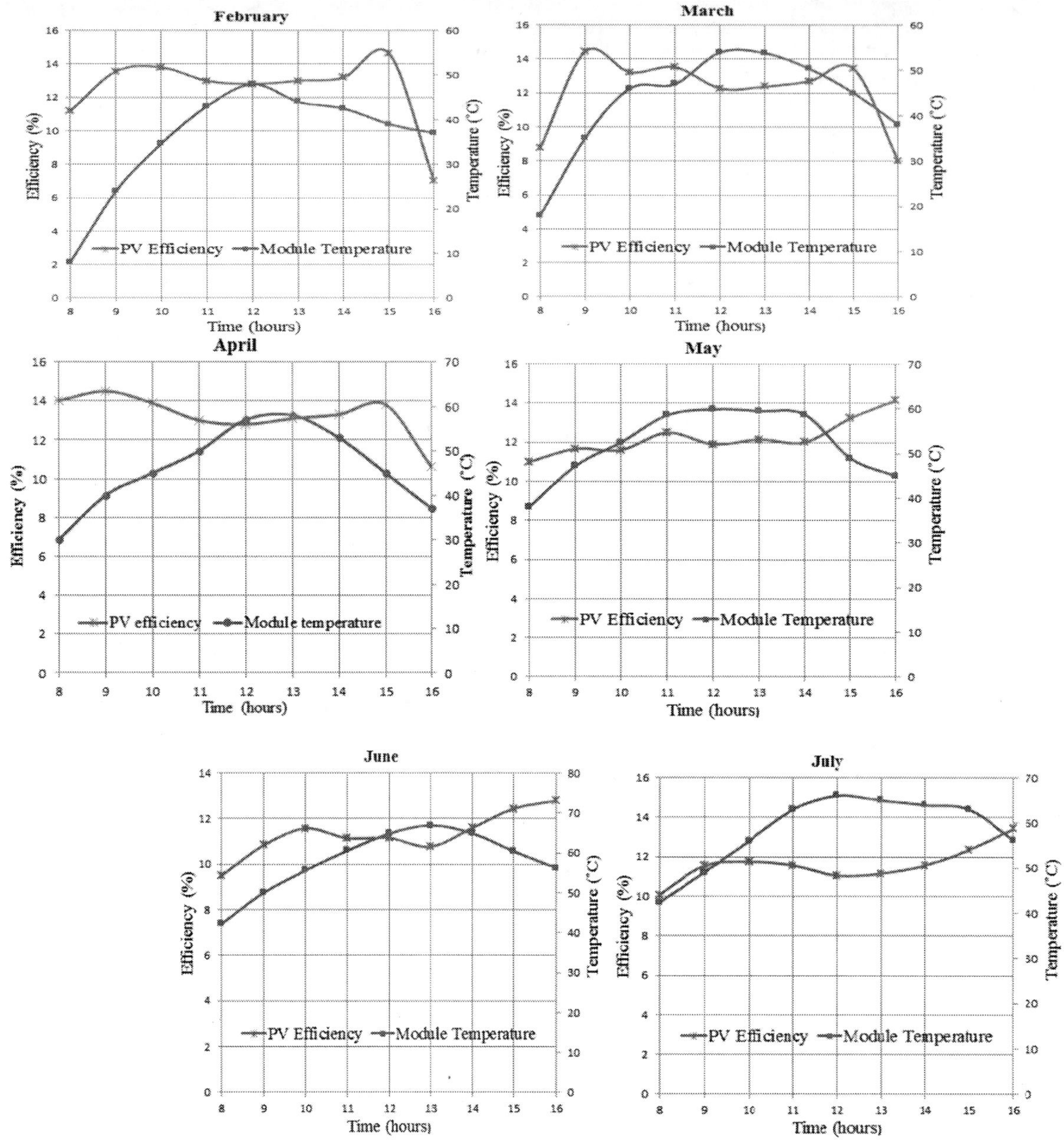

Figure7: Daily average PV efficiency and module temperature

IOP Conf. Series: Materials Science and Engineering **765** (2020) 012015 doi:10.1088/1757-899X/765/1/012015

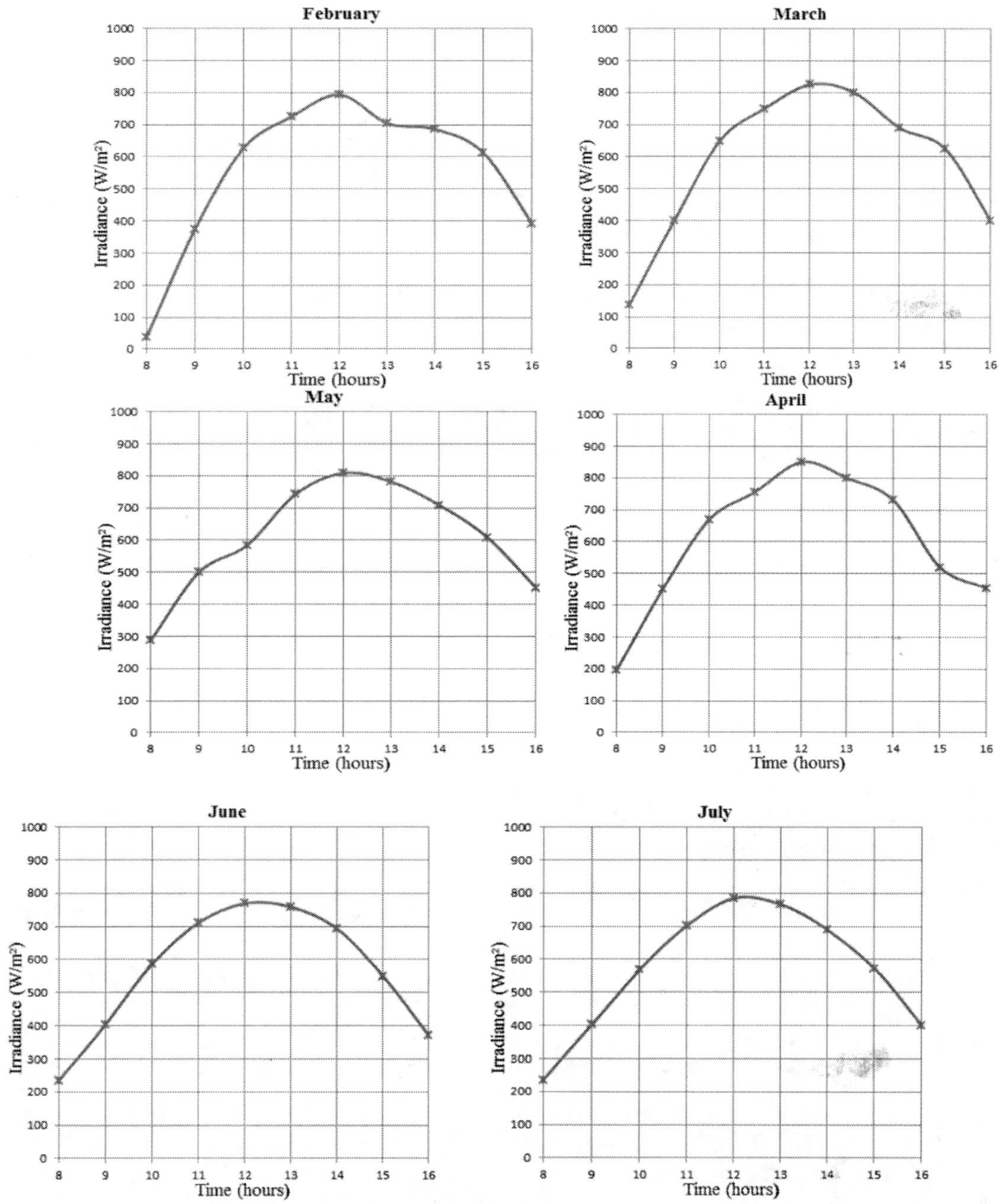

Figure 8: Average daily of irradiance

From figure 8, it can be seen that the maximum incident irradiance for the six months is during the time between 12:00 midday and 01:00 pm while the minimum irradiance is at 08:00 am.
The minimum average daily irradiance is at February, where it reaches about 560 W/m^2 due to the cloud and rain during the days in this month. while the maximum is at May where it reaches about 615 W/m^2, then the irradiance decreased gradually during June and July. This is because there is an error in omitting of solar panels which leads to loss of absorbing the solar energy in the summer season, especially (June and July). In order to get high radiation in the summer when the sun is high, the tilt angle of the system must be less than 40°, and azimuth must be 0° toward south exactly.. The

IOP Conf. Series: Materials Science and Engineering **765** (2020) 012015 doi:10.1088/1757-899X/765/1/012015

optimum tilt and azimuth must be 30° and 0°, respectively according to PVsyst software as in figure 9.

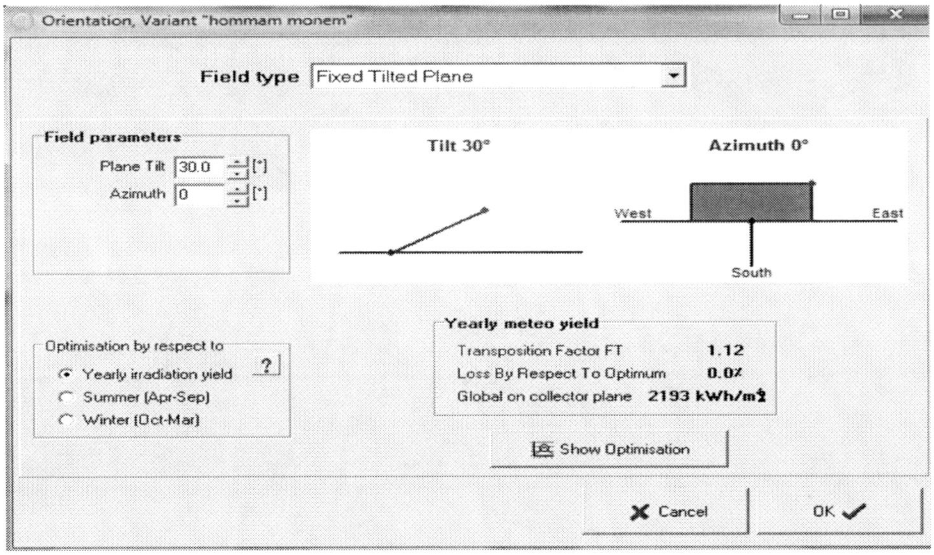

Figure 9: Optimum orientation of PV array from PVsyst software

7. The effect of dust accumulated on PV

Figure 10 shows the state of modules after 54 days exposure to the different weather condition without cleaning for interval (from 9/5/2019 to 1/7/2019). In order to calculate the decrease in PV power due to the accumulation of dust and other pollutants on the panels. we took the data for a whole day before cleaning the modules and after cleaning them with water. All the modules were cleaned with water on 2 / 7 / 2019 as shown in figure 11.

Figure 10: The modules are soaked with dust after 54 day (1/7/2019)

IOP Conf. Series: Materials Science and Engineering **765** (2020) 012015 doi:10.1088/1757-899X/765/1/012015

Figure 11: The PV modules after cleaning (2/7/2019)

From figure 12; it can be seen that the average daily output power in dirty module case is decreased by 7.88 % compared to the output power in clean module case.

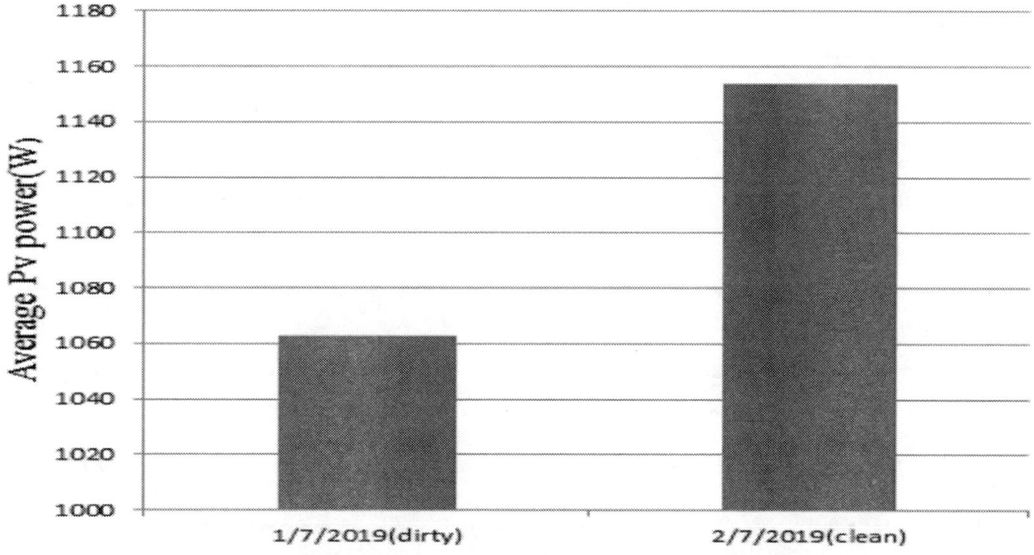

Figure 12: Average Power in dusty and clean modules

Table 4 presents the reduction in power due to the dust with time exposure for previous studies that conducted by various authors, compared with our work results. from the table It is noted that the percentage of reduction in power in this work is acceptable compared with the rest of the results, except the result of ref. 5 that it is conducted in desert area (Saudi Arabia), and also the exposure period is longer.

Table 4: Effect of dust in previous studies and present study

authors	Time exposure	Power reduction
Kaldellis et al.[4]	Four weeks	5%
Adinoyi and Said[5]	Six months	50%
Appels et al.[6]	Five weeks	4%
Ali et al. [7]	Three months	16%
Present work	54 day	7.88 %

8. Conclusions

This paper presents the performance of an off-grid 2.5 kW_p under various environment condition (temperature, dust, shading, cloud and rain).

From the results; we can observe that the performance of the PV system is strongly depend on the following:

1- the ambient temperature: where the increase of it leads to decrease in efficiency of PV system from 13.05% to 11.01 during July.

2- the dust: where this factor affecting the performance of the PV system, that it is reduced by 7.85% for 54 days.

3- the shading: where this factor has a significant effect on the performance of the PV system that it is reduced by 51.8% during February.

9. References

[1] Bloomberg New Energy Finance and Lighting Global, an innovation of the World Bank Group in cooperation with Global Off-Grid Lighting Association 2016, *Report Off-Grid Solar Market Trends Report.*

[2] Milosavljevic´ D D, Pavlovic´ T M and Piršl D S 2015 *Elsevier J. of Renewable and Sustainable Energy Reviews* 44 pp.423–435.

[3] Lorenzo E, Araujo G, Cuevas A, Egido M, Minano J and Zilles R 1994 *Handbook on Solar Electricity Engineering of Photovoltaic Systems* Earthscan Publications Ltd.

[4] Kaldellis J, Kokala A and Kapsali M 2010 *J. Fresenius Environmental Bulletin* **19** (12) pp.2864-2872.

[5] Adinoyi M J and Said S A M 2013 *J. Renewable Energy Elsevier* **60** pp. 633-635.

[6] Appels R, Lefevre B, Paesen R, Beerten A, Driesen J and Poortmans J 2012 (USA) *IEEE Proc. Conf. 38th Photovoltaic Specialists.*

[7] Ali H M, Zafar M A, Bashir M A, Nasir M A, Ali M and Siddiqui A M 2017 *J.Thermal Science* **21**(2) pp. 915-923.

A Hardware Implementation of Neuro-PMD Model classifier based FPGA

Wisam H. Ali[1] , Sammar Jaafar Ismail[2] , Reem Jaafar Ismail[3]
Dept. of Electrical Engineering, Dept. of Electromagnetic Engineering, [1,2] Dept. of Electrical Engineering- University of Technology, Dept. of Electromagnetic Engineering- University of Technology, [3] Dept. of Computer Science- Cihan University
30088@uotechnology.edu.iq, 50240@uotechnology.edu.iq, reem.jafar@cihanuniversity.edu.iq

Abstract. A classifier (polarization mode dispersion) has been introduced in this article. The suggested NeuroPMD classifier uses various types of fiber optics, and it's deviation effect properties to handle light transmission using Polarization Mode Desperation (PMD). The suggested classifier receives an unspecified signal with many percentage points, classifies it base on stored signals and shows the reliability of the received signal. The proposed system is built using VHDL, modeled by the Xilinx package (ISE 9.2i), the proposed NeuroPMD is implemented on the Spartan-3A XC3S700A kit. simulation and Implementation behavioral results demonstrate that the proposed model satisfies the specified operational requirements and gives solutions to overcome the scattering problems in the optical systems.

Keywords: artificial intelligent AI; PMD; classifier; VHDL; signal deviation

1. Introduction

The machine has recently been the main tool for simulating the performance of fiber optics such as numerical calculation for visual device planning[1,2]. This computer updates each new generation faster and thus improves the speed of operation necessary to model multiple tasks so that many programs can be introduced with a barely visible delay[2]. Because of this requirement, this allows the researcher to improve the system and at the same time see the outcome in the form of graphics with a single motion[3].

The designer must now reflect the method, which means that instead of using software engineering to simulate the design of the fiber optics network, the production of fiber optics can be used to implement a new generation of artificially smart computers with optical transmission rather than electrical transm ission[4,5]. The purpose of this paper is to design a VHDL model for a neural network classified technology based on a network of artificial intelligence to link photosynthesis. This technique depends on the optical polarization.

2. Neuro Optics

One of the most important features in the development of electronic circuits is high-efficiency communication and less data transmission loss. The optical connection for the artificial communication circuit where optics (instead of electronic) transmission offers an advantage in the model in terms of understanding, strong connection and flexibility[5]. Optic's connection for the neural memory is a public method that takes the interest of Scientifics for many years [6]. It can be beloved that the transmission of light in the fiber optics transmitter is more efficient than the transmitter that operates according to the electronic signal to transmit the signal along the track. The simplicity in the design and the high-efficiency connection is the major backbone for the employment of the optical connection in the intelligent circuit design.

The optical connection value for the artificial intelligent circuit which operates as the classifier is shown to be [7, 8]:

$$K_n = J_{pd} \frac{\sin(x)}{\cos(x)} \tag{1}$$

where:

K_n is the optical weight connection, J_{pd} is photocurrent, x can be calculated from equation (2).

.

$$x = K_1 \frac{V_n - V_{n+1}}{2} \tag{2}$$

K_1 is constant, V_n is the initial input voltage for generating the optical current (n).

The proposed Neuro-PMD classifier is designed to satisfy the high-speed transmission, output efficiency, fast implementation, and high performance.

3. Polarization analysis

The polarization mode dispersion is a new method to control the deviation of the optical signal to put the signals on the right target [8]. The major serious problem is the light scattering and the highest fading of the optical signals; make a question to find the answer to maintain this problem. Fig (1) shows the method to polarize the signal toward the wanted optical frequency (e.g. filter the wanted signal from the unwanted signal) [9].

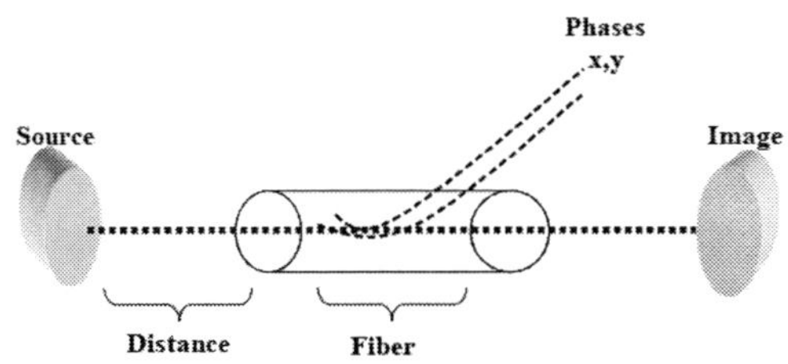

Figure 1. Proposed signal deviation block diagram

"Fig. 1" is in use to be under test in order to show the characteristics of the modeling of fiber coupling for more simulation software like MATLAB and Optics Software for Layout and Optimization (OSLO). The proposed VHDL model is implemented to give a finite parameter to such program analysis. The program must have many features, which are the type of the parameter source (laser and diode), the distance between source and fiber (it is a very important analysis for the signal transmission; it includes spacing and dispersion), and the type of fiber. The entire previous characteristic is required in optical design but the most important thing is the deviation of the optical signal in fiber, as illustrated in "Fig. 1", so that the phase (x, y) is substantial in the connection for fiber coupling [10].

4. The proposed model

The proposed model has to classify the communication signals according to the circuit optical link instead of electronic connectivity. This classifier takes the required signals and stored it in the form of digital signals, and ranked them according to desired requirements. Therefore, it will recognize the unknown input signal based on the stored signal and display the type of unknown signal. "Fig. 2" shows the block diagram of the proposed model. In this figure, the model consists of input signals to insert the communication signals to model, artificial memory to store the signals, and the output signals to display the type of the recognized signals. All parts are connected through an optical fiber, which operates depending on polarization mode dispersion. The pin assignments of the model are illustrated in "Fig. 3", it contains data pins to input the unknown signals to be recognized, write pin (wr) to enable input data to be entered to model, the display recognized signal pins (a0), and the matching pins (a1, a2, a3, and a4) to display the percentage of compatibility with the other of the stored signals.

IOP Conf. Series: Materials Science and Engineering **765** (2020) 012016 doi:10.1088/1757-899X/765/1/012016

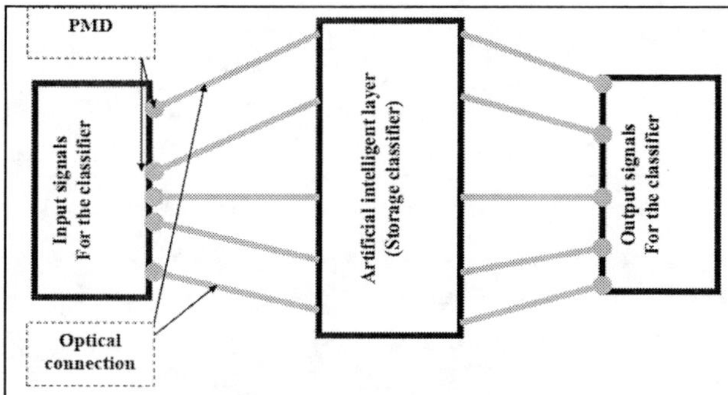

Figure 2. Block diagram of the proposed model.

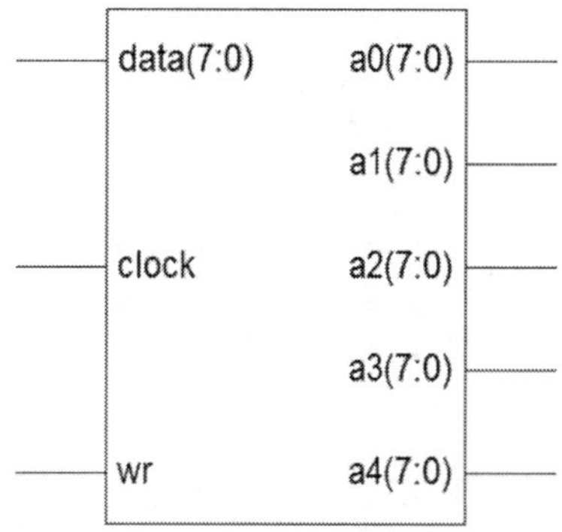

Figure 3. The pin assignments of the proposed model.

The proposed model has a major advantage over the traditional models in that it accepts multiple bits signal instead of one bit.

5. The VHDL model

The VHDL model of the proposed intelligent classifier is built using the Xilinx ISE 9.2i package and implemented using the Spartan-3A XC3S700A FPGA kit. This model consists of an entity declaration, processes, and sub-model. The entity declaration is used to declare the signals of the classifier. The process is assigned to store the designed signals, read the unknown communication signals, recognized the unknown signals, displayed the type of signals that match with the stored signals and showed the compatibility with the other stored signals. This model classified the unknown signals according to artificial intelligent optical weight. The sub-model is designed to receive the signal sequentially and prepares them for the classifier in the parallel form. "Fig. 4" shows the flowchart of the proposed classifier.

In this paper, four different signals have been selected and non-specific as shown in the "Table. 1", where the signals are (x, x1, x2, and x3), all the signals are dissimilar. After building the program in a VHDL language, the signals have been introduced to the program so that the signals are stored. Subsequently, when the real operation of the light classifier is examined, the signal must be inserted sequentially at the beginning of the program. As is clear in "Fig. 5" the (a0) indicates that the signal input is identical with x signal after the implementation of the program and lists the compatibility with the other of the signals in (a1, a2, a3). The previous events are demonstrated in "Figs. 6-8" whenever the input signal is (x1, x2, and x3) are simulated via the program. When the unknown signal is inserted to the classifier, it has not been compatible with any of the stored signals as illustrated in "Fig. 9". "Fig. 10" shows the implementation of the proposed classifier using the Spartan-3A XC3S700A FPGA kit.

TABLE I. THE FOUR DIFFERENT SIGNALS

Signal name	**Optical weight**							
X	*1*	*11*	*4*	*3*	*6*	*13*	*8*	*5*
	14	*7*	*15*	*12*	*10*	*16*	*9*	*2*
X1	*13*	*10*	*12*	*9*	*3*	*1*	*6*	*14*
	5	*8*	*16*	*7*	*4*	*15*	*2*	*11*
X2	*12*	*5*	*1*	*9*	*14*	*2*	*8*	*15*
	10	*3*	*11*	*16*	*13*	*7*	*6*	*4*
X3	*16*	*10*	*1*	*14*	*4*	*2*	*8*	*9*
	15	*3*	*7*	*13*	*5*	*12*	*11*	*6*

IOP Conf. Series: Materials Science and Engineering **765** (2020) 012016 doi:10.1088/1757-899X/765/1/012016

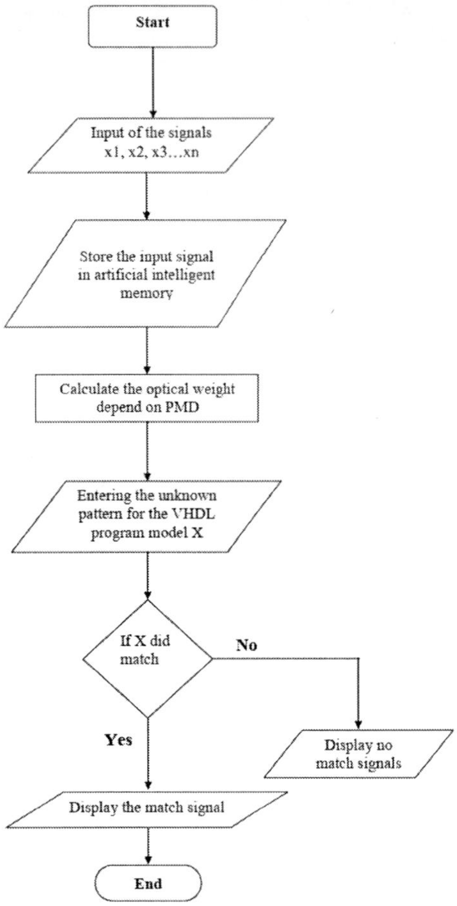

Figure 4. The flowchart of the proposed classifier

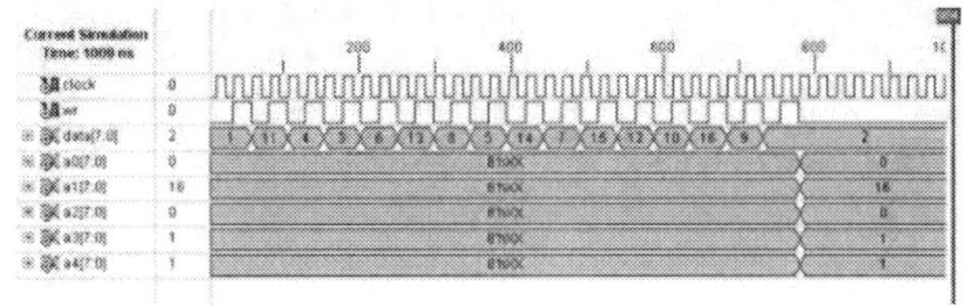

Figure 5. Classifier simulation results when a signal match with x.

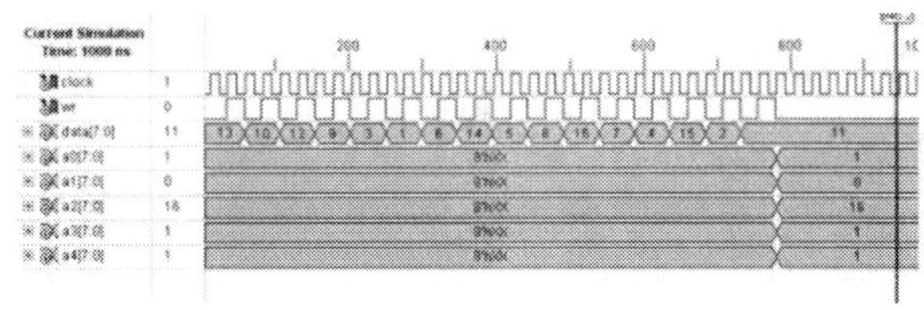

Figure 6. Classifier simulation results when a signal match with x1.

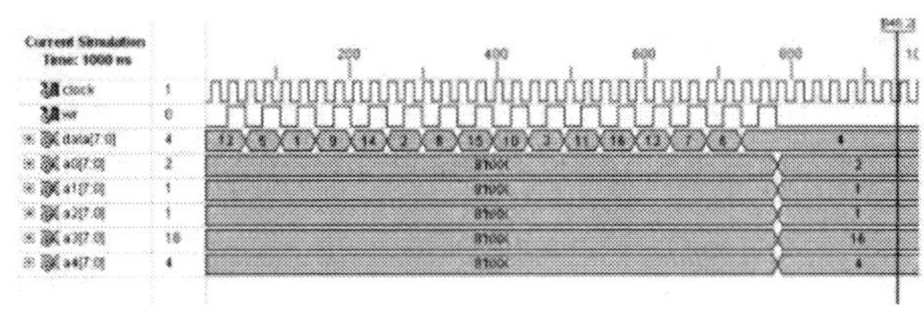

Figure 7. Classifier simulation results when a signal match with x2.

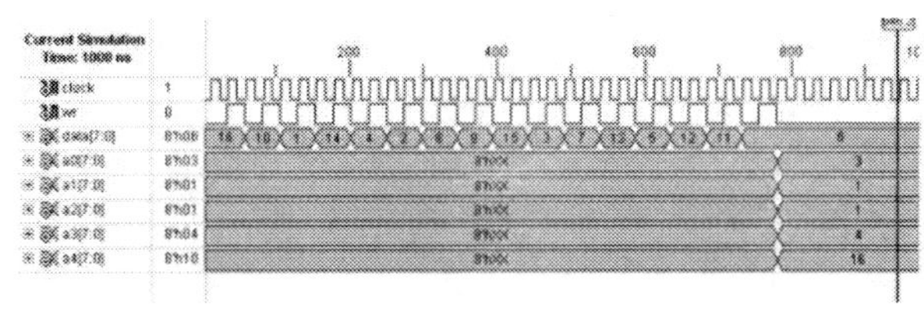

Figure 8. Classifier simulation results when a signal match with x3.

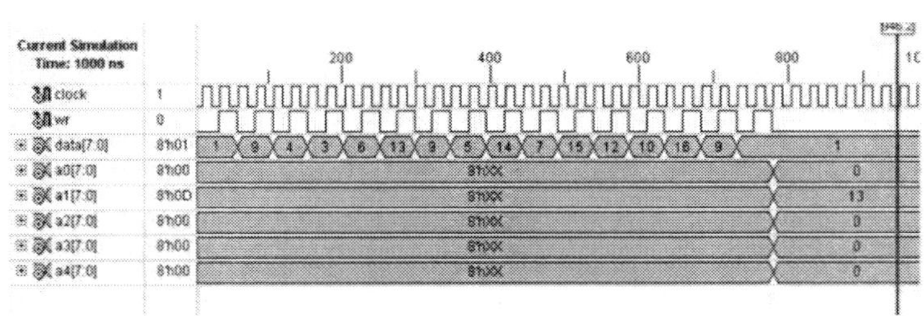

Figure 9. Classifier simulation results when the signal mismatch any signal.

Figure 10. Implementation of the proposed classifier using the Spartan-3A XC3S700A FPGA kit.

6. Conclusions

Most of the traditional classifier receives the signals with one bit. These processes determine the type of signal within the classifier as well as make the classification process need more time to classify the signals. So it was necessary to design a classifier that works with multi-bit signals and high-speed classification. In this paper, a VHDL model of a proposed artificial intelligent optical classifier has been presented. The classifier receives multi-bit signals, classifies them according to optical weight storage, which depends on the polarization mode dissipation to transmit the signals, display the classified signal in high-speed, and lists the compatibility with the others signal. Furthermore, this model's high efficiency has a signal connection and fewer losses. This classifier operation is based on the PMD so the scattering of the optical signal is fading. The proposed model is very useful in encryption and smart storage for multiple signals.

7. References

[1]. J. M. Howard, "Optical Design Using Computer Graphics," APPLIED OPTICS, Vol. 40, No. 19, 2001, pp. 3225 – 3231.

[2]. Lois Mignard-Debise, “Tools for the paraxial optical design of light field imaging systems”, thesis, Université de Bordeaux, 2018

[3]. R. John Koshel and A. Gupta, "Characterization of Light Pips For Efficient Transfer Of Light," Society of Photo-Optical Instrumentation Engineers, 2005.

[4]. F.P. Zen1, B.E. Gunara1, W. Hidayat1, Z.A. Thalib2, H. Zainuddin2, J. Aminuddin1," Application of Artificial Neural Network in Jitter Analysis of Dispersion-Managed Communication System,", Nonlinear Sciences / Pattern Formation and Solitons, arXiv:nlin/0407032, 14 Jul 2004.

[5]. Kazuki Otago, et. al. “Air Mounted Eyepiece Optical See-Through HMD Design with Aerial Optical Functions”, AH2018, February 7–9, 2018, Seoul, Republic of Korea.

[6]. L. Meflah, B. Thomsen, J. Mitchell, P. Bayvel, G. Lehmann, S. Santoni, B. Bollenz," Advanced Optical Performance Monitoring for Dynamically Reconfigurable Networks," Proceeding 10th European Conference on Networks and Optical Communications (NOC'05), 01-JUL-2005, pp. 554- 561.

[7]. R. Ramachandran and N. Gunasekaran, " Optical Implementation of a Two-dimensional Bipolar Hopfield Model Neural Network," Proc.Natl.Sti, Vol.24, No.1, 2002, pp.73-78.

[8]. Chaitra.P," Hardware Implementation Of Artificial Neural Networks Using Back Propagation Algorithm On Fpga" ICESMART, Vol. 05 No.04, 2016,pp211-214

[9]. A. Romariz and K. Wagner, "Optoelectronic Implementation of a FitzHugh-Nagumo Neural Model," Proceeding of Neural Information Processing Systems 15, Key: CiteULike: 423198, 9 December 2003.

[10]. J.M. Zurada, "Introduction to Artificial Neural Systems", Jaico Publishing House, 1996.

[11]. E. Ip and J. M. Kahn, "Digital Equalization of Chromatic Dispersion and Polarization Mode Dispersion," JOURNAL OF LIGHTWAVE TECHNOLOGY, VOL. 25, NO. 8, AUGUST 2007, pp.2033-2043.

[12]. L. Yan, M. C. Hauer, Y. Shi, X. S. Yao, P. Ebrahimi, Y. Wang, A. E. Willner, and W. L. Kath, " Polarization-Mode-Dispersion Emulator Using Variable Differential-Group-Delay (DGD) Elements and Its Use for Experimental Importance Sampling," JOURNAL OF LIGHTWAVE TECHNOLOGY, VOL. 22, NO. 4, APRIL 2004, pp. 1051-1058.

[13]. J. P. Gordon and H. Kogelnik, " PMD fundamentals: Polarization mode dispersion in optical fibers," PNAS, vol. 97, no. 9, April 25, 2000, pp.4541–4550.

Obstacle Avoidance Method for Highly Redundant Robotic Arms

Firas S. Hameed[1, 2], Dr. Hassan M. Alwan[1], Dr. Qasim A. Ateia[1]

1 University of Technology, Mechanical Engineering Department, Baghdad, Iraq.

2 Email: firassubhyhameed@gmail.com

Abstract

Obstacle avoidance is an important concept to be considered when a robotic system is installed in an environment. Obstacles within the working space of a robotic system can prevent the robot from performing the tasks assigned to it properly; hence, the designer of the robotic system must program the robot to follow an emergency strategy that enables it to avoid any contact with a probable obstacle anywhere anytime within the working space of the robot. For mobile robots this task is relatively easier to accomplish since this type of robots have the flexibility to change their routes and take alternatives that are free of obstacles. For fixed robotic arm manipulators the situation is much more complex. In this paper we are only concerned with obstacle avoidance techniques and present a real-time obstacle avoidance method for fixed robotic arms that combines the benefits of both global and local techniques in that it can be used for both static and dynamic environments with a reduced computational effort.

Key words: Robotics, Inverse Kinematics, Obstacle Avoidance, Critical Points, Triangulization, Echo Sensors, Arduino Board.

Introduction

Obstacle avoidance techniques are classified into two categories, namely; global and local techniques. Global techniques are applied on the robotic arm joint space in such a way that they guarantee a collision-free path following performance within the robotic arm's working environment. The drawback of such techniques is that they assume a static working environment in which all the obstacles are fixed and pre-located resulting in a minimized computation effort to avoid them.

Local techniques, on the other hand, are applied when the environment is not static, i.e. the existence of the obstacles is not known in advance. These methods rely on a set of sensors to locate the proximity of those obstacles to the mechanism. These methods require high computational effort to re-adjust the configuration of the robotic arm while trying to avoid an obstacle.

Many researchers were interested in the aspect of obstacle avoidance for robotic manipulators. D. Puiu and F. Moldoveanu [1] presented a joint trajectory planning strategy for a redundant manipulator. The controller modifies the joint configuration as a moving object (an obstacle) is getting closer to the robotic arm. The object proximity was detected by a visual system around the robotic manipulator. E.J. Solteiro Pires et al [2] used a multi-objective genetic algorithm (MOGA) to optimize the configurations of the robotic arm such that the optimized configuration ensures obstacle avoidance while performing the task. Jing Liu et al [3] used Cylindrical Bounding Box to simplify the geometry of the obstacle and, by using optimization, they can consider the major axis of the box and its radius as the dimensions of the obstacle to avoided. Tse-Ching Lai et al [4] proposed an obstacle avoidance method based on Non Uniform Rational B-splines (NURBS). A safe distance between the end-effector of robot arm and obstacle is considered to avoid the collision. Tianjian Hu et al [5] developed what is termed a backward quadratic search algorithm (BQSA) as another option for solving IK problems in obstacle avoidance. The BQSA detects possible collisions based on the root property of a category of quadratic functions, which are derived from ellipse-enveloped obstacles and the positions of each link's end-points. The algorithm executes a backward search for possible obstacle collisions, from the end-effector to the base, and avoids obstacles by utilizing a hybrid IK scheme, incorporating the damped least-squares method, the weighted least-norm method and the gradient projection method. In this paper, an obstacle avoidance method was applied on highly redundant robotic arms. The method relays mainly on the existence of a solution to the inverse kinematics problem for that type of robotic arm.

Description of the Method

In a previous work [6] we presented a new method to solve the inverse kinematics problem for redundant robotic arms. It is based on the idea of configuring the robotic arm to adjust its joints on a selected polynomial such that all the links are aligned on the polynomial starting from the base up to the end effector. When the robotic arm works in a free environment the second order polynomial of the form of equation (1) is the best choice based on which the links of the robotic arm are to be configured. This is demonstrated in Figure 1.

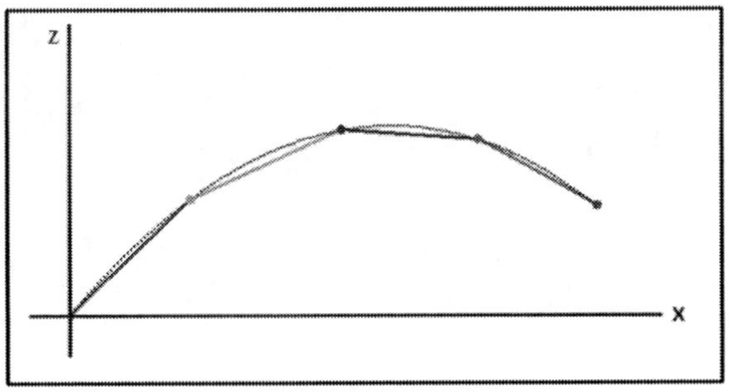

Figure (1): Robotic arm links (the colored lines) are fit on a second order polynomial (the dotted curve) in an obstacle-free working space.

$$Z(X) = a_o + a_1 X + a_2 X^2 \quad (1)$$

When an object gets closer to the arm, a set of sensors installed at certain locations on the robot body, will measure the distance of the robotic arm from that object. Based on the information acquired by the sensors the controller determines a number of "critical" points that lie closest to the arm referenced to the location of the detected obstacle. These critical points locate inflexion points on the suggested curve hence determine the overall shape (the degree) of the configuration polynomial that enables the robotic arm to avoid a possible collision with the object. Figure 2 shows the effect of one object proximity to the robotic arm. Point A represents a critical point that suggests a third-order polynomial expressed in equation (2) as the best configuration of the robotic arm that enables it to avoid this obstacle.

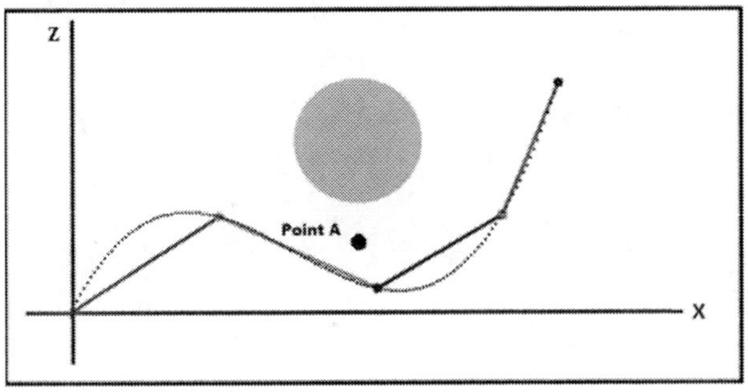

Figure (2): Working Space with One Obstacle. A third order polynomial is generated based on the position of the critical point A.

$$Z(X) = a_o + a_1 X + a_2 X^2 + a_3 X^3 \quad (2)$$

If the number of objects that are close to the robotic arm increases, the degree of the configuration polynomial increases as well. Figure (3) shows a fourth-degree configuration polynomial used to configure the robotic arm as

IOP Conf. Series: Materials Science and Engineering **765** (2020) 012017 doi:10.1088/1757-899X/765/1/012017

it gets close enough to two objects simultaneously. The sensors generate a critical point for each object and the degree of the polynomial is decided accordingly by the controller.

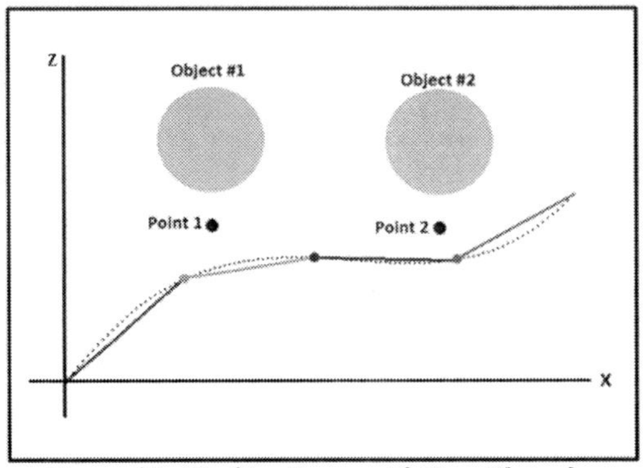

Figure (3): Working Space with Two Obstacles.

Let's take an example to best describe how the method works. Consider the working space shown in Figure (4) in which a robotic arm is to follow a semicircular 3D path that happens to pass a region in which a column (an obstacle) is suspended from the ceiling. As long as the obstacle is far enough (at regions R1 and R3), the second order polynomial is used to solve the inverse kinematics of the robotic arm. Once the obstacle gets closer to the arm by entering region R2, the sensors locate the critical point (which will be defined later) and trigger the controller to switch into using the third order polynomial that enables the robot to avoid the collision. Figure (5) shows the three regions R1, R2 and R3 that the robotic arm should move into while performing the task. The obstacle is represented as a solid black circle inside the region R2.

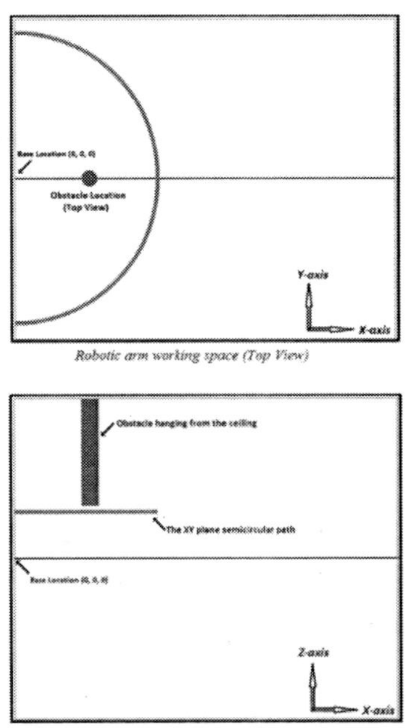

Figure (4): Top and side views of the robotic arm working space.

The position profile of the robotic arm as it tracks the path is shown in Figure (6). The figure represents a record of angle variations of each link which is derived from solving the inverse kinematics problem of the robotic arm during the motion. Since the path chosen is a semicircular and centered at the base, the joint angles of all the links, but not the base, are constant within R1 and R3.

As Figure 6 shows, the position profile of each link is smooth and continuous within each of the defined regions. However, at the instants of transition between the adjacent regions, the profile shows discontinuities that are large enough to cause very high torque variations of all the joints at these transition instances. These high torque spikes can cause damage to one or more joints due to high inertia forces resulting from sudden change of the links' angular position in a short period of time not to mention the possibility of losing the arm stability during the motion.

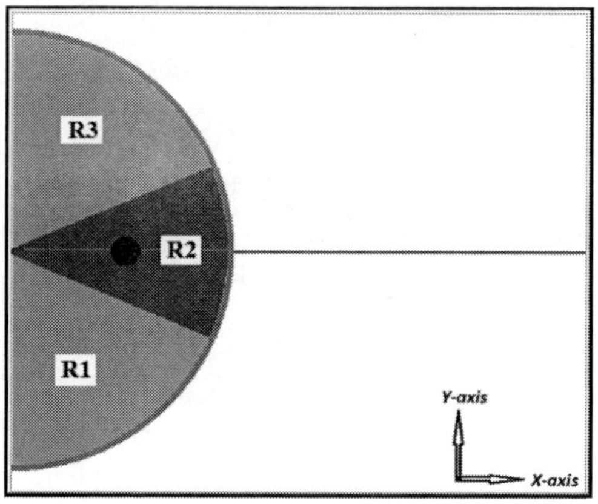

Figure (5): Working space regions. R1 and R3 are the regions where the arm is far from the obstacle. R2 is the region of obstacle proximity to the robotic arm.

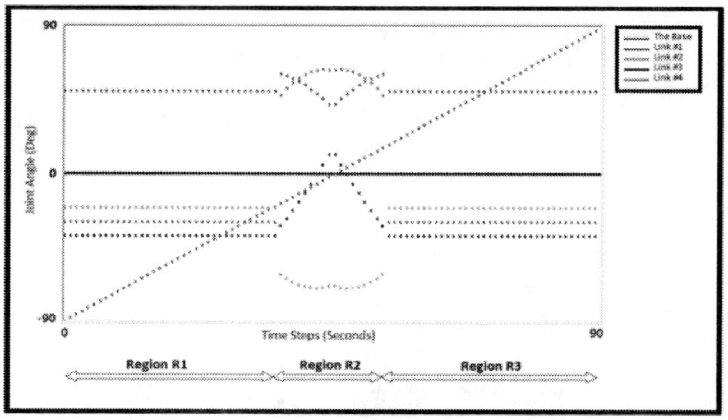

Figure (6): Position profile of joint angle of each link while the robot tracks the given semicircular path.

In order to guarantee a stable performance of the robotic arm during its motion, the controller must deal with those profile discontinuities generally in one of two ways:

1- The controller can increase the time step at the instant of any region-to-region transition in such a way that the speed of the corresponding motor(s) is reduced. This can highly lower the acceleration at the joint which, accordingly, lowers the inertia forces at the joints. This approach is followed when the task assigned to the robotic arm requires that the path should be preserved.
2- The position profile can be linearized to get rid of the discontinuity points and re-gain a smooth position profile which guarantees a stable arm performance during the motion. However, following this linearizing process will definitely affect the end-effector positioning which makes it impossible for the robotic arm to follow the path assigned to it. This approach is followed when the task is path-independent. Figure (7) shows the modified position profile after applying the linearizing process.

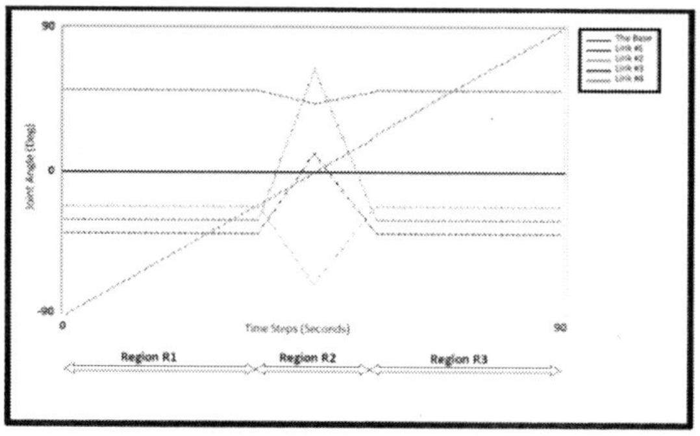

Figure (7): Linearized position profile

To linearize the position profile in R2, we make use of the fact that the variations in all of the joint angles are related to the angle of the base θ_o. In the example, R2 is divided into two sub regions, the first starts at $\theta_o = -20^o$ and ends at $\theta_o = 0^o$(in the direction of the obstacle). The second part starts at $\theta_o = 0^o$ and ends at $\theta_o = +20^o$. Let $\theta_{-20}^{(n)}$ be the displacement of the nth joint at $\theta_o = -20^o$ and $\theta_{20}^{(n)}$ be the displacement of the nth joint at $\theta_o = 20^o$. Also let $\theta_{Obs}^{(n)}$ be the displacement of the nth joint at that value if the base angular position in the direction of the obstacle (which equals to zero in our example). The linearization equation is:

$$\theta^n = \frac{(\theta_{obs}^n - \theta_{-20}^n)}{20}\theta_o + \theta_{obs}^n \quad (3)$$

$$\theta^n = \frac{(\theta_{20}^n - \theta_{obs}^n)}{20}\theta_o + \theta_{obs}^n \quad (4)$$

Equation (3) is valid for $-20 \leq \theta_o \leq 0$, and equation (4) is valid for $0 \leq \theta_o \leq 20$. n=1 to 4 for both equations.

The effect of the linearization process on the original semicircular path is shown in Figure (8) below. Since the modification, or the linearization, process is applied only for the points within region R2, the effect of the process appears only in that region. The other regions did not change in shape after the linearizing process was applied.

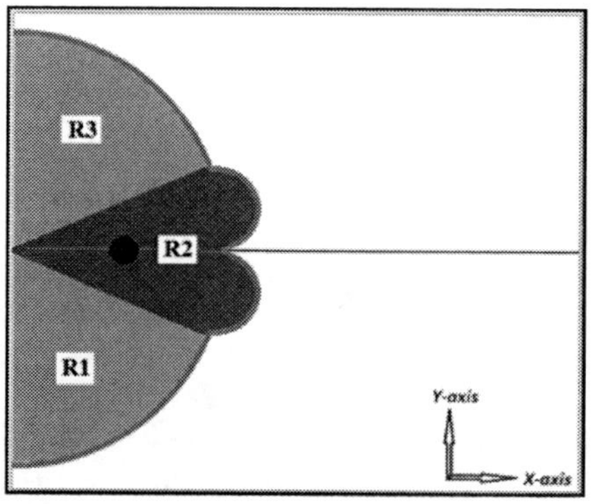

Figure (8): The semicircular path after applying the linearizing process.

Detection of an Obstacle in 3D Space

The ultrasound echo sensor with effectual angle of 15° is the most commonly used for distance measurement. It works by sending a sequence of ultrasound waves directed in front of the sensor and if an object exists it will reflect them back to the sensor. The time period between the instant of sending the ultrasound waves and that when they reflect back to the sensor is a measure of the distance since the sound speed is previously known. The distance measured by a single sensor does not give information about the location of the object in the 3D space, which makes the sensor useful only as a proximity indicator. Due to the effectual angle of 15°, the sensor view is a cone whose head lies on the sensor itself with its base lying away from the sensor.

In order to locate an object in the 3D space, we propose a method in which a set of sensors located at previously known points referenced to a certain origin making simultaneous measurements for the same obstacle and use the information to locate the object in the 3D space. Consider the coordinate system shown in Figure (9).

In Figure (9) the points A, B and C are the locations of three echo sensors that constitute the 3D distance measurement set. The plane X = 0 of the frame coincides with the body of the robot and the points A(0, d_1, $-d_2$), B(0, $-d_1$, $-d_2$) and C(0, d_1, d_2) are the locations of three distance sensors and are referenced to one origin point O(0, 0, 0). Because of the small value of the sensors' effectual angle the distances d_1 and d_2 must be chosen carefully such that all the three sensors make a successful measurement.

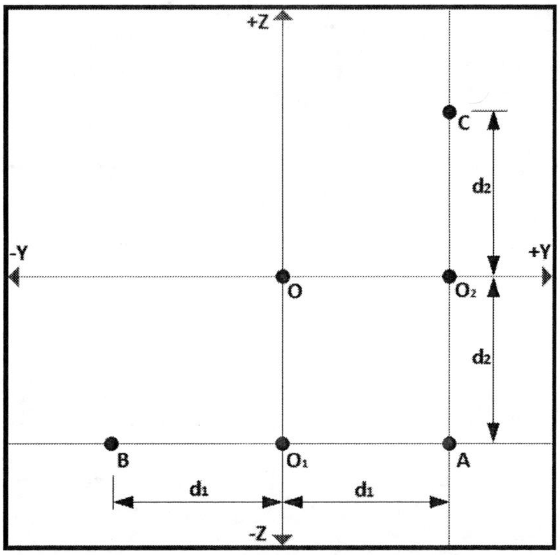

Figure (9): The frame on which the echo sensors are fixed.

Point P(X, Y, Z) is the 3D spatial position of an object detected simultaneously by the three sensors and it is required to calculate X, Y and Z in order to know where that objects is located with respect to the robotic arm. Choosing point O as the reference for the measurement has the advantage that the distance OP is the true distance between the obstacle and the link on which the sensors are attached regardless of the orientation of that link with respect to the obstacle.

In order to make the calculations easier, let $d_1 = d_2 = d$. When an object gets close enough from the sensors, each sensor will "see" the obstacle from its own angle, which means that each sensor will measure a different distance from the other two in the set. The object allocation procedure begins as follows:

1- The sensor at A measures the distance $AP = R_a$.
2- The sensor at B measures the distance $BP = R_b$.
3- The sensor at C measures the distance $CP = R_c$.
4- Now, point P(X, Y, Z) is a common point for three spatial scalene triangles, namely, Δ APB, Δ APC, and Δ BPC. See Figure (10).

5- For the scalene APB of which AB is the base, the median that starts at P and ends at O_1(0, 0, -d) which is the middle point of AB has a length found by the equation [7]:

$$R_1 = \overline{PO_1} = \frac{\sqrt{2R_a^2 + 2R_b^2 - 4d^2}}{2} \quad (5)$$

6- For the scalene APC with AC is the base, the median that starts at P and ends at O_2(0, d, 0), the middle point of AC, has a length given by the equation:

$$R_2 = \overline{PO_2} = \frac{\sqrt{2R_a^2+2R_c^2-4d^2}}{2} \qquad (6)$$

7- For the scalene BPC with BC is the base, the median that starts at P and ends at O(0, 0, 0), the middle point of BC, has a length given by the equation

$$R_3 = \overline{PO} = \frac{\sqrt{2R_b^2+2R_c^2-8d^2}}{2} \qquad (7)$$

8- Since R_1 represents the distance between point O_1 and point P, the following expression for the R_1 also applies, that is:

$R_1^2 = X^2 + Y^2 + (Z + d)^2$

Or,

$$R_1^2 = X^2 + Y^2 + Z^2 + 2dZ + d^2 \qquad (8)$$

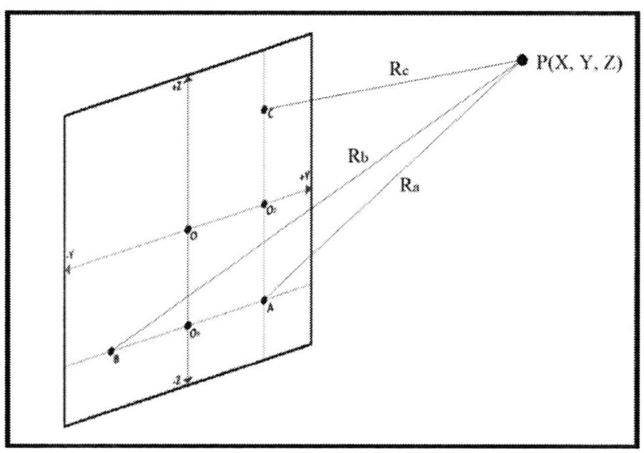

Figure (10): The frame on which the echo sensors are fixed in a 3D view.

9- Similarly for R_2, which is the distance between point O_2 and point P, whose length can be expressed as:

$R_2^2 = X^2 + (Y - d)^2 + Z^2$

Or,

$$R_2^2 = X^2 + Y^2 + Z^2 - 2dY + d^2 \qquad (9)$$

10- Since R_3 originates at O(0, 0, 0), its length can be expressed as:

$$R_3^2 = X^2 + Y^2 + Z^2 \qquad (10)$$

11- Equation (10) can be used to simplify equations (8) and (9) yielding:
$R_1^2 = R_3^2 + 2dZ + d^2$, and

$R_2^2 = R_3^2 - 2dY + d^2$

from which

$$Y = \frac{R_3^2 + d^2 - R_2^2}{2d} \quad (11)$$

$$Z = \frac{R_1^2 - R_3^2 + d^2}{2d} \quad (12)$$

And by using equation (10),

$$X = \sqrt{R_3^2 - Y^2 - Z^2} \quad (13)$$

This way, the exact position of the object, represented by point P(X, Y, Z), is determined. When the distance between O(0, 0, 0), which is located on the robot body, and the object becomes smaller than a predefined threshold, a critical point is created. A *critical point* is defined as that point located half way along the line connecting point P on the object and the local origin of the sensors set O, (See Figure 10).

The creation of a critical point is the trigger for the kinematics of the robot to turn from the parabolic configuration (Eq 1) to the third-degree-polynomial (Eq 2) passing through those three points (the base of the robot, the created critical point and the required end effector position) to avoid hitting the object while the robotic arm is in motion.

In certain cases, however, the size of the obstacle may be too large compared to the size of the measuring set (like when the end effector moves towards a wall for example). In such cases, the measurement set will saturate because all the sensors will measure the same distance making the above equations inapplicable. Then, the only alternative for the controller to avoid hitting that obstacle is to "freeze" or stop the robotic arm when it gets dangerously close to that kind of obstacles.

Conclusion

Obstacle avoidance is an emergency strategy taken by the robotic arm to prevent any chance of collision between a robotic arm and an obstacle that happens to enter the working space of the robot. Many techniques are available to accomplish obstacle avoidance each with its own advantages and disadvantages. The method presented in this work enables a highly redundant robotic arm to reconfigure its overall geometrical shape passing around the obstacle and continue performing the task safely. A set of three echo sensors in certain arrangement where used to make the distance measurement. The sensors are connected to an Arduino board as the micro controller.

References

[1] D. Puiu and F. Moldoveanu (2011) "Real-time Collision Avoidance for Redundant Manipulators". 6th IEEE International Symposium on Applied Computational Intelligence and Informatics, May 19–21, 2011.

[2] E.J. Solteiro Pires, P.B. de Moura Oliveira and J.A. Tenreiro Machado (2007) "Manipulator Trajectory Planning Using a MOEA". Applied Soft Computing 7 (2007) 659–667.

[3] Jing Liu, Ruimin Liu, Xin Shen and Linmin Meng (2018) "Research on Obstacle Avoidance of Space Manipulators Based on Cylindrical Bounding Box Model". Proceedings of 2018 IEEE International Conference on Mechatronics and Automation August 5 - 8, Changchun, China

[4] Tse-Ching Lai, Sheng-Ru Xiao, Hisayuki Aoyama, and Ching-Chang Wong (2017) "Path Planning and Obstacle Avoidance Approaches for Robot Arm". Proceedings of the SICE Annual Conference 2017, September 19-22, 2017, Kanazawa University, Kanazawa, Japan

[5] Tianjian Hu, Tianshu Wang, Junfeng Li and Weiping Qian (2016) "Obstacle Avoidance for Redundant Manipulators Utilizing a Backward Quadratic Search Algorithm". International Journal of Advanced Robotic Systems, 2016

[6] Firas S. Hameed, Hasan M. Alwan and Qasim A. Ateia (2019) "Novel Approach to Solve the Inverse Kinematics Problem for a Multi-Degree-of-Freedom Robotic Arm. Journal of Engineering and Applied Sciences (2019 Volume 14 Issue 13).

[7] Nathan Altshiller-Court, "College Geometry: An Introduction to the Modern Geometry of the Triangle and the Circle" Dover Publications, Inc. Mineola, New York 2007.

[8] Mehmet Ismet Can Dede, Omar W. Maaroof and Enver Tatlicioglu (2016) "A New Objective Function for Obstacle Avoidance by Redundant Service Robotic Arms". International Journal of Advanced Robotic Systems (2016).

[9] Haobin Shi , Jialin Chen, Wei Pan, Kao-Shing Hwang and Yi-Yun Cho (2018) "Collision Avoidance for Redundant Robots in Position-Based Visual Servoing". IEEE SYSTEMS JOURNAL (2018).

IOP Conf. Series: Materials Science and Engineering **765** (2020) 012018 doi:10.1088/1757-899X/765/1/012018

Design and implementation of a single-side flat synchronous linear motor model

Raghda'a Ahmed [1] **Farag Mahel Mohammed** [2] **Jamal Abdul-Kareem Mohammed** [3]

Department of Electromechanical Engineering / University of Technology / Baghdad-Iraq
[1] Email: gadaahmad460@yahoo.com
[2] Email: drfaragmahel@yahoo.com
[3] Email: jamalshwesh@gmail.com

Abstract. In recent decades, the demand for a linear motor has increased in various applications, due to its ability to develop a linear motion directly. One of the significant problems facing the operation of the linear motor is the presence of force ripple caused by the irregular magnetic field of the permanent magnets. Accordingly, a developed single-side permanent magnet synchronous flat linear motor (PMSLM) model has been designed based on minimizing its force ripple produced. A Maxwell ANSYS simulation is performed to study and enhance the dynamic characteristics and performance of the model and then verifying it with experimental measurements. Multi-objective optimization had been achieved through reducing the size of permanent magnets and reaction rail at specific rates with maintaining the required thrust and reducing the cogging force. The results of the improved model design showed that the maximum value of the force ripple is reduced by about 81.13% compared with the primary model at a smaller ripple coefficient of 0.22%. The comparison between the experimental measurements and numerical analysis showed a good agreement and accuracy of the analytical results.

1. Introduction

More than a hundred years ago, industry rotary motors with a traditional mechanism are still dominated to get linear motion [1]. Thus, the need for linear motors (LMs) is required to compensate those rotary ones for generating the linear motion directly. Furthermore, the LM has several attractive features such as simple structure, easy maintenance, and high acceleration/deceleration [2]. Though, in slotted permanent magnet (PM) linear motors, there is an undesirable phenomenon, known as detent (or cogging) force [3], produced due to slotting and finite length of the moving parts. This force is generated as a result of mutual attraction between the magnets and iron core of the motor with respect to the position of the translator relative to the magnets [4].

Thus, different methodologies had been applied in the literature to enhance motor performance and achieve the best results by reducing the force ripple, such as structural optimization or control method [5]. Optimizing structures include the tantamount to adjust the size of some critical structures, such as PMs modification [6], stator shape, and size in addition to the PM or stator skewing angle [7] likewise the employment of the fraction number of slots per pole [8]. Furthermore, the adjustments of the air gap and slot opening length are applicable methods [5]. Among control methods, the most popular one was applied by involving a feed-forward current compensation method [9]. Although these techniques had reduced the cogging forces effectively, some methods either increased fabrication difficulties. Therefore, this paper suggests an optimization method to estimate the optimum dimensions of the single-sided 12/10, flat type permanent magnet synchronous linear motor PMSLM based on regular steps with the aid of the Maxwell ANSYS simulation. The optimum dimensions required are the magnet length, magnet width, magnet height, air gap length, motor length at z-direction and the motor ends' width.

2. 2-D Analysis and optimization method

2.1. Structure and operating principle

The PMSLM is mainly composed of a moving platform and a stator base. Three-phase windings are mounted on the mover's core with tablet cover assembled on the moving platform. The moving platform translates along with a pair of linear guides represented by four linear rolling bearings while the stator base is located along the moving path with mounted permanent magnets.

Generally, the production of the thrust in the PMSLM is influenced by electrical and magnetic sources. Electrical source normally depends on current feeding the motor coils. According to the Lorentz law [10], if a conductor carrying a current of I(A) is placed in a magnetic field with flux density B (Wb/m^2), an induced electromagnetic force *F* (N) on a section of conductor with a length of L (m) will be generated. The time-space distribution of the magneto-motive force (*F* or *MMF*) of a symmetrical three-phase winding with distributed parameters fed with a balanced system of currents can be expressed as [3]:

$$F(x,t) = \frac{1}{2}\sum_{v=1}^{\infty} F_{mv} \left\{ \begin{array}{l} \left[\sin\left(\omega t - v\frac{\pi}{\tau}x\right) + (v-1)\frac{2\pi}{3}\right] \\ + \left[\sin\left(\omega t + v\frac{\pi}{\tau}x\right) - (v+1)\frac{2\pi}{3}\right] \end{array} \right\} \tag{1}$$

Where F_{mv} is the magnitude of the vth harmonic of the armature *MMF* and v is the harmonic number. The generated electromagnetic power (P_{elm}) and thrust force (F_{dx}) could be obtained as follows [11]:

$$P_{elm} = \frac{m_1[V_1(R_1\cos\delta + X_{sd}\sin\delta) - E_f R_1]}{(X_{sd}X_{sq} + R_1^2)^2} \times \left[V_1(X_{sq}\cos\delta - R_1\sin\delta)(X_{sd} - X_{sq}) + E_f(X_{sd}X_{sq} + R_1^2) - E_f X_{sq}(X_{sd} - X_{sq})\right] \quad \text{(W)} \tag{2}$$

Where,

$$E_f = 2\pi f N_1 k_w \Phi_f \quad \text{(V)} \tag{3}$$

$$F_{dx} = \frac{P_{elm}}{v_s} \quad \text{(N)} \tag{4}$$

$$v_s = 2\tau f \quad \text{(m/s)} \tag{5}$$

Where, υ_s is the synchronous velocity, depends only on the frequency f and pole pitch τ. To verify the results of design calculation, maximum thrust force be determined for the single-sided PMSLM model. The motor has rated values of the input power of 402.7W, the linear velocity of 1m/s, the thrust force of 350N at m_{load} = 17kg, and the current of 4.3A. The dimensions listed in Table 1 show the stator and the mover specification for the designed PMSLM model, which is used as an initial model for the optimization process.

Table 1. Basic specification of the PMSLM designed model

Parameters	Symbols	Quantities
Conductor		70 turns/Slot 0.5 (mm^2,105°C Insulation temperature Copper
Air gape	g_1	1 mm
Motor active		288mm

length		
Motor active width		95mm
Translator		1010 steel Material [12]
	b_{11}	14mm
	h_{11}	17.2mm
	h_{14}	3mm
	b_{14}	4.4mm
	t_1	24mm
		12 slots
Stator		1010 steel Material
	hc_1	8mm
	τ	28.8
		10 poles
PM Material		NdFeB Class N42SH (B_r = 1.2T, μ_r = 1.05) [3]

2.2. 2-D Numerical field analysis

The 2-D finite element method FEM is an effective method for numerical calculation of electromagnetic field distribution. The commercial FE package based on Maxwell ANSYS has been used for this purpose depending on Maxwell equations [11]. The nonlinear material properties and structural details are taken into consideration. Maxwell ANSYS solution is divided into a load and no-load transient magnetic solution stages where its time period for every load step is 0.0016sec with 0.0001 nonlinear residual and backward Euler time integration solving methods. Whereas, the single solved curve during 2τ is composed of 36 steps.

Meshes generation with the triangular element shape, the element sizes adjusted by the user to adequate the analysis, depending on the physics of the model structure. The model is designed with three meshing regions; each one contains 1200 elements with 5.76mm maximum element length. The triangular element shape of modeling 2-D geometries is well suited for automatic mesh generation as shown in Fig. 1 (a). Figure 1(b) shows the results of the field distribution. As noted, the flux lines move in a transverse direction performing closed loops as it should. The maximum flux density B_{max} equals 3.5T and *MMF* of 426A.T.

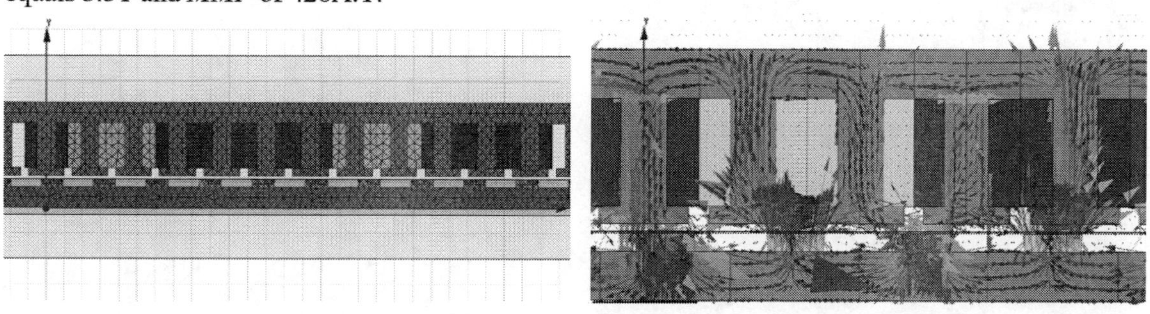

(a) The optimum model meshing (b) Magnetic field density as a vector

Figure 1. FEA of semi-closed core of the LM at position = 4.8 mm, time = 0.0048 sec (1m/s)

2.3. Optimal design

The optimization had accomplished based on the designed model demonstrated in Table. 1. Maxwell ANSYS optimization has based on two cases in order to predict best dimensions that lead to lowest cogging force at no load case. The first iteration of Table 2(a) has affirmed that the maximum cogging force has decreased by 73.9% while the second one has achieved with another set of constant values as it's displayed where the cogging force results have decreased by 77.43%. As a result, the second iteration offers the best percentage value of cogging force reduction in the first case of optimization. Otherwise, ends width of the motor translator affect on the variation of the cogging forces, therefore, the second case should be applied according to Table 2(b) depending on the best results obtained from the first case represented by the second iteration. The results of 2.625mm the translator end length led to a reduction in cogging force by 81.13%.

Although the area of the translator end section has decreased, it is still suitable for the linkage flux passing and did not cause flow dispersion out of the core metal. As well as taking several possibilities separately for high reaction rail (h_{cl}) and showing that the best value is the designed value itself. Table 3 illustrates the final results of optimization and Figs. (2 and 3) show the thrust force (at load case) and cogging force (at no-load case) before after the improvement. Figure 4 shows The Flowchart of the optimization procedure.

Table 2. (a) First and (b) Second case of optimization procedure (All dimensions in mm)

(a)

Parameters / Iterations	g_1	h_M	ω_p & L'_1
First	1	5	24.32 - 95 18.66 - 95 12.32 - 95 24.32 - 85 18.66 - 85
Second	1.5	3	12.32 - 85 24.32 - 75 18.66 - 75 12.32 - 75

(b)

ends length e_m (Iterations)	Constant Parameters
0.25	g_1= 1.5
1.4375	h_M = 3
2.625	ω_P = 12.32
3.8125	L'_1 = 75
5	L_M= L'_1 = 75

Table 3. Basic specification of the PMSLM optimized model (All dimensions in mm)

Parameters	g_1	L'_1	L_M	h_M	e_m	ω_P	h_{cl}
Quantityes	1.5	75	75	3	2.625	12.32	8

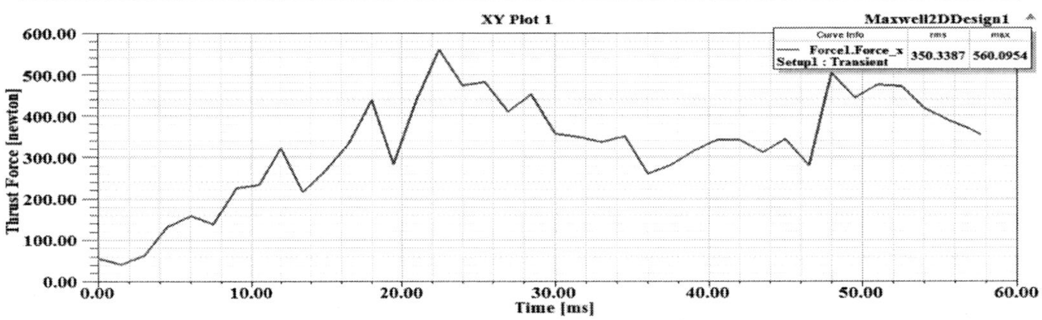

Figure 2. Primary model output thrust

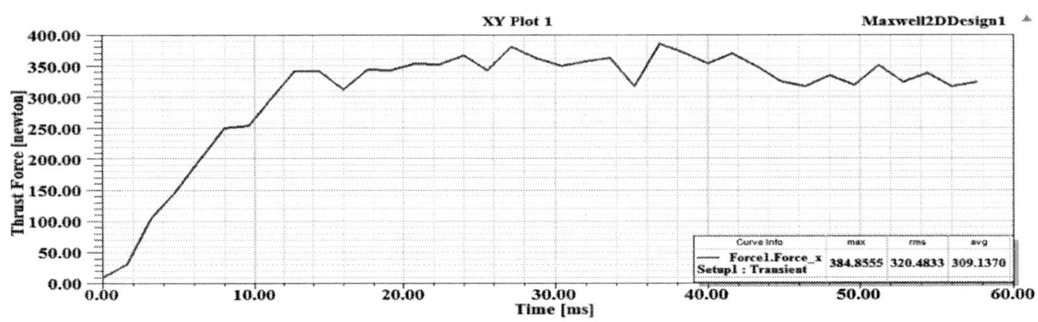

Figure 3. Optimized model output thrust

The size of the translator is smaller and therefore the motor needs less thrust force proportional to the weight difference with preserving the thrust force requires for carrying the desired load, as proposed in Figs. 2 and 3. The rms value of the optimum thrust force is found to be equal to

IOP Conf. Series: Materials Science and Engineering **765** (2020) 012018 doi:10.1088/1757-899X/765/1/012018

320.4833N at the mean value of the rms motor currents I_a =3.555A at m_{load} = 17kg and optimum $m_{translator} \approx$ 4.75kg with the same velocity at a frequency, f=17.36Hz.

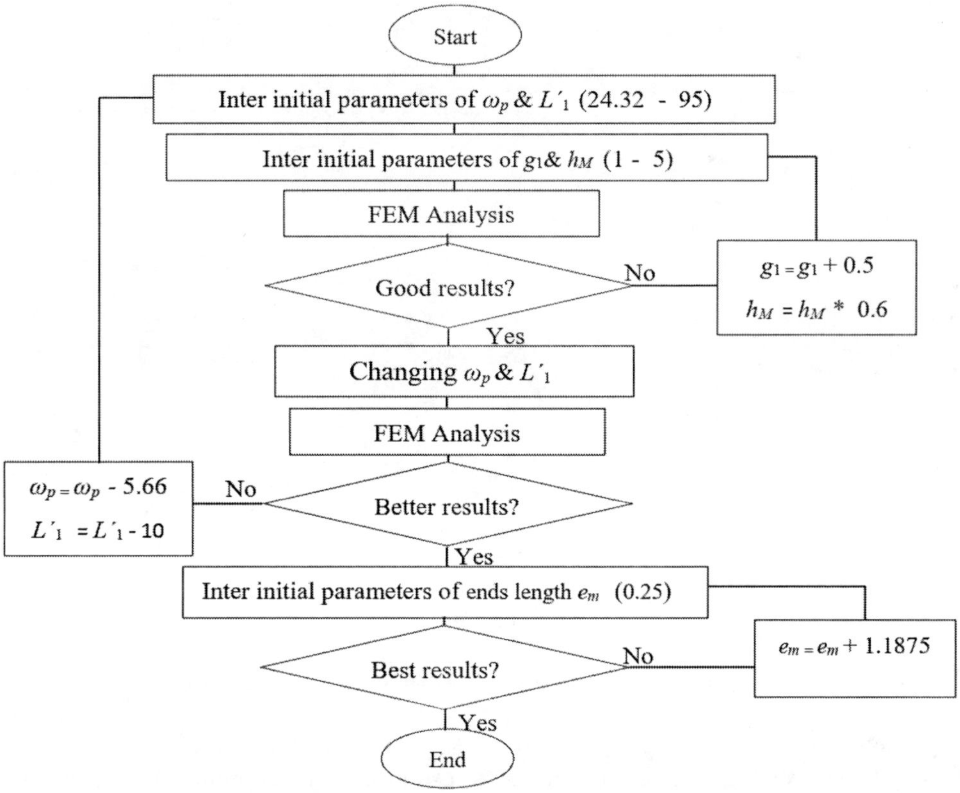

Figure 4. Flowchart of the optimization procedure

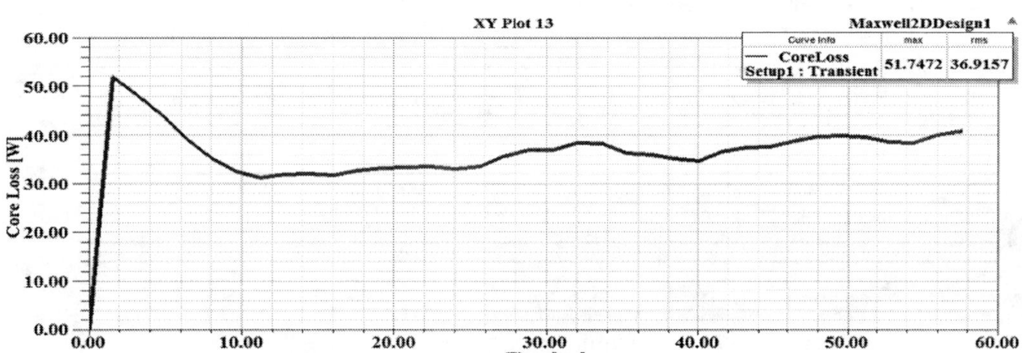

(a)

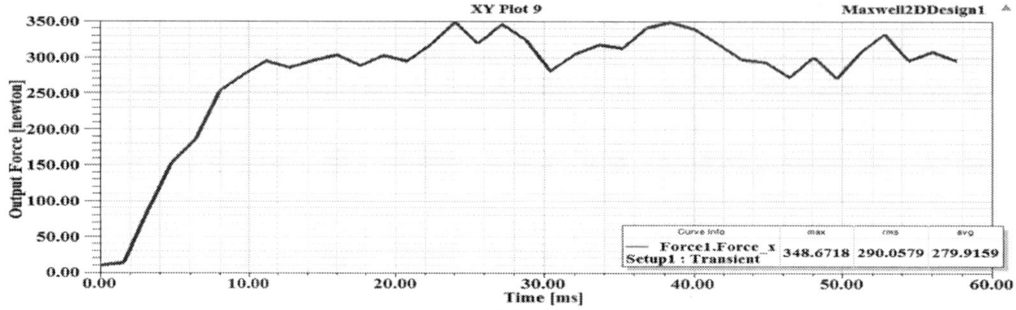

(b)

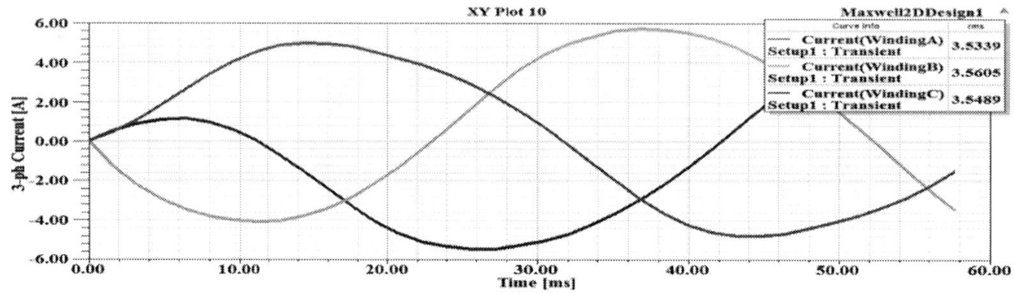

(c)

Figure 5. (a) Iron losses of the optimized laminated core, (b) Output thrust in this case, and (c) 3-phase currents

In another case where the core loss component is taken into consideration, it will be equal to 36.9W as illustrated in Fig. 5(a), where these analyses have achieved at low speed with the laminated core and the output thrust force in Fig. 5(b) is found to be 290.059N at a mean value of rms motor currents of I_a = 3.547A in Fig. 5(c) where the core loss affects the machine flux linkages and its reduction causes decreasing the thrust.

3. PMSLM model implementation and discussion

The manufacturing of the proposed motor is a long process comprising different stages. Starting with core sheets shown in Fig. 6 provided according to the designed model, and manufacturing it using CNC wire-cut KNUTH smart DEM. This process will produce 115 pieces with 0.65mm sheet thickness as shown in Fig. 7.

Figure 6. Electrical single sheet after manufacturing

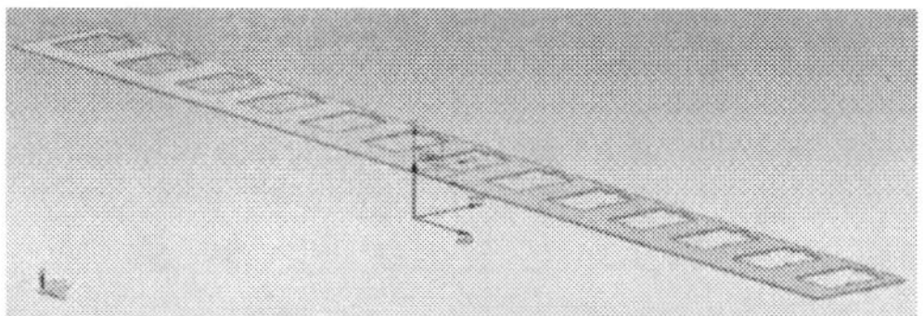

Figure 7. A single sheet was drawn by the NX

115 pieces of the core laminations have been cut and arranged by using three iron guides as shown in Fig. 5, then they were pressed by the machine vice to weld all sheets together as one block using MIG welding.

(a) Lamination Stack (b) Linear motor translator

Figure 8. Prototype experimental translator

F-Class fiber insulator has used with 25μm thickness with 0.95mm diameter (copper) winding coil has been selected to accomplish a single layer concentrated winding, while for the stator, the permanent magnets have been mounted on the reaction rain surface. Fig. 8(b) shows the completely wounded translator. Both the insulation resistance test and the series test had been carried out on the translator.

Finally, all the parts were assembled along with trying to adjust the air gap as possible so as to form a PMSLM model prototype shown in Fig. 9.

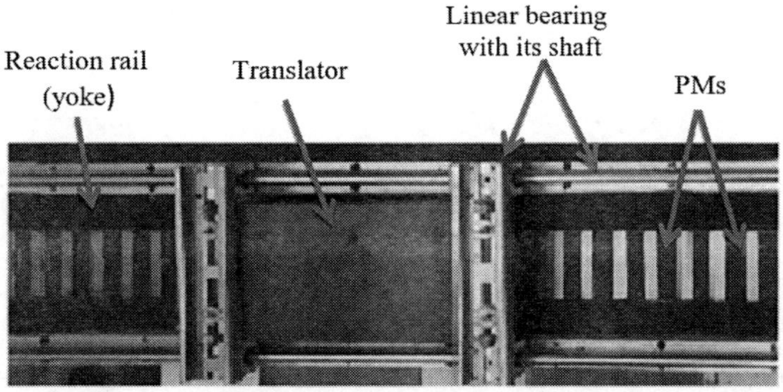

Figure 9. PMSLM prototype

Figure 10(a) shows the device screen and the result of phase A with 0.705m/s at 12.25Hz, note that B and C phases have the same results with 120° shifting. Likewise, Fig. 10(b) illustrates the FE analysis of the back *emf* at the no-load case at the same speed. As noticed, the value of the experimental back *emf* is 21.8% larger than the FE analytical value since the Maxwell ANSYS neglects the ends of the motor coils at 2-D analysis and this is responsible for back *emf* increment in the experimental model case.

IOP Conf. Series: Materials Science and Engineering **765** (2020) 012018 doi:10.1088/1757-899X/765/1/012018

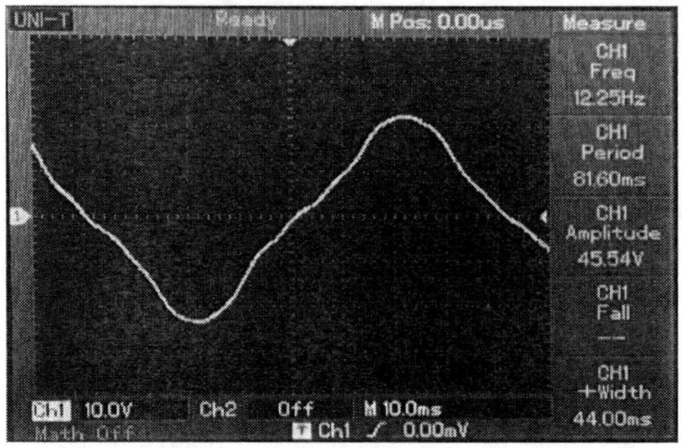

(a) An experimental test of phase A back *emf*

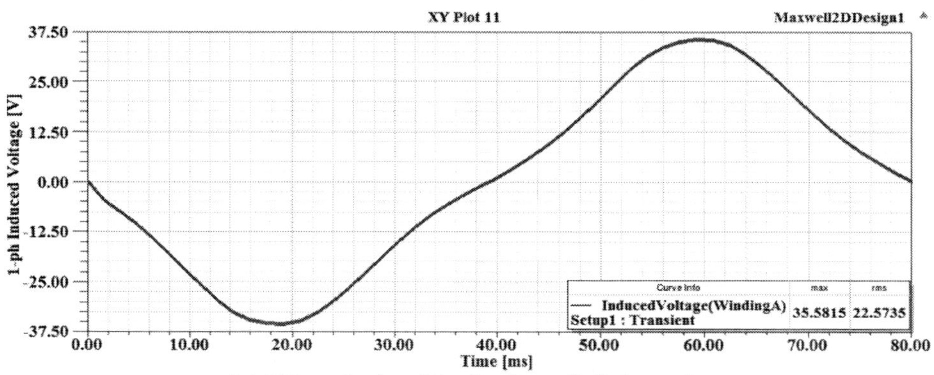

(a) FE analysis of the back *emf* of phase A.

Figure 10. Back *emf* measurement by two methods

Fig. 11 shows the comparison between the three cases of cogging forces that have been tested. The analysis had accomplished using FEM solving the cogging force of the optimized design as shown in Fig. 11 (a), while the experimental test has accomplished using a force meter with pull and push sensors as shown in Fig. 11(b). The case (b) presents a slightly lower value of the cogging force because of the error resulting from the air gap length. The translator frame in the prototype was made of steel and aluminum, it was deformed severely related to the great normal force, and thus the air gap comes to be uneven.

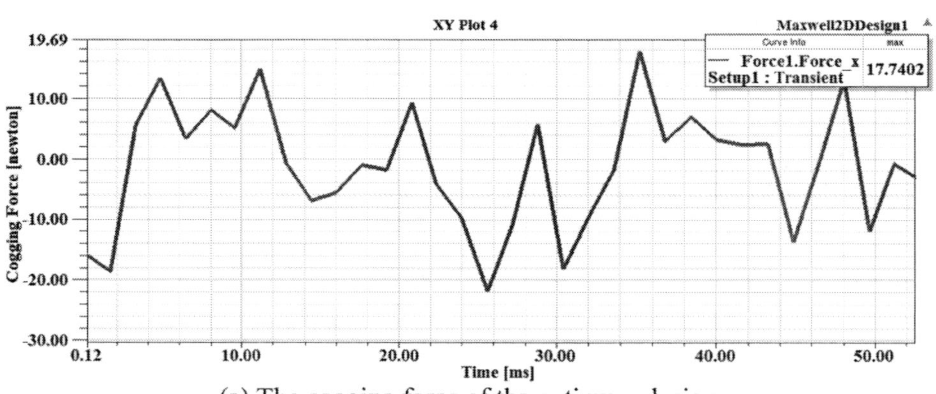

(a) The cogging force of the optimum design

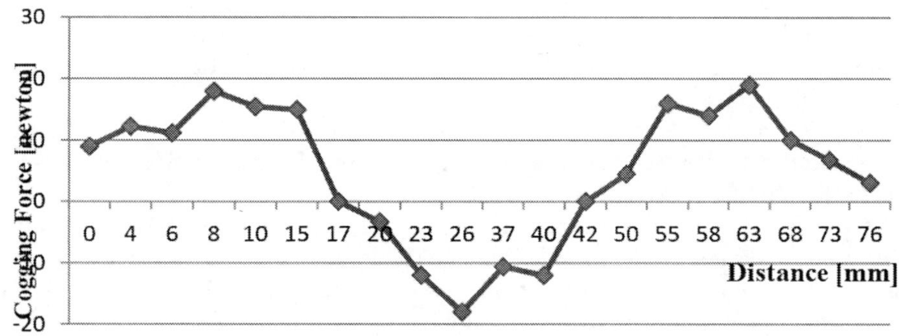

(b) The cogging force of the experimental prototype
Figure 11. Cogging forces at the no-load case

Thrust force constants, force ripple coefficient and the efficiency percentage results are illustrated in Table 4. The greater value of the force constant ($k_F = F_{dx}/ I_a$) leads to least thrust ripple k_r and highest efficiency which represent the best model. Variant results were observed in the model's $k_{F\text{-RMS}}$ values. This disparity is located due to a few reasons whereas two of them are convergent, unlike the rest. The convergence of the first two models' values is related to the error between the analytical and numerical equations while the increment of the third model is due to applying the optimization method in order to enhance the performance which is the goal of this paper. On the other hand, loss consideration in the fourth model causes the $k_{F\text{-RMS}}$ value to decrease. However, the ripple coefficient is equal to 0.223 which is considered an acceptable value to the practical designs in comparison with the rest models that offer high efficiency or better ripple coefficient but remain a virtual design. Finally, the improved (optimum) model has been built and tested experimentally which has low efficiency as expected, since practical models differ from the theory, related to the existence of electrical and mechanical losses.

Table 4. Force constant values, ripple coefficient and efficiency

Models	$k_{F\text{-Max}}$ (N/A)	$k_{F\text{-Min}}$(N/A)	$k_{F\text{-RMS}}$(N/A)	I_a(A)	k_r	Efficiency %
Analytical model	......	......	81.395	4.3	......	86.91
Simulated model	129.65	59.814	81.096	4.32	0.93	86.92
Optimal model	108.257	88.963	90.15	3.555	0.22	83.57
Optimal model with losses	98.279	76.082	81.775	3.547	0.281	75.64
Experimental model	......	......	63.636	3.3	......	60

4. Conclusions

In this paper, an improved design of a three-phase flat PM excitation SLM with a single-side type of 10/12 pole-slot combination has been designed and modeled using 2D and 3D FEA simulation whereas ANSYS package has been employed. The following conclusions are drawn:

- The design achieves optimum motor performance with minimum thrust ripple coefficient of 0.22 and thrust force of 320.4833N at rms load current of 3.555A where $k_{F\text{-RMS}}$ of 90.15.
- The optimized motor gives a thrust ripple twice less than the original model.
- The core losses of the optimum model causes a thrust ripple coefficient of 0.281 and thrust force of 290.057 N at rms load current of 3.547A where $k_{F\text{-RMS}}$ of 81.775.
- The experimental back *emf* is greater than the FEA value by 6.79%, while the experimental cogging force is smaller than the FEA value by 7.41%.

5.References

[1] Velázquez R, Hernández H, Gutiérrez C A and Rodrigo P 2015 An analytical and experimental study of ultrasonic linear motors *Tehnički vjesnik.* **22** 1057-1063

[2] Hamzehbahmani H 2011Modeling and simulating of single side short stator linear induction motor with the End Effect *J. o. Elec. Eng.* **62** 302-308

[3] Yang L, Fazhi S, Xiaofeng Y, Yue D and Jiubin T 2018 Intelligent measurement and compensation of linear motor force ripple: a projection-based learning approach in the presence of noise *Meas. Sci. Technol.* **29** 1-12

[4] Gieras J F, Piech Z J and Tomczuk B Z 2000 *Linear Synchronous Motor-Transportation and Automation Systems 2nd Edition* (Florida: CRC Press, Taylor & Francis)

[5] Wang M, Li L and Pan D 2015 Detent Force Compensation for PMLSM Systems Based on Structural Design and Control Method Combination *IEEE Trans. O. Indus. Elec.* **62** 6845-6854

[6] Abdalla I, Ibrahim T and Mohd Nor N 2014 Optimization of iron-cored linear motor with trapezoidal magnet using finite-element analysis *IEEE 2014 5th Int. Con. on Intelligent and Advanced Systems (ICIAS)* 1 - 6

[7] Ou J, Liu Y, Qu R and Doppelbauer M 2018 Experimental and theoretical research on cogging torque of PM synchronous motors considering manufacturing tolerances *IEEE Trans. O. Indus. Elec.* **65** 3772-3783

[8] Kwon Y S and Kim W J 2016 Detent-force minimization of double-sided interior permanent magnet flat linear brushless motor *IEEE Trans. O. Mag.* **52** 1-8

[9] Zhu Y W, Jin S M, Chung K S and Cho Y H 2009 Control-based reduction of detent force for permanent magnet linear synchronous motor *IEEE Trans. O. Mag.* **45** 2827-2830

[10] Shin-ichi M 2017 Torque Measurement Method Based on the Lorentz Force Law for Miniature DC Motor *IEEJ Trans* **13** 1-8

[11] Toliyat H A and Kliman G B 2004 *Handbook of Electric Motors, 2nd Edition, Revised and Expanded* (Florida: LLC, Taylor & Francis Group)

[12] Juanjuan P, Liang Y, Kuo H and Zongxia J 2016 Influence of Stator Parameters on Output Performance of Linear Switchedhju Reluctance Motor *IEEE/CSAA Int. Conf. on AUS* yc1186-1191

Adaptive Control Using NARMA-L2 Model and Electric Circuit for Condenser of Thermal Power Plant

Hosham Salim Anead[1] ,Khalid Faisal Sultan[1] ,Raheel Jawad[1*]
[1] University of Technology, Electromechanical Engineering Department, Baghdad, Iraq *Corresponding author e-mail: 51262@uotechnology.edu.iq

Abstract. Design and implantation electric circuit for enhancement performance of steam power plant and artificial neural networks technique used to control of condenser. Artificial neural network has been applied to control of the important variables of condenser in Al–Dura power plant in Baghdad such as pressure, temperature. in this study, applied Simulink model in Matlab program (v 2014 a) by using artificial neural network toolbox. the model of condenser of neural network by using NARMA to generated data and train network and using back propagation algorithm for training neural network. The method of control by ANN is off line. the electric circuit of pressure sensor and temperature sensor consist of transformer, dc bridge and voltage regulator. Experimental data of actual power plant obtained from al-dura power plant. comparing results of modelling neural network and electric circuit with experimental data of actual power plant. the results shown the maximum deviation between the ANN and electric circuit with experimental data is less than 4%. artificial neural network can be used in many industrial and engineering application.

Keywords: Artificial Neural Network, NARMA, Condenser, Steam Power Plant, Pressure, Temperature

1. Introduction

Condenser (atmospheric) of steam power plant that's working under the ranking cycle is a simply heat exchanger and the function of a condenser is to convert gaseous form exhaust steam into liquid by cooling mediums such as water/air and then pumped back into the boiler. To operation steam power plant strongly under minimum cost and maximum efficiency the condenser must be operation at optimum condition. The studies and literature review of condenser and artificial neural network as: [1] presented to control of cooling water flow rate, pressure and temperature of the condenser which these variables affected of performance of thermal plant, and thus on energy efficiency and its specific heat rate of the coal fired plant. The study applied when the plant runs at full load. The results suggest some approaches for the active control of backpressure operating the condenser at optimum cooling water mass flow rate to increase energy efficiency. [2] present a model of heat exchanger for analytical of mass transfer and heat in a flue gas condensing. The analytical model used to validate a pilot-scale heat exchanger. The result shows a good agreement of experimental result with analytical model results. [3] present a study of using a ANN for control of changeable-speed air conditioner by developed a dynamic model of a steam condenser. The model is based on the assumptions of constant liquid and vapour volumes and total condensation. The operating condition is taken from the static balance of a power plant. The neural network controller is feed forward network trained with two hidden layer using a model reference technique. Training network by back-propagation algorithm. The results obtained the responses of the model reference is similar with neural network control system and compared the results with using a fixed speed conventional of air conditioner, ANN control system consumed. [4] applied artificial neural network to determine exit water temperature of a condenser at various operation conditions and compared with the water temperature exit by experimental work. Back propagation algorithm used for training and learning neural network. Input of ANN include surface area, inlet temperature, flow

IOP Conf. Series: Materials Science and Engineering **765** (2020) 012019 doi:10.1088/1757-899X/765/1/012019

rate of water, steam temperature, enthalpy difference and steam flow rate. Experimental results and measurements obtained by engineering equation solver used as a set of input. The results shown the deviations of ANN models are less than 0.12 and shows the change of exit water temperature with the change of input parameters. [5] presented investigates task -blind and task -this method is reminiscent of time series prediction, but the network would also utilize information on the current state of the system. The controller design and apply it to the control of an actual air-cooled condenser and compare its performance with the one of the conventional controller. [6] presented PID neural network to get satisfactory control effect for the advantages of PID , capabilities of learning and nonlinear propose a deaerator water level of condenser decoupling control strategy founded on PID ANN which integrates PID and ANN by establishing proportional neuron, integral neuron and derivative neuron corresponding to proportional, integral and derivative, also used a method of choosing initial weights and learning coefficient from experience of PID to enhance the convergence performance of PID ANN . The results of simulation model show that the PID ANN decoupling control strategy can meet the requirements of multivariable system decoupling control. It is more successful in condenser water level and deaerator water level decoupling than PID control strategy.
The aim of this research is to control of variables pressure and temperature of condenser of al-dura power plant in Baghdad which affected of maximum generated power and plant performance.

2. Neural Network Control System

ANN applied to simulate their biological counterparts. The first simple model of a neuron knows by Mcculloch and Pitts at 1943. [7]. ANN are applied in many fields of science and technology. They provide an alternative approach to the simulation of complex systems [8, 9, 10 11]. Neural networks have been applied in many engineering applications because they are ability and easy to learn and cost effective. Neural network is applied to fit curves through given data without being provide a predetermined function [12-16]. ANN is a flexible for developing environment and capable to take decision and learned the proper function of control shown in Figure 1, the single artificial neuron consists of input layer, output layer, weighted connection between neurons and activation function as a single model shown in Figure 2.

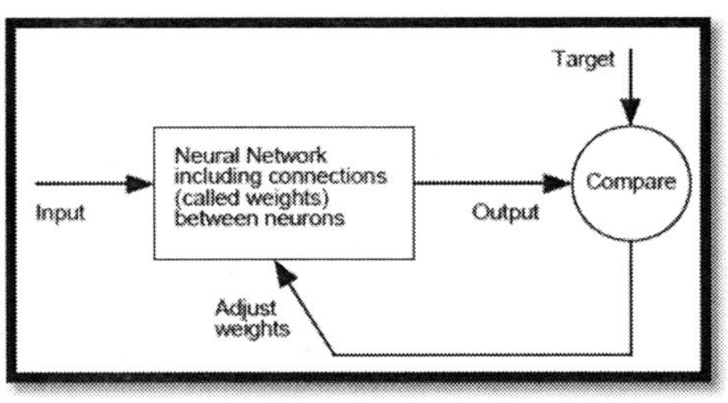

Figure 1. Schematic of an Artificial Neural Network

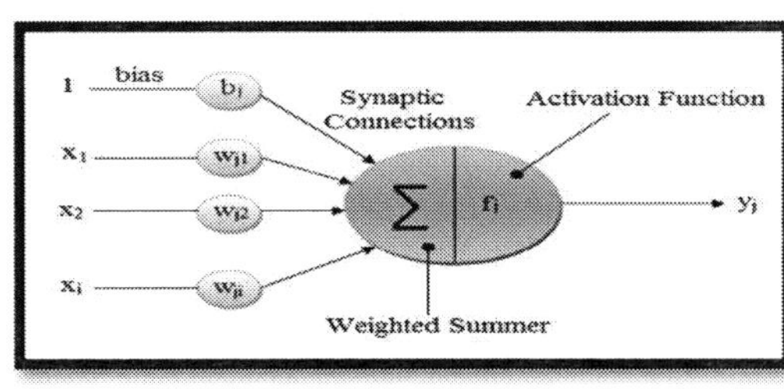

Figure 2. Basic Element of Artificial Neural Network

3. Experimental Devices

Experimental devices of temperature and pressure sensor. The equivalent circuit this pressure signal is processed to be transfer to the interface unit. In interface unit the signals are processed to enter into the computer after which to take control of an application that has been designed with the system requirements, and consequently offers a message to the operator for you to make the right selection. Pressure sensor consist of electric transformer which capable of increasing or decreasing the voltage and current levels of their supply, DC bridge which it uses to convert signal to continuous voltage signal to obtain logic value of the signal represented by way of one and zero, where in the one represents five volts and zero represents zero volts. the continuous voltage signal is inserted to the voltage regulator type 7805 (5 volts or zero volts). Its purpose is to supply a stable voltage and used as stabilizer may be restricted to ensure that the output as definite limits. The analog signal converted to a digital signal when sensor pressure read high. Figure 3 shown the diagram of sensing pressure circuit and Figure 4 show the photograph of sensing pressure circuit.

IOP Conf. Series: Materials Science and Engineering **765** (2020) 012019 doi:10.1088/1757-899X/765/1/012019

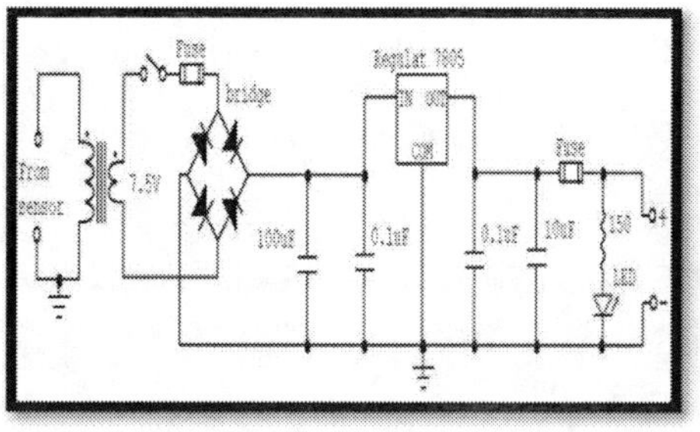

Figure 3. The Diagram of Sensing Pressure Circuit

Figure 4. The Photograph of Pressure Circuit

Temperature sensor which consists of a comparative voltage, after changing the temperature signal to voltage it's far in comparison with the threshold value (80°c) which has been almost set. The temperature increases more than 80°C or decrease less than 80 °C, signal may be sent to the control system through the equivalent circuit to make the right decision. This limit of temperature makes a condenser in steam power plant have been working in steady state. Figure 5 shown the diagram of temperature sensor circuit and Figure 6 shows the photograph of temperature sensors circuit.

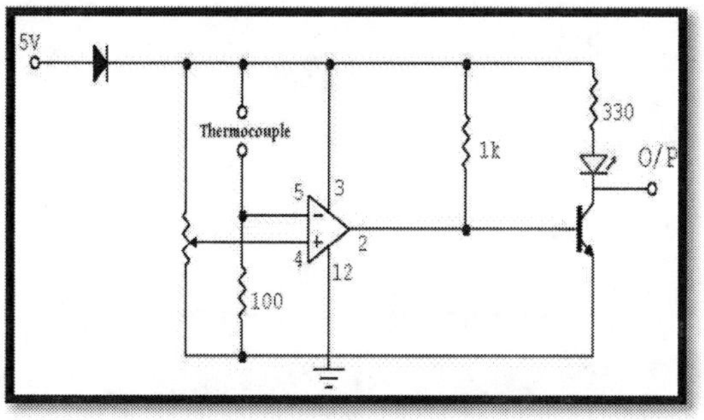

Figure 5. The Diagram of Temperature Sensor Circuit

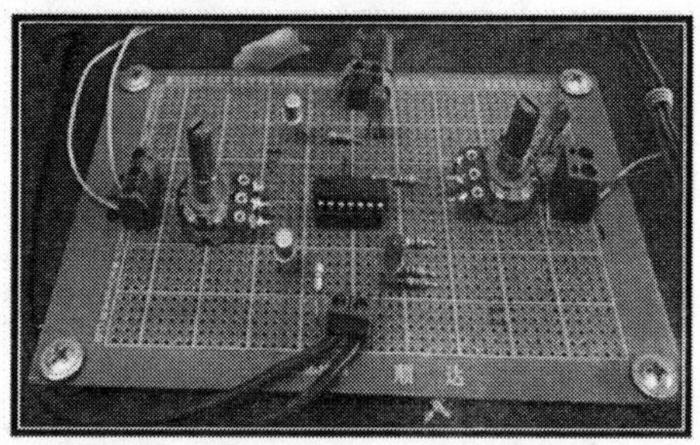

Figure 6. The Photograph of Temperature Sensors Circuit

4. Condenser Model Using Neural Network

The kind of controller of neural network used in this study is the nonlinear autoregressive moving average (NARMA) [11] are designed and employed to control of pressure and temperature of condenser of steam power plant. NARMA implemented in the ANN tool-box of simulation of Matlab. In system identification of ANN model of the plant is developed. Controller block of NARMA show in Figure 7.

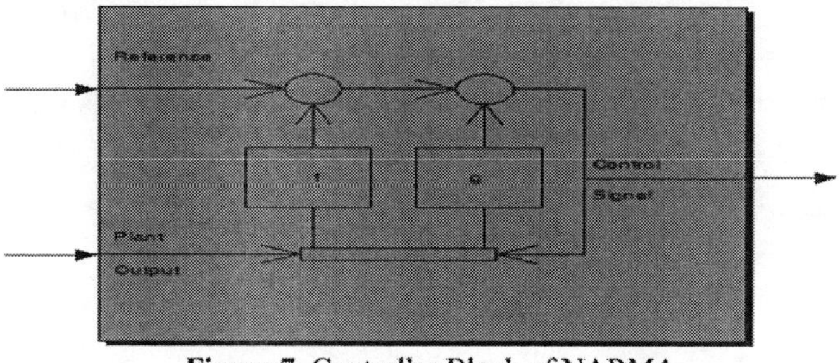

Figure 7. Controller Block of NARMA

IOP Conf. Series: Materials Science and Engineering **765** (2020) 012019 doi:10.1088/1757-899X/765/1/012019

The building block diagram of plant identification for the NARMA of condenser model working by adjusted parameters for generating data by insert minimum and maximum values for the plant input and output, minimum and maximum interval values as 0.1 and 1 seconds. The size of the hidden layer the number of delayed plant inputs and outputs, the sampling interval and finally the training function is Trainlm. The training network then selected by the response of the resulting plant model was displayed. Separate plots for validation data training data and testing data that's shown in Figures 8-12.

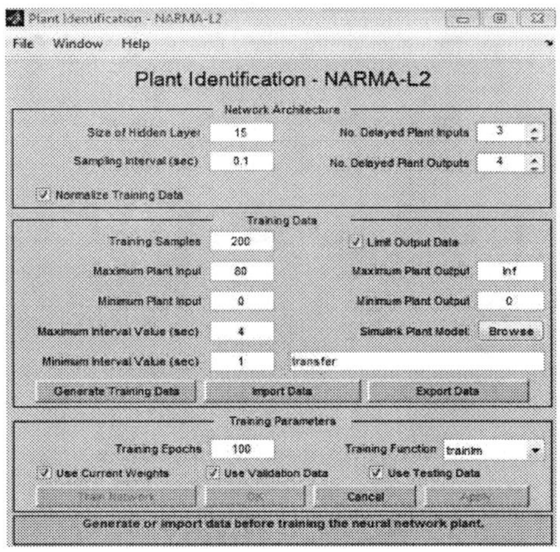

Figure 8. Generate Data of Plant Identification of NARMA

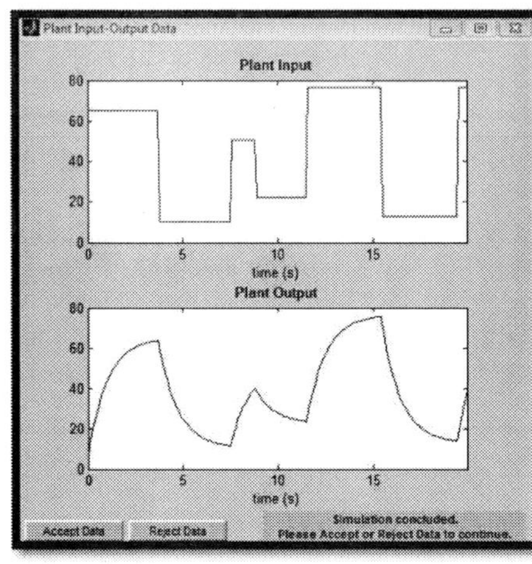

Figure 9. Accept Data of Input \ Output Data

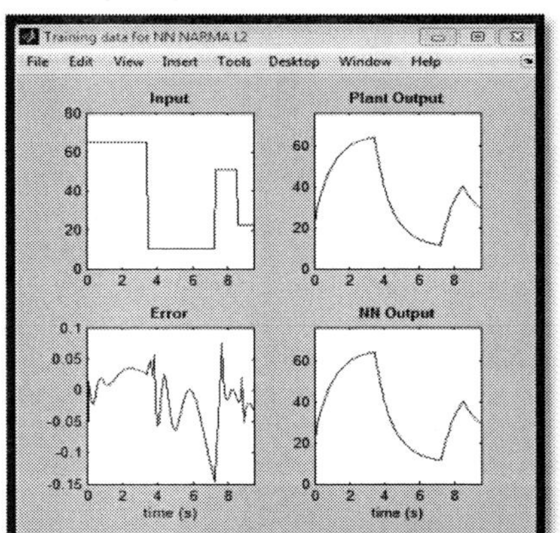

Figure 10. Validation Data of Neural Network

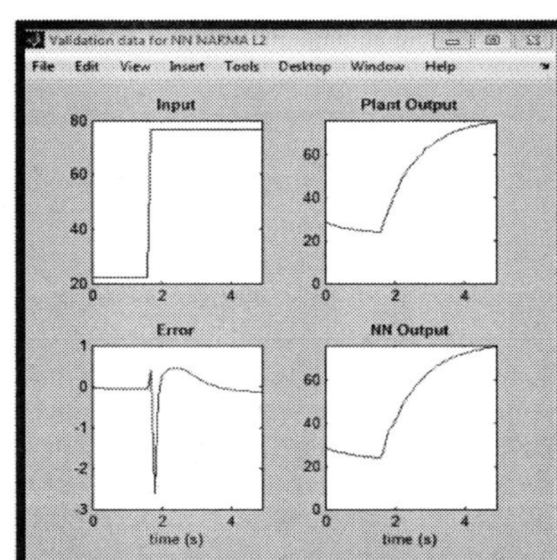

Figure 11. Training Data of Neural Network

IOP Conf. Series: Materials Science and Engineering **765** (2020) 012019 doi:10.1088/1757-899X/765/1/012019

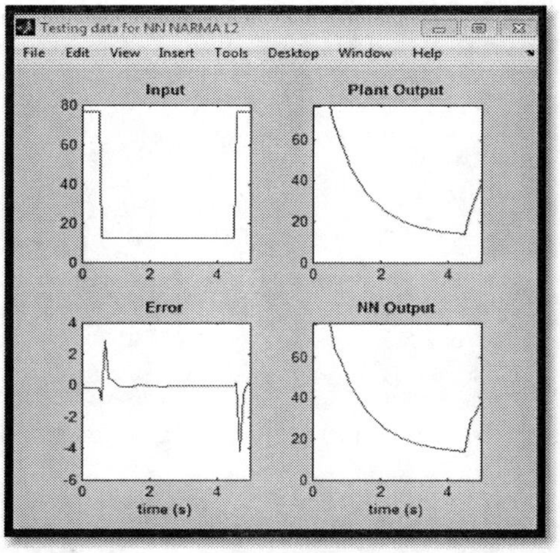

Figure 12. Testing Data of Neural Network

5. Results and Discussion

Figure 13 shows the model developed in Simulink \ Matlab to control of pressure and temperature of condenser at optimum value to enhancement of plant efficiency Figure 14 shows the output of pressure and temperature of condenser of power plant and compared these results with output of experimental device. The experimental results of an electrical device shown in Figure 15, this figure explains the practical pressure signal that measured from electric circuit that allowed to work the condenser under the normal condition, and shown the temperature signal measured practically, represent the perfect temperature to work condenser. The results obtained good agreement between NARMA result and experiment device.

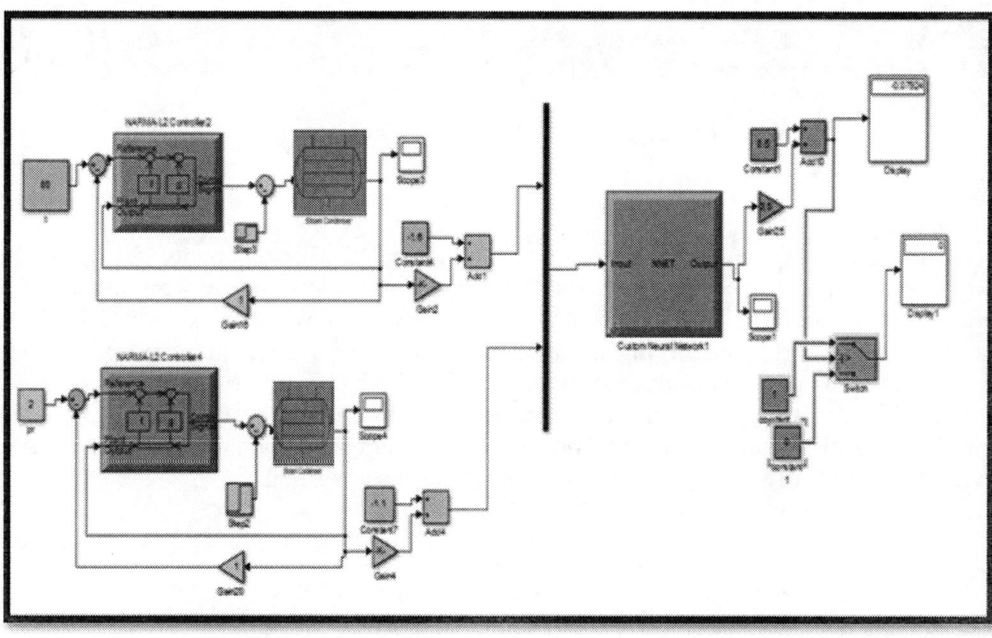

Figure 13. Simulation Model of Condenser in Power Plant

IOP Conf. Series: Materials Science and Engineering **765** (2020) 012019 doi:10.1088/1757-899X/765/1/012019

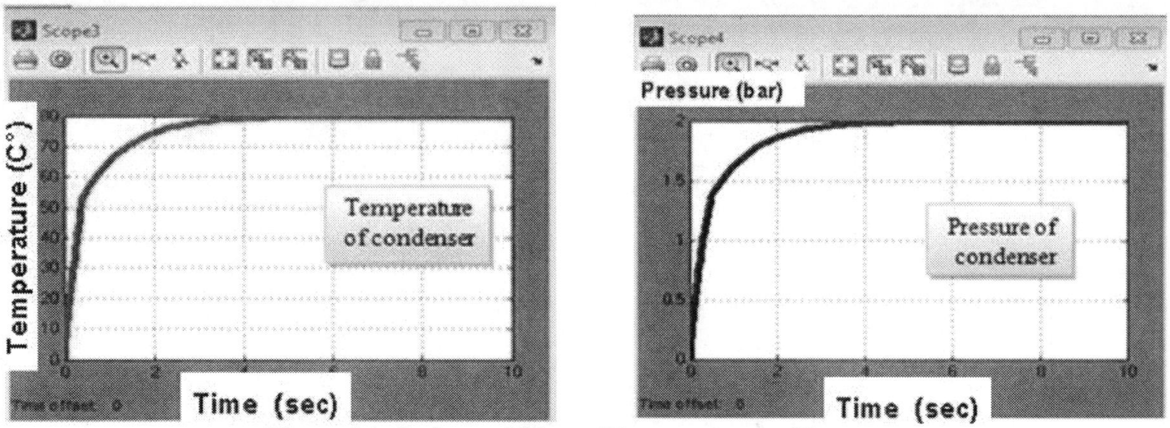

Figure 14. Optimum Value of Temperature and Pressure

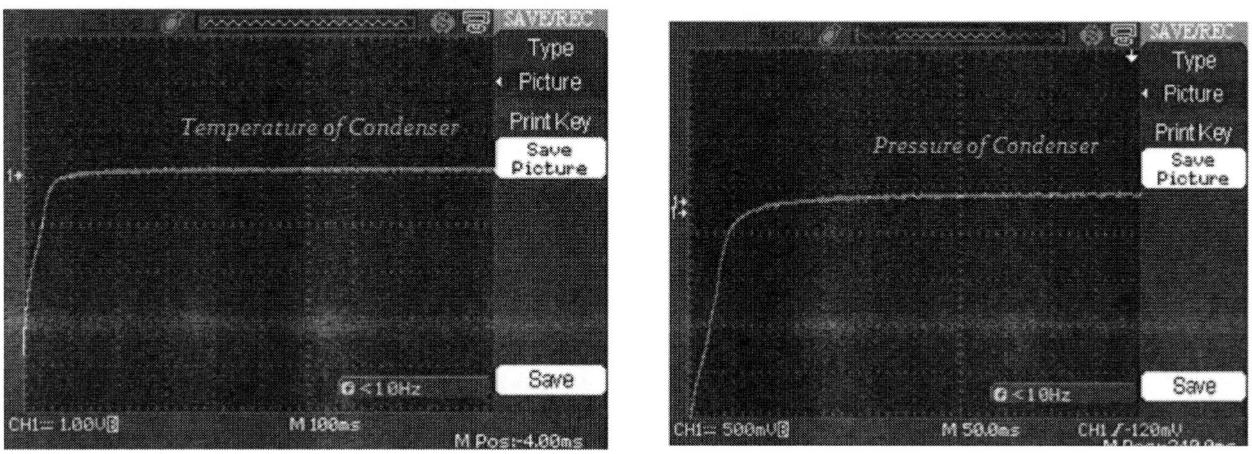

Figure 15. Experimental Signals of Temperature and Pressure

6. Conclusion

This paper applied artificial neural network technique to control of condenser for improvement the performance and design electric circuit for this purpose. This study shows the inputs of pressure and temperature of condenser. The trained NN model can approach desired values with height accuracy. And reach for normal condition of working steam plant by increase expansion ratio of steam and thus increase efficiency. The ANN model can advance desired values with height accuracy. When compared the results were used in Matlab program built with experimental circuit that's designed for control. Show a good agreement of signals pressure and temperature between them. This model gives high accuracy and these rough models are usually used in adaptive control of nonlinear systems.

References

[1] V. Haldkara, A. Sharmaa, R. Ranjana And V.K. Bajpaib."An Energy Analysis of Condenser",International Journal of Thermal Technologies01, Vol.3, No.4. December 2013

[2] K. Jeong, M. Kessen, H.Bilirgen, E. Levy. "Analytical Modeling of Water Condensation In Condensing Heat Exchanger",ELSEVIER International Journal of Heat And Mass Transfer (53) 2361-2368. 2010

[3] W. Thambanchacheep and S. Kuntanapreed."Amodel Reference Neural Network Controlfor A Variable-Speed Air Conditioner" Research And Development Center For Intelligent Systems Faculty Of Engineering King., Bangkok, THAILAND 10800

[4] H. Jasim ,H. Jaffa, "Effect Of Operation Conditions On Exit Water Temperature Of Condenser (Atmospheric) By Using Neural Network ". Journal of Engineering and Development, Vol. 15, No. 3, ISSN 1813-7822. September 2011

[5] G. Henze, R. Hindman. "Control of Air-Cooled Chiller Condenser Fans Using Clustering Neural Networks". ASHRAE. TRANSACTIONS- AMERICAN SOCIETY OF HEATING REFRIGERATING AND AIR CONDITIONING ENGINEERS. 2002.

[6] Wang Peng ,Meng Hao,Dong Peng And Dai Ri-Hui " Decoupling Control Based On Pid Neural Network For Deaerator And Condenser Water Level Control System" **IEEE** . **ISSN:** 1934-1768. 14 September 2015

[7] Roland S.Burns," Advance Control Engineering", Butterworth-Heinemann, Published Oxford . 2001.

[8] D. Pallares, F. Johnsson, (2006) Macroscopic Modelling of Fluid Dynamics in Large-Scale Circulating Fluidized Beds, Progress In Energy And Combustion Science 32 (5) 539–569.

[9] C. Zhao, L. Dun, W. Zhou, X. Chen, D. Zeng, T. Flynn, D. Kraft," Coal Combustion Characteristics On an Oxy-Cfb Combustor with Warm Flue Gas Recycle, In: Proceedings of The 21st International Conference On Fluidized Bed Combustion, 2012. Pp. 3–6.

[10] U. Kesgin, "Genetic Algorithm and Artificial Neural Network for Engine Optimization of Efficiency and Nox Emission, Fuel 83 (7) ,2004, 885– 895.

[11] Resat Celikel And Omur Aydogmusnarma-L2 Controller For Single Link Manipulator. IEEE Published In: 2018 International Conference On Artificial Intelligence And Data Processing (IDAP) 24 January 2019.

[12] A.G. Shempelev, V.M. Sushchikh, P.V. Iglin,"On The Results of Comparison of the Calculated and Normative Characteristics of Condensers Steam Turbines in A Wide Range of Steam Loads", Energetic,2015 Pp. 60-64.

[13] Xiao Fan ; Zhou Youbin ; Ruan Lin ; Zhou Kunpeng ; Wang Tao ; Cao Kan ; Rao Yuzestudy "On Transient Reactive Power Characteristics Of New-Generation Large Synchronous Condenser .IEEE 2018 China International Conference On Electricity Distribution (CICED) 31 December 2018.

[14] Zekun Yang ; Xiaoping Shiresearch "On Vacuum Control For Condenser With Double Compensation Loops: IEEE 2017 IEEE International Conference On Information And Automation (ICIA) Date Of Conference: 18-20 23 October 2017

[15] A.Shempelev ; P. Iglin ; N. Tatarinovaon Condenser Mathematical Model Method Introduction Into Steam Turbine Unit Mathematical Model. IEEE 2017 International Conference On Industrial Engineering, Applications And Manufacturing (ICIEAM)Date Of Conference: 16-19 May 2017.Date Added To IEEE Xplore: .

[16] Shempelev ; P. Iglin ."On Mathematical Model Of Condenser Equipped With Water-Jet Ejector . IEEE 2018 International Conference On Industrial Engineering, Applications And Manufacturing (ICIEAM)Date Of Conference. 06 June 2019

Modelling and Controlling of position for electro-pneumatic system using Pulse-Width-Modulation (PWM) techniques and Fuzzy Logic controller

Wisam Essmat Abdul-Lateef [1], Glebov Nikolay Alexeevich[2],Naseer Hamed Farhood [1], Ahmed H. Khdir[1], Dina H. Shaker[1]

[1]Electromechanical Engineering Department -University of Technology-IRAQ
[2]South-Russain State Polytechnical University – Russia
Corresponding e-mail : 50110@uotechnology.edu.iq

Abstract. The main objective of this paper is to design a low cost electro-pneumatic module which can be used in the industrial automation equipment's , robotic systems and mechatronic applications. Electro-pneumatic module includes four low-cost on/off solenoid valves and double acting cylinder. A Pulse width modulation (PWM) technique was used, which enable the model to correct the control of piston position and then to enable the system to follow the required trajectories accurately in many applications. Thus, it can be a good alternative at a lower cost than the using of proportional valves in the electro-pneumatic applications. A Fuzzy logic algorithm with PD control on the model in Matlab/Simulink environment to adjust the control values was applied. The simulation results and response of model were compared with the results from a previous practical with the similar model and controller but with a proportional valve. Finally, the simulation results showed a good accuracy with the practical results.

Keywords: Electro- pneumatic; Fuzzy Logic; PD control; Pulse width modulation (PWM); solenoid valves.

1. Introduction

The modern industrial applications require that the human works in different environments operation ranging from the industrial, nuclear, surgery to space exploration...etc. These environments may be highly risk to the humans and it is difficult to reach, or tiring and boring operations, in addition to the thermal or radiological environments that limits from the presence and work of human. But the automation systems and the robotic systems had enabled the humans to works in these environments without need to presence of connection or direct human presence, and these leads to a significantly reduce in humans accidents [1]. So, at present in the many industrial applications, the most of human workers are replaced by automation systems or robotic systems.
From the above, it found that the robotic systems and automation systems are an ideal solution for the many problems and difficulties in different fields.
Nowadays, the electro-pneumatic actuators have become an important component in robotic systems and industrial automation systems and widely used in these fields.

These electro- pneumatic actuators offer a lot advantages compared with the others actuators, where this type of actuators can to combine between the advantages and benefits for two types of actuators are the pneumatic and electrical actuator. The pneumatic system in the magnetic and electrical fields is considered an inert, so, this feature has made the pneumatic actuator use in many critical applications. in addition to that, this system is characterized by simplicity, reliability and cleanliness, with the low costs in configuration, work, maintenance and safety, as well as it has the ability to work in environments of aggressive, and at extreme temperatures (low and high).

While, the electric system can provide the ability to control in the elements of system and thus the ability to control in the positioning processes. However, the poor damping and large non-linear factors due to mechanical friction and highly air compressibility makes it difficult to control in the pneumatic systems, especially to precise control in the piston position [2, 3].

The electro- pneumatic systems have a widely application with the robotic systems as well as in the industrial automation field due to for characteristics that have been mentioned previously, but the positions identifying process remain the most important in these applications. For the positions identifying processes, the pneumatic valves use to control in air pressure inside the cylindrical chambers thus controls the positioning process .The control valves in electro-pneumatic system which used to control the flow of inputs (air) in the actuator can be classified into two common types. The first type is a proportional valve, which have a high control accuracy and linear behavior with functions operate as follows: the armature takes some position between the open and closed position entirely according to the electrical signal, and this considered is a very suitable for the most applications of industrial automation. But they are expensive because of their characteristics and its complex structure. A proportional valve represents the most expensive of the system components, where the proportional valve cost sometimes is close to the same cost of all other components in the system, this leads to reduce its use in some applications [4].

The solenoid valves (on / off) can be used instead of using the proportional valves because of these valves are very cheap, and by using these types of valves can be built and developed an electro-pneumatic system at low cost. But the mechanism work of this type of valves takes either the open position entirely or closed according to the electric driving, and this is not appropriate with the work requirements in the industrial automation field and its applications. To overcome of this problem, a Pulse Width Modulation (PWM) technique is used, where that the PWM technique allows to systems using the solenoid valves to have a similar behavior to the behavior of systems that using a proportional valve but at much lower costs[5,6].

The method used to control and to overcome from the difficulties in electro-pneumatic systems depend on the artificial intelligence techniques .Where use fuzzy logic , which is one of the artificial intelligence techniques branches, and the fuzzy logic can to adjust the system parameters automatically and improve the performance of system elements. In this search fuzzy PD controller is applied on the model in Matlab/Simulink environment for the purpose to control and adjust the values of electro- pneumatic system [7,8].

2. Modelling of electro - pneumatic system

The pneumatic systems can be modeled mathematically using the following differential equations: equation of motion, through this equation can be describe the piston load dynamics, forces balance, friction compensation, two equations for the rate of pressure change in the single chambers, equation of the solenoid valves dynamics and standard equation for the mass flow.

The equation of motion for the piston rod load was derived according to the Newton's second law and can be expressed as shown in Eq.1

$$(M_L + M_p)\ddot{x} + \beta\dot{x} + F_f + F_L = P_1A_1 - P_2A_2 - P_aA_r, \quad (1)$$

Where:

$\ddot{x}$ –piston acceleration; ML – is the external load mass ; MP– mass of piston and rod; $\dot{x}$– piston velocity; β – coefficient of viscous friction; Ff –friction force; FL – the external force; P1&P2 – are the pressure inside the behind and front chambers respectively; Pa – absolute ambient pressure; A1&A2 – are the piston area in the behind and front chambers respectively; Ar – is the cross-sectional area of the rod; Ps– the supply pressure[9].
The change of pressure rate for the behind and front chambers are presented in the Eq.2 and 3.

$$\dot{P}_1 = \frac{RT}{V_{o1}+A_1x}(\dot{m}_{in}\alpha_{in} - \dot{m}_{out}\alpha_{out}) - \alpha\frac{\pm A_1P_1}{V_{o1}+A_1x}\dot{x} \tag{2}$$

$$\dot{P}_2 = \frac{RT}{V_{o2}+A_2(L-x)}(\dot{m}_{in}\alpha_{in} - \dot{m}_{out}\alpha_{out}) - \alpha\frac{\pm A_2P_2}{V_{o2}+A_2(L-x)}\dot{x} \tag{3}$$

Where:
R – is the universal gas constant; T - is the air temperature; min, mout - mass flow at the inlet and outlet of cylinder cavity respectively; $\dot{x}$ - piston velocity; αin = 1.4,αout= 1;α = 1,2; L - stroke length of the cylinder; V1,V2 - inactive cylinder volume for the behind and front chambers respectively; A1, A2- effective piston area of cylinder for the behind and front chambers respectively.
The standard equation for mass flow rate($\dot{m}_v$) through the orifice of the effectivearea for solenoid valves(Av)can be written as in Eq. 4 [10.11].

$$\dot{m}_v = \begin{cases} C_f * A_v * C_1 * \frac{P_u}{\sqrt{T}} & \text{if } \frac{Pd}{Pu} \leq Pcr \\ C_f * A_v * C_2 * \frac{P_u}{\sqrt{T}}\left(\frac{P_d}{P_u}\right)^{\frac{1}{k}}\sqrt{1-\left(\frac{P_d}{P_u}\right)^{(k-1)/k}} & \text{if } \frac{Pd}{Pu} > Pcr \end{cases} \tag{4}$$

Where:
$\dot{m}_v$–mass flow rate through the solenoid valvesorifice; Cf–discharge coefficient, non-dimensional; Pu – upstream pressure; Pd – downstream pressure; Pcr – is the critical value of pressure =0.528; C1, C2 – constant coefficients; K – is the specific heat ratio.
In this system, four of low-cost on/off solenoid valves (2\2) ,were used as a command elements for the pneumatic system, and the effective area (Av) for solenoid valves determine by using Eq 5.

$$Av_{=d^2*\frac{\pi}{4}} \tag{5}$$

Where: d = diameter of thesolenoid valves[5].
The electro- pneumatic system proposed and modeled in this paper is shown in Fig.1

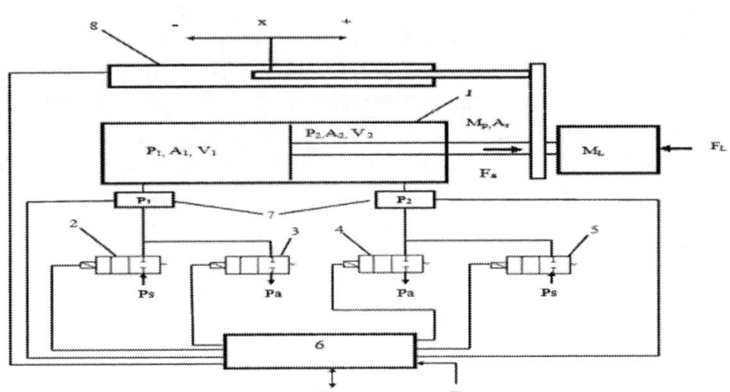

Figure1 The electro-pneumatic system with identification for each component

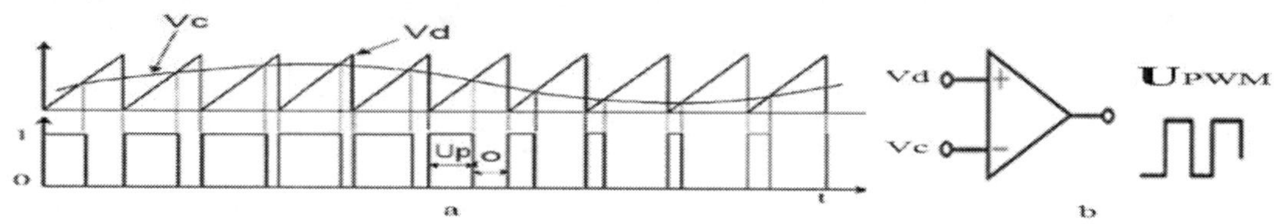

2.1. *Pulse Width Modulation (PWM) technique*

The pulse width modulation (PWM) technique is used to approximate or obtain to the required signal (multi-level or continuous) for real binary signals, a key point to operate the many devices used in a various fields. The principle of PWM, as the name indicates, it is the change of signal pulse width. When using the technique of pulse width modulation, the signal frequency and capacity remain constant. The PWM signal can be achieved by comparing the continuous control signal (Vc) with the high frequency carrier wave (Vd), as shown in Fig. 2 (a and b) [12,13].

Figure 2. Concept of PWM modulation Signal

The mathematical description of the PWM signal can be given by the following relation:

$$U_{PWM}(t) = \begin{cases} U_p & \text{for} \quad V_c(t) \geq V_d(t) \\ 0 & \text{for} \quad V_c(t) < V_d(t) \end{cases}$$

Where:
UPWM– input voltage (signal),

$$U_{PWM}(t) = \begin{cases} Up & -\text{switch is open} \\ 0 & -\text{the switch is off} \end{cases}$$

2.2. *Fuzzy Logic controller*

The controller system is designed and achieved according to the fuzzy logic controller. A fuzzy logic controller was chosen because existence of many independent variables that effect on the behavior of electro - pneumatic system such as (temperature, damping and large non-linearity due to mechanical friction, high air compression and cylinder vibrations, etc.). Thus, the fuzzy logic seems enough to solve the control problem in the electro-pneumatic systems, this is because of the fuzzy logic is a very flexible and strong tool for dealing with the systems it have properties like inaccurate data, incomplete mathematical models and non-linearity behaviors, through this controller getting the minimum error in position and good behaviours of electro-pneumatic systems [14,15].

3. The mathematical modelling of Electro-pneumatic Module with fuzzy PD controllers and PWM

The modelling of system is an important step for any form of the control systems design, where the modelling helps to understand the system requirements. Through using the modelling and numerical simulations, good information on the dynamic behavior of system can be obtained before it is actually perform. As above mentioned, the data derived from modelling helps end users to determine their requirements, development the system to meet these requirements and design a good and efficient controller with high durability and accuracy.

The fuzzy logic controller of PD with pulse width modulation (PWM) were developed and tuned by Simulink Toolbox of Matlab-Simulink package, which depends on the piston motion equations of pneumatic cylinder, rate of the pressure change in the cylinder cavities and the mass of air flow rate. This model have a two inputs are the error (e) and error change rate (de), and the output is the position of pneumatic cylinder piston (x). Seven memberships functions are used in this design and Table 1 shows the full database of fuzzy logic rules.

Table 1. Rule base of fuzzy logic PD controller

e \ de	N_B	N_M	N_S	Z	P_S	P_M	P_B
N_B	N_B	N_B	N_B	N_B	N_B	N_B	N_B
N_M	N_B	N_B	N_B	N_M	N_M	N_M	N_S
N_S	N_B	N_M	N_S	N_S	Z	N_S	P_S
Z	N_B	N_M	N_S	Z	P_S	P_M	P_B
P_S	N_S	P_S	Z	P_S	P_S	P_M	P_B
P_M	Z	P_S	P_M	P_M	P_B	P_B	P_B
P_B	P_B	P_B	P_B	P_B	P_B	P_B	P_B

The model is developed using a pulse width modulation technique (PWM), the input signal for each solenoid valve is the voltage of pulse width modulation (PWM). The signal frequency in this model was 50 Hz, this means that the solenoid valves open and close every 20 ms at running. Therefore, through the solenoid valves that control by PWM, a multi-functional electro-pneumatic system can be developed at an appropriate price and good performance. Fig.3 shows a block diagram of the PWM in Simulink Toolbox of Matlab-Simulink package. Fig. 4 shows the comparator generates a modulated PWM signal. The simulation was carried out using the following parameters for the MP-9C pneumatic robot manipulator are: piston stroke - 0.15 m, piston diameter - 0.042 m, rod diameter - 0.022 m. The piston motion was controlled by 4 solenoid valves of type (2\2).

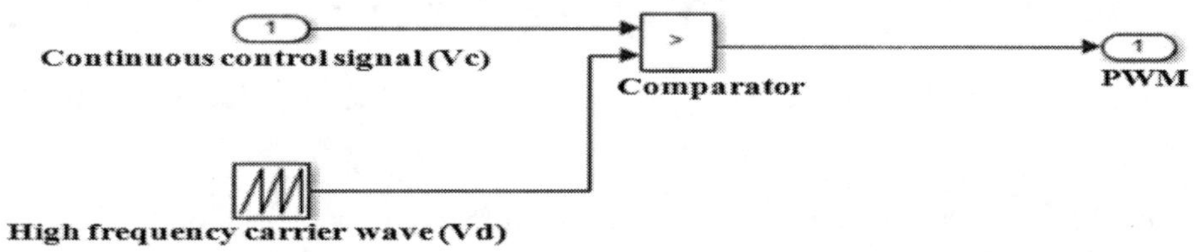

Figure 3. Block diagram of the PWM in Simulink Toolbox of Matlab-Simulink package

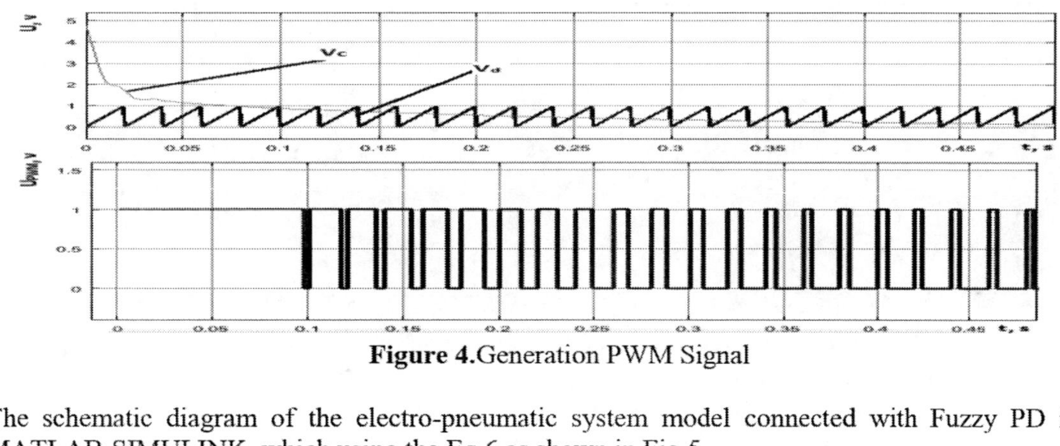

Figure 4. Generation PWM Signal

The schematic diagram of the electro-pneumatic system model connected with Fuzzy PD in the MATLAB-SIMULINK, which using the Eq.6 as shown in Fig.5

$$u(k) = K_p . e(K) + K_D . \frac{de(K)}{T} = K_p . e_p(K) + K_D . e_D(K) \quad (6)$$

IOP Conf. Series: Materials Science and Engineering **765** (2020) 012020 doi:10.1088/1757-899X/765/1/012020

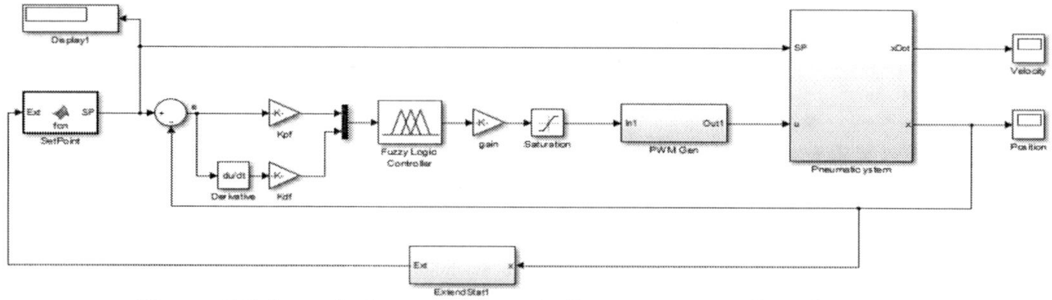

Figure 5. Schematicelectro-pneumatic fuzzy PD controller with PWM

Figure 6 shows the membership functions of the input variables (e) and (de), and Figure 7 shows the membership function of the output variable (x).

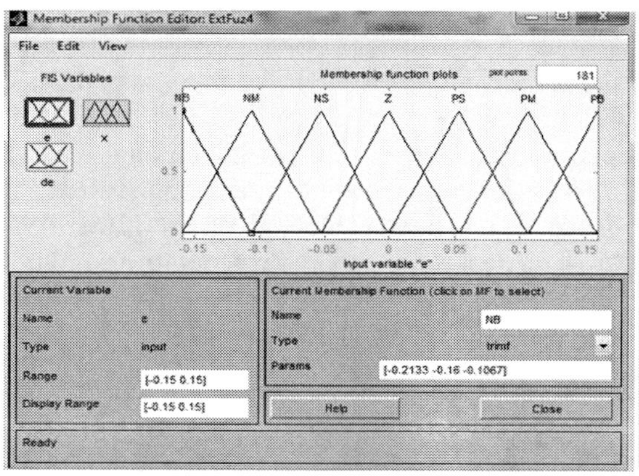

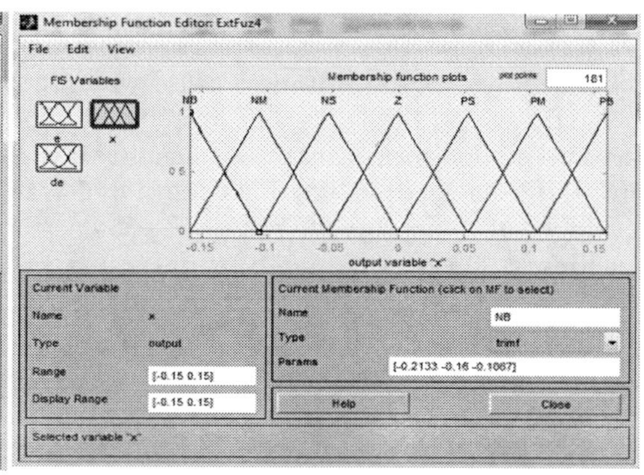

Figure 6. Error input membership function for- **Figure 7**.Position input membership function

4. Results

After the modelling and simulations for the electro-pneumatic model based on PWM which controlled by the fuzzy PD at different strokes, the model results showed that the control strategy presented is a good and stable enough as shown in Figures 8 (a) and (b).

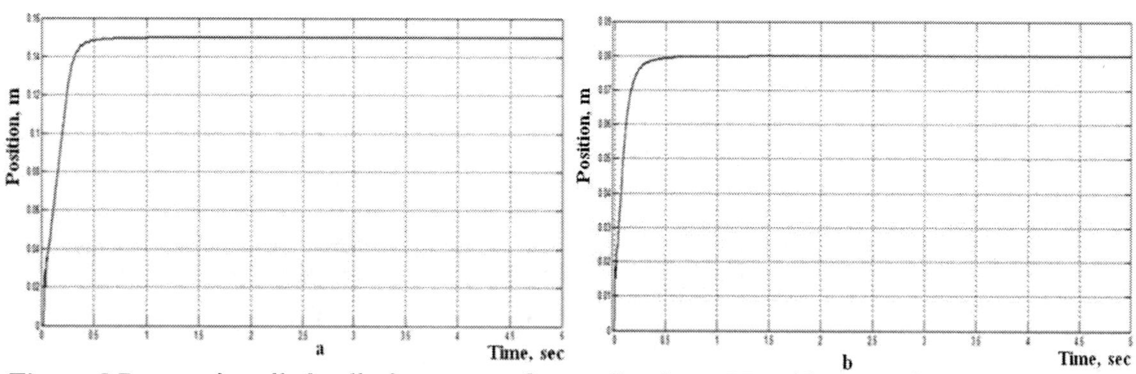

Figure 8.Pneumatic cylinder displacement of controller fuzzy PD with PWM for (a) stroke of 0.15 m, (b) stroke of 0.08 m

These figures represent the model results with a fuzzy PD based on PWM during the stroke 0.15 m and during the stroke 0.08 m, respectively.

The model results are compared with the results taken from [16], for a similar model using the same of controller unit but with a proportional valve. It was found that the results were a good and very close as shown in Table (2).

Table2. The comparison of positioning errors for the model results and the results of previous experimental work

Stroke, (m)	Position errors of fuzzy PD, (m) (Experimentally results)	Position errors of fuzzy PD with PWM, (m) (Simulations results)	Percentage Error
0,15	0,00052	0,00054	2·10-5
0,14	0,0005	0,00052	2·10-5
0,12	0,00048	0,0005	2·10-5
0,1	0,00046	0,00050	4·10-5
0,08	0,00044	0,00047	3·10-5

5. The discussion and Conclusions

The model of the electro-pneumatic system was proposed , designed and developed in this search. This model includes a double acting pneumatic cylinder with four cheap (2\2) on–off solenoid valves. As it is known, the solenoid valves works only in the opening and closing position, this is not suitable for applications in the modern industrial systems in general and mechatronic systems in particular. Therefore, to enhance the behaviour of solenoid valves and obtain on a similar behaviour to the proportional valves behaviour used in electro -pneumatic systems, the PWM technique was proposed and used. In this search , the PWM technique work at 50,Hz frequencies with pressure 5 bar .To obtain the required piston position for the system, the fuzzy PD controller was used.The response and results were a good as shown in figure (9) , and very close when compared it with the results from the previous practical [16] for a similar model use the same of controller unit but with a proportional valve.Where the difference was very little, about (2·10-5 ~ 4·10-5), and as shown in table 2.

References

[1] Ioannis Tsitsimpelisa, C. James Taylora, Barry Lennoxb, Malcolm J. Joycea.A review of ground-based robotic systems for the characterization of nuclear environments.Progress in Nuclear Energy . 2019, 111(109–124). www.elsevier.com/locate/pnucene.

[2] Fei Zhao, Yonggang Ma and YuansongSun.Application and Standardization Trend of Maintenance and Inspection Robot (MIR) in Nuclear Power Station. 3rd International Symposium on Mechatronics and Industrial Informatics . ,(ISMII 2017), ISBN: 978-1-60595-501-8

[3] Shadi Mohammad Munshi. Analysis, investigation and design of flexible universal pneumatic industrial manipulators involving cartesian and joint control in the basis of economic feasibility and appropriate technology. Degree of doctor of philosophy. Thesis, University of new south Wales, 2012.

[4] Minh-Quyen LE. Development of bilateral control for pneumatic actuated teleoperation system // Département Méthodes pour l'Ingénierie des Systèmes – Laboratoire Ampère Ecole doctorale: Electronique, Electrotechnique et Auto-matique (EEA) Spécialité: Automatique Industrielle thesis Submitted on 13 Jul 2012.

[5] Cajetinac S., Seslija D., Aleksandrov S., Todorovic M. PWM Control and Identification of Frequency Characteristics of a Pneumatic Actuator using PLC Controller // ELECTRONICS AND ELECTRICAL ENGINEERING, AUTOMATION, ROBOTIC , SISSN 1392 – 1215, 2012. № 7(123).

[6] Sean Hodgson, Minh Quyen Le, Mahdi Tavakoli, Minh TuPham..Improved Tracking and Switching Performance of an Electro-Pneumatic Positioning System. This paper appears in Mechatronics,2012, vol. 22, no. 1, pp. 1-12.
[7] BikasMondal, Md. ArifBillah, Bhaskar Roy and RohitSaha. Performance comparison of conventional PID and fuzzy logic controller in the field of over headed water level control system. International Journal of Computer Sciences and Engineering.Volume-4. Aug 2016, pp. 76-81.
[8] Ali A. Tlisov, Yuri V.Mitrishkin. Adaptive control system for pipeline valve pneumatic actuator. Proceedings of the 13th IFAC Symposium on Information Control Problems in Manufacturing .Moscow, Russia .2009. PP373-378.
[9] YusWitdarko, NursigitBintoro, BandulSuratmo and Budi Rahardjo. Modelling on mechanical cassava flour pneumatic drying. AGRITECH, Vol. 36, № 3, Agustus 2016, pp. 362-368.
[10] Xinhua Zheng, Lili Xie and Lizhuo Liu. Stability analysis of pneumatic cabin pressure regulating system with complex nonlinear characteristics. Journal of Control Science and Engineering.Volume 2015, Article ID 871621, pp.1-8.
[11] Rahmat M.F., Sunar N.H. and SyNajibSy Salim, MasturaShafinaz Zainal Abidin, MohdFauzi A.A. and Ismail Z.H. Reaiew on modelling and controller design in pneumatic and actuator control system. International journal on smart sensing and intelligent systemc Vol. 4, No 4. December 2011, pp. 630-661.
[12] A.Mahendran,K.Krishnakanth,S.A.Ramesh. Modelling and Simulation of Electro-Pneumatic Actuator Controller using Pulse Width Modulation. South Asian Journal of Engineering and Technology Vol.2, № 22 (2016) - C. 79 – 88.
[13] ErlendHelgeland. Sliding Mode Control of an Electro-Pneumatic Clutch Actuator. Master thesis, Norwegian University of Science and Technology, May 2008.
[14] Pradeep Rohilla and Abhishek PratapSingh.To Control Nonlinear Pneumatic System using Fuzzy Logic Controller.International Journal of Engineering Research and Technology (IJERT). Vol. 5, Issue 04, April-2016, pp. 491- 494.
[15] Mustafa Can Buken. Fuzzy controller design for lateral movement of a boeing 747 aircraft. Thesis of undergraduate program in Istanbul Bilgi University Computer Science Department. 2012. - 70 p.
[16] WisamE.Abdul -Lateef. Adaptive electric- pneumatic modules of mechatronic control systems in the technological equipment's and robots with pneumatic actuator. Degree of doctor of philosophy, 2017. Thesis South-Russian University (NPI) - Department of Mechatronics and Automation .Russia/ Rostov-on-Don.

IOP Conf. Series: Materials Science and Engineering **765** (2020) 012021 doi:10.1088/1757-899X/765/1/012021

Performance Improvement of Photovoltaic Module Using an Air-Cooling Micro Finned Heat Sink

Ibtisam A. Hasan

Sahar R. Faraj

Israa A. Mohammad

Electromechanical Engineering Department, University of Technology, Baghdad, Iraq

Corresponding author's e-mails: 50255@uotechnology.edu.iq

dr.sahar_alsakini@yahoo.com

israaalrkaby@gmail.com

Abstract: In this work, improved PV panel was fabricated by replacing the tedlar layer with aluminum alloy micro pin fin heat sink to increase the heat transfer conduction. Air flow system was designed and implemented to apply force convection on the heat sink, which named MPFHS-PV/T integrated system, to decrease the PV module temperature. The integrated system was tested on 25Jun 2019 in Baghdad climate and the results recorded as flowing: decreasing the PV module temperature by 14.65% along with electrical performance improvement by 13% for the output power and 13.32% for efficiency. The daily averages of electrical efficiency .1102%, thermal efficiency 40.94% and overall thermal efficiency 51.9%.

Keywords: Micro, Heat sink, Photovoltaic module, Air-cooling system, Performance improvement, Micro finned heat sink.

Symbol	Meaning	Unit
AB	The Fins Base Area	m^2
I_s	Solar Radiation.	W/m^2
I_{mp}	Maximum Current of The PV Module at the Maximum Power	A
Vmp	The Maximum Voltage of PV Module at Maximum Power	V
η_e	Electrical Efficiency	______
η_o	Overal Thermal Efficiency	______
η_{th}	Equivalent Thermal Efficiency	______

1. Introduction

A photovoltaic module represents a band of PV cells arranged together to convert solar energy to electric power, by investing the behavior of semiconductor with solar radiation. High-efficiency photovoltaic panels become a necessity in today high power demands globally as many countries moving toward renewable energy sources. One of the main limitations to photovoltaic panel's efficiency is temperature toleration, which has a tremendous impact on it, which will lead to deteriorating performance and eventually stop working [1]. The need for thermal management in PV panels becomes imperative to emit this unwanted heat away. Various types of heat sinks have been designed to address this issue in PV panel, radiators, and printed circuit boards. One particular solution which is micro fins has received extensive research. Micro fins aim to increase the heat transfer through the extension of the exchange surface area [2]. Fins cross section can be a square, elliptical, cylindrical, and rectangular or any other shape. Fins attached to the base of a heat sink to increase the heat transfer area, and gather as many fins on the base of the heat sink as possible to provide extra heat transfer area [3]. But for the case of natural fluid flow, increasing fin density can cause obstruction to the fluid flow during fins and reduce heat transfer, [4]. Also, higher fin density at forced fluid flows may lead to increase power consumption, which results in high power consumption, thus reducing the system efficiency, [6]. For these reasons, Engineers have to determine a proper fins density and fins array arrangement that optimally work for a specific system. This leads to an abundance of possible arrangements for the fins.

PV panel cooling system utilizes water and air which is a thermal unit appended to the rear side of the module. However, most preferred is air because of the minimal use of the material and low-cost operation in spite of its thermal physical properties [5]. The thermal performance of an air circulation system to remove heat is spacious used due to air properties such as low viscosity, small density, and a little power fun consumption.

Many research carried out and various technologies utilized to improve the performance of PV panels. The related studies of the heat sinks with various geometries and arrangements of fins, with heat sinks effect on the cooling process and heat flow are reviewed. As well as, it included the outcomes of using micro fin heat sink as a useful technique for cooling and different flows such as free and forced convection. Peles et al., 2005 [6] investigated the phenomenon of pressure drop also heat transfer for flow over the micro pin fins heat sink. The result was that quite low thermal resistances were obtainable at using the pin fin heat-sink. Abel Siu-Hoetetal., 2007[7] carried on an experimental study for heat transfer and pressure drop features for a heat sink of micro pin fin. The measurement of temperature distribution and pressure drop were used to estimate averaged local Nusselt number (heat transfer coefficient) and the average friction factor. The found results were the coefficient of Heat transfer had a higher number near the heat sink entrance and decreased with the flow direction. Mahmoud et al.2011 [8] studied the effects of fin's high for the range of 0.25 - 1.00 mm in thermal performance of copper heat sinks. The results were the rates of convective coefficient improved when decreasing fin height or increasing fin spacing.

The micro-fins heat sink is our focus of an investigation, as a model of micro-fins with a duct attached to the rear surface of the panel developed that allows studying the effect of micro-fin as a cooling system on panel electrical output at a different velocity of fluid flow. The main objective of this work is to reduce PV panel temperature by employing a micro fins heat sink, and estimating the improvement in heat transfer carried by the introduction of micro fins with forced convection of airflow in the cooling channel.

2. Methodology:

For writing the equation of energy balance for the suggestion improved PV module with a thermal collector system, these assumptions are used: 1. A system is supposed to be at nearly a steady state for any time during their execution, 2. whole thermal characteristic for materials are supposed to be independent on temperature [9]. 3. The EVA (Ethylene-vinyl acetate) transmissivity is close to 100 % [10].

The formulated equation of energy balance is:

$$\text{Solar energy incident on panel} = \begin{Bmatrix} Energy\ loss\ from\ upper\ surface + \\ Output\ electricenergy\ + \\ Heat\ separated\ to\ thermal\ collator\ channel \end{Bmatrix} \quad (1)$$

The electrical efficiency of a module is calculated as follow:

$$\eta e = \frac{I_{mp}V_{mp}}{I_{s\times AB}} \quad (2)$$

Thermal efficiency is defined as follows [11]:

$$\eta_{\mathrm{th}} = \frac{\Delta \mathrm{Hairflow}}{AB\times I_s} \quad (3)$$

While, the overall thermal efficiency (η_o) can be obtained by the following equation [12]:

$$\eta_o = \eta_e + \eta_{th} \quad (4)$$

3. System Description

The PV panel has been developed by replacing the Tedlar layer with a micro finned plate, Figure 1 shows the fined plate which was fabricated by CNC machine. The cooling system depended on forced convection for the micro pin fin heat sink (MPFHS). MPFHS was enveloped by an isolated channel (a Perspex plastic channel) for flowing the cooling fluid (air) as a forced convection. This system will name MPFHS-PV/T system and its schematic is presented in Figure 2. It consists of a glass layer, EVA layer, cell layer, EVA layer and a heat sink of micro pin fins as well as an airflow channel. All these components have been described in detail with their physical properties and dimensions in Table 1. The schematic of the submitted MPFHS-PV/T integrated system is displayed in Figure 3. The experiment was performed on the site of Baghdad/Iraq (The latitude of Baghdad, Iraq is 33.3, and the longitude is 44.36), during Jun 2019.

Figure 1: Micro Finned Plate Heat Sink

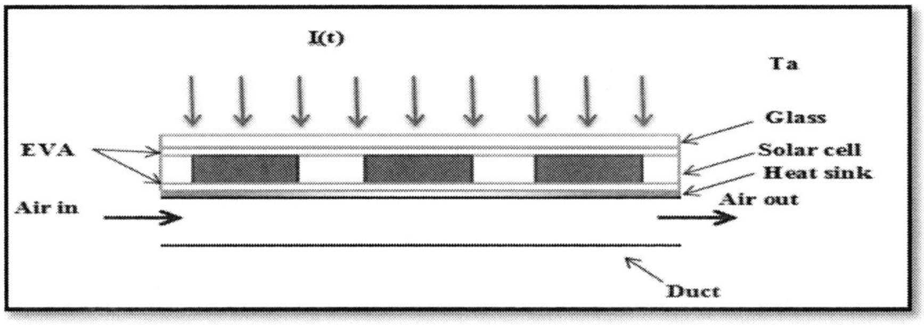

Figure 2: View of Cross-Sectional M PFHS-PV/T System

Table 1: Dimensions and thermal properties of the proposed PVT module component.

Layer	**Thickness (t), (M)**	**Thermal conductivity (K), (W/MK)**	**Absorptivity (α)**	**Reflectivity (A)**	**Transmit t-ance (τ)**	**Backin g factor (bc)**	**Emissivity (ε)**
Cover glass	0.0032	1.0	0.05	0.05	0.9		0.91
EVA(1)	500×10^{-6}	0.35			1		
PV cell	180×10^{-6}	148	0.85	0.15		0.9	0.9
EVA(2)	500×10^{-6}	0.35					
Heat Sink Base	0.003	78					
Tested heat sink Geometry	Fin length (L_{fin}),mm	Fin width (d_f),mm	Fin spacing (S),mm	Fin number (N_{fin})	Kfin w/mk	$AB\ m^2$	
	1	0.9	1.5	29150	195	0.1444	
Channel Geometry	Channel height(Hch) (m)	Channel length (Lch) (m)	Channel length (Wch) (m)	Material			
	0.02	0.38	0.38	Perspex plastic			

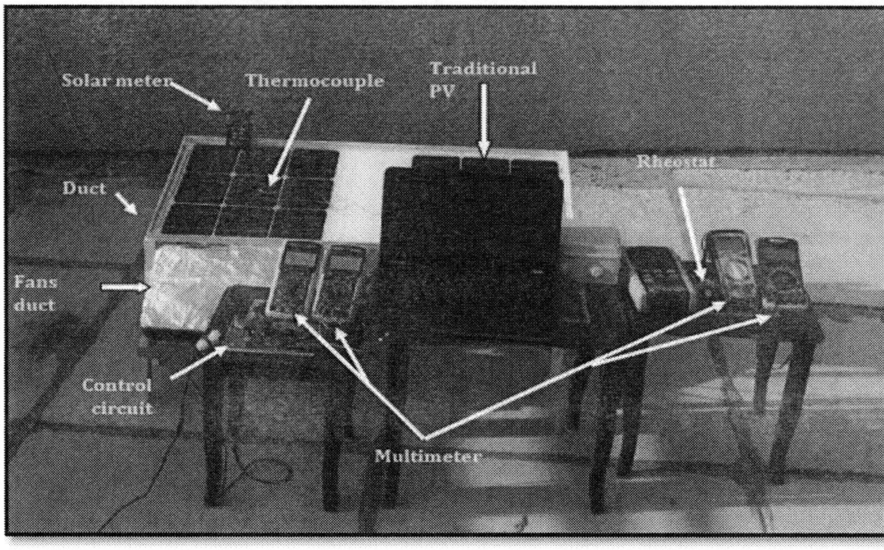

Figure 3: Experimental setup

The photovoltaic panels used in the experimental workare two units of 21.4W, monocrystallinetype, manufactured in AL-Mansoor Factory/ Ministry of Industry. The layers that constructed this two panel are shown in Figure 4.

Each half-hour, the value of solar radiation, cell temperature, inlet, outlet fluid temperature, the current and voltage related to maximum power and ambient temperature by temperature recorder have measured. Measurements were taken on Jun 25 in Baghdad's Weather conditions and at a fixed flow rate of 4.5 m/s in the cooling duct. The measurements started at 8:00AM and ended at the 14:30 PM.

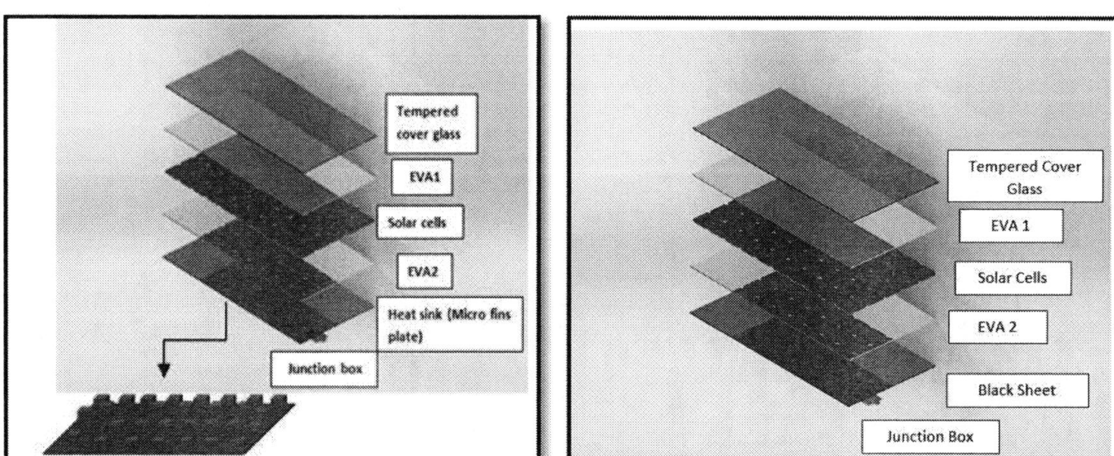

Figure 4: The Layers of Two Panels Constructed.

4. Results and Discussion

The PV module temperature during June may reach more than 70°C, due to the influence of hot climate in Iraq. The improved PV module has been tested on Jun -19- 2019 and its performance result has been compared with that of the traditional module and without any cooling system or under free convection effect. Figure 5 shows hourly temperatures distribution of improved PV module along with traditional module, ambient temperature and radiation intensity. There was an expected enhancement in average operating temperature about 3.2% due to the existence of micro finned plate at the back side of improved module.

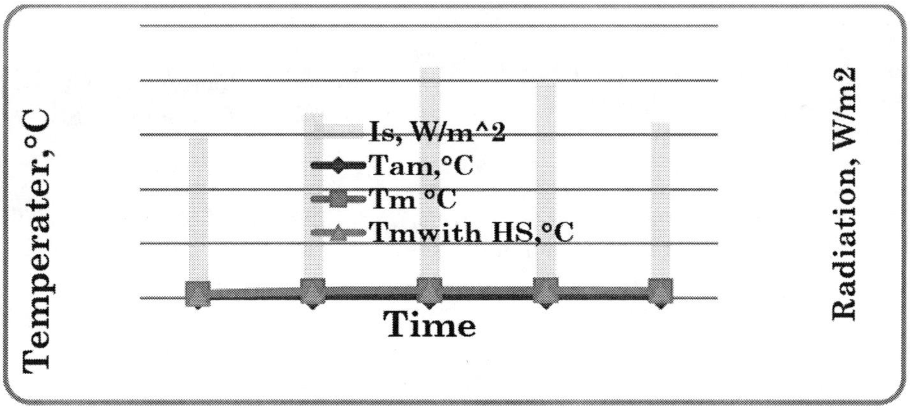

Figure 5: Hourly Temperatures of Improved PV Module Along with Traditional Module and Ambient Temperature

The experimental performances of MPFHS-PV/T integrated system were conducted at Baghdad city on Jun 25 2019 from 8:00 AM to 14:30 PM. The airflow velocity in cooling channel was determined at 4.5 m/s due to its tangible effect on PV module temperature. These data are measured by using solar power meter, digital thermometer and wind speed device, respectively. Figures 6 show hourly solar radiation, wind speed, ambient temperature along the day time on Jun 25.

Figure 7 presents a comparison of hourly PV modules temperatures during the day of measuring with and without cooling. The temperature for the non-cooling solar module varies from minimum values of 53.6 ◦C at 8 AM to maximum values of 72.4. ◦C at 12:30 pm, whereas for the cooling module it varies from a minimum values of44.5 ◦C at 8 am and a maximum value of 65.3 ◦C at12 pm. This figure reveals that the use of the cooling system caused to decrease in the average daily operating temperature of two modules about 13.5.

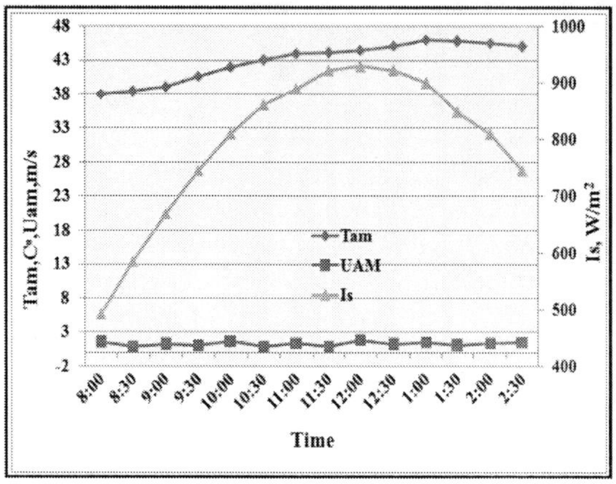

Figure 6: Climatic condition on Jun 25 2019, modules in Baghdad city temperature

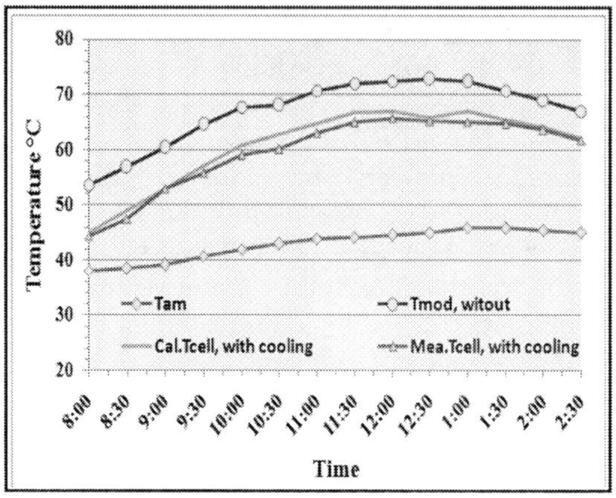

Figure 7: Hourly temperatures for PV in a different state along with ambient

Figures 8 illustrate output voltage during the testing period. It was found that it was an enhanced of 6.01% in average daily output voltage. This enhances had been achieved due to a reduction in operating temperature of module. Voltage increasing has a positive influence on the productive power and electrical efficiency of improved module, as represented in Figures 9 and 10. The percentage increases are 13% and 13.32%, for output power and efficiency at 12:00 PM respectively.

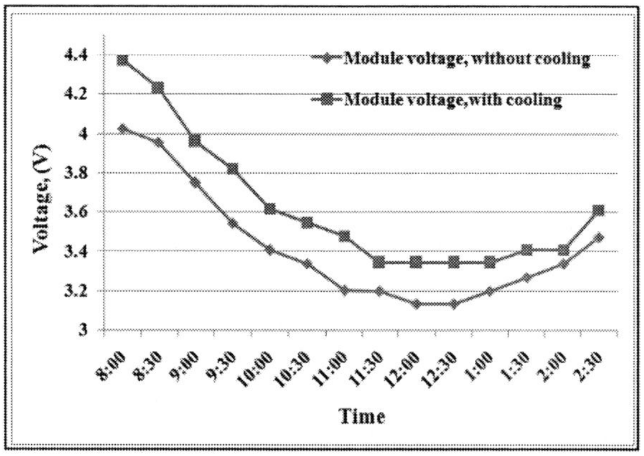

Figure 8: Hourly maximum voltage cooling for cooling and not cooled PV module

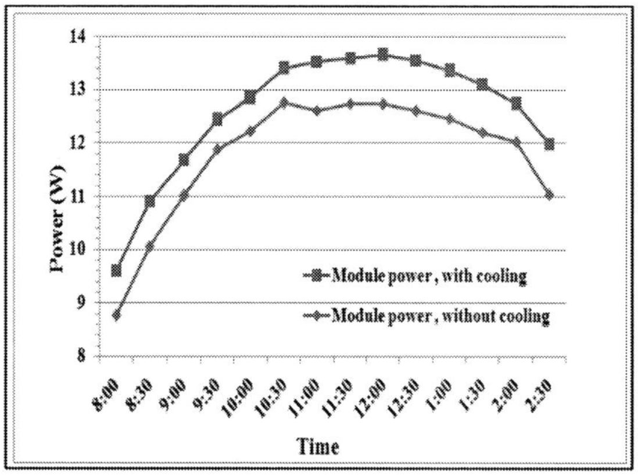

Figure 9: Hourly maximum power for and not cooled PV module

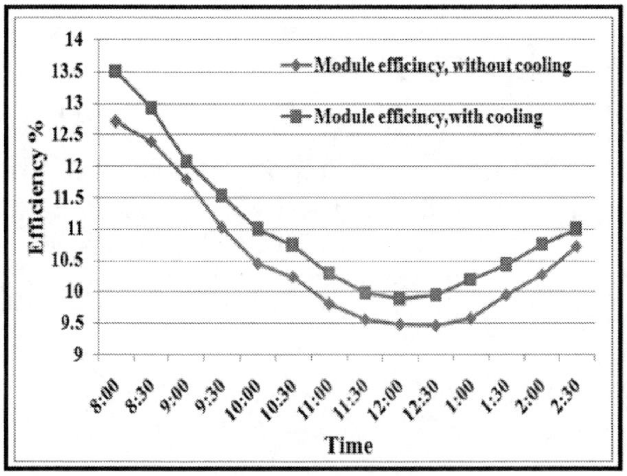

Figure 10: Hourly maximum efficiency for cooling and not cooled PV module

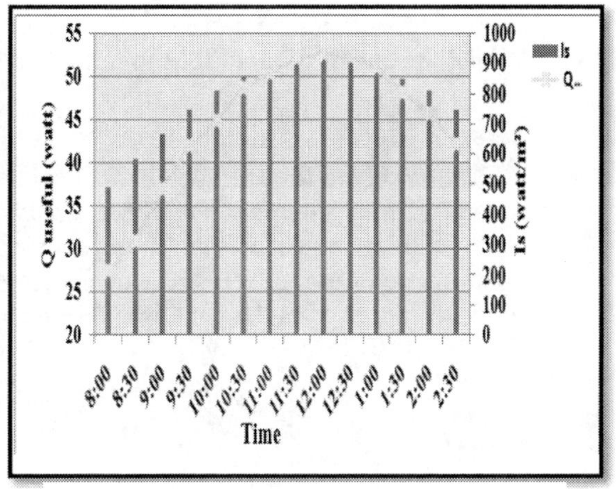

Figure 11: Hourly Heat Gain of MPFHS-PV/T integrated system

Figure 11 shows the hourly variations of heat energy gained and solar radiation. Thermal energy gained and solar radiation is connected directly to each other, which was the reason for the observed change in airflow temperature difference along with enhancement in module temperature efficiencies. The thermal efficiency variations for each hour during measuring date have been depicted in Figure 12. Solar radiation, panel's area, and airflow velocity played a main role in evaluating the thermal efficiency. The multiple efficiencies of the MPFHS-PV/T system which converts solar energy into electrical and thermal energy are shown in Figure 13. Cooling the PV module not only recaptures the excess thermal energy, but it also keeps its electrical efficiency during the time operation in hot climatic condition. The daily averages of electric, thermal along with overall thermal efficiency were found to be 11.02%, 40.94% and 51.9% respectively. The air flow rate extracts thermal energy from the Micro fined heat sink at the panel back side also fairly enhances its electrical efficiency.

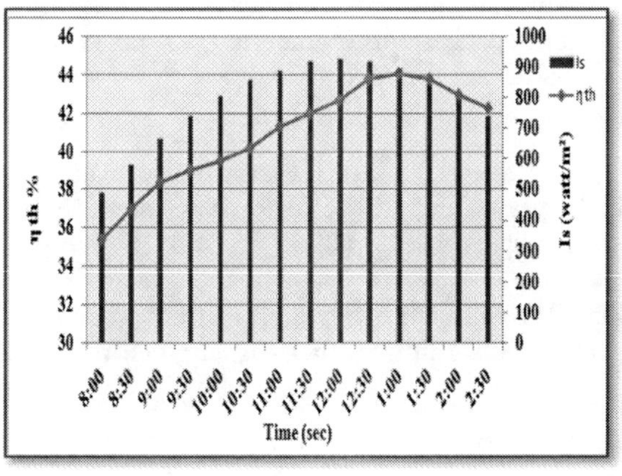

Figure 12: Hourly Thermal Efficiency with of MPFHS-PV/T system

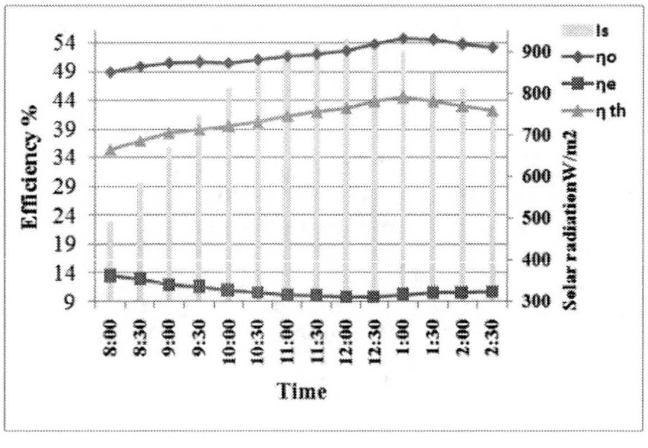

Figure 13: Hourly electric, thermal along overall efficiency of MPFHS-PV/T integrated

IOP Conf. Series: Materials Science and Engineering **765** (2020) 012021 doi:10.1088/1757-899X/765/1/012021

5. Conclusion

The improved PV module and airflow channel to collect the excess are one of the solutions to tempered module temperature rising. The improved PV module and airflow channel to collect the excess heat are one of the solutions to tempered module temperature rising.

A prototype for a single channel solar collector with micro bin fins heat sink has been made and the performance outputs have been analyzed. The use of the proposed cooling system has the potential to achieve an increase in output power of 13%, and 13% for efficiency when compared with traditional module results. While the decreasing in average daily temperature of module was 14.65%. The daily averages of electrical, thermal and overall thermal efficiency were found to be 11.02%, 40.94% and 51.9% respectively.

References

[1] Hasan, I.A., 2018. Enhancement the Performance of PV Panel by Using Fins as Heat Sink. Engineering and Technology Journal, 36(7 Part (A) Engineering), pp.798-805.

[2] Micheli, L., Sarmah, N., Luo, X., Reddy, K.S. and Mallick, T.K., 2013. Opportunities and challenges in micro-and nano-technologies for concentrating photovoltaic cooling: A review. Renewable and Sustainable Energy Reviews, 20, pp.595-610.

[3] Popovici, C.G., Hudişteanu, S.V., Mateescu, T.D. and Chereches, N.C., 2016. Efficiency improvement of photovoltaic panels by using air cooled heat sinks. Energy Procedia, 85, pp.425-432.

[4] Micheli, L., Reddy, K.S. and Mallick, T.K., 2016. Thermal effectiveness and mass usage of horizontal micro-fins under natural convection. Applied Thermal Engineering, 97, pp.39-47.

[5] Khan, M., Ko, B., Alois Nyari, E., Park, S. and Kim, H.J., 2017. Performance evaluation of photovoltaic solar system with different cooling methods and a bi-reflector PV system (BRPVS): an experimental study and comparative analysis. Energies, 10(6), p.826.

[6] Yu, X., Feng, J., Feng, Q. and Wang, Q., 2005. Development of a plate-pin fin heat sink and its performance comparisons with a plate fin heat sink. Applied thermal engineering, 25(2-3), pp.173-182.

[7] Peles, Y., Koşar, A., Mishra, C., Kuo, C.J. and Schneider, B., 2005. Forced convective heat transfer across a pin fin micro heat sink. International Journal of Heat and Mass Transfer, 48(17), pp.3615-3627.

[8] Siu-Ho, A., Qu, W. and Pfefferkorn, F., 2007. Experimental study of pressure drop and heat transfer in a single-phase micropin-fin heat sink. Journal of Electronic Packaging, 129(4), pp.479-487.

[9] Tonui, J.K. and Tripanagnostopoulos, Y., 2007. Improved PV/T solar collectors with heat extraction by forced or natural air circulation. Renewable energy, 32(4), pp.623-637.

[10] Wu, J., Zhang, X., Shen, J., Wu, Y., Connelly, K., Yang, T., Tang, L., Xiao, M., Wei, Y., Jiang, K. and Chen, C., 2017. A review of thermal absorbers and their integration methods for the combined solar photovoltaic/thermal (PV/T) modules. Renewable and Sustainable Energy Reviews, 75, pp.839-854.

[11] Chiu, H.C., Hsieh, R.H. and Jang, J.H., 2013. Numerical analysis on the heat transfer of heatsink with micro-pin-fins. ECS Transactions, 52(1), pp.759-764.

[12] Matsumoto, N., Tomimura, T. and Koito, Y., 2014. Heat transfer characteristics of square micro pin fins under natural convection. Journal of Electronics Cooling and Thermal Control, 4(03), p.59.

IOP Conf. Series: Materials Science and Engineering **765** (2020) 012022 doi:10.1088/1757-899X/765/1/012022

The optimum thickness of paraffin wax insulator filling a double glazed window

J M Jalil and S M Salih

Electromechanical Engineering Department, University of Technology, Iraq.

Email: jalalmjalil@gmail.com
Salih0499@gmail.com

Abstract. A numerical three-dimensional investigation is performed to estimate the optimum doping content (i.e. thickness) of the PCM filling a double glazed window. For the summer weather of Baghdad, Iraq. Finite Volume method combined with the enthalpy method is utilized to acknowledge the conduction related with phase change problems within the wax by using a constructed FORTRAN (F90) program. Based on the constructed numerical model, the transient heat transfer characteristics of the insulation were calculated and analyzed in detail. The obtained results indicate that optimum doping content of wax filling the window can be estimated, The required thickness of wax paraffin (melting point 40 °C) to keep temperature within the required range were as follows: May 2 cm, June 3 cm, July 3.5 cm, August 3.5 cm, September 3 cm and October 1.5 cm.

Key Words; thermal insulation, finite volume, wax paraffin, phase change

Nomenclature:
A: Area (m^2)
a, b : Coefficient in the discretization equation
B: 'Bottom' neighbor of grid P
b: Control-volume face between P and B
Cp: Specific heat (kJ/kg °C)
Cp_s: Specific heat of solid phase (kJ/kg °C)
Cp_l: Specific heat of liquid phase (kJ/kg °C)
H: Enthalpy (kJ/kg)
e: Control-volume face between P and E
E: 'East' neighbor of grid P
L: Latent heat (kJ/kg)
i, j, k: Unit vector
k: Thermal conductivity (W/ m °C)
l: Liquid
N: 'North' neighbor of grid P
n: Control-volume face between P and N
P: Grid point
q: Heat generation (W/m^3)
$\bar{q}$: Average heat generation
S: 'South' neighbor of grid P

s: Control-volume face between P and S
s: Solid
T: Temperature (°C)
t: Time(s)
T: 'Top' neighbor of grid P
t: Control-volume face between P and T
Tm: Melting temperature (°C)
V: Volume (m^3)
W: 'West' neighbor of grid P
w: Control-volume face between P and W
x, y, z: Cartesian coordinates(m)
ρ: Density (kg/m^3)
$\Delta x, \Delta y, \Delta z$: The width of the control volume in x, y, z directions
$\delta x, \delta y, \delta z$: The distance between two adjacent grid points in x, y, z direction
λ: Diffusion coefficient
Δ: Difference
∂ : Partial derivative
ΔT: Temperature range (°C)
u, v, w: Velocity in x, y, z directions

Introduction

In nowadays, the Phase Change Material (PCM) is a possible option for reducing the energy consumption and for increase the thermal comfort in buildings. The use of PCM in building applications provides the potential to increase the indoor thermal comfort for occupants due to the reduced indoor temperature fluctuations and lower global energy consumption. The possibility to integrate the PCM into the material of construction for cooling and heating the buildings gained the interest of researchers from all the world because the PCM have a high latent heat, meaning it is capable of storing and release high amounts of heat during its melting and solidifying process at a particular temperature [1].

Hasan et al. investigated the validation of installing layers of PCM as insulation Layers with different thickness located in walls and ceiling of a room located in Kut, Iraq. The acquired results shown a considerable amount of reduction in the indoor temperature and cooling load which led to saving in electricity consumption. Furthermore, **Xie et al**. investigated numerically the transient heat transfer characteristics of combined multilayer thermal insulation materials combined with two types of PCM. Goia et al. designed and built a test facility (glazing prototype) containing DGU_PCM (double glazed window with PCM) and DGU_GG (without PCM) and observed the performance of both of them within six months of the year focusing on the thermal comfort aspect and under real working condition. For the majority of the time, the DGU_PCM performance was much better. however, the interior surface temperature of DGU_PCM was greater than DGU_GG which is a promising feature in winter but it must be avoided in summer. **Ismail et al.** studied numerically and experimentally the thermal efficiency of two types of window, traditional and composite (filled with PCM or air). The experimental investigation was carried out by means of spectrophotometry on both types. The reflectance test indicated that the reflectance of the single glass window was 12 % to 13 %, however, the reflectance of the window with PCM was 7 %. **Gowreesunker et al.** analyzed numerically and experimentally the thermal and optical aspect of a double glazed unit with PCM compared to the usual double glazed unit. The obtained results illustrated that the transmittance during a fast phase change is unstable, however, throughout solid and liquid phases the visual transmittance rates are 40% and 90% respectively. Moreover, during the phase change process, the utilized PCM enhanced the thermal mass. **Liu et al.** demonstrated numerically the effect of semi-transparent property and zenith angle on the thermal performance of double glazed roof filled with PCM, the thickness of the PCM was also studied. The acquired results shown that the semi-transparent property, zenith angle and the thickness of the PCM have an enormous effect on the thermal performance of the double glazed roof. **Li et al.**

investigated numerically the thermal performance of the double glazed unit filled with PCM with different thermophysical parameters of PCM. The effects of specific heat capacity, density, thermal conductivity, latent heat and melting temperature of the PCM were studied. The obtained results indicated that all of the mention properties affect significantly the performance of the double glazed unit. **Li et al.** experimentally investigated the thermal behaviour of triple –pane window (TW) + PCM and compare it with (TW) and (DW + PCM). The obtained results indicate that the inside surface temperature of the TW + PCM was lower than the DW+PCM and TW by 2.7 and 5.5 °C, respectively. Additionally, minimized the heat entering the building by 16.6 % and 28% throughout the summer sunny day. **Liu et al.** developed a numerical model to determine the optical and thermal performance of glazed roof incorporate with PCM. In addition, the effect of air convection (h wind), PCM thickness and melting temperature were also studied, in the northeast China climates. Results obtained shows that regarding the internal surface temperature of the glazed roof the influence of the air convection should be considered, and the recommended PCM thickness and melting temperature were 12-20 mm and 16-18 °C, respectively.
The aim of this paper is to estimate the optimum thickness of the paraffin wax filling a double glazed window under the summer condition of Baghdad.

2. Numerical Formulations

Stefan problems were used to represent the heat conduction problems with phase change. This problem is addressed by the enthalpy transforming method; this method was offered to convert the energy equation into a non-linear equation with a single dependent variable enthalpy (H). The benefits of enthalpy method are that it can solve the problem when it's formulated in a fixed region, in addition to temperature, this method deals with the enthalpy as a dependent variable and separates the energy equation into a set of equations that include both temperature and enthalpy.
The analysis of the model in 3-D is regarding to Cao [11].
The energy equation is:

$$\frac{\partial}{\partial x}\left(k\frac{\partial T}{\partial x}\right)+\frac{\partial}{\partial y}\left(k\frac{\partial T}{\partial y}\right)+\frac{\partial}{\partial z}\left(k\frac{\partial T}{\partial z}\right)+\overline{q}=\rho\frac{\partial H}{\partial t} \quad (1)$$

In addition, the state equation [11]

$$\frac{dH}{dT}=C_P \quad (2)$$

Where the specific heat case is constant for each phase and the change of phase happens at a single temperature [11],[12].

$$T=\begin{cases} T_m + {}^{H}/_{C_{ps}} & H\le 0 \quad (Soild\ phase) \\ T_m & 0<H<L \quad (Phase\ change) \\ T_m + {}^{(H-L)}/_{C_{pl}} & H\ge L \quad (Liquid\ phase)\end{cases} \quad (3)$$

For the above relation, $H = 0$ was selected according to the phase change material (PCM) in their solid state to temperature T_m.
Also, the "Kirchhoff" temperature is presented as [13]:

$$T^*=\int_{T_m}^{T}k(\eta)d\eta=\begin{cases}k_s(T-T_m), & T<T_m\\ 0, & T=T_m\\ k_l(T-T_m), & T>T_m\end{cases} \quad (4)$$

Utilizing Equation (3) and the definition illustrated in equation (4) [11].

$$T^* = \begin{cases} {k_s H}/{C_{ps}}, & H \leq 0 \\ 0, & 0 < H < L \\ {k_l (H-L)}/{C_{pl}}, & H \geq L \end{cases} \tag{5}$$

Now, enthalpy function is introduced as [11}:

$$T^* = \lambda(H)H + S(H) \tag{6}$$

In Kirchhoff temperature's terms, adjusting equation (1) and replacing equation (6), Provides [14]:

$$\rho\frac{\partial H}{\partial t} = \frac{\partial}{\partial x}\left(\frac{\partial(\lambda H)}{\partial x}\right) + \frac{\partial}{\partial y}\left(\frac{\partial(\lambda H)}{\partial y}\right) + \frac{\partial}{\partial z}\left(\frac{\partial(\lambda H)}{\partial z}\right) + p + q \tag{7}$$

With

$$p = \frac{\partial}{\partial x}\left(\frac{\partial S}{\partial x}\right) + \frac{\partial}{\partial y}\left(\frac{\partial S}{\partial y}\right) + \frac{\partial}{\partial z}\left(\frac{\partial S}{\partial z}\right) \tag{8}$$

And for $\lambda = \lambda(H)$ *and S=S (H)*, the above equation exhibits the control-volume Finite-difference. Integrating the equation over the control volumes as [14]:

$$\begin{aligned} \iiint_{\Delta V} \rho\frac{\partial \mathrm{H}}{\partial \mathrm{t}}\Delta V = & \iiint_{\Delta V} \frac{\partial}{\partial x}\left(\frac{\partial(\lambda H)}{\partial x}\right)\Delta V + \iiint_{\Delta V} \frac{\partial}{\partial y}\left(\frac{\partial(\lambda H)}{\partial y}\right)\Delta V + \iiint_{\Delta V} \frac{\partial}{\partial z}\left(\frac{\partial(\lambda H)}{\partial z}\right)\Delta V \\ & + \iiint_{\Delta V} \frac{\partial}{\partial x}\left(\frac{\partial S}{\partial x}\right)\Delta V + \iiint_{\Delta V} \frac{\partial}{\partial y}\left(\frac{\partial S}{\partial y}\right)\Delta V + \iiint_{\Delta V} \frac{\partial}{\partial z}\left(\frac{\partial S}{\partial z}\right)\Delta V + \iiint_{\Delta V} q\Delta V \end{aligned} \tag{9}$$

By means of an explicit scheme, the time variation term becomes [11],

$$\iiint_{\Delta V} \rho\frac{\partial H}{\partial t}\Delta V = \rho\Delta V\left(\frac{H_p - H_p^{\circ}}{\Delta t}\right) \tag{10}$$

Now, let the coefficient of H_N, H_S, H_E, H_W, H_T, *and* H_B be a_N, a_S, a_E, a_W, a_T and a_B, respectively. Writing equation (10) in the familiar standard form [14]:

$$a_p H_p = a_N H_N + a_S H_S + a_E H_E + a_W H_W + a_T H_T + a_B H_B + b \tag{11}$$

With, H_p^o representing the old value of H at P (grid point), the coefficients' values are [14]:

$$a_p = a_N + a_S + a_E + a_W + a_T + a_B$$

$$a_E = \frac{\Delta t}{\rho\Delta V}\frac{\lambda_E A_x}{\delta x_e}, a_W = \frac{\Delta t}{\rho\Delta V}\frac{\lambda_W A_x}{\delta x_w}, a_N = \frac{\Delta t}{\rho\Delta V}\frac{\lambda_N A_y}{\delta y_n}, a_S = \frac{\Delta t}{\rho\Delta V}\frac{\lambda_S A_y}{\delta y_s}, \ a_T = \frac{\Delta t}{\rho\Delta V}\frac{\lambda_T A_z}{\delta z_t}, a_B = \frac{\Delta t}{\rho\Delta V}\frac{\lambda_B A_z}{\delta z_b} \tag{12}$$

$$\begin{aligned} b = & -[a_N + a_S + a_E + a_W + a_T + a_B - 1]H_p^o + b_N S_N + b_S S_S + b_E S_E + b_W S_W + b_T S_T \\ & + b_B S_B - b_p S_P + \frac{dt}{\rho}q\Delta \end{aligned} \tag{13}$$

Where, $b_p = b_E + b_W + b_S + b_N + b_T + b_B$

$$b_E = \frac{\Delta t}{\rho \Delta V}\frac{A_x}{\delta x_e}, b_W = \frac{\Delta t}{\rho \Delta V}\frac{A_x}{\delta x_w}, b_N = \frac{\Delta t}{\rho \Delta V}\frac{A_y}{\delta y_n}, b_S = \frac{\Delta t}{\rho \Delta V}\frac{A_y}{\delta y_s}, \quad (14)$$

The last procedure is similar to liquid and solid regions.

3. Validation of the numerical model

A room with the dimensions of 1m*1m*1m was manufactured. The walls and ceiling were made of PVC sandwich panel (5 cm thickness) containing a window in the south side of it with the dimensions of 50*50 cm, in addition, the ground was made of wood with a thickness of 2 cm. the thickness of the PCM filling the glazed unit was 1.7 cm. The properties of the wax are given in table 1. The solar radiation of the sun was simulated by using 4 halogen lamps which placed 70 cm away from the window. Figure 1 illustrates the specifics details of the double glazed window doped (filled) with PCM, which showcase the incident solar radiation and the portion of it that was transmitted to the room. In addition, heat transfer processes (thermal convection and radiation) occur on the boundary of the interior and exterior surfaces, respectively.
The numerical model is validated by means of this experimental room, under the environmental condition of Baghdad (extremely hot summer) in 25/6/2018. Figure 2 shows the comparison between the experimental and numerical results regarding the interior surface temperature of the double glazed window (filled with PCM) and the indoor temperature of the room, respectively. Which displays a quite good agreement between them.

4. Results and discussion

The constructed FORTRAN computer program solved the energy equation with phase change in the layer of paraffin wax filled the double glazed window. At x = 0, the solar radiation of Baghdad weather was applied (monthly average radiation was taken for each month). As shown in figure 3 (a).
In order to evaluate the performance of the double glazed window (including the wax), two criteria were selected, temperature decrement factor (d_{pcm}) and internal surface temperature of the double glazed window. Where the performance of the glazed unit is accepted if the temperature decrement factor is low and internal surface temperature is within the accepted range (i.e. above the melting temperature of the wax by 4 to 5 °C or below it).

$$d_{pcm} = \frac{T_{pcm,max} - T_{pcm,min}}{T_{a,max} - T_{a,min}} \quad (15)$$

Where ($T_{pcm,max}$) and ($T_{pcm,min}$) are the maximum and minimum internal surface temperature of the double glazed unit respectively, on the other hand, ($T_{a,max}$) and ($T_{a,min}$) are the maximum and minimum ambient temperature [10]. Figure 3 (b) illustrate the ambient temperature levels for Baghdad in 2018. The solution was applied for summer months starting from May to October 2018. The results were plotted as follows: Figure 4 (a) for May, figure 4 (b) for June, figure 4 (c) for July, figure 4 (d) for August, figure 4 (e) for September and figure 4 (f) for October, respectively.
Figure 4 (a) shows the internal surface temperature of the unit with different PCM's thicknesses in May. The tested thicknesses were 2.5 cm, 2 cm, and 1.5 cm. It can be seen that the 1.5 cm thickness could not keep the internal surface temperature of the window within the accepted range, however, thicknesses 2 and 2.5 cm would successfully maintain the temperature within the accepted range. Moreover, the temperature decrement factor was 0.73, 1.15 and 1.53, respectively. Figure 4 (b) shows the internal surface temperature of the unit in June. The required thickness to keep the temperature within the accepted range was 3 cm, in which temperature decrement factor was 1. Furthermore, figure 4 (c) shows the internal surface temperature of the unit in July. The required thickness was between 3.5 cm and 4 cm, and the temperature decrement factor was 0.95 and 0.76 respectively. Also, for August the required thickness was between 3 and 3.5 cm (temperature decrement factor was 1 and 0.74, respectively), as shown in figure 4 (d). Also, for September the required thickness was 3 cm, in

which temperature decrement factor was 0.92, as shown in figure 4 (e). However, figure 4 (f) illustrates October data, in which the required thickness was between 1.5 and 2 cm to achieve the accepted temperature and temperature decrement factor (i.e. 1.2 and 0.6, respectively).
Consequently, figure 5 illustrate the variation of the required thickness of the wax for each month (summer period). Which demonstrate the required thickness of the PCM to ensure reliable performance from the window. And it's start from 2 cm (in May) and reaches the maximum in July (3.5 cm) and returns back to 1.5 cm in October. Additionally, the recommended thickness for the whole summer period of Baghdad is 3 cm.

5. Conclusions

1. For the summer weather of Baghdad, the required thickness of the double glazed window filled with paraffin wax (melting point 40 °C) was as follows: May 2 cm, June 3 cm, July 3.5 cm, August 3.5 cm, September 3 cm, October 1.5 cm.
2. For the whole summer weather of Baghdad the required thickness is 3 cm.

References

[1] Socaciu L, Pleşa A, Ungureşan P and Giurgiu O 2014 Review on phase change materials for building applications. *Leonardo Electronic Journal of Practices and Technologies* **25** 179-194.

[2] Hasan I, Hadi O, and Ahmed O 2018 Experimental investigation of phase change materials for insulation of residential buildings *Sustainable cities and society* **36** 42-58.

[3] Xie T, He L and Tong X 2016 Analysis of insulation performance of multilayer thermal insulation doped with phase change material. *International Journal of Heat and Mass Transfer* 102 934-943.

[4] Goia F, Perino M and Serra V 2013 Improving thermal comfort conditions by means of PCM glazing systems. *Energy and Buildings* **60** 442-452.

[5] Ismail K. and J. Henrıquez 2002 Parametric study on composite and PCM glass systems. *Energy conversion and management* **43.7** 973-993.

[6] Gowreesunker L, Stankovic B, Tassou A and Kyriacou A 2013 Experimental and numerical investigations of the optical and thermal aspects of a PCM-glazed unit. *Energy and Buildings* **61** 239-249.

[7] Liu C, Zhou Y, Li D, Meng F, Zheng Y and Liu X 2016 Numerical analysis on thermal performance of a PCM-filled double glazing roof. *Energy and Buildings* **125** 267-275.

[8] Li D, Li Z, Zheng Y, Liu C, Hussein K and Liu X 2016 Thermal performance of a PCM-filled double-glazing unit with different thermophysical parameters of PCM. *Solar Energy* **133** 207-220.

[9] Li S, Sun G, Zou, K and Zhang X 2016 Experimental research on the dynamic thermal performance of a novel triple-pane building window filled with PCM. *Sustainable Cities and Society* **27** 15-22.

[10] Liu C, Wu Y, Bian J, Li D and Liu X 2018 Influence of PCM design parameters on thermal and optical performance of multi-layer glazed roof. *Applied energy* **212** 151-161.

[11] Cao Y, Amir F and Won C 1989 A numerical analysis of Stefan problems for generalized multi-dimensional phase-change structures using the enthalpy transforming model. *International journal of heat and mass transfer* **32.7** 1289-1298.

[12] Norton T, Delgado A, Hogan E, Grace P and Sun W 2009 Simulation of high pressure freezing processes by enthalpy method. *Journal of food engineering* **91** 260-268.

[13] Cho H and Sunderland E 1969 Heat-conduction problems with melting or freezing. *Journal of Heat Transfer 91*(**3**) 421-426.

[14] Abdulmunem R and Jalal M 2018 Indoor investigation and numerical analysis of PV cells temperature regulation using coupled PCM/Fins. *International Journal of Heat and Technology* **36** 121-1222.

[15] Dinçer İ, Midilli A and Kucuk H 2014 *Progress in Sustainable Energy Technologies*. Creating Sustainable Development Vol. 2 (Springer).

Table 1. Thermal properties of materials

Material	Melting temp. (°C)	Density (Kg.m^{-3})	Thermal cond. (W.m^{-1}.K^{-1})	Specific heat (J.kg^{-1}.K^{-1})	Latent heat (KJ.kg^{-1})
Glass	--	2500	0.96	840	--
Paraffin wax [15]	40	880	0.2	1800	174.12

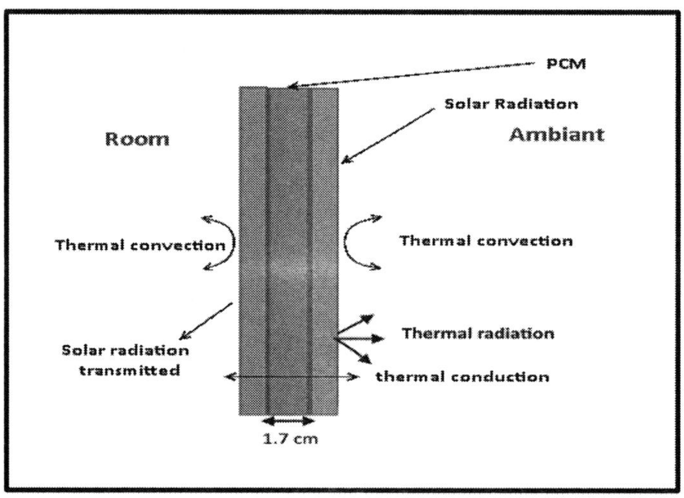

Figure 1. Double glazed window filled with paraffin wax

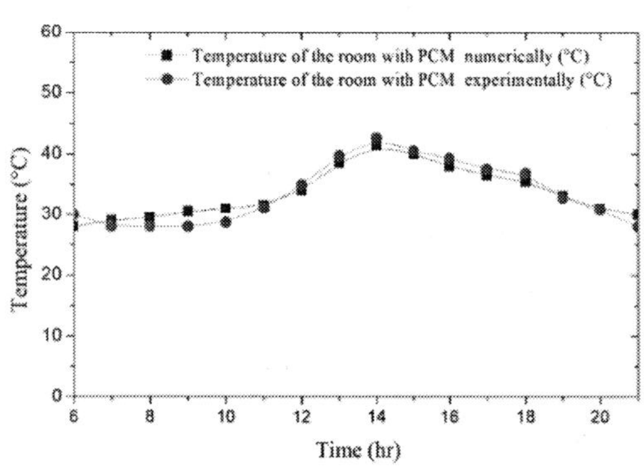

(a)

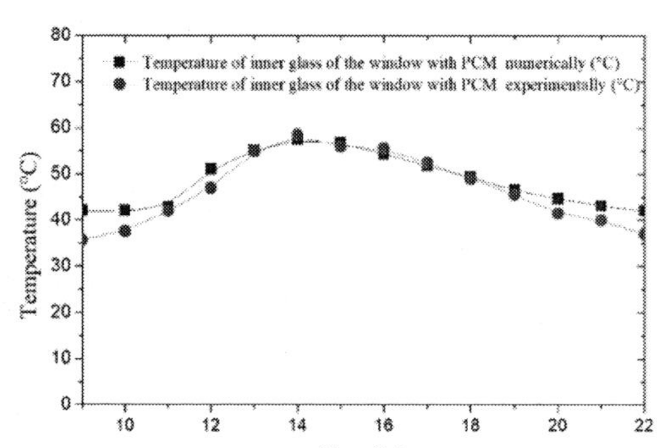

(b)

Figure 2. Comparison between experimental and numerical results (a) Room temperature (b) Internal surface temperature of the double glazed window with 1.7 cm wax thickness

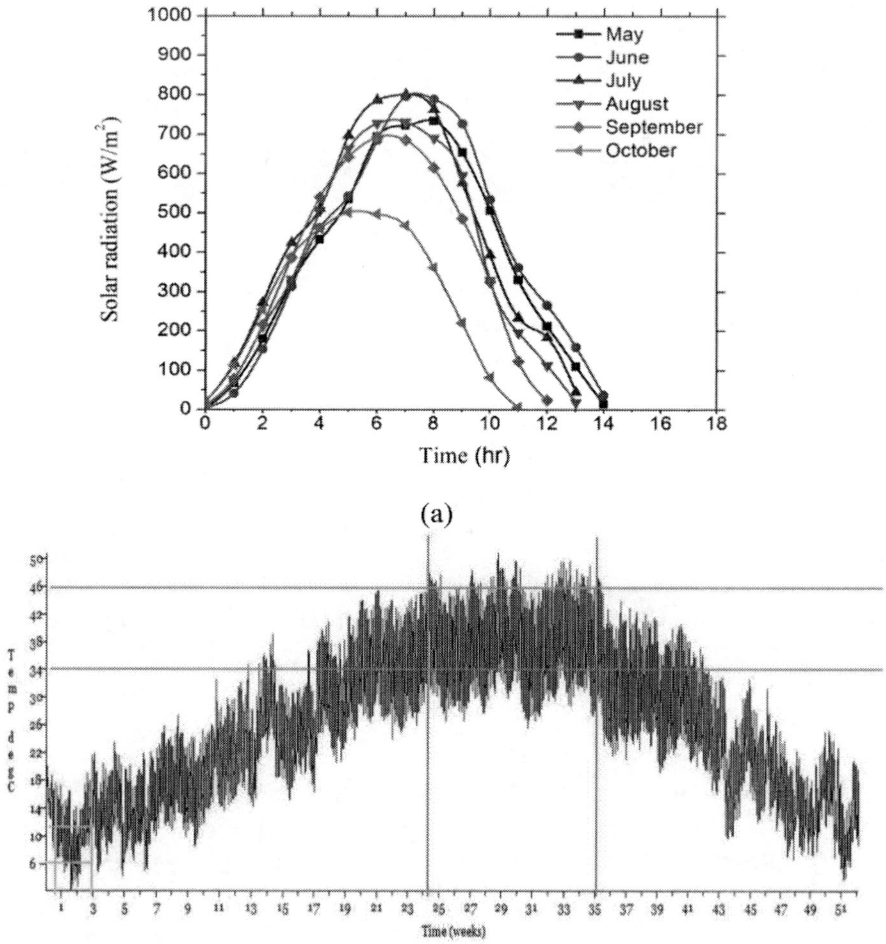

(b)

Figure 3. (a) The monthly average solar radiation in the summer of 2018 (b) Temperature levels for Baghdad in 2018.

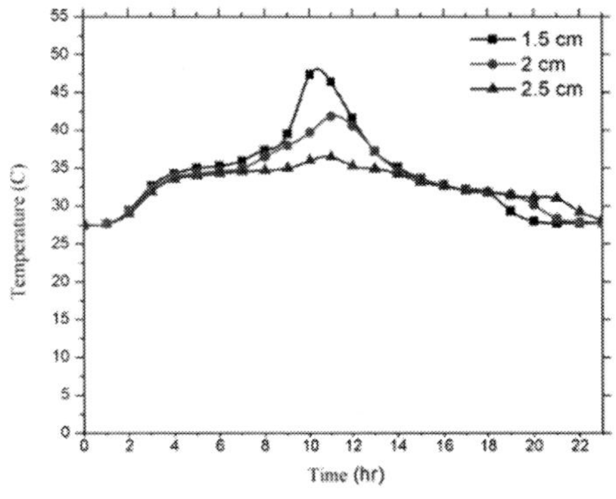

(a)

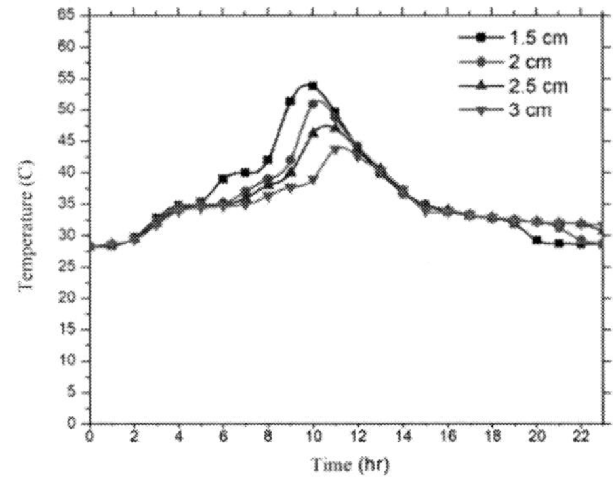

(b)

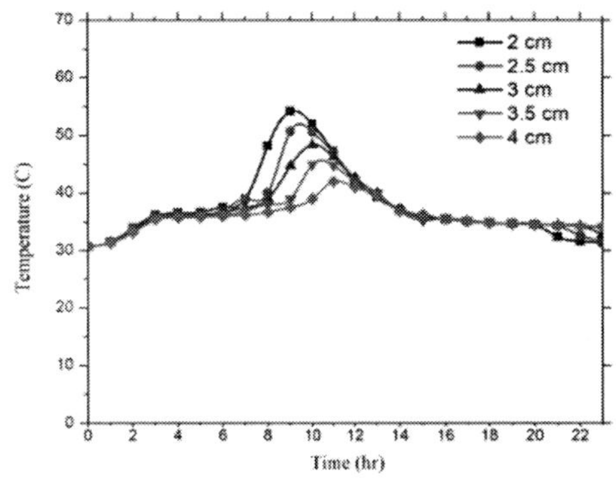

(c)

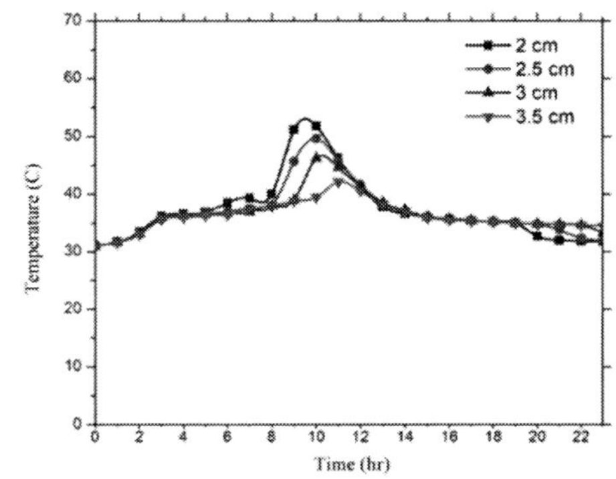

(d)

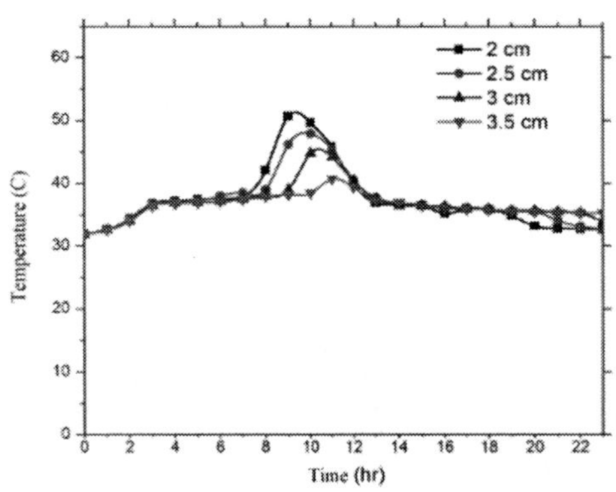

(e)

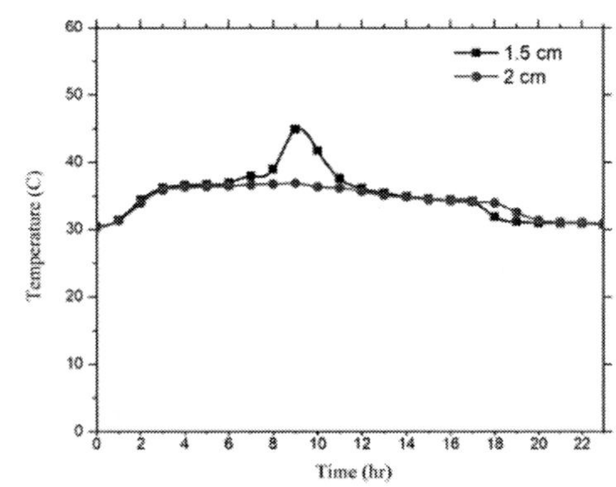

(f)

Figure 4. Temperature of the inner surface of the double glazed window with different PCM thicknesses at (a) May (b) June (c) July (d) August (e) September (f) October

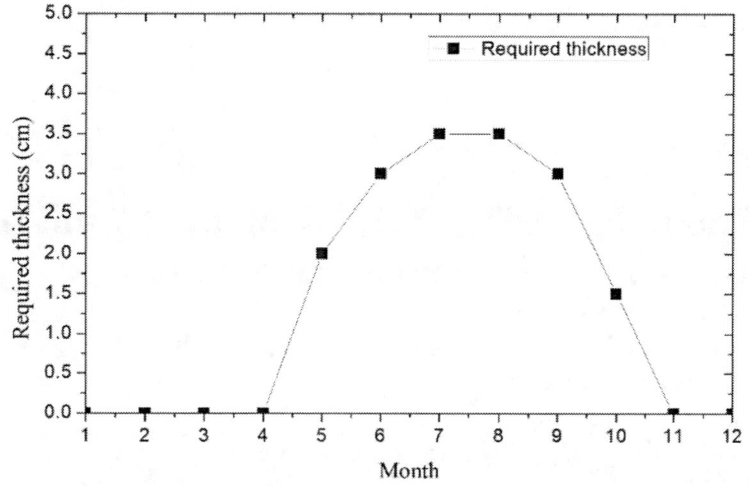

Figure 5. The variation of the required thickness of the paraffin wax filled double glazed window for the summer weather of Baghdad

Optimized manufacturing of the small dual wind turbine used to generate electricity in central Iraq areas

Muhannad Z. Khalefa

Shereen F. Abd Alkarim

Rasha S. Salih

Electromechanical Engineering Department, University of Technology, Baghdad, Iraq

corresponding author's e-mail : 50119@uotechnology.edu.iq

Abstract. Start your Due to the importance of generating electricity from wind turbines, single rotor wind turbine and dual rotor wind turbine. There was a need to design and manufacture the turbine system operating in both types mentioned above and have advantages to increase the generation of electric power. The aim of this work was to generate electrical power at a wind speed of up to 1m/s and more to achieve this wind speed and direction of the wind were adopted from tables of the Zerbatiya Station of the Wasit Governorate (Geographically located in central Iraq). The simulation was used to generate rotational speed for front and rear blades from 1m/sec to 13m/sec. A new idea was used to induce the dual small wind turbine to generate electrical power at low speed (1m/ sec). In this case, there is latent energy in the surrounding wind turbine, but it is not that helps to rotate the wind turbine. To apply this idea the starter motor is used. Passive magnetic bearing supports have been used to reduce friction of the rotors this will reduce losing of wind energy. Brakes are the mechanical devices which aimed to slow or shut down the machine. They are also designed to prevent moving after stopping. Finally, the Genetic Algorithm (GA) was used within the program for the purpose of obtaining the best amount of power generated. All the above were for the purpose of determining the best position for blades at different speeds to generate electricity by using Permanent Magnetic Synchronous Generator (400W).Results of this paper: (a)-The starter motor proved good performance when using it. (b)- Using passive magnetic bearing proved effect on increase the power generation. (c)- Mechanical brakes work to stop the system within a period of time and not immediately as this will negatively affect the performance of the system.

Keywords: Dual wind turbine, passive magnetic bearing, the starter motor, passive magnetic bearing, mechanical brakes.

1. Introduction

The single-rotor wind turbines (SRWT) suffer high aerodynamic loss deficiency in the blade root region (approximately bottom 25%). This particular region of the blade is primarily designed for maintaining the structural integrity of the blade. It causes high aerodynamic losses and thus reduces the overall efficiency of a turbine, [1]. A modern design of dual rotor wind turbines (DRWT) is suggested to mitigate the losses by introducing another rotor. The (DRWT) design is able to harness more wind energy by minimizing root loss in the vicinity of the roots of the main rotor blades for the same airflow, it also causes much larger dynamic wind load operating on the dual rotor wind turbine model, and higher velocity errors following the (DRWT) model, compared to the (SRWT) case, [2, 3]. Wind turbines have to be able to stop in case of failure of critical components or if wind speed goes over a critical limit. If the wind turbine does not brake, the rotation of the wind turbine may lead to loss in the whole structure, [4, 5]. Studied an experimental to investigated the aeromechanics and the properties of double rotor wind turbines (DRWTs) in either co-rotating or counter-rotating configuration, which compared with conventional single-rotor wind turbine (SRWT) model illustrating basic physics to optimize the wind turbine design in order to increase energy productivity and best durability [6]. Showed that the strongest correlation was between power and speed of the air stream. While incidence angles of the two rotors also affected the turbine's power, no such effect was observed for changes in the distance between the rotors. Field tests confirmed the findings and observations made in the wind tunnel [7] . they found that the stiffness is improved by stacking the magnet in the special combination of different radial cum axial direction of magnetization. The best configuration based on this analysis is fabricated and tested in the vertical axis wind turbine [8]. Showed the efficiency of the wind turbines or the energy output can be increased by reducing the cut-in-speed and/or the rated-speed by modifying and redesigning the blades. The problem is tackled by identifying the optimization parameters such as annual energy yield, power coefficient, energy cost, blade mass, and blade design constraints such as physical, geometric, and aerodynamic. The present paper provides an overview of the commonly used models, techniques, tools and experimental approaches applied to increase the efficiency of the wind turbines [9]. Designed dual wind small turbine with used a device to reverse the rotor rotation to study multiple cases to achieve the best results and performance of the separation mechanism [10]. This paper shows the manufactured small wind turbine at different cases, single rotor wind turbine and dual rotor wind turbine. The most important advantage of using a starter motor is to generate an internal torque for the system and this torque is concentrated in the middle of rotor wind turbine. Passive magnetic bearing used to reduce friction to take advantage of wind energy without loss. A mechanical brakes performance for the front and rear rotors was tested at different operating cases as well as at different speeds.

2. Analysis of power produced from dual wind turbine

According to Betz theory, the maximum wind power is nearly 16/27 of the obtainable energy in the wind while less wind axial speed by two-thirds through a singular rotor disc. Whatever, the process of wind turbines is transforming lower than 40% of wind energy to electrical energy. Thus, approximately 60% of probable wind energy evades without being used. In fact, the energy in the wake of one rotary wake is large. Some of this energy might be extracted by constituting the other rotor in the wake. Since rotation behind first rotor rotates in reverse direction to the rotating direction of rotor. The output power of the turbine rotor and the wind kinetic energy per unit time are given as follows, [11].

$$P_T = T_m \cdot \omega \tag{1}$$

$$P_W = \frac{1}{2} \rho \cdot V_0^3 \cdot A \tag{2}$$

This limit is known as the Betz limit. So, the maximum power which the turbine can produce is expressed as follows.

IOP Conf. Series: Materials Science and Engineering **765** (2020) 012023 doi:10.1088/1757-899X/765/1/012023

$$P_W = \left(\frac{16}{27}\right)\left(\frac{1}{2}\right)\rho \, . \, V_0^3 \, . \, A \tag{3}$$

The rotor power coefficient C_p is defined as the ratio between the output power of the rotor and the dynamic power of the air as shown below.

$$C_P = \frac{Actual \;\; Electrical \;\; Power \;\; \text{Pr}oduced}{Wind \;\; Power \;\; \text{int}\,o \;\; Turbine}$$

$$C_P = \frac{P_T}{P_W} = \frac{T_m \, . \, \omega}{\left(\frac{1}{2}\right)\rho \, . \, V_0^3 \, . \, A} \tag{4}$$

The power coefficient is a nonlinear function of the tip speed-ratio λ, which depends on the rotation speed of the shaft and wind velocity, [12].

$$\lambda = \frac{V_{Tip}}{V_0} = \frac{R.\omega}{V_0} \tag{5}$$

3. Experimental work

In this section, explained the practical aspect of manufacturing a wind power system in two cases, single and dual wind turbine blades, which include different secondary operations states. The experimental work can be divided into three main sections, namely the mechanical components section, the electrical components section and the last section is the control of the whole system and the extraction of all the values to be desired.

3.1. Mechanical Components

The mechanical components can be divided into three parts, the first part is front part to be placed in the direction of the expected wind, the second part is the rear part and third part is gearbox and three-phase permanent magnet synchronous generator, as shown in Figure 1. The front and rear parts is divided into, movable wing and non-movable wing. All specifications and dimensions of the components will be handled according to their location as follows.

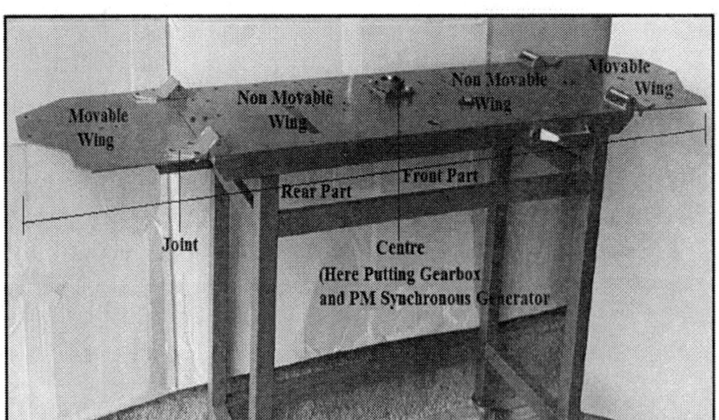

Figure 1: Simplified layout of the dual wind turbine

3.1.1 The Front Part

(a)-Movable Wing

The movable wing is important because it includes a new idea to be accomplished in this work, it contains:
i-Set of blades (three blades) with each Model Number:NE-M3, The total diameter is about (165cm) that made from (PVC) and variable thickness as shown in Figure 2-a and hub that the blades carry be with that dimensions an internal diameter of (16 mm) as shown in Figure 2-b. Thickness of blade tip 3mm and other side 4mm, thickness in the chord 4mm and other side 8mm and final in the root of blade 14.5mm.

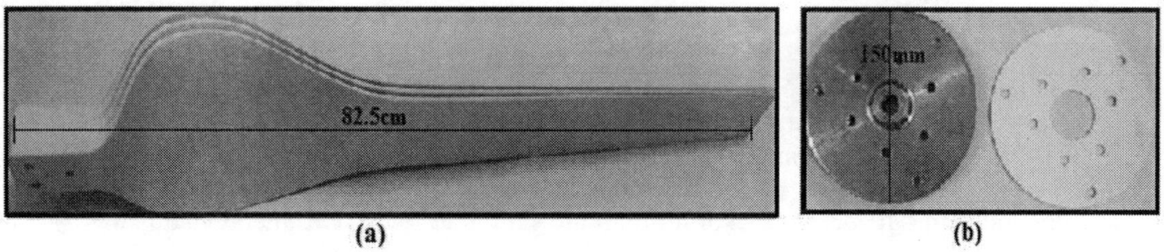

Figure 2: Wind turbine blades used in this work
(a)-Front View of Blades (b)- The Hub of Blades

ii- The movable wing base that dimensions (32cm x 40cm x 0.4cm).
iii- Figure (3) explains all that will be mentioned in (iii) to (v). Two ball bearings, that internal diameter (25mm), model P205, W/ Grease Zerk. Pillow, with manufactured bases of aluminum with dimensions (128 x 40 x 29 mm).
iv- The shaft length 575 mm, consist of three parts with diameters (16mm, 25 mm and 28 mm), made of mild steel material.
v- The movable wing shall be fixed with the non-movable wing of the front and rear parts by the help of two joints, it characterized by high durability and high tolerance to lift the moving part with full flexibility.

Figure 3: Movable wing.
(1)-Blade, (2) - Hub of blade, (3)-Bush, (4)-Ball bearing internal diameter (25mm), (5)-Shaft, (6)-Wing base, (7)-Arm for lifting, (8)-Joint, (9)-Hollow shaft with slots, (10)-Cardan joint, (11)- Ball bearing internal diameter (20mm), (12)- Simulation Device.

IOP Conf. Series: Materials Science and Engineering **765** (2020) 012023 doi:10.1088/1757-899X/765/1/012023

(b)-Non Movable Wing

Non movable wing with dimensions 52cm x 40cm x 0.4cm, as shown in Figure (4), it contains the following components:

i- Two types of ball bearings that internal diameter (20mm). The first type of ball bearing will be supported the horizontal front and rear shafts and the second type of ball bearing for supported the shaft of gearbox.
ii- Shaft with length 410 mm and diameter 20 mm is made of low carbon steel.

3.1.2 The Rear Part

(a)- Movable Wing

Contains the following components:
i- A sets of blades a symmetric with the front part.
ii- Hub that the blades carry which Featured they can be tightened in two cases, which match the angles between the front and rear parts, as well as when the deviation between blades of the front and rear. All components of this part are similar to which was mentioned in the front part.

(b)- Non Movable Wing

The end of the shaft of the front set and the shaft of the rear set involves gear from type (helical) number of teeth (15) that inner diameter (16mm) as shown in Figure 3. Figure 4 shows that these two gears they give motion to horizontal gear number of teeth (33) that inner diameter (20mm) and material (low carbon steel) which is connected to a vertical shaft with total length 225mm this shaft contain variable diameters as follows from beginning to end of length (12mm, 13mm, 16mm, 20mm, 23.5mm, 24mm,12mm).

Figure 4: Nacelle of this work during manufacture.

The non-movable wing contains the following components :

i- Two ball bearing that internal diameter (20mm) with manufactured bases of aluminum with dimensions (128*40*29mm).
ii- The total splined shaft length 210 mm, consist four parts with diameters (16mm, 20 mm ,21mm and 16 mm), made of low carbon steel material,
iii- Gear that connected shaft for the rear part with the center gear of the system number of teeth (15), material (Steel) and the inner diameter (16mm).

IOP Conf. Series: Materials Science and Engineering **765** (2020) 012023 doi:10.1088/1757-899X/765/1/012023

3.1.3 The Mechanical Brake

System of the brake is type Antilock Brake System (ABS) as shown in Figure 5, is installed in the front part used in this work consisting of a mechanical part and the other electric part (motor), components of the mechanical part are:
a-Rotary disk manufacture from stain steel with thickness 2mm, inner diameter 20mm and outer diameter 160mm.
b- Caliper assembly and brake pads: The disc caliper is further classified as floating and fixed caliper. This types of brake use only a single piston to squeeze the brake pad against the rotary disk.
c- Beam and hollow rod No. (1) using for supported wire with caliper assembly and brake pads and beam and hollow rod No. (2) using for supported another component.
d-Motor and Limit switch will discuss it later.
e- Two steel wire with diameter 0.4 cm used to pull the arm for move the piston of brake.
Operating principles of the brake as following:

i- By feeding the computer with the information of the amount of wind speed specified previously installed in the program when the arrival of this speed will work brake system and when the speed of arrival without it will stop brake work and the rotor wind turbine works well.
ii- To force brake performance by using a steel wire, a limit switch is used to reach close. The limit switch is at the bottom and vice versa.

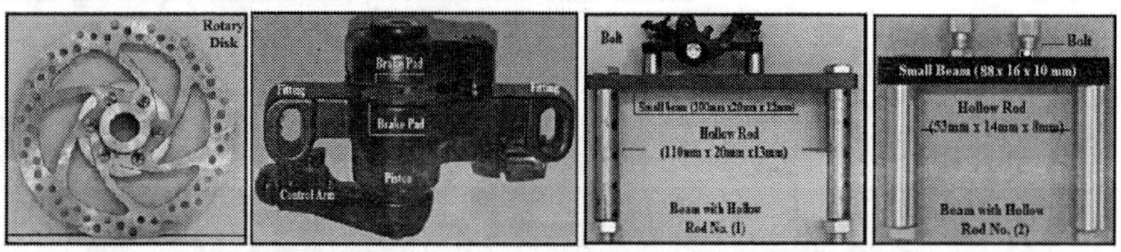

Figure 5: Components of the mechanical brake.

3.2 Electrical Components

There is a state of compatibility between the work of the electrical components in the operation of all the mechanical parts of the dual wind turbine system in this work, details will be mentioned as follows:

3.2.1 Starter Motor Used in Dual Wind Turbine for Internal Rotation

Current wind turbines have a high inertia so that they must be provided by a starter motor in order to get them running in low winds (wind speed is about V_{min}=1m/sec). The principle of the starter motor is the rotation of dual wind turbine by turning the ignition switch, the starter motor is energized throughout the magnetizing solenoid, causing a rod to be pushed out .The gear meets the flywheel and the starter turns. This increases the speed of the turbine. Starter Motor type New Starter 10455513 as shown in Figure 6 (a,b,c) and specifications are reported in the Table 1, [13].

Figure 6: Starter motor with their support.
(a)- Complete Structure of The Starter Motor, (b)- Starter Motor Without Coil and Solenoid (c)- Interlock The Starter Motor With Small Flywheel

IOP Conf. Series: Materials Science and Engineering **765** (2020) 012023 doi:10.1088/1757-899X/765/1/012023

Table 1 Specifications of the starter motor,[13].

Specifications		Value
Direct Drive	Teeth	10T
	Voltage	12V
	Rotation	CCW
Delco No.		10455513
Bosch		SR5044X
Brand		URQS
Item Weight		4.5 kg
Manufacturer Part Number		6744
No. of teeth		10

3.2.2 Passive Magnetic Bearing

The passive magnetic bearing was used in dual wind turbine in this work as the number one in each rotor instead of ball bearing to know the performance of the system would improve and thus the efficiency of electric generation would be more, type of magnetic bearing from Neodymium Magnets (These Neodymium Iron Boron magnets are also known as "Rare Earth" or NdFeB magnets), as shown in Figure 7. Magnetic bearings are manufactured of a typical passive magnetic bearing as following descriptions in Table 2, [14]:

Table 2 Data of the passive magnetic bearing, [14].

Descriptions	Value
Outer Diameter	5.08 cm
Inner Diameter	2.54 cm
Thick Ring	2.54 cm
Material	Nickel (Ni) Plated (silver-like shine).
Grade	N48 - Extremely powerful magnetic material.
Pull Force is approximately	24.25 N against a 2.54 cm steel plate.
These rings are axially magnetized, so the strong North/South poles are on the flat sides. That makes them easier to stack into a cylinder than to connect side by side.	-

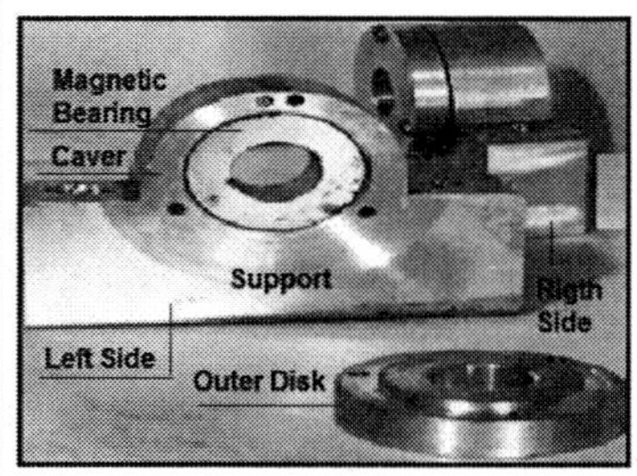

Figure 7 Two passive magnetic bearing with their support.

3.2.3 The Electrical Motor for The Mechanical Brake

An electric motor type ZGB37R is used for the purpose to complete the work of the mechanical brakes for two rotors of dual wind turbines used in this work, as shown in Figure 8 and specifications are reported in the Table 3.

Table 3 Specifications of the motor ZGB37R.

Specifications	Value
Motor diameter	32.8 mm
Gear box diameter	36.8 mm
Total height	51.5 mm (without bearing)
Output shaft	6 mm (flat position is 5.35 mm)
Output shaft length	21.5 mm (measured from the panel)
Weight	159 g
Voltage	12 V
Speed	6.5 rpm (motor speed 3000 rpm)
Current	0.02 A
Voltage	24 V
Speed	13 RPM (motor speed 6100 rpm)
Current	0.03 A
Reduction ratio	approximately 1:460
Torque	15 N.m

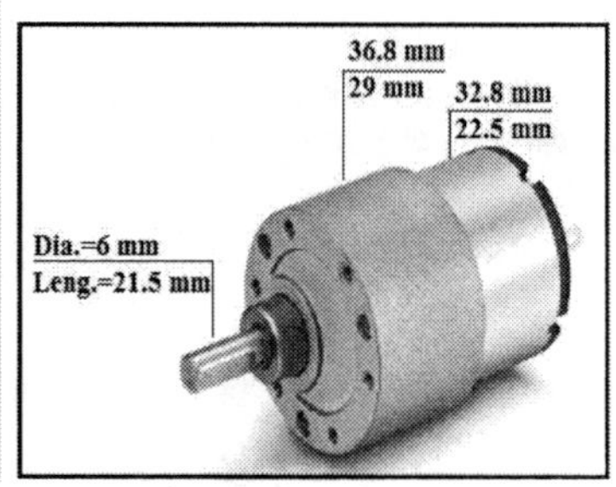

Figure 8 The used motor for the mechanical brake.

Figure 9 shows the manufactured system, with the parts mentioned above with the base.

Figure 9 The installation of mechanical parts.

4. Results and discussion

4.1 Production of Wind Speed by Simulation Motor

The operation of the manufactured system in all studied cases was done by simulations by using the stepper motor mentioned in chapter four. Stepper motor rotates the front rotor and so the whole system will rotate in a manner similar to that happens in the presence of natural wind. The main goal of the system as mentioned previously is to take advantage of wind energy when the wind speed between (1-2m/sec) and more. In order to ensure normal operation of the simulation, the speed of the motor was changed to suit the wind speed (1-13m /sec) as shown in Table 4.

Table 4: Effect of Simulation Motor Speeds on The Dual Wind Turbine With Radius of Blade is 0.825 m.

No.	Rotational Speed of Motor (r.p.m)	Wind Speed	
		m/sec	km/hr
1	11	0.950	3.421
2	13	1.123	4.043
3	16	1.382	4.976
4	18	1.555	5.598
5	20	1.727	6.220
6	23	1.987	7.153
7	25	2.159	7.775
8	27	2.332	8.397
9	29	2.505	9.019
10	32	2.764	9.952
11	34	2.937	10.574
12	36	3.110	11.196
13	39	3.369	12.129
14	41	3.542	12.751
15	48	4.146	14.928
16	52	4.492	16.172
17	59	5.097	18.350
18	71	6.133	22.082
19	82	7.084	25.503
20	93	8.034	28.924
21	105	9.071	32.656
22	116	10.021	36.078
23	128	11.058	39.810
24	139	12.008	43.231
25	151	13.045	46.963

4.2 Produce the power of the wind by theoretically analysis

The efficiency of the wind turbine is determined by the amount of energy obtained from the wind, which in turn depends mainly on several parameters: (i) Wind speed (ii) radius of wind turbine (iii) Other less important parameters. It was found the amount of energy obtained from the manufactured system by focusing on the wind speed (V_0) between (1-13m/sec) and the air density is (1.25kg/m^3). Note that the radius of the turbine is (0.825m) and from the application of Equations (2) and (3), find that the amount of the wind power generated is shown in Table 5.

Table 5: Theoretical power obtained from front rotor.

Wind Speed (V_0) (m/sec)	Rotated in The Same Direction	
	Wind Power (Watt) into Turbine (P_W) Front Rotor By Eq.(2)	Wind Power (Watt) into Turbine (P_W) Front Rotor By Eq.(3)
1	1.335	0.788
2	10.680	6.301
3	36.045	21.266
4	85.440	50.409
5	166.875	98.456
6	288.360	170.132
7	588.735	347.350
8	603.520	356.076
9	973.215	574.196
10	1335.000	787.650
11	1776.885	1048.362
12	2306.880	1361.059
13	2932.995	1730.467

4.3 Single Wind Turbine

When construct a wind turbine system, it should be placed in the direction of the expected wind and this is known from the meteorological department. Table 6 shows in detail the operation of the single-rotation wind turbine when the speed (simulation speed) ranges from less than 1m/sec to more than 13m/sec. Since the stepper motor which is used for simulation operate in the mode of pulses, this means that when the measured speed is 11rpm, the system will give the actual number of readings for the voltages and current as well as the generated electrical energy. A software program was developed to read the genetic algorithm to obtain the best actual generation values. The genetic algorithm was programmed by (visual studio 2015).

To illustrate the importance of the change in the values of the generated power in Table 6 includes the actual values and the Genetic Algorithm (GA) for the voltages, the currents and powers. Figure 10, which includes three curves, the maximum values and minimum values generated during the operation of the system, found that the generated power when calculating by the Genetic Algorithm (GA) values is limited between these values.

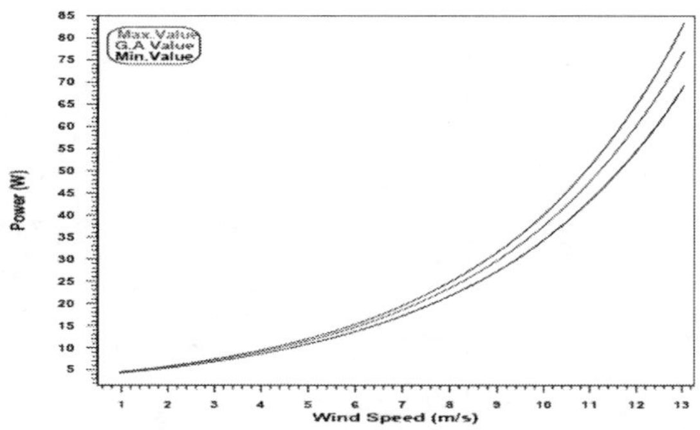

Figure 10: Electric power generated at different wind speeds

Table 6: All results from operation single wind turbine with non tilt angle.

Simulation		Actual Values			Genetic Algorithm (GA)		
Rotational Speed of Front Rotor	Wind Speed (m/sec)	Voltages (Volt)	Current (A)	Power (W)	Voltages (Volt)	Current (A)	Power (W)
11	0.950	12.41	0.224	2.784	12.51	0.222	2.778
11	0.950	12.25	0.215	2.634			
13	1.123	12.41	0.276	3.428	12.36	0.273	3.380
13	1.123	12.2	0.262	3.201			
16	1.382	12.25	0.346	4.248	12.25	0.333	4.081
16	1.382	12.05	0.289	3.482			
18	1.555	12.2	0.355	4.339	12.20	0.336	4.106
18	1.555	11.9	0.332	3.960			
20	1.727	12.2	0.403	4.925	12.20	0.403	4.925
20	1.727	11.85	0.391	4.633			
23	1.987	12.15	0.488	5.938	12.15	0.4545	5.522
23	1.987	12.25	0.444	5.448			
25	2.159	12.2	0.510	6.222	12.10	0.499	6.042
25	2.159	12	0.494	5.928			
27	2.332	12.2	0.568	6.929	12.15	0.546	6.645
27	2.332	12.1	0.550	6.664			
29	2.505	12.3	0.603	7.428	12.20	0.583	7.113
29	2.505	12	0.597	7.173			
32	2.764	12.2	0.680	8.296	12.15	0.666	8.096
32	2.764	12.15	0.652	7.931			
34	2.937	12.2	0.744	9.078	12.15	0.693	8.427
34	2.937	12.1	0.599	7.256			
36	3.110	12.36	0.719	8.888	12.15	0.719	8.737
36	3.110	11.95	0.711	8.501			
39	3.369	12.2	0.803	9.807	12.15	0.795	9.666
39	3.369	12.05	0.779	9.387			
41	3.542	12.3	0.801	9.859	12.15	0.810	9.844
41	3.542	11.85	0.757	8.982			

48	4.146	12.15	0.957	11.639	12.15	0.957	11.639
48	4.146	12.1	0.905	10.961			
52	4.492	12.15	1.038	12.620	12.15	1.047	12.726
52	4.492	11.95	1.005	12.016			
59	5.097	12.15	1.216	14.776	12.10	1.178	14.260
59	5.097	11.85	1.191	14.114			
71	6.133	12.15	1.508	18.331	12.10	1.448	17.525
71	6.133	11.95	1.388	16.587			
82	7.084	12.2	1.707	20.833	12.10	1.637	19.819
82	7.084	12.1	1.620	19.608			
93	8.034	12.15	1.916	23.291	12.15	1.9169	23.291
93	8.034	12.05	1.857	22.384			
105	9.071	12.15	2.253	27.380	12.10	2.075	25.108
105	9.071	12.10	2.052	24.838			
116	10.021	12.15	2.612	31.746	12.15	2.341	28.452
116	10.021	12.1	2.193	26.545			
128	11.058	12.36	3.862	47.739	12.56	3.536	44.412
128	11.058	12.3	2.992	36.801			
139	12.008	12.76	4.844	61.811	12.76	4.369	55.753
139	12.008	12.71	3.898	49.555			
151	13.045	13.06	6.642	86.745	13.01	6.289	81.821
151	13.045	12.96	5.422	70.279			

4.4 Dual Wind Turbine

(i)- The Rotation of The Front and Rear Rotors in The Same Direction

Table 7 show all the results obtained from the operation of the system when they rotate in one direction. Found the best value of the power generated at the rotation speed is about 151 rpm and these all values extracted are from the genetic algorithm.

Table 7: Results from The Dual Wind Turbine Rotation in The Same Direction.

Simulation		Genetic Algorithm (GA)		
Rotational Speed of Front Rotor (r.p.m)	Wind Speed (m/sec)	Voltages (Volt)	Current (A)	Power (W)
11	0.950	12.25	0.236	2.892
13	1.123	12.2	0.265	3.235
16	1.382	12.25	0.316	3.873
18	1.555	12.41	0.355	4.414
20	1.727	12.25	0.395	4.841
23	1.987	12.36	0.454	5.618
25	2.159	12.25	0.494	6.052
27	2.332	12.25	0.533	6.536
29	2.505	12.20	0.573	6.991

32	2.764	12.36	0.625	7.732
34	2.937	12.30	0.664	8.175
36	3.110	12.20	0.711	8.679
39	3.369	12.20	0.770	9.402
41	3.542	12.30	0.801	9.859
48	4.146	12.41	0.938	11.645
52	4.492	12.36	1.016	12.565
59	5.097	12.25	1.153	14.129
71	6.133	12.2	1.403	17.118
82	7.084	12.25	1.603	19.637
93	8.034	12.36	1.837	22.716
105	9.071	12.2	2.119	25.860
116	10.021	12.2	2.760	33.681
128	11.058	12.46	3.944	49.142
139	12.008	12.76	5.198	66.334
151	13.045	13.11	6.674	87.498

(ii)- The reverse rotation direction of the front and rear rotors

Table 8 shows the power generated by the system if the rotation of the front rotor reverses the rotation of the rear rotor. The following is found from the operation of the system in this case:

A - The work of the system is more stable, especially when the motor simulation used in this work is rotating at high speed when the position of the front blades match the position of the rear blades. But when the opposite position (non match) will generate mechanical vibration in the system and this is not desirable.

B - The results listed in the following table taken in the case of mismatching in the position of the front and rear blades as well as their reverse rotation. It was found that the generated power is lower than the previous case due to the resistance of the rear blades rotation.

C. In all previous cases, the rotation of the gearbox rotor is only in one direction for the purpose of increasing the torque.

Figure 11 shows a comparison between the results obtained from the two cases according to the direction of rotation and the positions of the front and rear blades. It was found that one-way rotation is better for generating power.

Table 8 Power From Dual Wind Turbine in Reverse Rotation Case.

Simulation		Genetic Algorithm (GA)		
Rotational Speed of Front Rotor (r.p.m)	Wind Speed (m/sec)	Voltages (Volt)	Current (A)	Power (W)
11	0.950	12.2	0.215	2.623
13	1.123	12.25	0.265	3.248
16	1.382	12.41	0.319	3.966
18	1.555	12.25	0.351	4.310
20	1.727	12.36	0.386	4.780
23	1.987	12.36	0.449	5.557
25	2.159	12.25	0.494	6.052
27	2.332	12.30	0.527	6.492

IOP Conf. Series: Materials Science and Engineering **765** (2020) 012023 doi:10.1088/1757-899X/765/1/012023

29	2.505	12.25	0.573	7.020
32	2.764	12.25	0.632	7.746
34	2.937	12.20	0.679	8.285
36	3.110	12.30	0.703	8.656
39	3.369	12.20	0.770	9.402
41	3.542	12.25	0.827	10.139
48	4.146	12.36	0.877	10.842
52	4.492	12.15	1.016	12.351
59	5.097	12.36	1.065	13.171
71	6.133	12.20	1.403	17.118
82	7.084	12.20	1.620	19.770
93	8.034	12.41	1.561	19.374
105	9.071	12.25	1.874	22.959
116	10.021	12.15	2.292	27.853
128	11.058	12.30	3.264	40.147
139	12.008	12.66	4.991	63.196
151	13.045	13.01	6.257	81.404

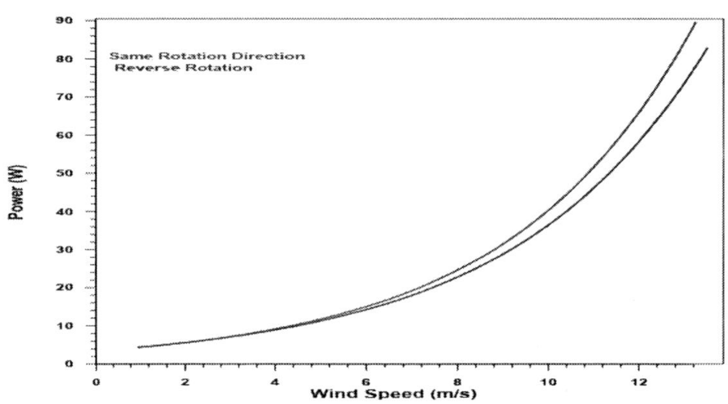

Figure 11: Power from dual wind turbine in different rotation directions.

4.5 The advantages of using starter motor

The most important advantage of using a starter motor is to generate an internal torque for the system and this torque is concentrated in the middle of dual rotor wind turbine. The following results were obtained: When the dual rotor wind turbine is stopped, starter motor is operated for 3 seconds and the front and rear blades will rotate at a speed of more than 25rpm. This corresponds to the previous Tables to a wind speed of 2m/sec.

It was found that the best cases of starter motor work, which give the desired goal, are in the following two cases:

i- single rotor wind turbine

ii- dual rotor wind turbine, When the front and rear blades rotate in one direction and these blades are in matched position.

4.6 The effect of magnetic bearing

The reasons for the use of this type of bearing, which are alternative to ball bearing, are that they have low friction and do not need lubrication this was previously mentioned. In this work, only two passive magnetic bearings were used, as follows:

i- They were placed on the front movable wing. When testing the single wind turbine it was found that the power generated was more than the power generated when ball bearings is used as shown in Figure 12, because the friction in movable wing is low, as shown in Figure 13.

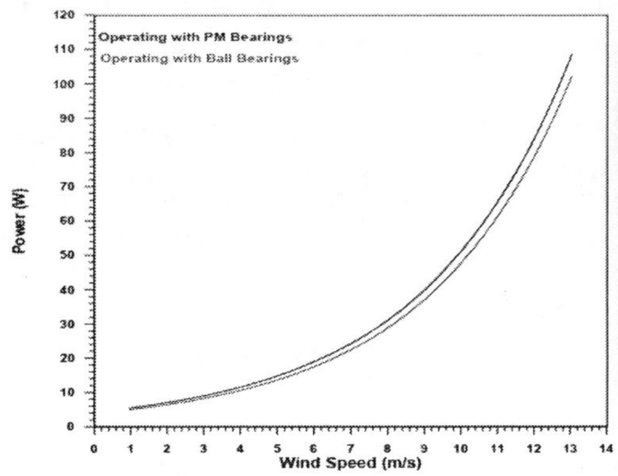

Figure 12: Comparing between operating ball bearings and PM bearings.

Figure 13: Passive magnetic bearings

5.Conclusions

From the results of this work the following conclusions can be obtained:

1- It was found that the operation of the system in the case of single rotor wind turbine will generate a little power at speed 1m/sec, but will multiply this power at speed 2m/sec. The value of GA is considered the best value, the best power value obtained at 151rpm is 81.821W.

2- Dual wind turbine operates in two cases: The best results of DWT rotation in one direction done with in position of the blades identical and inverse rotation when the blades position is not identical.

3- The starter motor proved good performance with external source of wind for the system.

4- Using one passive magnetic bearing in each side had a minimum effect on the power generation

5- Mechanical brakes work to stop the system within a period of time and not immediately as this will negatively affect the performance of the system.

References

[1]- Pardalos M. P., Steffen R., Mario V. F. P., Niko A.I and Vijay P 2013 Handbook of wind power systems Springer Berlin, Heidelberg.

[2]- Zhenyu W., Ahmet O., Wei T., Anupam S. and Hull H.2015 An experimental investigation on the wake characteristics behind a novel twin-rotor wind turbine 33rd Wind Energy Symposium, Kissimmee, Florida, Published by the American Institute of Aeronautics and Astronautics.

[3]- Xiang L. and Shuangxia N.2016 A novel contra-rotating power split transmission system for wind power generation and its dual mppt control strategy IEEE Transactions on Power Electronics, Vol.32, No.9.

[4]- Kenny L. S.2017 A numerical investigation of dual-rotor horizontal axis wind turbines using an in-house vortex filament code (DR HAWT) Thesis of Master's in Applied Science in Mechanical Engineering, Ottawa-Carleton Institute for Mechanical and Aerospace Engineering, Department of Mechanical and Aerospace Engineering, Carleton University, Ottawa, Ontario.

[5]- Margrét O. O.2014 A general description and comparison of horizontal axis wind turbines and vertical axis wind turbines Thesis of Master's in Mechanical Engineering, Faculty of Industrial Engineering, Mechanical Engineering and Computer Science, University of Iceland, Iceland.

[6]-Ahmet O., Wei T. and Hui H.2016 Experimental investigation on the wake characteristics and aeromechanics of dual-rotor wind turbines Journal of Engineering for Gas Turbines and Power (ASME), Vol.138, No.4.

[7]- Leszek Romański, Jerzy Bieniek, Piotr Komarnicki, Marcin Dębowski and Jerzy Detyna 2016 Operational tests of a dual-rotor mini wind turbine Maintenance and Reliability, Vol.18 No.2, pp 201–209.

[8]- Micha Premkumar T., Seralathan S. and Mohan Thangaraj 2017 Design and Analysis of a Permanent Magnetic Bearing for Vertical Axis Small Wind Turbine ", Energy Procedia, Vol. 117, pp 291–298

[9]- Shafiqur Rehman, Md. Mahbub Alam , Luai M. Alhems and Mujahid Rafique M. 2018 Horizontal axis wind turbine blade design methodologies for efficiency enhancement Energies, 11, 506.

[10]. Muhannad Z. Khalefa, Shereen F. Abd Alkarim and Rasha S. Salih 2109 effect of the tilt angles on the power generation of the small manufactured dual wind turbine equipped by separation mechanism and rotation reversing device Thi_Qar University Journal for Engineering Sciences , Vol.10, No.1.

[11]- Peter E. Jenkins, Abdalfadel Younis and Yuxuan Chen 2017 Design and analysis of a dual rotor turbine with a shroud using flow simulations Journal of Power and Energy Engineering 05(04), DOI: 10.4236/jpee.

[12]- Peter F. Odgaard, Chris Damgaard and Rasmus Nielsen 2008 On-Line Estimation of Wind Turbine Power Coefficients Using Unknown Input Observers Proceedings of the 17th World Congress The International Federation of Automatic Control Seoul, Korea, July 6-11.

[13]- Murugesan V. M., Chandramohan G, Senthil Kumar M., Rudramoorthy R., Ashok Kumar L., , Suresh Kumar R, Basha D. and Vishnu Murthy K. 2012 an overview of automobile starting system faults and fault diagnosis methods ARPN Journal of Engineering and Applied Sciences, Vol. 7, No. 7.

[14]- Aravind R. and Ashwin K. S, 2017 Magnetic levitation wind turbines, Imperial Journal of Interdisciplinary Research (IJIR) ,Vol.3, No. 3.

The hybrid Photovoltaic-Thermal double-pass solar system for drying applications

Abdullatif A.Jadallah*[1], Mohammed K.Alsaadi[2], Saifalddeen A.Hussien[3]

Electromechanical Engineering Department, University of Technology, Iraq.

Email: abdullateef.aljad@gmail.com
50055@uotechnology.edu.iq
Saifalddeen1994@gmail.com

Abstract. In this paper, is devoted to evaluating the performance of the double-pass hybrid Photovoltaic-Thermal (PVT) solar system proposed for drying purposes theoretically and experimentally as well as the system is designed, fabricate and modeled in order to simulate the productivity of this system. The hot air extracted from the PVT collector can be used as a heat source for the drying application. The critical parameter such as temperature distribution, useful heat gain, electrical power, and thermal efficiency are computed using MATLAB 2015b program built for this purpose. Results show that the higher output fluid temperature was 63°C at a lesser mass flow rate which 0.017 kg/s and the maximum efficiencies of electrical, thermal and overall at a higher mass flow rate which reached 12.65%, 56.73%, and 85% respectively at mass flow rate 0.031 kg/s. In addition, the optimum electrical power and thermal energy reached 50.57 W and 389.37 W at 0.031 kg/s.

Key Words; photovoltaic cell, thermal collector, performance, dryer, heat gain

Nomenclature

A_m	PV module area	m^2
A_p	absorber plate area	m^2
A_f	fin area	m^2
A_b	backplate area	m^2
A_{gd}	area of dryer glasses	m^2
A_c	cross-section of channel	m^2
a	ideality factor of diode	
c_{pa}	specific heat of the air	$kJ/kg.k$
F. F.	fill factor	
G_t	total solar radiation	W/m^2
$G_{t,r}$	total solar radiation at STC	W/m^2
h_r	radiative heat coefficient	$W/m^2\,k$
h_c	convective heat coefficient	$W/m^2\,k$
h_w	convective heat by wind	$W/m^2\,k$
H_f	high of fin	m
k_g	glass thermal conductivity	$W/m\,k$

k_p	plate thermal conductivity	$W/m\,k$
k_b	back thermal conductivity	$W/m\,k$
k_a	air thermal conductivity	$W/m\,k$
L_p	plate length	m
L_f	fin length	m
$\dot{m}_a$	air mass flow rate	kg/s
t_p	plate thickness	m
t_f	fin thickness	m
t_g	glass thickness	m
T_C	cell temperature	°C
$T_{C,r}$	cell temperature at STC	°C
T_p	absorber plate temperature	°C
T_b	backplate temperature	°C
T_f	fluid temperature	°C
T_d	dryer chamber temperature	°C
T_a	ambient temperature	°C
T_s	sky temperature	°C
t_{bd}	dryer insulated thickness	m
U_t	top losses heat coefficient	$W/m^2\,k$
U_b	back losses heat coefficient	$W/m^2\,k$
V_d	dryer volume	m^3
v_f	fluid velocity	m/s
Wp	plate width	m
W_b	backplate width	m
μ_w	air viscosity	
μ_a	viscosity of air	
ρ_a	density of air	
ε_c	cell emissivity	
ε_p	plate emissivity	
α_c	cell absorptivity	
α_p	plate absorptivity	
τ_g	glass transitivity	
β_c	packing factor	
STC	standard test condition	

1. Introduction

The modern (PVT) solar system technologies are a special design of solar collectors that consists of a flat plate thermal collector with a PV collector which attached to the top of the thermal collector. The PVT solar system is used for generating electrical and thermal energy at the same time. The useful heat output from this system can be used for drying applications as dryer and also cause an improvement in electrical efficiency by cooling the back PV cells by blowing air to extract heat from back PV cells with thermal collector and produce high air temperature output applied dryer chamber. The PVT solar dryer system is one of the clean and renewable technologies because of self-generated electricity without using traditional sources; hence, the dependence upon fossil fuel and global warming problems are reduced [1]. A lot of previous works of scholars have studied theoretically and experimentally of hybrid PVT solar system technologies for drying applications.

Yuting et al. reviewed various types of PVT systems, using different kinds of working fluids. They also presented the utilization and improvement of the PV/T systems in a different application [2]. **Ahmed et al.** developed and predicted an indirect type of solar dryer that operates at both forced and natural convection modes in the different weather conditions [3]. **Sumit et al**. designed and fabricated a hybrid PVT air collector connected with a dryer and analyzed a thermal model of this system. The performance of the dryer chamber under forced convection mode was better for crops than natural convection mode [4]. **Abhay et al**. developed an indirect solar dryer system for drying crops. It was found the indirect dryer is more efficient than an open sun dryer [5]. **Essalhi et al.** developed an

IOP Conf. Series: Materials Science and Engineering **765** (2020) 012024 doi:10.1088/1757-899X/765/1/012024

indirect solar dryer system for drying pear. The results showed the higher output temperature reached 57°C when solar radiation 900 $\mathrm{W/m^2}$ and the thermal efficiency of the drying room was 11.11% [6]. **Neha el al.** developed a thermal model for the proposed translucent photovoltaic thermal with a thermoelectric cooler collector. It was found that a lesser mass flow rate causes higher in temperature of outlet air [7]. **Ehsan et al.** analyzed experimentally a solar dryer with the storage of thermal energy and the hot output air is supplied the dryer chamber with 50°C [8]. **G.N.Satyender and Prashant** compared between the airflow patterns, single and double pass airflow and found that the double-pass is more efficient than a single pass because more heat extraction can be obtained from a double pass of counterflow [9]. **Jin-Hee et al.** designed a hybrid PVT air-based solar system and analyzed the electrical and thermal performance with experimental results. The output warm air from the PVT collector could be used as a heat supply in the buildings [10]. **Ahmad et al.** developed thermal modeling of dual-pass solar collector in two cases, fins and without fins. Results showed that in the case of fins, the solar system is more cost-effective compared to the solar system without fins [11].
In this work, we will focus on design, fabrication, and simulation of the hybrid PVT solar system for drying purposes. The PVT system is locally unused for drying; therefore, the whole PVT solar system proposed in this work is designed and utilized for drying applications. Besides, analysing the performance of hybrid PVT with the solar dryer has been made to check out from the match between the theoretical and experimental results. In addition, adding fins and double pass technologies are utilized to improve thermal performance. The general schematic shape of the system in this study is shown in figure (1).

2- Methodology

The hybrid PVT system with the solar dryer was modeled. The electrical characteristics of the PV module depend on a single diode have been carried out by using the MATLAB SIMULINK 2015b and the electrical model has presented to calculate the electrical power, electrical efficiency and fill factor. In addition, the energy balance equations were applied to the different sections of the PVT system components with solar dryer room to evaluate the temperature of PV cell, absorber plate, backplate, output fluid, and dryer chamber respectively and estimate the useful heat gain, thermal efficiency, and overall efficiency.

2.1 Theoretical Work

2.1.1 Electrical Model

The electrical model of the PV module can be accomplished by analyzing the equations built on the single diode equivalent circuits of the PV module which presented in [12]. At a constant cell temperature (25°C) and incident radiation (1000 $\mathrm{W/m^2}$), the mathematical expressions of the PV module are expressed as:

$$I = I_{pv} - I_D \tag{1}$$

Whereas:
I: load current.
I_{pv}: light current.
I_D: saturation current.
The load current in Eq. (1). can be described as:

$$I = \frac{G_t}{G_{t,r}}(I_{sc,r} + \alpha_I(T_C - T_{C,r})) - I_{s,r}\left(\frac{T_c}{T_{c,r}}\right)^3 e^{\left[\left(\frac{q\,E_g}{a\,K}\right)\left(\frac{1}{T_{C,r}}-\frac{1}{T_C}\right)\right]}\left[e^{\left(\frac{V+IR_S}{a\frac{KT_C}{q}}\right)} - 1\right] \tag{2}$$

Where:
shunt current.I_{sh} :
R_s: series resistance.
R_{sh}: shunt resistance.
E_g: bandgap energy.

q: electron charge.
k: Boltzmann's constant.
α_I: current temperature coefficient.
I_{mp}: current at the maximum power point.
V_{mp}: voltage at the maximum power point.
The electrical efficiency and fill factor can be calculated from Eq. (3, 4) [12].

$$\eta_E = \frac{P_E}{G_t A_m} = \frac{I\,V}{G_t A_m} \tag{3}$$

$$F.F. = \frac{I_{mp} V_{mp}}{I_{sc} V_{oc}} \tag{4}$$

2.1.2 Thermal Model

The energy balance equation of thermal modelling of each PVT part component was carried out for assuming the quasi-steady-state condition has considered, the airflow through the duct is uniform, the electrical losses of PV cells and heat losses from edges of the system are negligible, and the heat transfer coefficients by convection mode from absorber and back plates to fluid are equal. All heat transfer coefficients by convection and radiation distributions, top losses, and bottom losses in the hybrid PVT system and in the dryer chamber are illustrated in figure (2) and (3). Energy balance equations for each part of the PVT system in watts unit are obvious in the block diagram which illustrated in figure (4). The energy equation of the PVT system with the dryer room states the input and output energies and taking into consideration the heat losses from the PVT and dryer system. The energy balance equations in each part of the hybrid PVT and dryer chamber are expressed below:

1. Energy Balance of Solar cells of PV module:

According to Eq. (5), the input energy is represented by the amount of solar radiation that is absorbed by the solar cells on the left side. On the right side, the output energies from PV cell comprises the heat transfer by radiation between solar cell and thermal collector, the heat transfer by convection between the solar cell and airflow, and the electrical productivity. The heat losses include the losses by radiation between the solar cell and ambient air, the top losses by convection between the top cell surface and ambient air.

$$G_t\,\alpha_c\,\tau_g = h_{r,c-p}\left(T_c - T_p\right) + U_{t,c-a}\,(T_c - T_a) + h_{r,c-sky}\left(T_c - T_{sky}\right) + h_{c,c-f1}\left(T_c - T`_{f1}\right) + h_{r,c-a}\,(T_c - T_a) + G_t\,\beta_c\,\tau_g\,\eta_e \tag{5}$$

From Eq. (5), the cell temperature expression can be written as:

$$T_C = \frac{G_t\,\tau_g(\alpha_c - \eta_e\beta_c) + (\frac{1}{h_w} + \frac{t_g}{k_g})^{-1}\,T_a + h_{r,c-sky}\,T_{sky} + h_{r,c-p}\;T_p + 0.5\,h_{c,c-f1}\left(T_{f1} + T_{f2}\right)}{(\frac{1}{h_w} + \frac{t_g}{k_g})^{-1} + h_{r,c-p} + h_{c,c-f1} + h_{r,c-sky}} \tag{6}$$

The mathematical expressions of heat transfer coefficients such as (h_w, $h_{c,c-f1}$, $h_{r,c-sky}$, $h_{r,p-b}$, and, $h_{r,c-p}$) are illustrated in [13].

2. Energy Balance of Fluid Flow in The Upper Pass:

The left side in Eq. (7) shows the output energy or heat gain from the PVT system as hot air in the upper channel. The input energies in the right side contain the heat transfer between the back surface of solar cell to airflow and heat transfer between solar collector and airflow.

$$\dot{m}_a c_{pa}(T_{f2} - T_{f1}) = h_{c,p-f`1}(T_p - T`_{f1})A_{pf} + h_{c,c-f`1}(T_c - T`_{f1})A_m \quad (7)$$

From Eq. (7), the airflow temperature in the upper channel can be expressed as Eq. (8).

$$T_{f2} = \frac{(\dot{m}_a c_{pa} - 0.5A_{pf}h_{c,p-f`1} - 0.5A_m h_{c,c-f`1})T_a + A_m h_{c,c-f`1}T_c + A_{pf}h_{c,p-f`1}T_p}{(\dot{m}_a c_{pa} + 0.5A_{pf}h_{c,p-f`1} + 0.5A_m h_{c,c-f`1})} \quad (8)$$

Whereas:

$$T`_{f1} = \frac{T_{f1} + T_{f2}}{2} \quad (9)$$

3. Energy Balance of Absorber Plate:

The left side in Eq. (10) shows the input energies to solar collector and the output energies in the right side are contains the convective heat transfer between the absorber plate and airflow in upper and lower channel, the convective heat transfer between the backplate and airflow in lower channel, and heat transfer by radiation between absorber plate and backplate.

$$G_t(1 - \beta_c)\,\alpha_p\,\tau_g^2 A_p + h_{r,c-p}(T_c - T_p)A_p = h_{c,p-f`1}(T_p - T`_{f1})A_{pf} + h_{c,p-f`2}(T_p - T`_{f2})A_{pf} + h_{r,p-b}(T_p - T_b)A_b \quad (10)$$

From Eq. (10), the plate temperature can be expressed as:

$$T_p = \frac{(G_t(1-\beta_c)\,\alpha_p\,\tau_g^2 A_p + h_{r,c-p}T_c + h_{r,p-b}T_b)A_p + 0.5A_{pf}(h_{c,p-f`1}(T_{f1}+T_{f2}) + h_{c,p-f`2}(T_{f2}+T_{f3}))}{A_p(h_{r,p-b} + h_{r,c-p}) + 0.5A_{pf}(h_{c,p-f`1} + h_{c,p-f`2})} \quad (11)$$

Where:

$$A_{pf} = A_p + A_f \quad (12)$$

4. Energy Balance of Fluid Flow in the Lower Pass:

The left side in Eq. (13) shows the output energy or heat gain from the PVT system as hot air in the lower channel. The input energies on the right side contain the heat transfer by convection between the solar collector to airflow and heat transfer between backplate and airflow.

$$\dot{m}_a c_{pa}(T_{f3} - T_{f2}) = h_{c,p-f`2}(T_p - T`_{f2})\,A_{pf} + h_{c,b-f`2}(T_b - T`_{f2})\,A_b \quad (13)$$

$$T_{f3} = \frac{(\dot{m}_a c_{pa} - 0.5A_{pf}h_{c,p-f`2} - 0.5A_b h_{c,b-f`2})T_{f2} + A_{pf}h_{c,p-f`2}T_p + A_b h_{c,b-f`2}T_b}{(\dot{m}_a c_{pa} + 0.5A_{pf}h_{c,p-f`2} + 0.5A_b h_{c,b-f`2})} \quad (14)$$

Whereas:

$$T`_{f1} = \frac{T_{f1} + T_{f2}}{2} \quad (15)$$

5. Energy Balance of Back Plate:

Heat transfer by radiation represents the input energy on the left side from collector to backplate. The back losses and output energy are illustrated in right side.

$$h_{r,p-b}(T_p - T_b) = U_b(T_b - T_a) + h_{c,f`2-b}(T_b - T`_{f2}) \quad (16)$$

By arranging the equation Eq. (16), the backplate temperature can be expressed as:

$$T_b = \frac{h_{r,p-b}\,T_p\,A_p + (\frac{1}{h_w}+\frac{t_b}{k_b})^{-1}\,T_a\,A_b + 0.5\,h_{c,b-f2}\,A_b(T_{f2}+T_{f3})}{(\frac{1}{h_w}+\frac{t_b}{k_b})^{-1}\,A_b + h_{r,p-b}\,A_p + h_{c,b-f2}\,A_b} \tag{17}$$

6. Energy Balance of Dryer Chamber:

The output useful heat gain from the PVT system considers the input energy to solar dryer on the left side. Besides, from Eq. (18), the solar radiation that passes from the glass walls of solar dryer from all directions. The output energies contain the top and back losses from glass covers and back dryer respectively.

$$\dot{m}_a\,c_{pa}(T_{f3} - T_d) + \tau_g(G_E A_E + G_W A_W + G_N A_N + G_T A_T) = U_{td}\,A_g(T_d - T_a) + U_{bd}A_b(T_d - T_a) \tag{18}$$

From Eq. (18), the dryer temperature can be written as:

$$T_d = \frac{\dot{m}_a c_{pa} T_{f3} + (G_E+G_W+G_N+G_T)\tau_g\,A_g + \left(\frac{1}{h_w}+\frac{t_{gd}}{k_{gd}}\right)^{-1} T_a\,A_g + \left(\frac{1}{h_w}+\frac{t_{bd}}{k_{bd}}\right)^{-1} T_a\,A_{bd}}{\dot{m}_a c_{pa} + (\frac{1}{h_w}+\frac{t_{gd}}{k_{gd}})^{-1}\,A_g + (\frac{1}{h_w}+\frac{t_{bd}}{k_{bd}})^{-1}\,A_{bd}} \tag{19}$$

Where $(G_E,\ G_W,\ G_N,\ G_T)$ are represented the value of solar radiation from east, west, north, and top of the dryer room which measured experimentally by the solar power meter. The air has thermo physical properties such as density, viscosity, specific heat capacity, Prandtl number, and thermal conductivity, also, the heat transfer coefficients by forced mode. These properties are significant parameters to solve the thermal model of the PVT system and solar dryer. All expressions of these properties are illustrated in [14]. The rate of useful heat gain that carried away from the PVT system by fluid in watt unit and the thermal and overall efficiencies are described as [15]:

$$Q_u = \dot{m}_a c_{pa}(T_{fo} - T_{fi}) \tag{20}$$

$$\eta_{th} = \frac{Q_u}{G_t A_m} = \frac{\dot{m}_a c_{pa}}{G_t A_m}\,(T_{fo} - T_{fi}) \tag{21}$$

$$\eta_{overall} = \eta_{th} + \frac{\eta_E}{0.36} = \frac{1}{G_t A_m}\left[\dot{m}_a c_{pa}(T_{fo} - T_{fi}) + \frac{I\,V}{0.36}\right] \tag{22}$$

2.1.3 Computer Model

The design parameters of the electrical model such as voltage, current, and power were solved by using the MATLAB program in an iteration manner and displayed the behavior of the PV module. The flow chart of the PVT system program is shown in figure (5). The heat transfer coefficients by convection and radiation were calculated theoretically by MATLAB, and these coefficients need initial trails of temperatures and electrical efficiency, moreover, the properties of air are evaluated. By solving the energy balance equations, the temperature distributions have been solved by built program numerically. The MATLAB program needs input parameters to generate, which contain the value of mass flow rate of air, the design dimensions of the hybrid PV/T system with dryer room, and condition weather parameters such as solar radiation, ambient temperature, and wind speed. Finally, it has been conducted the temperatures of the PVT system, electrical efficiency, heat gain, thermal efficiency, and overall efficiency.

2.2 Experimental Work

The system consists of 32 solar cells of a monocrystaline silicon that connects in series as the PV module and surrounds double glass covers from the top and below of PV cells as shown in figure (6) and the dimensions of each glass cover such as length, width, and thickness are 120 cm, 56 cm, and 1.75 cm respectively. The electrical characteristics such as voltage, current, and power of PV module are demonstrated in table. (1). The components of the proposed PVT solar system with the dryer chamber are illustrated in the figure (7). The solar collector locates directly under the PV module with dimensions (110 cm*54 cm*0.1 cm) and the backplate is in parallel with the thermal collector with dimensions (120 cm*54 cm*0.1 cm). The backside of the PVT system is covered by insulating layer to prevent heat leakage and decrease back losses. The thermal solar collector is made from pure aluminum metal and contains 20 fins in the upper and lower surface. The rectangular fin type is utilized in this study with dimensions (100 cm*3 cm*0.1 cm) in order to increase the surface area and heat transfer coefficient by convection between the airflow and solar collector. The double pass counterflow technique is utilized to extract more heat from the system. A DC fan used for blowing air into the PVT system, the fan is generated by the PV module and the table (3) shows the power consumption by DC fan which depends on the amount of airflow rate. The solar dryer system is connected to the PVT system by a duct that carries the hot air output from the hybrid PVT system to the dryer chamber. The solar dryer volume in this work is (52cm*42cm*60cm) that consists of a wooden frame, glass covers, and, chimney for exhausting air. The PVT solar system faces south and tilted at 33° to collect more radiation intensity. The experimental work starts with measure the solar radiation, ambient air, and wind velocity by utilizing solar power meter, temperature sensor, and speed meter. Further, the DC fan is operated by the PV module and set the speed of fan at a convenient mass of airflow by using voltage regulator to control in fan speed. Further, the thermal and electrical behavior of PVT system with dryer solar system is measured by utilizing the measurement devices which are illustrated in figure (8). Repeat the measuring process each hour and record the output readings.

3. Results and Discussion

The experimental results were taken on, 19th of June in 2019 for Baghdad, Iraq. The statistical analysis for quantifying the degree of identity between the theoretical and the experimental results of hybrid PVT solar system have been depended to evaluate the root mean square of percentage deviation (e) and linear coefficient of correlation (r) with based on the equations of approach study which presented by [15], and these values in this work are illustrated in Table 2. Figure (9) illustrates the effect of solar intensity on the open-circuit voltage and short circuit current and can be noticed that the current has more sensitive at changing solar radiation than voltage. For this reason, the electrical power dropped because of reduction current. Figure (10) shows the effect of solar cell temperature on the open-circuit voltage and electrical power and can be found that voltage affected by changing the cell temperatures. Hence, electrical power decreased when the voltage decreased. Figure (11) shows the effect of solar radiation on the open-circuit voltage and electrical power and can be found that electrical power affected by changing solar radiation. Hence, electrical power decreased when solar radiation increased. Figure (12) clarifies the hourly variations of solar intensity with time in the summer day. Figure (13) shows the hourly change of inlet temperature which measured experimentally with time in the summer day. Figures (14, 15, 16, 17, and, 18) demonstrate the comparisons between the theoretical and the experimental results of useful heat gain and temperatures of PV cell, absorber plate, backplate, and output fluid at the same input fluid temperature. Figure (17). slows that the output fluid temperature when the mass flow rate increased because the fluid at high velocity cannot extract further heat from PV cell and absorber plates and maximum temperature is obtained at 0.017 kg/s and 1100 W/m^2, and it was 63°C when the input fluid temperature was 44°C. It is obvious from figure (18) that the useful heat gain is directly proportional to the mass flow rate according to Eq. (21), the maximum value of useful heat gain reached 389.375 W at 0.031 kg/s when the solar radiation 1100 W/m^2. The root mean square of percentage deviations (e) and linear coefficients of correlation (r) to approve the agreement between these results are obvious in Table 2. Figure (19) represents the variation of electrical power (P_E) in each hour of the day at different solar radiation and mass flow rate. It is evident that electrical power is increased by increasing the solar intensity because of more electrons are emitted from solar cells and generate more current. Further, the change of mass flow rate effects on the electrical power which increases when the mass flow rate increases due to the cell temperature decreased and that cause increasing voltage and electrical power, and the maximum power is found at 1100 W/m^2 and 0.031 kg/s. Figure (20) states the change of electrical efficiency (η_E) with time. The electrical efficiency mainly depends on solar radiation and cell temperature, therefore, the increase of mass flow causes raise in electrical efficiency by extracted heat from PV cell and reduction cell temperature. The electrical efficiency is decreased in spite of the electrical power is increased due to the electrical efficiency depends significantly on cell temperature that causes drop in efficiency but the electrical power continues in increasing when solar radiation increases. The maximum electrical efficiency is 12.6% at 750 $W/m^2$and 0.031 kg/s at 10:00 p.m. Figure (21) demonstrates the hourly change of thermal efficiency and indicates that the maximum thermal efficiency is 56.73% when solar radiation 1050 W/m^2 at 0.031 kg/s. Figure (22) displays the hourly change of overall efficiency and found that the maximum overall efficiency is about 85% when solar radiation 750 W/m^2 at 0.031 kg/s because of the electrical efficiency at 750 W/m^2 is high.

4. Conclusion

A mathematical model of the electrical and thermal performance was presented in this study; also, the predicting of results were simulated by utilizing MATLAB program. The comparison between the theoretical and experimental results has been proven. The double pass technology and adding fins were applied in the PVT system for two reasons, firstly, to raise the heat extraction from the absorber plate and secondly, to increase the surface area to transfer further heat from plate to fluid. The maximum temperature difference between input and output airflow was 19°C at a low mass flow rate 0.017 kg/s which considers acceptable temperature to supply the solar dryer. In addition, the results indicated that the maximum electrical and thermal efficiencies of the PVT system were 12.65% and about 56.73%, respectively, at a higher mass flow rate about 0.031 kg/s.

References

[1] Nurul Syakirah Nazri, Ahmad Fudholi, Bardia Bakhtyar, Chan Hoy Yen, Adnan Ibrahim, Mohd Hafidz Ruslan, Sohif Mat, Kamaruzzaman Sopian. Energy economic analysis of photovoltaic-thermal-thermoelectric (PVT-TE) air collectors. Renewable and Sustainable Energy Reviews 92 (2018) 187–197.

[2] Yuting Jia, Guruprasad Alva, Guiyin Fang. Development and applications of photovoltaic–thermal systems: A review. *Renewable and Sustainable Energy Reviews* 102 (2019) 249–265.

[3] Ahmed Zoukit, Hicham El Ferouali, Issam Salhi, Said Doubabi, Naji Abdenouri. Takagi Sugeno fuzzy modeling applied to an indirect solar dryer operated in both natural and forced convection. *Renewable Energy* 133 (2019) 849e860.

[4] Sumit Tiwari, Sanjay Agrawal, G.N. Tiwari. PVT air collector integrated greenhouse dryers. Renewable and Sustainable Energy Reviews 90 (2018) 142–159

[5] C.K.K. Sekyere, F.K. Forson, F.W. Adam. Experimental investigation of the drying characteristics of a mixed-mode natural convection solar crop dryer with back up heater. *Renewable Energy* 92 (2016) 532e542.

[6] Essalhi Hajar, Tadili.Rachid, Bargach.Mohammed.Najib. Conception of a Solar Air Collector for an Indirect Solar Dryer. Pear Drying Test. *Energy Procedia* 141 (2017) 29–33

[7] Neha Dimri, Arvind Tiwari, G.N. Tiwari. Thermal modelling of semitransparent photovoltaic thermal (PVT) with thermoelectric cooler (TEC) collector. *Energy Conversion and Management* 146 (2017) 68–77

[8] Ehsan Baniasadi, Saeed Ranjbar, Omid Boostanipour. Experimental Investigation of the Performance of a Mixed-mode Solar Dryer with Thermal Energy Storage. *Renewable Energy* (2017).

[9] Satyender Singh, Prashant Dhima. Thermal performance of double pass packed bed solar air heaters – A comprehensive review. *Renewable and Sustainable Energy Reviews* 53 (2016) 1010–1031.

[10] Jin-Hee Kim, Se-Hyeon Park, Jun-Tae Kim. Experimental performance of a photovoltaic-thermal air collector. Energy Procedia 48 (2014) 888 – 894. [11] Liu C, Wu Y, Bian J, Li D and Liu X 2018 Influence of PCM design parameters on thermal and optical performance of multi-layer glazed roof. *Applied energy* **212** 151-161.

[11] Ahmad Fudholi, Kamaruzzaman Sopian, Mohd Hafidz Ruslan, Mohd Yusof Othman. Performance and cost benefits analysis of double-pass solar collector with and without fins. *Energy Conversion and Management* 76 (2013) 8–19

[12] Sangram Bana, R.P. Saini. A mathematical modeling framework to evaluate the performance of single diode and double diode based SPV systems. *Energy Reports* 2 (2016) 171–187

[13] Wei Chang, Yunfeng Wang, Ming Li, Xi Luo, Yongbao Ruan, Yongrui Hong, Shaobo Zhang. The theoretical and experimental research on thermal performance of solar air collector with finned absorber. *Energy Procedia* 70 (2015) 13 – 22

[14] Ahmad Fudholi, Kamaruzzaman Sopian, Mohd Yusof Othman, Mohd Hafidz Ruslan, B. Bakhtyar. Energy analysis and improvement potential of finned double-pass solar collector. *Energy Conversion and Management* 75 (2013) 234–240.

[15] Karima E. Amori, Hussein M. Taqi Al-Najjar. Analysis of thermal and electrical performance of a hybrid (PV/T) air based solar collector for Iraq. *Applied Energy* 98 (2012) 384–395.

Table 1. Specifications of designed PVT solar system.

Parameter	Value	Parameter	Value
V_{oc}	21.55 V	β_C	0.74
I_{sc}	5.33 A	τ_g	0.95
A_m	0.66 m^2	α_c	0.88
P_{mp}	84.3 W	t_g	0.0035 m
V_{mp}	16.99 V	α_p	0.8
I_{mp}	4.962 A	k_p	200 w/m.k
F.F.	0.734	k_g	0.8 w/m.k
No. of cells	32	t_f	0.0009 m
Lp	1.1 m	t_p	0.001 m
L_b	1.2 m	t_b	0.001 m
L_f	1 m	t_f	0.0009 m
H_f	0.03 m	ε_g	0.9
Wp	0.54 m	ε_p	0.9
W_b	0.54 m	No. of fins	20

Table 2. Produced power by DC fan.

Mass Flow of Air (kg/s)	Consumed Power (W)
0.017	6.78 W
0.023	6.5 W
0.032	7.5 W

Table 2. Values of (e%) and (r) at a different mass flow rate.

	T_C	T_p	T_b	T_{fo}	Q_u
At m=0.017kg/s					
e %	2.3	3.2	2.8	1.7	8.3
r	0.992	0.988	0.987	0.995	0.996

At m=0.023kg/s					
e %	1.9	2.3	2.7	2.19	10.7
r	0.995	0.983	0.995	0.993	0.996
At m=0.031kg/s					
e %	2.3	2.6	3.02	1.7	10.4
r	0.978	0.989	0.992	0.995	0.997

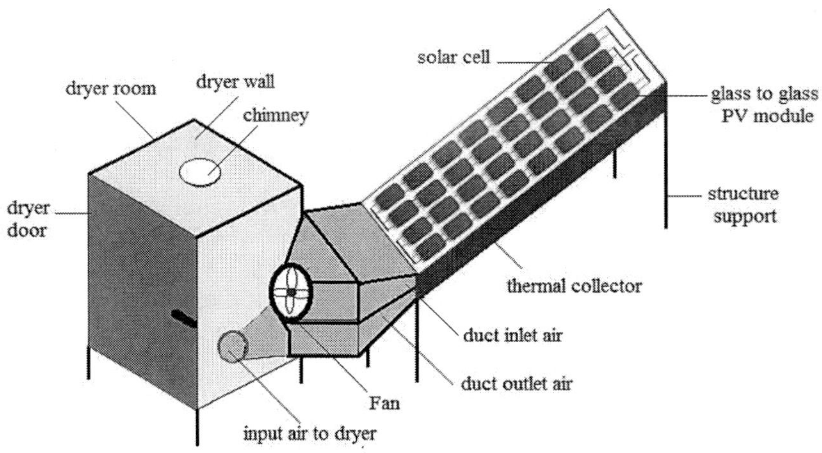

Figure 1. Schematic diagram of the PVT solar system with dryer.

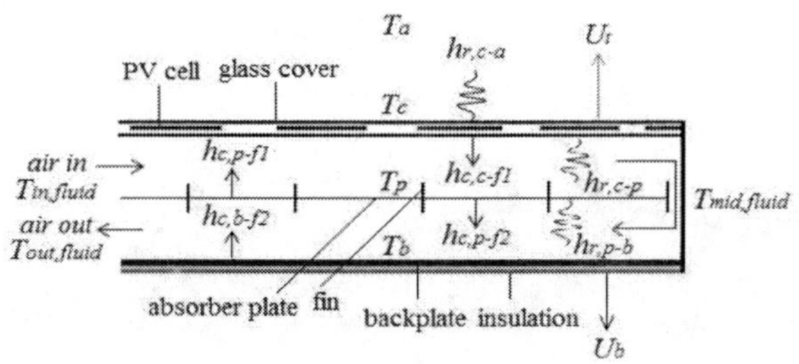

Figure 2. Convection and radiation heat transfer coefficients distribution in the PVT system.

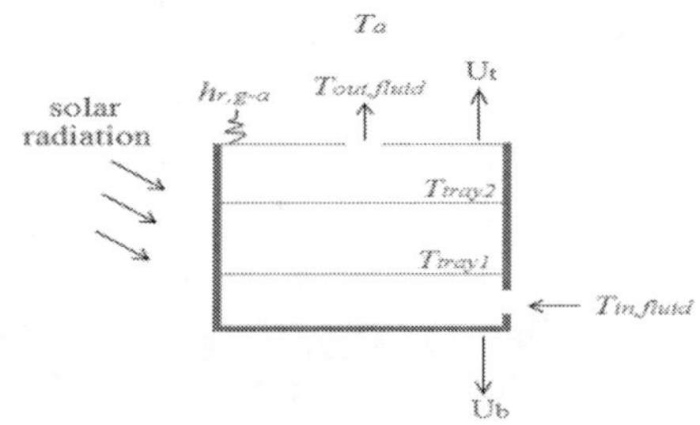

Figure 3. Convection and radiation heat transfer coefficients distribution in the solar dryer.

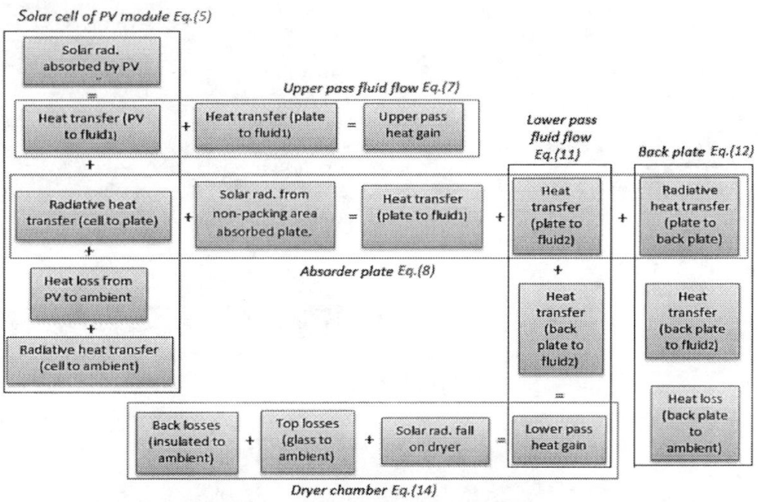

Figure 4. The block diagram of energy balance equations.

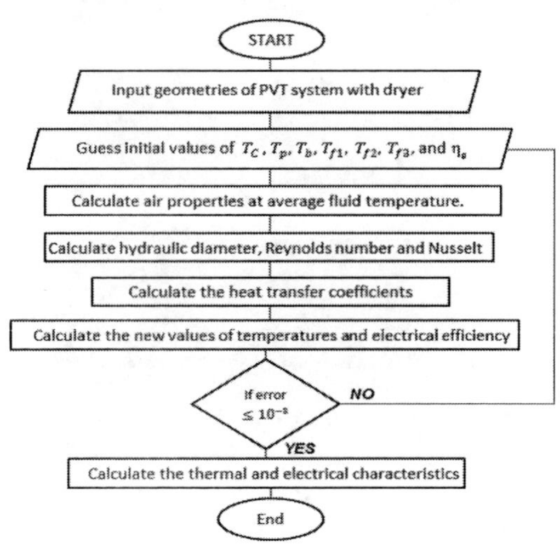

Figure 5. Flow chart of the PVT system.

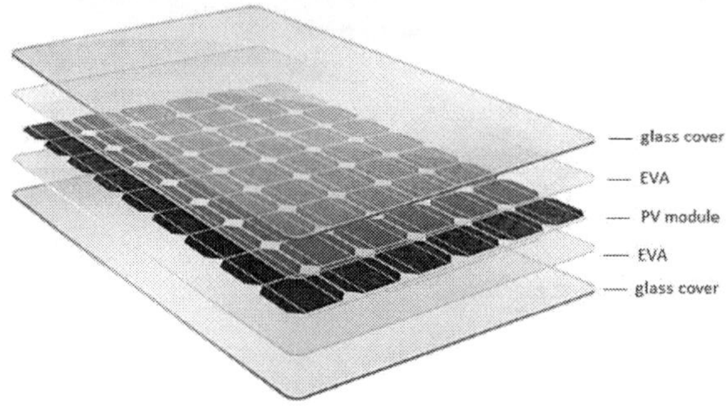

Figure 6. The double glass PV module component.

(a) **(b)**

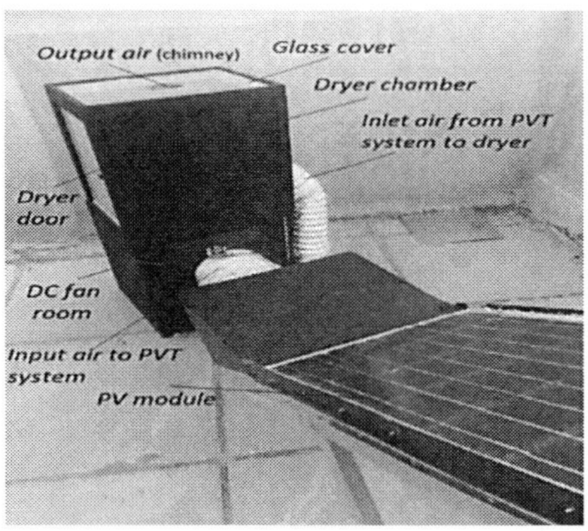

Figure 7. the hybrid PVT solar system With dryer.

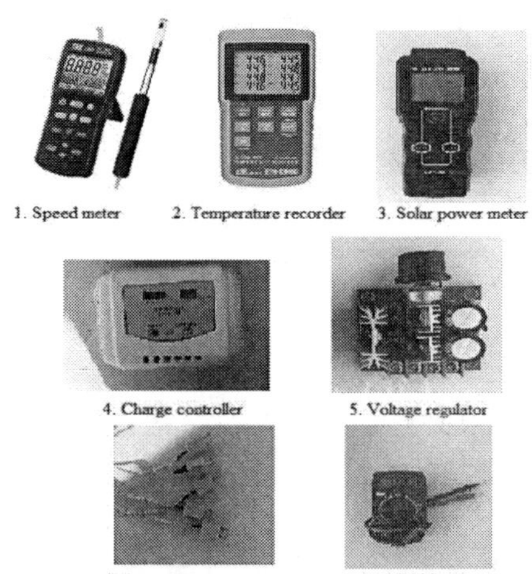

Figure 8. Measurements devices.

IOP Conf. Series: Materials Science and Engineering **765** (2020) 012024 doi:10.1088/1757-899X/765/1/012024

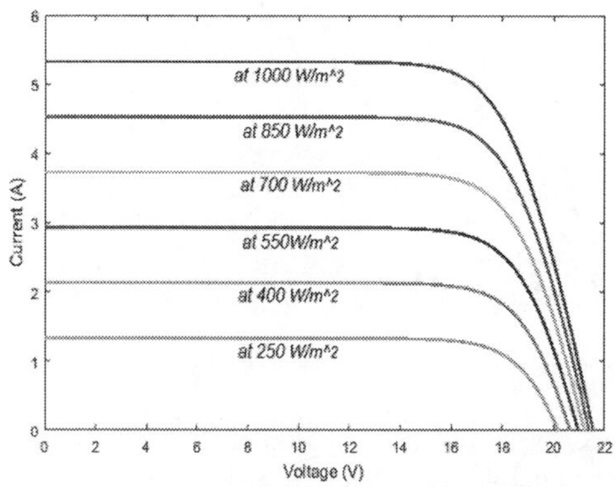

Figure 9. Effect of solar radiation on voltage and current.

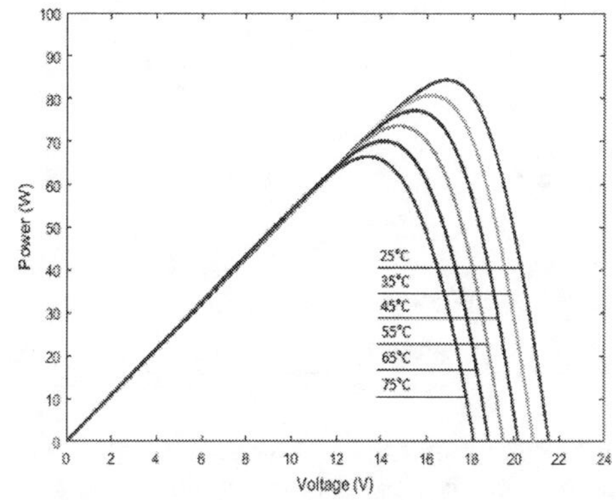

Figure 10. Effect of cell temperature on voltage and power.

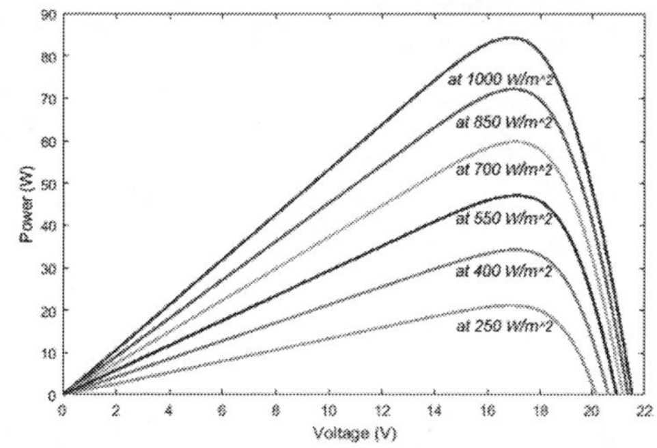

Figure **11**. Effect of solar radiation on voltage and power.

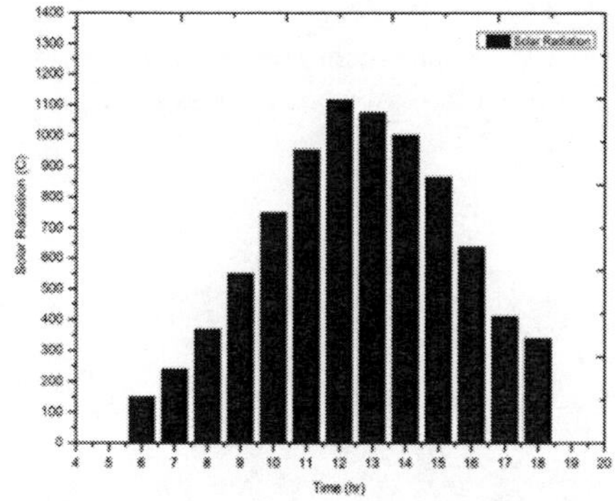

Figure 12. Hourly variations of solar intensity.

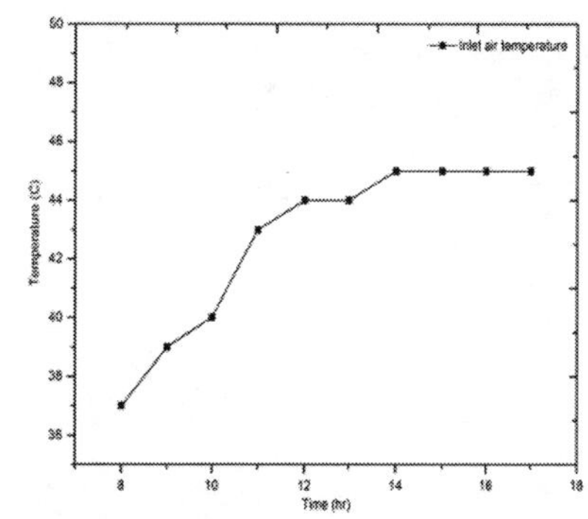

Figure 13. Hourly variations of inlet temperature.

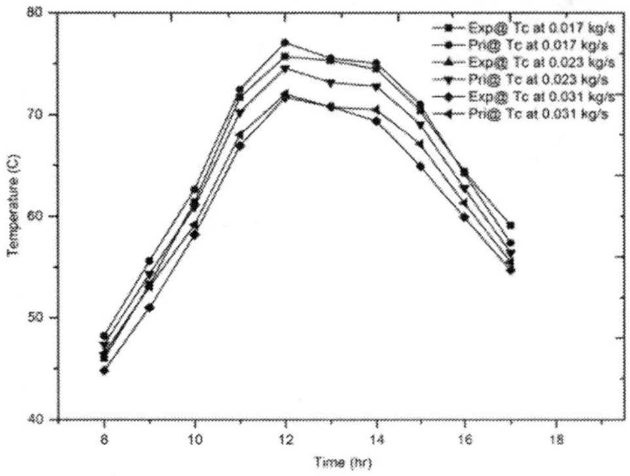

Figure 14. Experimented and predicted cell temperature at several mass flow rates.

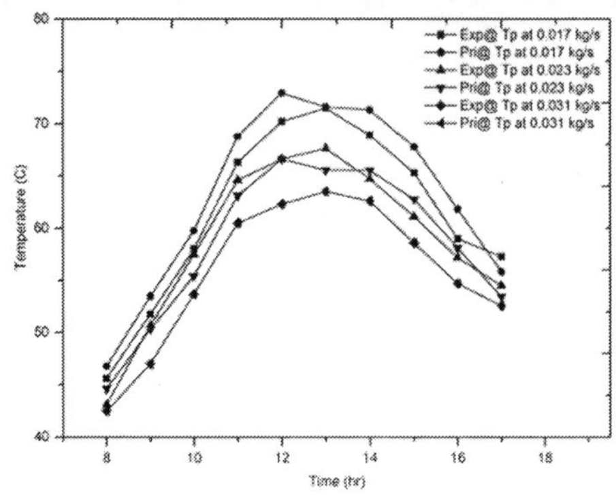

Figure 15. Experimented and predicted absorber plate temperature at several mass flow rates.

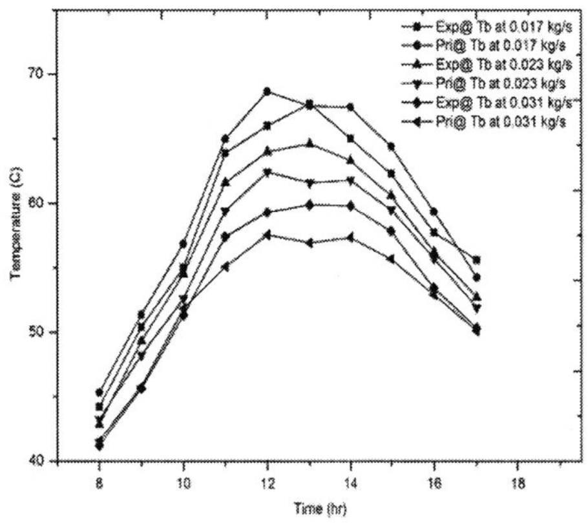

Figure 16. Experimented and predicted backplate temperature at several mass flow rates.

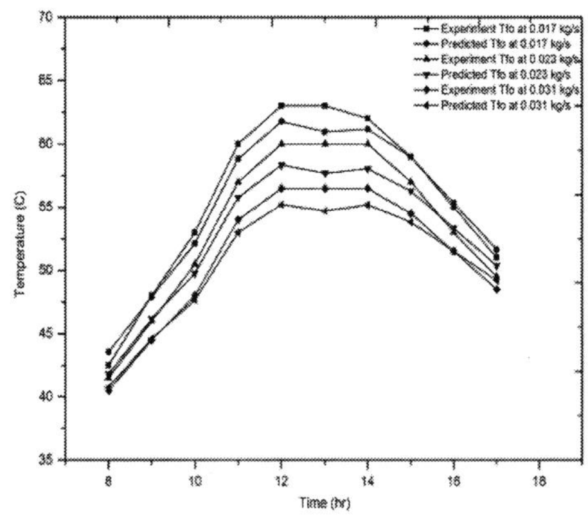

Figure 17. Experiment and predicted output fluid temperature at several mass flow rates.

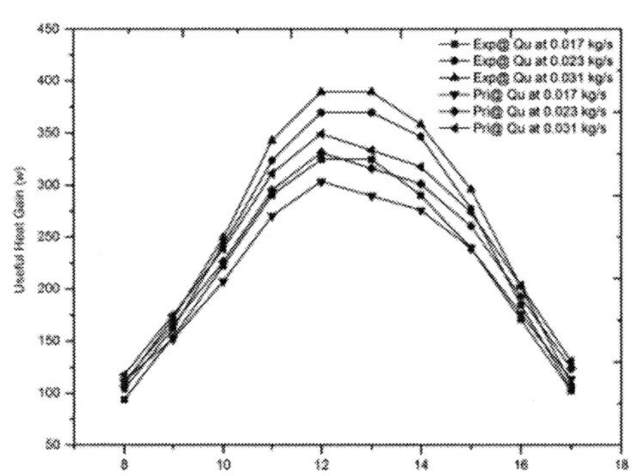

Figure 18. Experimented and predicted useful heat at several mass flow rates.

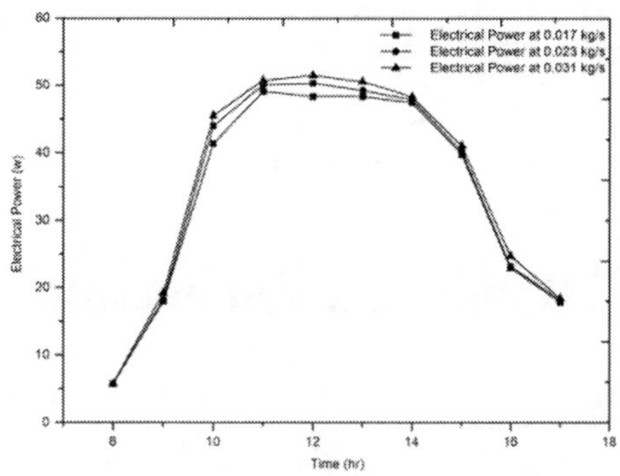

Figure 19. Variation of electrical power with time at a different mass flow rate.

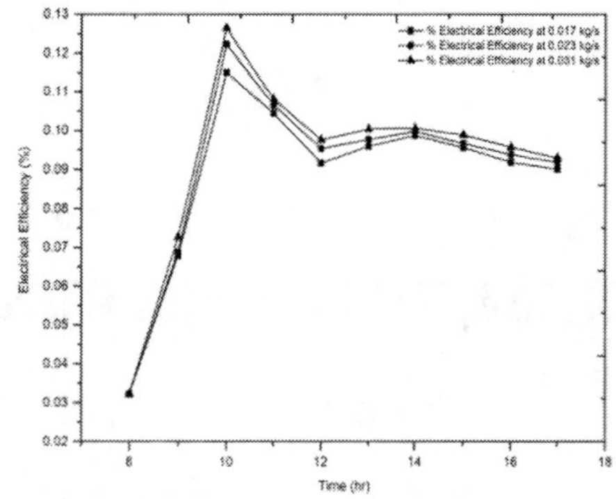

Figure 20. Variation of electrical efficiency with time at a different mass flow rate.

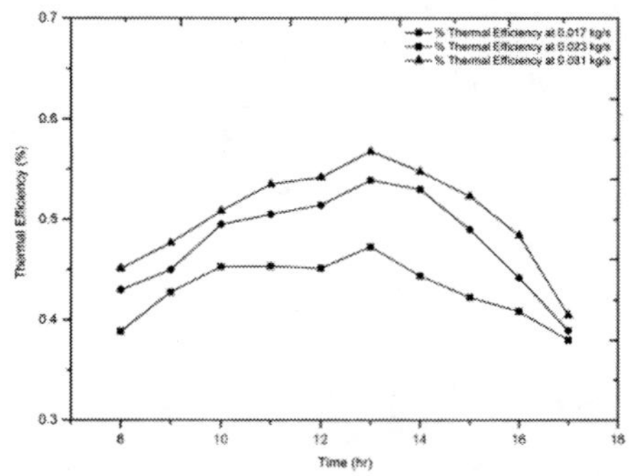

Figure 21. Effect of increasing mass flow rate on thermal efficiency.

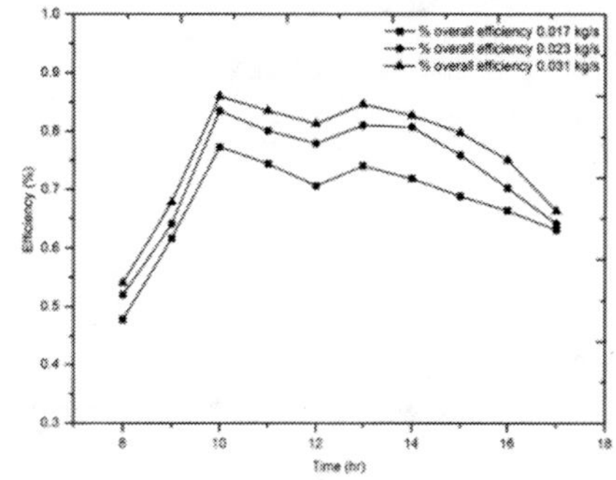

Figure 22. Variation of overall efficiency with time.

Palm Olein with Jatropha Curcas Oil Blend as a Renewable Bio -Fluid

Mohammed Hassan Jabal

Hussain Saad Abd

A. A. Gaddoa

Ibrahim Falih Kadhim

Syahrullail Samion

Ameer S.Shajar

Abdulwahhab A.Hardan

Electromechanical Engineering Department, University of Technology, Baghdad, Iraq

corresponding author's e-mail : 50124@uotechnology.edu.iq

Abstract. Many bio-fluids have been introduced to replace petroleum-based products one of which is vegetable oil which has characteristics that are friendly to the environment. Based on its characteristics, it is turning into a significant bio fluid source. Vegetable oil has a great advantage over other forms of bio-oils which is its wild availability as a renewable source. Furthermore, when used in engines, vegetable oil based lubricants have the ability to decrease hydrocarbon emissions as well as carbon dioxide. Using the pure vegetable oil with no additions and using certain blending ratios are the two methods of using vegetable oil as a bio-fluid. This paper focuses on conducting an investigation on the influence of using the normal load on the performance characteristics for the blending of two types of vegetable fluid, and making a comparison between this and a commercial mineral fluid using the four ball tribotester, palm olein with Jatropha curcas oil which were mixed in (RB60/J40) ratio. All conducted experiments were conformed to ASTM D4172. It has been noticed from the obtained results that RBD palm olein and Jatropha curcas oil Blend has the ball bearings wear scar and frication coefficient less than the commercial fluid. Based on that, it is concluded that the RB60/J40 blend is more efficient than neat RBD palm olein and commercial mineral fluid as well.

Keywords: RBD palm olein, jatropha curcas, applied normal load, anti-wear, anti-friction, parameter of flash temperature.

1. Introduction

An efficient method to boost petroleum production has been found by engineers and scientists since the beginning of the 19th century which led to cheap petroleum based lubricants

IOP Conf. Series: Materials Science and Engineering **765** (2020) 012025 doi:10.1088/1757-899X/765/1/012025

productions. These cheap lubricants caused the effect of greenhouse and global warming issue. The need to biodegradable products has increased due to the increased health awareness and restricted resources of petroleum. A lot of efforts were put to fight pollution especially that produced by petroleum-based lubricants. It has been found that every year the environment was polluted by dumping about 12 million tons of lubricant wastes [1]. Oil products of biodegradable have become an effective substitute to regular lubricants due to the pollution of ecological. Furthermore, other bio-oils such as animal fat and vegetable oil are being used as alternatives to mineral- based oil. It has been discovered that ancient Egyptians had used vegetable oils in making their monuments [2]. This makes it safe to say that using vegetable oil in the industry is applicable. Promoting vegetable oil as a lubricant is based on its features which are first, it is biodegradable and second, it is less toxic than other petroleum based oil. Additionally, vegetable oil is an easy renewable source to produce and work with. Also, an investigation has found that vegetable oils have better lubricating ability compared to the synthetic oils and mineral oil when used between two moving metals. This lubricating ability is due to that fact that vegetable oil has a huge quantity of unsaturated and polar ester elements which influence the sliding movement of the two metals [3]. On top of that, vegetable oil has a long of fatty acids that have superior fundamental limits lubricant properties.

Several researchers have reported that the wear rate is still high even in the case of low friction coefficient associated with using vegetable oils due to their ability to increase friction coefficients. This makes vegetable oils have quality lubricating abilities. An investigation was conducted to monitor the chemical attack of the vegetable oil fatty acid on the surface [4]. It has been found that during sliding movements, the metallic soap film is wiped away. Palm oil has been tested by many researchers for applications that are engineering related such as using it as a diesel engine fuel [5, 6]. Palm oil was studied as a hydraulic fluid [8]. Also, palm oil characteristics were investigated as a metal forming lubricant by Syahrullail and his co-workers [7, 8]. Studying the pure palm oil as a bio-fluid can be divided to many categories. The first category includes using one hundred percent palm oil as a test lubricant [9, 10]. The second category includes suing emulsion of palm oil [11, 12]. The third category includes using additive mixed with palm oil [13, 14]. The forth category includes using palm oil as an additive [15]. The obtained results were promising and showed great potential which can utilized in engineering applications. Though, the oxidation of vegetable oil among other factors has to be taken into account. This paper aims to explore the RBD palm olein friction coefficient and wear performance and Jatropha curcas oil friction coefficient and wear performance blending with various normal loads using four ball tribotester.

2. Experimental Method

2.1. Equipment

A four-ball wear tester was utilized for this research. This tester was initial defined by [16] already having attained the standing of a long-time establishment within the elementary investigation of lubricants characteristics. This device utilizes four balls, one ball on the highest and 3 balls at rock bottom. The 3 balls are placed inflexibly in a sphere pot comprising the lubricant subjected to the test and pushed against the highest ball. The highest ball is created to rotate at the wanted speed whereas rock bottom balls force the highest ball to stay in place. The up described tester is shown in Figure 1, while the necessary parts such as the ball bearings, collect, and the oil cup assembly are shown in Figure 2. Acetone was used to clean components surfaces prior to executing each experiment. (100,200 to 600 Newton) normal loads were applied under (1200 rpm) as a rotary velocity for a time period of 60 minutes and 75°C lubricant temperature in this investigation.

Figure 1: Photo of Four-ball tribotester machine

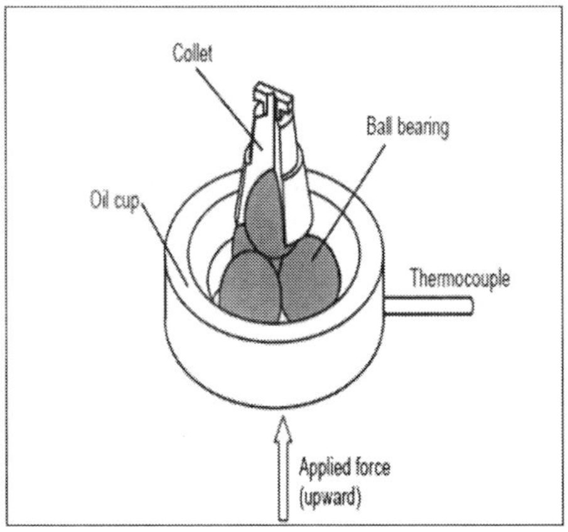

Figure 2: Schematic sketch of four-ball tribotester apparatus

2.2. *Materials*

In the study, the four ball are made of AISI E-52100 chrome alloy steel. The specifications of the ball are as followed: a 12.7 mm diameter; extra polish grade 25; hardness 64–66 HRC (Rockwell C Hardness). For each experiment, A new cleaned with acetone and dried with a fresh lint-free industrial wipe set of balls were used.

2.3. *Lubricants*

RBD palm olein and jatropha curcas were used in this study as lubricants. The two oils were mixed in 60% of RBD palm olein and 40% of Jatropha curcas oil. The obtained results from each experiment were compared to the results obtained from conducting the same experiment using commercial mineral fluid (SAE 40). The experiment was designed to test 10 ml of fluid each time.

2.4. *Test procedures*

In order to prop the steel balls, all components including the four balls and the sphere pot were completely cleaned by acetone and dried utilizing a fresh lint free industrial wipe to assure no residuals of any solvent are remained before putting together the parts and pouring the lubricant into the pot. To fix the three bottom ball in position, a torque wrench was used to tighten the ball bearings. After that, the highest ball placed and fixed in the collector and bounded onto the spindle. Finally, the fluid being tested was poured into the assembled pot. The oil level was observed to fill all the gaps within the cup assembly. A non-friction disc in the four ball machine was used a platform to fix the pot component on. Then, the load was slowly applied to dodge any shock loading. After that, the tribotester created in heater was used to heat up the lubricant to 75 °C. After this temperature was

IOP Conf. Series: Materials Science and Engineering **765** (2020) 012025 doi:10.1088/1757-899X/765/1/012025

achieved, the drive motor was started to rotate the highest ball at the wanted speed. After 60 minutes, the heating was stopped and the pot was removed from the machine.

2.5. *Wear scar diameter*

Commonly known, the greater the diameter of the wear scar, the worse the wear. Therefore, measuring the diameter of wear scar of the three ball of each test was performed to examine and find out the lubricity performance of each tested lubricant. To evaluate and determine the wear scar for all the three balls individually, high resolution computer software using the obtained photomicrograph was used.

2.6. *Torque of friction and coefficient of friction*

A specific data acquisition was used to recode the friction torque which was observed to increase fast at first then after 5 to 10 min, it became a steady-state condition. Also, in the steady state case, the torque of friction average was recorded and the coefficient of friction was determined according to (IP-239) which is shown in the Equation below.

$$\mu = \frac{T\sqrt{6}}{3Wr} \tag{1}$$

where μ = friction coefficient.
T = the frictional torque in kg mm.
W= the applied load in kg.
r= 3.67 (the distance from the centre of the contact surface on the lower balls to the axis of rotation).

This method was used by [17-19] and [9]. A computer was used to record and calculate the frictional torque and the coefficient of friction automatically.

2.7. *Flash temperature parameter*

a particular number known as the flash temperature parameter (FTP) which is used to determine the critical flash temperature. At this temperature and at given conditions, a lubricant would fail. The FTP can be determined by Equation (2). The FTP shows low breaking dawn possibility of lubricant film [18]. High value of FTP expresses high lubricant performance.

$$FTP = \frac{W}{(WSD)^{1.4}} \tag{2}$$

Where:
W= the applied load in kg
WSD= the wear scar diameter in mm.

3. Result and discussions

An investigation and a characterization of the Jatropha curcas oil with refined, bleached & deodorised palm olein (RBD) mixing effects under various loads. The obtained results which were also compared with 100% commercial lubricant give a deeper understanding of ball bearings worn surfaces utilizing oil analysis for example WSD, COF and FTP.

3.1. Density and kinematic viscosity

The unit of mass per volume is the fluid density. To determine the density of the lubricants specified earlier in this paper, an experiment has been conducted. The fluid resistance defined as kinematic viscosity. Shear stress or the tensile stress can deform the fluids also known as the fluid internal friction. To determine viscosity and fluidity, a viscometer was used which has a rotating spindle a specific speed. The rotation speed of the spindle encounters the viscosity or fluidity after dipping the spindle into the lubricant could be measured. Table 1 shows that as the lubricant temperature raises, the viscosity declines but the fluidity increases. It is easier for particles to move in lubricants with more fluid. Therefore, the fluids viscosity is influenced by the fluids temperature.

Table 1: fluid samples viscosity (mm^2/sec)

Temp(°C)	RB60/J40	RB 100%	ENG 100%
40	28.02	33.85	102.50
75	9.26	14.25	23.22
100	5.54	9.85	12.07

For lubricating oil, viscosity is the most significant property. The lubricant's viscosity is between 5.54 and 12.07mm^2/sec at 100°C. Whereas, it is between 28.02 and 102.50 mm^2/sec at 40°C. Prior to conducting the test, comparison between the viscosity of the of RBD palm olein and the blend of RBD palm olein and Jatropha curcas oil (RB60/J40) and the commercial mineral oil at 40, 75 and 100°C is present in Figure 3. The neat RBD palm olein viscosity is higher than blend of Jatropha curcas oil with refined, bleached & deodorised palm olein (RB60/J40) viscosity. However, at all temperature tests, it is less than viscosity of 100% commercial lubricant.

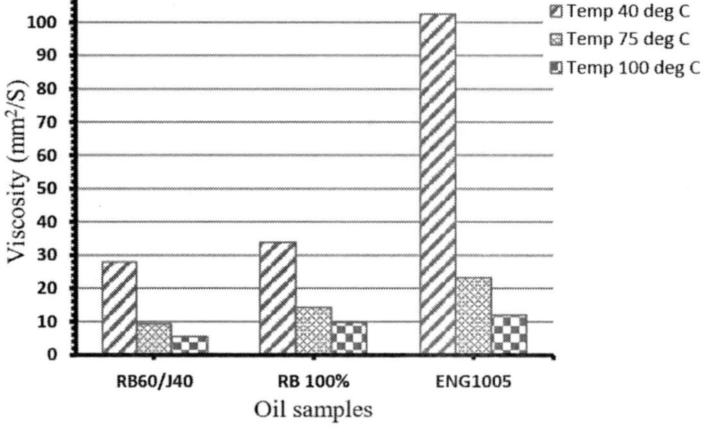

Figure 3: Oil samples viscosity at 40, 75 and 100°C.

IOP Conf. Series: Materials Science and Engineering **765** (2020) 012025 doi:10.1088/1757-899X/765/1/012025

3.2. Wear scar diameter

By a personal computer operating optical & scanning electron microscope, the diameter of wear area on the three ball bearings was monitored and determined. Figure 4 shows the wear scar diameter average values. The data of all lubricant samples presented in this figure indicates that wear scar diameter (WSD) increases with the increment of the normal loads. Thus, by reducing the WSD, the blend of RBD palm olein (RB60/J40) and Jatropha curcas oil operated as an anti-wear additive.

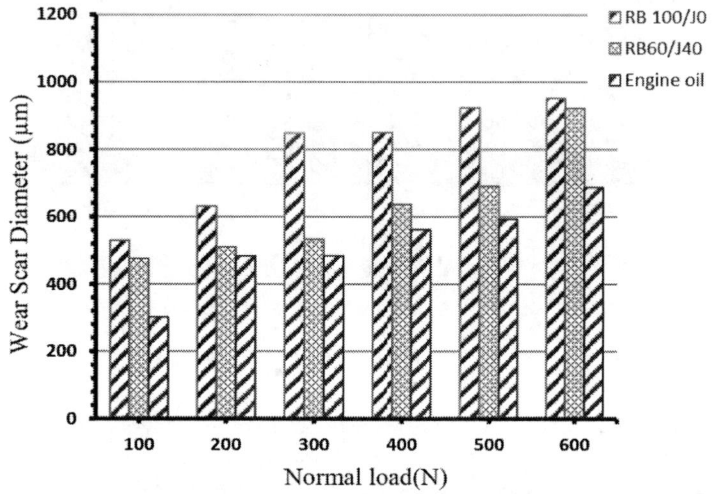

Figure 4: Wear Scar Diameter (µm) vs. Normal load (N)

3.3. Friction Torque

The mixture of refined, bleached & deodorised palm olein with Jatropha curcas oil (RB60/J40) friction performance was evaluated under 100 to 600 normal loads, 1200 rpm rotational speed, and 75°C, oil temperature for 60 min. the obtained friction torque results are shown in Figure 5 which shows that the friction torques of all oil mixtures used in this research increase in a similar pattern as normal load raises. The torque of friction was quickly raised in the starting phase of the tests and became a steady-state while heading to the last phase of the experiments. The steady-state condition of the friction torque indicates that the film of lubricant between ball bearings was unchanging with no lubricant layer breaking down. Nevertheless, at 600N, the friction torque increased along the process because of the breakdown of the lubricant film causing great friction at the metal to metal contact area.

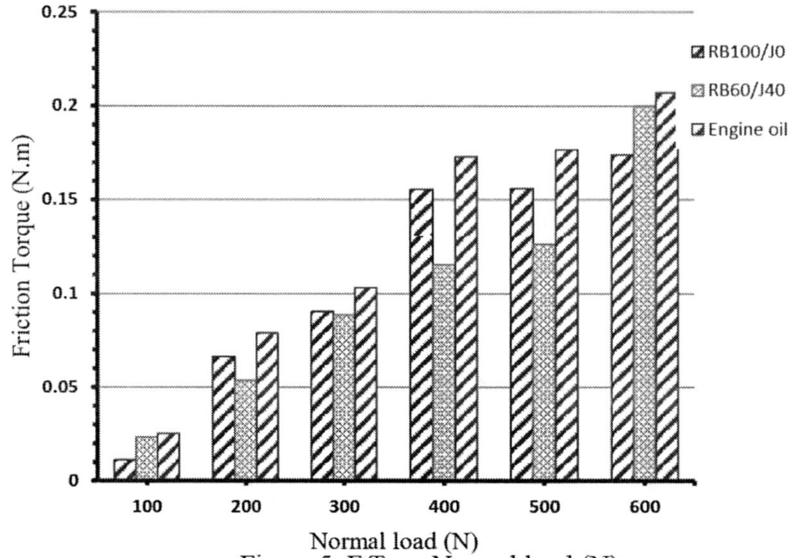

Figure 5: F.T vs. Normal load (N)

3.4. Coefficient of friction

The friction coefficients for all tested lubricants under the specified conditions were determined and plotted as illustrated in Figure 6. An analysis of the plotted results concludes that for all specified samples, the friction coefficients are less than those for 100% commercial lubricant. Also, at 100N normal load for the pure refined, bleached & deodorised palm olein, the least friction coefficient is (0.025126) also, for the mixture of the Jatropha curcas oil with RBD palm olein is (0.051319). Upon that it is concluded that 60% of refined, bleached & deodorised palm olein blended with 40% of Jatropha curcas oil gives the highest lubricating ability in comparison to the other tested lubricants.

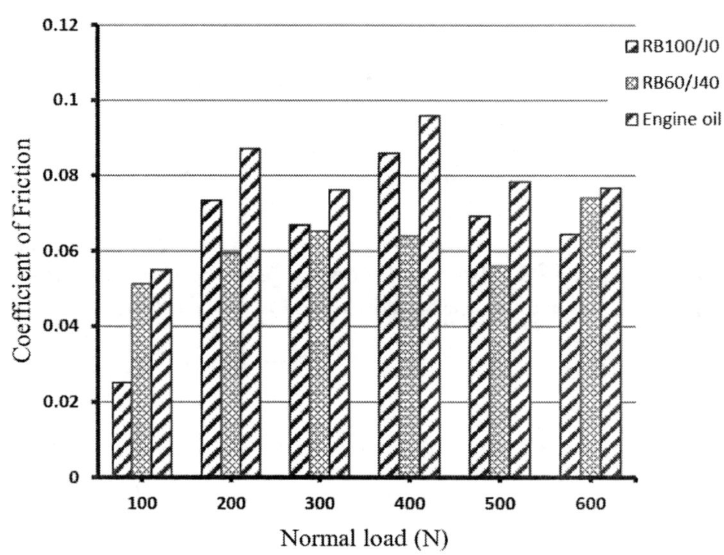

Figure 6: friction coefficients under different normal load

IOP Conf. Series: Materials Science and Engineering **765** (2020) 012025 doi:10.1088/1757-899X/765/1/012025

3.5. Flash temperature parameter

Figure 7 shows the results obtained from calculating and plotting the flash temperature parameter for the three tested lubricants using various normal loads.

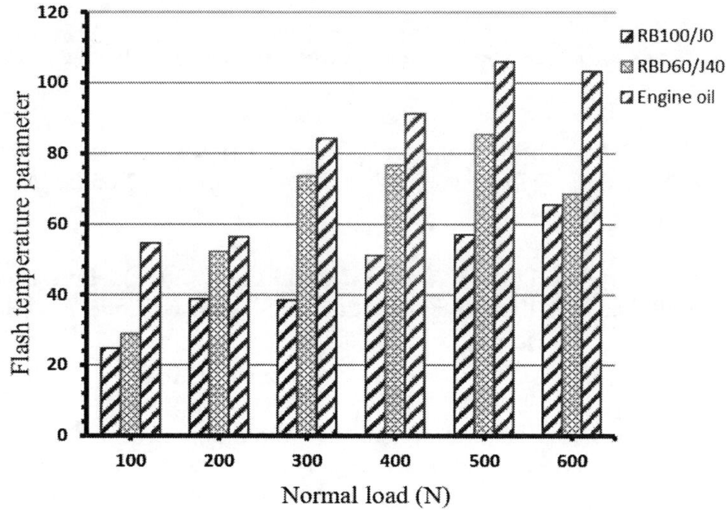

Figure 7: FTP vs. Normal load (N)

As shown in Figure 7, the greatest FTP For the neat refined, bleached & deodorised palm olein which is (65.458) happened at 600N normal load and (85.46) for palm olein (RB60/J40) at 500N. Thus, it can be concluded that when blending 60% refined, bleached & deodorised palm olein oil with 40% of Jatropha curcas oil, it boosts the lubricity performance and also reduces breakdown of lubricant film comparing to pure RBD palm olein lubricant.

4. CONCLUSIONS

A four ball tribotester was used to execute the tribological evaluation of the RBD palm olein and Jatropha curcas oil (RB60/J40). The experiment time was set to 60 min and the rotating speed of the top ball was 1200 rpm. A set of normal loads was used which are 100, 200, 300,400,500 and 600N. The collected data illustrated that (RB60/J40) had less friction coefficient in comparison to the other two tested lubricants. It also had less wear scar diameter under all normal loads. However, the RBD PS had a tendency to create larger wear scar due to the production of non-reactive detergent. From the observation on the surface topography of the worn surface, the rough surface (deep valley of asperities) that formed helped to create an oil reservoir of the RBD PS, and prevented metal-to-metal contact.

References

[1] Delgado, M.A.; Quinchia, L.A; and Galegos, C.(2010). Viscosity modification of different vegetable oil with EVA copolymer for lubricant applicants. Industrial Crops and Products, 32(1),607–612.

[2] Nosonovsky, M.(2000). Oil as a lubricant in the ancient Middle East. Tribology Online, 2(2), 44–49.

[3] M. Hassan et al.,(2016), The Tribological Characteristics of RBD Palm Olein with Jatropha Oil Blend Using Four-Ball Tribotester with Different Normal Loads, Applied Mechanics and Materials, Vol. 819, 499-503.

[4] Golshokouh,I.;Syahrullail,S.;Ani,F.N.;and Masjuki.H.H.(2013). Investigation of Palm Fatty Acid Distillate Oil as an Alternative to Petrochemical Based Lubricant.MPOB Journal,7(1),72-80.

[5] Mohammed Hassan Jabal, Abdulmunem R. Abdulmunem and Hussain Saad Abd (2019) Experimental investigation of tribological characteristics and emissions with nonedible sunflower oil as a bio lubricant, Journal of the Air & Waste Management Association, 69:1, 109-118, DOI: 10.1080/10962247.2018.1523070.

[6] Ahmed, H. I., Hassan, X. M., & Abdul, C. A. (2011). Performance changes for window type air-conditioner unit chemically treated to work in an industrial petrochemical environment. Engineering & Technology Journal, 29, 430-447.

[7] Wan Nik, W.B.; Ani, F.N.; and Masjuki, H.H.(2005). Thermal stability evaluation of palm oil as energy transport media, Energy Conservation and Management, 46 , 2198-2215.

[8] Syahrullail, S.; Nakanishi, K.; and Kamitani, S.(2005). Investigation of the effects of frictional constraint with application of palm olein oil lubricant and paraffin mineral oil lubricant on plastic deformation by plane strain extrusion. Journal of Japanese Society of Tribologists,50(12),877–885.

[9] Ing, C.T.; Mohammed Rafiq; A.K.; Azli, Y.; and Syahrullail, S.(2012). The effect of temperature on the tribological behavior of RBD palm stearin. TribologyTransactions, 55(5), 539–548.

[10] Hassan, M.;Syahrullail, S.; and Ani, F. N.(2016a).The tribological characteristics of the cactus and mineral oil blends using four-ball tribotester. Jurnal Teknologi, 78 (9-2), 33-38.

[11] Jabal M.H.; F.N. Ani; and S, Syahrullail.(2014). The Tribological Characteristic of the Blends of RBD Palm Olein with Mineral Oil using Four-ball Tribotester. Jurnal Technology, 69(6),11-14.

[12] Maleque, M.A.; Masjuki, H.H.; and Haseeb, A.S.M.A.(2000).Effect of mechanical factors on tribological properties of palm oil methyl ester blended lubricant. Wear, 239(1),117–125.

[13] Jabal, M. H., and Khalefa, M. Z. (2018). Tribological Characteristics Evaluation of Mustard Oil Blends. Journal of Engineering, 24(3), 1-12.

[14] Hassan, M.; Ani, F.N.; and Syahrullail, S. (2016). Tribological Performance of Refined, Bleached and Deodorised Palm Olein Blends Bio-lubricants. Journal of Oil Palm Research, 28(4), 510-519.

[15] Masjuki, H.H.;and M.A. Maleque.(1997).Investigation of the anti-wear characteristics of palm oil methyl ester (POME) contaminated with lube oil using a Four-ball machine of IP239 standard. Wear, 206,179-186.

[16] Fattah, F. A., Ariff, M. M., Hassan, J. M., and Nasir, A. F. (2017). The Tribological Characteristics of Rubber Seed Oil With Mineral Oil Blends. Malaysian Journal Of Industrial Technology (MJIT).

[17] Mohammed Hassan, Farid Nasir Ani and S. Syahrullail. (2016).Tribological Features of Refined, Deodorized, and Bleached Palm Olein with Mineral Oil Blend. Tribology Transactions, 59:4, 671-678.

[18] Hussain, A., Mehdi, S., Ali, A., Adeel, M., Jabal, M., & Ani, F. (2018, June 30). Investigation of tribological characteristics of castor oil with mineral oil blends. JOURNAL OF ENGINEERING AND APPLIED SCIENCES, 37(1). https:// doi.org/https: //doi.org /10.25211 /jeas.v37i1.2688.

[19] M. H. Jabal, E. O. Hassoun, A. R. Abdulmunem, F. N. Ani (2019).Experimental Investigation Of The Neat Rubber Seed Oil As A Hydraulic Fluid Using Four-Ball Tribotester. Journal of Engineering Science and Technology.Vol. 14, No. 5 . 2670 – 2680.

Experimental and Numerical Study of a New Corrugated and Packing Solar Collector

Jalal M. Jalil

Majida K. Ahmed

Hussein A. Idan

Electromechanical Engineering Department, University of Technology, Baghdad, Iraq

corresponding author's e-mail : 50124@uotechnology.edu.iq

Abstract. The thermal performance of a new solar air collector with corrugated packing is studied in this work. A new innovate shape was suggested in this paper to improve the thermal performance of the solar air collector. Mathematical models are developed to investigate the thermal performance of the collector and the results are verified by experiments. Effects of the operating parameters, solar radiation intensity and inlet air velocity are studied to optimize the thermal performance of the collector. Comparisons are conducted between the experimental and numerical results and good agreement was found. The new innovate shape shows better performance than other corrugated shapes and the absorber and air gained higher temperatures. The maximum temperature of air inside duct at the exit point was 69.1 C° at solar heat flow of 725 W/m² and inlet velocity 1.0 m/s. The new shape shows better performance than other corrugated absorber.

Keywords: Solar air collector; Corrugated packing; Thermal performance; Computational fluid dynamics

Nomenclature

As	South direction area heat transfer (m2)	An	North direction area heat transfer (m2)
Ae	East direction area heat transfer (m2)	Aw	West direction area heat transfer (m2)
At	Top direction area heat transfer (m2)	Ab	Bottom direction area heat transfer (m2)
qs	South direction conduction heat transfer (W)	Tgab	Temperature of gab (K)
qn	North direction conduction heat transfer (W)	To	Ambient temperature (K)
qe	East direction conduction heat transfer (W)	Tf	Air temperature inside duct (K)
qw	West direction conduction heat transfer (W)	Ti	Inlet air temperature (K)
qt	Top direction conduction heat transfer (W)	Te	Exit air temperature (K)
qb	Bottom direction conduction heat transfer (W)		
qsolar	Solar energy (W/m2)		
u,v,w	Velocity of x, y and z components (m/sec)		Greek letters
		μ	Dynamic viscosity of fluid (kg/m sec)
Cp	Specific heat at constant pressure (J/kg K)	αg	absorptivity of the glass
kg	Thermal conductivity of glass cover (W/m K)	ρ	Density of fluid (kg/ m3)

kw Thermal conductivity of absorbing wall (W/m K)
hi Heat transfer coefficient of air inside duct (W/m2 K)
ho Heat transfer coefficient due to wind (W/m2 K)
hgab Heat transfer coefficient of air between absorber plate and glass cover (W/m2 K)

1. Introduction

Solar energy is clean and their applications are increasingly utilized in buildings and in drying crops in the world. One of the main effective methods to enhance the convective heat-transfer rate is to increase the heat-transfer surface area and to increase the turbulence inside the channel by utilizing fins or corrugated surfaces and many studies have been carried out on this topic. For example, Ali and Hanaoka [1] studied experimentally the effects of the operating parameters on laminar flow forced-convection heat transfer for air flowing in a channel having a V-corrugated upper plate heated by radiation heat flux while the other walls are thermally insulated. Karim and Hawlader [2] studied experimentally the thermal performance for v-groove solar air collector for drying applications. In addition, they investigated both experimentally and theoretically in an effort to improve the performance of conventional air heaters. In both studies displayed the results indicate better thermal efficiency for a v-corrugated collector compared to a flat plate collector. Effects of operating variables on the thermal performance have been investigated. The results show that the temperature of the fluid at the exit of the collector decreases with flow rate resulting in an increase of efficiency due to decreased thermal losses to the environment [3]. [4] Investigated numerically the effects of wavy surface and thermal radiation heat transfer over an inclined wavy solar collector. The mathematical model presented is based on the energy transfer phenomenon within the various components of the collector. The transfer equations are discretized using the finite difference method. The proposed approach indicates that the wavy surface increases in general the performance of the solar collector.

Double-pass counter flow v-grove collector is considered one of the most efficient solar air-collectors. A mathematical model is developed for this type of collector by Karim, Perez and Amin [5]The simulation results proven the difference between the predicted and experimental results is, at maximum, approximately 7% which is within the acceptable limit considering some uncertainties in the input parameter values to allow comparison. To make a better use of the jet impingement, a GTC (glazed transpired solar collector) with perforating corrugated plate is developed by Zheng and his co-workers [6, 7]. The results indicate that the GTC with perforating corrugated plate is applicative enough for its advantages in economy and thermal performance in rural areas of cold regions. Also they studied the thermal performance of a novel solar air collector with metal corrugated packing in the buildings of cold regions. Mathematical models are developed to investigate the thermal performance of the collector and the results are verified by experiments. The hydraulic analysis is conducted experimentally to study the pressure drops of the air flows in the corrugated packing. The results indicate that the metal corrugated packing solar air collector is more appropriate to be used in the rural buildings of cold regions for its advantages of large heat transfer area, high heat transfer coefficient and good economic performance. The thermal performance of three types of solar air-heaters under several configurations and operating conditions are analyzed, measured and compared by [8]. All the analytical and experimental results show that, although the thermal performance of the type 2 heater is just slightly superior to that of the type 1 heater, both of these cross-corrugated solar air-heaters have a much superior thermal performances to that of the flat-plate one. An investigation was conducted on the thermal performance of a solar air collector with a v-groove absorber [9]. The results show that the v-groove collector has considerably superior thermal performance to the flat-plate collector.

By analyzing the finite element method the nonlinear behavior of steel plate shear walls with corrugated plates is investigated under lateral pushover loading conditions [10]. The results of this study have demonstrated that in the wall with constant dimensions, the trapezoidal plates have higher energy dissipation, ductility and ultimate bearing than sinusoidal waves, while decreasing the steel material consumption. An experimental investigation on the absorber surface of the collector whose shape was made up to provide better heat transfer surfaces was presented by [11].

This paper introduced mathematical and experimental work to investigate the solar collector performance with new design of absorber plate and different operation parameter cases of the city of Kut in Iraq.

2. Methodology and divided as a sub-suction as follows

2.1. Mathematical analysis

Figure. 1 displays the model for the solar collector. Air enters the collector from the opening that is in first end for any duct in this collector with an inlet temperature which is assumed equal to the uniform ambient air temperature. Hot air exits from the other side (second end) of the duct at outlet temperature. The collector has an eleven simple rectangular shape (ducts) connected together in special method and glass cover on one side with collector walls are isolated on the other sides. The mathematical formulation of the airflow problem is ruled by basic conservation of mass, momentum and energy equations. In the present study, the flow is laminar according to Reynolds number values (Re = 2138 for inlet velocity 1 m/s and Re = 2875 for inlet velocity 1.3 m/s). Reynolds number was calculated according to equation (1) and for the inlet air velocity range from 1.0 to 1.3 m/s and hydraulic diameter 0.038 m calculated according to equation (2), noting that all other properties were extracted from the standard air tables at the average temperature.

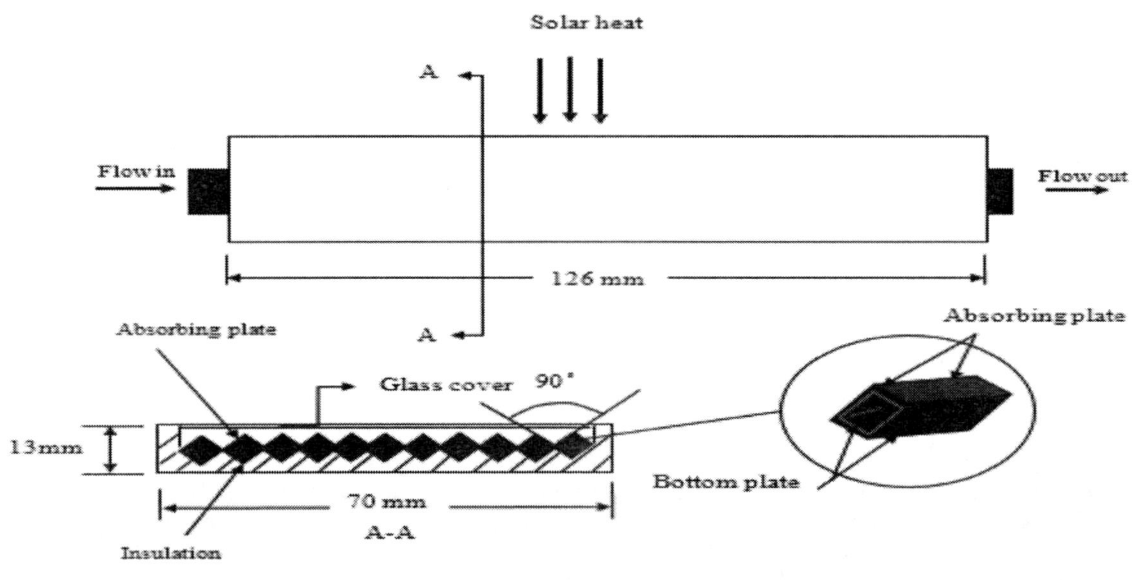

Figure 1: The Schematic diagram of solar collector

$$Re = \frac{\rho V D_h}{\mu} \qquad (1)$$

Where D_h is the hydraulic diameter of duct,

IOP Conf. Series: Materials Science and Engineering **765** (2020) 012026 doi:10.1088/1757-899X/765/1/012026

Dh= 4 A/p = (4 W*H)/2(W+H) (2)

Where A and P, are the area and perimeter of the duct

The following assumptions have been made:

1- Three-dimensional of conservation equations.
2- Steady state incompressible flow.
3- Conduction three-dimensional heat transfer through glass cover and absorber plate.
4- Air flow in the duct has been considered laminar according to the value of Reynolds number.
5- All properties are evaluated at an average temperature.

2.2. Governing Equations (Laminar Flow)

For solar collector, laminar 3-D Cartesian Coordinates Navier Stokes equations (continuity, momentum and energy) were solved by finite volume (SIMPLE algorithm) using build in program (Fortran 90), and the graphs were plotted using Tecplot software. The 3-D laminar Navier Stokes equations with forced convection are Awbi [12]:

I-Continuity equation

$$\frac{\partial}{\partial x}(\rho u) + \frac{\partial}{\partial y}(\rho v) + \frac{\partial}{\partial z}(\rho w) = 0 \qquad (3)$$

II-Momentum equation

u-Momentum (x-direction)

$$\frac{\partial}{\partial x}(uu) + \frac{\partial}{\partial y}(uv) + \frac{\partial}{\partial z}(uw) = -\frac{1}{\rho}\frac{\partial P}{\partial x} + v\left(\frac{\partial^2 u}{\partial x^2} + \frac{\partial^2 u}{\partial y^2} + \frac{\partial^2 u}{\partial z^2}\right) \qquad (4)$$

v-Momentum (y-direction)

$$\frac{\partial}{\partial x}(vu) + \frac{\partial}{\partial y}(vv) + \frac{\partial}{\partial z}(vw) = -\frac{1}{\rho}\frac{\partial P}{\partial y} + v\left(\frac{\partial^2 v}{\partial x^2} + \frac{\partial^2 v}{\partial y^2} + \frac{\partial^2 v}{\partial z^2}\right) \qquad (5)$$

w-Momentum (z-direction)

$$\frac{\partial}{\partial x}(uw) + \frac{\partial}{\partial y}(vw) + \frac{\partial}{\partial z}(ww) = -\frac{1}{\rho}\frac{\partial P}{\partial z} + v\left(\frac{\partial^2 w}{\partial x^2} + \frac{\partial^2 w}{\partial y^2} + \frac{\partial^2 w}{\partial z^2}\right) \qquad (6)$$

III-Energy equation

$$\frac{\partial}{\partial x}(uT) + \frac{\partial}{\partial y}(vT) + \frac{\partial}{\partial z}(wT) = \frac{k}{\rho\, C_p}\left(\frac{\partial^2 T}{\partial x^2} + \frac{\partial^2 T}{\partial y^2} + \frac{\partial^2 T}{\partial z^2}\right) \qquad (7)$$

2.3. Boundary Conditions

The mesh used was 51*21*21, and the number of iteration was 5000. The program was already tested for consistency in previous study.

Table 1: boundary conditions

Variable	**Wall**	**Inlet**	**Outlet**
Velocity	u, v, w = 0	$u = u_{in}$ v = w = 0	$\left[\frac{\partial u, v, w}{\partial n} = 0\right]$

IOP Conf. Series: Materials Science and Engineering **765** (2020) 012026 doi:10.1088/1757-899X/765/1/012026

Temperature	$\left[\frac{\partial T}{\partial n} = 0\right]$	$T = T_{in}$	$\left[\frac{\partial T}{\partial n} = 0\right]$

2.4. Heat Conduction

The energy is exchanged through solid wall. For steady state, plane, the heat flow in the x or y or z direction may be calculated by Fourier equation Holman [13]:

$$q = -kA\frac{\partial T}{\partial x} \qquad (8)$$

The total heat flow at any point in the material is the result of the (q) at that point. In this section, an analytical approach can be used for the solar collector walls. For glass cover shown in Figure.2.

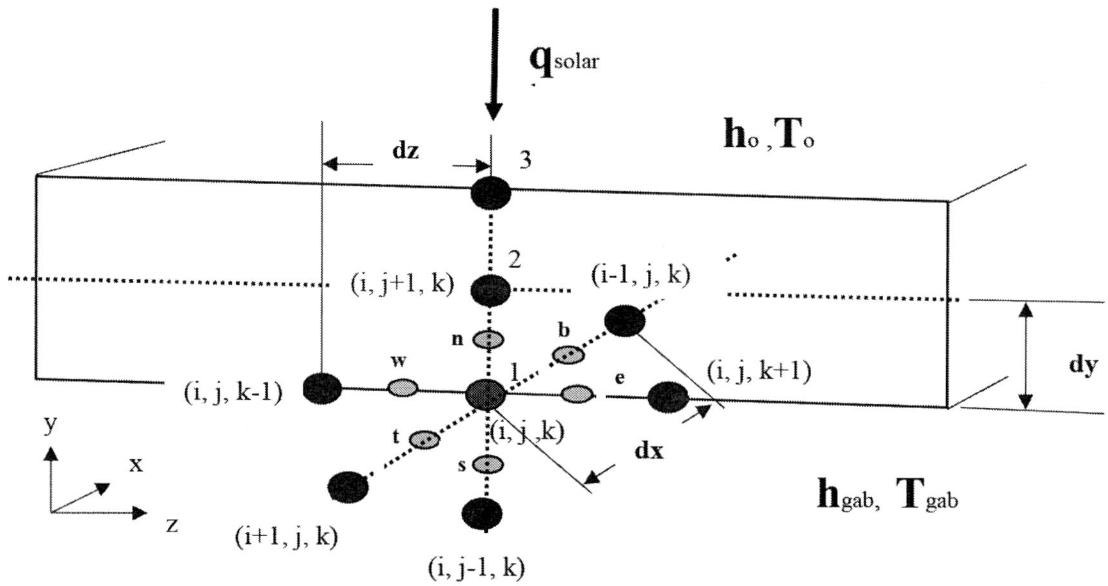

Figure: 2 Energy balance for node (1) on the glass

$$\sum q_{\text{in all face}} = q_{\text{storage}} \qquad (9)$$

Energy balance for node (1) on the glass cover.

$$\sum q_{\text{in all faces}} = q_s + q_n + q_e + q_w + q_t + q_b + q_{\text{solar }1}/4 \qquad (10)$$

$q_{\text{solar }1}$= the heat absorbed by glass cover from the solar radiation incident = $q_{\text{solar}} \times \alpha_g$

$$h_{gab}\, A_s\left(T_{i,j,k} - T_{gab}\right) + \frac{k_g A_n}{\Delta y}\left(T_{i,j,k} - T_{i,j+1,k}\right) + \frac{k_g A_e}{\Delta z}\left(T_{i,j,k} - T_{i,j,k+1}\right) + \frac{k_g A_w}{\Delta z}\left(T_{i,j,k} - T_{i,j,k-1}\right) + \frac{k_g A_t}{\Delta x}\left(T_{i,j,k} - T_{i+1,j,k}\right) + \frac{k_g A_b}{\Delta x}\left(T_{i,j,k} - T_{i-1,j,k}\right) + q_{\text{solar }1} A_n/4 = 0 \qquad (11)$$

Where

$A_s = \Delta x\, \Delta z$, $A_n = \Delta x\, \Delta z$ }

$$A_e = \left(\frac{\Delta y}{2}\right)\Delta x \quad , \quad A_w = \left(\frac{\Delta y}{2}\right)\Delta x$$
$$A_t = \left(\frac{\Delta y}{2}\right)\Delta z \quad , \quad A_b = \left(\frac{\Delta y}{2}\right)\Delta z \tag{12}$$

Energy balance for node (2) on the glass cover.

$$\sum q_{\text{in all faces}} = q_s + q_n + q_e + q_w + q_t + q_b + q_{solar\,1}/2 \tag{13}$$

$$\frac{k_g A_s}{\Delta y}\left(T_{i,j,k} - T_{i,j-1,k}\right) + \frac{k_g A_n}{\Delta y}\left(T_{i,j,k} - T_{i,j+1,k}\right) + \frac{k_g A_e}{\Delta z}\left(T_{i,j,k} - T_{i,j,k+1}\right) + \frac{k_g A_w}{\Delta z}\left(T_{i,j,k} - T_{i,j,k-1}\right) + \frac{k_g A_t}{\Delta x}\left(T_{i,j,k} - T_{i+1,j,k}\right) + \frac{k_g A_b}{\Delta x}\left(T_{i,j,k} - T_{i-1,j,k}\right) + q_{solar\,1} A_n/2 = 0 \tag{14}$$

Where

$$\left.\begin{aligned} A_s &= \Delta x\,\Delta z \quad , \quad A_n = \Delta x\,\Delta z \\ A_e &= \Delta y \Delta x \quad , \quad A_w = \Delta y\,\Delta x \\ A_t &= \Delta y \Delta z \quad , \quad A_b = \Delta y\,\Delta z \end{aligned}\right\} \tag{15}$$

Energy balance for node (3) on the glass cover.

$$\sum q_{\text{in all faces}} = q_s + q_n + q_e + q_w + q_t + q_b + q_{solar\,1}/4 \tag{16}$$

$$\frac{k_g A_s}{\Delta y}\left(T_{i,j,k} - T_{i,j-1,k}\right) + h_o A_n\left(T_{i,j,k} - T_o\right) + \frac{k_g A_e}{\Delta z}\left(T_{i,j,k} - T_{i,j,k+1}\right) + \frac{k_g A_w}{\Delta z}\left(T_{i,j,k} - T_{i,j,k-1}\right) + \frac{k_g A_t}{\Delta x}\left(T_{i,j,k} - T_{i+1,j,k}\right) + \frac{k_g A_b}{\Delta x}\left(T_{i,j,k} - T_{i-1,j,k}\right) + A_n q_{solar\,1}/4 = 0 \tag{17}$$

Where h_o the convection heat transfer due to wind, is given by McAdams [14] as:

$$h_o = 5.7 + 3.8v \tag{18}$$

$$\left.\begin{aligned} A_s &= \Delta x\,\Delta z \quad , \quad A_n = \Delta x\,\Delta z \\ A_e &= \left(\frac{\Delta y}{2}\right)\Delta x \quad , \quad A_w = \left(\frac{\Delta y}{2}\right)\Delta x \\ A_t &= \left(\frac{\Delta y}{2}\right)\Delta z \quad , \quad A_b = \left(\frac{\Delta y}{2}\right)\Delta z \end{aligned}\right\} \tag{19}$$

Similar approach was used for absorber and the number of mesh was 51*3*21. The conduction in glass and absorber was performed in subroutines within main program, which solve the governing equations for duct flow simultaneously.

2.5 Experimental Works

The experimental rig was built to study air flow and heat transfer in corrugated packing solar collector shown Figure. 3. The aim was to investigate the impacts of the change inlet air velocity and input power on air flow and heat transfer in the system. It consists of eleven ducts, glass cover and wood box all these as solar collector. The solar collector dimensions are 126 cm length, 65 cm width and 13 cm height. The one duct is manufactured of Aluminium plate with length 126 cm, width 4 cm and 4 cm height. All these ducts are assembled in special technique. The glass cover consists of glass windows with dimensions of 65 cm width and length of 126 cm and a thickness of 0.4 cm. A gap was created between the glass cover and the absorbing surface with thickness of 1.5 cm from the top of the wave and this gap was constant throughout the test. The absorbent surface formed from the assembly of aluminum ducts, so that it gives a crispy surface at an angle 90 ° (V-grooved).

Figure: 3 Corrugated packing solar collector

2.6. Experimental Apparatus

Figure. 4 shows a schematic diagram of the experimental apparatus, and a photograph of experimental apparatus is shown in Figure. 5, which consists of:
1. Blower. 2. Pipe. 3. Control Valve. 4. Transition Piece. 5. Test Main Section 6. Solar Simulator. 7. Voltage Regulator. 8. Measuring Units. a- Sensors, Arduino and lap top (temperature measurement). b- Solar Power Meter. c- Hot Wire Anemometer. d- Voltmeter. 9. Electrical Control.

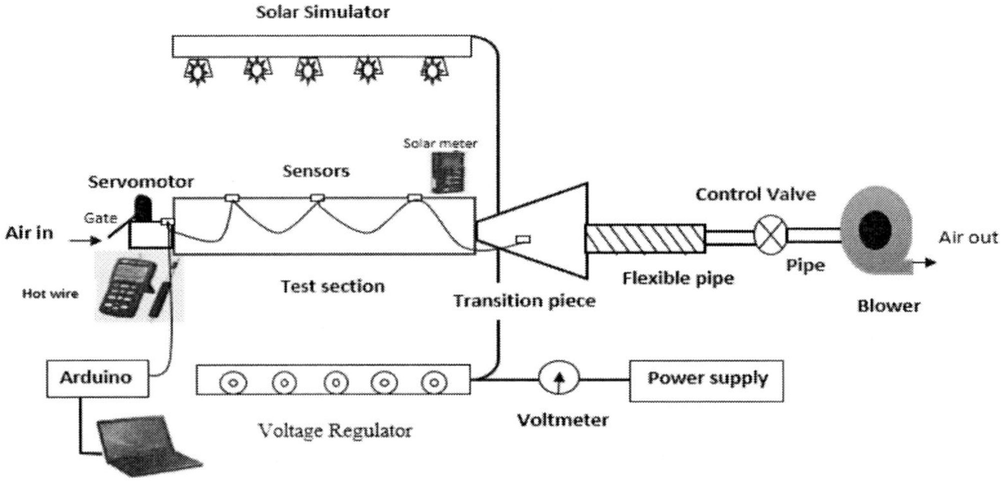

Figure: 4 Schematic diagram of the experimental apparatus

Figure: 5 Photograph of experimental apparatus

3. Result and discussions

The results have been calculated numerically and experimentally. The numerical results represent three-dimensional, laminar flow forced convection inside solar collector. Different parameters were discussed and studied numerically to exhibit the performance of a solar collector such as air velocity (1.0, 1.1, 1.2 and 1.3 m/s) and solar heat flux (400, 540, 640 and 725 W/m²). The experimental part also, exhibited the performance of a solar collector under the same conditions that were utilized in numerical part. The temperature change was compared through a channel for numerical and experimental testing.

Figure. 6 shows the comparison between experimental and numerical results. The comparison seems acceptable. The numerical results are higher than experimental results because of the losses through walls.

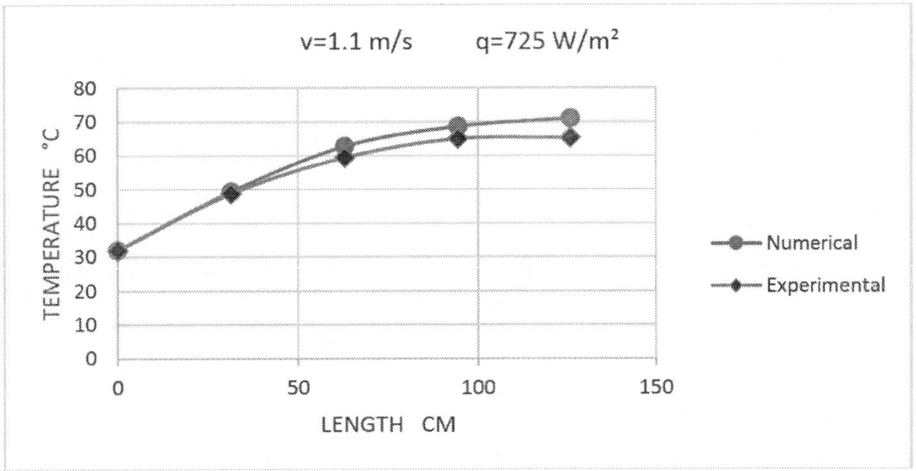

Figure: 6 Comparison between experimental and numerical results

Figure. 7 shows the comparison with Ali and Hanaoka (2002). Ali and Hanaoka (2002) used upper V-corrugated plate with no fins. Our new shape shows higher temperatures for absorber and air because both surfaces (top and bottom) transfer heat to the air. The fins transfer the heat quickly between upper and bottom surfaces.

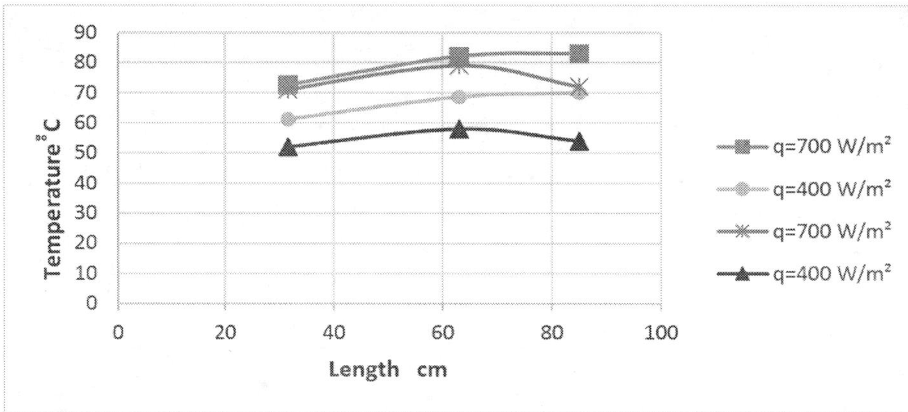

Figure: 7 Comparison with Ali and Hanaoka (2002) results at velocity = 0.65 m/s

Figure. 8 shows the variation in air temperature measured through the channel section in the various solar heat flows (400, 540, 640 and 725 W / m²) and air velocity is a constant input. All of these curves in the figure shows an increase in air temperature through the channel. Improved heat exchange zone and enhanced heat absorption contributed to the enhancement of thermal gains of the V-corrugated collector. V-corrugated collector benefited from its absorption of greater capacity of solar heat flux in increasing temperature.

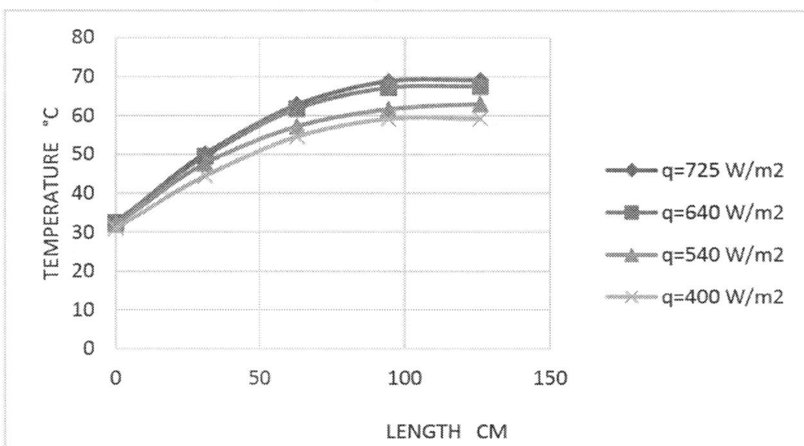

Figure: 8 .Change the air temperature through the channel at different solar heat fluxes and inlet velocity = 1.0 m/s.

Figure. 9 shows the variation of measured air temperature along the duct at different inlet air velocities (1.0, 1.1, 1.2 and 1.3) m/s and solar heat flux was constant at 725 W/m². Normally, all these curves in figure demonstrates an increase in the air temperature through the duct. However, the temperature decreases with an increase in inlet velocity from 1.0 to 1.3 m/s. For sample at 725 W/m², the maximum temperature of air inside the duct was in the exit point about 69.1 C° at 1.0 m/s and the minimum was about 61.8 C° at 1.3 m/s.

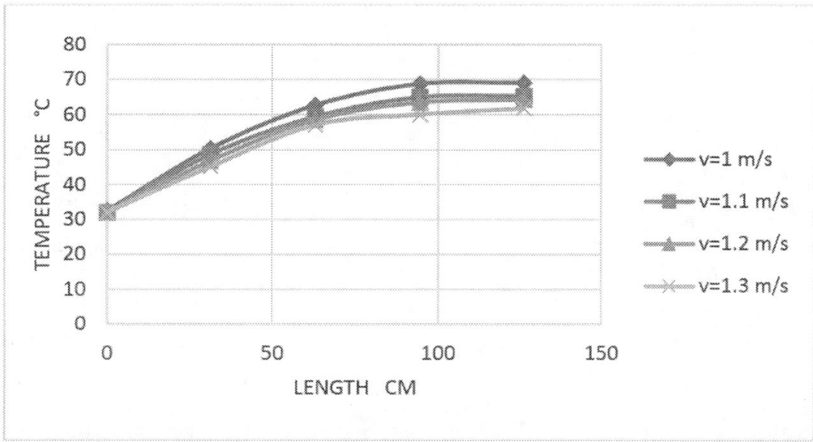

Figure: 9. Air temperature change through the channel at different inlet velocities and solar heat flow =725 W/m²

Figure. 10 shows the variation of absorber plate temperature with solar heat flux at different inlet air velocities (1.0, 1.1, 1.2 and 1.3 m/s). The absorber plate temperature increases with an increase in solar heat flux. Figure. 11 shows the variation of absorber plate temperature with inlet velocity at different heat fluxes (400, 540, 640 and 725 W/m²). The absorber plate temperature decreases with an increase in inlet air velocity.

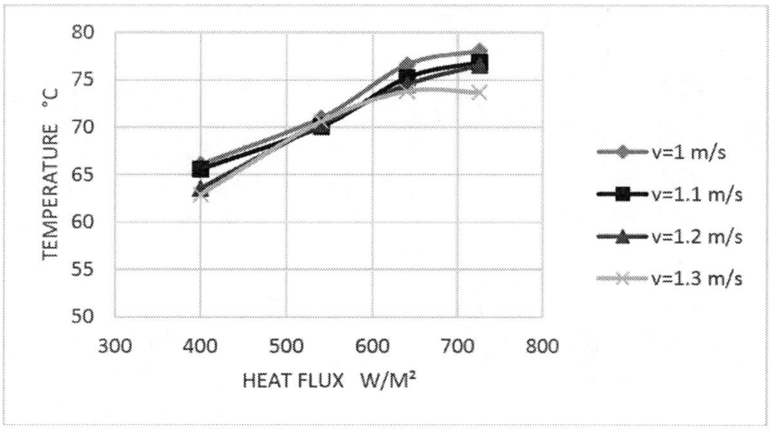

Figure: 10. Temperature variation of absorption plate with heat flow at different inlet velocities.

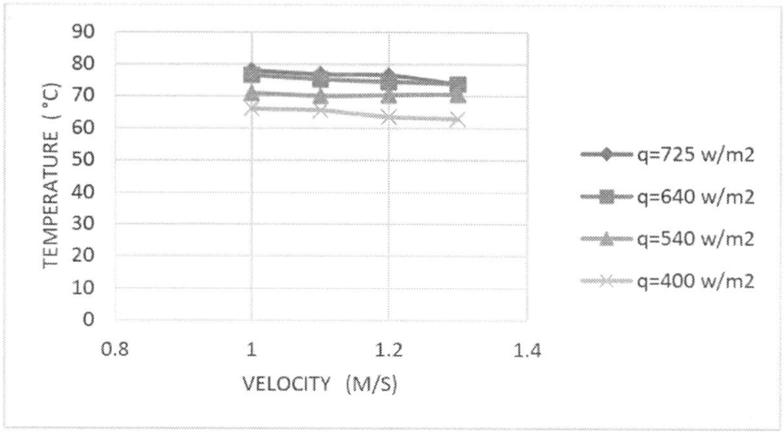

Figure: 11. Variation of absorption plate temperature with inlet velocity at different heat flux.

The effect of solar heat flux and inlet air velocity on glass cover temperature were similar to its effects on absorber plate temperature as shown in Figures.12 and 13.

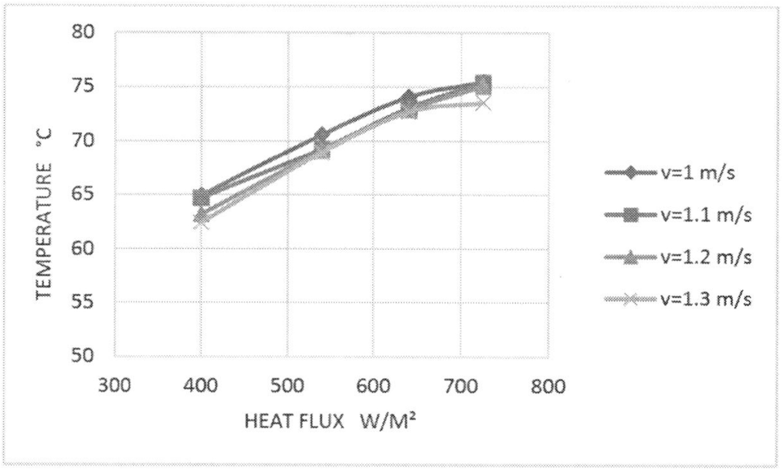

Figure: 12. Variation of glass cover temperature with heat flux at different inlet air velocities.

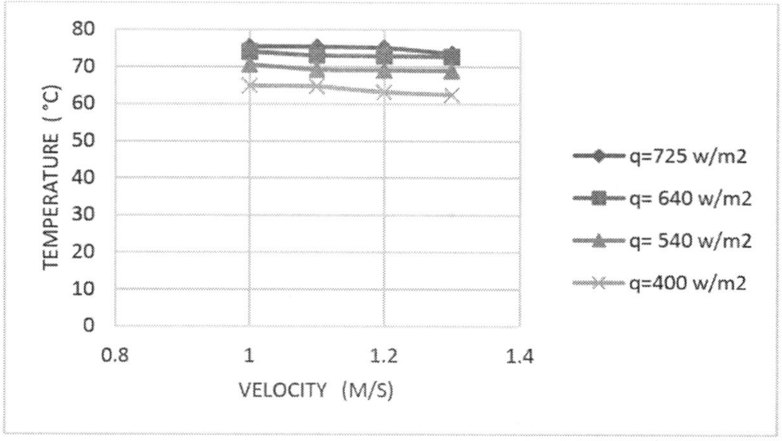

IOP Conf. Series: Materials Science and Engineering **765** (2020) 012026 doi:10.1088/1757-899X/765/1/012026

Figure: 13. Variation of glass cover temperature with inlet velocity at different heat flux.

Figure. 14 shows the isothermal contours inside duct in solar collector for 400 W/m^2 solar heat fluxes and 1.0 m/s inlet air velocities. It was observed that temperature begins to increases along duct, while in the radial direction, the high temperature at the channel walls begins to decline towards the center of that channel. Figure. 15 shows the velocity vector inside duct for 400 W/m² and inlet velocity = 1.0 m/s. The velocity was decreases towards the duct walls until it becomes zero at the walls, while at the center of the channel at its maximum value (boundary layer development).

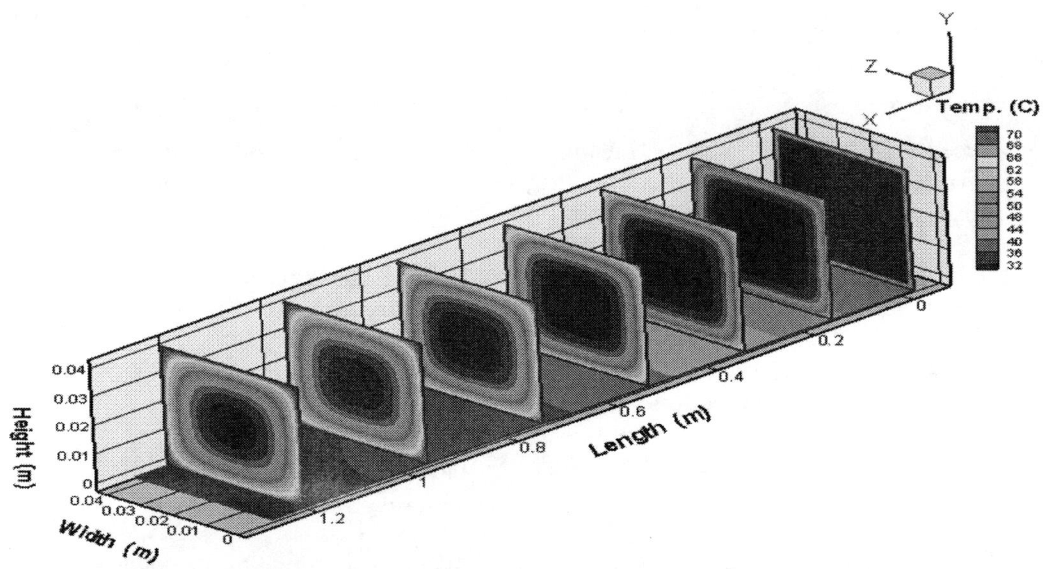

Figure: 14. Isothermal contours through the duct at solar flux = 400 W/m² and inlet velocitv = 1.0 m/s

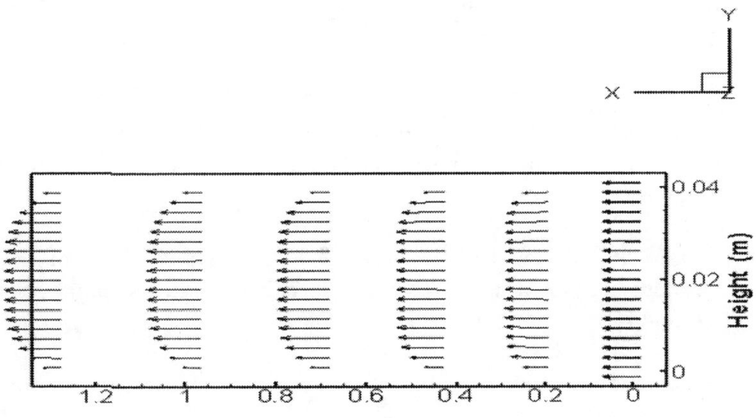

Figure: 15. Flow field of air through the duct for solar heat flow = 400 W/m² and inlet velocity=1.0 m/s

4. CONCLUSIONS

1- The maximum temperature of air inside duct at the exit point of the duct was 69.1 C° at solar heat flow of 725 W/m² and inlet velocity 1.0 m/s, and the minimum was about 52.6 C° at solar heat flux of 400 W/m² at inlet velocity 1.3 m/s.
2- Good agreement was found when compared between numerical and experimental results with deviation of about 4-11% for air temperature (at exit), 1.2-8% for absorber plate temperature (at center) and 1-6.7% for glass cover temperature (at center).
3- The new shape shows better performance than other corrugated absorber. The new shape separates the flow into many parallel duct flows where walls of the ducts work as fins.

References

[1] Ahmed Hamza H. Ali, Yutaka Hanaoka (2002). Experimental study on laminar flow forced-convection in channel with upper V-corrugated plate heated by radiation, Int. J. Heat and Mass Transfer 45 2107–2117.

[2] Md Azharul Karima, M.N.A Hawlader (2006). Performance evaluation of a v-groove solar air collector for drying applications, Applied Thermal Engineering 26, 121–130.

[3] Karima Md Azharul, Hawlader M.N.A (2006). Performance investigation of flat plate, v-corrugated and finned air collectors, Elsevier Journal, Energy 31, 452–470.

[4] Mayouf Si Abdallah,"Effects of the Wavy Surface on a Solar Collector(2016). Advances in Energy and Environmental Science and Engineering, ISBN: 978-1-61804-338-2.

[5] M.A. Karim, E. Perez, Z.M. Amin, (2014). Mathematical modelling of counter flow v-grove solar air collector, Renewable Energy 67, 192-201

[6] Wandong Zheng, Bojia Li, Huan Zhang, Shijun You, Ying Li, Tianzhen Ye, (2016). Thermal characteristics of a glazed transpired solar collector with perforating corrugated plate in cold regions, Energy 109, 781-790

[7] Wandong Zheng, Huan Zhang, Shijun You, Yindan Fu and Xuejing Zheng (2017). Thermal Performance Analysis of a Metal Corrugated Packing Solar Air Collector in Cold Regions, Applied Energy 203, 938–947.

[8] Wenfeng Gao, Wenxian Lin, Tao Liu, and Chaofeng Xia (2007). Analytical and experimental studies on the thermal performance of cross-corrugated and flat-plate solar air heaters, Applied Energy 84, 425–441.

[9] Tao Liu , Wenxian Lin , Wenfeng Gao , Chuanxu Luo , Ming Li , Qinhong Zheng & Chaofeng Xia (2007). A Parametric Study on the Thermal Performance of a Solar Air Collector with a V-Groove Absorber, International Journal of Green Energy, 4:6, 601-622.

[10] Yadollahi Y., Pakar I. and Bayat M (2015)..Evaluation and Comparison of Behavior of Corrugated Steel Plate Shear Walls, Latin American Journal of Solids and Structures 12, 763-786.

[11] S. Li, H. Wang, X. Meng, X. Wei, (2016). Comparative study on the performance of a new

IOP Conf. Series: Materials Science and Engineering **765** (2020) 012026 doi:10.1088/1757-899X/765/1/012026

solar air collector with different surface shapes, Applied Thermal Engineering, S1359-4311 (16) 33866-2

[12] Awbi, H. B, Ventilation of Buildings, (2003) London.

[13] Holman J. P., Heat Transfer,(2004) McGraw-Hill, Inc.

[14] McAdams, W. H, Heat Transmission, (1994) 3rd ed., New York: McGraw-Hill.

Heat Transfer Enhancement in Air Duct Flow By Micro-Channel Experimental And Numerical Investigation

Jalal M. Jalil

Ghada A. Aziz

Amjed A. Kadhim

Electromechanical Engineering Department, University of Technology, Baghdad, Iraq

corresponding author's e-mail : 50124@uotechnology.edu.iq

Abstract. The heating of air flow through micro-channel was studied experimentally and numerically to examine the improvement in the thermal performance achieved by micro-channel. Numerically, laminar incompressible 3D steady-state Navier Stokes equations were solved by Finite volume method. Experimentally, a rig was built to investigate the cooling of air flow through micro-channel for different velocities and electric powers heater. The micro-channels block (length = 0.1, width = 0.05, height = 0.005 m) was manufactured from copper metal with 10 rectangular micro-channels (length = 0.1, width = 0.001, height = 0.0005 m). The performance of the microchannel was evaluated through exit air temperature value. The studied parameters numerically and experimentally were air velocities inside micro-channels (1 to 20 m/s) and heater powers (1 to 5 W). The comparison between numerical and experimental results was acceptable and reached 3% as maximum. Also the results were compared with other investigator results. High heat transfer coefficients values were achieved by micro-channel, reaches maximum value of 150 W/m² K at velocity air 20 m/s and heater power 5 W.

Keywords: rectangular channel, micro-channel, CFD, 3D simulation, heat transfer coefficient.

NOMENCLATURE

cp	specific heat J/(kg K)
D_H	hydraulic diameter (m)
H_{ch}	channel height (m)
$H_{H.E}$	exchanger height (m)
K	thermal conductivity (W/m K)
L	channel length (m)
$M^{\cdot}$	mass flow rate (kg/s)
q	heat transfer rate (W)
T	temperature (K)
t_s	separating wall thickness (m)
u	x-component velocity (m/s)
V	y-component velocity (m/s)

W z-component velocity (m/s)
W_{ch} channel width (m)
A cross-sectional area of the flow
P wetted perimeter of the cross-section
ε heat exchanger effectiveness
μ dynamic viscosity (Pa.s)
ρ density (kg/m$^{3)}$
α= H/W aspect ratio
T_{out}, T_{in} outlet and inlet air Temperature, K)
υ kinematic viscosity of air (m^2/s)
X axial coordinate (m)
Y horizontal coordinates (m)
Z vertical coordinates (m)
cv Control Volume
FVM Finite Volume Method

1. Introduction

Over the past years there have been an increased investigations in micro-channel flow, that has included all manufacturing and industry part; from space, medical application and the car world, it have accuracy and small design of this work tend to have small good cooling.

Vladimir Glazar et al [1] Authors' main findings ,numerical analysis of fluid flow and heat transfer in the heat exchanger with microchannel coil (MCHX)., 3D mathematical model has been defined and appropriate numerical simulation of heat transfer has been performed , as it gives more precise results. Isaev et al [2] Authors presented Enhancement of heat transfer in laminar air flow in narrow channels, with designing mini- and micro-channels of microelectronics cooling devices, compact miniature heat exchangers, air capacitors, radiators. Vortex heat transfer is considered using the computational domain of the narrow micro channel of dimensionless height 1, width 6 and length 4 at periodic boundary conditions. An oval-trench dimple of dimensionless width 1 and length 4.5 is located in the center of the heated wall at an angle of 45° to incoming flow. To solve the Navier–Stokes equations and the energy equation, . The Reynolds number is equal to 103. Sun, X. Y., et al [3] presented two types of DCHEs, desiccant coated microchannel heat exchanger (DMHE) and desiccant coated fin-and-tube heat exchanger (DFHE), are manufactured and tested. The heat transfer, mass transfer and pressure drop performances of two DCHEs are analyzed and compared by experimental results. DMHE shows lower heat capacity, higher heat and mass transfer capacity compared with DFHE. Mushtaq et al. [4] numerically studied the effect the size and shape of channels on the performance of counter-flow micro-channel in heat exchanger and used water as a cooling liquid. They reached that if the size of channels decreasing, both the effectiveness of heat exchanger and pressure drop will increase and they mention that the decision of increase or decrease the size of channels will depends on the application. Mathew and Hegab [5] analyzed theoretically performance of parallel flow microchannel in heat exchanger subjected to constant external heat transfer. The equation predicts temperature distributions as well as effectiveness for the heat exchanger. Moreover, the model equation used when the fluids subjected to equal or unequal external heat transfer. Costaschuk et al. [6] investigated experimentally flow of water through aluminum rectangular microchannel (hydraulic diameter equals 169 μm and a Reynolds number range from 230 to 4,740). The average Poiseuille number for the flows was 86.4, which is in good agreement with the theoretical value of 86.9. The average critical Reynolds number was found 2,370. Li and Wang [7] examined experimentally fluid flow and heat transfer of six array of louvered fins for a range of Reynolds no 400–1600 and subsequently developed. Sabbah et al. [8] mentioned that the calculation of heat transfer in Microchannels becomes difficult with complexity of the geometry of the Microchannels, requiring

3d analysis of heat transfer in both liquid and solid phases. CFD models were implemented in order to investigate and optimize the fluid flow and heat transfer in microchannel heat sinks. Ryu et al. [9] parametrically studied the procedure to improve the work of corrugated louvered fins. They have worked fin pitch, louver pitch, and louver angle as the three most important parameters. The JF factor of the resulting optimum model was improved 14–32% compared to the base model for Reynolds numbers in the range $0 \leq Re\ Lp \leq 500$. Jang and Chen [10] used the conjugate gradient method to numerically investigate the thermal and hydraulic performance of a louvered fin heat exchanger with a variable louver angle for nine face velocities. Xia et al. [11] investigated the improved heat transfer effect in microsized channels. A numerical model has been set using CFD platform, hot flow is squeezed into the micro-channel and the heat is gradually transferred into the cold surface, top and bottom surfaces of the channel with rectangular area cross-sections. They reached that, heat transfer value per unit effective area is higher if the inlet area was decreased, and new design of heat exchanger is proposed based on these extra micro effects. Shinde and Lin [12] investigated the heat transfer and pressure drop of compact heat exchangers with louvered fins and flat tubes was conducted within a low air-side Reynolds number range of $20 < Re < 200$. They reached that better represented by two equations in two flow regimes (one for Reynolds = 20 – 80 and one for Reynolds = 80 – 200) than single regime equation in the format of power-law. Liu et al. [13] investigated experimentally the phase distribution of two-phase slug flow in horizontal multi-parallel micro channels, which was consisted of a header with hydraulic diameter of 0.667 mm and three parallel channels with hydraulic diameter of 0.5 mm, all of them have rectangular cross-sections. The inlet velocities of gas and liquid were from 0.130 to 1.47 m/s and from 0.128 to 0.900 m/s, respectively. They found that the phase distribution characteristics of two-phase flow in parallel channels highly depended on the inlet gas slug length and the inlet real velocity, a uniform phase distribution was achieved at high real velocity with short gas slugs. Shen et al. [14] investigated analytically the optimal location of internal vertical bifurcation integrated with a micro channel heat sink and the corresponding rectangular smooth micro channel is compared with those with internal vertical bifurcation. The simulation indicated that a clear inflection point of pressure gradient may prevail with the presence of internal vertical bifurcation. It was also found that the best thermal performance and the proposed optimal design of internal vertical bifurcation thermal performance without any pressure drop penalty.

To add more data on micro-channel literature, this work was organized to study micro-channel air heating experimentally and numerically. Numerically, laminar incompressible 3D steady-state Navier Stokes equations will solve by Finite volume method. Experimentally, a rig will built to investigate the heating of air flow through micro-channel for different velocities and electric powers heaters. The performance of the microchannel will evaluate through exit air temperature value.

2. Numerical Studies

In this study, the following assumptions have been made:

1- Three-dimensional
2- Steady-state and incompressible flow.
3- Conduction three-dimensional heat transfer in block of micro-channels.
4- Laminar flow (inlet Reynolds number is between is 500 to 2069).
5- All properties are evaluated at an average temperature.

The equations that describe the flow of air inside the micro-channel (continuity, momentum, and energy) were solved by finite volume method (SIMPLE algorithm) using Fortran 90 language (build in program). These equations are as follows [15, 16, and 17].
Continuity equation

Continuity equation

$$\frac{\partial}{\partial x}(\rho u) + \frac{\partial}{\partial y}(\rho v) + \frac{\partial}{\partial z}(\rho w) = 0 \tag{1}$$

Momentum equations

X direction (U Momentum)

$$\frac{\partial}{\partial x}(uu) + \frac{\partial}{\partial y}(uv) + \frac{\partial}{\partial z}(uw) - \frac{1}{\rho}\frac{\partial P}{\partial x} + v\left(\frac{\partial^2 u}{\partial x^2} + \frac{\partial^2 u}{\partial y^2} + \frac{\partial^2 u}{\partial z^2}\right) \tag{2}$$

Y direction (V Momentum)

$$\frac{\partial}{\partial x}(vu) + \frac{\partial}{\partial y}(vv) + \frac{\partial}{\partial z}(vw) = -\frac{1}{\rho}\frac{\partial P}{\partial y} + v\left(\frac{\partial^2 v}{\partial x^2} + \frac{\partial^2 v}{\partial y^2} + \frac{\partial^2 v}{\partial z^2}\right) \tag{3}$$

Z-direction (W Momentum)

$$\frac{\partial}{\partial x}(uw) + \frac{\partial}{\partial y}(vw) + \frac{\partial}{\partial z}(ww) = -\frac{1}{\rho}\frac{\partial P}{\partial z} v\left(\frac{\partial^2 w}{\partial x^2} + \frac{\partial^2 w}{\partial y^2} + \frac{\partial^2 w}{\partial z^2}\right) \tag{4}$$

Energy equation

$$\frac{\partial}{\partial x}(uT) + \frac{\partial}{\partial y}(vT) + \frac{\partial}{\partial z}(wT) = \frac{k}{\rho\, C_p}\left(\frac{\partial^2 T}{\partial x^2} + \frac{\partial^2 T}{\partial y^2} + \frac{\partial^2 T}{\partial z^2}\right) \tag{5}$$

The boundary conditions for the block and air flow are shown in the Table 3. The boundary conditions for air are shown in Table 1

Table 1 Boundary Conditions

Variable	Wall	Inlet	Outlet
Velocity	u, v, w = 0	$u = u_{in}$, v = w = 0	$\left[\frac{\partial u, v, w}{\partial n} = 0\right]$
Temperature	$\left[\frac{\partial T}{\partial n} = 0\right]$	$T = T_{in}$	$\left[\frac{\partial T}{\partial n} = 0\right]$

By solving the above governing equations, the velocities and temperature distribution is determined in the air flow.

IOP Conf. Series: Materials Science and Engineering **765** (2020) 012027 doi:10.1088/1757-899X/765/1/012027

For the block, one-tenth of the block was studied (symmetry) by solving energy equation in the block as shown in Fig. 1. The energy equation was solved by thermal resistances. For example,

Node 1,

$$q_w = -\frac{k_w A_w}{\Delta z}\left(T_{i,j,k} - T_{i,j,k-1}\right) \tag{6}$$

$$q_e = h_{in} * Ae * (T_{i,j,k} - T_{air}) \tag{7}$$

$$q_n = -\frac{k_w A_n}{\Delta y}\left(T_{i,j,k} - T_{i,j+1,k}\right) \tag{8}$$

$$q_s = -\frac{k_w A_s}{\Delta y}\left(T_{i,,j,} - T_{i,j-1,k}\right) \tag{9}$$

$$q_t = -\frac{k_w A_t}{\Delta x}\left(T_{i,j,k} - T_{i+1,j,k}\right) \tag{10}$$

$$q_b = -\frac{k_w A_b}{\Delta x}\left(T_{i,j,k} - T_{i-1,j,k}\right) \tag{11}$$

$$C_S = \frac{kA_s}{\Delta y}, \quad C_e = h_{in} * Ae\,, \qquad C_t = \frac{kA_t}{\Delta x}$$

$$C_n = \frac{kA_n}{\Delta y}, \qquad C_w = \frac{kA_w}{\Delta z}, \qquad C_b = \frac{kA_b}{\Delta x} \tag{12}$$

$$C = C_s + C_n + C_e + C_w + C_t + C_b \tag{13}$$

$$T_{(I,J,K)} = T_{(i,j,k)} * (1 - w) + w * \left[c_s\, T_{(i,j-1,k)} + c_n\, T_{(i,j+1,k)} + c_e\, T_{air} + c_w\, T_{(i,j,k-1)} + c_t\, T_{(i+1,j,k)} + c_b\, T_{(i-1,j,k)} + q_{(i,j,k)}\right]/C \tag{14}$$

The energy equation for the block was solved through a subroutine within the main program which solves Navier-Stokes equations in air flow simultaneously.

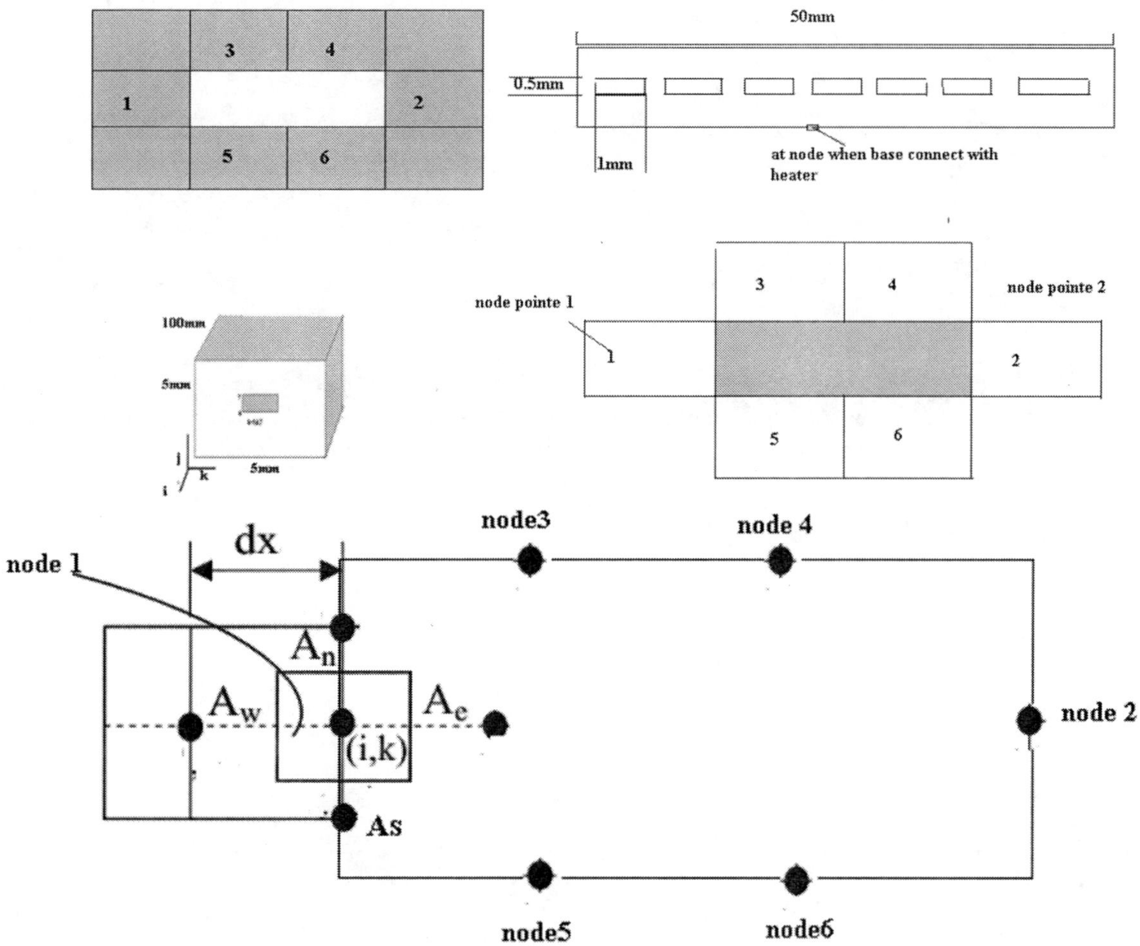

Figure 1: Block of micro-channels nodes surrounding micro-channel

3. Experimental Studies

A schematic diagram of the test facility (block) is shown in Figs. 2, 3 and 4, which was made from copper. The overall dimensions are 50 mm wide x 5 mm high x 100 mm long. The block has ten micro-channels. The dimensions of single rectangular micro-channel are 1 mm wide x 0.5 mm high x 100 mm long (Tables 1 and 2). The working fluid is air; the inlet temperature was maintained at 26°C for all tests by control circuit. Air was drawn from the room through a blower and then through micro-channel equipped with a programmable variable speed drive (digital). Electric heater, with PID controllers, was installed in the bottom of the block of the microchannels. The electric power heater values vary between 1 to 5 W, while the velocity of the air inside the micro-channel varies between 2 and 20 m/s.

The hydraulic diameter,

$$D_h = \frac{4A}{P} = \frac{4 * W_{ch} * H_{ch}}{2(W_{ch} + H_{ch})} . \qquad (15)$$

IOP Conf. Series: Materials Science and Engineering **765** (2020) 012027 doi:10.1088/1757-899X/765/1/012027

Reynolds number,

$$Re = \frac{\rho \, v \, D_h}{\mu} \qquad (16)$$

To calculate the convective heat transfer coefficient of the flow,

$$Q = m^{\cdot} * C_p * (T_{ain} - T_{aout}) \qquad (17)$$

The averaged heat transfer coefficients were defined as

$$h = \frac{Q}{A_s * (T_w - T_b)} \qquad (18)$$

Where $A_s = (2H + 2W)\,L$. The channel wall temperature, Tw, is assumed to be uniform and equal to the average of the readings from the six thermocouples located in the copper block.

$$T_b(\text{Average bulk mean temperature}) = \frac{T_{in+}T_{out}}{2}$$

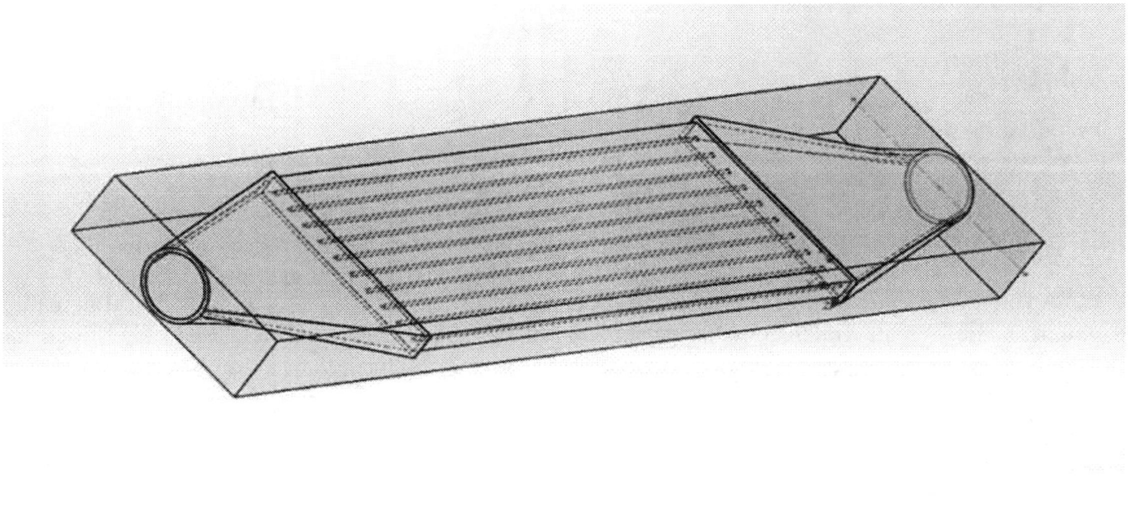

Figure 2: Schematic of Microchannels

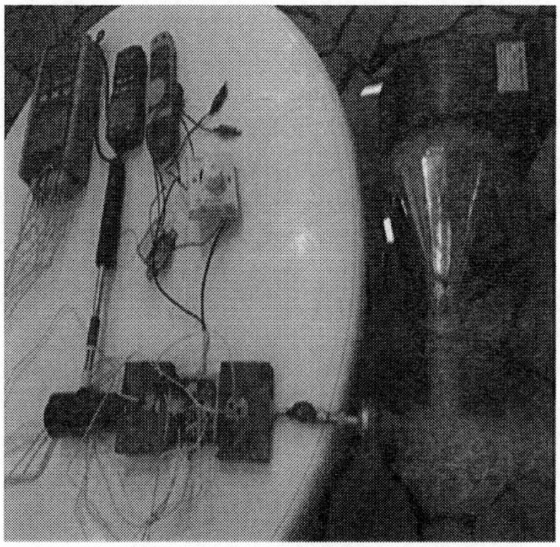

Figure 3 Photo of the experimental rig

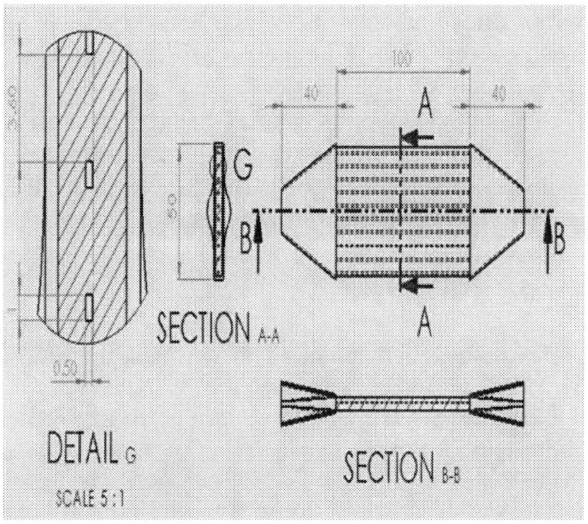

Figure 4 Geometry of micro-channels b

Table 2 Dimensional geometry of the micro-channel

No of channels	L_{ch}	w_{ch}	H_{ch}
10	100 mm	1 mm	0.5 mm

Table 3 Dimensional geometry of the micro-channels block

length	width	thickness
100 mm	50 mm	5 mm

4. Results and Discussions

In the experimental part, the studied parameters were; air velocity (1 to 20 m/s) and the heater powers (1 to 5 W). Fig. 5 shows the variation of air outlet temperature with inlet air velocities for various heater powers. The temperature decreases as the inlet air velocities increases from 2 to 20 m/s. The temperature increases as heater power increases. The micro-channel shows great performance; the temperature of the air starts with 26 °C at inlet and reaches 80 °C at outlet with heater power 5.5 W in only 10 cm length. Fig. 6 shows the variation of heat transfer coefficient with various inlet air velocities at variable heater powers. The heat transfer coefficient increases with increases of velocity reaching value of 150 W/m^2 K which is very high value for air to achieve. Fig. 7 shows variation of heat transfer coefficient with Reynolds number for different heater powers. The figure shows that the flow is laminar, although the flow is laminar, the heat transfer coefficient reaches 150 W/m^2 K. This value usually cannot be achieved even with turbulent flow for air. Fig. 8 shows the variation of air outlet temperature with inlet air velocity for various power heaters (numerical results). Fig. 9 shows the comparison between the experimental and numerical results of the outlet air temperature with inlet air velocities for various heater powers. The comparison was good. The numerical value was higher than experimental value because of small losses which usually ignored in numerical solution

IOP Conf. Series: Materials Science and Engineering **765** (2020) 012027 doi:10.1088/1757-899X/765/1/012027

(insulated walls). Fig. 10 shows that the flow field of air through micro-channel at power = 5 W, U_{ch} = 5 m/s, the boundary layer was built even in small length (10 cm). Fig. 11 shows the isothermal contours of air flow inside micro-channel .The developing of thermal boundary layer was clear in three distances. The temperature reaches 80 °C as average at the end for velocity 2 m/s and power 5 W. The outlet air temperature is nearly uniform in cross-section which means that the heat reaches the core of the flow (main advantage of the micro-channel). Fig. 12 shows comparison of the present results with Ref [18] results for the variation of Nusselt numbers with Reynolds number. The present study fits well with Ref [18] results.

5. Conclusions

The conclusions that can be drawn from the present study are:

1- Good agreement with deviation of about 2-3% between numerical and experimental results for outlet temperature.

2- The outlet air temperature is nearly uniform in cross-section.

3- The outlet air temperature from micro-channel was increased as velocity decreased.

4- CFD is a powerful tool to analyze and model to predict the temperature and velocity distribution in the flow of the microchannel and also to predict temperature of the body of the microchannel.

7. The value of heat transfer coefficient was reached high value 150 W/m² K, which cannot be reached without micro-channel.

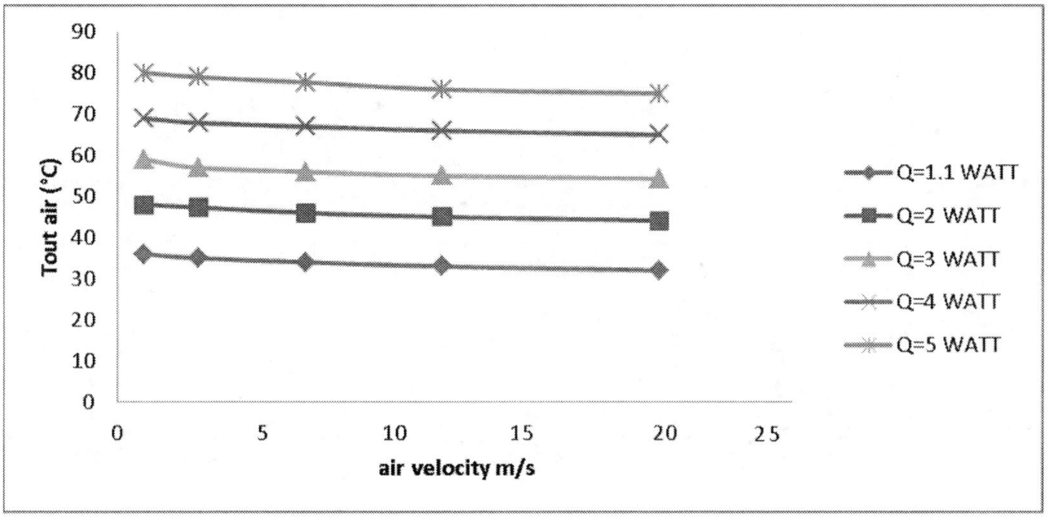

Figure 5: Variation of air outlet temperature with inlet air velocities for various power heaters

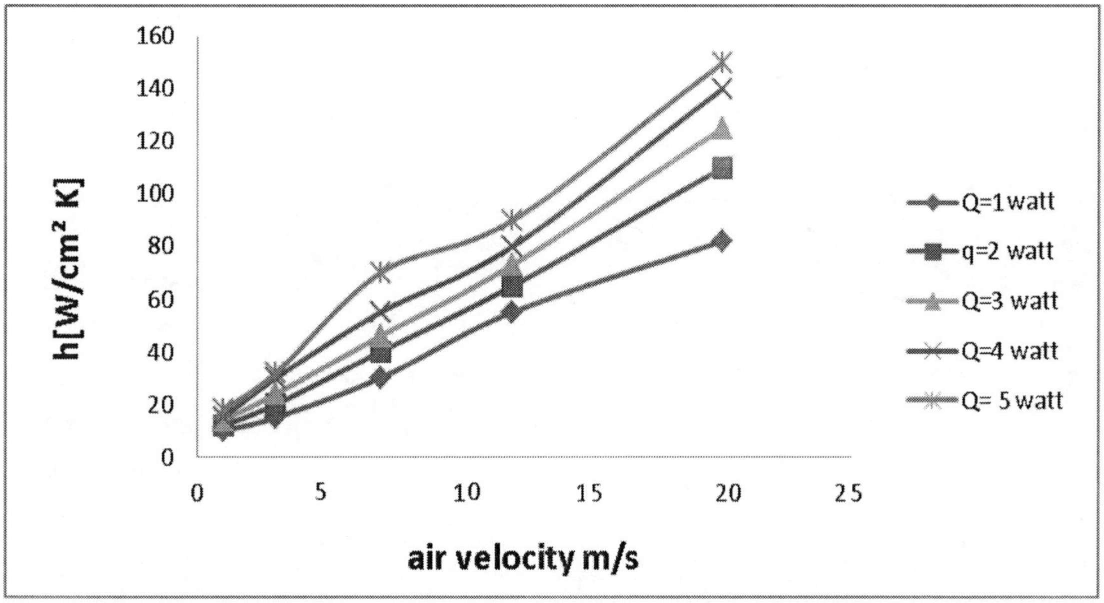

Figure 6: Variations of heat transfer coefficient with inlet air velocities at different power heaters

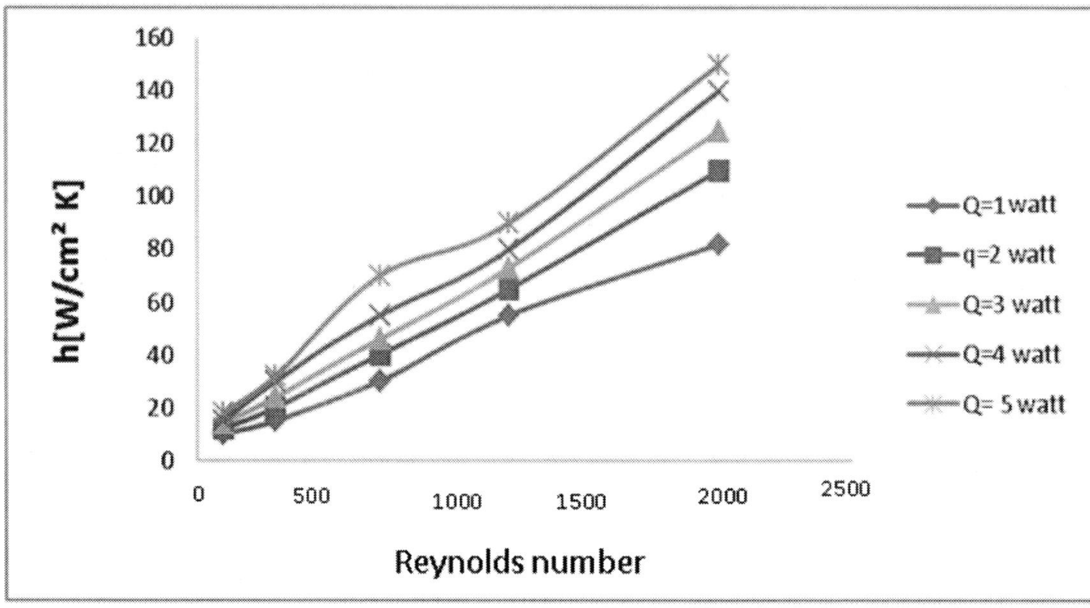

Figure 7: Variation of heat transfer coefficient with Reynolds number for different power heaters

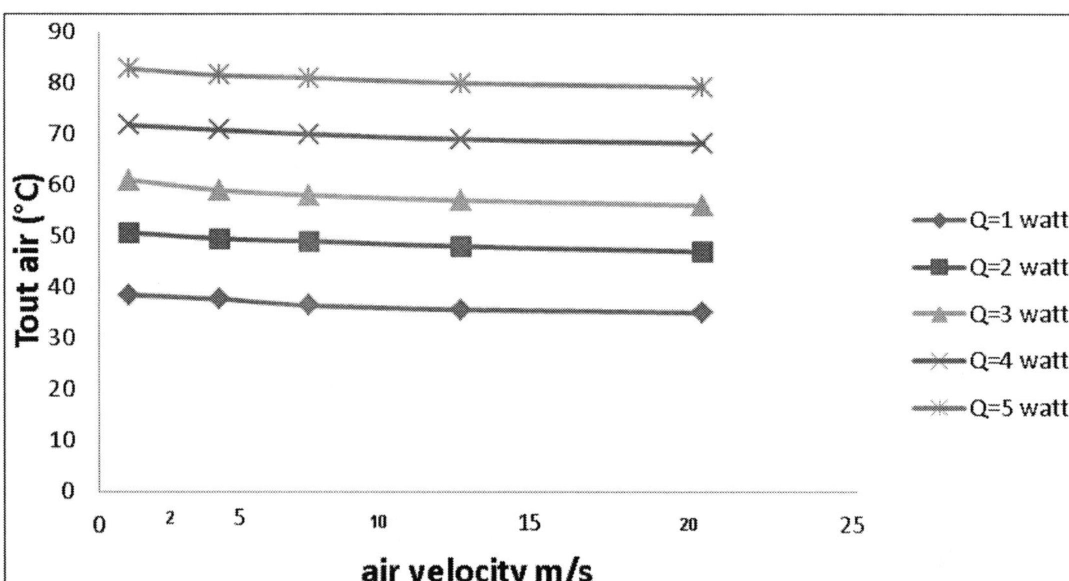

Figure 8: Numerical variation of air outlet temperature with inlet air velocities for different power heaters

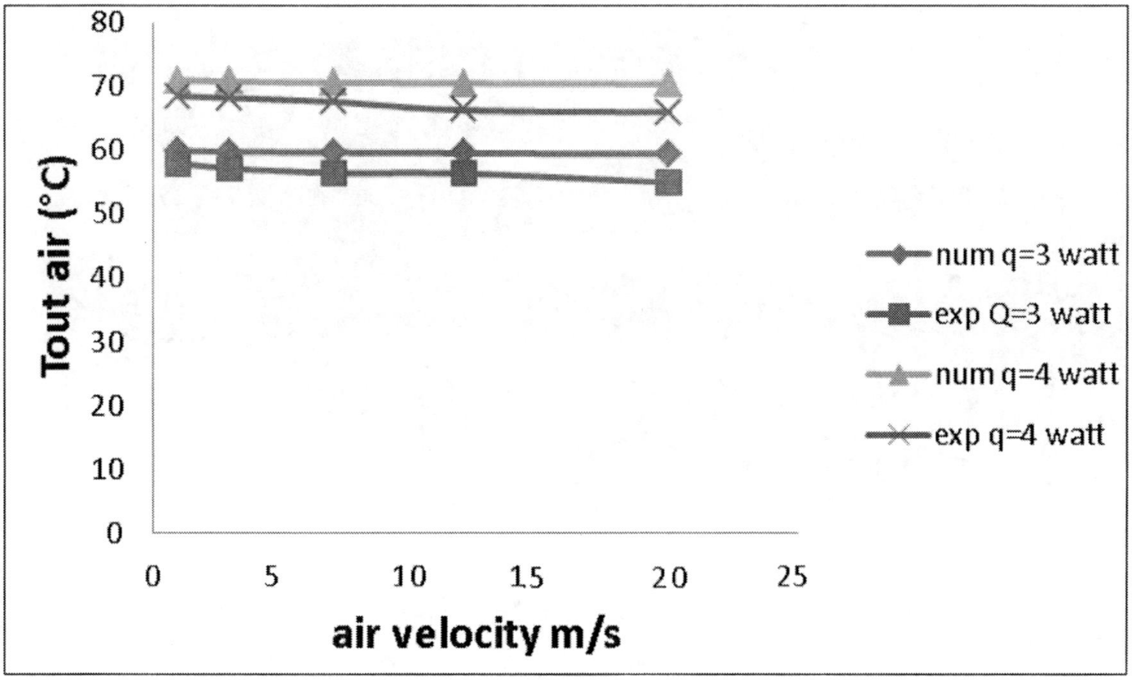

Figure 9: Comparison between the experimental and numerical results of the variation of air outlet temperature with inlet air velocities for 3 and 4 W power heaters

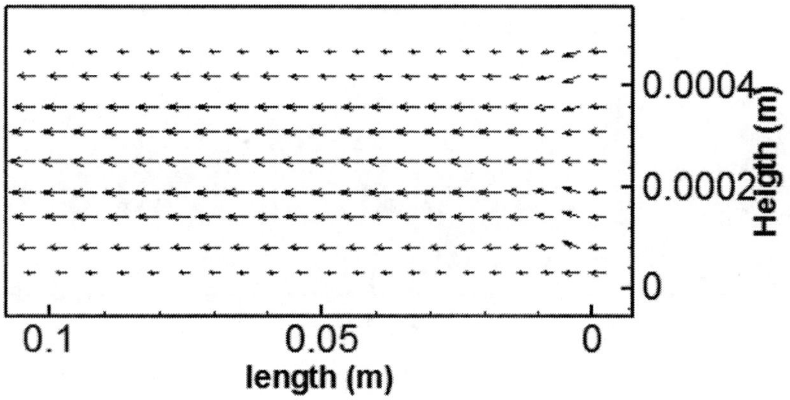

Figure 10: Flow field through micro-channel at power = 5 W, U_{ch} = 5 m/s

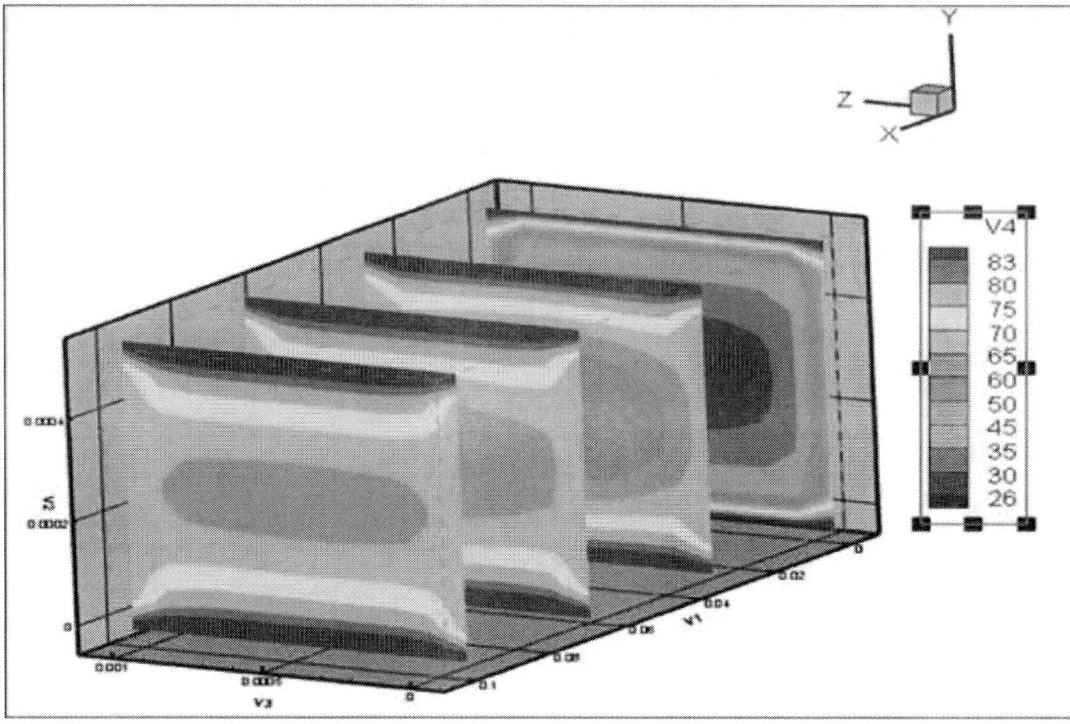

Figure11: Isothermal contours of air flow at power = 5 W, U_{in}= 5 m/s)

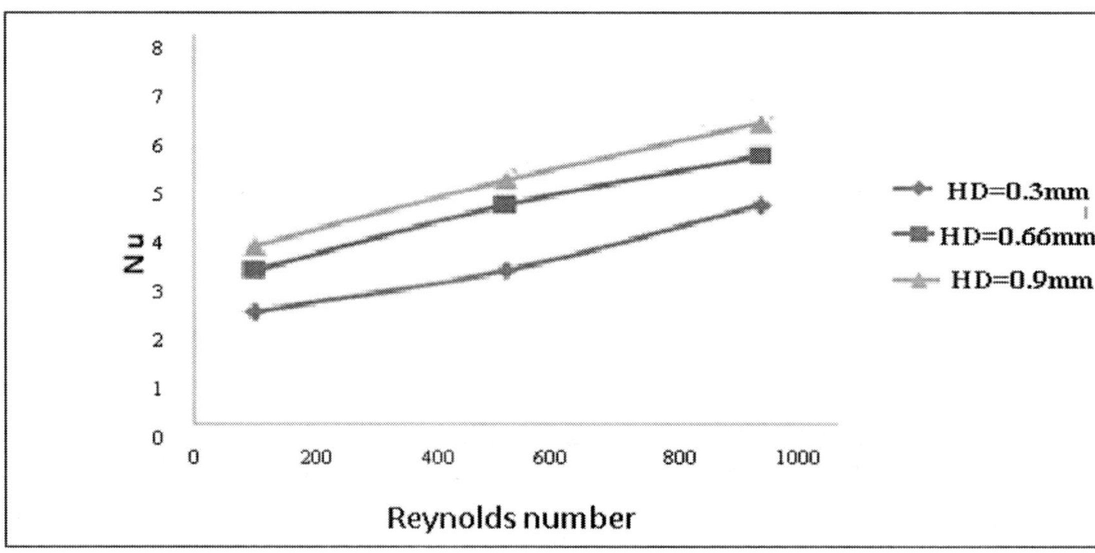

HD = 0.3 and 0.9 mm [18], 0.66 mm present study

Figure 12: Comparison between numerical Nusselt number with Reynold number with different hydraulic diameters of present and other research results [18],

References

[1] Glazar, V., Trp, A., Lenic, K., & Torbarina, F. (2019). Numerical analysis of heat transfer in air-water heat exchanger with microchannel coil. In E3S Web of Conferences (Vol. 95, p. 02004). EDP Sciences.

[2] Isaev, S. A., Leontiev, A. I., Milman, O. O., Popov, I. A., & Sudakov, A. G. (2019). Influence of the depth of single-row oval-trench dimples inclined to laminar air flow on heat transfer enhancement in a narrow micro-channel. International Journal of Heat and Mass Transfer, 134, 338-358.

[3] Sun, X. Y., Dai, Y. J., Ge, T. S., Zhao, Y., & Wang, R. Z. (2019). Heat and mass transfer comparisons of desiccant coated microchannel and fin-and-tube heat exchangers. Applied Thermal Engineering, 150, 1159-1167..

[4] Hasan, M. I., Shkarah, A. J., Ali, M. S., & Rageb, A. M. A. (2010). Using of Microencapsulated Phase Change Material Suspension to Enhance the Performance of Microchannel Heat Exchanger. Thi-Qar University Journal for Engineering Sciences, 1(2), 1-12.

[5] Mathew, B., & Hegab, H. (2010). Application of effectiveness-NTU relationship to parallel flow microchannel heat exchangers subjected to external heat transfer. International Journal of Thermal Sciences, 49(1), 76-85.

[6] Costaschuk, D., Elsnab, J., Petersen, S., Klewicki, J. C., & Ameel, T. (2007). Axial static pressure measurements of water flow in a rectangular microchannel. Experiments in fluids, 43(6), 907-916.

[7] Li, W., & Wang, X. (2010). Heat transfer and pressure drop correlations for compact heat exchangers with multi-region louver fins. International Journal of heat and mass transfer, 53(15-16), 2955-2962.

[8] Sabbah, R., Farid, M. M., & Al-Hallaj, S. (2009). Micro-channel heat sink with slurry of water with micro-encapsulated phase change material: 3D-numerical study. Applied Thermal Engineering, 29(2-3), 445-454.

[9] Ryu, K., Yook, S. J., & Lee, K. S. (2014). Optimal design of a corrugated louvered fin. Applied thermal engineering, 68(1-2), 76-79.

[10] Jang, J. Y., & Chen, C. C. (2015). Optimization of louvered-fin heat exchanger with variable louver angles. Applied Thermal Engineering, 91, 138-150.

[11] Xia, L., & Chan, Y. (2015). Investigation of the enhancement effect of heat transfer using micro channel. Energy Procedia, 75, 912-918.

[12] Shinde, P., & Lin, C. X. (2017). A heat transfer and friction factor correlation for low air-side Reynolds number applications of compact heat exchangers (1535-RP). Science and Technology for the Built Environment, 23(1), 192-210.

[13] Liu, Y., Sun, W., & Wang, S. (2017). Experimental investigation of two-phase slug flow

distribution in horizontal multi-parallel micro-channels. Chemical Engineering Science, 158, 267-276.

[14] Shen, H., Wang, C. C., & Xie, G. (2018). A parametric study on thermal performance of microchannel heat sinks with internally vertical bifurcations in laminar liquid flow. International Journal of Heat and Mass Transfer, 117, 487-497.

[15] Awbi, H. B. (2002). Ventilation of buildings. Routledge..

[16] Versteeg, H. K., & Malalasekera, W. (2007). An introduction to computational fluid dynamics: the finite volume method. Pearson education.

[17] Patankar, S. V., Minkowycz, W. J., & Sparrow, E. M. (1980). Series in computational methods in mechanics and thermal sciences. McGraw-Hill Book Company, New York, 1-197.

[18] Sahar, A. M., Wissink, J., Mahmoud, M. M., Karayiannis, T. G., & Ishak, M. S. A. (2017). Effect of hydraulic diameter and aspect ratio on single phase flow and heat transfer in a rectangular microchannel. *Applied Thermal Engineering*, *115*, 793-814.

IOP Conf. Series: Materials Science and Engineering **765** (2020) 012028 doi:10.1088/1757-899X/765/1/012028

The Effect of Mass Flow Rate on the Thermal Performance for Solar Collector with Compound Parabolic Concentrator During the Summer Season

[1]**Asst.Prof. Dr. Talib Z. Farge**
[1]**Asst.Prof. Dr.Abduljabbar Owaid H.**
[2]**Asst.Lec. Mina Abdulkareem Nsaif**

[1]**Electromechanical Engineering Department, University of Technology, Baghdad, Iraq**
[2] **AL-Nisour University College, Baghdad, Iraq**
corresponding author's e-mail : 50270@uotechnology.edu.iq

Abstract

Present study investigated the effect of the summer season, on the thermal performance (heat energy and thermal efficiency) of the solar collector with CPC without cover, by designing a proper solar collector with compound parabolic concentrator (CPC). It was implemented a solar collector with a cylindrical receiver diameter 1.5 inches and an acceptance angle of 45°with concentration ratio of 2.613. The surface reflector was made from stainless steel an aperture area of 0.87 m^2. The orientation of the solar collector at 150° was set as (south-east) direction and 45° inclined angle at Baghdad city. The experimental results showed that the performance of the solar collector was improved significantly as the water circulation mass flow rate increased. The percentage increase in the heat energy and thermal efficiency are (69.95 and 68.68) % resectivelly in case of the mass flow rate of 300 kg/hr with respect to 60 kg/hr.

Key word: colar coolector ,compound parabolic concentrator, concentration ratio.

Introduction

It is well known that the solar collector is a special kind of a heat exchanger which is normally converted the solar radiation heat to the heat energy of the various transport fluids like air, water and oil which is usually flowing across the collector. The circulating fluid could be used as directed or circulate way, through a storage tank to be used in need **[1]**. In fac, the solar collectors with CPC are non-imaging types, which concentrate the solar radiation rays and then reflect them on the absorber surface (receiver surface) **[2]**. Various applications utilize solar collectors during the summer season. Some of them are normally used to supply the hot water in the hotels and hospitals for laundry and bathing applications in addition to the milk dairies (pasteurization, condensation and cleaning), leathers processing industry (drying and tanning), metal finishing industry (phosphating and degreasing) and swimming pool heating.The solar collectors also could be used to pre-feed hot water to the boilers to generate a steam of a steam turbine power plant. Several studies investigated various. types of solar, solar collector. **Yong Kim, et al. (2007) [3]**, improved the performance of an evacuated tube solar collector by using CPC tracking with single axis. The performance results heat flux, outlet water temperature and collector efficiency were critically compared with and without using tracking. It was found that the collector efficiency with tracking was much higher by about 14.9% than that without tracking solar collector.**Santos- Gonzalez, et.al (2011) [4]**, studied the solar collector with CPC experimentally and numerically. Reasonable comparison was made between the numerical and experimental results. The area of the The collector was equal of 1.33 m^2. Acceptance angle about 30° and the concentration ratio about 3.5. The result confirmed that the inlet water temperatures were varied from 70° C to 30° C and the water mass flow rates changed from 0.25 kg/s to 0.05 kg/s. The comparison results showed an acceptable agreement.

Singiresu S. Rao, et.al (2013) [5], optimized a solar collector with CPC and compared with the results of the optimum flat plate. The optimization was carried with three objectives. The first one to maximization of the average solar energy through the annual, the second to maximize the lowest solar energy through the month and the third objective to minimize the cost. The comparison objective to minimize the cost. The comparison between the optimum results of the solar collectors with the flat plate and with the CPC showed a significant reduction of cost per a unit ratio of energy. A full scale of solar collector with a CPC system with an area of (1x0.87) m^2 was designed by Wisut Chamsa-ard et.al (2014) [6]in order to examine the thermal efficiency of the evacuated tube solar collector with CPC. the effect of acceptance angle tube diameter and absorbing on the performance of compound parabolic concentrator solar collector have been experimentally investigated by [8, 9].they concluded that the best thermal of the solar collector with CPC were at acceptance angle of 45° and at largest diameter of absorbing tube .

The objective of the present work is to investigate the performance of the solar collector with CPC for differents circulation of mass flow rate during the summer season.

Experimental Work

Figure (1) shows the major components of the solar collector with CPC, which represented by a solar collector system that consists of CPC reflector surface and cooper pipes coated by black thermal dye, plywood, Reflector holder (polyurethane foam), frame, circulating pipes, valves, circulating water pump, thermal storage tank and measuring devices. The dimension of solar collector with CPC was (100x87) cm. All readings were taken from 9 A.M to 1:30 P.M.

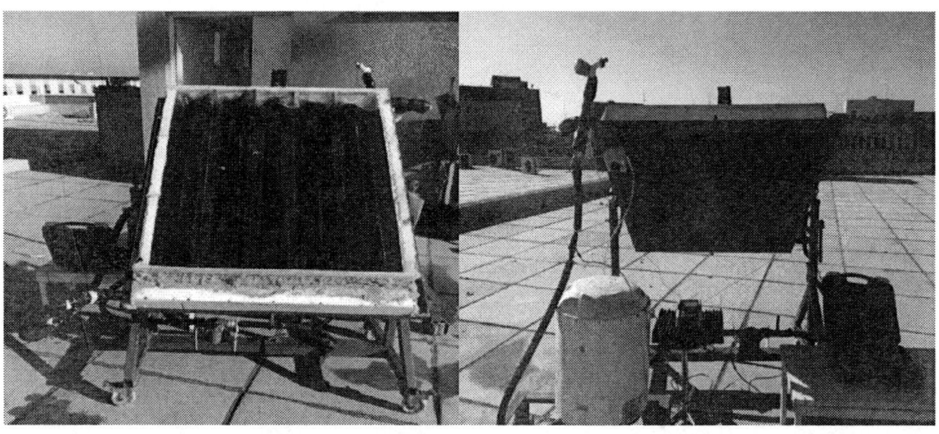

(a) Front view **(b)** Back view

Figure (1): The test rig system (full- scale solar collector), (100x87) cm

Compound parabolic concentrator

The surface curvature of the compound parabolic concentrator constructs from two curves. First curve from A to B and the second curve from B to C, as shown in Figure (2). There are three independent variables (r,θ_{ac} and θ), where :

$R = f(r, \theta_{ac}, \theta)$

For a certain surface of the compound parabolic shape the values of radius of the absorber tube (r) and the acceptance angle (θ_{ac}) kept constant and the value of the angle (θ), which its values change from zero to the ($\frac{3\pi}{2} - \theta_{ac}$) on the y-axis. It can be seen that the curves of the CPC are symmetric about the y-axis and can obtained from the following equations **[7]**.

a) The equation of the first part of the curve from A to B is:

$$D = r.\theta \quad for\ (|\theta| \le \theta_{ac} + \frac{\pi}{2}) \qquad (1)$$

b) The equation of the second part of the curve from B to C is:

$$D = r.\frac{\theta+\theta_{ac}+\pi/2-\cos(\theta-\theta_{ac})}{1+\sin(\theta-\theta_{ac})} \quad for\ (\theta_{ac} + \frac{\pi}{2} \le |\theta| \le \frac{3\pi}{2} - \theta_{ac}) \qquad (2)$$

Where: R is the distance from point D on the absorber tube to point S on the surface of CPC. The concentration ratio of the CPC is a function of the acceptance angle given by following equation **[7]** .

$$CR = \frac{1}{\sin(\frac{1}{2}\theta_{ac})} \qquad (3)$$

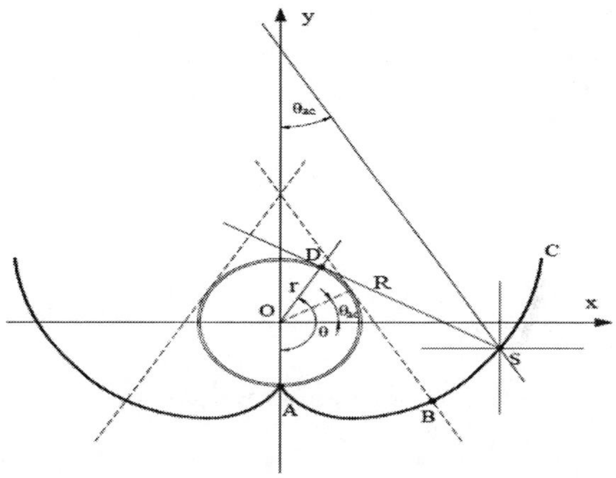

Figure (2): The CPC shape **[7]**

Calculation Procedures

The calculation of the solar collector with CPC performance was made as follows:

a) The volume flow rate of water ($\dot{v}_w$)

The volume flow rate of water was measured by flow meter, where the three values of volume flow rate were chosen (60 and 300) (L/hr)
So the mass flow rate can be determined by the equation (4):

$$\dot{m}_w = \frac{\rho_w\, x\, \dot{v}_w}{3600} \qquad (4)$$

b) Heat energy received by the water in the collector ($\dot{Q}_w$) is:

$$\dot{Q}_w = \dot{m}_w \,.\, C_{p_w} \,.\, \Delta T_w \qquad (5)$$

c) The efficiency of solar collector

$$\eta_{s.c} = \frac{\dot{Q}_w}{A.I} x100 \qquad (6)$$

d) Heat loss energy Q_L

To calculate heat energy loss from the solar collector without cover to the surrounding by using free convection only, the following equations were used [10]: The film temperature as follows:

$$T_f = \left(\frac{T_a + T_s}{2}\right) + 273 \quad (7)$$

And to find the coefficient of thermal volume expansion:

$$\beta = \frac{1}{T_f} \quad (8)$$

The correction equations of kinematic viscosity, thermal conductivity and Prandtl number, by using the following linear interpolation:

$$v = v_1 + \frac{T_f - T_1}{T_2 - T_1} \cdot (v_2 - v_1) \quad (9)$$

$$k = k_1 + \frac{T_f - T_1}{T_2 - T_1} \cdot (k_2 - k_1) \quad (10)$$

$$p_r = p_{r_1} + \frac{T_f - T_1}{T_2 - T_1} \cdot (p_{r_2} - p_{r_1}) \quad (11)$$

Then to calculate the Grashof number, Nusselt number, heat transfer coefficient and heat lost uising the following equations:

$$G_r = \frac{g \cdot \beta \cdot (T_s - T_a) . d^3}{v^2} \quad (12)$$

$$N_u = C\,[G_r . P_r]^n \quad (13)$$

Where the values of C=0.53 and n= 0.25 for the value of Gr Pr in the range between (10^4-10^9)[10] .

$$h = \frac{N_u \cdot K}{d} \quad (14)$$

$$Q_L = h . A . (T_s - T_a) \quad (15)$$

Result and Discussion

Figures (3-7) were clearly shown the corresponding performance results for the solar collector with CPC, without cover and during the summer season (summer time) for different mass flow rate along the daytime from 9 a.m. to 1:30 p.m.Figures (3 and 4) display the temperature variation with respect to time for different water circulation mass flow rate (60 and 300) kg/hr. It can clearly noticed that the difference between the inlet and outlet of the water temperature through the collector was increased significantly, when the mass flow rate decreased. Further , the temperature gradient for the water circulation mass flow rate of 300 kg/hr provide a higher value than the water circulation mass flow rate of 60 kg/hr , since the solar radiation intensity were approximately equal for the two cases, and have a higher value at 12 p.m. It was found that the heat lost from the solar collector which was increased with the lower water circulation mass flow rate. The reason for that attributed to the lower values of heat transfer coefficient as well as to the temperature difference between the receiver surface temperature and ambient temperature, as it is shown in Figure (7). It should be noted that the receiver tube had a long time to loss the heat to the surrounding in the case of lower water circulation mass flow rate.Figures (5, 6) show the comparison performance results (heat energy and collector efficiency) of the solar collector with CPC for different water circulation mass flow rate through the collector. It can be seen that the heat energy and thermal solar collector efficiency, increase significantly, as the water circulating mass flow rate increase. The reason for that was attributed to the higher water temperature gradient as well as the higher mass flow rate, since the heat energy is a function of mass flow rate and temperature difference. Accordingly, the heat lost in the case of the higher mass flow rate was lower than the lower mass flow rate as shown in Figure **(7)**. It was clearly observed that the percentage of the heat energy and thermal solar efficiency increase in the water circulation mass flow rate of 300kg/hr with respect to that of 60 kg/hr were (69.95 and 23.79%) respectively at 12 p.m.

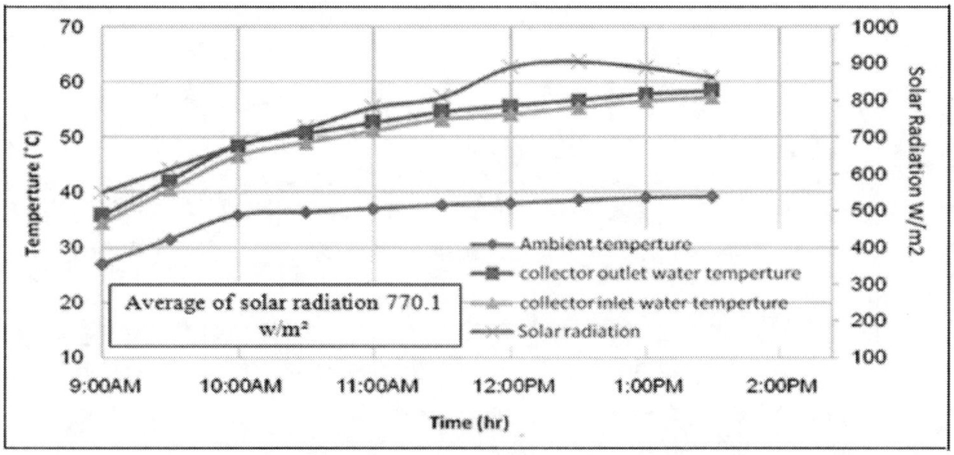

Figure (3): Temperature distribution with respect to the time with $m^{\cdot}$ = 60 kg/hr during 11/6/2015.

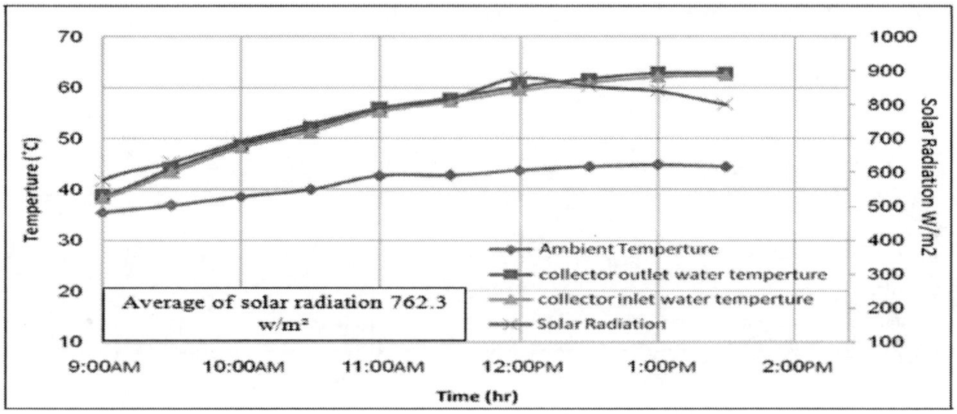

Figure (4): Temperature distribution with respect to the time with $m^{\cdot}$ = 300 kg/hr during 14/6/2015.

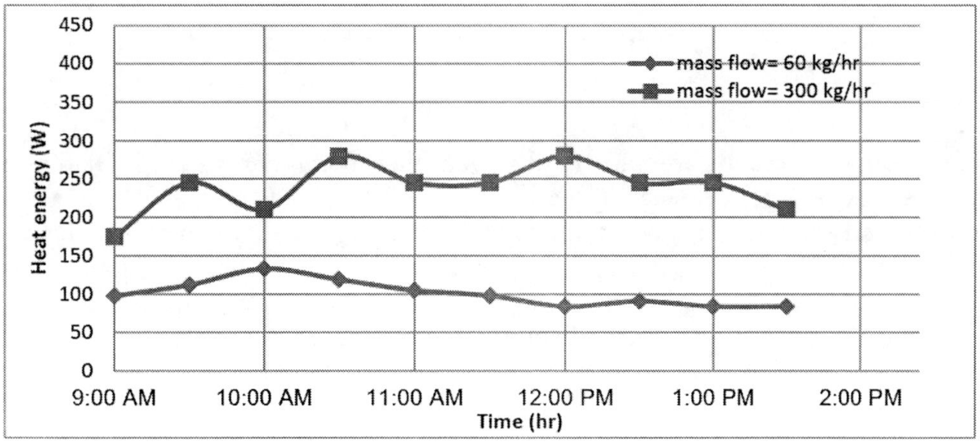

Figure (5): Comparison Heat Energy for two mass flow

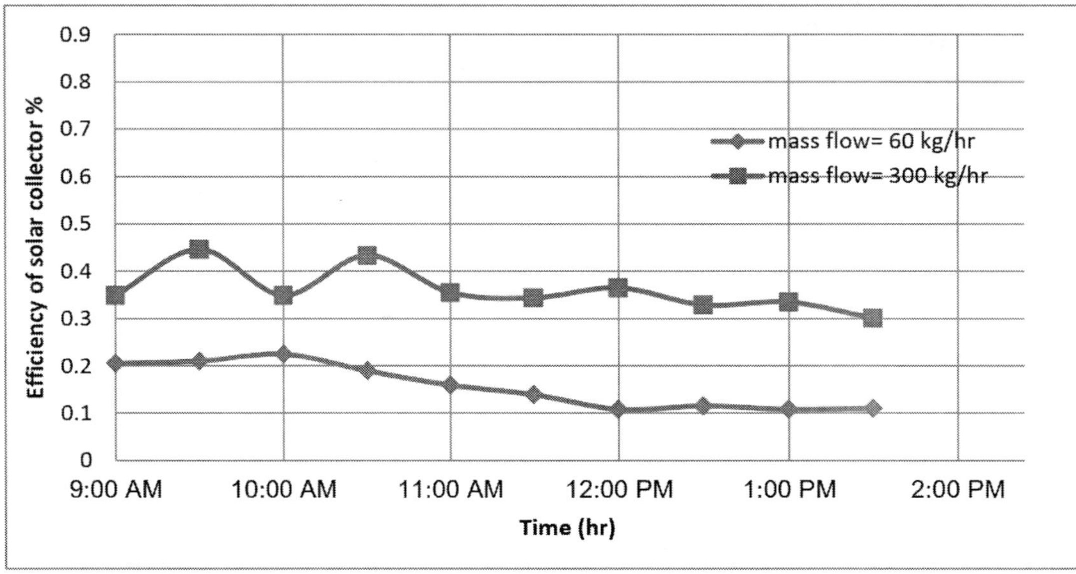

Figure (6): Comparative Efficiency of solar collector for two mass flow

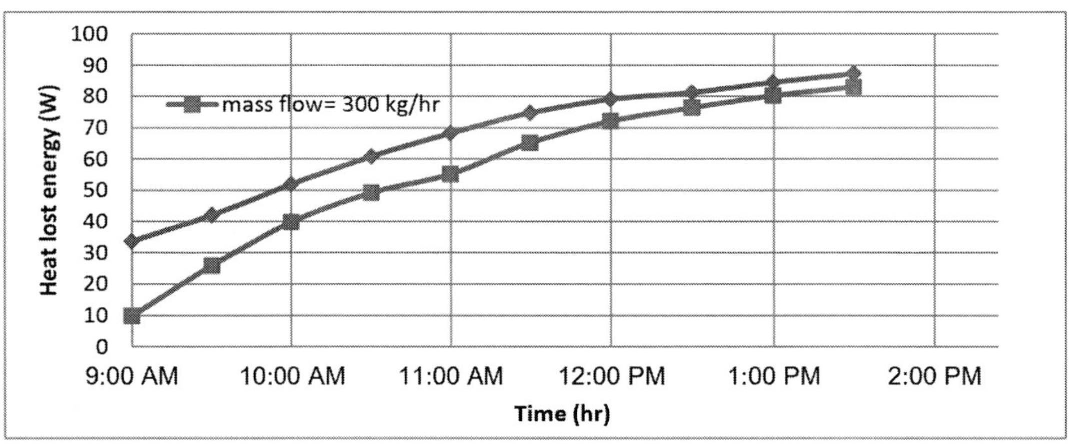

Figure (7): Comparison heat lost energy results for two mass flow

6. Conclusion

1. The experimental results are clearly observed that the higher water circulation mass flow rate was given the best performance of the solar collector with CPC.
2. The comparison performance results for heat energy and thermal efficiency of the solar collector with cpc show a better performance in the case of water circulation mass flow rate of 300 Kg/hr than of 60kg/hr .The percentage of the heat energy and thermal efficiency are (69.95 and 68.68) % respectively in case of water circulation mass flow rate of 300kg/hr.
3. The heat lost from the solar collector was decreased significantly with increasing the water circulation mass flow rate.

List Of Symbols:

SYMBOL	TITLE	UNIT
A	Area of the Collector	m^2
C_{pw}	Specific heat capacity of water	J/kg.K
d	Absorber tube diameter	m
G_r	Grashof number	....
g	Gravitational acceleration	m/s^2
h	Heat transfer coefficient	$W/m^2.k$
I	Solar intensity	W/m^2
K	Thermal conductivity	W/m.k
m	Water mass flow rate	kg/s
Nu	Nusselt number	....
Pr	Prandtl number	....
Q_L	Heat losses	W
r	Absorber tube radius	m
T	Temperature	K
T_F	Film temperature	K
β	coefficient of thermal volume expansion	1/K
θ	Incidence angle	Degree
Θ_{ac}	Acceptance angle	Degree
v	Kinematic viscosity	m^2

References

[1] S. Kalogirou," Solar thermal collectors and applications .Progress in energy and combustion science 30 (3), p. 231-295, (2004).

[2] D.K. Patel and P. K. Brahmbhatton "Analysis and performance investigations a solar air heater with compound parabolic concentrator" international journal on recent and innovation trends in computing and communication, Vol.3, (2015).

[3] Y. Kim, G. Han, &T. Seo "An evaluation on thermal performance of CPC solar collector" International Communications in Heat and Mass Transfer VOL. 35 ,NO.4 (2008) 446–457.

[4] S.González et.al. "Development and experimental investigation of a compound parabolic concentrator" International journal of energy research, Vol. 36,No. 12, pp. (1151–1116), (2011).

[5] S. S. Rao, H. G. Lee & Y. Hu,, "Optimal Design of Compound Parabolic Concentrator Solar Collector System" Journal of Mechanical Design, Vol. 136,No.9, (2013).

[6] W. Chamsa-ard, S. Sukchai, S. Sonsaree, &C. Sirisamphanwong, "Thermal Performance Testing of Heat Pipe Evacuated Tube with Compound Parabolic Concentrating Solar Collector by ISO 9806 –1",
Energy Procedia, 56, pp. (237 – 246), (2014).

[7] J. Blanco Gálvez And S. Malato Rodríguez "Solar Photochemistry Technology", Solar Energy Conversion and Photoenergy Systems – Vol. II (2005).

[8] Z. Talib , O. Abduljabbar and A. Mina "Expermental Study On The Effect Of The Acceptance Angle Of The Compound Parabolic Concentrator On The Performance Of The Solar Collector" ,journal of al-Nisour university college ,vol.4,2018.

[9] Z.Talib, A.Ali and A. Hiba" Expermental Study The Effect of Absorbing Tube Diameter The Performance of Compound Parabolic Concentrators Solar Collector "Journal of Engineering and Sustainable Development Vol.22 No.4,july 2018.

[10] Holman JP, "Heat Transfer", Ninth Edition ,2002.

IOP Conf. Series: Materials Science and Engineering **765** (2020) 012029 doi:10.1088/1757-899X/765/1/012029

Design and Implementation of Water Distillation System Using Solar Energy

Assist. Prof. Dr. Talib Z. Farge[1] Lec. Dr. Mohammed Jawad Mohammed [2] Nuha Ali Jassim[3]
[1,2,3] University of Technology / Electromechanical Engineering Department Baghdad - Iraq
50270@uotechnology.edu.iq[1] 50024@uotechnology.edu.iq[2] nuhajassim5@gmail.com[3]

Abstract. An investigation of the solar distillation using a solar collector with a graduated absorbed surface has been utilized in order to improve the solar still performance. The solar collector is contained on graduated absorbing surface with 25 steps. Absorbing surface made from copper plate and dye by a thermal black pigment to increase the solar absorption. The top of the solar collector is covered by a glass with 4mm of thickness. While, the other side surfaces of the solar collector are made from the transparent plastic with 6mm of thickness. Also, Aperture area of the solar collector was 0.4526 m2. Experiments had been carried out in Baghdad (33.3° N latitude) during May in 2019 to investigate the performance of the still. The results appeared the maximum water production that reached up to 486 ml at the time interval between 12.30 and 1.00 pm, while the total thermal solar collector efficiency was 36%. The proposed method succeeded to reduce the TDS from 679 ppm to 39 ppm. for inlet water and outlet pureed water respectively, and the volume distilled water production was 3696 ml for the time from 7.30 am to 4.00 pm.

1. Introduction

Water is one of the resources that potentially useful to all living beings. Often, water sources are brackish containing dissolved solids and a harmful bacterium, and therefore cannot be used for drinking. Distillation is one of the processes that can be used for water purification. Desalination refers to the process of removing salt and other minerals from water. The reason to distil the water in order to extract the fresh water from salt water which is suitable for human consumption or irrigation. Most of the researches in distillation were focused on developing the cost-effective ways of providing fresh water for human uses. Therefore, the solar distillation one of the most popular methods for water purification in the villages and far territory from the source of water purification nowadays [1-2].

2. Literature review

Various researches are carried out to improve the performance of the still. The basin area of the still, free surface area of water, inlet temperature of water, wind velocity, solar radiation and depth are some of the factors that effect on the productivity of the solar still. Abujazar et al. [3] done the efficiency evaluation of an inclined copper-stepped solar still experimentally. They pointed out that using of the inclined trays in stepped solar still increased the freshwater productivity up to 605 mL/m2l, wick type evaporator still and the efficiency of the still are improved by reducing the radiation losses from the basin. Dubey et al. [4] designed a solar still with separated condensing chamber to enhance the rate of condensation and hence the productivity. Their performances in terms of energy and exergy efficiencies were compared with the conventional still. Dimri, et al. [5] done the theoretical and experimental analysis of a solar still with a flat plate collector under various condensing materials. Rashidi et al. [6] investigated the effect of Nano fluid on the productivity of a stepped solar still numerically. Also, they have optimized the geometry of steps. The results showed an enhancement of 22% hourly productivity. Nagarajan et al. [7] investigated the performance of an inclined solar still with and without baffles. The results showed that the freshwater productivity of the solar still with baffles was 1.68 times the freshwater productivity of the solar still without baffles. Suleiman [8] studied the effect of water depth on productivity, and their experimental results showed that a higher productivity of 6.7 L/day was obtained for a low water depth. Saettone et al. [9] designed and studied a stepped solar still with internal reflective walls, and attached a set of heat pipes. This technique has enhanced the solar still performance by 32.4%. Velmurugan, et al. [10] worked with an industrial effluent in a fin type single solar still and a stepped solar still separately. Velmurugan and Srithar [11] used asponge cubes in the still and acquired about 57.8% more yield than the conventional still. Zurigat and Abu-Arabi [12] modelling of the system along with the performance analysis was compared. Finally, the aim of present work is to investigate the performance of the new type of solar water distillation system using a gradual absorbed surface collector.

3. Experimental work

The experimental setup comprises of a transparent plastic box with dimensions of 73cmx62cmx45cm for length, width and high respectively with an inclination angle of 38.8° as shown in Figure 1. The top of the box was covered by a glass with 4mm of thickness. Therefore, the solar energy would be reaching the surface of absorbing plate through all aspects and from the top of the collector box. The surface of absorbing plate was design as shown in Figure 2. The surface of absorbing plate was made from copper plate with thickness of 1mm. The base area of the aperture area of the solar collector was 66cmx62cm, the maximum height was 30 cm and minimum height was 10 cm. Absorbing surface was reducing gradually from the maximum height to minimum height by 25 steps. Each step had a dimension of 2 x1 x62cm for width, height and length, respectively. As shown in Figure 3. The solar collector contained on three water ports. The first one used for flow of the brackish water into the solar collector, while the other two ports were used for water flow outside the solar collector (one for distilled water and the other for saline water).

Figure 1. Solar distillation system

Figure 2. Absorbing plate

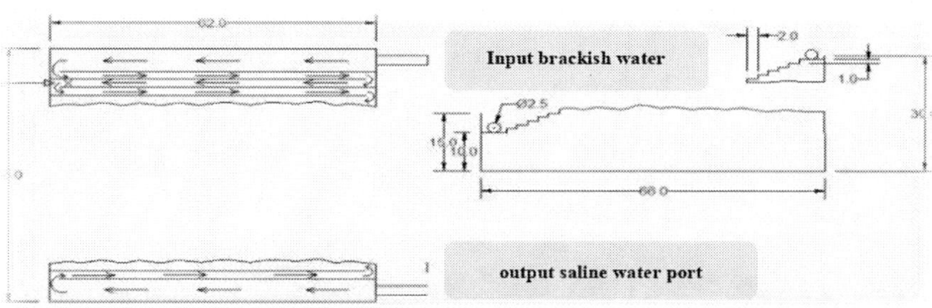

Figure 3. Surface of the absorbing plate sketch

A storage tank with capacity of 250 L used to supply the solar collector with brackish water. A gate valve was fitted on the water storage tank in order to control the mass flow rate of water that enters the solar distillation collector. The procedure of the solar distillation work is through the flow of brackish water inside the solar collector when the water is circling around the steps of the gradient surface. Then, the water will be evaporated due to heat absorption from the absorption plate. Finally, the evaporated water condenses as a result of its contact with surface of the glass and slips down to get out of the distilled water port. The remaining water will be very salty and coming from of the saline water port. The inlet and outlet water temperatures were measured by three thermocouples type- K (copper – constantan) with accuracy of ±0.5°C. Solar radiation intensity was measured by using a solar power meter type TES1333

4. Theory and calculations

Heat of evaporation can be calculated as follows:

$$Q_d = mh_f \quad (1)$$

Where
h_f Refers to enthalpy of evaporation of water and equals to 417460 J/kg [13].
Also, the mass of distilled water equal to:

IOP Conf. Series: Materials Science and Engineering **765** (2020) 012029 doi:10.1088/1757-899X/765/1/012029

$$md = V\rho \quad (2)$$

Where, ρ refers to the water density and equals to $1000\ kg/m^3$ .The heat energy of the distilled water is:

$$Q_d = \frac{V}{1000} * h_f \quad (3)$$

The heat power of evaporation with respect to time interval is:

$$\dot{Q}_d = \frac{Q_d}{t} \quad (4)$$

The efficiency of the solar collector for distilled water is:

$$\eta d = \frac{\dot{Q}_d}{AI} * 100\% \quad (5)$$

Where, A is aperture Area of the solar collector equal to $0.4526m^2$.If the heat energy of saline water that outside the saline port use in other applications, so the heat power of the saline is:

$$\dot{Q}_s = \dot{m} c_p (Ts - Tb) \quad (6)$$

Where, $C_p$equal to 4180 J/kgK [13]. And the saline water flow through saline water port is:

$$\dot{m} = \frac{m}{0.5 \times 3600} \quad kg/sec \quad (7)$$

And, the total heat supplied by the solar collector is:

$$\dot{Q}_{total} = \dot{Q}_d + \dot{Q}_s \quad (8)$$

The total efficiency of the solar collector in case of using the heat power of saline water in other applications is:

$$\eta_{total} = \frac{\dot{Q}_{total}}{AI} * 100\% \quad (9)$$

5. Results and discussions

Table (5.1) shows the reading data and the obtained results from 7.30 am to 4.00 pm at a time interval of 30 minutes for each reading at the volume flow rate 2.05 of ml/s. All these reading have taken in centre of Baghdad city with east into west with angle of tilted of 33°and along the day of 19 May 2019.

Table (5-1) measurements and results data along the day of 19 May 2019.

Time (hr)	I [W/m2]	T_b [°C]	T_s [°C]	T_d [°C]	Tair [°C]	Vol [ml]	$\dot{Q}_d$ [W]	$\dot{Q}_s$ [W]	Q_{total} [W]	η_d%	η_{total} %
8.00	567.7	30	54	39	31	5	1.16	8.36	9.52	0.45	3.71
8.30	688.9	31	57	41	32	20	4.64	12.1	16.74	1.49	5.37
9.00	765.8	32	59	45	33	50	11.6	15.69	27.29	3.35	7.87

9.30	816.6	33	60	46	34	81	18.79	18.85	37.64	5.08	10.18
10.00	852.3	34	62	48	35	125	28.99	22.71	51.7	7.52	13.4
10.30	899.7	36	64	50	36	162	37.57	25.98	63.55	9.23	15.61
11.00	903.3	39	66	52	37	217	50.33	28.22	78.55	12.31	19.21
11.30	915.6	39	67	55	38	275	63.78	32.54	96.32	15.39	23.24
12.00	926.4	41	69	58	39	340	78.85	36.28	115.13	18.81	27.46
12.30	938.9	43	71	60	40	405	93.93	38.97	132.9	22.1	31.27
1.00	945.6	45	73	62	41	486	112.71	42.25	154.96	26.34	36.21
1.30	932.1	44	72	59	41	375	86.97	45.53	132.3	20.62	31.36
2.00	897.3	42	69	57	40	320	74.22	47.1	121.32	18.28	29.87
2.30	856.1	40	67	55	39	285	66.1	50.11	116.21	17.1	29.99
3.00	759.4	39	65	51	38	240	55.66	51.29	106.95	16.19	31.12
3.30	622.7	38	62	48	38	200	46.38	50.16	96.54	16.46	34.25
4.00	521.2	36	58	45	37	110	25.51	48.55	74.06	10.81	31.39

Figure 4 shows the solar radiation intensity and the temperature distribution (inlet brackish water, outlet distilled water, outlet saline water and ambient air) along the day time from 8.00 am to 4.00 pm. All the different temperatures of the component of solar still and ambient air temperature increased with time until reacheing the maximum values of 1.00 pm due to increaseing in solar radiation intensity which reached the maximum values of 1.00 pm, where in this case, an increasing in the absorb solar radiation by the collector exceeded the losses to the atmosphere. After 1.00 pm, the temperature decreased due to decreasing in the solar radiation intensity. Where, in this case, the losses from solar collector became larger than the absorb solar radiation.

Also Figure 4 shows that the curve of solar radiation decreased sharply after 1.00 pm due to the cloud that appeared at that time. While, the temperature of distilled water and saline water decreased gradually and this due to absorbed plate worked as a heat storage.

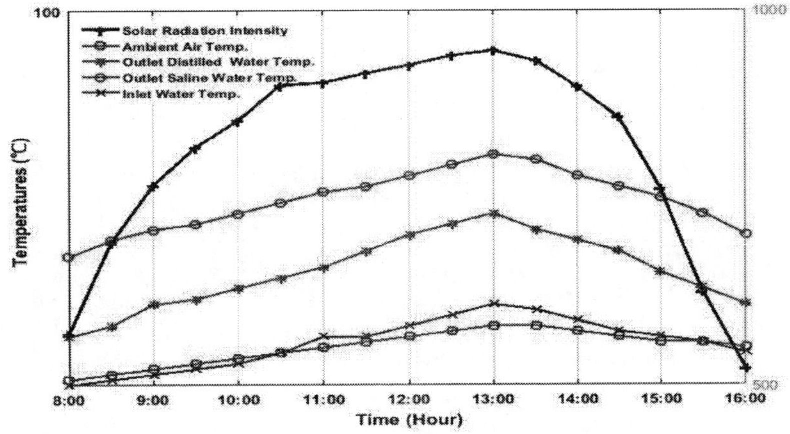

Figure 4. The temperatures and solar radiation intensity distribution with respect to time

Figure 6 shows the thermal efficiency of the solar distillation collector along the intervals of the day time from 7.30 am to 4.00 pm, where, the time intervals was equal to 30 minutes. Figure 6 shows that both of the distilled water efficiency and total efficiency are increased towards the day time to 1.00 pm, and then decreased due to behaviour of the solar radiation intensity along the day. Also, Figure 6

IOP Conf. Series: Materials Science and Engineering **765** (2020) 012029 doi:10.1088/1757-899X/765/1/012029

shows that the total thermal efficiency greater than the distilled water efficiency due to higher outlet water temperature which is used in other thermal applications . The highest value of the distilled water efficiency and total thermal efficiency are 26% and 36% respectively for time interval between 12.30 pm and 1.00 pm due to highest solar radiation intensity at that time.

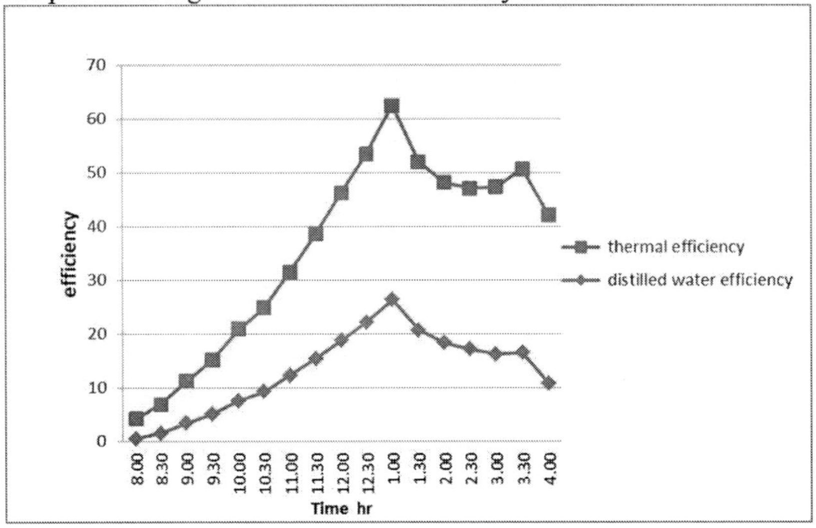

Figure 5. Productivity of distilled water with time

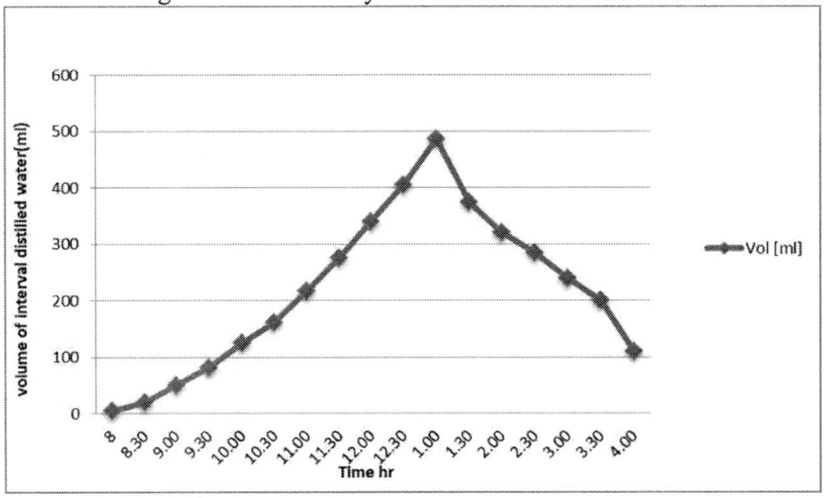

Figure 6. variation of solar collector efficiency with respect to time

6. Conclusion

An investigation of the solar distillation using a solar collector with a graduated absorbed surface has been utilized in order to improve the solar still performance. Experiments had been carried out in Baghdad (33.3° N latitude) during May in 2019 to investigate the performance of the still. The following conclusions are extracted from this investigation:

1. The values of solar distillation collector efficiency and production of distilled water reached the maximum amount at 1.00 pm.
2. The volume of distilled water production by the solar collector was 3696 ml, which equivalent to 8166 ml/m^2 through the time between 7.30 am to 4.00 pm.
3. The maximum values of efficiency of the solar collector for both of the distillation and total thermal are 26% and 36% respectively.

IOP Conf. Series: Materials Science and Engineering **765** (2020) 012029 doi:10.1088/1757-899X/765/1/012029

Nomenclature

Symbol	Description	Units
A	Aperture Area of The Solar Collector	m2
Cp	Specific Heat Capacity	J/kgK
hf	Enthalpy of Evaporation of Water	J/kg
I	Solar Radiation Intensity	W
m	Mass of Water	kg
$\dot{m}$	Distilled Water Mass Flow Rate	Kg/sec
Q	Heat Energy	J/kg
$\dot{Q}$	Heat Power	W
T	Temperature	°C
t	Time	sec
V	Distilled Volume	ml
ρ	The Water Density	Kg/m3
η	Efficiency	-

Subscripts

Symbol	Meaning
b	Brackish
d	Distilled
s	Saline

References

[1] Farge, T. Z., Sultan, K. F., & Ahmed, A. M. (2017). Experimental Study of the Performance Water Distillation Device by Using Solar Energy. Engineering and Technology Journal, 35(6 Part (A) Engineering), 653-659.

[2] B. Pushpa, K. S. Chendra, K. P. Reddy, "A study on solar still with separate Condenser" International Refereed Journal of Engineering and Science (IRJES), Volume 4, Issue 6 ,PP.94-98,June, 2015

[3] Abujazar, M. S. S., Fatihah, S., Lotfy, E. R., Kabeel, A. E., & Sharil, S. (2018). Performance evaluation of inclined copper-stepped solar still in a wet tropical climate. Desalination, 425, 94-103.

[4] Dubey R., Rai A.K. (2017). Comparative performance of a double condensing chamber solar still with a conventional solar still. International Journal of Mechanical Engineering and Technology (IJMET).: 8(11), 106-112.

[5] Dimri, V., Sarkar, B., Singh, U., & Tiwari, G. N. (2008). Effect of condensing cover material on yield of an active solar still: an experimental validation. Desalination, 227(1-3), 178-189.

[6] Rashidi, S., Bovand, M., Rahbar, N., & Esfahani, J. A. (2018). Steps optimization and productivity enhancement in a nanofluid cascade solar still. Renewable Energy, 118, 536-545.

[7] Nagarajan, P. K., El-Agouz, S. A., DG, H. S., Edwin, M., Madhu, B., Sathyamurthy, R., & Bharathwaaj, R. (2017). Analysis of an inclined solar still with baffles for improving the yield of fresh water. Process Safety and Environmental Protection, 105, 326-337.

[8] Tarawneh, M. S. K. (2007). Effect of water depth on the performance evaluation of solar still. JJMIE, 1(1).

[9] Saettone, E., Valencia-Tovar, Y., & Gómez-de-la-Torre-Gastello, A. (2017). Preliminary overview and evaluation of a stepped solar distiller with internal reflective walls and borosilicate vacuum tubes. Desalination, 413, 136-143.

[10] Velmurugan, V., Kumar, K. N., Haq, T. N., & Srithar, K. (2009). Performance analysis in stepped solar still for effluent desalination. Energy, 34(9), 1179-1186.

[11] Velmurugan, V., & Srithar, K. (2007). Solar stills integrated with a mini solar pond—analytical simulation and experimental validation. Desalination, 216(1-3), 232-241.
[12] Zurigat, Y. H., & Abu-Arabi, M. K. (2004). Modelling and performance analysis of a regenerative solar desalination unit. Applied thermal engineering, 24(7), 1061-1072.
[13] Keenan, J. H. (1969). Steam tables: thermodynamic properties of water, including vapor, liquid, and solid phases (English units).

Photovoltaic Thermal (PV/T) Collector Performance Evaluation as Electricity Generation and Hot Air Supplier For Partially Covered with PV Modules

Jalal Mohammed Jalil, Ahmed Abdulqader Hussein, Anwar Jabbar Faisal

Electromechanical Engineering Department, University of Technology

E-mail: jalalmjalil@gmail.com , ahmedabdulqaderhussein@gmail.com , anwarjabarfaisal@gmail.com

Abstract. This study experienced the performance of air biased PV/T collector partially covered with glass at various values of irradiance (400, 600, 800,1000) W/m^2 and constant ambient temperature. A three various configurations have been demonstrated in this work, case A (glass at inlet portion), case B (glass at mid portion) and case C (glass at outlet portion). The investigation has been performed in term of outlet air temperature and electrical power. A numerical model was developed via using the computational fluid dynamic (CFD) program. 3D steady state, turbulent forced convection model is utilized to solve Navier Stokes and energy equations of air flowing inside the duct. The results compared with the experimental measurements that carried out from indoor conditions using solar simulator, a strong agreement has been achieved between experimental and numerical outcomes. It has been observed that case C (glass at outlet portion) is more favorable in electrical and thermal gain, which generates 44.3°C hot air and 26.6 W electric power, while the maximum outlet air temperature and electric power reach 39°C, 25.4 W and 43°C, 21.05 W for configurations A and B respectively. Furthermore, the analysis of PV/T system concluded that the PV/T system should be operate at moderate air flowrate (0.013 kg/s) which is the specific or optimum flowrate.

Keywords: PVT air collector; PV/T partially covered with PV; Thermal efficiency; Thermal Analysis

1. Introduction

PV/T systems is one of the most popular renewable energy resource that facing the energy crisis all around the world [1]. Its allow us to utilize the solar radiation efficiently and reduce the usage of expensive fossil fuel consumed [2].The electrical power generated from photovoltaic system depends on the cells temperature and the solar radiation incident directly on it. PV system can converts 6-16% of solar incidence into electricity at 25°C. The remaining 80% of this incident is converted to heat and acted as losses dissipated to the environment. The heat dissipation is not desirable and must be converted into useful energy such as forced air or water circulation with backside of the PV surface. This sophisticated system is recognized as hybrid photovoltaic thermal systems, these systems able to generate electric energy as well as heat energy simultaneously. The generated heat is used for various application, such as the hot water/air can be used for heating space, solar dryers and solar stills. Solar cells power

Nomenclature	
A_x	Area in the direction of east and west (m^2)
A_y	Area in the direction of tob and bottom (m^2)
A_z	Area in the direction of north and south (m^2)
k_{ap}	Thermal conductivity of the absorber plate (W/m.K)
T_{af}	Bulk temperature of air flowing (°C)
α_{ap}	Absorber plate absorptivity
τ_g	Transmissivity of glass surface
U, V, W	Velocities in X,Y,Z directions (m/s)
x, y, z	Cartesian coordinates m
V_f	Air following velocity (m/s)
i, j, k	Indices which indicate positions in the (x, y, z) direction
ρ	Density of air (kg/m^3)
G	Solar radiation
h_{in}	Internal heat transfer coefficient (W/m.K)
υ_e	Effective kinematics viscosity(m^2/s)
Γ	Diffusion coefficient (N.s/m^2)

and efficiency is a leaner decreasing function of temperature this mean when the operating cells temperature is high, the electrical power becomes low so that this circulation of fluid below the panel will act as coolant to the PV cells, enhance its performance and helps to maintain the PV system in a good condition during its operation [3]. In this system, the working fluid used to extracting the heat from PV cells is the air. The system is consist of a basic air channel placed at the back of the PV cells. PV modules is enclosed in a duct and air is circulated in channel by fans. A part of energy generated is required to operate the fans, therefore it must be considered when designing the duct. Optimum channel high and mass flow rate of air are the most effecting parameters in this type [4]. PV/T system is partially covered by PV module, glass is used in the non-packing area of PV modules because it has a superior thermal properties. The irradiance is transmitted through the glass and then fall into the black painted absorber plate. Moreover, the thermal energy related with pack surface of PV module is transferred by convection to the absorber plate for extra heating. The fluid at a lower level of PV modules gets the thermal energy and flow in the upward direction [5]. Many experiences and researches on different types of solar PV/T collector had been studied, to analyze the behavior of PV/T system and factors effected on it. **Al-Imam et al. (2015) [6]** Analyzed the performance of water biased photovoltaic thermal (PV/T) collector with energy storage into Phase Change Material (PCM) and Compound Parabolic Concentrator (CPC) to get better solar radiation by concentrate it into the absorber plate of hybrid collector. The total useful energy is the sum of heat transferred to the phase change material (PCM) and heat carried by water. The results shows that for a clear day the overall efficiency varied between 55% and 63% and for semi-cloudy day is varied between 46% and 55%. **Saidan et al. (2016) [7]** Investigated the negative effect of dust accumulation on PV models as a result of scattering the solar radiation from PV modules by dust. The experiment has been carried out in Baghdad - Iraq for three photovoltaic solar models. The result showed the maximum current reduced from 6.9% to 16.4% and this value depend on the period of the cell exposing to the dust. **Tiwari et al. (2016) [8]** Designed a photovoltaic thermal system for greenhouse. Studying the effect of packing factor, absorptivity, mass flow rate of air and transmissivity on the system. Electrical energy, thermal energy and overall thermal energy have been evaluated. It was observed that the system could be used for another application like electrical production, enhancing biogas production, space heating, the crop cultivating and crop drying, etc. **Tonui and Tripanagnostopoulos (2008) [9]** Studied the air cooling of photovoltaic thermal collectors by natural flowing when a slim mineral sheet or fins hanging at the mid side attached to the back surface of air-channel to enhance heat transfer from PV cell to air stream. Over and above, develop a numerical model to validate experimental data. The work improve that the system efficiency for every parameter evaluated with the fin system giving better performance than the thin metal sheet system but both enhancing heat transfer from PV module whereas the maximum outlet air temperatures for fin system was 58°C and 60°C for thin metal sheet system. **Ooshaksaraei (2016) [10]** Designed and fabricated four air-based photovoltaic thermal collectors with four different packing factor namely: double-path parallel flow design, double-path, counter flow, double-pass, returning flow and single-path design. The experiment done indoor using the solar simulator, the result showed that single-path design is better than the other in electricity generation and for heat generation, double-path parallel flow design is the better one, its showed the maximum overall efficiency of 51% - 67%, while the single-path design showed minimum overall efficiency of 28% - 49% for packing factor equal (0.7). **Mishra and Tiwari (2013) [5]** studied the influence of coverage rate with PV module on the effectiveness of water biased PV/T system at constant collection temperature, with two configuration are: case A (PV/T system partially covered with PV model) and case B (PV/T system fully covered with PV model). It's seen that the thermal gain is better for case A due to lesser area of the collector covered with PV in case A than that for case B. Electrical gain in case B is higher than case A due to larger area covered with PV module in case B than that for case A. This work will investigate the electrical and thermal optimization for air based PV/T collector partially covered with glass at three different configurations and investigate the specific flow rate and optimum glass configuration that combination with PV modules.

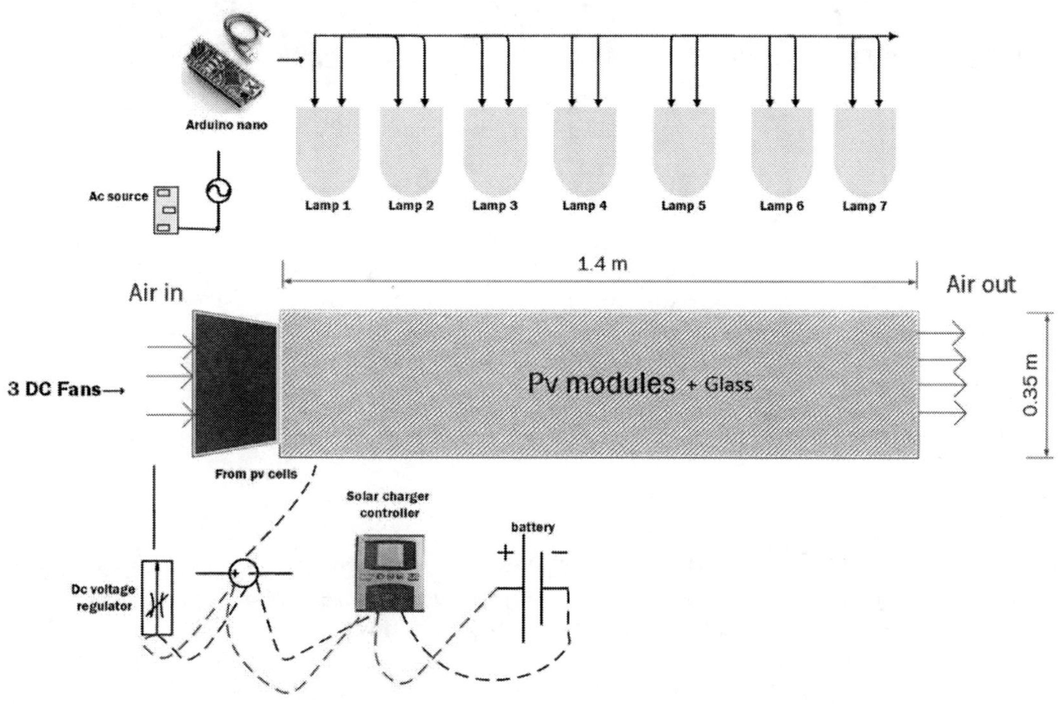

Figure 1: Sketch of the experimental rig.

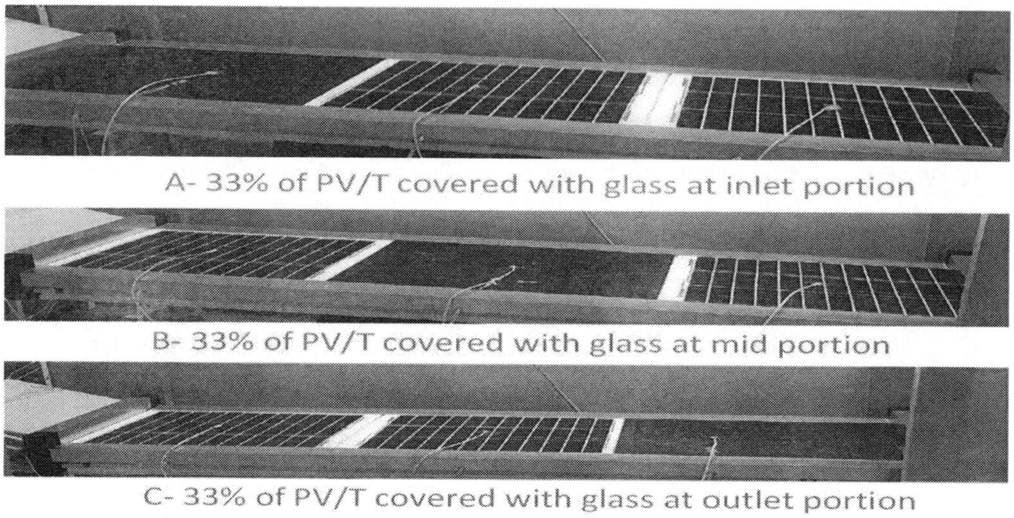

Figure 2: The photograph for three cases of PV/T system.

Table 1: Specification of PV modules.

Two (20 watt) solar cells	
Dimensions	47cm * 35cm
Current at P max	1.1 Amp
Voltage at P max	17.5 V
Short circuit current (Ish)	1.2 Amp.
Open circuit voltage	20 V

2. System Description

A sketch of the experimental rig is illustrated in Fig.(1), and Fig.(2) shows the photograph of three cases for PV/T system that carried out in this work, which is mainly consists of tow pieces of solar cells and one piece of glass attached to the aluminum absorber plate, a three DC fans for forced cooling system, DC voltage regulator for control the flowrate. Solar simulator with microcontroller for tungsten control, sensors and measuring Units. Other important components are solar charge controller and battery for power storage. Air is the working medium and supplied to the thermal channel by three DC fans each of 3 W capacity. The experiment was situated indoor to preserve ambient temperature and the intensity of solar radiation and to avoid rabidly change in outdoor condition. The detail description and dimensions of PV solar cells specification is illustrates in Table (1).

3. Numerical Modeling

This study includes mathematical formulation, which govern partial differential equations (PDEs) to depict turbulent fluid flow field and heat transfer in a PV/T system. The hybrid PV/T system, which studied numerically, has full scale of experimental duct. The duct has 0.35 m width and 1.4 m length. The cross-section of the duct is (0.35 * 0.03) m^2. A mesh size of (62 * 22 * 22) elements involved in numerical solution of PV/T system. The mathematical model of the airflow problem is governed by basic conservation of momentum, mass and energy balance equations. Turbulent flow is considered in the present work (the Reynolds number values of the air inlet is 6133). The following assumptions have been considered:

(i) Assumption has been considered steady state with negligible side and bottom losses, body forces and air leakage.
(ii) Conversation equation of three-dimensional model.
(iii) Forced convection of operation and the streamline of air through the duct is uniform.
(iv) Three-dimensional heat transfer conduction through PV modules, glass and absorber plate.
(v) Turbulent airflow has been considered in the channel.

3.1. Governing Equations

The flow of air inside the PV/T system is turbulent with, 3-D, incompressible and steady state. Navier Stokes equations (NSEs) (continuity, momentum and energy) are resolved by control volume (CV) in Cartesian Coordinates and after simplification the equations are as follows [11]:

(i) The Continuity Equation

$$\frac{\partial}{\partial x}(U) + \frac{\partial}{\partial y}(V) + \frac{\partial}{\partial z}(W) = 0 \qquad (1)$$

(ii) The Momentum Equation
(a) x-Direction

$$\frac{\partial}{\partial x}(UU) + \frac{\partial}{\partial y}(UV) + \frac{\partial}{\partial z}(UW) = -\frac{1}{\rho}\frac{dP}{dx} + \frac{\partial}{\partial x}\left(\nu_e \frac{\partial U}{\partial x}\right) + \frac{\partial}{\partial y}\left(\nu_e \frac{\partial U}{\partial y}\right) + \frac{\partial}{\partial z}\left(\nu_e \frac{\partial U}{\partial z}\right) + \frac{\partial}{\partial x}\left(\nu_e \frac{\partial U}{\partial x}\right) + \frac{\partial}{\partial y}\left(\nu_e \frac{\partial V}{\partial x}\right) + \frac{\partial}{\partial z}\left(\nu_e \frac{\partial W}{\partial x}\right) \qquad (2)$$

(b) y-Direction

$$\frac{\partial}{\partial x}(UV) + \frac{\partial}{\partial y}(VV) + \frac{\partial}{\partial z}(VW) = -\frac{1}{\rho}\frac{dP}{dx} + \frac{\partial}{\partial x}\left(\nu_e \frac{\partial V}{\partial x}\right) + \frac{\partial}{\partial y}\left(\nu_e \frac{\partial V}{\partial y}\right) + \frac{\partial}{\partial z}\left(\nu_e \frac{\partial V}{\partial z}\right) + \frac{\partial}{\partial x}\left(\nu_e \frac{\partial U}{\partial y}\right) + \frac{\partial}{\partial y}\left(\nu_e \frac{\partial V}{\partial y}\right) + \frac{\partial}{\partial z}\left(\nu_e \frac{\partial W}{\partial y}\right) \qquad (3)$$

(c) z-Direction

$$\frac{\partial}{\partial x}(UW) + \frac{\partial}{\partial y}(VW) + \frac{\partial}{\partial z}(WW) = -\frac{1}{\rho}\frac{dP}{dx} + \frac{\partial}{\partial x}\left(\nu_e \frac{\partial W}{\partial x}\right) + \frac{\partial}{\partial y}\left(\nu_e \frac{\partial W}{\partial y}\right) + \frac{\partial}{\partial z}\left(\nu_e \frac{\partial W}{\partial z}\right) + \frac{\partial}{\partial x}\left(\nu_e \frac{\partial U}{\partial z}\right) + \frac{\partial}{\partial y}\left(\nu_e \frac{\partial V}{\partial z}\right) + \frac{\partial}{\partial z}\left(\nu_e \frac{\partial W}{\partial z}\right) \qquad (4)$$

(iii) The Energy Equation

$$\frac{\partial}{\partial x}(UT) + \frac{\partial}{\partial y}(VT) + \frac{\partial}{\partial z}(WT) = \frac{\partial}{\partial x}\left(T_e\frac{\partial T}{\partial x}\right) + \frac{\partial}{\partial y}\left(T_e\frac{\partial T}{\partial y}\right) + \frac{\partial}{\partial z}\left(T_e\frac{\partial T}{\partial z}\right) \quad (5)$$

k- *ε model* is used to simulate the turbulent flow numerically [12]. This model has couple of equations, which join between turbulent kinetic energy (k) equation and turbulence dissipation rate (*ε*). All partial differential equations can be prescribed in the following general equation [11]:

$$\frac{\partial}{\partial x}(\rho U\emptyset) + \frac{\partial}{\partial y}(\rho V\emptyset) + \frac{\partial}{\partial z}(\rho W\emptyset) = \frac{\partial}{\partial x}\left(\Gamma_\emptyset\frac{\partial \emptyset}{\partial x}\right) + \frac{\partial}{\partial y}\left(\Gamma_\emptyset\frac{\partial \emptyset}{\partial y}\right) + \frac{\partial}{\partial z}\left(\Gamma_\emptyset\frac{\partial \emptyset}{\partial z}\right) + S_\emptyset \quad (6)$$

In this equation, the left side is represent the convection terms expressions, while the right side represent diffusion and source expressions.

3.2. Thermal Analysis of Solid Walls

3.2.1. Absorber Plate :

Thermal energy is barge to the blackened Absorber Plate from the direct radiation passed through glass cover in the non-packing area of PV module and fall on it. In addition by the connection between back surface of PV module and the absorber plate due to temperature difference between them, the air flowing in the channel get heated by the thermal energy brought from absorbing plate to the following air [13-14]. At node 3 of absorber plate, the energy balance is shown in Fig.(3) and represented as following:

$$q_e + q_w + q_t + q_b + q_n + q_s + q_{flux} = 0 \quad (7)$$

$q_e, q_w, q_t, q_b, q_s, q_n$ are the heat transfer by conduction to the east, west, top, bottom, north and south of the node respectively which can be given by [15]

$$q_e = k_{ap}A_x\frac{\partial T}{\partial x} = \frac{k_{ap}\,A_x}{\Delta x}\left(T_{ap}(i,j,k) - T_{ap}(i+1,j,k)\right) \quad (8)$$

$$q_w = k_{ap}A_x\frac{\partial T}{\partial x} = \frac{k_{ap}A_x}{\Delta x}\left(T_{ap}(i,j,k) - T_{ap}(i-1,j,k)\right) \quad (9)$$

$$q_t = k_{ap}A_z\frac{\partial T}{\partial z} = \frac{k_{ap}A_z}{\Delta z}\left(T_{ap}(i,j,k) - T_{ap}(i,j,k+1)\right) \quad (10)$$

$$q_b = k_{ap}A_z\frac{\partial T}{\partial z} = \frac{k_{ap}A_z}{\Delta z}\left(T_{ap}(i,j,k) - T_{ap}(i,j,k-1)\right) \quad (11)$$

$$q_n = q_{conv}(ap\_af) = h_{in}\,A_y\left(T_{ap}(i,j,k) - T_{af}\right) \quad (12)$$

$$q_s = k_{ap}A_y\frac{\partial T}{\partial y} = \frac{k_{ap}A_y}{\Delta y}\left(T_{ap}(i,j,k) - T_{ap}(i,j-1,k)\right) \quad (13)$$

$q_{conv(ab-af)}$ is the heat transferred by convection from absorber plate to the flowing air

$$h_{in} = 2.8 + 3v_f \quad (14)$$

Where h_{in} is the coefficient of heat transferred by convection between the absorber plate and air flowing inside the collector [16]. Where v_f is the air following velocity.

d

an

$$q_{flux} = (\tau_g\alpha_{ab})A_{yG} \quad (15)$$

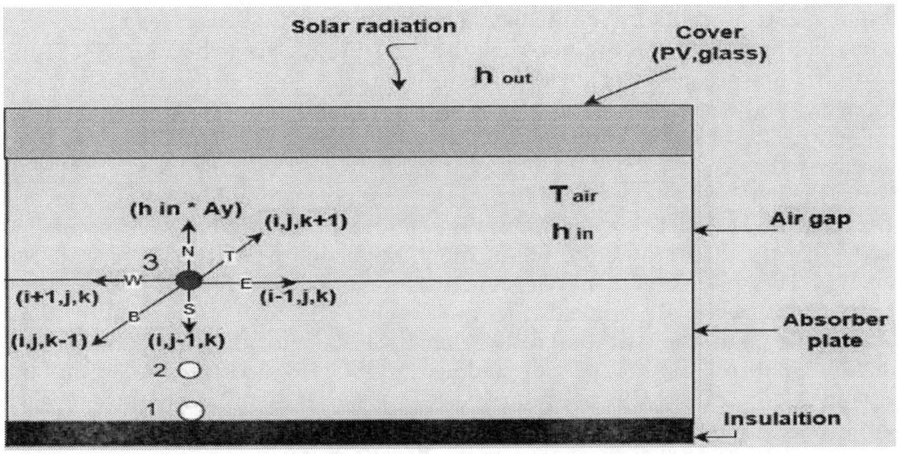

Figure 3: The energy balance of absorber plate.

for the portion covered by glass where a_{ab} is the absorber plate absorptivity and equal 0.8 [17], τ_g is the transmissivity of glass surface and equal 0.9 [18], and

$$q_{flux} = 0 \tag{16}$$

for the portion covered by PV modules. Eq.(7) can be represented as:

$$\frac{K_{ab}A_x}{\Delta x}(T_{ab}(i,j,k) - T_{ab}(i+1,j,k)) + \frac{K_{ab}A_x}{\Delta x}(T_{ab}(i,j,k) - T_{ab}(i-1,j,k)) +$$

$$\frac{K_{ab}A_z}{\Delta z}(T_{ab}(i,j,k) - T_{ab}(i,j,k+1)) + \frac{K_{ab}A_z}{\Delta z}(T_{ab}(i,j,k) - T_{ab}(i,j,k-1)) +$$

$$h_{in}A_y(T_{ab}(i,j,k) - T_{amb}) + \frac{K_{ab}A_y}{\Delta y}(T_{ab}(i,j,k) - T_{ab}(i,j-1,k)) + A_y q_{flux} = 0 \tag{17}$$

Suppose:

$a_e = a_w = \frac{K_{ap}A_x}{\Delta x}$

$a_t = a_b = \frac{K_{ap}A_z}{\Delta z}$

$a_s = \frac{K_{ap}A_y}{\Delta y}$

$a_n = h_{in}A_y$

$a_p = a_e + a_w + a_t + a_b + a_n + a_s$ these terms are substituted in Eq. (17) to give:

$$T_{ab}(i,j,k) = [a_e * T_{ab}(i+1,j,k) + a_w * T_{ab}(i-1,j,k) + a_t * T_{ab}(i,j,k+1) + a_b * T_{ab}(i,j,k-1) + a_n * T_{amb} + a_s * T_{ab}(i,j-1,k) + a_y * q_{flux}]/a_p \tag{18}$$

The absorber plate temperature of any point can be calculated by Eq.(18)

4. Results and Discussion

For the three cases, 0.33% of PV/T area is covered with glass at three different position (A- inlet portion, B- mid portion, C- outlet portion), in order to increase heat gain. Figure (4) reveals the variation of air temperature along the distance of PV/T system at different values of irradiance for the three configurations. Figure (5) illustrates the variation of absorber plate temperature along the distance of PV/T system at different values of irradiance for the three configurations.

Figure (6) shows the variation of (PV cell and glass) temperature along the distance of PV/T system at different values of irradiance for the three positions. From Fig.(4) one can notice that the maximum temperature of outlet air was obtained at configuration C reached 44.3 °C at 1000 W/m^2 and this sharp increase in outlet air temperature is due to the direct radiation fall into glass at outlet portion and restricts the heat gain in this position without make this radiation to passing parallel with PV modules. It was also observed that there is gradual increase in air temperature in case B due to the existence of glass at mid portion make the air get heated by direct radiation in the mid side moreover, to the thermal energy acquired by convection with PV modules back surface at outlet portion. Minimum outlet air temperature was obtained at case A because of that the direct radiation fall in the inlet portion when the cooling air was supplied. From Fig.(5) one can observe that the configuration A have convergent temperatures along the distance of absorber plate, while configuration B have the higher temperature along the distance of absorber plate at mid portion rather than the last end of absorber plate because of the solar radiation fall directly in mid portion and obscured by PV modules in the front and last end of plate, configuration C have the optimum rising in absorber plate temperature reached 60 °C. The maximum PV modules temperature achieved at configuration A reached 70 °C at 1000 W/m^2 due to the existence of PV module at the last side of PV/T air system and get heated by heat transferred from the hot air below it as well as the thermal energy gained from the direct radiation as illustrates in Fig.(6a). The PV cell and glass temperature have the same raising in temperature in configuration B due to the existence of glass in mid portion it will withdraw thermal energy from PV modules that connected with it from two direction, as shown in Fig. (6b). For configuration C Fig. (6c) shows that the maximum PV modules temperature was 63 °C at 1000 W/m^2 witch is the less among the other configuration as a result of existence the PV modules at cooling side of PV/T system. Additionally, the reduction in area of PV cell is lead to reduction on electric power generated also, the effect of the position of the glass on the electric power is depicted in Fig.(7) it's obvious that the configuration B (glass at mid portion) have the lowest generation due to high cell temperature because of receive the radiation from two direction. Furthermore, the electrical power is augmented in configuration C (glass at outlet portion) due to cooling effect and restricted the direct radiation at last end of PV/T system. To express the effect of cooling air more in partially covered PV/T with glass at outlet portion Fig(8) shows the variation of outlet air temperature with electrical power generated depending on the air flowrate variation and at constant irradiance (600 W/m^2). It is clear that the optimum air flowrate was (0.013 kg/s) when the electric power and outlet air temperature were 18 W and 37°C. Thermal efficiency of PV/T system partially covered with glass at three different configurations analyzed in Fig.(5.9), one can observe from this figure that case C have the maximum thermal efficiency reached 48.7% due to the effect of direct radiation fallen in the glass and hot air flowing in the outlet portion lead to high thermal energy, in contrast to the case A which the direct radiation from the glass fallen in to inlet portion when the cooling air was supplied. In contours Figures, the three dimensional plot for absorber plate and air temperature is plotted at airflow of 0.013 kg/s for configuration C (partially covered with glass at outlet portion), Figs.(10) and (11) shows isotherm contours of absorber plate and air respectively. The results clearly show that the initial increase in the temperature of absorber plate edges is due to the connection between PV modules and absorber plate in this point and for air, fig (11) reveals that the upper layers of air that attached to the back surface of PV modules have got higher temperature and at last portion of PV/T system this increase will be at the upper and lower layers when the glass is existence and the radiation fallen in the absorber plate directly. The PV modules, plate and air temperatures obtained by experimental work has been compared with the numerical results and plot in Fig.(12) for the case C glass at outlet portion, a strong agreement has been achieved between the experimental and numerical outcomes. Fig. (13) shows the flow field of three-dimensional PV/T system at air flowrate equal 0.013 kg/s, this figure shows the creation of boundary layers, its observed that the thickness of boundary layers is small near the wall.

IOP Conf. Series: Materials Science and Engineering **765** (2020) 012031 doi:10.1088/1757-899X/765/1/012031

5. Conclusion

On the basis of study the hybrid PV/T air collector with different configurations the following points have been concluded:

(i) PV/T system partially covered with different glass configurations concluded that placing the glass at outlet portion give the higher electrical and thermal gain reached 26W and 44°C so that its prefer over other two configurations (inlet portion and mid portion).

(ii) Optimizing the electrical power and outlet air temperature with mass flow rate, it is advised to operate the PV/T collector at moderate flow rate (0.013 kg/s) to obtain the required hot air and electrical power.

(iii) Glass cover located at mid portion (case B) will maximize PV modules temperature hence, the electrical power generated will be worst in this configuration amongst other two configurations.

References

[1] D. Atheaya, A. Tiwari, G. Tiwari, and I. Al-Helal, *Analytical characteristic equation for partially covered photovoltaic thermal (pvt) compound parabolic concentrator (cpc),Solar Energy, vol. 111, pp.176–185, 2015.*
[2] *A. Saxena, V. Tirth, and G. Srivasatava, "Thermal performance studies of a low cost hybrid solar air heater."*
[3] *Kalogirou, Soteris A " Solar energy engineering: processes and systems," Elsevier Inc., 1st Ed, 2009.*
[4] *M. I. Hussain and G. H. Lee, "Experimental and numerical studies of a u-shaped solar energy collector to track the maximum cpv/t system output by varying the flow rate," Renewable energy, vol. 76, pp. 735–742, 2015.*
[5] *R. Mishra and G. Tiwari, "Energy and exergy analysis of hybrid photovoltaic thermal water collector for constant collection temperature mode," Solar Energy, vol. 90, pp. 58–67, 2013.*
[6] *M. Al Imam, R. Beg, M. Rahman, and M. Khan, "Performance of pvt solar collector with compound parabolic concentrator and phase change materials," Energy and Buildings, vol. 113, pp. 139–144, 2016.*
[7] *M. Saidan, A. G. Albaali, E. Alasis, and J. K. Kaldellis, "Experimental study on the effect of dust deposition on solar photovoltaic panels in desert environment," Renewable Energy, vol. 92, pp. 499–505, 2016.*
[8] *S. Tiwari, J. Bhatti, G. Tiwari, and I. Al-Helal, "Thermal modelling of photovoltaic thermal (pvt) integrated greenhouse system for biogas heating," Solar Energy, vol. 136, pp. 639–649, 2016.*
[9] *J. Tonui and Y. Tripanagnostopoulos, "Performance improvement of pv/t solar collectors with natural air flow operation," Solar energy, vol. 82, no. 1, pp. 1–12, 2008.*
[10] *P. Ooshaksaraei, K. Sopian, S. H. Zaidi, and R. Zulkifli, "Performance of four air-based photovoltaic thermal collectors configurations with bifacial solar cells," Renewable energy, vol. 102, pp. 279–293, 2017.*
[11] *H. B. Awbi, "Ventilation of building," E and FN Spon, 1991.*
[12] *B.T. Launder, D.B. Spalding, "Lecturers in Mathematical models of turbulence," Department of Mechanical Engineering, imperial college of Science and Technology, London, England, 1972.*
[13] *G. Tiwari, I. Al-Helal et al., "Analytical expression of temperature dependent electrical efficiency of n-pvt water collectors connected in series," Solar Energy, vol. 114, pp. 61–76, 2015.*
[14] *G. T. Shyam, "Analysis of series connected photovoltaic thermal air collectors partially covered by semitransparent photovoltaic module,"Solar Energy, vol. 137, pp. 452–462,2016.*
[15] *J. P. Holman, Heat Transfer-Si Units-Sie. Tata McGraw-Hill Education, 2002.*
[16] *S. A. Kalogirou, Solar energy engineering: processes and systems. Academic Press, 2013.*
[17] *S. Dubey, S. Solanki, and A. Tiwari, "Energy and exergy analysis of pv/t air collectors connected in series," Energy and Buildings, vol. 41,no. 8, pp. 863–870, 2009.*
[18] *B. Widyolar, L. Jiang, and R. Winston, "Spectral beam splitting in hybrid pv/t parabolic trough systems for power generation," Applied Energy, vol. 209, pp. 236–250, 2018.*

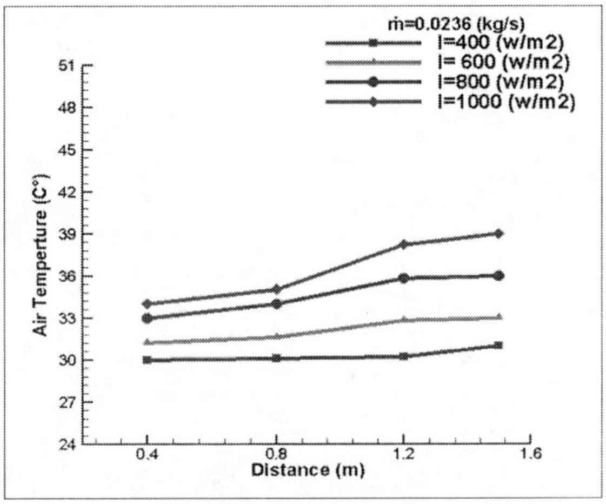

(a) Glass at inlet portion

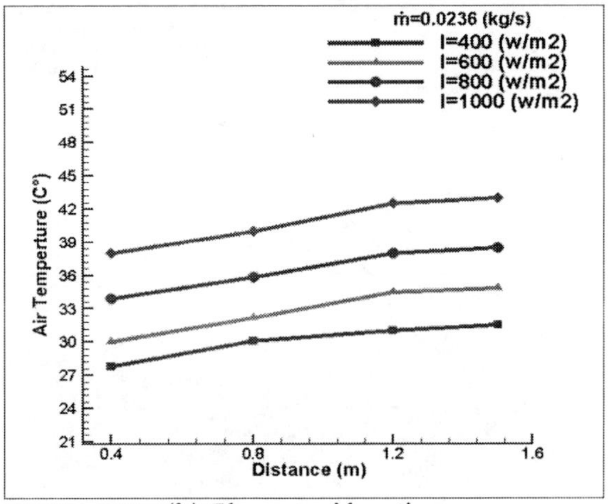

(b) Glass at mid portion

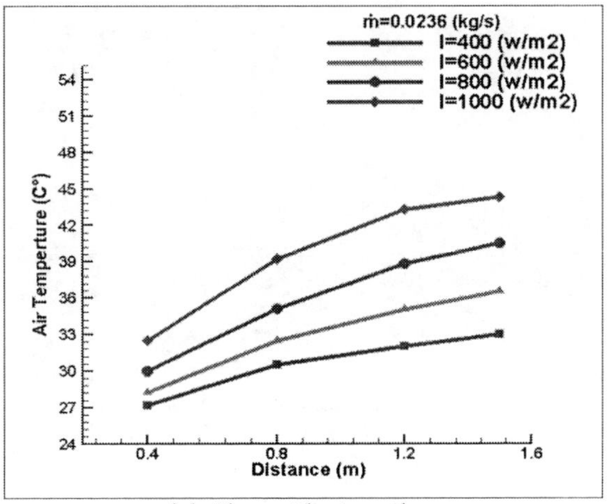

(c) Glass at last portion

Figure 4: Experimental variation of air temperature along the distance of PV/T collector for the three cases at different values of solar radiation.

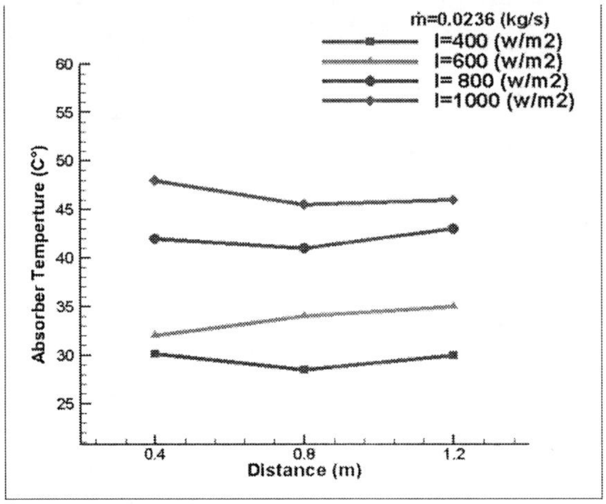

(a) Glass at inlet portion

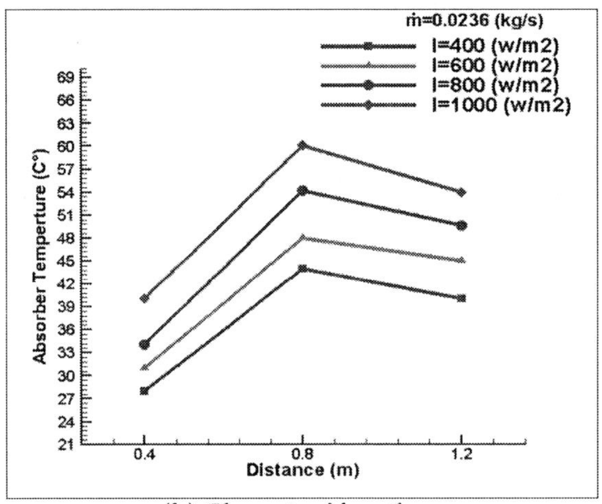

(b) Glass at midportion

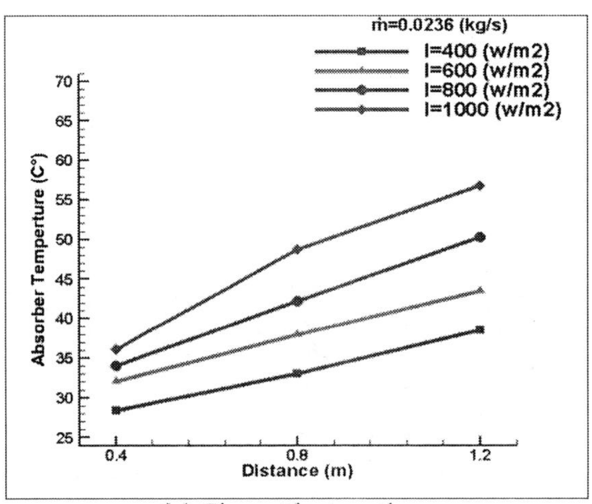

(c) Glass at last portion

Figure 5: Experimental variation of absorber plate temperature along the distance of PV/T collector for the three cases at different values of solar radiation.

IOP Conf. Series: Materials Science and Engineering **765** (2020) 012031 doi:10.1088/1757-899X/765/1/012031

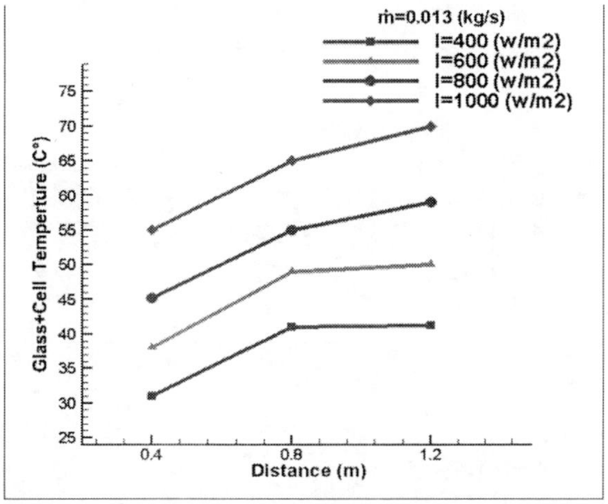

(a) Glass at inlet portion

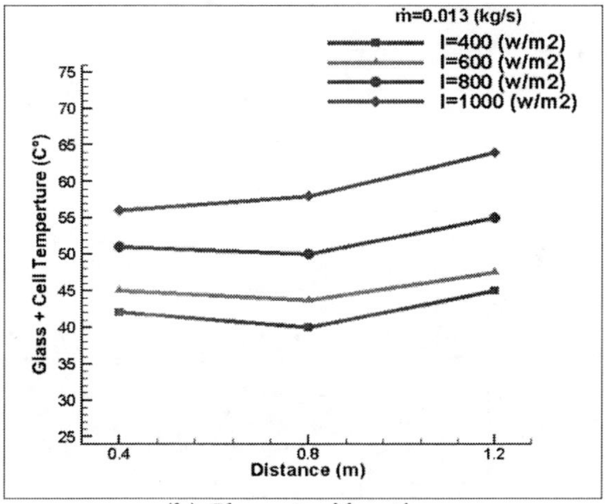

(b) Glass at midportion

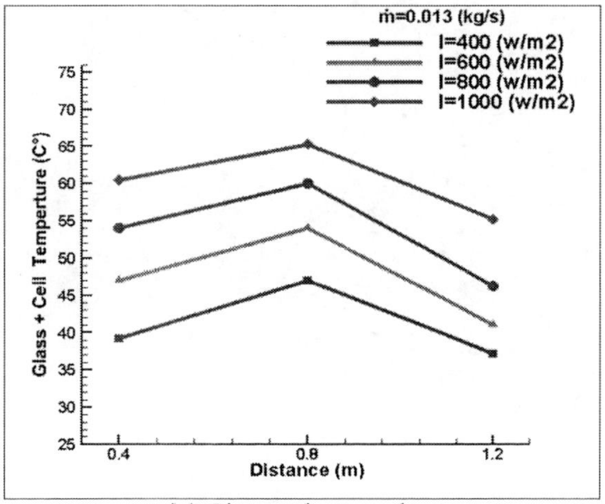

(c) Glass at last portion

Figure 6: Experimental variation of the PV cell and glass temperature along the distance of PV/T collector for the three cases at different values of solar radiation.

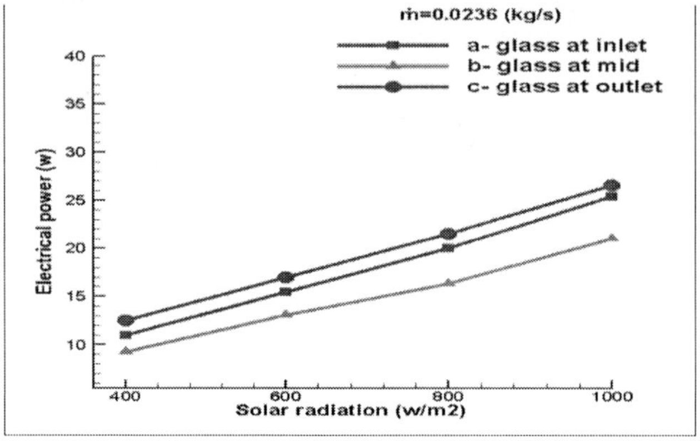

Figure 7: Experimental variation of electrical power generated with respect to the irradiance variation at different glass configuration.

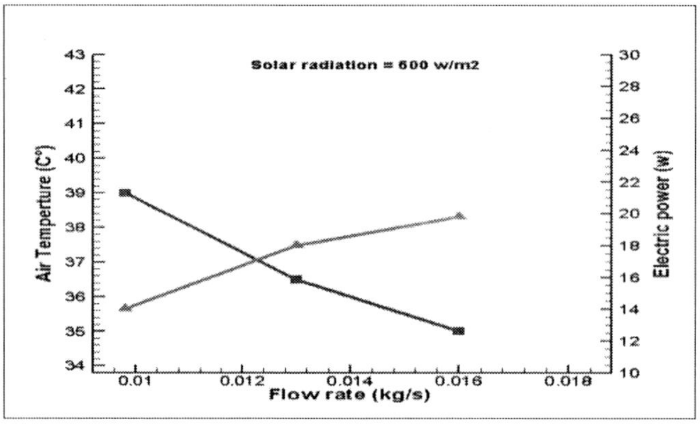

Figure 8: Variation of electrical power and outlet air temperature with respect to air flowrate variation for configuration C (glass at outlet portion) at constant solar radiation (600 W/m^2).

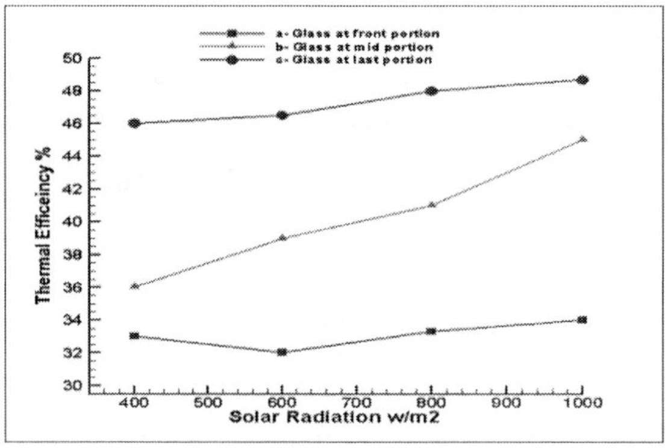

Figure 9: Experimental variation of thermal efficiency with respect to the irradiance variation at different glass configurations.

IOP Conf. Series: Materials Science and Engineering **765** (2020) 012031 doi:10.1088/1757-899X/765/1/012031

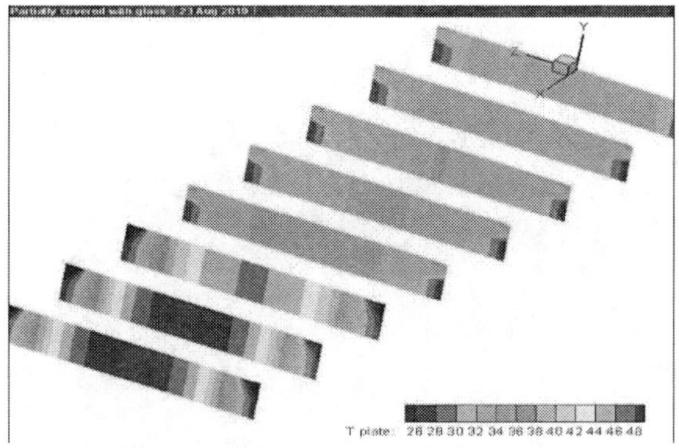

Figure 10: Isotherm contours of absorber plate in PV/T system for configuration C (glass at outlet portion).

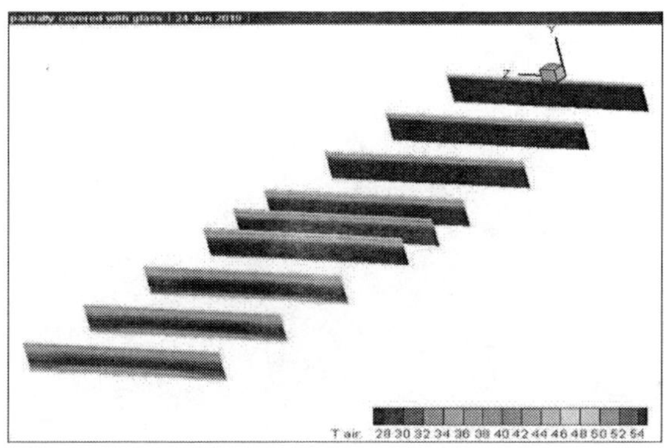

Figure 11: Isotherm contours of air in PV/T system for configuration C (glass at outlet portion)

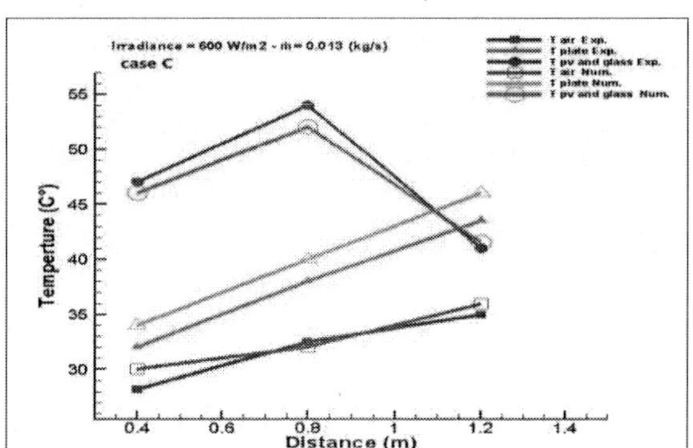

Figure 12: Comparison between the experimental and numerical results for the configuration C (glass at outlet portion)

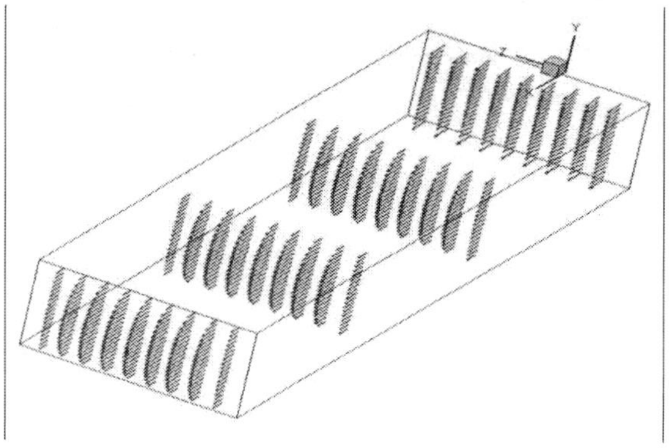

Figure 13: Velocity vector (u, v and w) at flowrate of 0.013 kg/s.

Outdoor testing of an evacuated tube closed two phase thermosyphon solar water heater charged with Nano- fluid

Mahmoud Maustafa Mahdi

Ameer Abed Jaddoa

Ibrahem Falih Kadhim.

Wisam Asmet

Electromechanical Engineering Department, University of Technology, Baghdad, Iraq

corresponding author's e-mail : 50087@uotechnology.edu.iq

Abstract. This paper reports the performance of two types of evacuated tube heat pipe solar water heater systems that have been experimented outdoors under the local meteorological conditions of the capital of Iraq. Tests were conducted using two systems made of the same materials but with different design specifications. To build each of the two trail system, a single evacuated tube to two phase thermosyphon is used to make the investigation of the experiments simpler. With the absence of loading conditions, experiments with the first system were conducted. Graphene water Nano-fluid (GW) was used as the working fluid in this system. The water storage tank system capacity was varied, to analyze the performance for different storage capacity. The system was operated with a flat reflector plate and without this reflector. A modified second system was built with a longer evaporator section for higher performance and with tank that enables hot water removed from it. The hot water from the storage tank was withdrawn intermittently and continuously. Graphene water (GW),Graphene acetone(GA) and distilled wate(d-W) were used as the working fluids in this second system. Energy stored at the water storage tank and overall efficiencies were calculated and found to be comparable to those obtained using natural circulation solar water heating systems. A comparison between the obtained Results and other works was made. The comparison showed that the present systems were just as good in performance. An improvement in the overall efficiency was found when operating the system under load conditions. This improved value is about 10% higher than that without load condition.

Keywords: Solar two phase thermosyphon, Solar Evacuated tube water heater, Two phase thermosyphone with Nano fluid,

IOP Conf. Series: Materials Science and Engineering **765** (2020) 012032 doi:10.1088/1757-899X/765/1/012032

1. Introduction

Since the eighteenth century, scientists and engineers have been interested in the use of solar energy for water heating systems. (CTPT) is able with a very small temperature gradient to transfer high rates of heat [1]. Compared to the more conventional flow absorber, a (CTPT) absorber has many advantages in a s solar collector system such as superior freeze tolerance, less parasite pumping demands, a thermal diode benefit, where heat loss is minimized when the collector temperature is less than ambient [2]. The evaporator end of a heat pipe was first inserted in a flat plate collector by Bienert and Wolf using one of the early solar water applications [3]. A number of experiments were carried out to include the heat pipe in collectors [4, 5, 6] but the results were not certain or promising [4]. Bairamov and Toiliev [7] conducted experiments on two type solar water heating systems, namely, with and without heat pipe. Experimental results on a model converted into a solar collector with and without a heat pipe, shows that at the end of water heating period in summer, the water temperature in the storage tank of the System with a heat pipe is (10 - 11 °C) higher than that without a heat pipe. They mentioned, the diode effect being the cause. Akyurt [8], designed, and manufactured heat pipes incorporated into a prototype solar water heater developed for this purpose, and was tested under actual insolation conditions. It was observed that the heat pipes perform satisfactorily as a heat transfer element in solar water heaters. Number of the heat transfer factors of a heat pipe absorber array connected to a common manifold were analyzed by Hull [9]. It was shown that the number of absorbers and the heat transfer rate at the condenser were key factors in determining array efficiency. Georgier [10] presented a mathematical simulation for a vacuum solar collector with a heat pipe. Hammead [11] described the performance of a flat plat solar collector cooled by a set of heat pipes designed and manufactured locally to work at low temperature conditions equal to that of the flat plate solar collector. Chun, et.al [12] conducted a series of tests on solar domestic hot water systems manufactured with heat pipes. Results showed interesting performance data stemming from the difference in the working fluids, presence of a wick, and other various design parameters associated with the collection and utilization of solar energy. Ismaill and Abogderah [13] carried out a comparative study between theoretical prediction and experimental results of a flat plate solar collector with heat pipes. Lima and Palanyana [14] described the experimental behavior of a flat plate solar collector with heat pipes and selective surfaces. It was observed that the thermal degradation of energy absorbed and the climate conditions influence the overall energy loss coefficient. However, their losses are comparable to the other devices produced for the same purpose. Corliss et.al.[15] conducted a detailed study of a heat pipe augmented passive heating system with different kinds of heat pipes made of different pipe materials, with different working fluids and collector configurations. Lu , et al.[16] fabricated an open thermysyphon system with an evacuated tub collector to reach a high temperature .They concluded that the evaporating heat transfer coefficient was increased by about 30% compared with deionized water . Senthil kumar et al. [17] designed a two identical solar collectors with a flat surface water was the working fluid in first collector, while CNT-water Nano fluid was in the second system. They concluded that the collector charged with Nano fluid is better compared with that with purr water. Nh.Mujawar and S.M.Shaikh [18] designed and developed an evacuated heat pipe solar water heating systems using Nano fluid .It was found a mark rise in the collector efficiency using nano fluid. Sumehh S. Jagtap et al [19] made an experimental investigation of evacuated tube two phase thermosyphon charged with copper oxide Nano fluids. They found that the collector charged with Cuo is 20% more than purr water in term of transfer capacity and heat absorption.

Most of the above studies were conducted without removing the hot water from the storage tank of the solar heat pipe system. In this study a complete solar water heater, utilizing an evacuated tube two phase thermosphon, was developed and tested under actual insolation conditions. Two systems were tested outdoors. The performance of both systems was investigated experimentally. Experiments were carried out on the first system with different amounts of water in the storage tank and with and without a flat reflector. Experiments on the second system were conducted with and without hot water removal out of the storage tank. Hot water withdrawal was continuous and intermittent. The objective of the present work is to investigate the performance of an evacuated tube two phase thermosyphon solar water heater under various operating parameters with and without load.

1.1 System Description

Experiments were carried out with two solar water heater systems made of the same materials but with different design specifications. To build each of the two trail system, a single evacuated tube to two phase thermo syphon is used to make the investigation of the experiments simpler .The first system was basically a solar water heater utilizing the evacuated tube heat pipe technology with no hot water removed from its storage tank. This system consisted of Pyrex glass evacuated tube of (100 cm) length and (5cm) outer diameter. The heat pipe was 22 mm in outer diameter a 20 mm inner diameter, and 1270 mm Jong. The heat pipe was installed inside the glass tube. A wick is built of enfolded screen wick of 100 mesh size. The length of the evaporator is 1000 mm, the length of the adiabatic is 70 mm, and the length of the condenser is 200 mm. part of the heat pipe called the condenser portion is directly inserted into the storage container as shown schematically in Fig.1 Graphen – Water GW was used as the working fluid in this CTPT. The system was operated with a flat reflector plate, and without this reflector. With the reflector the collector receives radiation heat from the back as well as the top of the absorber surface. This effect was pioneered by Owens-Illinois [16]. It is achieved by using a highly reflective aluminium sheet surface behind the assembly to illuminate the rear surface, as shown in Fig.1.

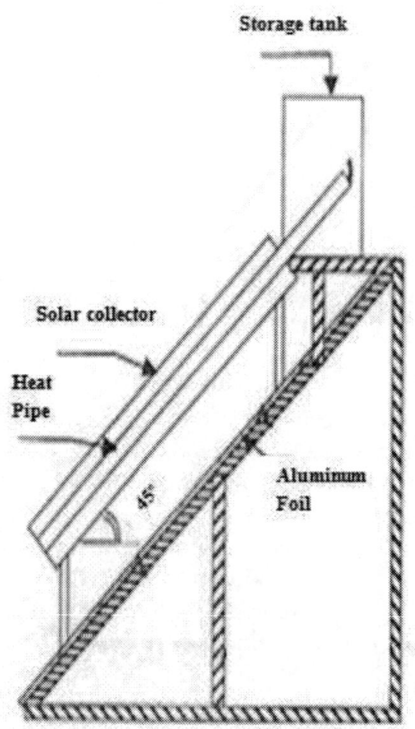

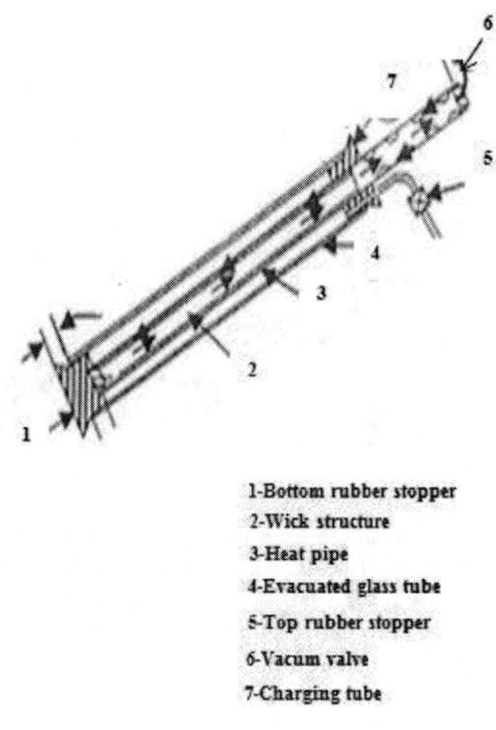

Figure 1: Schematic diagram of the solar collector heat pipe Collector details.

Figure 2: Evacuated tube heat pipe solar.

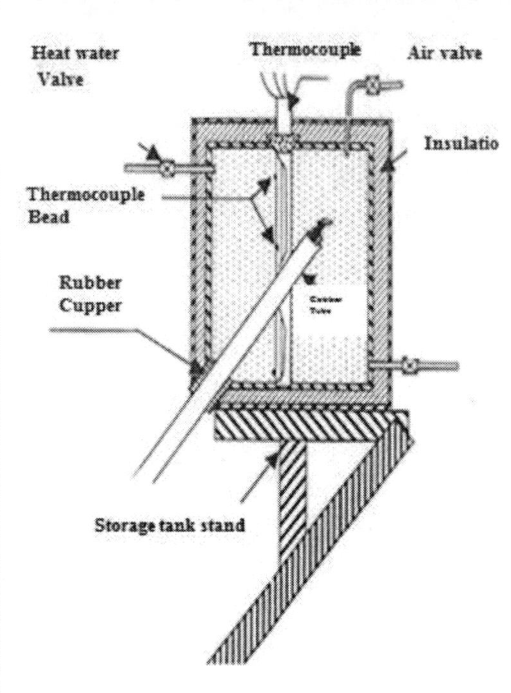

Figure 3: Cross sectional diagram of the storage tank

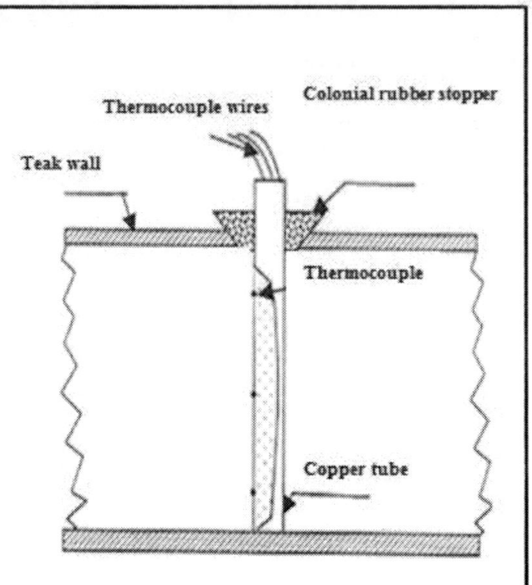

Figure 4: Schematic diagram of a centreline probe

2.Manufacture and Assembly

The solar collector tested in the present trial system is shown in Fig.1. A conventional tubing and piping with a 22 mm nominal diameter was chopped to (1270 mm) length. Ends caps were of copper and were mechanically modified in a way allows them to go inside the pipe. The condenser end cap was drilled providing the necessary access for the charging tube. All CTPT components were cleaned thoroughly before assembly with a series of chemical and mechanical processes to remove all possible contaminates. A rig was specifically built to charge the heat pipe on. After filling operation, the filling tube was crimped. Then, the excess tube was clipped off and brazed which would lead to producing a heat pipe. Paint was used to cover the area exposed to sun of the evaporator section. The evaporator section was then installed within a glass envelop forming a collector with its ends closed by two conical rubber stoppers as shown in Fig.2. The stoppers were modified by a machine in a way that allows them to fit in the evacuated glass tube. The top end was drilled providing the necessary access holes. The centre hole was used for the insertion of the heat pipe tubing, while the other was used for drain tube. The drain tube has a vacuum shut-off valve to subject the glass tube to a vacuum after coupling it with vacuum pump. The complete solar collector was supported by a steel stand with a tilt angle of 45°C facing south as shown in Fig.1. The water storage tank of the system was fabricated from a galvanized steel of 0.6 mm thickness and measuring 0.11m * 0.15m*0.325m in size, and it can contain 5L of water. Fig.3 shows a schematic cross-sectional diagram of the water storage tank with tank accessories. To minimize heat losses to the environment, the storage container of the water was surrounded with 3cm thick glass wool insulation placed between the tank outer surface and the outer covering made of aluminium sheet with 1 mm thickness. The storage tank included a hole in the lower part through which the condenser portion of the heat pipe was inserted directly through a rubber stopper. The mains cold water inlet and hot water outlet of the storage tank were position as shown in Fig.3. Cold mains water enters from the bottom side of the storage tank and hot water leaves from the upper opposite side through a 5mm copper tube. The copper tube was drilled for a temperature sensor.

The water storage tank was supported by a steel stand Fig. 3. The water storage tank was instrumented with three thermocouples were located inside the storage tank and two in the inlet and outlet for water temperature measurement. The three thermocouples were distributed into a vertical centreline probe, to determine the distribution of the temperature distribution along the height of the tank. The probe shown in Fig.4 was made of a copper tube 5mm in diameter with holes drilled in its outer surface used for inserting the thermocouples wires. The complete solar collector system was supported by a steel stand with a tilt angle of 45° above the horizontal facing south. This inclination angle was chosen to provide approximately the optimum performance of the heat pipe and highest heat transport rate [17]. In addition, it symbolizes a decent wintertime tilt angle for solar collectors in the capital of Iraq where the latitude angle is 33°N. Fig. (4a) typical solar radiation intensity during a sunny day.

3.Test Procedure

Experiments were carried out with two solar water heater trial systems with different design specifications. With the purpose of gaining usual everyday data with perfect skies, test measurements were conducted on sunny days. The trials were conducted in May, June, July, and August 2019. The experimental preparation and test procedure of each system is described below:

3.1 Solar CTPT system.

The performance of the first system was conducted outdoors. The solar collector which consisted of a single evacuated tube CTPT inclined at 45°C toward south. G.W was used as the working fluid in the CTPT. The system was tried in steady – state conditions wherein the solar intensity and ambient temperature were assumed constant for a window of time. This window was usually 30 minutes for an operating from 16:00 p.m. till 24 p.m.

At the onset of each experiment the storage tank was filled with a specified quantity of fresh water, the glass tube was cleaned thoroughly; the thermocouple connections were checked for contact. The measuring instruments were switched on. For this system the temperature distribution of the water inside the storage tank was determined by a probe containing three thermocouples. The mean tank temperature was determined from the average experimental temperature data. The same experimental procedure was repeated for each of the different storage capacities and the two different configurations (with and without solar reflector). The temperatures of the storage tanks at the beginning and end of each operation were used to determine the overall collected heat. In this case from 7:30 or 8:00 a.m. till 17:00 p.m. The overall. Stored heat was given by [18] as:

$$Q_{ove} = M_1 CW(T_{mf} - T_{ms}) \quad (1)$$

While the overall or bulk efficiency of the system was counted using the following equation;

$$\eta_{ove} = \frac{M_1 CW(T_{mf} - T_{ms})}{\int_0^t A_a I(t)dt} \quad (2)$$

3.2 Solar CTPT system with hot water removal.

In the same manner as in the first trial system, the second trial system was tested out- doors. The system was tested with flat reflector behind the collector and with a fixed quantity of 5 kg of water in the storage tank. The system was tested with and without hot water withdrawn from the storage tank and for different patterns, of hot water-removal. Three working fluids were used with this system namely GW,GA and dw.For the case of water withdrawal from the storage tank, the load was connected to the storage tank such that the cold inlet water entered at the bottom of the tank and the hot water outlet was taken from the top of the tank. The mass flow rate of load water was regulated by a valve at the outlet of the storage tank. The load mass flow rate was measured by timing the water collected in a vessel of 1 litter capacity. The inlet and outlet temperatures of the load were noted simultaneously with the other data. Two types of water withdrawal programs were employed. In the first test, the water was withdrawn intermittently. Three patterns of hot water removal were used with this type of test with two different CTPT working fluids. Limited trials were carried out forcing a big load on the system in a small period of time to determine the response of the solar collector. Load water in the range from 2.5 litres to 3 litres was removed rapidly from the tank at a 10 minutes period. In this case the average temperature of the water withdrawn from the tank and the temperature of mains water entering the tank were recorded in addition to the other measurements. The overall heat or bulk heat collected by the system with intermittent load conditions was estimated and compared for the different CTPT working fluids which were G.W and G.A. In the second type of tests, the water was withdrawn continuously from the storage tank and for seven different load mass flow rates of water through the tank. These flow rates were (0.5 [for, 1 I /hr, 2 fpr, 3 1 fbr, 4 I for, 5 I or, and 6 / /hr). In each case, the load usually started at 9:00 a.m. and continued until 17:00 p.m. The overall or bulk heat collected by the system with load conditions was estimated from the following equation [20];

$$Q_{ove.l} = M_t CW\left(T_{mf} - T_{ms}\right) + \int_0^t m\ load\ CW(T_{out.L} T_{in,L})\, dt \quad (3)$$

Further simplifications were done to the equation to generate the following form;

$$Q_{ove} = M_t CW\left(T_{mf} - T_{ms}\right) + M_l Cw(\bar{T}_{out.L} - \bar{T}_{in,L}) \quad (4)$$

Where;

$$M_l = \Sigma_{mL}^{.i} \Delta t^i \quad (5)$$

$$(T_{out}^{-L} - T_{in}^{-L}) = \frac{\Sigma^{i=k} m_L^i (T_{out,L}^i - T_{in,L}^i) \Delta t^i}{\Sigma_{mL}^{.i} \Delta t^i} \quad (6)$$

Where the System Overall or Bulk Efficiency was determined using the Following Equation;

$$\eta_{ove} = \frac{M_t CW(T_{mf} - T_{ms}) + M_L Cw(\bar{T}_{out.L} - \bar{T}_{in,L})}{\int_0^t I(t) A_a dt} \quad (7)$$

4. Discussion of the results:

Experiments were conducted with the aim of evaluating the behaviour of an evacuated tube CTPT solar water heater system. Two systems with different specifications were experienced for the following condition:

Condition	First system	Second system
CTPT working fluid	Graphen-water	Graphen-acetone, Graphen-water, d-water
Collector inclination angle	45 degree	45 degree
Evaporator length	1000 mm	1200 mm
Condenser length	200 mm	200 mm
Storge capacity	2.5,3,4 and 5 kg	5 kg
Weather capacity	Clear sky	Clear sky
Load condition	No load	Continuous and intermittent load

The effect of the various operating factors on the performance of the above systems will be discussed in the following sections. The first experiments were conducted, on the first system, without the reflector. These experiments were 'carried out in the same weak such that the weather conditions were considered similar. Fig.5 shows a comparison of the increase in the mean tank temperature in the month of May for different storage tank capacity. The trend of variation is similar. The temperature of the mean tank was found to rise with time and reach its highest magnitude at mid-day and then slightly decreases. The difference in temperature is due to the stored water. Also, the figure shows that the maximum mean tank temperature is 65°C for 2.5 kg with a daily temperature increase of 33°C, whereas this maximum is only 55°C for the 5kg storage capacity with a daily temperature rise of 25°C. The maximum mean tank temperature for the other storage capacities lies between those two values. Also it is observed from Fig.5 that, the maximum temperature for water occurs at earlier times with the lower storage capacity. A number of experiments had been carried out through the same month and conditions and it would be tedious and repetitive to show all such results m the form of performance curves similar to those discussed above. An example of this is Fig.6. When a flat reflector is used behind the collector of the solar water heater system as shown in (Fig.1), a slight increase in the maximum mean tank temperature is noticed for 4 and 5 kg storage capacity when compared to that without reflector as shown in Fig.7.

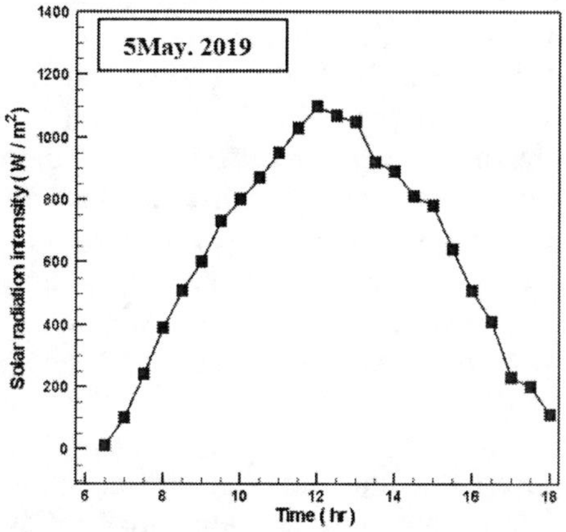

Figure 4a: Typical solar radiation intensity During a sunny day

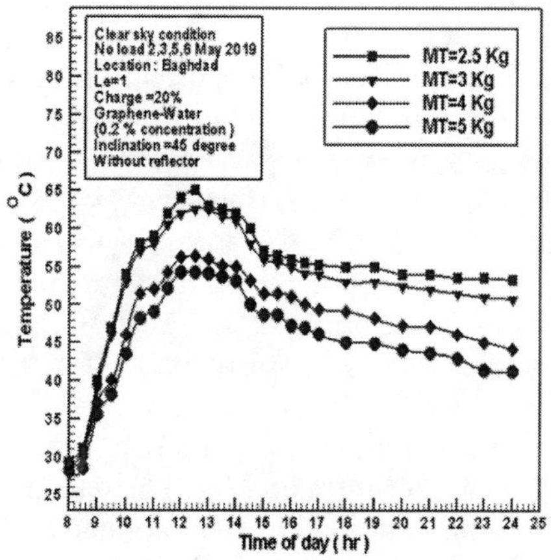

Figure 5: The variation of tank Temperature with storage capacity

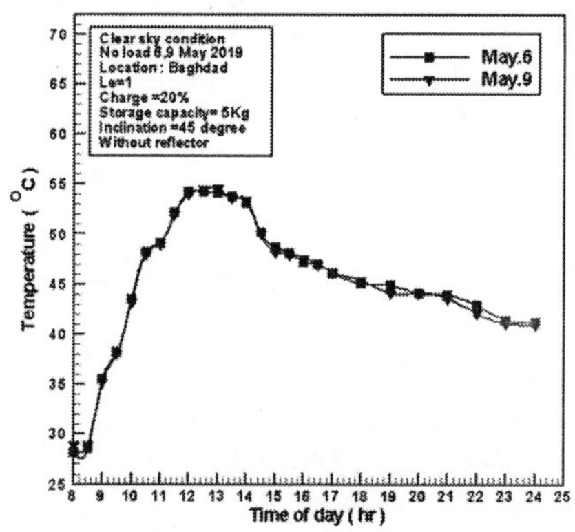

Figure 6: The variation of tank Temperature for different days

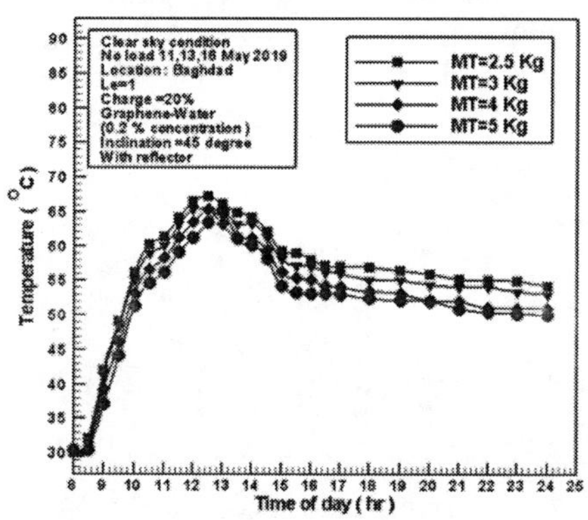

Figure 7: The variation of tank Temperature with storage capacity

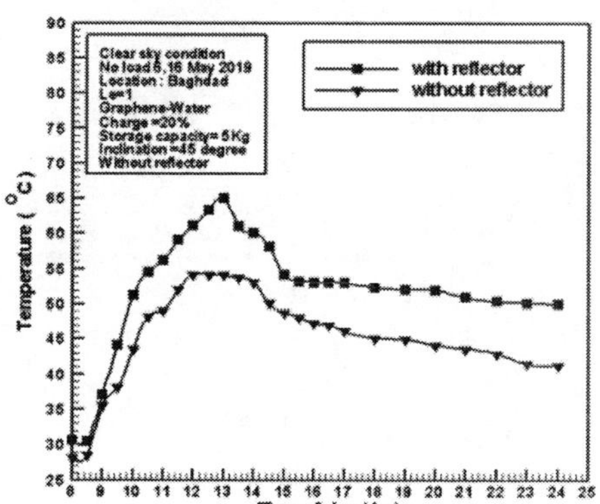

Figure 8: The variation of tank Temperature with and without reflector

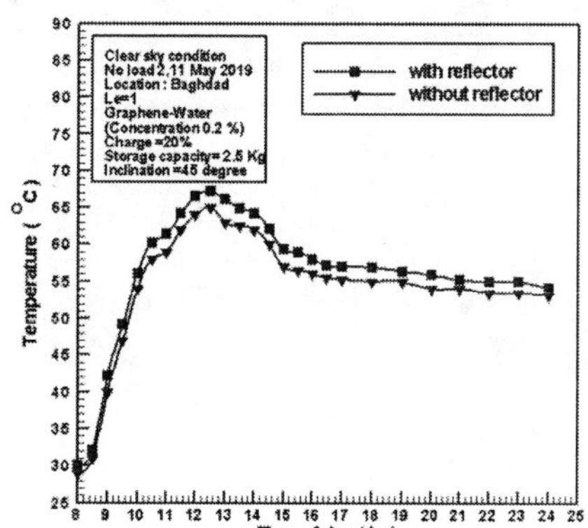

Figure 9: The variation of tank Temperature with and without reflector

This increase in temperature is approximately 4°C for 4kg storage and 6°C for 5kg at approximately 12.00 noon. In order to assess the effect of the flat reflector, Figs.8 and 9 are plotted for each of two storage capacities of 2.5 and 5 kg respectively, It is clear that for the system with the flat

IOP Conf. Series: Materials Science and Engineering **765** (2020) 012032 doi:10.1088/1757-899X/765/1/012032

reflector at 5kg storage capacity shows a higher mean tank temperature than without the reflector. At 2.5 kg, Fig.9, an approximately similar maximum mean tank temperature at 12.00 noon. This is because the maximum possible temperature attainable with this system is limited by the storage even with using the flat reflector. The distribution of tank temperature is a tremendously complex fiction of the delivered energy, the mixing effect and energy losses that the body of water causes at various temperatures. Heat transfer inside the storage tank is essentially controlled by convection phenomena that cause temperature redistribution. Fig.10 illustrations the temperature differences at three positions within the storage tank, of Skg water capacity, at several times on a standard clear day in May. The natural convection mixing effect within the storage tank is proven and illustrated well and the tank content becomes nearly at the same temperature. Once the mean tank temperature and its rate of increase have been determined, the heat stored at the storage tank can then be calculated employing equation 1 and also the bulk or overall efficiency employing equation 2. The overall heat stored was taken in the period between 9.00 a.m. and 19.00 p.m. Fig.11 shows the daily bulk heat stored as mathematically counted from equation 1. The incident solar radiation intensity was determined by the method of Faber and Morrison [18]. The data points represent actual values. The connecting lines have no physical significance other than connecting the data points. Values of overall heat stored range between 580 kJ for 2.5 ke and 667 kJ for the 5 ke water. The system without flat reflector and 5kg water storage capacity provided a 2.5% higher heat stored than with 4kg storage, 4% higher than 3kg storage and 7% higher than 2.5 kg storage water. This increase in overall heat stored was 4, 9 and 11% for the case in which the system operated with flat reflector. As the system operates with a flat reflector is more advantageous in clear climates. .The overall daily system efficiency for the operating period is calculated from equation 2 and range from 45 to 50%depending on the system configuration and storage tank capacity as displayed in Fig,-12.

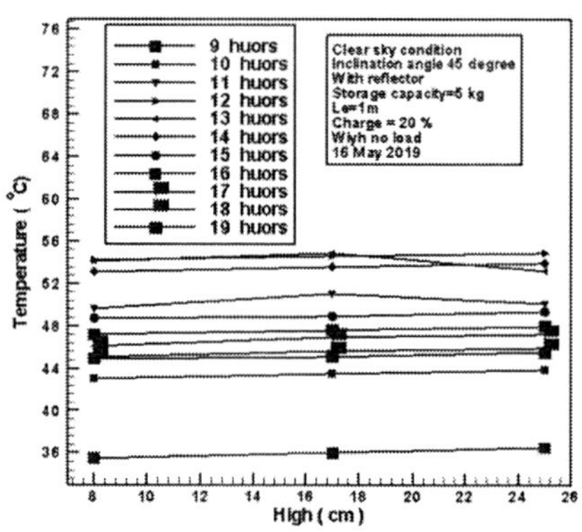

Figure 10: Temperature variation of water within the storage tank at different times

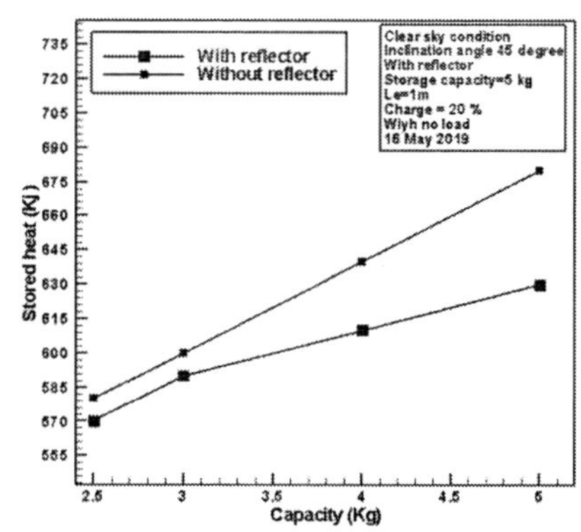

Figure 11: Overall heat storage as a function of storage capacity

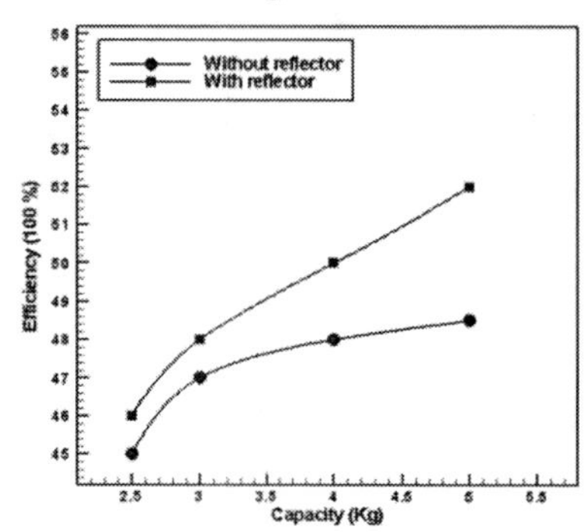

Figure 12: Overall bulk efficiency as a function of storage capacity

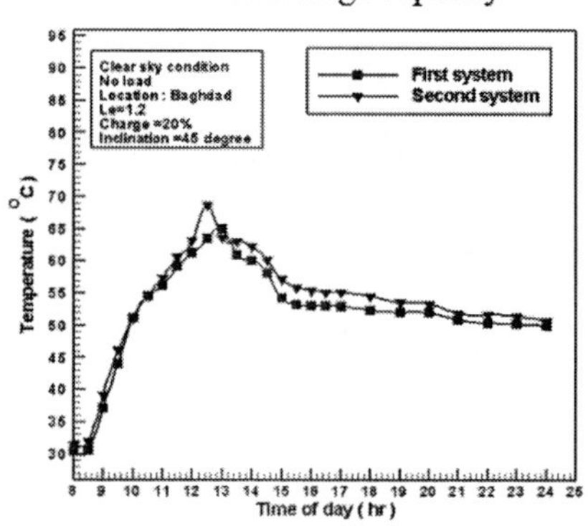

Figure 13: The variation of tank temperature for various system configuration

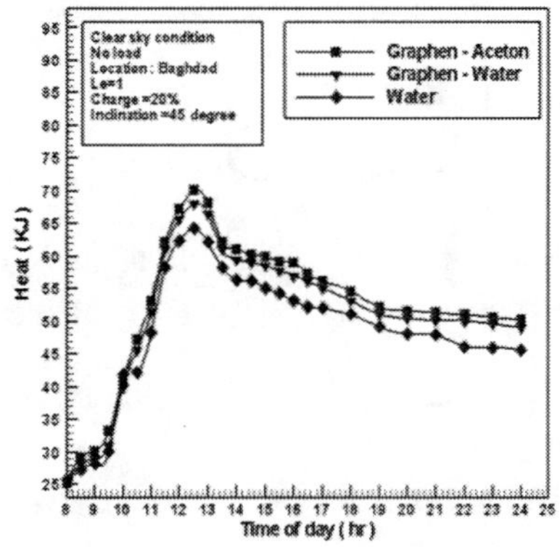

Figure 14: The variation of tank temperature for various working fluid

Firstly, the CTPT of the second solar water heater system was charged with G.W. in order to compare its results with the "system. Fig.13 shows the variation of mean tank temperature for the two solar water heater systems. The second system shows a significant increase in the mean tank temperature, the temperature difference is approximately 6°C at 12.00 noon. This improvement is due to the longer evaporator length (larger collector area) of the second system, which enables it to collect and transfer more energy to the storage tank. In order to assess the effect of the working fluid, Fig.14 shows the variation of the mean tank temperature during the operation period with no load condition for the various working fluids. A maximum temperature difference of 8°C between G.A. and d.W is observed at 12.00 noon. This is because G.A. has a larger value of the overall heat | transfer coefficient than G.W. and d.W. [17]. At the end of operating period, the final storage tank temperature was 53.8°C for G.A., 52.6°C for G.W and 49.2°C for d.W. Fig.15 shows the variation of calculated bulk heat stored during the operation period for the three working fluids. For the effect of load condition on the system performance to be shown, trials were conducted while withdrawing hot water from the storage tank during the operation period. The hot water withdrawn was taken intermittently and continuously. The first load pattern of hot water withdrawn from the storage tank is shown in Fig.16 was suggested by reference [19]. Fig.19 shows the variation on mean tank temperature, load outlet and load inlet temperature during the operation period. The system is observed to operate at a lower temperature than that without load. Also Fig.19 shows that the mean tank temperature and the load outlet temperature are increasing throughout the period between 8.00 a.m. and 14.00 p.m., which indicates that the energy transferred to the water in the storage tank, is higher than that carried out by the load water. A maximum temperature difference. Of 2°C between meat tank temperature and load outlet temperature is also observed. These temperatures start decreasing after 14.00 p.m. This is because the useful energy transferred to the water storage tank is lower than the heat carried away with the load water. The overall bulk efficiency was calculated to be 53% for this continuous load condition. This value is 5.6% higher than the value for no-load condition. It can be concluded that the overall bulk efficiency improved by operating the system under load conditions. The second load pattern of hot water withdrawal is shown in Fig.17 which was suggested by reference [20]. Fig.20 shows the variation of mean tank temperature with load condition. The load was taken for an hour and half starting from 8:00. Till 9.30 am. At 12.00 noon water was withdrawn at intervals of 1/4 an hour starting from 12.00 noon till 13.30 p.m. It is clearly observed from Fig.20, that the mean tank temperature does not increase smoothly between 12.00 noon and 14.00 p.m. This is of course is due to the introduction of fresh cold water to the tank. During the period between 14.00 p.m. and 17.00 p.m., it is noted that the fluctuation in mean tank temperature diminishes because of the large interval of time between loading and unloading. However, the rate of increase of the mean tank temperature is small compared to that obtained with no load condition.

IOP Conf. Series: Materials Science and Engineering **765** (2020) 012032 doi:10.1088/1757-899X/765/1/012032

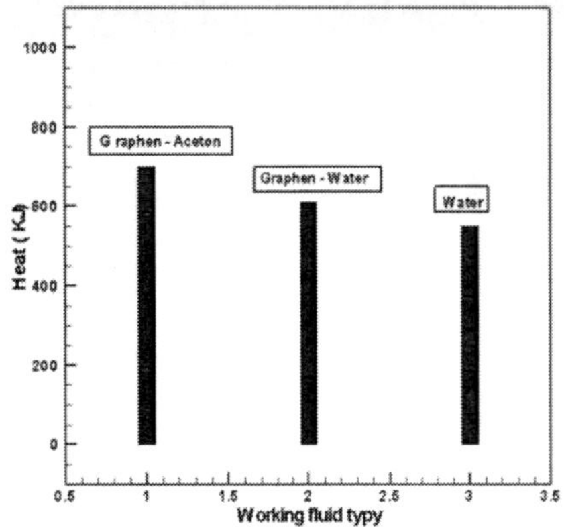

Figure 15: Overall stored heat for different Working fluids

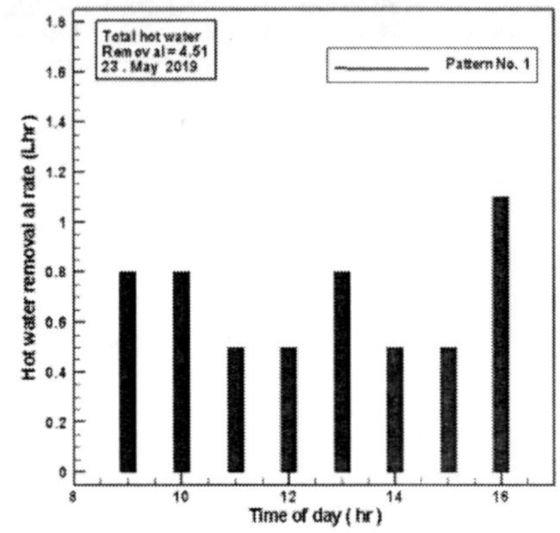

Figure 16: Hot water use profile

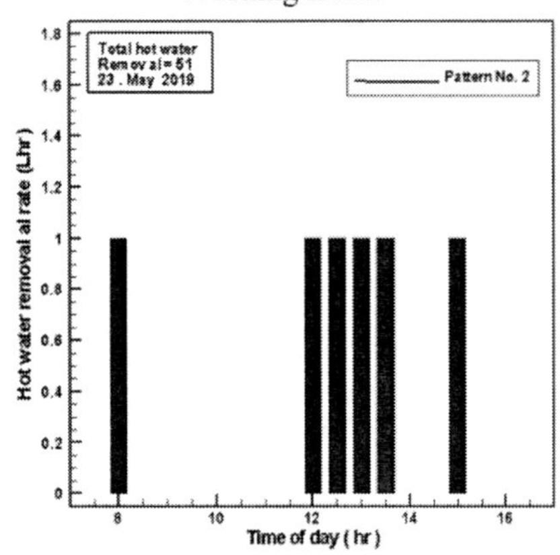

Figure 17: Hot water up profile

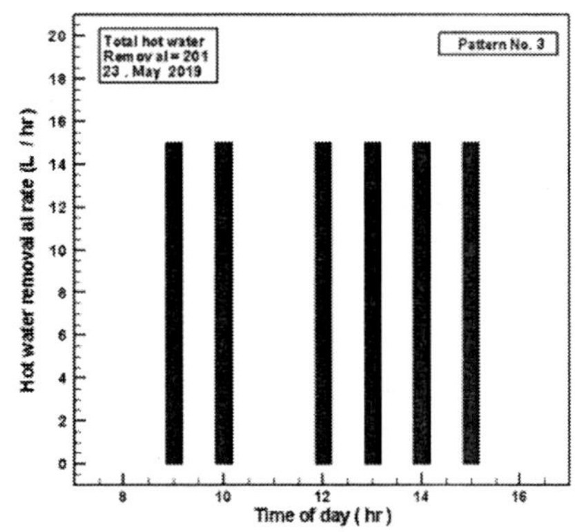

Figure 18: Hot water up profile

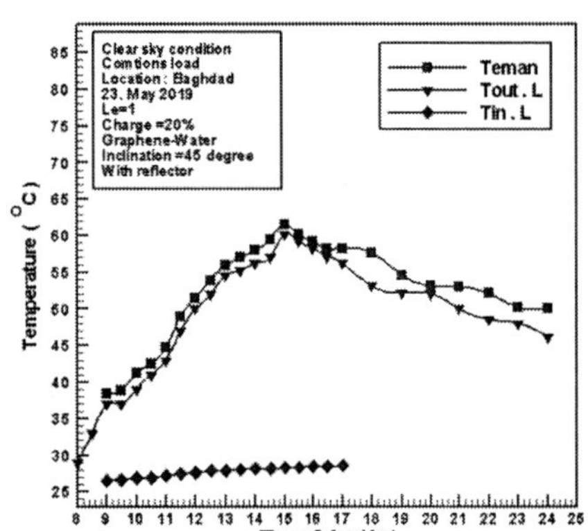

Figure 19: The variation of mean tank temperature inlet and Outlet temperatures during the operating period

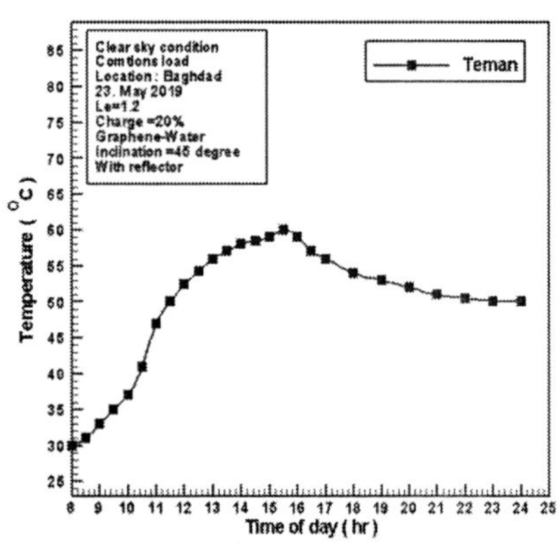

Figure 20: The variation of tank temperature During the operating period

Fig.21 shows the variation of mean tank temperature during the operation period: with a 3rd pattern of intermittent load condition as shown in Fig.18. A relatively large volume of hot water was withdrawn from the storage tank at relatively short periods of time. It is seen in Fig.21, that the increase in the mean tank temperature is not uniform between 9.00 a.m. and 14.00 p.m. The system shows a good response to the large load variation of hot water withdrawal. Also, it is observed that the rate. Of increase of mean tank temperature is lower compared to that obtained with no load condition.

Fig.22 shows a comparison of the overall bulk heat collected in the load patterns and that collected with no load condition. The heat stored is observed to be higher with loading. Loading pattern no.2. Appears to give the best results in term of heat stored. The preceding discussion shows that the variation in system performance is mainly dependent on the flow rate of load water. However, this effect is not very clear. Therefore, It was decided to test the system with differing constant continuous loads. Hot water withdrawal was carried out between 9.00 a.m. and 17.00 p.m. in June and July 2004. at the 2 flow rates of 0.5, 1, 2,3, 4, 5 and 6 Z /hr. All three working fluids were used with this continues loading. Figs.23, 24 and 25 show the system performance for each of the three working fluids with the seven loading conditions. These figures show that a higher loading resulted in the system operating at a lower temperature. Consequently, the hot water removal rate also affects the overall heat collected. It is observed that, G.A. shows a higher mean storage tank temperature than G.W. and d.W., resulting in a higher overall heat stored. This is clear depicted in Fig.26. However, for water, there is a trade of between the benefit of large hot water removal rate and loss of performance due to low overall heat transfer coefficient of the CTPT as we will see in the following discussion. It is seen from Fig.26, that in the range 0.5 to 4 l /hr all working fluids show a significant increase in the overall heat collected. This is because increasing the rate of hot water removal resulted in decreasing the storage tank temperature, which in turn decreases the heat losses from the storage tank. However, for water and for rates higher than 4 l /hr a substantial decrease in overall bulk heat occurs. Whereas, G.W. and d.W maintain approximately constant values. This behaviour is due to the effect of the overall heat transfer coefficient of the CTPT with these fluids. The reduction in the overall bulk heat, for water, is thought to be caused by the lower overall: heat transfer coefficient of the CTPT at lower operating temperature. G.W. and d.W render approximately: constant overall heat transfer coefficient with variation in operating temperature [17]. Therefore, there is an optimum value of hot water removal rate with water at. Which the overall bulk heat is maximum. Fig.27 shows the variation of calculated. Overall daily system efficiency as a function of hot water removal rate for the experimental working fluids. The rate-of hot water removal at which the maximum system efficiency occurs is 4 different for different working fluids. It is observed that the overall bulk system efficiency follows the variation of the bulk heat collected by the system. Fig.27 shows that, for water, the overall bulk efficiency reaches its maximum value of 61.2 at approximately 3 to 4 l/ /hr hot water removal rate. It is also observed that at high hot water removal rate there .is a slight decrease in the overall efficiency resulting from the reduction in the bulk heat as was shown in Fig.26. . However, the bulk efficiency of G.W. and d.W are less sensitive to the variation of hot water removal rate with maximum values of 55% for G.W and 47% for d.W at approximately 5 l /hr hot water removal rate. In this study, the mean tank temperature distribution is similar in trend to the works of Akyurt [8] and Chun [12]. As illustrated in Fig.28. However, the peak value is vary which is caused by the difference in design requirements of the present work as compared to that employed by the above workers, direct comparison between mean tanks temperature alone is not really valid. As a result of the above discussion, comparison can be made only for the overall bulk efficiency of the current system with those deduced from the available literature. Table 1 below shows the overall bulk efficiency obtained with the conditions listed for this work and the works of Akyurt [8] and Chun [12].

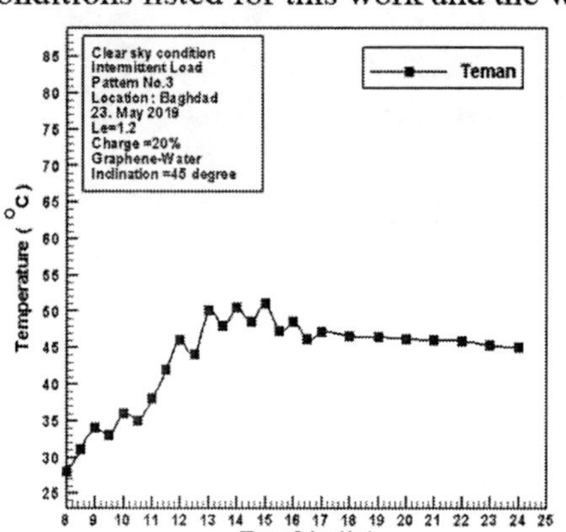

Figure 21: The variation of tank temperature During the operating period

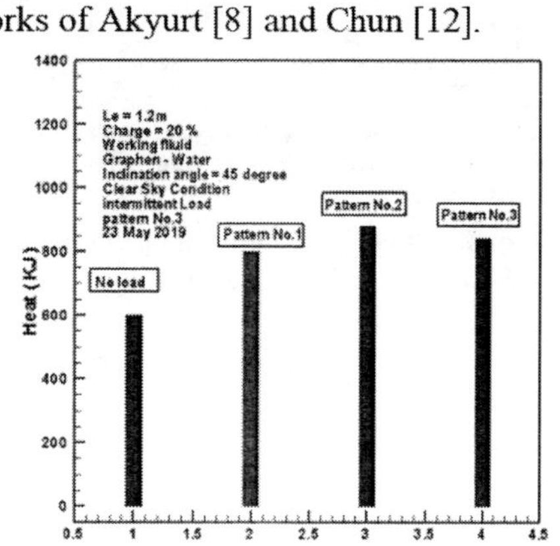

Figure 22: Overall heat stored for Different load conditions

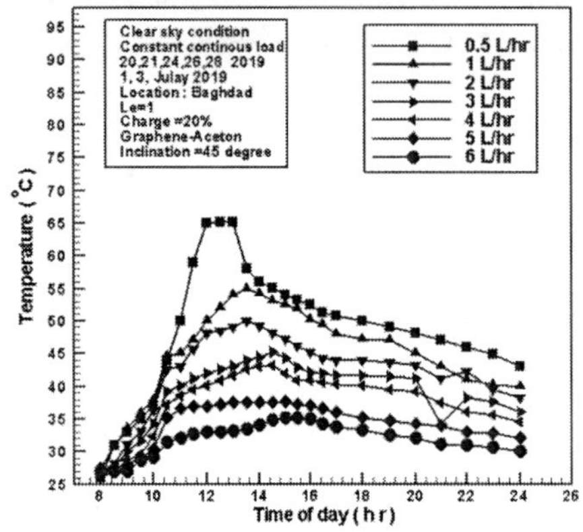

Figure 23: The variation of tank temperature with for various hot water removal rates

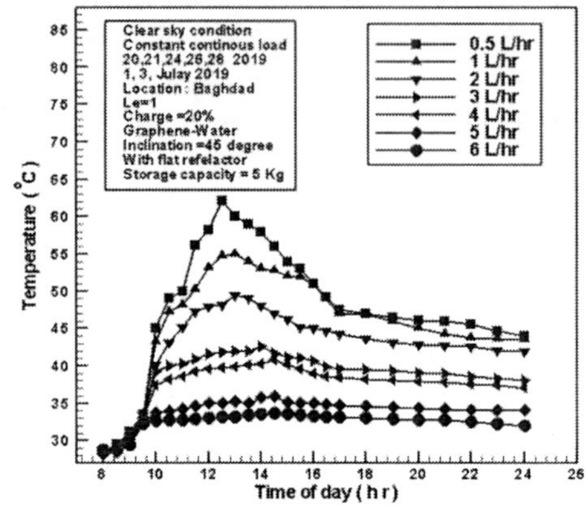

Figure 24: The variation of tank temperature with for various hot water removal rates

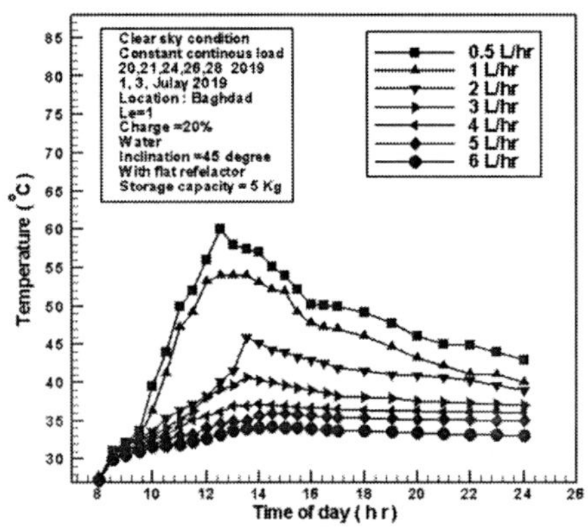

Figure 25: The variation of tank temperature with for various hot water removal rates

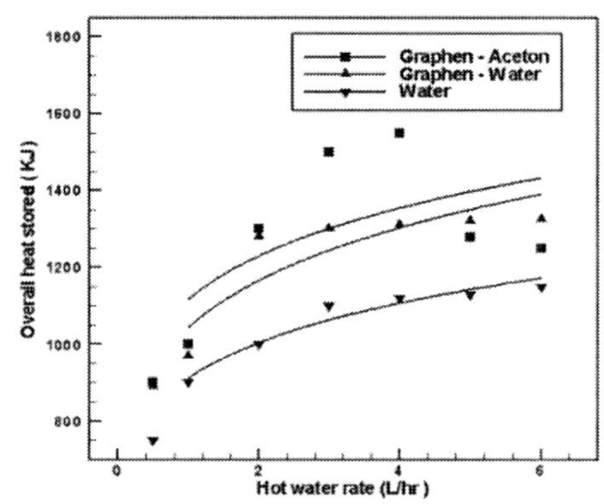

Figure 26: The effect of hot water removal rate on theoverall heat various working fluid

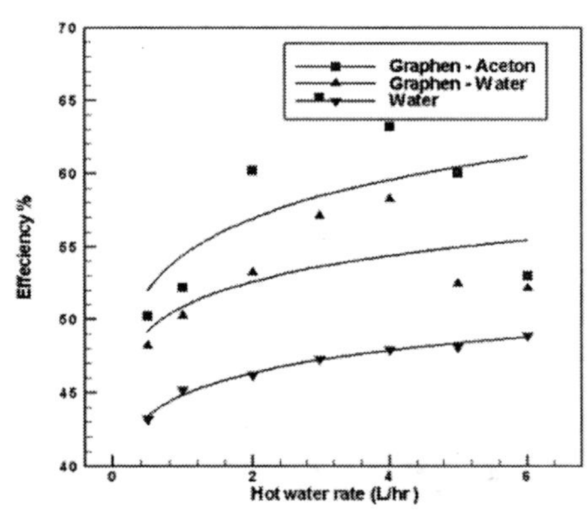

Figure 27: The effect of hot water removal rate on the overall bulk efficiency for various working fluid

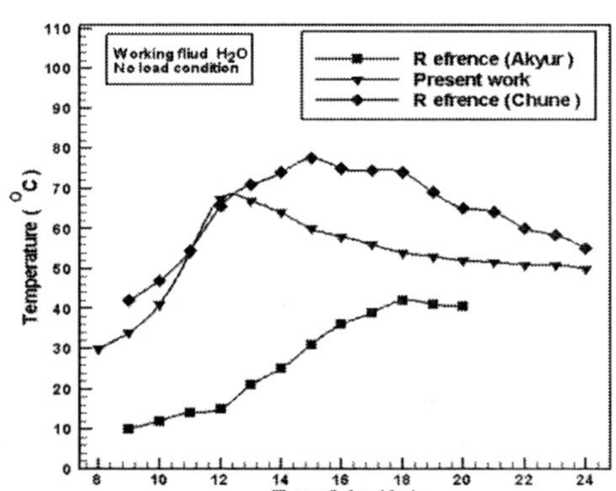

Figure 28: variation of tank temperature for various investigator

Table (1) overall bulk efficiency of some solar heat pipe system

	Load condition	Design specifications				overall bulk efficiency (%)
		Collector type	Le (m)	Le (m)	Working fluid	
Present work	No load	Evacuated Tube CTPT	1.2	0.2	G.W	48%
	Continuous	Evacuated Tube CTPT	1.2	0.2	G.A	61%
Reference [8]	No load	Flat Plat	2.1	1.2	Ethanol	52%
Reference [12]	No load	Panel Type Flat Plat	1.7	0.2	Ethanol	45%

5. Conclusions

Several essential conclusions can be drawn from this study on the single evacuated tube solar water heater system.

1. The maximum value of the mean tank temperature and the overall bulk heat stored depend mainly on the storage capacity.
2. The larger evaporator length system, the higher maximum mean storage temperature and a higher bulk heat stored is obtained.
3. For no load condition, the system performance was found to be is similar for both G.A. and G.W.working fluid, whereas d.W shows relatively a slight reduction in system performance.
4. Through the operation period, the system with constant continuous load has shown unchanging temperature distribution in comparison with intermittent load condition.
5. The system has a good response and it is sensitive to the large variation in the hot water removal rates.
6. The overall bulk heat stored of the system improves with operating under load condition.
7. In the range of study, the overall bulk heat collected of such system. Depends on the rate of hot water removal.
8. The performance of the system charged with water is maximized when the daily hot G.A. removal rate is approximately equal to 3-4 W/hr.

6. Nomenclature

Aa	Absorber area [m^2]	η	Efficiency %
Cw	Specific heat of water [kJ/kg °C]	mf	End temperature **(°C)**
I	Solar radiation intensity[W/m^2]	ms	Start temperature **(°C)**
T	Temperature [°C]	L	Load
Tmean	Mean Temperature [°C]	out,L	Outlet temperature of load water **(°C)**
Mt	Mass of water in the storage [kg]	in,L	Inlet temperature of load water
m	Mass flow rate [kg/s]	ove.	Overall (bulk)
Q	Heat collected [kJ]		
GW	Graphen - water		
GA	Graphen – aceton		
d-W	Distilled Water		
CTPT	Closed tow phase thermosyphon		

References

[1] Dunn. P.D. and Reay, D.A. Heat Pipes.3rd Edition :Pergamon Press; 1982.

[2] Bairamov, R. and Toiliev, K. Heat Pipes In Solar Collector. Proc. 4th Int. Heat Pipe Conference, London,UK, .

[3] Pergamon Press, 1982, pp 47-34.

[4] Bienert, W. B. and Wolf, D. A. Heat Pipe In Flat Solar Collectors. ASME Publ. Paper, 76-WA/SOL-12, 1976.

[5] Dietrich, B. and Gehrke, K. Analysis, Design and Thermal Performance of Flat Plate Solar Collector. Am. Sect. Int. Sol. Energy Soc., 1977, Section 1, Sess.A.2, pp 4.

[6] Ortabasi, U. and Buhl, W. M. internal Cusp Reflector For an Evacuated Tabular Solar Thermal Collector. ,Solar Energy 1980; 25: 67-78.

[7] Ernst, D. M. and Eastman, G. Y. Cost Effective Solar Collector Using Heat Pipes. P.5 Paper, Thermo core Inc., Leola, Pennsylvania, 1978.

[8] Akyurt, M. Development of Heat Pipes For Solar Water Heaters, Solar Energy 1984; 32(5): 625-631.

[9] Hull, J. R. Comparison of Heat Transfer in Solar Collectors with Heat Pipe Versus Flow-Through Absorber. ASME, Journal Of Solar Energy Engineering, 1987;109: 253-258.

[10] Georgier, V. Mathematical Model of a Vacuum Solar Collector With eat Pipe and Flat Absorber. Breunstofi- Warme Kraft, 1993;45(12): 527-534.

[11] Hammead, M. Experimental Study of The Performance of A Solar Collector Cooled By Heat Pipes. Renewable Energy, 1995;6(1): 11-15. 2.

[12] Chun, W, Kang, Y.H. Lee, Y. S. An Experimental Study of The Utilization Of Heat Pipes For Solar Water Heaters. Applied Thermal Engineering, 1999;19: 807-817.

[13] Ismail, K. R. and Abogderah, M. M. Performance of a Heat Pipe Solar Collector. Journal of Solar Energy - Engineering- Transaction of ASME, 1998;120(1): 51-59.

[14] Lima, P. and Palanyana, S. Flat Plate Solar Collector with Heat Pipes and Selective Surfaces. Latin American Applied Research. 2000;30(3):249-251.

[15] Gorliss, J. M. Evaluation of Heat Pipe Application For Passive Solar Systems. DOE Report, Dep. of Energy, Washington, D.C., 1979.

[16] Lu L., Liu., Xiao H. Thermal performance of an open thermosyphon using nanofluids for high-temperature evacuated tublar solar collectors part1:indoor experiment . solar energy 2011, 85:379-387.

[17] Senthil Kumar R., Manimarn R., Ramadoss K., Shankar N. Experimental analysis of nano fluid-charged solar water heater by solar tracking system. Archives of applied science research 2012, 4 : 2582-2590.

[18] N.H.Mujawar and S.M.Shaikh Thermal performance investigation of evacuated tube heat pipe solar collector with nanofliuid IJESRT, PP.824,Vo.5(12), 2016.

[19] Sumedh S. Jagatap, et al , Performance analysis of evacuated tube heat pipe solar wather heating system using copper oxide nanofluids,IJARIIE ,Vol.3, Issue-2,2017.

[20] Lunde, P. J. Solar Thermal Engineering. Joon Wiléy and Sons, Inc., 1980.

[21] Mahmoud, M. Mahdy. Experimental Investigation. of Low Heat Flux Heat Pipes with and without Adiabatic Section for Solar Application. Ph. D. Thesis, Baghdad-Iraq, 2005.

[22] Faber, E. A. Morrison, C. A. Clear-Day Design Values of Solar Energy, ASHRAE GRP 170, Application of Solar Energy For Heating and Cooling of Buildings, Edited By Jordon, R. C. and Liu, B.Y. H., 1977.

[23] Tzafestass, G. A. Spyridonos, A. V and Koumoutsos, N. G. Finite Difference Modeling, Identification and Simulation of a Solar Water Heater. Solar Energy, 1974; 16:25-31.

[24] Yong, M. F. Bergquam, J. B. The Performance of a; Thermosyphon Solar Demostic Hot Water System with Hot Water Removal. Solar Energy, 1984;32(5): 655-658.

IOP Conf. Series: Materials Science and Engineering **765** (2020) 012033 doi:10.1088/1757-899X/765/1/012033

Analysis of Vector Controlled Variable Speed DFIG for Wind Turbines

Talib Z. Farge

Ali H. Numan

Ghasaq M. Hussein

Electromechanical Engineering Department, University of Technology, Baghdad, Iraq

Correspondence Author Ali H. Numan: 50059@uotechnology.edu.iq

Abstract. Double Fed Induction Generators (DFIG) has been widely used for the past two decades in large wind farms. This paper presents control and analysis of the doubly fed induction generator (DFIG) connected to variable speed horizontal axis wind turbine (HAWT). The detailed wind turbine (WT) model, DFIG dynamic model in d-q - synchronous reference frame, power electronic converters, and controllers are presented. The reactive and active power of the DFIG can be precisely decoupled by controlling the rotor side converter (RSC). The field-oriented control (FOC) strategy and the proportional-integral (PI) controllers were used to control the RSC of the DFIG. Furthermore, a space vector pulse width modulation (SVPWM) was utilized to generate pulsing signals required to drive three phases DC-AC converter (inverter). The proposed control method of DFIG based on HAWT was verified by simulation in MATLAB-SIMULINK software. The obtained simulation results showed the capability and effectiveness of the proposed control method.

Keywords: WT, DFIG, SVPWM, Vector Control.

1. Introduction

For the previous period, the demands of energy have been increasing gradually, electrical power and environmental issues especially and this has become a challenging problem for the world. Besides, pollution is gradually growing parallel with the energy request while sources of conventional energy such as fossil fuels are rapidly depleting. Over the last few decades, scholars have been directing studies to improve the energy efficiency. This has led to the finding of many alternatives as renewable energies to generate electricity [1].

It is well known that clean energy can be generated by various environmental sources such as solar radiation, wind movement, water flow and biological wastes. However, such energy is almost unexpansive and handy with no clear environmental pollutants. The power generated by wind being soundly alternative of those produced by classic fuels and also has no impacts on global warming. Furthermore, such energy is produced by transferring kinetic power to electric form by using well designed turbines [2]. Obviously, wind power was applied in mill circulation and ship navigation particularly in case of absence of classic energy [1]. It is a replacement of fossil fuels because of minor influence upon the environment and free to use.

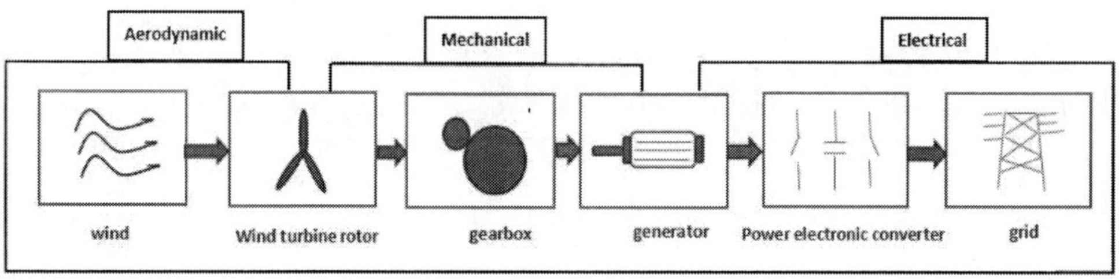

Figure 1: The topology of the entire wind power converting system.

Figure 1 describes topology of the entire wind power converting system. WECS generates electric energy via applying the wind intensity to operate electricity generators. The transforming of incoming air power in to the electric power occurs at two stages: the extraction device where the wind turbine rotor hunts the wind power speed by aerodynamically blades resulting into rotating mechanical energy to operate the generator rotor [3]. The electricity generator transfers the rotating mechanical energy into electric. A gearbox can be used for controlling the speed of rotation of the turbine rotor matching a generator. Electric power is transmitted to the grid by a transformer. The linking of the turbine to the grid is probable at different voltage levels with a known level rate of 600 - 700 V. Where the first stage consists of extracting tool, power electronics converters can also be used for better power extraction and variable speed working of the wind turbine [3]. The wind generator converts mechanical energy rotating inputs into electrical energy outputs.

The doubly feed induction generator (DFIG) presented in Figure 2, shows that DFIG, or WRIG are the common terms to be used in defining the electric machine with a stator in a cylindrical shape with a number of slots(typically36–48) in the inner face which contains 3-phase windings, producing an air gap magnetic field in the with either 2 or 3 pole pairs. The magnetic field generated via the rotor and stator windings necessity is turn at the identical speed nevertheless with shifting to some degrees to create the torque by a machine. By means of the revolving portion of the machine(rotor) to provide it, it should have a slip ring (three-slip rings). Slip rings are needed to maintenance, cost, as well as efficiency [4].

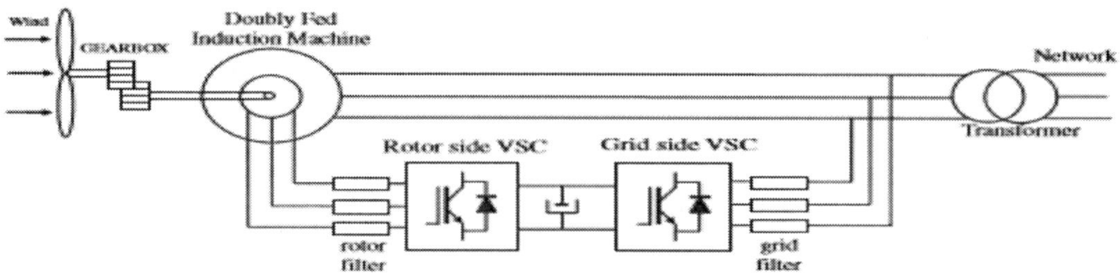

Figure 2: The DFIG schematic diagram.

DFIG is applied for linking towards the network by means of a variable-pace wind generator. DFIG has a sum of rewards associated with fixed speed generators, consisting that the machine can work in sub-synchronous, synchronous, and super-synchronous mode, control of active power in the same time with the reactive power control, improved energy detention and mechanical stress decreasing with low cost by the advance of the power converters. The bi-directional buck-boost DC-DC converter is mediated between the fixed DC Link voltages. Rotor side converter (RSC) is linked to rotor to control active power in addition to reactive power of DFIG 's stator. While, grid side converter (GSC) is linked with network for controlling the DC Link voltage [5].

2. Wind Turbine Aerodynamic Model

Wind turbines generate electric energy by extracting the wind energy in order to move an electrical generator. The swept area of the blades extracts the kinetic energy from the WT according to equation (1), and produces the moving air power (Pair) according to equation (3).

$$E = ½\, m\, v^2 \tag{1}$$

Where $m = \rho V$ and ρ is density of air particles (nearly$1.225\ kg/m3$), V is the volume with $= Ax$, and A is the blades swept area equals to:

$$A = \pi R^2 \tag{2}$$

The moving air power is equal to

$$Pair = dE/dt$$

$$= ½\, \rho\, \pi\, R^2\, v^2\, dx/dt$$

$$= ½\, \rho\, \pi\, R^2 v^3 \tag{3}$$

P_{air} indicates the existing energy in the wind but the energy actually moved to the rotor of the wind turbine ($P_{wind\ turbine}$) is decreased by Cp according to equations 4 and 5. Where Cp is the wind power coefficient and according to Betz limit

$$Cp = Pwind\ turbine\ /\ Pair \tag{4}$$

$$Pwind\ turbine\ = \tfrac{1}{2}\,\rho\,\pi\,R^2 v^3 Cp \tag{5}$$

The torque and power coefficient can be represented via analytical terms related to tip speed ratio (λ) and the pitch or blade angle (β). The frequently expression applied and improved to various WTs, are given by equation below:

$$Cp(\lambda, \beta) = c1\ (c2\ /\ \lambda i - c3\ \beta - c4)e^{-(c_5/\lambda_i)} + c6\ \lambda \tag{6}$$

Where $c_1$to c_6 are equal to 0.517, 116, 0.4, 5, 21 and 0.0068 respectively for HAWT with a variable speed of wind [6]. And (λ_i) is given via the following equation:

$$\lambda i = \frac{1}{\frac{1}{\lambda+0.08\beta} - \frac{0.035}{\beta^3+1}} \tag{7}$$

The ratio of tip speed is calculated by the following equation:

$$\lambda = \frac{wR}{V} \tag{8}$$

Cp and λ are dimensionless and applied for defining the action of the rotor of wind turbine at different sizes Cp which depend on geothermal parameters and manufacturer, typically Cp is introduced being related with two limits ratio of tip speed (λ) and bitch angle (β) such as shown in Figure 3 [7] [8].

Finally, the rotor of WT produces a torque according to the following equation:

$$\mathrm{Tw} = \mathrm{Pwind\ turbine}\ /w \tag{9}$$

The block scheme for wind turbine calculations is shown in Figure 4:

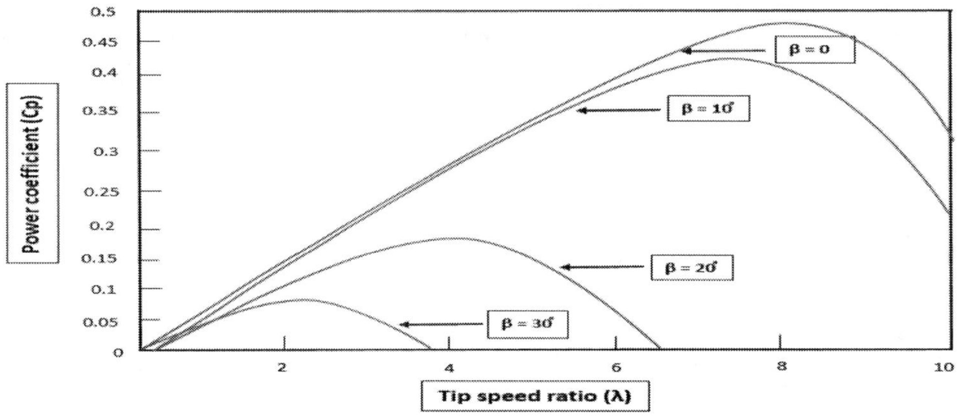

Figure 3: The power coefficient Cp (λ, β).

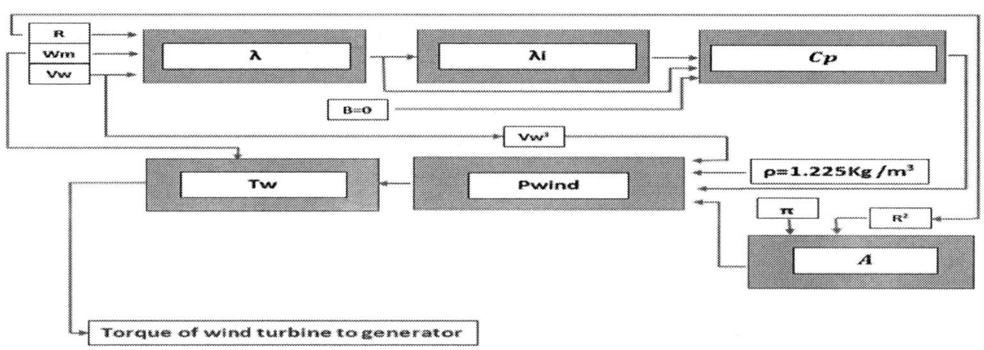

Figure 4: Block scheme of wind turbine calculations.

3. Dynamic Model Of DFIG

DFIG is a predictable induction generator with wound-rotor. Generator stator is linked to network directly through a transformer, and the rotor is linked to AC-DC-AC voltage source convertor. AC-DC-AC converters includes the rotor side converter (RSC) and the grid side converter (GSC) which linked via DC link. Representation of DFIG is shown in Figure 5[9].

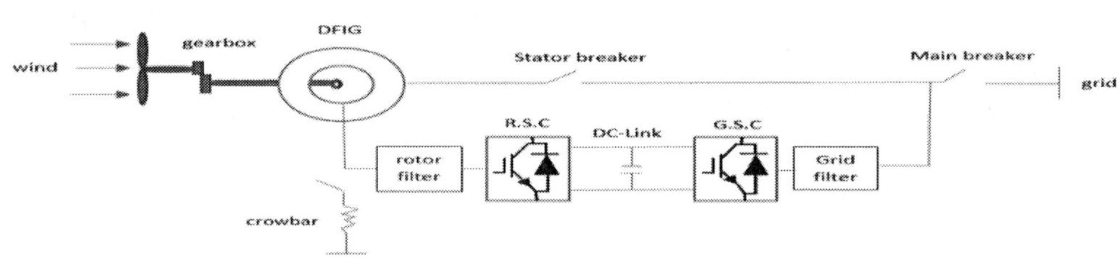

Figure 5: The representation of the DFIG.

IOP Conf. Series: Materials Science and Engineering **765** (2020) 012033 doi:10.1088/1757-899X/765/1/012033

In an AC machine, rotor windings move regarding the stator windings because of this, the dynamic action of an AC machine is much hard [9].

Then the Park's and Clarke's transformation will be used to transform three phase signals (abc) to a (αβ0) stationary reference frame according to the following matrices [10]. And shown in Figure 6:

$$[fdqo] = Tdqo\,(\theta)[fabc] \qquad (10)$$

f denotes to the voltages, magnetic flux linkages, electric currents or electric charges and the transformation matrix is shown below

$$Tdq0(\theta) = 2/3 \begin{bmatrix} cos\theta & \cos(\theta - 120°) & \cos(\theta + 120°) \\ sin\theta & \sin(\theta - 120°) & \sin(\theta + 120°) \\ 1/2 & 1/2 & 1/2 \end{bmatrix}$$

$$[f\alpha\beta 0] = T\alpha\beta 0[fabc] \qquad (11)$$

The transformation matrix is shown below:

$$T\alpha\beta 0 = 2/3 \begin{bmatrix} 1 & -1/2 & -1/2 \\ 0 & \sqrt{3}/2 & -\sqrt{3}/2 \\ 1/2 & 1/2 & 1/2 \end{bmatrix}$$

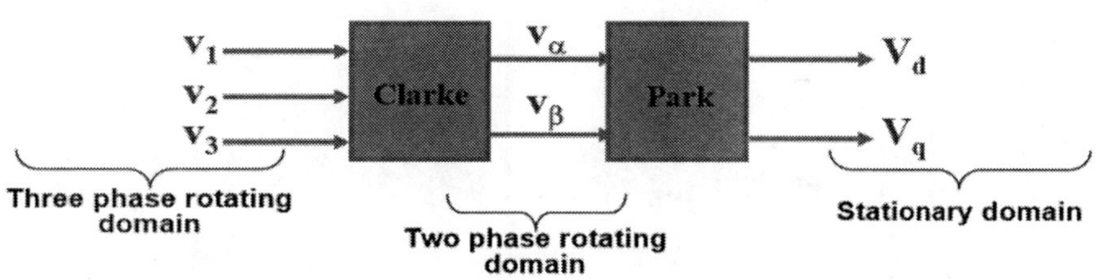

Figure 6: presented Park's and Clarke's transformation reference frames

In the dynamic modeling of DFIG, the instantaneous stator and rotor voltages are presented in equations below[11]:

$$V_{as}(t) = R_s\, i_{as}(t) + d\psi_{as}(t)/dt \qquad (12)$$
$$V_{bs}(t) = R_s\, i_{bs}(t) + d\psi_{bs}(t)/dt \qquad (13)$$
$$V_{cs}(t) = R_s\, i_{cs}(t) + d\psi_{cs}(t)/dt \qquad (14)$$

$$V_{ar}(t)=R_r\, i_{ar}(t) + d\psi_{ar}(t)/dt \tag{15}$$
$$V_{br}(t)=R_r\, i_{br}(t) + d\psi_{br}(t)/dt \tag{16}$$
$$V_{cr}(t)=R_r\, i_{cr}(t) + d\psi_{cr}(t)/dt \tag{17}$$

The currents of the stator and rotor can be represented in αβ components such as follows:

$$d/dt \begin{bmatrix} i\alpha s \\ i\beta s \\ i\alpha r \\ i\beta r \end{bmatrix} = (1/\sigma\ Ls\ Lr).\ M. \begin{bmatrix} i\alpha s \\ i\beta s \\ i\alpha r \\ i\beta r \end{bmatrix} + (1/\sigma\ Ls\ Lr).\ N. \begin{bmatrix} v\alpha s \\ v\beta s \\ v\alpha r \\ v\beta r \end{bmatrix} \tag{18}$$

Where

$$M = \begin{bmatrix} -Rs\ Lr & \omega m\ Lm^2 & Rr\ Lm & \omega m\ Lm\ Lr \\ -\omega m\ Lm^2 & -RS\ Lr & -\omega m\ Lm\ Lr & Rr\ Lm \\ Rs\ Lm & -\omega m\ Ls\ Lm & -Rr\ Ls & -\omega m\ Lr\ Ls \\ \omega m\ Ls\ Lm & Rs\ Lm & \omega m\ Lr\ Ls & -Rr\ Ls \end{bmatrix}$$

$$N = \begin{bmatrix} Lr & 0 & -Lm & 0 \\ 0 & Lr & 0 & -Lm \\ -Lm & 0 & Ls & 0 \\ 0 & -Lm & 0 & Ls \end{bmatrix}$$

After finding $i_{\alpha s}$, $i_{\beta s}$, $i_{\alpha r}$, $i_{\beta r}$, the fluxes expanded in αβ components are obtained according to the following expressions:

$$d/dt \begin{bmatrix} \psi\alpha s \\ \psi\beta s \\ \psi\alpha r \\ \psi\beta r \end{bmatrix} = \begin{bmatrix} \frac{-Rs}{\sigma Ls} & 0 & \frac{RsLm}{\sigma LsLr} & 0 \\ 0 & \frac{-Rr}{\sigma Lr} & 0 & \frac{RsLm}{\sigma LsLr} \\ \frac{RrLm}{\sigma LsLr} & 0 & \frac{-Rr}{\sigma Lr} & -\omega m \\ 0 & \frac{RrLm}{\sigma LsLr} & \omega m & \frac{-Rr}{\sigma Lr} \end{bmatrix} . \begin{bmatrix} \psi\alpha s \\ \psi\beta s \\ \psi\alpha r \\ \psi\beta r \end{bmatrix} + \begin{bmatrix} v\alpha s \\ v\beta s \\ v\alpha r \\ v\beta r \end{bmatrix} \tag{19}$$

Where σ is the leakage coefficient, which can be obtained from

$$\sigma = 1 - Lm^2/Ls.Lr \tag{20}$$

The electromagnetic torque calculation according to the following equation:

$$Tem = \frac{3}{2}\ P\ Lm\ Im\ \{\vec{\imath s}.\vec{\imath r}^*\} = \frac{3}{2} P\ Lm\ (i\alpha r.i\beta s - \alpha s.\ i\beta r) \tag{21}$$

Finally, to obtain the dynamic and responsive powers of the stator and rotor and torque via αβ model, the following expressions will be used [12]:

$$Ps = \frac{3}{2} Re\ \{\vec{vs}\,.\,\vec{is}^*\} = \frac{3}{2}(v\alpha s.\, i\alpha s + v\beta s.\, i\beta s) \quad (22)$$

$$Pr = \frac{3}{2} Re\ \{\vec{vr}\,.\,\vec{ir}^*\} = \frac{3}{2}(v\alpha r.\, i\alpha r + v\beta r.\, i\beta r) \quad (23)$$

$$Qs = \frac{3}{2} Im\ \{\vec{vs}\,.\,\vec{is}^*\} = \frac{3}{2}(v\beta s.\, i\alpha s - v\alpha s.\, i\beta s) \quad (24)$$

$$Qr = \frac{3}{2} Im\ \{\vec{vr}\,.\,\vec{ir}^*\} = \frac{3}{2}(v\beta r.\, i\alpha r - v\alpha r.\, i\beta r) \quad (25)$$

The detailed block scheme for DFIG calculations is shown in the Figure 7:

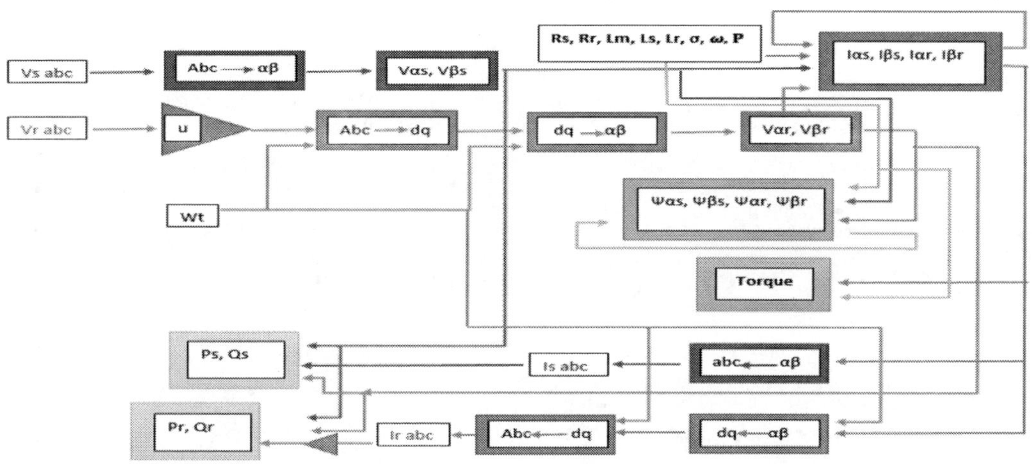

Figure 7: The detailed block scheme of DFIG calculation

4. Field Oriented Control (FOC)

The main impartial of FOC or vector control (VC) is for finding and keeping an angular association between the space vector current of the stator and one internal field vector [13].

The first and the standard selection for controlling DFIM is the flux orientation. It summarizes approximating of the flux of the machine and then aligning the d axis reference frame to it [14].

IOP Conf. Series: Materials Science and Engineering **765** (2020) 012033 doi:10.1088/1757-899X/765/1/012033

The modeling of FOC contains the external loop of the rotor speed of the generator which is the slower loop for comparison between the rotor generator speed and the reference speed which is needed as a constant value. The internal loop is the dq rotor currents which is the fastest loop where the two loops enter to proportional- integral controller (PI) for controlling them via the PI parameters (K_P& K_I). And the outer loop, the comparator block measures the error value between the desired value ω_r and the variable value of the rotor speed from generator contentiously. In case of the turbine speed is higher or lower than the rated speed, the comparator measures the error and sends it to PI controller to adjust it via reset of the value of K_P and k_I to keep the rotor speed at speed level which produces maximum Cp in order to increase the efficiency of WT [15]. Figure 8 which represents the two control loops and their controllers.

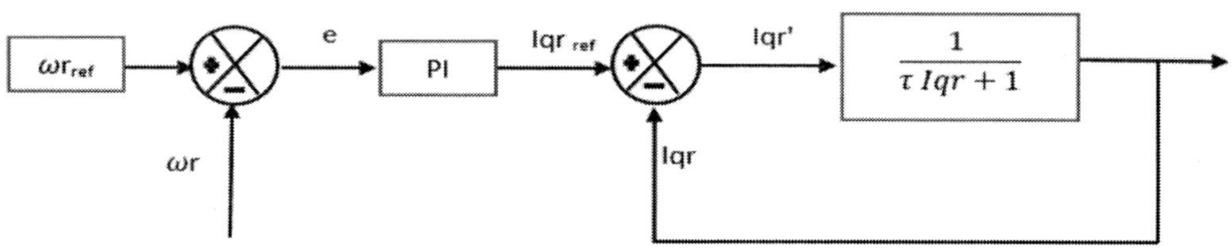

Figure 8: The representation PI controller loops

To obtain the values of K_P and K_I, the following equations were used.

$$K_P = 2Iqr\ /\ \omega ref - \omega r \tag{26}$$

$$K_I = (1/\ \omega ref - \omega r)\ *(2\ Iqr/\ (\omega ref - \omega r)\text{-}K_P)\ *(d\Delta\omega/\ dt) \tag{27}$$

Figure 9 represents the FOC modeling and PI controllers

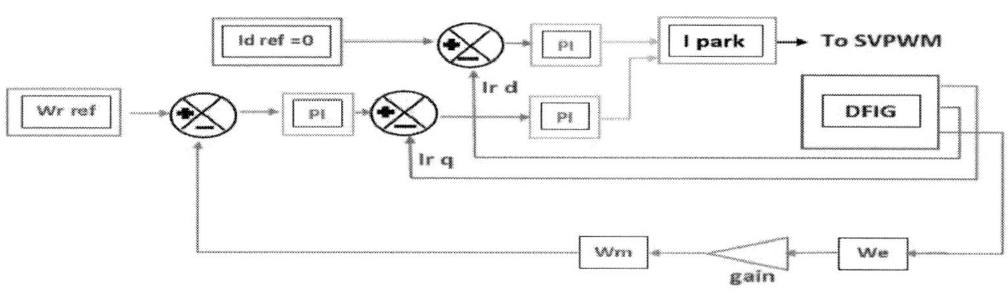

Figure 9: The representation of FOC and PI controller

IOP Conf. Series: Materials Science and Engineering **765** (2020) 012033 doi:10.1088/1757-899X/765/1/012033

5. SVPWM Model

The variable speed of wind turbine and constant frequency technology are widespread in wind power generation system where the space vector control technology is presented in DFIG system for decoupling controller of active power in addition to reactive power by using IGBT as the switching element together with SVPWM [16].

PWM of 3-phase (VSI) voltage source inverter is presented in Figure 10. For formation the output using six switches of power from S1 to S6, controlling on it via switching variables a, a', b, b', c and c'.

In case of the upper transistor is switched on that means that a, b or c is 1, then the matching lower transistor is switched off and this means that a', b' or c' is 0. So, the cases of the upper transistors (on and off) S1, S3 and S5 are applied to regulate the voltage output [17] [18].

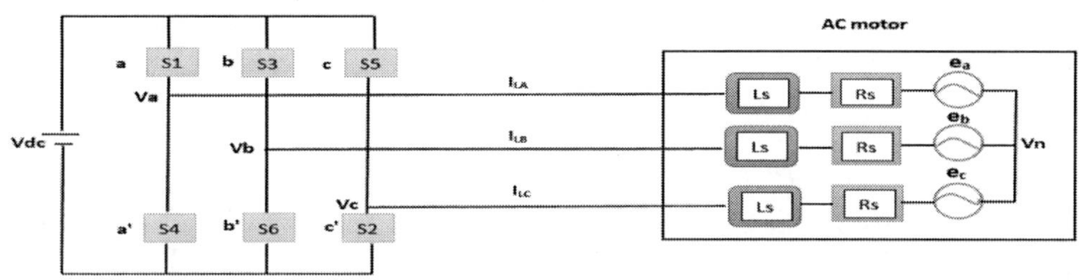

Figure 10: 3-phase PWM-VSI

6. Simulation Model

Figure 11 shows the aerodynamic modeling of wind turbine. The wind profile generated by signal builder and enters to the wind turbine as a variable wind speed simulation to rotate wind blades. The pitch angle of the blades enters the mask as a constant block. Inside the wind turbine mask, the representation of all equations of aerodynamic of wind which used to generate the torque of wind turbine which required to operate the DFIG, then the sub-system block of DFIG which contains the simulation of the equations of doubly fed induction generator to generate required electrical power. The outputs of this sub-system are the currents, fluxes, active and reactive powers of both stator and rotor of the generator and the electrical angular speed and the electromagnetic torque.

To control on the speed of DFIG, the PI controller block used as three loops where the outer loop was the speed controlling and the other two loops was to rotor current control. Also, the SVPWM block was presented to generate the pulses to operate the inverters of the RSC.

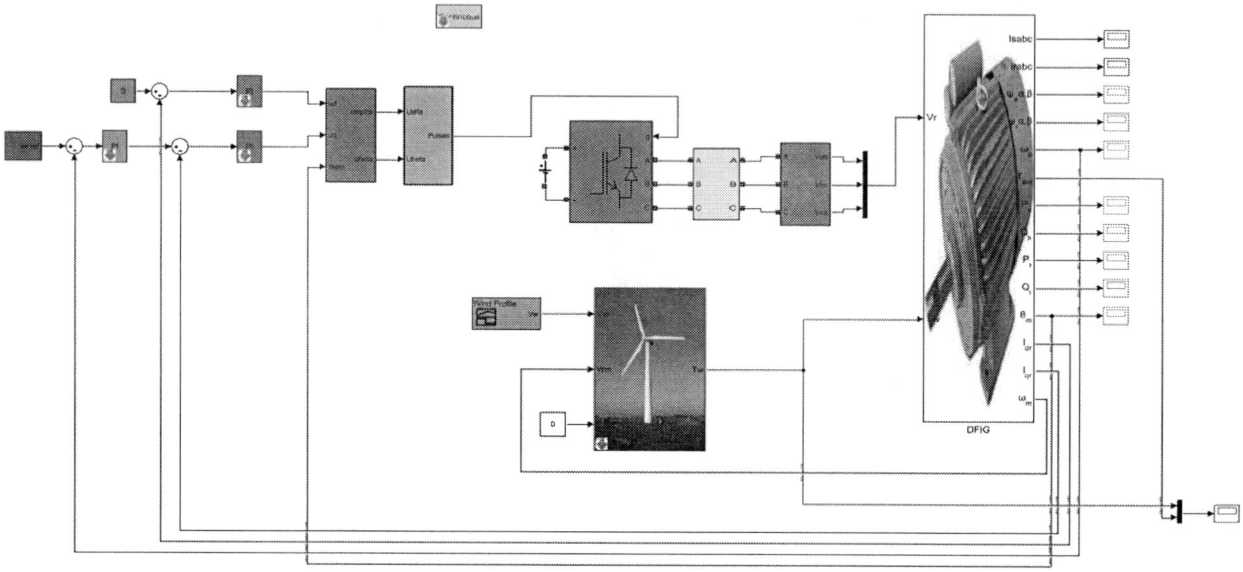

Figure 11: The DFIG based WT simulation model

This simulation model was built by using DFIG with the parameters which presents in following table at variable speed of wind for 4sec operation time.

Table 1: The Input Parameters of DFIG

Parameter	Value
Stator voltage	208 V
Rotor voltage	104 V
Supply frequency	50 Hz
Pole pair	2
No. of stator winding	516 turns
No. of stator winding	264 turns
Stator resistance	12.6 ohm
rotor resistance	16.7 ohm
Stator inductance	0.376 H
rotor inductance	0.38 H
Magnetizing inductance	0.352 H
Moment of inertia	0.0016 $kg.m^2$
Bitch angle	0^0

IOP Conf. Series: Materials Science and Engineering **765** (2020) 012033 doi:10.1088/1757-899X/765/1/012033

7. Results and Discussions

The velocities of the wind as step signal which vary from 5m/sec for time from 0 to 1 sec., 11.25m/sec for time from 1 to 2 sec., 17.5m/sec for time from 2 to 3 sec. and 8m/sec for time from 3 sec. to 4 sec. As shown in Figure 12, they were used as an input data to the modeling.

Inside the DFIG model, Figure 13 represents the 3-phases instantaneous voltages of the stator with respect to time as a sine wave signal with phase shifting by120° and varying according to the velocity of the wind which is transformed to αβ components.

Figures 13 &14 shows the 3-phases instantaneous voltages of the stator and rotor with respect to time as a sine wave signal with phase shifting by120° and varying also according to the velocity of the wind which is transformed to αβ components.

Figures 15 & 16 indicate that the voltages of the stator in αβ components with respect to time after being transformed it via Clarke mask to facilitate the simulation it by computing program.

Figures 17 & 18 represent the voltages of the rotor in αβ components with respect to time after being transformed to dq components via Park mask and then transformed to αβ with angular position wt of the stationary and rotating reference frame.

Figures 19 & 20 represent the currents of the stator in αβ components, which varied according to wind velocity, increasing with wind velocity and decreasing with it.

Figures 21 & 22 represent the currents of the rotor in αβ components, which also increased and decreased with wind velocity.

Figures 23 & 24 show that the three phase currents of the stator and rotor respectively.

Figures 25 & 26 represent the fluxes of the stator and rotor in αβ components respectively, which increased by increased in wind velocity.

Figure 27 indicates that the electrical angular frequency of generator which increases with wind velocity by enhancing the rotation speed of the shaft.

Figure 28 represents the electromagnetic torque of the generator which was decreased with increasing of the wind speed.

Figures 29 & 30 represent an active and reactive power of the stator, which increasing after the speed of the wind reaches its rated value.

Figures 31 & 32 indicate that an active and reactive power of the rotor, which were increased after the speed of the wind reaches the rated value.

Figure 33 represents the torque of wind and torque of generator where torque of wind had a constant value while the generator torque varied with wind velocity.

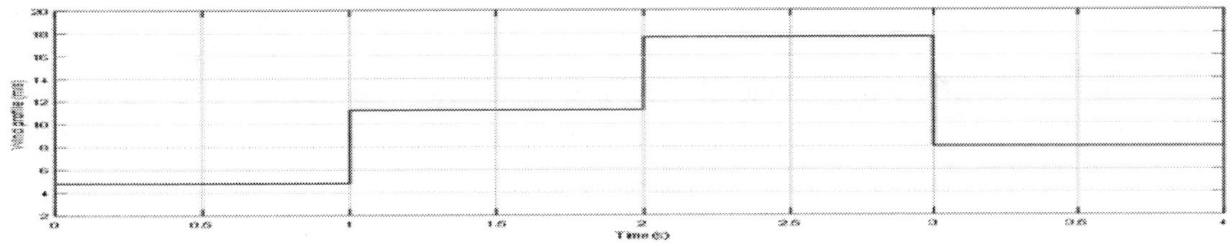

Figure12: The wind velocity profile

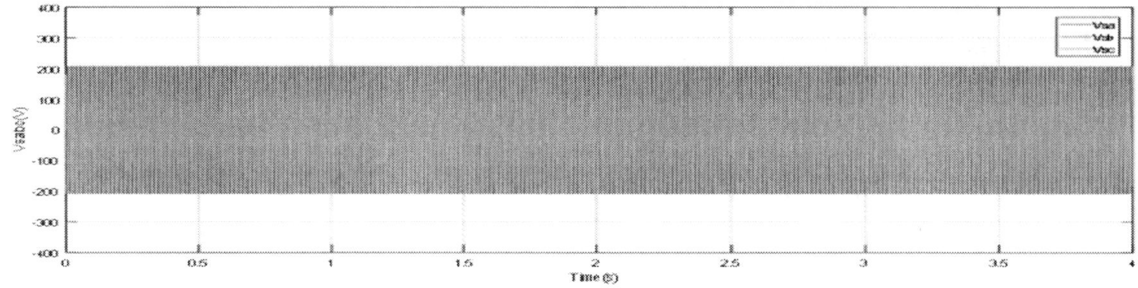

Figure 13: The 3-phases instantaneous voltages of the stator

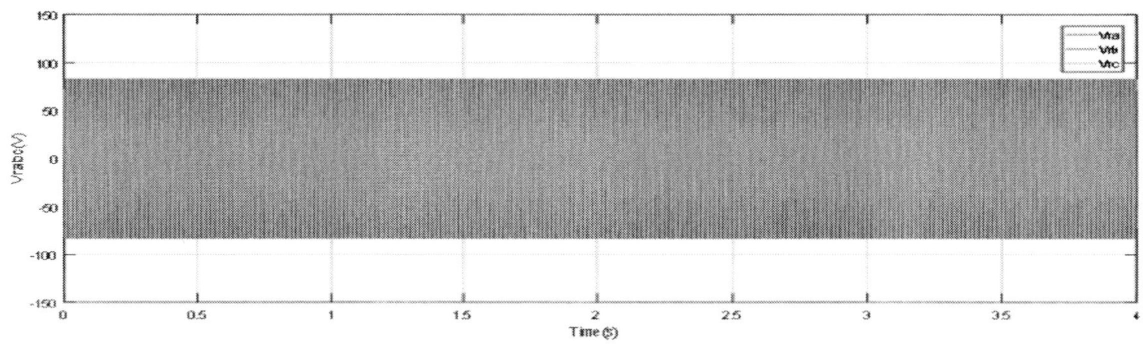

Figure 14: The 3-phases instantaneous voltages of the rotor

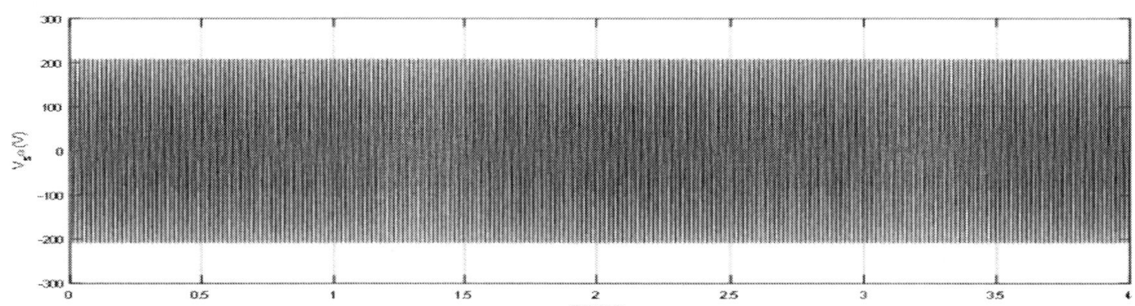

Figure 15: The voltages of the stator in α component

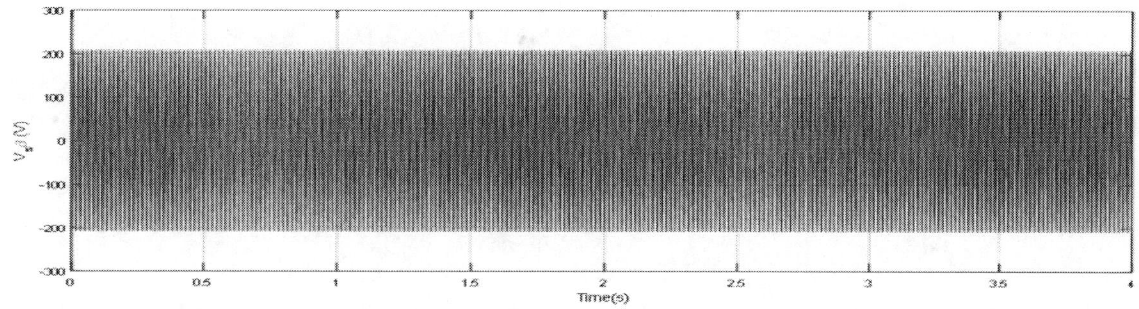

Figure 16: The voltages of the stator in β component

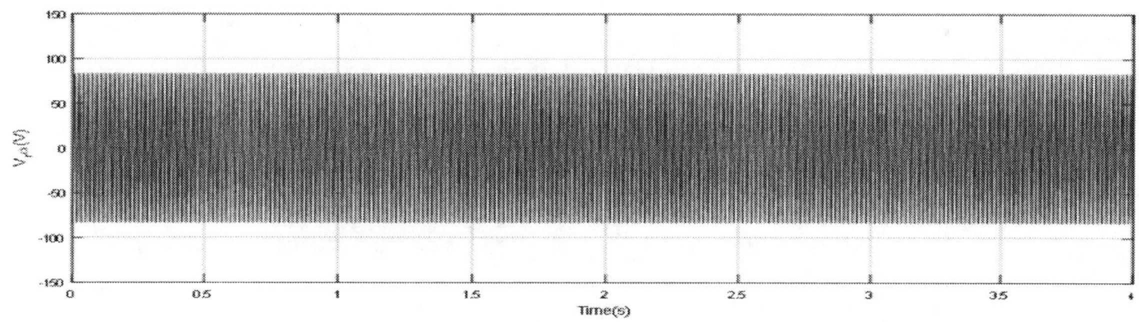

Figure 17: The voltages of the rotor in α component

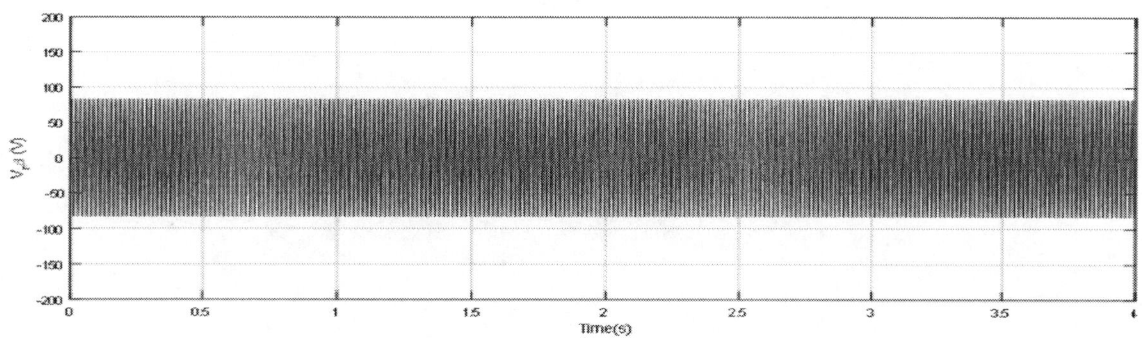

Figure 18: The voltages of the rotor in β components

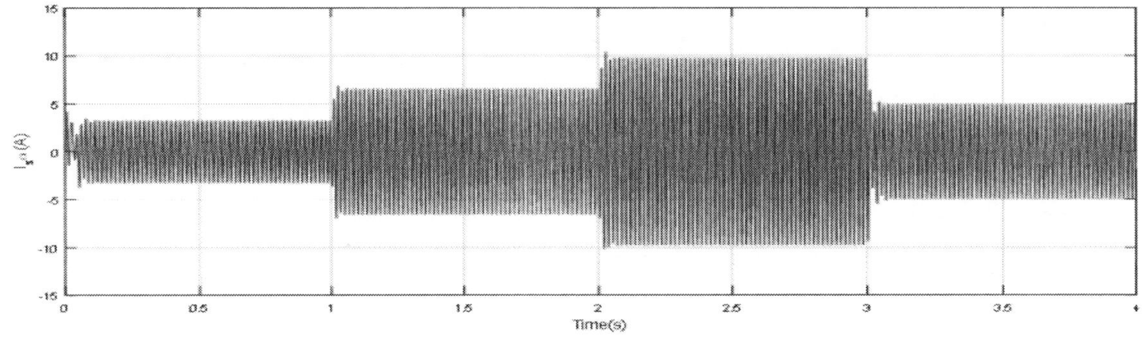

Figure 19: The currents of the stator in α components

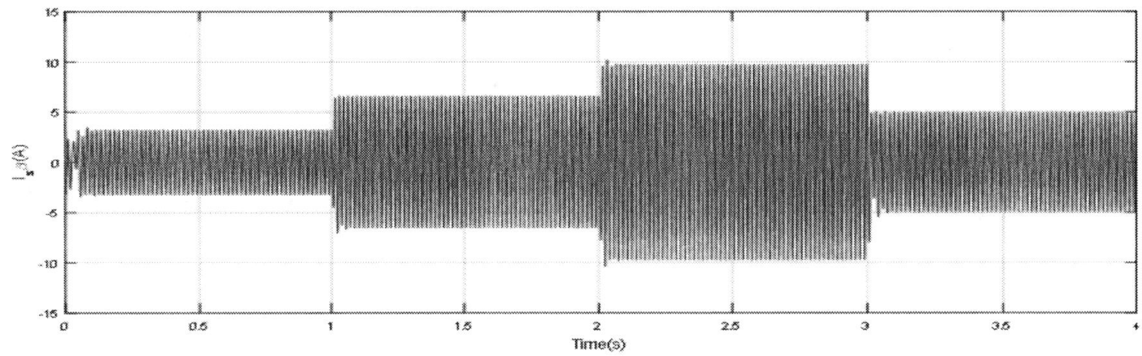

Figure 20: The currents of the stator in β components

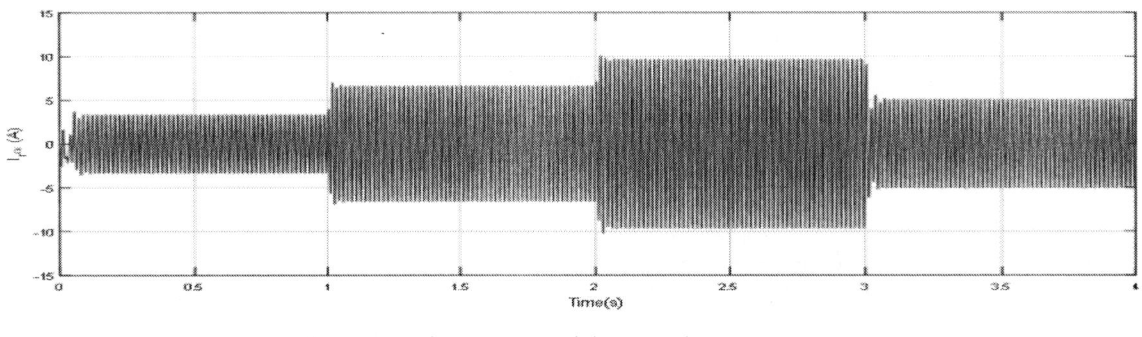

Figure 21: The currents of the rotor in α components

IOP Conf. Series: Materials Science and Engineering **765** (2020) 012033 doi:10.1088/1757-899X/765/1/012033

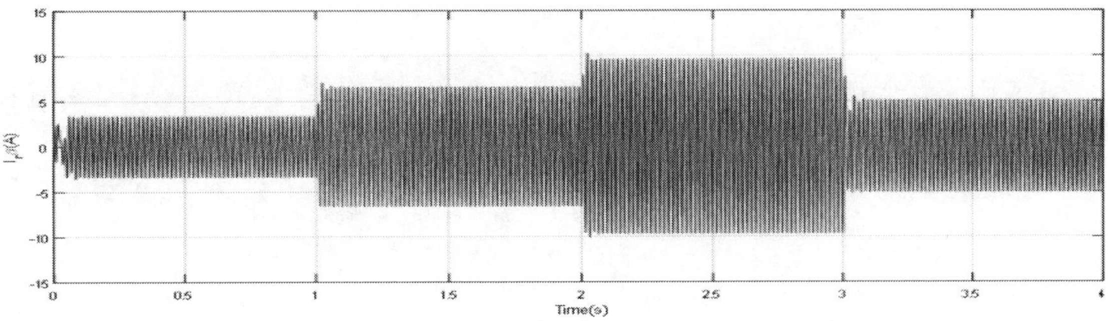

Figure 22: The currents of the rotor in β components

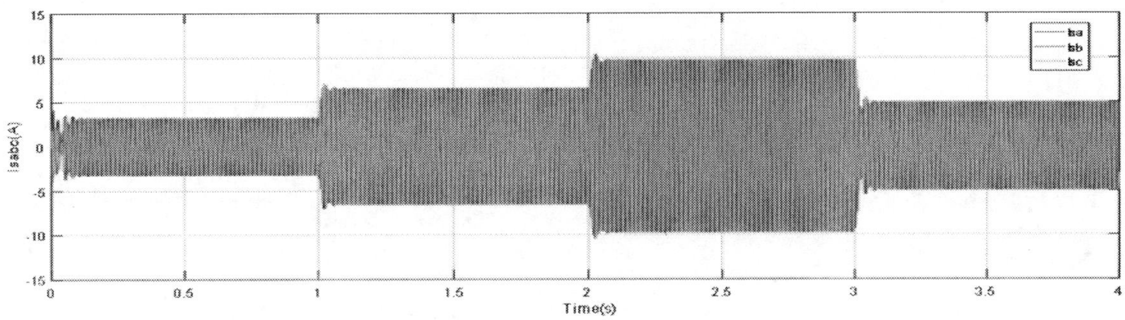

Figure 23: Three-phase stator currents

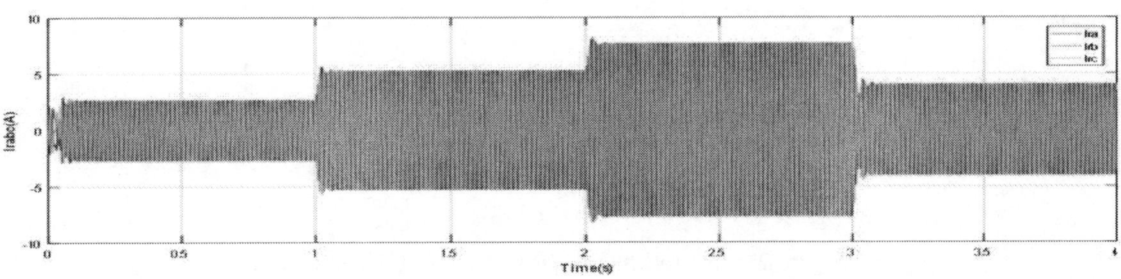

Figure 24: Three-phase rotor currents

IOP Conf. Series: Materials Science and Engineering **765** (2020) 012033 doi:10.1088/1757-899X/765/1/012033

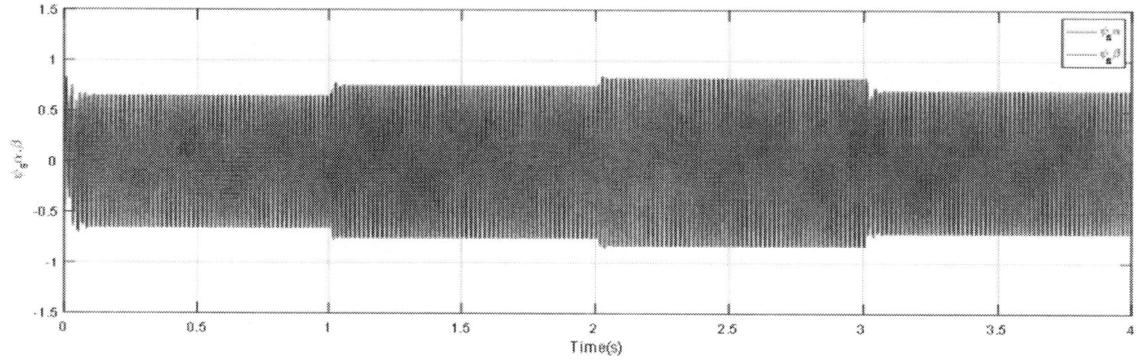

Figure 25: $\alpha\beta$ stator flux

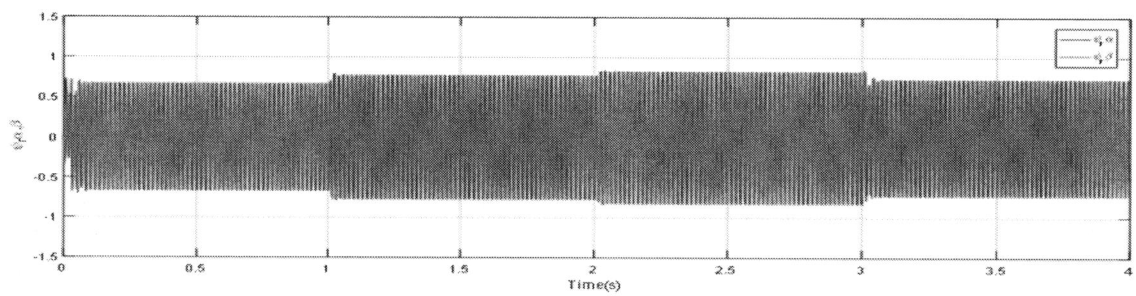

Figure 26: $\alpha\beta$ rotor flux

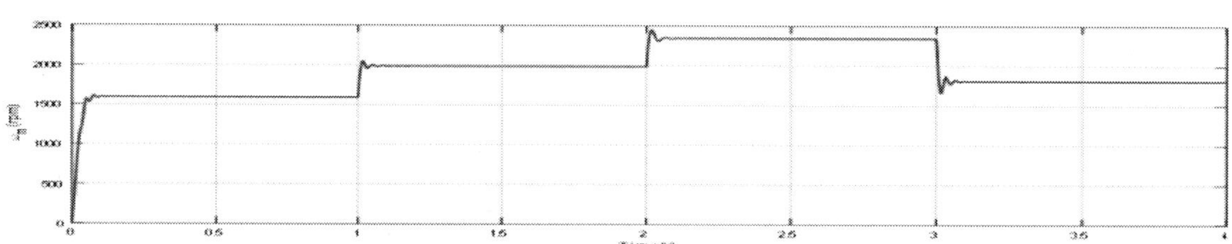

Figure 27: The mechanical shaft speed of the generator

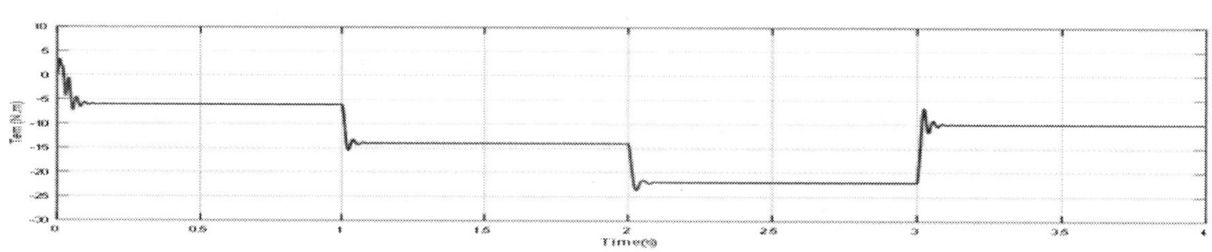

Figure 28: Electromagnetic torque of the generator

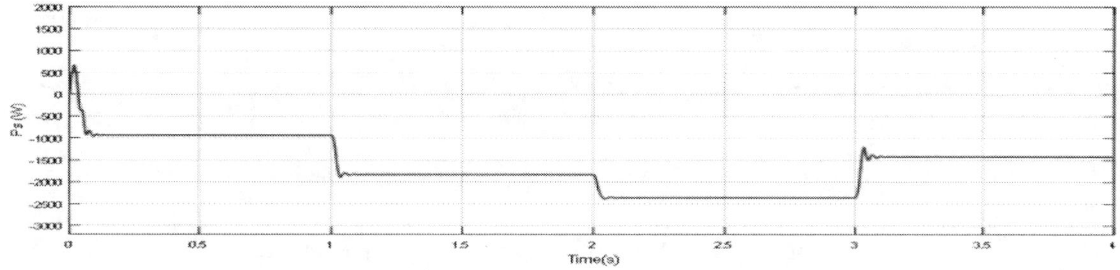

Figure 29: Stator active power

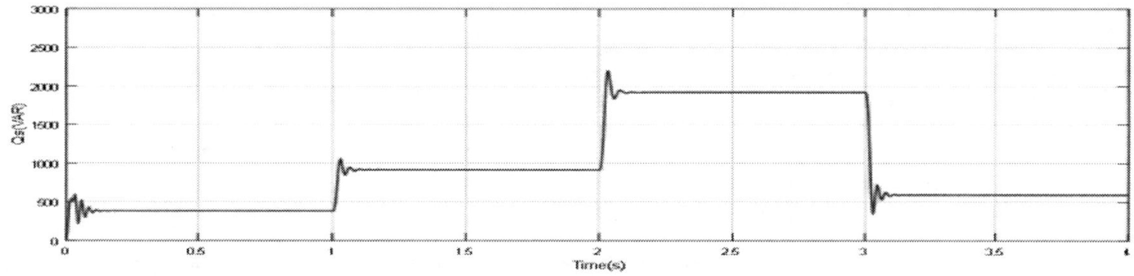

Figure 30: Stator reactive power

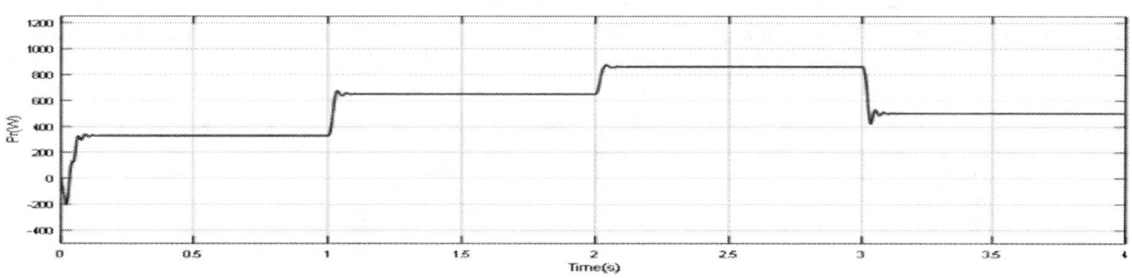

Figure 31: Rotor active power

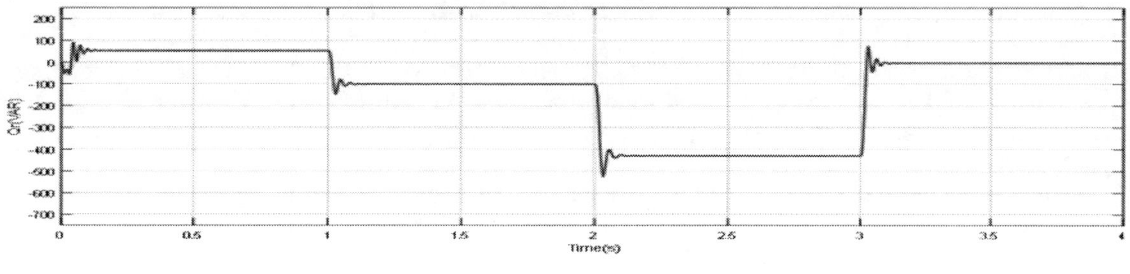

Figure 32: Rotor reactive power

IOP Conf. Series: Materials Science and Engineering **765** (2020) 012033 doi:10.1088/1757-899X/765/1/012033

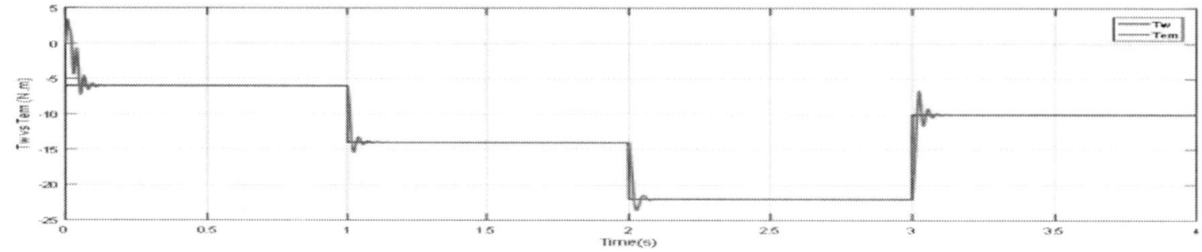

Figure 33: WT torque vs electromagnetic toque response

8. Conclusions

In this modeling, control and analysis of doubly fed induction generator (DFIG) connected to variable speed horizontal axis wind turbine (HAWT) was analyzed and several remark conclusions was obtained from the simulation results as summarized follows:

1) By using of the MATLAB/Simulink, the aerodynamic model of WT under different wind speed values over time was obtained and give accurate results.
2) By using of the MATLAB/Simulink, the steady state and dynamic performance of the DFIG was modeled, simulated and decoupled controlled on the reactive and active powers by controlling rotor side converter (RSC) by various control methods such as FOC and SVPWM.
3) By using of the MATLAB/Simulink, the modeling of AC-DC and DC-AC power electronics converters was implemented and generating pulsing of the inverters by using SVPWM.
4) Field oriented control was applied and investigated and efficient result as one of the power control strategies was used in induction machines.

REFERENCES

[1] BoroumandJazi G., Saidur R., Rismanchi B. and Mekhilef S.,(2012). "A review on the relation between the energy and exergy efficiency analysis and the technical characteristic of the renewable energy systems," Renewable and Sustainable Energy Reviews, VOL. 16, PP. 3131-3135.

[2] Sarkar M. R., Julai S., Tong C. W., Chao O. Z. and Rahman M., (2015)." Mathematical modelling and simulation of induction generator based wind turbine in MATLAB/SIMULINK", University of Malaya, VOL. 10, NO. 22.

[3] Yaramasu V. and Wu B., (2017). "Model predictive control of wind energy conversion systems".
[4] Sanchez E. N., (2016). "Doubly Fed Induction Generators Control for Wind Energy", Mexico.

[5] https://www.google.iq/search?hl=en&site=imghp&tbm=isch&source=hp&biw=1040&bih=622&q=The+wind+energy+conversion+system+topology+&oq=The+wind+energy+conversion+system+topology+&gs_l=img.12...3229.3229.0.4628.4.4.0.0.0.0.253.253.2-1.1.0....0...1.2.64.img.3.0.0.0.H9HFEIeABP4#imgrc=s1ma7TWZTK9RgM:

[6] Slootweg J.G. Polinder H. and Kling W. L., (2008)."dynamic modeling of wind turbine with doubly fed induction generator", IEEE, vol.1, pp.644-649.

[7] Kaneko A., Hara N. and Konishi K., (2012)."Model predictive control of DFIG-based wind turbines", Osaka Prefecture University, Japan.

[8] Fortmann J., (2015) "Modeling of Wind Turbines with Doubly Fed Generator System", Duisburg, Germany, 2014.Alhamzi M. M., "modeling of doubly fed induction generator using MATLAB/SIMULINK", California State University, Los Angeles.

[9] Masaud T., (2013)."Modeling, Analysis, Control and Design Application Guidelines of Doubly Fed Induction Generator (DFIG) for Wind Power Applications".

[10] http://read.pudn.com/downloads169/sourcecode/embed/779893/Part%204%20Clark&Park.pdf.

[11] Mannah M. A., Hammoud A., Haddad A., Elias R. and Bazzi Hassan, (2016)."Modeling and Real Time Simulation of A DFIG-Based Wind Energy Conversion System", Beirut, Lebanon.

[12] Alhamzi M. M., (2015). "modeling of doubly fed induction generator using MATLAB/SIMULINK", California State University, Los Angeles.

[13] Kazmierkowski M.P., Krishnan R. and Blaabjerg F., (2002). "Control in Power Electronics: Selected Problems", Academic Press, San Diego.

[14] Wang S. and Ding Y., (1993). "Stability Analysis of Field Oriented Doubly- Fed Induction Machine Drive Based on Computer Simulation", Vol. 21, pp. 11-24.

[15] Tadesse T., (2016). "Study of Doubly Fed Induction Generator Control Under Grid Fault Conditions", Addis Ababa University.

[16] Cheng J. and Wang W., (2013)."Simulation Study of Space Vector Pulse Width Modulation in Wind Power Generation System", China.

[17] Blaabjerg F, Liserre M, Ma K. Power electronics converters for wind turbine systems. IEEE Transactions on industry applications. 2011 Dec 23;48(2):708-19.

[18] Hamdi L. Q., (2007). " Design of Indirect Field-Oriented Voltage Controller Using Space Vector PWM Inverter for Three-Phase Induction Motor Drive", Ph.D., University of Technology, Al-Rasheed College of Engineering and Science.

IOP Conf. Series: Materials Science and Engineering **765** (2020) 012034 doi:10.1088/1757-899X/765/1/012034

BIM2BEM for room heating-cooling time estimation after turning on the HVAC systems

Zahra Pezeshki[1,*], **Ahmad Darabi**[1], **Mohammad Moghiman**[2]

1 Shahrood University of Technology, Faculty of Electrical and Robotics Engineering, Shahrood, Iran

2 Ferdowsi University of Mashhad, Department of Mechanical Engineering, Mashhad, Iran

E-mail: tejaratemrooz@gmail.com*

Abstract. The complexity of real constructions, with irregular geometry, inhomogeneous materials, variable morphology, alterations and damages, poses numerous challenges in digital modeling and simulation of structural performances under different types of actions. Recent developments in Building Information Modeling (BIM) have introduced advanced modeling capabilities. This paper presents a new methodology using BIM2BEM for the estimation of heating and cooling time after we turn on the HVAC systems. The generation of the BIM is carried out with information of a case study model in Autodesk REVIT software. Then, BIM is turned into a Finite Element Model (FEM) in COMSOL Multiphysics software and Building Energy Modelling (BEM) is obtained in winter and summer which preserves irregularities and anomalies, such as verticality deviation and variable thickness. Afterwards, the cooling and heating time of the whole room after turning on the cooling and heating appliances was calculated and compared with reality results.

1. Introduction

Architectural firms are moving from CAD to BIM and today relationship between BIM and BEM is a solution to develop HVAC systems, because BIM is an advantageous method to produce efficiency solutions for energy and reduce cost and time during the design of the building [1, 2, 3], so BIM and BEM are able to work efficiently and improve sustainability [3, 4, 5, 6, 7], but researchers have shown that a few papers have utilized BIM as an accurate database for BEM [8]. Studies have shown that in winter, cities have greater temperature than rural areas and this phenomenon causes the air conditioners to work higher and release carbon dioxide into the air which is harmful to the climate and environment [9]. Some studies have proposed methods for avoiding overheating as well as decrease the loss [10, 11]. Another study on heating time have shown when the heating time is shorter, the heat transfer is better. Also, researchers believe that decrease of cooling time is more important and has a great impact on its system performance [12, 13]. But this issue is different in heating systems such as radiators. In radiators, the surface is very important in heating. If the radiator surface is small, the heating temperature is very low. Also, the heating is very high when the radiator surface is too large. So, the size of radiator is a necessary component for suppliers and customers [14]. Generally, one of the important things in studies is data accuracy specifically in simulation studies to know how much these simulations are near to reality. In this paper, according to before studies that we have done for BEM simulation for HVAC systems in COMSOL Multiphysics, we want to calculate and estimate the cooling and heating time after turning on the systems. So, we followed BIM2BEM with: 1) BIM definition in Autodesk REVIT software, 2) BEM definition in COMSOL software, and 3) Conducting tests to depict and validate BIM2BEM approach. This paper has three section. After introduction, it is begun with BIM2BEM analysis in section 2. Then, in section 3, experiments are performed and the results are shown. Finally, conclusion and future work are presented.

2. BIM2BEM analysis

In this research, we use the case study model in Mashhad, Iran, as known Toos Arman Star Apartment Hotel. So, at first, the Computer-Aided Design (CAD) layout is introduced into the Autodesk REVIT software, and the unit geometry (unit 502) is manually drawn in REVIT (Figure 1) [3, 6].

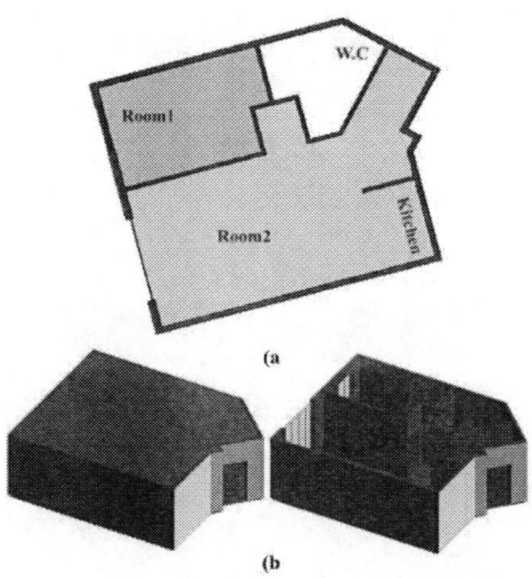

Figure 1. (a) a view of room 1 and room 2 and (b) a view of the building geometry of unit 502 drawn in REVIT [3, 6].

Then, the REVIT model is created in COMSOL Multiphysics. The unit has two rooms and for cooling and heating the unit, it is used a fan-coil and a radiator for room 1 and a fan-coil and two radiators for room 2. Thus laminar and turbulent processes are done (Figures 2, 3) [3, 6].

The total process of BIM2BEM has been shown in Figure 4 [3, 8, 15].

2.1. *Parametric study and synthetic data*

Engineers usually integrate the local heat flux density over total heat transfer area to calculate the overall heat transfer rate. Thus the total transferred heat rate is as below (Eq. 1) [14]:

$$Q = \int_0^A k \cdot (\vartheta_2 - \vartheta_1) dA \tag{1}$$

Where, ϑ is the heat-transfer coefficient, A is cross-sectional area, and k is the thermal conductivity coefficient. Also, energy balance equations 2 and 3 are employed to determine the temperature difference in the surface area dA [14]:

$$\delta Q = m_1 \cdot d\vartheta_1 = m_1 \cdot C_{p1} \cdot d\vartheta_1 \tag{2}$$

$$\delta Q = -m_2 \cdot d\vartheta_2 = -m_2 \cdot C_{p2} \cdot d\vartheta_2 \tag{3}$$

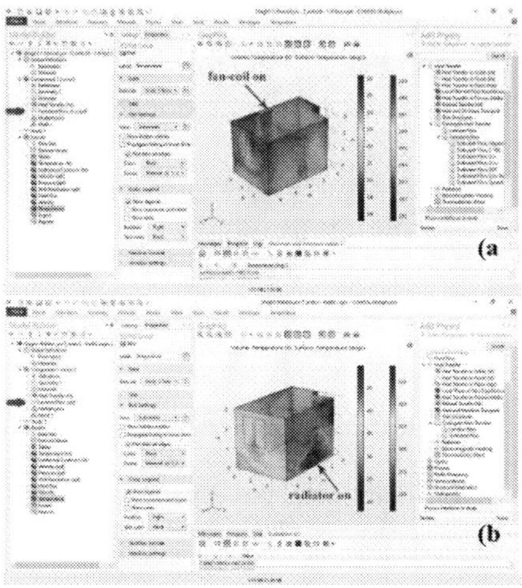

Figure 2. (a) turbulent flow (for fan-coil); and (b) laminar flow (for radiator) (room 1) [3, 6].

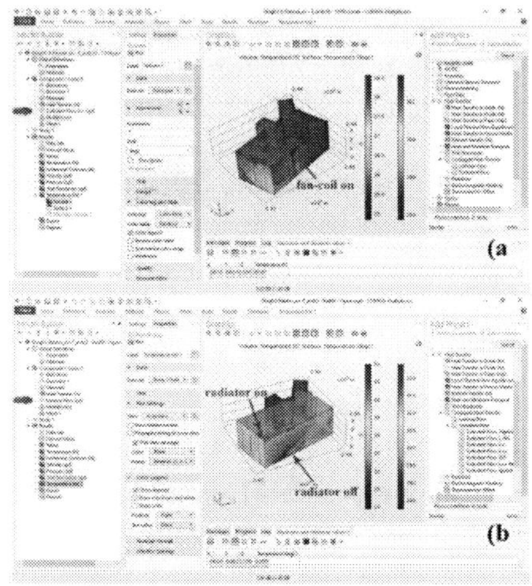

Figure 3. (a) turbulent flow (for fan-coil); and (b) laminar flow (for radiator) (room 2) [3, 6].

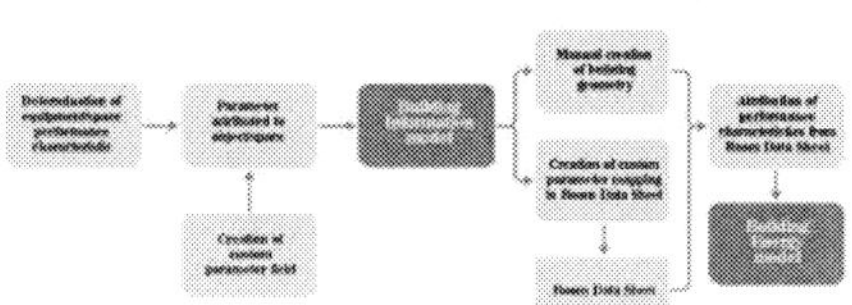

Figure 4. Bringing information to BIM, definition, collection and documenting information for BEM [3, 8, 15].

Where m is mass, ϑ is overall heat-transfer coefficient, and C_p is the specific heat at constant pressure. The temperature change, $\vartheta_2 - \vartheta_1$ is replaced with $\Delta\vartheta$. Thus, this variation can be determined by the change of the fluid temperatures (Eq. 4).

$$d\Delta\vartheta = d\vartheta_2 - d\vartheta_1 = -\delta Q \cdot \left[\frac{1}{m_1 \cdot C_{p1}} + \frac{1}{m_2 \cdot C_{p2}}\right] \quad (4)$$

From equations 2, 3 and 4, we have (Eq. 5),

$$Q = m \cdot C_p \cdot \Delta\vartheta = \rho \cdot V \cdot C_p \cdot \Delta\vartheta \quad (5)$$

Where V is the volume of the rooms. Here, the space of the rooms has been filled by air, so ρ and C_p is equal to 1.2 and 1003.5, respectively. All the information of room 1 and room 2 for calculation of the Q have been given in Table 1.

Table 1. The information of room 1 and room 2.

Room information	$\rho(kg/m^3)$	$V(m^3)$	$C_p(J/(kg°K))$	$m \cdot C_p(J°K)$
Room 1	1.2	39.92	1003.5	48071.664
Room 2	1.2	114.7	1003.5	138121.74

IOP Conf. Series: Materials Science and Engineering **765** (2020) 012034 doi:10.1088/1757-899X/765/1/012034

For calculation of the power of P for fan-coil and radiators, it obtained from tables [16]. The characteristic of powers is given in Table 2.

Table 2. Characteristic of radiator and fan-coil powers [16].

Characteristic	$P(kCal/h)$	$P(W)$
Fan-coil (Model 400)	2730	3166.8
Radiator with 11 blades	1430	1658.8
Radiator with 12 blades	1560	1809.6

3. Experiments and results

After turning on the cooling and heating appliances, i.e. fan-coil and radiator, the cooling and heating time of the whole room 1 and room 2 is as follow (Eq. 6):

$$t = Q/P \tag{6}$$

In this study, two methods are considered for calculation of the temperature difference, which obtains as follow (Eqs. 7,8):

$$\Delta\vartheta = \vartheta_{amb} - ((\vartheta_{Max} + \vartheta_{min})/2) \tag{7}$$

$$\Delta\vartheta = \vartheta_{amb} - \sum_{i=1}^{n} \frac{\vartheta_i}{n} \tag{8}$$

Where, ϑ_{amb} is ambient temperature, ϑ_{Mx} and ϑ_{min} is the maximum and minimum of temperature which obtain after turning on the fan-coil and radiators from COMSOL Multiphysics, and n is the number of temperature points. Thermal comfort is 25°C. This study is tested and evaluated for two days in the year (the warmest day in summer and the coldest day in winter) [6]. In summer we turn on the fan-coil when the ambient temperature is 27, 28 and 29°C. The fan-coil is tuned in desired temperature, 23°C, and the velocity of fan-coil wind is 1.016 m/s. The cooling time is calculated for each of the rooms, and the time is calculated with both equations 7 and 8 (Table 3).

The chart of fan-coil $t - \vartheta_{amb}$ of room 1 for two equations are given as Figure 5.

In winter we turn on the radiators when the ambient temperature is 5°C. The radiators is tuned in temperatures 50, 55, 60 and 65°C. The room 1 has one radiator and room 2 has two radiators with 11 and 12 blades. The radiator in room 1 is tuned in temperatures 50, 55, and 60°C, because this room is smaller than room 2 and for room 2, the radiator temperature 65°C is also added. The heating time is calculated for each of the rooms, and the time calculated from both equations 7 and 8 (Table 4).

IOP Conf. Series: Materials Science and Engineering **765** (2020) 012034 doi:10.1088/1757-899X/765/1/012034

Table 3. Room cooling time estimation after turning on the fan-coil.

Characteristic	The fan-coil has been tuned in desired temperature, 23°C with velocity of 1.016 m/s						
Room1	ϑ_{amb}(°C)	Eq. 9			Eq. 10		
		$\Delta\vartheta(°K)$	$Q(J)$	$t(minute)$	$\Delta\vartheta(°K)$	$Q(J)$	$t(minute)$
	27	2.501	120227.232	0.63	1.969	94689.675	0.50
	28	3.158	151810.315	0.80	2.537	121966.900	0.64
	29	6.953	334242.280	1.76	5.643	271285.026	1.43
Room2	27	1.989	274655.08	1.45	0.972	134243.557	0.71
	28	2.689	371340.298	1.95	1.247	172296.551	0.91
	29	6.061	837086.805	4.41	2.819	389371.074	2.05

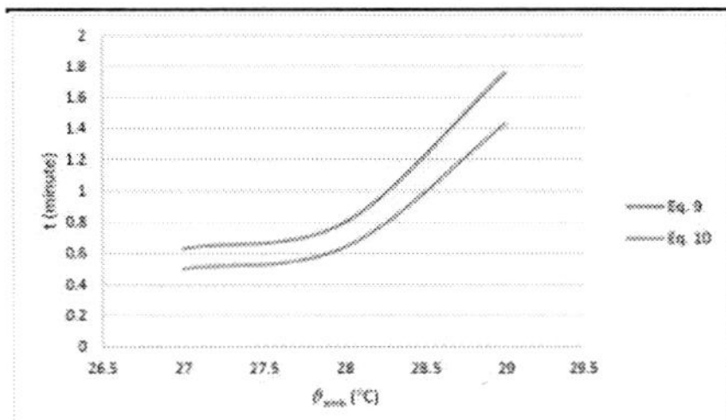

Figure 5. The chart of fan-coil $t - \vartheta_{amb}$ for room 1.

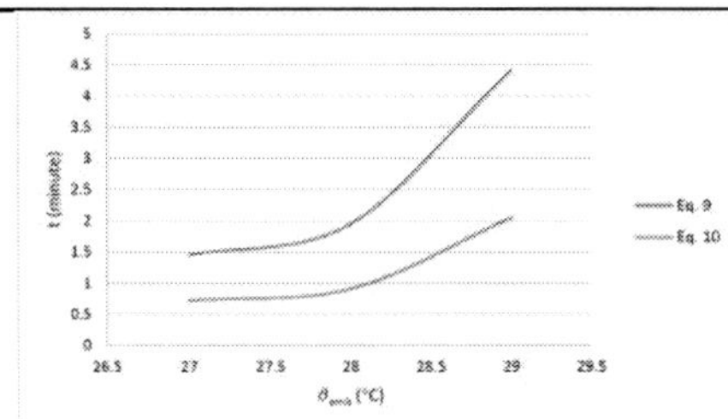

Figure 6. The chart of fan-coil $t - \vartheta_{amb}$ for room 2.

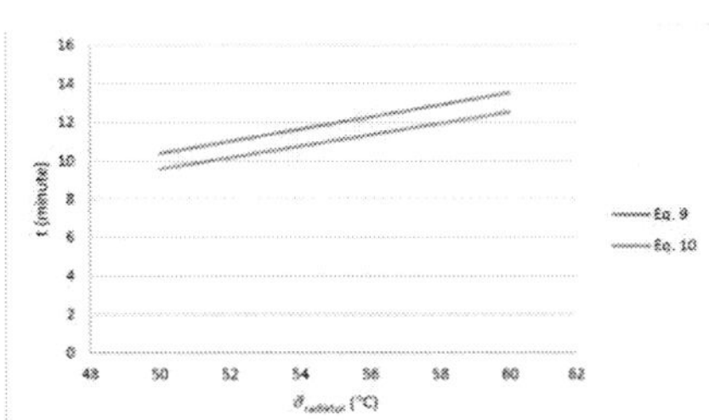

Figure 7. The chart of 11 blades radiator $t - \vartheta_{radiator}$ for room 1.

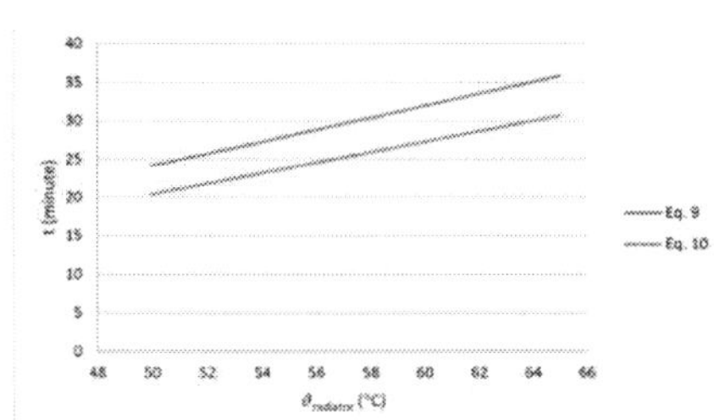

Figure 8. The chart of 11 blades radiator $t - \vartheta_{radiator}$ for room 2.

The charts of 11 blades radiator $t - \vartheta_{radiator}$ of room 1 and room 2 for two equations 7 and 8 are given as Figures 7 and 8.

The charts of 12 blades radiator and 11&12 blades $t - \vartheta_{radiator}$ of room 2 for two equations 7 and 8 are given as Figures 9 and 10.

According to the cooling experiments, the size of the room is larger, the temperature difference is smaller and the cooling time is higher. Also, with equation 10, the cooling time is lower than the equation 9, and fan-coil requires more time for cooling the rooms when the ambient temperature is high. According to the heating experiments, the temperature difference and heating time of the rooms are higher when the radiator temperature is being more and more. Also, the temperature difference with equation 10 is lower than the equation 9. The temperature difference is lower when the size of the room is larger, but the heating time is higher. The radiator type, i.e. 11 blades or 12 blades, does not have an effect on time heating, and the location of the radiator at room is important accordance with Table 4 and

IOP Conf. Series: Materials Science and Engineering **765** (2020) 012034 doi:10.1088/1757-899X/765/1/012034

Table 4. Room heating time estimation after turning on the radiators.

Characteristic	The radiator has been tuned in temperatures, 50, 55, 60 and 65°C							
Room1	$\vartheta_{radiator}$(°C)		Eq. 9			Eq. 10		
			$\Delta\vartheta(°K)$	$Q(J)$	$t(minute)$	$\Delta\vartheta(°K)$	$Q(J)$	$t(minute)$
		50	21.48	1032555.307	10.37	19.79	951395.619	9.56
		55	24.75	1189557.362	11.95	22.87	1099367.997	11.05
		60	28.01	1346439.237	13.53	25.94	1247142.122	12.53
Room2	11	50	17.36	2398345.893	24.10	14.67	2026735.699	20.36
	blades	55	20.18	2787020.470	28.002	17.15	2368701.092	23.80
	radiator	60	22.99	3175487.863	31.91	19.62	2709929.674	27.23
	on	65	25.80	3563817.135	35.81	22.09	3050429.021	30.65
Room2	12	50	18.11	2501591.894	23.04	17.69	2443973.331	22.51
	blades	55	21.01	2901730.575	26.73	20.51	2832508.598	26.09
	radiator	60	23.91	3301869.256	30.41	23.32	3220378.953	29.66
	on	65	26.80	3701938.875	34.10	26.12	3607572.773	33.23
Room2	11&12	50	17.29	2388677.372	11.48	16.37	2260791.241	10.86
	blades	55	20.08	2773415.478	13.33	19.02	2626685.549	12.62
	radiators	60	22.86	3158084.524	15.18	21.66	2991868.041	14.38
	on	65	25.65	3542684.509	17.02	24.30	3356350	16.13

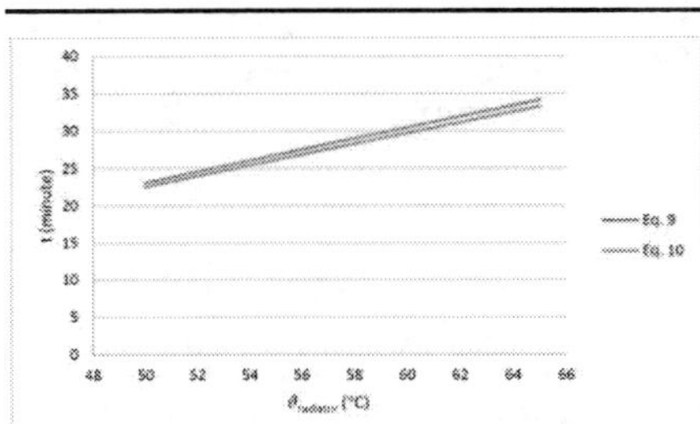

Figure 9. The chart of 12 blades radiator $t - \vartheta_{radiator}$ for room 2.

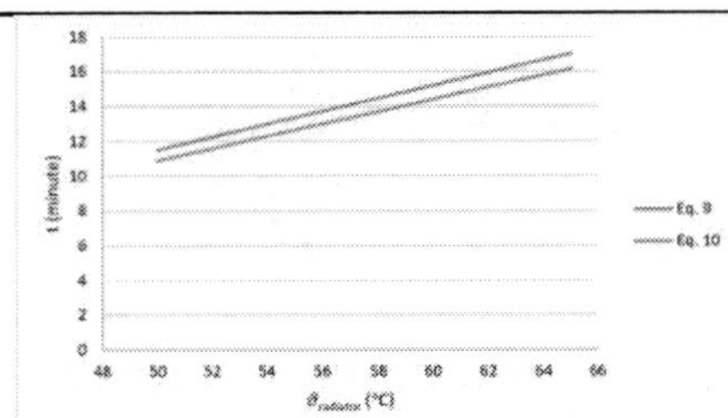

Figure 10. The chart of $t - \vartheta_{radiator}$ for room 2 when both radiators are on.

Figures 8 and 9. It is considered that the heating time of room 2 with 11 blades radiator is lower than the heating time of that with 12 blades radiator. At last, according to the Table 4, when both radiators are on in room, the heating time is lower. Therefore, it is concluded when the number of radiators is more, the heating time will be lower.

3.1. Evaluation

For evaluation of this study to see these results are near to reality or not, we collected data for cooling and heating time from the unit. For cooling time, we turn on the fan-coil in room 1. The fan-coil was tuned in desired temperature, 23°C, and the ambient temperature was 29°C. For heating time, we turn on the radiator in room 1. The radiator was tuned in 60°C, and the ambient temperature was 5°C. All the results are shown in Table 5.

Table 5. Room heating time estimation after turning on the radiators.

Characteristic	Time from Eq. 9 (minute)	Time from Eq. 10 (minute)	Time from Reality (minute)
Fan-coil cooling time	1.76	1.43	2
Radiator heating time	13.53	12.53	14

According to the Table 5, the estimation results are very near to the reality results.

4. Conclusion

This paper tries to estimate the time of heating-cooling after turning on the HVAC systems to see the results are near to reality or not. To estimate the time, all the information was obtained from the BIM, BEM and the fan-coil and radiator catalogs. For this purpose, BEM was used for the acquiring the temperature data from COMSOL Multiphysics software. The experimental results demonstrated the good results in contrast to reality. Another work, BEO for energy saving, is under further development.

Acknowledgments

This paper is related to Zahra Pezeshki's future PhD studies that she wants to begin. The authors express their sincere gratitude to the anonymous reviewers for their constructive comments and helpful suggestions, which have led to substantial improvements of this paper. This case study was carried out on one of the Samen Municipality projects in Mashhad, Iran and in collaboration with Khorasan Razavi Construction Engineering Organization (IRCEO) in Mashhad, Iran.

References

[1] G. Spiridigliozzi, L. Pompei, C. Cornaro, L. De Santoli, and F. Bisegna, "BIM-BEM support tools for early stages of zero-energy building design" *IOP Conf. Series: Materials Science and Engineering* 2019; vol 609, 072075, pp. 1-6, https://doi.10.1088/1757-899X/609/7/072075.

[2] Z. Pezeshki, and S.A.S. Ivari, "Applications of BIM: a brief review and future outline" *Archives of Computational Methods in Engineering* 2018; vol 25, no 2, pp. 273-312.

[3] Z. Pezeshki, "Energy Modeling for Estimation of Indoor Temperature of the Building on the BIM Platform", MSc thesis of Faculty of Electrical and Robotics Engineering, Shahrood University of Technology, Shahrud, Iran. 2018; pp. 1166 (http://shahroodut.ac. ir/en/thesis/thesis.php?thid=TK658.) https://doi.org/10. 13140/RG.2.2.21522.09924.

[4] W.S. Jeong, J.B. Kim, M.J. Clayton, J.S. Haberl, and W. Yan, "Translating Building Information Modeling to Building Energy Modeling Using Model View Definition" *The Scientic World Journal* 2014; vol 2014, pp. 1-21.

[5] E. Kamel, and A.M. Memari, "Review of BIM's application in energy simulation: Tools, issues, and solutions" *Automation in Construction* 2019; vol 97, pp. 164-180.

[6] Z. Pezeshki, A. Soleimani, and A. Darabi, "Taguchi method to determine the best location for cooling and heating appliances" *Proceedings of the Institution of Civil Engineers Energy* 2019; pp. 1-69, https://doi.org/10.1680/jener.18.00024.

[7] L.M. Khodeir, and A.A. Nessim, "BIM2BEM integrated approach: Examining status of the adoption of building information modelling and building energy models in Egyptian architectural firms" *Ain Shams Engineering Journal* 2018; vol 9, Issue 4, pp. 1781-1790.

[8] Z. Pezeshki, A. Soleimani, and A. Darabi, "Application of BEM and using BIM database for BEM: a review" *Journal of Building Engineering* 2019; vol 23, pp. 117, https://doi.org/10.1016/j.jobe.2019.01.021.

[9] S.P. Singh, M. Katiyar, and S. Agarwal, "An Overview of Passive Cooling Techniques in Buildings Design Concepts" *Journal of Building Construction* 2019; vol 1, no 2, pp. 1-6.

[10] O. Dudina, and A. Dudin, "Optimization of Queueing Model with Server Heating and Cooling" *Mathematics* 2019; vol 7, no 9, 768, pp. 1-14. https://doi.org/10.3390/math7090768.

[11] P. Puls, C. Lange, and R. chsner, "Hybrid cooling towers in a free cooling application - modelling and field measurement verification" *Chemical Engineering & Technology* 2019; vol 42, no 9, pp. 1871-1878.

[12] L. Ming, Y. Yanguang, L. Wanqing, and L. Zhi-hui, "Study of Heating Time of test model and Application in Low Density Wind Tunnel Using Infrared Thermography", 2014; http://dx.doi.org/10.21611/qirt.2014.048.

[13] M. Xia, X.P. Chen, H.Y. Li, and Z.H. Gan, "Cool down time optimization of the Stirling cooler" *IOP Conf. Series: Materials Science and Engineering* 2017; vol 278, no 012043, pp. 1-7. http://doi.10.1088/1757899X/278/1/012043.

[14] P. Bckh, and T. Wetzel, *Heat Transfer Basics and Practice* 2012; Springer-Verlag, Berlin, Heidelberg. https://doi.org/10.1007/ 978-3-642-19183-1.

[15] T. Gerrish, K. Ruikar, M. Cook, M. Johnson, and M. Phillip, "Using BIM capabilities to improve existing building energy modelling practices", *Eng. Constr. Archit. Manag.* 2017; vol 24, no 2, pp. 190208.

[16] M. Moghiman, *Air Conditioning and Central Heating Engineering* 2016; Ferdowsi University of Mashhad, Mashhad, Iran, 9th edition, pp. 1-535, 9789643860134 (In Persian).

IOP Conf. Series: Materials Science and Engineering **765** (2020) 012035 doi:10.1088/1757-899X/765/1/012035

Impacts of non-homogeneous nanofluid approach and orientation angle on convection heat transfer within 3D wavy cavity

Ammar I. Alsabery[a,b,*], Adel H. Ayaal[c] and Ishak Hashim[b]

[a]Refrigeration & Air-conditioning Technical Engineering Department. College of Technical Engineering, The Islamic University, Najaf, Iraq
[b]Department of Mathematical Sciences, Faculty of Science & Technology, Universiti Kebangsaan Malaysia, 43600 UKM Bangi, Selangor, Malaysia
[c]Department of Electromechanical Engineering, University of Technology, Baghdad, Iraq

*Corresponding author's: alsabery_a@ukm.edu.my

Abstract. Steady laminar natural convection and heat transfer inside a tilted 3D wavy cavity filled by water-Al_2O_3 nanofluid does examined numerically by applying the finite element technique. An isothermal curved heater is located at the right surface of the chamber while the left surface remains in a steady cold temperature. The top and bottom horizontal surfaces are held adiabatic. The governing equations are formulated in the non-dimensional model using the two-component non-homogeneous equilibrium model through transporting phenomena in nanofluids connecting among the influences of Brownian diffusion and thermophoresis. The results of the dimensionless buoyancy-driven parameter, inclination angle, nanoparticle volume fraction and number of undulations were examined. It is shown that the inclination angle, the wavy shape of the surface plus the features of nanofluid are outstanding control parameters to the rate of the heat transfer and the distribution of the fluid flow.

1. Introduction

The significance of natural convection including heat transfer within enclosed mediums own recognizing in many engineering applications toward its diverse applications in solar thermal collectors, ventilation, cooling about electronics and buildings, containment building cooling, exchange of heat, storage tanks, heat exchangers, and dual-pane windows, and various others [1]. A numerical investigation on the natural convection heat transfer in a differentially heated 3D cubical cavity was explained by Fusegi et al. [2]. Wang et al. [3] investigated the problem of 3D natural convection heat transfer numerically in a cubical cavity with high Rayleigh numbers. They concluded that the flow motion turned to be fully turbulent when the Rayleigh number was approached to 10^9. Gibanov and Sheremet [4] examined the effect of various heat source configurations on the steady natural convection heat transfer into a 3-D cubical hollow employing the finite difference method. They discovered that the addition of Rayleigh number points to the heat transfer augmentation concerning the trapezoidal chamber. Very recently, [5] demonstrated the unsteady natural convection and heat transfer within a 3-D cubical cavity with the presence of trapezoidal heater. They concluded that the increase of the heater length led to an apparent reduction in the average Nusselt number.

Within the field of thermal science, the heat transfer enhancement is the most critical process that engineers are widely attempting for, due to the enormous interests in industry for controlling such procedures [6]. Several concepts have investigated the area of enhancing the heat transfer, an essential technique and innovative idea toward this domain is combining nanometer-sized particles into the core fluid; as a result, the heat transfer is improved. Adding those nanometer-sized solid particles within the variety of sizes 10-50nm into the base liquid is named nanofluid. Nanofluid regards to active liquid that holds suspended nanoparticles sized smaller than 100nm concerning conventional heat transfer, such as, water, ethylene glycol, and oil [7]. Khanafer et al. [8] described an excellent comprehensive investigation on the natural convection heat transfer in a 2D square cavity that partially occupied by nanofluids. Applying the finite volume method, Al-Rashed et al. [9] described the effect of an external magnetic field on natural convection heat transfer in a cubical cavity filled with nanofluid (carbon nano-tube-water).

Another concept that used for improving the thermal transmission is the corrugated borders of the cavity. Such a topic is widely investigated by many researchers in many engineering applications, such as nuclear component, micro-electronic devices, electrical and solar collectors, etc. [7,10]. A recent work on the free convection and heat transfer inside a wavy hollow that loaded with water-Al_2O_3 nanofluid and having an inner solid block was reported by Hashim et al. [11]. They noticed by using the two-phase nanofluid model that the convection heat transfer in the wavy cavity is improved by adding nanoparticles as well as the number of oscillations.

According to the given literature and the best of our knowledge, there are no studies on the laminar free convection inside an inclined 3D wavy chamber filled by water-Al_2O_3 nanofluid. Thus, the present numerical work aims to study the impacts of non-homogeneous nanofluid approach and orientation angle toward convective heat transfer within a 3D wavy chamber. Indeed, this sort of complex geometry performs a primary role in numerous engineering applications, such as solar energy (solar thermal collector), cooling of electronic devices, heat exchangers, and many others.

2. Methodology

2.1 Mathematical Analysis

The steady three-dimensional (3D) natural convection inside a wavy-walled chamber with length L, width $L/2$ and thickness $L/2$, as described in Fig. 1. The right wavy surface is taking the following form $1 - A(1 - \cos(2N\pi\,\mathbf{V}))$. An isothermal curved heater is located at the right surface of the chamber while the left surface remains in a steady cold condition. Those top and bottom horizontal surfaces are taken adiabatic. The chamber borders remain impermeable; the space inside the wavy region is loaded with water-Al_2O_3 nanofluid. The Boussinesq approximation holds applicable. Through the attention of those hypotheses, continuity, momentum and energy equations of the convection flow can be addressed in the dimensionless scheme as [7]:

2.1.1 Governing Equations

$$\nabla \cdot \mathbf{V} = 0, \tag{1}$$

$$\mathbf{V} \cdot \nabla \mathbf{V} = -\nabla P + \frac{\rho_f}{\rho_{nf}} \frac{\mu_{nf}}{\mu_f} \nabla^2 \mathbf{V} + \frac{(\rho\beta)_{nf}}{\rho_{nf}\beta_f} \frac{1}{\Pr} Ra \cdot \theta, \tag{2}$$

$$\mathbf{V}\cdot\nabla\theta=\frac{(\rho C_p)_f}{(\rho C_p)_{nf}}\left(\frac{k_{nf}}{k_f}\frac{1}{\Pr}\nabla^2\theta+\frac{D_B^*}{\Pr\cdot Le}\nabla\varphi^*\cdot\nabla\theta+\frac{D_T^*}{\Pr\cdot Le\cdot N_{BT}}\frac{\nabla\theta\cdot\nabla\theta}{1+\delta\theta}\right), \tag{3}$$

$$\mathbf{V}\cdot\nabla\varphi^*=\frac{D_B^*}{Sc}\nabla^2\varphi^*+\frac{D_T^*}{Sc\cdot N_{BT}}\cdot\frac{\nabla^2\theta}{1+\delta\theta}. \tag{4}$$

where **V** explains the non-dimensional velocity vector (U,V,W).

The governing equations of Eqs. (1)-(4) are moulded into dimensionless forms by employing the following non-dimensional variables:

$$X,Y,Z=\frac{x,y,z}{L},\ \mathbf{V}=\frac{\mathbf{v}L}{\nu_f},\ P=\frac{pL^2}{\rho_{nf}\nu_f^2},\ \varphi^*=\frac{\varphi}{\phi},$$
$$D_B^*=\frac{D_B}{D_{B0}},\ D_T^*=\frac{D_T}{D_{T0}},\ \delta=\frac{T_c}{T_h-T_c},\ \theta=\frac{T-T_c}{T_h-T_c}. \tag{5}$$

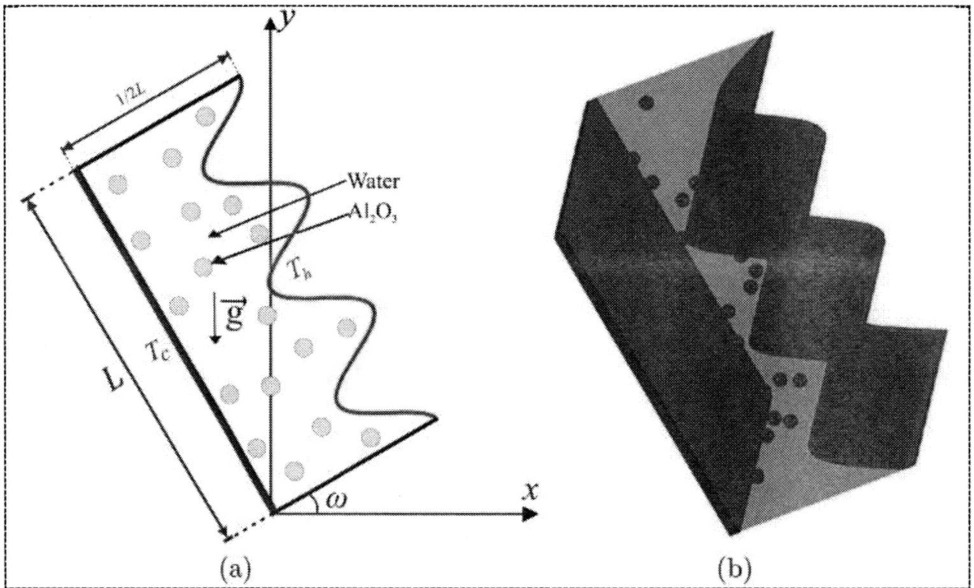

Figure 1. Physical design regarding convection inside a tilted wavy chamber (a) 2D-xy-plane including the coordinate system and (b) 3D cavity.

2.1.2 Boundary Conditions

The corresponding conditions concerning the dimensionless boundary for Eqs. (1)-(4) do provided by:

$$U=V=0, \text{ toward all the surfaces of the cavity.} \tag{6}$$

On the right hot surface:

$$\mathbf{V}=0,\ \frac{\partial\varphi^*}{\partial\mathbf{V}}=-\frac{D_T^*}{D_B^*}\cdot\frac{1}{N_{BT}}\cdot\frac{1}{1+\delta\theta}\frac{\partial\theta}{\partial\mathbf{V}},\ \theta=1, \tag{7}$$

On the left cold surface:

$$\mathbf{V}=0,\ \frac{\partial\varphi^*}{\partial\mathbf{V}}=-\frac{D_T^*}{D_B^*}\cdot\frac{1}{N_{BT}}\cdot\frac{1}{1+\delta\theta}\frac{\partial\theta}{\partial\mathbf{V}},\ \theta=0, \tag{8}$$

On the the adiabatic top surface:

$$\mathbf{V}=0,\ \frac{\partial\varphi^*}{\partial\mathbf{V}}=0,\ \frac{\partial\theta}{\partial\mathbf{V}}=0, \tag{9}$$

On the the adiabatic bottom surface:

$$\mathbf{V}=0,\ \frac{\partial\varphi^*}{\partial\mathbf{V}}=0,\ \frac{\partial\theta}{\partial\mathbf{V}}=0, \tag{10}$$

The used thermophysical properties concerning the nanofluid are: heat capacitance ($(\rho C_p)_{nf}$), effective thermal diffusivity (α_{nf}), effective density (ρ_{nf}), the thermal expansion coefficient (β_{nf}), thermal conductivity ratio $\left(\frac{k_{nf}}{k_f}\right)$ and dynamic viscosity ratio $\left(\frac{\mu_{nf}}{\mu_f}\right)$ which can be explained as the following [19, 15]:

$$(\rho C_p)_{nf}=(1-\phi)(\rho C_p)_f+\phi(\rho C_p)_p, \tag{11}$$

$$\alpha_{nf}=\frac{k_{nf}}{(\rho C_p)_{nf}}, \tag{12}$$

$$\rho_{nf}=(1-\phi)\rho_f+\phi\rho_p, \tag{13}$$

$$(\rho\beta)_{nf}=(1-\phi)(\rho\beta)_f+\phi(\rho\beta)_p \tag{14}$$

$$\frac{k_{nf}}{k_f}=1+4.4\mathrm{Re}_B^{0.4}\mathrm{Pr}^{0.66}\left(\frac{T}{T_{fr}}\right)^{10}\left(\frac{k_p}{k_f}\right)^{0.03}\phi^{0.66}, \tag{15}$$

$$\frac{\mu_{nf}}{\mu_f}=1/\left(1-34.87\left(d_p/d_f\right)^{-0.3}\phi^{1.03}\right). \tag{16}$$

The local Nusselt number on the right wavy surface may now be defined as follows:

$$Nu_{nf}=-\frac{k_{nf}}{k_f}\left(\frac{\partial\theta}{\partial\mathbf{V}}\right)_{\mathbf{V}=0}. \tag{17}$$

Lastly, the average Nusselt number evaluated on the right wavy surface and it is provided by:

$$\overline{Nu}_{nf}=\int_0^1 Nu_{nf}\,d\mathbf{V}. \tag{18}$$

2.2 Simulation Method and Validation

The dimensionless governing equations of Eqs. (1)–(4) controlled by the adopted boundary conditions of Eqs. (6)–(10) are determined by the Galerkin weighted and finite element techniques. The computational region does discretized toward small triangular components, as outlined by Fig. 2. Triangular Lagrange finite elements about several forms are employed to each concerning the flow variables inside the computational area. Residuals regarding to any conservation equation hold achieved through replacing the approximations toward the dimensionless governing equations. For explaining the nonlinear expressions into equations of momentum, the algorithm of Newton-Raphson iteration is applied. The convergence concerning the solution does appropriate if the relative error of each variable meets the following convergence criteria:

$$\left|\frac{\Gamma^{i+1}-\Gamma^{i}}{\Gamma^{i+1}}\right| \leq \eta,$$

where i denotes the iteration number and η stands for the convergence criterion. In this numerical work, the convergence criterion was set at $\eta = 10^{-6}$.

Toward verifying the developed computational code, Fig. 3 explains a comparison of the used nanofluid models for the most pair influenced properties (thermal conductivity and the dynamic viscosity) between the present results and the experimental data of Chon et al. [12] and Ho et al. [13] and numerical one provided by Corcione et al. [14].

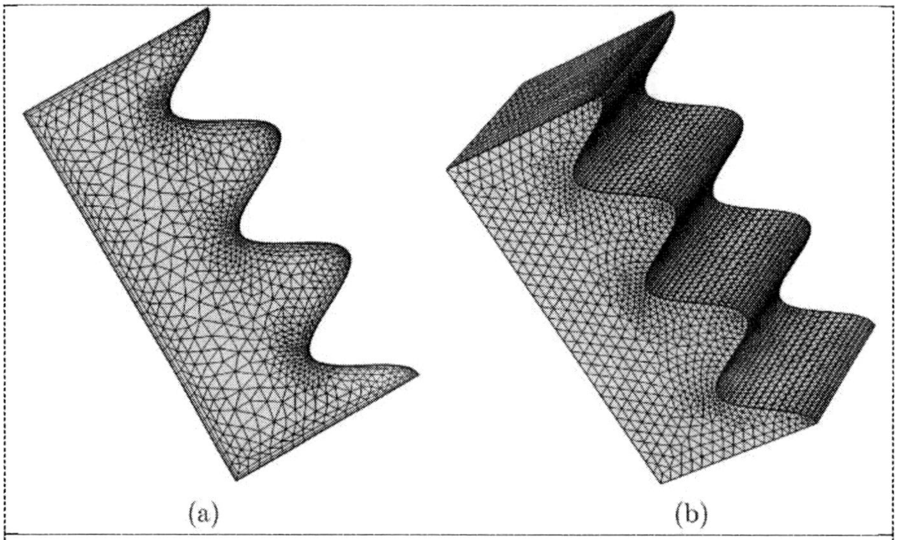

Figure 2. Grid-points distribution for (a) two-dimensional XY-plane and (b) three-dimensional cavity.

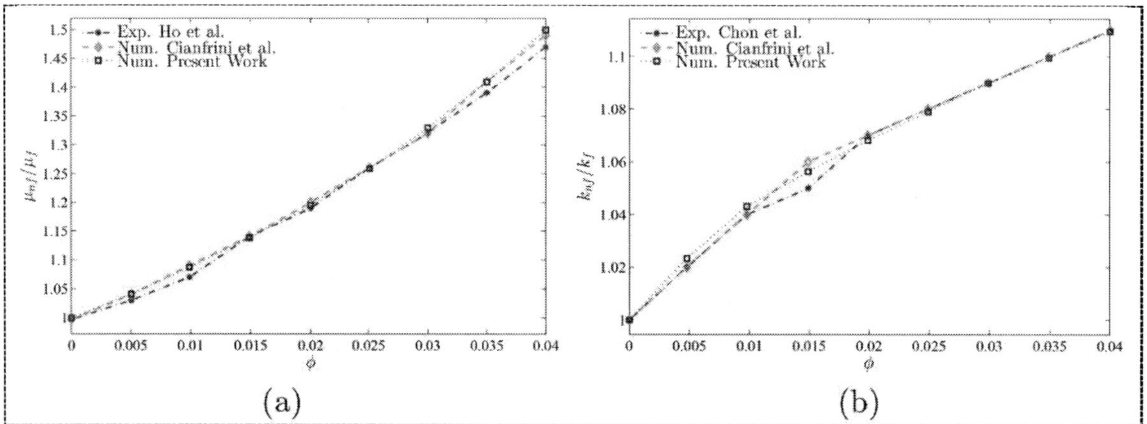

Figure 3. Validations concerning (a) thermal conductivity ratio of the present work among Chon et al. [12] and Corcione et al. [14] and (b) dynamic viscosity ratio by Ho et al. [13], Corcione et al. [14] and the present work.

3. Results and Discussion

This section of the work explains computational data of the stream function (streamlines), temperature profile (isotherms) and nanoparticles concentration for the inclination angle ($0 \leq \Omega \leq 120°$) and undulations number ($0 \leq N \leq 4$). The inclination angle in this study is applied to the considered cavity not to the flow only for indicating the clear effects of the orientation angle. Effects of other parameters, such as the dimensionless buoyancy-driven parameter (Rayleigh number) ($10^4 \leq Ra \leq 10^7$) and nanoparticle volume fraction ($0 \leq \phi \leq 0.04$) are presented. The thermophysical characteristics of water and solid Al_2O_3 phases are described in Table 1.

Table 1. Thermo-physical properties of water with Al_2O_3 nanoparticles at $T = 310K$.

Physical properties	**Fluid phase (water)**	**Al_2O_3**
C_p, (J/kg K)	4178	765
ρ, (kg/m^3)	993	3970
k, (W m^{-1} K^{-1})	0.628	40
$\beta \times 10^5$, (1/K)	36.2	0.85
$\mu \times 10^6$, (kg/ms)	695	-
d_p, (nm)	0.385	33

Figure 4 describes the 3D plot of differences regarding the flow, temperature field, and distribution of nanoparticles for several orientation angle at $Ra = 10^6, \phi = 0.02$ and $N = 3$. Generally, the warm fluid close to the wavy surface produces two streamlines cells with the absence of the orientation angle ($\Omega = 0$). Affected by the buoyancy forces, which is resulting in the imposed temperature gradient, streamlines cells appear in the anticlockwise direction, as clearly shown in Figure 4. Isotherms are highly destroyed within the middle of the cavity and next to the wavy wall, while vertical shape dominants on the isotherms patterns. Nanoparticles concentration show a strong behaviour on the top of the cavity due to the buoyancy forces. Imposing a low value of Ω tends to influence the fluid flow and temperature distribution. The second cell of the streamlines vanishes, and the flow circulation reduces. The isotherms patterns less destroyed within the centre of the cavity.

Figure 5 shows the 2D plot of differences regarding the flow, temperature field, and distribution of nanoparticles for several orientation angle at $Ra = 10^6, \phi = 0.02$ and $N = 3$. The anti-clockwise vortex (positive values of Ψ) of the streamlines takes place within the middle of the wavy chamber (please see the value of the streamlines colour bar). The strength of the flow circulation decreases with an increment of Ω. Rising the orientation angle up to 90° leads to a strong influenced on the flow and temperature field. Multiple streamlines cells (anti-clockwise and clockwise vortexes) appear next to the wavy surface, which indicated the changing of the heating system. The heating system, in this case, show a hot temperature from the top wavy surface and a cold one from the bottom surface. Due to that, the isotherms patterns occur with almost horizontal lines except the area next to the wavy surface. The concentration of nanoparticles show a clear propagation within the cavity which evaluated by the orientation angle (Ω) for $Ra = 10^6, \phi = 0.02$ and $N = 3$. Indicate the influence of the Brownian motion of thermophoresis. At the high orientation angle ($\Omega = 120°$), the direction of the fluid heat flow changes to the clockwise. The stream function takes a form of single clockwise vortex at the middle of the wavy chamber. The strong concentration of nanoparticles observed at the lower segment of the wavy.

Figures 6 and 7 explain variations of 3D and 2D plots of the streamlines, isotherms, and nanoparticle distribution for various undulations number (N) at $Ra = 10^6, \Omega = 30°$ and $\phi = 0.02$. For vertical flat surfaces ($N = 0$), the flow motions appear with a singular anti-clockwise vortex that drags diagonally within the centre of the chamber. Isotherms patterns observed with low distortion and almost a horizontal shape with the middle of the cavity. Weak concentration of nanoparticles is noted within the distribution evaluated by the orientation angle (Ω) for $Ra = 10^6, \phi = 0.02$ and $N = 3$ in XY-plane. With the considered cavity with very low propagation. Appling number of undulations on the right surface enhances the flow motions temperature distributions and nanoparticles distribution.

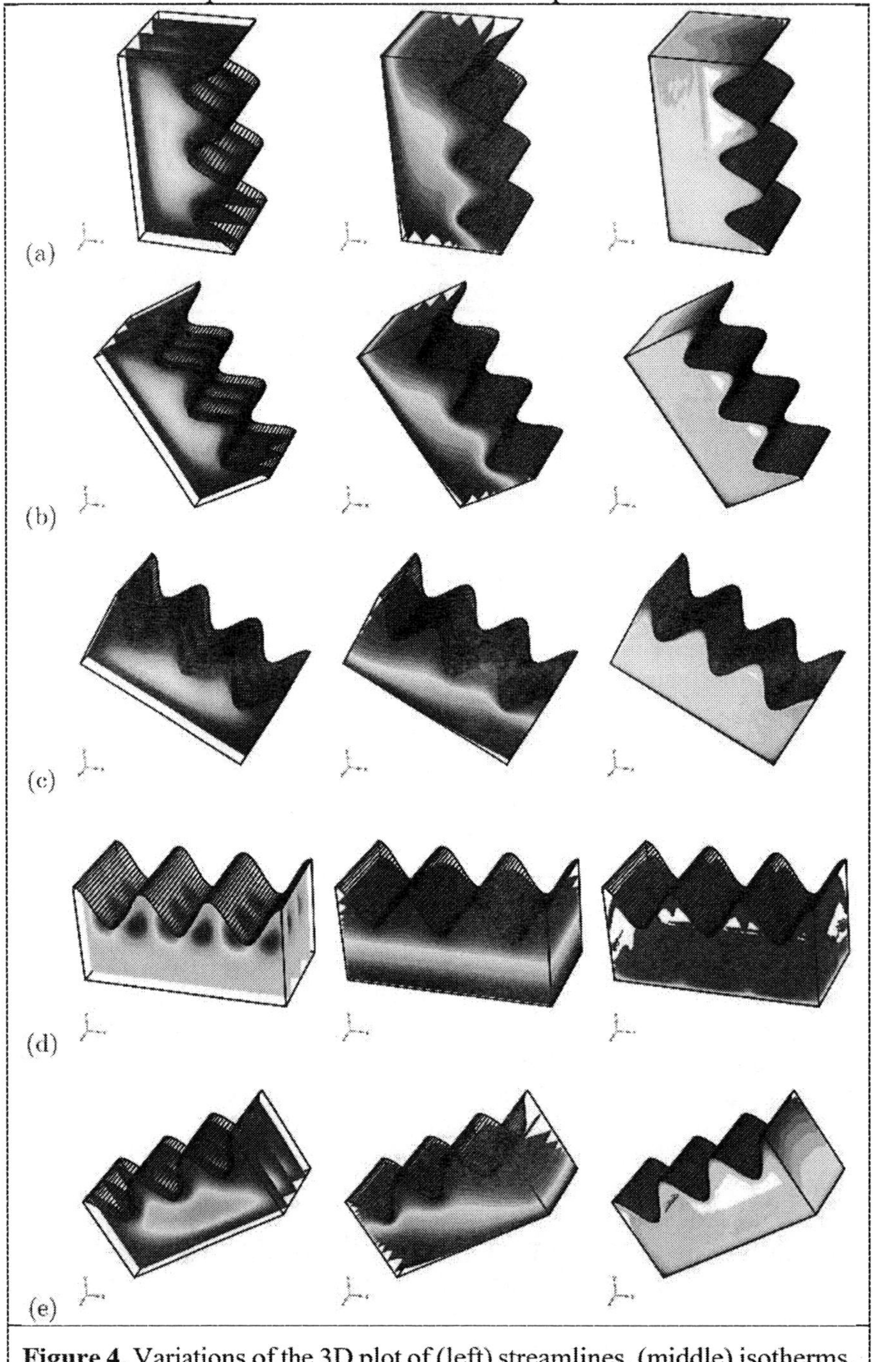

Figure 4. Variations of the 3D plot of (left) streamlines, (middle) isotherms, and (right) nanoparticle distribution evaluated by the number of undulations (N) for $Ra = 10^6, \Omega = 30°$ and $\phi = 0.02$.

The anti-clockwise vortex of the streamlines shrinks vertically due to the narrow space inside the cavity. A substantial enhancement is recognized toward the force of the flow circulation by the increasing of N; please see the values of the streamlines color bar in Figure 7. The isotherms patterns are highly destroyed, especially nearby to the wavy surface influenced by the intense thermal gradient. Increasing the number of undulations is to enhance the concentration of nanoparticles. Strong concentration of nanoparticles was indicated at the top of the cavity and near to the wavy surface. Besides, a high propagation of nanoparticles concentration is noted within the cavity. The strength of the nanoparticles increases from 0.035 to 0.054, with an increment of the number of undulations, as shown in Figures 7(c) and 7(d). The strength of the flow circulation shows a continuous enhancement with the rising of N which affected by the nanofluid intensity, and as a result, the fluid flow circulation enhances.

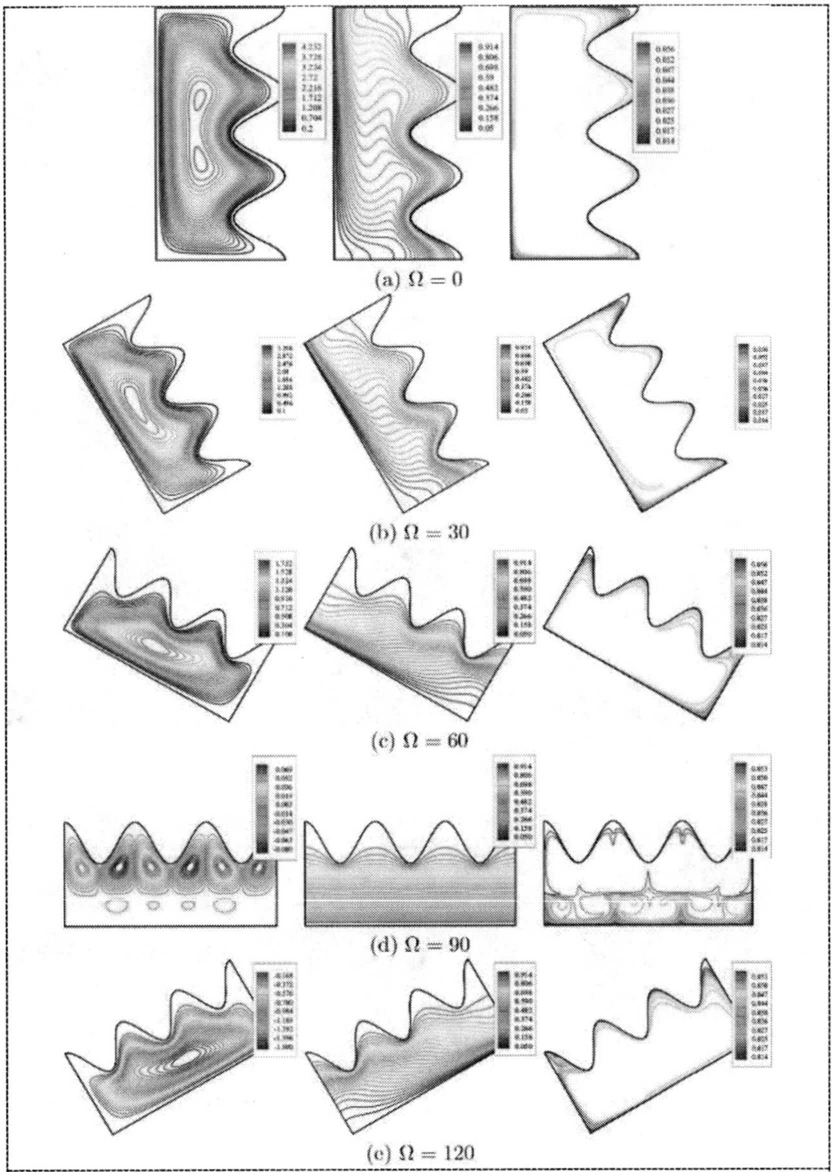

Figure 5. Variations of the 2D plot of (left) streamlines, (middle) isotherms, and (right) nanoparticle distribution evaluated by the orientation angle (Ω) for $Ra = 10^6, \phi = 0.02$ and $N = 3$ in XY-plane.

Figure 8 demonstrates differences of the local Nusselt number (Nu_{nf}) through the wavy surface (S) for diverse (a) Ω and (b) N at $Ra = 10^6$ and $\phi = 0.02$. The increasing of the orientation angle is enhanced the local heat transfer, as presented in Fig. 8(a). The strongest enhancement on the heat transfer is noted on the lower and upper corners of the wavy surface due to the large temperature gradient. At the lower corner from the curved wall, lower and higher amounts of Ω (0 and 120°) display a substantial augmentation in the local heat transfer. While the growth of the orientation angle up to 60° revels the rising on the local heat transfer at the upper corner of the wavy surface. Figure 8(b) shows a robust growth of the local heat transfer with the rise of the number of undulations. This due to the growth of the temperature gradient, which indications in the maximizing local heat transfer.

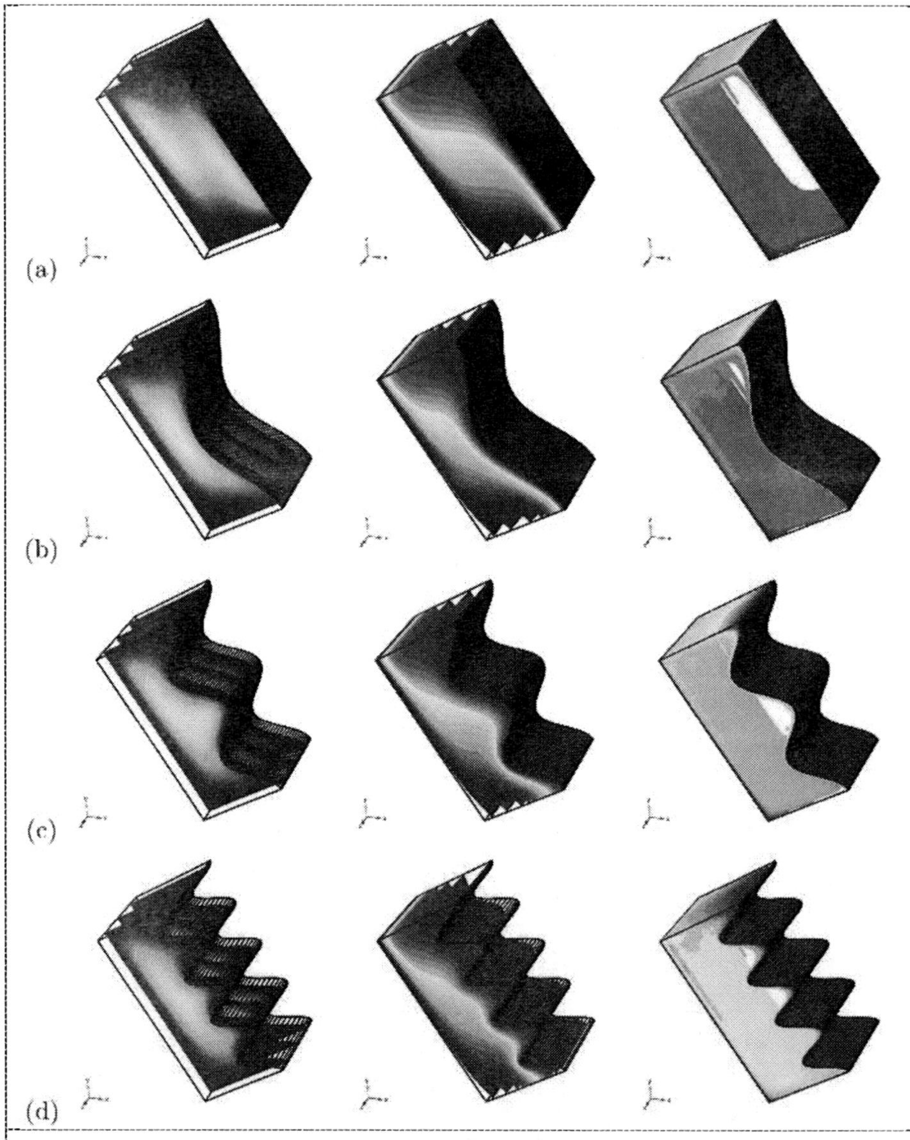

Figure 6. Variations of the 3D plot of (left) streamlines, (middle) isotherms, and (right) nanoparticle distribution evaluated by undulations number (N) for $Ra = 10^6, \Omega = 30°$ and $\phi = 0.02$.

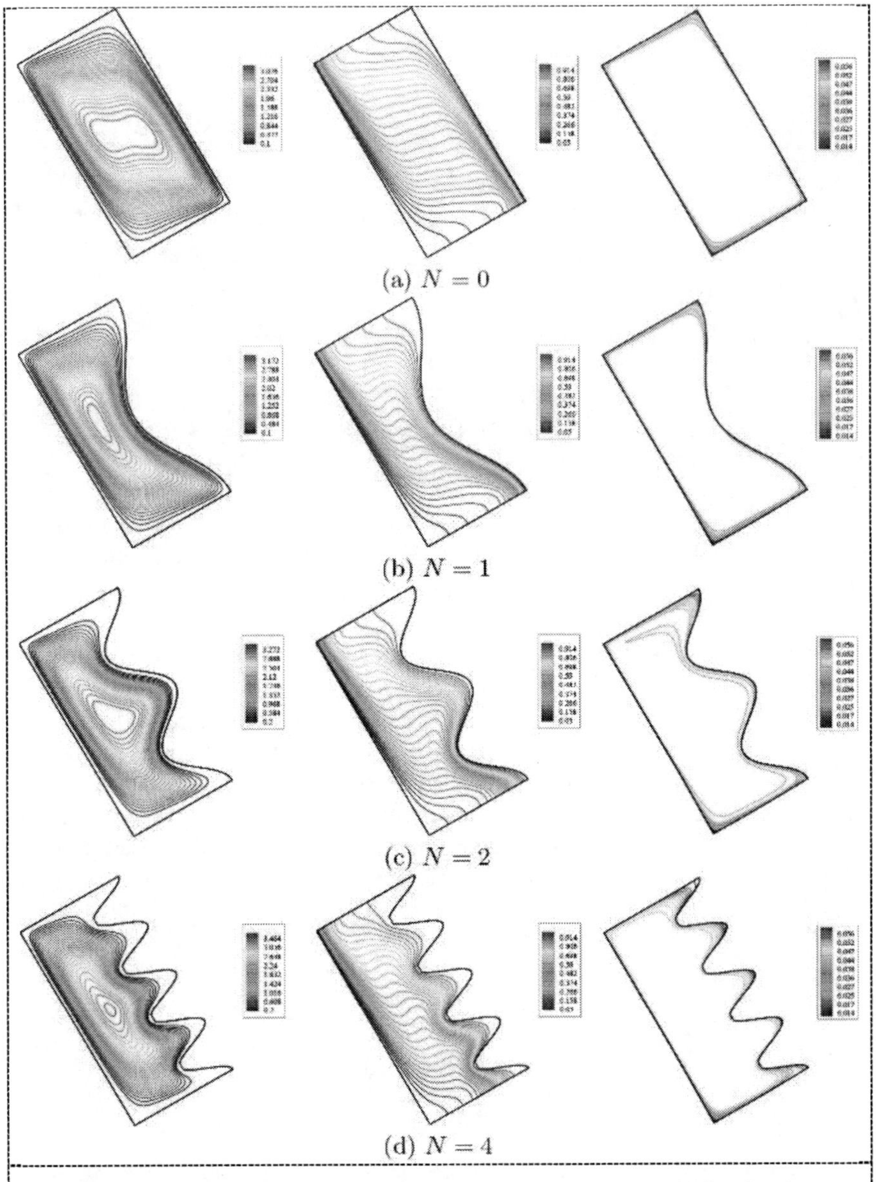

Figure 7. Varieties of the 3D plot of (left) streamlines, (middle) isotherms, and (right) nanoparticle distribution evaluated by undulations number (N) for $Ra = 10^6, \Omega = 30°$ and $\phi = 0.02$ in XY-plane.

Figure 9(a) clarifies the enhancement of the heat transfer among the increasing of the dimensionless buoyancy-driven parameter (Rayleigh number). The is because of the increasing of the thermal gradient following the rising of Rayleigh number. Increasing the number of undulations tends to increase the drag effect on the wavy segments, which raises the energy plus as a result; the rate of the heat transfer increments. This enhancement is distinctly obtained in Figure 9(b). The average Nusselt number seems with an improvement by sinusoidal plots with the increasing of the number of undulations. Also, an optimal enhancement is noted on the average Nusselt number with the increasing of the nanoparticle volume fraction (ϕ) up to 0.03.

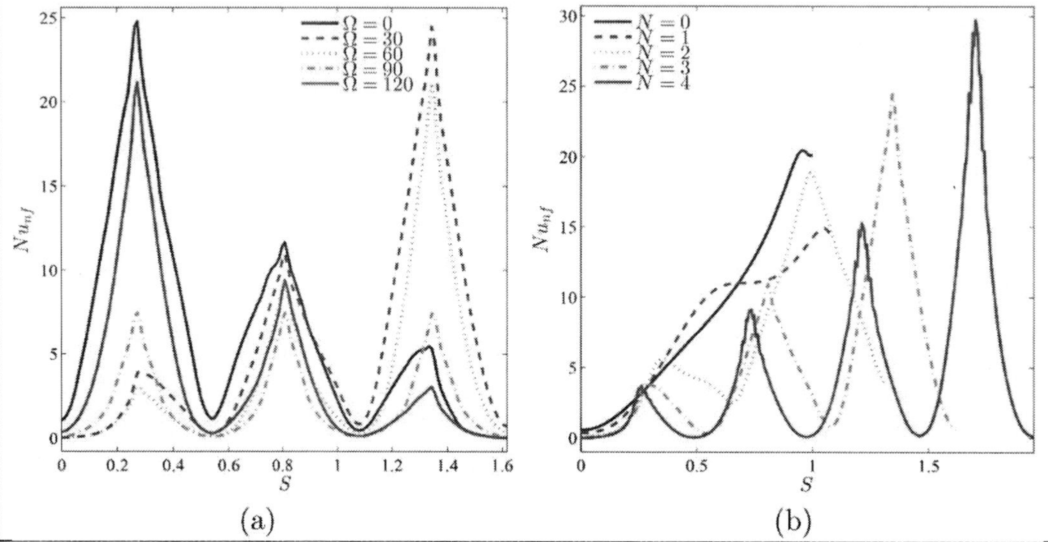

Figure 8. Differences of the local Nusselt number relates to the wavy surface (S) for different (a) Ω and (b) N at $Ra = 10^6$ and $\phi = 0.02$.

Figure 10 reports the variations of the heat transfer rate interface with the orientation angle for different (a) nanoparticle volume fraction and (b) number of undulations at fixed Rayleigh number ($Ra = 10^6$). The average Nusselt number shows a reducing function with the increasing of the orientation angle up to 90°. However, rising Ω more than 90° indicates an enhancement on the thermal gradient furthermore as a result; convective heat transfer raises. The convection heat transfer growths by the extension of ϕ due to the higher thermal conductivity of the nanoparticles. Additionally, the rise in the number of undulations revels a substantial enhancement on the overall heat transfer which observed with the ascension of the orientation angle more than 90°, as clearly shown in Figure 10(b).

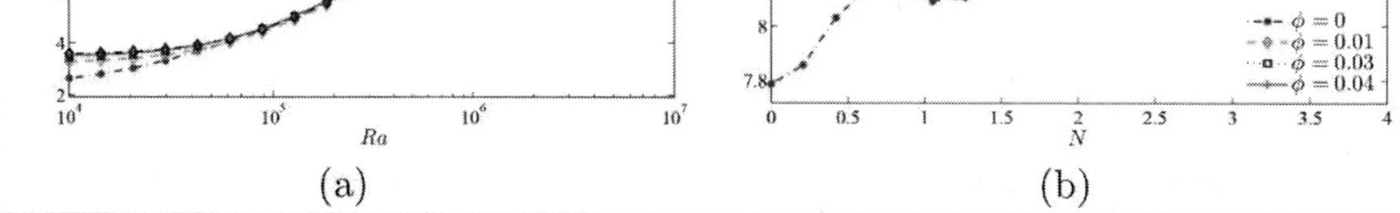

Figure 9. Differences of $\overline{Nu}_{nf}$ with (a) Ra and for various N, and (b) N for various ϕ at $\Omega = 30°$.

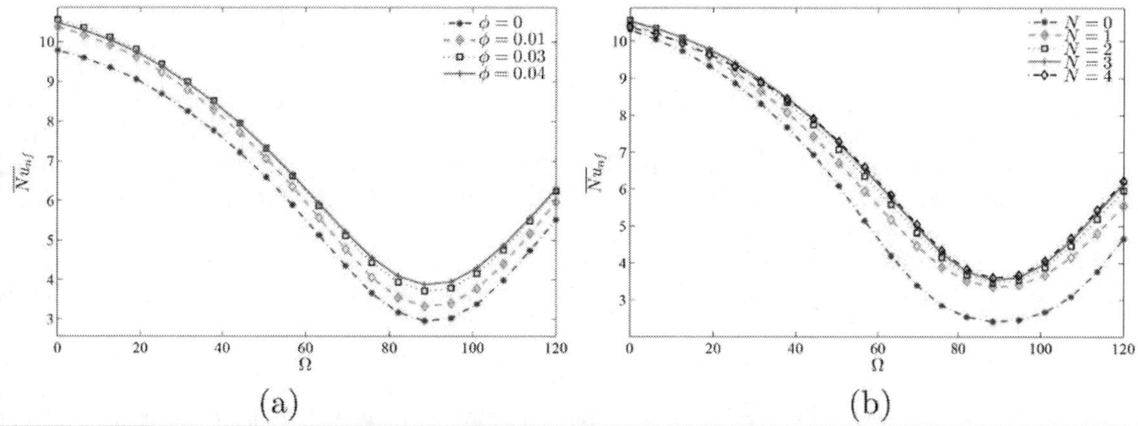

Figure 10. Differences of $\overline{Nu}_{nf}$ with Ω for various (a) ϕ and (b) N at $Ra = 10^6$.

4. Conclusions

The present study considered the steady laminar natural convection and heat transfer inside a tilted 3D wavy chamber packed by water-Al_2O_3 nanofluid. The governing equations are written in the dimensionless form by employing the two-component non-homogeneous approach toward conveying phenomena in nanofluids and solved numerically using the finite element method. It is determined that the intensity concerning the streamlines, isotherm patterns and nanoparticles distribution boosted with an increment of the number of undulations. The strongest enhancement on the local heat transfer is noted on the lower and upper corners of the wavy surface due to the large temperature gradient. The average Nusselt number appears with a clear reduction with the increasing of the orientation angle up to 90°. However, rising Ω more than 90° indicates an augmentation on the convection heat transfer. The overall heat transfer is increased by the addition of the nanoparticle volume fraction up to 0.03 under the influence of the number of undulations. Furthermore, it is shown that the inclination angle, the wavy shape of the surface including the features of nanofluid are outstanding control parameters for the rate of the heat transfer plus the distribution of the fluid flow.

Acknowledgments

The work was supported by the Universiti Kebangsaan Malaysia (UKM) research grant DIP-2017-010.

References

[1] S. Ostrach, "Natural convection in enclosures," Journal of Heat Transfer, **110**(4b), pp.1175–1190, 1988.
[2] Fusegi, T, Hyun, J.M, Kuwahara, K, and Farouk, B, 1991, A numerical study of three-dimensional natural convection in a differentially heated cubical enclosure, International Journal of Heat and Mass Transfer, **34**(6), pp.1543-1557.
[3] Wang, P., Zhang, Y. and Guo, Z., 2017. Numerical study of three-dimensional natural convection in a cubical cavity at high Rayleigh numbers. International Journal of Heat and Mass Transfer, 113, pp.217-228.
[4] Gibanov, N.S. and Sheremet, M.A., 2018. Natural convection in a cubical cavity with different heat source configurations. Thermal Science and Engineering Progress, **7**, pp.138-145.
[5] Gibanov, N. and Sheremet, M.A., 2019. Effect of trapezoidal heater on natural convection heat transfer and fluid flow inside a cubical cavity. International Journal of Numerical Methods for Heat & Fluid Flow, **29**(4), pp.1232-1248

[6] Alsabery, A.I, Yazdi, M.H, Altawallbeh, A.A, and Hashim, I, 2018, Effects of nonhomogeneous nanofluid model on convective heat transfer in partially heated square cavity with conducting solid block, Journal of Thermal Analysis and Calorimetry. https://doi.org/10.1007/s10973-018-7789-3.

[7] Alsabery, A.I, Sheremet, M.A, Chamkha, A.J, and Hashim, I, 2019, Impact of nonhomogeneous nanofluid model on transient mixed convection in a double lid-driven wavy cavity involving solid circular cylinder, International Journal of Mechanical Sciences, **150**, pp.637-655.

[8] Khanafer, K, Vafai, K, and Lightstone, M, 2003, Buoyancy-driven heat transfer enhancement in a two-dimensional enclosure utilizing nanofluids, International Journal of Heat and Mass Transfer, **46**(19), pp.3639-3653.

[9] Al-Rashed, A.A., Kalidasan, K., Kolsi, L., Aydi, A., Malekshah, E.H., Hussein, A.K. and Kanna, P.R., 2018. Three-dimensional investigation of the effects of external magnetic field inclination on laminar natural convection heat transfer in CNT–water nanofluid filled cavity. Journal of molecular liquids, 252, pp.454-468.

[10] A. Shenoy, M. Sheremet, and I. Pop, Convective Flow and Heat Transfer from Wavy Surfaces: Viscous Fluids, Porous Media, and Nanofluids. CRC Press, 2016.

[11] Hashim, I, Alsabery, A.I, Sheremet, M.A, and Chamkha, A.J, 2019, Numerical investigation of natural convection of Al_2O_3-water nanofluid in a wavy cavity with conductive inner block using Buongiorno's two-phase model, Advanced Powder Technology, **30**(2), pp.399-414.

[12] Chon, C.H, Kihm, K.D, Lee, S.P, and Choi, S.U, 2005, Empirical correlation finding the role of temperature and particle size for nanofluid (Al_2O_3) thermal conductivity enhancement, Applied Physics Letters, **87**(15), p.153107.

[13] Ho, C.J, Liu, W.K, Chang, Y.S, and Lin, C.C, 2010, Natural convection heat transfer of alumina-water nanofluid in vertical square enclosures: an experimental study, International Journal of Thermal Sciences, **49**(8), pp.1345-1353.

[14] Corcione, M, Cianfrini, M, and Quintino, A, 2013, Two-phase mixture modeling of natural convection of nanofluids with temperature-dependent properties, International Journal of Thermal Sciences, **71**, pp.182-195.

Numerical Investigation of Thermal Performance for Air Solar Collector with Multi Inlets

Jalal M. Jalil

Adel H. Ayaal

Abdulwahhab A.Hardan

Electromechanical Engineering Department, University of Technology, Baghdad, Iraq

Corresponding author's e-mail : 50249@uotechnology.edu.iq

Abstract. Over the years, there has been rigorous modification of solar collectors for the thermal needs of the era to be efficiently served. A variety of design innovations have paved their way into the invention of novel ways of securing more solar energy as useful heat. Primarily, this study investigates the effects of multi air inlets on the transfer of heat between air flow and the absorber of flat plate collector. Computational Fluid Dynamics was employed to simulate a flat plate collector with 1, 2 and 3 inlets to evaluate the thermal performance with more than one inlet. The results showed that the 3 inlets case has the highest exit temperature followed by 2 inlets and then 1 inlet. Also higher exit air temperature can be obtained for higher heat flux and slower inlet velocity.

Keywords: Keywords: multi-inlets; CFD; solar collector; flat plat collector; heat transfer.

Nomenclature

T	Temperature (K)
ΔT	Wall temperature difference.
U	Velocity in x direction (m/s)
V	Velocity in y direction (m/s)
W	*V*elocity in z direction (m/s)
x	Coordinate (m)
y	Coordinate (m)
z	Coordinate (m)

Greek symbols

Φ	Dependent variable used in discretization equation.
Γ_{Φ}	Diffusion coefficient used in discretization equation.
ρ	Density of the fluid (kg/m³)
$\partial_x, \partial_y, \partial_z$	grid space (m)

1. Introduction

Through utilization of solar energy, solar air heaters are employed in a variety of applications, space heating and crop drying for instance, requiring heat of low to moderate temperatures below 60 °C. The main heaters are the single pass with rear duct, front duct, double duct and double pass [1] . Ambient temperature, solar radiation, and inlet temperature strongly affect the performance of the solar collector. There has been no substantial progress for cooling systems despite previous research in FPCS focusing on heating applications. In most of the research in this field, inlet temperatures were within range of ambient temperature in solar heating systems [2]. In actuality, the thermal efficiency of a simple solar air collector is found to be very poor. This is due to the low convective heat transfer coefficient of air as well as the thermos-physical property of air. Consequently, viscous sub-layer formation appears on the absorber which is resistant to the heat transfer rate. In light of this, many techniques are available, active or passive, to enhance the efficiency of the solar collector. A multi air flow inlet could be used which results in the enhancement of thermal heat transfer. The enhancement is achieved through increasing the temperature difference between the collector and fluid (inlet air) [3].

A number of researchers are aiming to refine the air heater by incorporating the flat plate collector with packed bed and energy storage systems. The performance of FPC was investigated by Zhang et al. [4]. Their findings were used to verify a mathematical method used to calculate the FPC collector's thermal performance. Geometrically speaking, the width of the collector was 1m, 2m in length; with a thickness of 0.065m. Researchers found that the mass flow rate was a central component impacting performance and outlet temperature. The temperature of the inlet ranged between 21.1 and 45.1 °C. The ambient temperature was between 15.5 and 23.9 °C, and the outlet temperature was in the scope of 37 to 55.4 °C. Mohammad et al [5] investigated the effects of flow rate (m˙) and inlet temperature (Tin) on thermal efficiency (hth) of flat plate collectors (FPC). Computational Fluid Dynamics (CFD) was used to stimulate an flat plate collectors (FPC), and experimental data derived from literature was used to authenticate the results. An examination of the high and low level flow rates in addition to the inlet temperatures (298 to 373 K) was conducted in the FPC. At 298 K and 370 K, there was a thermal efficiency of 93% and 65% respectively. The thermal efficiency of a double pass solar air heater was researched by Gonzalez et al., [6]. The inlet's air temperature rose to 35°C with a solar radiation of around 900 W/m2. It was concluded that the average daily efficiency is 34%. A study by Kareem et al [7] presents the performance of a forced convective multi-pass solar air heating collector (MPSAHC). Guillermo et al [8] brought forth the notion of flexible operation while designing the solar thermal utility systems for low temperature processes. The purposes of the design were: (A), Supply of the process's thermal needs (minimum required temperature and heat duty) and (B), Inflation of the operating time during the day. This method showed that the network structure is defined by altering the mass flow rate and the inlet temperature of the working fluid to obtain the smallest collector surface area. The study conducted by Foued et al. [9] examined the solar performance of the double flow and found that double pass operation increases solar collector efficiency. Using a single pass solar with double flow enhances the thermal system's performance which may be higher than the single flow model based on the same rate of flow (Ozgen et al.) [10].

Gleaning from previous research in solar air collector, there are no studies related to use of multiple air inlets. In this paper, the influence of multiple air inlets (1, 2 and 3) on thermal performance of air stream flowing inside collector with different locations will be studied with different velocities and solar flux.

IOP Conf. Series: Materials Science and Engineering **765** (2020) 012036 doi:10.1088/1757-899X/765/1/012036

2. Numerical Formulation

The collector has one, two or three inlets. Glass cover on the top and collector walls are insulated on the other sides. The mathematical formulation of the airflow problem is governed by basic conservation of mass, momentum and energy equations. In the present study, the flow is turbulent according to Reynolds number values, (Re > 2800).
Assumptions:
1- Three-dimensional (3D) of conservation equations.
2- Steady state incompressible flow.
3- All the properties are calculated at an average temperature

2.1. Governing Equations

Numerically solving the governing 3D elliptic PDEs (partial differential equations) allowed the predications of turbulent flow field duct to be obtained. FORTRAN program was used to conduct simulations. Using the Boussinesq approximation, the fluid flow in a duct equations were calculated. The following form usually applies to generic partial differential equations, otherwise known as transport equations, for turbulence scales k, ε, temperature, momentum, and continuity:

$$\frac{\partial}{\partial x}(\rho U\phi)+\frac{\partial}{\partial y}(\rho V\phi)+\frac{\partial}{\partial z}(\rho W\phi)=\frac{\partial}{\partial x}\left(\Gamma_\phi\frac{\partial\phi}{\partial x}\right)+\frac{\partial}{\partial y}\left(\Gamma_\phi\frac{\partial\phi}{\partial y}\right)+\frac{\partial}{\partial z}\left(\Gamma_\phi\frac{\partial\phi}{\partial z}\right)+S_\phi$$

Convection terms are displayed by the three expressions shown on the left, whereas on the right hand side, the four expressions show source expression and diffusion. The diffusion coefficient Γe, (diffusivity) is stated in the following equation:

$$\Gamma_e=\frac{v_e}{\sigma_e}+\frac{v_t}{\delta_t}$$

Here, the effective Prandtl number is σe and this includes the turbulent dynamic viscosity as well as the turbulent diffusion coefficient. Three kinds of inlet (one, two and three) were studied. The duct length was 1.5 m. The cross section area was 0.8 * 0.04 m2 and the absorber area is 1.5 * 0.8 m2.

2.2. Boundary conditions

At the outlet, inlet, and walls, the conditions of boundary applied to the collector must be categorised. Specification details are shown below. Furthermore, All scalars and vectors variables (U, V, W, T, k and ε) arrangements' estimates can be seen, as well as the inlet's conditions of boundary for the CFD domains in the following equation:

$$U(0,y,z)=U_{in}\,,\quad V(0,y,z)=W(0,y,z)=0\,,\quad T(0,y,z)=T_{in}$$

A change was applied in the values (1.5, 2.0 and 2.5 m/s) of inlet air velocity U. Setting the normal gradients to zero is common practice and the outlet boundary condition can be presented as follows:

$$\frac{\partial \phi(x, y, z)}{\partial x} = 0$$

For the absorber and glass, 3d energy equation in solid was used to calculate the temperature distribution in the class and absorber. Three cases will be examined with the same mass flow rate. Case 1 with one inlet, case 2 with two inlets and case 3 with three inlets.

3. Result and discussions

The current program used in this paper was validated for different cases in previous studies so it used in this paper without validation. Fig. 1 shows the flow field for three cases, 1, 2 and 3 inlets. The building of the boundary layer is clear in the three plots. In one inlet, the boundary layer was built early, while in two inlets, the building of the boundary layer was delayed in the lower surface because of the new fresh air inlet. In three inlets, more delay was observed in the lower surface because of two new fresh inlets in the second and third inlet.

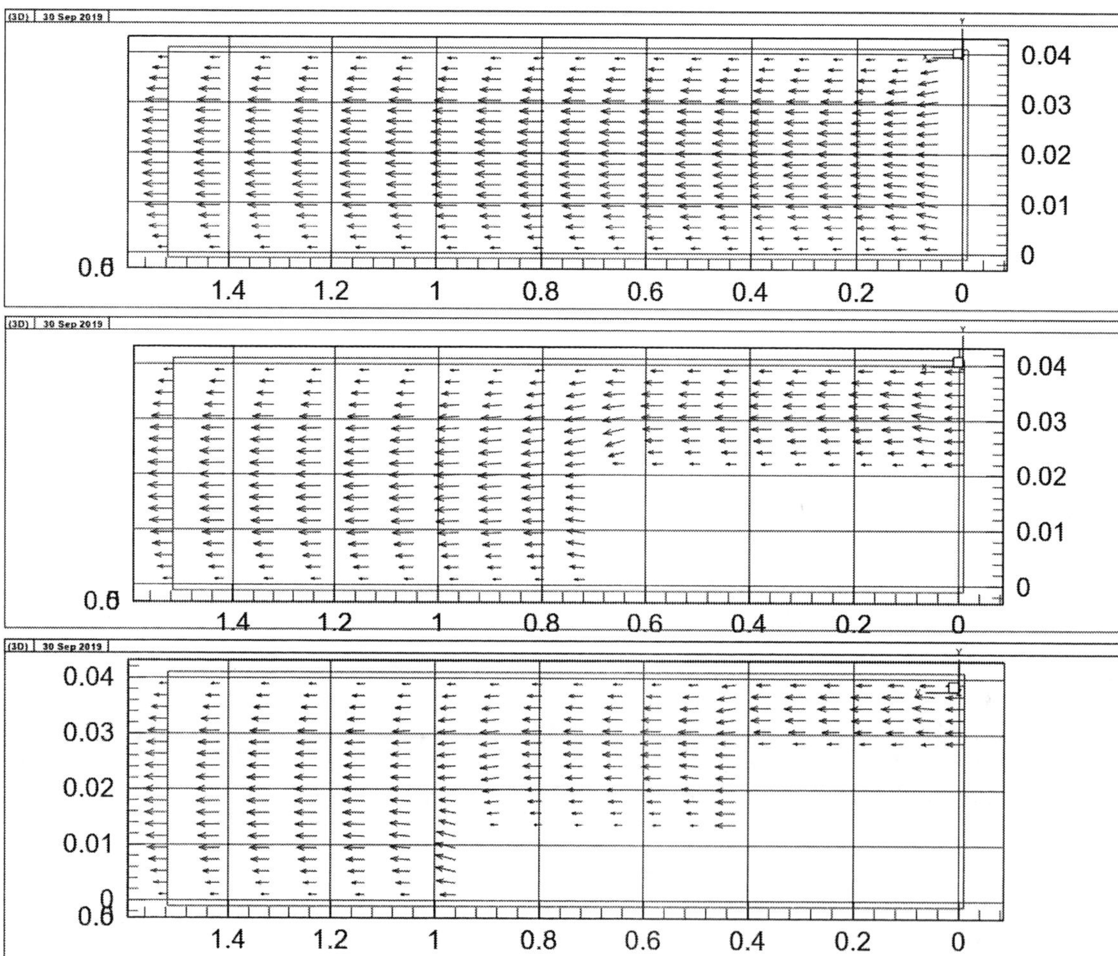

Figure 1 Flow Field for 1, 2 and 3 inlets for 1.5 m/s.

Fig. 2 indicates the variation of temperature along the length of the 2 inlets for three different velocities 1.5, 2.0 and 2.5 m/s. The behaviour is clear, the temperature increases along the length until the position of the second inlet where the temperature decreases sharply because of the new fresh air, then it increases with high rate till the end of the length. Lower velocity 1.5 m/s will usually have highest temperature than the higher velocity.

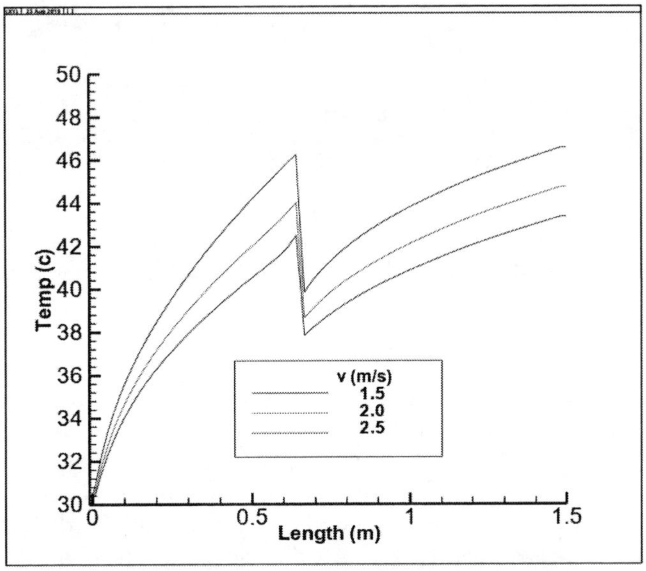

Figure 2 Temperature variation along the length of the collector for different velocities (1.5, 2, 2.5 m/s) at 700 W/m2 and 2 inlets.

Fig. 3 shows the variation of temperature along the length for 2 inlets and 1.5 m/s for three different values of solar irradiations 500, 700 and 900 W/m2. The behaviour is similar to Fig. 2, the temperature increases until the starting of the second inlet where the temperature decreases sharply because of the entrance of the fresh air. Higher solar irradiation usually has highest temperature so 900 W/m2 will have highest exit temperature than 500 and 700 W/m2.

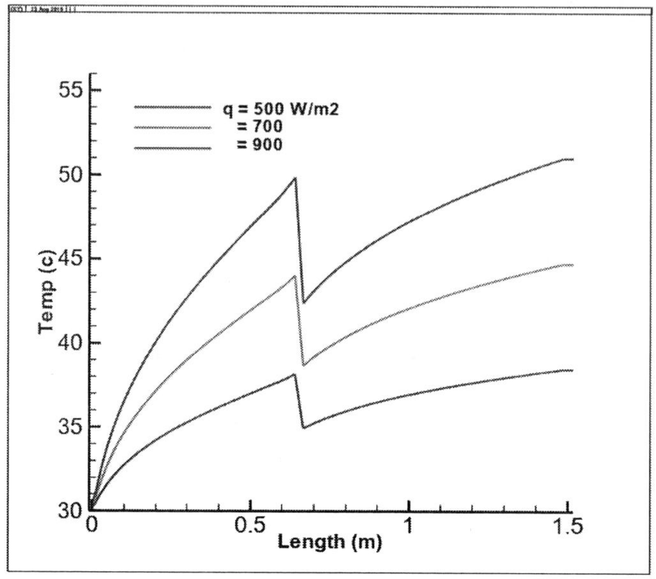

Figure 3 Temperature variation along the length of the collector for 1.5 m/s and different solar flux (500, 700 and 900 W/m2).

Fig. 4 shows the variation of temperature along the length for 1.5 m/s and 500 W/m2 for three different cases (1, 2 and 3 inlets). The purpose of this research was appeared in this figure. In the beginning, 3 inlets case has highest temperature then 2 inlets case and then 1 inlet case because of the small value of flow rate of the three inlets case comparing to two and one inlets cases. Low value of mass flow rate means more heat can diffuse into the core of the flow. In each new inlet, a new fresh air will disturb the boundary layer and allows more heat can diffuse into the core of the flow and the temperature increases sharply. At the end, 3 inlet case has highest exit temperature the other two cases. Similar behaviour in Fig. 5 for 2.0 m/s and 700 W/m2 and Fig. 6 for 2.5 m/s and 900 W/m2.

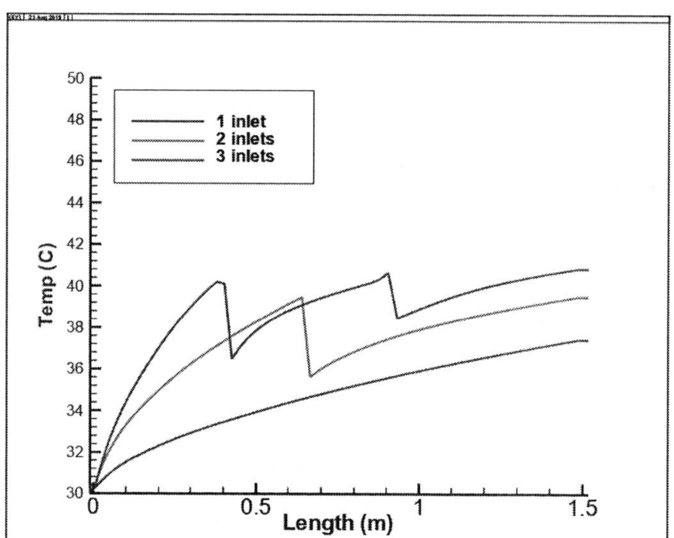

Figure 4 Temperature variation along length of the collector for the three cases (1, 2 and 3 inlets) for 1.5 m/s and 500 W/m2.

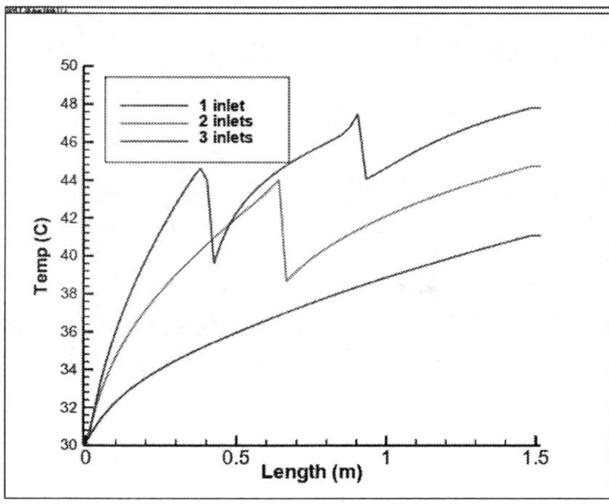

Figure 5 Temperature variation along length of the collector for the three cases (1, 2 and 3 inlets) for inlet velocity 2.0 m/s and 700 W/m2.

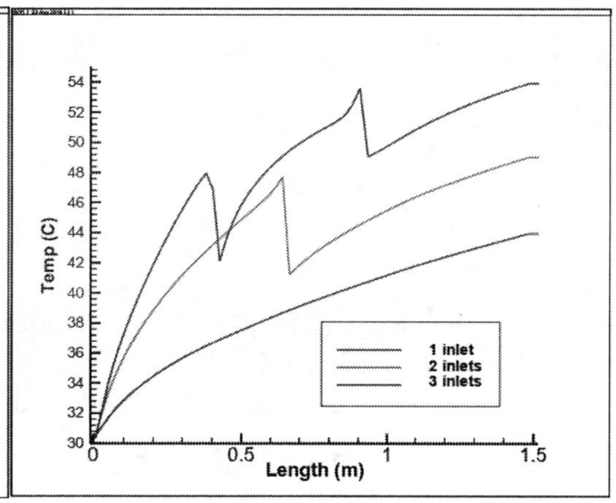

Figure 6 Temperature variation along length of the collector for the three cases (1, 2 and 3 inlets) for inlet velocity 2.5 m/s and 900 W/m2.

4. CONCLUSIONS

1. Air temperature at the exit is increases with higher solar flux and lower velocity.
2. Divided the main inlet into two or three inlets will improve thermal performance of the collector so higher exit air temperature will be obtained.

References

[1] Irfan K. and Emre T. (2006)'Experimental Investigation of Solar Air Heater with Free and Fixed Fins: Efficiency and Exergy Loss" International Journal of Science & Technology Volume 1, No 1, 75-82,.

[2] Calise, F.; d'Accadia, M.D.; Palombo, A.; Vanoli, L. (2013).Dynamic simulation of a novel high-temperature solar trigeneration system based on concentrating photovoltaic/thermal collectors. Energy, 61, 72–86.

[3] Shashank D. Bhujade1, R. S. Shelke, , (2017).Studies on Shapes of Turbulators on the Absorber Plate of Solar Air Heater System International Journal of Engineering Science and Computing, Volume 7 Issue No.12.

[4] Zhang, D.; Li, J.; Gao, Z.;Wang, L.; Nan, J. (2016). Thermal performance investigation of modified flat plate solarcollector with dual-function. Appl. Therm. Eng. 108, 1126–1135.

[5] Mohammad A., Ben H., Andrew H. and Dominic O. (2018). Determining the Effect of Inlet Flow Conditions on the Thermal Efficiency of a Flat Plate Solar Collector, Fluids, 3, 67; doi:10.3390/fluids3030067.

[6] Gonzalez, S..Thermal, (2012). Evaluation and Modeling of Double-pass solar Collector For Air Heating. Conference, opportunities, limit & Need Towered an environmental responsible, 7-9.

[7] Kareem MW, Habib K, Ruslan MH, Saha BB, (2017). Thermal performance study of a multi-pass solar air heating collector system for drying of Roselle (Hibiscus sabdariffa), Renewable Energy, doi: 10.1016/j.renene.2016.12.099.

[8] Guillermo M., Amanda L., Juan R. and Martín P., (2019). Incorporating the Concept of Flexible Operation in the Design of Solar Collector Fields for Industrial Applications Energies, 12, 570; doi:10.3390/en12030570.

[9] Foued, C., Noureddine, M., Abdelhafid, B., & Said, B. (2013). Thermal Efficiency Analysis of a Single Flow Solar Air Heater with Different Mass Flow Rates In A Smooth Plate. Frontiers in Heat and Mass Transfer. http://dx.doi.org/10.5098/hmt.v4.1.3006

[10] Ozgen, F., Esen, M., & Esen, H. (2009). Experimental Investigation of Thermal Performance of a Double-Flow Solar Air Heater Having Aluminium Cans. Renewable Energy, 34, 2391-2398. http://dx.doi.org/10.1016/j.renene.2009.03.029.

Numerical Investigation of Thermal Performance of Micro-Pin Fin with Different Arrangements

Jalal M. Jalil, Ahmed H. Reja, Amro M.Hadi
Electromechanical Engineering Department, University of Technology, Iraq, Baghdad
E-mail: jalalmjalil@gmail.com

Abstract. The impact of micro-pin fins on heat transfer performance in this study is analyzed numerically. Navier-Stokes equations are solved by steady-state three dimensions turbulent forced convection. Various numbers of micro-pin fins with in-line configuration and different micro-pin fins configurations such as staggered configuration additionally to the in-line configuration are explored. In addition, comparison between micro-pin fins duct and smooth duct is studied. Results illustrate that the thermal performance was enhanced by 7.9 - 9.3% with increasing the number of micro-pin fins from 902 to 1312 and by 10.2 - 11.1% with increasing the number from 902 to 1640. Also, displays that the heat transfer performance in the staggered configuration is slightly better than the in-line configuration by 2.6 - 4.2% with the same micro-pin fins number. Moreover, the micro-pin fins duct gave an increase of 30.24% in heat transfer performance relative to the smooth duct.

Keywords: Micro-pin fins, Micro cooler, Micro heat sink, Passive cooling, Micro cooling techniques.

Nomenclature:
x, y, z: Cartesian coordinates.
ρ: *Density of air.*
U, V, W: Velocities in X,Y,Z directions.
Γ: *Diffusion coefficient.*
φ: *General dependent variable.*
T_p: *Plate temperature.*
i, j, k: Indices which indicates positions in the (x,y,z) directions.
A: Area.
W: Relaxation factor.

1. Introduction

Micro-pin fins are a passive technique used to improve the heat transfer efficiency of mechanical or electronic constructions. Micro-pin fins are used to dissipate heat produced by mechanical or electronic devices into a fluid environment, often air or liquid coolant, to keep the optimum temperature [1]. In the domain of passive cooling devices, micro-pin fins are the most differentiated technique that can intervene with fluid and dissipate the boundary layer. In addition, small working fluid inventory requirements [2] with low cost. Micro-pin fins use in semiconductor industry and to reduce the temperature from projector and LED. Also to minimize the temperature of graphic chip and high power laser [3],[4] . There are various researchs in the field of micro-pin fins, **Mei et al. [5]** researched the impact of tip clearance in micro-reactor on both pressure drop and thermal efficiency of micro-reactor with micro-pin fins numerically and experimentally. This study was done to improve hydrodynamic efficiency at small numbers of Reynolds. The results illustrated that the pressure drop was enhanced when the tip clearance increased. It was also discovered that thermal efficiency has a inverse relationship with tip clearance. Moreover, the comparison between micro-pin fins with and without tip clearance displayed that the thermal efficiency of the hc / hf ratios equal to 1.1 or 1.2 micro-pin fins was better than that of the Reynolds number without tip clearance between 33 and 350. **Ndao et al. [6]** investigated Experimentally four distinct forms (square, circular, elliptical, hydrofoil) were used to show the impact of both distinct micro-pin fins forms and cross-flow area on the heat efficiency of jet impingement with micro-pin fins. As for Re, the working fluid (R134a) Re ranged from 8000 to 80000. Results have shown that the cross-flow area heat enhancement impact can be seen at elevated Re, but the main impact in the experiment was on the distinct micro-pin fin forms, and this was obvious when contrasting micro-pin circular and square fins with other micro-pin fin configurations. **Yeom et al. [7]** experimentally and numerically analyzed the thermal performance impacted by the use of both active and passive methods in a single channel. Active component depicted by a piezo-electric temporary agitator used to produce turbulent air using a blade of oscillation. As far as active method is concerned, eleven samples with two instances of distinct dimensions (H250D400) and (H400D250) were used, representing a passive cooling method. Results indicates that the operation of micro finned surface and agitator improved thermal transfer efficiency in various proportions. In the case of the use of a 920 Hz agitator with micro-pin fins with dimensions (H400D250), the heat transfer coefficient was increased by almost 230 percent compared to the non-agitator flow rate of 40-60 LPM. Almost 250 percent enhanced in the plain wall as well. **Guan et al. [8]** experimentally tested the effect of three different shapes of micro-pin fins, elliptical shape diamond shape, and circular hydrophobic shape on both pressure drop and friction factor, using ultra-pure water as a working fluid. Four specific contact angles (99.5, 119.5, 151.5) were used in this study. **Hua et al. [9]** experimentally investigated the impact on flow resistance, friction factor (FF) and pressure drop (ΔP) of different micro-pin fin shapes and sizes when high-pressure nitrogen flows through it. **Qian et al. [10]** studied the impact of a non-uniform micro-pin fins array (NMPFAR) contrast to the impact of a uniform micro-pin fins array (MPFAR) on both thermal performance and flow field was examined numerically and experimentally. In addition, the effect of distinct (NMPFAR) in three cases with three types of diameter (two, four, eight) on heat transfer performance was evaluated numerically. **Chiu et al. [4]** numerically and experimentally analyzed the impact of micro-pin fins on the cooling effectiveness of the liquid. Aluminum models of the micro-pin fins board have been manufactured to evaluate both the thermal performance and the flow features. The effects of both pressure drop and geometrical parameters on thermal performance are also studied. **Jia et al. [11]** studied the impact of micro-channel integration with micro-pin fins on both fluid flow and heat efficiency at Re between 147 and 637. Four instances of cone-shaped micro-pin fins are evaluated in the upper, center, downstream, and sub divided across the micro-channel. The main factors are the relative diameter of the fins, relative fin space, and relative fins height. **Naphon and Wiriyasart [12]** investigated experimentally the effect of many parameters such as pulsating flow, nanofluids, and magnetic field on both flow characteristics and thermal performance of micro-finned tubes comparison with smooth tubes. All Parameters illustrated direct relationship with the thermal performance. **Li et al. [13]** studied the effect of integrating the micro pin fins and piezo electric fan on flow and heat transfer performance of heat sink experimentally. Significant factors are taken into considerations in this study like vibrating fan's Reynolds number, the ratio of fan tip to the heated surface, and Re of channel flow. **Ambreen et al. [14]** showed numerically the impact of two parameters (distinct forms of micro-pin fins and nano fluid) on the heat transfer performance of the micro-pin fins surface with the introduction of a discrete phase model (DPM). **Kewalramani et al. [15]** analyzed experimentally and numericallythe hydrothermal performance by applied constant heat flux by a heater from the base on elliptical micro pin fin heat sink. The area of the heat sink is 11 mm length and 10 mm width contains twenty-five elliptical micropinfins(five columns and five rows) having 2.02 mm major axis and 1.84 mm minor axis with 130 μm height.

In the present study, the impact of different micro-pin fins numbers and configurations on the thermal performance and flow field are studied numerically. In addition, the impact of the micro-pin fins duct and smooth duct on the heat transfer performance are compared numerically.

2. Mathematical Model and Numerical Analysis

The governing three-dimensional partial differential equations were numerically analyzed to predict the behavior of heat transfer and turbulent flow through the micro-pin fins duct. The governing equations that categorize heat transfer and fluid flow in the micro-pin fins are focused on mass, momentum and energy conservation formulas.

2.1. General Formula for Governing Equations

The general form for all partial differential equations is :

$$\frac{\partial}{\partial x}(\rho U\varphi) + \frac{\partial}{\partial y}(\rho V\varphi) + \frac{\partial}{\partial z}(\rho W\varphi) = \frac{\partial}{\partial x}(\Gamma_\varphi \frac{\partial \varphi}{\partial x}) + \frac{\partial}{\partial y}(\Gamma_\varphi \frac{\partial \varphi}{\partial y}) + \frac{\partial}{\partial z}(\Gamma_\varphi \frac{\partial \varphi}{\partial z}) + S_\varphi \quad (1)$$

The first term on the left is the convection term while the right term is the diffusion and source expressions

2.2. Thermal Analysis of Solid Walls

Solid walls with micro-pin fins is emulated with a steady-state3-dimensions model. The goal of any heat system emulation is to assess the impact of distinct factors on device results under varying working conditions. The technique of energy equilibrium is implemented to small components with 6 faces. This method states that summing in and out of heat is equivalent to zero. **Fig. 3** displays the energy balance of node 2.

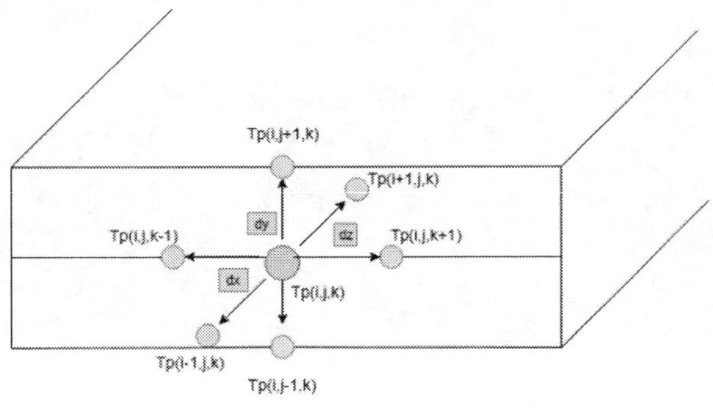

Figure 1: Energy balance at node 2

$$T_p(i,j,k) = T_p(i,j,k) \times (1-w) + w \times (A_e \times T_p(i+1,j,k) + A_w \times T_p(i-1,j,k) \quad (2)$$
$$+A_n \times T_p(i,j+1,k) + A_t \times T_p(i,j,k+1) + A_b \times T_p(i,j,k-1) + q) \div A_p$$

2.3. *Analyzing of Flow and Physical Model*

To analyze the impact of different micro-pin fins numbers on the heat transfer performance, three model with various dimensions are analyzed. The length (L), width (W), and height of the duct are 200 mm , 100 mm, and 10 mm respectively. **Fig. 1** displays the distribution of the micro-pin fins throughout the duct of all three model. The dimensions of the three models are shown in **Table 1**

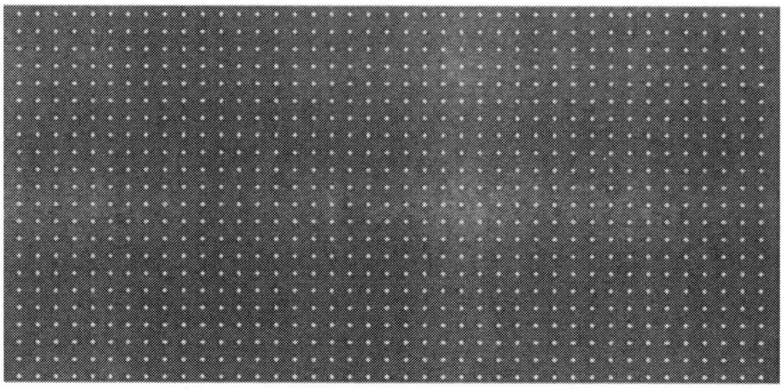

(a) 902 MPFs

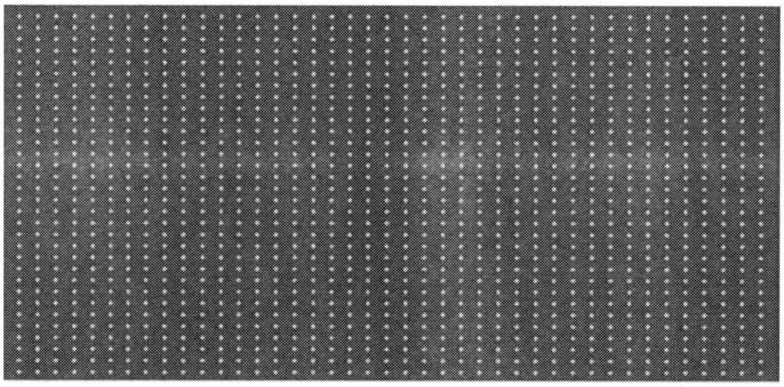

(b) 1312 MPFs

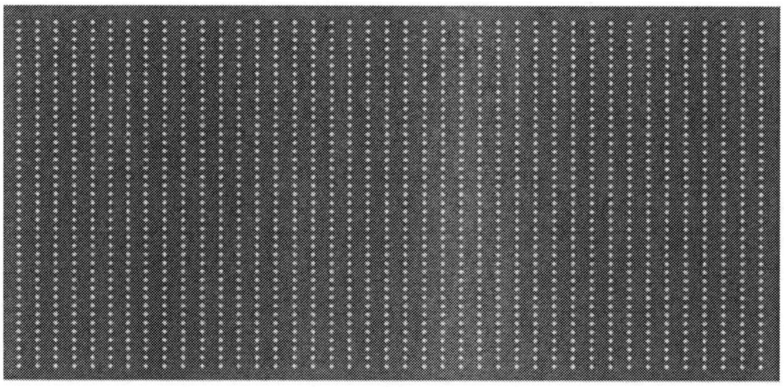

(c) 1640 MPFs

Figure 2: Comparison between the results of different numbers of in-line micro-pin fins

As for the comparison between the staggered and the in-line configurations, two simulation models was established with same duct dimensions (200 mm * 100 mm * 10 mm)

Different in-line micro-pin fins numbers	Diameter (mm)	SL (mm)	ST (mm)	Height (mm)
902 micro-pin fins	0.875	3.96	3.78	10
1312 micro-pin fins	0.875	3.96	2.27	10
1640 micro-pin fins	0.875	3.96	1.52	10

Table 1: The dimensions of different micro-pin fins numbers models

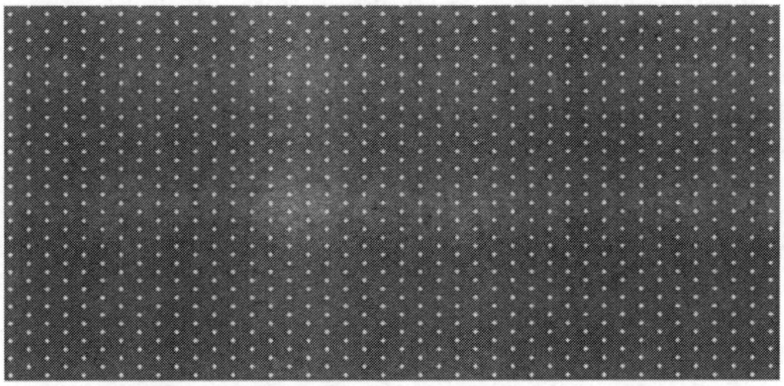

(a) Staggered configuration

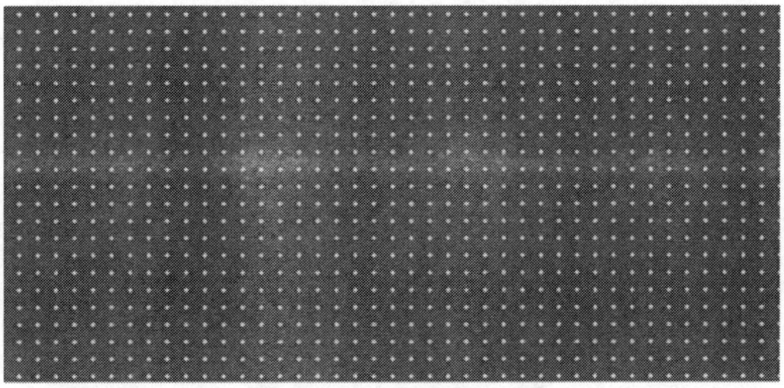

(b) In-line configuration

Figure 3: Simulation models of the two configurations

and same micro-pin fins numbers (924 micro-pin fins). In **Fig. 2** the two configurations are shown. The same longitudinal distance (3.96 mm), transverse distance (3.78 mm), height (10 mm), and diameter (0.875 mm) are considered for both staggered and in-line configurations.

From the right side, air joins the duct at an inlet temperature that is considered to be equal to the ambient temperature (45°C) and passes through the all various micro-pin fins arrangements. The hot air is coming out of the right side at the output temperature. The duct's walls are isolated from all sides while the steady-state 3-D model calculates the temperature inside the duct. The mathematical definition of fluid flow issues is controlled by the fundamental conservation of volume, momentum and energy equations. The previous assumptions were created, considered turbulent flow in the current work :

- 3-dimensional conservation equations.
- Incompressible Steady-state flow.
- 3-dimensional conduction of heat transfer through the micro-pin fins plate.

- All features are evaluated at an average temperature.
- The heat is transferred by forced convection
- The body forces are negligible.

3. Results and Discussion

3.1. *Effect of Different Micro-Pin Fins Numbers with In-line Configuration*

In this section, the impact of distinct in-line configuration on the heat transfer efficiency of micro-pin fins is investigated. The test was done for three different airspeeds and heat fluxes. The effect of three different numbers on the thermal performance in case of 2 m/s and 250 Watt is displayed in **Fig. 4(a)**. The results illustrated that the heat transfer performance had a direct relationship with the number of micro-pin fins, it was enhanced by 8.4% when increasing the number of micro-pin fins from 902 to 1312, and nearly 10.2% when increasing from 902 to 1640 micro-pin fins. In **Figs. 4(b)** and **4(c)** the effect of various micro-pin fins numbers are displayed for two cases 3 m/s with 250 Watt and 4 m/s with 250 Watt respectively. Relative to the results obtained, the thermal performance was improved by 9.3% and 7.9% respectively when increasing the number of micro-pin fins from 902 to 1312 while nearly 11.1% and 8.8% when increasing from 902 to 1640. The thermal performance enhanced with increasing of micro-pin fins number because both surface area and obstruction of air were increased.

3.2. *Comparasion between In-line and Staggered Micro-Pin fins Configurations*

In this test, three cases of airspeeds and heat fluxes were applied for both in-line and staggered arrangement which displayed in **Fig. 2**. The air temperature variation throughout the micro-pin fins duct at 1 m/s with heat flux 250 Watt is displayed in **Fig 5(a)** where the air entered the duct with temperature 45°C and was rising gradually but not equally because at the staggered arrangement the temperature difference is slightly higher than the in-line arrangement. The air exit at 71.1°C in staggered arrangement while in the in-line arrangement exit at 68.1°C which means that the thermal performance enhanced by 4.2%. In the case of 2 m/s and 250 Watt, the thermal performance improved nearly 2.8% as shown in **Fig. 5(b)**. In addition, **Fig. 5(c)** exhibits the temperature difference in the case of 3 m/s and heat flux 250 Watt for both two arrangements where the enhancement in thermal performance reached 2.6%. The thermal performance for three different airspeeds was shown in the above cases and it was evident that the heat transfer performance of the staggered configuration was slightly higher than that of the in-line configuration. Due to an increase in the effective contact region between the airflow and the micro-pin fins in the staggered configuration, this temperature distinction between the two configurations occurred. Also in the staggered configuration, the obstruction of air is more than that in the in-line configuration which means ensuring better heat transfer. Moreover, dissipation of the boundary layer is more in the staggered system because of the propagation of the micro-pin fins throughout the plate. The explanation for the slight difference in temperature between the staggered configuration and the in-line configuration in this analysis is the small distance of the sample segment Additionally, the difference in temperature between the two configurations decreased due to the same reason when rising air speed.

3.3. *Comparison Between Micro-Pin Fins Duct and Smooth Duct*

Thermal performance for the in-line micro-pin fins duct and smooth duct are compared numerically in this section. The air enterd with a temperature 38°C in both different ducts. In the conditions of 2 m/s airflow and 250 Watt heat flux, the air exit at a temperature 65.46°C in the micro-pin fins duct while exit at 50.26°C in the smooth duct. Relative to the results obtained, the thermal performance was enhanced by 30.24% with 902 micro-pin fins duct compared with the smooth duct. **Fig. 6** illustrated the temperature variation along the two ducts.

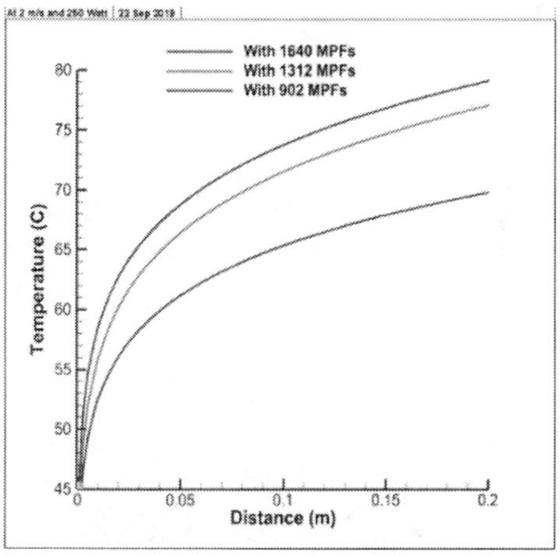

(a) At 2 m/s and 250 Watt

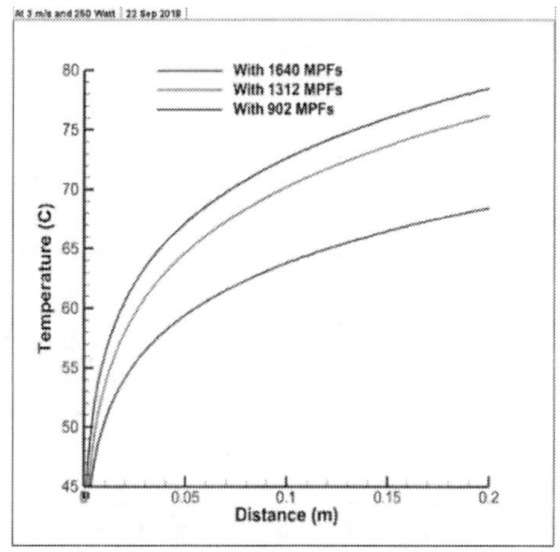

(b) At 3 m/s and 250 Watt

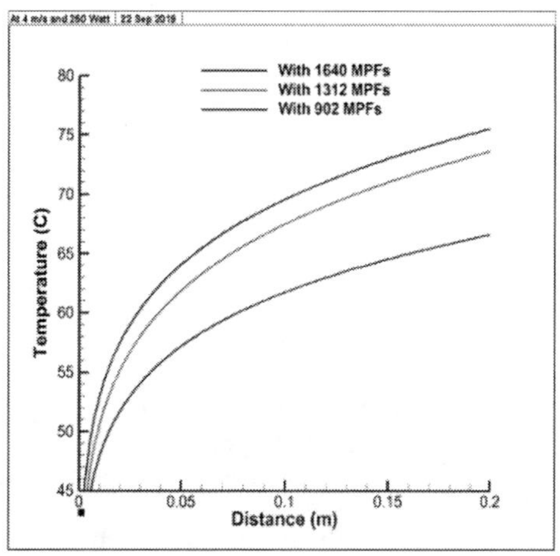

(c) At 4 m/s and 250 Watt

Figure 4: Comparison between the results of different numbers of in-line micro-pin fins

4. Temperature Contours

The temperature contours are plotted in two-dimensional form to show the temperature distribution along the micro-pin fins duct. In the tests, the air reaches the duct with almost 45°C and the temperature of the solid buildings is calculated before. **Fig. 7** displays the temperature contours of the distinct micro-pin fins numbers 902, 1312, and 1640 for 2 m/s and 250 Watt. In addition, **Fig 8** shows the temperature contours of staggered and in-line systems in the same conditions in the case of 1 m/s and 250 Watt.

5. Results of Analysis for Flow Field

Fig. 9(a) illustrates the distribution of the two-dimensional airflow of the in-line micro-pin fins in the situation of 2 m / s. The velocity vector in the x-direction (u) and z-direction (w) are shown in the figure. In the center, the velocity is greater than the other points close the walls. There are circulations near the micro-pin fins locations with a straight flow between the micro-pin fins. As for staggered arrangement, air obstruction is greater than in-line structure, which causes more fluid field circulation. **Fig. 9(b)** Displays the airflow allocation along the

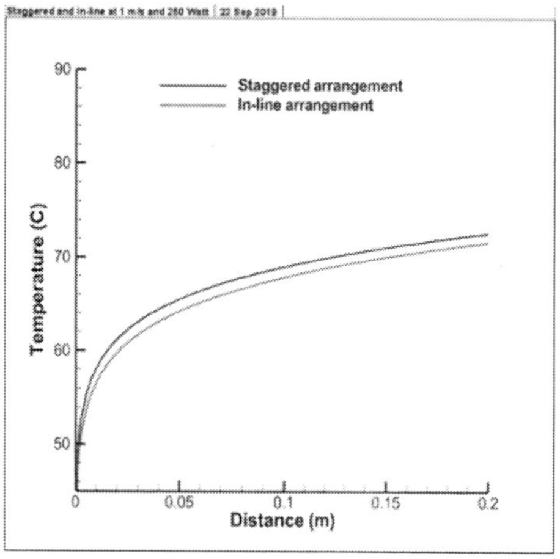

(a) At 1 m/s and 250 Watt

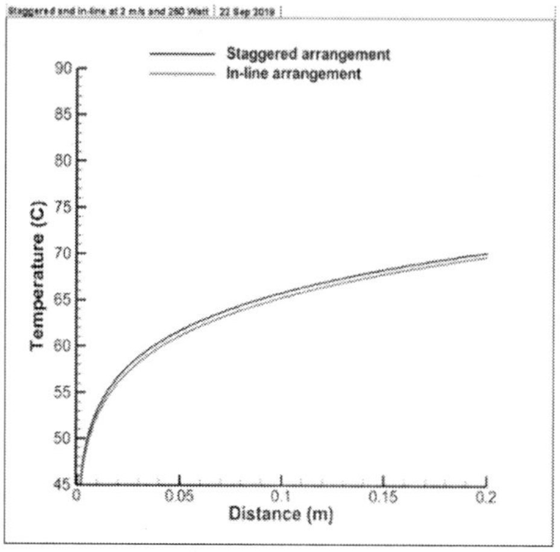

(b) At 2 m/s and 250 Watt

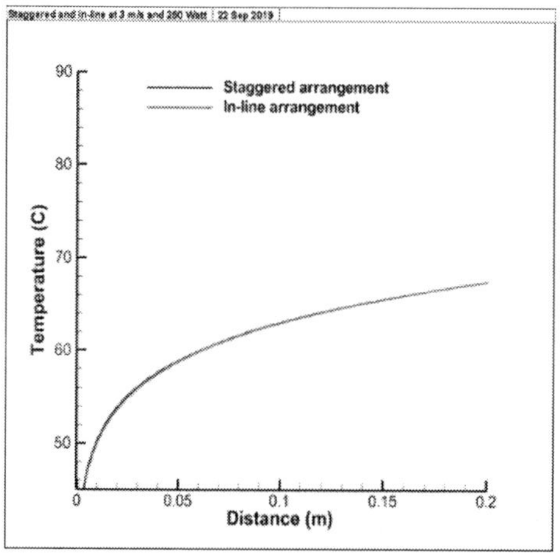

(c) At 3 m/s and 250 Watt

Figure 5: Comparison between the results of staggered and in-line configurations

duct.

6. Conclusions

The main conclusions of the present job can be described in the following notes :

(i) The thermal performance enhanced with increasing of micro-pin fins number from 902 to 1312 by 7.9 - 9.3% and from 902 to 1640 by 10.2 - 11.1%.

(ii) In the staggered configuration, the heat transfer performance is slightly higher than the in-line configuration by nearly 2.6 - 4.2% with the same number of micro-pin fins.

(iii) In addition, the micro-pin fins duct increased the thermal transfer performance by 30.24% compared to the smooth duct.

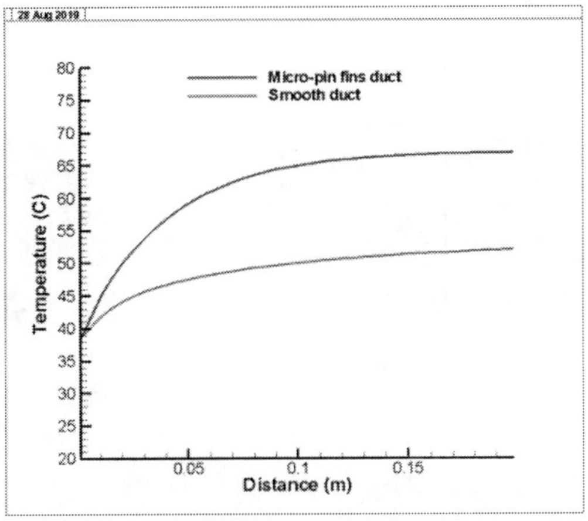

Figure 6: Comparison between the temperature variation inside the micro-pin fins duct and the smooth duct

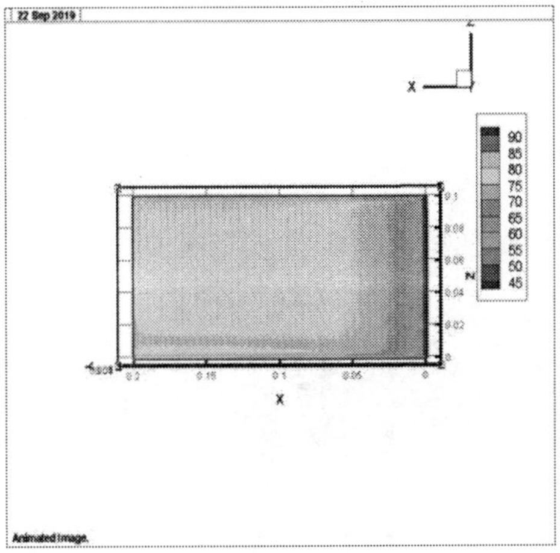

(a) 1640 MPFs Watt

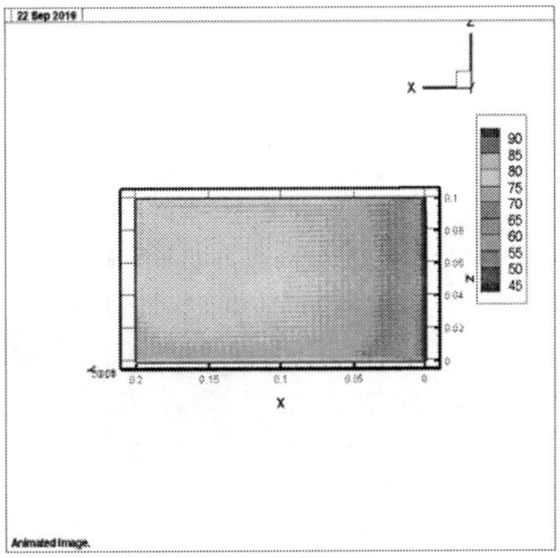

(b) 1312 MPFs Watt

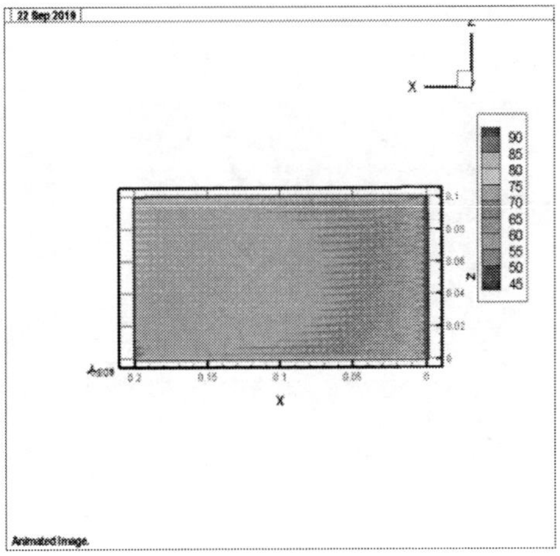

(c) 902 MPFs Watt

Figure 7: Temperature distribution of different micro-pin fins numbers at 2 m/s and 250 Watt

IOP Conf. Series: Materials Science and Engineering **765** (2020) 012037 doi:10.1088/1757-899X/765/1/012037

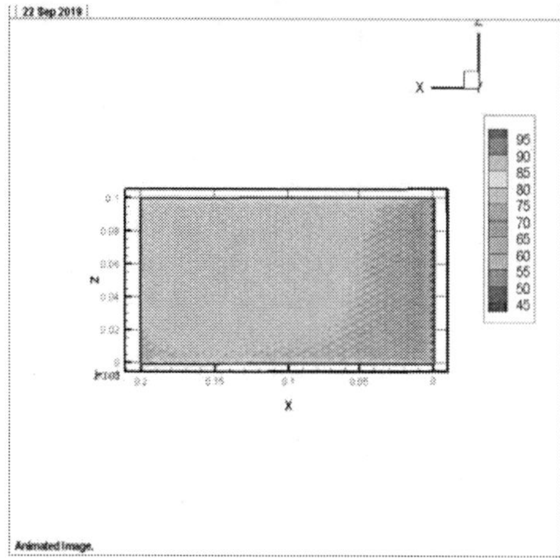

(a) Staggered configuration

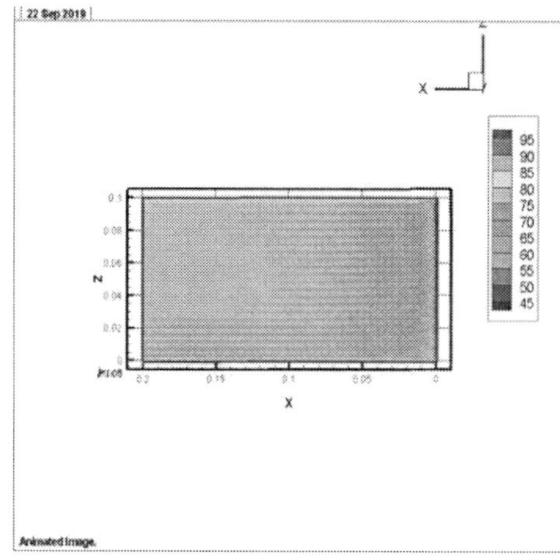

(b) In-line configuration

Figure 8: Temperature distribution of different configurations at 1 m/s and 250 Watt

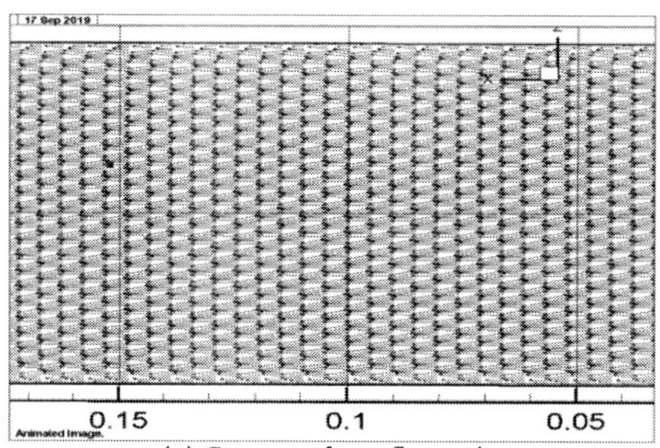

(a) Staggered configuration

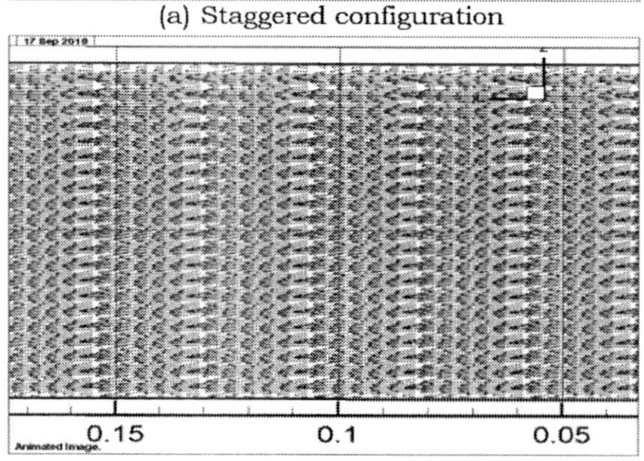

(b) In-line configuration

Figure 9: Flow field vectors of (u) and (w) for different configurations

References

[1] Oguntala G, Abd-Alhameed R, Sobamowo G, Abdullahi HS. Improved thermal management of computer microprocessors using cylindrical-coordinate micro-fin heat sink with artificial surface roughness. Engineering science and technology, an international journal. 2018 Aug 1;21(4):736-44.
[2] Wan W, Deng D, Huang Q, Zeng T, Huang Y. Experimental study and optimization of pin fin shapes in flow boiling of micro pin fin heat sinks. Applied Thermal Engineering. 2017 Mar 5;114:436-49.
[3] Falsetti C, Magnini M, Thome JR. A new flow pattern-based boiling heat transfer model for micro-pin fin evaporators. International Journal of Heat and Mass Transfer. 2018 Jul 1;122:967-82.
[4] Chiu HC, Hsieh RH, Wang K, Jang JH, Yu CR. The heat transfer characteristics of liquid cooling heat sink with micro pin fins. International Communications in Heat and Mass Transfer. 2017 Aug 1;86:174-80.
[5] Mei D, Lou X, Qian M, Yao Z, Liang L, Chen Z. Effect of tip clearance on the heat transfer and pressure drop performance in the micro-reactor with micro-pin–fin arrays at low Reynolds number. International Journal of Heat and Mass Transfer. 2014 Mar 1;70:709-18.
[6] Ndao S, Peles Y, Jensen MK. Effects of pin fin shape and configuration on the single-phase heat transfer characteristics of jet impingement on micro pin fins. International Journal of Heat and Mass Transfer. 2014 Mar 1;70:856-63.
[7] Yeom T, Simon TW, North M, Cui T. High-frequency translational agitation with micro pin-fin surfaces for enhancing heat transfer of forced convection. International Journal of Heat and Mass Transfer. 2016 Mar 1;94:354-65.
[8] Guan N, Jiang G, Liu Z, Zhang C, Ding N. The impact of contact angle on flow resistance reduction in hydrophobic micro pin fins. Experimental Thermal and Fluid Science. 2016 Oct 1;77:197-211.
[9] Hua J, Li G, Zhao X, Li Q, Hu J. Study on the flow resistance performance of fluid cross various shapes of micro-scale pin fin. Applied Thermal Engineering. 2016 Aug 25;107:768-75.
[10] Qian M, Mei D, Yi Z, Feng Y, Chen Z. Fluid flow and heat transfer performance in a micro-reactor with non-uniform micro-pin-fin arrays for hydrogen production at low Reynolds number. International Journal of Hydrogen Energy. 2017 Jan 5;42(1):553-61.
[11] Jia Y, Xia G, Li Y, Ma D, Cai B. Heat transfer and fluid flow characteristics of combined microchannel with cone-shaped micro pin fins. International Communications in Heat and Mass Transfer. 2018 Mar 1;92:78-89.
[12] Naphon P, Wiriyasart S. Experimental study on laminar pulsating flow and heat transfer of nanofluids in micro-fins tube with magnetic fields. International Journal of Heat and Mass Transfer. 2018 Mar 1;118:297-303.
[13] Li XJ, Zhang JZ, Tan XM. Effects of piezoelectric fan on overall performance of air-based micro pin-fin heat sink. International Journal of Thermal Sciences. 2018 Apr 1;126:1-2.
[14] Ambreen T, Kim MH. Effect of fin shape on the thermal performance of nanofluid-cooled micro pin-fin heat sinks. International Journal of Heat and Mass Transfer. 2018 Nov 1;126:245-56.
[15] Kewalramani GV, Hedau G, Saha SK, Agrawal A. Study of laminar single phase frictional factor and Nusselt number in In-line micro pin-fin heat sink for electronic cooling applications. International Journal of Heat and Mass Transfer. 2019 Aug 1;138:796-808.

IOP Conf. Series: Materials Science and Engineering **765** (2020) 012038 doi:10.1088/1757-899X/765/1/012038

Control on Temperature of Hybrid Nanofluid in Evacuated Tube Solar Collector Using Smart Curtain

Hosham S. Anead, Khalid F. Sultan and Ahmed M. Salman
Electromechanical Engineering Department, University of Technology,Baghdad,Iraq.
E-mail: Hoshams.2000@yahoo.com, ksultan61@yahoo.com, ahmedm.salman90@gmial.com

Abstract.
This research was revealed control on temperature hybrid nanofluid in evacuated tube solar collector system, where the hybrid nanofluid utilized as working fluid in ETC for enhancement the thermal performance for heating device. The smart curtain was used for purpose shadow the evacuated tube solar collector for controlling on temperature of the hybrid nanofluid in heating or cooling conditions. In addition, Adaptive neuro fuzzy inference system (ANFIS) a way of artificial intelligence was applied to control on the shadow the evacuated solar tube collector. Where the main principle of the curtain is to control the polarization of the sun radiation, the curtain work depends on three parameters were applied in ANFIS technique, two parameters are input variable like as (temperature of hybrid nanofluids (inlet, outlet) and ambient temperature, and one output parameter represents by (distance parameter).

Keywords— Adaptive Neuro Fuzzy Inference System, Fuzzy Logic Controller, Temperature of Hybrid Nanofluid.

1. Introduction

The future is the renewable energy. It protects the environment and enables more energy to be produced. Solar power is a significant form of power. Currently, solar collectors like flat panel, evacuated tube, parabolic trough, dish collectors are used to collect heat from solar energy [1, 2, 3] .Hybrid nanoparticles are the most desirable means of improving the efficiency of heat transfer devices. The hybrid nanofluid is composed of mixing hybrid nanoparticles with the working fluid [4, 5, 6, 7, 8, 9]. Adaptive network-based fuzzy inference system is an adaptive network that enables the implementation of neural network topology in conjunction with fuzzy logic.

ANFIS system involves elements of the traditional fuzzy system, Also the calculation on each level was executed by a layer of hidden neurons and the learning capacity of the neural network is provided to improve system knowledge. ANFIS is able to handle complicated, non-linear, dynamic systems.in additional it implements a Takagi-Sugeno fuzzy inference system [10] and has a five-layered. Analyzing the mapping relationship between the input and output data. ANFIS can determine the ideal distribution of membership functions using either a back-propagation algorithm by itself or in combined with a least-square technique similar to ANNs. ANFIS network is trained based on experience of supervisor learning to achieve a specific target output from a particular input. Jang initially introduced the ANFIS methodology in the early 1990 [11], its applications to the modeling of thermal energy systems is somewhat recent. Several studies using ANFIS for modeling, control, performance prediction in energy systems [12, 13, 14, 15].

IOP Conf. Series: Materials Science and Engineering **765** (2020) 012038 doi:10.1088/1757-899X/765/1/012038

2. Neuro-Fuzzy system Layers (ANFIS Architecture)

Hybrid (ANFIS) is one of the Artificial Intelligence methods that belong to one of the adaptive networks, functionally equivalent to the fuzzy systems. It is assumed that there are two inputs x, y and output z in the fuzzy inference system. According to figure(1) ANFIS implements a fuzzy inference system from Takagi Sugeno [11] and has five layers as shown in Figure 1 Sugeno fuzzy model has the following rules of first order as follows:

Rule 1: If x is A_1 and y is B_1 Then F1=P_1 x + q_2 y + r_1

Rule 2: If x is A_2 and y is B_2 Then F2=P_2 x + q_2 y + r_2

Layer1: Fuzzification layer each node I in layer 1 is a node-function adaptive node. Parameters in this layer: parameters of the premise (or precedent).

$$O_{1,i} = \mu_{A_1}(x) \quad i = 1, 2$$

$$O_{1,i} = \mu_{B_{i-2}}(x) \quad i = 3, 4$$

$$\mu_A(x) = \frac{1}{1 + [(x - c_i)^2/a_i^2]^{b_i}}$$

Where: $\mu_A(x)$ and $\mu_{B_{i-2}}$ are the membership function levels for the fuzzy sets A_i and B_i, respectively, and (a_i, b_i, c_i) are the membership function parameters that can change the membership function shape.

Layer 2: rule layer is a fixed node marked () with the output of which all incoming signals are produced:

$$O_{2,i} = w_i = \mu_{A_i}(x) \times \mu_{B_i}(x) \quad i = 1, 2$$

Layer3: normalization layer, the nodes of which are labelled N. Its function in the following process is to normalize the weight function:

$$O_{3,i} = \overline{w_i} = \frac{w_i}{w_1 + w_2} \quad i = 1, 2$$

Layer 4: defuzzification layer with adaptive nodes. The output equation is w (px+qy+r), where the linear parameters or so-called consequent node parameters are denoted by pi, qi and ri. The defuzzy relationship between this layer's input and output may be defined as:

$$O_{4,i} = \overline{w_i} f_i = \overline{w_i}(p_i x + q_i y + r_i)$$

Where: $\overline{w_i}$ is the normalized firing strength from layer 3, and (p_i , q_i , r_i) are the consequent variables set of the node.

Layer 5: summation neuron the single node in this layer is a fixed node labelled 5,1 (overall output), which computes the overall output as the aggregation of all incoming signals

$$O_{5,1} = \textbf{overall output} = \sum \overline{w_1} f_i$$

When premise parameters are fixed:

$$F = \frac{w_1}{w_1 + w_2} f_1 + \frac{w_2}{w_1 + w_2} f_2 = \overline{w_1} f_1 + \overline{w_2} f_2$$

Which is linear in the consequent parameters (p_1 , q_1 , r_1 and p_2 , q_2 , r_2).

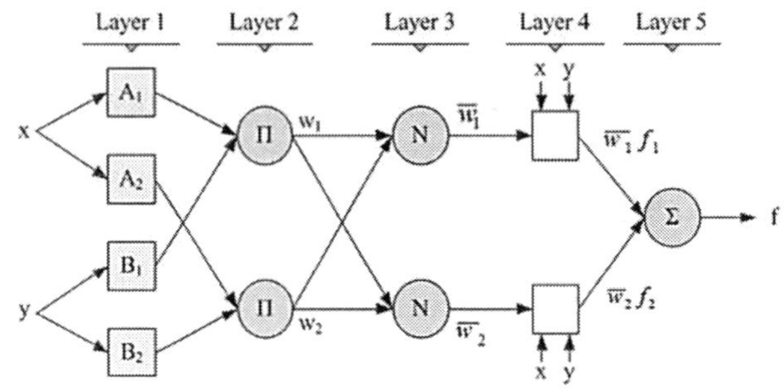

Figure 1. Takagi Sugeno layers.

3. Evacuated tube solar collector with shadow system

In this work was added a shadow system to the ETCS In order to contribute to the development of the system performance of the solar heater and the control of the temperature of the fluid hybrid as shown in Figure 2 and 3. The curtain contributes to covering the solar heater system where it works for controlling the temperature of the hybrid nanofluid present in the heater. The curtain moves up and down by mechanical and electrical parts depending on the set conditions that control the movement. This curtain is made of (Flex) white rubber material and is considered one of the strongest printing materials because it is characterize d by resistance to tension, pressure and weather. Figure 4 and 5 illustrate the curtain which is mounted on frame and the ETC with shadow system.

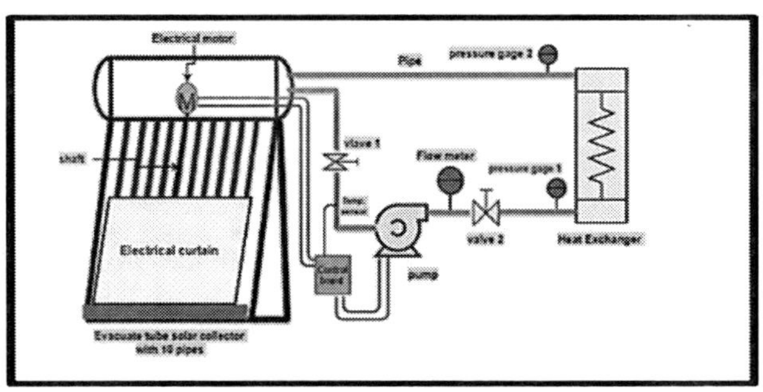

Figure 2. Schematic diagram of the ETSC with shadow system.

Figure 3. The solar heating system used in the experiments with shadow system.

Figure 4. The mounted curtain on frame.

Figure 5. ETC with shadow system.

4. Components used in shadow the solar collectors

Here we explain the hardware equipment response on movement the curtain. The curtain system consists of a mechanical part and an electrical part. Where a mechanical part consists of the outer frame base is made of metal iron, as show in Figure 6 the goal of this framework is the installation of curtain, stepper motor and shift. From another hand the electrical part consists of a stepper motor and a power supply in addition microcontroller (arduino). In Figure 7 show the control board involve the electrical part. It has also been linked to some sensors such as temperature sensors (thermocouple sensor) as well as sensors for distance (ultra sonic) to the control circuit Figure 8 and 9 .

Figure 6. The mainframe with curtain.

IOP Conf. Series: Materials Science and Engineering **765** (2020) 012038 doi:10.1088/1757-899X/765/1/012038

Figure 7. Control board.

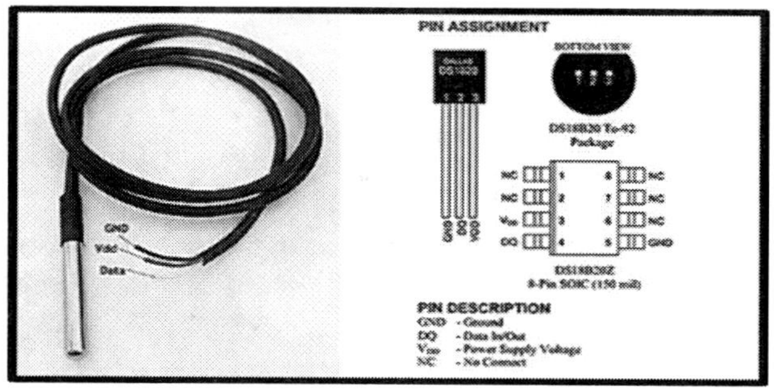

Figure 8. DS18B20 Temperature Sensor.

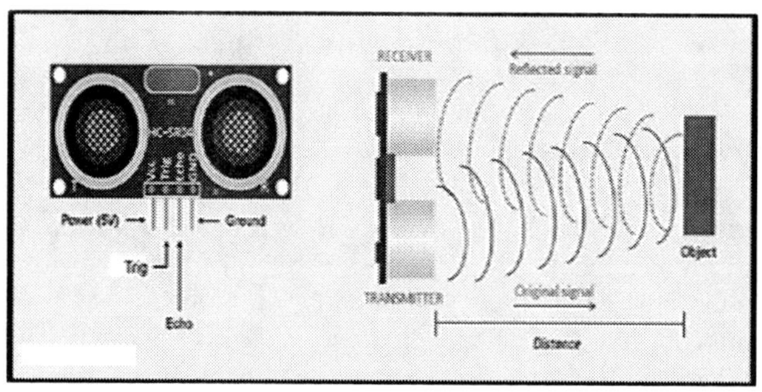

Figure 9. work Ultrasonic Sensor.

5. Design the sugeno fuzzy rules by using ANFIS

Fuzzy logic tries to simulate our decision-making and is capable of providing precision findings depending on unknown and indefinite information. Mamdani and Sugeno were most popular fuzzy inference designs. Usually both designs may be described as (if the antecedent is the consequence) to all fuzzy designs, the antecedence is the same, But the result of each template has distinct types. At Sugeno fuzzy design two factors are regarded as fuzzy inputs such as T_{hf} and T_a. A typical Sugeno fuzzy rule in the proposed ANFIS model with two inputs T_{hf} and T_a, and one output Dist. can be shown as If T_{hf}=x and T_a=y, then Dist.= $f(T_{hf},T_a)$, where x and y are the membership function of data entered T_{hf} and T_a, respectively. At Sugeno fuzzy design, f is a crisp function, frequently a linear input factor feature as:

$$f(T_{hf},T_a) = pT_{hf} + qT_a + r \tag{1}$$

The suggested classification of ANFIS, as shown in Figure 10, involves three main components: Fuzzification, Sugeno Fuzzy Inference Engine, and Defuzzification. As noted above, for each sample n, two input variables: Thf and Ta are considered. ANFIS goal is to calculate the distance for each sample n, Dist. (n), according to data entered. Fuzzifier switching the normalized inputs to linguistic fuzzy parameters utilizing the membership functions. The membership functions for data entered could be viewed in Figure 11 The fuzzy sets employed to fuzzy of Each entry parameter that can be categorized into three membership features: (Low, Medium and High).

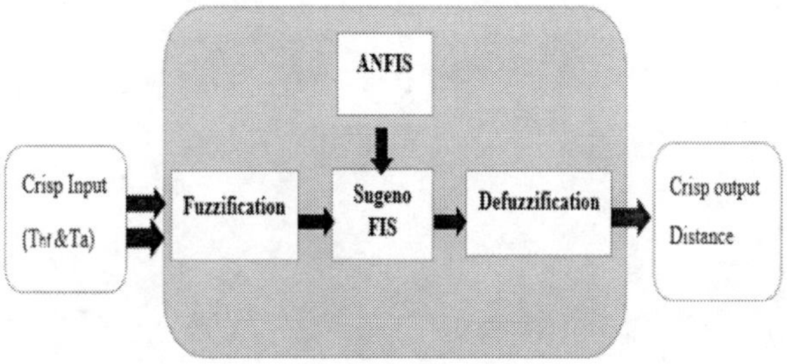

Figure 10. Suggested classification of ANFIS.

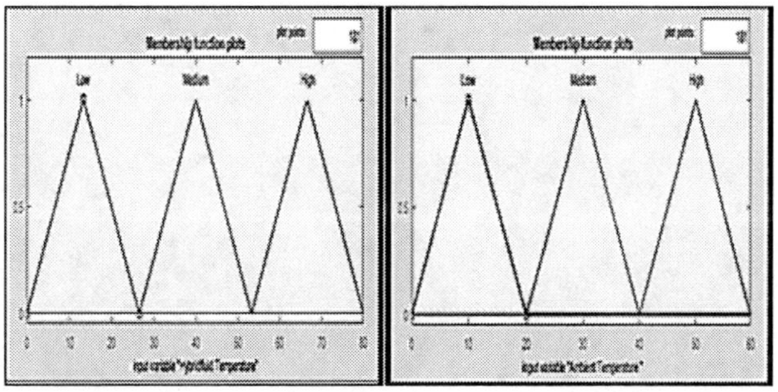

Figure 11. Two inputs of memberships function.

According to the range of the data entry for the sample n , several of the fuzzy rules will fire. Then, weighted average defuzzification method was applied to create the final crisp output Dist(n) from the aggregation of the fired outputs $out_k(n)$, k=1,2,..,9 depend on Equation 2 , in which, w_k is the minimum of the memberships of the data entered T_{hf} and T_a within the Rule k of Equation 2 .

$$Dist(n) = \frac{\sum_{k=1}^{9}(w_k \times out_k)}{\sum_{k=1}^{9} w_k} \quad (2)$$

In fuzzy system, the number of fuzzy memberships for every data entry is 3 (Low, Medium, and High). Therefore, the number of fuzzy if-then rules is 3 3 = 9, where every rule own 3 constant variables (p_k, q_k and r_k). Hence, the total amount of Manageable variables in the fuzzy system is 3 9 = 27.This is a Sugeno fuzzy design for stepping motor control. We apply the technique of ANFIS to predict the membership function factors and the resulting functions. Of each linguistic variable, we utilized a fuzzy design of 9 rules and 3 membership functions.

6. ANFIS editor

The ANFIS Editor contains on fuzzy tool box can be utilize it for generate constructing, and optimizing Sugeno-types adaptive neural fuzzy inference systems. The following Figure 12 Showed the name of each input factor on the left and the output on the right. Figure 13 Shows ANFIS models structure was developed with 2 inputs, 1 output and 9 fuzzy rules. Figure 14 Revealed the relationship between the error and number of epoch .From another hand The Rule Editor allows automatic construction of the rule sentence as indicated in the FIS Editor by to the description of the input and output factor specific with the FIS Editor As shown in Figure 15. in addition the Rule Viewer shown in Figure 16 Displays a road map of the whole fuzzy inference process. Furthermore the Surface Viewer shown in Figure 17 with two-input one-output schemes as can generate three-dimensional charts that MATLAB can adjust to manage .

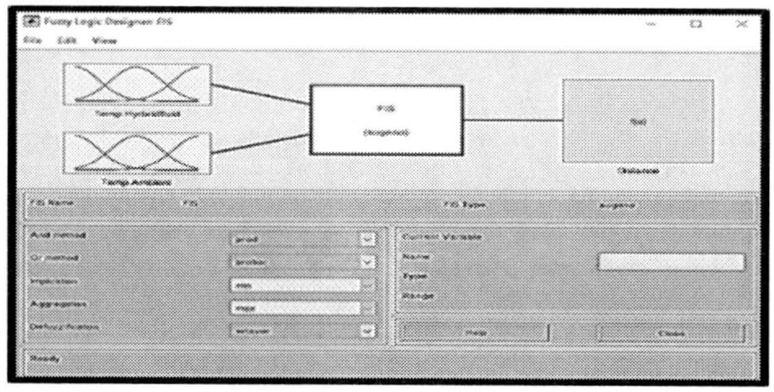

Figure 12. Input/Output membership function .

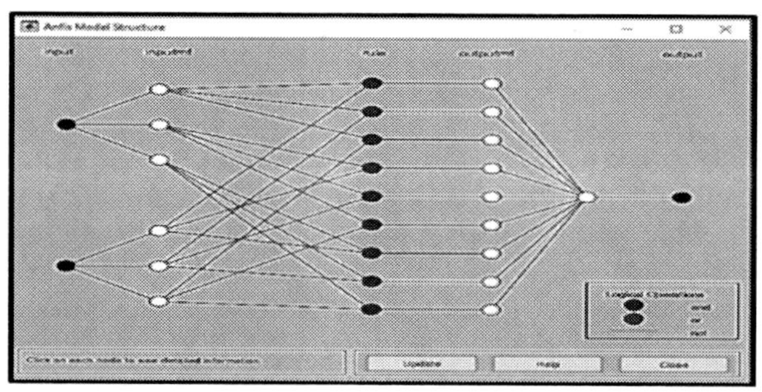

Figure 13. ANFIS structure.

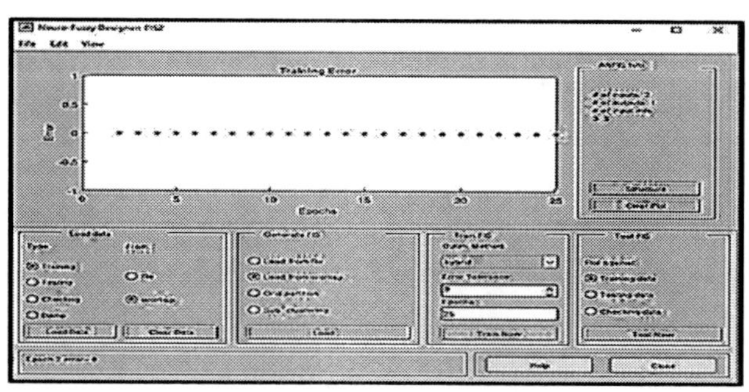

Figure 14. (FIS) editor.

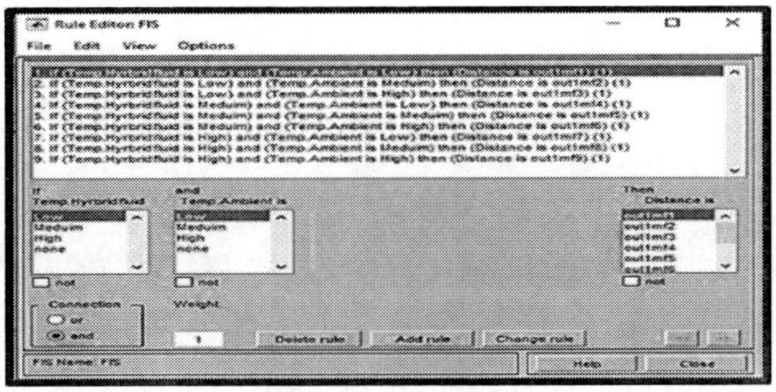

Figure 15. Rule editor.

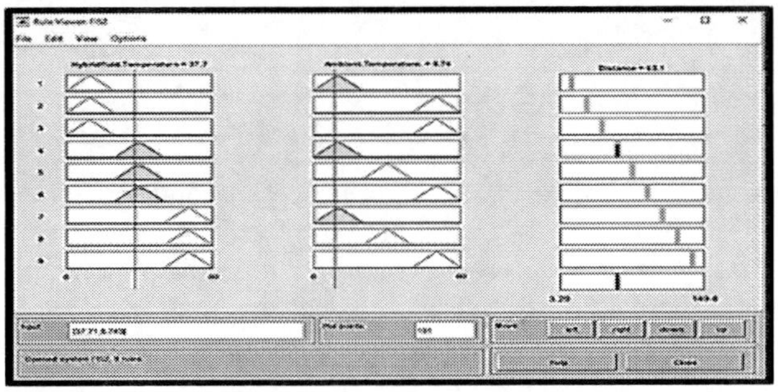

Figure 16. Rule viewer .

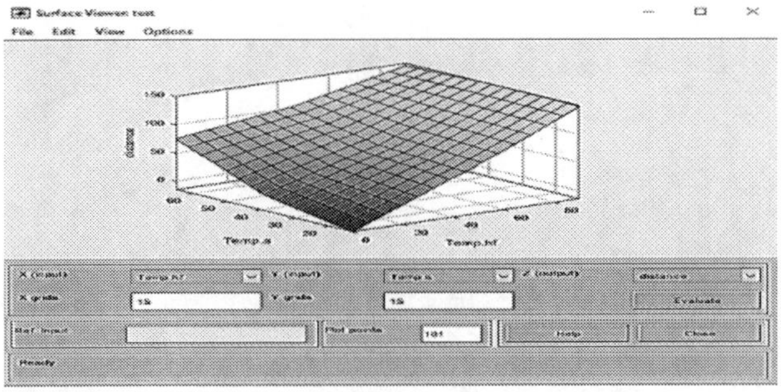

Figure 17. Surface viewer.

7. Simulation Results

After obtaining the standard real data of parameters such as (temperature of hybrid nanofluids, ambient of temperature and distance) from the solar collector. Adaptive Neuro-Fuzzy Inference System (ANFIS) method was utilized for forecasting the output of distance of electrical curtain. After completing the training process for Multifunction's (Triangular - Gaussian - Bell - Trapezoidal) was applied for training the parameters findings show best linear functions were obtained by the number of epoch and number of the layer. Table 1 shows the Training Error of the (inlet and outlet) Temperature hybrid nanofluids with shadow in ETC.

Figure 18 explain the stepping motor reaction to a sequence of step input signals. We show in Figure 19 the findings of the application of the ANFIS procedure with test data (240 samples were used). The

Table 1. Training Error of the (inlet and outlet) Temperature hybrid nanofluids with shadow in ETC

Multifunction's Type	No. of Layers	No. of Iterations	Average Error	
			Constant	
Triangular	6	40	0.00227698	
Trapezoidal	6	40	**0.00237731**	
Bell shapped	6	40	0.00225577	
Symmetric Gaussian function	6	40	0.00227093	
			Linear	
Triangular	6	40	0.00222149	Best
Trapezoidal	6	40	0.00228354	
Bell shapped	6	40	0.0022579	
Symmetric Gaussian function	6	40	0.0022612	

training error was explained in Figures 20. In the process of shading the ETC collector the electric curtain starts to work based on the temperature of the hybrid nanofluid, where the maximum height of the curtain is approximately 140 centimeters when the nanofluid temperature is 90°C, The curtain is in a fully open state at low temperatures around 5°C, at various mass flow rate and volume concentration.

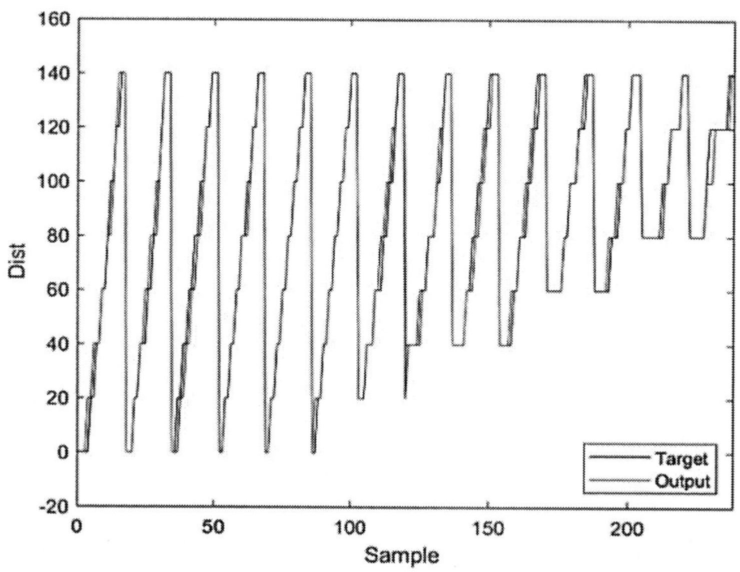

Figure 18. Response of the stepping motor to a sequence of step input signals.

IOP Conf. Series: Materials Science and Engineering **765** (2020) 012038 doi:10.1088/1757-899X/765/1/012038

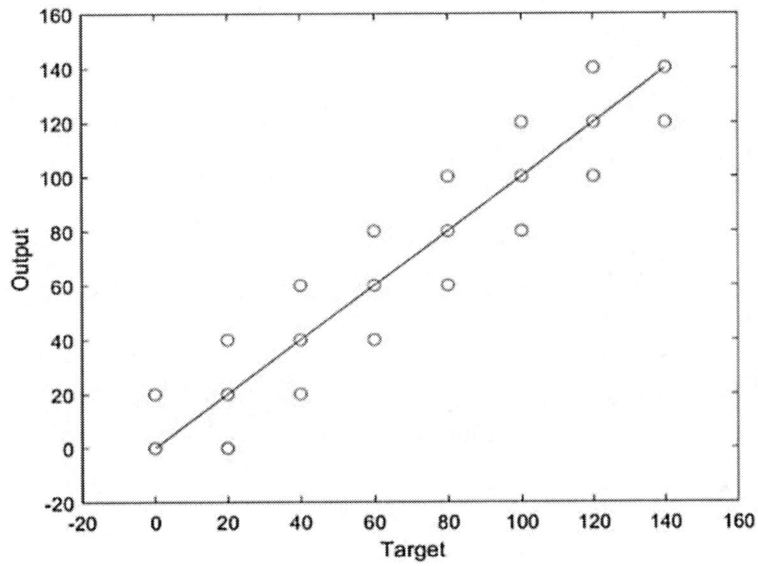

Figure 19. Findings of applying the ANFIS methodology.

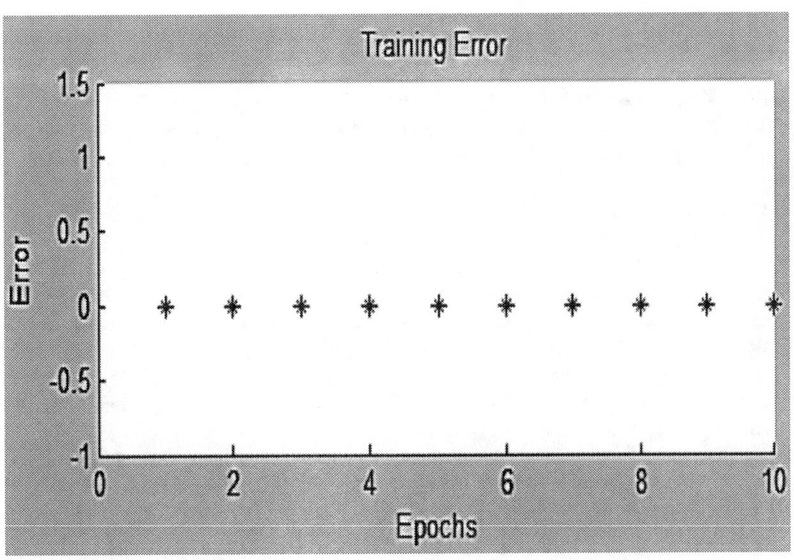

Figure 20. Training Error and Training data with FIS.

8. Conclusion

(i) ANFIS technique was for forecasting the output distance parameter of the smart curtain.

(ii) ANFIS a technique is applied to designing the fuzzy rule base of the smart system into control. In specific.

(iii) The ANFIS approach has been applied to construct a Sugeno fuzzy model to regulate the stepper motor drive.

(iv) The role of the electric curtain is to control the temperature of the hybrid fluid and increase the efficiency of the ETC.

(v) Applying ANFIS method for simulation the ETC parameter and evaluated the performance of these parameters.

(vi) ANFIS is a perfect option for realize the desirable precision and efficiency in controlling stepping motor.

References

[1] Kalogirou S. The potential of solar industrial process heat applications. Applied Energy [Internet]. Elsevier BV; 2003 Dec;76(4):33761. Available from: http://dx.doi.org/10.1016/s0306-2619(02)00176-9.

[2] Morrison GL, Tran NH, McKenzie DR, Onley IC, Harding GL, Collins RE. Long term performance of evacuated tubular solar water heaters in Sydney, Australia. Solar Energy [Internet]. Elsevier BV; 1984;32(6):78591. Available from: http://dx.doi.org/10.1016/0038-092x(84)90253-6.

[3] Sabiha MA, Saidur R, Mekhilef S, Mahian O. Progress and latest developments of evacuated tube solar collectors. Renewable and Sustainable Energy Reviews [Internet]. Elsevier BV; 2015 Nov;51:103854. Available from: http://dx.doi.org/10.1016/j.rser.2015.07.016.

[4] Huminic G, Huminic A. Hybrid nanofluids for heat transfer applications A state-of-the-art review. International Journal of Heat and Mass Transfer [Internet]. Elsevier BV; 2018 Oct;125:82103. Available from: http://dx.doi.org/10.1016/j.ijheatmasstransfer.2018.04.059.

[5] Mehta S, Chauhan KP, Kanagaraj S. Modeling of thermal conductivity of nanofluids by modifying Maxwells equation using cell model approach. Journal of Nanoparticle Research [Internet]. Springer Science and Business Media LLC; 2010 Dec 14;13(7):27918. Available from: http://dx.doi.org/10.1007/s11051-010-0167-0.

[6] Masuda H, Ebata A, Teramae K, Hishinuma N. Alteration of Thermal Conductivity and Viscosity of Liquid by Dispersing Ultra-Fine Particles. Dispersion of Al2O3, SiO2 and TiO2 Ultra-Fine Particles. Netsu Bussei [Internet]. The Japan Society of Thermophysical Properties; 1993;7(4):22733. Available from: http://dx.doi.org/10.2963/jjtp.7.227.

[7] AZWADI, CS Nor; ADAMU, I. M. Turbulent force convective heat transfer of hybrid nano fluid in a circular channel with constant heat flux. J. Adv. Res. Fluid Mech. Therm. Sci, 2016, 19 (1): 1-9..

[8] SULTAN, Khalid Faisal. Experimental evaluation of the thermal performance in the solar nanofluid heating system by using cupper and titanium oxide. Iraqi journal of mechanical and material engineering, 2015, 15 (4): 272-287.

[9] Toghraie D, Chaharsoghi VA, Afrand M. Measurement of thermal conductivity of ZnOTiO2/EG hybrid nanofluid. Journal of Thermal Analysis and Calorimetry [Internet]. Springer Science and Business Media LLC; 2016 Apr 13;125(1):52735. Available from: http://dx.doi.org/10.1007/s10973-016-5436-4.

[10] Takagi T, Sugeno M. Fuzzy identification of systems and its applications to modeling and control. IEEE Transactions on Systems, Man, and Cybernetics [Internet]. Institute of Electrical and Electronics Engineers (IEEE); 1985 Jan;SMC-15(1):11632. Available from: http://dx.doi.org/10.1109/tsmc.1985.6313399.

[11] Jang J-SR. ANFIS: adaptive-network-based fuzzy inference system. IEEE Transactions on Systems, Man, and Cybernetics [Internet]. Institute of Electrical and Electronics Engineers (IEEE); 1993;23(3):66585. Available from: http://dx.doi.org/10.1109/21.256541.

[12] Viswavandya M, Sarangi B, Mohanty S, Mohanty A. Short Term Solar Energy Forecasting by Using Fuzzy Logic and ANFIS. Computational Intelligence in Data Mining [Internet]. Springer Singapore; 2019 Aug 18;75165. Available from: http://dx.doi.org/10.1007/978-981-13-8676-3-63.

[13] Olatomiwa L, Mekhilef S, Shamshirband S, Petkovi D. Adaptive neuro-fuzzy approach for solar radiation prediction in Nigeria. Renewable and Sustainable Energy Reviews [Internet]. Elsevier BV; 2015 Nov;51:178491. Available from: http://dx.doi.org/10.1016/j.rser.2015.05.068.

[14] Jovi S, Anii O, Marseni M, Nedi B. Solar radiation analyzing by neuro-fuzzy approach. Energy and Buildings [Internet]. Elsevier BV; 2016 Oct;129:2613. Available from: http://dx.doi.org/10.1016/j.enbuild.2016.08.020.

[15] Bassam A, May Tzuc O, Escalante Soberanis M, Ricalde L, Cruz B. Temperature Estimation for Photovoltaic Array Using an Adaptive Neuro Fuzzy Inference System. Sustainability [Internet]. MDPI AG; 2017 Aug 15;9(8):1399. Available from: http://dx.doi.org/10.3390/su9081399.

IOP Conf. Series: Materials Science and Engineering **765** (2020) 012039 doi:10.1088/1757-899X/765/1/012039

Blood flow through the circulatory system element affected by double stenotic lesions.

A F Tazyukova[1], F Kh Tazyukov[2], Hind D. Salman[3] , E R Kutuzova[4], A G Kutuzov[5], S V Yushko[6],

[1]Chemical Cybernetics Department, Kazan National Research Technological University, Kazan, K. Marx st., 68, 420015, Russia
[2]Mechanical Engineering Department, Kazan National Research Technological University, Kazan, K. Marx st., 68, 420015, Russia
[3]University of Technology – Iraq , 50260@uotechnology.edu.iq
[4]Control actuators system department, Kazan National Research Technological University, Kazan, K. Marx st., 68, 420015, Russia
[5]Chemical Cybernetics Department, Kazan National Research Technological University, Kazan, K. Marx st., 68, 420015, Russia
[6]Rector of the Kazan National Research Technological University, Kazan, K. Marx st., 68, 420015, Russia

four

Abstract. Destruction of the elements of the blood circulatory system may in many cases cause many dangerous diseases. Despite many medical diagnostic tools being used in medical laboratories, mathematical modeling at present can be successfully used for predicting, setting the diagnosis, forming coherent picture of the disease, and determining further treatment. One of the most dangerous pathologies of cardio-vascular system is stenosis. This disease is a consequence, and cause of many dangerous conditions. Due to small characteristic size of the element of the circulatory system (1-3 mm) and moderate values of Reynolds number (0.1-300), non-Newtonian properties of blood appear at the flow in the affected cells of the circulatory system. The rheological constitutive relation FENE - P is used for the flow modeling which predicts rheological properties of the blood such as anomalous viscosity and effects of elasticity. This paper examined the effect of the rigidity of the walls of blood vessels and investigated the effect of double - stenotic lesions on blood flow. An element of the circulatory system with a double - stenotic lesions with 50 % narrowing was chosen as the model case. As a result of the simulation, streamlines were obtained, which are the distribution of the principal stresses difference (PSD) in the most significant places of the affected elements of the circulatory system.

1. Introduction

Atherosclerotic lesions of vessels are some of the most common vascular diseases and are categorized as chronic dangerous pathologies. Atherosclerosis is a condition associated with formation of narrowing plaques on the vessel's walls. As a result, blood circulation, composition and properties of blood flow as well as the nutrition of organs and tissues do not run as usual. There are many reasons for this pathology: lack of physical activity, injury, quality of food, age, family history and the presence of other diseases [1-4].

At present, medical science developed numerous ways to perform diagnosis of stenosis such as ultrasound of the heart, magnetic - resonance angiography and computed tomography to name a few. Clinicians performing the diagnostics sometimes utilize several methods at the same time in order to have a complete picture of the disease. Mathematical modeling is another promising method utilized in diagnostic and therapeutic surgery [5-6, 14]. Building different model cases allows to apply an individual approach to each particular case, forecast and minimize possible risks and choose the most effective treatment.

Creation of a mathematical model is difficult due to the complex internal structure of blood, nature of the blood flow, the structure of the vessel walls, and localization of the affected area of the circulatory system. In addition, limited amount of research has been done on mechanical properties of the formed elements in the blood flow. This is one of the reasons for the lack of consensus among researchers on necessity of considering non-Newtonian properties of the blood. Some scientists believe that the non-Newtonian properties should be considered only for calculations of small vessels (less than 0.5 mm [7]) when the value speed and tangential stresses are small so for relatively large elements of the circulatory system simulations carried out with Newtonian or power-law fluid models [8].

Since the end of last century scientists have undertaken attempts to simulate blood flow as a viscoelastic fluid using various models for the components of the circulatory system with a diameter of more than 1 mm [9]. Research [10] presents the results of an experiment, aimed at comparing the flow of Newtonian and viscoelastic liquids during the course of the flow in the branching element of a circulatory system. Glycerol solution was used as a Newtonian fluid while the diluted solution of polyacrylamide was used as a non-Newtonian. Three-dimensional channel of 1 mm diameter was used as a branching element of the circulatory system. The experiment was carried out for values of Reynolds number 498 to 951 and showed, that the stream lines for two fluids are different. A stagnant area formed in the channel in the case of a non-Newtonian fluid. This effect didn't observe in the case of Newtonian fluid. Studies [11-13] discuss modeling of the flow of blood in various kinds of elements of the circulatory system with pathologies (stenosis, aneurysms, and etc.) using Newtonian and non-Newtonian fluid models. A comparison of the results shows, that the representation of the blood as a non-Newtonian fluid allows to catch those effects, which arise in the flow of blood, but not reproduced by Newtonian fluid.

One of the most important characteristics of the blood flow in the affected elements of the circulatory systems is normal and shear stresses. Normal stresses arise where the blood flow meets a restriction or expansion of vessels, separation, as well as rotation of the flow. Relatively high values of normal stress may lead to the damage of the formed elements of the blood, leading to deformation of the vessels walls and formation of plaques. High normal stresses are the consequence of pressure increasing in the blood, it is provoking factor for heart attacks.

There is a number of rheological constitutive relations predicting the behavior of such complex liquids as blood. One of these relations is rheological constitutive relation FENE – P [6]. The advantage of this model is its ability to predict characteristics of fluids such as those specific to blood. It can be noted, that this constitutive relation is used to simulate the flows of dilute polymer solutions too.

The purpose of this work is to achieve mathematical simulation of the flow of blood (a composition which contains protein macromolecules) in a stenosis affected elements of the circulatory system and discuss the influence of these macromolecules configurations on the flow picture in the vicinity of the affected element of the blood circulatory system.

2. Mathematical model

The schematic representation of the element of the circulatory system is shown in Fig.1. The model of an element of the circulatory system is a 2D channel with symmetrical double stenosis and the characteristic values are given in the table below.

Table 1. Symbols and accepted values

Title	Symbol	Value (mm)
The width of the channel	H	3
The length of stenosis	d	6

The width of the gap	h = 0,5H	1.5

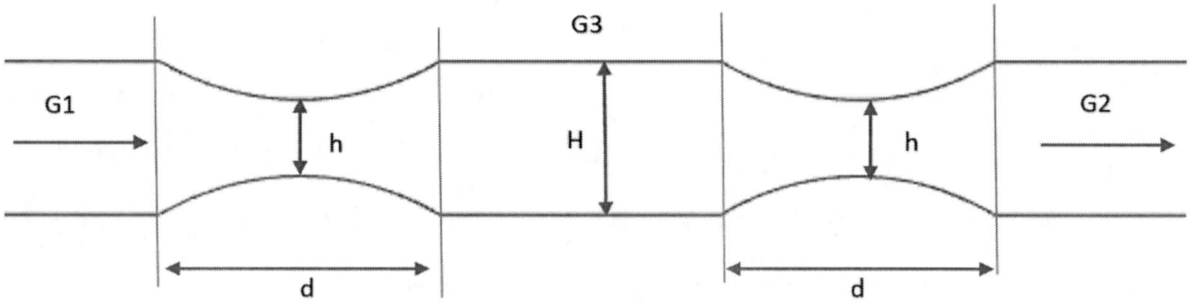

Figure.1 The shape of the affected areas of the blood circulatory system

It is assumed that the blood is an incompressible viscoelastic fluid with a given average density (ρ = 1000 kg / m^3), dynamic viscosity at zero shear rate (η_0 = 0.005 Pa·sec), and a stress relaxation time λ = 0.0189 sec.

In this paper, the FENE-P model (Finitely Extensible Nonlinear Elastic by Peterlin) is selected. It predicts the following properties: viscosity anomaly, the presence of stress relaxation and the dependence of the longitudinal viscosity on the longitudinal strain rate. These properties are also shown when the blood flows through the affected element of the circulatory system and underlie the occurrence of negative effects. Based on this, blood is considered to be a non-Newtonian fluid. This viscoelastic model is based on the representation of macromolecules in the form of flexible dumbbells. It is assumed that in the stream, such dumbbells can stretch relatively more to their original length and change orientation. While conducting an analogy with blood components, protein macromolecules were considered as flexible dumbbells.

The movement of blood through the circulatory system is described by the equations of motion and continuity [6, 14]:

$$\rho\left(\frac{\partial \vec{v}}{\partial t}+\vec{v}\cdot\vec{\nabla}\vec{v}\right)=-\vec{\nabla}p+\vec{\nabla}\cdot\tilde{\tau}, \tag{1}$$

$$\vec{\nabla}\cdot\vec{v}=0. \tag{2}$$

The total stress can be written as a sum as follows:

$$\tilde{\tau}=\tilde{\tau}^{p}+\tilde{\tau}^{s}. \tag{3}$$

Rheological constitutive relation can be written as follows:

$$\tilde{\tau}^{p}=\frac{\eta^{p}}{\lambda}\left[\frac{\tilde{A}}{1-\frac{tr(\tilde{A})}{3L^2}}-\frac{\tilde{I}}{1-\frac{1}{L^2}}\right], \tag{4}$$

$$\frac{\tilde{A}}{1-(tr\tilde{A})/(3L^2)}+We\overset{\nabla}{\tilde{A}}=\frac{\tilde{I}}{1-1/L^2}. \tag{5}$$

$$\overset{\nabla}{\tilde{A}}=\frac{\partial\tilde{A}}{\partial t}+\vec{v}\cdot\vec{\nabla}\tilde{A}-\vec{\nabla}\vec{v}\cdot\tilde{A}-\tilde{A}\cdot(\vec{\nabla}\vec{v})^{T}. \tag{6}$$

where $\vec{v}$ - is the velocity vector; ρ - the density of the blood; λ - the characteristic stress relaxation time; $\eta^0 = \eta^S + \eta^P$ - the blood viscosity at zero shear rate; η^P - the dynamic viscosity of non-Newtonian liquid component at zero shear rate; η^s - the dynamic viscosity of the blood plasma; $\tilde{\tau}^P$ - the non-Newtonian stress component; $\tilde{\tau}^S$ - the Newton stress component; $\tilde{A} = \frac{3\langle \vec{Q}\vec{Q} \rangle}{Q_{eq}^2}$ - the configuration tensor, where $\vec{Q}\vec{Q}$ - is the dyadic product of configuration vectors; $\langle \vec{Q}\vec{Q} \rangle = \int_{-\infty}^{\infty}\int_{-\infty}^{\infty}\int_{-\infty}^{\infty} \vec{Q}\vec{Q} \cdot P_N(\vec{Q}) \cdot d\vec{Q}d\vec{Q}d\vec{Q}$ - the averaging over the ensemble, where $P_N(\vec{Q})$ - the probability that a randomly chosen protein macromolecule has a predetermined size, which is in the range of $\vec{Q}$ to $\vec{Q} + d\vec{Q}$; $\tilde{D} = \frac{1}{2}\left(\vec{\nabla}\vec{v} + \left(\vec{\nabla}\vec{v}\right)^T\right)$ - is the deformation rate tensor, where $(\cdot)^T$ - is the transposition procedure.

Standard procedure for transforming equations into dimensionless form provide the following dimensionless variables:

$$We = \frac{\lambda U}{l}, \quad Re = \frac{\rho U l}{\eta^0}, \quad \beta = \frac{\eta^s}{\eta^0}, \quad L^2 = 3\left(\frac{Q_0}{Q_{eq}}\right)^2,$$

where U - characteristic velocity; l - characteristic linear scale; Q_{eq} - he length of the configuration vector of the protein macromolecule in equilibrium; Q_0 - the maximum possible length of the configuration vector; We-Weissenberg number; Re - Reynolds number; L^2 - is the parameter characterizing the degree of unraveling of the coil of the protein macromolecule.

2.1. Boundary conditions

At the channel entrance (boundary G1, fig.1) the constancy of the horizontal velocity component and the equality to zero of the vertical velocity components are set as follows:

$$U_x = const, \qquad U_y = 0. \tag{7}$$

To form a steady-state velocity profile, and prevent the influence of a given velocity profile in the input section on the flow in the central part of the channel, the length of the input part of the channel is selected as 10 widths. With this choice, as shown by test calculations, the fluid flow in the central part of the channel does not depend on the specification of the velocity profile in the inlet section.

At the output of the channel (boundary G2, fig.1) "soft" conditions are set:

$$\frac{\partial U_x}{\partial x} = 0, \quad U_y = 0, \quad \frac{\partial \tau_{xx}}{\partial x} = \frac{\partial \tau_{xy}}{\partial x} = \frac{\partial \tau_{yy}}{\partial x} = 0. \tag{8}$$

Boundary conditions (8) mean that the flow at the output is assumed to be steady with uniform velocity and stress profiles. According to test calculations, the selected length of the output part as 20 width is also suitable to fulfill condition (8).

On the solid walls of the channel (boundary G3, fig.1), no-slip condition are set, according to which the horizontal and vertical velocity components are zero:

$$U_x = 0, \qquad U_y = 0. \tag{9}$$

2.2. Initial conditions at the initial time in the entire flow field

The fluid flow at the initial time is so, that

$$U_x = 0, \qquad U_y = 0. \tag{10}$$

The problem was solved numerically, using the control volume method based on the integrated OPENFOAM 4.0 platform.

The decision was made for a semi - explicit time scheme. The stability was ensured by satisfying the Courant criterion [14].

3. Results and discussion

In numerical simulations, blood behavior is described using two fluid models: Newtonian and viscoelastic. By implementing a model of a viscoelastic fluid, a blood flow with pathologies of various "complexity" is represented. The degree of fluid "complexity" in this paper is determined by the value of parameter L^2, which characterizes the degree of unraveling of the protein macromolecules. For comparison, two limiting cases are considered, where $L^2 = 50$ and $L^2 = 700$, and the rest of the model parameters ($We = 1$, $Re = 100$, $\beta = 0.1$) are assumed as the same in value. In the model case, the walls of the circulatory system element are rigid, which is typical for vessels affected by stenosis. Such assumption takes place due to the following possible factors: with age, the walls lose their elasticity and permeability, which can be both a consequence and one of the causes of metabolic disturbances

Obtaining results for Newtonian fluids is possible when setting the Weissenberg number value, $We = 0.01$.

Fig. 2 shows the streamlines for non-Newtonian fluid at various values of the parameter L^2 and Newtonian fluid in the affected element of the circulatory system.

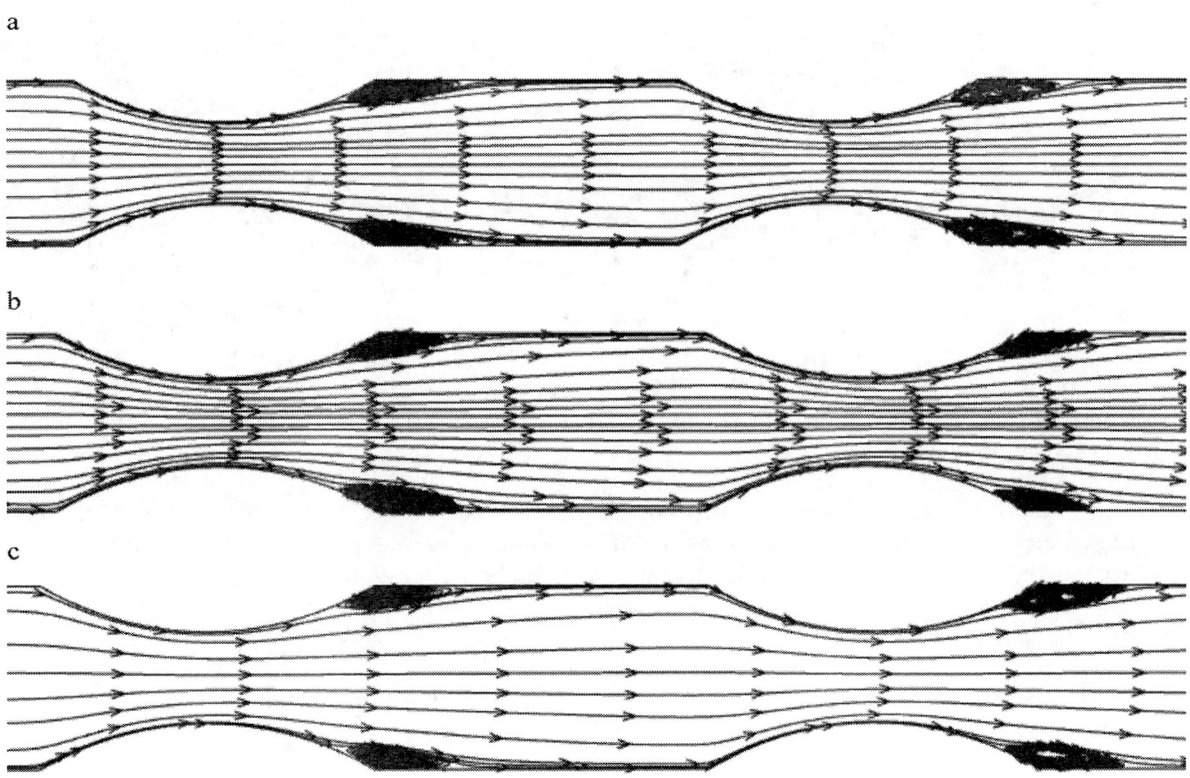

Figure 2. Streamlines. $L^2 = 50$ (a); $L^2 = 700$ (b) and Newtonian fluid (c)

The formation of stagnant areas is primarily associated with the form of the circulatory system element under consideration and the presence of special fluid properties. Near the walls there is a rapid change in speed of the flow velocity, while the change in speed in the central part of the artery is insignificant. In places where there is a change in the shape of the walls (in this case, the area where the stenosis and artery wall connect), the flow structure changes and stagnant areas form. The size of the stagnant areas depends on the shape, degree of narrowing and the length of stenosis.

Figure 3 shows the superposition of streamlines on the isolines of the principal stress difference. Evidently, that the influence of L^2 (Fig. 3 (b)) leads to the formation of a "tube" of oriented macromolecules from the central part of the stenosis along the stream in the stenotic element of the circulatory system. This "tube" of oriented macromolecules blocks the stagnant regions. Such "tube" prevents the penetration of fluid from the central part of the channel into a stagnant region. As a result, the geometric dimensions of stagnant areas are reduced.

In the discussed case, the influence of the "tube" on the size of the stagnant region is not significant, however, a comparison of these graphs (Fig. 3) demonstrates that for a Newtonian fluid, the emerging stress state is associated with a change in the shape of the channel.

a

b

c

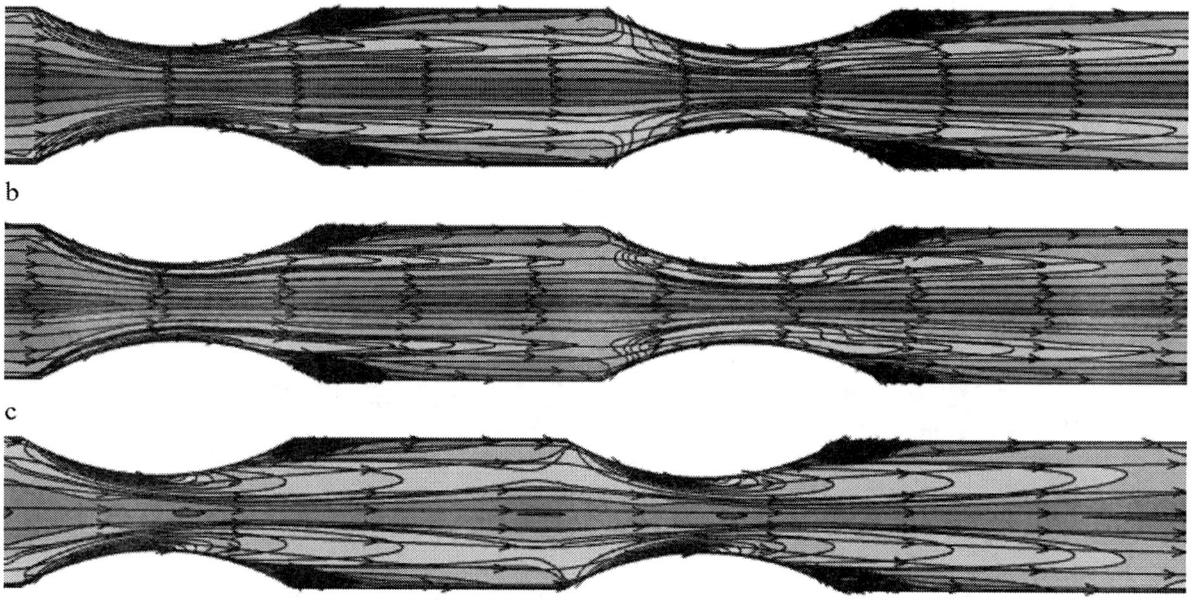

Figure 3. The distribution of principal stress difference in the artery to $L^2 = 50$ (a), $L^2 = 700$ (b) and a Newtonian liquid (c)

Fig. 3 (b) it is clearly distinguishable that the stagnant regions have different sizes. In this case, asymmetry is detected in the distribution of the isolines of the principal stress difference. This asymmetry is a consequence of differences in the orientation of protein macromolecules.

4. Conclusion

Using numerical modeling, models of blood flow in an artery with double stenotic plaques were created. The possibility of obtaining the flow behavior and predicting the occurrence of stagnant areas and other possible negative effects allows to complement the clinical picture of the disease and predict the consequences. Despite the apparent similarity of streamlines, it should be noted that two limiting model cases (Newtonian fluid and viscoelastic fluid at $L^2 = 700$) are significantly different from each other. The presence of protein macromolecules in the elements of the circulatory system affects the flow pattern in such a way that the symmetric form of the flow becomes asymmetric and areas of increased stress values appear in the flow.

A change in the parameter L^2 (increase to 700) significantly affects the nature of the flow too. Despite the decrease in stagnant areas, a "tube" of oriented macromolecules forms inside the circulatory system element. Thus, a 50% narrowing of the flow is retained in the channel outside the stenosis. The presence of asymmetry is explained by the combined influence of normal stress and altered conformation of protein macromolecules.

Thus, it may be concluded that the presence of a second stenotic lesion has a noticeable effect on the flow structure, and significantly increases the load on the human cardiovascular system.

5. References

[1] Chang R. W., Goodney P. P., Baek J. H. et al. Long-term results of combined common femoral endarterectomy disease. J Vasc Surg. 2008; 48 (2): 362-367.

[2] Dosluoglu H. H., Lall P., Cherr G. S., Role of simple and complex hybrid revascularization procedures for symptomatic lower extremity occlusive disease / J Vasc Surg. 2010; 51 (6): 1425-1435

[3] Adams R. J., Alberts G., Alberts M. J. et al. Update to the AHA / ASA recommendations for the prevention of stroke in patients with stroke and transient ischemic attack // Stroke. - 2008. - Vol. 39. - №5. - P. 1647-1652.

[4] Brott T. G., Hobson R. W., Howard G. et al. Stenting versus endarterectomy for treatment of carotid artery stenosis // N. Engl. J. Med. - 2010. - Vol.363, №1. - P.11-23

[5] Birchall D., Zaman A., Hacker J., Davies G., Mendelow D. Analysis of haemodynamic disturbance in the atherosclerotic carotid artery using computational fluid dynamics // Eur. Radiol. - 2006. - vol. 16. - P.1074-1083. DOI: 10.1007 / s00330-005-0048-6

[6] Tazyukov F. Kh., Kutuzova E. R., Snigerev B.A., Garifullin F.A., Layek G.C., Sharafeev A.Z. Blood flow peculiarities in vessels bifurcation. Research journal of pharmaceutical, biological and chemical sciences, ISNN: 0975-8585 - 2016 - v. 7 (6) - pp. 2750-2760.

[7] Mandal S. M., Mukhopadhyay S., Layek G. C. Pulsatile flow of shear-dependent fluid in a stenosed artery // Theoretical and Applied Mechanics. - 2012. - V. 39. - number. 3. - P. 209-231.

[8] Huh H. K., Ha H., Lee S. J. Effect of non-Newtonian viscosity on the fluid-dynamic characteristics in stenotic vessels // Exp. Fluids. - 2015. - Vol. 56, iss. 167. DOI: 10.1007 / s00348-015-2037-0

[9] Cho Y.I., Kensey K. R. Effects of the non-Newtonian viscosity of blood on flows in a diseased arterial vessel. Part 1: Steady flows // Biorheology. - 1991. - V. 28. - №. 3-4. - S. 241-262.

[10] Moravec S., Liepsch D. Flow investigation in a model of a three human artery with Newtonian and Non-Newtonian fluids, Biorheology, 20, 745, 1983

[11] Perktold K., Peter P., Resch M. Pulsatile non-Newtonian blood flow simulation through a bifurcation with an aneurysm, Biorheology, 26, 1011-1030, 1989

[12] Perktold K. Non-Newtonian blood flow simulation and wall shear stress in an arterial bifurcation, Proc. 2nd Int. Symp. On biofluid Mechanics and Biorheology, 1989, 685-694, Munich W., Germany

[13] Stefan H., Brandstatter W., Bachler G., Pucher R. Comparison of Newtonian and non-Newtonian blood flow in stenotic vessels using numerical simulation, Proc. 2nd Int. Symp. On biofluid Mechanics and Biorheology, 1989, 695-701, Munich W., Germany

[14] Tazyukov F. Kh. Kutuzova E. R., Snigerev B.A., Garifullin F.A. Blood flow in branched vessels with aneurysm. Russian journal of biomechanics, ISNN: 2409-6601 - 2018 - v. 22 - pp. 345-360.

IOP Conf. Series: Materials Science and Engineering **765** (2020) 012040 doi:10.1088/1757-899X/765/1/012040

Innovative Design of a Ventilated Modern Greenhouse Powered by PV/T Solar System

Abdullateef A.Jadallah*[1], Mohammed K.Alsaadi[2], Zayn A.Hafidh[3]

Electromechanical Engineering Department, University of Technology, Iraq.

Email: abdullateef.aljad@gmail.com
50055@uotechnology.edu.iq
Zayn.ahmed100@gmail.com

Abstract: In this paper , a modern design is proposed for ventilating and conditioning of greenhouse integrated thermal solar system in summer season in (33.3°N, 44.4 °E) Baghdad, Iraq with respect to East-West orientation. The proposed system involves photovoltaic panels and counter flow double pass (with fins). The photovoltaic solar thermal system is fixed on the roof of single span greenhouse. Experimental work was carried out during summer season. The results show the ventilation process changes the air inside temperature of the greenhouse to make it equal to the temperature required by the plants. The calculated root mean square error is 2.04 and correlation factor is 0.993 for the inside air temperature and 4.98, 0.98 respectively for the ground temperature.

Keywords: Ventilation process, BIPVT, Greenhouse ventilation, Greenhouse, Solar energy.

Nomenclature:

A_d, A_p, A_g	Door,Plant,Ground areas (m^2)	$P_{(T)}$	Partial vapor pressure at saturation (pa)
C_a	Air specific heat	S_t	Total solar radiation falling on the greenhouse cover (W)
E_v	Heat transfer through ventilation	τ	Transmissivity of the greenhouse cover (-)
F_{PR}	Shape factor between plant and greenhouse room (-)	α_p	Plant absorptivity of solar radiation (-)
h_a	Convective heat transfer coefficient between the greenhouse floor and inside air (W m^{-2}°C^{-1})	α_g	Ground absorptivity of solar radiation (-)
h_d	Heat transfer coefficient from the greenhouse door to the ambient air (W m^{-2}°C^{-1})	ρ	Reflectivity of the ground(-)
M_p	Total mass of plant (kg)	ε	Emissivity (-)

Abbreviations:

I.A.T Inside air temperature (C)
G.T Ground temperature (C)

1. Introduction

The framework of the glasshouse is made from glass, it used to provide appropriate environment for specific type of plants in the specific season, other times the heat must be pulled out using of

IOP Conf. Series: Materials Science and Engineering **765** (2020) 012040 doi:10.1088/1757-899X/765/1/012040

ventilating or cooling system. The air inside room temperature is a significant factor which affects the plants and varies with ambient temperature, wind speed and solar radiation falling on the greenhouse **[1]**.

Photosynthesis is a process which diverts the energy of sun into the nutrition by the plants (i.e., carbohydrates). This process is achieved in suitable circumstances by nature and artificially in glasshouse **[2]**. Greenhouse shape and orientation are important parameters that affect the amount of the total absorbed solar radiation. According to (**Mobtaker** and **Ajabshirchi**, 2016) the single span is acceptable quantum of solar radiation than other shapes of the greenhouses, such as Quonset, Arch, Vinery, uneven span and even span under the same climate conditions **[3]**. The two main elements that have higher cost which entered in glasshouse construction are the cooling and heating units. Heating process is conventionally supplied by burning diesel, liquid petroleum or fossil fuel that increases emission in the environment **[4]**. Recently, researches have showed that the photovoltaic systems with or without solar thermal units is satisfied solution in the greenhouse application specially for the rural, desert and distant zones. The solar energy is cheap, clean and zero emission technology. Besides, the maintenance is lower compared with conventional generators, The employing of photovoltaic panels on the top of greenhouse's roof increases the productivity of land and reduces utilized space **[5]**. One of using of the glasshouses on the planet is drying which is defined as the process of removing the moisture out of the crops after harvesting season to economize the agricultural product **[6]**. Sun drying, reduces the moisture in the crops by the 16% in 24 h whereas 17 h has been taken in the hothouse **[7]**. The outlet temperature depends on the solar radiation which considered a main effect, moreover the ambient temperature and mass flux, increases with time from 8 AM to 1 PM then reducing because of the decay of the solar radiation **[8]**. On the search data of air heating systems of (**A. El-Sebaii** and **S.Aboul-Enein**), the outlet temperature is higher in v-corrugated plate than the flat plate and more efficient by 11%-14% **[9]**. Between the words of (**E. Mashonjowa** and **F. Ronsse**) showed that the high investment project for the greenhouse in Zimbabwe were imported from Europe also few of them were successful due to the climatic differences conditions and the lack of experiences of the farmers to control the modern types of the greenhouse. With ventilated greenhouse with sensors devices which distributed in the room of the glasshouse, the thermometer sensed the thermal behavior for cover. Under the sunny condition was the difference 10 C between the measured and numerical results and 2 C in night **[10]**. (**C. Kittas** and **M. Karamanis**) explained the solar radiation is the main paramter to calculate the air temperature under the cover and the forced ventilation improve the conditions inside the greenhouse and reduce the overheating **[11]**. (**M. Z. Jacobson** and **V. Jadhav**) cleared up the tilting and tracking are used in all countries with all latitude to increase the solar radiation that incident on the rooftop of photovoltaic system **[12]**. (**A. Arbel** and **M. Barak**) used a combination between the forced ventilation and fogging system which based on providing water in small possible drops that can enhance the mass and heat exchange between air and water with accepted conditions 28 C in the air inside temperature **[13]**. In the present paper, an innovative design for a single span greenhouse is built also the temperatures distribution inside the greenhouse is investigated with or without the ventilation process.

2. Materials and Dimensions

A wooden structure of single span greenhouse is built with dimensions of (120cm*44.5cm*70cm). The walls contain 4mm glass cover , while the roof consists of photovoltaic solar panel (120cm*54cm) with packing factor 0.66 and 7mm thickness as shown in **Fig.1**.The boundaries of the glass layer and PV that are attached with wooden structure are covered with silicon to prevent leakage of air or water. With four small fans are fixed at the top of the north wall to support the ventilation process.

Fig.1. The present design of the single span greenhouse

3. Solar Radiation of the Glasshouse

The hourly incident solar radiation was measured using of solar power during summer season in Baghdad, Iraq. The Single span glasshouse is divided into five sections (west wall, east wall, south roof, south wall and north wall). The model is prepared for a clear day also counter (reverse) flow double pass photovoltaic thermal system has been used to dry the crops in summer

The total radiation available on the green house is given by:

$$(1) S_t = A_{SR} I_{SR} + A_{SW} I_{SW} + A_{NW} I_{NW} + A_{WW} I_{WW} + A_{EW} I_{EW}$$

where Asw, Asr, Aww, Aew, Anw are the Area of south wall, south roof, west wall, east wall and north wall respectively (m^2). The refraction index ' n ' and extinction coefficient K are used for calculating transmitted solar radiation falls through the cover of greenhouse.

(**Duffie** and **Beckman**, 1991) proposed an expression for refraction radiation transported from air with refraction index n_i considered as (1) to glass with refraction index (1.526) **[14]**.

4. Thermal Modeling of the Greenhouse:

4.1 Glasshouse plants:

According to **Fig.2** the energy balance that determines the plants temperature is:

$$\alpha_p(1-\rho)\tau S_t(1-F_t) = M_p C_p \frac{dT_p}{dt} + h_{pr} A_p (T_p - T_R) + h_r A_p (T_p - T_R) \quad (2)$$

where: (T_p, T_R) the plant and air inside temperatures (C) respectively. Also (F_t) the ratio of the transmitted solar radiation falling on walls/ roof inside the greenhouse to the total transmitted solar radiations inside the greenhouse at the same time. Total solar fraction is taken from references. In Addition to the expression h_r , h_{pr} and h_p has analyzed in [1].

The Convective and evaporative heat transfer coefficient from the plant to the inside air ($W\ m^{-2} C^{-1}$) is showed in equation (3a) :

$$h_{pr} = h_p + \frac{0.016 \times h_p [P(T_p) - \gamma P(T_R)]}{T_p - T_R} \quad (3a)$$

The Convective heat transfer coefficient between plant and inside air (W $m^{-2}C^{-1}$) is:

$$h_p = 2.8 + 3(V) \tag{3b}$$

Also the radiation heat transfer coefficient between plant and inside air (W $m^{-2}C^{-1}$) is:

$$h_r = F_{PR}\varepsilon\sigma(T_p^2 - T_R^2)(T_p + T_R) \tag{3c}$$

4.2 Glasshouse floor:

The energy balance which calculates the ground temperature is:

$$\alpha_g(1-\alpha_p)(1-\rho)\tau S_t(1-F_t) = -KA_g\frac{\partial T}{\partial x}_{x=0} + h_a A_g(T_{x=0} - T_R) \tag{4a}$$

The expression in the steady state condition of the conduction thermal energy in the ground is

$$-KA_g\frac{\partial T}{\partial x}_{x=0} = h_b A_g(T_{x=0} - T_\infty) \tag{4b}$$

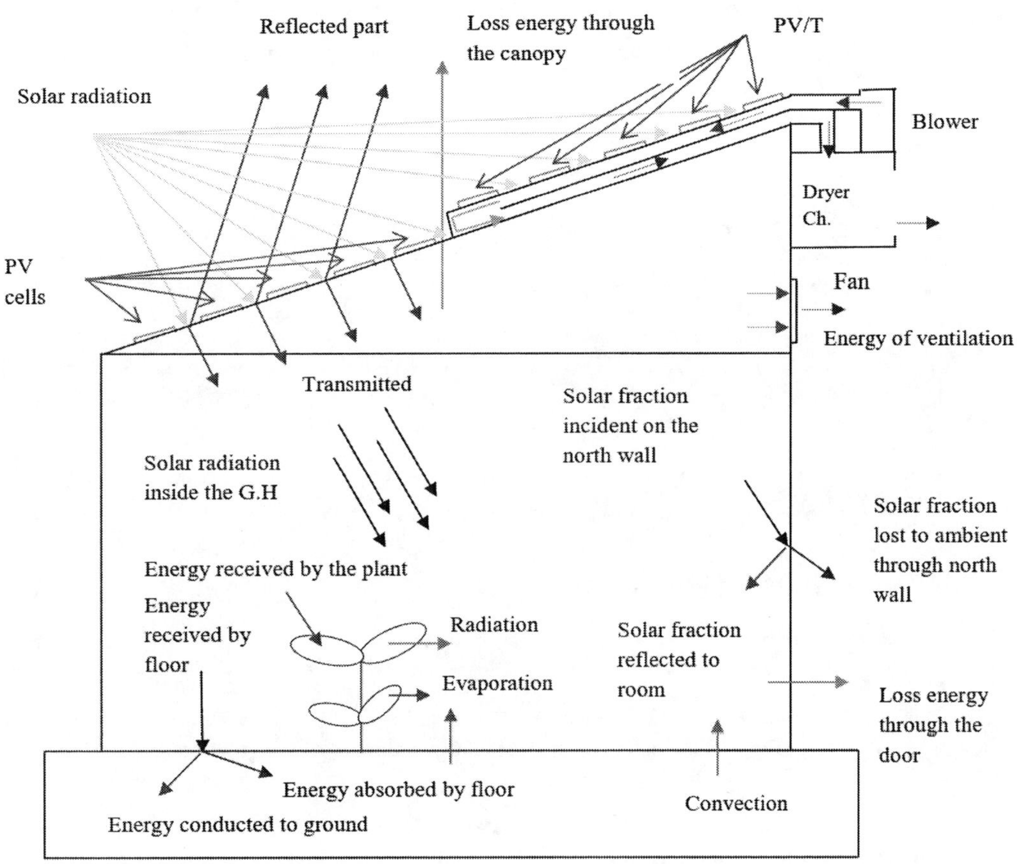

Fig.2. Working principle of a greenhouse with solar radiation

where: (h_b) bottom heat transfer coefficient between the greenhouse floor and the ground beneath (W $m^{-2}{}^{o}C^{-1}$). And (T_∞) the temperature which is constant after certain depth of ground also is considered equal to the underground temperature beneath the glasshouse floor (T_o). Hence the equation (4a) can be written according to the **Fig.3**:

IOP Conf. Series: Materials Science and Engineering **765** (2020) 012040 doi:10.1088/1757-899X/765/1/012040

$$\alpha_g(1-\alpha_p)(1-\rho)\tau S_t(1-F_t) = h_b A_g(T_{x=0}-T_\infty) + h_a A_g(T_{x=0}-T_R) \tag{5}$$

4.3 Glasshouse air

As shown in Fig.3 the energy balance that carried out to reach the air inside temperature is:

$$A_p h_{pr}(T_p - T_R) + A_p h_r(T_p - T_R) + h_a A_g(T_{x=0} - T_R) + (1-\alpha_g)(1-\alpha_p)(1-\rho)\tau S_t(1-F_t) + \rho F_t\, \tau S_t(1-\rho) = m_a C_a \frac{dT_R}{dt} + U_t A_c(T_R - T_a) + h_d A_d(T_R - T_a) + E_v \tag{6}$$

Where: (U_t) the total heat transfer coefficient of the glasshouse. According to equation (6) the inside air temperature can be simplified as:

$$T_R = \frac{\left[\rho F_t + \left((1-\alpha_g)(1-\alpha_p) + \alpha_{ge} H_G\right)(1-F_t)\right](1-\rho)\tau S_t + Z T_a + A_p(h_{pr}+h_r)T_p + A_g h_b T_o H_G \pm E_v}{Z + A_p(h_{pr}+h_r)T_p + A_g h_b H_G}$$

Where; (7)

$$Z = A_c U_t + h_d A_d + h_b A_g \tag{8a}$$

$$H_G = \frac{h_a A_g}{h_b A_g + h_a A_g} \tag{8b}$$

$$\alpha_{ge} = \alpha_g(1-\alpha_p) \tag{8c}$$

$$P_{(T)} = \left[exp\left(20.386 - \frac{5132}{T}\right)\right] \times 133.3 \tag{8d}$$

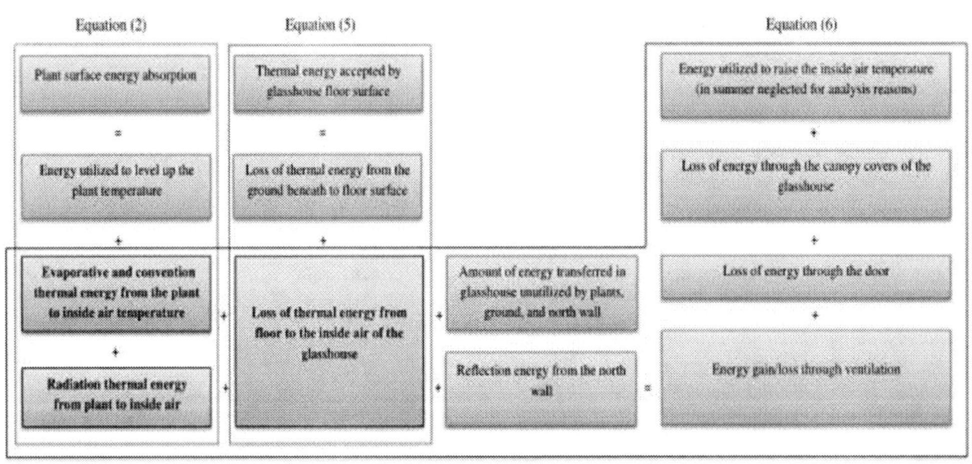

Fig.3. Schematic diagram of energy balances of the greenhouse

A computer program is written in MATLAB is built to carry out the necessary determinations to calculate the inside air temperature in each hour of the day in addition to mathematical simplifications $(Z, H_G, \alpha_{ge}, P_{(T)})$ prepared as constants in (table 1) :

Table 1

Constants used for the mathematical thermal model

Bottom heat transfer co.(h_b)	1 W $m^{-2}°C^{-1}$	Plant specific heat (C_p)	4190 J $Kg^{-1}°C^{-1}$
Cover area (A_c)	1.9 m^2	Reflectivity of the ground (ρ)	1
Door area (A_d)	0.09 m^2	Shape factor (F_{PR})	1
Emissivity (ε)	0.7	Stefane-Boltzman constant (σ)	5.67*10^-8 W $m^{-2}K^{-4}$
Ground area (A_g)	0.54 m^2	Total mass of plant (M_p)	5 kg
Ground absorptivity (α_g)	0.75	Underground temperature (T_o)	17 C
Plant area (A_p)	0.54 m^2	Upper heat transfer co. (h_a)	10 W $m^{-2}°C^{-1}$
Plant absorptivity (α_p)	0.4	Velocity (v)	0.5 m S^{-1}

5. Statistical Analysis

The accuracy of the modules is determined with different indicators contains root mean square error (RMSE) and correlation coefficient (r) were computed according to data series **[15, 16]**.

$$RMSE = \left[\frac{1}{K}\sum_{i=1}^{K}\left(e_{i,c} - e_{i,m}\right)^2\right]^{0.5} \tag{9}$$

$$r = \frac{\sum_{i=1}^{K}(e_{i,c}-e_{c,ave})(e_{i,m}-e_{m,ave})}{\sqrt{\left[\sum_{i=1}^{K}(e_{i,c}-e_{c,ave})^2\right]\left[\sum_{i=1}^{K}(e_{i,m}-e_{m,ave})^2\right]}} \tag{10}$$

6. Experimental Details

Single span glasshouse is built with East-West orientation and (0.54m^2 floor area), at Baghdad (33.3°N latitude and 44.4 °E longitude) Iraq. A door is located the east-wall of the greenhouse wit screw to serve the mechanism of opening and closing. The south roof consists of photovoltaic panel with 7mm thickness, packing factor 0.66 and 33.2° as slope angle. The length, width and height of greenhouse are 1.16m, 0.47m, and 0.74m respectively. The totals cover area of the glasshouse is (four walls and one roof) 1.9m^2. Four square holes is opened in the top of the north wall to fix the fans sequentially .The hourly solar radiation was measured in solar meter. In addition the inside air and ground temperatures were measured with thermocouples. One sensor is fixed outside the glasshouse to sense the ambient temperature and three inside the glasshouse under shade (0.3m, 0.5m, and 0.7m) far from the door.

7. Discussion and Results

Every wall or roof receives solar intensity according to the sun movement, for example the East-Wall gained a solar radiation from 5 AM – 12 AM then starting to decrease due to movement of sun which reaches to the middle of south wall and roof of the greenhouse. Because of the convergent measured values that is sensed between (15/7/2019) and (16/7/2019) then the measured values for the solar radiation and ambient temperature in (15/7/2019) is decided to be the reference for this study. Also Fig.4 illustrates the hourly solar radiation available on each part of the greenhouse in summer season at 33.3°N (15/7/2019).

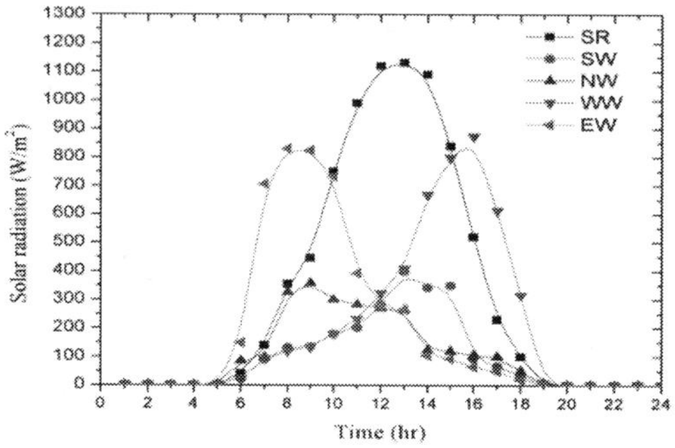

Fig.4. Hourly solar radiation available on each wall and roof of the greenhouse

It can be notice from **Fig.5** that the relationship between the total solar radiations which is gained by the greenhouse and the ambient temperature. The ambient temperature increases from 6 AM to 2 PM when the sun is perpendicular on the earth. The total solar radiation is effects largely because of the exposed area of glass to sun light. At first hours of the light, the rays of sun effected on the east wall of the greenhouse then reaching to south wall and roof which are the biggest areas among the other walls of greenhouse that exposed to sun (the top value of total solar radiation at 1 PM is 730 w). So, the amount of the total radiation increases according to increment in area of glass which gained the sun rays.

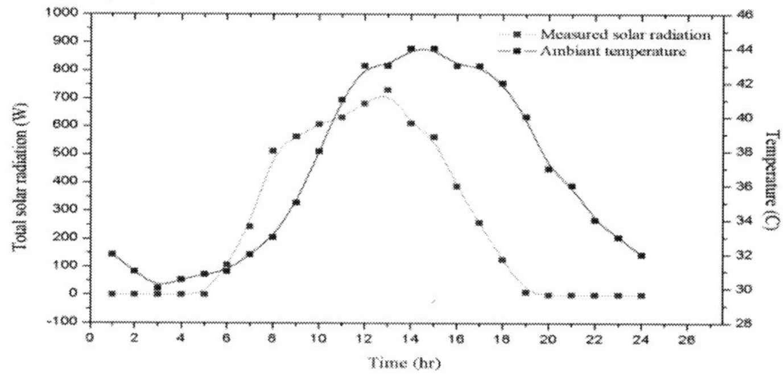

Fig.5. The relationships between the total solar radiation and the ambient temperature with respect to time

(I.A.T) is influenced by the solar radiation gained in each hour of the day. Positive relationship is available between (I.A.T) and the total solar radiation. **Fig.6** shows the experimental and calculating values of the (I.A.T) during the day (15/7/2019) without consideration of ventilation and (16/7/2019) with it.

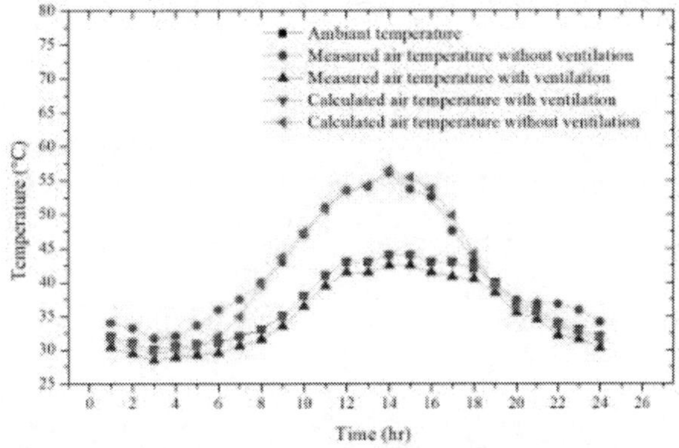

Fig.6. The measured and calculated computation of the (I.A.T) during the day (15/7/2019) addition to (16/7/2019) the process of ventilation.

It is noted that the convergence between the measured and determined (I.A.T) without ventilation and closer to what is known as congruence between the ambient temperature and calculated (I.A.T) with ventilation because of the fan pumped an air supplied from inside of the green house. Four fans worked from 8AM-6PM and one fan is worked from 6PM-8AM due to that can explain the difference between measured (I.A.T) with ventilation and calculated (I.A.T) with ventilation in most of hours in day.

The experimental and calculated ground temperatures are illustrated in **Fig.7** with and without ventilation. G.T is affected by the falling radiation from the walls, the inverted radiation from the north wall and underground temperature which are the reasons of the high value of G.T. The range of G.T is expanded between 8 AM – 5 PM because of the increment of total solar radiation which is passed through the glasshouse. In addition to the cover of the south roof with packing factor 0.66 not permeable to huge amount of solar radiation to enter from the roof to the inside. So the south wall is responsible to increment of G.T.

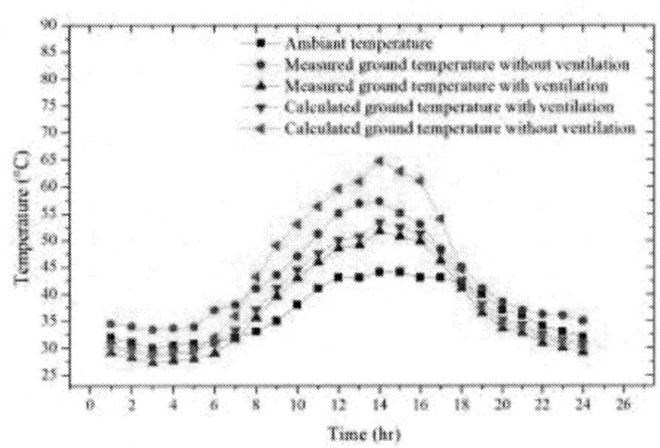

Fig.7. The measured and calculated determination of G.T during the day (15/7/2019) addition to (16/7/2019) the process of ventilation.

8. Conclusion

The total solar radiation of glasshouse is affected by the solar radiation that is provided by the sun with respect to the radiation fall on the area. Also the increment of (I.A.T) depends on the total solar radiation and ambient temperature. (G.T) is influenced by the total solar radiation, reflected radiation from the north wall, the underground temperature and (I.A.T). Further, the innovative design with large scale the purpose of the counter (reverse) flow double pass photovoltaic solar thermal system dries the crops that converts the hot air from the PVT to small dryer room that includes the harvest fruits.

References:

1-V.P. Sethi, "On the selection of shape and orientation of a greenhouse: thermal modeling and experimental validation", Sol. Energy 83 (2009) 21e38.
2-R.D. Singh, G.N. Tiwari, "Thermal heating of controlled environment greenhouse: a transient analysis", Energy Convers. Manage 41 (2000) 505e522.
3-H.G Mobtaker ,Y.Ajabshirchi,S.F.Ranjbar, M.Matloobi, "Solar energy conservation in greenhouse: Thermal analysis and experimental validation", Renewable Energy 96 (2016) 509-519.
4-L.Chai, C.Ma, J.Q.Ni , "Performance evaluation of ground source heat pump system for greenhouse heating in northern China" , bio systems engineering 111 (2012) 107-117
5-R.H.E.Hassanien , M.Li , W.D.Lin , "Advanced applications of solar energy in agricultural greenhouses", Renewable and Sustainable Energy Reviews 54 (2016) 989–1001.
6-R.K.SAHDEV, A.K.DHINGRA, M.KUMAR, "A comprehensive review of greenhouse shapes and its applications", Front. Energy, DOI 10.1007/s11708-017-0464-8.
7-A.Elkhadraoui, S.Kooli, I.Hamdi, A.Farhat, "Experimental investigation and economic evaluation of a new mixed mode solar greenhouse dryer for drying of red pepper and grape", Renewable Energy 77 (2015) 1-8.
8-M.A.Karim, M.N.A.Hawlader, "Performance evaluation of a v-groove solar air collector for drying applications", Applied Thermal Engineering 26 (2006) 121–130.
9-A.A.El-Sebaii , S.Aboul-Enein, M.R.I. Ramadan, S.M. Shalaby, B.M. Moharram, "Investigation of thermal performance of-double pass-flat and v-corrugated plate solar air heaters", Energy 36 (2011) 1076-1086.
10- E. Mashonjowa, F. Ronsse , J.R. Milford, J.G. Pieters, "Modelling the thermal performance of a naturally ventilated greenhouse in Zimbabwe using a dynamic greenhouse climate model", Solar Energy 91 (2013) 381–393.
11- C. Kittas, M. Karamanis, N. Katsoulas, "Air temperature regime in a forced ventilated greenhouse with rose crop", Energy and Buildings 37 (2005) 807–812.
12- Mark Z. Jacobson, Vijaysinh Jadhav, "World estimates of PV optimal tilt angles and ratios of sunlight incident Upon tilted and tracked PV panels relative to horizontal panels", Solar Energy 169 (2018) 55–66.
13- A. Arbel, M. Barak, A. Shklyar, "Combination of Forced Ventilation and Fogging Systems for Cooling Greenhouses", Biosystems Engineering (2003) 84 (1), 45–55.
14-J.A.Duffie,W.beckmen, "Solar Engineering of Thermal Processes".
15- M. Zarzo, P. Marti, "Modeling the variability of solar radiation data among weather stations by means of principal component analysis", Appl. Energy 88 (2011) 2775e2784.
16-O.O. Ajayi, O.D. Ohijeagbon, C.E. Nwadialo, O. Olasope, "New model to estimate daily global solar radiation over Nigeria", Sustain Energy. Tech. Assess. 5(2014) 28e36.

Control Strategy on Ducting Vertical Axis Wind Turbine Based on Permanent Magnet Synchronous Generator

Abdullateef A. Jadallah

Sahar R. Farag

Jinan D. Hamdi

Electromechanical Engineering Department, University of Technology, Baghdad, Iraq

corresponding author's e-mail : abdullateef.aljad@gmail.com

Abstract. Control methodology for a standalone vertical axis wind turbine with variable speed with a PMSG is studied in this research. Where a simplified control strategy adopted for the generator side converter in order to harness ultimate power was investigated and achieved employing Sim power dynamical system simulating code. The controller is able to maximize output of varying speed vertical axis turbine during variable wind. vector control scheme is utilized to control the load side PWM inverter. This is to sustain the inverter's amplitude output voltage. It was noticed that, the controller can keep the load voltage well at a load and kept constant by changing the load status. Generating system having this proposed control statement is viable for the installation of the small-scale wind turbine. This wind turbine system is characterized by its standalone status and variable speed. Such types of wind turbines are suitable for rural areas power supply. Results reveal that controller runs very well and depicts very well dynamical and a good steadily performance.

Keywords: Drive train, inverter, LC filter, tip speed ratio, vertical axis wind turbine, pitch angle.

1. Introduction

The capability of changing wind turbines speed may be greatly enhanced by using a PM synchronous with a direct motor. PMSG received a lot of interest in applying with wind turbine **[1]**. PMSG have. self-exciting, that permits running at a. high factor of power and good efficiency **[2]**. It works in at higher relatively factor of power due to the miss-magnetization current which may be more effective over induction machines. For production of ultimate power gained from the variable wind velocities; running of the wind turbine conversion system generator is essential. It needs a proven controlling scheme. Optimal tracking of torque and produced power is a well-known control strategy as it serves to provide optimal harvesting of wind energy **[3-5]**. These strategies do not appropriate because of its high cost for micro-scale turbines. Many sources of wind energy have been fitted in remote villages and isolated islands **[6]**. In recent years, wind power generation systems present a focus point in the search for renewable energy resources. This is largely because of the quick progress in the capacity of wind turbines, further the advancement of power electric and their applications in wind power production. For the isolated settlements settled far away from the utility grid, a sensible trend to independent power generating consists employing turbines having storages to initiate complete system **[7]**. Many studies have been conducted focusing on establishing the energy capturing privileges associated with changeable wind speed operation. Xabier Juankorena et.al 2009 **[8]** demonstrated a procedure for initial regulating of the variable-speed vertical axis wind turbine.

IOP Conf. Series: Materials Science and Engineering **765** (2020) 012041 doi:10.1088/1757-899X/765/1/012041

Basic management is the method by which the unsteadiness or fluctuations between the power produced and the energy required are stabilized. The system involves of the linear change of the power generated with the network frequency, and if the network frequency rises, the power would generated be reduced and so on. This proposed methodology is included by varying the power, and the speed curve of the wind turbine according to the network frequency, hence changing the percent of the ultimate available power added into the network. It is obtained by running the turbines at different rotating speeds than the ideal, with uncontrolled angle. This study recommends that the primary method of pitch regulation in depend on of wind speed and do not require controlling the angle of pitch. Rajveer Mittal 2010 **[9]** proposed a battery storage system for different speed of turbines using (PMSG). Off-network wind systems inherently have storage because of the wind availability doesn't provide the requested energy demand continuously. If the load request is fewer than the power generated by the wind. Thus, the overstocking power is consigned in the batteries. K.Kalyan Raj.et.al 2012 **[10]** literate and compared various methods and schemes for the stability of voltage of SEIG insulated variable speed drops. Their concluded that power electronics transformers ensure promoting results on voltage control. Arshad Nawaz.et.al2013 **[11]** studied the regulation method of output voltage of independent wind turbines run at a variable wind speed. This method bases upon the voltage regulating utilization for the vibration voltage of the wind turbine driven at variable speed. The voltage regulation is provided for utility. The battery system is also suggested for a system that can ensure power when the wind regulator voltage is fallen below threshold value because of wind absence or low wind speed. Samanvorakij and Kumkratug 2013 **[12]** this study presented of PMSG modelling. It includes of wind turbine, PMSG, the angle of pitch control and the AC/DC/AC converter model. Permanent magnet synchronized generator and converter are created in (d-q) model. Results of simulation showed that the angle of pitch controls actively with at high wind speeds, to set the power coefficient for generator.

2. Duct and Wind Turbine Model

The flow is considered as one dimensional, steady and incompressible, so the velocity and pressure at out let of convergent duct calculated by **[13]**:

$$V_{out}=V_{in}A_{in}/A_{out} \;...\; (1)$$

$$P_{out} = P_{in} - \rho_{in} V_{in} (V_{out} - V_{in}) \;...(2)$$

For divergent duct wind turbine, the increment in wind velocity at inlet of the divergent duct is referred to the aspect area ratio **[14]**.

$$V_{rotor} = A_{out}/A_{in} * V_{\infty} * (1 - a) \;...\; (3)$$

$$p_{rotor} = p_{\infty} - (1/2)\rho \; (V_{\infty}^2 - V_{rotor}^2) \;...(4)$$

Combining the convergent part with the divergent duct via the test section which include the VAWT rotor present the third configuration that is the convergent-divergent ducting system. Figure (1) shows all case of ducted wind turbine for variable wind speed input.

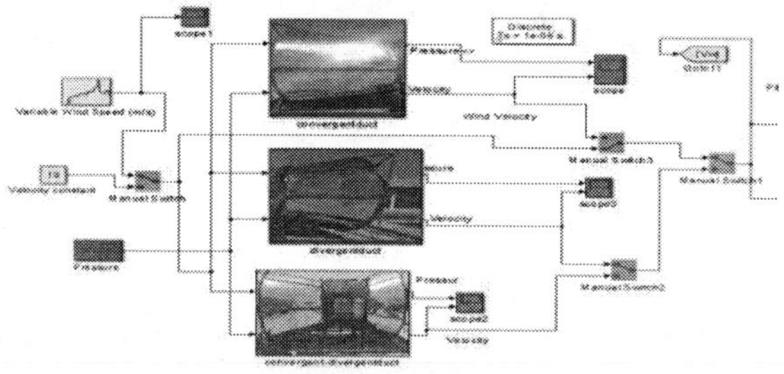

Figure 1:Matlab simulink ductimg system

3.Model of Wind Turbine

The wind turbine model shown in figure (2) describes the aerodynamic model that extract power from the wind in the form of mechanical energy which is fed to the generator via a shaft. The aerodynamical power is given by the relation below **[15]**.

$$P_m = \frac{1}{2}\rho A V^3 Cp \ldots (5)$$

We shall use a generic equation of as proposed by **[15]**. The equation is expressed below as

$$C_{P(\beta.TSR)} = 0.22\left(\frac{116}{TSR_i} - 0.4 * \beta - 5\right) * \exp\left(\frac{-12.5}{TSR_i}\right) + 0.0068\, TSR_i \ldots (6)$$

And

$$\frac{1}{TSR_i} = \frac{1}{TSR + 0.08\,\beta} - \frac{0.035}{1+\beta^3} \ldots (7)$$

The relation between the TSR and the angular speed of the wind rotor is given as:

$$\omega = \frac{TSR * V}{R} \ldots (8)$$

While, the mechanical torque in terms of mechanical power is given by the equation below:

$$T_m = \frac{P_m}{\omega} \ldots (9)$$

Subsisting for P from (5) and from (8) into (9), the mechanical torque will be:

$$T_m = \frac{1}{2}\rho A R V^2 C_t \ldots (10)$$

Where the torque coefficient is given below:

$$C_t = \frac{Cp}{TSR} \ldots (11)$$

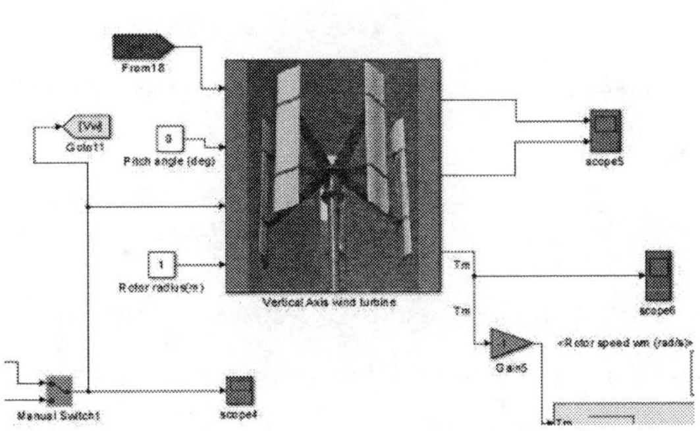

Figure 2:Matlab simulink of VAWT

4. Modeling and Control of the Variable vertical axis Turbine

4.1 General

IOP Conf. Series: Materials Science and Engineering **765** (2020) 012041 doi:10.1088/1757-899X/765/1/012041

The control frame of a PMSG installed wind turbine operating with variable speed shown in figure (3) which involve the wind turbine- vertical axis. PMSG with single switch and (3) phase switch-ode rectifier and also a vector controlled (PWM) voltage source inverter. PMSGs do not appropriate for implementing as it is variability in amplitude and the frequency because variability of wind speed. Constant DC voltage will be desired for a direct usage, storing or converting to AC by employed inverter. The proposed single switch 3- phase rectifier includes a 3-phase diode bridge rectifier and also DC to DC converter. The output of the switch-mode rectifier may be controlled via managing the duty cycle of the active switch at any the air speed to produce ultimate power gained from the wind rotor and hence to provide the demand consequentially.

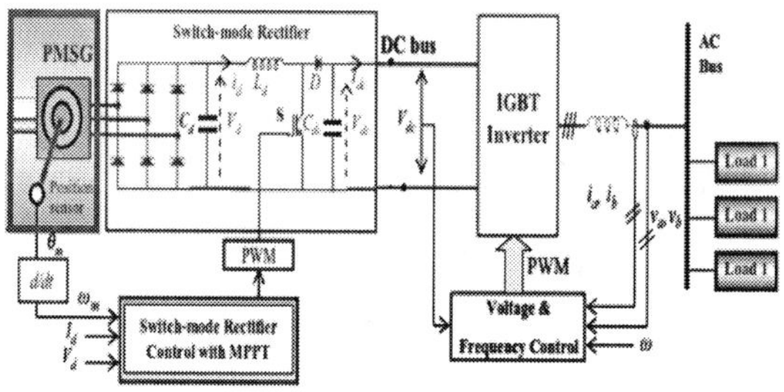

Figure 3: Control Structure of a PMSG based standalone variable speed VAWT [16].

4.2 Control of Switch-Mode Rectifier

Figure (4) shows the frame of the suggested control policy of the switch mode rectifier. Control top was aimed to control the duty cycle of switch (S) presented in figure (3) for utilizing the power maximum from the variable wind speed vertical wind turbine and transmit the power to the demand load. The program of control consists the following steps:

- Measuring of the speed of the generator,
- Obtain the reference torque considering the equation bellow:

$$K = \frac{1}{2}\rho A(\frac{R}{TSR})^3 \mathrm{Cp} \text{ ...(12)}$$

$$T_g^* = Kw^2 \text{ ...(13)}$$

The reference torque is inherently used to compute the value of DC reference current via measuring the output voltage of the rectifier v,V_d as it may give as:

$$I_d^* = T_g^* * \frac{w^2}{V_d} \text{ ....(14)}$$

The difference occurred between the based DC-current and the DC-current from measurement employed to change the duty cycle of the switch in order to manage the output and generated torque got from the switch-mode rectifier via a PI controller. **[16].**

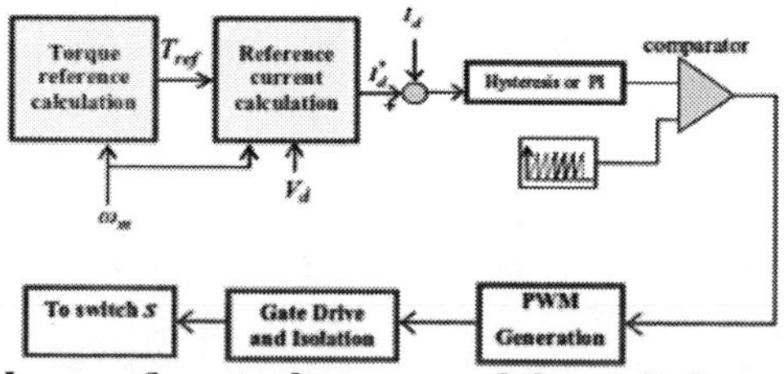

Figure 4: Scheme of control process of the switch-mode rectifier [16]

4.3 Control Load Side of Inverter

The supply side converter is targeted to tune the frequency and voltage. The output voltage must be controlled in term of frequency and amplitude as there is non- grid coexist standalone wind energy conversion system. The control frame for standalone control mode includes of voltage output controller, current controller and dc link voltage controller. Voltage output controller is utilized to control the output-voltage among transient load or fluctuation of speed of wind. DC-link controller voltage is utilized to balance the dc link voltage. The dc-voltage PI controller keeps voltage referred to the referencing value. The PI controller is utilized to modulate the voltage output and current in the internal control loops so the dc voltage controller in the outer-loop. In order to requite the cross-coupling impact to keep the output filter into the rotating reference frame. The compensating terms are then added as illustrated in the following figure.

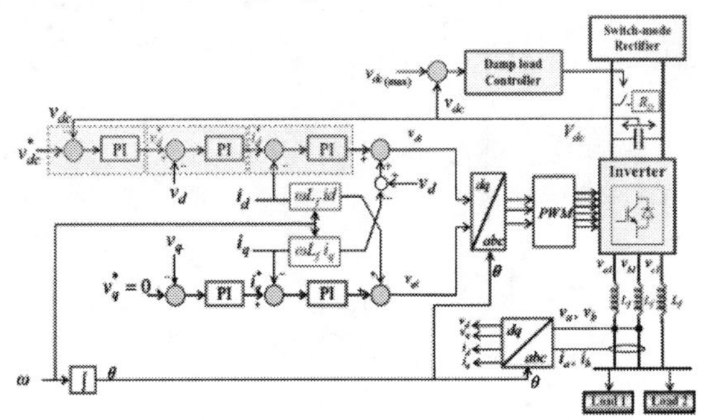

Figure 5. Structure of the vector control for the standalone mode of operation [16]

Figure (6) shows the **MATLAB** Simulink module of duct wind turbine based on permanent magnet synchronous generator. This model shows a direct driven generator by wind turbine shaft.

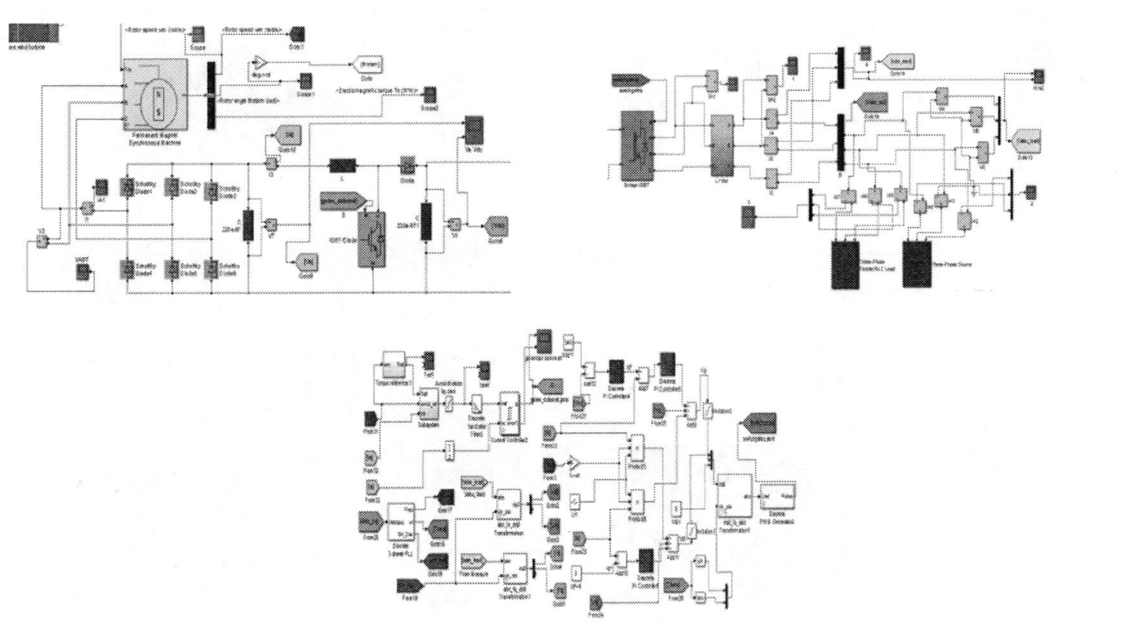

Figure 6: Wind turbine system Simulink model

5. Results and Discussion of Wind Turbine Simulink Model:

In Simulink work, for a wind turbine having 0.5 m rotor radius, zero and section pitch angle, the following results were achieved: Figure (7) shows the use of variable wind speed as an input for ducted wind turbine- vertical axis, variation in the wind speed is used to describe the process of control of the generated power got from the generator of the change and acceptance of wind speed in

IOP Conf. Series: Materials Science and Engineering **765** (2020) 012041 doi:10.1088/1757-899X/765/1/012041

case of ducted and non-ducted wind turbines as well. That the variation in rotor speed according to variable wind speed has reached maximum value about 32 rad/sec of rotor speed. This fluctuation in rotation speed entering the generator lead to variable voltage output as presented in figure (8). Figure (9) shows produced mechanical torque by the wind turbine while the electromagnetic torque was represented in figure (10). It was noticed that the generator torque has a similar behavior of mechanical torque but negative sense as the mean of input generator torque. Figures (11) and (12) depicts the voltage and current at generator terminal respectively. DC-DC converter is utilized to control the voltage. So, the Dc voltage output from the 3-phase diode rectifier depends upon the input voltage which has a varying level. As demonstrated in figure (13) the converter controls the voltage to be approximately constant vs. figure (14). Line to line voltage output from DC-AC converter was illustrated in figure (15) and after filter line to line voltage output becomes presented in figure (16). Figures (17 and 18) illustrate the output voltage (3-phase) from the wind turbine system.

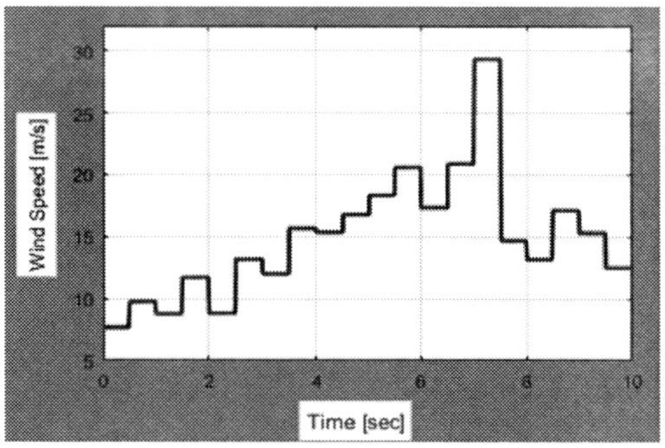

Figure (7) Wind speed as a function of time

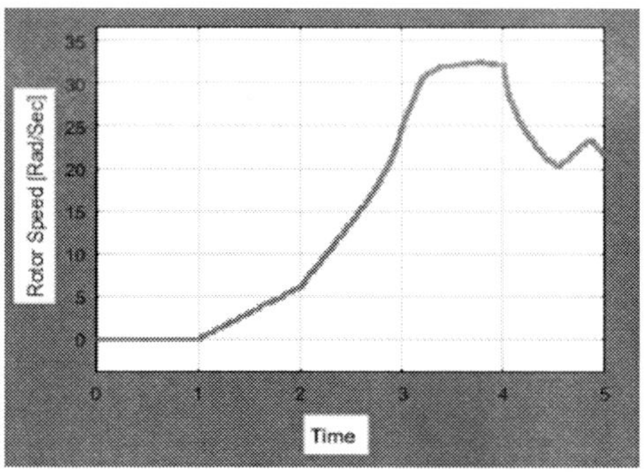

Figure (8) Rotor speed as a function of time

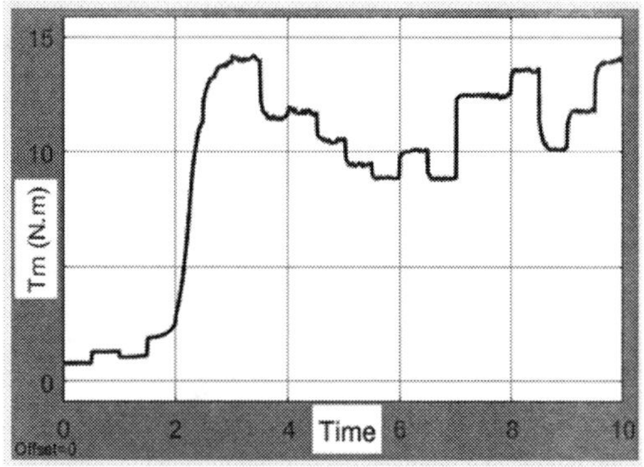

Figure (9) Turbine output torque as a function of time

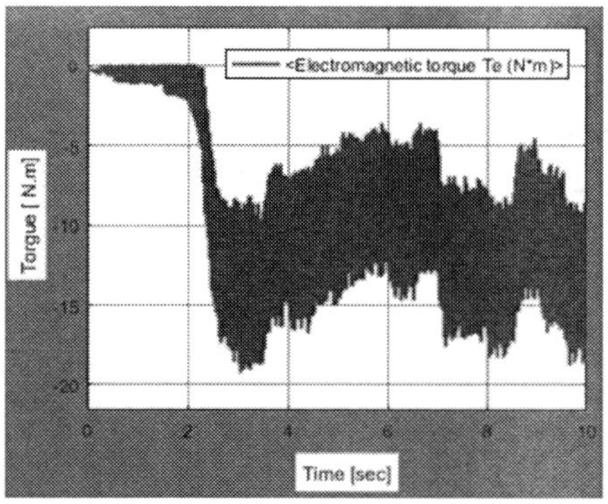

Figure (10) PMSG electromagnetic torque as a function of time

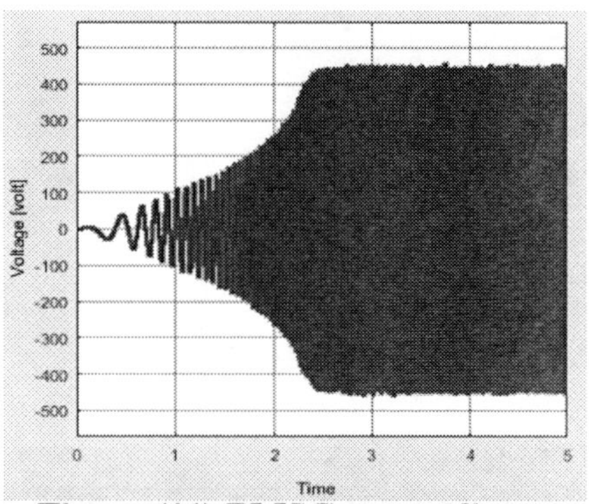

Figure (11) PMSG output line to line voltage

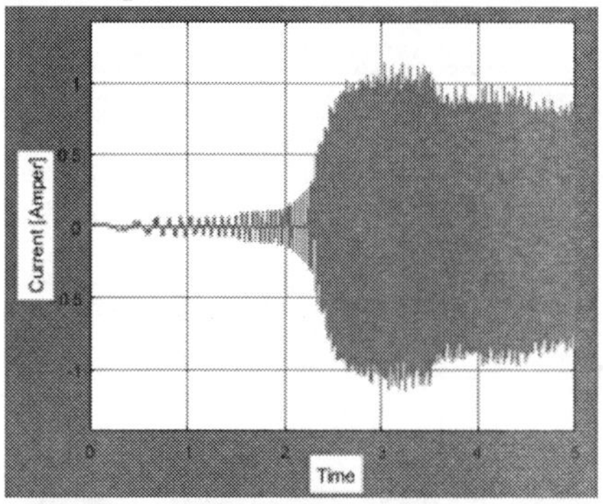

Figure (12) PMSG output line to line current

IOP Conf. Series: Materials Science and Engineering **765** (2020) 012041 doi:10.1088/1757-899X/765/1/012041

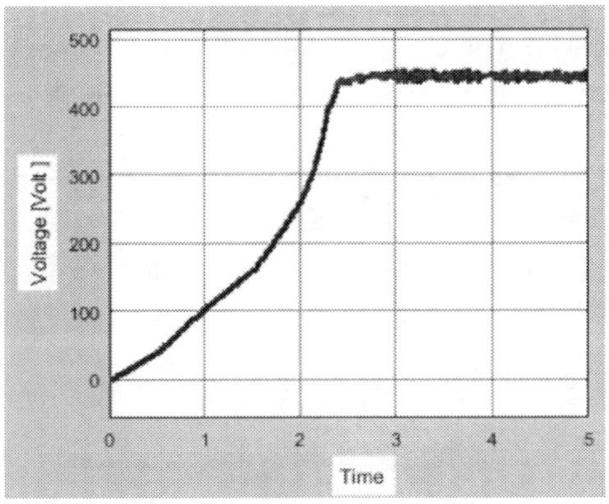

Figure (13) Rectifier output voltage

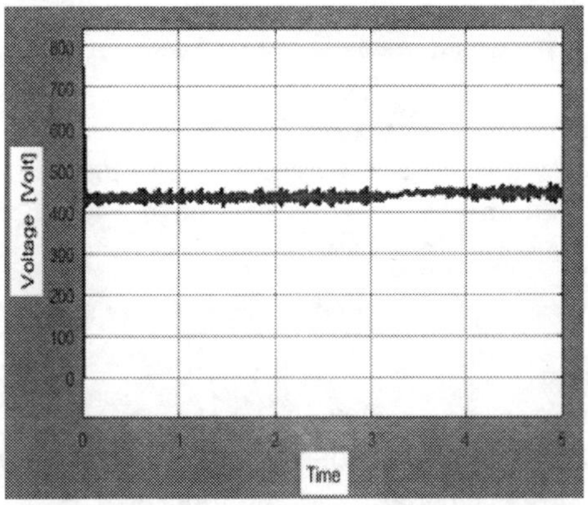

Figure (14) Controlled Dc voltage

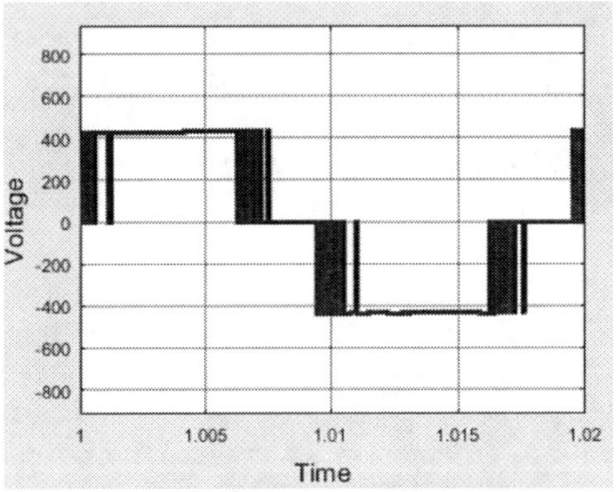

Figure (15): Line to line inverter output voltage before filtering

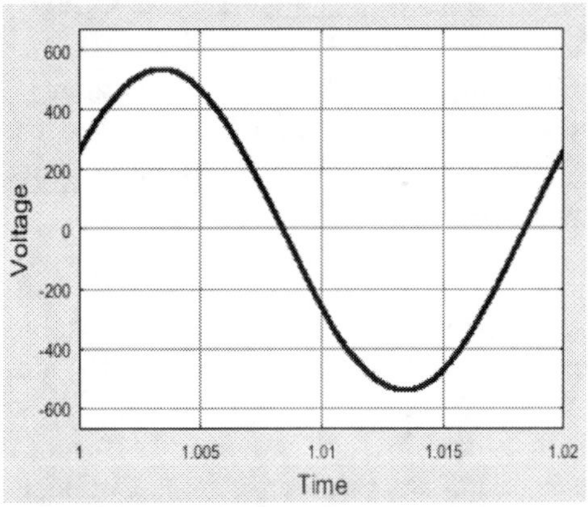

Figure (16): Line to line inverter output voltage after filtering

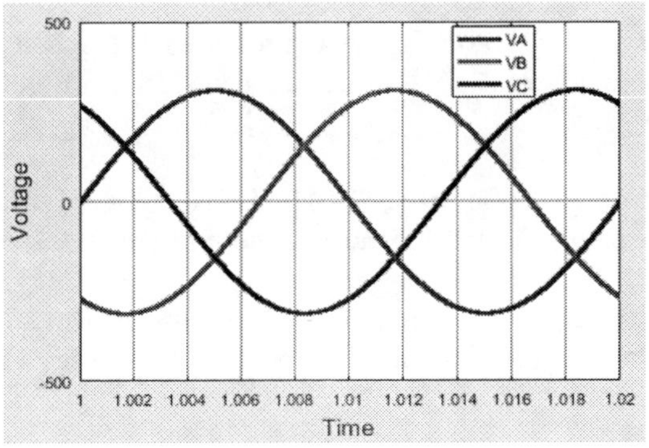

Figure (17) Wind turbine system AC output voltage

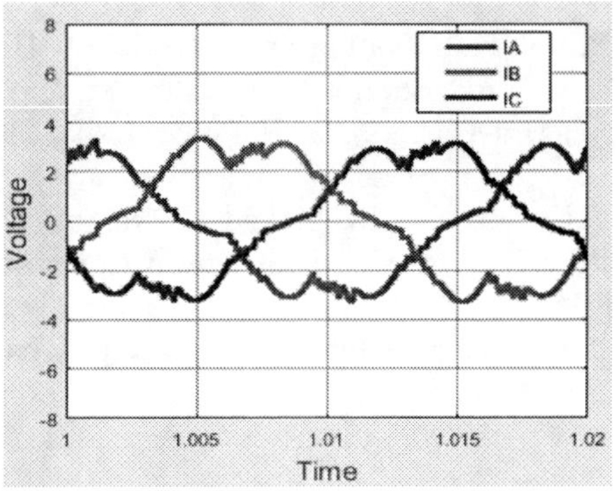

Figure (18): Wind turbine system AC output current

IOP Conf. Series: Materials Science and Engineering **765** (2020) 012041 doi:10.1088/1757-899X/765/1/012041

6. Conclusion

In this paper, the Darrius vertical axis wind turbine was modelled and simulated and conducted using **MATLAB/Simulink**. The modeled system with all subsystems has specified and investigated for validation. In the modeling process of the mechanical transmission, gear ratio is neglected which leads to a reduction in losses in the system. The algorithm of control for the grid side converter along the LC filter was developed in order to regulate the voltage of grid side with specified power quality.

References

[1] M. E. Haque "Control of a Stand-Alone Variable Speed Wind Turbine with a Permanent Magnet Synchronous Generator" M1-8 ember, IEEE, PP, (2008).

[2] T. F. Chan, L. L. Lai "Permanent-Magnet Machines for Distributed Generation: A Review in Proc" IEEE power engineering annual meeting, (2007).

[3] N. Eswaramoorthy, M. Siva Ramkumar, G. Emayavaramban, A. Amudha, S. Divyapriya, M. Sivaram Krishnan, D. Kavitha " A Control Strategy for A Variable Speed Wind Turbine with A Permanent Magnet Synchronous Generator Using Matrix Converter with SVPWM" ISSN: 2277-3878, Volume-8, Issue-1S4, June)2019.

[4] Anjana Jain1, S. Shankar1 and V. Vanitha2" Power Generation Using Permanent Magnet Synchronous Generator (PMSG) Based Variable Speed Wind Energy Conversion System "Journal of Green Engineering, Vol. 7 4, 477–504(2018). doi: 10.13052/jge1904-4720.742

[5] M. Chinchilla, S. Arnaltes, and J. C. Burgos "Control of Permanent-Magnet Generators Applied to Variable Speed Wind-Energy Systems Connected to The Grid" IEEE Trans" On Energy Conver, vol 21, no. 1, March, (2006).

[6]. S. R. Hadian-Amrei and H "Iranmanseh. Novel Direct Power Control for Compensating Voltage Unbalance and Load Fluctuations in PWM Rectifiers" ACSE Journal, vol. 6, no. 4, Dec, (2006).

[7] Ankit Kumar Singh.et.al "Modeling and Control of Grid Connected Variable Speed PMSG Based Wind Energy System" Atlantis Press, April (2013).

[8] X. Juankorena, I. Esandi, J. Lôpez, and L. Marroyo "Method to Enable Variable Speed Wind Turbine Primary Regulation" (2009).

[9] R. Mittal, K. S. Sandhu, and D. K. Jain "Battery Energy Storage System for Variable Speed Driven PMSG for Wind Energy Conversion System" Joint International Conference on Power Electronics, Drives and Energy Systems & Power India, (2010).

[10] K. Kalyan, E. Swati, and C. Ravindra, "Voltage Stability of Isolated Self Excited Induction Generator (SEIG) for Variable Speed Applications using Matlab / Simulink" no. 3, (2012).

[11] Arshad Nawaz, Muhammad Naeem Arbab "Voltage Regulation of Variable Speed Wind Turbine using MATLAB/Simulink" International Journal of Engineering and Advanced Technology (IJEAT) ISSN: 2249 – 8958, Volume-2, Issue-5, June (2013).

[12] S. Samanvorakij and P. Kumkratug " Modeling and Simulation PMSG based on Wind Energy Conversion System in MATLAB/SIMULINK " Proc. of the Second Intl. Conf. on Advances in Electronics and Electrical Engineering — AEEE (2013) doi:10.3850/ 978-981-07-5939-1_15.

[13] Patrick H, Oosthuizen , William E. Carscallen "Introduction to compressible fluid flow" 2nd Edition (2013).

[14] Geurts B, Simao Ferreira C, Van Bussel G "Aerodynamic analysis of a vertical axis wind turbine in a diffuser" Carlos Simão Ferreira (2015).

[15] B.O.Omijeh, C. S. Nmom, E. Nlewem. " Modeling of a Vertical Axis Wind Turbine with Permanent Magnet Synchronous Generator for Nigeria" International Journal of Engineering and Technology Volume 3 No. 2, February, (2013).

[16] M. Haque, M. Negnevitsky & K. Muttaqi, "A Novel Control Strategy for A Variable Speed Wind Turbine with A Permanent Magnet Synchronous Generator" IEEE Industry Applications Society Annual Meeting, (2008).

IOP Conf. Series: Materials Science and Engineering **765** (2020) 012042 doi:10.1088/1757-899X/765/1/012042

MPPT Control for Grid Connected Wind Energy Conversion System Based Permanent Magnet Synchronous Generator (PMSG) and Five-Level Neutral Point Clamped Converter

Youssef Errami, Abdellatif Obbadi, Smail Sahnoun

Laboratory: Electronics, Instrumentation and Energy- Team: Exploitation and Processing of Renewable Energy, Department of Physics, Faculty of Science, University Chouaib Doukkali, Eljadida, Morocco

errami.emi@gmail.com

Abstract. This paper presents nonlinear control of a Wind Energy Conversion System (WECS) based Permanent Magnet Synchronous Generator (PMSG) and Five-Level Neutral Point Clamped Voltage Source Converter (5L-NPC-VSC). For WECS, the Maximum Power Point Tracking (MPPT) is a very important requirement in order to maximize the effectiveness. Therfore, MPPT approach and grid-connected control methods are discussed. On the other hand, the proposed control techniques are based on Sliding Mode Approach (SMA) and Vector Control (VC) to achieve the MPPT, to keep the dc link voltage constant and to the control active and reactive powers. The stability of the regulators is obtained using Lyapunov analysis. An extensive simulation study, using MATLAB/Simulink, is conducted on a 4-MW wind farm composed of two 2 MW PMSG based wind turbines.

1. Introduction

At the present time, Permanent Magnet Synchronous Generator (PMSG) is widely used in Wind Energy Conversion System (WECS), due to its advantages such as excellent power factor, simple control, low noise, low losses and highly efficient operation [1-2]. On the other hand, power electronic converters are principally used as the interface between the WECS and the power grid to achieve high performance and effectiveness when connected to the electric network. Diverse topologies have been employed for the WECS based PMSG such as PWM rectifiers, boost converters with diode rectifier and back to back Voltage Source Converters (VSC) [3]. In addition, the Maximum Power Point Tracking (MPPT) is a very important requirement in order to maximize the effectiveness to increase the annual energy yield of the WECS [4].

The consistent and efficient operations of the WECS depend on the control methods. Hence, numerous control approaches for the WECS based PMSG have been considered, for both the generator and the grid-side VSCs, in several researches [5]-[18]. Jaramillo-Lopez et al. and Wu et al. [5-6] proposed novel RBF neural network methods and cloud model to extract maximum power of the WECS based PMSG. In [7], Das et al. presented a novel $H\infty$ based robust control system to control active and reactive power flow between the WECS and the electrical network. In [8], Errami et al. developed control strategies for integrated control of wind turbine based PMSG. In [9], backstepping control approach was proposed to compensate the parameter uncertainties of the WECS based PMSG. Moreover, a Proportional Integral Control (PIC) model was presented in [10], and a MPPT method was presented in [11]. On the other hand, sliding mode (SM) control method is used as one of the most attractive nonlinear controllers for the WECS, because such controller owns excellent anti-disturbances ability, also it has low sensitivity with respect to uncertainty, dynamic performance and good robustness [12-13]. The SM is one of the powerful control approaches for systems with uncertainties and unknown trouble [14-15].

In this paper, the SM approach and the Vector Control (VC) are employed for controlling the wind farm system based PMSG and Five-Level Neutral Point Clamped Voltage Source Converter (5L-NPC-VSC) (Fig. 1). The control of the rectifiers by the SM ensures the MPPT. The control of the 5L-NPC-VSC is based on a VC. The grid side inverter sustains a constant DC-link voltage and control the active and reactive powers that the WECS exchanges with the electrical power. Also, it is used to achieve unity power factor (UPF).

This paper is organized as follows. Section 2 presents description of the system dynamic model. In Section 3, the proposed control system is discussed. Section 4 presents and discusses the simulation results. Several conclusions are summarized in section 5.

2. Proposed system architecture

2.1. Wind turbines

The mechanical power extracted from the wind is defined as [16]:

$$P_{turbine} = \frac{1}{2}\rho A_{turbine} C_p(\lambda_t, \beta_t) v_w^3 \tag{1}$$

where ρ is the air density, $A_{turbine}$ is the area swept by the rotor blades (in m^2), v_w is the wind speed and $C_p(\lambda_t, \beta_t)$ is the turbine power conversion coefficient. λ_t and β_t are tip speed ratio (TSR) and pitch angle, respectively.

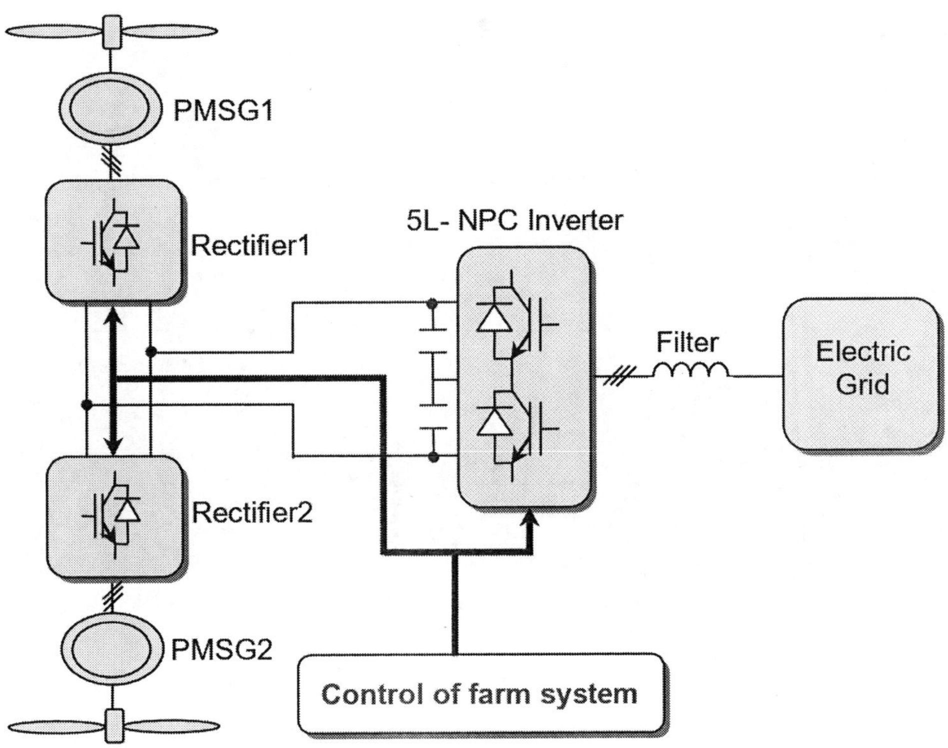

Fig. 1. Structure of the wind farm system.

IOP Conf. Series: Materials Science and Engineering **765** (2020) 012042 doi:10.1088/1757-899X/765/1/012042

The expression for $C_p(\lambda_t, \beta_t)$ is defined as [17]:

$$C_p(\lambda_t, \beta_t) = \frac{1}{2}\left(\frac{116}{\lambda_g} - 0.4\beta_t - 5\right)e^{-\left(\frac{21}{\lambda_g}\right)}$$
$$\frac{1}{\lambda_g} = \frac{1}{\lambda_t + 0.08\beta_t} - \frac{0.035}{\beta_t^3 + 1} \tag{2}$$

To extract MPP from wind, the C_p must be set in maximum value ($C_{p-\max} = 0.4104$). So, the TSR is equal to optimal value ($\lambda_{t-opt} = 8.1$) and $\beta_t = 0$. The TSR is:

$$\lambda_t = \frac{\omega_{PMSG} R_t}{v_w} \tag{3}$$

where ω_{PMSG} is the angular velocity of the generator, R_t is the radius of blades. The mechanical torque T_t is given with:

$$T_{turbine} = \frac{P_{turbine}}{\omega_{PMSG}} \tag{4}$$

Fig. 2 shows the relationship between the power coefficient C_p and tip speed ratio, so to achieve the MPPT, the optimum velocity is obtained when the TSR is set in $\lambda_{t-opt} = 8.1$. Then, the power extracted from the turbine will be at its maximum value:

$$P_{turbine\_\max} = \frac{1}{2}\rho A_{turbine} C_{p-\max}\left(\frac{R_t \omega_{PMSG-opt}}{\lambda_{t-opt}}\right)^3 \tag{5}$$

Consequently, the PMSG velocity reference is:

$$\omega_{PMSG-opt} = \frac{v_w \lambda_{t-opt}}{R_t} \tag{6}$$

For each wind velocity, the MPPT algorithm can be completed by adjusting the speed of the generator to maintain an optimal λ_{t-opt} [17]. Fig. 3 shows the curves of extracted power and maximum extracted power of the turbine at different rotor speeds.

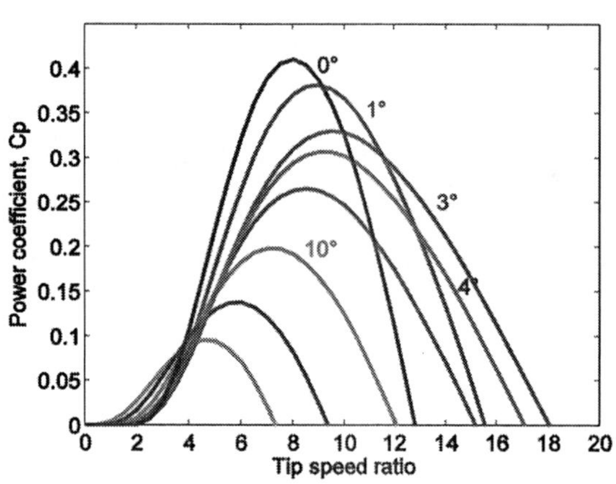

Fig. 2. Characteristics of C_p for various values of the pitch angle β

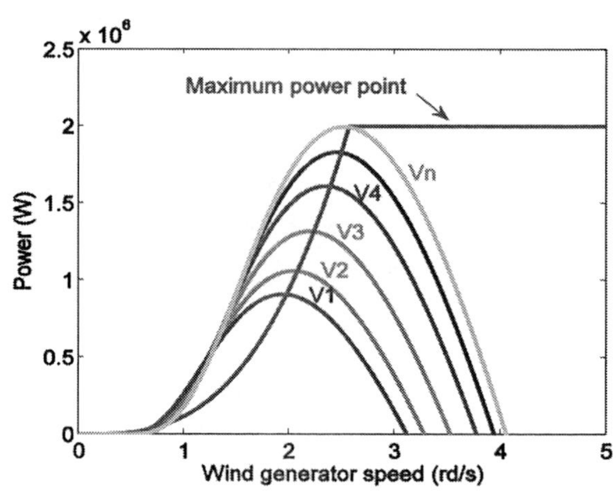

Fig. 3. Wind generator power curves at various wind speed

On the other hand, to avoid excessive rotor acceleration under high-velocity wind conditions, pitch angle control is activated to limit the output power and keep the PMSG speed at the reference value $\omega_{PMSG-opt}$. The structure of the typical pitch control system is shown in Fig. 4.

2.2. *Modelling of generator*

The voltage equations of a PMSG in dq reference frame are expressed as [18]:

$$\begin{bmatrix} v_{md} \\ v_{mq} \end{bmatrix} = \begin{bmatrix} R_g + p.L_s & -\omega_e L_s \\ \omega_e L_s & R_g + p.L_s \end{bmatrix} \begin{bmatrix} i_d \\ i_q \end{bmatrix} + \begin{bmatrix} 0 \\ \omega_e \psi_f \end{bmatrix} \tag{7}$$

Where v_{mq}, v_{md} are the stator voltages in the dq frame, i_q, i_d are the stator current in the dq frame, R_g is the stator resistance, L_s is the in inductance of q-d axis and ψ_f is the permanent magnetic flux. The electrical rotating speed of the generator, ω_e is defined as:

$$\omega_e = p_n \omega_{PMSG} \tag{8}$$

where p_n is the number of pole pairs of the generator, ω_g is the mechanical angular velocity. If the PMSG is assumed to have equal ($L_q = L_d = L_s$), the expression of the electromagnetic torque of generator can be described as:

$$T_{ge} = \frac{3}{2} p_n \left[\psi_f i_q \right] \tag{9}$$

The dynamic equation of the wind turbine system is described by:

$$J \frac{d\omega_{PMSG}}{dt} = T_{ge} - T_t - F_v \omega_{PMSG} \tag{10}$$

where J is the moment of inertia, F_v is the viscous friction coefficient, T_t is the mechanical torque developed by the wind turbine.

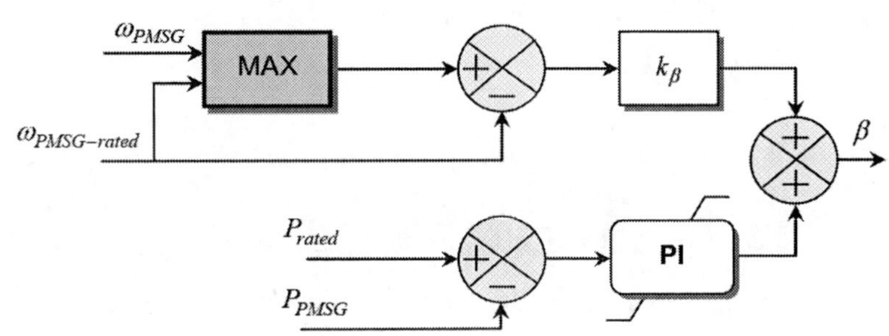

Fig. 4. Pitch angle controller.

3. Wind farm control

3.1. Proposed Sliding Mode control of PMSG

The following sliding surface is defined:

$$S_{\omega-PMSG} = \omega_{PMSG-opt} - \omega_{PMSG} \tag{11}$$

based on (10),

$$\frac{dS_{\omega-PMSG}}{dt} = \frac{d\omega_{PMSG\_opt}}{dt} - \frac{1}{J}(T_t - T_{ge} - F_v\omega_{PMSG}) \tag{12}$$

The regulator is employed to track a $\omega_{PMSG-opt}$ such as [19]:

$$S_{\omega-PMSG} = \frac{dS_{\omega-PMSG}}{dt} = 0 \tag{13}$$

Also, the following function will be used to guarantee the sliding mode conditions:

$$u_{\omega-PMSG} = k_{\omega-PMSG}\,\mathrm{sgn}(S_{\omega-PMSG}) \tag{14}$$

where $k_{\omega-PMSG}$ is a constant that is positive and $\mathrm{sgn}(S_{\omega-PMSG})$ is a sign function defined as [20-21]:

$$\mathrm{sgn}(S_{\omega-PMSG}) = \begin{cases} 1 & S_{\omega-PMSG} \succ \gamma \\ \dfrac{S_{\omega-PMSG}}{\gamma} & \gamma \geq |S_{\omega-PMSG}| \\ -1 & -\gamma \succ S_{\omega-PMSG} \end{cases} \tag{15}$$

Therefore, using equations (9-10) and (11-15), we can obtain the following command:

$$i_{q-ref} = \frac{2}{3p_n\psi_f}(T_{ge} + J\frac{d\omega_{PMSG-opt}}{dt} + F_v\omega_{PMSG} + Jk_{\omega-PMSG}\,\mathrm{sgn}(S_{\omega-PMSG})) \tag{16}$$

Then the control surfaces for the currents are chosen:

$$S_{q-PMSG} = i_{q-ref} - i_q \tag{17}$$

$$S_{d-PMSG} = i_{d-ref} - i_d \tag{18}$$

where i_{q-ref} and i_{d-ref} are the reference of current components. In addition, the regulators are employed to track the reference of current components such as:

$$S_{q-PMSG} = \frac{dS_{q-PMSG}}{dt} = 0 \tag{19}$$

$$S_{d-PMSG} = \frac{dS_{d-PMSG}}{dt} = 0 \tag{20}$$

From (7) and (17-20), the commands to track i_{q-ref} and are given by:

$$v_{q-ref} = L_s\omega_e i_d + \omega_e\psi_f + R_g i_q + L_s\frac{di_{q-ref}}{dt} + L_s\,k_{q-PMSG}\,\mathrm{sgn}(S_{q-PMSG}) \tag{21}$$

$$v_{d-ref} = -L_s\omega_e i_q + R_g i_d + L_s k_{d-PMSG}\,\mathrm{sgn}(S_{d-PMSG}) \tag{22}$$

where $k_{d-PMSG} \succ 0$ and $k_{q-PMSG} \succ 0$. The function of Lyapunov is chosen [22]:

$$\Re_{Generator} = \frac{1}{2}S^2_{\omega-PMSG} + \frac{1}{2}S^2_{d-PMSG} + \frac{1}{2}S^2_{q-PMSG} \tag{23}$$

$$\frac{d\Re_{Generator}}{dt} = S_{\omega-PMSG}\frac{dS_{\omega-PMSG}}{dt} + S_{d-PMSG}\frac{dS_{d-PMSG}}{dt} + S_{q-PMSG}\frac{dS_{q-PMSG}}{dt} \tag{24}$$

IOP Conf. Series: Materials Science and Engineering **765** (2020) 012042 doi:10.1088/1757-899X/765/1/012042

Using equations (7), (11-12), (16-18) and (21-22), (24) becomes:

$$\frac{d\Re_{Generator}}{dt} = -k_{\omega-PMSG}\, S_{\omega-PMSG}\, \text{sgn}(S_{\omega-PMSG}) - k_{d-PMSG}\, S_{d-PMSG}\, \text{sgn}(S_{d-PMSG}) - k_{q-PMSG}\, S_{q-PMSG}\, \text{sgn}(S_{q-PMSG}) \tag{25}$$

Consequently:

$$\frac{d\Re_{Generator}}{dt} = -k_{\omega-PMSG}\left|S_{\omega-PMSG}\right| - k_{d-PMSG}\left|S_{d-PMSG}\right| - k_{q-PMSG}\left|S_{q-PMSG}\right| \prec 0 \tag{26}$$

The trajectory is maintained on the sliding surfaces. Therefore, the objectives of the control strategy of the PMSG are ensured. Fig.5. presents a scheme of each PMSG control using the SM technique.

3.2. *Proposed Vector Control of 5L-NPC*

The 5L-NPC is used:
1) To control the dc-link voltage to be stable under wind speed variations;
2) To synchronize the ac power generated by the WECS with the power grid;
On the other hand, the grid active and reactive powers are expressed as [23-24]:

$$P_{grid} = \frac{3}{2} V_{grid} i_{d-grid} \tag{27}$$

$$Q_{grid} = \frac{3}{2} V_{grid} i_{q-grid} \tag{28}$$

So, the grid active and reactive powers are controlled through direct and quadrature electric network current components, respectively. The proposed control system is shown in Fig. 6. As seen, the system uses the d- and q-axis PI current controllers. The d- and q-axis current commands are generated by the PI regulators and compared with their actual values. In addition, the q-axis current reference $i_{q-grid-r}$ is produced by the reactive power Q_{grid-r} according to (28) and it sets to zero to achieve Unity Power Factor Control (UPFC).

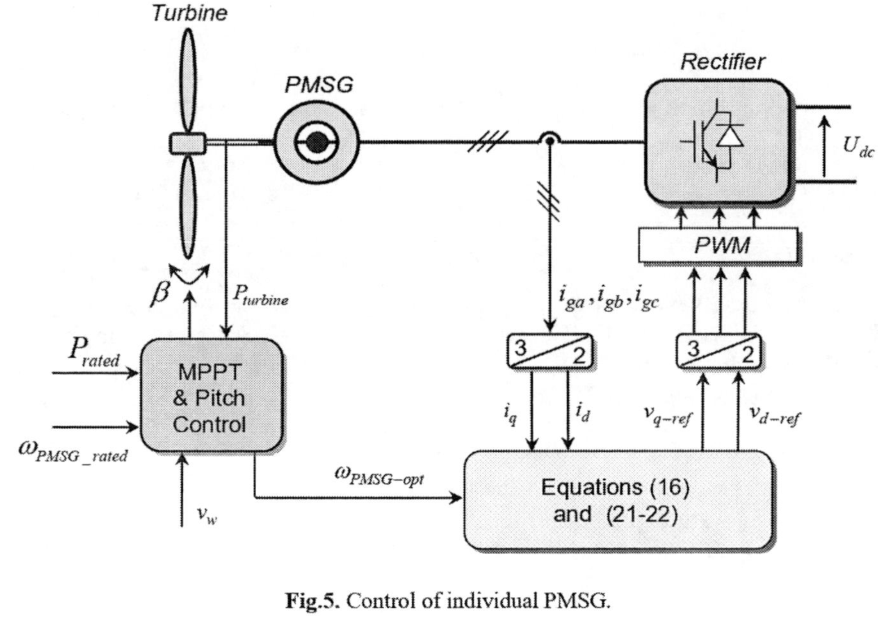

Fig.5. Control of individual PMSG.

The d-axis current reference $i_{d-grid-r}$ is provided by dc-bus voltage regulator to control the active power injected into the grid.
The DC link voltage and the grid side 5L-NPC can be modelled by [25-28]:

$$C\frac{dU_{dc}}{dt}=\frac{P_{recifiers}}{U_{dc}}-i_{5L-NPC} \tag{29}$$

$$e_{5L-NPC-d}=L_f\frac{di_{d-grid}}{dt}+R_f i_{d-grid}-\omega L_f i_{q-grid}+V_{grid} \tag{30}$$

$$e_{5L-NPC-q}=L_f\frac{di_{q-grid}}{dt}+R_f i_{q-grid}+\omega L_f i_{d-grid} \tag{31}$$

where C is the dc link equivalent capacitor; $P_{recifiers}$ is the total power at the output of the rectifiers; i_{5L-NPC} is the converter line current; U_{dc} is dc-link voltage; $e_{5L-NPC-d}$ and $e_{5L-NPC-q}$ are the direct and quadrature components of the voltage at the output of the 5L-NPC; V_{grid} represents the component in the d-axis of the network voltage; L_f and R_f are the resistance and inductance of filter, respectively.
To compensate the cross-coupling effect equations (30-31), the decoupling voltages are added to the current regulator outputs. Then, we use: $e_d^{'}$ et $e_q^{'}$ as:

$$e_q^{'}=L_f\frac{di_{q-grid}}{dt}+R_f i_{q-grid} \tag{32}$$

$$e_d^{'}=L_f\frac{di_{d-grid}}{dt}+R_f i_{d-grid} \tag{33}$$

Consequently, the 5L-NPC q-axis d-axis voltage components can be given by:

$$e_{NPC-d}=e_d^{'}-\omega L_f i_{q-grid}+V_{grid} \tag{34}$$

$$e_{5L-NPC-q}=e_q^{'}+\omega L_f i_{d-grid} \tag{35}$$

So, the control voltages of the 5L-NPC can be obtained by the use of two correctors PI such:

$$e_d^{'}=K_P(i_{d-grid-ref}-i_{d-grid})+K_I\int(i_{d-grid-ref}-i_{d-grid})dt \tag{36}$$

$$e_q^{'}=K_P(i_{q-grid-ref}-i_{q-grid})+K_I\int(i_{q-grid-ref}-i_{q-grid})dt \tag{37}$$

Fig. 6 shows the control block diagram of 5L-NPC based on the above strategy. Besides, the decoupling voltages $V_{grid}-L_f\omega i_{q-grid}$ and $L_f\omega i_{d-grid}$, are added to guarantee the decoupling of i_{q-grid} and i_{d-grid} components.

4. Simulation analysis

In order to validate the effectiveness of the proposed methods, simulation studies are carried out on a grid connected wind farm based PMSG. The parameters of the WECS are given in the Appendix. In addition, during the simulation, for each generator side converter, the d-axis command current component, i_{dr}, is set to zero. Also, for the 5L-NPC, $Q_{NPC-ref}$, is set to zero to ensure the unit power factor. The dc-link voltage reference and the electric network frequency value are $U_{dc-ref}=1800V$ and 50 Hz, respectively. The performances of the wind farm are given in Fig. 7 and Fig. 8. The wind speeds of two wind turbines are depicted in Fig. 7(a). From Fig. 7(c), the pitch angles are controlled depending on the wind velocity. So, at the wind velocities less than the rated wind velocity,

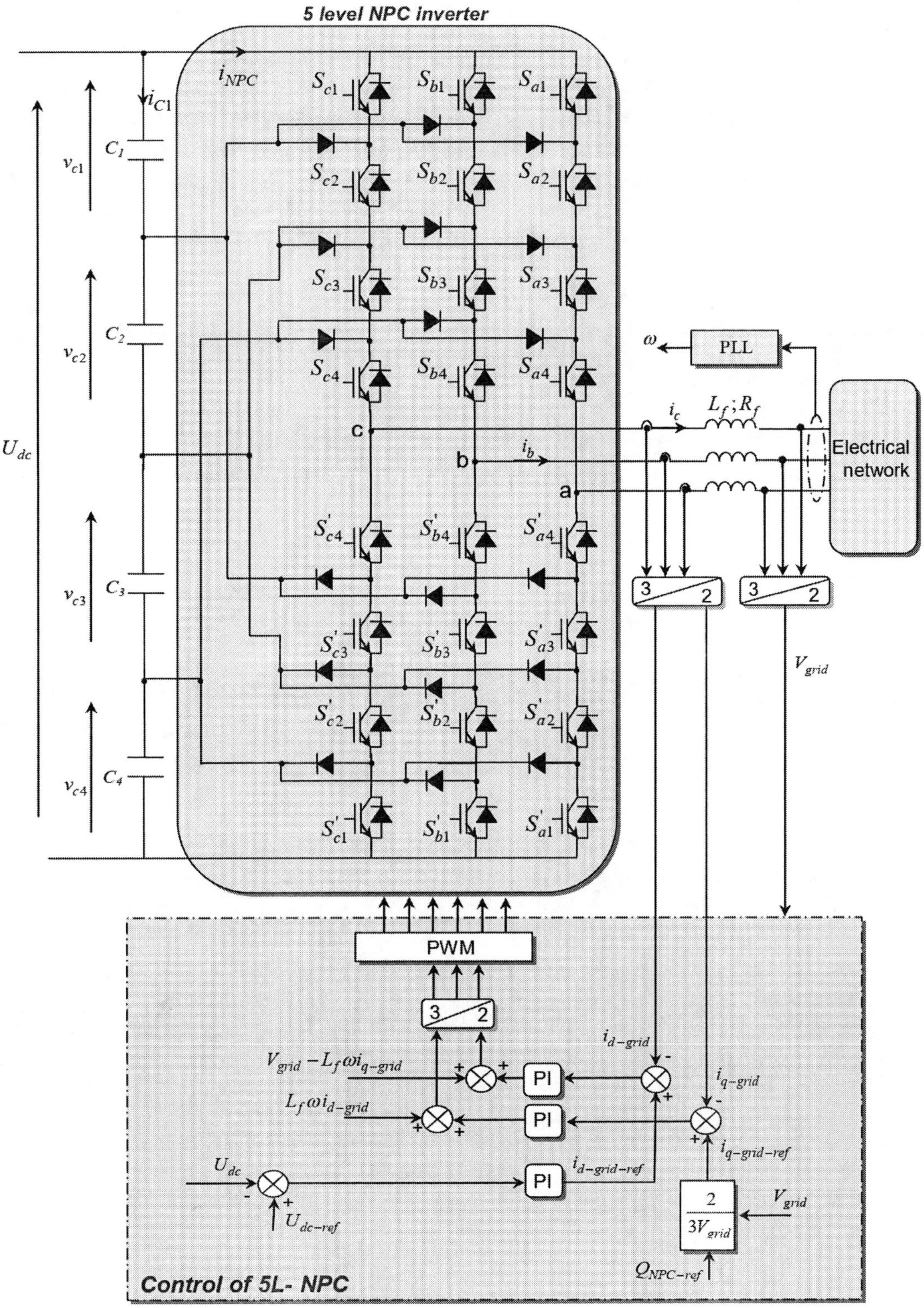

Fig.6. Control of five level NPC inverter.

IOP Conf. Series: Materials Science and Engineering **765** (2020) 012042 doi:10.1088/1757-899X/765/1/012042

the pitch angles are fixed at 0° and the power performance coefficients of the turbines are fixed at their maximum value, around 0.41. On the contrary, at higher wind speeds, the pitch angles increase and the coefficients of power decrease to limit the PMSG velocities. Consequently, rotational velocities and power extracted remains constant at the design limits. As shown in Fig. 7(e), the mechanical velocity of the PMSG1 controlled with the proposed strategy tracks its reference, obtained from the MPPT algorithm, successfully with rapid response. In Fig. 8(a), the WECS dc link voltage and the external reference U_{dc-ref} are depicted together. The dc link voltage is regulated toward its reference of 1800V.

This proves the effectiveness of the established controller systems. The output voltage of the NPC inverter is depicted in Fig. 8(b). Fig. 8 (c-d) shows the variation and the magnified view of the three-phase current and voltage of the electricity network. It can be seen that, the unit power factor is ensured because the reference of the reactive power has been set to zero.

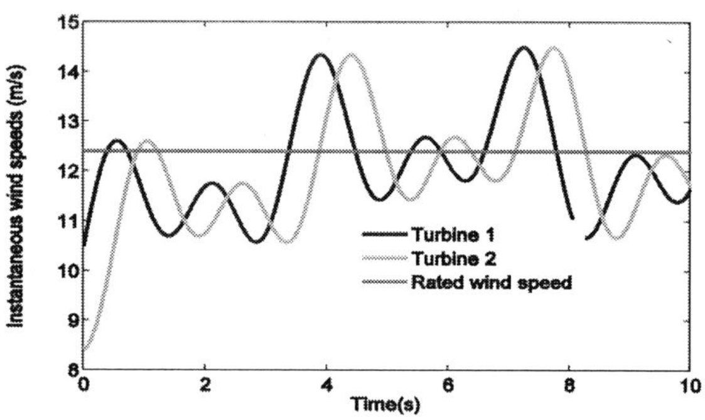

(a) Instantaneous wind speeds (m/s).

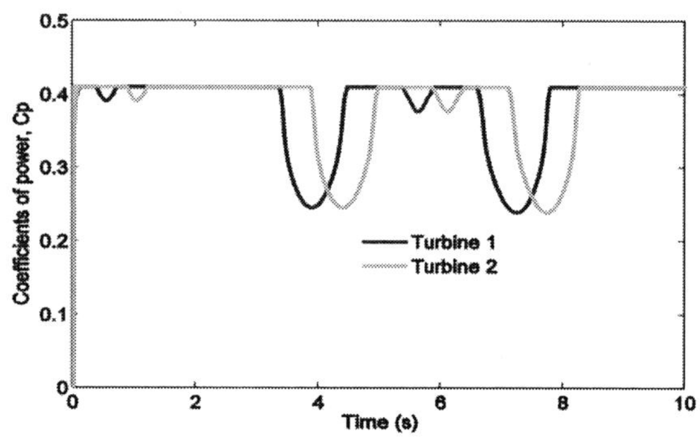

(b) Coefficients of power, C_p

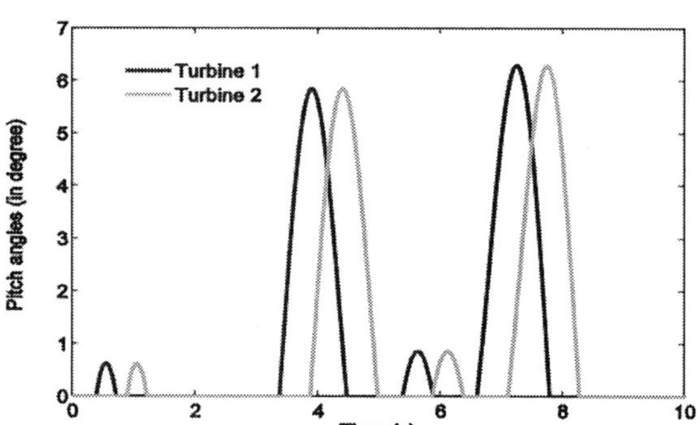

c) Pitch angles β_t *(in degree).*

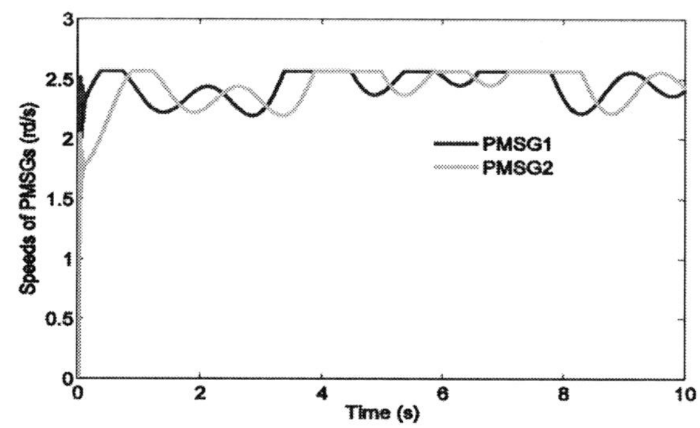

(d) Speeds of generators (rd/s).

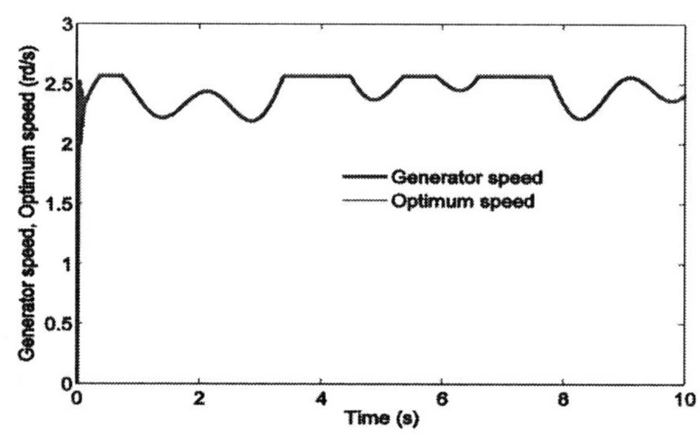

(e) Speed of PMSG1 (rd/s).

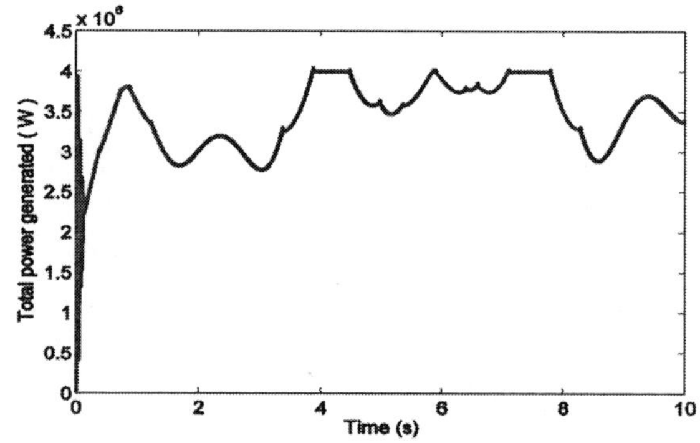

(f) Total power generated (W).

Fig. 7. Performances of wind farm during wind speed variations (1).

IOP Conf. Series: Materials Science and Engineering **765** (2020) 012042 doi:10.1088/1757-899X/765/1/012042

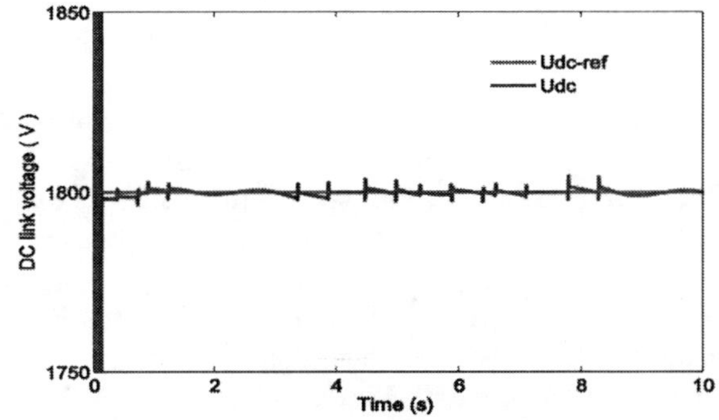

(a) DC link voltage (V)

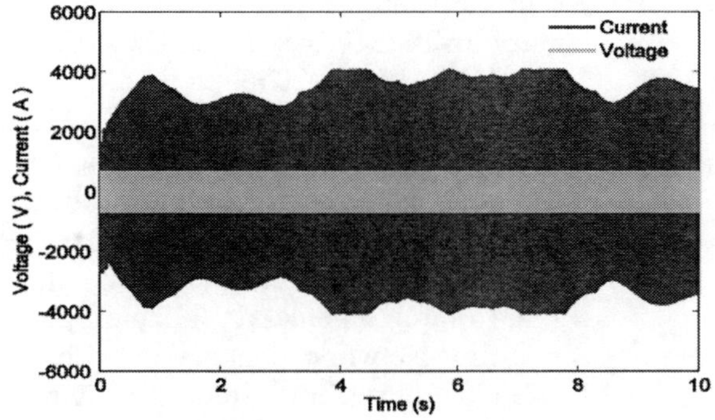

(c) Voltage and current of grid phase.

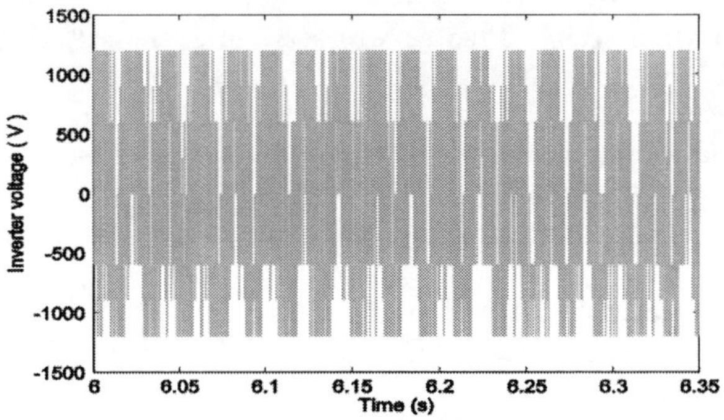

(b) *Inverter voltage (V).*

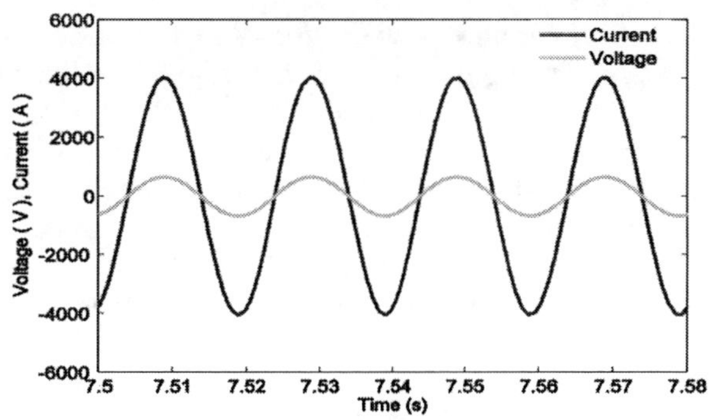

(d) Voltage and current of grid phase. *(zoom)*

Fig. 8. Performances of wind farm during wind speed variations (2).

5. Conclusion

This paper has proposed the MPPT control for grid-connected wind farm based PMSG and 5L- NPC. Nonlinear control SM and the VC have been combined to achieve the MPPT, to keep the dc link voltage constant and to control the active and reactive powers. The theoretical sufficient conditions have been driven in terms of stability based on Lyapunov stability theory. Also, Pitch Control strategy has been employed to prevent wind turbine damage in case of high wind speed. The effectiveness of the proposed strategies has been demonstrated by simulation results for a 4-MW wind farm based PMSG. The simulation results demonstrate that combined control method shows very good dynamic and stable state performance and works very well.

Appendix

TABLE I. PARAMETERS OF THE PERMANENT MAGNET SYNCHRONOUS GENERATOR

Parameter	Value
P_r nominal power	2000 (KW)
ω_{PMSG} nominal mechanical velocity	2.57 (rd/s)
R stator resistance	8 (mΩ)
L_s stator q -axis and d -axis inductance	0.3 (mH)
ψ_f permanent magnet flux	3.86 (Wb)

IOP Conf. Series: Materials Science and Engineering **765** (2020) 012042 doi:10.1088/1757-899X/765/1/012042

References

[1] Arumugam Vinodkumar, Mani Prakash, Young Hoon Joo, "Impulsive observer-based output control for PMSG -based Wind Energy Conversion System ," IET Control Theory & Applications, Vol. 13, No. 13, pp. 2056 - 2064, 2019.
[2] Mahmoud A. Soliman ; Hany M. Hasanien ; Haitham Z. Azazi ; E. E. El-Kholy ; Sabry A. Mahmoud , "An Adaptive Fuzzy Logic Control Strategy for Performance Enhancement of a Grid-Connected PMSG-Based Wind Turbine," IEEE Transactions on Industrial Informatics, Vol. 15, No. 6, pp. 3163 - 3173, 2019.
[3] Youssef Errami, Mohammed Ouassaid, Mohamed Cherkaoui, Mohamed Maaroufi, " Sliding Mode Control Scheme of Variable Speed Wind Energy Conversion System Based on the PMSG for Utility Network Connection," Chapter, Book Title: Advances and Applications in Sliding Mode Control systems, Studies in Computational Intelligence book Series, Volume 576, 2015, pp 167-200, DOI 10.1007/978-3-319-11173-5_6.
[4] Chun Wie, Zhe Zhang, Wei Qiao, Liyan Qu, " An Adaptive Network-Based Reinforcement Learning Method for MPPT Control of PMSG Wind Energy Conversion Systems," IEEE Transactions on Power Electronics, Vol. 31, No. 11, pp. 7837 - 7848, 2016.
[5] Jaramillo-Lopez. F, Kenne. G, Lamnabhi-Lagarrigue. F, " A novel online training neural network-based algorithm for wind speed estimation and adaptive control of PMSG wind turbine system for maximum power extraction," Renewable Energy, 2016, 86:38-48.
[6] Z.Q. Wu, W.J. Jia, C.H. Wu, " Maximum wind power tracking based on cloud RBF neural network," Renewable Energy, 2016, 86:466-472.
[7] Das and B. Subudhi , " A H∞ robust active and reactive power control scheme for a PMSG-based wind energy conversion system ," IEEE Transactions on Energy Conversion, Vol. 33, No. 3, pp. 980-990, 2018.
[8] Errami, Y., M. Ouassaid, and M. Maaroufi, " Modeling and variable structure power control of PMSG based variable speed wind energy conversion system, " Journal of Optoelectronics and Advanced Materials 15.11-12 (2013): 1248-1255.
[9] Marwa Ayadi, Babil Derbel, "Nonliear adaptive backstepping control for variable-speed wind energy conversion system-based permanent magnet synchronous generator," International Journal of Advanced Manufacturing Technology, 92(1-4): 39-46, 2017.
[10] Muljadi, E., Butterfield, C.P, " Pitch-controlled variable-speed wind turbine generation ," IEEE Transactions on Industry Applications, Vol. 37, No. 1, pp. 240–246, 2011.
[11] Tan, K., Islam, S. , "Optimum control strategies in energy conversion of PMSG wind turbine system without mechanical sensors," IEEE Transactions on Energy Conversion, Vol. 19, No. 2, pp. 392–399, 2004.
[12] Slotine, J.E., Li, W , " Applied Nonlinear Control," Prentice Hall, New Jersey, 1991.
[13] Utkin, V.I. , " Sliding mode control design principles and applications to electrical drives," IEEE Transactions on Industrial Electronics, No. 40, No. 1, pp. 23–36, 1993.
[14] Li, S., Du, H., Yu, X., "Discrete-time terminal sliding mode control systems based on Euler's discretization ," IEEE Transactions on Automatic Control, Vol. 59, No. 2, pp. 546 - 552 , 2014.
[15] Li, S., Zhou, M., Yu, X, , "Design and implementation of terminal sliding mode control method for PMSM speed regulation system," IEEE Transactions on Industrial Informatics, Vol.9, No. 4, pp. 1879–1891, 2013.
[16] Y.Errami, M.Hilal M.Benchagra, M.Ouassaid, M.Maaroufi, " Nonlinear Control of MPPT and Grid Connected for Variable Speed Wind Energy Conversion System Based on the PMSG", Journal of Theoretical and Applied Information Technology, Vol. 39, No.2, pp.205- 217, May 2012.
[17] Youssef Errami, Abdellatif Obbadi, Smail Sahnoun, Mohammed Ouassaid, and Mohamed Maaroufi, "Performance evaluation of backstepping approach for wind power generation system-based permanent magnet synchronous generator and operating under non-ideal grid voltages", Int. J. Power and Energy Conversion, Vol. 10, No. 4, 2019. DOI: 10.1504/IJPEC.2019.10024034.

[18] Ngoc Dat Dao, Dong-Choon Lee, Sangmin, "A Simple and Robust Sensorless Control Based on Stator Current Vector for PMSG Wind Power Systems", IEEE Access, Vol. 7, pp. 8070 – 8080, 2018.
[19] Evangelista, C., P. Puleston, F. Valenciaga, L. M. Fridman, " Lyapunov-Designed Super-Twisting Sliding Mode Control for Wind Energy Conversion Optimization", IEEE Transactions On Industrial Electronics 60, pp. 538-545, 2013.
[20] Youssef Errami, Mohammed Ouassaid, Mohamed Cherkaoui, Mohamed Maaroufi, "Variable Structure Sliding Mode Control and Direct Torque Control of Wind Power Generation System Based on The PM Synchronous Generator, " Journal of ELECTRICAL ENGINEERING, Vol. 66, No. 3, 2015, pp. 121–131. DOI: 10.1515/jee-2015-0020.
[21] Errami, Y.; Maaroufi, M.; Ouassaid, M., "Maximum Power Point Tracking of a Wind Power system based on the PMSG using Sliding Mode Direct Torque Control, " Renewable and Sustainable Energy Conference (IRSEC), 2013 International, vol., no., pp. 228-233, 7-9 March2013, doi: 10.1109/IRSEC.2013.6529706.
[22] Errami, Y.; Maaroufi, M.; Ouassaid, M., "Variable Structure Direct Torque Control and grid connected for wind energy conversion system based on the PMSG," in Complex Systems (ICCS), 2012 International Conference on , vol., no., pp.1-6, 5-6 Nov. 2012. doi: 10.1109/ICoCS.2012.6458524.
[23] Y. Errami, M. Ouassaid, M. Maaroufi, " Control of Grid Connected PMSG Based Variable Speed Wind Energy Conversion System, " International Review on Modelling and Simulations (I.RE.MO.S), Vol. 5, No. 2, pp. 655-664, April 2012. ISSN: 1974-9821.
[24] Errami, Y., M. Benchagra, M. Hillal, M. Ouassaid, M. Maaroufi, "MPPT Strategy and Direct Torque Control of PMSG Used for Variable Speed Wind Energy Conversion System, " International Review on Modelling and Simulations (I.RE.MO.S), Vol. 5, No. 2, pp. 887- 898, April 2012. ISSN: 1974-9821.
[25] Youssef Errami, Mohammed Ouassaid, Mohamed Maaroufi, " Modelling and Optimal Power Control for Permanent Magnet Synchronous Generator Wind Turbine System Connected to Utility Grid with Fault Conditions, " World Journal of Modelling and Simulation, Vol. 11 (2015) No. 2, pp. 123-135.
[26] Errami, Youssef; Ouassaid, Mohammed; Maaroufi, Mohamed, "Variable structure control for Permanent Magnet Synchronous Generator based wind energy conversion system operating under different grid conditions," Complex Systems (WCCS), 2014 Second World Conference on , vol., no., pp.340,345,10-12Nov.2014. doi: 10.1109/ICoCS.2014.7060996.
[27] Errami, Y.; Maaroufi, M.; Cherkaoui, M.; Ouassaid, M., "Maximum power point tracking strategy and direct torque control of permanent magnet synchronous generator wind farm," in Complex Systems (ICCS), 2012 International Conference on , vol., no., pp.1-6, 5-6 Nov. 2012. doi: 10.1109/ICoCS.2012.6458520.
[28] Y. Errami, M. Ouassaid, M. Maaroufi, "Control scheme and power maximization of permanent magnet synchronous generator wind farm connected to the electric network", International Journal of Systems, Control and Communications (IJSCC), Vol. 5, No. 3/4, pp. 214 – 230, 2013.

IOP Conf. Series: Materials Science and Engineering **765** (2020) 012043 doi:10.1088/1757-899X/765/1/012043

Intensification of heat transfer in the cooling systems of radio electronic equipment under free and forced convection

A.A. Lopatin[1], Ali Kamil Sebur[2], Israa S.Ahmed [2] ,R.A. Gabdullina[1], A.R. Biktagirova[1]

[1] Federal State Budgetary Educational Institution of Higher Education «Kazan National Research Technical University named after A.N. Tupolev–KAI», Department of Jet Engines and Power Plants, 420111, Kazan, st. K. Marx, 10.

[2] Electromechanical Engineering Department, University of Technology, Baghdad, Iraq

[1] aalopatin@kai.ru, arbiktagirova@kai.ru, rozzy94@mail.ru
[2] 50278@uotechnology.edu.iq, 50112@uotechnology.edu.iq.

Abstract. This paper considers the issues of heat transfer intensification in the air cooling system of heat-loaded elements of radio-electronic equipment based on split fins. The results of the study of heat transfer in free and forced convection cooling systems for heat-loaded elements of electronic equipment are presented. The thermal efficiency of systems based on smooth and partially split ribs is compared; the efficiency of split ribs is 16-20%. The dependences of thermal resistance on thermal load are determined. The results of a comparison of mathematical and physical modeling are provided. The influence of the method of supplying heat to split ribs is estimated. Optimization of the work area is presented.

Keywords: heat transfer, heat transfer intensification, cooling systems, partially split ribs, heat flow,

1. Introductions

Modern electronic and power equipment, a part of aviation equipment, is characterized by high heat generation, the absolute values of which can exceed 300-800 W [1]. Such values result from, first of all, by the rapid development of radio electronics and the nanoindustry, contributing to an increase in productivity and simultaneous miniaturization of microchips and other elements of modern electrical equipment.

One of the primary reasons for the failure of the expensive elements of radio-electronic equipment in aviation equipment is the failure to comply with thermal operating conditions. In addition to the problem of heat removal, issues related to the optimization of the weight and size characteristics of cooling and thermal stabilization systems also come to the fore. Evidently, the existing problems with the removal of high-density heat fluxes cannot be addressed using a universal cooling method, since each heat-loaded element carries its own individual operating characteristics and parameters, due to specific design features and operating conditions of the equipment, power supply, the requirements for ensuring environmental and fire safety, etc.

Depending on the design and shape of the components that require cooling, various solutions are used for organizing permanent heat removal. With regard to power transistors and diodes of medium power, it is possible to use pin fin heat sink with a solid or dissected base. The cooling of such surfaces is usually carried out under conditions of natural convection by the surrounding air. These systems are disadvantaged because of their low thermal efficiency (heat transfer coefficient $\alpha \leq 10$ W / m^2 ° C).

In the current thematic scientific literature, detailed studies on this subject are presented [2-4], simultaneously, however, discussions are quite acute regarding the possible ways of intensifying heat transfer. The significant interest in the issues related to the intensification of heat transfer at surfaces of various shapes and sizes as applied to the cooling systems of electronic equipment is evidenced by the presence of a large number of publications in modern scientific journals, both in Russia and abroad. The issues of heat transfer intensification in technical systems are also described in a significant number of works [5-11]. One of the most effective methods for increasing heat transfer in air cooling systems is split or partially split finning. [12,13]. The use of ribs as a method of intensifying heat transfer leads to an increase in the value of the heat transfer coefficient by tens of times. This method of intensification implies a wide variety of different types of finning: longitudinal, transverse, rolling, spiral and many others [14, 15].

Due to the practical impossibility of extensively considering in one work the entire range of problems associated with the intensification of heat transfer in conditions of free and forced convection, the authors decided to focus on the practical significant topics - the cooling system of a continuous X-ray flaw detector. With a power of up to 190 W for glass x-ray tubes, the maximum possible operating conditions is a temperature of 80 ^{0}C.

2. Methods

To study the effectiveness of the proposed method of heat transfer intensification, a comparative analysis of sections with smooth and sections with split ribs was carried out (Figure 1,2). The ribs are made of M1 brand copper and are arranged concentrically. The influence of the tightness conditions during the alternate lowering of the casing pipe on the working section, as well as the optimal opening angle of the "petals" of the ribs were determined experimentally, the thickness of the boundary layer and the criterion equations are also analytically determined. The influence of the method of heat input, the determination of the optimal number of cuts of the edge of the rib and the temperature distribution along the surface of the rib are calculated numerically.

To analyze the possible optimization parameters for split finning, a number of numerical studies were conducted to determine the most effective method for supplying heat and identifying the number of sections of the edge of the rib, and the geometry of the working section.

In order to determine the most effective way for supplying heat to the split finning, modeling was carried out under the condition of heating the entire lateral edge of the rib (100 mm), as well as heat transfer through a sector 30, 50 and 70 mm long. The temperature of the upper part of the rib of the working section with a maximum length of 100 mm is slightly higher than the temperatures with the other three methods of heat input, however, the apparent difference is insignificant and falls into the deviation error range of 5%.

IOP Conf. Series: Materials Science and Engineering **765** (2020) 012043 doi:10.1088/1757-899X/765/1/012043

Figure 1. The appearance of the work area with solid ribs.

Figure 2. The appearance of the working area with split ribs.

3. Experimental Research

The experimental bench included a heater simulating isothermal heating, a working section located on the surface of the heater, and instrumentation. To study forced convection, an additional blower was installed. The working section consisted of copper fins located along an axial pattern. A sketch of the working area is shown in Figure 3.

In the process of research, a comparison was made of the heat transfer in the area with smooth and split ribs. On each edge of the section with a split finning, three dissections are plotted. To study the effectiveness of the proposed method of heat transfer intensification, a comparative analysis of sections with smooth and sections of split ribs was conducted. The ribs are made of M1 brand copper and are concentrically arranged. The influence of the tightness conditions during the alternate lowering of the casing pipe on the working section, as well as the optimal opening angle of the "petals" of the ribs were determined experimentally, the thickness of the boundary layer and the criterion equations are analytically determined. "Petal" refers to a segment of a cut rib. The influence of the method of heat input, the determination of the optimal number of cuts of the edge of the rib and the temperature distribution along the surface of the rib are calculated numerically.

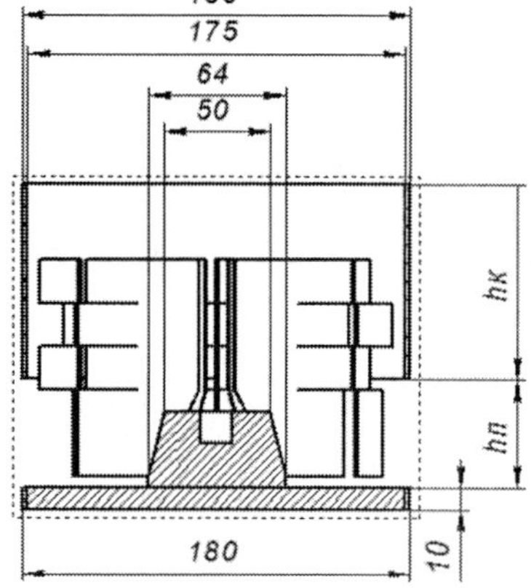

Figure 3. A sketch of the working area, front view

4. Results

Figure 4 shows the dependence of heat transfer in the case of the forced movement of the coolant in cooling systems created on the basis of partially split (index "p") and flat ("p") ribs. It can be observed that at moderate Reynolds numbers (Re <5000), the difference in heat transfer between the two systems is insignificant. However, at Re> 10000, a considerably sharp increase in Nur begin starts and at Re ≈ 22000, the intensification reaches 1.4 times (Nur / Nup ≈ 1.4 is the ratio of the Nusselt number in the section with split finning to the Nusselt number with flat edges), which is in accordance with the results, presented in Gortyshov Yu.F. [15]. Thus, it can be noted that the working section with partially split ribs features a higher thermal efficiency than a system with flat ribs. Figure 5 indicates the dependence of the differential pressure on the working areas with flat ribs and partially split fins. It is observed that in the entire range of Reynolds numbers there exists a slight excess of hydraulic resistance at the working section with partially split fins.

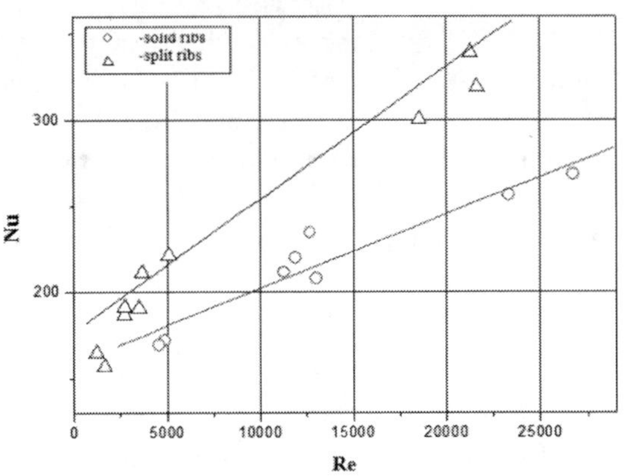

Figure 4. Dependence of the Nusselt number on the Reynolds number:
○, Δ - Smooth and split edges, respectively

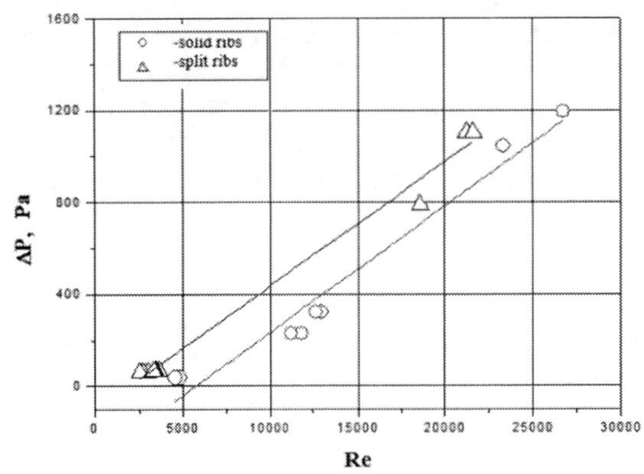

Figure 5. Pressure drop P (hydraulic resistance) in areas with smooth (○) and split (Δ) ribs

Figures 6 and 7 Indicate the results of heat transfer studies for smooth and split ribs at different heights of lifting the casing, as well as at different angles of the opening of the ribs.

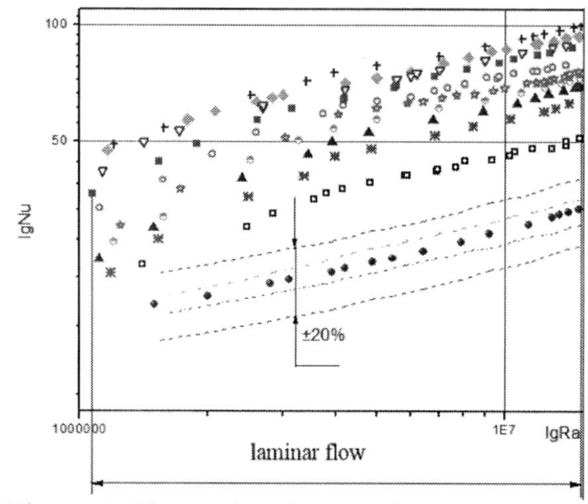

Figure 6. The study of convective heat transfer surfaces. Casing lifting height: - there is no casing, ⁎ - hp = 90mm, ▲ - hp = 80mm, ◦ - hp = 70mm, ⋆ - hp = 60mm, ○ - hp = 50mm, + - hp = 40mm, ◆ - hp = 30mm, ■ - hp = 10mm, ▪ -hp = 20mm, ● - smooth rib, calculation according to known dependencies: ⋯ - Mikheev M. A. formula, - - - Isachenko V.P. formula

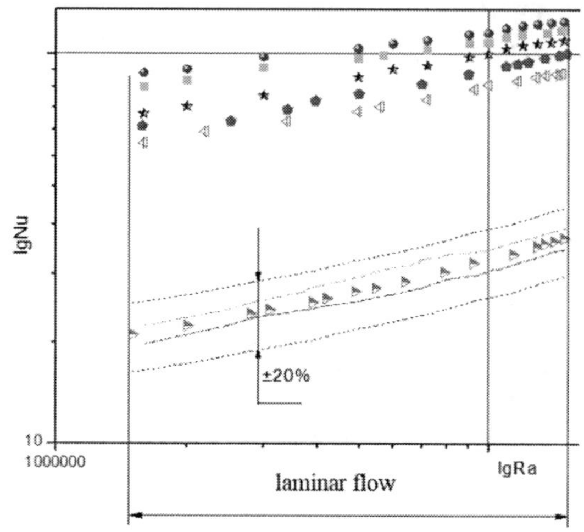

Figure 7. The effect of the angle of extension of the ribs on the heat transfer in comparison with a smooth rib. The opening angle of the ribs: ★ - 10 °, ■ - 20 °, ● - 30 °, ⬟ - 45 °, ◁ - 60 °, ▷ - smooth rib, calculation according to the known dependencies: ⋯ -Mikheev M.A. formula, - - - V.P. Isachenko

Zero height of the casing corresponds with the base of the working area. As a comparison of the experimental results with the data from studies by renowned authors, Fig. 3 indicates the calculation lines by M.A. Mikheev, formula (1), Written E.N. [16] and V.P. Isachenko, formula (2), Lopatin A.A. [12] for free convection:

$$Nu = 0,46 * Ra^{0,25} \tag{1}$$

$$Nu = 0,63 * Ra^{0,25} \tag{2}$$

The maximum values of heat transfer are attained when the height of the casing is increased by 40 mm and the opening angle of the ribs is 30 °. The presence of these maxima is explained by the formation of the self-traction effect in the cavity of the constraining casing and the opening of the boundary layers. With a decrease in the height of raising the casing, the amount of working air entering the casing region decreases and the air surrounding the ribs heats up. With an increase in the lifting height of the casing pipe, the area of ribs inside the pipe decreases, which impedes the intensity of the process. When the opening angle of the ribs is less than 30°, the boundary layers close together, which significantly worsens the heat transfer process. At the opening angles of the ribs by 45 ° and 60°, the diameter of the circle of the base of the working section decreases, thereby negatively affecting the overall intensification process.

IOP Conf. Series: Materials Science and Engineering **765** (2020) 012043 doi:10.1088/1757-899X/765/1/012043

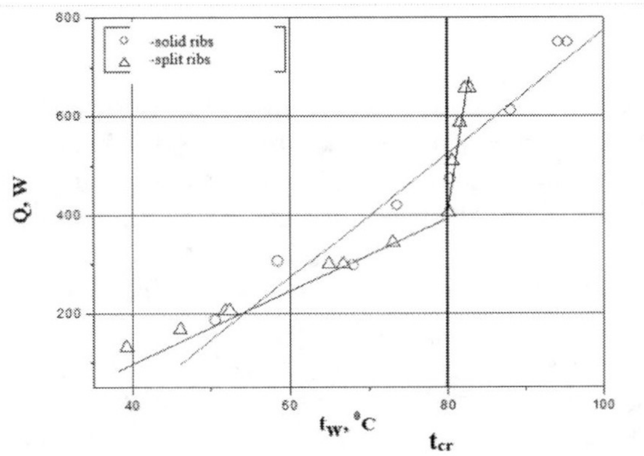

Figure 8. The dependence of the supplied heat flux on the temperature of the anode of the x-ray tube

The results of calculating the temperature distribution over the surface of the split fin are indicated in Figure 9. The results of optimizing the geometric characteristics of the cooling system are shown in Figure 10.

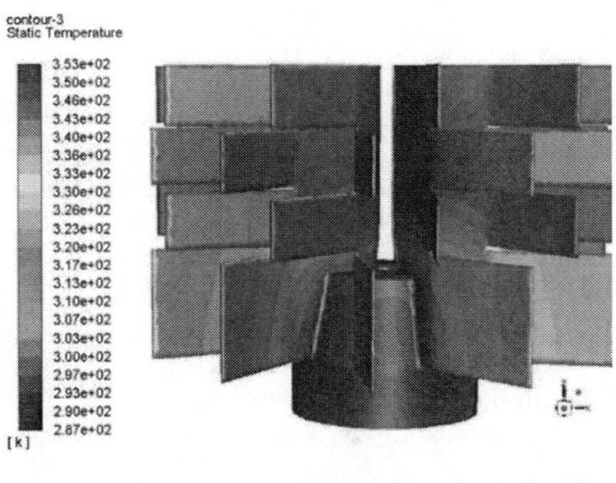

Figure 9. Temperature distribution along the rib, power 188.8 W

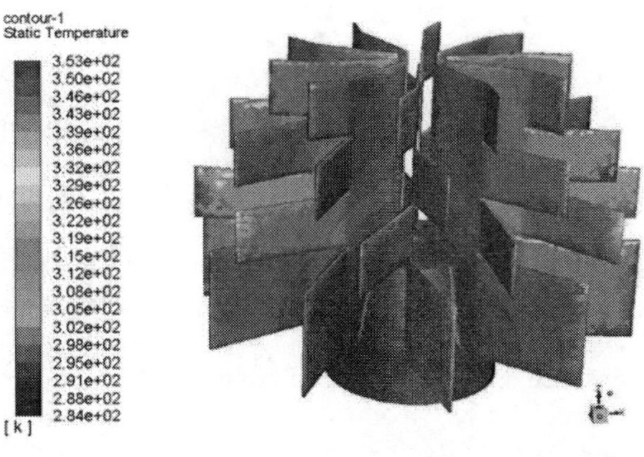

Figure 10. Thermal calculation of the upgraded model

5. Discussion

As a result of a joint consideration of the thermal and hydraulic efficiency of air cooling systems, it can be noted that the achieved heat exchange intensification using partially split ribs can be explained by the combined influence of two indicated factors: firstly, the presence of dissections of the heat-exchange surface does not allow the formation of a complete boundary layer along the length of the rib and secondly, due to the constant stalling of the flow from the initial and final edge of the rib, some additional turbulence occurs of flow.

The primary parameter characterizing the quality of cooling systems is thermal resistance - the ratio of the temperature difference on the cooled surface to the amount of heat flowing through it (° C / W). However, as previously noted, due to a significant increase in thermal efficiency at Re> 20,000 (Figure 5), partially split ribs are more preferable for use in forced convection cooling systems for continuous X-ray flaw detectors.

The operating temperature range of cooling systems with external air blowing is indicated in Fig. 8. It can be indicated that a system based on flat and split edges at tcr = 80 ° C can divert ≈500 W, which is not sufficient for ensuring the stable operation of the device. The maximum possible operating conditions should be understood by tcr. Therefore, systems built using partially split ribs can be use adopted d for flaw detectors of medium and low power ($Q \leq 600$ W). The low heat flux removed by the air cooling system, created on the basis of partially split fins, can be explained by some coupled factors that have a decisive influence on the heat transfer process as a whole. Therefore, with all of the above-mentioned advantages of partially split fins, it should be noted that in the absence of an increase in the heat transfer area, additional turbulization caused by the dissection of the heat exchange surfaces, and, consequently, the mixing of the flow, leads primarily an increase, though insignificant, in the hydraulic resistance. Moreover, the increase in heat transfer is not leading and, accordingly, does not have a decisive influence on the process as a whole.

An analysis of the results of numerical modeling of the temperature distribution in the working area facilitates determining the areas of fins with maximum and minimum heat transfer, which subsequently became the basis for further optimization.

The resulting model of split ribs allows one to remove the necessary heat fluxes, power up to 190 watts, and also provides a reduction in the mass of the working area by 16.5%.

6. Conclusion

As a result of the work, studies of heat transfer intensification in natural and forced convection using split ribs were carried out and presented. The effectiveness of using split finning is indicated; during the experiment, the most optimal geometric parameters of the working section were identified. Criteria are obtained for calculating the heat transfer of systems with axial split finning. It can also be assumed that the intensity of heat transfer processes will depend not only on the mode, but also on the geometric parameters of the partially split finning, including the depth of the cut, step and shape of the “petals”, etc. To confirm the above assumptions, it is necessary to conduct a series of refinement experiments. As a result of physical and numerical simulations, data were obtained indicating the absence of the influence of the length of the edge of the rib through which heat is supplied to the heat transfer, as well as illustrating the possibility of optimizing mass and size characteristics, which leads to a saving of 16.5% in mass.

The aim of this study is the verification of experimental data in numerical simulation programs and the topological optimization of the resulting model. Creating rib geometry by modeling based on topological optimization allows you to purposefully adapt the component and the cooling system as a whole. Solutions that cannot be manufactured industrially or can only be created with disproportionate effort can also be produced as separate parts using additive manufacturing processes. However, it is also necessary to consider the structural nuances associated with the production process.

References

[1] Lopatin A.A., Schelchkov A.V., "Intensification of heat transfer in cooling systems of continuous x-ray apparatus.", In the mat. RNKT-5. M .: Publishing House of MPEI, 2008, v.

6, p. 113 - 116.
[2] Yarkaev M.Z., Al-Janabi A.Kh.A., Gilmanov A.Kh., Gortyshov Yu.F., Popov I.A., Schelchkov A.V., " A comparative study of heat exchangers with intensification of heat transfer", Bulletin of Kazan State Technical University. A.N. Tupolev. - 2013. - No. 2-2. - pp. 73-79.
[3] Popov I.A. "Hydrodynamics and heat transfer of external and internal free-convection vertical flows with intensification." Intensification of heat transfer: monograph / Ed. ed. Yu.F. Gortishova. - Kazan: Center for Innovative Technologies, 2007. - 326 p.
[4] Gortyshov Yu.F., Popov I.A., Usenkov R.A. "Heat transfer of free-convection flows in the presence of surface intensifiers", News of Higher Education Institutions. Aircraft technology. - 2003. - No. 3.- ss. 29-32.
[5] Vasiliev V.Ya., "Rational intensification of convective heat transfer by dissection of long smooth channels", Bulletin of the Moscow Aviation Institute. - 2010 .-- T.17. - No. 3. - pp. 143-152.
[6] Popov I.A. "Hydrodynamics and heat transfer of external and internal free convective vertical flows with intensification.", Intensification of heat transfer: monograph, Ed. ed. Yu.F. Gortyshova. - Kazan: Center for Innovative Technologies. - 2007 .-- 326 s.
[7] Dreitser G.A., Lobanov I.E., "Ultimate intensification of heat transfer in pipes due to artificial turbulization of the flow for gaseous coolants with variable thermophysical properties", Moscow Aviation Institute Bulletin. - 2005. - T.12. - No. 3. - ss. 18-25.
[8] Gortyshov Yu.F., Popov I.A., Usenkov R.A., " Heat transfer of free-convection flows in the presence of surface intensifiers", News of Higher Education Institutions. Aircraft technology. - 2003. - No. 3.- ss. 29-32.
[9] Baldwin, C. S., Bhavani, S. H., Jaeger, R. C., "To ward Optimized Enhanced Sur faces for Passive Immersion Cooled Heat Sinks", IEEE Trans. Comp. Hy brids, Manufact. Technol., 23 (2000), pp. 70-79.
[10] Gortyshov Yu.F., Popov I.A., Olimpiev V.V., Schelchkov A.V., Kaskov S.I. "Intensification of heat transfer. Thermohydraulic efficiency of promising methods of heat transfer intensification in the channels of heat exchange equipment", Center for Innovative Technologies Publishing House. 2009. Kazan. 530.
[11] Vasiliev V. Ya.,"Efficiency of the process of rational intensification of convective heat transfer in channels with discrete turbulators", Herald of the Moscow Aviation Institute. 2010.Vol. 17. No. 3. P. 153-162.
[12] Lopatin A.A. "Heat transfer and resistance in cooling systems of electric power equipment with split finning", Bulletin of Kazan State Technical University named after A.N. Tupolev. - 2013. - No. 2-1.- ss. 187-190.
[13] Lopatin A.A., Nikolaeva D.V., "Investigation of the capabilities of split fins in cooling systems of electronic equipment", Problems and prospects for the development of aviation, land transport and energy ANTE-2015: International Scientific and Technical Conference, October 19-21, 2015. Collection of reports. Kazan: Brig Publishing House. - 2015 .-- pp. 566-592.
[14] Fraas A., Otsisik M., " Calculation and design of heat exchangers", M .: Atomizdat, 1971, 356 p.
[15] Gortyshov Yu.F., Olimpiev V.V.," Heat exchangers with intensified heat transfer ", Kazan: Kazan Publishing House. state tech. University, 1999.176 s.
[16] Written E.N., " New effective developed heat exchange surfaces for solving problems of energy and resource conservation", Prom. heat engineering. , 2007. - T.29. , No. 5. , pp. 7-16.

IOP Conf. Series: Materials Science and Engineering **765** (2020) 012044 doi:10.1088/1757-899X/765/1/012044

Intelligent Nero modelling methods for PV panel system

Ibtisam Ahmed Hasan, ,Mohammed Jawad Mohammed, Fatima Attya Lafta

Electromechanical Engineering Department, University of Technology- Iraq,

50024@uotechnology.edu.iq

Abstract. Nonlinear system identification modelling for the temperature of photovoltaic (PV) panel has been conducted in this work. In the beginning, an experimental work has been extracted from previous work in order to collect the input (ambient temperature, humidity, irradiance and wind speed) and output (PV module temperature) parameters. Then, Neural Network time series and Adaptive Neuro-Fuzzy Inference System (ANFIS) models represented as system identification method to predict the temperature of PV panel as an output for the system. Both of modelling methods verified using mean square error (MSE). The effectiveness of all methods has been compared to know which method is the batter. Finally, the achieved results stated that the ANFIS method recorded the lowest MSE of $2.2627*10^{-7}$ compared with NARX method which recorded of 5.078. ANFIS technique proved that will be able to use it in the control process in future.

Keywords: Nonlinear Autoregressive Exogenous (NARX), ANFIS, System identification modelling , PV panel.

1-Introduction

In the last years, employment of solar energy has been increased in many countries due to increasing in the fuel prices and confined traditional resources. Photovoltaic (PV) conversion is the direct transformation of sunlight into electricity without any heat engine to interpose [1]. Using the solar energy reduces the environmental damage that results from burning of fuel used in power generation plants. The sun radiates an enormous amount of energy through nuclear fusion process. During one second, the sun converts about four million tons of the mass which mostly consists of the helium and hydrogen into energy producing solar radiation and neutrinos [2]. The amount of sunlight available is one factor to take into account when considering using solar energy. The most important characteristic of a photovoltaic panel is the Conversion efficiency, the solar radiation that is not converted into electricity is almost entirely transformed into heat. The efficiency is decrease while the temperature of PV cell rises [3]. Various factors can significantly affect PV electrical performance during operation, one of these factors is PV temperature which has great impact on PV performance particularly in hot weather such as summer season in Iraq where PV temperature reaches to 80 °C which leads to significantly decreasing in PV performance. In fact, more than 80% of the radiation falling on a PV cells is converted to thermal energy or reflected Cooling system is substantial in order to overcome problem at hot environment and high irradiation. Generally, there are several cooling techniques for cooling the PV panel and overcome PV temperature rising problem and enhance its performance such as air-cooling and water-cooling [4]. Studied different cooling methods trying to enhance the

performance of PV system, used: back, front and duel cooling. The results showed the dual cooling reduced PV temperature by 18% while front cooling and back cooling reduced PV temperature by 11% and 2% respectively. [5] Improved PV array efficiency experimentally by spraying water on front side of PV. The output hot water was used as heating water for houses, buildings etc. The result showed the electrical power of each one panel increased by 0.354 w/degree. [6] Used method of passive heat exchange as back cooling for PV panels. The result showed increasing in electrical and thermal efficiency that reached to 11.4% and 41.6% respectively, in a sunny day.
[7] Designed a micro-heat pipe array to cool a PV panels. It was found that the cooling by water is more effective than cooling by air. [8] Suggested ducts that arranged in parallel with inlet and outlet channels on the back side of PV module. The results showed the temperature reduced from 68°C to 38°C and increased the efficiency from 8.6% to 12.5%. [9] found that using the air cooling increased the electrical efficiency by 2.6% and the output power 8.4% due to reduction in cell temperature about 4.7°C while by using water cooling, the cell temperature reduced by 8°C. This led to increase the PV efficiency by 3% and the output power by 13.9%. [10] Considered the appropriateness of artificial neural networks for 1 day ahead solar power generation estimating. [11] Estimating the everyday sun based radiation with two dynamic artificial neural networks (Feed forward Time Delay Neural Network and NARX) was proposed. As indicated by the creators, the two models had a palatable presentation. Built up an ANFIS model to foresee the ideal measuring coefficient of PV frameworks dependent on land arranges. The outcomes acquired were contrasted and broke down and counterfeit neural systems (ANN) demonstrating that ANFIS model displays the most exact outcomes.[12] Created an ANFIS model to anticipate month to month sun based worldwide radiation for PV framework estimating. The model showed a flat out rate blunder of 0.48, which yielded better outcomes contrasted with different strategies, for example, ANN.

The entirety of the past referenced investigations had a fundamental job in the improvement of this paper. It was created in three phases: right off the bat, the information assortment; also, NARX models and, at long last, ANFISE models with examination between two diverse system expectation models: NARX and ANFISE systems.

2- PV Experimental Work

In order to predict the actual PV panel model based module temperature as an output, an experimental design has been obtained from [13]. The system input consisted on radiation intensity, ambient temperature, wind speed and relative humidity. The experimental environment was in Baghdad city under coordinates of 33.33°N latitude, 44.39°E longitude, and 34 m elevation. two panels used in this work which are; one, the panel is type RNQ – 20D and made from a Mono-crystalline solar cells while, the other panel is type JP 01 that made from Poly crystalline solar cells. Many devices employed to collect input and output data such as digital multi-meter, regulator and battery. While, Thermocouple type k was used to obtain the panels temperature using a digital data recorder. The specifications of the used solar panels can see in Table (1).

Table (1): Technical specification of the PV panel. type poly crystalline. (b) type mono-crystalline

Model	**SR-100S**
Maximum power (Pm)	10W
Open circuit voltage(V∞)	22V
Short circuit current (Isc)	0.64A
Peak voltage (Vm)	17V
Peak current (Im)	0.59A
Short circuit current temperature coefficient(KI)	2.1mA/°C
Open circuit voltage temperature coefficient(KV)	-0.79V/°C
No.of serious cells (Ns)	36
No.of parallel cells (Np)	1
Dimensions	285*350*mm

Table (2): Technical specification of the PV panel type mono-crystalline

Maximum power (Pm)	**SR-100S**
Maximum power(Pm)	20W
Open circuit voltage(V∞)	21.6V
Short circuit current (Isc)	1.23A
Optimum operating voltage (Vmp)	17.5V
Optimum operating current (Imp)	1.14A
temperature coefficient of (Pmax)	-0.23%/°C
temperature coefficient of(V∞)	-0.33%/°C
temperature coefficient of(Isc)	0.05%/°C
Max system voltage	600 VDC (UL)
Fire Rating	Class C
Weight	2.2kg/4.8lb
Dimension	345*470*25mm
STC	Irradiance=1000W/ m, T=25°C, Am=1.5

A theoretical model for the thermal performance of PV panel with heat sink as a cooling system was formulated. The proposed cooling system was consisted of an array of pin fins as a heat sink. The model also included the determination of the suitable dimensions of fins and arrangement. A mathematical model was built and solved using MATLAB package in order to calculate the thermal characteristic of PV panel with heat sink. In experimental part, three passive methods applied and tested in order to reduce the operating temperature of solar module.

3-NARX Model

Prediction is a kind of dynamic filtering, in which past values of one or more time series are used to predict future values. A neural framework is a movement of figuring that endeavours to see shrouded associations in a great deal of data through a methodology that imitates the way wherein the human personality works. Neural frameworks can conform to developing input; so the framework makes the best result without hoping to redesign the yield criteria. Dynamic neural systems, which incorporate tapped postpone lines, are utilized for nonlinear separating and expectation. The model nonlinear autoregressive with exogenous (Outer) inputs (NARX) is proposed in [14]. NARX foresee arrangement y (t) given (d) past estimations of arrangement y (t) and another outer arrangement x (t), which can be single or multidimensional. The condition that models the NARX arrange conduct for time arrangement expectation is appeared in Equation (1).

$$\mathbf{y(t) = f(x(t-1), x(t-2), \ldots, x(t-d), y(t-1), y(t-2), \ldots, y(t-d))} \quad (1)$$

In this investigation, we use NARX with one contribution for the temperature of sun based module time arrangement at time t-1, y(t-1), and another contribution with exogenous information at time t−1, x(t−1), to give a solitary yield information y(t), comparing to the estimation of the temperature of solar module one step forward. Figure (1) shows this architecture.

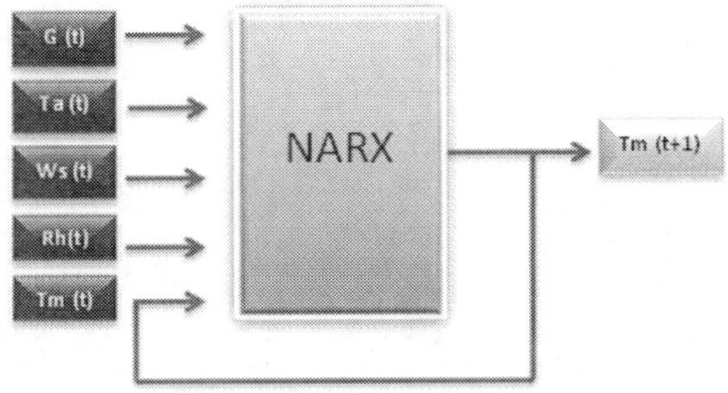

Fig. (1): Dynamic for PV temperature forecasting (NARX model)

4-ANFIS Model

ANFIS is a multilayer arranges mix of ANN learning calculations and a fluffy deduction framework (FIS) to delineate info parameters to yield esteem.

ANFIS has a system type structure like that of a neural system, which maps contributions through information enrolment works and related parameters. Figure (2) represent the ANFIS design with x and y sources of info and F yield. A1 and A2 are the fluffy participations that fuzzified input x. Likewise B1 and B2 are the fluffy enrolments that fuzzified input y [14].

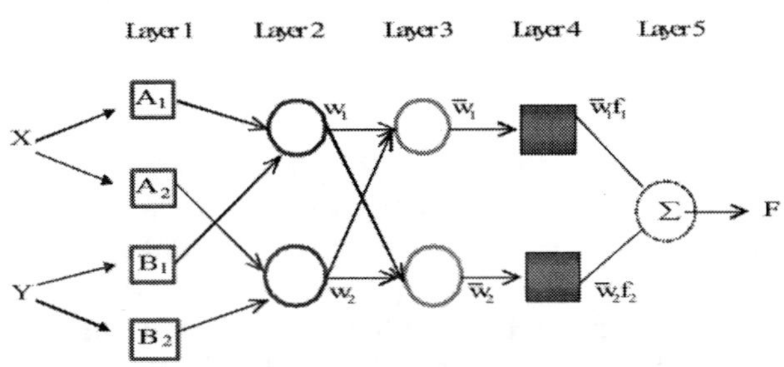

Fig. (2): ANFIS Structure

ANFIS model is created by five-layer engineering (Figure 2), where every one of the hubs in a similar layer have a comparable capacity [15]. Layer 1 is made out of participation work hubs. This layer changes over the crisped contribution to fluffy qualities, giving the contributions to the accompanying layers. The yield of this layer is given by:

$$O_{1,1} = \mu Ai(x) \quad (2)$$

$$O1, 2 = \mu Bi(y) \quad (3)$$

Where μ is the participation work, O1, I is the enrolment esteem for the fresh information sources x and y, and the subscripts 1 and I speak to the layer number and hub number, individually. Layer 2 contains the standard hubs. The hubs in this layer compute the terminating quality of a standard wi as:

$$O2, i = wi = \mu Ai(x)\ \mu Bi(y) \quad (4)$$

In layer 3, every hub figures the proportion of the I-the standard's terminating solidarity to the total of all guidelines terminating quality ($\overline{w}_i$):

$$O3, i = \overline{w}_i = \frac{w_i}{w_1+w_2} \quad (5)$$

IOP Conf. Series: Materials Science and Engineering **765** (2020) 012044 doi:10.1088/1757-899X/765/1/012044

Layer 4 figures the standard yields dependent on the ensuing parameters (Equations (2) and (3)) and the yield of the past layer:

$$\mathbf{O_{4,i} = \bar{w}_i\, f_i = \bar{w}_i(p_i\, x + q_i\, y + r_i)} \qquad \textbf{(6)}$$

At long last, layer 5 processes the general yield adding every one of the information sources sign to get the yield model worth.

$$\mathbf{O_{5,i} = \textstyle\sum_i \bar{w}_i\, f_i = \frac{\sum_i w_i\, f_i}{\sum_i w_i}} \qquad \textbf{(7)}$$

5- Model approval

The confirmation procedure of the outcomes is considered as one of the significant strides to quantify the exhibition of numerical calculations. Model approval was utilized to check model estimation execution through estimating and looking at genuine information (input-yield) acquired by the test arrangement with the anticipated information (yield) created by SI. Mean square blunder (MSE) was utilized to approve the dynamic conduct execution of the PV temperature by contrasting the real and anticipated information esteems. The type of mean square mistake was spoken to as:

$$\boldsymbol{\varepsilon(t) = \frac{1}{N} \textstyle\sum_{t=1}^{N} y(t) - \hat{y}(t)^2} \qquad \textbf{(8)}$$

Where $y(t)$ is the actual output and $\hat{y}(t)$ is the predicted output. [15]

6-Results and Discussion

Approach containing steps is utilized to appraise the temperature of the PV cluster. The system input consisted on radiation intensity, ambient temperature, wind speed and relative humidity, figure (3) show these variables, whilst the system output was temperature of PV panel and figure (4) illustrates the output. Representing dynamic data: 1650 time steps of 4 elements.

At using NARX system, the range selected for the preparation, testing, and approval sets was 70%, 15%, and 15%, individually, of the time arrangement utilized as information sources. Firstly, it was fixed number of neuron (Nn) at 10 and change number of delay (Nd) from 2_10, each MSE is detailed in the table (3) at this table found the best MSE at Nd=10. Secondly, it was fixed number of delay at 10 and change number of neuron from 2 _10. In table (3), it was found the best MSE at Nn=9, So, it was taken 9 ,10 for Nn and Nd, respectively, as show in table (4). From this table, it was found the MSE for all NARX system was 5.078. Figure (5) show the block diagram of the system, figure (6) illustrates a flowchart of the methodology employed, figure (7) displays predict the error and figure (8) offers the actual and predict temperature of PV panel.

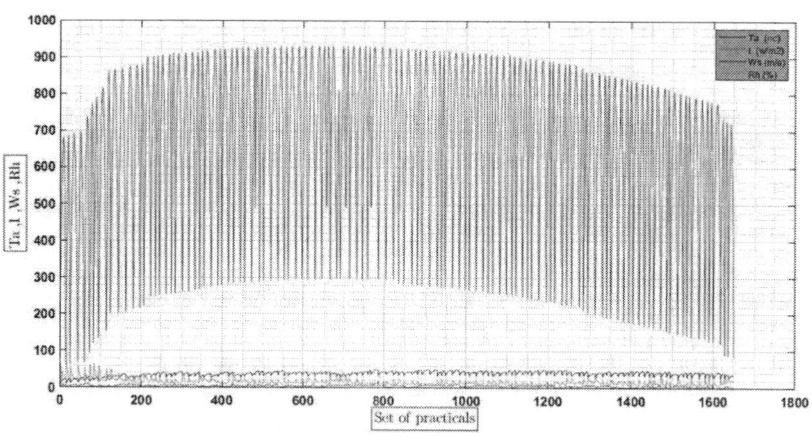

IOP Conf. Series: Materials Science and Engineering **765** (2020) 012044 doi:10.1088/1757-899X/765/1/012044

Fig. (3): illustrates the input data

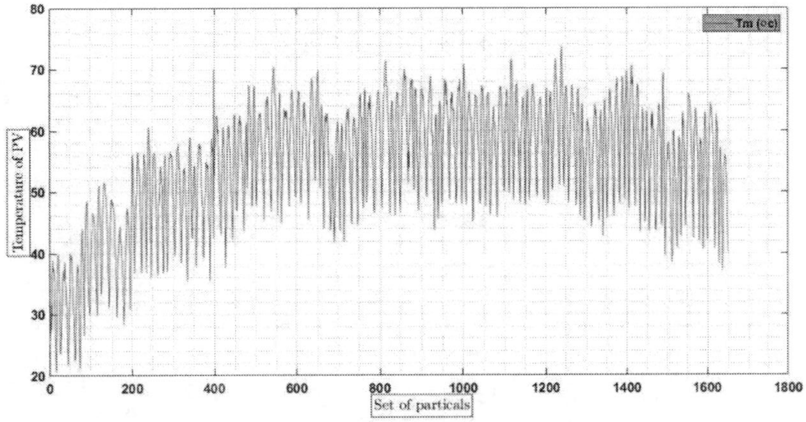

Fig. (4): illustrates the output data

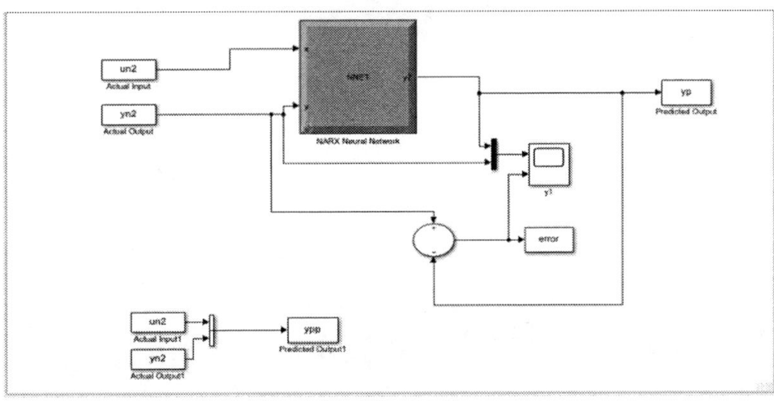

Fig. (5): Block diagram

IOP Conf. Series: Materials Science and Engineering **765** (2020) 012044 doi:10.1088/1757-899X/765/1/012044

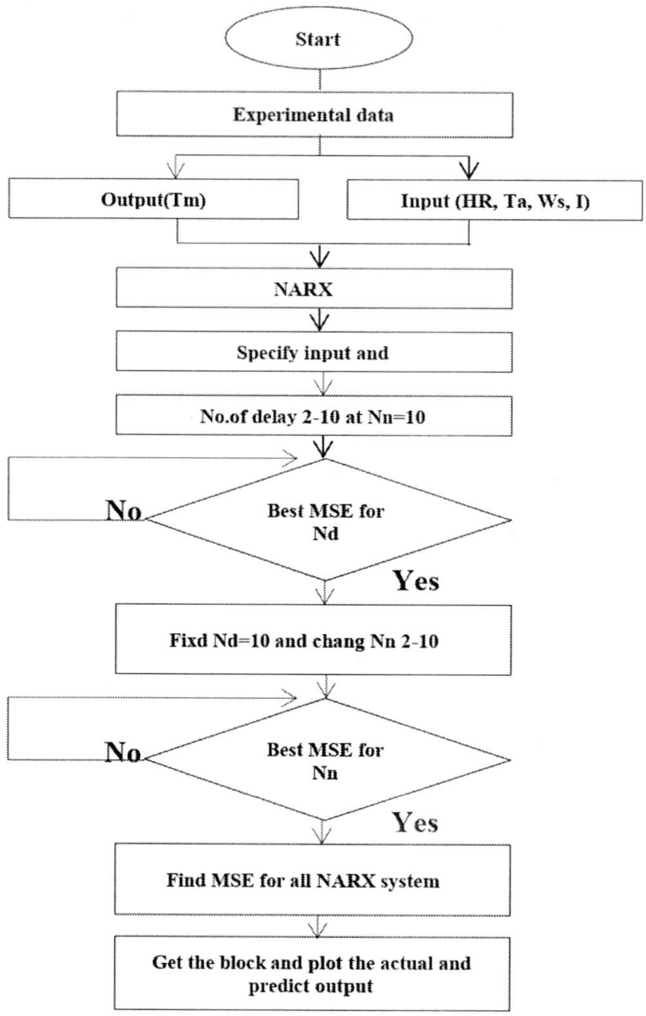

Fig. (6): NARX methodology flowchart.

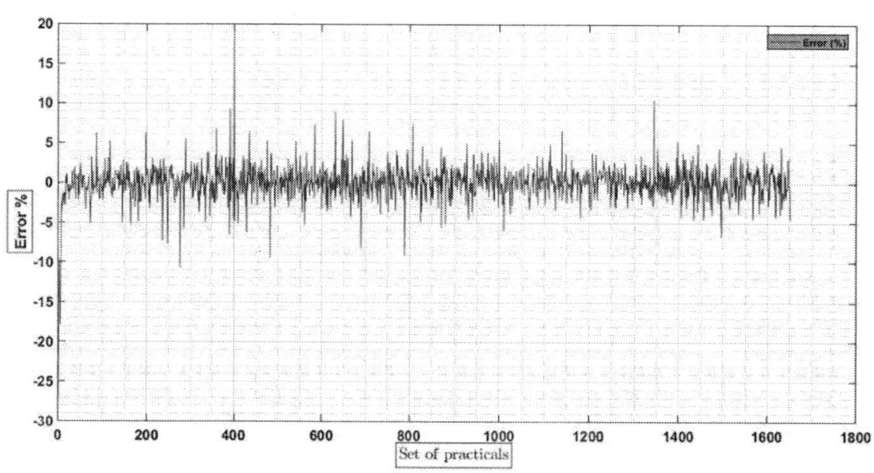

Fig. (7): illustrates predict the error

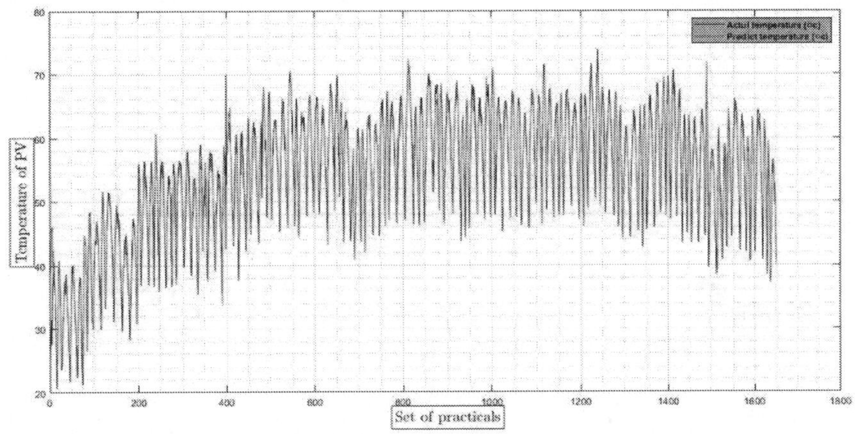

Fig. (8): illustrates actual and predict temperature of PV panel.

Table (3) - Training and predicting error results at fixed: a- Nn=10, b- Nd=10

STEP ONE						STEP Two					
No.of Neuron	No.of delay	MSE Training	MSE Validation	MSE Testing	MSE	No.of Neuron	No.of delay	MSE Training	MSE Validation	MSE Testing	MSE
10	2	4.371	3.540	3.750	4.66	2	10	4.550	4.614	4.47	5.69
10	3	3.206	4.822	4.450	4.19	3	10	3.710	3.640	4.45	5.36
10	4	3.954	4.146	4.968	5.03	4	10	4.360	7.770	8.78	5.93
10	5	3.750	3.560	5.230	5.05	5	10	3.440	6.970	3.50	5.23
10	6	2.870	4.59	7.290	5.o3	6	10	3.840	5.860	7.40	6.49
10	7	2. 369	8.022	3.850	4.86	7	10	3.410	5.430	5.77	6.86
10	8	3.280	3.760	5.905	4.68	8	10	3.290	4.450	6.73	5.25
10	9	3.540	4.080	4.620	6.30	9	10	2.970	5.590	5.50	5.11
10	10	3.032	8.050	4.280	4.01	10	10	2.930	4.240	5.36	6.57

Table (4) - Training and predicting error results

Number of Neurons	Number of delay	MSE Training	MSE Validation	MSE Testing	Mean Squared Error
9	10	3.085	4.718	6.172	5.078

Preparing of the ANFIS model is practiced through a progression of ages playing out an improvement procedure to limit the contrasts between test information and reproduced yield,. Figure (9) show a flowchart of the methodology employed For our implementation, data for training and checking were 850and 800, respectively, the best zero-request Surgeon ANFIS engineering contains an info layer with five hubs, two summed up ringer melded enrolment capacities per each information hub, the ages were 10, a standard layer with 32 hubs, a yield participation work with 32 consistent hubs, and one yield relating to TPV,Sim Figure (10)., Figure (11) it shows predict error percentage ,while ,figure(12) displays actual and predict temperature of PV panel. At MFs=2, MSE=2.2627e-07. However, MSE=0.1072 at MFs=3, as show in table (5).The plan was change the MFs from 2_10, but it was noticed that when MFs increase, the error increase too, so, the best choice was MFs=2 .

At a comparison between the Mean Square Error (MSE) which predicted by ANFIS and NARX models, in table (3) the MSE is 5.078 while the MSE in table (4) is 2.2627e-07 and 0.1072 that is mean the forecast delivered by the ANFIS method is relatively better .

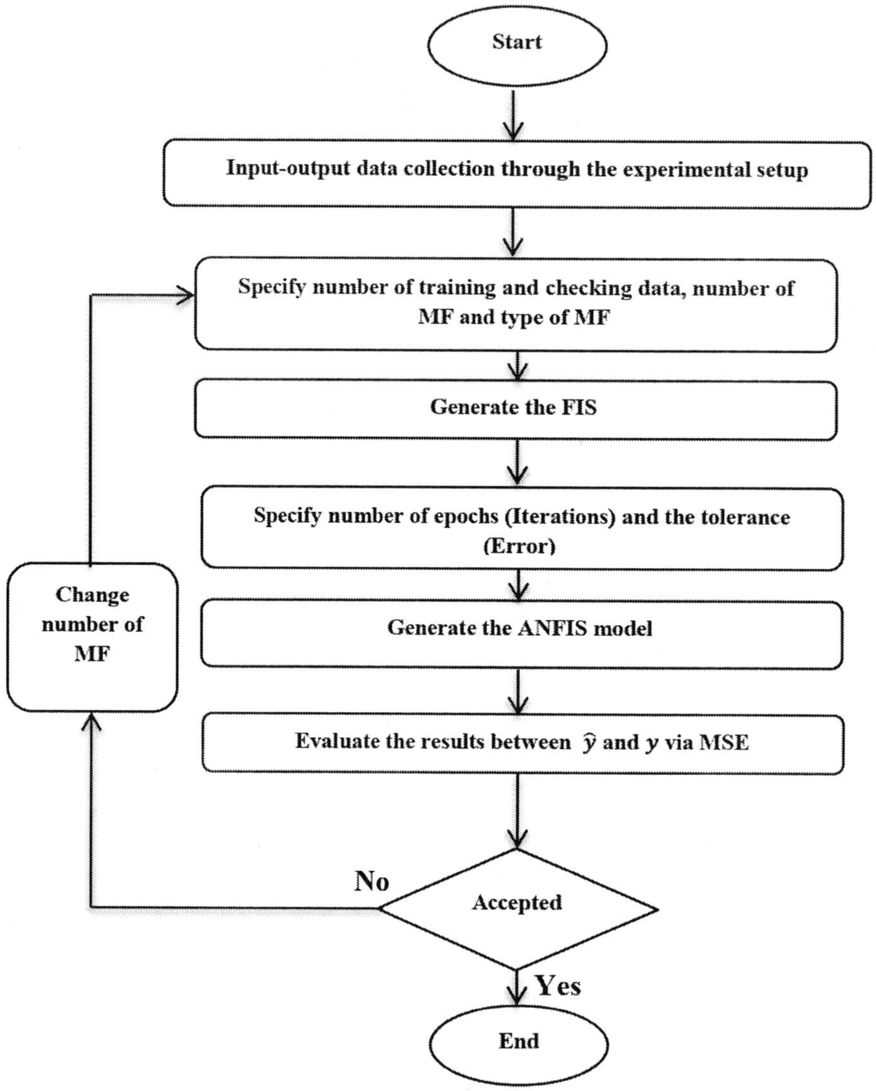

Fig. (9): ANFIS methodology flowchart.

IOP Conf. Series: Materials Science and Engineering **765** (2020) 012044 doi:10.1088/1757-899X/765/1/012044

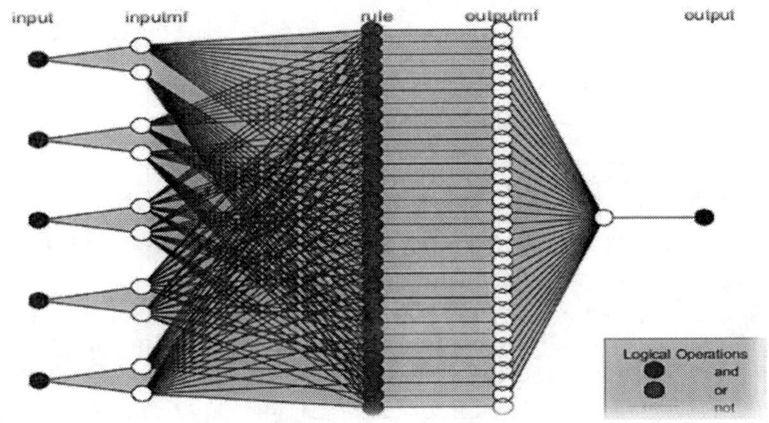

Fig. (10): ANFIS engineering model for estimation of PV temperature.

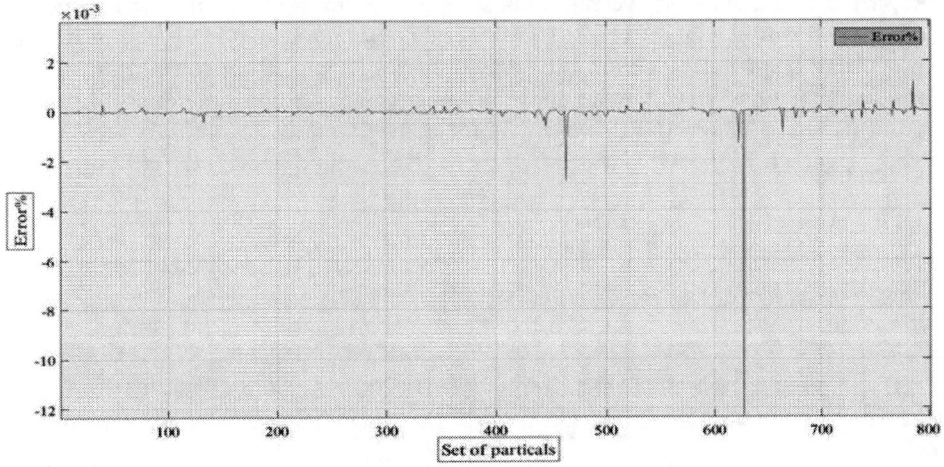

Fig. (11): Shows predict error

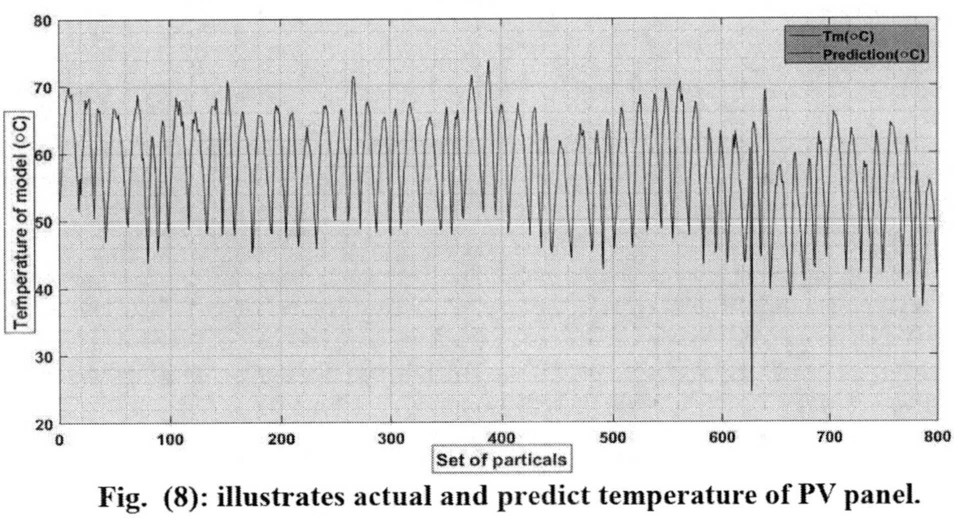

Fig. (8): illustrates actual and predict temperature of PV panel.

Table (5) - Results of the ANFISE model

Number of MFs	Minimal Training RMSE	Minimal Checking RMSE	Mean Squared Error	Number of nods	Number of fuzzy rules
2	2.1e-05	4.59e-04	2.2627e-07	92	32
3	7.8e-05	3.27e-01	0.1072	524	243

7- Conclusions

The thermal model utilized in foreseeing the temperature of PV module is significant as the module temperature influences its capacity yield. The thermal model where utilized in anticipating the module temperature of PV module working in Baghdad ecological conditions. A careful inquiry was performed dependent on the considered information parameters to choose the arrangement of the ideal mix of data sources that has the most impact on expectation of PV temperature. The NARX and ANFIS models were built by using 1650 data base composed of radiation intensity, ambient temperature, wind speed and relative humidity's were used to compare the two models. In the first model No.of neuron was changed from 2_10 until reached to the lower MSE, the best result fixed and change No.of delay from 2_10 .Whilst, the second model depended on selection the No.of MFs and membership function type. It very well may be construed from the outcomes that with respect to foreseeing the PVtemperature ANFIS based figure conveys better results when contrasted with the NARX based gauge, where the MSE was 5.078 by NARX method and 2.2627e-07 by ANFIS method.

References

[1] Parida, B., Iniyan, S., Goic, R., 2011. A review of solar photovoltaic technologies. Renewable and Sustainable Energy Reviews, Vol.15, pp.1625-1636.

[2] S.A. Kalogirou, Solar energy engineering: processes and systems, 2nd Edition, Academic Press, 2014

[3] C. G. Popovicia, S. V. Hudişteanua, T. D. Mateescua and N. Cherecheş "Efficiency improvement of photovoltaic panels by using air cooled heat sinks" , Energy Procedia, Vol. 85, pp. 425-432, 2016.

[4] A. Hachicha, C. Ghenai and A. Kadir, "Enhancing the Performance of a Photovoltaic Module Using Different Cooling Methods," World Academy of Science, Engineering and Technology. International Journal of Energy and Power Engineering, vol. 9, no. 9, pp. 1068-1071, 2015.

[5] S. M. Salih, O. I. Abd and K. W. Abid, "Performance Enhancement of Based on Water Spraying Technique," SciencePG journals. International Journal of Sustainable and Green Energy, vol. 4, no. 3-1, pp. 8-13, 2014.

[6] Y. Ibrahim, N. Y. Hassan, M. Momoh and M. G. Mahmud, "Design and Thermos phonic Solar Photovoltaic Thermal Water Heating System," International Organization of Scientific Research. Journal of Applied Physics (IOSR-JAP), vol. 7, no. 2, pp. 88-96, 2015.

[7] Xiao Tang, Zhenhua Quan, Yaohua Zhao, “Experimental investigation of solar module cooling by a novel micro heat pipe array”, Energy and Power Engineering, Vol. 2, No. 3, pp. 171-174, 2010

[8] H.G. Teo, P.S. Lee and M.N.A. Hawlader, "An active cooling system for photovoltaic modules", Applied Energy, Vol. 90, pp.309-315, 2012.

[9] Tang X., Quan Z. and Zhao Y., "Experimental Investigation of Solar Panel Cooling by A Novel Micro Heat Pipe Array", Energy and Power Engineering, Vol.2, PP.171-174, 2010.

[10] Chaouachi, A.; Kamel, R.M.; Ichikawa, R.; Hayashi, H.; Nagasaka, K., 2009).

[11] Di Piazza, Di Piazza, & Vitale, 2013).

[12] Mohanty, S. ANFIS based Prediction of Monthly Average Global Solar Radiation over Bhubaneswar (State of Odisha). Int. J. Ethics Eng. Manag. Educ. 2014, 1, 2–6

[13] Duha, A. Attar .Cooling Enhancement of PV Panel Using Heat Sink and Wood Wool Pad with Temperature Prediction by Partials Swarm Optimization.2018

[14] Mohammad, M. J., & Darus, I. Z. M. (2014, April). Nonlinear system identification of vortex induced vibration on pipe cylinder. In Proceedings of the 14th International Conference on Robotics, Control and Manufacturing Technology (ROCOM'14), Kuala Lumpur, Malaysia (pp. 23-25).

[15] Mohammed, M. J., Ahmed, M. K., & Abbas, B. A. MODELING AND CONTROL OF HORIZONTAL FLEXIBLE PLATE USING PID-CS CONTROLLER. August 2019 .

IOP Conf. Series: Materials Science and Engineering **765** (2020) 012045 doi:10.1088/1757-899X/765/1/012045

Modeling and Investigation Performance of Small-Scale PMSG based HAWT with FOC

Talib Z. Farge

Ali. H. Numan

Zaineb Tahir

Electromechanical Engineering Department, University of Technology, Baghdad, Iraq

Correspondence Author Ali H. Numan: 50059@uotechnology.edu.iq

Abstract. This paper focused on modeling and simulation of small-scale horizontal axis wind turbine (HAWT) based on permanent magnetic synchronous generator (PMSG) with filed oriented control (FOC) circuit connected to 3-phase full bridge rectifier by MATLAB software. FOC system developed for generator side converter to regulate mechanical speed and electromagnetic torque at variable wind speed. Space vector pulse width modulation (SVPWM) had been implanted in FOC system to turn on six-IGBT switches and validate pure and steady-state of output voltage. Results showed optimum performance coefficient was 0.48 and tip speed ratio 8.1 at zero pitch angle, also when pitch angle increase by 5^ocoefficient performance decreased to 0.35. The simulated waveforms of the mechanical speed, and electromagnetic torque were contained two case, before and after applied FOC. With FOC, the results shown faster and smooth response to the faster change in wind speed.

Keywords: HAWT, PMSG, FOC, and SVPWM.

1. Introduction

Variable speed wind turbine system has many features as compared with fixed speed wind turbine system such as low cost, small size, high efficiency, high power production, high power quality and, maximum power can be extracted at variable wind speed. There are many types of generators use in wind turbine systems. Permanent Magnet Synchronous Generator (PMSG) its common applied in modern wind turbines because of low weight, small size, high pole number, gear-less, less maintenance and, compact structure. Therefore it is an excellent applicant for this purpose. Power delivered to grid by variable speed wind turbines depend on wind speed and rotor speed of generator [1]. In the generating side, there are many control strategies used to regulate speed and extract maximum power of variable speed wind turbines. Felid oriented control (FOC) is a popular and modern control method because archive optimal utilization of wind energy [2].

A literature review represent ideas and researchers efforts in dealing with theoretical (simulation) and experimental of wind energy conversion system based on PMSG. P. Thayumanavan et al. [3] presented maximum power point tracking (MPPT) method based on sensor-less FOC to control speed and torque of PMSG. Electromagnetic torque had been control by set i_d to zero and control by i_q. B. Lahfaoui et al. [4] modeling and simulation of MPPT based on PMSG. MPPT connected DC boost convertor. Perturb and observe algorithm used in MPPT. B. Jain et al. [5] reviewed control methods of PMSG based wind energy conversion system connected to grid. Control methods included direct and vector methods. Direct torque and field oriented control method used to control speed and torque in machine side, while direct power control and voltage oriented control used in grid side. This paper focused as follows, section 2 presents modeling aerodynamic HAWT. In section 3, presents modeling of PMSG, MPPT based FOC and, VOC.

2. Aerodynamic Modeling of HAWT

The kinetic energy available from wind energy for x distance [6]:

$$\mathrm{U} = \frac{1}{2} m u^2 = \frac{1}{2} (\rho_a A x) u^2 \tag{1}$$

Wind power is derivative of kinetic energy:

$$P_w = \frac{dU}{dt} = \frac{1}{2} \rho_a u^2 A \frac{dx}{dt} = \frac{1}{2} \rho_a A u^3 \tag{2}$$

$$\rho_a = 3.485 \frac{P}{T_a} \tag{3}$$

Air density ρ_a equal 1.225 kg/m^3 at stander conditions (101.3kPa, 288K).

$$P_m = C_P \left(\frac{1}{2} \rho_a A u^3 \right) = C_P P_w \tag{4}$$

Betz coefficient (C_P) is factor prove that ideal turbine can't extract more than 0.593 from the wind power.

$$\lambda = \frac{r_m \omega_m}{u} \tag{5}$$

$$C_P = b_1 \left(\frac{b_2}{\lambda_i} - b_3 \beta - b_4 \right) exp(\frac{-b_5}{\lambda_i}) + b_6 \lambda \tag{6}$$

$$\lambda_i = (\frac{1}{\lambda + 0.08\beta} - \frac{0.035}{\beta^3 + 1})^{-1} \tag{7}$$

For variable wind speed, $b_1 = 0.5176, b_2 = 116, \; b_3 = 0.4, b_4 = 5, b_5 = 21, b_6 = 0.0068$
Modeling shaft of wind turbine described in the following equation:

$$J_m \frac{d}{dt} \omega_m + B_m \omega_m = T_e - T_m \tag{8}$$

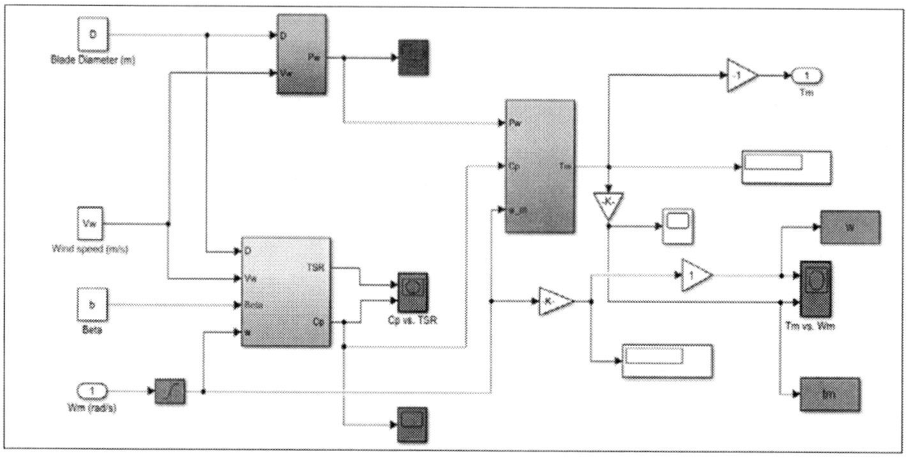

Figure 1: Simulation of HAWT.

Table 1: Parameters of wind turbine

Rated power	220W
Rated wind speed	11.5m/s
Blade diameter	1.3m
Rated R.P.M	600 r.p.m

3. Electrical Modeling

3.1. PMSG modeling

PMSG rotor contained poles of permanent magnet (PM) generate magnetic field. Figure 2 shows equivalent circuit of 3-phase PMSG has 2-poles with *abc*-axis of stator. Current induced by rotor divide into two quadrature, q-axis and d-axis. Rotor position angle θr is angle made angle between d-axis and a-axis [7]. Following basic equations of PMSG modeling:

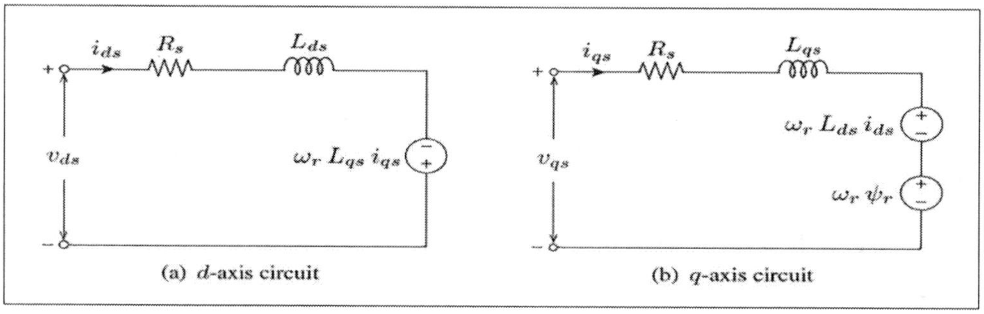

Figure 2: Equivalent circuit of PMSG.

IOP Conf. Series: Materials Science and Engineering **765** (2020) 012045 doi:10.1088/1757-899X/765/1/012045

$$v_{ds} = r_s i_{ds} + L_d \frac{di_{ds}}{dt} - \omega_r L_q i_{qs} \quad (9)$$

$$v_{qs} = r_s i_{qs} + L_q \frac{di_{qs}}{dt} - \omega_r L_d i_{ds} + \omega_r \psi_r \quad (10)$$

Active power P_{dq} and reactive power Q_{dq} in dq-frame

$$P_{dq} = \frac{3}{2}\left(v_{ds} i_{ds} + v_{qs} i_{qs}\right) \quad (11)$$

$$Q_{dq} = \frac{3}{2}\left(v_{qs} i_{ds} - v_{ds} i_{qs}\right) \quad (12)$$

Electromagnetic torque:

$$T_e = \frac{3P_p}{2}\left(\psi_{ds} i_{qs} - \psi_{qs} i_{ds}\right) = \frac{3P_p}{2}\left(\psi_r i_{qs} - (L_{ds} - L_{qs}) i_{qs} i_{ds}\right) \quad (13)$$

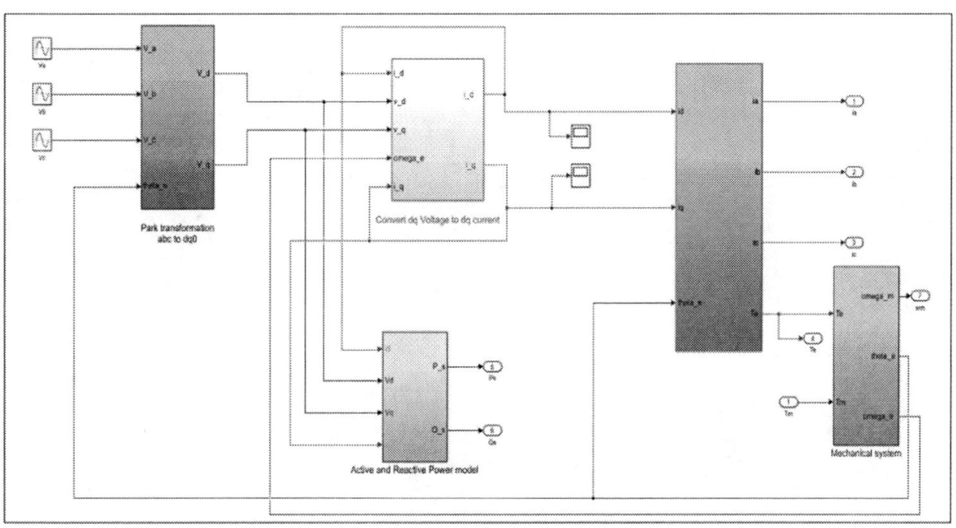

Figure 3: Modeling of PMSG.

3.2. Modeling of flied oriented control

Because of fluctuation of wind speed that effect on speed of generator and power production field oriented control (FOC) use in generator side. By set $i_{ds}^{*} = 0$, maximum electromagnetic torque can be achieve. By quadrature current (i_{qs}^{*}) can be controlled by electromagnetic torque [8].

IOP Conf. Series: Materials Science and Engineering **765** (2020) 012045 doi:10.1088/1757-899X/765/1/012045

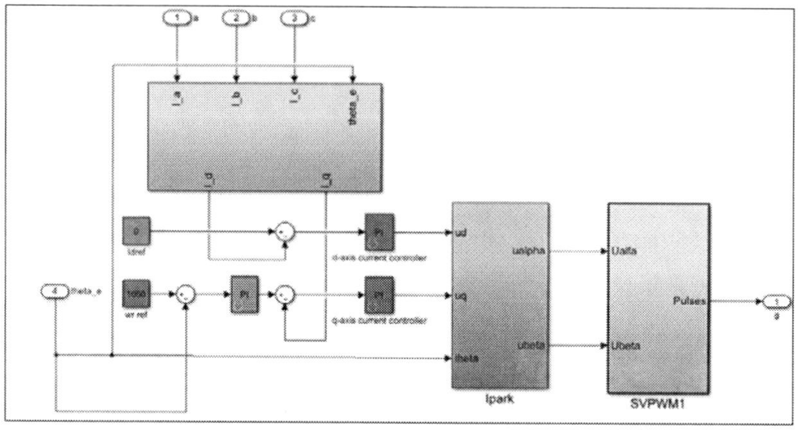

Figure 4: FOC simulation.

4. Results and Discussions

Wind turbine based on PMSG system was simulated by MATLAB/Simulink 2017b. Figure 5 shows the coefficient performance with tip speed ratio, optimum performance coefficient was 0.48 at tip speed ratio 8.1 while pitch angle was 0°. It is clear from the figure when pitch angle increase by 5^ocoefficient performance decreased to 0.35 at tip speed ratio 9.3. Mechanical torque effected mainly by pitch angle, when pitch angle fixed at 0° maximum torque was 3.85N.m at mechanical speed 600r.p.m. When pitch angle increased by 5° mechanical torque drop to 2.64N.m at mechanical speed 320.3r.p.m.

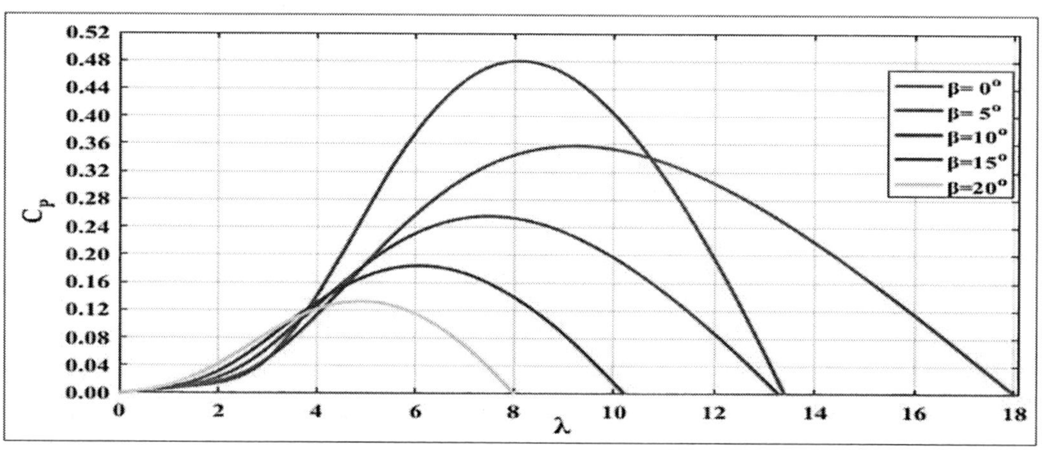

Figure 5: Performance coefficient vs. tip speed ratio.

IOP Conf. Series: Materials Science and Engineering **765** (2020) 012045 doi:10.1088/1757-899X/765/1/012045

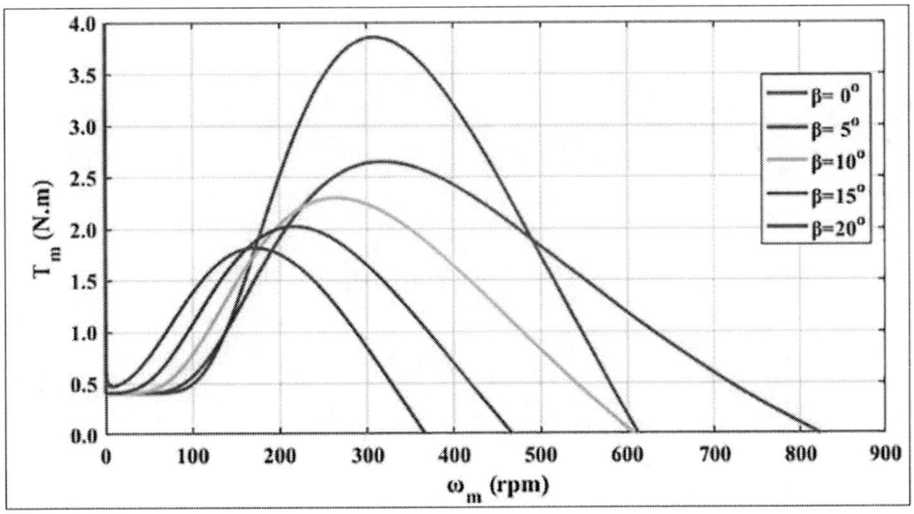

Figure 6: Mechanical torque vs. mechanical speed.

Three air speeds (11.5, 8 and, 3 m/s) were selected for operation simulation of wind turbine system for 1sec as shown in figure (7). Output current is directly proportional to the air speed. At 11.5, 8 and, 3m/s the output current was 4.81, 3.15, 1.6 A, respectively as figure 8. Mechanical rotational speed effected mainly by wind speed fluctuation, this is clearly shown by the figure 9. With FOC we could get the curve without fluctuation and stable speed. Also, electromagnetic torque effected mainly fluctuation of wind speed, FOC system cancelled these fluctuations and get stable electromagnetic torque as shown figure 10.

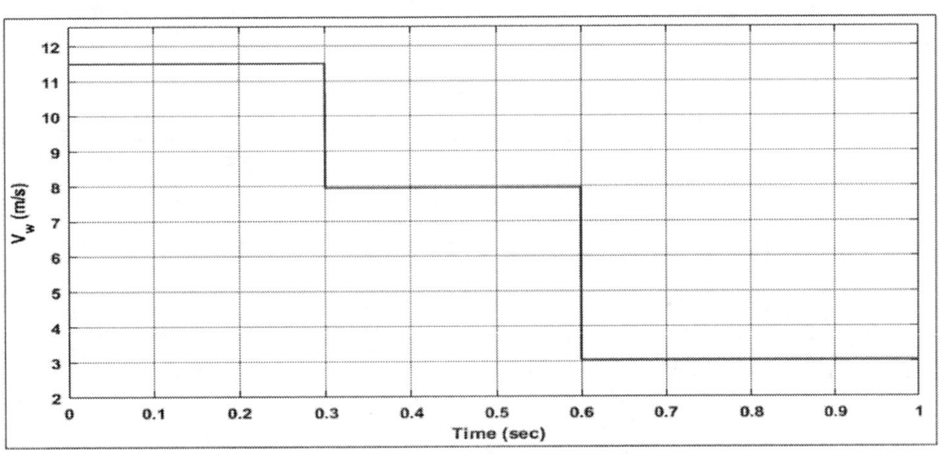

Figure 7: Input wind speed.

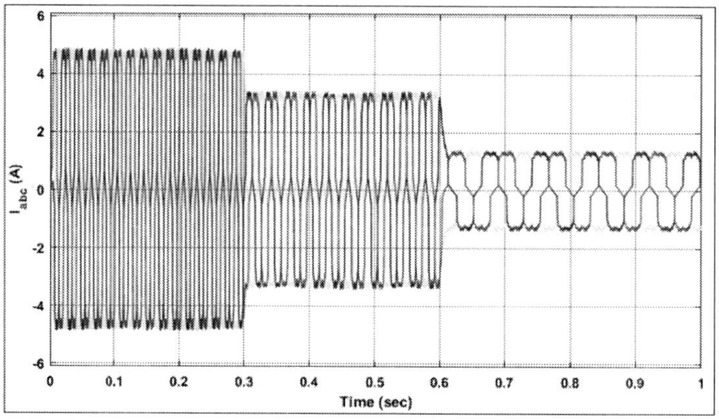

Figure 8: Output current of PMSG

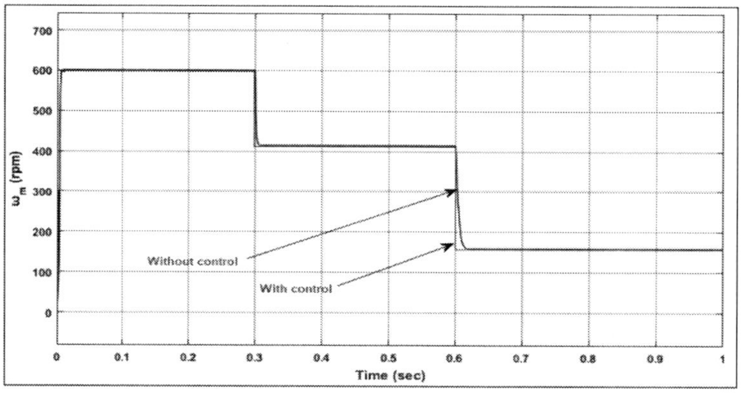

Figure 9: Mechanical speed with and without FOC

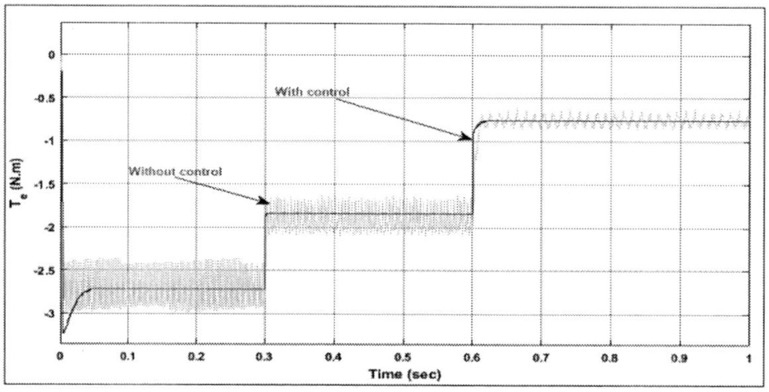

Figure 10: Electromagnetic torque with and without FOC

IOP Conf. Series: Materials Science and Engineering **765** (2020) 012045 doi:10.1088/1757-899X/765/1/012045

5. Conclusions

This paper presented modeling and simulation of HAWT based on PMSG with filed oriented control system had been simulated by MATLAB/Simulation. Wind turbine system has rated 220W. To achieve stable operation of PMSG at variable wind speed the flied oriented control system was applied. Maximum mechanical torque and optimum performance coefficient were obtained at fixed pitch angle (0°). Steady-state of electromagnetic torque and mechanical speed had been achieved under variable wind speed when used field oriented control.

Nomenclature

A	Cross-section area	m^2
$b_1, b_2, b_3, b_4, b_5, b_6$	Constants of solution aerodynamic differential equations	
B_m	Viscous friction	N.m.s
C_P	Performance coefficient	
g	Standard acceleration of gravity	m/s^2
i_{as}, i_{bs}, i_{cs}	Machine phase *abc* stator voltage	A
i_{ds}, i_{qs}	PMSG stator currents in *dq*-frame	A
J_m	Momentum inertia of rotor shaft	Kg.m^2
L_q, L_d	PMSG rotor inductance in *dq*-frame	H
L_a	Armature inductance	Wb
m	Air mass	Kg
P	Air pressure	Pa
P_w	Wind power	W
P_m	Mechanical power	W
P_p	Pair poles	
P_{dq}	Active power of PMSG in synchronous *dq*-reference frame	W
Q_{dq}	Reactive power of PMSG in synchronous *dq*-reference frame	VA
r_m	Maximum radius blade of wind turbine	m
r_s	Resistance of PMSG stator windings	Ω
T_m	Mechanical torque	N.m
T_e	Electromagnetic torque	N.m
T_a	Air temperature	°C
t	Time	sec
v_{as}, v_{bs}, v_{cs}	Machine phase *abc* stator voltages	V
v_{ds}, v_{qs}	Machine stator voltages in *dq* frame	V
U	Kinetic wind energy	J
u	Wind velocity	m/s
x	thickness of air parcel in x-direction	m
ρ_a	Air density	Kg/m^3
λ	Tip speed ratio	
β	Pitch angle	degree
ω_r	Electrical angular speed	rad/s
ω_m	Mechanical angular speed	rad/s
ω	Angular electrical frequency	rad/s
ψ_r	Peak flux linkage of permanent magnetic	Wb
$\psi_{as}, \psi_{bs}, \psi_{cs}$	Machine stator flux linkages in dq-frame	Wb

References

[1] Neeraj Pareta, Naveen Sen. (2014) Modelling and Simulation of PMSG Based Wind Energy Conversion System. International Journal of Emerging Research in Management & Technology; pp43-52.

[2] Ankush Mudholkera, P. M. Menghalb and, A. Jaya Laxmic, (2015) "SVPWM Based Converter for PMSG Based Wind Energy Conversion System," *Elsevier, Procardia Computer Science*, vol. 70, pp 676 – 682.

[3] P. Thayumanavan, R. Muthu and, J. Sankararaman, (2014) "Sensor-less Field Oriented Control of Wind Turbine Driven Permanent Magnet Synchronous Generator Using Flux Linkage and Back EMF Estimation Methods," *Research Journal of Applied Sciences, Engineering and Technology*, vol.20, pp 4303-4312.

[4] B. Lahfaoui, S. Zouggar, B. Mohammed, Mohamed L. Elhafyani, (2017) "Real time study of P&O MPPT control for small wind PMSG turbine systems using Arduino microcontroller," *Elsevier, Energy Procedia*, vol.111, pp 1000– 1009.

[5] Bhavna Jain, Shailendra Jain and, R. K. Nema, (2015) "Control strategies of grid interfaced wind energy conversion system: An overview," Elsevier, Renewable and Sustainable Energy Reviews vol. 47, pp 983–996.

[6] Gary L. Johnson, "Wind Energy System,(2006) " *Prentice Hall PTR, University of Wisconsin – Madison.*

[7] V. Yaramasu and, B. Wu, (2017) "Model Predictive Control of Wind Energy Conversion System," *John Wiley & Sons.*

[8] Muamer M. Shebani and Tariq Iqbal, (2017) "Dynamic Modeling, Control, and Analysis of a Solar Water Pumping System for Libya," *Hindawi, Journal of Renewable Energy,* pp 1-13.

Experimental and Numerical Assessment on S-shaped Diffuser performance with different Turbulence Intensity

Raed Jessam[1]

Hussain H. Al-Kayiem[2]

Mohammad S. Nasif,[2]

Ameer Abed Jaddoa Albuali[1]

[1]Electromechanical Engineering Department, University of Technology, Baghdad, Iraq

[2]Mechanical Engineering Department, University of Technology, Petronas, Malaysia

corresponding author's e-mails : 50097@uotechnology.edu.iq, hussain_kayiem@utp.edu.my, mohammad.nasif@utp.edu.my

Abstract. The turbulence structure in air intake S-shaped diffuser is proven to be influencing parameter on the diffuser performance. In the present research work, experimental and numerical investigations have been undertaken to explore the effect of inlet turbulence intensity level on the performance of air intake S-shaped diffuser. Detailed measurements including pressure and velocity at the inlet and outlet planes and static pressure on the top and bottom walls were taken at three pre-selected Reynolds numbers of 4.8×104, 6.4×104 and 7.5×104. The predicted corresponding turbulence intensities at the experimented three Re, were 4.16%, 3.15% and 2.8%. ANSYS-FLUENT 15 software with Standard k-ε turbulence model has been used for numerical simulations. Numerical results of static pressure recovery, total pressure loss coefficient and wall static pressure recovery have been compared with the experimental results. The experimental results indicate that increasing Reynolds number at the inlet of S-shaped diffuser resulting in slight decrease in measured turbulence intensity. Also, as Re increased from 4.8x104 to 7.5x104, the static pressure recovery increased by 9% and the total pressure loss coefficient was reduced by 4.9%.

Keywords: distortion coefficient; CFD; S-shaped diffuser; turbulence intensity; static pressure coefficient.

Nomenclature

AR	area ratio ($AR=Ae/Ai$)	P_{ws}	wall static pressure (N/m^2)
Ae	cross section area of outlet plane	P_{wsi}	wall inlet static pressure (N/m^2)
Ai	cross section area of inlet plane	Re	Reynolds number
D_h	hydraulic diameter (mm)	P_t	total pressure (N/m^2)
C_{PR}	static pressure recovery coefficient	$P_{t\,ave\,o}$	outlet average total pressure (N/m^2)
C_{WPR}	wall static pressure coefficient	R_c	radius of curvature (mm)
C_{TL}	total pressure loss coefficient	U_{ave}	average velocity (m/s)
$T.I$	turbulence intensity	ρ	density of air (kg/m^3)
P_s	static pressure (N/m^2)	μ	viscosity of air (kg/m. s)
P_{si}	inlet static pressure (N/m^2)	AIP	aerodynamic inlet plane
P_{ti}	inlet total pressure (N/m^2)		

1. Introduction

The S-shaped diffuser is one of the widespread research topics in mechanical engineering due to the imperative applications for guiding flow at the aero-engine inlet, gas turbine inter-connecting-duct and engine exhaust. Therefore, S-shaped diffuser represented the main component of breathing aero-engines. The design of S-shaped diffusers is more complicated than regular diffusers since additional pressure losses are resulting from the complicated flow phenomena consisting of flow separation and circulating bubble, in high bending walls. The experimental and numerical investigation is complicated with the added challenges in predicting boundary layer growths and flow separation under the interactions between the secondary counter rotating flows and the adverse pressure gradient regions [1].

The main issue that is taken concern in the numerical simulation is the modeling of the turbulence. Wenbiao and Xiaocui [2] used SST turbulent model to get the optimum design of S-shaped diffuser. Xiao *et al.* [3] applied a modified k–ω turbulence model to investigate transonic flow features of an S-duct. Gerolymos *et al.* [4] adopted Reynolds Stress Model to analyze the flow field structure of a dual S-duct. Saha *et al.* [5] studied flow characteristics of the S-duct based on RNG k–ε turbulence model. Zhang *et al.* [6] used k–ε turbulence model to design an S-duct. Lee and Kim [7] were studied the sensitivity of the S-duct with different turbulence models. Zhang *et al.* [8] investigated multidisciplinary design strategy of an S-duct, which used SST k–ω turbulence model for numerical simulation. Gopaliya *et al.* [9-10] investigated S-shaped diffuser of 90°/90° turn with rectangular inlet (Aspect ratio = 2) and semicircular outlet (Aspect ratio = 2) by the effect of offset. The obtained show that the increase in Reynolds number (Re) leads to increase in coefficient of static pressure gain (C_{PR}).

Gavin *et al.* [11] investigated the Allison 250 gas turbine S-shaped diffuser. The calculated Allison 250 S-shaped diffuser flow features were then compared to those in a constant diameter S-shaped tube with the same inlet cross-sectional area and offset.

Gavin *et al.* [12] have investigated numerically six variations of the S-shaped diffuser geometry of the Allison 250 gas turbine with range of turbulence intensity ($T.I$) at one Re and with range of Re at one turbulence intensity. The qualitative results show that, the pressure coefficient variations for the three inlet $T.I$ values considered are qualitatively similar. The quantitative results show that the highest inlet $T.I$ returns slightly higher-pressure coefficients.

The present study focuses on the turbulent flow in the air intake S-shaped diffuser with different inlet $T.I$ and different inlet Re. An experimental test rig is constructed and used to collect data that

IOP Conf. Series: Materials Science and Engineering **765** (2020) 012046 doi:10.1088/1757-899X/765/1/012046

have been used for validation. ANSYS-FLUENT15 with Standard *k-ε* turbulence model has been adopted as the investigation tool. The main performance parameters of S-shaped diffuser in terms of (C_{PR}), total pressure loss coefficient (C_{TL}) and wall static pressure coefficient (C_{WPR}) have been predicted to permit exploration of sensitivity of these parameters with the different inlet *T.I* and *Re*.

2. Experimental set up

The geometry of the S-shaped diffuser used in this study is shown in figure 1. It was designed based on area ratio AR of 3.1. The radius of curvature $R_c = 280$ mm, the turning angle $\theta = 45°$. The total length of the diffuser $L = 526$ mm, and the hydraulic diameter $D_h = 65$ mm. The modeled diffuser has a constant area conduit at the inlet with 65 mm long x 65 mm height, for smooth air inflow, and constant area conduit at the outlet with 65 mm long x125 mm height.

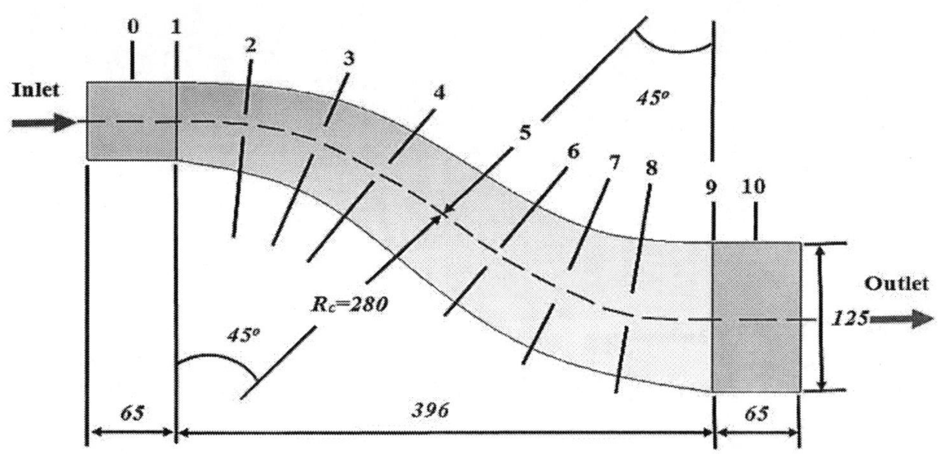

Figure 1: Geometry of the modeled S-shaped diffuser and plane of interest (all dimensions in mm)

An 1100 mm duct length which has same diameter of inlet S-shaped diffuser has been used to connect the air blowers with the S-shaped diffuser as shown in figure 2.

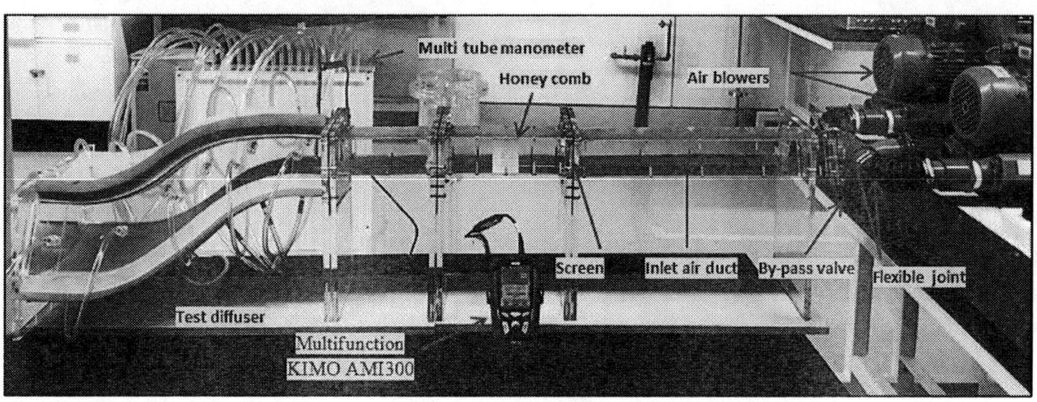

Figure 2: S-shaped diffuser Test Apparatus

IOP Conf. Series: Materials Science and Engineering **765** (2020) 012046 doi:10.1088/1757-899X/765/1/012046

Three inlet velocities have used for investigation 14.4 m/s,15.8m/s and 17.4 m/s, i.e three *Re* of 4.8×10^4, 6.4×10^4 and 7.5×10^4 obtained depending on the inlet hydraulic diameter of S-shaped diffuser. Also, it has been used for measuring the *T.I* experimentally at the inlet plane of S-shaped diffuser by interface the datalogger with PC as shown in figure 3. These inlet velocities and inlet turbulence intensities used an inlet boundary conditions for CFD simulation.

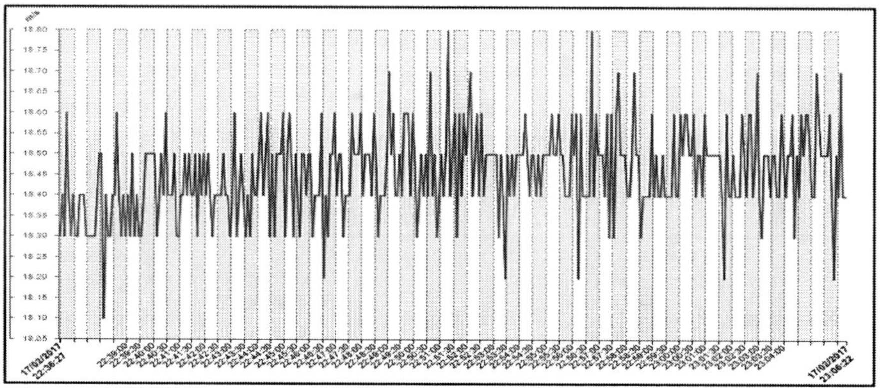

Figure 3: Measurement of *T.I* by velocity fluctuation with time at the inlet of S-shaped diffuser.

3. Performance parameters

The main important parameters are using for evaluation the performance of S-shaped diffuser in the current study are:

3.1. Wall static pressure coefficient CWPR

The wall static pressure coefficient is represented by the following equation:

$$C_{WPR} = \frac{P_{WS}-P_{WSi}}{\frac{1}{2}\rho U^2_{ave\ i}} \qquad (1)$$

3.2. Static pressure recovery coefficient CPR

It represents the magnitude by which kinetic energy has been converted into pressure energy due to diffusing action at any location along the S-shaped diffuser, it is represented by the equation:

$$C_{PR} = \frac{P_S-P_{Si}}{\frac{1}{2}\rho U^2_{avi}} \qquad (2)$$

IOP Conf. Series: Materials Science and Engineering **765** (2020) 012046 doi:10.1088/1757-899X/765/1/012046

3.3. Total pressure loss coefficient CTL

It is a measure of how much total pressure is lost as a proportion of the mean inlet dynamic pressure due to viscous forces and turbulent mixing, this defined by the equation

$$C_{TL} = \frac{P_{ti} - P_t}{\frac{1}{2}\rho U_{avi}^2} \quad (3)$$

4. Numerical Approach

4.1. Mesh generation

The S-shaped diffuser used in the experiments was modeled by modular design ANSYS 15 for CFD simulation, a mesh generation by ANSYS - FLUENT 15. Figure 4 shows the unstructured mesh along the S-shaped diffuser.

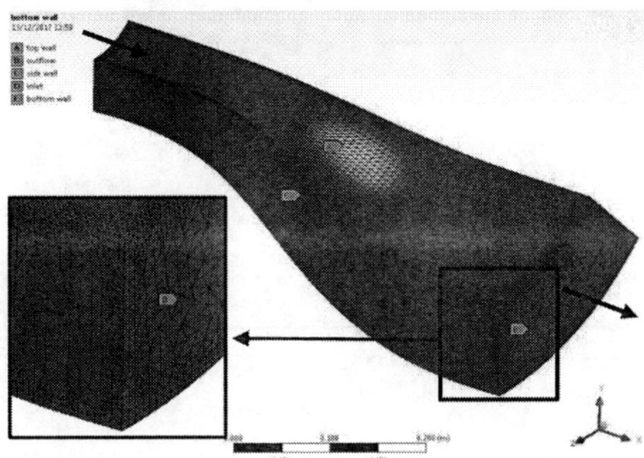

Figure 4: Unstructured mesh of S-shaped diffuser

4.2. Mesh independency checking

When doing analysis of any CFD problem using software or code the first important and necessary thing is checking the mesh dependency, because of the solution has a great dependency on the fineness of meshing. In this study, Different meshes have been checked for S-shaped diffuser for $Re = 6.4\times10^4$ and $T.I = 3.15$, with performance parameter C_{WPR} on the top and bottom walls as shown in figure 5-A and 5-B, respectively.

IOP Conf. Series: Materials Science and Engineering **765** (2020) 012046 doi:10.1088/1757-899X/765/1/012046

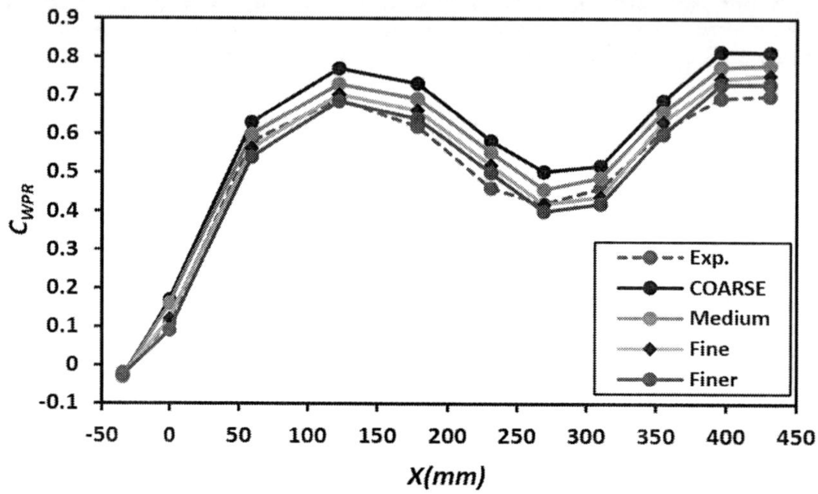

A. Top wall

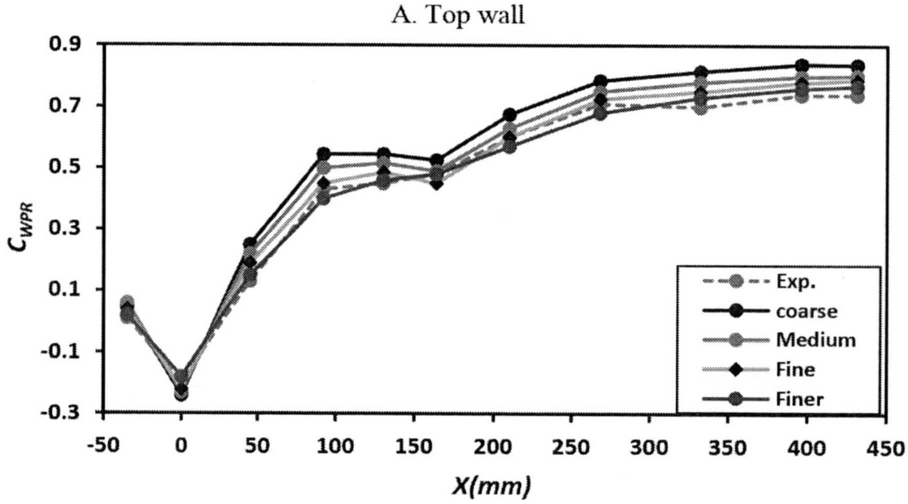

B. Bottom wall

Figure 5: Calculated C_{WPR} with different grid resolution: A. top wall, B. bottom wall

Refining the mesh from coarse to medium, results in a significant change of the C_{WPR} values as much as 11% at the top wall and 12% at the bottom wall. A further refinement of the mesh results in an average variation of the C_{WPR} values of 5.5% and 7% at the top and bottom walls, respectively. This variation drops to 1.4% at the top wall and 2.6% bottom wall when fine mesh is compared to the finer mesh which has 119548 volume cells. This means that with fine mesh results were almost independent of the computational mesh, and the mesh could be used with certainty for validation. The important features of these meshes are summarized in table 1.

Table 1. Grid independency check.

Mesh No.	No. of volume cells	C_{WPR} of top wall		C_{WPR} of bottom wall	
		CFD	Error %	CFD	Error %
coarse	19404	0.81	11	0.84	12
medium	82042	0.77	5.5	0.80	7
fine	119548	0.74	1.4	0.77	2.6
finer	124572	0.73	-	0.75	-

4.3. Boundary conditions

For the CFD simulations of S-shaped diffuser, same values of inlet velocity profile and *T.I* as found experimentally are used as input for velocity inlet and *T.I* boundary conditions.. Enhanced wall treatment was adopted for near wall treatment. No slip boundary condition was specified at the walls of S-shaped diffuser. Finally, all simulation cases were performed with standard air properties at the temperature 30°C, as below:
Inlet Reynolds number: (4.8×10^4, $6.4\times10^{4,}$ 7.5×10^4)
Inlet Turbulence intensity: (4.16 %, 3.15 %, 2.8 %)
Outlet gauge pressure: (0 Pa)
Air density: (1.1649 kg/m^3)
Air viscosity: (1.868×10^{-5} kg/ (m s))

4.4. Governing equations

The governing equations for continuity and momentum, three–dimensional, steady state, Reynolds Average Navier Stokes (RANS) equations were numerically solved turbulent incompressible flows of air [13]. The 3D Standard k–ε model for simulating the turbulence quantities has been adopted. The enhanced wall treatment was opted to cope with high complex three-dimensional near wall flow phenomena of diffusers. The inlet boundary conditions were selected like those of the experimental inlet conditions. Near-wall modeling is performed with enhanced wall treatment method to address the boundary layer formed during grid generation.

4.5. Validation

The calculated C_{WPR} using the *Re* of 6.4×10^4 and *T.I* of 3.15% at the inlet of the S-shaped diffuser, and the ANSYS-FLUENT Standard k-ε, RNG k-ε and Realizable k-ε turbulence models are compared with the experimental data in figure 6. Very good quantitative and qualitative agreement can be observed between the experimental data and Standard k-ε turbulence model data along the length of S-shaped diffuser. An observation of the validation results from table 2 the predicted results by Standard k-ε model was better agreement with the experimental results compared to the RNG k-ε and Realizable k-ε turbulence models. The Standard k-ε model predicts C_{WPR} as 0.73 and 0.75 on the top and bottom wall, respectively, which is closer to the experimental results of 0.69 and 0.74. This is correspondent to NG & Birk (2013) [1] and Gopaliya *et al.* [10].

IOP Conf. Series: Materials Science and Engineering **765** (2020) 012046 doi:10.1088/1757-899X/765/1/012046

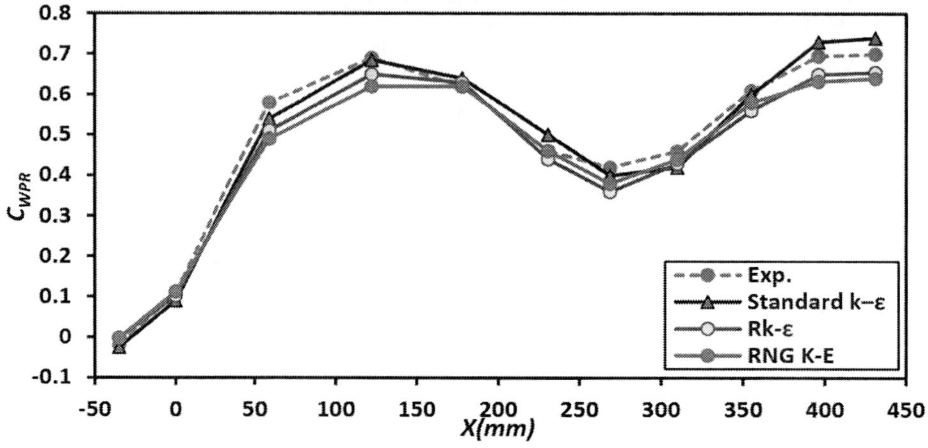

(A)

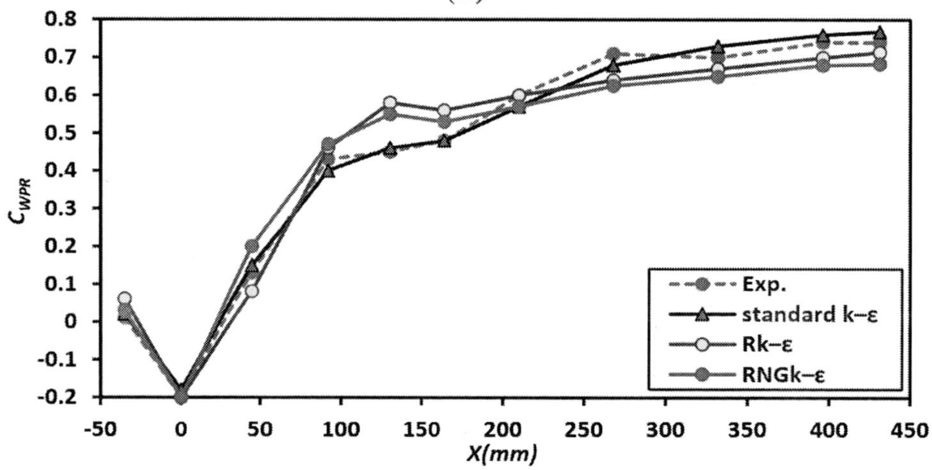

(B)

Figure 6: Comparison of experimental values of C_{WPR} with different turbulence models: A. Top surface, B. Bottom surface.

Table 2: Comparison of C_{PR} values of experimental study with values of CFD turbulence models

Type of study	Top wall		Bottom wall	
	C_{WPR} value	Deviation %	C_{WPR} value	Deviation %
CFD- R k-ε	0.65	5.8	0.7	5.4
CFD- RNG k-ε	0.63	8.7	0.68	10.8
CFD- Stand. k– ε	0.73	5.8	0.75	2.5
Experimental	0.69	-	0.74	-

5. Results and discussion

5.1. Flow field structure in the S-shaped Diffuser

The development of flow velocity along the S-shaped diffuser at $Re = 6.4\times10^4$ and $T.I. = 3.15$ is showing in figure 7. After the inlet plane, the longitudinal velocity profile trend to distort towards the bottom surface due to the concave shape of bottom wall of diffuser in addition to high curvature at the first bend. While the flow near the top surface is decelerated until the first bend at X=120 mm approximately. After the first bend, the flow will be accelerating with the top wall and decelerate with bottom wall until the outlet plane after the second bend. This behavior of flow due to the secondary flow which is resulting from increase flow separation zone at the bottom surface.

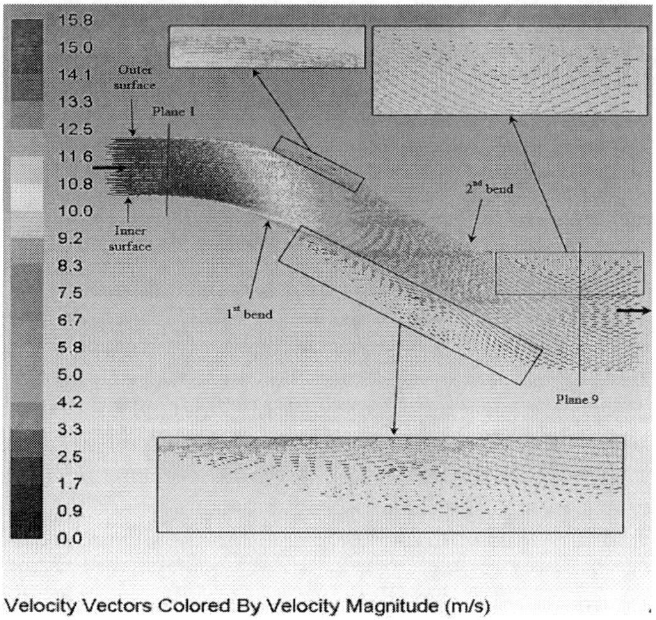

Figure 7: Development of the flow field structure in the S-shaped Diffuser by velocity vectors

Figure 8 describe the development of flow velocity by velocity vectors on the top and bottom walls at $Re = 6.4\times10^4$ and T. I = 3.15%. The flow velocity decreased on the top wall until the first bend at X = 120mm, after that the flow velocity increased until the second bend at X = 270mm and then decreased in the direction of outlet due to the increase of cross-sectional area and back flow as illustrated in figure 8-A. On the bottom wall of S-shaped diffuser the flow on the first bend undergoes a larger acceleration and hence diffusion due to the high bending of wall, as shown in figure 8.B. This part of the flow was found to be responsible for a significant part of the overall duct total pressure loss caused by a flow separation zone at the bottom surface.

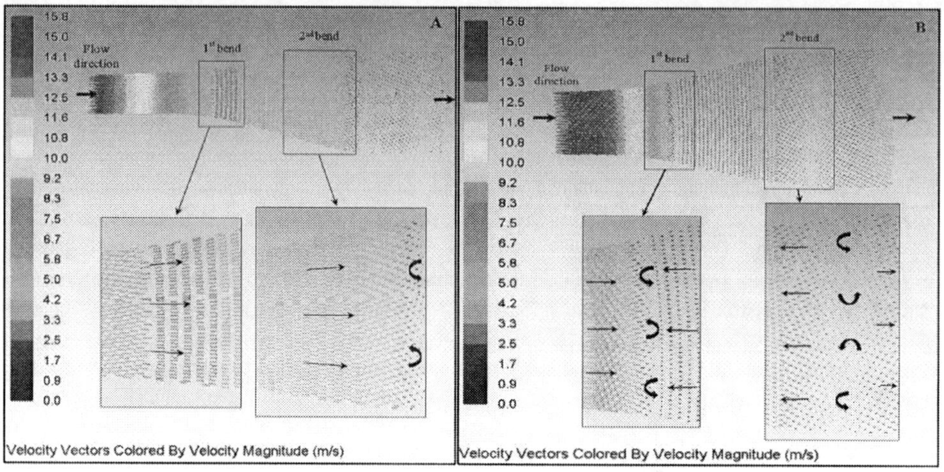

Figure 8. S-shaped diffuser with development of velocity vector at Re 1 = 6.4×104 and T. I1 = 3.15% on the: A. top wall (top view), B. bottom wall (top view).

5.2. Wall Static Pressure Recovery CWPR

The C_{WPR} is one of the performance parameters which can be determined by measuring wall static pressure on the top and bottom walls along the length of S-shaped diffuser. figure 9 presents the distribution of C_{WPR} on the top wall for S-shaped diffuser. The trend of C_{WPR} for experimental and CFD results is similar by recovering until the first bend. The maximum value of C_{WPR} from experimental and CFD investigation at the first bend is 0.75 and 0.62, respectively at $X \approx$ 150 mm with *Re* 3. After that, the C_{WPR} dropped to be around 0.38 and 0.42 for experimental and CFD results, respectively at $X \approx$ 270mm. This drop resulting from the accelerating flow between the first and second bend due to high curvature. The flow separation zone indicated at the second bend at $X \approx$ (240 -320) mm for S-shaped diffuser. After the second bend, the C_{WPR} is recovering to the outlet plane for both diffusers due to the decelerating flow resulting from expansion cross sectional area.

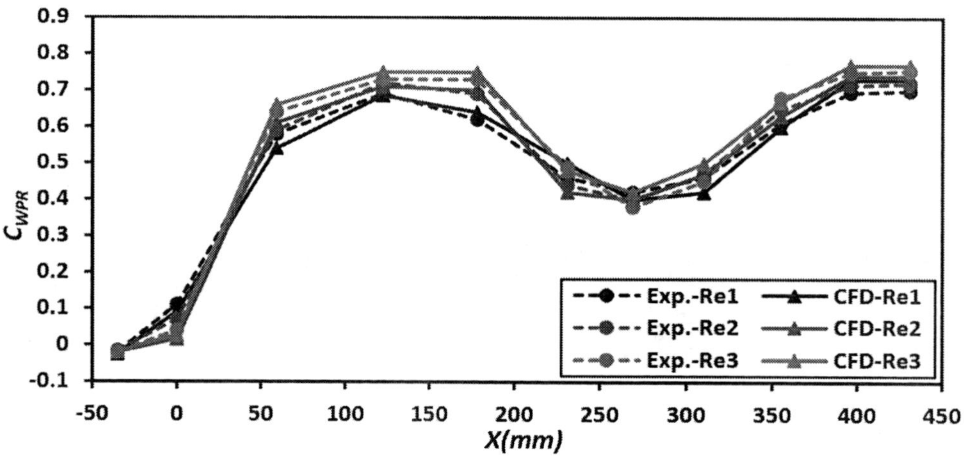

Figure 9: Distribution of CWPR along the top wall with, Re 1 = 5.8×104, T. I1 = 4.16%, Re 2 = 6.4×104, T. I2= 3.15%, Re 3= 7.5×104, T. I3= 2.8%

The trend of C_{WPR} along the bottom wall showed a continuous increase after an initial decrease at the start of the first bend as shown in figure 10. This initial decrease was observed by other researchers Sullerey *et al.* [14] and Asghar *et al.* [15]. This drop in C_{WPR} indicates an acceleration of the flow, while the rise in static pressure signifies a strong adverse pressure gradient. During the first bend, the bottom surface of the diffuser is the region where low momentum fluid exists. The C_{WPR} value after this drop increased due to high diffusion resulting from the expansion of cross-sectional area of S-shaped diffuser and may be due to the formed separation bubble as indicated approximately between X =90mm-160 mm.

Generally, the behavior of C_{WPR} at the bottom wall has an opposite trend to that of the top wall. The increase of Re from 4.8×10^4 to 7.5×10^4 leading to increase C_{WPR} on the top wall from 0.69 to 0.75 and from 0.73 to 0.77 for the experimental and CFD investigation, respectively. While the C_{WPR} was increase on the bottom wall from 0.74 to 0.81and from 0.76 to 0.83 for experimental and CFD investigation. In addition, the experimental results were lower in both top and bottom walls as compared to the CFD results and there is a deviation between the experimental and CFD results may be due to surfaces finish where in the experiments the S-shaped diffusers were made using wood be different from the surface finish assumed in the CFD simulations.

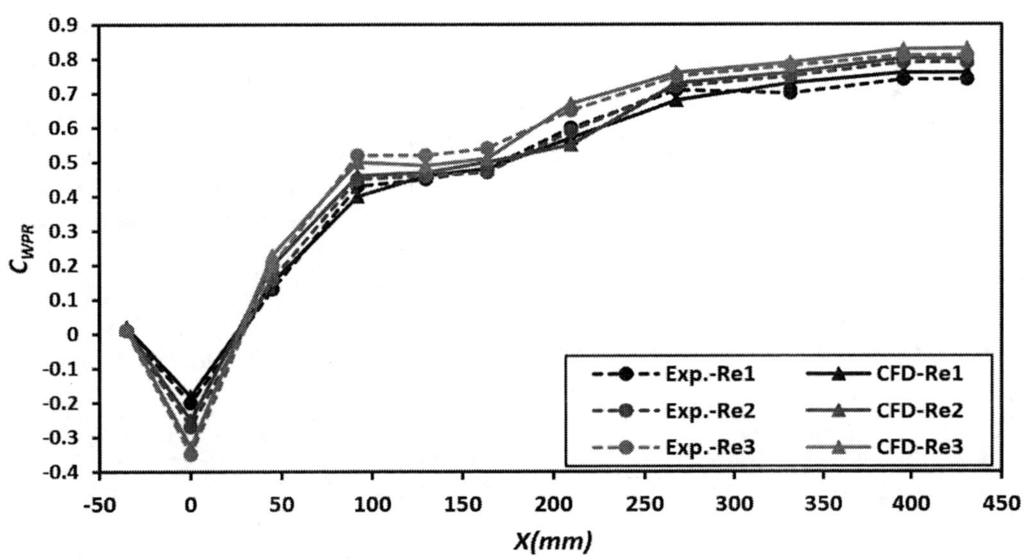

Figure 10: Distribution of CWPR along the bottom wall with, Re 1 = 5.8×104, T. I1 = 4.16%, Re 2 = 6.4×104, T. I2= 3.15%, Re 3= 7.5×104, T. I3= 2.8%

5.3. Static Pressure Recovery CPR

Figure 11 shows the variation of static pressure recovery coefficient C_{PR} along the length of the S-shaped diffuser. C_{PR} is computed, experimentally by test rig measurements and numerically with CFD, with different Re and $T.I.$ The C_{PR} values computed by CFD at $Re = 4.8\times10^4$ increase in the initial phase up to plane 3 around $X\approx$ 120mm is observed. After that it drops due to flow separation until flow reattachment occurs somewhere before plane 7 at $X\approx$ 240mm and then it increases steadily till the exit plane. At the same case with experimental measurements the trend of C_{PR} similar to the C_{PR} by

CFD with some deviations in the location or length of the separated flow region. C_{PR} increase in the initial phase up to around $X \approx$ 100mm and then it decreases due to flow separation. The flow separation occurs early with experimental measurements and flow reattachment occurs at $X \approx$ 205mm. This trend of C_{PR} can be observed with the $Re = 6.4 \times 10^4$ and $Re = 7 \times 10^4$, for the experimental and numerical investigations.

In general, the C_{PR} increases with increasing from $Re_1 = 4.8 \times 10^4$ to $Re_3 = 7.5 \times 10^4$ from 0.71 to 0.78 and from 0.75 to 0.8 for experimental and numerical investigation. This improvement of C_{PR} may be resulting from increase the mixing level between the core flow of high kinetic energy with the flow of low kinetic energy close to the bottom wall.

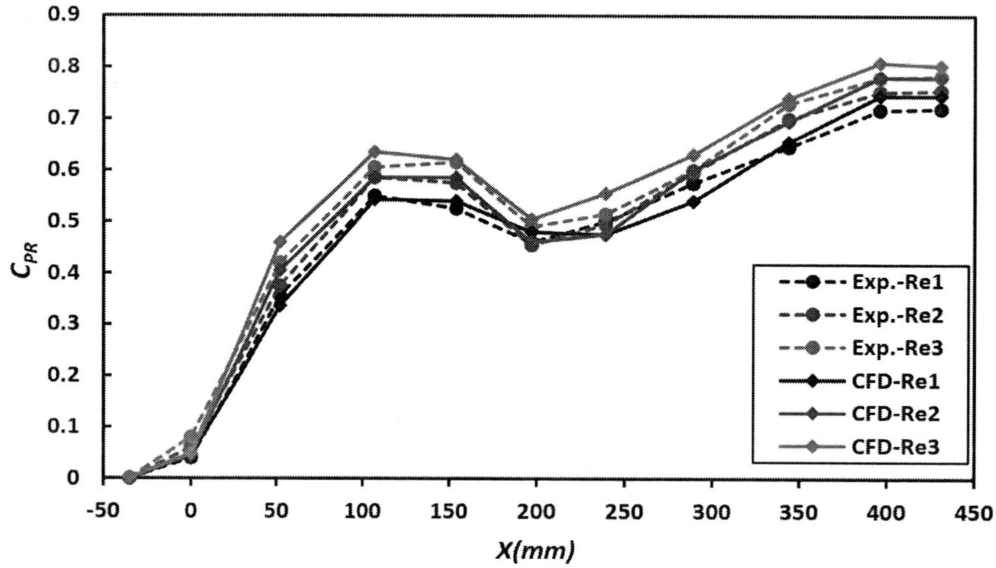

Figure 11: Comparison of experimental and numerical results of CPR along the length of S-shaped diffuser for various Re.

5.4. Total pressure loss coefficient, CTL

Figure 12 shows that the C_{TL} decreases with increase the Re along the centerline length of S-shaped diffuser. The experimental results of C_{TL} reduced from 0.28 to 0.263 and the CFD results reduced from 0.257 to 0.245 as illustrated in Fig. 12. This reduction be due to the reduction of flow separation zone which is resulting from increase the mixing level between the flow of high kinetic energy with the flow of low kinetic energy close to the bottom wall as mentioned previously.

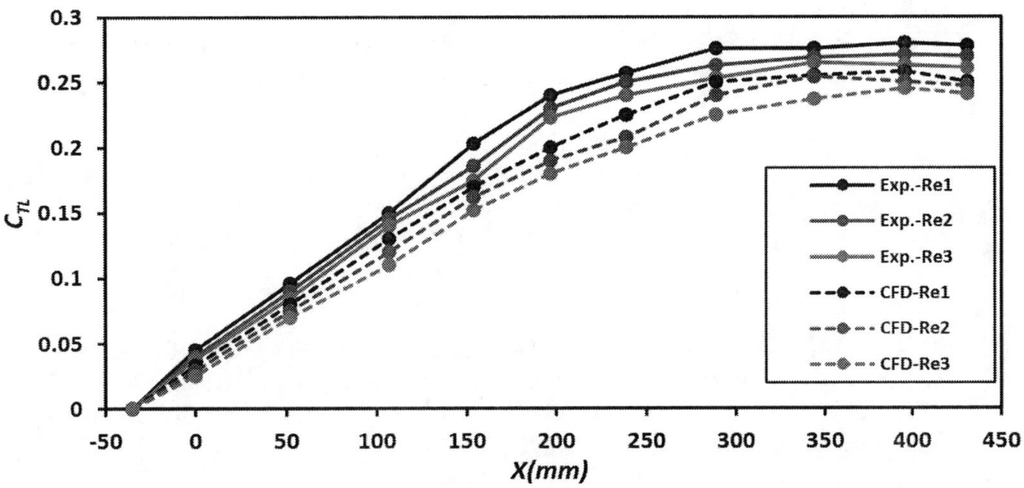

Figure 12. Comparison of Experimental and numerical results of CTL along the length of S-shaped diffuser for various Re.

There is a deviation between the experimental results and the CFD results of C_{TL} with the investigated range of *Re*, the higher value of 7.6% with Re_2 and the lower value of 6% with Re_1, as in Table 3. This deviation may be due to the small fluctuation of inflow during the experimental test compared to constant inflow assumed with CFD simulation, thereby the values of C_{TL} are affected by the inflow conditions.

Table 3. Experimental and numerical results of CTL along the length of S-shaped diffuser.

Re	C_{TL} value		
	Experimental	CFD	Deviation%
4.8×10^4	0.28	0.257	6
6.4×10^4	0.271	0.25	7.6
7.5×10^4	0.263	0.245	6.8

6. Conclusions

Experimental and numerical studies have been conducted on S-shaped diffuser with range of *Re* from 4.8×10^4 to 7.5×10^4. The following conclusions have been drawn from current investigations:

- Increase *Re* from 4.8×10^4 to 6.4×10^4 and 7.5×10^4 experimentally resulting in slightly reduction in the measured inlet *T.I* as, 4.16%, 3.15% and 2.8%, respectively.
- Increase in *Re* from 5.8×10^4 to 7×10^4 resulting in increase in the wall static pressure recovery, C_{WPR} for both top and bottom walls, as demonstrated by the experimental and numerical results.
- Increase in *Re* 4.8×10^4 to 7.5×10^4 resulting in increase in static pressure recovery coefficient for both CFD results from 0.75 to 0.8 (6.2%) and experimental results from 0.71 to 0.78 (9%), respectively. Also, a difference has been realized between the experimental and the CFD

pressure recovery coefficient results around 5.6%, 4% and 2.5% at 4.8×10^4 to 6.4×10^4 and 7.5×10^4 Reynolds numbers, respectively

- The total pressure loss coefficient C_{TL} is decreasing from 0.28 to 0.263(6.4) for CFD results and from 0.257 to 0.245(4.9) for experimental results as *Re* values increasing from 5.8×10^4 to 7×10^4.

References

[1] Ng, B. C., & Birk, A. M. (2013, June). Experimental and CFD Study of a Rectangular S-Bend Passage with and Without Pressure Recovery Effects. In ASME Turbo Expo 2013: Turbine Technical Conference and Exposition (pp. V06BT38A008-V06BT38A008). American Society of Mechanical Engineers

[2] Gan, W., & Zhang, X. (2016). Design optimization of a three-dimensional diffusing S-duct using a modified SST turbulent model. Aerospace Science and Technology.

[3] Xiao, Q., Tsai, H. M., & Liu, F. (2003). Computation of Transonic Diffuser Flows by a Lagged k-? Turbulence Model. Journal of propulsion and power,19(3), 473-483.

[4] Gerolymos, G. A., Joly, S., Mallet, M., & Vallet, I. (2010). Reynolds-stress model flow prediction in aircraft-engine intake double-S-shaped duct. Journal of Aircraft, 47(4), 1368-1381.

[5] Saha, K., Singh, S. N., Seshadri, V., & Mukhopadhyay, S. (2007). Computational analysis on flow through transition s-diffusers: effect of inlet shape. Journal of aircraft, 44(1), 187-193.

[6] Zhang, W. L., Knight, D. D., & Smith, D. (2000). Automated design of a three-dimensional subsonic diffuser. Journal of Propulsion and Power, 16(6), 1132-1140.

[7] Lee, B. J., & Kim, C. (2007). Automated design methodology of turbulent internal flow using discrete adjoint formulation. Aerospace Science and Technology, 11(2-3), 163-173.

[8] Zhang, J. M., Wang, C. F., & Lum, K. Y. (2008). Multidisciplinary design of S-shaped intake. In 26th AIAA Applied Aerodynamics Conference (p. 7060).

[9] Gopaliya, M. K., Kumar, M., Kumar, S., & Gopaliya, S. M. (2007). Analysis of performance characteristics of S-shaped diffuser with offset. Aerospace Science and Technology, 11(2-3), 130-135.

[10] Gopaliya, M. K., Goel, P., Prashar, S., & Dutt, A. (2011). CFD analysis of performance characteristics of S-shaped diffusers with combined horizontal and vertical offsets. Computers & fluids, 40(1), 280-290.

[11] Lee, G. G. W., Allan, W. D. E., & Boulama, K. G. (2012, June). Numerical and experimental analysis of the airflow inside an A250 diffuser tube. In ASME Turbo Expo 2012: Turbine Technical Conference and Exposition (pp. 1703-1712). American Society of Mechanical Engineers.

[12] Lee, G. G., Allan, W. D., & Boulama, K. G. (2013). Flow and performance characteristics of an Allison 250 gas turbine S-shaped diffuser: Effects of geometry variations. International Journal of Heat and Fluid Flow, 42, 151-163.

[13] Ahmed.K. A., "Turbulence Modelling "in Introduction to Computational Fluid Dynamics(CFD)2008: USM, 2008.

[14] Sullerey, R. K., Chandra, B., & Muralidhar, V. (1983). Performance comparison of straight and curved diffusers. Defence Science Journal, 33(3), 195-203.

[15] Asghar, A., Stowe, R.A., Allan, W.D. and Alexander, D., 2017. Entrance Aspect Ratio Effect on S-Duct Inlet Performance at High-Subsonic Flow. Journal of Engineering for Gas Turbines and Power, 139(5), p.052602.

IOP Conf. Series: Materials Science and Engineering **765** (2020) 012047 doi:10.1088/1757-899X/765/1/012047

Lead-free perovskite and double perovskite solar cells.

Aqel Mashot Jafar[1], Kawther A. Khalaph[2] and Amar Moula Hmood[3]

[1]Solar Energy Research Center, Renewable Energy Directorate, Higher Education and Scientific Research Ministry/ Baghdad/ Iraq.

[2]Medical College, Ibn Sina University of Medical and Pharmaceutical Sciences, Baghdad, Iraq.

[3]Materials Researches Directorate, Higher Education and Scientific Research Ministry, Baghdad, Iraq

1. aqeel@scbaghdad.edu.iq, 2. Kawthar.ali1104a@csw.uobaghdad.edu.iq, 3.dr.amarmoula@scbaghdad.edu.iq.

Abstract. The toxicity and instability of lead- halide perovskite materials in the solar cells are yet demands must be resolved to vie to find alternatives of free lead-based halide perovskites and double perovskite materials. Here we report on the experimental synthesis of a new family of lead-free inorganic halide double perovskites Cesium copper antimony chloride ($Cs_2CuSbCl_6$) and perovskite Cesium copper chloride (CsCuCl3) as a based sensitizer of the solar cells and studies of its characterize and optical properties of the Photo-Voltaic (PV) device. The configuration structural of PV devices is (FTO/TiO_2/$CsCuCl_3$ or $Cs_2CuSbCl_6$/Cu_2O/Al electrode). The double perovskite $Cs_2CuSbCl_6$ and Perovskite $CsCuCl_3$ are successfully synthesized, performed of structural investigations by X-ray diffraction, and depicted of optical properties via optical absorption of UV−vis measurement. The perovskite ($CsCuCl_3$) belongs to the space grope P6122 and the Hexagonal structure phase. The strongest peeks of the XRD diffraction pattern of the double perovskite $Cs_2CuSbCl_6$ are taken at the reflection points (-201), (-202), (-3-11), (220), (-223) and (-5-34). From UV−vis measurements obtained values of a direct gap of both ($Cs_2CuSbCl_6$) and ($CsCuCl_3$) are (1.5 and 2.2) eV, respectively. The optimum power conversion efficiency (PCE) of the PV device which used $Cs_2CuSbCl_6$ as a sensitizer is 0.94%, the filling factor is 24%, the open-circuit voltage is 582.5 mV and the short circuit current was 2.656 mA. The measurement is carried out under one sun intensity illumination (1000 W/m^2).

Key Words: Perovskite, Double Perovskite, Free-Lead.

1. Introduction

Free- Lead Perovskite of inorganic halides materials with a chemical formulas, ABX_3 and A_2BCX_6, where A represents a monovalent non-organic molecule, B and C are metal of a positive valence group, and X are an halogen anions [1]. These materials have recently attracted of a tremendous amount of interest in the solar cells society. The free lead perovskite solar cells based on

IOP Conf. Series: Materials Science and Engineering **765** (2020) 012047 doi:10.1088/1757-899X/765/1/012047

this materials are stilled under investigation and research [2]. Previous studies refer that major part of materials in this family have a stabilization phase anti decomposition and exhibit flexible synthesis of optical properties with gaps in the range from infrared to ultraviolet radiation. These unique features, coupled with low cost solutions, make them an ideal candidate as new generation optical absorbers [3]. In previous literature, efforts has been made by researchers to understand how structural Characteristics and optical properties affect the performance of photovoltaic (PV) devices [4] ,[5]. The structure and synthesis of the photovoltaic thin films were investigated by analysis X-ray diffraction (XRD). The morphology of the surface was studied by EDAX electron microscope [6]. The spectral analysis of the thin films in solar cell, the energy band gap and range energy bands of HOMO, LUMO were studied [4], [7].Here, in the current work, the Free-Lead perovskite materials such as $CsCuCl_3$ and double perovskite materials as $Cs_2CuSbCl_6$ were investigated as absorption layer in the perovskite photo-voltaic devices, which were not previously studied according to the researcher's survey of the literature studies as absorption layer in photovoltaic device. We try to investigate the performance of this materials as the thin film for absorption or harvesting the light in the photovoltaic device of perovskite solar cells. In addition, the studying investigates of the structural characteristics and optical properties of this material.

2. Basic theory :

The perovskite material CsCuCl3 have a direct optical band gap[6] . The allowed direct transition of optical band gap was expressed accordance to the equation (1) [8].

$$hv * \alpha = B * \left[hv - E_g\right]^{1/2} \dots (1)$$

Where, v is the light frequency, h is the Planck constant, Eg is the optical energy gap, B is empirical constant and (α) is the absorption coefficient which expressed as the equation (2) according to the Beer-Lambert law for optical absorption of the thin film [9].

$$\alpha = 2.303\,\frac{A}{t} \dots\dots (2)$$

Where A is the Absorbance of a film which have thickness (t), The relation between the Extinction Coefficient (ko) and the Absorption coefficient (α) is shown in the equation (3) :

$$K_o = \alpha * \lambda/_{4\pi} \dots (3)$$

Where λ is expressed the wave length of the incident light . The relation between the Refractive Index N_o and the Extinction Coefficient K_o is expressed in the equation (4) :

$$N_o = \left[\frac{4R}{(R-1)^2} - K_o^2\right]^{1/2} - \frac{(R+1)}{(R-1)} \dots\dots (4)$$

Where the R is the reflectance of the film. The relations between the Extinction Coefficient (K_o) with the Real and the Imaginary of the Dialectic Constants (C_r and C_I) as shown in the equations 5 and 6[10], respectively [10].

$$C_R = N_O^2 - K_O^2 \dots\dots (5)$$

$$C_I = 2 * K_O * N_O \dots\dots (6)$$

Where the No is represented to the refractive index. The Power Conversion Energy (PCE) and Full Factor (F.F.) of a solar cell are calculated by using equations (7) and (8) [11] ,[12], [13]:

$$PCE = \frac{V_{Oc} \cdot I_{SC} \cdot FF}{P_{in} * A_{SC}} \dots\dots (7)$$

$$F.F. = \frac{I_m V_m}{I_{sc} V_{oc}} = \frac{P_m}{I_{sc} V_{oc}} \dots\dots\dots (8)$$

Where, (V_{OC}), (I_{SC}), (I_m) , (V_m) and (A_{sc}) are the open circuit voltage, short circuit current, Maximum current, Maximum Voltage and active area of the solar cell, respectively. The power conversion efficiency (PCE) can be improved by increasing the photo-generated current (I_{ph}) which correlated of the Quantum Efficiency (QE) of PV device. QE is defines that the number of charge carrier collected of the solar cell divided on the number of incident photons while I_{ph} only takes into account the absorbed photons. QE expresses with the following formula [14] :

$$Q.E. = \frac{h * C}{q * \lambda} * \frac{I_{ph}}{P_{in}} \ldots\ldots\ldots (9)$$

where c is the light speed, h is Planck's constant, q is the electronic charge, λ is the wavelength, and P_{in} is the power incident on the PV device.

3. Characterization of samples:

Crystal structure and composition studied via X- Ray Diffraction using (XRD Shimadzu 6000, Cu-Kα) for coating film sample. The surface morphology of sample and the formations of the perovskite crystals studied via the Scanning Electron Microscope (SEM) (Bruker Nano GmbH, Germany). Light I-V Measurements are recorded as the Test Reports via the Photovoltaic measurements system which composed of Oriel I−V test station and an Oriel Solar simulator. The solar simulator is type AAA for spectral performance which have uniformity of irradiance, and temporal stability. A xenon lamp with a power 450 W equipped the solar simulator. The output power is adjusted to match (1000 W m^{-2}). I−V curves are calculated via a computer which connected with an external bias to the PV cell and measuring Iph with a Keithley model 2400 digital source meter. The transmittance and absorption of film coated are analyzed via (SPECTRO UV/Vis Double Beam (UVD-3500) Labomed ,Inc) in the wavelength range of (190 - 1200) nm. The temperature of the substrates of deposition processes is monitored via an infrared temperature indicator in the any run of the experimental. Quantum Efficiency measurement system contains of monochromatic at (300-900)nm, reflector halogen lamp (75)W, and Kethley to record I_{ph}.

4. Experimental:

4.1. *Materials and solvents:*

Copper chloride $CuCl_2$ (99%) and cesium chloride CsCl (99%) were purchased from Fluka AG. Glass substrates of Fluorine-doped Tin Oxide (FTO) was employed as electrodes Transparent Conductive Oxide (TCO) which purchased of Visiontek, sheet Resistance 8 Ω/□ at dimensions (2x2) cm^2. Dioxide Titanium TiO_2 (99%) nano-particles> 30 nm of powder was origin of China. Copper oxide Cu_2O (99%) was purchased from BDH. N,N Di Methyl Formamide (DMF) (99.9% anhydrous) was purchased from Sigma Aldrich. Hydrochloric acid HCl (35.4%) was purchased from Fluka Chemika.

4.2. *Preparation of Perovskite precursor solution:*

The cesium copper chloride Perovskite precursor solution was synthesized by dissolving 0.134 g (1 mmol) of copper chloride ($CuCl_2$) in 2.0 ml of hydrochloric acid at room temperature and completed to 10 ml via adding DMF to obtain the sample A. Following by, 0.168 g (1 mmol) of cesium chloride was dissolved in 3 ml of hydrochloric acid at room temperature and completed to 10 ml via adding DMF to obtain the sample B. After complete dissolution of the materials as precursors solutions, two samples, A and B, were mixed in the Becker on the starrier and increased the temperature at 70°C with 30 minutes time [6].

precursor solution of double perovskite Cs2CuSbCl6 was prepared of the reaction of the $SbCl_3$ (228 mg, 1 mmol) and $CuCl_2$ (98 mg, 1 mmol) were dissolved in 10 mL DMF. Then, CsCl (336 mg, 2 mmol) was dissolved in 2 mL of hydrochloric acid and completing to 10mL by adding DMF, following by adding the Cs precursor solution to the previously made solution of Sb and Cu of salts precursor. Black precipitates were obtained by ultrasound sonication of the reaction mixture for 30 s to obtain $Cs_2CuSbCl_6$ precursor solution [15] , [4].

IOP Conf. Series: Materials Science and Engineering **765** (2020) 012047 doi:10.1088/1757-899X/765/1/012047

4.3. Preparation of Cu2O precursor solution:

Precursor solution of Copper oxide is prepared via dissolving 0.143g (1mmol) of Cu2O in 8ml of 1-butanol and adding 2ml of deionized water on the hot starrier at 50°C temperature and 10 min time.

4.4. PV Device fabrication:

Preparation of compact layer TiO_2 of photo-electrode is a first fabricated of PV device by employment Aerosol Assisted Chemical Vapor Deposition (AACVD) technique via using Ultrasonic Atomizer (402AI) as deposition system doing with ultrasonic frequency (1.5MHz) [13],[12].The precursor solution, which preparation of Nano particles> 20 nm of TiO_2 powder dispersed via Ultra-sonically frequency in ethanol solvent ,is sprayed on per-heated FTO at 450 °C, the time of deposition is 1h. The hot substrates are left to cold at 50°C. Following by, the perovskite precursor solution of $CsCuCl_3$ or $Cs_2CuSbCl_6$ are deposited by drop casting on (FTO/ TiO_2) substrates and annealing at 100°C. Following by depositing Cu_2O precursor solution by drop casting process and annealing at 100°C. Then, put a foil of Al with area 0.1cm^2 as a counter electrode on Cu_2O layer to obtain two samples of Free-Lead Perovskite Solar Cells (FLPSCs) with the configurations (FTO/ TiO_2/$CsCuCl_3$/Cu_2O/Al) and (FTO/ TiO_2/$Cs_2CuSbCl_6$/Cu_2O/Al) as shown in figure (1) which depicts the structure thin films of Photo-Voltaic (PV) device.

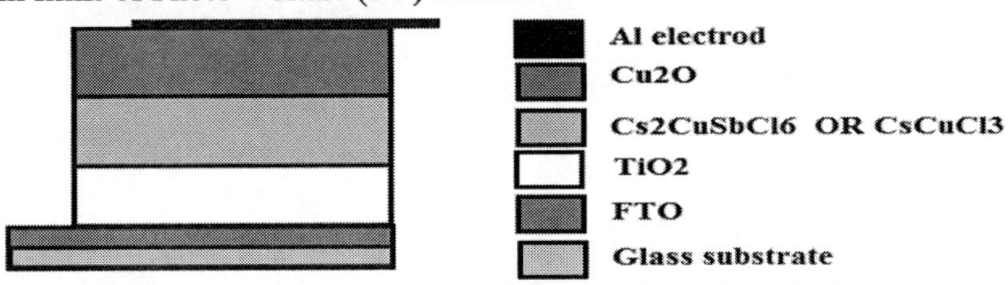

Figure (1). structure thin films of configurations FLPSCs.

5. Results and Discussion:

5.1. Structural Studies:

Figure(2-a) depicts the XRD pattern of TiO_2 thin film (Red line), which achieved by depositing of TiO_2 precursor solution via using AACVD on preheated glass substrate at 400 °C and 1 hour deposited time. The reflections are indicated to the tetragonal TiO_2. Reflections positions (black bars) of (101), (004), (204) and (301) corresponding with tetragonal structure are indicated to TiO_2 Anatase system by PDF # 211272 for comparative. Reflection position of (210) corresponding with tetragonal structure is indicated to TiO_2 Rutile system by PDF # 211276. Figure (2-b) depicts XRD pattern of the perovskite structure of $CsCuCl_3$ thin film sample which prepared via drop casting process on glass substrate followed by annealing at 100 °C. XRD pattern, red curve, of $CsCuCl_3$ perovskite is exhibited the Hexagonal structure phase, belonged the space grope P6122 and can be confirmed by PDF # 180349 at reflections positions of (100), (103), (110), (202), (204), (300) and (222). Figure (2-c) shows the patterns XRD of the thin film of $Cs_2CuSbCl_6$ depositing on glass substrate which was preparation by drop casting process Followed by annealing to 100 °C at 15 min. The XRD patterns of the sample preparation appear with red curve line in the figure (2-c), but the black bars are denoted to the double perovskite $Cs_4CuSb_2Cl_{12}$ [15]. The strongest peeks of pattern lines of the sample preparation are taken at the reflection points (-201), (-202), (-3-11), (220), (-223) and (-5-34). Figure (3-a,b) shows the high-resolution Scanning Electron Micrographs (SEM) images of thin film of CsCuCl3 and $Cs_2CuSbCl_6$, respectively, which have been deposited by drop casting process on glass substrates, followed by annealing at 100 °C temperature and 15 min deposited time. The (SEM) image clearly show the transformation from particles to aggregations structure in the sample of $CsCuCl_3$, as can be seen from these micrograph (a), but the structure of particles are nearly crystals in shape with reasonably uniform size distributions size (1- 2)μm of SEM image of $Cs_2CuSbCl_6$ sample. The scale bars of the images are 10 μm, scanning with high voltage 10kV and magnification 5 kx.

IOP Conf. Series: Materials Science and Engineering **765** (2020) 012047 doi:10.1088/1757-899X/765/1/012047

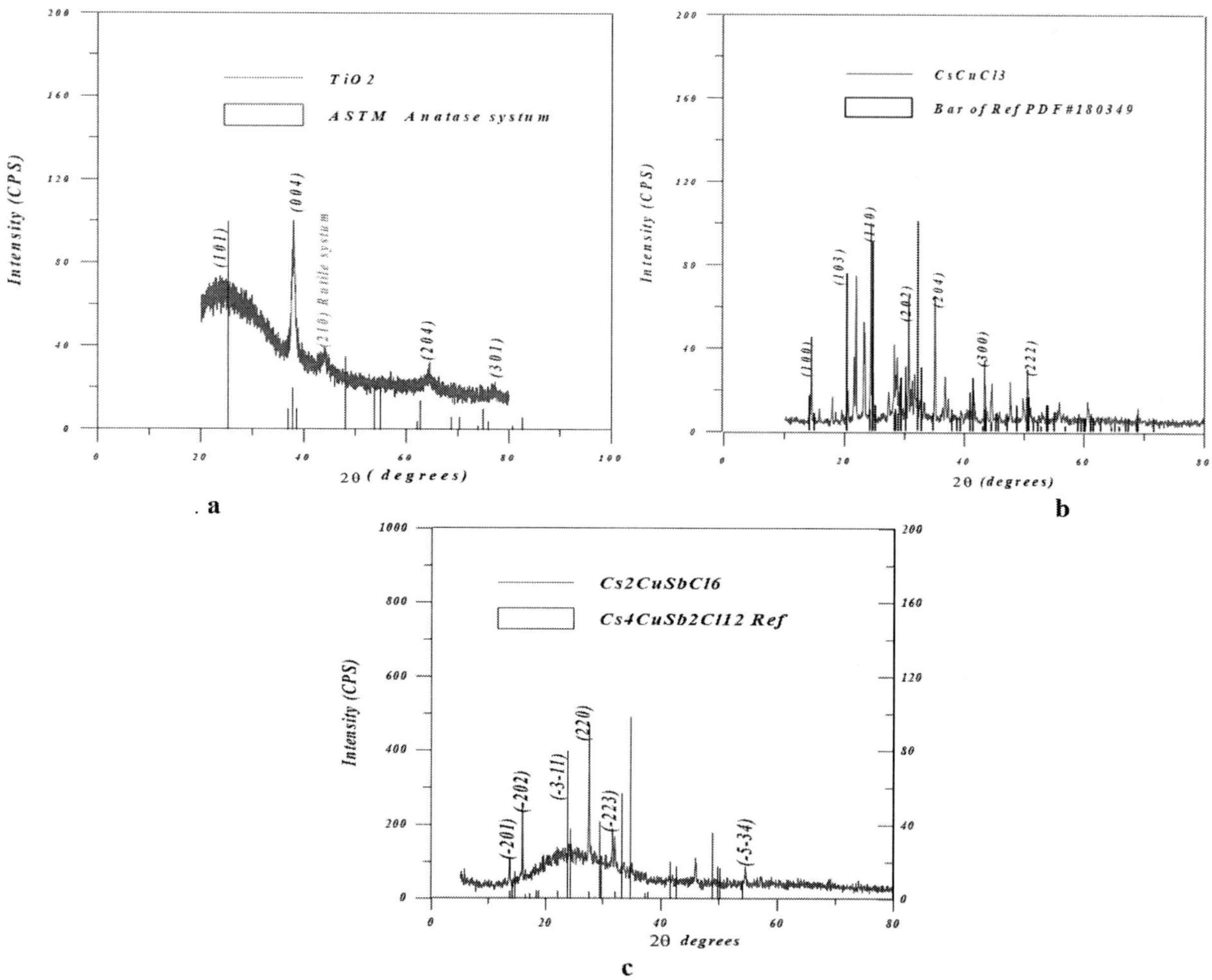

Figure (2).XRD of thin film of (a) TiO_2 (b) $CsCuCl_3$ (c) $Cs_2CuSbCl_6$.

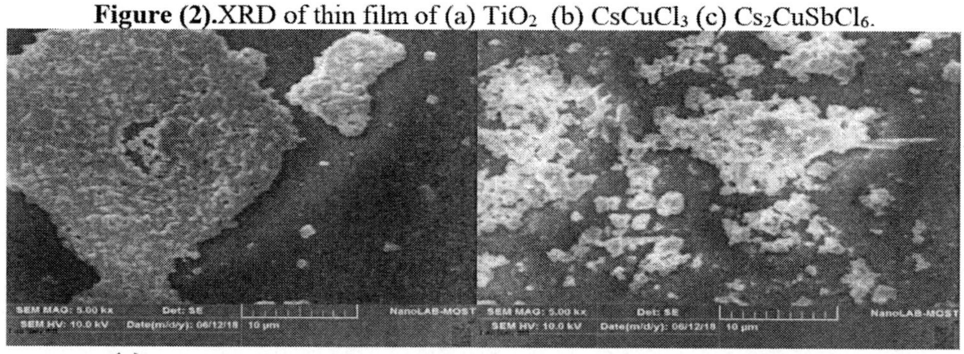

(a) **(b)**

Figure (3). Top SEM images thin film of (a)$CsCuCl_3$ (b) $Cs_2CuSbCl_6$.

5.2. Optical Studies:

Figure(4-a) is illustrated the absorbance of Pb-free halide perovskite thin films and it clears more absorbance of the sample $Cs_2CuSbCl_6$ than other sample $CsCuCl_3$. Both of the samples are analyzed at the wavelength range (300-1100) nm employing a (UV/VIS Double Beam SPECTRO (UVD-3500) Labomed, Inc.) and blank glass substrates as the reference samples.

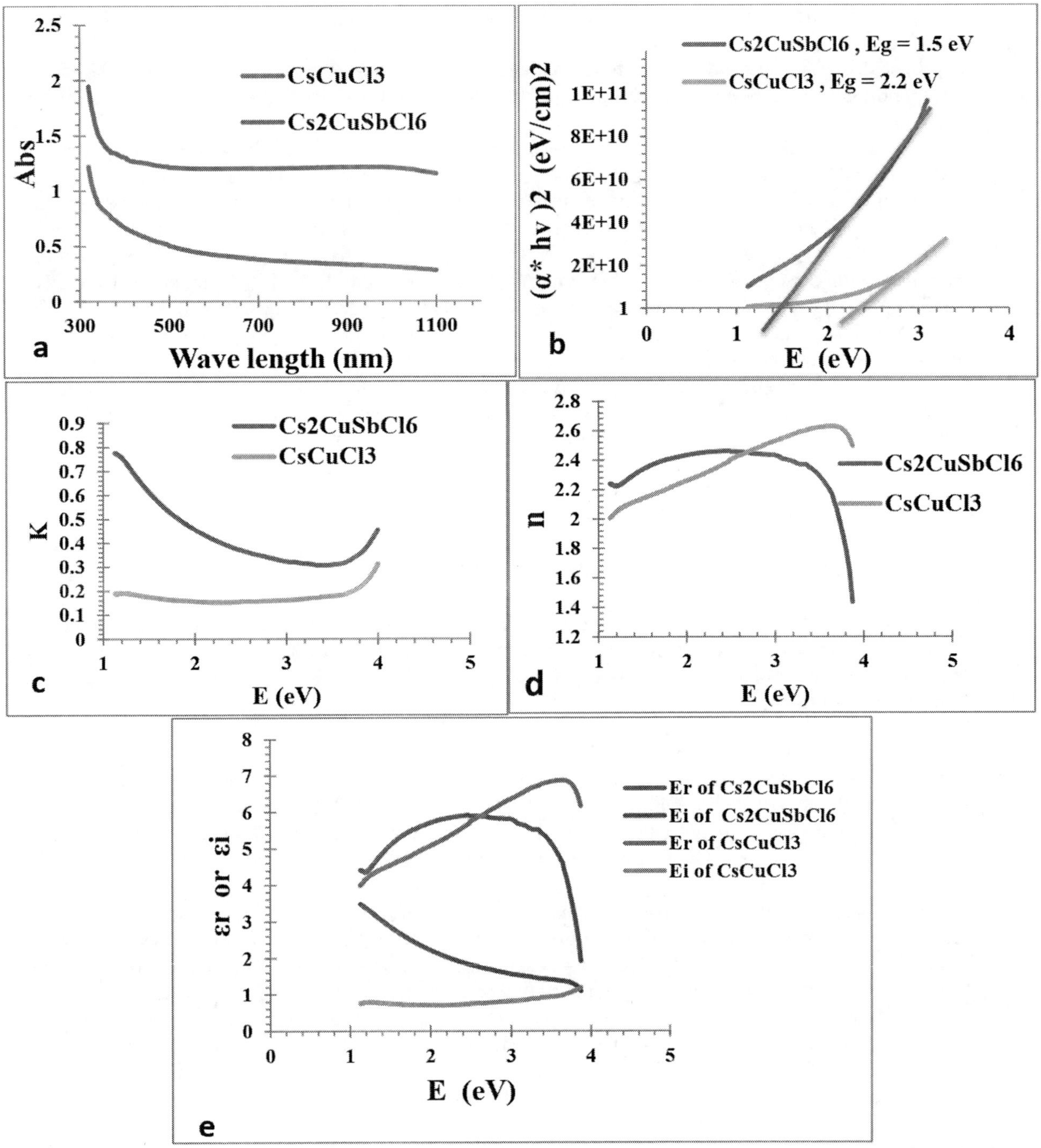

Figure (4). Optical properties of the thin film of $Cs_2CuSbCl_6$ and $CsCuCl_3$. (a) Absorbance (b)Optical Energy gap (c) extinction coefficient (d) Refractive Index and (e)Real and Imaginary parts of dielectric constant.

IOP Conf. Series: Materials Science and Engineering **765** (2020) 012047 doi:10.1088/1757-899X/765/1/012047

Plots of photon energy (hv) versus $(\alpha hv)^2$ for the thin films of deposited peroveskite materials on glass substrates are estimated of Absorption charts of the thin film which have the thickness (400nm) as shown in Figure 7(b). Pb-free halide perovskites materials $CsCuCl_3$[6], and Cs2CuSbCl6[4] have a direct optical band gap energy (E_g) which is determined of both samples by fitting the absorption data to the direct transition equation (4). The values of optical band gaps are estimated by extrapolating the linear part of the curves $(\alpha hv)^2$ as a function of the energy, E , intercept with the E axis at $\alpha = 0$. The E_g values of $CsCuCl_3$ and $Cs_2CuSbCl_6$ via data UV-Vis are estimated (2.2 and 1.5)eV which corresponding to the reported of References of [6], [4] , [16], respectively, as depicted in figure 4-b and table 1.

Figure (4-c) is depicted the extinction coefficient (k_o) to the free-lead halide perovskite films of $Cs_2CuSbCl_6$ and $CsCuCl_3$ materials. The extinction coefficient (k_o) refers to the amount of attenuation of the electromagnetic wave as it passes through the physical medium, and expresses the basis of value that determined by electromagnetic wave interactions with the medium. In figure (4-c), It can be observed clearly that the best interactions of electromagnetic wave of the visible light absorption with $Cs_2CuSbCl_6$ than other material, $CsCuCl_3$. Figure (4-d) is depicted the Refractive Index (n_o) to the free-lead halide perovskite films of $Cs_2CuSbCl_6$ and $CsCuCl_3$ materials. The refractive index (n) is a function of reflectivity (R) and extinction coefficient (K_o) as the basis of the equation (4). According of the Figure (4-d), the spectral refractive index and its energy change of the visible range were estimated about (2 to 2.6) of the $CsCuCl_3$ film but estimated (2.3 to 2.4) of the $Cs_2CuSbCl_6$ film at the visible range of the energy. Figure (4-e) depicts the real εi and imaginary εr parts of the dielectric constant as a function of the photon incident energy to the free-lead halide perovskite films of $Cs_2CuSbCl_6$ and $CsCuCl_3$ materials. The interaction between light and material charges is due to the absorption of energy in the material, and thus polarization of the medium's charges. This polarization is usually described by the dielectric constant complex (ε), Real and Imaginary parts, as shown of equations 5 and 6. The behavior of εi is similar to refractive index because the smaller value of ko^2 comparison of n^2, while εr is mainly depends on the k_o values, which are related to the variation of absorption coefficient. The imaginary part represents the absorption associated of radiation by free charges carriers [17]. The imaginary part of the dielectric constant is a measure less than of real part of dielectric constant ($\varepsilon r > \varepsilon i$), as shown in figure (4-e), the imaginary part of the dielectric constant is a measure of absorption of the energy of the radiation failing by the atoms of the material.

5.3. Performance of P.V. Devices Studies::

Figure (5) depicts the report of the measurements by the Photovoltaic measurements system of I-V curve of FLPSCs with configurations (a): (FTO/ $TiO_2/Cs_2CuSbCl_6/Cu_2O$/Al) and (b): (FTO/ $TiO_2/CsCuCl_3/Cu_2O$/Al). The measurement is carried out under intensity illumination (1000 W/m^2), active area of solar cell is 0.4 cm^2 and sweeping voltages in the scan-direction with a scan rate of s = 50 mV/s. Figure (5-a) is illustrated I-V curves of photovoltaic device that employs of $Cs_2CuSbCl_6$ as absorption layer and have the best efficiency of performance FLPSCs, (PCE = 0.94%). All the parameters of the performance FLPSCs are inserted in the table (1). The best efficiency is achieved to the FLPSC which have $Cs_2CuSbCl_6$ as a sensitizer due to the best optical properties, higher absorption and lower optical energy gap, than other sensitizer or absorption layer $CsCuCl_3$, as shown in figure (5) and table (1). R_s and R_{sh} in the table (1) are expressed the reverse slop of the (I-V) curve at points V_{oc} and I_{sc}, respectively, and determined the F.F. values [18]. It can to be observed the small values of F.F. of the solar cells (F.F.=24%), due to the low values of R_s and high values of R_{sh} of PV devices as shown in table (1). The PV devices which have typical F.F. must to be have very low value of R_s and very high value of R_{sh}.

IOP Conf. Series: Materials Science and Engineering **765** (2020) 012047 doi:10.1088/1757-899X/765/1/012047

Table 1. Performances parameters of FLPSCs have active area ($0.4cm^2$).

FLPSCs	E_g(eV)	V_{oc}(mV)	I_{sh}(mA)	V_m(mV)	I_m(mA)	P_m(μW)	FF	E_{ff}%	R_s(Ω)	R_{sh}(Ω)
$Cs_2CuSbCl_6$	1.5	582.5	2.656	302.0	1.241	374.95	0.24	0.94	147.8	787.95
$CsCuCl_3$	2.2	361.7	1.790	196.4	0.787	154.57	0.24	0.39	341.8	862.92

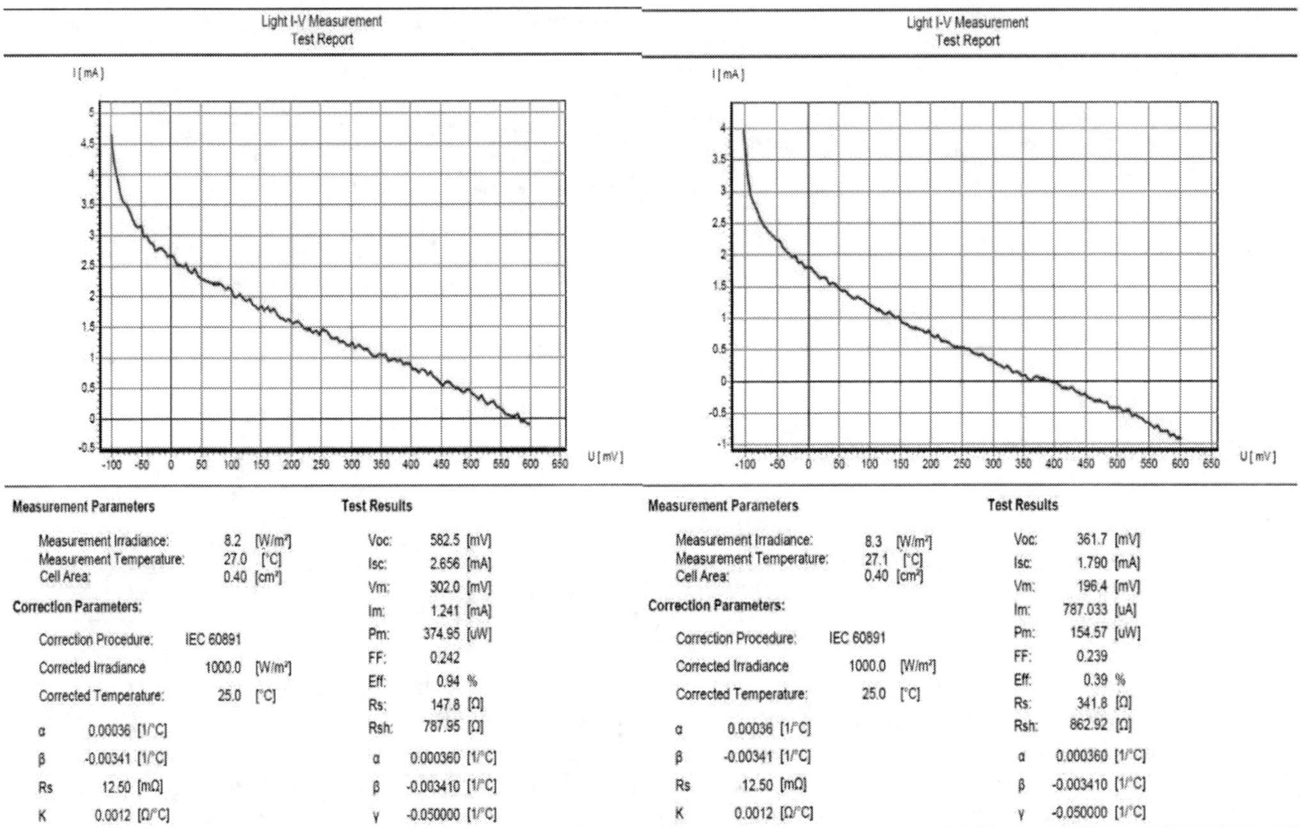

(a) ***(b)***

Figure 5. I-V curve of FLPSCs have (a): $Cs_2CuSbCl_6$ and (b): $CsCuCl_3$ as absorption of PV device.

The data of the Quantum Efficiency (Q.E.) in the wave length range (300-900)nm is recorded by Quantum Efficiency measurement system which contains of monochromatic at (300-900)nm, reflector halogen lamp (75)W, and Kethley to record Iph , as shown in figure (6-a). The output light power of the reflector halogen lamp at the wavelength range (300-900)nm, input power (P_{in}) of the solar cell sample, is calculated via employing power solar meter and monochromatic as shown in figure (6-b).

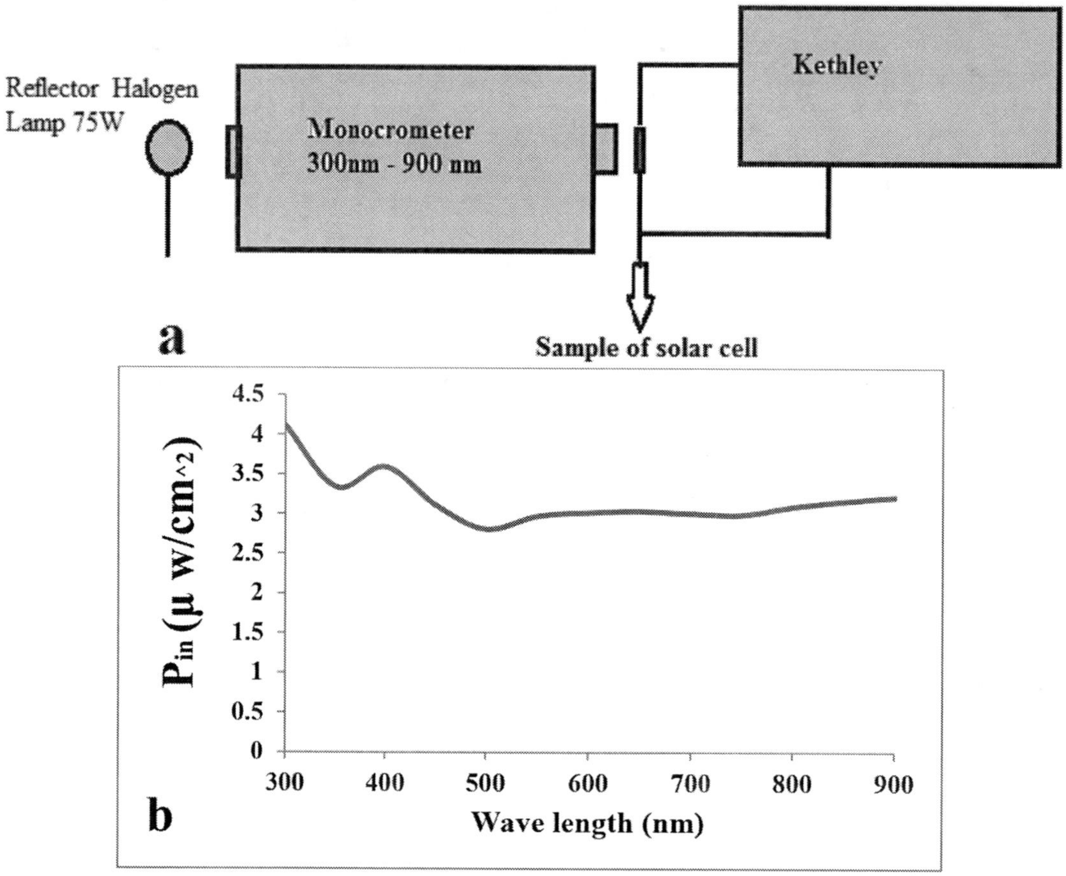

Figure 6. depicts (a): Quantum Efficiency measurement system and (b): Input power on PV device at wave length range (300-900)nm.

Figure (7) depicts the (Q.E.) of the PV device with configurations (a): (FTO/ $TiO_2/Cs_2CuSbCl_6/Cu_2O/Al$) and (b): (FTO/ $TiO_2/CsCuCl_3/Cu_2O/Al$). It is clearly visible in Figure (7-a) that the Q.E. is observed to be increased with wavelength in the range of 350-750 nm. It is reached to maximum at wave length 500 nm, beyond this maximum decreased rapidly and found minimum at the wavelengths 300 nm and 900 nm. The decreasing rapidly of the Q.E. curve of peak at wave length 500nm is observed that may be attributed to the decreased rapidly of behavior in the absorption curve as shown in figure (7-a). The increased of the Q.E. curve at the visible range (300-900) in figure (7-a) confirms the increased efficiency of the PV device that have $Cs_2CuSbCl_6$ as sensitizer, comparative with the other PV device as shown in figure (5). It is clearly of the Figure (7-b) that the Q.E. curve of PV device which have CsCuCl3 as sensitizer is observed to be decreased gradually with increased the wavelength in the range of 350-750 nm. It is reached to maximum Q.E. at 300 nm, decreased gradually and found minimum Q.E. at the wavelengths 650 nm. This behavior of Q.E. curve is similar to absorption curve of the $CsCuCl_3$ sensitizer as shown in figure (7-b).

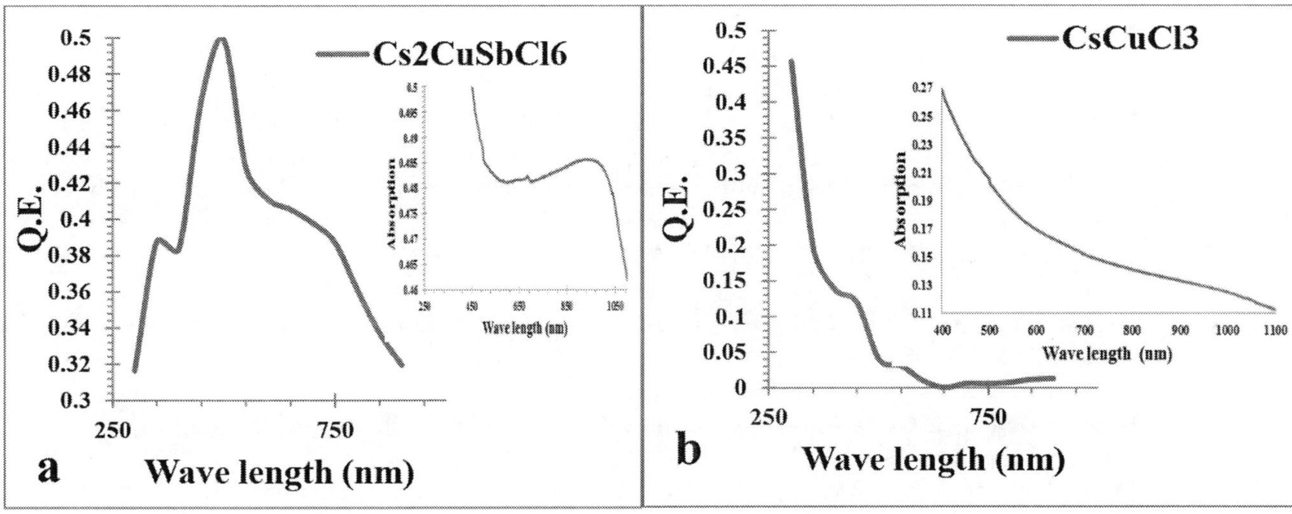

Figure 7. depicts Q,E. at wave length range (300-900)nm of PV device have (a): Cs2CuSbCl6 (b): CsCuCl3 as sensitizer.

6. Conclusion :

In summary, through a combined studies of optical and structural properties of a new family of free-lead halide double-perovskite, we have observed that its correlated of the P.V. devices performance. Perovskite $CsCuCl_3$ and double perovskite $Cs_2CuSbCl_6$ were successfully synthesized and then, FLPSCs with construction structural ($FTO/TiO_2/$ $Cs_2CuSbCl_6$ OR $CsCuCl_3$ $/Cu_2O/Al$ electrode) have been successfully fabricated. The empirical results give that an optimum PCE = 0.94 % can be achieved by using double pereovskite, $Cs_2CuSbCl_6$, as harvesting light layer in FLPSC, while PV device which have $CsCuCl_3$ as a sensitizer is given (PCE= 0.39%) . It is attributed that the PCE of PV device was correlated with the absorption curve of sensitizer material and the quantum efficiency curve of PV device Performances of the wave length spectrum. The optical characterization of the double perovskite $Cs_2CuSbCl_6$ and perovskite $CsCuCl_3$ indicating a direct gap semiconductor.

7. Acknowledgements :

The authors would like to thank Ministry of Science and Technology, Renewable Energy Directorate, Baghdad-Iraq, and Materials Research Directorate, Baghdad-Iraq, for their support in the present work.

References

[1] F. Giustino and H. J. Snaith, "Toward Lead-Free Perovskite Solar Cells," *ACS Energy Lett.*, vol. 1, no. November, p. 1233–1240, 2016.

[2] S. F. Hoefler and G. Trimmel, "Progress on lead-free metal halide perovskites for photovoltaic applications : a review," *Monatshefte für Chemie - Chem. Mon.*, vol. 148, no. 5, pp. 795–826, 2017.

[3] A. Jain, O. Voznyy, and E. H. Sargent, "High-Throughput Screening of Lead-Free Perovskite-like Materials for Optoelectronic Applications," 2017.

[4] Brenda Vargas, E. Ramos, E. Pérez-Gutiérrez, J. C. Alonso, and and D. Solis-Ibarra, "A Direct

Bandgap Copper–Antimony Halide Perovskite," *J. Am. Chem. Soc.*, no. 139, p. 9116–9119, 2017.

[5] Xingang Zhao *et al.*, "Design of Lead-Free Inorganic Halide Perovskites for Solar Cells via Cation-Transmutation," *J. Am. Chem. Soc.*, vol. 139, no. 7, pp. 2630–2638, 2017.

[6] Muhammad Aamir, "SYNTHESIS OF METAL HALIDE PEROVSKITE MATERIALS FOR LIGHT HARVESTING APPLICATIONS," Allama Iqbal Open University, Islamabad, 2018.

[7] P. Karuppuswamy *et al.*, "Role of a hydrophobic scaffold in controlling the crystallization of methylammonium antimony iodide for efficient lead-free perovskite solar cells," *Nano Energy*, vol. 45, no. December 2017, pp. 330–336, 2018.

[8] A. M. Jafar, K. Al-amara, T. M. Lafta, M. H. Mahmod, and I. M. Abood, "Fabrication and Characterization of P-Type Co-Doped Tin Oxide Nano-Films," *Elixir Int. J.*, vol. 65, no. June 2013, pp. 19799–19803, 2013.

[9] A. M. Jafar, K. Al-Amara, F. L. Rashid, and A. I. K. Fayyadh, "Fabrication and Characterization of Fluorine-Doped Tin Oxide Transparent Conductive Nano-Films," *Int. J. Innov. Res. Eng. Sci.*, vol. 6, no. 2, pp. 49–58, 2013.

[10] M. N. Al-Jabery, Q. A. Abbas, and Hamid S. Al-Jumaili, "Study the optical properties of CuInS 2 non stoichiometric thin films prepared by chemical spray pyrolysis method," *Iraqi J. Phys.*, vol. 10, no. 19, pp. 70–75, 2012.

[11] A. M. Jafar, M. H. Suhail, and Falah I. Mustafa, "Organolead Iodide Perovskite Solar Cells (OPSC)," *AASCIT J. Energy*, vol. 2, no. 6, pp. 81–87, 2016.

[12] M. H. Suhail and A. M. Jafar, "Fabrication and Characterization of Organolead Halide Peroviske Solar Cell," *Elixir Int. J.*, vol. 98, no. September 2016, pp. 42709–42713, 2016.

[13] A. M. Jafar, M. H. Suhail, F. M. Al-Attar, and and M. K. Kalaf, "ORGANOLEAD HALIDE PEROVSKITE SOLAR CELLS," in *32st European Photovoltaic Solar Energy Conference and Exhibition*, 2016, pp. 1296–1302.

[14] S. Chander, A. Purohit, A. Nehra, S. P. Nehra, and M. S. Dhaka, "A Study on Spectral Response and External Quantum Efficiency of Mono-Crystalline Silicon Solar Cell," vol. 5, no. 1, pp. 1–4, 2015.

[15] N. Singhal, R. Chakraborty, P. Ghosh, and A. Nag, "Low-Bandgap Cs 4 CuSb 2 Cl 12 Layered Double Perovskite : Synthesis , Reversible Thermal Changes , and Magnetic Interaction," pp. 2085–2092, 2018.

[16] G. Volonakis *et al.*, "Lead-Free Halide Double Perovskites via Heterovalent Substitution of Noble Metals," 2016.

[17] C. Kittel and H. Y. Fan, *Introduction to Solid State Physics Eighth Edition*. 2005.

[18] G. Sissoko and S. Mbodji, "A Method to Determine the Solar Cell Resistances from Single IV Characteristic Curve Considering the Junction Recombination Velocity (Sf)," *International Journal of Pure and Applied Sciences and Technology*, vol. 6, no. 2. pp. 103–114, 2011.

An Experimental Investigation of Performance of Hybrid PV/T-PCM System with Control System

Hashim A. Hussein

Ali H. Numan

Zahraa Aqeel

Electromechanical Engineering Department, University of Technology, Baghdad, Iraq

Correspondence Author Ali H. Numan: 50059@uotechnology.edu.iq

Abstract. In this paper we investigated experimental performance of designed hybrid PV/T with phase change material (PCM) of paraffin wax type based thermal storage material. Paraffin wax embedded with aluminum matrix foam to enhancement the physical properties. Designed hybrid system consists of three main parts, glass to glass PV module, thermal collector and, control system. Thermal results indicated the optimum layer thickness of PCM was 3cm because discharge time of thermal energy storage was longer of 6 hours. Also, optimum temperature of difference air, absorber plate, glass to glass PV module and, paraffin wax were 8.6°C, 80°C, 44°C and, 55°C at air speed of 0.8m/s, respectively.

Keywords: PCM, glass to glass PV module, hybrid PV/T.

1. Introduction

The use of conventional energy sources in energy production over the past years has resulted in increased carbon dioxide emissions [1]. This has led to an increase in global warming, which has resulted in significant changes in the climate of the globe. Renewable energies express the optimal solution to get rid of the problems of environmental pollution caused by traditional energy sources as they are inexhaustible, clean and constantly renewed. There are many types of renewable energy sources such as solar energy, wind energy, geothermal energy, hydro energy, tidal wave energy …etc. Solar energy is one of the most renewable energy sources scattered around the world. It uses sunlight as an input for energy production by equipment [2].

Solar energy is used in the production of thermal energy and electrical energy. Solar collectors are used to convert solar radiation into thermal energy that can be used to heat air or heat water. PV panels are also used to convert solar radiation into electrical energy. Hybrid energy systems combine more than one source of power generation. Hybrid photovoltaic/ thermal (PV/T) system is used to generate both thermal and electric power. Thermal energy is invested in heating domestic water or air. Phase change materials (PCMs) organic commonly used for thermal storage in solar collectors. S. Tiwari et. al. [3], designed hybrid PV/T system (agricultural crop drier) , investigate experimental and theoretical performance under different parameters such as, PV cell temperature, solar radiation and, crop temperature. MATLAB/Simulink software was used to verify the theoretical performance of the system. Results showed theoretical and experimental of thermal energy were 1.92kWh and 2.03kWh, respectively. A. Tabook et. al. [4] designed hybrid PV/T system consisted of designed glass to glass PV module and designed thermal collector. Glass to glass PV module allows the passage of radiation to the thermal collector unlike what is found in conventional PV modules used in hybrid PV/T systems. Abdulrahman S. Mahmood et. al. [5] designed and fabricated solar air collector with thermal storage material (paraffin-wax based PCMs). At day, results showed difference between outlet and input temperature was 1.5-6.5°C. At night, results showed solar air heater was given for 5 hours heat energy.

In this paper we have designed a hybrid PV/T-PCMs air-based with control system to generate and store thermal and electrical energy. Glass to glass PV module and solar thermal collector have been designed, also paraffin-wax based PCM had been used to store excess heat during the day. Control system had been used to control the outlet air temperature.

2. Experimental Work

2.1. Specification of designed system

Hybrid PV/T-PCM air-based system consist of container $0.6\times1m^2$(inner layer coated with aluminum plate 2mm thickness and an outside covered with wood fiber has thickness 4mm as insulation), glass to glass PV module as cover, diverge duct, converge duct, absorber plate, DC fan, iron support frame, DC battery, control system based on Arduino microcontroller and charge controller as shown in figure 1. The distance between absorber plate and bottom was 40mm filled with paraffin-wax embedded with aluminum foam matrix to increase the melting temperature of paraffin-wax. Area of inlet and outlet duct $0.54\times0.075m^2$. Glass to glass PV module used as cover of solar collector. Glass to glass PV module consists of two layers of transparent glass granulated heat pressed where solar cells are placed inside those layers and specifications are shown as in Table 1.

Table 1: Specification of designed glass to glass PV module

Max. power	67.5 W
Max. voltage	14.15 V
Max. current	4.77 A
Open circuit voltage	18.27 V
Short circuit current	5.14 A
Cells in series /parallel	28 / 1

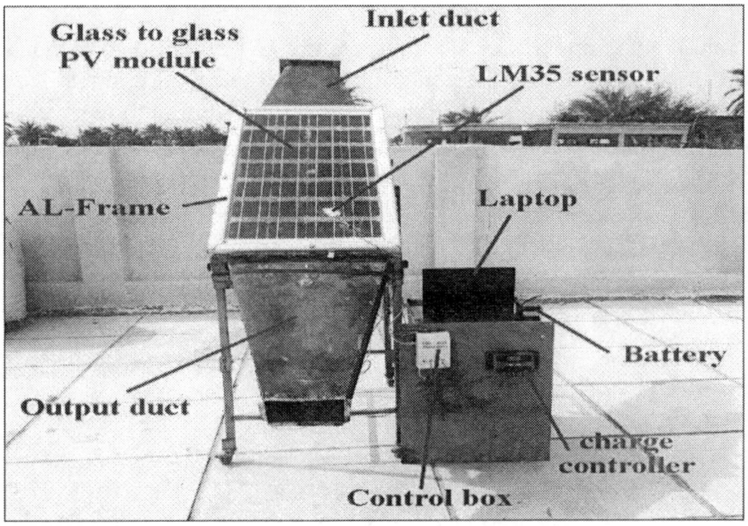

Figure 1: Photographic image of test rig of hybrid PV/T-PCM system.

2.2. Control and measurement process

The control circuit consists of a microcontroller (Arduino-UNO), DC motor driver (IBT-2) and, temperature sensors (LM35). The main function of control circuit to control the speed of DC fan and control the outlet air temperature. Five LM35 sensors were fixed on hybrid system, such as absorber plate, outlet duct, glass to glass PV module, paraffin wax and, ambient temperature. During the day, the fan is fed by the power generated from the glass to glass PV module through DC driver. During the night the DC fan is fed by the battery attached to the PV panel via the charging regulator.

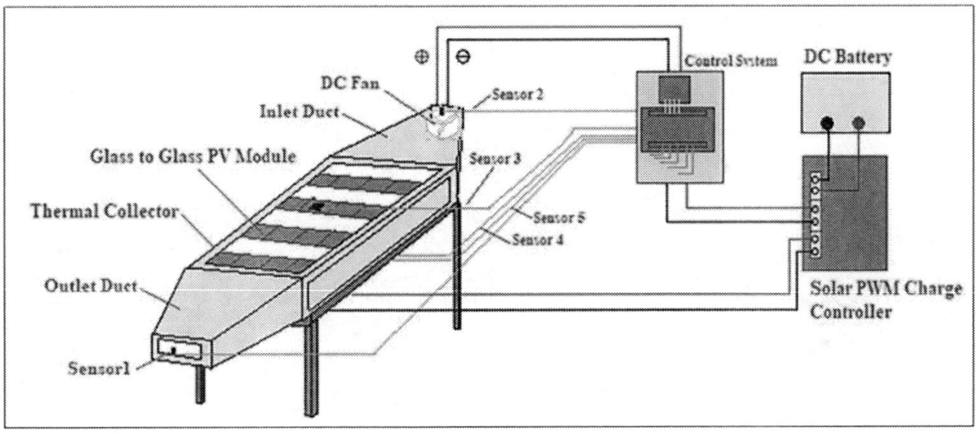

Figure 2: Schematic diagram of control system of hybrid PV/T-PCM system

Solar power meter (TES-1333) had been used to measure the intensity of sunlight on hybrid system with accuracy $\pm$ 10W/m^2 and resolution 0.1W/m^2. Also, wind speed meter (GM8908 type) used to measure the speed of inlet air. Rheostat load had been used to measure the current and voltage load and to calculate maximum power of glass to glass PV module.

3. Results and Discussion

Figures 3, 4 and 5 show maximum difference temperature (ΔT) of air data recoded at 21st March-2019. Output temperature of air measured at difference air speed to select optimum air speed. Three Thickness of paraffin wax had been used to investigate which thickness is optimum to give long discharge time of heat energy after sunset. The best thickness of PCM was 3cm at air speed of 0.8m/s. At thickness 3cm, the discharge time of heat energy given 6 hours.

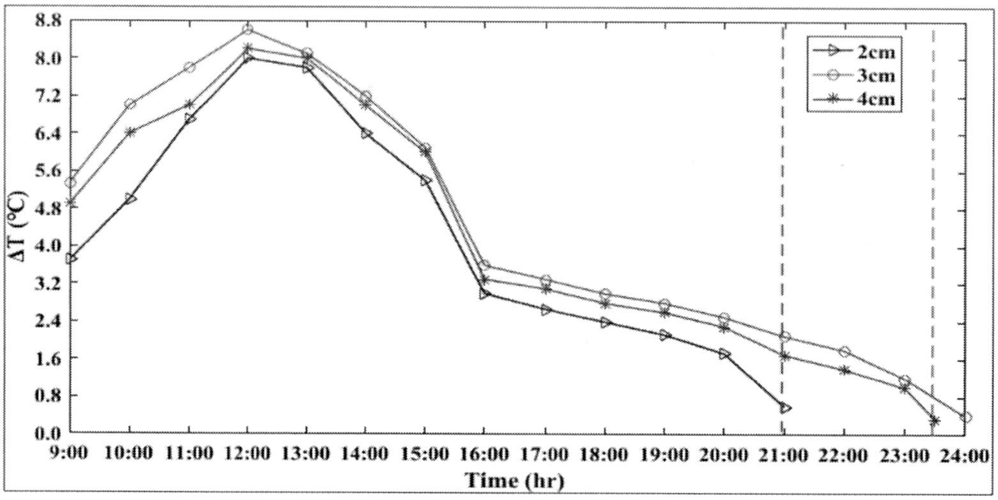

Figure 3: Difference air temperature at air speed 0.8m/s

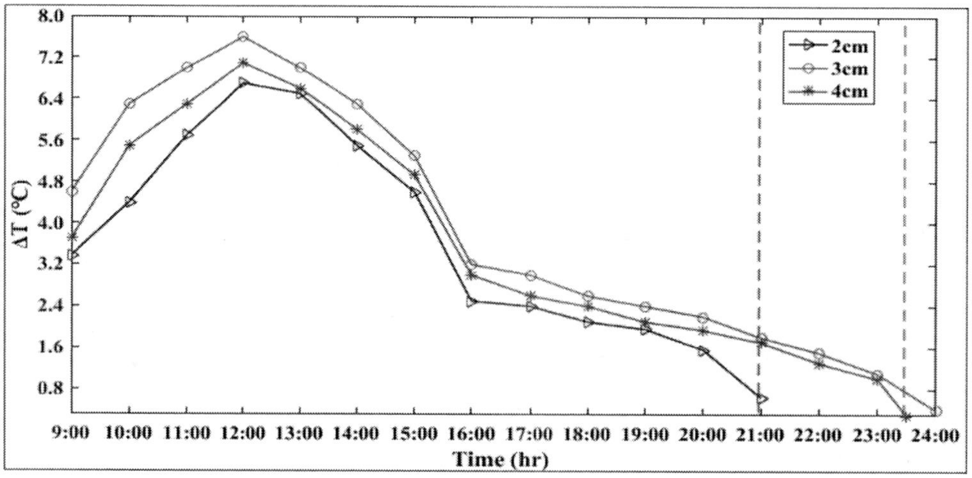

Figure 4: Difference air temperature at air speed 1m/s

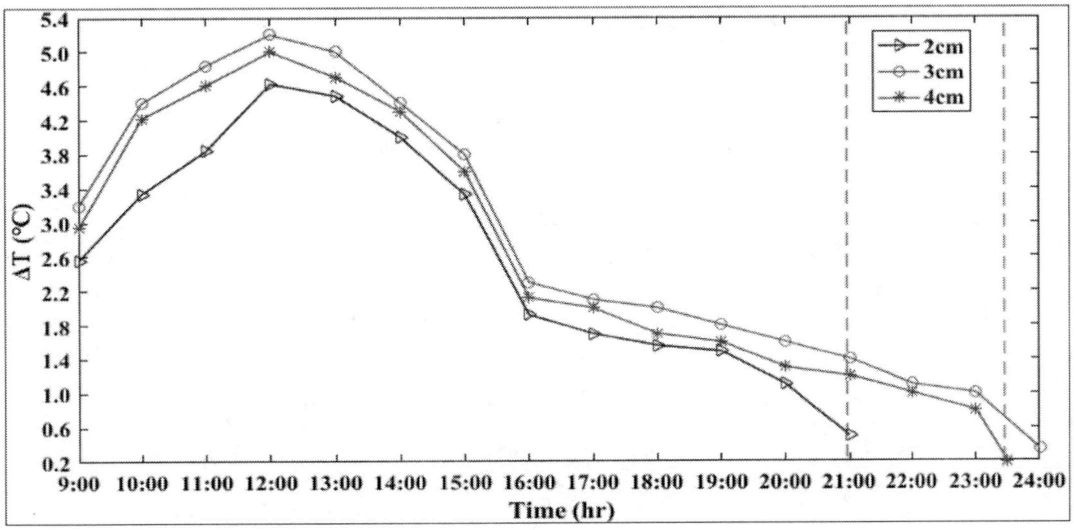

Figure 5: Difference air temperature at air speed 1.5m/s

Absorber plate temperature, cell temperature of glass PV module and paraffin wax temperature were recoded as shown in figures 6, 7 and 8, respectively. The optimum temperatures were at thickness of 3cm and air speed 0.8m/s.

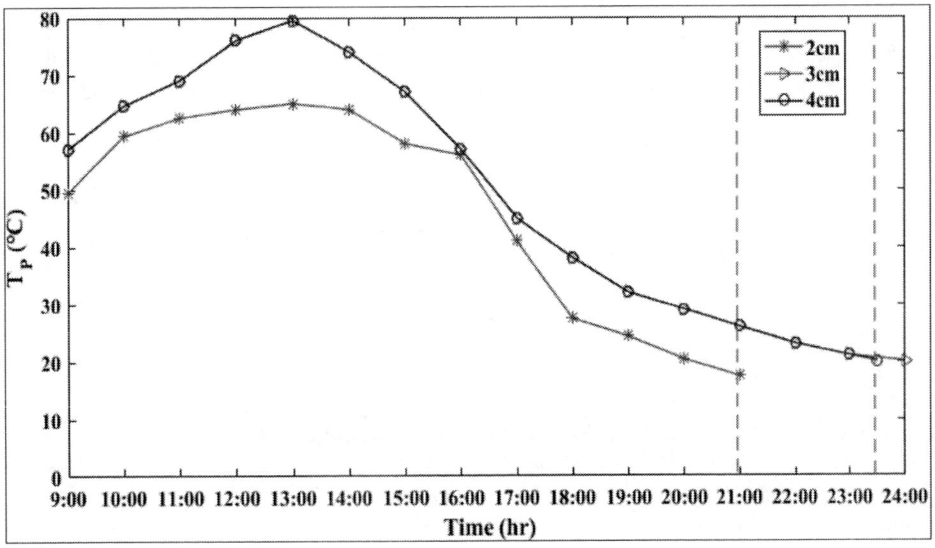

Figure 6: Absorber plate temperature at air speed0.8m/s

IOP Conf. Series: Materials Science and Engineering **765** (2020) 012049 doi:10.1088/1757-899X/765/1/012049

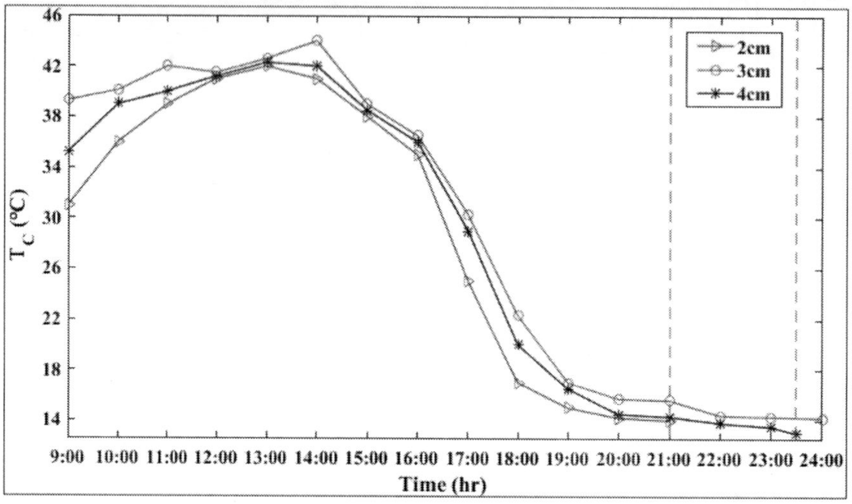

Figure 7: Cell temperature at air speed of 0.8m/s

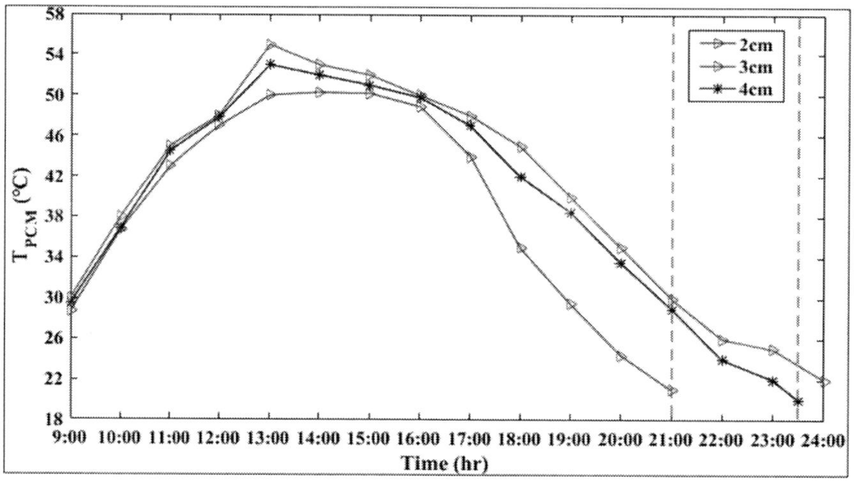

Figure 8: PCM temperature at air speed of 0.8m/s

4. Conclusion

The main points of this work can be concluded as the following:

1. Designed of hybrid PV/T-PCM system represents development in hybrid PV/T systems because it generates and storage thermal and electrical energy.
2. Designed glass to glass PV module improvement in generation and storage of thermal energy.
3. Using paraffin wax with aluminum matrix foam enhance thermal performance of hybrid PV/T system.
4. Use the control system given optimum output air temperature and long discharge time.

References

[1] Sumath S, Kumar LA, Surekha P. Solar PV and Wind Energy Conversion Systems an Introduction to Theory. Modeling with Matlab/Simulink, and the role of soft Computing Techniques, ISSN. 2015:978-3.

[2] Kalogirou, Soteris, "Solar energy engineering: processes and systems," Elsevier, 2009.

[3] Tiwari S, Tiwari GN, Al-Helal IM. Performance analysis of photovoltaic–thermal (PVT) mixed mode greenhouse solar dryer. Solar Energy. 2016 Aug 1;133:421-8.

[4] Tabook MA, Othman MY, Sopian K, Ruslan MH. Double glass PV cell efficiency in PV/T combination water and air heating flat plate collector system. J of Novel Appl Sci. 2014;3:1629-34.

[5] Mahmood AS. Experimental Study on Double-Pass Solar Air Heater with and without using Phase Change Material. Journal of Engineering. 2019 Jan 31;25(2):1-7.

Improving the Performance of the Pump Station in Pipe Line Transportation System Using PLC Controller and Remote Monitoring

Hashim A. Hussein

Abdzahraa J. aleeby

Zahraa M. Mahdi

Electromechanical Engineering Department, University of Technology, Baghdad, Iraq

E-mail: hashim171967@gmail.com, abdjb2003@yahoo.com

Abstract. In industries, analysis and control of pressure and flow rate control is very difficult, this leads to factory closings and heavy maintenance through PLC and SCADA implementation. Automatic control plays an important role in the continuous operation of the system. In this project, the actual pump performance is tested in terms of pressure, efficiency, and flow rate of the pump for the operating system using a pipeline by crossing pressure in the pipeline with a length of 2 m by connecting pressure sensing equipment. The global control unit SIMATIC S7-1214 as the main decision-making unit that uses this data to make the required decisions .Thus the operation and stopping of the pump as it asks the details related to all other correct information to SCADA to monitor and control the parameters in the pipeline system of the oil pumping station. Then it performs continuous supervision of oil pump station pipeline in order to allow solving any problem and thereby regulating the control system in the structure consists of three layers. These include the first layer of field devices second remote terminal units, and third domain controllers. The signals are sent from the devices via the transmitters to the dedicated PLC boards in the second layer. The central level of SCADA contains a high-speed computer to supervise or operate the station remotely in order to display information through the LABVIEW screens that was showed the final results of pressure and flow rate by operating the system with a voltage rated by Plus With Modulation at a value of 2.73 V for both light and heavy crude oil. It was showed that the pressure value is 0.22bar and the value of the flow rate is 0.25 L/M for heavy oil. While the light is 0.23bar for pressure and the flow rate is 0.652L/M. In addition, to conducting several experiments that show that an increase in the value of the voltage obtained using heavy oil by a value of 3.77 V the pressure value becomes 0.35 bar the flow rate is at 4.78L/M, while light oil has shown the results are that the rated voltage value is 3.77 V, the pressure value is 0.33bar , and the flow rate value through the tube becomes 5.17 L/M. Thus, the greatest value of the voltage 3.77 V , The result when the flow rate through the pipeline increases rapidly, and the pressure decreases, the pipeline passes through the pipeline faster

Keywords: SCADA ; PLC ;Oil pumps stations; Pressure sensor .

IOP Conf. Series: Materials Science and Engineering **765** (2020) 012050 doi:10.1088/1757-899X/765/1/012050

1. Introduction

Oil is formed on the ground of the earth as a result of the remains of animals and plants as a result of the conversion of organic materials in geological conditions under the surface of the earth that led to the formation of crude oil and natural gas, and thus leads to the transfer of large quantities of crude oil between the production site and refining and consumption points and thus be transferred across lines. The pipes for transporting heavy oil from the Qayara refinery and light oil came from the Dora refinery and made a comparison between the two types to control the pressure and flow rate using PLC by conducting a signal transmission SCADA and a linear process with time delay{1.2.3.4} represented by LABVIEW in order to get rid of the labor cost and monitor line safety Oil pipelines while ensuring effective transport safety and assurance [1] can alter the flow rate that can be sensed through monitoring and control using LABVIEW demonstrated differences in pressure value [2] in order to improve control performance by analyzing pressure signals at various points during a period Transport [4] Indirectly maintained fluids are transported as flow rates and pressure stability may enhance production rate as well as safety of oil transport without human intervention and reduce losses around the pipeline [4] therefore it is an innovative model of pipeline control system using the LABVIEW platform which includes Information about the change in pressure status indicating the continuous flow of any liquid in the long-range transport pipeline system [5].

LABVIEW with SCADA has the ability to remotely control distinct parameters such as pressure and flow in the oil pipeline and the result is continuous control in order to regulate the flow rate in the destination in real time [6]. Morsi (2014) [7] generally relies on the control system Conventional (DCS) to provide all the functions of controlling equipment and operation, but it does not provide sufficient conditions for the discovery of problems. Therefore, it was suggested that the SCADA / PLC system be used to control a complete oil refinery instead of conventional control via DCS, thus avoiding wasting manpower and material resources to increase the safety of workers using SCADA / PLC in order to perform the examination time which means observer System, and the discovery of problems by recording and storing the amount of data in addition to the screens SCADA more realistic than the DCS user and the operator can monitor the process from all over the world. Priyanka (2018) [8] suggested that industries need more reliable control, power, efficiency and flexibility, as the Fuzzy-PID controller is designed to automatically regulate the flow rate of petroleum products by controlling multiple pressure signals in long transmission. Fuzzy-PID performance is compared to the PLC-based Cascade-PID console. FUZZY-based and Cascade-PID based PLC controllers are applied for experimental setup, and performance analyzes are verified empirically. Real-time data from pressure and flow is monitored through the SCADA screen using data logging. Therefore experimental analysis has shown that the Fuzzy-PID-based PLC controller maintains a steady state with the minimum installation time is 23 seconds and the minimum exceeds 24.36% compared to the PLC-based Foreman PLC. Fuzzy PID controller provides fair response with minimum error rate of 18.92%

2. Methodology

Identify and classify equipment in the crude oil pumping control and monitoring station using PLC and connected to the supervisory control program and obtain data for a sample from the oil tank with the pump. A case study suggested an understanding of the level of oil in a pipeline pumping station and a switch of support for the design of crude oil pumping station [9].

3. Modelling of the Oil Pump

The study of the characteristics of the oil pump to determine the quality of the product when sending the product through continuous succession in one pipeline. The flow rate and pressure are controlled by PLC and liquid control through the pipeline of SCADA displays as shown in the figure 1. Oil pump station pipeline is measured by piezoresistive and flow rate sensor. Piezoresistive pressure sensor measures the pressure of gases and liquids on the manufacturing scale where pressure sensors are manufactured. In bulk case, the M12 connectors and cable connectors are used as the standard for electrical conductivity (20 mA) or three wires with voltage output as output signal current for obtaining a voltage of 0.5 to 4.5 V. Measuring the flow rate with the help of a flow rate sensor of the type Turbine flow sensor. The flow rate transmitter will convert the fluid flow into an analog signal and the load will be monitored through PLC connected to SCADA to display the results on the LABVIEW screen.

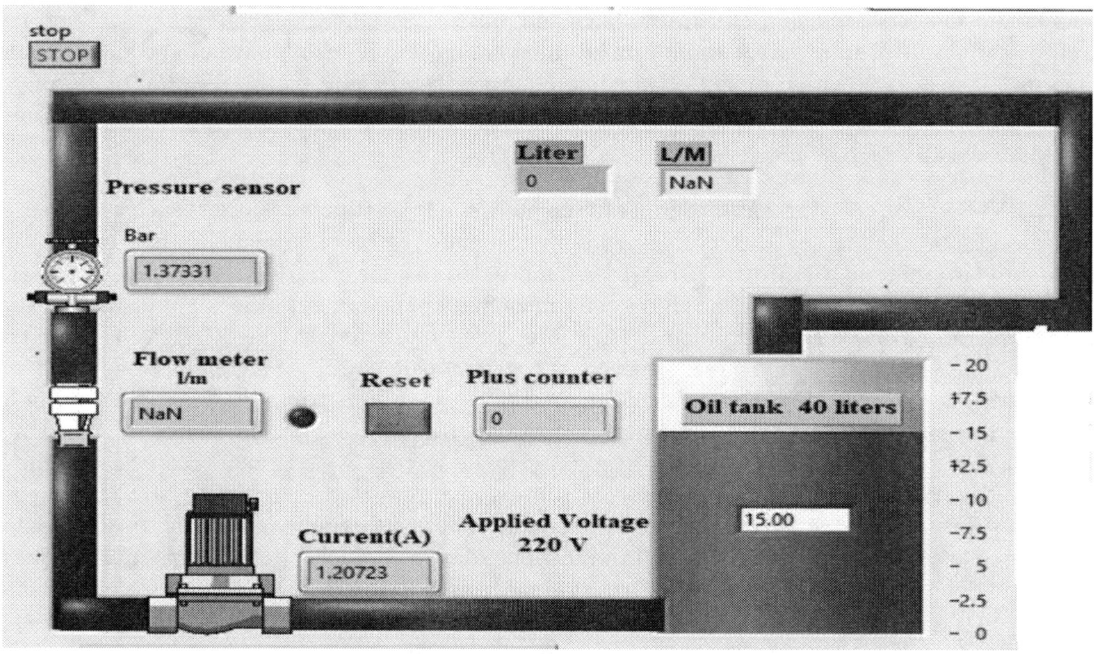

Figure 1: SCADA screen for observation and controlling of flow and pressure.

4. Implementation

This work implemented a system to control the parameter such as pressure and flow rate of the pump petroleum products using PLC and SCADA. control information is given to SCADA from HMI.SCADA converted into a control signal in order to be processed in PLC considered as a control signal to operate the final control [10].

5. PLC (programmable logic controller):

IOP Conf. Series: Materials Science and Engineering **765** (2020) 012050 doi:10.1088/1757-899X/765/1/012050

The main objective is to monitor and control the pressure and flow rate of the system in the pipeline automatically using an integrated SCADA system and PLC to connect it [14]

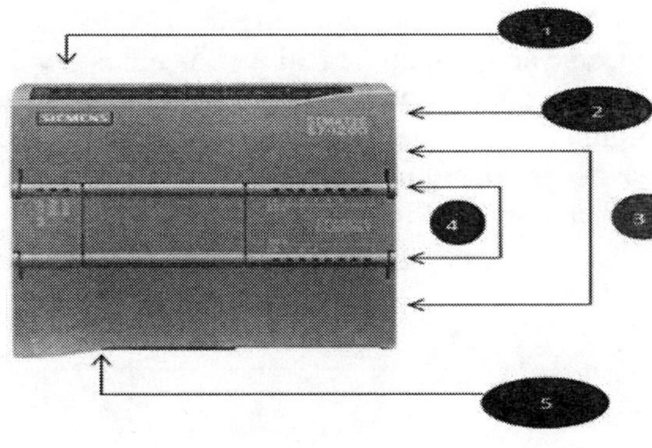

Figure 2: PLC components (① Power connector② Memory card slot under top door③ Removable user wiring connectors behind the doors④ Status LEDs for the on-board I/O⑤ PROFINET connector (on the bottom of the CPU)

Therefore, pressure and flow are transmitted through the pipeline and thus readings appear on SCADA screens so PLC is the continuous scanning of the program and includes the following Procedures:

Procedure 1: Test input status:

the information obtained from the processor is storage in memory for using in the next procedures

Procedure 2: Program implementation:

the input is taken from the previous procedure and action is taken to activate some output and the result can be postponed and stored in memory for recovery later

Procedure3 : Checking and Correction of output status from 1&2:

PLC inspect the output signals and adjusts it by making changes to the inputs that were checked during the first step and executing the program in the second step and then the third step is performed PLC returns the beginning of the cycle and repeat these steps continuously

Scan time = Execution time procedure 1+ Execution time procedure 2+ Execution time procedure 3 [2].

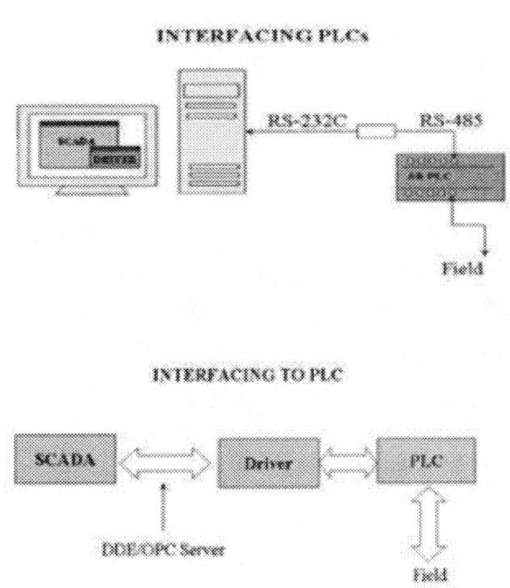

Figure 3: Interfacing PLC

6. SCADA System

Supervisory control and data acquisition are used to control the factory or device and it may be automatic or through commands that work and data is obtained first by a PLC programmable logic controller and then the data transfer is processed again to the central SCADA system [2] includes LABVIEW is a development program mainly used for virtual machines and gets the information that makes up the program from several units that allow the creation of windows applications or graphical user interfaces or connectivity or automatic control or device simulation in a loop [12] Simulation can be used with Lab VIEW real and a unit of time income and output the real world, and is fed through the PLC where the signal is given to the PLC block with the desired set point and can change its value in the foreground panel on the basis of the type of liquid transported in the pipeline. The output signal that corresponds to the opening of the control valve is considered to regulate the flow rate and control it [13]as shown in the diagrams for flow control and pressure control in oil pipelines.

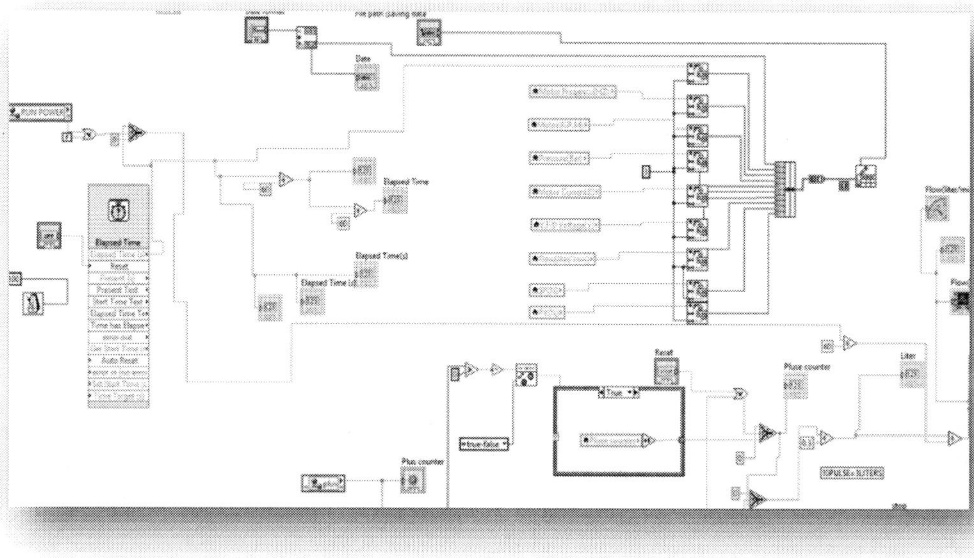

Figure 4: Monitoring and flow and pressure in oil piping with control

7. Results and Discussion

Simulations were carried out in detail in the experimental preparation section of the performance monitoring of the pump, pressure and flow control in the oil pipeline, the transmission system .The open loop system is obtained using the LIABVIEW control unit and the block control of the Lab VIEW platform is developed where the graph is given Control signal for control valve as shown in Figure 2. Performance analysis. VIEWLAB performance-based PLC controls are exploration of working range of 520 l / h at specified time t = 0-20 minutes Corresponding performance indicators are clearly calculated by PLC control in a short period of time in about 20 and maintaining a stable state with the use of different types straight oil, This ensures high speeds to enhance the operating condition of the oil pipeline transportation system. The same analyzes are performed with different operating points such as 230, 520 and 600 l / h to ensure the durability of the open-loop controller. The simulated responses that have reached different operating points provide better performance with the same settings for different operating points. PLC control settings showed strong response when operating point is deflected. The control valve opens and closes for control. In order to obtain the desired flow rate of the oil pipeline system by giving simulation and experimental results through LABVIEW-based PLC controllers and by simulating the results the performance in the observation increases 20.64%. A valve to obtain the desired flow rate at the destination

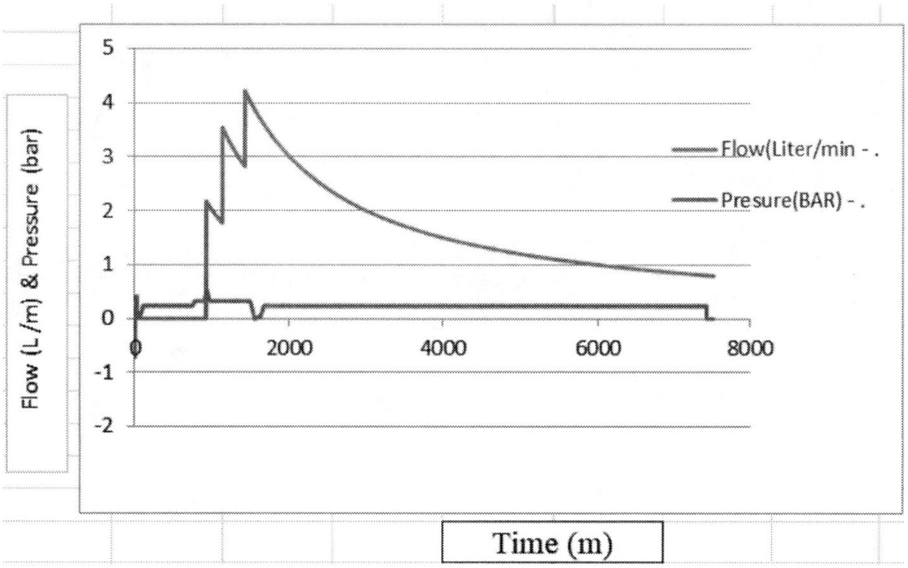

Figure 5: Pressure & flow rate for 20% of heavy oil for pump

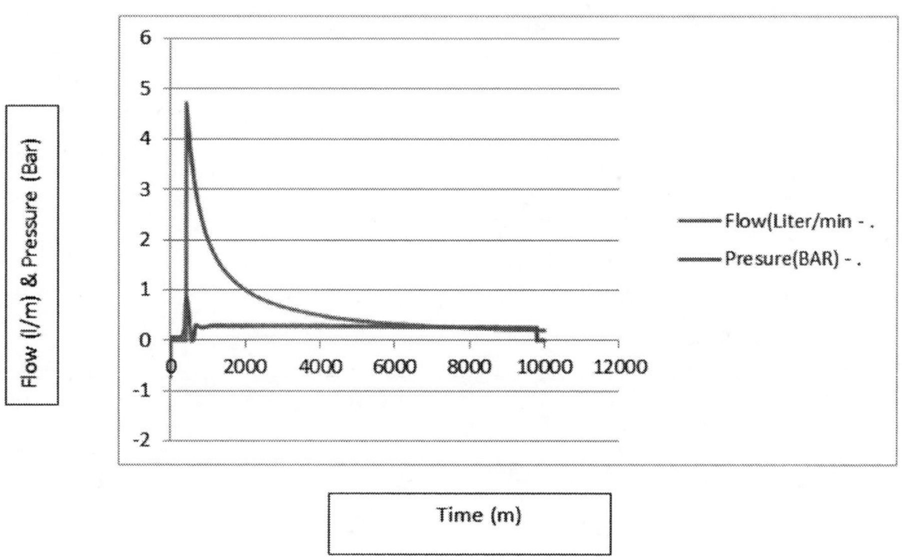

Figure 6: Pressure &flow rate for 20% of heavy oil for pump

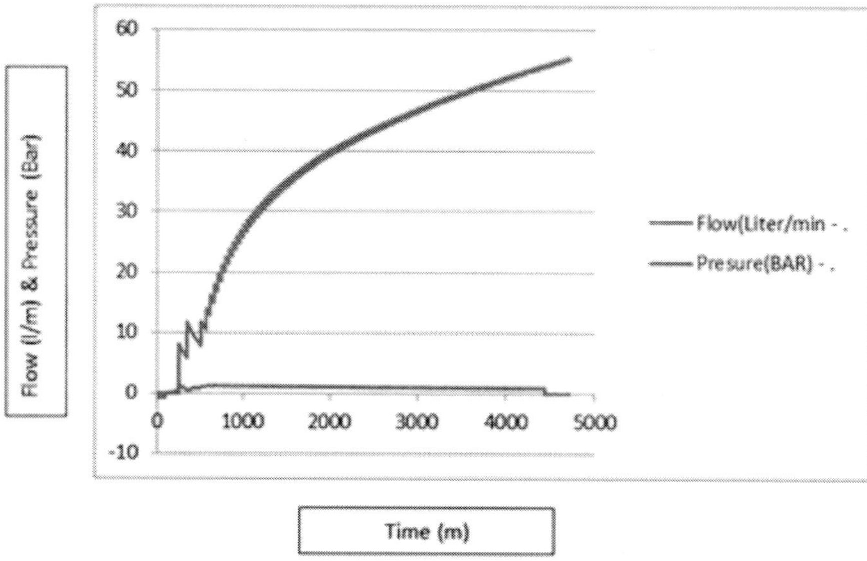

Figure 7: Pressure & flow rate for 40% of heavy oil for pump.

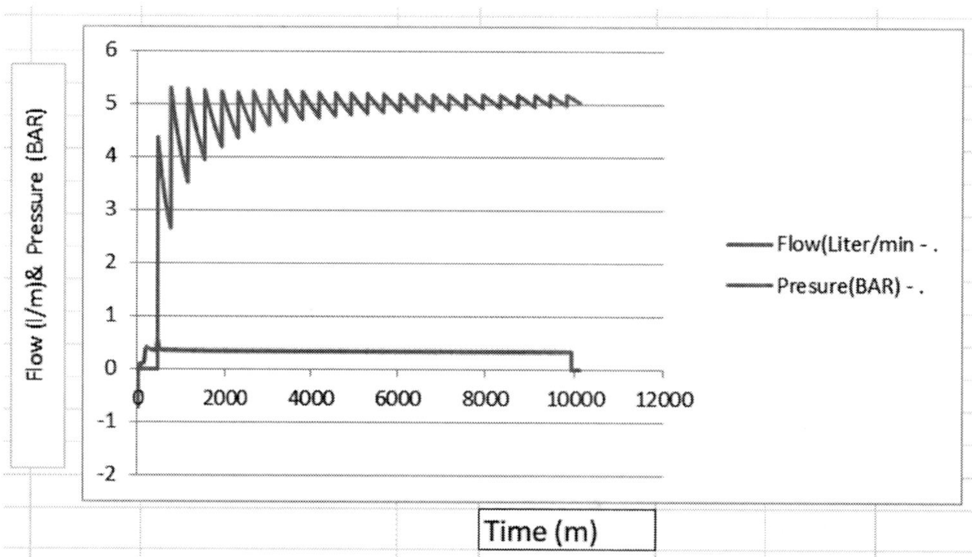

Figure 8: Pressure &flow rate for 40% of light oil for pump

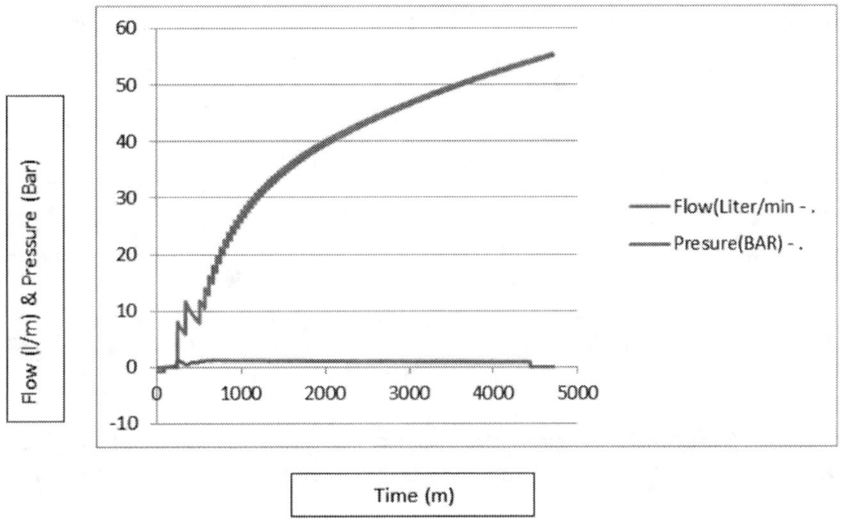

Figure 9: Pressure & flow rate for 60% of heavy oil for pump

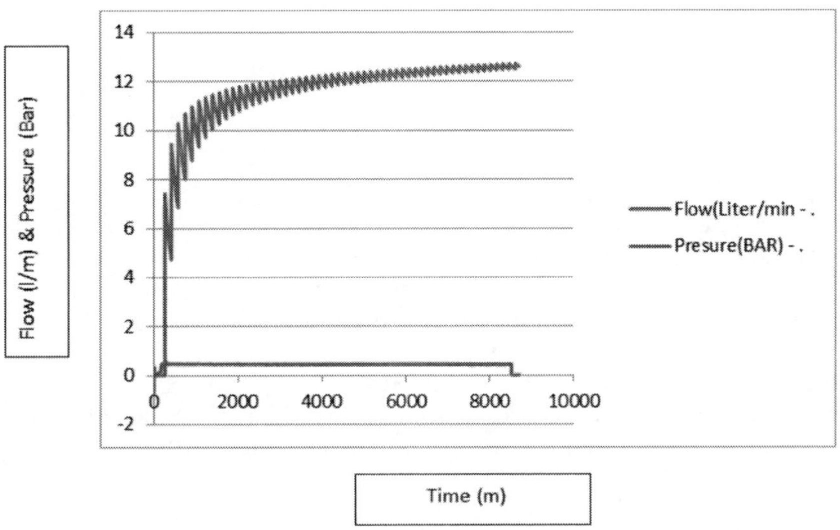

Figure 10: pressure & Flow rate for 60% of light oil for pump

8. Conclusion

This research presented a new plan to meet real time requirements and high data transmission needed for remote monitoring of pipelines as the scheme relies on the SCADA framework for the remote control to monitor oil pipelines and control the transportation of oil and petroleum products through the pipeline on several hundreds to thousands of miles are done using PLC to monitor various physical parameters control valves and pumps along the pipeline to maintain oil flow through the pipeline to be installed and maintained easily devices are easily connected and replaced. the exchange technology was used to create data in order to obtain information in real time remote testing and analysis can be carried out and improvement has been done through service-based SCADA is a promising and profitable option for improving the oil pipeline, and with this the increased voltage value from 2.73 to 4.65 has led to an increase in the flow rate and consequently an increase in pressure reaching from 0.25 to 0.546 bar and an increase in the flow rate from 0.22 to 5.64 L/M for the heavy oil and increased voltage value from 2.73 to 4.65 has led to an increase in the flow rate and consequently an increase in pressure reaching from 0.23 to 0.43 bar and an increase in the flow rate from 0.657 to 15.37 L/M for the heavy and light oil .

References

[1] Zhang LB, Wang ZH, Liang W. Remote and intelligent diagnostic methods on leakage fault for crude oil pipelines. In2006 International Pipeline Conference 2006 Jan 1 (pp. 683-688). American Society of Mechanical Engineers Digital Collection.

[2] Venuprasad A, Narayan YS. Automatic monitoring and controlling of pressure using PLC and SCADA. International Journal of Innovative Research in Science, Engineering and Technology. 2016;5(7):2347-6710.

[3] Deshmukh RN, Zope PH, Suralkar SR. Generation of Electricity by Renewable Energy Sources & Transmission of Energy Production Units using PLC & SCADA. International Journal of Advanced Research in Computer Engineering & Technology (IJARCET). 2012 Nov;1(9).

[4] Mobayen S. Chaos synchronization of uncertain chaotic systems using composite nonlinear feedback based integral sliding mode control. ISA transactions. 2018 Jun 1;77:100-11.

[5] Shamsuzzoha M, Skogestad S. The setpoint overshoot method: A simple and fast closed-loop approach for PID tuning. Journal of Process control. 2010 Dec 1;20(10):1220-34.

[6] He JB, Wang QG, Lee TH. PI/PID controller tuning via LQR approach. Chemical Engineering Science. 2000 Jul 1;55(13):2429-39.

[7] Morsi I, El-Din LM. SCADA system for oil refinery control. Measurement. 2014 Jan 1;47:5-13.

[8] Priyanka EB, Maheswari C, Thangavel S. Remote monitoring and control of LQR-PI controller parameters for an oil pipeline transport system. Proceedings of the Institution of Mechanical Engineers, Part I: Journal of Systems and Control Engineering. 2019 Jul;233(6):597-608.

[9] Elawad MA, Hamad MM, Fadl Elmola MF, Ahmed HA. CRUDE OIL PUMP STATION MONITORING AND CONTROLLING DEPEND ON PLC BASED (Doctoral dissertation, Sudan University of Science and Technology).

[10] Ibrahim MA. An automation system for Alsahafa Pumping Station By Using (PLC and SCADA) Khartoum State for Water Corporation (Doctoral dissertation, Sudan University of Science and Technology).

[11] Zhang LB, Wang ZH, Liang W. Remote and intelligent diagnostic methods on leakage fault for crude oil pipelines. In2006 International Pipeline Conference 2006 Jan 1 (pp. 683-688). American Society of Mechanical Engineers Digital Collection.

[12] Husin H, Azizi A, Husna A. An overview of viscosity reducers in heavy crude oil production. Chemeca 2014: Processing excellence; Powering our future. 2014:1246.

IOP Conf. Series: Materials Science and Engineering **765** (2020) 012050 doi:10.1088/1757-899X/765/1/012050

[13] Yasin G, Bhanger MI, Ansari TM, Naqvi SM, Ashraf M, Ahmad K, Talpur FN. Quality and chemistry of crude oils. Journal of Petroleum Technology and Alternative Fuels. 2013 Mar;4(3):53-63.

[14] Bolton W. Programmable logic controllers. Newnes; 2015 Mar 6.

IOP Conf. Series: Materials Science and Engineering **765** (2020) 012051 doi:10.1088/1757-899X/765/1/012051

Utilizing the Computation Intelligence Technique for Boiler Pressure Controlling

Hosham S.Anead
Khalid F. Sultan
Ayat Ahmed

Electromechanical Engineering Department, University of Technology, Baghdad, Iraq

corresponding author's e-mails : Hoshams.2000@yahoo.com, Ksultan61@yahoo.com
Ayat.alniamey@yahoo.com

Abstract. This paper aims to detect a damage in boiler of steam plant by using intelligence technique to improve the boiler performance. Experimental information is obtain by use an industrial boiler operating at AL-Dura power plant. This manuscript presents is to compare intelligence technique for boiler of steam power plant PID,ANFIS,PSO technique . Designing various pressure controllers to keep the boiler operating in normal conditions and enhance the efficiency. Those parameters are regulated through building a model and implementing in MATLAB, based on steam power plant requisite and the aims of the control. Results have shown that the PSO algorithm is optimal controlled and offers superior power plant performance from the PID controller. PSO, ANFIS and PID were implemented in AL–Dora power plant in Baghdad. Those techniques have been extensively utilized in numerous industry applications.

Keyword: ANFIS, PSO, control, PID.

1. Introduction:

Most of the electricity formed as of thermal power plant. An important component of thermal plant is a steam generation (boiler), it is a closed container which is made of high-quality steel where the steam is produced from the water by applying thermal from combustion of fuel such as coal, oil or gas. When a fuel is burning a surface of boiler is heating , hot gases generated in heating surface and exposed on one side and water or steam on the other. The steam collected over the surface of the water has been taken from boiler via superheater and afterwards, appropriate pipes to drive turbine for the electric generation or for some purpose in industrial application, The performance of steam power plant determine by performance of boiler[1] .Many of experimental and numerical researches were discussed on the boiler characteristics in several decades as S.Hosham (2018),(2019). which applied ANN for control of pressure, pressure, air\fuel ratio, flame, water level, gas available of boiler, this technique give agood agreement, and when compare between PID,ANN,the result give ANN a best performance than PID [2] [3].

Rahul (2011) This paper discussed the 2 approaches, which are the PID controller and the fuzzy logic controller compared between them, the best results have been shown in the case of the use of the fuzzy logic controller. Maximal overshoot of approximately 9.35% for the fuzzy controller in comparison with 47.3% identified by the PID controller. The settling time which has been measured for the fuzzy controller at 7.18s and 10.14 in the PID controller, which has been utilized for controlling parameters like the pressure, which changes due to the changes of the circumstances [4].

Then Pranitachavan (2015), discussed controlling of superheated steam pressure in the thermal power plants with the use of the fuzzy logic controller. The fuzzy logic controller was advanced with the use of the PIC 16F-871 micro-controller which produces the desirable signals. MATLAB programming with the fuzzy control has been utilized for coping with issues that result from steam generators. Conventional controller PID can't treat with the non-linearities of the system. In addition to that, this research discusses comparisons between the fuzzy controller and the conventional controller. In the case where the comparison concluded that fuzzy logic control has higher flexibility and needs a smaller amount of time compared to the conventional approach [5] As well as H. Aygun1(2011) applied Particle Swarm Optimization based PID , FLC which are modern controllers, and classical PID controller are utilized to control the bed pressure of a circulating fluidized bed boiler and they are compared , Bed pressure is important parameters in a circulating fluidized bed boiler. Because the efficiency of the combustion enhances with rising the bed pressure and also harmful emissions are influenced with the bed pressure. Simulation results show that settling time in PSO-PID controller is lower than other controllers. PSOPID decreases overshoot like FLC but in classical PID controller overshoots are the biggest. In FLC there are no overshoot [6].

And D.Thangavelusamy (2015) Design and accomplishment of PI/PID controllers Based Particle Swarm Optimization apply to boiler element. Designing Coordinated PID Controller and tuning the parameter with the use of the (PSO), the Results obtained that controller provides more sufficient performance for each of the servo and the regulatory responses.

This study aims at controlling variables pressure of boiler of AL-Dora power plant in Baghdad that impacted maximal produced power and plant efficiency with the use of intelligent techniques and made a comparison between them [7].

2. Intelligent Control Techniques

2.1. Particle Swarm Optimization

PSO is a method for optimizing difficult numerical function by recruit or metaphor the social behavior of fish schools and flocks of birds. The most important PSO algorithm is founded by simplifying the social simulation model. The algorithm was planned the first time to emulate birds search for food which is simulated to be a cornfield vector. The bird would find food through united cooperation among birds Then, expand to contain multidimensional search. In PSO algorithm, each particle in swarm appear to be as a result of the trouble and it is resolute by its velocity and position.

For numbers of variable (n) in optimization task, a flock of particles are placed in the equal number of dimensional of search space at random selection positions and velocities in order to know their best magnitude for determining (Pbest) the greatest speed value and location in the specified dimensional space throughout the random movement of each part and at each iteration. Or, the velocity of both particles is fitted as to its own flying test and another particle flying experiment.

IOP Conf. Series: Materials Science and Engineering **765** (2020) 012051 doi:10.1088/1757-899X/765/1/012051

The particle can be represented [8], as:

$$x_i = (x_{i,1}, x_{i,2}, x_{i,3}, \dots \dots . . x_{i,n}) \quad (1)$$

In n-dimensional space, the most excellent previous particle position as:

$$P_{best_i} = \left(P_{best_{i,1}}, P_{best_{i,2}}, \dots \dots . P_{best_{i,n}}\right) \quad (2)$$

The adapted velocity and location of both particles may be considered with the use of current distance and velocity from $(P_{best_{i,d}})$ to (g_{best_d}) as illustrated in following formula *as:*

$$V_{i,m}^{(It.+1)} = W * V_{i,m}^{(It.)} + c_1 * rand * \left(P_{best_{i,m}} - x_{i,m}^{(It.)}\right) + c_2 * rand * \left(g_{best_m} - x_{i,m}^{(It.)}\right) \quad (3)$$

$$x_{i,m}^{(It.+1)} = x_{i,m}^{(It.)} + v_{i,m}^{(It.)} \quad (4)$$

Where i=1, 2, 3,……, n

m=1, 2,3, ….., d

n stands for the Number of particles.

Iter. stands for the Iterations pointer.

d stands for the Dimension.

W stands for the Inertia weight factor.

$V_{i,m}^{(It.)}$ stands for the Particle (no. i) velocity at iteration It.

rand stands for a Random number between 0-1.

c_1, c_2 stand for the Acceleration constant.

P_{best_i} stands for the Best previous position of the particle.

$x_{i,m}^{(It.)}$ stands for the Particle i current position at iteration It.

g_{best_m} stands for the Best particle among all the particles in the population.

The integrated time weight square error (ITSE) is used for evaluating the coefficients.

$$ITSE = \int_0^t t.\, e^2(t)dt \quad (5)$$

IOP Conf. Series: Materials Science and Engineering **765** (2020) 012051 doi:10.1088/1757-899X/765/1/012051

2.2. The Adaptive Network Based Fuzzy Inference System (ANFIS)

The ANFIS neuro-fuzzy controller has been implement by Jang and employed Takagi – Sugeno – Kang (TSK) system fuzzy inference. Square nodes in ANFIS construction denotes the limitation sets of the functions of membership of TSK fuzzy system. The circular nodes can be considered static / non – modifiable and present operations like the produce or the max/min computations. A hybrid learning rule has been utilized for increasing the speed of the parameter adaption. This utilization of the sequential least square in forward by-pass for identifying successive parameter, and back – propagation in backward pass for setting basis parameters up[9].

2.3. PID Controller

Genetic algorithms applied to analyze a range of engineering problems. By implementing this type of algorithms in the tuning of the PID. The common strategy of the PID control was discovered as one of the potential approaches as a result of its little costs, simple maintenance also easy in the control designs. Basically, variables of the PID controller are made up of 3 distinct variables, which are: integral (ki), proportionality (kp), in addition to the derivative values (kd). A proper setting of those variables is going to enhance the dynamic system response, improve system stability, and eliminate steady state error [9]

The most important PID control system constriction has been illustrated in figure 1. The set point was changed, the error is going to be calculated between set point and actual output. Error signal, E(s), has been utilized for the production of proportionate, derivative, and integral actions, with resultant signals that have been weighted and summed for forming control signal, $U_C(s)$, that has been applied to the particular model. After that, obtaining new signal of the output. This new real signal is going to be emitted to controller, and again, error signal is going to be computed. New control signal, $U_C(s)$, is going to be transmitted to plant. This process is going to run in a continuous manner, to the point where the steady – state – error becomes near 0.

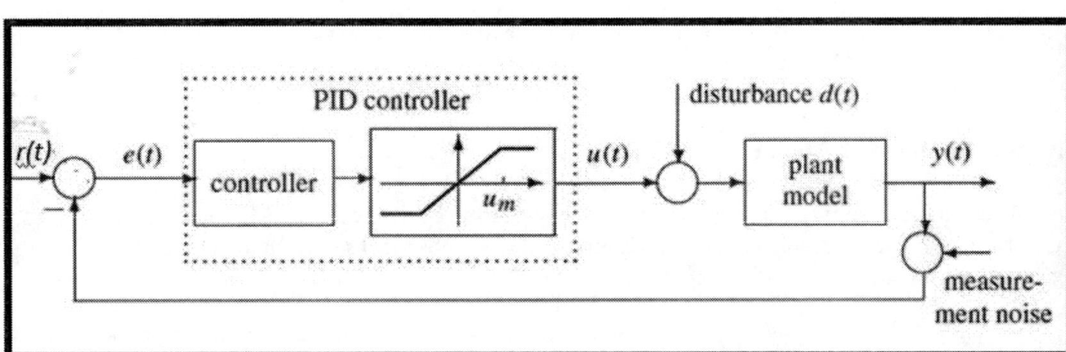

Figure 1: PID control system

IOP Conf. Series: Materials Science and Engineering **765** (2020) 012051 doi:10.1088/1757-899X/765/1/012051

3. Simulation Models of pressure

Pressure generate from boiler are similar to entering the turbine. In the case where the data attainment, shows that pressure must be <140 bar for avoiding the explosion.

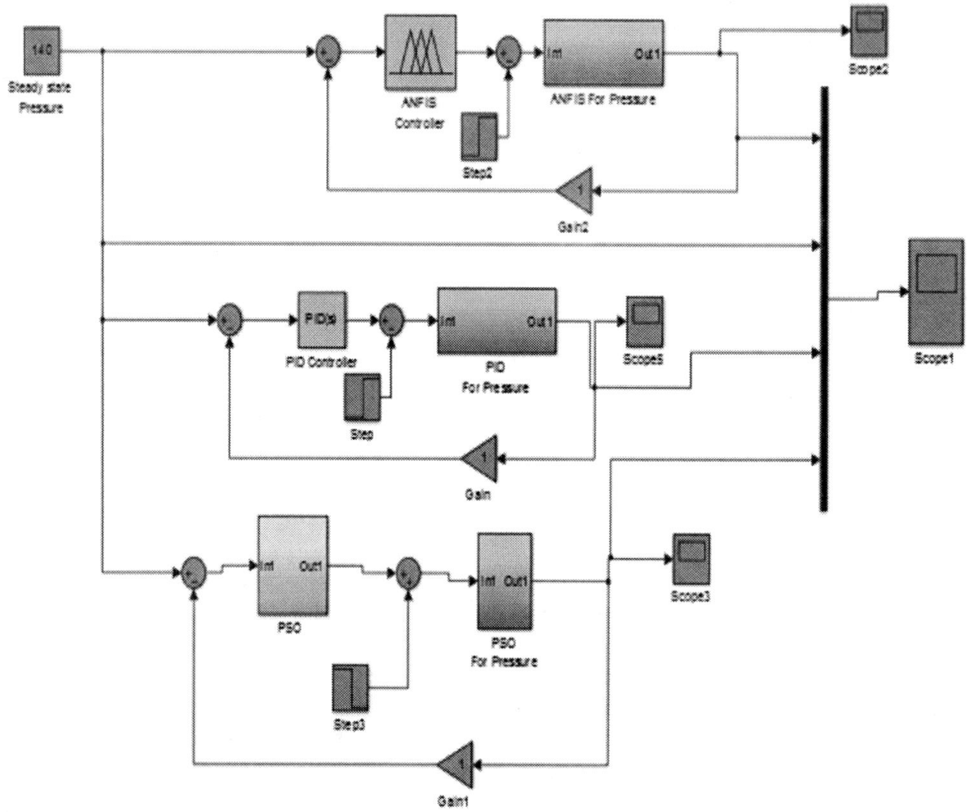

Figure 2: Simulation Models of Pressure

4. Results of Simulink Model

Figure 3 and table 1 shows the pressure increase with time of the boiler in AL – Dura power plant,using (PSO,ANFIS,PID) This result give a small setting time in PSO technique than ANFIS,PID. But ANFIS is less overshoting comparter other methods

IOP Conf. Series: Materials Science and Engineering **765** (2020) 012051 doi:10.1088/1757-899X/765/1/012051

Table 1: Results of compartion of intelligent technique of pressure

Parameter Pressure	Rise Time	Settling Time	Peak	Peak Time	Overshoot
PSO	**0.85**	**2**	**160**	**30**	**20**
ANFIS	**3.4**	**4**	**135**	**30**	**5**
PID	**2.6**	**18**	**180**	**30**	**40**

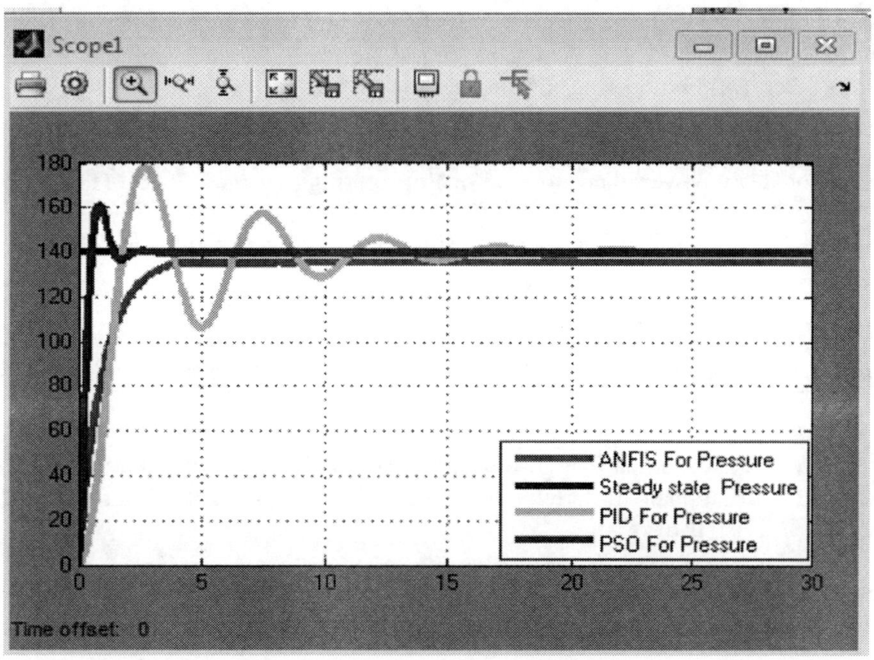

Figure 3: The pressure of the boiler.

5. Conclusion

This paper applied to Computation intelligence technique to control of pressure of steam generation. In this study, a case study of the process control which takes the boiler was implemented with the use of the PSO, PID, and ANFIS. Initially, a mathematical system model has been developed and a traditional PID controller, PSO, ANFIS have been applied in it. Through comparing with PSO, PID, ANFIS controller, it has been shown that this approach enhanced the dynamic system efficiency in a more sufficient manner. PSO can be considered as the optimal one, due to the fact that it has given adequate performances and is known for its sufficient robustness. The conventional algorithm of the PSO. It utilizes only primary mathematical operators, and provides sufficient outputs in the noisy, static, and continuously varying environments. All those benefits result in implementing PSO in a wider range of the research areas. Lately, PSO approaches have been applied with success in numerous engineering and real-world issues as well.

Abbreviation

PID	Proportional-integral-derivative
ANFIS	Adaptive Neuro Fuzzy Inference System
PSO	Particle Swarm Optimization

References

[1] Raja AK, Srivastava AP, Dwivedi M. Power plant engineering, 2006 New Age International (P) Ltd. Publishers, New Delhi.:110.

[2] Anead HS, Sultan KF, Abd-Kadhum RJ. Evaluation and Improvement Performance of a Boiler in a Thermal Power Plant Using Artificial Neural Network. Engineering and Technology Journal. 2018;36(6 Part (A) Engineering):656-63.

[3] Salim H, Sultan KF, Jawad R. Comparison between PID and artificial neural networks to control of boiler for steam power plant. Journal of Engineering Sciences. 2019;6(1):E10-5.

[4] Malhotra R, Sodhi R. Boiler flow control using PID and fuzzy logic controller. IJCSET. 2011 Jul 1;1(6):315-9.

[5] Chavan P. Superheated steam temperature control in thermal power plant using fuzzy logic controller. International Journal of Advance Foundation And Research In Science & Engineering. 2015;1.

[6] Aygun H, Demirel H. Comparison of PSO-PID, FLC and PID in a circulating fluidized bed boiler. In2011 7th International Conference on Electrical and Electronics Engineering (ELECO) 2011 Dec 1 (pp. II-376). IEEE.

[7] Thangavelusamy D, Ponnusamy L. Optimization Techniques Applied to Boiler Turbine Unit. Indian Journal of Science and Technology. 2015 Oct;8:26.

[8] Allaoua B, Gasbaoui B, Mebarki B. Setting up PID DC motor speed control alteration parameters using particle swarm optimization strategy. Leonardo Electronic Journal of Practices and Technologies. 2009 Jan;14:19-32.

[9] Roland S.Burns,"Advance Control Engineering", Butterworth-Heinemann,Published,2001.

IOP Conf. Series: Materials Science and Engineering **765** (2020) 012052 doi:10.1088/1757-899X/765/1/012052

Influence of Cryogenic Temperature (CT) on Tensile Properties and Fatigue Behavior of 2024-Al2O3 Nanocomposites

M N Abdulridah[1], A D Assi [2] and H J Al-Alkawi [3]

[1] Electrical Engineering Department, University of Technology/ Iraq- Baghdad.

[2] Middle Technical University/ Institute of Technology/ Iraq- Baghdad

[3] Electromechanical Engineering Department, University of Technology/ Iraq- Baghdad.

E-mail:11122@uotechnology.edu.iq

Abstract. The stir casting process used to manufacture the AL/ Al_2O_3. AA2024 with matrix material alumina Al_2O_3 at different weight percentage ratio 0.3, 0.5, 0.7 and 0.9%. Were fabricated by using cryogenic temperature (CT). It was observed that the maximum enhancement is occurred at nanocomposite containing 0.9% Al_2O_3 which shows improvement percentage of 9.9%, 14.63% and 14.28% for ultimate strength, yield stress and Brinell hardness respectively and the ductility is reduced by 34.57% compared to 0% by weight AA2024. It clearly shows that all the mechanical properties are improved by increasing weight percentage of Al_2O_3 but the best properties are obtained at 0.9% Al_2O_3 nanocomposite. The fatigue S-N curve tests were done for all the specimens having average of roughness between 0.2 to 0.5Mm.The fatigue endurance limit of the 0.9% wt. Al_2O_3 composite into two cases without (CT) and with (CT) is taken at 107 cycles. The fatigue limit at 10^7 cycles increases from 83 MPa for without (CT) to 89.6 MPa with (CT) resulted in 7.95% fatigue limit improvement.

1. Introduction

Among the metal matrix composites particle-reinforced aluminum matrix composites have attracted substantial interest among researchers as engineering materials due to their mechanical properties like strength to weight ratio and improved hardness, superior to those of general aluminum alloys. Some of the typical applications of these MMCs are bearings, automobile pistons, cylinder liners, piston rings, connecting rods, sliding electrical contacts, turbo charger impellers, space structures, etc.[1].

Different methods are used to fabricate MMCs one of them is stir casting method, it is the most commonly used due to have advantages such as the ability to mass production, lower processing cost, simplicity in operation, easier control over matrix composite, improve particle matrix bounding [2].

Al-alkawi et al. [3] used *MMCs* of 7075 AA with different wt% of Al_2O_3 at average size of 10nm, were manufactured using stir casting method. The results showed that when adding particles of Al2O3 the mechanical properties of nanomaterial were improved such as Brinell hardness, yield stress (σy), ultimate strength (σu) and ductility. Maximum improvement in hardness BHN, yield stress and ultimate strength were observed by weight percentage 0.2 Al_2O_3. While the lowest value is obtained of ϵ % at weight percentage 0.2 Al_2O_3.

Kumar K. et al. [4] used Electromagnetic stir casting process and mechanical stir casting combined with a process of Hybrid Casting. AA 2024 reinforced with 1 % wt. Al_2O_3 nanocomposite was manufactured. The composite was tested by microscopy (SEM) and tensile testing. The results of tensile strength has been improved by 43% from Al2024 with 1 % wt. Al_2O_3 nanocomposite when compare it with the Al2024 alloy.

Ashwath p. et al, [5] studied the manufacture of aluminum alloy AA-SiC and AA- Al_2O_3 with weight percentage (3, 6 and 9) of SiC and Al_2O_3 with 10 μm average particle size. Aluminum alloy 2024 reinforced with 6% wt. of Al_2O_3 samples were showed improved hardness results, stress-strain

IOP Conf. Series: Materials Science and Engineering **765** (2020) 012052 doi:10.1088/1757-899X/765/1/012052

and strength behaviour. Aluminum alloy 2900 reinforced with 6 wt% Al_2O_3 showed good formability and ductility properties closer to 2024AA. The results showed that frictional coefficient behaviour of AA 2900 with 6 wt% Al_2O_3 was better than of AA 2024.

S.A. Sajjadi et al. [6] fabricated Micro composites and nanoparticles (A356/ Al_2O_3) with a different percentage of particles wt. using two melting techniques such as compo-molding and stir- molding. Pressure tests, stiffness, hardness and tensile were conducted to determine the mechanical properties of compounds. The results of the study showed that mechanical properties (ultimate tensile, yield strength, hardness and compression strength) are improved when added alumina (nano, mico).

Gao S. et al [7] studied deep cryogenic treatment used to AA5A06 at temperature (-155C) for (4H,8H,10H) respectively. The results showed that after the use of cryogenic treatment (CT), the mechanical properties of 5A06 were significantly improved. The main object of this work is to study the mechanical properties of Al_2O_3 particle-reinforced 2024 aluminium alloy composites and investigate the effect of cryogenic temperature on mechanical properties of the same material.

2. Experimental Work

Metal matrix composites materials (MMCs) are two or more materials mixed with different properties, one material is called matrix - phase , the other one is called reinforcement material, in this study AA 2024 with chemical composition at table (1), and reinforcement material aluminum oxide Al_2O_3 with chemical composition at table (2) are used.

Table 1. Chemical Composition of AA 2024

Component	% Si	% Fe	% Cu	%Mn	% Mg	%Cr	% Zn	%Ti	% other	% Al
Standard [8]	Max. 0.5	Max. 0.5	3.8-4.9	0.3-0.9	1.2-1.8	0.1	0.25	0.15	Max 0.15	Ba.l
Measured	0.48	0.46	4.2	0.52	1.48	0.8	0.11	0.11	-	Bal.

Table 2. Chemical Composition of Al_2O_3

Chemical Composition	α -Alumina	Fe_2O_3	TiO_2	CaO	Other magnetic material
Reference [9] (wt.%)	±93	±0.8	±1.8	±1.1	±0.2

2.1 Mixing

The stir casting process used to manufacture the AL/Al_2O_3. AA2024 with matrix material alumina Al_2O_3 at different weight percentage ratio 0.3, 0.5, 0.7 and 0.9%. Were fabricated by using stir casting method. This method involves several steps that can be summarized as follows [2]:

1. Preparing the specimens by cutting into cubes (1-2) cm3 and washed it with alcohol, then washed with steam of hot air at a temperature 100°C after that dry parts.
2. Heated parts by an electric heater at 200°C.
3. Lift the oven covers and carry the resulting parts of the heating process, then close the oven cover tightly. Then, withdrawn air from the oven using vacuum.
4. After pumping the furnace with Argon gas, it is heated to 800°C
5. The alumina (Al_2O_3) are heated to 200°C.
6. The nanoparticles are added to the molten AA2024 through the gas pump.
7. Raised the temperature of furnace is over melting temperature of Al. about 850°C then cooled it less 650°C.

8. Four minutes are designed for stirring and with stirring speed 450 rpm.
9. Mixing is heated to 850±10°C
10. Finally, pour the molten into appropriate molds to obtain an aluminum rod.

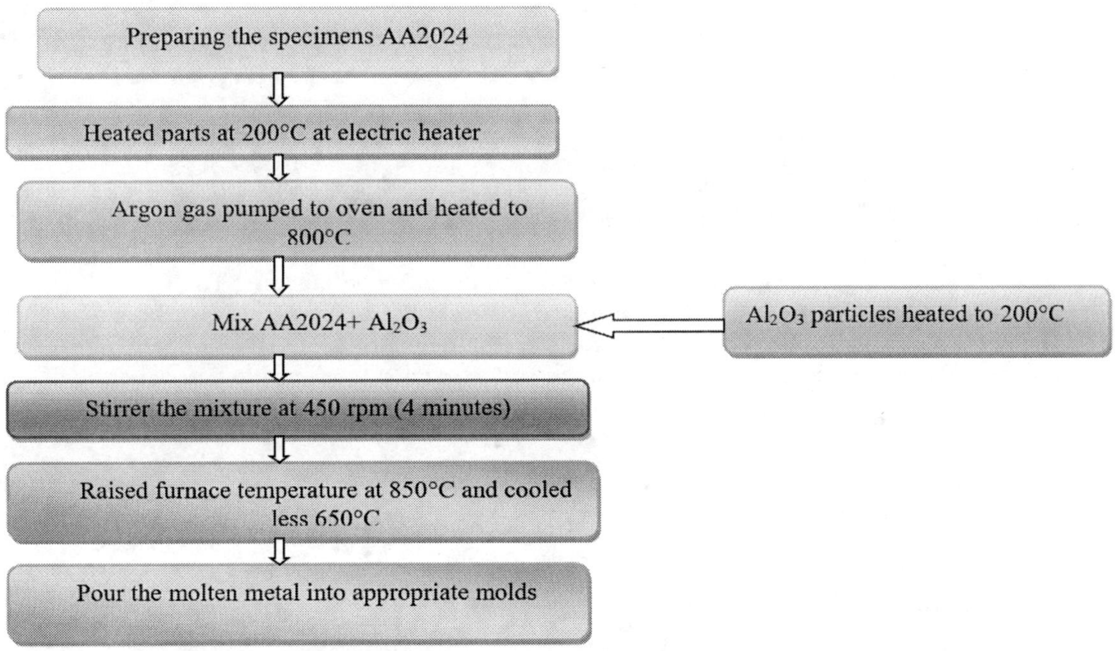

Figure 1. Block diagram of stir casting process

2.2 Cryogenic Treatment

Deep cryogenic treatment (DCP) is a processing that can be used to improve the mechanical properties of components, this process have several advantages to improve fatigue resistance, hardness and grain size. The block diagram of DCP represents in figure (2). DCP was applied to samples of 2024Al/Al_2O_3 nanocomposite that fabricated at several step:

1. Wash the samples several times with methanol then dried the samples
2. Used cooling instrument type SKL500u
3. The sample put in vertical position in holder of the cooling chamber
4. The samples was slowly cooled to avoid thermal shock by using to cooling chamber
5. The cooling process till the samples reach at (-160°C) had been taken at 4 hour.
6. Hold the samples at (-160°C) for 24 hour
7. The samples reheated slowly without thermal shock to ambient temperature
8. The sequence of this treatment was carried out in vacuumed environment.

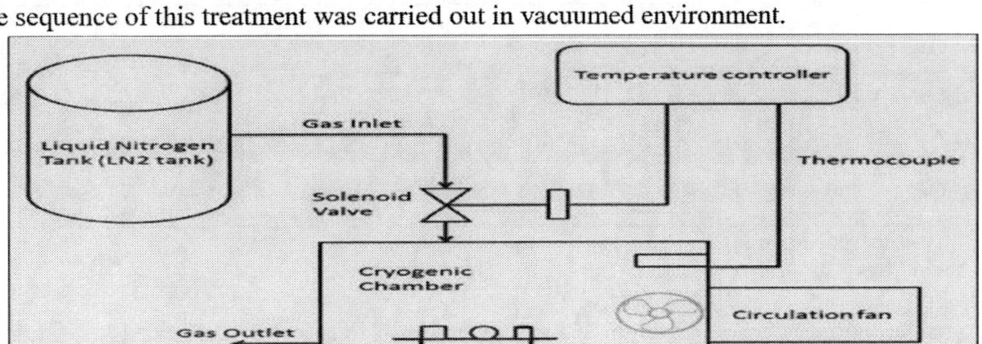

Figure 2. Schematic diagram of deep cryogenic treatment processing [10].

IOP Conf. Series: Materials Science and Engineering **765** (2020) 012052 doi:10.1088/1757-899X/765/1/012052

2.3 Sample dimensions and test apparatus

Figure (3) shows the fatigue specimen According to America Standard Specification (ASTM) which means the (American Society Testing and Materials).

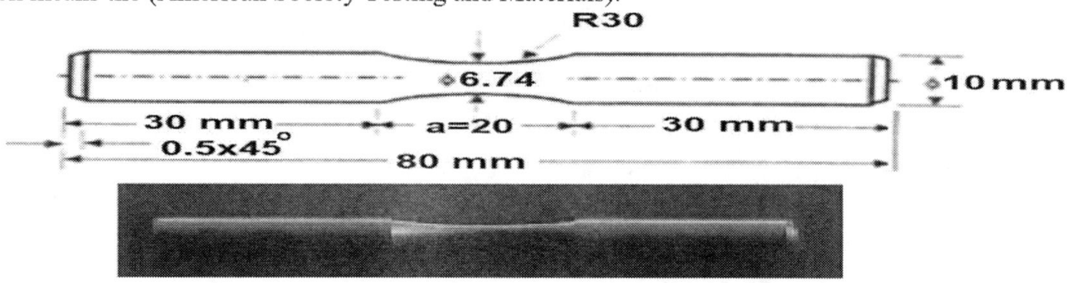

Figure 3. The Fatigue Specimen According to ASTM

A rotational-type fatigue test device was used (Scheck pun rotary bending machine) as shown as in figure (4) and with a speed of 1420 rpm so as to test the matrix material alumina Al_2O_3 specimens. 24 specimens were tested for each level of stress and extraction to eliminate the scattering that occurs in fatigue test, this is the usual context in fatigue tests.

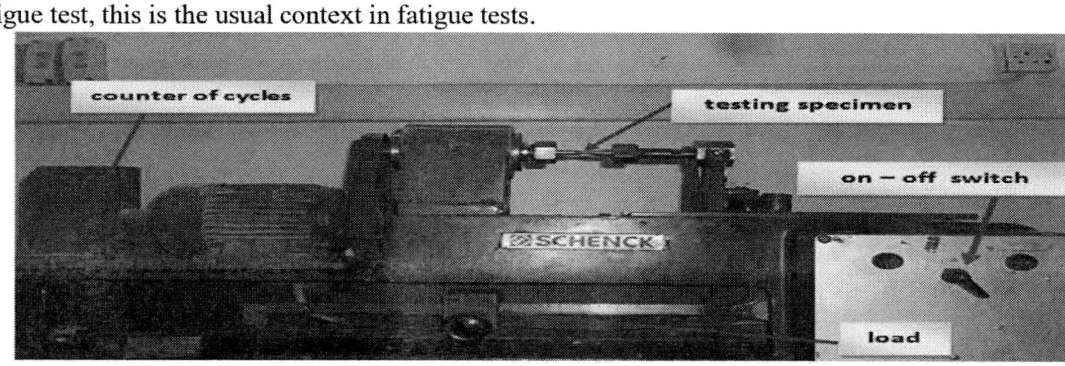

Figure 4. The Fatigue Test Device Type (Scheck pun rotary bending machine)

3. Results and discussion

3.1 Mechanical Properties

Table (3) and figure (6) gives the tensile strength and mechanical properties results with different weight percentage of the reinforced material under to cases of testing (RT) and (CT).

Table 3. Mechanical Properties of 2024 Al. alloy with different amount of Al_2O_3.

Material	Without (CT)					With (CT)				
	*UTS (MPa)	**YS (MPa)	BH	E(GPa)	∈ %	*UTS (MPa)	**YS (MPa)	BH	E(GPa)	∈ %
AA2024	182	82	58	68.4	11	186	87	63	69	10.7
0.3% Al_2O_3	188	86	60	69	10	201	90.5	65	70	8.4
0.5% Al_2O_3	192	89	62	69.7	9.2	207	91	68	70.3	8
0.7% Al_2O_3	196	91	64	70	8.6	211	95	70	70.6	7.3
0.9% Al_2O_3	200	94	66	70.2	8.1	215	98	72	71	7

*(UTS) Ultimate Tensile Strength
**(YS) Yield Stress

Mechanical properties can be determined by performing a tensile test during which a homogenous standardized test sample is subjected to a controlled the applied load (P) and deformation (ΔL).

Considering the initial dimensions of test sample, cross-section area (A_o) and initial length (L_O) then the average linear stress (σ) can be calculated by:

$$\sigma = \frac{P}{Ao} \quad (1)$$

$$\text{And strain } \in = \frac{Lf - L_0}{L_0} = \frac{\Delta L}{L_0} \quad (2)$$

Where L_f is the final length of failure

The stress (σ) and stain ($\in$) can be plotted to outline the engineering stress-strain curve. Fig.(5) shows the stress- strain experimental curve of 0.9wt.% Al_2O_3 in both (RT) and (CT).

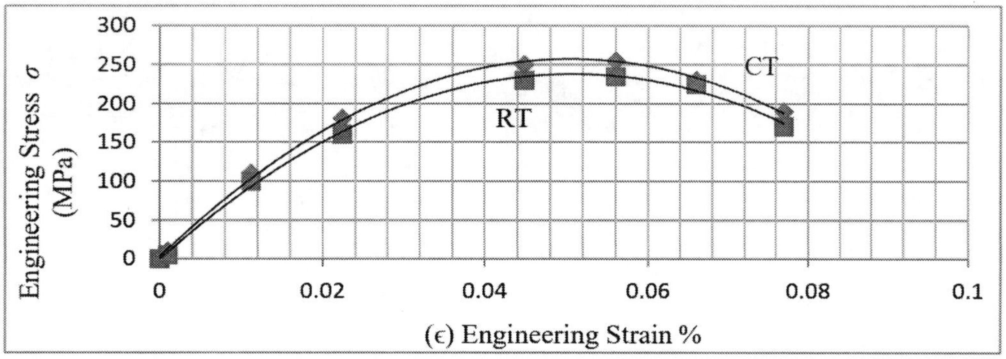

Figure 5. Engineering Stress-Stain curve for 0.9 wt.% Al_2O_3 composite at (RT) and (CT)

From the results presented in table (3), it is seen that there is a significant improvement in mechanical properties and hardness after (CT) treatment and the maximum enhancement is occurred at nanocomposite containing 0.9% Al_2O_3 which shows improvement percentage of 9.9% and 14.63% for UTS and YS respectively and the ductility is reduced by 34.57% compared to 0% by weight AA2024. It clearly shows that all the mechanical properties are improved by increasing weight percentage of Al_2O_3 but the best properties are obtained at 0.9% Al_2O_3 nanocomposite show fig(6). These improvement may be coming from reinforced particles are uniformly distributed on AA2024 metal matrix and this parameter is a great drawback in producing metal matrix composite [11]. The thermal mismatch between matrix and Al_2O_3 is the main factor for increasing the density of dislocation leading to increase the composite mechanical properties [12].

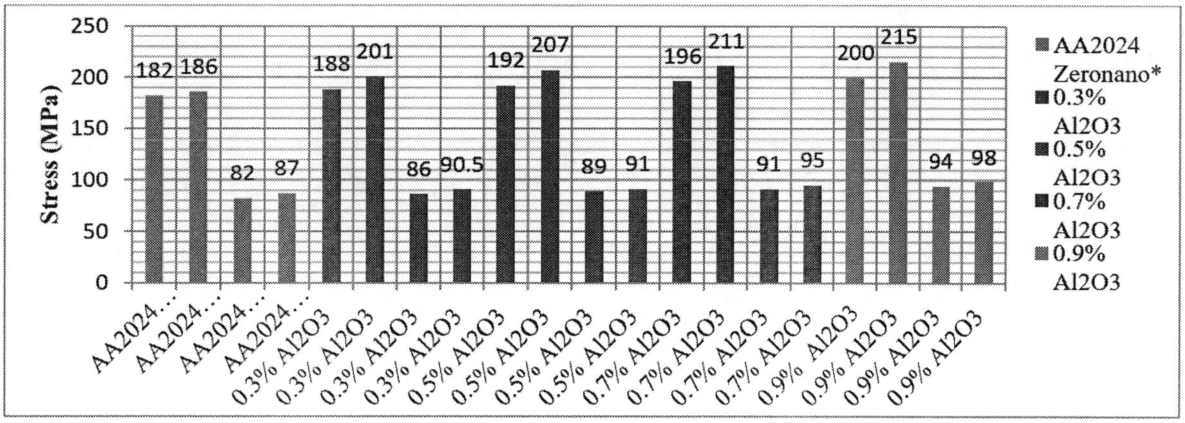

Figure 6. Mechanical Properties (UTS and YS) of 2024 Al. alloy with different percentage of Al_2O_3 at (RT) and (CT).

IOP Conf. Series: Materials Science and Engineering **765** (2020) 012052 doi:10.1088/1757-899X/765/1/012052

Modulus of elasticity (E) defined as (Stress/ Strain) and it is measured of the stiffness of material (E $\frac{A_0}{L_0}$) and it is used for calculating structural deflections. Ref [13] mentioned that the modulus of elasticity is slightly high at low temperature compared to ambient temperature and it drops at elevated temperature.

It is observably determined that (E) increases when the temperature decreases. Generally (E) under (CT) is greater than that of modulus of elasticity at ambient temperature. For composite containing 9 wt.% Al_2O_3 the value of (E) increased from 70.2 to 71GPa without and with (CT) respectively showing 1.139% improvement.

The variation of (E) verses the amount of Al_2O_3 under ambient temperature and (CT) can be illustrated in fig. (7).

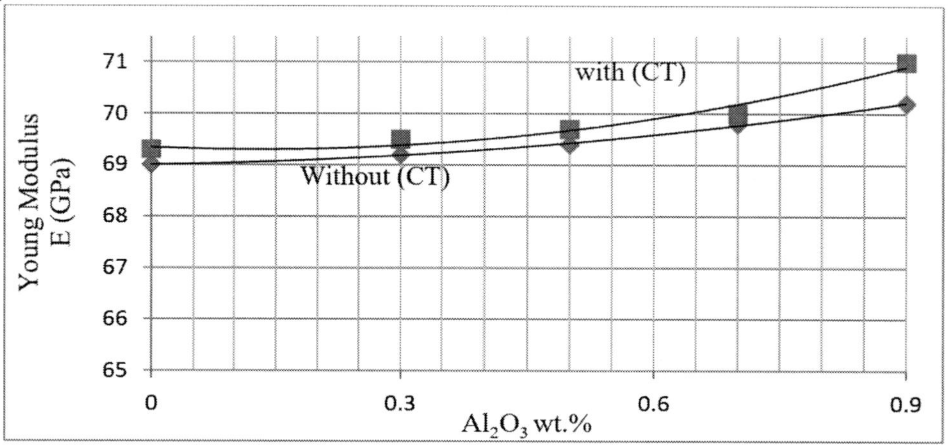

Figure 7. Young Modulus (E) against wt.% of Al_2O_3 for room temperature (RT) and (CT).

The ductility of aluminum alloys depends on temperature and their crystal structure. Aluminum and its' alloys are ductile even at low temperatures. This property provides using aluminum alloys in cryogenics applications. The experimental results revealed that the ductility of matrix and composites reduced at (CT) compared to (RT) as shown in fig. (8)

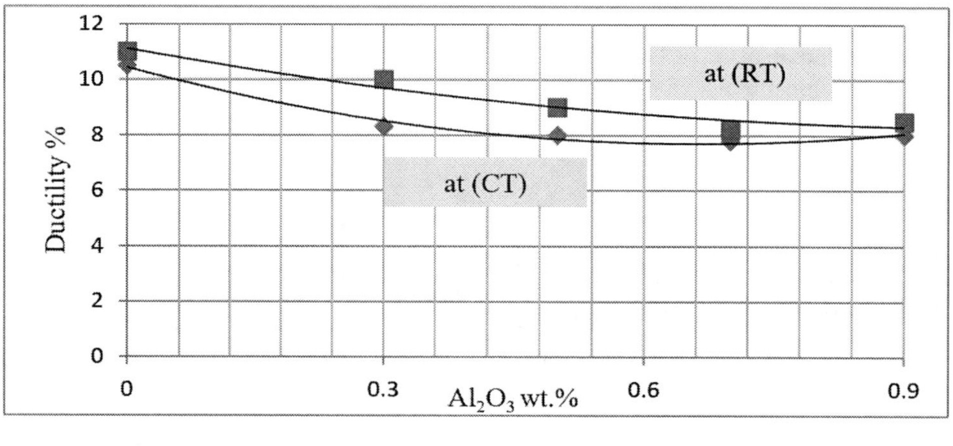

Figure 8. Ductility VS wt. % of Al_2O_3 at (RT) and (CT).

It is clear that the behaviour of AA2024 changes greatly when tested at (CT) compared to (RT) behaviour. This finding is greed well with Ref [14].

IOP Conf. Series: Materials Science and Engineering **765** (2020) 012052 doi:10.1088/1757-899X/765/1/012052

3.2 Fatigue S-N Curve Results

There are many factors that influence the S-N curve performance which are (temperature, surface treatments, loading ratio, cycling frequency and type of reinforcement) [15].

The fatigue S-N curve tests were done for all the specimens having average of roughness between 0.2 to 0.5Mm. four applied stresses were selected depend on the mechanical properties and for each stress level there are three specimens were tested with stress ratio (R= -1),12 specimens were examined without (CT) and 12 specimens with (CT) and the result have been listed in table (4) and graphically displayed in form of constant S-N curves as shown in fig.(9). The power law formulas which describe the fatigue behaviour have good correlation factor (R^2) as given in table (4) which indicated that the experimental results are well explained by power law equation. The fatigue endurance limit of the 0.9% wt. Al_2O_3 composite into two cases without (CT) and with (CT) is taken at 10^7 cycles. . The fatigue limit at 10^7cycles increases from 83 MPa for without (CT) to 89.6 MPa with (CT) resulted in 7.95% fatigue limit improvement.

Table 4. Fatigue S-N curves Results of AA2024

Without (CT)				With (CT)		
Spe. No.	Nano wt%	Applied stress (MPa)	Nf (cycles)	Spe. No.	Applied stress (MPa)	Nf (cycles)
1			980	13	140	2600
2			1020	14		3100
3		140	1200	15		1800
4		120	35600	16	120	82800
5			38000	17		90000
6	0.9%		41800	18		78800
7		100	1008000	19		1460000
8			1120000	20	100	1825000
9			1256000	21		2120000
10		90	3680000	22	90	5680000
11			4860000	23		4860000
12			3380000	24		3980000
$\sigma_f = 205N_{fav}^{-0.056}$ $R^2 = 0.9865$				$\sigma_f = 221N_{fav}^{-0.056}$ $R^2 = 0.9764$		
Fatigue endurance limit (MPa)						
83				89.6		

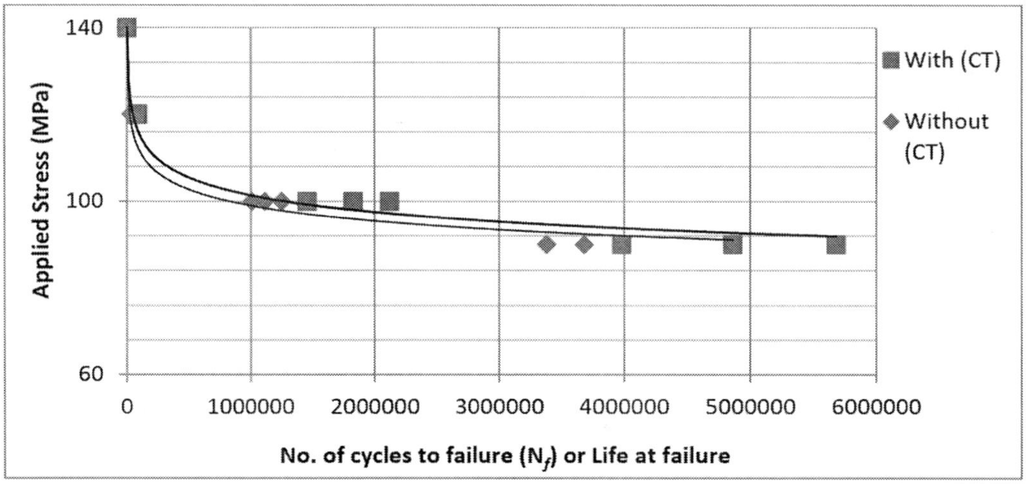

Figure 9. S-N curves results with (CT) and without (CT).

4. Conclusion

1- The effect of (CT) treatment on mechanical properties, hardness and fatigue behavior of AA 2024 matrix and nanocomposites has been examined through experimental investigation.
2- The stir casting technique was successfully adopted in the fabrication of AA 2024 -Al_2O_3 nanocomposites.
3- Mechanical properties and hardness of the AA2024-Al_2O_3 nanocomposites increased as compared to that of matrix AA2024 at room temperature.
4- The (CT) treatment show an increase in mechanical properties and hardness but the ductility of matrix and composites decrease significantly with increase the amount of Al_2O_3. The improvement in (UTS), (YS) and hardness due to (CT) treatment were increased by 9.9% ,14.63% and 14.28%, respectively while percentage elongation decrease with (CT) treatment.
5- It has been demonstrated that the (CT) treatment is an effective technique to improve the fatigue constant S-N curve of 0.9% Al_2O_3 composite.

5. References

[1] Purohit, R., Rana, R.S. and Verma, C.S., 2012. Fabrication of Al-SiCp composites through powder metallurgy process and testing of properties. *International Journal of Engineering Research and Applications*, *2*(3), pp.420-437.

[2] Hussain J.M. Al-Alkawi , Maha N. Abdulridah and Raad H. Majid,2019. Fatigue Strength of Nanocomposite Under High Temperature. *Journal of Engineering and Applied Sciences*, **14** (14), PP 4742-4746.

[3] Al-Alkawi, H.J.M., Al-Rasiaq, A.A. and Al-Jaafari, M.A.A., 2017. Mechanical Properties of 7075 Aluminum Alloy Matrix/Al2O3 Particles Reinforced Composites. *Engineering and Technology Journal*, *35*(3 Part (A) Engineering), pp.239-245.

[4] Kumar, K., Verma, D. and Kumar, S., 2014, December. Processing and tensile testing of 2024 Al matrix composite reinforced with Al_2O_3 nano-particles. In *5th International & 26th All India Manufacturing Technology, Design and Research Conference (AIMTDR 2014)*.

[5] Ashwath, P., Joel, J., Xavior, M.A. and Kumar, H.P., 2018. Effect of SiC and Al_2O_3 particles addition to AA 2900 and AA 2024 MMC's synthesized through microwave sintering. *Materials Today: Proceedings*, *5*(2), pp.7329-7336.

IOP Conf. Series: Materials Science and Engineering **765** (2020) 012052 doi:10.1088/1757-899X/765/1/012052

[6] Sajjadi, S.A., Ezatpour, H.R. and Parizi, M.T., 2012. Comparison of microstructure and mechanical properties of A356 aluminum alloy/Al_2O_3 composites fabricated by stir and compo-casting processes. *Materials & Design*, ***34***, pp.106-111.
[7] Gao, S., Wu, Z.S., Jin, P.F. and Wang, J.J., 2011. Study on microstructure and properties of 5A06 aluminum alloy welded joint by deep cryogenic treatment. *In Advanced Materials Research* (Vol. **314**, pp. 927-931). Trans Tech Publications.
[8] *International Aiioy Designation and Chemical Composition Limit for Wrought Aluminum and Wrought Aluminum*, 2017. registration recorded series, pp3.
[9] Kok, Metin. 2005. Production and mechanical properties of Al_2O_3 particle-reinforced 2024 aluminium alloy composites. *Journal of Materials Processing Technology* **16**, no. 3, pp.381-387.
[10] Bolobov, V. I., and Thanh BinhLe. 2018. Influence of deep cryogenic treatment on structure and wear resistance of materials of hydraulic breaker chisels. In IOP Conference Series: *Materials Science and Engineering*, vol. **327**, no. 4, pp. 042016. IOP Publishing.
[11] Raghavendra M.J., Praveen K., Arun R., Arjun S. 2017. A study on microstructure and characterization of aluminum 7075 metal matrix reinforced with silicon carbide particles using stir casting method. *IJRET*, Vol.**6**, issue 12.
[12] Mazahery, A., Abdizadeh, H. and Baharvandi, H.R., 2009. Development of high- performance A356/nano-Al_2O_3 composites. *Materials Science and Engineering: A*, ***518***(1-2), pp.61-64.
[13] Roylance, D., 2001. Stress-strain curves. *Massachusetts Institute of Technology study, Cambridge.*
[14] Ekin, J.W. 2006 Experimental techniques for low temperature measurements *Contemporary Physics*, ***49***(3), pp.228-229.
[15] Degrieck, J. and Van Paepegem, W., 2001. Fatigue damage modeling of fibre-reinforced composite materials. *Applied mechanics reviews*, **54**(4), pp.279-300.

Characterization of Ni-Ti-Ag Shape Memory Alloy

Khansaa D. Salman[1]

Laith J. saud[2]

Suad A. Shihab[1]

[1]Electromechanical Engineering Department, University of Technology, Baghdad, Iraq
[2] Control and Systems Engineering Department, University of Technology, Baghdad, Iraq

corresponding author's e-mail : soad.ahmdshehab@gmail.com@uotechnology.edu.iq

Abstract. In this paper, specimens of Ni-Ti shape memory alloy incorporating different percentages of Ag nanoparticles are constructed and three of its properties are studied which are density, porosity, and electrical conductivity. The specimens are constructed using powder technology. The construction process involved three main steps; mixing, compacting, and sintering. Before conducting the tests necessary to reveal the alloy properties mentioned above which represent the focus of this work, the constructed ternary Ni-Ti-Ag alloy specimens were first tested to check some of its characteristics and prove that the constructed alloy is a shape memory alloy. These tests included the Differential Scanning Calorimeter (DSC) test, the X-Ray Diffraction (XRD) test, the Scanning Electro-Microscopy (SEM) test, and the Shape Memory Effect (SME) test. The results of these tests reflected the homogeneous distribution of the Ag nanoparticles in the Ni-Ti matrix and indicated the formation of the austenite (Ti_2Ni) and martensitie (Ti 002) phases, and that the alloy has a very good SME (89.9% for the 10 wt.% percentage of Ag nanoparticles). Next to this step, porosity test and electrical resistance test were carried out. The results and related calculations for these tests showed that increasing the weight percentage of Ag nanoparticles decreases the porosity and increases the density as well as the electrical conductivity..

Keywords: Shape memory alloy, Ni-Ti matrix, Ag nanoparticles, microstructure, physical properties.

1. Introduction

Nickel –Titanium alloy has been considered for many decades as an important alloy in many industrial applications because of its shape memory effect (SME) [1]. A smart alloy is an important intelligent material (smart material) which can remember its original shape. The technical name of smart alloys is "Shape Memory Alloys (SMAs)". A SMA is defined as an alloy that exhibits special properties of pseudo elasticity and shape memory effect (SME) [2]. The Nickel–Titanium (Ni-Ti)alloy was first developed in the1960s. This alloy was named Nitinol as an acronym for the composing

IOP Conf. Series: Materials Science and Engineering **765** (2020) 012053 doi:10.1088/1757-899X/765/1/012053

elements and the location of commencing research [3].Many methods exist for preparing shape memory alloys such as: melting, heat treating, powder metallurgy, thermal spray, and thin film fabrication [4]. Powder metallurgy is considered as a new technique used to fabricate shape memory alloys. The work done so far using this technique is limited as it is new, but, there is an increasing interest in it due to its advantages [5]. Powder metallurgy is used to produce a near net shape memory alloy. There are two types of powder metallurgy to manufacture Ni-Ti SMAs. The first is raw powder sintering, while the second is alloying powder sintering. Raw powder sintering improves the homogeneity of Ni-Ti SMAs [6]. The Ni-Ti matrix got high attention in engineering applications. The addition of ternary or even quaternary elements to the binary Ni-Ti alloy is used to alter specific properties. Ag element is considered as an important element to obtain new ternary functional material for industry. Ag is to replace either Ti or Ni atoms in the Ni-Ti matrix [7].The aim of this work is to prepare an Ni-Ti-(Ag nanoparticles) shape memory alloy with improved characteristics and properties.

2. Experimental procedures

2.1 Materials Used

In the present work, Ni and Ti powders (manufactured by Sky Spring Nano Materials - USA) are used with grain size of 40 μm for Ni and 50 μm for Ti. Ag nanoparticles with grain size of 25nm is used as an alloying element to produce the ternary Ni-Ti-Ag alloy. The weigh percentages of Ag nanoparticles as well as for Ni and Ti are summarized in Table 1 for the five specimens made in this work.

Table 1: The weight percentages of powders in the made specimens.

	Wt.% of powders		
	Ag (nm)	Ni(μm)	Ti (μm)
Specimen 1	0	50	50
Specimen 2	3	50	47
Specimen 3	5	50	45
Specimen 4	7	50	43
Specimen 5	10	50	40

2.2 Preparation of the specimens

The powder mixture consists specific portions of Ni, Ti and Ag nanoparticles with 1% volume fraction of acetone. Acetone plays the role useful as a lubricant [8]. The powders were mixed by an electrical mixer with a rotating speed of about 350 rpm for 2 hrs. After the mixing process, the powder mixture was cold pressed (compacted) by uniaxial direction pressure at 600 MPa to obtain a green compact in the form of a cylindrical specimen with 10mm height and 10mm diameter.Afterwards, the green compacts were sintered at 600°C for 6 hrs. With a heating rate of 1.666°C/min. in an electrical furnace provided with argon.

2.3 Microstructure and Phases Formation Tests

After preparation of the specimens, a microstructure test for all of the specimens was carried out using Scanning Electron Microscopy (SEM). This test aimed at determining the effect of compacting pressure and sintering temperature on the microstructure of the constructed shape

IOP Conf. Series: Materials Science and Engineering **765** (2020) 012053 doi:10.1088/1757-899X/765/1/012053

memory alloy. The second test carried out was the X-Ray Diffraction test. The purpose of this test was revealing phases' formation before and after adding Ag nanoparticles. The Differential Scanning Calorimeter (DSC) test was done to obtain the starting and finishing temperatures for phases transformation. The weight of the specimens used ranges between 5 and 10 mg, and are put in aluminum crucibles. The specimens are heated between -50°C and + 200 °C with scan rate of 5°C/min. Cooling the specimens was done through pumping liquid nitrogen.

2.4 Shape Memory Effect Test

The shape memory effect for the specimens was determined by compressing the specimen by 0.06 % of its original length [9] with a compression rate of 1mm/min. This step was followed by heating the specimen in a furnace at 120 °C for 5 Mins and then cooling it in the air. The shape memory effect was calculated using the following equation [10].

SME%=(L_2-L_1)/(L_0-L_1)× 10 1

Where:
L_0: the original length of the specimen (mm).
L_1: the length of the specimen after compression (mm).
L_2: the length of the specimen after heating and cooling (mm).

2.5 Physical Properties Tests

The specimens' physical properties tests done in this work included porosity test and four-point probe resistance test. The porosity was found using the following equation [11].

$$\text{Porosity\%} = [1\text{-}(\frac{\text{actual density}}{\text{theoretical density}})] \times 100 \quad \text{.........................} \; 2$$

The actual density for each of the samples was obtained by dividing its weight by its volume, and the theoretical density was obtained using the equation 2:

$$\rho_{th} = \sum_{i=1}^{n} wt_1 * \rho_1 + wt_2 * \rho_2 + wt_3 * \rho_3 + \cdots + wt_n * \rho_n \text{.......} \; 3$$

Where:
ρ_th: Theoretical density of specimen (g/cm3).
Wt. : weight percentage (%).
ρ1, 2, 3, ..., n: density of elemental powders (g/ cm3).

$$\rho_e = R * \frac{A}{L} \quad \text{..........................} \; 4$$

$$\sigma = \frac{1}{\rho_e} \quad \text{..........................} \; 5$$

Where:
e: electrical resistivity.
R: Specimen electrical resistance.

A, L: Area and length parameters for the cylindrical specimen.
σ : Specimen electrical conductivity.

3. Results and Discussion

3.1 Microstructural analysis

The microstructure of the constructed alloy specimens, that contain different percentages of Ag nanoparticles, was achieved using SEM test. The images obtained through this test are given in figure 1 The images depict the homogeneous distribution of Ag nanoparticles in the Ni-Ti matrix with little amount of pores. Due to their low melting temperature, the Ag nanoparticle melted and filled the interstitial voids between the particles of the Ni-Ti matrix. It could be observed that Ag particles surrounded those of the Ni-Ti matrix and that there is no inter diffusion between them. This agrees with ref. [12]. The shape and distribution of the created pores depend extremely on the sintering temperature, and the coalesce goodness of Ag nanoparticles with those of Ni-Ti depends on this temperature.

Figure 1: Images obtained with SEM test for the tested specimens.

3.2 X-Ray Diffraction (XRD) analysis

The results of the XRD analysis are shown in figure 2. The results indicated the existence of austenite (Ti2Ni) and martensite (Ti 002) phases using JCPDS – Card file as shown for Ti (96-900-8518), Ag(96-9008694), Ni (96-2100647), Ti 002(96-900-8518).

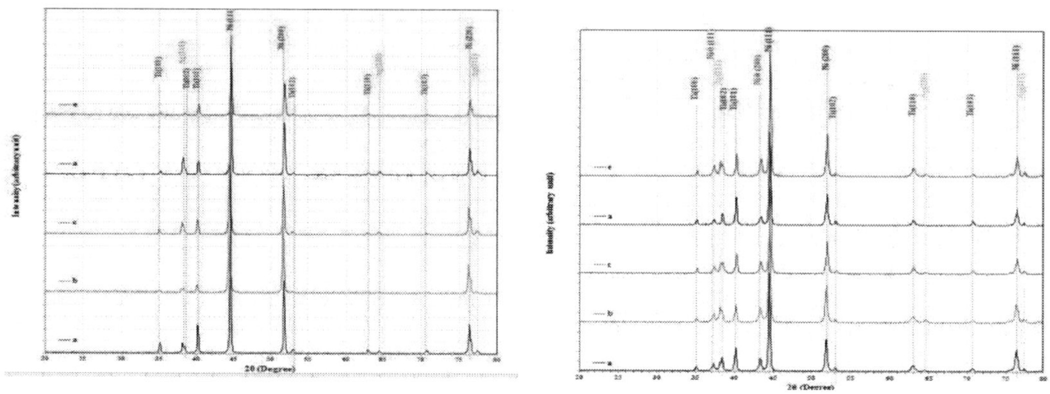

a- XRD before sintering **b- XRD after sintering**

Figure 2: X-Ray Diffraction test results.

3.3 Differential Scanning Calorimeter (DSC) test results

The DSC test was done for all of the specimens to obtain the phases transformation temperatures using DSC devic (TA-1A) made in Iran. The test was conducted at Polymer Engineering Department, (Amirkabir University of Technology/ Tehran- Iran). These temperatures represent the starting and the finishing temperatures of the austenite phase (As, Af) and the starting and the finishing temperatures of the martensite phase (Ms, Mf). The outcome of this test is given in figure 3. The results indicate the presence of peak temperatures for heating as well as cooling cases. This emphasis the transformation between austenite martensite phases. The results obtained here revealed that the constructed Ni-Ti-Ag alloy is a shape memory alloy.

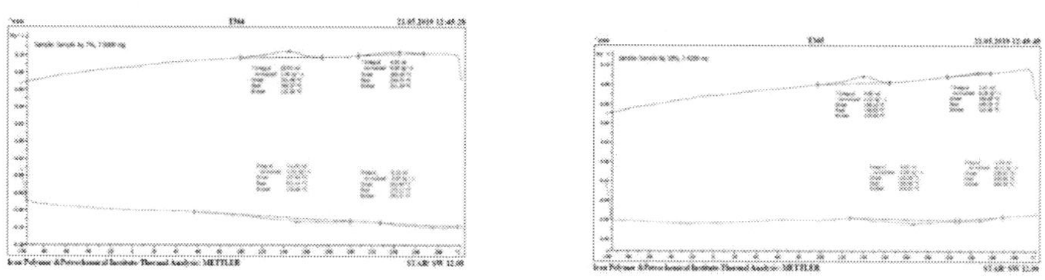

Figure 3: Result for the DSC test for the specimens with 7 and 10 wt.% of Ag nanoparticles

3-4 Shape Memory Effect

The value of SME for the specimen with 10 wt.% Ag nanoparticles was calculated according to procedure of section 2.4 and it was found to be 89.9%.

3-5 Density and porosity results

Regarding density and porosity test, the results obtained indicated that the density increases and porosity decreases with the increase of Ag percentage. The added Ag particles fill the voids between the other particles, and during compression, the applied pressure arranges the powder particles and gets them close together, and causes more contact points between them. This leads to forming localized plastic deformation and then the density increases and porosity decreases [13]. The effect of increasing Ag nanoparticles on density and porosity is depicted in the graphs of figures 4 and 5.

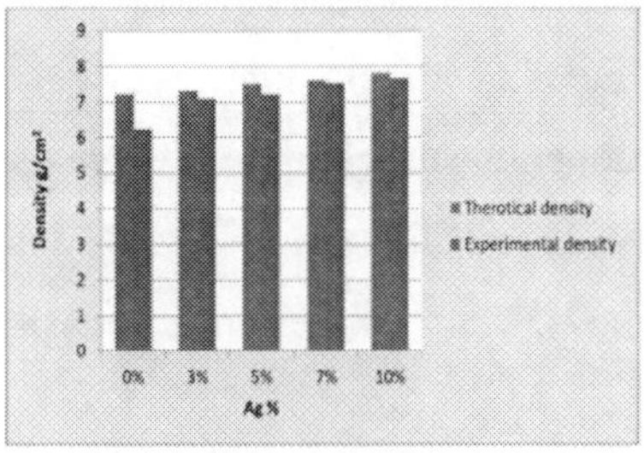

Figure 4: Density verses Ag. percentage.

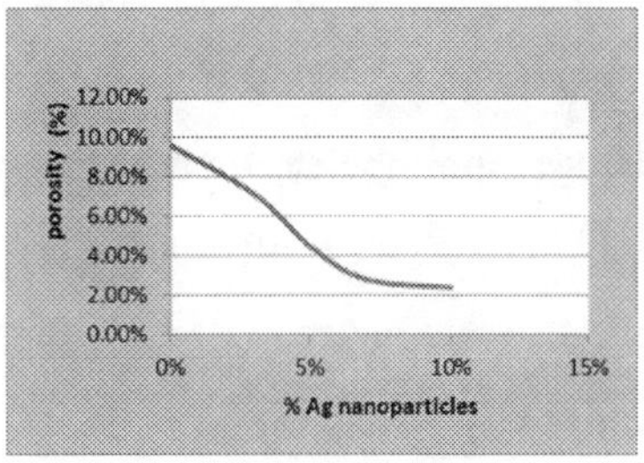

Figure 5: Porosity verses Ag percentage.

Also, after sintering necks are created between particles of the powders and afterward these necks coalesced together causing an increase in the density and a decrease in the porosity.

3-6 Electrical conductivity

The electrical conductivity for each specimen was found depending on its measured resistance using the procedure stated in section 2.4. The measurements results obtained with the values of the calculated conductivity are listed in table 2, with two charts in figures 6 and 7 being given for depicting the effect of adding the Ag part and for comparison purposes. The results indicate that the resistivity decreases, and so the conductivity increases, as the percentage of Ag in the alloy increases.

Table 2: The resistivity and conductivity values for the five alloy samples considered.

Ag percentage	Electrical Resistivity (×10-6 Ω.Cm)	Electrical Conductivity (Ω.Cm)-1
0 wt.%	80.7	12391
3 wt.%	73	15872
5 wt.%	50.7	19723
7 wt.%	43.2	22355
10 wt.%	40.5	24691

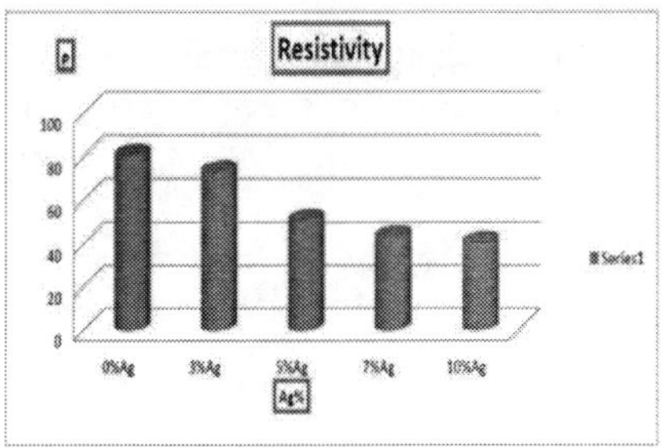

Figure 6: Resistivity versus Ag percentage chart.

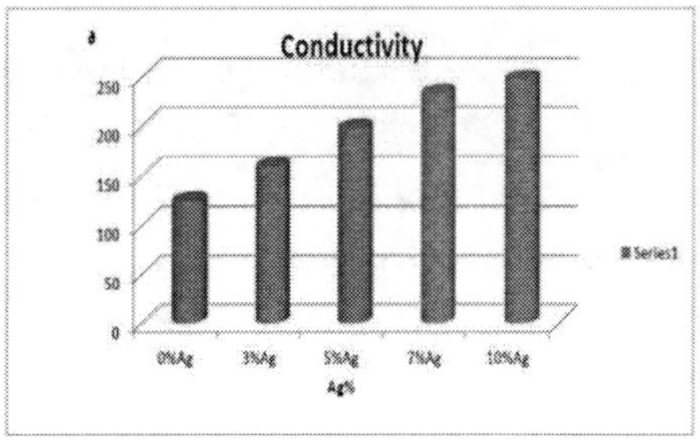

Figure 7: Conductivity versus Ag percentage chart.

IOP Conf. Series: Materials Science and Engineering **765** (2020) 012053 doi:10.1088/1757-899X/765/1/012053

Conclusions

This paper was about composing and analyzing a ternary alloy with Ni-Ti matrix as the base and Ag nanoparticles as the addend. The results of this work showed that the density and electrical conductivity increases with increasing the amount of Ag in the alloy, while the porosity decreases. The 10 wt.% percentage of Ag gave the best value for the shape memory effect which was about 89.95%. The results of scanning electron microscopy revealed the homogenous distribution of Ag nanoparticles in the Ni-Ti matrix with little amount of pores. While the X-ray diffraction results depicted the creation of Austenite phase (Ti_2Ni) and martensite phase (Ti 002). These phases are responsible for the shape memory effect.

The main contribution of this work relies in using nanoparticles of Ag powder as well as trying different Ti percentages with keeping fixed percentage of Ni. The outcomes of this study are hoped to be a contribution to the related literature and the world of SMA applications considering that the study used powder technology in construction and used nanoparticles of Ag, as well as taking into account samples with different percentages of Ag.

References

[1]K. Otsuka, X-Reu, "Physical metallurgy of Ti-Ni based shape memory alloy", Prog. Material Sci-50 Pf., pp. 511-678, 2005.

[2]Raul B. Perez Socz, Vicente Recart, Maria Lino, Oscar A. Ruano and Tose San Juan, "Advanced shape memory Alloys Processed by power Metallurgy", Advanced engineering materials, vol.2, no.1-2,2000,pp.49-53, 2000.

[3] N. M. Dawood, M. Juber, "Studied the Effect of Cooling Rate on The Phase Transformation Behavior and Hardness of Ni-Ti Shape Memory Alloy", No.(2)/Vol. 23,No.2, 2015.

[4]CN aresh, PSC Bose, CSP Rao," Shape Memory alloys: a state of art review", IOP Conf. Series: Materials Science and Engineering 012054, 149 (2016).

[5] R. M. GERMAN and K.S. CHURN, " Sintering Atmosphere Effects on the Ductility of W-Ni-Fe Heavy Metals", Metallurgical Transactions A, Vol. 15A, 1984-747.

[6] M. Bram, A. Ahmed – Khanlou , A. Heckmann , B. Fuchs, "Powder metallurgical Fabrication processes for Ni-Ti shape memory alloy parts", Article in materials science and Engineering A 337 (1-2): 254-263, 2002.

[7] Giberto H. T. Alvares da Silva, Jorge Otubo,"Desiging Ni-Ti Ag Shape Memory Alloy by Vacuum Arc Rcmelting: First practical Insights on Melting and Casting", Shape Memory. Super elasticity, ASM International 2018.

[8] Bozzone S., "Solid Oral Dosage Forms Powder Blending", 1kev Meeting, 2001.

[9] Marek Novoltny, "Shape memory alloys (SMAs)" , 2008

.

[10] Jianfeng Wan and Xing Huang ,"Effect of Nitrogen Addition on Shape Memory Characteristics of Fe-Mn-Si-Cr alloy", Material Transaction, Vol.43, pp. 920-925, 2002.

[11] J.T. Al- Haidary and Shadi Al-Khatib, "Manufacturing and characterization of dental shape memory alloy", Materials Science

[12] M. Bitzer, M.Bram, H. P. Buchxemer and D. Stover "Phase Transformation Behavior of Hot Isostatically Pressed NiTi", Journal of Materials and Engineering A,Vol. 419, pp. 45-49. 2006.

[13] Bolton W., 1988, Engineering Materials Technology, Butter worth Heinemann Oxford

.

Experimental study to increase the strength of the adhesive bond by increasing the surface area arrangement

K. A. Ameen[1]
M J Al-Dulaimi [1]
Hussain Abdulaziz Abrahem [2]
H A Hasan [1]

[1]Department of Air Conditioning and Refrigeration Engineering Technologies, Al Esraa University College
[2] Electromechanical Engineering Department, University of Technology, Baghdad, Iraq

corresponding author's e-mail : mustafa.ingenium@gmail.com

Abstract . This research represents an experimental and numerical tensile study to investigate the effect of adding epoxy and both sides of two specimens on the shear stress at which the failure of the bond occurs. Also, the effect of the of epoxy bond arrangements on the toughness and the area of contact were investigated. Fifteen specimens in three arrangement s are used for the tensile test. In the first arrangement (case 1) the epoxy bond is placed in between the steel specimens. In the second arrangement (case 2), the epoxy bond is placed between the steel specimens and it extended 5 cm outward from both sides. In the third arrangement, the epoxy bond is placed between the steel specimens and it extended 10 cm outward from both sides. These arrangements used to increase the contact area of the bond. The single Lap – Simple bond type is used for this study. The experimental and numerical results show a god agreement. The arrangement of the epoxy bond is found to have a strong impact on the mechanical behavior. The shear stress increase by 500 % in case of arrangement 3 .

Keywords: increasing strength, Adhesive bond, Epoxy , surface area

1. Introduction

Over the last decades , composite materials have been used significantly in all engineering and became a dominant class of materials for wide range of applications , like spacecrafts and building structure . Also the military industries have been using composite material widely for high performance applications where the cost is secondary . Technological developments in composite materials have been accompanied with development in adhesive bonds . The bonded joints have become a complementary of the traditional mechanical fasteners in composite materials and metallic structure . The use of adhesively bonded joints are very common in such structures due to the longer service life , improved load distribution , lower machining cost and less complexity . in addition to that, the recent

IOP Conf. Series: Materials Science and Engineering **765** (2020) 012054 doi:10.1088/1757-899X/765/1/012054

advances in the filed of epoxies and adhesives lead to higher temperature sustainable adhesives which increases the integrity of the structure under severe operating environment .

There are many types of the adhesive bonds . Focusing on our study , the adhesive bonds can be classified according to their configurations into four groups

1.1 Single Lap

This configuration is very simple and active. There are three types of it (Simple, Beveled and Radlused) as shown in figure (1-1a). The single Lap – Simple type is chosen for this study.

1.2 Double Lap

This configuration has three types (Simple, Beveled and Radlused) as shown n figure (1-1b).

1.3 Scarf

(Single Taper, Double Taper and Increased Thickness) as shown in figure (1-1c),

1.4 Strip

This configuration has three types (Single Strip, Double strip and Beveled Strip). as shown in figure (1- 1d),

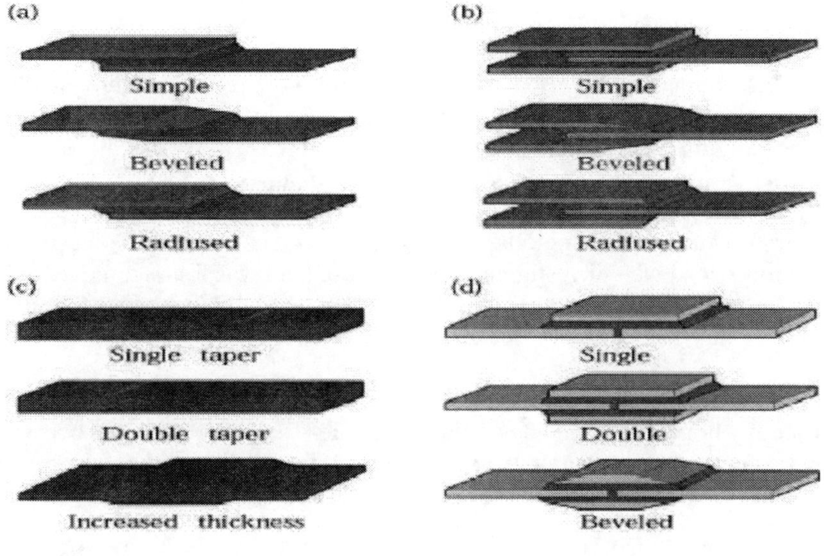

Figure 1 . Configurations of adhesive bonds

IOP Conf. Series: Materials Science and Engineering **765** (2020) 012054 doi:10.1088/1757-899X/765/1/012054

2. Literature Review

In recent years , more attention has been paid to the use of adhesive – bonded joints. The use of adhesive bonding offers additional advantages over the traditional jointing methods .Theses advantages are fatigue and corrosion resistance , lighter weight , crack retardation , good damping and cost saving [1].
Mechanical performance of adhesive bonded joints depends on a number of parameters such as: material properties and joint geometry (adhesive thickness, adherend thickness), surface roughness etc.

Many experiments have been conducted to investigate the relation between the adhesive layer thickness and its shear strength [2].Numerous experiments have been conducted to determine the strength and the fracture behavior of material pieces adhesively jointed for different bond geometries . These experiments investigated the effect of bond geometry on the crack path propagation and the assessment of fracture initiation criteria [3]. In another work , a simplified model has been developed to evaluate the stress distribution in the adhesive layer of double -lap joint (DLJ) under a cyclic tensile lap- shear load .The adhesive material was assumed to follow the constitutive relation of elastic perfectly shear stress -strain [4] . The effect of adhesive bond thickness on the mechanical behavior of polyurethane adhesive for automotive applications [5] . The effect of adhesive bond thickness on the fractural toughness was tested for thickness range from 0.2 to 2 mm .specimens were fabricated from hard tool steel . The fractural toughness was found to increase with thickness of the adhesive bond . It increases by 20% for thickness between 1 and 2 mm and linearly between 0.2 and 1 mm . Another experimental work was conducted to investigate the effect of surface roughness of different adhered sheets on the strength of the adhesive bond[6] .The adhered sheets were from wood and Aluminum. Sand papers and emery papers were used to obtain different degree of roughness sheets . The results revealed that there is a direct effect of roughness on the strength of the adhesive bond .For the wood , the adhesive bond strength was found to decrease with the increase of the roughness. For the Aluminum the roughness has an optimum value because the strength of adhesive bond increase first and then tend to decrease with the roughness increase . Measurements were conducted to investigate the effect of specimen geometry on strength of single lap adhesive bond . The effects of adhesive bond length and thickness on the failure load were investigated . The failure load was found to increase linearly with the increase of bond length for bond thickness of 1.5 mm . The bond thickness was found to have a variant effect on the failure load . The optimum thickness was 0.5 mm .

The works from literature reveal that there is not enough contribution in the field of studying the effect of the epoxy bond arrangements on the mechanical behavior. The present work represents an experimental and numerical investigation to examine the effect of adhesive bond length and arrangement in term of the contact area on the shear stress and toughness.

3. Experimental Method

In the current study, specimens used for the tests are made from Steel and with one type of epoxy. The chemical composition for steel used in this study as in the table (1).

Table 1. The chemical composition for steel

Symbol	Element	(%) Concentration	Symbol	Element	(%) Concentration
Mg	Magnesium	0.023	Cu	Copper	**0.4061**
Al	Aluminum	0.0053	Zn	Zinc	**0.0063**
Si	Silicon	0.0019	As	Arsenic	**0.01143**
P	Phosphorus	0.0012	Zr	Zirconium	**0.05**
S	Sulfur	0.002	Nb	Niobium	**0.0013**
Ti	Titanium	0.0019	Mo	Molybdenum	**0.124**
V	Vanadium	0.0016	Ag	Sliver	**0.001**
Cr	Chromium	0.1776	Cd	Cadmium	**0.00069**
Mn	Manganese	0.678	Sn	Tin	**0.001**
Fe	Iron	97.13	Sb	Antimony	**0.0012**
Co	Cobalt	0.296	W	Tungsten	**0.0075**
Ni	Nickel	0.1639	Pb	Lead	**0.0014**
C	**Carbon**	**1**			

The Epoxy used in this study is a low strength adhesive paste. When fully cured, the adhesive has good chemical resistance and good sustainability e at high temperatures. It is suitable for bonding a wide variety of metals, ceramics, glass rubbers, rigid plastics and other materials and is widely used in many industrial applications where resistance normal environments.

The properties of the materials used in this study are shown in table (2). The properties of Steel and adhesive were obtained with web material databases.

Table 2. Material properties at room temperature

Material Properties	Adhrends Steel	Adhesives Epoxy
(Ultimate tensile strength (MPa	1000*690	60
(Ultimate shear strength (MPa	1000*80	50
(Young's modulus (MPa	1000*200	12

IOP Conf. Series: Materials Science and Engineering **765** (2020) 012054 doi:10.1088/1757-899X/765/1/012054

3.1. Specimen

Three specimens of different arrangements were used in this study for tensile test as shown in figure. The number of specimens used in this study (5) for each configuration as shown in figure 3. The test device used in this study is universal test device as shown in figure 4.

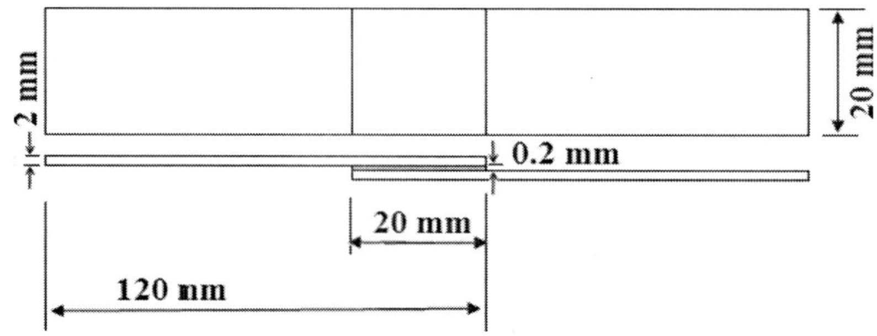

(a)

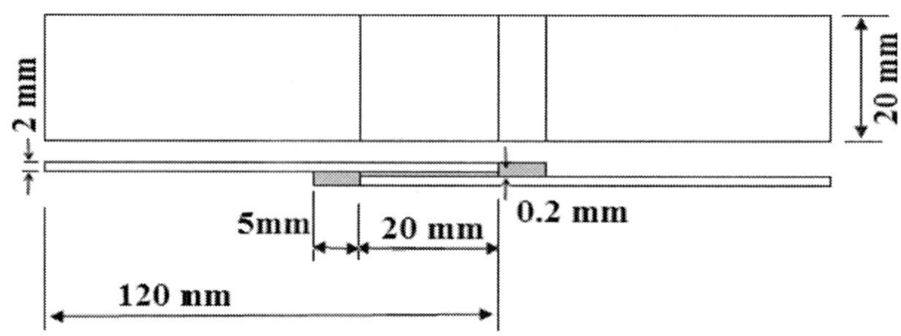

(b)

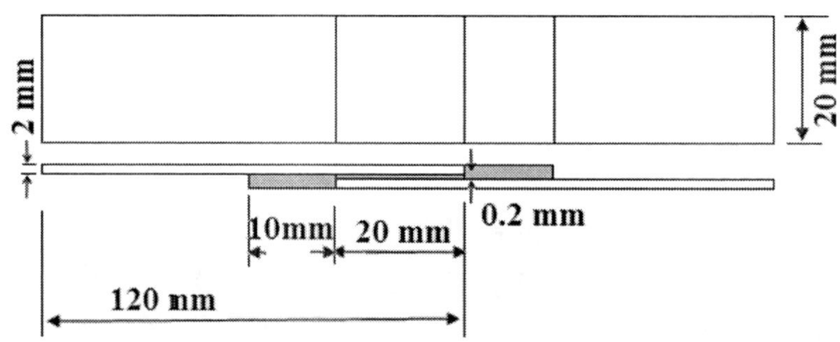

(c)

Figure 2. Specimens arrangements

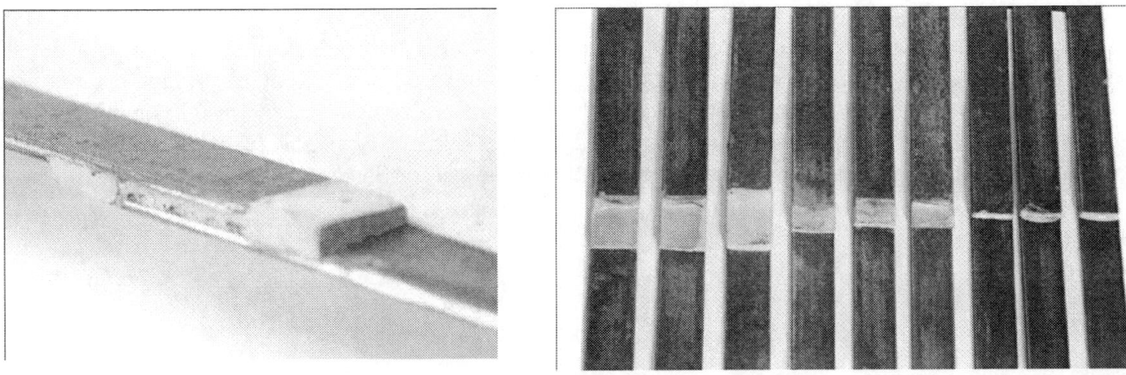

Figure 3. Specimens used in this study

Figure 4. Universal test device

IOP Conf. Series: Materials Science and Engineering **765** (2020) 012054 doi:10.1088/1757-899X/765/1/012054

4. Numerical Work

The two dimension problem is analyzed by the finite element method using the APDL program ANSYS. In this study the specimens is represented by plane element (Plane182), this element used for planes stress structure model. As shown in Figure (5). The element has two degree of freedom, all of them translations (u and v). After defining the element type, the material properties must be defined. The material properties used are listed in Table (2) for Steel and Epoxy.

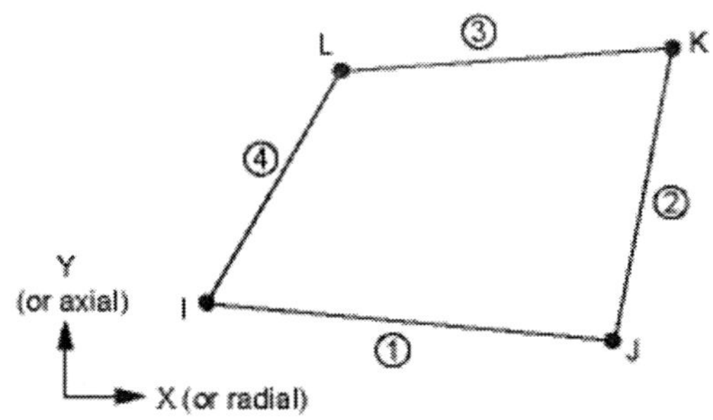

Figure 5. Plane182 element

After limited the element type and the material properties, used the rectangular element and mesh the all areas of specimen (the steel and epoxy) as shown in Figure (6). For two case.

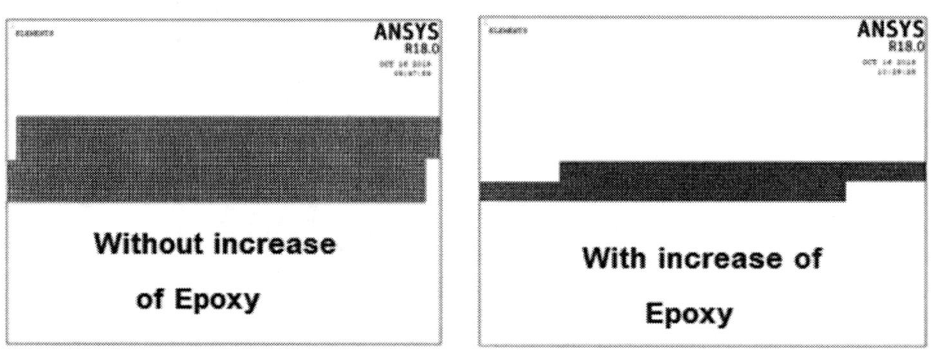

Figure 6. Meshing of the Specimens

5. Result and discussions

Results from experimental and numerical investigation of the effect of epoxy bond arrangement on the shear stress , toughness and elastic region are presented .

Figure 7 represent a comparison of the shear stress from experimental and numerical results from ANSYS. A god agreement can be seen between the numerical and experimental results since the average error is less than 17 % . The shear stress increase as the contact area of the epoxy increases due different arrangements . Figures 8 and 9 show the contours of stress distribution for arrangement 1 and 2 .

Also, the contact area was found to play dominant effect on the toughness. The toughness was found to increase with the area of contact as shown in Figure 10.The toughness increased by 5.7 times in case of arrangement. 3 as compared with arrangement 1 . The relation between elastic region and area of contact is shown in Figure 11. The elastic region changes non linearly with the area of contact

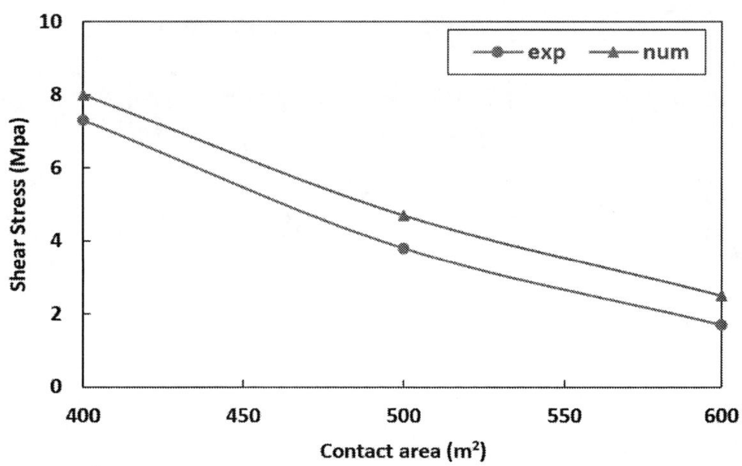

Figure 7. The Comparison between the Experimental and numerical results

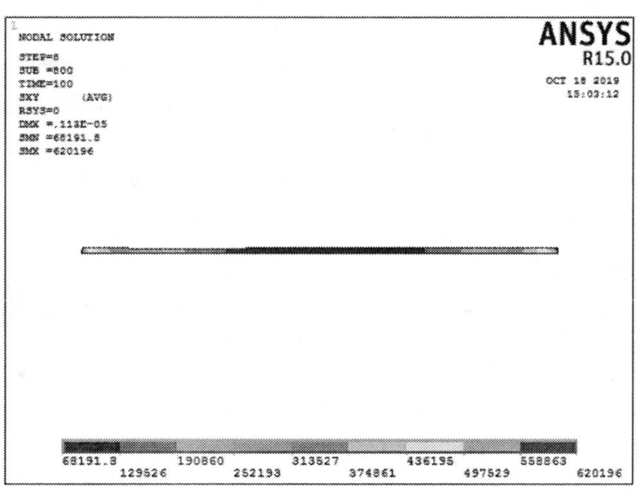

Figure 8. The Shear Stress Dis. On the resin (arrangement 1)

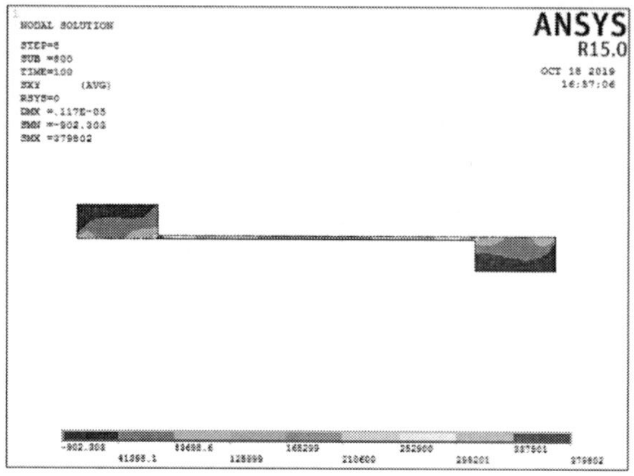

Figure 9. The Shear Stress Dis. On the resin (arrangement 2)

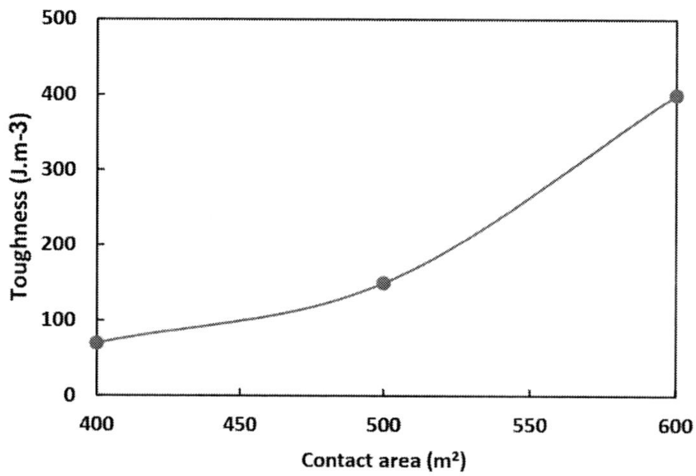

Figure 10. The relation between the Toughness and contact area

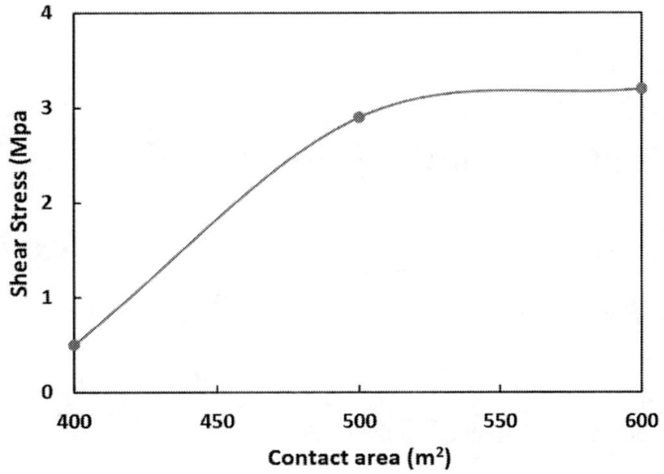

Figure 11. The relation between contact area and elastic region

6. CONCLUSIONS

The present work is an experimental tensile study of the effect of epoxy bond contact area on the shear stress of the failure, the toughness and the elastic region of the bond. Fifteen specimens of three arrangements were used of tests. It was found that the contact region has a significant influence on the shear stress, toughness and elastic region.

References

[1] J. Yeon, Y. Song, K. K. Kim and J. Kang, Materials Journal, 27 July 2019.

[2] M. Afendi, Thesis, U.V. of Tsukuba, July 2011.

[3] Da Silva M, Adams D, Int. J. Adth Adhes, 22727, 2007 .

[4] Cheuk P, Tong L, Hwang C, Baker A, Chalkley P, 2002. *.109 57 .Comp. Struct*

[5] M. Banea , L Silva , C Raul , 2014. Journal of *.91 .Adhes*

[6] S. Budhe, A. Ghumatkar, N. Birajdar and M.D. Banea., Springerm Applied Adhesion Science, Dec. 2015.

[7] A. Ghumatkar, S. Budhe, R.Sekhar, M.D. Banea, S. de Barros., Lat. Am. J. solids struct., Vol. 13, No.,13, Rio de Janeiro, Dec. 2016.

[8] M. Muller, P. Hrabe, R. Choteborsky, D. Herak., Res. Agr. Eng., 52, Vol. *.2006 ,(1)*

IOP Conf. Series: Materials Science and Engineering **765** (2020) 012055 doi:10.1088/1757-899X/765/1/012055

Evaluation the Sensitivity of Bone Natural Frequency as a Diagnosis Tool to Identify Bones Integrity

Shaker S Hassan[1]

Ameen M Al Juboori[2]

[1]Mechanical Engineering Dep., Technology University, Baghdad, Iraq

[2]Biomedical Engineering Dep., AL Mustaqbal University Collage, Babil, Iraq

Abstract. This study explain the reason behind the low sensitivity of the natural frequency for the changes in bone health. The mechanical properties of the bone were reduced by placing the bone in boiling water for 2 hours and the mass of the bone was reduced by placing the bone in diluted acidic medium for 24 hours. The natural frequency of the bone samples (10 samples) was calculated in three cases (fresh, after boiling, and after acidic treatment). The natural frequency was calculated theoretically by using finite element analysis based upon CT-Scan DICOME images and experimentally by using impulse test. The difference between the results in the two methods was less than 15%. The boiling process caused a reduction in the natural frequency while the acidic treatment caused an increase in the natural frequency compared with the natural frequency of the fresh condition. The mutual influence of these two component cannot be separated. This mutual influence will reduce the sensitivity of the natural frequency as diagnosis factor for the changes in the health condition of the bone.

Keywords: Bone, finite element, modal test, natural frequency.

1. Introduction

Evaluation of bone integrity requires certain diagnosis procedures that are mostly radioactive, very expensive, and time consuming. The investigation of alternative diagnosis procedure was the main concern for the researchers. Several fields of research were explored to invent nonradioactive test device and frequency analysis was one of those fields of study. The natural frequency was the first factor to be investigated for the used as diagnosis tool to identify the physiological health condition of the bone. Natural frequency evaluated by many researchers by different methods for animal and human bones in two ways (in-vitro and in-vivo).

This field of study started in the early 1970, when Thompson et al. [1] conducted several experiments on dog radii to extract the resonant frequency. The aim of work was to investigate the mechanical properties of the bone by taking into consideration the effect of twist in the principle axis,

IOP Conf. Series: Materials Science and Engineering **765** (2020) 012055 doi:10.1088/1757-899X/765/1/012055

shear deformation, and viscoelastic behavior. Jurist and Kianian [2] calculated the resonant frequency of the ulna bone. Three models were used to illustrate the effect of the boundary condition, skin, and accelerometer weight effect. William P. Doherty et al. [3] investigated the efficiency of the resonant frequency as an evaluation factor for the bone integrity. The results showed that the resonant frequency sensitivity to the changes occur in the bone is lower than that for the mass and stiffness. Thompson et al. [4] presented an in-vivo experimental study to investigate the resonant frequency of ulna bone. The arising problem in the in vivo test is the effect of the soft tissue, which can influence the results due to the heavy damping. This damping will require higher energy load waves to insure the excitation of the system, but this increase in the energy will involve pain and discomfort to the patient. Viano et al. [5] presented a theoretical and experimental study for the human femur to investigate the elastic property and bone mineral density change due to aging. Timothy K. Hight [6] investigated the natural frequency of the tibia bone by using finite element model to calculate the natural frequency by taking into consideration the influence of irregularity in shape and different boundary conditions at proximal and distal ends. Kahlil et al. [7] investigate the resonant frequency of the human femur bone for different modes of shape (transverse in the maximum and minimum inertia plane, torsional, and longitudinal). G. Vander Perre et al. [8] conducted in vivo and in vitro tests to evaluate the structural dynamics properties of the bone (natural frequency, mode shape, and damping ration). P. Cornelissen et al. [9] conducted a study on the effect of the soft tissues and joints on the structural dynamic properties of human tibia. The results showed that soft tissue has almost the same influence on the vibration results in both methods. Cornelissen, M. et al. [10] conducted a study focusing on the effect of the boundary conditions on the calculated natural frequency of human tibia. J. C. Misra and S. C. Samanta [11] investigated a mathematical model for the transverse vibration of the human tibia. G. Van Der Perre et al. [12] presented a study to evaluate the mechanical properties in group of patient infected by osteoporosis. Béatrice Couteau [13] studied the frequency analysis of human femur bone taking into consideration the material properties and the geometrical explicitly. Pinar Arpinar et al. [14] presented a study to evaluate the bone strength by using vibration analysis and dual energy x-ray absorptiometry Bekir Bediz [15] conducted an in vivo and in vitro study to evaluate the structural dynamic properties change of human tibia by using vibration analysis. G. Campoli et al. [16] conducted a study to explain the effect of bone shape and density distribution on the resonant frequency of the bones. All the studies conducted by using the frequency analysis provides important information on the factors influencing the values of the natural frequency for different modes of vibration. The behavior of the natural frequency change due to the effect of bone diseases was the main goal of investigation to invent a new diagnosis procedure to distinguish between the normal and abnormal health conditions.

2. Finite Element Analysis

Bones have a very complicated shape and building 3-dimensional modal is not easy by using traditional ways. In addition, mechanical properties and density values change along the bone. In order to overcome those problems CT-Scan DICOME (Digital Imaging and Communications in Medicine) were used to build the three-dimensional modal (Figure 1). Those images has the size of 512×512 pixels. CT-scan device type GE Medical Systems\Discovery CT 750 HD was used and the images were taken with the same setting (KVP 120, slice thickness 2.5 mm, and tube current 120 mA).

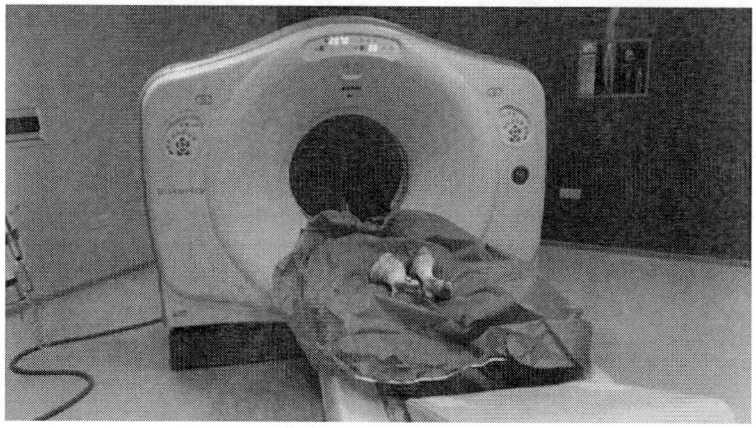

Figure 1: CT-Scan test for the bones.

The DICOME images was used to build the three-dimensional modal by using MIMICS software and the meshed modal produced by using TET-4 elements by using 3-matic software. The material assignment was done by using MIMICS software again based upon the gray-scale value for each pixel (Figure 2). The modal analysis for the meshed modal was solved by using ANSYS software version 14.5. The natural frequency for different modes of vibration were evaluated (Figure 3). This analysis will be repeated in three cases fresh, after boiling, and after acidic treatment.

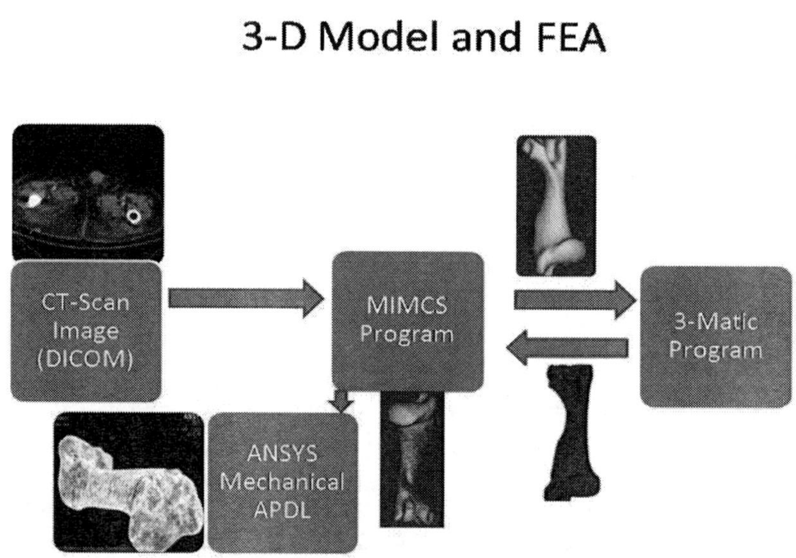

Figure 2: Flow chart for the construction of the FEA model.

IOP Conf. Series: Materials Science and Engineering **765** (2020) 012055 doi:10.1088/1757-899X/765/1/012055

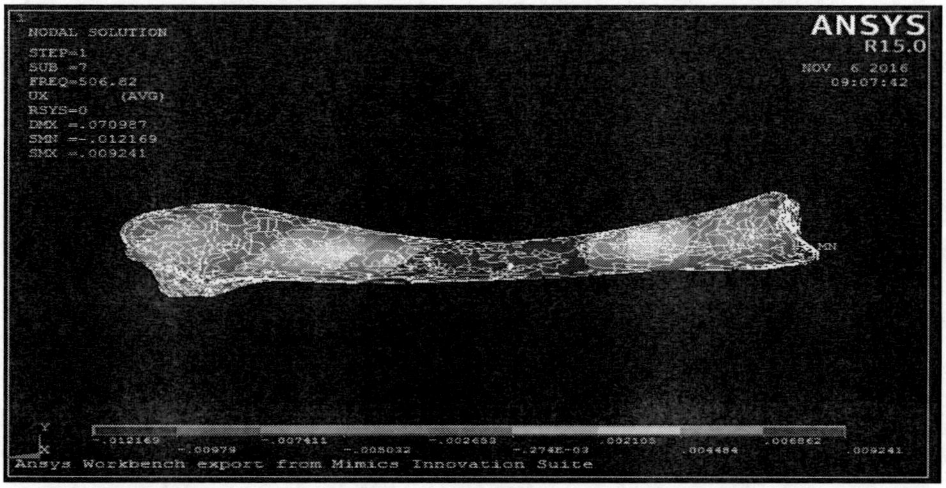

Figure 3: Modal analysis solution by using ANSYS software version 14.5.

3. Material and Experimental Analysis

In this study, two bovine bones were selected (Humerus and Tibia). These bones are different in length and weight. They were collected from different animals. The bones were subjected to severe condition to produce an influence on their organic and inorganic materials to reduce its stiffness and mass.

Viscoelastic behavior is exhibited due to the presence of the collagen fibers [17]. Therefore, the bone is placed inside boiling water to affect the organic materials (collagen) inside the bone matrix without influencing the inorganic materials. This treatment will disturb the bone matrix, and will results erosive pits on the surface, the protein will be lost, crystallizing increased, porosity increased as result of collagen lost [18]. The bones were submerged into boiling water for three hours.

In order to simulate the calcium efflux from the bones, the bone will be placed inside diluted sulfuric acid. Sulfuric acid will work as reducing agent and react with the salts (mainly with the calcium). This chemical reaction will significantly reduce the calcium content and the mass of the bone will be reduced [19].

The natural frequency will be extracted experimentally by using impulse test. This test is conducted by using impulse hammer, accelerometer, and dynamic signal analyzer. The bone was placed on soft sponge in order to simulate the free-free boundary conditions. Figure 4 shows the experiment setup. Impact hammer was used to excite the bone by tapping along the bone and in different locations. Minimum excitation force was 140 N. Although the duration of the excitation force was 0.25 s, it provides enough energy to excite the system. The excitation was recorded by channel 1, and the response was recorded by using channel 2. Both of the excitation and the response were stored in the software for further processing (FRF). This experiment was repeated for all the bones samples in different conditions (fresh, boiled, and sulfuric acid treatments).

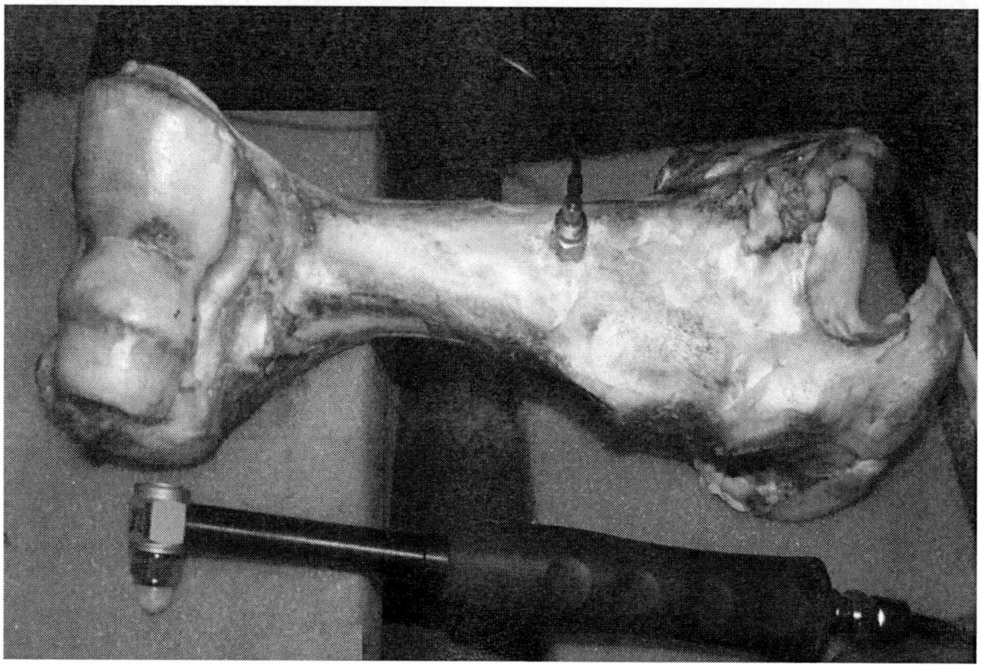

Figure 4: The modal test For Fresh bone.

4. Result and discussions

Tables (2 & 3) illustrate the natural frequencies of the bones (Humerus & Tibia) before and after the treatments. These tables demonstrate the results obtained from finite element analysis and experimental part, discrepancy between the results obtained from the two methods, and percentage of decrease in the natural frequency due to boiling treatment, and percentage of increase in the natural frequency due to the acidic treatment effect.

Table 1: Natural frequency (Hz) of Humerus.

No	Fresh			Boiled			%↓	H2SO4			%↑
	Exp	**FEA**	**Disc**	**Exp**	**FEA**	**Disc**		**Exp**	**FEA**	**Disc**	
1	941	1090	14.2%	911	939	-3.0%	2.6%	1052	1131	-9.2%	10.5%
2	5365	5549	3.5%	5120	5022	2.4%	4.3%	5536	5355	3.5%	3.3%
3	5654	5938	4.9%	5485	5272	3.7%	3.0%	6210	5575	10.0%	8.9%
4	6022	6110	1.4%	5713	5629	1.5%	5.3%	6684	6465	3.2%	9.7%
5	6734	6733	-0.1%	6436	6083	5.4%	4.5%	7056	7062	-0.2%	4.5%
6	7001	7210	2.8%	6753	6871	-1.9%	3.7%	7372	8075	-9.5%	5.0%
7	7110	7264	2.0%	6953	7440	-6.8%	2.2%	7556	8427	-11.1%	6.0%
8	7953	8040	1.2%	7730	8112	-4.7%	2.6%	8469	9286	-9.6%	6.2%

IOP Conf. Series: Materials Science and Engineering **765** (2020) 012055 doi:10.1088/1757-899X/765/1/012055

Table 2: Natural frequency (Hz) of Tibia Bovine bone.

No	Fresh			Boiled			%↓	H2SO4			%↑
	Exp	FEA	Disc	Exp	FEA	Disc		Exp	FEA	Disc	
1	721	789	8.3%	671	681	-1.0%	3.5%	759	668	-13.3%	4.8%
2	4152	4257	2.5%	3712	3767	1.5%	10.6%	4662	4248	-9.8%	11.1%
3	5372	5183	-3.7%	5245	5327	1.6%	2.3%	5519	5462	-1.0%	2.6%
4	6052	6151	1.6%	5345	5610	4.8%	11.7%	6595	6770	2.7%	8.1%
5	7274	7323	0.6%	6862	6688	-2.7%	5.7%	7475	8262	9.7%	2.6%
6	7981	8205	2.8%	7671	7412	-3.6%	4.0%	8500	8612	1.4%	6.0%
7	8833	8415	-5.1%	8140	8265	1.8%	8.0%	9462	8873	-6.8%	6.7%

The agreement between the finite element analysis and the experimental results is good (less than 15%). The tables above show clearly that the natural frequency changes after boiling. The natural frequency value decreased in all samples after treatment. The amount of this decrease is almost the same for the two samples (200-300 Hz). This converging can be related to stability of treatment condition for all samples. The natural frequencies of the bone after the acidic treatments are higher than the fresh bone natural frequency. The increase in natural frequency is between 400-700 Hz.

The natural frequency of the bones after the acidosis treatment is higher than the boiling treatment natural frequency. The increase is in the interval 700-1000 Hz. A conclusion can be achieved from these results the calcium efflux has a greater impact on the natural frequency than the collagen transformation.

Although the finite element analysis provides us with the ability to investigate unlimited number of modes of vibrations, but the restriction was in the experimental part of the work. Modes of vibration, which were extracted from the experimental work, are compared with their counterparts in the finite element analysis results. Some frequencies were dropped, because it was difficult to capture their counterpart in one of the three cases (fresh, boiled, and acidosis treatment).

5. CONCLUSIONS

The sensitivity of the natural frequency to the changes in the bone health conditions was relatively low. The explanation done by the researcher relate the low sensitivity to the mass and stiffness change. In this study, the bone stiffness was decreased by disorder the organic materials inside the bone matrix. This decrease in the stiffness results a decrease in the natural frequency of the bone. The mass of the bone was decreased by the chemical reaction and this decrease results an increase in the natural frequency of the bone. The impact of the mass on the natural frequency is greater than the stiffness and the bone length has a minor effect on the bone natural frequency. The influence of these two factors (stiffness and mass) cannot be separated in the real life and the superposition of their opposite effect will reduce the change in the natural frequency when the bone health condition of the bone is changed.

IOP Conf. Series: Materials Science and Engineering **765** (2020) 012055 doi:10.1088/1757-899X/765/1/012055

References

[1] Delgado Thompson G. A. Anliker M. and Young D.R. "Determination of Mechanical Properties of Excised Dog Radii from Lateral Vibration Experiment". 1970, J.R. CAMERON (ED) CONF-700515 (Chicago), Oak Ridge: USA.

[2] Jurist J. M. and Kianian. "Three Models of Vibration Ulna". Journal of Biomechanics 1973, vol 6, pp 331-342.

[3] William P. Doherty, Edwin G. Bovill, Edward L. Wilson. "Evaluation of use resonance frequency to characterize the physical properties of human long bones". Journal of Biomechanics, November 1974, Vol 7, Issue 6, pp 559-561.

[4] Gerald A. Thomson, Donald R. Young, and David Orne. "In Vivo Determination of Mechanical Properties of Human Ulna by Means of Mechanical Impedance Test: Experimental results and Improved Mathematical Model". Journal of Medical and Biological Engineering, 1976, vol 14, pp 253-262.

[5] D. Viano, U. Helfenstein, M. Anliker, and Ruegsegger. "Elastic properties of Cortical Bone in Female Human Femur". Journal of Biomechanics, 1976, vol 9, pp 703-710.

[6] T. K. Hight. "Natural frequency analysis of a human tibia". Journal of Biomechanics, 1980, vol 13, issue 2, pp 139-146.

[7] T. B. Khalil, D. C. Viano, L. A. Taber. "Vibrational Characteristics of the Embalmed Human Femur". Journal of sound and Vibration, 1981, vol 75, issue 3, pp 417-436.

[8] G. Vander Perre, R. Van Audekercke, M. Martens and J. C. Mulier, "Identification of in Vivo Vibration Modes of Human Tibia by Modal Analysis", ASME Journal of Biomechanical Engineering, 1983, vol 105, pp 244-248.

[9] P. Cornelissen et al. "Assessment of tibial stiffness by vibration testing in situ—II. Influence of soft tissues, joints and fibula". Journal of Biomechanics, 1986, vol 19, issue (7), pp 551-561.

[10] Cornelissen, M., Cornelissen, P., Van der Perre, G., Christensen, A. B., Ammitzboll, F., Dyrbye, C., "Assessment of Tibial Stiffness by Vibration Testing in Situ – I. Identification of Mode Shapes in Different supporting Conditions", Journal of Biomechanics, 1987, v. 20, pp. 333-342.

[11] J. C. Misra and S. C. Samanta. "A MATHEMATICAL ANALYSIS OF THE Vibration Characteristics of Human Tibia". Computers & Mathematics with Applications, Elsevier, 1988, vol 16, no 12, pp 1017-1026.

[12] G. Van Der Perre and G. Lowet. "In Vivo Assessment of Bone Mechanical Properties by Vibration and Ultrasonic Wave Propagation Analysis". Elsevier Science Inc. 1996, Bon, vol 1, Supplement, pp 29S-35S.

[13] Béatrice Couteau, Marie-Christine Habotho, Robert Darmana, Jean-Claude Brignola, Jean-Yves Arlaud. "Finite element modelling of the vibrational behavior of the human femur using CT-based individualized geometrical and material properties". Journal of Biomechanics, 1998, vol 31, pp 383-386.

[14] Pinar Arpinar, Beyza Simsek, Onder Cem Sezgin, Gulin Birlik, and Feza Korkusuz. "Correlation between mechanical vibration analysis and dual energy X-ray absorptiometry (DXA) in the measurement of in vivo human tibial bone strength". Technology and Health Care 13 (2005) 107–113.

[15] Bekir Bediz, H. Nevzat Ozguven, Feza Korkusuz. "Vibration measurements predict the mechanical properties of human tibia". Clinical Biomechanics 25 (2010) 365–371.

[16] G. Campoli, N.Baka, B.L.Kaptein, E.R.Valstar, S.Zachow, H. Weinans, A.A.Zadpoor. "Relationship between the shape and density distribution of the femur and its natural frequencies of vibration". Journal of Biomechanics 47 (2014)3334–3343.

[17] S. Viguet-Carrin, P. Garnero and P. D. Delmas, "The role of collagen in bone strength". Osteoporosis International, vol. 17, pp. 319-336, 2006.

[18] M. Todoh, S. Tandano, Y. Imari. "Effect of Heat Denaturation of Collagen Matrix on Bone Strength". 13th International Conference on Biomedical Engineering, Volume 23 of the series IFMBE Proceedings pp 2034-2037, 2008.

[19] Sascha Kopic, John P. Geibel, Gastric Acid, Calcium Absorption, and their impact on Bone Health. Physiological Reviews, Vol. 93 no. 1, 189-268, 2013.

IOP Conf. Series: Materials Science and Engineering **765** (2020) 012056 doi:10.1088/1757-899X/765/1/012056

Compression strength, dielectric and magnetic properties of new aluminium matrix hybrid nanocomposites

Hussain A. Aljawad [1], H. J. M. Alalkawi [2] and Ghada Adel Aziz [3]
Department of Electromechanical Engineering, University of Technology, Baghdad, Iraq
[1]hussein94aljawad@gmail.com
[2]alalkawi2012@yahoo.com
[3]50070@uotechnology.edu.iq

Abstract. Nanocomposites are materials fabricated from two or more materials with different mechanical and electrical properties. Combining these materials produces a new designed material with new and better properties compared to the individual components. In recent years nanocomposites have been developed and employed almost in all industries. The current study deals with fabricating a hybrid nanocomposite (when there are a minimum of three materials, the composite is called as hybrid composite). Pure aluminium as the base matrix. Iron oxide Fe_2O_3 (alpha) and aluminium oxide Al_2O_3 (alpha alumina). Fe_2O_3 weight percentage (wt.%) is varied (1.5 , 2.5 and 5 % by weight) and the wt % of Al_2O_3 is held constant (2 wt.%). The new designed nanocomposite was produced using Powder Metallurgy (P/M) method. This method has been widely used for fabricating aluminium matrix nanocomposites (AMNCs), due to it is low costs and gives high accuracy as well as the ease of using. The matrix used was aluminium powder with an average particle size of (60μm) with 99.6% purity and Fe_2O_3 (99% purity and 30 nm particle size) and Al_2O_3 with (99.5% purity and 14-20 nm particle size). The experimental results revealed that the microstructure images of composites showed uniformly distributed of Fe_2O_3 and Al_2O_3 in aluminium matrix. The maximum compressive strength is 152 MPa in nanocomposite containing (1.5 Fe_2O_3 + 2 Al_2O_3) wt.%. The improvement percentage was recorded to be 30% for CS. The real and imaginary components of relative permittivity decrease with increase of frequency. The hysteresis curve of nanocomposites showed that the magnetic properties were improved for all the nanocomposites but the better response of magnetic properties was found with the nanocomposite of (1.5% Fe_2O_3 + 2% Al_2O_3) content .

1.Introduction

Aluminium metal matrix nanocomposites (AMMNCs) are the most common in various industries such as automotive, aerospace, electronics and power transmission lines, because of the high ratio of strength to weight and good electrical properties. The main properties on which the applications of aluminium is based are the high mechanical strength, low density and relatively good electrical conductivity [1].

A. Tan, et al. [2] (2016) , investigated the effect of the hybrid reinforced materials (micro SiC + nano TiB_2) with various content on the microstructure and mechanical properties of the aluminium using powder metallurgy method. The experimental results revealed that the mechanical properties are improved with increase in content of TiB_2 nanoparticles reinforcement. The best improvement was observed in composite including (5 TiB_2 + 10 SiC wt%). Ferreira, et al. [3, 4] (2016 , 2018), designed and fabricated Al. alloy (AMCs) reinforced with different percentages of nano-Fe_3O_4

magnetic particles using powder metallurgy method. Microstructural, electromagnetic and mechanical properties were studied. Compression test results showed the effect of Fe_3O_4 on the impact resistance of the nanocomposites. In addition to, the results were observed good improvement in magnetic properties with increase wt% of Fe_3O_4 particles. S. J. Vijaya , et al. [5] (2018), studied the effect of the hybrid nanoreinforced materials (CNTs + Fe_2O_3) with various contents on the electromagnetic properties of the aluminium using modelling techniques to analyze the material properties. The results revealed significant improvement in the electromagnetic properties of hybrid nanocomposites compared to the base alloy. It was found that as the vol.% Fe_2O_3 increased, the magnetic flux density value as well as energy increased. But the magnetic field strength decreased with the increase in the vol.% Fe_2O_3 particles.

The aim of present work is to fabricate of an aluminium metal matrix nanocomposites (AMMNCs) based on powder aluminium with 60 μm using powder metallurgy (P/M) route. The microstructure, compressive strength, permittivity, dielectric loss and magnetic permeability testing are attempt to be make using different samples sizes of AMMNCs hybrid (Al_2O_3 14-20 nm and Fe_2O_3 30nm). The major objective of the current work is the preparation of the composites containing (0 , 1.5 Fe_2O_3 + 2 Al_2O_3 , 2.5 Fe_2O_3 + 2 Al_2O_3 and 5 Fe_2O_3 + 2 Al_2O_3) wt%.

2.Materials and methods

2.1. Materials and nanocomposites fabrication

In this work nano Fe_2O_3 and Al_2O_3 particles were reinforced pure aluminium. The size of nano Fe_2O_3 (α) was 30 nm while the Al_2O_3 (α) particles varies from 14-20 nm. The manufacturing of nanocomposites will be using powder technology method.

The aluminium powders were mixed with the nanoparticles to prepare nanocomposites with (1.5, 2.5, and 5) wt.% of Fe_2O_3 and 2 wt.% of Al_2O_3 for each percent reinforcement, these powders are added to alcohol to be mixed well in two steps : First, mixing by magnetic field for 5 min. Second, mixing by (ultrasonic –cleaner) for 20 min, In order to increase the homogeneity between the particles of the material. These mixed powders were dried at 70 °C for 15 min, after that the powders have been added 5% of (pva) as a lubricant to reduce friction during compressive, then leave dry at room temperature. The powder mixtures were compressed at room temperature under uniaxial press at 250 MPa. The samples were sintered at 600 °C under argon gas. By two steps, first step at 300°C for (1 h) and second step at 600 °C for (2 h). The total weight per case (added percentage) was 25 g. Table 1 shows the percentage and weight details.

Table 1. Mixture rule of composites.

Al (g)	Fe_2O_3 (g)	Al_2O_3 (g)	Total (g)	Fe_2O_3 wt%	Al_2O_3 wt%
25	0	0	25	0	0
24.125	0.375	0.5	25	1.5	2
23.875	0.625	0.5	25	2.5	2
23.25	1.25	0.5	25	5	2

2.2. Experimental measurements

During microstructure analysis, proper preparation of the specimen surface requires a small sample of the nanocomposite selected, prepared and manufactured. Then polishing in addition to coating samples with gold and palladium spray (for 135 seconds), to discover accurate content and obtain the best accuracy. After preparing the microstructure specimens , they were tested by Field Emission Scanning Electron Microscope (FESEM), using test device type (Cam Scan Mv 2300).

The mechanical properties of the manufactured specimens were obtained by using compression test machine model STM-50 according to standard ASTM E-4 with capacity 50KN. Cylindrical specimens of 14 mm height and 10 mm diameter with compression speed of 0.5 mm/min at room temperature were tested .

Base metal and composite material samples are subjected to electrical tests using (IVIUMSTAT.XR) device at room temperature with (1 V) voltage, (1 A) current avg. and frequency ranges between 10 KHz to 10 MHz . The electrical tests include dielectric loss (δ), real and imaginary relative permittivity (εx).

The magnetic test measures the magnetic properties of powders and the composites by using a vibrating sample magnetometer (VSM) device model (Meghnatis Daghigh Kavir Co., Iran) at a constant frequency. The test measure the magnetization against the magnetic field at room temperature. As the magnetic field is varied over a limited range, the magnetization of the powders is measured by the sense coils with a lock in amplifier.

3.Experimental results and discussion

3.1. Microstructure results

Figure 1 shows the FESEM images under 200 nm or under 800000X magnification respectively. These images show the microstructure of pure aluminium and nanocomposites containing (1.5 Fe_2O_3 + 2 Al_2O_3 , 2.5 Fe_2O_3 + 2 Al_2O_3 and 5 Fe_2O_3 + 2 Al_2O_3) wt.% revealing the presence of Fe_2O_3 and Al_2O_3 and homogenous dispersion in the Al matrix. The images contain Fe_2O_3 , Al_2O_3 concentrated at the grain boundaries of aluminium particles because the grain boundaries energy is high, attracting foreigner particles to it. The gray light which represents the Al matrix. Microstructure of nanocomposites reveals uniform distribution of the hybrid nanomaterials and less porosity along the grain boundaries. The nano ceramic Al_2O_3 is shown as white, while the nano Fe_2O_3 is dark.

IOP Conf. Series: Materials Science and Engineering **765** (2020) 012056 doi:10.1088/1757-899X/765/1/012056

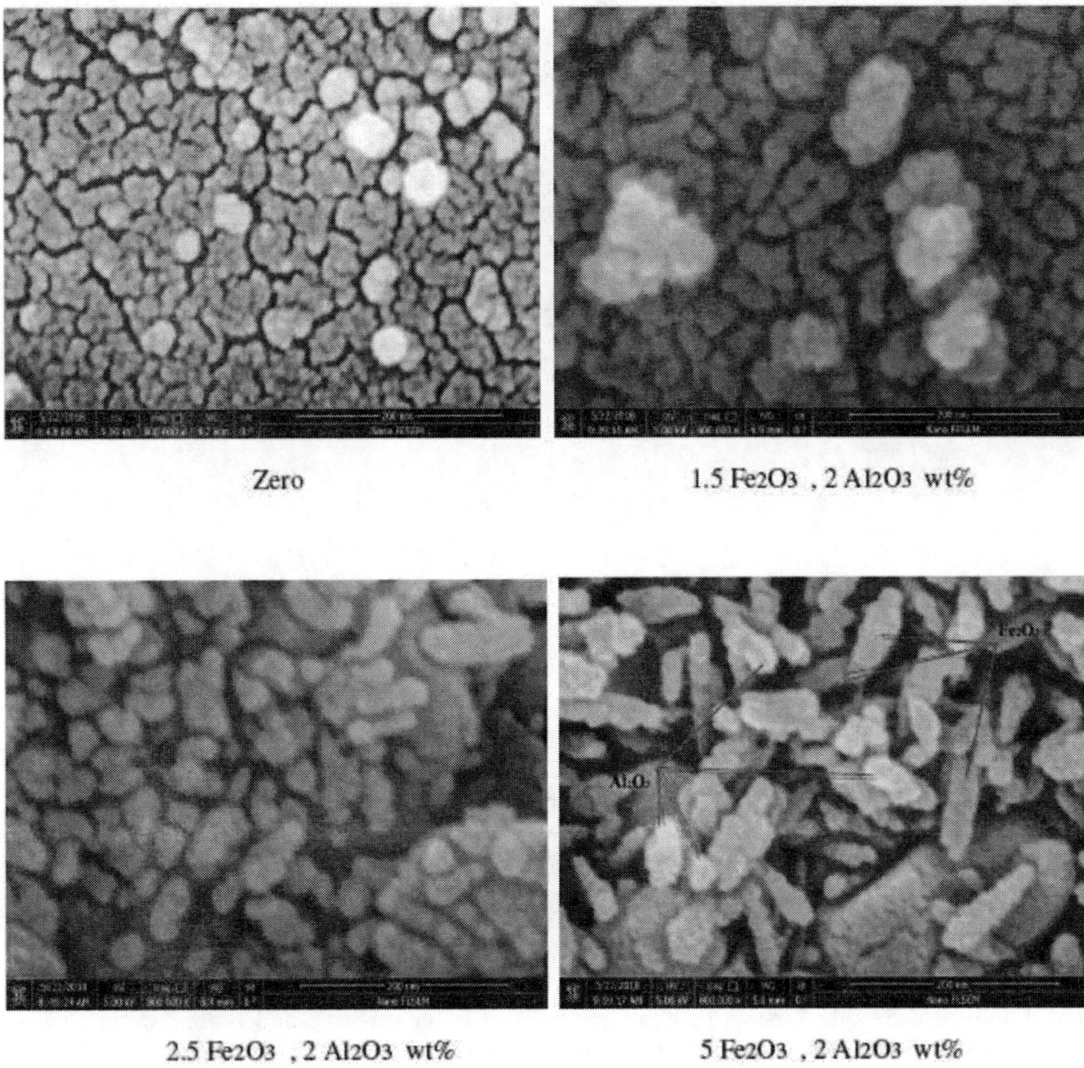

Figure 1. Shows the FESEM images under 200 nm.

3.2. Compression results

Figure 2 shows the compression engineering stress – strain curves of as-received and different weight percentage of Fe_2O_3 + Al_2O_3 nanomaterials. It can be seen that the compression strength of the nanocomposite containing (1.5 Fe_2O_3 + 2 Al_2O_3 wt%) is significantly enhanced compared to as-received and the other nanocomposites. The ultimate compression strength of the base metal with (1.5 Fe_2O_3 + 2 Al_2O_3 wt%) was improved by 30 % compared with the as-received metal. This finding is agreed with the results reported by [6]. The experimental results of compression tests of the nanocomposites containing different weight percentage are listed in table 2, which shows slightly enhanced of failure strain of as-received. For analysis the compression strength and failure strain, 12 specimens were tested with 0.5 mm/min test speed and the average results of three samples for each percentage were recorded.

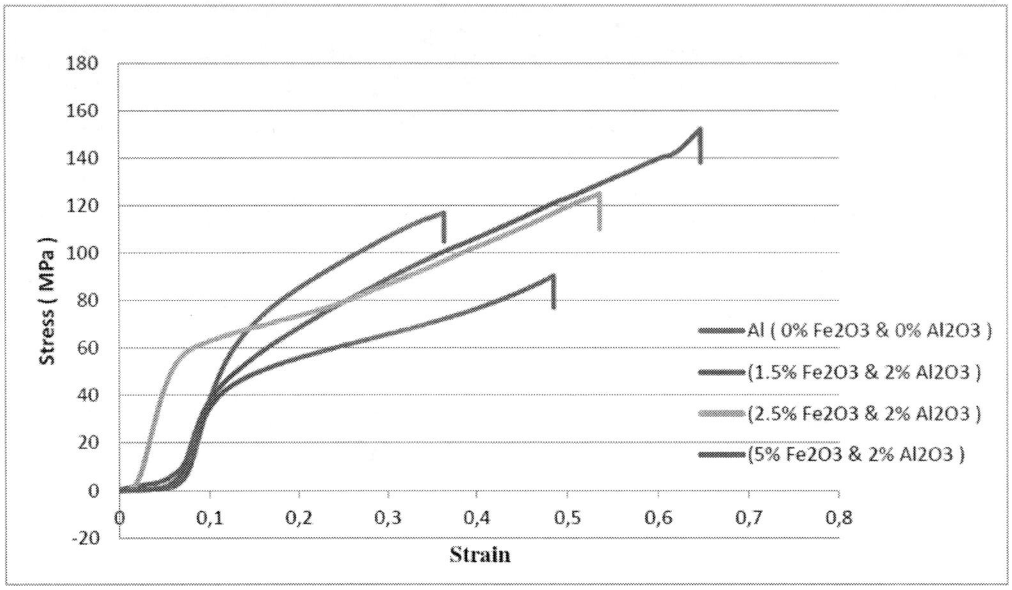

Figure 2. Shows the compression engineering stress – strain curves.

Table 2. Experimental compression testing at room temperature (RT).

Composites	Compression strength (MPa)	Failure strain
As-received	117	0.3635
Al + (1.5% Fe_2O_3 + 2% Al_2O_3) wt.%	152	0.6467
Al + (2.5% Fe_2O_3 + 2% Al_2O_3) wt.%	125	0.5362
Al + (5% Fe_2O_3 + 2% Al_2O_3) wt.%	90	0.4853

Figure 2 and table 2 show that the compression strength (CS) and failure strain (FS) of the nanocomposite were improved with increasing the Fe_2O_3 + Al_2O_3 weight percentage. The CS was obtained to be 117 MPa for as-received and 152 MPa at (1.5% Fe_2O_3 + 2 % Al_2O_3) nanoreinforced materials, with 30% improvement factor compared to the base metal. The main reasons for above improvement can be attributed to many factors, the incorporation of harder Fe_2O_3 + Al_2O_3 particles for strengthening the nanocomposite resulted in hard nature of nanocomposite. The proper interfacial bounding between the pure aluminium with the harder nanoparticles, in addition to that, the nanoparticles were homogeneous and uniformly distributed into the base metal. Thus, the CS and FS are improved [4].

3.3. Electrical results (Permittivity)

Electrical properties of the pure aluminium and three nanocomposites were analyzed by testing carried out on the computerized universal testing device type (IVIUMSTAT.XR). By measuring the real, imaginary and static permittivity values of the metal according to Debye Dispersion theory, the relative permittivity can be described [7] :

$$\varepsilon r = \varepsilon' \pm j\varepsilon'' \quad \ldots\ldots (1)$$

Where : εr is the relative permittivity
ε' is real part of permittivity
ε'' is imaginary part of permittivity

$$\text{When}: \varepsilon' = \varepsilon_\infty + \frac{\varepsilon_s - \varepsilon_\infty}{1+\omega^2\tau^2} \quad \ldots\ldots (2)$$

$$\& \quad \varepsilon'' = \frac{\varepsilon_s - \varepsilon_\infty}{1+\omega^2\tau^2} \quad \ldots\ldots (3)$$

Where : ε_∞ is Dielectric constant at high frequency
ε_s is Dielectric constant at low frequency
$\omega = 2\pi f$, is Angular frequency of the electric field
τ is Relaxation time .

The dielectric loss angle (tan δ), is taken as measure of dielectric loss and is known also as loss tangent.

$$\tan\delta = \frac{\varepsilon''}{\varepsilon'} \quad \ldots\ldots\ldots (4)$$

Figure 3 shows the relation of real and imaginary parts of permittivity against frequency (log Hz). Figure 4 shows a comparison of dielectric loss and relative permittivity properties of hybrid nanocomposites with different amount of hybrid reinforcements (a, b, c, d). The relative permittivity (εx), volume resistance (Rx) and dielectric loss (tanδ) were measured to test the electrical properties of the hybrid nanocomposites materials. Figure 3 shows the electrical properties of as-received and hybrid nanocomposites for different amount of $Fe_2O_3 + Al_2O_3$ against log frequency (log Hz). These properties revealed a nonlinear relation by adding the hybrid amount of nanoparticles materials.

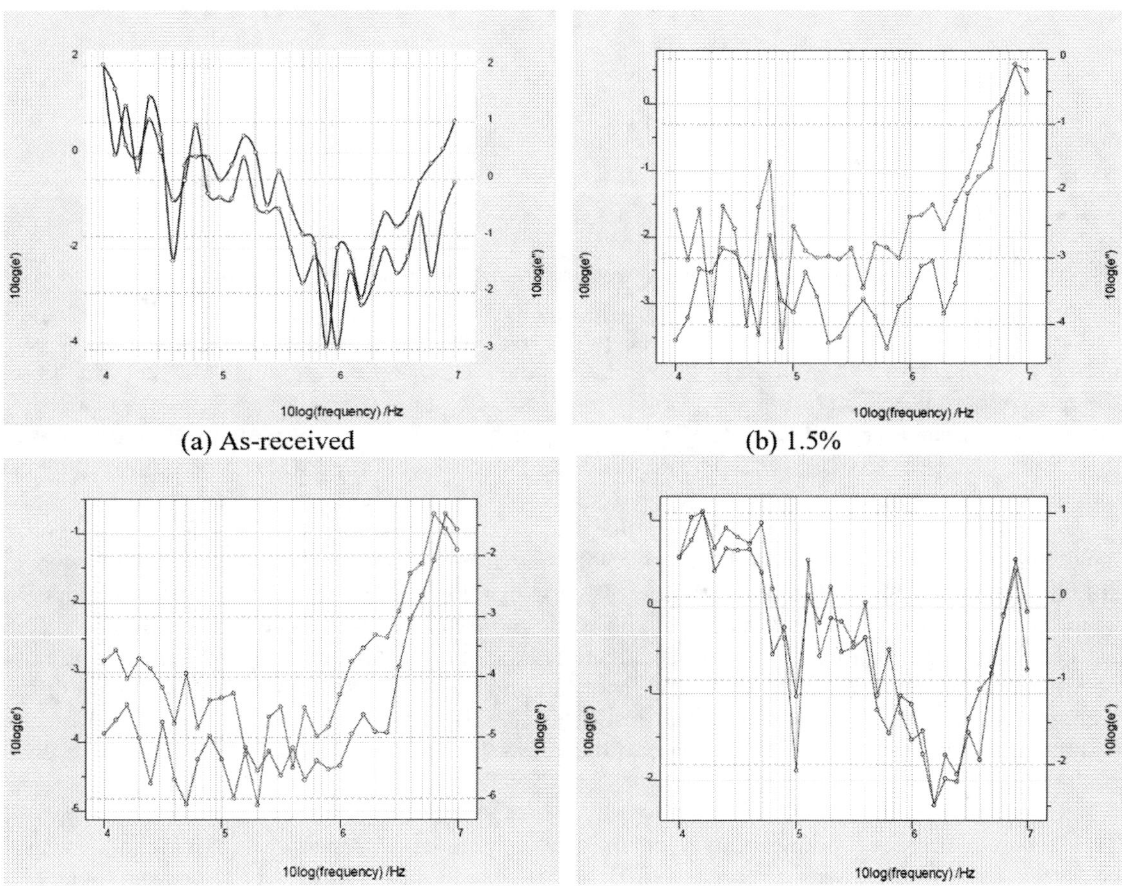

(a) As-received (b) 1.5%

(c) 2.5% (d) 5%

Figure 3. Shows the curves of the real and imaginary parts of permittivity

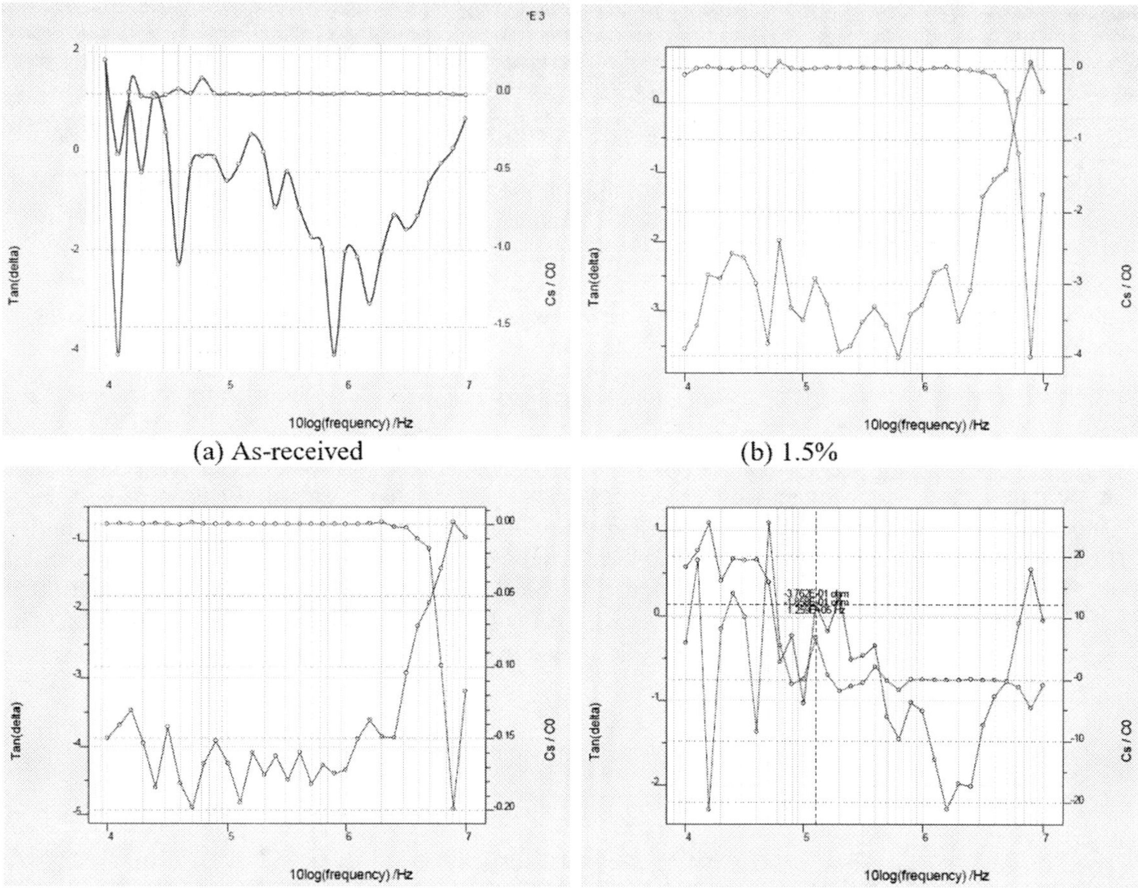

(a) As-received (b) 1.5% (c) 2.5% (d) 5%

Figure 4. Shows the curves of relative permittivity and dielectric loss angle

Where Ɛ' (black line) is real part of permittivity and Ɛ" (blue line) is the imaginary part of permittivity in figure 3. Tan (delta) (black line) is the dielectric loss angle and Cs/Co (blue line) is the relative permittivity in figure 4.

The aluminium doped 1.5% Fe_2O_3 + 2 % Al_2O_3 having higher values of Ɛ' , Ɛ" and tanδ = Ɛ"/Ɛ' than those undoped (as-received). The electrical and mechanical properties can be improved by using one reinforced material [8].

It is observed from above curves that the relative permittivity values for as-received and nanocomposites varies in random behaviour with increase of frequency and then at high frequency it slightly decrease. This is true and agreed well with what concluded by Ref. [9] . Many attempts have been carried out on the improvements of dielectric properties using unique nanoreinforced material. Z.Li et al [10] manufactured Al. doped SiC nanopowder and they observed that the increasing in wt% of SiC leads to increasing in the dielectric loss. Aluminium – SiC nanocomposite containing different amount of SiC was fabricated and electrically tested. They deduced that the increase of electrical conductivity leads to significantly increase of dielectric loss [11].

3.4. Magnetic results

The absolute magnetic permeability (μabs) is main magnetic property of a material , which is concerning to magnetic susceptibility (χm). The magnetic susceptibility is a measurement of the amount of magnetization of a metal in reaction to an exterior magnetic field, as follows [12]:

$$\mu abs = (1 + \chi m)\, \mu o \qquad \ldots\ldots (5)$$

$$\chi m = \frac{M}{H} \quad \ldots\ldots (6)$$

where : (μo) is vacuum permeability, equal to $4\pi \times 10^{-7}$ henry/m.
(M) is The Magnetization , ampere per meter.
(H) is Magnetic field, ampere per meter.

According to equations (5 & 6), the value of Xm (magnetic susceptibility) can be obtained by apply the above equation. It is clear that the value of Xm is maximum for the nanocomposite containing 1.5% Fe_2O_3 + 2 % Al_2O_3. Table 3 shows the value of Xm and absolute magnetic permeability μabs with magnetic field (H) between (12000 to - 12000 A/m) for four types of composites. Maximum magnetization (M) occurred at the nanocomposite containing 1.5% Fe_2O_3 + 2 % Al_2O_3 which is equal to 0.11666 A/m. The improvement in (M) value equal to 26 % compared to the magnetization for pure aluminium. Figure 5 shows the variation of magnetic field (H) against magnetization (M).

Table 3. Magnetic properties of nanocomposites with different hybrid nanomaterials.

Composites	Xm	μabs (H/m)	Magnetization (M) A/m
Pure Aluminium	7.7×10^{-6}	1.254×10^{-6}	0.0927 to - 0.0927
Al + (1.5% Fe_2O_3 + 2% Al_2O_3) wt.%	9.73×10^{-6}	1.257×10^{-6}	0.11666 to - 0.11666
Al + (2.5% Fe_2O_3 + 2% Al_2O_3) wt.%	8.95×10^{-6}	1.256×10^{-6}	0.10733 to - 0.10733
Al + (5% Fe_2O_3 + 2% Al_2O_3) wt.%	8.79×10^{-6}	1.256×10^{-6}	0.10544 to - 0.10544

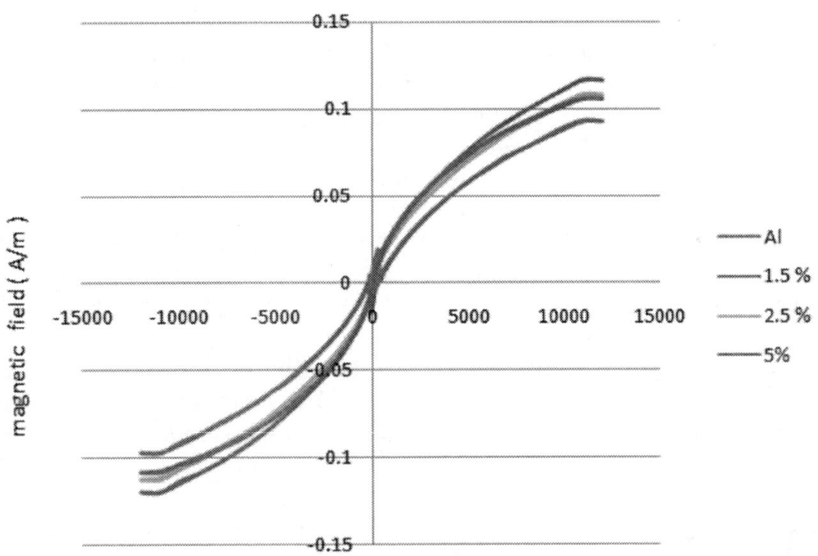

Figure 5. The variation of magnetic field (H) against magnetization (M)

Maximum μabs was found in (1.5% Fe_2O_3 + 2% Al_2O_3) nanocomposite which was equal to 1.257×10^{-6} H/m and reduced to 1.256×10^{-6} H/m for (5% Fe_2O_3 + 2% Al_2O_3) nanocomposite, while this value was recorded as 1.254×10^{-6} H/m for pure aluminum. The magnetization (M) is an intrinsic property depends on the distribution of the nanoparticles. The experimental results also show that the value of (M) for (1.5% Fe_2O_3 + 2% Al_2O_3) nanocomposite is better than that of pure aluminium and other nanocomposites. This can be attributed to relatively fine distribution of the nanoparticles for (1.5% Fe_2O_3 + 2% Al_2O_3) nanocomposite [13]. The enhancement in μabs may be due to the improvement of the permeability (μabs) of the nanocomposites was due to the enhanced of dielectric loss properties. The improvement of microstructure and mechanical properties resulted in improving the μabs induced by containing the hybrid nanomaterial [14].

4.Conclusions

From the experimental results that have been obtained , the following remarks can be listed below :

- From microstructure observation it would be concluded that the hybrid nanomaterials (Fe_2O_3 + Al_2O_3) could be uniformly distributed in pure aluminum. Adding Fe_2O_3 and Al_2O_3 nanoparticles to the aluminum matrix significantly improved their mechanical properties. The best improvement which observed in ultimate compression strength was 30 % at (1.5% Fe_2O_3 + 2% Al_2O_3) wt.%.
- Adding Fe_2O_3 and Al_2O_3 nanoparticles to the matrix significantly improved their electrical properties. In addition it was observed that the real and imaginary values of relative permittivity for all nanocomposites decreased with the frequency increasing at specific range.
- The increase in weight percentage of nanoparticles Fe_2O_3 and Al_2O_3 led to increase the magnetic properties of the nanocomposite studied. However, the best magnetic properties were found in nanocomposite containing (1.5% Fe_2O_3 + 2% Al_2O_3) wt.%. The hysteresis curves for the three nanocomposites with pure aluminium gave the measured magnetization value was equal to 0.0927 A/m for pure aluminium, 0.11666 A/m for 1.5% Fe_2O_3 + 2% Al_2O_3, 0.10733 A/m for 2.5% Fe_2O_3 + 2% $Al_2O_3$and 0.10544 A/m for 5% Fe_2O_3 + 2% Al_2O_3.

References

[1] John Dwight 2010 Aluminum Design and Construction *Taylor Francis Group* **fourth edition**

[2] Tan, Aoshuang, Jie Teng, Xiang Zeng, Dingfa Fu and Hui Zhang 2017 Fabrication of aluminium matrix hybrid composites reinforced with SiC microparticles and TiB_2 nanoparticles by powder metallurgy *Powder Metallurgy* **60** 1 66-72.

[3] Ferreira, L. M. P., E. Bayraktar, and M. H. Robert 2017 Magnetic and electrical properties of aluminium matrix composite reinforced with magnetic nano iron oxide (Fe_3O_4) *Advances in Materials and Processing Technologies* **2.1** 165-173.

[4] L. Ferreira, E. Bayraktar, I. Miskioglu and M.-H. Robert 2018 New magnetic aluminum matrix composites (Al-Zn-Si) reinforced with nano magnetic Fe_3O_4 for aeronautical applications *Advances in Materials and Processing Technologies* 1-12.

[5] Vijay, S. J., Alexandre Tugirumubano, Sun Ho Go, Lee Ku Kwac and Hong Gun Kim 2018 Numerical simulation and experimental validation of electromagnetic properties for Al-MWCNT-Fe_2O_3 hybrid nano-composites *Journal of Alloys and Compounds* **731** 465-470.

[6] Mohammed A. Taha , G.M. Elkomy, H. Abo Mostafa and El Said Gouda 2017 Effect of ZrO2 Contents and Ageing Times on Mechanical and Electrical Properties of Al–4.5 wt.% Cu Nanocomposites Prepared by Mechanical Alloying *Materials Chemistry and Physics*.

[7] P. Debye, Polar Molecules 1929 The Chemical Catalog Co., Reprinted by *Dover Publications* New York.

[8] O. Matarredona , H. Rhoads , Z.L. , J.H. Harvell , L. Balzono, D.E. Resasco 2003 dispersion of single walled carbon nanotubes in Aqueous solution of the Anionic surfactant NaDDBS *J. phys. Chem. B.* **Volume 107** 13357-13367.

[9] Dias C. J., R. Marat, P. Inanico and N. Mendes 2004 Transaction on dielectric and electrical insulation *IEEE* 11-35.

[10] Su, Xiaolei, W. zhou , X. su. F. Luo, D. zhu 2010 Improvement of permittivity of SiC with Al doping by combustion synthesis using Al_2O_3 *Journal of Alloys and Compounds* **492**.1-2 L16-L19.

[11] Jianlei K. , Peng J. . Fanyoug R. Wenbin G. 2016 conductivity dependent dielectric properties and microwave absorption of Al- doped SiC whiskers *J. alloys and compounds* **687** 227-231.

[12] Rajinder P. 2015 Electromagnetic, Mechanical, and Transport Properties of Composite Materials *Taylor & Francis Group* Canada.
[13] Kaw A. K. 2006 Mechanics of composite materials M *Taylor and francis Group* 2nd ed New York 40-46 .
[14] Y. Hirano, R. Hanaoka , N. osawa, K. Miyagi, Y. Fujita, Y. Kanamara 2016 Electrical and mechanical properties of nanocomposite material containing electrically dispersed MWCNTs *IEEE* **978-1** 6-16.

Thin-plate stress analysis solved by combining FEM functions and approximations of class C^0

[1] **Hussain Abdulaziz Abrahem,**

[1] **Enaam Obeid Hassoun**

[1] **Hind D. Salman,**

[2] **Serazutdinov M. N.**

[1]**Electromechanical Engineering Department, University of Technology – Iraq , 50008@uotechnology.edu.iq,50109@uotechnology.edu.iq, 50260@uotechnology.edu.iq**
[2]**Kazan National Research Technological University, Kazan, Russia, serazmn@mail.ru**

Abstract. The method for constructing functions with finite support is used, which combines the properties of FEM functions and approximations in the form of series with unknown coefficients. To illustrate the properties of the approximating function constructed, we present the results of the solution of the 2D theory of elasticity for a thin plate, shown on figure 1,*b*. The thickness of the plate is h. One facet of the plate (at $x = -a$) is fixed and the other one (at $x = -a$) unfixed; the remaining two are under shear stresses τ and an evenly distributed load q.The solution of the problem of determining the stress and strain in the plate can be obtained from the steady-state condition of the Lagrange functional by using the functions of class C^0 of the high-degree approximation for scientific calculations. The calculations show that the obtained solution has high accuracy, even in cases when the length of the rectangular domain is considerably greater than its width.

Keywords: theory of elasticity, FEM, high-degree approximation, degenerated domain solution.

1. Introduction

Variational methods are frequently used to solve problems in various fields of science and technology. To approximate the unknowns contained in the variational equations, finite functions of FEM may be used as well as functions in the form of series with unknown coefficients, imposed across the domain of definition of the solution. One of their advantages is that their usage allows achieving a relatively simple solution, and in order to increase the accuracy of the solution, a larger number of members in the series is used.

The advantages of FEM functions are determined by the fact that they are defined in subdomains; they are functions with finite support. While using FEM functions, it is possible to increase the accuracy of the solution by dividing the domain of definition of the solution into smaller elements as well as by increasing the degree of approximation. A number of well-known experts in the FEM field prefer using approximations of a high degree [1, 2, 3].

The finite elements—during whose construction, the basic functions are associated with unknown parameters in the nodes located mainly at the boundaries of the elements—are widely used nowadays. For several reasons, the most commonly used finite elements are those that are based on the use of low-approximation degree polynomials.

However, the use of high-approximation degree functions may smoothen contradictions and eliminate unwanted effects that arise in some cases. In particular, it relates to the calculation of thin shells [4]. This article presents a method for constructing functions with finite support, which combines the properties of FEM functions and approximations in the form of series with unknown coefficients. Some of the parameters, which are used as unknowns in these functions, determine the value of the function at the boundaries of the rectangle; others are coefficients of series entered in the domain. They are defined in a way that allows connecting functions determined in different rectangular domains. The distinctive feature of these functions is that they combine the properties of finite element approximation functions in the form of polynomials, trigonometric series, and others. The degree of approximation inside and on the boundary of the domain can be defined. Further, polynomial, trigonometric, and other functions can be used as basis functions.

Using the isoparametric mapping [5,6] of rectilinear quadrangles on the curvilinear, these functions can be used for obtaining solutions of theory of elasticity and the theory of thin shell [7] problems in the domains of different shapes. A system has a lot of algebraic equations with band structure while solving problems and determining the unknown coefficients of the approximating function. The method described in this article was used to construct one-dimensional functions with finite support and to calculate beam systems [8–10]. The formulas, on the basis of which calculations of the stress-strain state of structural elements were presented in the article, are also given in [11]. Another way of constructing functions with a finite carrier to solve three-dimensional problems is presented in [12].

2. Constructing functions of class C^0

Let's divide the domain of the solution Ω into rectangular subdomains Ω_i $(i=1,I)$, in each of which we introduce the dimensionless local coordinate system $O\xi\eta$ $(-1\le\xi\le1,-1\le\eta\le1)$ with the origin (beginning) in the center Ω_i (Fig. 1). We define the sides of the rectangle and the nodes in its corners as Γ_{ij} and Γ_{ij} $(j=1,4)$.

We define the function $u(\xi,\eta)$ in each of the subdomains Ω_i as

$$\begin{aligned} u(\xi,\eta)=u^i(\xi,\eta)=\sum_{m=1}^{M}u^i_{om}\,\varphi_{om}(\xi,\eta)+\sum_{n=1}^{N_1}u^i_{1n}\,F_{1n}(\xi,\eta)+\sum_{n=1}^{N_2}u^i_{2n}\,F_{2n}(\xi,\eta)+\\ +\sum_{n=1}^{N_3}u^i_{3n}\,F_{3n}(\xi,\eta)+\sum_{n=1}^{N_4}u^i_{4n}\,F_{4n}(\xi,\eta)+\sum_{j=1}^{4}u^i_j\,f_j(\xi,\eta). \end{aligned} \tag{1}$$

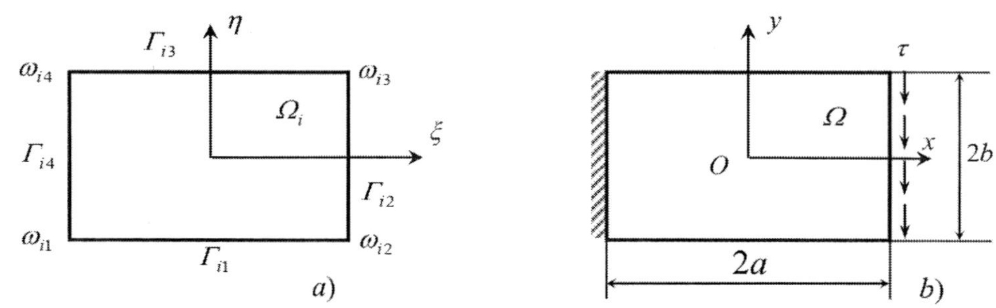

Fig. 1

Here, u^i_{0m}, u^i_{jn} – unknown coefficients; u^i_j – values $u(\xi,\eta)$ at the nodes ω_{ij}; $\varphi_{om}(\xi,\eta)$, $F_{jn}(\xi,\eta)$, $f_j(\xi,\eta)$ – basis functions, which satisfy certain conditions. Thus, functions $\varphi_{om}(\xi,\eta)$ must vanish at Γ_{ij} ($j=\overline{1,4}$); $F_{kn}(\xi,\eta)$ – equal to zero at Γ_{ij} if $k \neq j$, ($k=\overline{1,4}$, $j=\overline{1,4}$); $f_j(\xi,\eta)=1$ at the nodes ω_{ij} and equal to zero on the sides of the square Ω_i opposite to ω_{ij}.

If these conditions are met, placing the values $\eta=\pm 1$, $\xi=\pm 1$ into the formula (1), we get

$$
\begin{aligned}
u(\xi,-1) &= \sum_{n=1}^{N_1} u^i_{1n}\, F_{1n}(\xi,-1) + u^i_1\, f_1(\xi,-1) + u^i_2\, f_2(\xi,-1),\\
u(1,\eta) &= \sum_{n=1}^{N_2} u^i_{2n}\, F_{2n}(1,\eta) + u^i_2\, f_2(1,\eta) + u^i_3\, f_3(1,\eta),\\
u(\xi,1) &= \sum_{n=1}^{N_3} u^i_{3n}\, F_{3n}(\xi,1) + u^i_3\, f_3(\xi,1) + u^i_4\, f_4(\xi,1),\\
u(-1,\eta) &= \sum_{n=1}^{N_4} u^i_{4n}\, F_{4n}(-1,\eta) + u^i_4\, f_4(-1,\eta) + u^i_1\, f_1(-1,\eta).
\end{aligned} \tag{2}
$$

As seen from relations (1), (2), coefficients u^i_{om} determine values $u(\xi,\eta)$ only inside subdomains Ω_i, while on the boundaries of Γ_{ij}, values $u(\xi,\eta)$ depend on $u^i_{jn}F_{jn}(\xi,\eta)$ and $u^i_j f_1(\xi,\eta)$, $j=\overline{1,4}$. If we define $u(\xi,\eta)$ in subdomains Ω_i $(i=1, I)$ in such a way that at the boundaries of adjacent subdomain, it is expressed through the same values $u^i_{jn}F_{jn}(\xi,\eta)$, $u^i_j f_1(\xi,\eta)$, then we get a function $u(\xi,\eta)$ continuous in the transition from one subdomain Ω_i to another. Consequently, $u(\xi,\eta)$ is a function of class C^0 with finite support.

IOP Conf. Series: Materials Science and Engineering **765** (2020) 012057 doi:10.1088/1757-899X/765/1/012057

Please note that basis functions $\varphi_{0m}(\xi,\eta)$, $F_{jn}(\xi,\eta)$, $f_j(\xi,\eta)$ must satisfy the conditions of completeness. Let's have a look at one of the possible ways to select functions $\varphi_{om}(\xi,\eta)$, $F_{jn}(\xi,\eta)$, $f_j(\xi,\eta)$:

For some complete system of functions $\psi_m(\xi,\eta)$, $m=1,2,3,...$, using expressions

$$
\begin{array}{ll}
F_o(\xi,\eta)=(1-\xi^2)(1-\eta^2), & \\
F_1(\xi,\eta)=(1-\xi^2)(1-\eta)/2, & F_2(\xi,\eta)=(1+\xi)(1-\eta^2)/2, \\
F_3(\xi,\eta)=(1-\xi^2)(1+\eta)/2, & F_4(\xi,\eta)=(1-\xi)(1-\eta^2)/2, \\
f_1(\xi,\eta)=(1-\xi)(1-\eta)/4, & f_2(\xi,\eta)=(1+\xi)(1-\eta)/4, \\
f_3(\xi,\eta)=(1+\xi)(1+\eta)/4, & f_4(\xi,\eta)=(1-\xi)(1+\eta)/4,
\end{array}
\tag{3}
$$

We can write down the system of functions (1) in the following way:

$$
\begin{array}{l}
\varphi_{0m}(\xi,\eta)=F_o(\xi,\eta)\psi_m(\xi,\eta), \quad m=1,2,3,...; \\
F_{jn}(\xi,\eta)=F_j(\xi,\eta)\,\psi_k(\xi,\eta), \quad n=1,2,3,...;k(n)=1,k_{j1},k_{j2},... \quad .
\end{array}
\tag{4}
$$

It is easy to confirm that $\varphi_{om}(\xi,\eta)$, $F_{jn}(\xi,\eta)$ in the form (4) satisfy the required conditions. Have a look at the graphs representing the changes in some of the functions (3) in Fig. 2.

While selecting the value of indexes k_{jl} in (4), it's important to note that on the boundaries

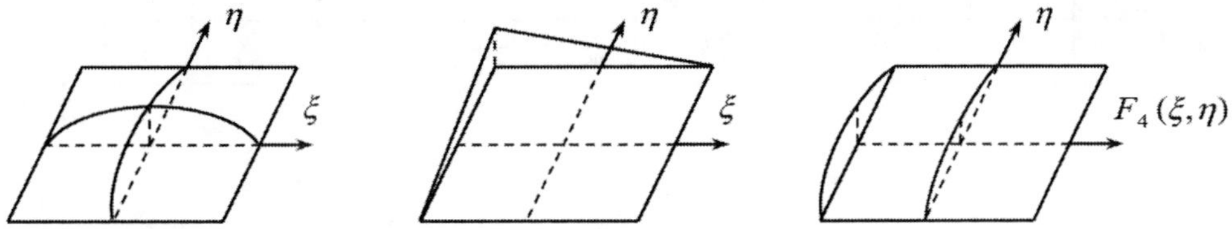

Fig. 2

Γ_{ij}, the function $F_{jn}(\xi,\eta)$ must be linearly independent.

Using power functions, we get the following:

$$
\begin{array}{l}
\psi_m(\xi,\eta)=\xi^k\eta^l, \quad k=0,1,2,...; \quad l=0,1,2,...; \\
F_{1n}(\xi,\eta)=F_1(\xi,\eta)\,\xi^{n-1}; \quad F_{3n}(\xi,\eta)=F_3(\xi,\eta)\,\xi^{n-1}; \\
F_{2n}(\xi,\eta)=F_2(\xi,\eta)\,\eta^{n-1}; \quad F_{4n}(\xi,\eta)=F_4(\xi,\eta)\,\eta^{n-1}.
\end{array}
\tag{5}
$$

IOP Conf. Series: Materials Science and Engineering **765** (2020) 012057 doi:10.1088/1757-899X/765/1/012057

In relations (5), exponents k, l are selected in a way for functions $\psi_m(\xi,\eta)$ form a complete polynomial of degree K.

` While using (3) – (5), expression (1) becomes a polynomial, degree and properties of which depend on the values M, N_j.

We believe $N_1 = N_2 = N_3 = N_4$. Opening brackets in (1) and grouping terms with same exponents, we present expression (1) in the form of a polynomial with unknown coefficients. Let the total number of these coefficients be K_c. Some of the members with a thus obtained function will form the complete polynomial of degree K_0, and the rest, power functions with unknown coefficients, with a degree greater than K_0. Let their total number be K_l.

It is obvious that K_0, K_c, K_l depend on K, N_1. Table 1 illustrates this dependence. You can see that with the increase of values K and N_1, K_0 increases as well.

Thus, based on the relations (3) – (5), functions with class C^0 finite support can be obtained in the form of a complete polynomial of degree K_0, to which a number of power functions with an exponent greater than K_0 is added.

Table 1

K	0				1				2				3			
N_1	0	1	2	3	1	2	3	4	2	3	4	5	3	4	5	6
K_c	9	13	17	21	15	19	23	27	22	26	30	34	30	34	38	42
K_0	2	3	4	4	3	4	5	5	4	5	6	6	5	6	7	7
K_l	3	3	2	6	5	4	2	6	7	5	2	6	9	6	2	6

3. Results of calculations.

To illustrate the properties of the approximating function constructed, we present the results of the solution of the 2D theory of elasticity for a thin plate, shown in figure 1,b. the thickness of the plate is h. One facet of the plate (at $x=-a$) is fixed, the other one (at $x=-a$) unfixed; the remaining two are under shear stresses τ and an evenly distributed load q.

The solution of the problem of determining the stress and strain in the plate can be obtained from the steady-state condition of the Lagrange functional:

$$\frac{1}{2}\delta\iint_{\Omega} \left\{ \frac{Eh}{1-\mu^2}\left(\frac{\partial u}{\partial x}+\frac{\partial v}{\partial y}\right)^2 + Gh\left[\left(\frac{\partial u}{\partial x}+\frac{\partial v}{\partial y}\right)^2 - 4\frac{\partial u}{\partial x}\frac{\partial v}{\partial y}\right]\right\} d\Omega = \\ = h\int_{-b}^{b} \left[\tau\,\delta\, v(-a,y)\right]dy + h\int_{-a}^{a}\left[q\delta\, u(x,b)\right]dx. \tag{6}$$

Here, u, v – displacements of the plate in the direction of axes Ox and Oy module, h – thickness of plate, E – Young's modulus, μ – Poisson's ratio, $G=E/2(1+\mu)$.

If the domain of the definition of the solution Ω is not divided into the rectangular subdomain, then $I=1$, Ω_1 occupies the entire domain Ω. We introduce the dimensionless coordinates $\xi = x/a$, $\eta = y/b$ and believe that $N_1 = N_2 = N_3 = N_4$. In this case, $u(x,y)$ and $v(x,y)$, using (1) can be written down as follows:

$$u(\xi,\eta) = \sum_{m=1}^{M} u_{om}\varphi_{om}(\xi,\eta) + \sum_{j=1}^{4} u_j f_j(\xi,\eta) + \sum_{k=1}^{4}\sum_{n=1}^{N_1} u_{kn}F_{kn}(\xi,\eta),$$
$$v(\xi,\eta) = \sum_{m=1}^{M} v_{om}\varphi_{om}(\xi,\eta) + \sum_{j=1}^{4} v_j f_j(\xi,\eta) + \sum_{k=1}^{4}\sum_{n=1}^{N_1} v_{kn}F_{kn}(\xi,\eta). \quad (7)$$

We believe that the approximating functions are defined in the form of (5).

The terms of the plate fixation will be satisfied, if the following is true for (7) formulas $u_{2n} = v_{2n} = 0$, $u_2 = u_3 = v_2 = v_3 = 0$.

Placing expression (7) into the condition (3), we obtain a system of algebraic equations of $N_c = 2K_c$ order for finding unknown coefficients in the series (7).

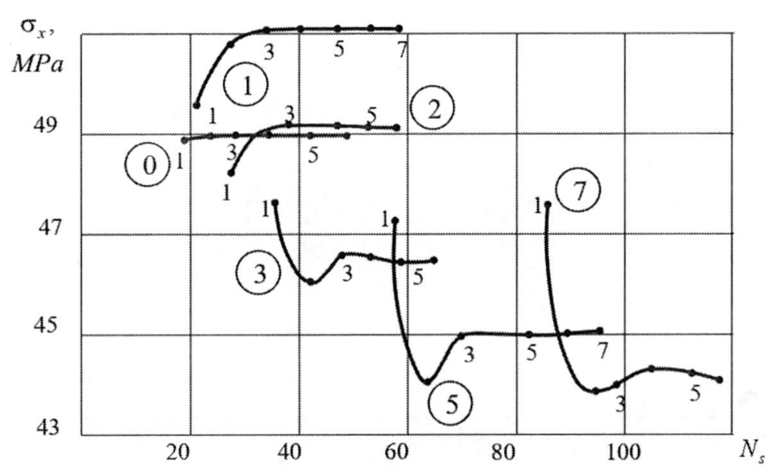

Fig. 3

The calculations assumed that the lengths of the sides of the plate were $a = b = 0.25\,m$, $E = 2\cdot 10^5$ MPa, $\mu = 0.3$, $h = 0.01\,m$, $q = 0$; shear stresses affecting facets of plates at $x = -a$ have constant thickness; height of the plates change in accordance to the law $\tau = \tau_0 (b^2 - y^2)/b^2$, $\tau_0 = 10$ MPa. Fig. 3 shows graphs of the changes in stress σ_x at the point $x = 0/125\,m$, $y = 0.25\,m$, depending on the used approximation. The numbers in the circles by the lines of graphs show degrees of K of complete polynomials inside Ω_1 used in the calculations. The numbers on the lines around the selected points indicate under which degrees of K_Γ of approximating polynomials on the boundaries of domain Ω_1 data of calculations were received ($K_{I'} = N_1 - 1$).

Here are the results of the calculations of the plate: for domain Ω divided into a number of subdomain Ω_i . Tables 2–4 present results obtained for $E = 2 \cdot 10^5$ MPa, $\mu = 0.3$, $a = 0.3\,m$, $\mathrm{b} = h = 0.01\,m$, $q = 0.1$ MPa, $\tau = 0$.

In this case, $\mathrm{b}/\mathrm{a} = 0.033$; therefore, the sizes of the sides of the domain Ω differ considerably and the plate can be seen as a beam fixed on one end. For the beam, the maximum stresses occur at $x = -a$ and are calculated according to the formula $\sigma_{\max}^{b} = \mathrm{M}_{z}^{\max} / \mathrm{W}_{z}$. Taking into consideration that $\mathrm{M}_{z}^{\max} = 2qh2a^2$ =180Hm, $\mathrm{W}_z = h(2b)^2/6 = 0{,}67 \cdot 10^{-6}\,\mathrm{m}^3$, we obtain $\sigma_{\max}^{b}$ = 270 MPa.

Table 2 shows the results of calculations in the case where subdomian Ω_1 feels the whole domain Ω ($I = 1$). We present the values of the maximum stress σ_{max} obtained using the relations (6), (7), (5) at different values of the number of members N_1, M in the series (7). In the table, $\Delta\sigma = \left|(\sigma_{\max} - \sigma_{\max}^{b}) \backslash \sigma_{\max}^{b}\right| \cdot 100\%$; the percentage difference between the value $\sigma_{\max}^{b}$ is calculated by the theory of rods and $\sigma_{\max}$ is found with the use of elasticity theory.

Table 2

M	N_1	σ_{max}, MPa	$\Delta\sigma$, %
2	4	270,17	0,063
4	6	270,51	0,18
6	8	271,14	0,42
8	10	272,22	0,82

Table 3

b_1, m	$\Delta b_1\,(b)$, %	$\Delta b_1\,(a)$, %	σ_{max}, MPa	$\Delta\sigma$, %
$2 \cdot 10^{-3}$	10	0.33	273,70	1,37
10^{-3}	5	0.17	272,66	0,81
$5 \cdot 10^{-4}$	2.5	$8.3 \cdot 10^{-2}$	271,42	0,5
$5 \cdot 10^{-5}$	0.25	$8.3 \cdot 10^{-3}$	270,61	0,22
10^{-7}	$5 \cdot 10^{-4}$	$1.7 \cdot 10^{-6}$	272,51	0,93

Table 3 shows the solution in the case when $I = 2$, domain Ω is divided into 2 parts Ω_1 and Ω_2 (Fig. 4,a). Domain Ω_1 on the figure 4, a is shaded, the length of the vertical side Ω_1 equals b_1. The length of the side Ω_2 equals $\mathrm{b}_2 = 2b - b_1$, the relative value b_1 calculated by the formula:

$\Delta b_1(b) = (b_1 / 2b)\cdot 100\%$, $\Delta b_1(a) = (b_1 / 2a)\cdot 100\%$. The calculations assumed $M = 4$, $N_1 = 6$.

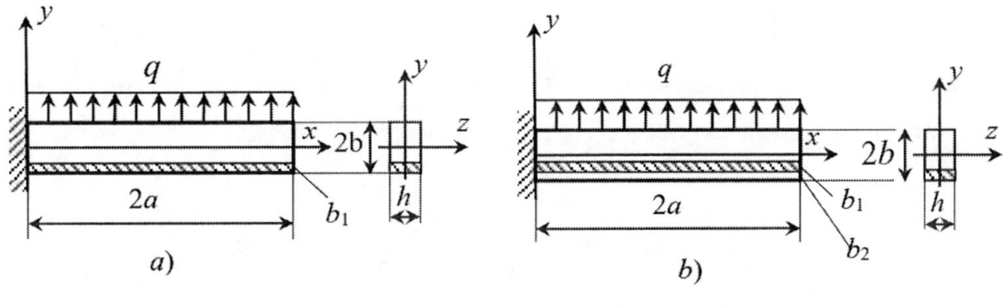

Fig.4

The results of calculations for domain Ω divided into 3 subdomains Ω_1, Ω_2, Ω_3 (Fig. 4,*b*) are shown in Table 4. Domain Ω_1 in Fig. 4, *b* is shaded, the length of the vertical side Ω_1 equals b_1. The length of the vertical side Ω_2 equals $b_2 = 5\cdot 10^{-3}\ m$. For Ω_3, the length of the vertical side $b_3 = 2b - b_1 - b_2$. We assumed $M = 4$, $N_1 = 6$.

Table 4

b_1, m	$\Delta b_1(b)$, %	$\Delta b_1(a)$, %	σ_{max}, MPa	Δ, %
$2\cdot 10^{-3}$	10	0.33	277,74	2,85
10^{-3}	5	0.17	277,17	2,65
$5\cdot 10^{-4}$	2.5	$8.3\cdot 10^{-2}$	276,84	2,53
$5\cdot 10^{-5}$	0.25	$8.3\cdot 10^{-3}$	276,51	2,41
10^{-7}	$5\cdot 10^{-4}$	$1.7\cdot 10^{-6}$	276,25	2,31

4. Discussions

As can be seen from the graphs shown in Fig. 3, for the given problem, starting from a certain value K_Γ, the increase of the order of approximation for unknown functions on the boundaries Ω_1 practically does not increase the accuracy of the solution. This conclusion is valid for the other case as well. Therefore, to construct the functions that have a high degree of approximation, it is important to use the degree of freedom inside the element as well as on its boundaries. It is possible to obtain a solution with high accuracy without breaking down the domain of the definition of the solution Ω into several subdomains Ω_i.

IOP Conf. Series: Materials Science and Engineering **765** (2020) 012057 doi:10.1088/1757-899X/765/1/012057

In some cases, it may be required to break down the domain of the definition of the solution into a certain amount of subdomains. The advantage of functions described here is that they allow using Ω_i that have significantly different sizes of sides while breaking down the domain into subdomains. The data in Tables 2, 3 and 4 shows that by using the functions with the finite support described here, the solution can also be obtained with high accuracy in the case where the sizes of subdomains Ω_i differ significantly. As seen from Tables 3 and 4, the solution is obtained with good accuracy in the cases where the length b_1 of one side of the subdomain Ω_1 is shorter than the length $2a$ of the other subdomain side by $1.7 \cdot 10^{-6}$ %. Therefore, the solution is obtained when a rectangular domain almost degenerates into a line.

5.Conclusions

Based on the results of the research presented, the following conclusions can be written:

1. The distinctive features of the approximating function $u(\xi,\eta)$ with finite support are that it is expressed through the four values at the nodes of the rectangular domain Ω_i and generalized parameters that determine the function inside and on the boundaries of Ω_i. To change the order of the approximation of the function, you can just change the number of these generalized parameters.
2. Due to its structure, the boundaries of the subdomains $u(\xi,\eta)$ is presented by functions $F_{jn}(\xi,\eta)$, which are independent of each other. Further, the functions are defined inside Ω_i. The required degree of approximation in each subdomain Ω_i can be reached by specifying a corresponding number of members of series M, N_j ($j=\overline{1,4}$). Moreover, in different subdomains, different M as well as different functions $\varphi_{om}(\xi,\eta)$ and even different $F_{jn}(\xi,\eta)$, $f_j(\xi,\eta)$ can be used if certain conditions are met.
3. As can be seen from the structure $u(\xi,\eta)$, various systems of functions can be used as the basis function. If necessary, in order to improve approximation, inside one or more subdomains Ω_i, various additional functions (e.g. singular)—depending on the properties of the chosen problem—can be introduced on the boundaries Γ_{ij} and at the nodes ω_{ij}.
4. While using the isoparametric mapping of rectilinear quadrangles Ω_i on curvilinear quadrangles, functions $u(\xi,\eta)$ can be used to obtain solutions in the domains other than rectangular.
5. Similar to the finite element method, to determine the unknown coefficients $u(\xi,\eta)$, a system of algebraic equations of the tape structure is obtained.

IOP Conf. Series: Materials Science and Engineering **765** (2020) 012057 doi:10.1088/1757-899X/765/1/012057

References

[1] G. Strang, G. Fix, *An Analysis of the Finite Element Method*, Wellesley-Cambridge Press, 2008.

[2] O.C. Zienkiewicz, K. Morgan, *Finite Elements and Approximation*, Courier Corporation, 2006.

[3] P. Solin, K. Segeth, I. Dolezel, *Higher-Order Finite Element Methods*, Chapman & Hall/CRC, 2004.

[4] K.-J. Bathe, E.L. Wilson, *Numerical Methods in Finite Element Analysis*, Prentice Hall, 1976.

[5] A.I. Golovanov, M.S. Kornishin, *Introduction* to the *Finite Element Method* in Statics of Thin Shells, Kazan, Fiz.-Tekh. Inst. KF AN SSSR, Kazan, 1989.

[6] K.-J. *Bathe Finite Element Procedures*, Prentice Hall, 1996.

[7] M.N. Serazutdinov, M.F. Garifullin, Shells of complex shape. *Soviet Applied Mechanics*, No. 27, 11, 1991, 1077–1082.

[8] M.N. Serazutdinov, M.N. Ubaidulloev, H.A. Abrahem, Calculation of reinforced loaded structures by the variational method. *Scientific and Theoretical Journal "Proceedings of Higher Educational Institutions. Construction"*, No. 7, 2010, S.118–123. (Russian copy)

[9] M.N. Serazutdinov, M.N. Ubaidulloev, H.A. Abrahem, Increasing the bearing capacity of reinforced loaded structures. *Journal "Structural Mechanics of Engineering Structures and Structures"*, No. 3, 2011, pp.23–30. (Russian copy)

[10] M.N. Serazutdinov, M.N. Ubaidulloev, H.A. Abrahem, The influence of mounting forces on the bearing capacity of reinforced rod systems. *Bulletin of Kazan Technological University*, No. 10, 2011, pp.116–124. (Russian copy)

[11] M.N. Serazutdinov, A method for constructing a compactly supported function of a high degree of approximation of class C0. *Bulletin of Kazan Technological University*, T. 19, No. 11, 2016, S.162–165. (Russian copy)

[12] F.S. Khayrullin, O.M. Sakhbiev, Calculation of orthotropic constructions by a variation method on the basis of three-dimensional functions with final carriers. *PNRPU Mechanics Bulletin*, No. 2, 2017, pp.195–207. doi:10.15593/perm.mech/2017.2.11.

The effect of crack length on the fracture parameters of an edge crack parallel to the material variation in a rectangular FGM plate

Haider Khazal Mehbes

Hassanein Ibraheem Khalaf

Ameen Ahmed Nasser

Mechanical Engineering Department, Basrah University, Basrah, Iraq

Abstract. In this study, the extended element free Galerkin method (XEFG) is used for crack analysis an edge crack parallel to the material variation in a rectangular functionally graded material (FGM) plate under traction loading. For the first time, the effective parameters of XEFGM are employed such as the sub-triangulation term for numerical integration, suitable influence domain, and enrichment functions for discontinues locations in the fracture analysis of different crack positions of an edge cracked FGM plate. In addition, the incompatible formulation is adopted to extract the stress intensity factors (SIFs). Good data and results are observed about the relationship between the crack lengths and SIFs with good verification during changing XEFG parameters. There is good agreement and stability in the results of SIFs through the sizes of J-integral rJ equal to 0.4 -0.8. In addition, the acceptance results for the size of the support domain are bound between dmax=1.52 to 2.

Keywords: extended element free Galerkin method; crack length, functionally graded materials; stress intensity factors

NOMENCLATURE

Symbol	Description	Unit
E	Modulus of elasticity (Young's Modulus)	MPa
G	Shear modulus	MPa
M	Interaction integral	N/m
N	number of nodes	....
a	The length of a crack	mm
b	Body force	N
d_{mI}	variation of the support domain	m
n	Number of nodes in the influence of domain	....
n_j	The unit outward normal to contour	....
r	Radial distance from a crack tip	m
rJ	Size of J-Inegral	m
$\bar{t}$	traction force	N
t	Time	s
u^h	The unknown trial approximation of displacement	m
$\bar{u}$	The prescribed displacements on the boundary	m
w	The weight function of an influence node	....

Ω	The global domain of the problem	....
Γ	Boundary of the global domain	....
δ	Kronecker delta	....
ϕ	Shape function	....
ν	Poisson's ratio	....
κ	Kolosov constant	....
ρ	Mass density	kg/m^3
λ	Lagrange multiplier	....

1. Introduction

In the recent decades, there are many applications for functionally graded materials (FGMs) in the engineering fields, involving military usages, space vehicle, electrical insulator, bio-medical, etc. According to the literature survey, the fracture analysis of FGMs was studied by many of researchers using the conventional and advanced numerical methods as the finite element method (FEM) [1-4], boundary element method [5], meshfree methods [6-10], and extended finite element method (XFEM) [11-15]. Most of the mentioned references focused on the employed of the numerical methods to extract the stress intensity factors at fixed crack tip position. The study of the effect of the crack length or crack size in certain materials not FGM was done by some of researchers as depicted in [16-19]. These studies were important to know the strength ranges of the related materials against the fracture, and to understand the critical fractures parameters and terms for the whole domain. However, the effect of the crack length on the fracture terms such as SIFs is unadulterated interest in this research. In the particular, lately, Khazal et al. [6] adopted an extended element free Galerkin method (XEFGM) for extract SIFs in one crack position in a rectangular FGM plate, therefore this paper will extend the investigation of the effect of the discontinuity length on the fracture parameters by using XEFGM. In addition, verification is employed between the results obtained by the present study and available references solutions in the literature. Hence, this study will report the effect of crack lengths on the fracture parameters of a functionally graded material numerically. The incompatible interaction integral method will use to determine the fracture parameters as SIFs. In addition, the effective parameters of XEFGM will employ such as the sub-triangulation term for numerical integration, suitable influence domain, and enrichment functions for discontinues locations in the fracture analysis of different crack positions of an edge cracked FGM plate.

2. Displacement and Stress Field Equations

Referring to Hook's law, the strain can be written as follow:

$$\varepsilon = C\sigma = \begin{bmatrix} C_{11} & C_{12} & C_{16} \\ C_{12} & C_{22} & C_{26} \\ C_{16} & C_{26} & C_{66} \end{bmatrix} \begin{Bmatrix} \sigma_{xx} \\ \sigma_{yy} \\ \sigma_{xy} \end{Bmatrix} \tag{1}$$

where C is the compliance matrix. In FGM as illustrated in Figure 1, the material traits such as the Young modules Y and the Poisson's ratio v, change referring to:

$$Y = Y(x_1, x_2) = Y(x), \upsilon = \upsilon(x_1, x_2) = \upsilon(x) \tag{2}$$

In addition, the singular stress field (σ) and displacement field in the terms of the standard angular functions at the crack location can be written as [20]:

$$\sigma_{11} = \frac{1}{\sqrt{2\pi r}}[K_I f_{11}^I(\theta) + K_I f_{11}^{II}(\theta)] \tag{3}$$

$$\sigma_{22} = \frac{1}{\sqrt{2\pi r}}[K_I f_{22}^I(\theta) + K_{II} f_{22}^{II}(\theta)] \quad (4)$$

$$\sigma_{12} = \frac{1}{\sqrt{2\pi r}}[K_I f_{12}^I(\theta) + K_{II} f_{12}^{II}(\theta)] \quad (5)$$

$$u=\{u_1, u_2\}T \quad (6)$$

$$u_1 = \frac{1}{G_{tip}}\sqrt{\frac{r}{2\pi}}[K_I g_1^I(\theta) + K_{II} g_1^{II}(\theta)] \quad (7)$$

$$u_2 = \frac{1}{G_{tip}}\sqrt{\frac{r}{2\pi}}[K_I g_2^I(\theta) + K_{II} g_2^{II}(\theta)] \quad (8)$$

$$G_{tip} = \frac{Y_{tip}}{[2(1+\nu_{tip})]} \quad (9)$$

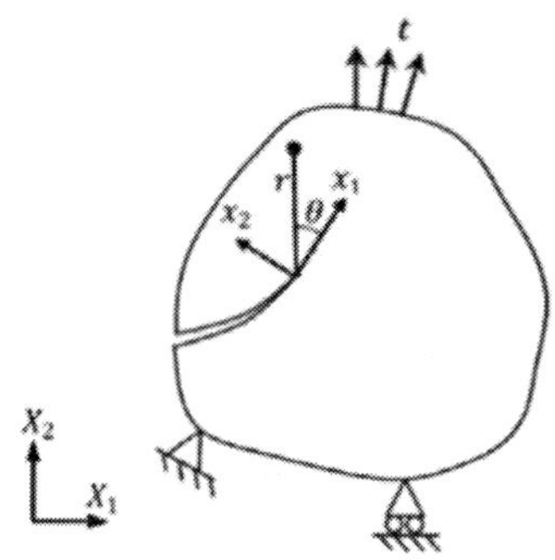

Figure 1: A cracked FGM domain.

3. XEFGM Procedure

Figure 2 depicts 2D problem of fracture problem that includes a crack $\boldsymbol{\Gamma_c}$. For this issue, the PD equation can be defined in the form of equilibrium equation as [6]:

$$\boldsymbol{L}^T\boldsymbol{\sigma} + \boldsymbol{b} = 0 \quad in\, problemd\; dmain\; \Omega \quad (10)$$

where L is the differential operator written as

$$L = \begin{bmatrix} \frac{\partial}{\partial x} & 0 \\ 0 & \frac{\partial}{\partial y} \\ \frac{\partial}{\partial y} & \frac{\partial}{\partial x} \end{bmatrix} \quad (11)$$

and $\boldsymbol{\sigma}, \boldsymbol{b}$ and $\boldsymbol{u}$ are the stress, body force, and displacement vectors, respectively. EFGM adopts moving least squares (MLS) technique [21] that put the shape functions to be worked totaly in the terms of distributed nodes. The weak form of the governing equation can be defined as[6]:

$$\int_\Omega (\boldsymbol{L}\delta\boldsymbol{u})^T(\boldsymbol{DLu})d\Omega - \int_\Omega \delta\, \boldsymbol{u}^T\boldsymbol{b}d\Omega - \int_{\Gamma_t} \delta\, \boldsymbol{u}^T\bar{\boldsymbol{t}}d\Gamma - \int_{\Gamma_u} \delta\, \boldsymbol{\lambda}^T(u-\bar{u})d\Gamma - \int_{\Gamma_u} \delta\, \boldsymbol{u}^T\lambda d\Gamma = 0 \quad (12)$$

where $\boldsymbol{\lambda}$ represents the Lagrange multiplier variable.

The displacement form that extract from the governing equation by applying XEFGM can be written as:

$$\boldsymbol{u}^h(\boldsymbol{x}) = \sum_{i=1}^{n} \phi_i(\boldsymbol{x})\boldsymbol{u}_i + \sum_{k=1}^{S_t} \phi_k \sum_{\alpha=1}^{4} E_\alpha(\boldsymbol{x})\boldsymbol{b}_k \tag{13}$$

where $\boldsymbol{b}_k$ represents extra DOFs that add crack tips, s_t is the nodes which the discontinuity in its support domain, and $E_\alpha(\mathbf{x})$ are the enrichment functions [6, 22]:

$$E(r,\theta) = \left(\sqrt{r}\cos\left(\frac{\theta}{2}\right), \sqrt{r}\sin\left(\frac{\theta}{2}\right), \sqrt{r}\sin\left(\frac{\theta}{2}\right)\sin(\theta), \sqrt{r}\cos\left(\frac{\theta}{2}\right)\sin(\theta)\right)$$
(14)

Discretization of Eq. (12) defines:

$$\begin{bmatrix} \boldsymbol{K} & \boldsymbol{Q} \\ \boldsymbol{Q}^T & \boldsymbol{0} \end{bmatrix} \begin{bmatrix} \boldsymbol{U} \\ \boldsymbol{\lambda} \end{bmatrix} = \begin{bmatrix} \boldsymbol{F} \\ \boldsymbol{q} \end{bmatrix} \tag{15}$$

where $\boldsymbol{K}$ represents the global stiffness matrix, $\boldsymbol{F}$ defines as the global force vector, $\boldsymbol{Q}$ and $\boldsymbol{q}$ are the Lagrange related terms for enforcement of the boundary conditions by the Lagrange multipliers $\boldsymbol{\lambda}$ [6]. In a particular, suitable election for the influence domain near a crack tip is adopted which is depended on the similar method used in FGMs crack examples by [6,23].

For the numerical integrations of the relevant equations, Gauss quadrature method is applied thru the background cell of XEFGM. The background cell that existed with a crack is divided at both sides into sub-triangles that edges are used to the crack surfaces. Gauss quadrature that uses a background structure and uniform nodal distribution are used in the modeling of the numerical examples. More details for the representation of the Gauss quadrature method and sub-triangles technique can be viewed in [23].

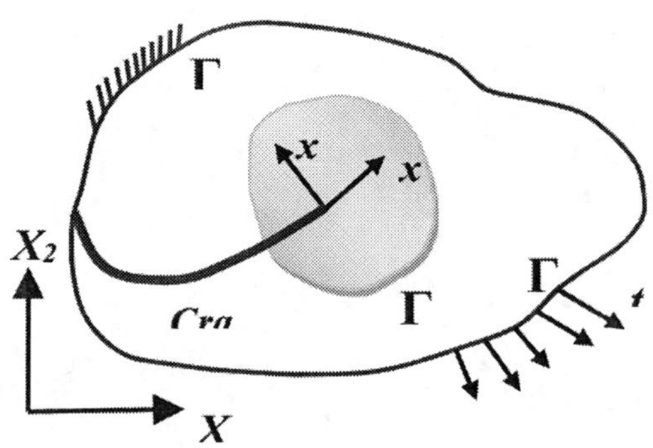

Figure 2: A 2D fractured domain.

4. Calculation of SIFs by the interaction integral method

The incompatibility formulation is followed for the auxiliary fields of cracked body [2,6]:

$$\sigma_{ij} = y_{ijkl}(x)\varepsilon_{kl}\ , \quad \varepsilon_{ij} \neq \frac{1}{2}\left(u_{i,j} + u_{j,i}\right)\ , \ \sigma_{ij,j} = 0 \tag{16}$$

The equivalent domain equation of the J-integral of Γ around the crack tip represents as (Figure 3):

$$J = \int_A \left(\sigma_{ij}u_{i,1} - w\delta_{1j}\right)q_{,j}\, dA + \int_A \left(\sigma_{ij}u_{i,1} - w\delta_{1j}\right)_{,j} q dA \tag{17}$$

where w represents the strain energy density:

$$w = \frac{1}{2}\sigma_{ij}\varepsilon_{ij} \tag{18}$$

The incompatibility formulation is implemented to find the mode I and II SIFs :

$$M = \int_A \left\{\sigma_{ij}u_{i,1}^{aux} + \sigma_{ij}^{aux}u_{i,1} - \frac{1}{2}(\sigma_{ik}u_{ik}^{aux} + \sigma_{ik}^{aux}u_i)\delta_{1j}\right\}q_{,j}\,dA + \int_A \left\{\sigma_{ij}\left(c_{ijkl}^{tip} - c_{ijkl}(x)\right)\sigma_{kl,1}^{aux}\right\}q dA \tag{19}$$

Therefore, J-integral can be written in terms of mode I and mode II SIFs:

$$J_{local} = (K_I^2 + K_{II}^2)/E_{tip} \tag{20}$$

$$J_{local}^{s} = \frac{\left(K_I + K_I^{aux}\right)^2 + \left(K_{II} + K_{II}^{aux}\right)^2}{Y_{tip}} = J_{local} + J_{local}^{aux} + M_{local} \tag{21}$$

Where Jlocal drived by Eq. (20), and J_{local}^{aux} represent by

$$J_{local}^{aux} = [(K_I^{aux})^2 + (K_{II}^{aux})^2]/Y_{tip} \tag{22}$$

and M_{local} is

$$M_{local} = 2(K_I K_I^{aux} + K_{II}K_{II}^{aux})\ /Y_{tip} \tag{23}$$

The mode I and mode II SIFs are extracted as gives:

$$K_I = M_{local}^{(1)}Y_{tip}\ /2\,, \qquad (K_I^{aux} = 1.0\,, K_{II}^{aux} = 0.0),$$

$$K_{II} = M_{local}^{(2)}Y_{tip}\ /2\,, \qquad (K_I^{aux} = 0.0\,, K_{II}^{aux} = 1.0). \tag{24}$$

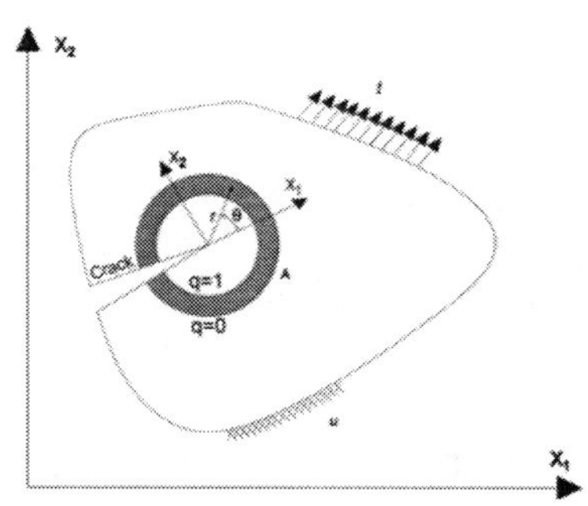

Figure 3: The model of contour at the tip of discontinuity .

5. Numerical case study

An edge crack FGM plate under traction load is adopted as shown in Figure 4. The change of the material properties is in the x1- direction according to the following equation:

$$Y(x1) = Yo\, e^{\alpha\, x1}, \alpha = \frac{1}{b}\ln\left(\frac{Yw}{Yo}\right) \tag{25}$$

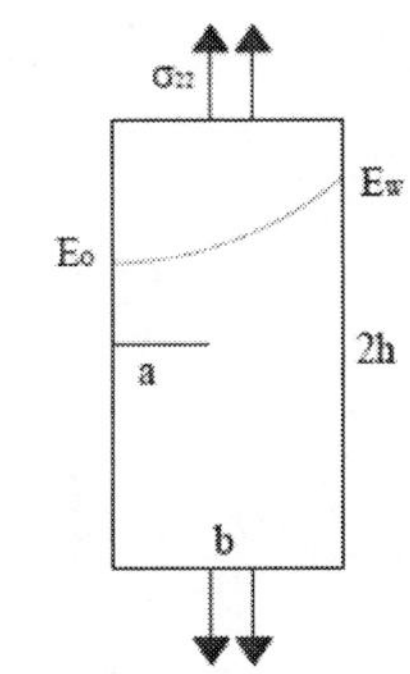

Figure 4: An edge crack FGM plate (b= 10 mm, 2h = 30 mm, v=0.25).

Where Y0=10000 MPa and Yw=50000 MPa are the Young's modulus at the left and right side of the plate, respectively. Circular support domain in term of cubic spline weight function with linear basis function are employed. Uniform distribution nodes, and background cells with Gauss quadrature term in each cell are adopted. Total number of nodes and background cells are 1624 and 1479 respectively. Total enrichment nodes in the crack tip are 64 node (16×4). The integration of the numerical equation is done by a 2×2 Gauss quadrature rule in all cell except at the crack sub-triangle cells is adopted by 13 Gauss nodes. Figures 5 and 6 represent the geometry and numerical model of XEFGM.

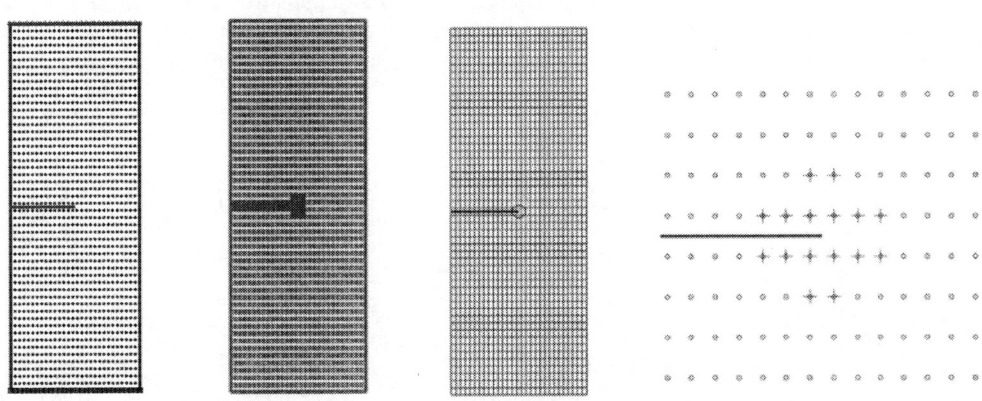

(a) (b) (c) (d)

Figure 5: (a) nodal distribution, (b) Gauss points, (c) background cell, and (d) enrichment nodes of the numerical model of an edge crack FGM plate (clack length a=5 mm).

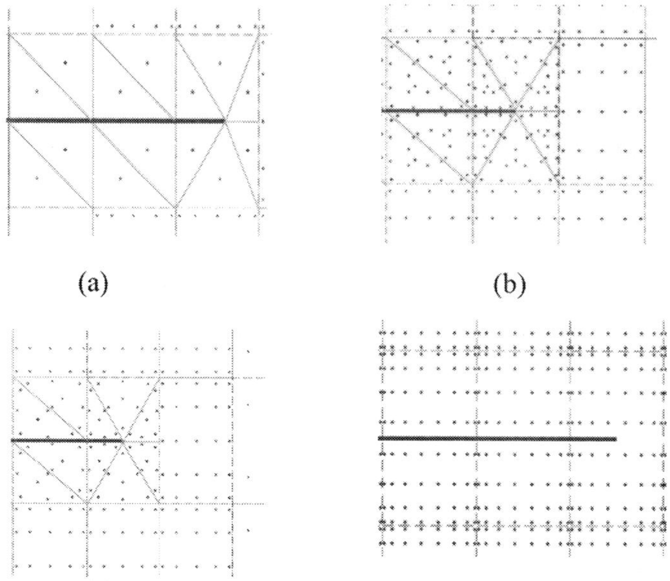

Figure 6. (a) 1 Gauss point for each proposed sub-triangle, (b) 13 Gauss points for each proposed sub-triangle, (c) 7 Gauss points for each proposed sub-triangle, and (d) 16 Gauss points of the standard approach of the numerical model in the crack region.

Table 1 compares the values of normalized SIFs in terms of crack lengths of the current work. For the verification issue, the work is compared with [5] at crack length equals to 4 mm because [5] have result only for this length only that used boundary element method. The error of the XEFG method with [5] can be viewed in Table 2. There is good agreement and stability in the results of SIFs through the sizes of J-integral rJ equal to 0.4 -0.8.

Table 1 Normalized SIFs (KI) of the present work in terms of the size of the J-integral domain (rJ) at size of support domain dmax=1.7 with 13 Gauss points for each sub-triangle.

a	$r_J = 0.4$	$r_J = 0.5$	$r_J = 0.6$	$r_J = 0.7$	$r_J = 0.8$	$r_J = 0.9$	$r_J = 1$	[5]
1	0.82388	1.0446	1.0446	1.0446	1.0502	1.0581	1.0768	-
2	0.70708	1.1816	1.1636	1.1479	1.1479	1.1598	1.1923	-
3	1.4034	1.4034	1.3818	1.3892	1.3892	1.3994	1.4646	-
4	1.7566	1.7566	1.7385	1.7681	1.7681	1.7873	1.8587	1.741
5	2.351	2.351	2.4041	2.4041	2.4041	2.4848	2.5095	-
6	3.5226	3.5226	3.5372	3.5206	3.5206	3.607	3.687	-
7	5.7538	5.7538	5.8018	5.7968	5.7968	5.9481	6.1907	-
8	9.6987	11.5217	11.8456	11.9466	11.9466	12.2571	12.7251	-
9	35.7161	42.3974	42.3974	42.3974	44.673	44.6849	46.7381	-

Table 2 The error of Normalized KI in comparing with [5] for r_J = 0.5, dmax=1.7, 13 Gauss points for each sub-triangle at a= 4 mm.

r_J	XEFG Normalized KI	Normalized KI [5]	Error %
0.4	1.7566	1.741	0.90
0.5	1.7566	1.741	0.90
0.6	1.7385	1.741	0.14
0.7	1.7681	1.741	1.56
0.8	1.7681	1.741	1.56
0.9	1.7873	1.741	2.66
1	1.8587	1.741	6.76

Table 3 shows the efficiency of the present technique (the sub-triangle technique) of the Gauss points distribution to calculate the correct SIF against standard approach. To further clarify Figure 6-a gives acceptance results for normalized KI at 1 Gauss point for each sub-triangle in comparing with standard approach 8×8 Gauss points. The best results can be determined at 7 and 13 Gauss points for each sub-triangle at the crack region as depicted in Table 4.

Table 3 Normalized SIFs (KI) of the present work in terms of various Gauss points in each cell in the crack region at dmax=1.7 with rJ=0.5.

a	Normalized KI at 1 Gauss point for each sub-triangle	Normalized KI at 7 Gauss points for each sub-triangle	Normalized KI at 13 Gauss points for each sub-triangle	Normalized KI at standard approach 4×4 Gauss points	Normalized KI at standard approach 8×8 Gauss points	Normalized KI [5]
1	1.0377	1.0441	1.0446	0.99138	0.99961	-
2	1.1422	1.1800	1.1816	1.0547	1.0805	-
3	1.3149	1.4005	1.4034	1.2883	1.3075	-
4	1.5835	1.7563	1.7566	1.5983	1.678	1.741
5	2.0243	2.3455	2.351	2.141	2.2965	-
6	2.9779	3.5006	3.5226	3.4958	3.5109	-
7	4.7191	5.7405	5.7538	6.0703	6.0671	-
8	9.9591	11.5894	11.5217	13.1671	12.613	-
9	56.7955	42.7133	42.3974	55.901	48.4083	-

Table 4 The error in Normalized KI in comparing with [5] for various number of Gauss points in each cell in the region of crack at crack length a=4 mm, and dmax=1.7 with rJ=0.5.

Gauss points distribution	XEFG Normalized KI	Normalized KI [5]	Error %
1 Gauss point for each sub-triangle	1.5835	1.741	9.04
7 Gauss points for each sub-triangle	1.7563	1.741	0.87
13 Gauss points for each sub-triangle	1.7566	1.741	0.89
standard approach 4×4 Gauss points	1.5983	1.741	8.19
standard approach 8×8 Gauss points	1.678	1.741	3.61

In addition, Figure 7 clears that the SIF increases when the crack length increases. The SIF at crack lengths equals to 9 mm is huge due to the plate reaches to the break as depicted in Figure 8.

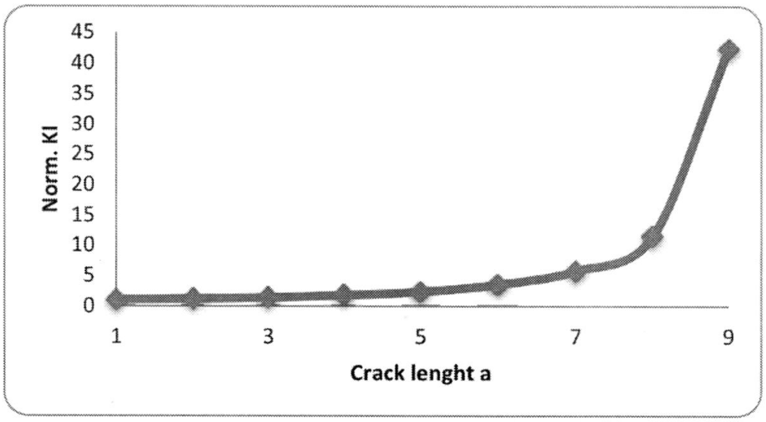

Figure 7: Values of normalized KI for various crack length at dmax=1.7, rJ=0.5, and 13 Gauss points for each sub-triangle.

Figure 8: The deform shape of an edge cracked plate at crack lengths (a=1 to 7 mm).

Figure (9) illustrates the result of SIFs in terms of a number of nodes and background cells. Therefore, optimal background cell has to be chosen to avoid unnecessary computational costs with giving the same accuracy. In addition, meshfree methed has been good agreement, convergence, and stability against change the total number of background cells.

Hence, Figure (*10*) is given to depict the influence of size of support domain on the precision of normalized KI. The acceptance results are bound between dmax=1.52 to 2. The best result of SIFs is at dmax=1.52 to 1.7. It is clear that when the size of the support domain more than 2, the results of SIFs is not well convergence. Finally, despite the change in properties of the material through the x-axis, as illustrated in Figure (10), exceedingly regular stress contours are extracted without any stress extrapolation or stress refine methods. It is also obviously cleared that the right side from the plate seems higher σ_{yy} stress, due to intense stiffness, where Yw/Y0 is more than 2.5.

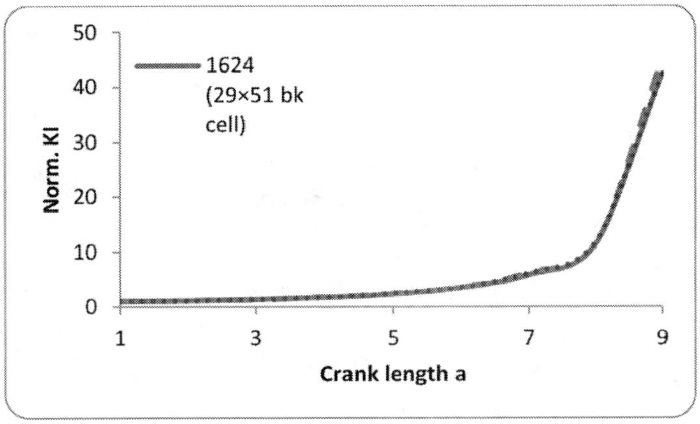

Figure 9: Values of normalized KI for various crack length at different number of nodes (dmax=1.7, rJ=0.5, and 13 Gauss points for each sub-triangle).

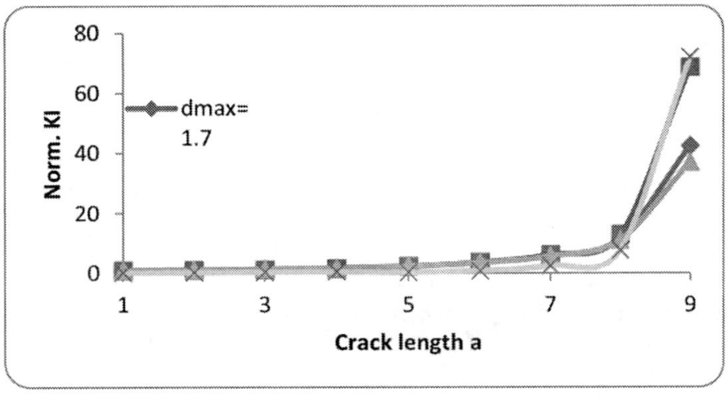

Figure 10: Values of normalized KI in terms of crack lengths at different sizes of support domain (rJ=0.5, and 13 Gauss points for each sub-triangle).

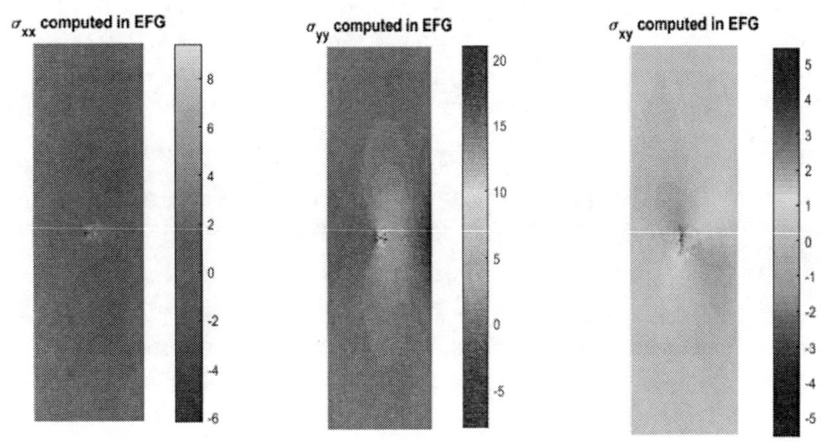

Figure 11: Stress contours of an edge cracked FGM plate under tensile loading (rJ=0.5, dmax=1.7, 13 gauss points for each sub-triangle)

6. CONCLUSIONS

This study reports the effect of crack lengths on the fracture parameters of a functionally graded material numerically and major conclusions of this study can explain in the following points:

1-The incompatible interaction integral method was used to determine the fracture parameters as SIFs.
2- The effective parameters of XEFGM were employed such as the sub-triangulation term for numerical integration, suitable influence domain, and enrichment functions for discontinues locations in the fracture analysis of different crack positions of an edge cracked FGM plate.
3- SIF in the plate increased when the crack length increased. The SIF at crack lengths equals to 9 mm is huge due to the plate reached to the break. In addition, in Gauss quadrature rule, the utilize of the sub- triangulation technique was given best results with good verification and stability with less integration points in comparing with the standard approach.
4- There is good agreement and stability in the results of SIFs through the sizes of J-integral rJ equal to 0.4 - 0.8.
5- The acceptance results for the size of the support domain are bound between dmax=1.52 to 2.

References

[1] J.-H. Kim and G.H. Paulino, On fracture criteria for mixed-mode crack propagation in functionally graded materials, Mechanics of Advanced Materials and Structures, vol.14, pp.227–244, 2007.

[2] J.-H. Kim and G. H. Paulino, Consistent formulations of the interaction integral method for fracture of functionally graded materials, Journal of Applied Mechanics, ASME, vol. 72, pp. 351-364, May 2005.

[3] J.-H. Kim and G.H. Paulino, Mixed-mode fracture of orthotropic functionally graded materials using finite elements and the modified crack closure method, Engineering Fracture Mechanics, vol.69 (14–16), pp.1557–1586, 2002.

[4] J. H. Kim, and G. H. Paulino, "Mixed mode crack propagation in functionally graded materials", Ph.D thesis, University of Illinois at Urbana-Campaign, 2000.

[5] X.W. Gao, Ch. Zhang, J. Sladek and V. Sladek, "Fracture analysis of functionally graded materials by a BEM" Composites Science and Technology 68 (2008) 1209-1215.

[6] H. Khazal, H. Bayesteh, S. Mohammadi, Sayyed Shahram Ghorashi, and Ameen Ahmed, An extended element free Galerkin method for fracture analysis of functionally graded materials, Mechanics of Advanced Materials and Structures, vol.23, pp.513-528, 2016.

[7] S. Garg and M. Pant, "Numerical simulation of adiabatic and isothermal cracks in functionally graded materials using optimized element-free Galerkin method", Journal of Thermal Stresses, Taylor and Francis 2017 .

[8] H. Khazal, and N. Saleh, "XEFGM for crack propagation analysis of functionally graded materials under mixed-mode and non-proportional loading", Mech. Adv. Mater. Struct., vol. 0,1-9 2018. Doi: 10.1080/15376494.2018.1432786.

[9] J. Sladek, V. Sladek, and C. Zhang, "A meshless local boundary integral equation method for dynamic anti-plane shear crack problem in functionally graded materials", Eng. Anal. Bound. Elem., vol. 29, pp. 334–342, 2005.

[10] W. Farouq, H. Khazal, and A. F. Hassan, "Fracture analysis of functionally graded material using digital image correlation technique and extended element-free Galerkin method", Optics and Lasers in Engineering, volume 121, October 2019, Pages 307-322

[11] J. Dolbow, "An extended finite element method with discontinuous enrichment for applied mechanics", Ph.D. thesis, Northwestern University, Evanston, IL, USA, 1999.

[12] H. Bayesteh and S. Mohammadi, "XFEM fracture analysis of orthotropic functionally graded materials, Composites, Part B 44, pp.8–25, 2013.

[13] S.S. Hosseini, H. Bayesteh and S. Mohammadi, Thermo-mechanical XFEM crack propagation analysis of functionally graded materials, Materials Science and Engineering, vol. 561, no.20, pp. 285–302, January 2013.

[14] J. Dolbow, An extended finite element method with discontinuous enrichment for applied mechanics, Ph.D. thesis. Northwestern University, Evanston, IL, USA, 1999.

[15] A. Bergara, J. I. Dorado, A. Martín-Meizoso and J. M. Martínez-Esnaola," Fatigue crack propagation at aeronautic engine vane guides using the extended finite element method (XFEM)", Mechanics of Advanced Materials and Structures, 11 Apr 2019.

[16] K. Ando, B. A. Kim, M. C. Chu, and S. Sato, "effects of crack length, notch root radius and grain size on fracture toughness of fine ceramics", FFEMS,Volume16, Issue9,September 1993,Pages 995-1006.

[17] J. Akbardoost, "Size and crack length effects on fracture toughness of polycrystalline graphite", Engineering Solid Mechanics 2 (2014) 183-192.

[18] H. Khazal, A. Hassan, W. Farouq, and H. Bayesteh, "Computation of Fracture Parameters in Stepwise Functionally Graded Materials Using Digital Image Correlation Technique," Materials Performance and Characterization 8, no. 1 (0): 344-354.

[19] A.Turnbull, and L.Wright,"modelling the electrochemical crack size effect on stress corrosion crack growth rate", corrosion science, volume 126, September 2017, Pages 69-77.

[20] B.N. Rao and S. Rahman, Mesh-free analysis of cracks in isotropic functionally graded materials, Engineering Fracture Mechanics, vol.70 pp.1–27, 2003.

[21] P. Lancaster and K. Salkauskas, Surfaces generated by moving least squares methods, Mathematics of Computation, vol. 37, pp. 141-158, 1981.

[22] M. Fleming, Y. Chu, Moran B, T. Belytschko, Enriched element-free Galerkin methods for crack tip fields, Int J Numer Methods Engng,1997;40:1483–504.

[23] S.S. Ghorashi, S. Mohammadi and S.R.S. Yazdi , "Orthotropic enriched element free Galerkin method for fracture analysis of composites" Engineering Fracture Mechanics-Volume 78, Issue 9, June 2011, Pages 1906–1927.

The working process of mixing Newtonian fluids in the channel of the multi-unit mixer diffuser-confusor type

[1]Israa S.Ahmed

[1]Khalid Faisal Sultan

[1]Ali Kamil Sebur

[2] A.A. Kurbangaleev

[2] S.V. Yushko

[2] F.Kh. Tazyukov

[2] F.R. Karibullina

[1]Electromechanical Engineering Department, University of Technology, Baghdad, Iraq
50112@uotechnology.edu.iq, 50278@uotechnology.edu.iq, 50084@uotechnology.edu.iq

[2] Kazan National Research Technological University , Russia.
ArturKurbangaleev@rambler.ru, tazyukov@mail.ru, faridafamily@mail.ru

Abstract The article presents the results of mathematical modeling of mixing the components of Newtonian fluids in the tubular channel of the mixer in the three-dimensional formulation. The mixing of components occurs under turbulent conditions, without taking into account the influence of mass forces and chemical interaction. The flow of the mixture is modeled in a tubular multi-channel diffuser-confuser type using the package "Fluent". The input of mixing components into the channel is made through jet nozzles. The quality of the resulting mixture is estimated by the mixing ratio. Using results of three-dimensional model the analysis of the mutual influence of the character of the current of the mixing components from the way of their giving in the channel is made: quantity and an arrangement on a channel rim of jet nozzles.

Keywords: Newtonian fluids, multi-channel diffuser-confuser, jet nozzles, three-dimensional model.

1. Introduction.

By the end of the 20th century, specialists and engineers in the design of tubular channels in mixer machines, widely used in chemical, petrochemical, food and other industries accumulated considerable experience in using these devices in different forms and types of tubular channels [1,4],

IOP Conf. Series: Materials Science and Engineering **765** (2020) 012059 doi:10.1088/1757-899X/765/1/012059

Fig. 1. It is known that the greatest practical efficiency in mixing the turbulent flows of liquids is represented by diffuser-confusor type channels [5,8].

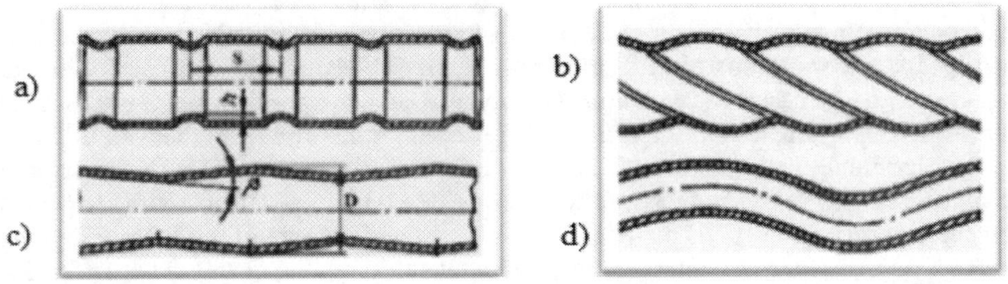

Figure 1: Different types of tubular channels: a) Channel with an annular projection, b) Pipe or a channel of the diffuser-confuser type, C) Spiral profiled channel, d) channel with a wave axis.

The hydrodynamics of the process in a multi-link channel of the diffuser-confusor type when mixing Newtonian fluids has been well studied by scientists such as Mukhametzyanova A. G., Takhavutdinov G. S., Danilov Yu. M., Garanin S. F., Petrov A. G., and so on. [6-11]. The tasks in these works were set as axisymmetric, however, with such a formulation of tasks it is impossible to simulate the input of components into the channel through jet nozzles, if in a real apparatus-mixer the supply of mixing components into the channel is carried out through a system of jet nozzles or nozzle heads. That is, the model in the axisymmetric formulation will not fully reflect the real picture of what is happening. The introduction of the mixture components into the channel in this simulation can be done only through the annular section. Therefore, the problem of mixing the components of liquids in a tubular multi-channel diffuser-confuser type with the input of components through the nozzles should be simulated only in a three-dimensional formulation [12], which will most adequately reflect the real mixing process.

2. Theoretical equations.

The aim of this work was to model a three-dimensional formulation for the problem of mixing two components of Newtonian fluids in a tubular multi-link channel of diffuser-confusor type with the introduction of the second component into the channel through jet nozzles. The mixing coefficient [13] (formula-1) was used for qualitative evaluation of the mixing process of components.

$$\gamma_a = 1 - \frac{1}{V}\iiint_V \left|C(x,y,z) - C_0\right| dV, \qquad (1)$$

where γ_a - the average volume mixing coefficient; C_0-the desired(specified) concentration; C -the resulting concentration; V -the specified volume; x,y,z -axis of the Cartesian coordinate system.

The mixing coefficient is associated with the rate of dissipation of the kinetic energy of turbulence, depending on the length of the confusor and diffusor part of the channel, the length of the entire tubular channel, the ratio of the cross-sectional diameter of the tubular channel and the diameter of the

IOP Conf. Series: Materials Science and Engineering **765** (2020) 012059 doi:10.1088/1757-899X/765/1/012059

inlet nozzles, the opening angle between the diffusor and confusor parts, the ratio between the thickness of the channel pipe in its widest and narrowest parts, as well as the properties and modes of mixing components.

A mathematical model based on the system of the Reynolds equations for turbulent mass transfer [14-16] (formula 2) was constructed for this problem. The channel of confuser - diffuser type (length Lк=1м. - 5м.) was used in the simulation. The opening angle of the diffuser was chosen so that the resulting unsteady micro separation of the flow of the resulting mixture from the wall did not cause a sharp increase in hydraulic resistances. The nozzles are located at right angles to the longitudinal axis of the channel section, the diameter of the nozzles dф=0,004 - 0,01 m. with respect to the diameter of the channel section 1: 10, dTp. = 0.04 m – 0.1 m. The flow rate of the mixture in the range from U_1= 0,3m/s – 1m/s, the speed of entering the components through the nozzles U_2= 1m/s-3M/s,

$$\begin{cases} \overline{\rho}\left[\frac{\partial \overline{u}_i}{\partial t}+\overline{u}_j\frac{\partial \overline{u}_i}{\partial x_j}\right]=-\frac{\partial \overline{p}}{\partial x_i}+\frac{\partial}{\partial x_j}\left(\mu\frac{\partial \overline{u}_i}{\partial x_j}\right)-\overline{\rho u'_j\frac{\partial u'_i}{\partial x_j}} \\ \frac{\partial \overline{u}_j}{\partial x_j}=0, \frac{\overline{\partial u'_j}}{\partial x_j}=0 \\ \frac{\partial C}{\partial t}+\overline{u}_j\frac{\partial C}{\partial x_j}=D\frac{\partial^2 C}{\partial x_{jj}^{\ 2}} \end{cases} \tag{2}$$

where D is the coefficient of mutual diffusion , $\overline{p}$ is the averaged pressure. $\overline{u}$ - Averaging-Favre velocities [17] (formula 3).

$$\overline{u}=\frac{1}{(t_1-t_2)}\int_{t_1}^{t_2} udt. \tag{3}$$

It is assumed that the mixing occurs without thermal interaction and without chemical reactions; the components of the mixture have no interfaces-free interpenetration; the mixture is considered to be a continuous medium with average, depending on local values, concentrations C(x,y,z), density and molecular viscosity (formula 4)[18]:

$$\rho=\frac{\rho_1\rho_2}{C\rho_1+(1-C)\rho_2}, \qquad \mu=\frac{\mu_1\mu_2}{C\mu_1+(1-C)\mu_2}. \tag{4}$$

The following turbulence model was used to close the system of equations (2): SST is a two - layer model proposed [14].

According to the boundary conditions for the tubular channel shown in Fig. 2 it is assumed that the velocity profiles, initial turbulence parameters and volume fractions of the components used are set in the input channel sections. $\varphi(x,y,z)$:

Input 1: $(u_1,\mu_1,C_1,k_1,\varepsilon_1)=\varphi(x,z)_1$, Input2 $(u_2,\mu_2,C_2,k_2,\varepsilon_2)=\varphi(x,z)_2$

In the output section, “soft” boundary conditions of the steady flow are set:

IOP Conf. Series: Materials Science and Engineering **765** (2020) 012059 doi:10.1088/1757-899X/765/1/012059

$$\left(\frac{\partial \phi}{\partial y}\right) = 0 \cdot$$

In the area of the walls, a "non - equilibrium" function was used-Non-Equilibrium Wall Functions-NEW F. on the walls themselves, the condition of "adhesion" - all components of the velocity are zero [19-21].

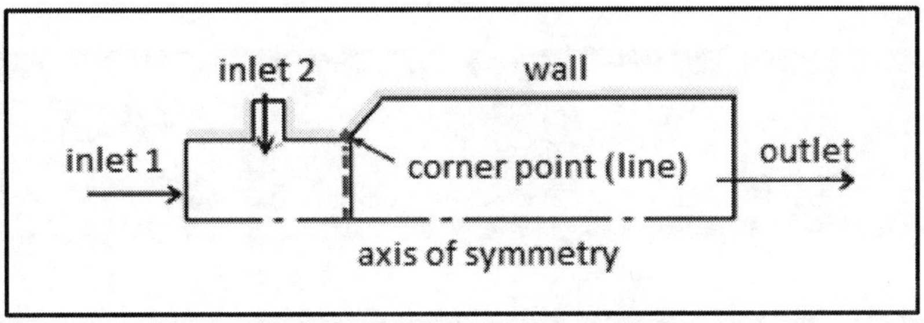

Figure 2: Initial section of the channel.

The task was solved with the help of the package "Fluent". The model of hydrodynamic mixing process is three-dimensional, the grid is adapted. A preliminary preparation of the grid area was carried out. In the area of the symmetry axis of the channel, the grid is sparse – "rough". Closer to the periphery, condensation was set, in the area of angular points, series decomposition was used to exclude non-physicality. The procedure of mirroring the grid area was also used to eliminate possible asymmetry of the flow [22]. The most detailed description of the adaptation and preparation of the grid to the problem is given in the article [23,24]. The number of nodes used was at least N = 150,000. The stability of the method was tested by the Courant condition. The convergence of the scheme was analyzed on thickening grids (by the mixing coefficient).

The number of nodes in the grid	The mixing coefficient y
50000	0.865
60000	0.843
70000	0.878
80000	0.905
90000	0.892
100000	0.928
110000	0.937
120000	0.944
130000	0.948
140000	0.949
150000	0.949

Table 1. The dependence of the coefficient - ya on the number of nodes in the grid.

3. Simulation Results

For Figure.3 based on the simulation results, the concentration field of the component in the tubular channel of the diffuser-confuser type, designed for mixing three components of liquids that do not enter into a chemical reaction with each other, is shown. The length of the diffuser-confusor tubular channel was equal to five transverse length scales. It is obtained that in this case the output mixing coefficient is equal to $\gamma_a = 0.949$.

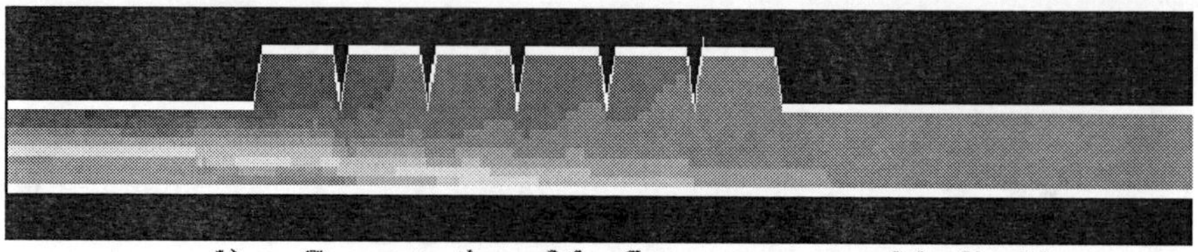

1) Concentration of the first component of the liquid

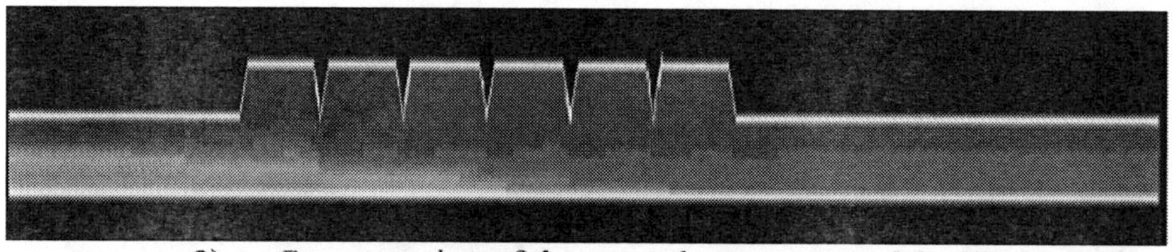

2) Concentration of the second component of the liquid

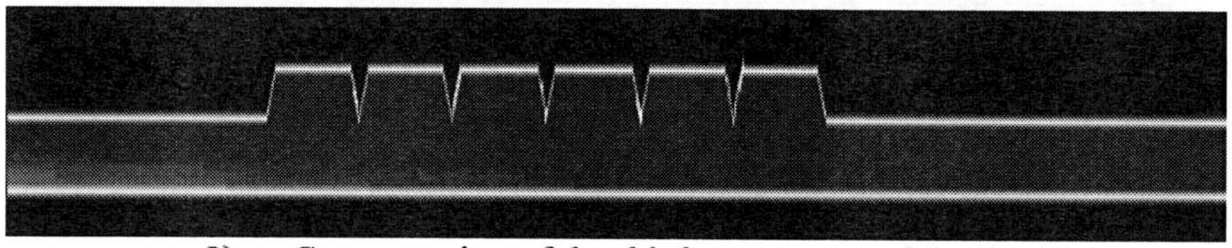

3) Concentration of the third component of the liquid

Figure 3: The distribution of the concentration of the liquid component.

When the relative length of the channel is up to 20 (the number of nodes is not less than N = 400 000), the mixing coefficient at the output increased to $\gamma_a = 0.975$, which indicates an increase in the quality of mixing Figure 4.

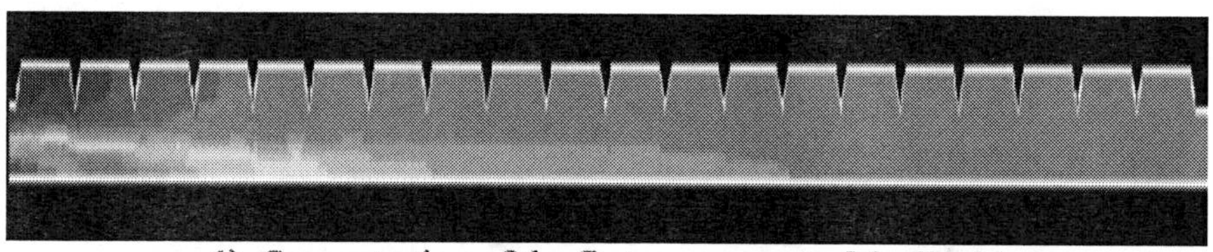

1) Concentration of the first component of the liquid

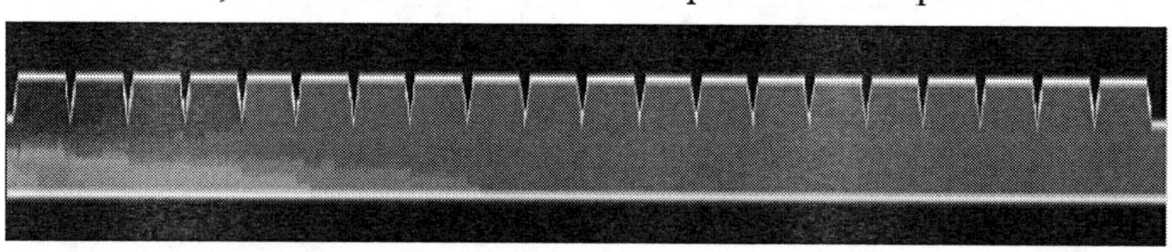

2) Concentration of the second component of the liquid

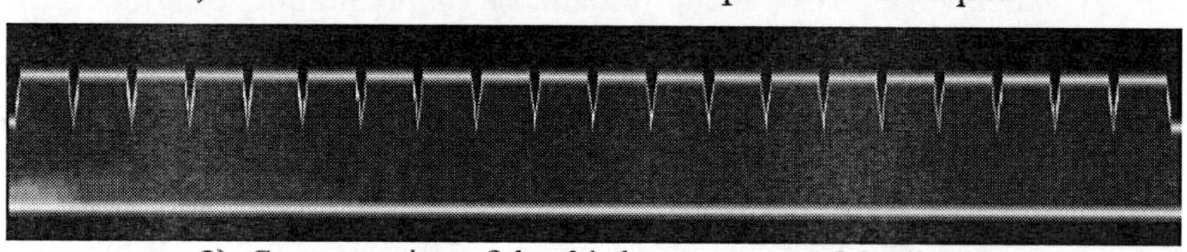

3) Concentration of the third component of the liquid

Figure 4: The distribution of the concentration of the liquid component.

The number of nodes in the grid	The mixing coefficient y
50000	0.675
70000	0.789
90000	0.826
110000	0.837
130000	0.849
150000	0.856
170000	0.875
190000	0.886
210000	0.893
230000	0.907
250000	0.918
270000	0.926
290000	0.934
310000	0.946
330000	0.955
350000	0.968
370000	0.974
390000	0.975
410000	0.975

Table 2. The dependence of the coefficient - ya on the number of nodes in the grid.

Figure.5 shows the concentration profiles in a tubular apparatus when mixing water (the first component) and ethyl alcohol (the second component) on the cross-section of the channel, where $z=L/r$ - is the number of the ratio of the scale of the length of the channel to the cross-section of its radius.

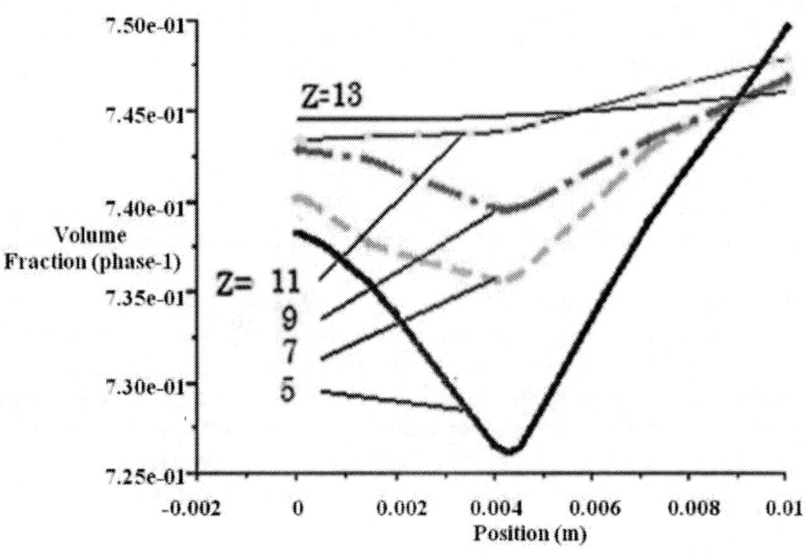

A)

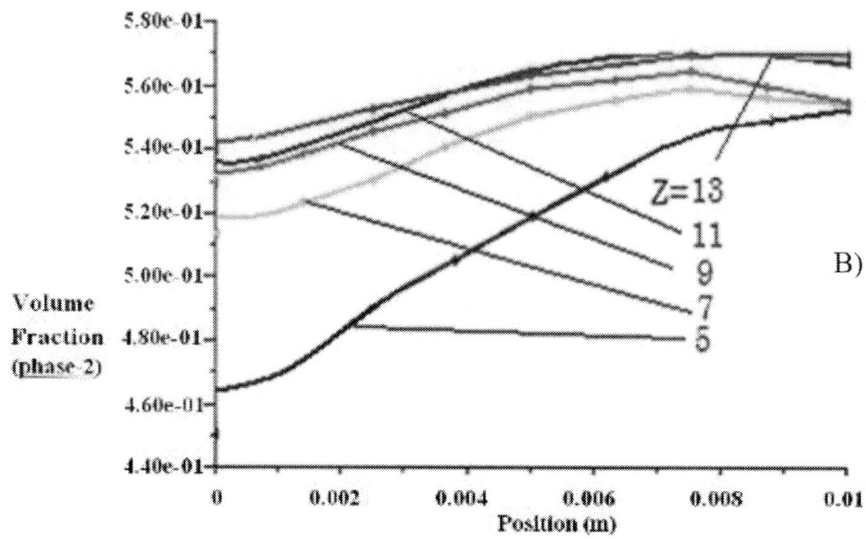

Figure 5: A) - Distribution of concentrations of the 1st component across the channel sections,

B) - Distribution of concentrations of the 2nd component across the channel sections

4. Conclusion.

As a result of the simulation, it is obtained that starting from approximately L/r=18, the concentration profiles in cylindrical apparatuses become almost identical regardless of the method of organization of the components supply. A similar pattern is observed with respect to the distribution of other hydrodynamic parameters.

As a rule, for tubular multi - link channels, the relative lengths z = 25-60 and in some cases even more are characteristic. Therefore, the flow in a larger area of the channel will not depend on the method of organization of the input of liquid components into it. Nevertheless, the main output variables affecting the efficiency of the channel (turbulence energy and its dissipation rate) will be strongly determined by the parameters of the injected fluid flows in the nozzle region, and it is possible to simulate the flow of mixing components through the nozzles into the channel only by a three-dimensional formulation of the problem [25]. That is, only in this case the obtained approximating expressions for the basic hydrodynamic parameters depending on the design and control variables should be used in the design of tubular channels, including the confuser-diffuser type. It can also be noted that before starting the process of modeling mixing in a tubular multi-channel diffuser-confusor type should first be analyzed: prepared and adapted to the flow area of the grid, to find out the mutual influence of the flow pattern of the mixture flow in the channel from the feed part of the nozzles of the channel, to calculate and analyze the provision of the necessary residence time of the components in the mixture for better mixing.

5. References

[1] Boyarinov A.I. ; "Optimization methods in chemical technology", A.I. Boyarinov, V.V. Kafarov ,M .: Chemistry. 1969.S. 568.

[2] Prochukhan Yu.A; "The influence of mixing methods on the nature of the course of ultrafast polymerization processes", Yu.A. Prochukhan, K.S. Minsker et al. , High Molecular Compounds 1988. V.30. No. 6. S. 1250.

[3] Schreiber A.A. ; "Turbulent flows of gas suspension", A.A. Schreiber, L.B. Gavin, V.A. Naumov, V.P. Yatsenko // Kiev. 1987.S. 240.

[4] Absi R. ; "Turbulent viscosity and velocity profiles in developed turbulent flows in channels", R. Absi , Bulletin of the Russian Academy of Sciences. Mechanics of fluid and gas. Moscow. - 2019. No. 1. S. 138-150.

[5] Mukhametzyanova A.G. ; "The movement of multiphase flows in tubular channels of a diffuser-confuser design", A.G. Mukhametzyanova, V.P. Zakharov, R.G. Takhavutdinov, G.S. Dyakonov, K.S. Minsker // Bulletin of the Bashkir University. 2002. No. 1. S. 28-31.

[6] Tahavutdinov R.G. ; " Calculation of mixing parameters of a two-phase flow in a diffuser-confuser tubular apparatus", R.G. Takhavutdinov, G.R. Dyakonov, A.G. Mukhametzyanova, K.S. Minsker, Collection of scientific and methodological articles "Topical issues of teaching in the reform of the educational process". Kazan. 2001. S. 135-150.

[7] Mukhametzyanova A.G. ; "Numerical calculation of mixing parameters of a two-phase turbulent flow in a diffuser-confuser tubular apparatus", A.G. Mukhametzyanova, R.G. Takhavutdinov, G.S. Dyakonov, V.P. Zakharov, K.S. Minsker ,Interuniversity. topics Sat scientific works. Kazan. KSTU. 2001. S. 4-13.

[8] Mukhametzyanova A.G. ; "Numerical calculation of the hydrodynamics of a two-phase turbulent flow in a diffuser-confuser apparatus", A.G. Mukhametzyanova, R.G. Takhavutdinov, G.S. Dyakonov // Scientific session of KSTU. Annotation posts. Kazan. 2001. S. 91.

[9] Danilov Yu.M. ; "Design technology for small-sized turbulent tubular apparatuses based on the results of numerical modeling", Yu.M. Danilov, A.G. Mukhametzyanova // Zh: Chemical and oil and gas engineering. 2011. No5. S.38-42.

[10] Garanin S.F., Kravets E.M., Pronina O.N., Stadnik A.L. ; "Numerical Modeling of Two-Dimensional Flow of a Nonhomogeneous Fluid in a Confined Domain", Fluid Dynamics. 2018. T. 53. № 1. C. 127-135.

[11] Petrov A.G. ; "The exact solution of the equations of axisymmetric motion of a viscous fluid between parallel planes when they approach and move apart", A.G. Petrov , Bulletin of the Russian Academy of Sciences. Mechanics of fluid and gas. Moscow. - 2019. No. 1. S. 58-67.

[12] Kurbangaleev A.A. ; " Numerical 3D modeling of mixing components in the MTA.", Yu.M. Danilov, A.A. Kurbangaleev, A.G. Mukhametzyanova, K.A. Alekseev // Bulletin of Kazan

Technological University. 2012. T15. No. 12. - S. 167–169.

[13] Kurbangaleev A.A. ; "Algorithm for 3D modeling the process of mixing liquids in tubular channels", A.A. Kurbangaleev ,Transactions of Academenergo.- ed. Kazan Scientific Center of the Russian Academy of Sciences. - 2017. - 2. - S. 34–47.

[14] Kurbangaleev A.A. ; "Method of 3D modeling the process of mixing Newtonian fluids in tubular channels of diffuser-confuser type", A.A. Kurbangaleev ,International Scientific and Research Journal - ed. Yekaterinburg. - 2016. - No. 12 (54). - Part 5. - December. - S. 16–21.

[15] DeardorffJW. ; "A numerical study of three-dimensional turbulent channel flow at large Reynolds numbers", JW. Deardorff, JournalofFluidMechanics1970. - cambridge.org 41, P. 453-480.

[16] Hoyas S., Jiménez J. ; " Reynolds number effects on the Reynolds-stress budgets in turbulent channels", S. Hoyas, J. Jiménez , PhysicsofFluids20, 2008.- aip.scitation.org 101511.

[17] Lunev V.V. ; "On the modification of the averaged Navier-Stokes equations", V.V. Lunev, Bulletin of the Russian Academy of Sciences. Mechanics of fluid and gas. Moscow. - 2019. No. 2. S. 134-144.

[18] Nikushchenko D.V. ; "Investigation of viscous incompressible fluid flows based on the FLUENT calculation complex", D.V. Nikushchenko ,Textbook SPbGMTU. St. Petersburg - 2005 94 sec.

[19] Aleksin V.A. ; " Application of the method of near-wall boundary conditions to an investigation of turbulent flows with longitudinal pressure gradients, V.A. Aleksin, Fluid Dynamics. 2017. T. 52. № 2. C. 275-287.

[20] Jimenez J. ; "Coherent structures in near-wall turbulence", J. Jimenez ,Journal of Fluid Mechanics, 842, P1.DOI: 10.1017.2018. P. 144.

[21] Gorbushin A.R., Zametaev V.B., Lipatov I.I. ; " Stationary secondary flow in a plane turbulent free stream", A.R. Gorbushin, V.B. Zametaev, I.I. Lipatov, Proceedings of the Russian Academy of Sciences. Mechanics of fluid and gas. Moscow. - 2019. No. 2. S. 99-111.

[22] Nikitin N.V., Popelenskaya N.V. ; " Characteristics of instability developing in a turbulent flow in a flat channel", N.V. Nikitin, N.V. Popelenskaya, Proceedings of the Russian Academy of Sciences. Mechanics of fluid and gas. Moscow. - 2019. No. 2. P. 72-93.

[23] Kurbangaleev A.A. ; "The process of preparing the grid area in 3D - modeling of a small-sized tubular apparatus (MTA) as a mixer in the Fluent software environment", A.A. Kurbangaleev, F.Kh. Tazyukov, G.N. Lutfullina // Bulletin of Kazan Technological University. - 2013. - T16. - No. 21. - S. 242–244.

[24] Kurbangaleev A.A. ; "Reducing computational errors in numerical 3D modeling of mixing in axisymmetric channels", Yu.M. Danilov, A.A. Kurbangaleev, Bulletin of Kazan Technological University. - 2012. - T15. - No. 12. - S. 161–163.

[25] Kurbangaleev A.A. ; " Design of small-sized tubular apparatus - MTA as a mixer of liquid components using 3D modeling", A.A. Kurbangaleev, F.Kh. Tazyukov, G.N. Lutfullina, Yu.M. Danilov // Bulletin of Kazan Technological University. - 2013. - T16. - No. 21. - S. 261–263.

INFLUENCE OF ROUGHNESS PARAMETERS ON THE HARDENING OF CAST STEEL PLATES

A.A. Lopatin[1]
Enaam Obeid Hassoun[2]
Hussain Abdulaziz Abrahem[3]
Ch. F. Eremeeva[4]
R.A. Gabdullina[5]

[1,4,5] KNITU - KAI named after AN Tupolev, Institute of Aviation, Land Transport and Energy, Kazan-Russia. aalopatin@kai.ru, cheremeeva@gmail.com, rozzy94@mail.ru

[2,3]University of Technology – Department of Electromechanical Engineering, Iraq. 50109@uotechnology.edu.iq, 50008@uotechnology.edu.iq

Abstract. This paper presents the experimental and numerical analysis of the effect of roughness parameters on the hardening of a 148 mm x 148 mm cast steel plate. The temperature distribution on the plate surface was measured at various points, and the infabstractrared (IR) camera was used to capture infrared images during the cooling process. In this study, the temperature distribution on the plate surface was evaluated at each pixel image, and the data was collected and recorded using the Therma CAM researcher 2001 software. The results show that the cooling rates on the smooth and rough sides were significantly high at 500 C. Also, it has been found that the total amount of energy is constant over the entire thickness of the sample

Keywords: Quenching, Metal Properties, Thermal stress, Jet & Nozzle, Leiden frost temperature, boiling curve

1. Introduction

Metal hardening is an important process in various technical applications. In many heat treatment processes, metal work pieces must be hardened in a certain way. The hardening process aims to control the properties of the metal, such as strength, hardness, flexibility. The main advantage of using this process is that it not only achieves the necessary qualities but also reduces the deformation of the work piece.

This study aimed to conduct an experimental investigation of the hardening of smooth and micro-structured surfaces obtained by applying micro-dispersed roughness using a spray nozzle installed in a vertical position, observing the heat transfer process and further analyzing the parameters that affect the temperature change [1].

When quenching products from steel alloys, it is very important to ensure the specified cooling rate of a part is maintained; otherwise, the already rebuilt atomic structure of the metal can go into an intermediate state. Meanwhile, cooling that is too rapid is also undesirable, as it can lead to the appearance of cracks on the part or its deformation or other defects such as described in the following:

1. The insufficient hardness of the hardened part is a consequence of low heating temperature, low shutter speed at operating temperature, or insufficient cooling rate. Such defects can be corrected by normalization, annealing with subsequent hardening, or the use of a more energetic quenching medium.

2. Overheating is associated with heating the product to a temperature significantly higher than the required heating temperature for quenching. Overheating leads to the formation of a coarse-grained structure, resulting in increased brittleness of steel. Such defects can be corrected by annealing (normalization) and subsequent quenching at the required temperature.

3. Burning occurs when steel is heated to very high temperatures, close to the melting temperature (1200–1300 °C) in an oxidizing atmosphere. Oxygen penetrates the steel, and oxides are formed along the grain boundaries. Such steel is fragile and impossible to fix.

4. The oxidation and decarburization of steel are characterized by the formation of scale (oxides) on the surface of parts and carbon burnout in the surface layers. Such defects are incorrigible. If the machining allowance allows, the oxidized and decarburized layer must be removed by grinding. To prevent such defects, heating the parts in ovens with a protective atmosphere is recommended.

5. Warping and the occurrence of cracks are consequences of internal stresses. During the heating and cooling of steel, volume changes are observed, depending on temperature and structural transformations (the transition of austenite to marten site is accompanied by an increase in volume up to 3%). The simultaneous change in the volume of the hardened part due to its various sizes and cooling rates over the cross-section leads to the development of strong internal stresses, which cause cracks and warping of the parts during hardening [3,4].

Cooling steel plates with a spray nozzle plays a key role in metal processing. In this study, the experiments used a spray nozzle 460.404 with a jet angle θ = 60 °, installed in a vertical position (Fig. 1)

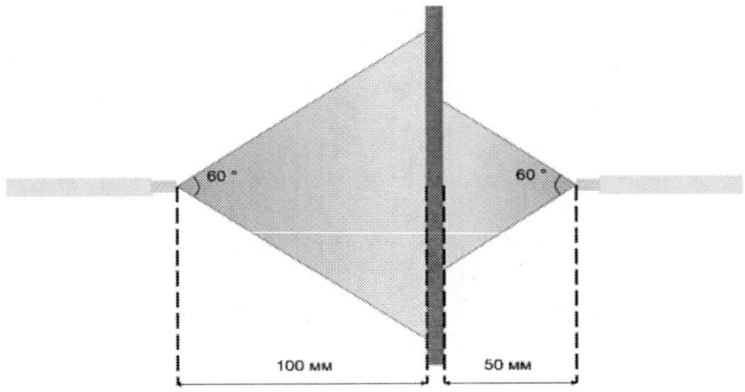

Figure 1: Front view illustrating spray cooling

In particular, in steel production, spray cooling is widely used in foundry and is also used to improve the final microstructure immediately after heat treatment. The scientific phenomenon is presented in Fig. 2. The heat transfer coefficient α is defined as $\alpha = q / (T_{surf} - T_{water})$, and the specific heat flux q passes through four boiling modes.

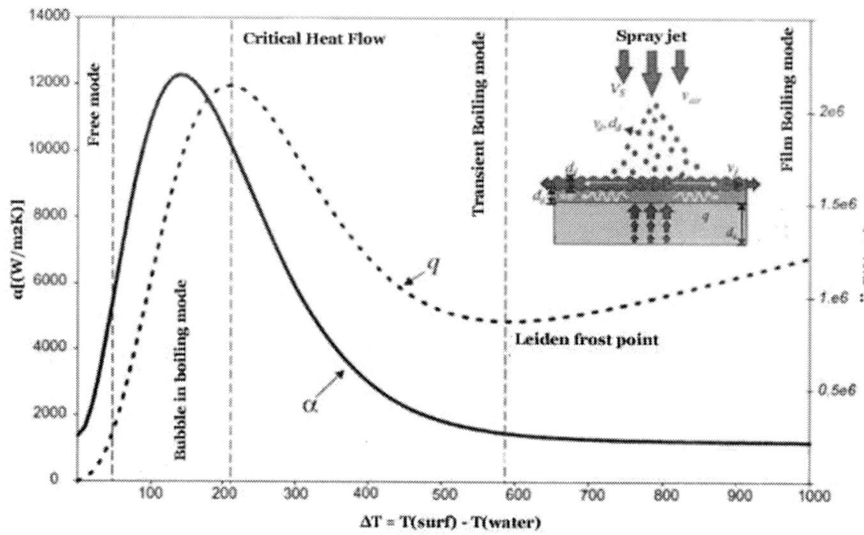

Figure 2: Spray cooling mechanism and boiling curve

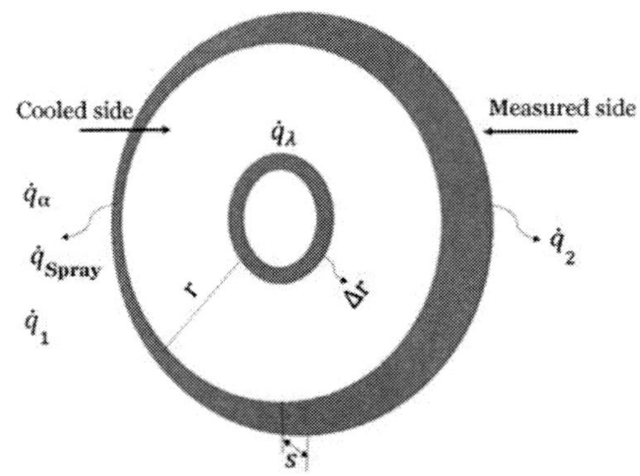

Figure 3: Heat transfer mechanism during spray nozzle quenching

At the very low-temperature gradient $\Delta T = T_{surf} - T_{water}$, the heat flux in the convective mode is released due to natural convection. With bubble boiling, vapor bubbles form jets until critical heat flux CHF is reached. In the film boiling mode, a stable vapor film is formed between the liquid film and the surface. The Leiden frost point characterizes the minimum heat flux density q at high temperatures. Below the Leiden frost point, the vapor film becomes unstable until the CHF has been reached again [2, 5].

2. Experimental Method

The main components of the experimental setup are presented in Fig. 4.

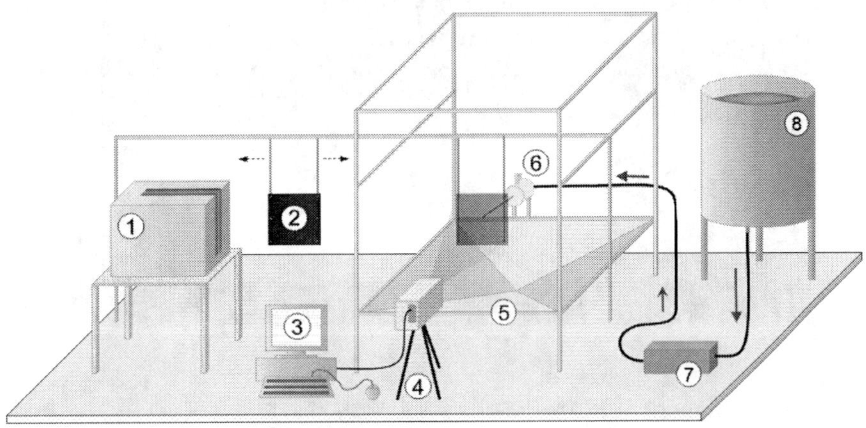

Figure 1: Test setup—Oven (1), Steel plate (2), Computer (3), Infrared camera FLIR SC3000 (4), Experimental cabin (5), Nozzle (6), Pump (7), Water tank (8).

The main component of this installation is the guide rail, which allows the insertion of a metal sample into the furnace for heating. As a first step in the process, the steel plate is heated in the furnace to a temperature of 520 °C. After heating, the stove is immediately installed in the cab with the help of a rail for cooling. The distance between the nozzle and the plate can be easily changed. Profiles allow configuring the test bench according to changes. A video camera records the process from the wetting site. The infrared (IR) camera FLIR SC3000 is used to record infrared images for the cooling process, on the other side of the steel plate. This site is covered with black graphite paint with an emissivity of 0.94. An IR camera measures temperature with a frequency of 150 Hz and a window of 240 × 80 pixels. The heat transfer mechanism is shown in Fig. 3. In all experiments, a steel plate with the dimensions of 148 mm x 148 mm and thickness s = 7 mm had an initial temperature $T_0 = 500$ ° C (± 3 ° C). Liquid cooling was performed using tap water having a temperature of 18 ° C and a volumetric flow rate of $V_0 = 1.2$ l / min.

3. Result and discussions

The temperature distribution of the plate was measured at 40 points to obtain the most accurate results. For the spray nozzle, the first measurement point was taken visually at the center of the spray cone. The next 39 measurement points were spaced at 2-pixel intervals, see Fig. 5. Then using software, the temperature curves were plotted over time at each point.

Curves were plotted until the temperature of the farthest measuring point from the center reached 80 °C. Ratio profile image temperature over time is presented in Fig. 6

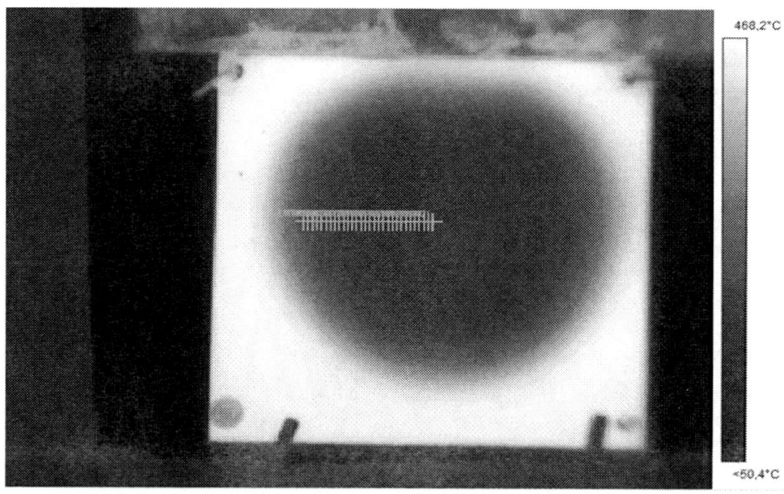

Figure 5: IR image of a steel plate heated to 500 ° C

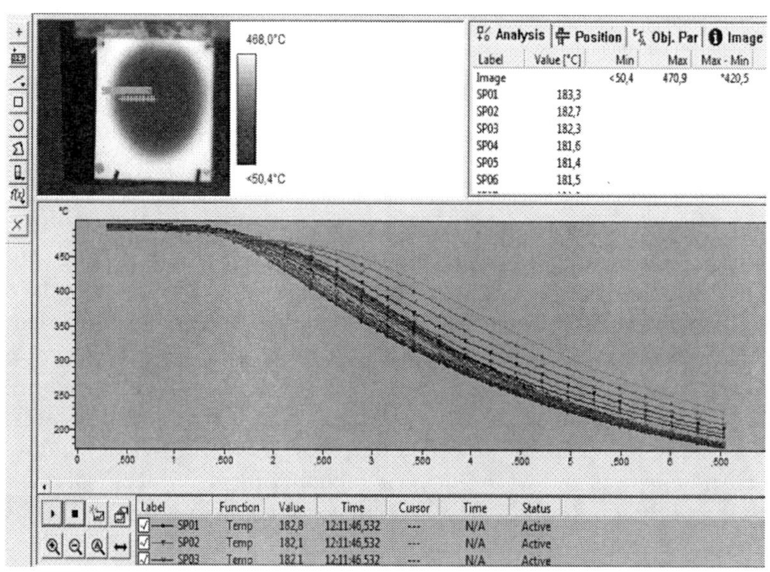

Figure 6: Graph of temperature versus time of the steel plate cooling process

The temperature curves over time for two points of smooth rough sides are presented in Fig 7. It can be seen from the results that the dashed lines represent the temperature curves for the rough side. In the case of a smooth side, the two lines of the mid and far points are close to each other. In other words, the cooling rate of the steel plate is practically independent of the location of the point. In the case of a rough surface, the two lines are more distinct.

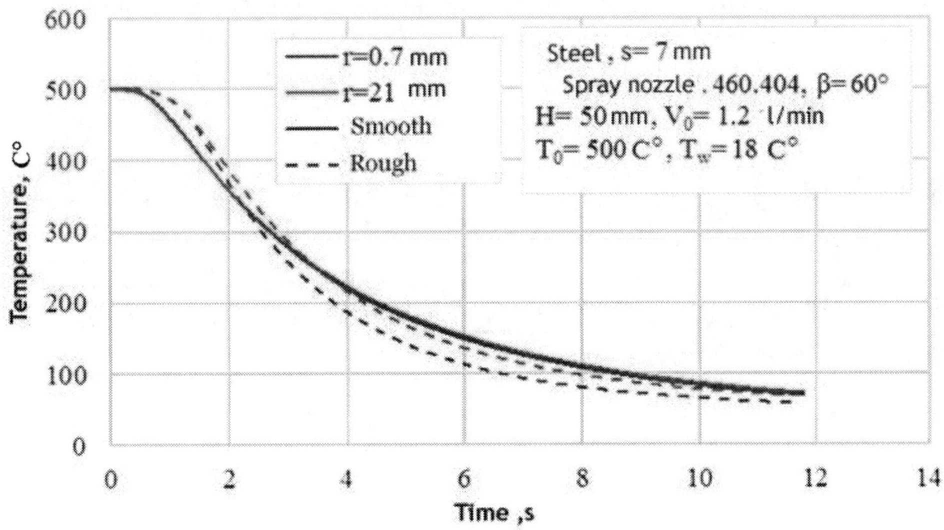

Figure 7: Comparison of temperature curves for different radii of smooth and rough surfaces (Steel, H = 50 mm, Tw = 18 C°, Spray nozzle 460.404, β = 60°)

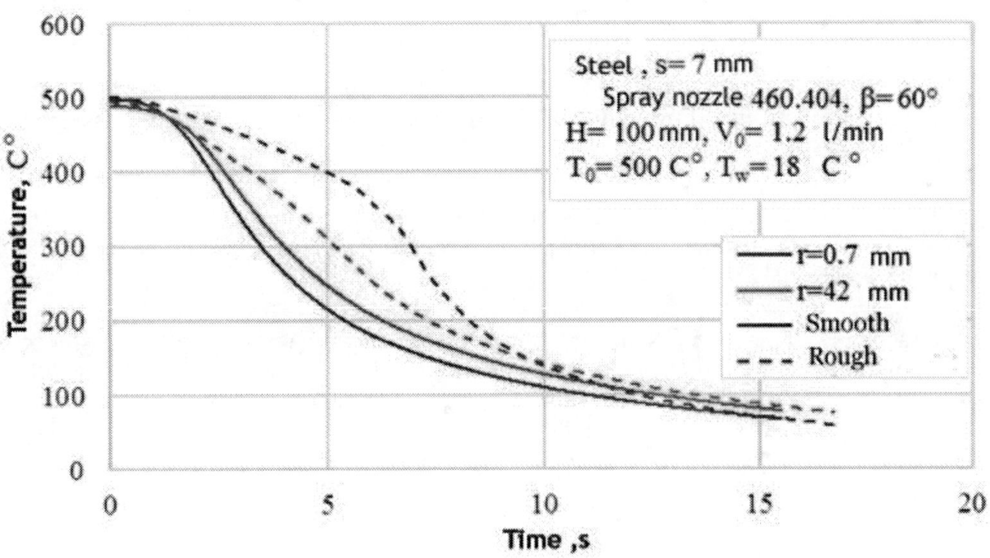

Figure 8: Comparison of temperature curves for different radii of smooth and rough surfaces (Steel, H = 100 mm, Tw = 18 °C, Spray nozzle 460.404, β = 60°)

Thus, cooling begins earlier on the plate with a smooth side; however, the roughness plate reaches the temperature of 100 °C faster.

Fig. 8 shows the results obtained during the cooling of steel at a distance of 100 mm from the nozzle to the plate. In this experiment, the water travels a distance twice as large as in the previous experiment. It is evident from the results that the cooling rates on the smooth and rough sides vary greatly. Starting at 500 °C, the smooth-side cooling is significantly faster than the rough-side cooling because the water reaching the plate does not have the adequate speed for spreading over a rough surface as fast as over a smooth surface. Water spreads rather slowly on uneven surfaces as well. However, 100 °C is achieved in two experiments at almost at the same time.

4. CONCLUSIONS

The effect of surface roughness during cooling was investigated, and a comparison was made between the variable parameters of a 148 mm x 148 mm steel plate. Operational parameters, such as the initial temperature of the sample, the temperature of the refrigerant (tap water), and the volumetric flow rate, remained unchanged during all experiments.

It was concluded that the plate with a smooth side has a higher heat transfer than that with a rough side due to the roughness on the surface of the plate and the difficulties of water spreading over the rough surface in an upright position.

References

[1] Eremeeva C. F., "Investigation of the influence of surface roughness on the cooling of cast steel plates by spray and full jet nozzles," Otto-von-Guericke University of Magdeburg, Master thesis, 2017.

[2] Wendelstorf J., Spitzer K. H., Wendelstorf R., "Spray water cooling heat transfer at high temperatures and liquid mass fluxes" International Journal of Heat and Mass Transfer, vol. 51, pp. 4902–4910, 2008.

[3] Smakulski P., Pietrowicz S., "A review of the capabilities of high heat flux removal by porous materials, micro channels and spray cooling techniques," Applied Thermal Engineering, vol. 104, pp. 636–646, 2016.

[4] Horacek, B., Kiger, K., Kim, J., "Single nozzle spray cooling heat transfer mechanisms," International Journal of Heat and Mass Transfer, vol. 48, no. 8, pp. 1425–1438, 2005.

[5] Bernardin, J., Mudawar, I., "The leidenfrost point: Experimental study and assessment of existing models," ASME Journal of Heat Transfer, vol. 121, pp. 894–903, 1999

Simulation of the analytical solution of lasing output power for Yb^{3+}: YAG thin disk laser regime in a fundamental mode operation

Raed Mahdi Salih[1]

Mudhir Shihab Ahmed[2]

[1]Al-khwarizmi Engineering College , University of Baghdad, Iraq

[2]College of education for pure science Ibn Al-Haitham, University of Baghdad, Iraq

corresponding author's e-mail : raedmahdi.rm@gmail.com

Abstract. A theoretical model is established to derivation equation of output Laser power for multi-pass pumping for a quasi-three level thin-disk laser, A numerical solution of this equation for Yb^{3+}:YAG thin-disk laser , which pumped at wavelength (940 nm) and lasing emission at wavelength (1030 nm) , regime in fundamental transverse mode (M^2=0.95) at room temperature were performed by calculated the variation of (Nt),(d),(Mp),(R2),(T) and (Pp) on the output Laser power and find the typical value of these parameters are used to obtain high output Laser power.

Keywords: thin disk laser, fundamental mode, continuous wave, output power.

1. Introduction

A Yb^{3+} as an ion that extracts from a simple electronic structure . Having two manifolds separated from (10000 cm^{-1}), The effects like excited state absorption (ESA), cross-relaxation, and upper transformation are absent. These processes, whose exist for another rare-earth ion usually used as lasant ions, can reduce the efficiency of the laser due to an alternative pathway for energy transfer [1,2]. a Yb^{3+} ion contain a low quantitative defect, which leads to reduced thermal load due to the energy mismatch between the wavelengths of pumping and lasing [2,3].

By considering Y3Al5O12 (YAG) as a host material, owns several acceptable qualities for medium-power laser applications. They were very mechanically strong [1,4].One of the charming features of Yb^{3+}: YAG is that it can contain very high concentrations of Yb^{3+}. Special laser architectures, such as a thin disk Laser (TDL)[4,5].

For example, in (TDL) design, the fracture limit is measured as an inverse of the thickness of the gain medium. So, maintaining a thin gain element while still owning enough rare-earth ion density from the ground to absorb the pump can provide excellent efficiency, good pump density and enhanced thermal management [5,6,7].

2. Thin-disk laser principle:

The (TDL) concept is a common suitable way when high efficiency, high power and excellent beam quality are required at the same time. Due to the ratio of a small volume to the surface of the area, The gain may be cooled with very high efficiency [9,10].

Also, the heat flow direction is parallel to the axis of the laser resonator, which, along within a very short optical path length during an active medium, reducing the effects of the thermal lens in addition to thermally induced deviation at orders of magnitude comparing with the laser rod power[9,11].

Thus, when the density of pump power is stable above a pump spot area, there are temperature gradients available within the volume of the pump, which leading a system easily scalable in the output power by increasing the diameter of a pump spot area at a constant pump density and this keeps the benefits of a pumping more efficiency. With improved thermal management [9,11].Figure (1) illustrate the principle of (TDL) design.

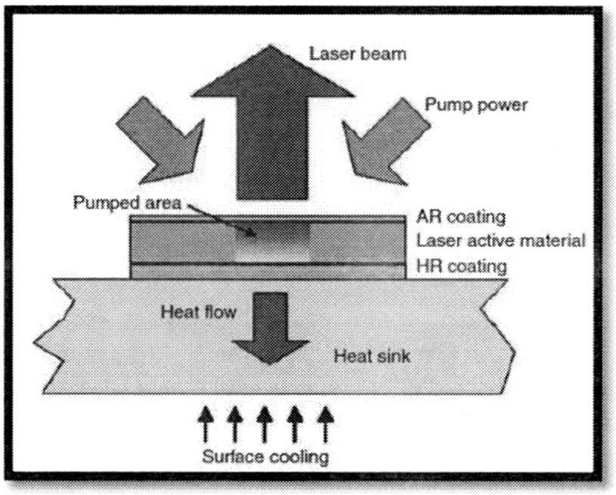

Figure 1: The principle of thin disk lasers design [12]

Because of its low quantum defect and broad absorption band, quasi-three-level Yb^{3+}:YAG has been studied extensively to develop high-efficiency and output power, pumped by laser diode (InGaAs) operation at (940 nm), when the lasing wavelength is (1030), with the quantum defect is (8.7%)[13,14]. However, the (TDL) scaled to a (kW) level whereas keeping close to the fundamental mode operation ($M^2<1$) is unrivalled and it is commonly limited to the output power about (1KW) [15].

The beam quality (M^2) in thin-disk laser described as [16, 17]:

$$M^2 = \left(\frac{a\,rp}{rfm}\right)^2 \tag{1}$$

Where (rp) is the pump spot radius, (rfm) is the fundamental mode radius and (a) is the indicates of the pump beam intensity and lasing mode intensity overlap, for a thin disk lasers (a=0.85) [18].

IOP Conf. Series: Materials Science and Engineering **765** (2020) 012062 doi:10.1088/1757-899X/765/1/012062

3. Analytical model:

The typical laser material for thin disk setup is Yb^{3+}:YAG, at room temperature a quasi- three-level laser material[19,20].The energy level system of Yb^{3+}:YAG consists of the stark split ($2F_{\frac{7}{2}}$) a ground state and ($2F_{\frac{5}{2}}$)) an excited state.

Because of the small energy splitting of each manifold, are assumed to be very small. This justifies the use of a two-manifold model, a ground state for energy (E1),and population (N1) and excited state for energy (E2) , and population(N2)[19,20] .

The energy levels of Yb^{3+} ion in YAG crystal at room temperature are shown in Figure (2).

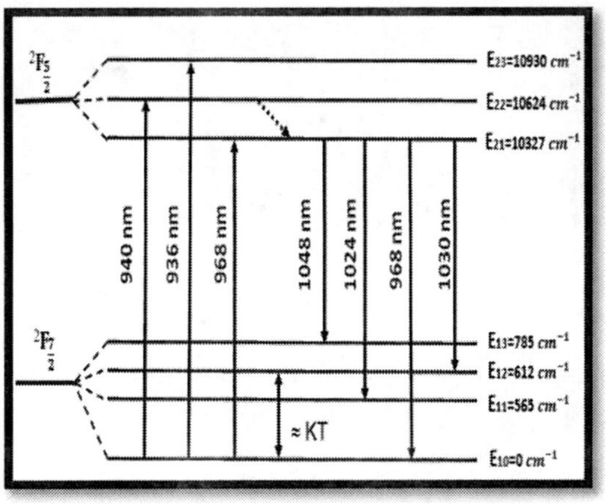

Figure 2: Energy level diagram for Yb^{3+}: YAG active

The rate equation for a ground state (N1) and excited state (N2), can be expressed as follows [22, 23, 24]:

$$\frac{dN_2}{dt} = \left[\sigma_{ap}(N_t - N_2) - \sigma_{ep}N_2\right]\frac{\eta_a \lambda_p}{\alpha d h c} Ip + \left[\sigma_{aL}(N_t - N_2) - \sigma_{eL}N_2\right]\frac{M_L \lambda L}{hc} I_L - \frac{N_2}{\tau} \quad (2)$$

$$\frac{dI_L}{dt} = \frac{M_L dc}{2L}\left[(\sigma_{aL} + \sigma_{eL})N_2 - \sigma_{aL}N_t\right]I_L - \frac{c}{2L}\left[-\ln(1 - T_r) - \ln(1 - \gamma)\right]I_L \quad (3)$$

$$N_t = N_1 + N_2 \quad (4)$$

Where (λp) is the pump wavelength , (λL) is the laser wavelength ,(τ) is the fluorescence lifetime of the upper manifold, (Nt) is the doping concentration , (Tr) is transmission, (ɣ) is the internal loss of the resonator ,(Ip) is the pump intensity,(IL) is the laser intensity, (Mp) is the number of pump beam passes through the active medium , (ML) is the number of laser beam through the resonator, (σap),(σep),(σep),(σeL) is absorption and emission cross section of pumping and lasing , at room temperature (T) respectively, (L) is the optical length of resonator , (d) is the crystal thickness, (h) is the Planck's constant and (c) is the velocity of an electromagnetic wave in the vacuum [24,25].

When the absorption coefficient (αp) is given by[25] :

$$\alpha p = \sigma apNt - \sigma p\ N2 \quad (5)$$

Where:

$$\sigma p = \sigma ap + \sigma ep \quad (6)$$

And the stokes efficiency (ηs) is defined as [25]:

$$\eta s = \frac{\lambda p}{\lambda L} \quad (7)$$

The round trip loss (δ) can be described as [24]:

$$\delta = -\ln(1-Tr) - \ln(1-\gamma) \quad (8)$$

Here:

$$Tr = 1-R2 \quad (9)$$

Where R2 is the output reflectivity.

The absorption efficiency (ηa) can be obtained [24]:

$$\eta a = Rp[1-\exp(1-Mp\ \alpha p\ d)] \quad (10)$$

Where (Rp) is the total reflectivity of the multi-pass pumping system.

In the steady-stare conditions ($\frac{dI_L}{dt} = 0$) , by solving eq.(3)to obtained the threshold population (N_{2th}) for lsaer operation.

$$N_{2th} = \frac{\sigma_{ap} N_t}{\sigma_L} + \frac{\delta}{M_L d \sigma_L} \quad (11)$$

Where: $\sigma_L = \sigma_{aL} + \sigma_{eL}$ (12)

When substituting the threshold condition (N2=N2th ,Ip=Ipth and IL=0) in eq.(2) the pump power at the threshold (P_{th}) can be described as :

$$P_{th} = \frac{hdcA_p}{\eta_a \tau} N_{2th} \quad (13)$$

Where (Ap) is the pump spot area is given by [25]:

$$Ap = \pi (rp)^2 \quad (14)$$

Here (rp) is the pump spot radius.

For the steady-state condition ($\frac{dN_2}{dt} = 0$) , solving eq. (2), the laser output power (Pout) can be obtained:

$$Pout = \frac{T_r}{\delta} \eta_a \eta_s (P_p - P_{th}) \quad (15)$$

IOP Conf. Series: Materials Science and Engineering **765** (2020) 012062 doi:10.1088/1757-899X/765/1/012062

4. Numerical model :

The numerical model consists three main steps at the beam quality factor is $(M^2 = 0.95)$. In the fist step, the Table (1) shows the value of all parameters for used to calculated (Pout) by matlab language.

Table (1) : the value of a Parameters for Yb^{3+}:YAG thin disk laser

Parameter	Value	Unit	Ref.
λP	$940X10^{-6}$	m	[22,25,26,27]
σap	$0.7506x10^{-24}$	m^2	[22,25,26,27]
σep	$0.1676x10^{-24}$	m^2	[22,25,26,27]
σp	$0.9182x10^{-24}$	m^2	[22,25,26,27]
λL	$1030X10^{-6}$	m	[22,25,26,27]
σaL	$2.1131x10^{-24}$	m^2	[22,25,26,27]
σeL	$0.1223x10^{-24}$	m^2	[22,25,26,27]
σL	$2.2354x10^{-24}$	m^2	[22,25,26,27]
Rp	0.77	—	[24]
ɣ	$1.7 X10^{-4}$	—	[28]
a	0.85	—	[18]
L	$50x10^{-3}$	m	[22,25]
ML	4	—	[25,29]
rfm	$195X10^{-6}$	m	[29]
h	$6.6205x10^{-34}$	J.sec	—
c	$3x10^{8}$	m / sec	—

In the second step, calculated the value of (Pout) where variation of the value of these parameters Nt [1,30],d [9,22,25],Mp [32],R2 [19,22],T [24,31] and Pp [25] respectively.

The range of values of these coefficient used in the numerical solution in this paper was selected from the references indicated at each coefficient.

5. Numerical results:

5.1. The effects of concentration (N):

The laser output power (Pout) increases with increasing the concentrations, reaching its highest value at (N=22.26x10^{26} $\frac{1}{m^3}$), after which it begins to decrease slightly despite the concentrations increasing, due to the indirect dependence of (Pout) on (N)

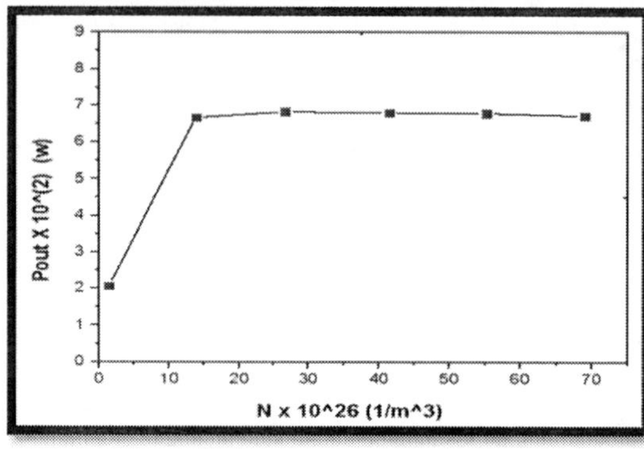

Figure 3: The laser output power (Pout) variations versus the concentration (N)

5.2. The effects of Disk thickness (d):

The laser output power (Pout) increases slightly by increasing the thickness of the laser thin disk and reaching its highest value at ($d=80x10^{-6}m$) and then decreasing slightly with increasing thickness.

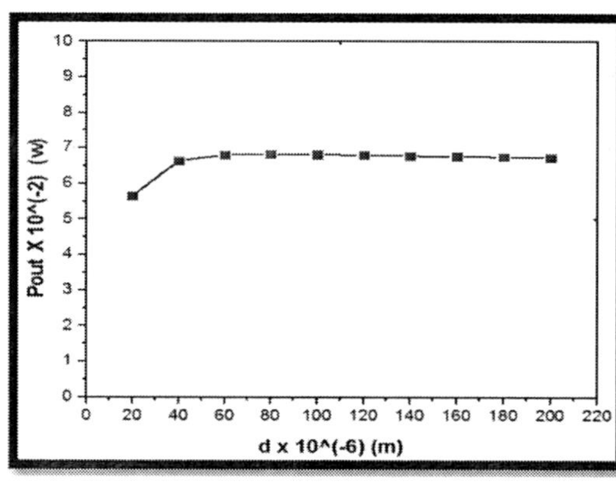

Figure 4: The laser output power (Pout) variations versus the thickness of thin disk laser (d)

5.3. The effects of number of pump beam passes through the active medium (Mp):

The laser output power (Pout) increases by increasing the number of pump beam passes through the active medium (Mp). This corresponds to the positive relationship between (Pout) and (Mp).

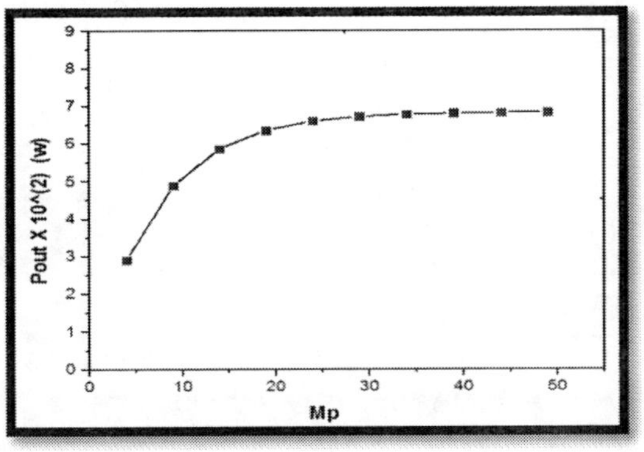

Figure 5: The laser output power (Pout) variations versus number of the pump beam passes through the active medium (Mp)

5.4. The effects of output reflectivity (R2):

The laser output power (Pout) is increased by increasing the reflectivity of the laser output mirror (R2), This corresponds to the positive relationship between (Pout) and (R2) .

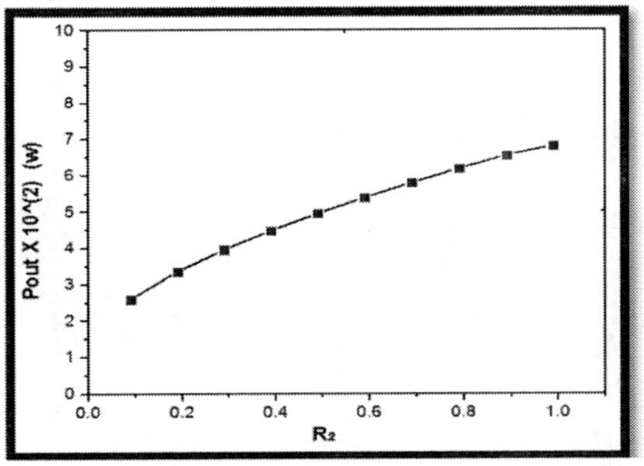

Figure 6 the laser output power (Pout) variations versus reflectivity of the laser output mirror (R2)

5.5. The effects of Temperature (T):

The ser output power (Pout) are slightly reduced by increasing temperature (T), due to the indirect dependence of (Pout) on (T).

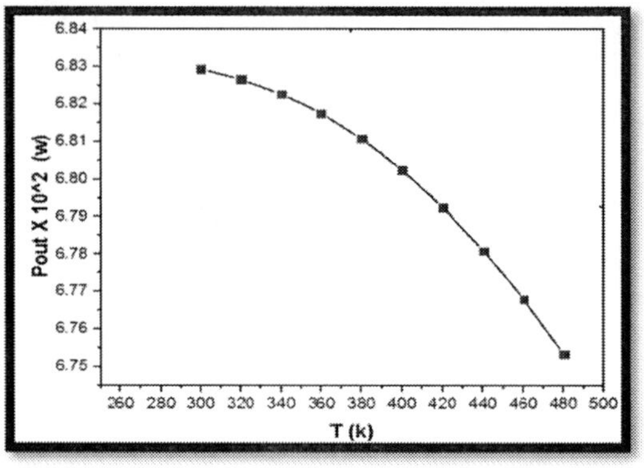

Figure 7: the laser output power (Pout) variations versus reflectivity of the temperature (T)

5.6. The effects of input pump power (Pp):

The laser output power (Pout) in a thin disk laser increase by increasing the input pump power (Pp), this corresponds to the positive relationship between (Pout) and (Pp).

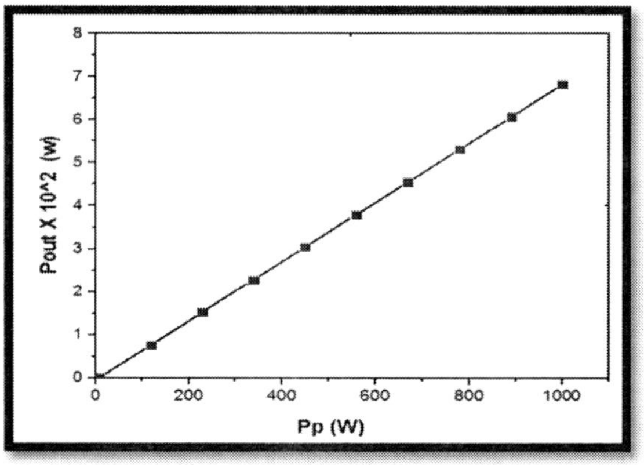

Figure 8 : The laser output power (Pout) variations versus the input pump power (Pp)

In the third step , finding the typical value of these parameters are used to obtained high lasing output power shows in Table(2).

Table (2): the typical values of parameters are used to obtained a low threshed power of laser , a high efficiency and high of output laser power

Parameters	Value	Unit
ηs %	**91.2621**	—
rfm	**195 x 10^{-6}**	m
rp	**223 x10^{-6}**	m
Nt	**26.66 x 10^{26}**	$\frac{1}{m^3}$
τ	**970 x10^{-6}**	Sec
d	**80 x10^{-6}**	m
Mp	**49**	—
R2	**0.98**	—
T	**300.15**	K^0
Pp	**1000**	W

6. CONCLUSIONS

In summary, a theoretical analysis is established to study the output lasing power for multi-pass pumping of Yb^{3+}:YAG thin-disk laser at a quasi-three level operation, In addition, we find a numerical solution of the equation of output Lasing power (pout) at the change of (Nt),(d),(Mp),(R2),(T) and (Pp) AT regime in a fundamental mode operation (M^2<1). Also we find, the (pout) is highest value at the concentration (26.66x10^{26} $\frac{1}{m^3}$) and thickness (80 x10^{-6} m), in addition, that the output Lasing power (pout) increases linearly by increasing the (Mp)),(R2) and (Pp) , but it is slightly reduced by increasing of the temperature , therefore we find the typical value of these parameters are used to obtain high output Laser power.

References

[1] M. Ramkumar, A.N. Akansu, " Some Design Issues for Robust Data Hiding Systems," *IEEE,* pp. 1528 - 1532, OCTOPER 1999.

[2] Ronak Doshi, Pratik Jain, Lalit Gupta, " Steganography and Its Applications in Security," *International Journal of Modern Engineering Research,* vol. 6, pp. 4634-4638, 2012.

[3] K. A. R. Yadav, "A New Method of Image Steganography using Last Three Bit Plane of Gray Scale Images," *Indian Journal of Science and Technology,* vol. 38, pp. 1-8, october 2017.

[4] Souvik Bhattacharyya, Gautam Sanya, " A Robust Image Steganography using DWT Difference Modulation (DWTDM)," *Computer Network and Information Security,* pp. 27-40, july 2012.

[5] "Wikipedia "color space" Wikipedia Free Encyclopedia," [Online]. Available: http://en.wikipedia. org/wiki/Main_page

[6] Levkowitz Haim, "color theory and modeling for computer Graphics, Visualization and Multimedia Application," *Kluwer Academic publishers,* 1997.

[7] "CIELAB color space," [Online]. Available: https://en.wikipedia.org/wiki.

[8] "Math | EasyRGB," IroGroup Limited, 2019. [Online]. Available: https://www.easyrgb.com/en/. [Accessed 4 2019].

[9] H. G. D. Al-Khoziay, Additive Nose Removed by Using Least Square Line Fitting, Baghdad: MSC, University Of Baghdad, College of Education for Pure Sciences / Ibn Al-Haitheam, 2004.

[10] M. Kamal and H. M. Abduljabbar, "The Effect of JPEG Compression Algorithm Attack on the LSB Steganography Method," *INFORMATION,* vol. 21, no. 5, pp. 1615-1624, 2018.

[11] H. L. Hussein, "Hiding Data in Color Image Using Least Significant Bits of," *Ibn Al Haitham Journal for Pure and Applied Science,* vol. 31, no. 2, pp. 193-198, 2018.

[12] F. A. Abdullatif and W. A. Shukur, "Blind Color Image Steganography in Spatial Domain," *Ibn Al Haitham Journal for Pure and Applied Science,* vol. 24, no. 1, pp. 1-9, 2011.

[13] M. Nosrati, R. Karimi and M. Hariri, "An Introduction to Steganography Methods," *World Applied Programming,* vol. 1, no. 3, pp. 191-195, August 2011.

[14] P. K. Saraswat and R. K. Gupta, "A Review of Digital Image Steganography," *Journal of Pure and Applied Science & Technology,* vol. 2, no. 1, pp. 98-106, 2012.

[15] W. Niblack, introduction to digital image processing - an introduction, 3 ed., vol. 4, Prentice- Hall , 1994, pp. 231-262.

Modeling finite element for stress state calculation in combined structures

[1] Hind D. Salman,[2] Ali Kamil Sebur,[3] Enaam Obeid Hassoun, [4] Sagdatullin M.K,

[5] Hussain Abdulaziz Abrahem.

[1]University of Technology – Department of Electromechanical Engineering, Iraq, 50260@uotechnology.edu.iq

[2]University of Technology – Department of Electromechanical Engineering, Iraq, 50278@uotechnology.edu.iq

[3]University of Technology – Department of Electromechanical Engineering, Iraq, 50109@uotechnology.edu.iq

[4]Kazan National Research Technological University- Department of Theoretical Mechanics and Strength of Materials, Kazan, Russia, ssmarat@mail.ru

[5]University of Technology – Department of Electromechanical Engineering, Iraq, 50008@uotechnology.edu.iq

Abstract. A method for constructing a transition finite element for orthotropic multilayer shells is developed in the article. Based on a three-dimensional finite element of the continuum, the family of the new finite elements is constructed, allowing joining elements of various structures in uniform element model. Calculation of the stabilizer of the plane on durability and stability for various cases of load is taken into account, along with the comparison of results of calculation with the data received in package ANSYS.
Keywords: transition finite element, stress state, orthotropic material, aircraft stabilizer

Introduction

While constructing physical models of multilayer orthotropic shells, various approaches are introduced, based on both different hypotheses for each layer of the shell [1-3], and on uniform hypotheses for all layers of a thin-walled structure [2,4-9]. In the first case, the order of the resolution system depends on the number of layers. In the second case, the order of the system does not depend on the number of layers, which opens up the possibilities for the effective use of FEM in the calculations of orthotropic multilayer shells. This orthotropic multilayer finite element (FE) was obtained based on a three-dimensional isotropic eight-node FE shell [10, 22], which consists discretizing three-dimensional

IOP Conf. Series: Materials Science and Engineering **765** (2020) 012063 doi:10.1088/1757-899X/765/1/012063

equations of the theory of elasticity in a curved coordinate system using some shell hypotheses. The application of such approaches to the calculation of shells of medium thickness is discussed in papers [11-18]. The introduction of orthotropy and multilayer into the design scheme of shells is described in publications [17-21].

To calculate combined structures and structures of substantially variable thickness by the finite element method, it is advisable to use both three-dimensional finite elements (FE) [10, 22] and shell ones. Various methods are used to dock them into a single calculation model. In [23], the creation and study of transitional three-dimensional FEs used in solving thermo elastic problems is described, in [24] transitional finite elements are used that connect quadratic three-dimensional elements of a continuous medium and elements of shells of the Timoshenko type into a single calculation model, which are chosen as nodal degrees of freedom with three projections of the displacement vector and two normal rotation angles; in [25], to calculate the stress-strain state of spatial structures, a combined model is used, where these conditions of elastic conjugation with shell elements are taken into account using the penalty method. In [26], the necessary conjugation conditions along the boundary between three-dimensional and shell FEs are realized by introducing the Lagrange multipliers into the original function, the parameters of which are excluded from the number of variable values at the assembly level of the structure. A review of some algorithms for creating transient FEs is given in [27].

The aim of this work is to develop a method to construct a three-dimensional finite element of a continuous medium, which allows modelling combined structures consisting of three-dimensional bodies and shells of medium thickness in a linear formulation. Typically, FE shells of medium thickness are constructed using degrees of freedom defined at the nodes of the median surface and including the rotation angles of a normal fiber. As a rule, FEs with angular degrees of freedom demonstrate good accuracy and are effective in calculating shells of small and medium thicknesses. However, it is very difficult to use them in modelling conjugations of shells with massive three-dimensional bodies, since it is necessary to express the nodal displacements of three-dimensional elements through the angles of rotation of the shell FEs. Therefore, the use of special shell elements, one of which is presented in this paper, allows us to simplify the procedure for joining elements modelling three-dimensional and shell substructures.

In this paper, we introduce approximations of the radius vector, covariant and contravariant basis vectors, metric tensors, displacements, strain and stress tensors. The technology of using the double approximation method for super convergence points and the technique of "lowering the approximation order" of lateral shear strains in a three-dimensional setting are described in detail in [10, 22]. The hypothesis of small compressive stresses is described in approximate form, the "simplified Hooke law", which relates the compressive stress to the compressive strain, is not used. The matrix of elastic constants for an orthotropic material is determined. The elastic characteristics and the winding angle can be generally different for each layer of the finite element. The assembly relationships of the stiffness matrix for a multilayer orthotropic FE are described.

Several test problems have been solved, and hence the efficiency of the proposed FE is tested and verified with solutions of other authors. The results show the acceptability of this technique to determine the stress-strain state of thin-walled structures. For illustration, the results of a static calculation of the stabilizer of a light-engine aircraft, consisting of multilayer shells, are presented.

1. Kinematic relationships

Within each element, a local coordinate system ξ_1, ξ_2, ξ_3 is introduced, which translates the curved box into a unit cube. We define a vector function of an arbitrary material point of a thin-walled structure.

$$\mathbf{r} = x^i\left(\xi_1, \xi_2, \xi_3\right)\vec{e}_i,$$

which would ensure the continuity of itself and its first derivatives, i.e. basis vectors, when crossing interelement boundaries. The projections of the radius vector are approximated as follows

$$x^i(\xi_1,\xi_2,\xi_3) = \sum_{t=1}^{8} x_t^i N_t(\xi_1,\xi_2,\xi_3),$$

where $N_t(\xi_1,\xi_2,\xi_3)$ are the functions of the shape of the FE shell.

To define covariant basis vectors

$$\mathbf{r}_k = \frac{\partial x^i}{\partial \xi^k}\mathbf{e}_i = r_k^i \mathbf{e}_i.$$

We write the components of the corresponding basis in the form

$$r_k^i = \sum_{t=1}^{8} x_t^i \frac{\partial N_t}{\partial \xi^k}.$$

The covariant basis is calculated by the formulas

$$\mathbf{r}^k = \frac{\partial \xi^k}{\partial x^i} = r^{k,i}\mathbf{e}_i.$$

Metric tensors in various bases characterizing the geometry can be represented using the following relations

$$g_{ij} = \mathbf{r}_i\mathbf{r}_j = \sum_m r_i^m r_j^m, \quad g^{ij} = \mathbf{r}^i\mathbf{r}^j = \sum_m r^{i,m} r^{j,m}.$$

We introduce approximations of the vector $\mathbf{u}$ and the gradient of the displacement vector $\mathbf{u}_{,i}$

$$\mathbf{u} = u^m(\xi_1,\xi_2,\xi_3)\mathbf{e}_m, \quad \mathbf{u}_{,i} = \mathbf{e}_m \sum_{s=1}^{8} u_s^m \frac{\partial N_s}{\partial \xi^i}.$$

The components of the strain tensor are defined as follows

$$\varepsilon_{ij} = \frac{1}{2}(\mathbf{u}_{,i}\mathbf{r}_j + \mathbf{r}_i\mathbf{u}_{,j}) = \sum_{r=1}^{8}\sum_{m}^{3} u_r^m E_{ij}^{r,m},$$

when

$$E_{ij}^{r,m} = \frac{1}{2}\sum_{s=1}^{8} x_s^m \left(\frac{\partial N_r}{\partial \xi^i}\frac{\partial N_s}{\partial \xi^j} + \frac{\partial N_s}{\partial \xi^i}\frac{\partial N_r}{\partial \xi^j} \right).$$

The technology of lowering the approximation order for transverse shear deformations ε_{i3}, $i = 1,2$ is described in detail in [10, 22].

Let us describe the technique of introducing a "decrease in the approximation order" for a three-dimensional finite element. Consider the shear strain $\varepsilon_{13}, \varepsilon_{23}$

$$E_{13}^{rm} = \frac{1}{128} X_t^m \left[\xi_r^1 \xi_t^3 \left(1+\xi_r^2\xi^2\right)\left(1+\xi_r^3\xi^3\right)\cdot \right.$$
$$\cdot\left(1+\xi_t^1\xi^1\right)\left(1+\xi_t^2\xi^2\right) + \xi_r^3\xi_t^1\left(1+\xi_r^1\xi^1\right)\cdot$$
$$\left.\cdot\left(1+\xi_r^2\xi^2\right)\left(1+\xi_t^2\xi^2\right)\left(1+\xi_t^3\xi^3\right)\right].$$

$$E_{23}^{rm} = \frac{1}{128} X_t^m \left[\xi_r^2 \xi_t^3 \left(1+\xi_r^1\xi^1\right)\left(1+\xi_r^3\xi^3\right)\cdot \right.$$
$$\cdot\left(1+\xi_t^1\xi^1\right)\left(1+\xi_t^2\xi^2\right) + \xi_r^3\xi_t^2\left(1+\xi_r^1\xi^1\right)\cdot$$
$$\left.\cdot\left(1+\xi_r^2\xi^2\right)\left(1+\xi_t^1\xi^1\right)\left(1+\xi_t^3\xi^3\right)\right].$$

The first "reduction of approximation" consists in eliminating the variability of these strains in thickness.

$$E_{13}^{rm} = \frac{1}{128} X_t^m \left[\xi_r^1 \xi_t^3 \left(1+\xi_r^2\xi^2\right)\left(1+\xi_t^1\xi^1\right)\cdot \right.$$
$$\left.\cdot\left(1+\xi_t^2\xi^2\right) + \xi_r^3\xi_t^1\left(1+\xi_r^1\xi^1\right)\left(1+\xi_r^2\xi^2\right)\left(1+\xi_t^2\xi^2\right)\right].$$

IOP Conf. Series: Materials Science and Engineering **765** (2020) 012063 doi:10.1088/1757-899X/765/1/012063

$$E_{23}^{rm} = \frac{1}{128} X_t^m \left[\xi_r^2 \xi_t^3 \left(1+\xi_r^1 \xi^1\right)\left(1+\xi_t^1 \xi^1\right) \cdot \right.$$
$$\left. \cdot \left(1+\xi_t^2 \xi^2\right) + \xi_r^3 \xi_t^2 \left(1+\xi_r^1 \xi^1\right)\left(1+\xi_r^2 \xi^2\right)\left(1+\xi_t^1 \xi^1\right) \right].$$

The second "reduction" is aimed at eliminating variability along coordinates ξ^1, ξ^2, deformations $\varepsilon_{13}, \varepsilon_{23}$, respectively.

$$E_{13}^{rm} = \frac{1}{128} X_t^m \left[1 + \left(\xi_r^2 + \xi_t^2\right)\xi^2 + \xi_r^2 \xi_t^2 \left(\xi^2\right)^2 \right] \cdot \left(\xi_r^1 \xi_t^3 + \xi_r^3 \xi_t^1\right).$$
$$E_{23}^{rm} = \frac{1}{128} X_t^m \left[1 + \left(\xi_r^1 + \xi_t^1\right)\xi^1 + \xi_r^1 \xi_t^1 \left(\xi^1\right)^2 \right] \cdot \left(\xi_r^2 \xi_t^3 + \xi_r^3 \xi_t^2\right).$$

The third modification involves defining strains $\varepsilon_{13}, \varepsilon_{23}$ as a linear approximation for ξ^1, ξ^2 at two points $\xi^1 = \pm 1$ and $\xi^2 = \pm 1$, respectively. To do this, consider the quadratic function $g(\xi)$

$$g(\xi) = \alpha_1 + \alpha_2 \xi + \alpha_3 \xi^2,$$

where $\alpha_1, \alpha_2, \alpha_3$ are - some coefficients. The linear approximation will have the form

$$\tilde{g}(\xi) = \frac{g(+1)+g(-1)}{2} + \frac{g(+1)-g(-1)}{2}\xi = \beta_1 + \beta_2 \xi.$$

By analogy, we write

$$E_{13}^{rm} = \frac{1}{128} X_t^m \left[\left(1+\xi_r^2 \xi_t^2\right) + \left(\xi_r^2 + \xi_t^2\right)\xi^2 \right] \cdot \left(\xi_r^1 \xi_t^3 + \xi_r^3 \xi_t^1\right).$$
$$E_{23}^{rm} = \frac{1}{128} X_t^m \left[\left(1+\xi_r^1 \xi_t^1\right) + \left(\xi_r^1 + \xi_t^1\right)\xi^1 \right] \cdot \left(\xi_r^2 \xi_t^3 + \xi_r^3 \xi_t^2\right).$$

It should be noted that taking lesser degrees in expressions for strains than those presented is not recommended, as this will lead to a reduced rank of the stiffness matrix, or in other words, to additional false rigid displacements.

A certain drawback of the scheme described above is the violation of the variational principle, as a result of which the conditions for the compatibility of deformations for truncated deformations are not satisfied, which leads to nonmonotonic convergence. Nevertheless, the choice of such approximations is expedient and justified from the standpoint of MSFE (moment scheme of finite elements) [16-18].

Hooke's generalized law is written as follows

$$\sigma_{ij} = D_{ijmn} \varepsilon_{mn}. \tag{1}$$

We introduce the hypothesis of small compression stresses

$$\sigma_{33} = 0. \tag{2}$$

Using relation (1), we write expression (2) in the following form

$$D_{33mn} \varepsilon_{mn} = 0. \tag{3}$$

We represent (3) as an expression

$$\begin{gathered} D_{3311}\varepsilon_{11} + D_{3312}\varepsilon_{12} + D_{3313}\varepsilon_{13} + D_{3321}\varepsilon_{21} + D_{3322}\varepsilon_{22} + \\ + D_{3323}\varepsilon_{23} + D_{3331}\varepsilon_{31} + D_{3332}\varepsilon_{32} + D_{3333}\varepsilon_{33} = 0 \end{gathered} \tag{4}$$

We express from (4) the compression strain

$$\begin{gathered} \varepsilon_{33} = -\left(D_{3311}\varepsilon_{11} + D_{3312}\varepsilon_{12} + D_{3313}\varepsilon_{13} + D_{3321}\varepsilon_{21} + \right. \\ \left. + D_{3322}\varepsilon_{22} + D_{3323}\varepsilon_{23} + + D_{3311}\varepsilon_{31} + D_{3332}\varepsilon_{32}\right) / D_{3333} \end{gathered} \tag{5}$$

Substituting (5) in (1), we get

$$\begin{aligned}\sigma_{ij}^{*} &= D_{ij11}\varepsilon_{11} + D_{ij12}\varepsilon_{12} + D_{ij13}\varepsilon_{13} + D_{ij21}\varepsilon_{21} + D_{ij22}\varepsilon_{22} + \\ &+ D_{ij23}\varepsilon_{23} + D_{ij31}\varepsilon_{31} + D_{ij32}\varepsilon_{32} - D_{ij33}\left(D_{3311}\varepsilon_{11} + \right. \\ &+ D_{3312}\varepsilon_{12} + D_{3313}\varepsilon_{13} + D_{3321}\varepsilon_{21} + D_{3322}\varepsilon_{22} + D_{3323}\varepsilon_{23} + \\ &\left. + D_{3311}\varepsilon_{31} + D_{3332}\varepsilon_{32}\right) / D_{3333}\end{aligned} \tag{6}$$

We write (6) in a simpler form

$$\sigma_{ij}^{*} = D_{ijmn}^{*}\varepsilon_{mn} = \left(D_{ijmn} - \frac{D_{ij33}D_{33mn}}{D_{3333}}\right)\varepsilon_{mn},$$

when

$$D_{ij11}^{*} = D_{ij11} - \frac{D_{ij33}D_{3311}}{D_{3333}},\ D_{ij12}^{*} = D_{ij12} - \frac{D_{ij33}D_{3312}}{D_{3333}},$$

$$D_{ij13}^{*} = D_{ij13} - \frac{D_{ij33}D_{3313}}{D_{3333}},\ D_{ij21}^{*} = D_{ij21} - \frac{D_{ij33}D_{3321}}{D_{3333}},$$

$$D_{ij22}^{*} = D_{ij22} - \frac{D_{ij33}D_{3322}}{D_{3333}},\ D_{ij23}^{*} = D_{ij23} - \frac{D_{ij33}D_{3323}}{D_{3333}},$$

$$D_{ij31}^{*} = D_{ij31} - \frac{D_{ij33}D_{3331}}{D_{3333}},\ D_{ij32}^{*} = D_{ij32} - \frac{D_{ij32}D_{3332}}{D_{3333}},$$

$$D_{3333}^{*} = D_{3333}.$$

The hypothesis of small compressive stresses recorded in this form makes it possible to explicitly introduce physical constants for both isotropic and orthotropic materials into the design scheme.

2. Orthotropy and layering

Among the many methods used to calculate layered composite shells, a special place is occupied by numerical methods, in particular, FEM. The main advantage of such algorithms is the possibility of calculating thin-walled structures of complex geometry with variable mechanical characteristics under the influence of various loads.

The generalized Hooke law for an orthotropic body has the following form:

$$\varepsilon_{11} = \frac{1}{E_1}\left(\sigma_{11} - \mu_{12}\sigma_{22} - \mu_{13}\sigma_{33}\right),\ \gamma_{23} = 2\varepsilon_{23} = \frac{\sigma_{23}}{G_{23}},$$

$$\varepsilon_{22} = \frac{1}{E_2}\left(\sigma_{22} - \mu_{23}\sigma_{33} - \mu_{21}\sigma_{11}\right),\ \gamma_{12} = 2\varepsilon_{12} = \frac{\sigma_{12}}{G_{12}},$$

$$\varepsilon_{33} = \frac{1}{E_3}\left(\sigma_{33} - \mu_{32}\sigma_{11} - \mu_{32}\sigma_{22}\right),\ \gamma_{31} = 2\varepsilon_{31} = \frac{\sigma_{31}}{G_{31}}.$$

Then the components of the elasticity matrix for an orthotropic material can be written as follows:

$$D_{1111}=\frac{1}{AE_2}\left(\frac{1}{E_3}-\frac{\mu_{23}^2}{E_2}\right), D_{2222}=\frac{1}{AE_3}\left(\frac{1}{E_1}-\frac{\mu_{31}^2}{E_3}\right),$$
$$D_{1122}=D_{2211}=\frac{1}{AE_3}\left(\frac{\mu_{31}\mu_{23}}{E_2}+\frac{\mu_{12}}{E_1}\right), D_{2323}=G_{23},$$
$$D_{2233}=D_{3322}=\frac{1}{AE_1}\left(\frac{\mu_{12}\mu_{31}}{E_3}+\frac{\mu_{23}}{E_2}\right), D_{1212}=G_{12}, \tag{7}$$
$$D_{1133}=D_{3311}=\frac{1}{AE_2}\left(\frac{\mu_{12}\mu_{23}}{E_1}+\frac{\mu_{31}}{E_3}\right), D_{3131}=G_{31},$$
$$D_{3333}=\frac{1}{AE_1}\left(\frac{1}{E_2}-\frac{\mu_{12}^2}{E_1}\right),$$

when

$$A=\frac{1}{E_1E_2E_3}(1-2\mu_{12}\mu_{23}\mu_{31}-\frac{E_1}{E_3}\mu_{31}^2-\frac{E_2}{E_1}\mu_{12}^2-\frac{E_3}{E_2}\mu_{23}^2).$$

Since in practice most materials have a layered structure with different mechanical characteristics in thickness, and generally not symmetrical, it is advisable to use a multilayer finite element. The thickness values of each layer h_k are more convenient to set in the form of their relative values,

$$\Delta_k=\frac{h_k}{h}, h=\sum_{k=1}^{N}h_k, \sum_{k=1}^{N}\Delta_k=1,$$

where N is the number of layers. If the thickness of each layer is the same, then $\Delta_k=1/N$. The interlayer coordinates by thickness inside the FE can be determined as follows

$$\zeta_1=-1, \zeta_2=-1+2\Delta_1, \zeta_3=-1+2\Delta_1+2\Delta_2,$$
$$..., \zeta_k=-1+2\sum_{k=1}^{n-1}\Delta_k, ..., \zeta_{N+1}=1.$$

When calculating the stiffness matrix, the Gauss-Legendre quadrature formula of the order $2\times2\times N$ where N is the number of layers used, is applied. Thus, the coordinates of the quadrature points in the direction of the axis ζ are determined in the middle of the thickness of each layer.

Then the components of the stiffness matrix can be calculated as follows:

$$D_{ab}^{rc}=g^{oi}g^{pj}D_{ijmn}E_{mn}^{ra}E_{op}^{cb}\sqrt{g}2\Delta_k,$$

when $r,c=1,8, a,b,i,j,m,n,o,p=1,3$.

The stress components are calculated at four quadrature points defined in the middle of each layer and then interpolated into nodes. Since a single-layer approximation is used when calculating the stress-strain state of thin-walled structures, acceptable accuracy is already achieved with the number of layers $N\geq3$.

3. Calculation of the stabilizer and elevator of a light-engine aircraft for strength

As an example, the calculation of the stress-strain state of the elevator and stabilizer of a light-engine aircraft is given. The stabilizer and elevator are a complex combined shell structure (Fig. 1), reinforced by root ribs and spars. When sampling the computational domain, both multilayer FE (for modelling the stabilizer and elevator sheathing) and a three-dimensional FE of a continuous medium (for modelling elements of root ribs and spars) are used.

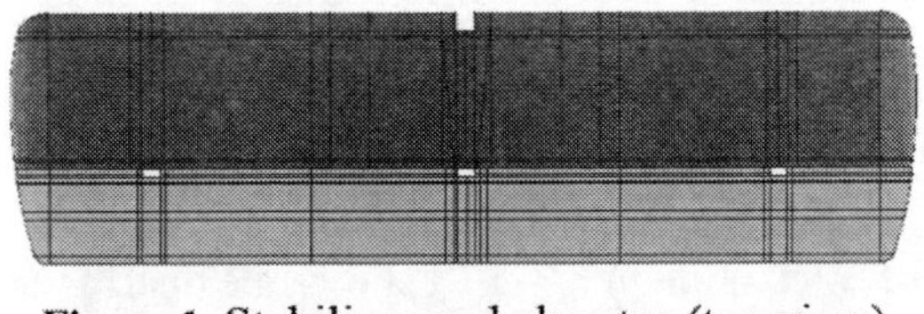

Figure 1. Stabilizer and elevator (top view).

Calculation of the stabilizer and elevator strength is carried out for balancing and maneuvering loads given in SNiPs, the application diagram of which is shown in Figure 2. Only symmetric cases of load application are considered.

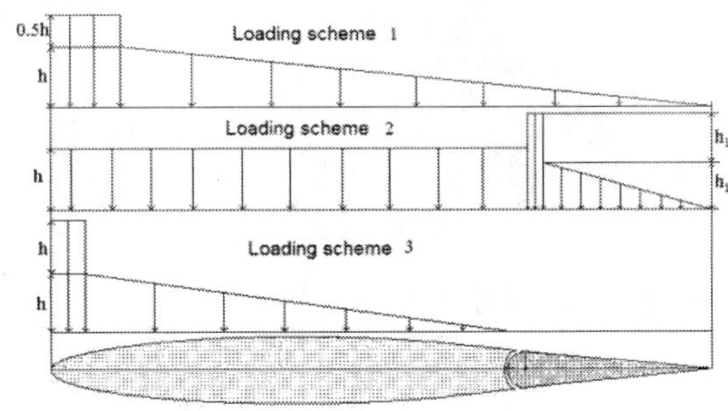

Figure 2. Loading schemes

The stabilizer mounting diagram for the fuselage is shown in Figure 3 (the elevator is attached to the stabilizer). Based on the symmetry of the design, only half of the stabilizer is shown. Fastening to the fuselage (for half of the stabilizer) is carried out at three points A , B and C (Fig. 3). A sliding hinge is realized at a point A , an ordinary hinge at a point B , and at point C a stabilizer is attached to the keel with a brace.

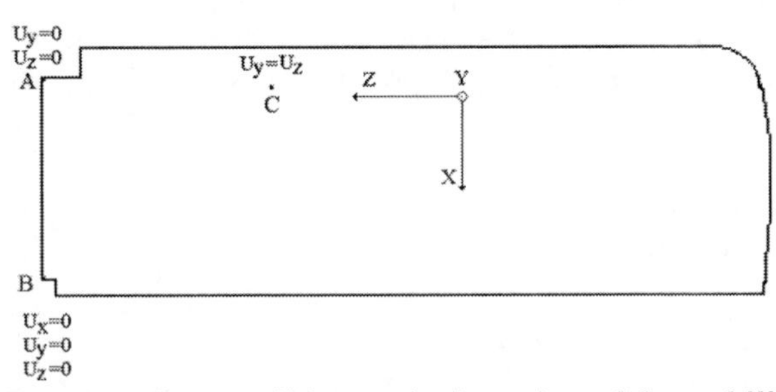

Figure 3. Scheme of kinematic fastening of the stabilizer

To verify the program, we compared the solution of the problem with the results obtained in the ANSYS program based on the finite element SHELL181. Table 1 shows a comparison of the reactions R_x , R_y and R_z occurring at the points of attachment of the stabilizer to the fuselage at the points A and B .

Table 1. Reactions (H) at the attachment points and transmitted to the fuselage from the stabilizer for three cases of loading.

№		R_X		R_Y		R_Z	
		progr	ansys	progr	ansys	progr	ansys
1	A	-	-	-5841	-5062	63,2	-9,2
	B	-524	-443	2164	1926	26,3	0
2	A	-	-	-9124	-8615	-21	-16
	B	-349	-337	11016	10497	-14	-3

3	A	-	-	-7811	-7223	-18	-13,8
	B	-414	-399	5642	5236	-37	-2

Analysis of the calculation results shows that the difference in the results of the calculation of force factors during the static calculation of a complex structure between those obtained by the developed method and in the ANSYS program does not exceed 15%.

To illustrate, the results of calculation in Figure 4 shows the distribution of the deflection W for half stabilizer constructions (in meters) for loading scheme 2.

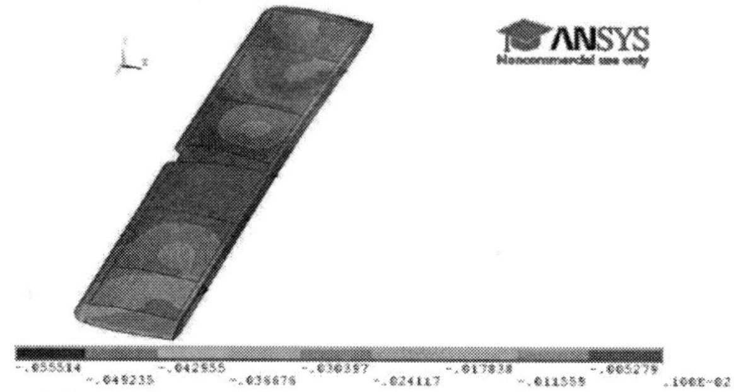

Figure 4. The picture of the deflections of half the stabilizer design for the loading scheme 2.

Based on the results obtained, it should be noted that the developed numerical method for studying the stress-strain state of orthotropic multilayer shells of complex geometry yields results that are in good agreement with the data obtained in other application software packages, in particular, in the ANSYS program. Therefore, based thereon, similar designs can be calculated and reliable results can be obtained.

Conclusion In this paper, we consider a technology for constructing a transitional finite element. The essence of the methodology is to impose shell hypotheses (the hypothesis of small compressive stresses, truncation of transverse shear deformations, “simplified Hooke's law”, reduced integration, etc.) on quadrature points, which are then interpolated into nodes at the junction of the element with the shell finite element. The nodes in which the transitional finite element is docked with massive three-dimensional elements are exempt from the imposition of shell hypotheses. Thus, smoothing of displacements within the finite element occurs, but does not guarantee smoothing of stresses within a given transitional FE, i.e., it is understood that such an operation should be applied to areas remote from the actual studied places of stress concentration. The criterion for assessing stresses in a given zone is the study of stress values while dividing this zone by various types of finite elements (three-dimensional, shell and transitional).

Thanks to this approach, it is possible to build a family of isotropic and orthotropic transition finite elements. Joining of single-layer and multi-layer finite elements is also possible.

References

[1] Bolotin V.V. *Mechanics of multilayer structures.* – Moscow, Engineering. – 1980. – 375p.
[2] Grigolyuk E.I. and Kulikov G.M. *Multilayer reinforced shells. Calculation of pneumatic tires.*- Moscow, Engineering. – 1988. – 288p.
[3] Soloviev S.S. Finite element model of a multilayer shell with anisotropic layers of variable thickness. – *University News. Aeronautical engineering* **4** – 1989. – Pp. 71-75.
[4] Alfutov N.A. and Popov B.G. *Calculation of multilayer plates and shells made of composite materials.*- Moscow, Engineering. – 1984. – 264p.
[5] Vasiliev V.V. *Mechanics of structures made of composite materials.* – Moscow, Eigineering. – 1988. – 270p.
[6] Grigorenko Y.M. and Vasilenko A.T. *Static anisotropic shells with finite shear stiffness.* – Kiev,

Scientific thought. – 1987. – 216p.

[7] Piskunov V.G. *Calculation of inhomogeneous shells and plates by the finite element method.* – Kiev, High school. – 1987. – 200p.

[8] Rickards R.B. *The finite element method in the theory of shells and plates*. – Riga, Knowledge. – 1988. – 284p.

[9] Rasskazov A.O. and Sokolovskaya I.I. *Theory and calculation of layered orthotropic plates and shells.* – Kiev, High school. – 1986. – 191p.

[10] Golovanov A.I. and Sagdatullin M.K. Three-dimensional finite element for the calculation of thin-walled structures. – *Scientific notes of Kazan State University* **151** No.3. – 2009. – Pp. 121-129.

[11] Perelygin O.A., Galyaviev Sh.Sh., Zaynullin R.Kh. and Berezhnoy D.V. The study of the strength of cylindrical shells with dents in the field of radial joints. – *Herald of the Kazan State Technological University* **1-2.** – 2001. – Pp. 75-77.

[12] Perelygin O.A., Tuykin N.M., Berezhnoy D.V. and Serazutdinov M.N. The study of the strength of cylindrical shells in the presence of removal or displacement of the edges of the welds. – *Herald of the Kazan State Technological University* **1-2.** – 2001. – Pp. 77-79.

[13] Berezhnoy D.V., Sagdatullin M.K. and Golovanov A.I. Multilayer orthotropic finite element of shells of medium thickness. - *Herald of the Saratov State Technical University* **57** No. 3. – 2011. – Pp. 9-19.

[14] Boorman Y.Z. and Soloviev S.S. Calculation of elastoplastic deformation of shells based on flow theory and FEM. – *Studies on the theory of plates and shells* **22.** – 1990. – Pp. 98-107.

[15] Parish H.A. Critical survey of the J-node degenerated shell element with special emphasis on thin shell application and reduced integration. – *Computer Methods in Applied Mechanics and Engineering* **20** No. 3. – 1979. – Pp. 323-350.

[16] Zienkiewicz O.C., Taylor R.L. and Too J.M. Reduced integration technique in general analysis of plates and shells. – *Int. J. Num. Meth. Eng* **3.** –1971. – Pp. 275-290.

[17] Golovanov A.I., Pesoshin A.V. and Tyuleneva O.N.- *Modern finite element models and methods for studying thin-walled structures.* – Kazan, KSU. – 2005. – 442p.

[18] Golovanov A.I., Tyuleneva O.N. and Shigabutdinov A.F. *The finite element method in the statics and dynamics of thin-walled structures.* – Kazan, KSU. – 2006. – 392p.

[19] Golovanov A.I. and Berezhnoi D.V. *The finite element method in the mechanics of deformable solids.* – Kazan, DAS. – 2001. – 301p.

[20] Popov B.G. *Calculation of multilayer structures by variational-matrix methods.* – Moscow, MSTU. – 1993. – 294p.

[21] Khechumov R.A., Keppler H. and Prokopyev V.I. *Application of the finite element method to structural analysis*. – Moscow, Association of Construction Universities. – 1994. – 353p.

[22] Berezhnoi D.V., Sagdatullin M.K. and Sachenkov A.A. Calculation of combined structures by the finite element method. – *Scientific and Technical Bulletin of the Volga Region* **4.** – 2012. – Pp. 13-16.

[23] Surana, Karan S. Three dimensional solid-shell transition finite elements for heat conduction. – *Comput. and Struct.* **26** No. 6. – 1987. – Pp. 941-950.

[24] Berezhnoi D.V., Krasnovsky I.Yu. Finite elements for structural analysis of substantially variable thickness. – *Statics and dynamics of structural elements of complex shape.* – 1990. – Pp. 30-36.

[25] Savula Ya.G. and Dubovik A.V. The use of a combined model for calculating the stress-strain state of spatial structures. – *Applied mechanics* **25** No. 9. – 1989. – Pp. 62-67.

[26] Adyasova N.M. Some questions of the calculation of nonlinear composite structures. – Applied *problems of strength and ductility* **1.** – 1975. – Pp. 124-135.

[27] Surana, Karan S. Transition finite elements for three-dimensional stress analysis. – *Int. J. Numer Meth. Eng.* **15** No. 7. – 1980. – Pp. 991-1020.

IOP Conf. Series: Materials Science and Engineering **765** (2020) 012064 doi:10.1088/1757-899X/765/1/012064

Characterization the wear behavior of Al-Ti alloy having different Ti ratios prepared by powder metallurgy

Nadia Ghadhanfer Hikmat

Sahib Mohammed Mahdi

Dalmn Yaseen Taha

Materials Engineering Department, Al-Mustansiriyah University, Baghdad, Iraq

corresponding author's e-mail : nadiaghadhanfer@gmail.com

Abstract. In this study, the Al-Ti alloy was fabricated by powder metallurgy method using three different Ti ratios (10wt%; 15wt%; and 20wt%Ti); and the wear behavior of these three alloys was investigated. The mixtures of powders were compacted at 146 MPa and sintered for 5 hours at 550- °C under vacuumed atmosphere. The samples were characterized using X-Ray Diffraction test (XRD), Scanning Electron Microscope test (SEM), microhardness and density tests. The tests of wear were achieved by employing a pin on disk machine under constant normal load 15N for four different sliding times 5, 10, 15, and 20 minutes. The results illustrate that the density and microhardness of Al-Ti alloy were increased as the Ti content increased. Also, it was observed that Al-Ti alloy with 20wt%Ti had the highest wear resistance and lowest weight loss.

Keywords: Wear Behavior; Powder Metallurgy; Al-Ti Alloy; Microhardness.

Introduction

Aluminium and its alloys are widely employed in aerospace and automotive sectors due to their easy formability and low density, whereas, titanium element is a material that has a special attention, owing to its good resistance to fatigue, low density and high strength [1]. By adding titanium into aluminium , high performance structural materials can be produced that are capable to use in many engineering industries and elevated temperatures applications [2]. This type of alloying can be achieved by various processes such as powder metallurgy, mechanical alloying, and melting [3-6]. Powder metallurgy is considered the most popular manufacturing method used in producing intermetallics, metallic or composites with small sizes and complex shapes. It can produce parts with very high precision and lower cost since the wasted material is little [7]. The mechanical and physical characteristics of powder metallurgy products are related closely to their density. Reducing the gradients of density in the components is a very important consideration if high mechanical performance and consistent response is required [8]. Improving the mechanical performance and tribological characteristics like wear behaviour of Al alloys can be achieved by forming a class of materials called intermetallic compounds. These materials are harder and stronger than the base alloy [9]. In general, the system of Al-Ti alloy has several intermetallic compounds, known as titanium aluminides. Among these compounds, $TiAl_3$ has a considerable interest because of its elevated temperature resistance, high wear resistance, excellent resistance to corrosion, good resistance to creep, low density and high hardness [10]. $TiAl_3$ intermetallic compound results from diffusion reaction that occurs between Ti and Al during sintering at a range of temperature about (500 to 600- °C) [11]. Tribology can be defined as the technology and science of contact or interacted surfaces as well as particles being in relative movement against each other [12]. This present work aims basically at developing a new material with enhanced wear characteristics; through adding titanium at three different percentages to aluminium and chooses the best ratio. The effect of $TiAl_3$ intermetallic compound, which formed during sintering, on the wear properties of Al-Ti alloy was investigated.

1. Experimental Method

1.1. Preparation of Al-Ti alloy

Al-Ti alloy was produced by powder metallurgy using Al powder from CDH Company with 99.8% purity, and Ti powder from Fluka Company with 99.8% purity added at three different weight percentages (10wt%; 15wt%; and 20wt%). The powders were mixed and blended for two hours using electrical parallel mixer with 70 rpm, as demonstrated in Figure1a. The mixed powders were compacted under 146 MPa pressure with 0.5 mm/min compaction speed; employing a double-action die (Figure1b), and producing cylindrical compacts (14mm diameter × 10mm height).

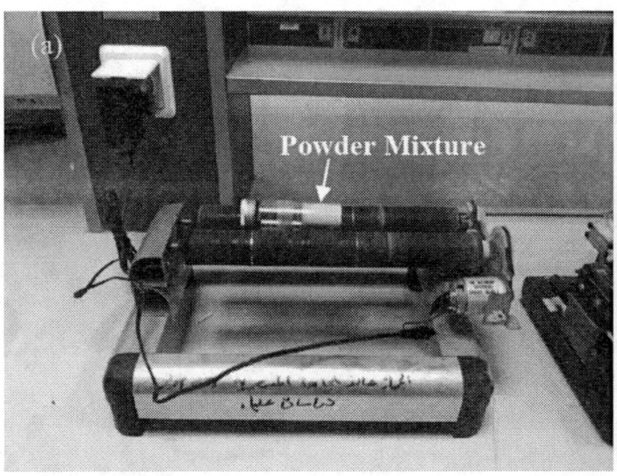

Figure 1a: Electrical parallel mixer

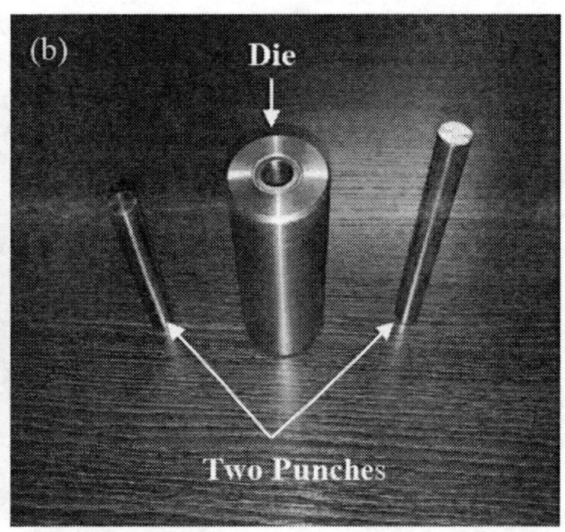

Figure 1b: Double-action die

These green compacts then subjected to sintering in a vacuumed atmosphere furnace, with 10^{-3} bar pressure of vacuum, at 550- °C for five hours, as illustrated in Figure 2.

Figure 2: System of vacuum tube furnace

1.2. Examination of Al-Ti alloy

The sintered specimens were etched using etching solution include 1ml HF, 2.5ml HNO_3, 1.5ml HCl, and 95ml distilled water for twenty seconds for Scanning Electron Microscope (SEM) examination. Archimedes' rule was employed to measure Al-Ti alloy density. The values of microhardness were measured by means of TH-715 device. The Al-Ti alloy examinations were achieved using X-Ray Diffraction (XRD) test by means of Shimadzu device (model 6000), Scanning Electron Microscope (SEM) test by means of VEGA3 LM-TESCAN. A pin on disk machine (as shown in Figure 3) was used to achieve the wear tests under constant normal load (15N) and (18 cm) diameter of the disk, with 490 revolution per minute (rpm) velocity of sliding, at four different sliding times: 5, 10, 15, and 20 minutes. The resulted worn surfaces were examined using optical microscope having 50X magnification. The specimens were weighted before and after every test using digital balance of ±0.0001 gm sensitivity, and the differences in weight were calculated. Equation (1) was used to determine the wear rates (W.R.) in (gm/cm) for each alloy [13]:

$$W.R. = \Delta W / \pi D N t$$
$$\Delta W = W1 - W2 \qquad (1)$$

Where:

W_1= Specimen weight before wear test (gm).

W_2= Specimen weight after wear test (gm).

D= Diameter of the used steel disk (cm).

N= Number of revolutions (rpm).

t= sliding time (min).

Figure 3: Pin on disk machine

2. Result and discussions

2.1. Results of XRD test

The examination of XRD was achieved to detect the intermetallic compound ($TiAl_3$). In a previous study made by Baisong et al. [14], it was reported that $TiAl_3$ intermetallic compound was resulted during sintering for five hours at 630 °C temperatures under 10MPa pressure. In this study, the XRD results of the three ratios of Al-Ti alloy show that Al was detected besides $TiAl_3$, which formed from a diffusion reaction that occurred between Ti and Al at 550- °C for 5 hours. This reaction includes [15]:

$$Ti + 3Al \rightarrow TiAl_3$$

Figure 4 illustrates XRD patterns of the three prepared Al-Ti alloys with different ratios of Ti.

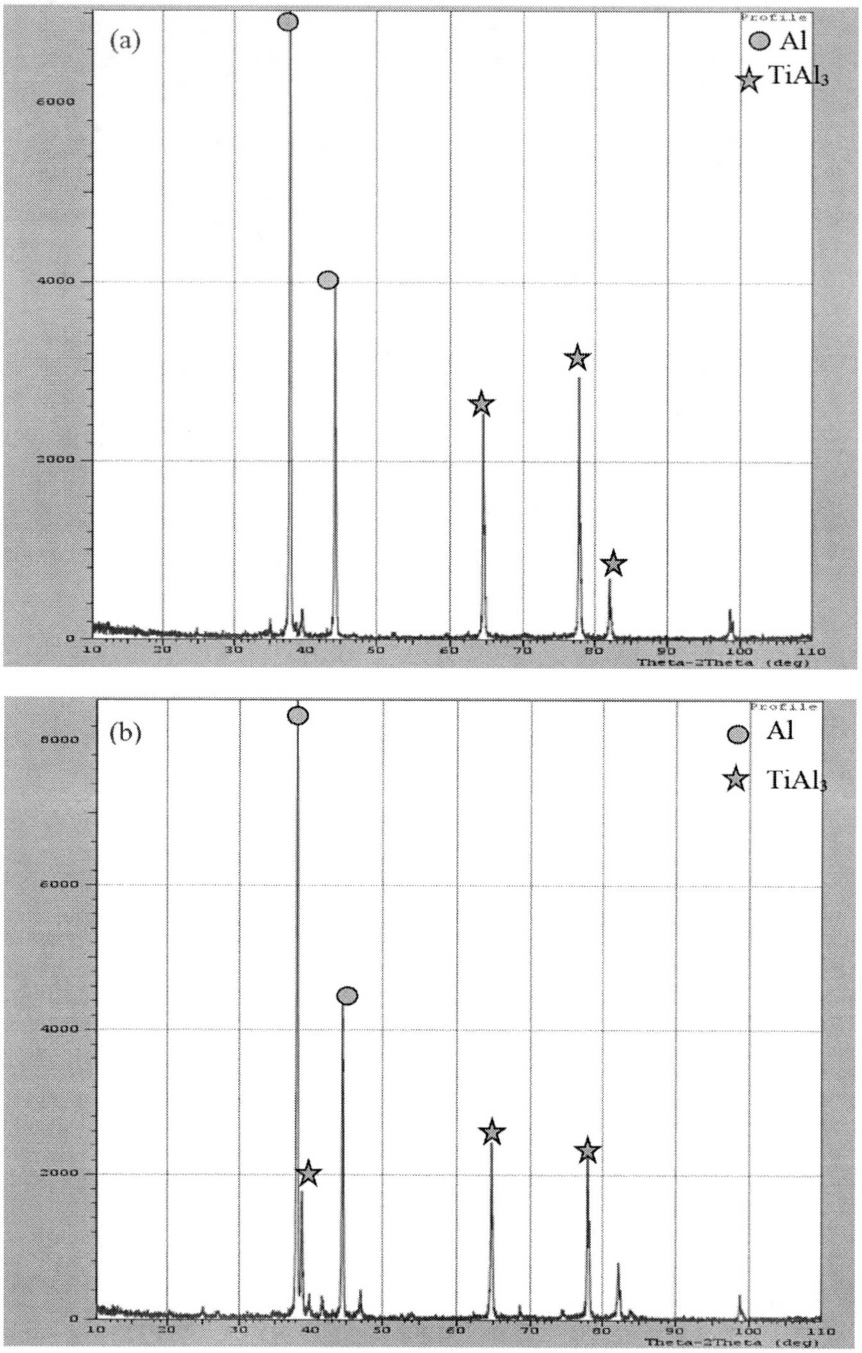
(a)
Profile
Al
TiAl3
6000
4000
2000
0
10
20
30
40
50
60
70
80
90
100
110
Theta-2Theta (deg)
(b)
Profile
Al
TiAl3
8000
6000
4000
2000
0
10
20
30
40
50
60
70
80
90
100
110
Theta-2Theta (deg)

IOP Conf. Series: Materials Science and Engineering **765** (2020) 012064 doi:10.1088/1757-899X/765/1/012064

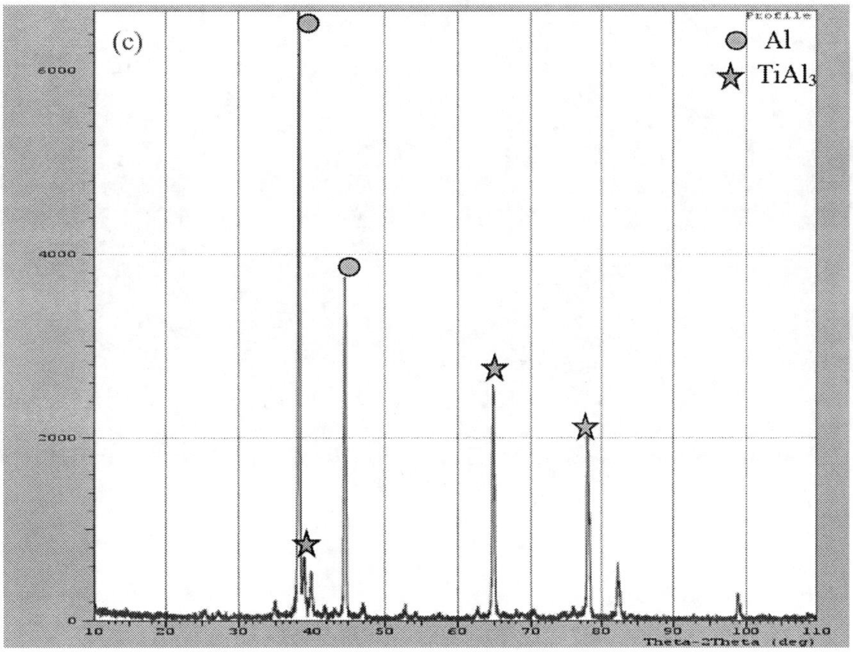

Figure 4: XRD pattern of a)10wt%Ti-Al; b)15wt%Ti-Al; c)20wt%Ti-Al

In Table (1), the angles of diffraction (2θ) and the intensities (I) of the X-ray beam are illustrated, and they referred to the formed intermetallic compound $TiAl_3$ in the three prepared Al-Ti alloys at approximately close angles but with different intensities.

Table 1: The data of XRD test results for $TiAl_3$ of the three Al-Ti alloys

Type of Al-Ti alloy	2θ (deg)	I
Al-10wt%Ti	64.5, 77.7, 81.9	41, 44, 11
Al-15wt%Ti	38.8, 64.7, 77.9	19, 32, 32
Al-20wt%Ti	39.9, 64.8, 78.0	6, 45, 39

2.2. SEM test results

From Figure 5, it is obvious that the addition of Ti into Al is done at dissimilar ratios (10w%; 15%; and 20%), and this is appeared in different distribution patterns in the microstructure of the prepared Al-Ti alloys. The SEM examination illustrates also the existence of $TiAl_3$ intermetallic compound, which resulted during the sintering from diffusion reaction between titanium and aluminum powders. The SEM results showed that the amount of $TiAl_3$ increases with increasing Ti content.

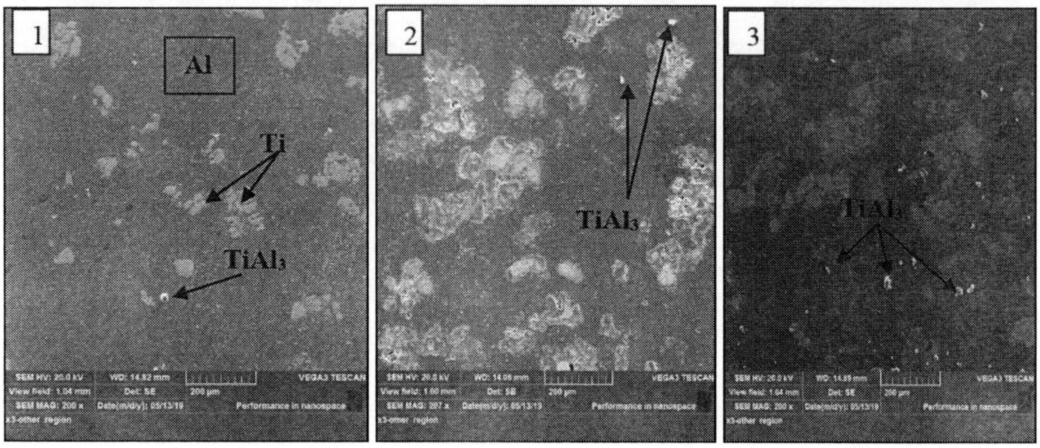

Figure 5: SEM images of Al-Ti alloys: 1)10wt%Ti; 2)15wt%Ti; 3)20wt%Ti

2.3. *Al-Ti alloys Density results*

The results show that 20wt%Ti-80wt%Al alloy has the highest value of bulk density about 2.989 gm/cm^3, compared to 10wt%Ti-90wt%Al alloy which has the lowest value of bulk density about 2.804 gm/cm^3, while 15wt%Ti-85wt%Al alloy has a moderate value of bulk density about 2.896 gm/cm^3, as shown in Figure 6.

This is back to the reason that Ti possesses higher density (about 4.506 gm/cm^3) than Al (which is about 2.7 gm/cm^3). Therefore; with increasing Ti ratio, the Al-Ti alloy bulk density would be higher [15].

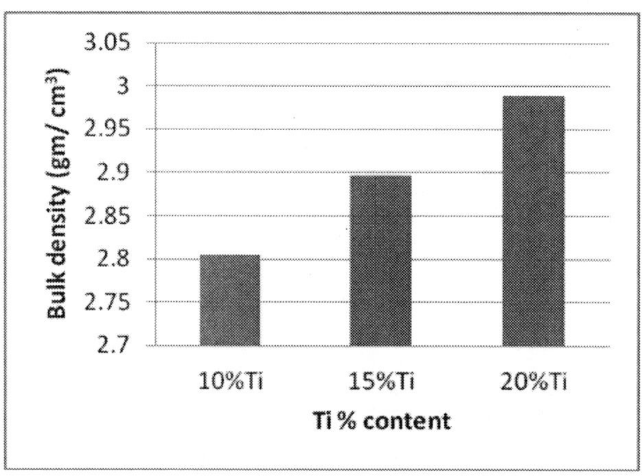

Figure 6: Bulk density of the three prepared Al-Ti alloys

2.4. *Al-Ti alloys microhardness results*

As the amount of titanium increase, the values of microhardness increases due to the increase in the generated $TiAl_3$ intermetallic compound, as shown in Figure 7.

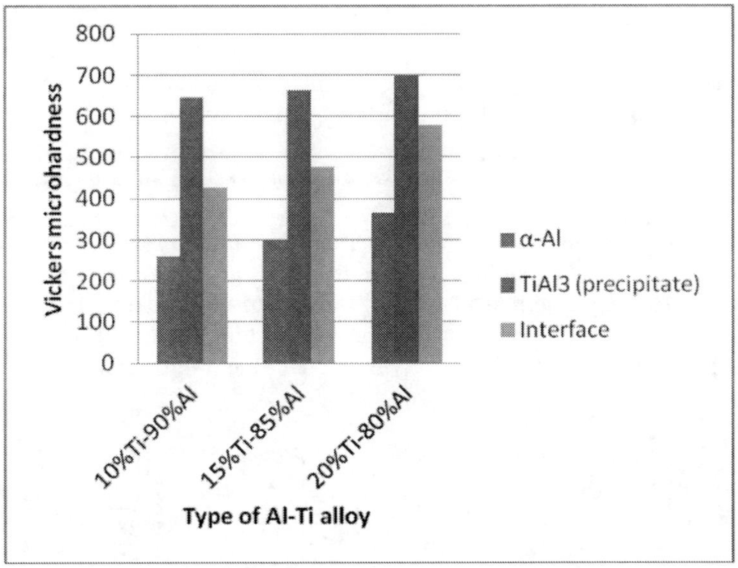

Figure 7: Al-Ti alloys microhardness results

2.5. Wear test results

The wear test results demonstrates that the highest wear rate was obtained in the Al-10wt%Ti alloy that possesses the lowest bulk density (2.804 gm/cm^3) and the lowest microhardness value (646.2 HV), while the lowest wear rate was obtained in Al-20wt%Ti that has the highest bulk density (2.989 gm/cm^3) and the highest microhardness value (698.9 HV) as shown in Figure 8.

The mechanism of wear is generally based on the hardness of the specimen which is considered as a key parameter that determines the amount of removed material, in which the sample is losing weight by sliding on a hard solid surface. The bright smooth regions with continuous grooves represents abrasive wear characteristics, while the dark regions with craters elongated in the direction of sliding represents the characteristics of adhesive wear [16].

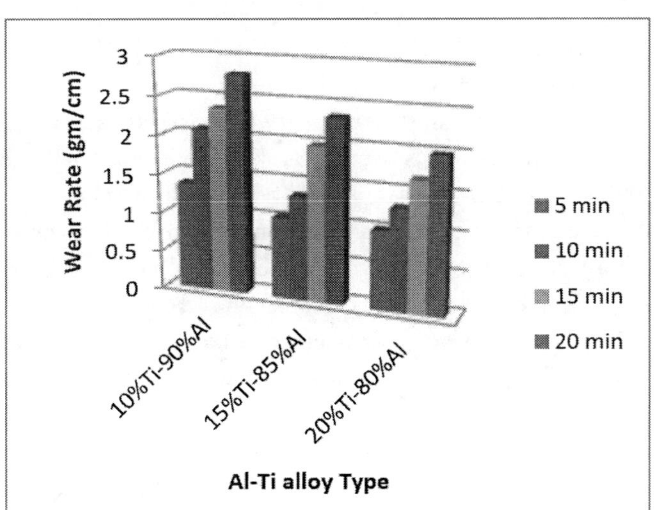

Figure 8: Al-Ti alloys Wear Rates under 15N load

Figure 9, reveals two types of wear mechanisms, but generally the results of optical microscope images showed that the adhesive wear tended to be the dominant type of wear mechanism at the Al-10wt%Ti alloy surface under the normal load 15N, which represents the dark regions in the image. While abrasive wear was more likely to appear as the predominant type of wear mechanism at Al-20wt%Ti alloy surface, which represents the bright regions, under the same normal load. In Al-15wt%Ti alloy, a combination of adhesive and abrasive wear mechanism was found at the worn surface.

These types of wear mostly appear clearer with the increase in sliding time. The adhesive wear can be attributed to the plastic deformation occurrence at the worn surface of Al-Ti alloy specimens which can leads to severe dislocations. These dislocations might cause cracks formation that would initiate and propagate in the direction of sliding as a result of the generated temperature between the sample and the disk with increasing time periods, and finally forms wear debris [9]. This plastic deformation could be reduced by increasing the alloy hardness. Therefore; the $TiAl_3$ plays an important role in improving the wear behavior of Al-Ti alloy by increasing the microhardness and decreasing the weight loss.

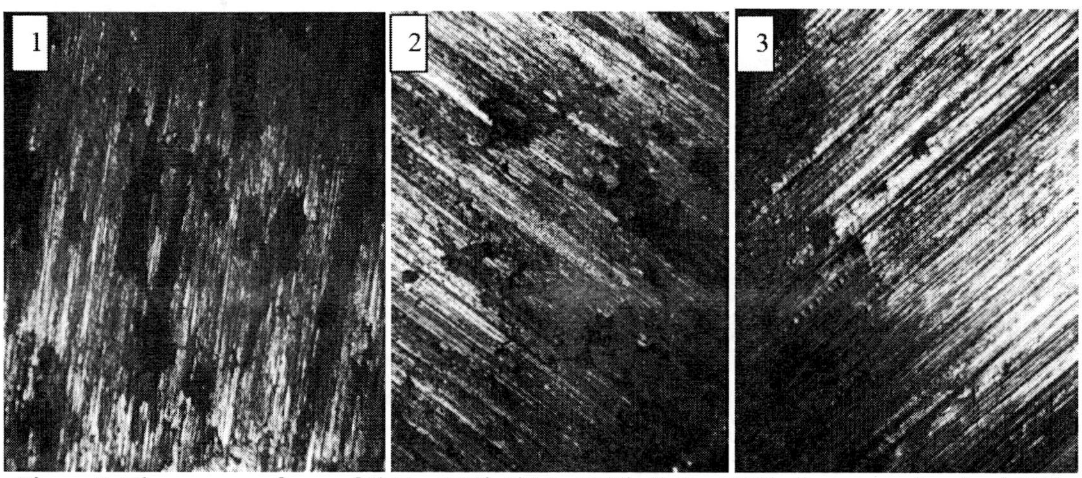

Figure 9: The worn surfaces of 1)10wt%Ti; 2)15wt%Ti; 3)20wt%Ti alloys tested under 15N load after 20min sliding time

3. CONCLUSIONS

1. $TiAl_3$ amount increases with increasing Ti content.
2. The values of microhardness and bulk density increases as the Ti weight percent increase.
3. Two types of wear mechanisms were observed on the worn surfaces of Al-Ti alloys that possess three different ratios of Ti, adhesive and abrasive wear mechanism.
4. The adhesive wear tends to be the predominant type of wear mechanism on the surface of Al-Ti alloy, having low Ti content; and it was observed that the adhesive wear increases with an increase in the sliding time and decreases with an increase in the Ti content.
5. In Al-10wt%Ti alloy, adhesive wear occurred due to plastic deformation that takes place on the worn surface and because of the low microhardness values of this alloy, the adhesive wear increases with an increase in the time of sliding. Further, abrasive wear is the predominant type of wear mechanism on the worn surface of Al-20wt%Ti alloy due to its high microhardness of this alloy which results from a higher content of $TiAl_3$, which helps to increase the wear resistance of the alloy and decrease the plastic deformation on the surface.

6. Al-15wt%Ti has a combined mechanism of wear which consists of adhesive and abrasive wear mechanisms, due to its moderate content of $TiAl_3$.

Acknowledgements

The authors acknowledge with gratitude the support of Al-Mustansiriyah University, and University of Technology, to achieve this work.

References

[1] D. Roy, S.S. Singh, B. Basu, W. Lojkowski, R. Mitra, I. Manna.(2009). Studies on wear behavior of nano-intermetallic reinforced Al-base amorphous/ nanocrystalline matrix In situ composite. Wear, 266 (11), 1113-1118.

[2] Sanjay Kumar Thakur, Manoj Gupta.(2007). Improving mechanical performance of Al by using Ti as reinforced. Composites Part A: Applied Science and Manufacturing, 38 (3), 1010-1018.

[3] Xiaoming Wang, Animesh Jha, Rik Brydson.(2004), In situ fabrication of Al_3Ti particle reinforced aluminium alloy metal-matrix composites, Materials Science and Engineering: A, 364(1), 339-345.

[4] A. Mandal, B.S. Murty, M. Chakraborty.(2009). Sliding wear behaviour of T6 treated A356-TiB_2 in-situ composites. Wear, 266 (7), 865-872.

[5] Tiezheng Li, Fengchun Jiang, Eugene A. Olevski, Kenneth S. Vecchio, Marc A. Meyers.(2007) Damage evolution in Ti6Al4V-Al_3Ti metal-intermetallic laminate composites. Materials Science and Engineering: A, 443(1), 1-15.

[6] I. C. Barlow, H. Jones, W. M. Rainforth. (2003). The effect of heat treatment at 500-655 °C on the microstructure and properties of mechanically alloyed Al-Ti-O based material. Materials Science and Engineering: A, 351, 344-357.

[7] Chuan-Yu Wu, Alan C.F. Cocks, Olivier T. Gillia, David A. Thompson.(2003). Experimental and numerical investigations of powder transfer. Powder Technology, 138 , 216-228.

[8] Chuan-Yu Wu, Luiza Dihoru, Alan C.F. Cocks.(2003). The flow of powder into simple and stepped dies. Powder Technology, 134 (1), 24-39.

[9] Özyürek Dursun, Tuncay Tansel, Evlen Hatice, Çiftci İbrahim.(2015). Synthesis, characterization and dry sliding wear behavior of in-situ formed $TiAl_3$ precipitate reinforced A356 alloy produced by mechanical alloying method. Materials Research, 18 (4), 813-820.

[10] C.L. Yeh, S.H. Su.(2006). In situ formation of TiAl-TiB_2 composite by SHS. Journal of Alloys and Compounds, 407 (1-2), 150-156.

IOP Conf. Series: Materials Science and Engineering **765** (2020) 012064 doi:10.1088/1757-899X/765/1/012064

[11] D. Özyürek, S. Tekeli, T. Tuncay, R. Yilmaz.(2012). The effect of synthesis time on wear behavior of Al-8%Ti alloy produced by mechanical alloying. Powder Metallurgy and Metal Ceramics, 51 (7), 491-495.

[12] Si-wei Zhang.(2004). Tribology of elastomers. book, Elsevier, First edition, ch. 1.

[13] François Cardarelli. (2008). Materials Handbook. A Concise Desktop Reference, book, Springer, 2nd Edition, 1019-1034.

[14] Baisong Guo, Song Ni, Rujuan Shen, Min Song. (2015). Fabrication of Ti-Al_3Ti core-shell structured particle reinforced Al based composite with promising mechanical properties. Materials Science & Engineering A, 639, 269-273.

[15] S. Çam, V. Demir, and D. Özyürek.(2016). Wear behaviour of A356/$TiAl_3$ in situ composites produced by mechanical alloying. Metals-open access metallurgy Journal, 6(2), 1-9.

[16] Xiang, Y. Z. Li, H. Zhang, Y. Huang, and H. Tang. (2012). Preparation and compressive behavior of porous titanium prepared by space holder sintering process. Journal of Procedia Engineering, 27, 768-774.

Investigation and characterization of coating and carburizing AISI 1011 steel

Diyar Abdulrazzaq Jabbar

Assist. prof. Dr. Zeyad Doshan Kadhim

Materials Engineering Department, Al-Mustansiriyah University, Baghdad, Iraq

corresponding author's e-mail : diyarabd19942017@gmail.com

Abstract. In this research work, the effect of coating layer thickness of nickel, which electrodeposited on low carbon steel AISI (1011), on mechanical properties has been studied. It also studied the mechanical properties of the carburized alloy at 950°C at the time of carburizing five hours. It was found that fatigue strength for this alloy increased as thickness of the nickel-coated layer increased. Also, the carburized specimen has a fatigue strength higher than the untreated metal. The metal that was treated with the carburizing process and nickel plating for 15 minutes has the best fatigue strength. The fatigue strength improvement against non-plated specimens was between 3.93% and 16.87 % . Furthermore, hardness of metal increased for specimens treated with electrodeposition, and it became 219.7 Hv for metal coated with nickel at 15 minutes and had a coated layer of 20.42 μm. The hardness value became 431.65 Hv for carburized alloy treated with plating for 15 minutes. However, there was an improvement in the surface roughness due to electroplating.

Keywords: fatigue strength, hardness, low carbon steel, coating, surface roughness

1. Introduction

Steel is a very important industrial material, as it has many more applications than any other engineering material [1]. Steel with carbon content less than 0.25 is termed as low carbon steel, while that with carbon content between 0.25% and 0.65% is known as medium carbon steel. If the content of carbon is high and ranges from 0.65% to 1.5%, then the steel is termed as high carbon steel [2]. Low carbon steel has high ductility and moderate strength and is used for its extensive fabrication properties for structural purposes, in buildings, bridges, ships, and cars [3]. Fatigue is the permanent, localized, and progressive structural change that occurs in metals and alloys subjected to fluctuating or repeated strains at values of stresses less than the tensile strength of metal [4]. Fatigue strength is an important mechanical property. However, the failure of fatigue starts on the metal surface [5]. Fatigue failure has four different stages: crack initiation, crack growth, crack propagation, and final rupture [6]. Most surface treatments produce compressive stresses in the metal surface, which reduce the probability of crack initiation and its expansion at the interface between the surface and core, thus increasing resistance to fatigue [7]. Carburization is an important and widely used process for surface hardening. This process I performed with the addition of carbon to the surface of the metal (low carbon steel) at a temperature between 850 and 950°C [8]. The nickel electroplating can be used for three purposes: decorative, functional, and electroforming. However, the appearance and other properties of the elecrodeposited nickel make this type of plating suitable for many applications by controlling the composition and the operating parameters of the plating solution [9].

IOP Conf. Series: Materials Science and Engineering **765** (2020) 012065 doi:10.1088/1757-899X/765/1/012065

2. Experimental Methods

A low carbon steel AISI (1011) has been used. The chemical analysis of this alloy was performed by the Thermo ARL 3460 optical emission spectrometer. The results can be summarized in the table below. This chemical analysis was carried out at 21 °C and moisture rate 44%.

Table (1) Chemical composition of low carbon steel AISI (1011)

Elements	C%	Si%	Mn%	P%	S%	Cr%	MO%	Ni%	Al%	Cu%	Fe%
Standard value	0.08-0.13	-	0.6-0.9	0.04 (Max)	0.05 (Max)	-	-	-	-	-	-
Actual value	0.0912	0.185	0.667	0.0233	0.0144	0.0652	0.0028	0.0338	0.0075	0.0457	Bal.

Fatigue specimens were prepared according to DIN, as illustrated in Figure (1). To get suitable and accurate dimensions for the fatigue specimens according to the standard dimensions of (HSM20 rotating bending fatigue machine), specimens were manufactured using the conventional lathe machine (Harrison 600, M350, EW700). Stress relief process was carried out for all specimens at 200 °C for three hours by using an eclectic furnace.

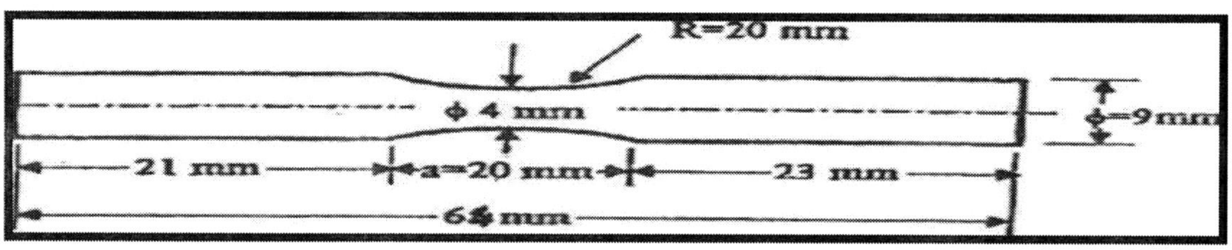

Figure 1: Fatigue test specimen according to DIN (50113), all dimension in mm

Fatigue samples were treated by using the pack carburization process as shown in Figure (2). Samples packed were with carburizing compounds in a metal box, which was sealed with clay and then placed in an electric furnace, where it was heated to 950°C and held for five hours. However, samples were quenched in oil. Table (2) shows the carburizing components used in the carburizing process.

Table (2): The carburizing components

Carburising component	charcoal	BaCO3	CaCO3	Na2CO3
wt %	80	10	9	1

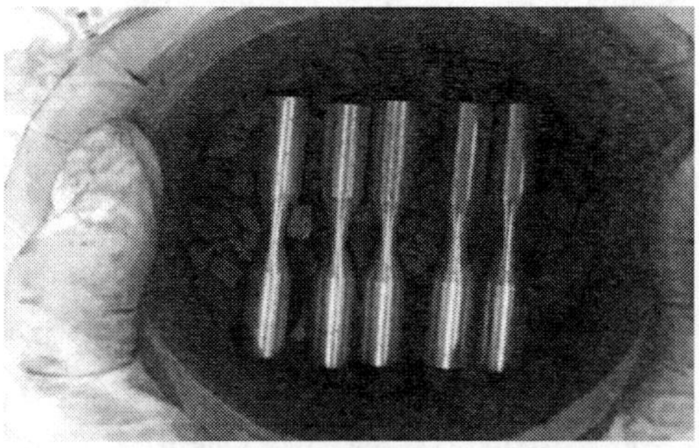

Figure 2: Carburizing box with fatigue specimens and carburizing components

Surface preparation of metals for electroplating was done in order to give good bonding of coating to the substrate and obtained finished parts. Surface preparation process consists of the following steps:

1- Alkaline cleaning: To remove fats from the outer surface.Sodium hydroxide is the best alkali in cleaners, so samples were immersied in hot solution of 75 gm/l NaOH and held for 5 minutes at 50°C.

2- Rinsing with water.

3-Acid cleaning: To immerse samples in dilute hydrochloric acid bath holding for two minute at room temperature, in order to remove oxides and rust from the surface.

4- Rinsing in water.

After the surface preparation process for coating, samples were coated with nickel by electrodeposition . The coating with nickel was done for three groups of samples with different coating times. Group one, specimens of the base metal was subjected to electroless nickel plating for five minutes. while ; the electroless nickel plating performed to the second group of specimens for ten minutes. However, group three consist of metal specimens subjected to electroless nickel plating for fifteen minutes. The nickel electroplating was done at 55 to 60 °C, current 7 A, voltage 3 volt, 4.5_5 pH, and the electroless nickel bath contained distilled water, 300 g/L of $NiSO_4$,30 g/L of NaCL, 40 g/L of H3-BO3, and polishing materials. After the nickel electroplating, the specimens were immersed in sodium dichromate composite in order to fixing color and prevent rust. After finishing the plating process, the specimens were dried. Ten samples of the selected alloy were manufactured as fatigue test specimens for each test except one sample for the purpose of the tensile test, as shown in Table (2). Figure (3) shows number of specimens for each group that is used for fatigue test.

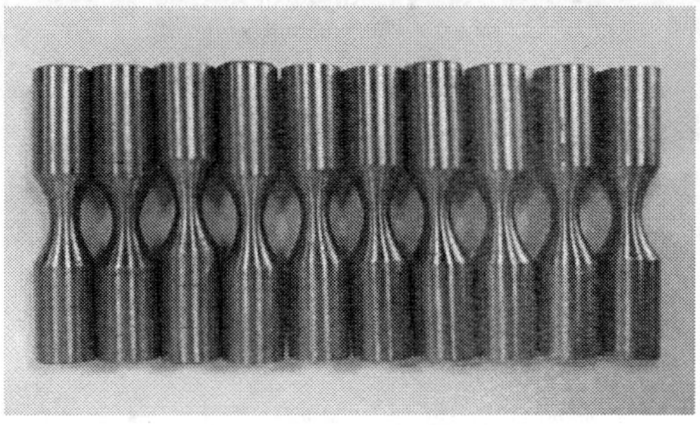

Figure 3: The sample of fatigue test according to DIN (50113)

IOP Conf. Series: Materials Science and Engineering **765** (2020) 012065 doi:10.1088/1757-899X/765/1/012065

Table (3) Categorization of the fatigue specimens

groups	*conditions*
A	*Ten untreated specimens*
B	*Ten specimens subjected to nickel coating process for 5 minutes time*
C	*Ten specimens subjected to nickel coating process for 10 minutes time*
D	*Ten specimens subjected to nickel coating process for 15 minutes time*
E	*Ten specimens subjected to carburizing process for five hour time at temperature* 950°C
F	*Ten specimens subjected to (carburizing process for five hour time at temperature* 950°C + *nickel coating process for 15 minutes time)*

2.1 Fatigue Test

The fatigue test machine of type (HSM20) rotating bending fatigue machine is used to do all fatigue tests, with the constant and variable amplitudes, as illustrated in Figure (4). The samples are exposed to an applied load from the right side of the perpendicular to the axis of the samples, developing a bending moment. The surface of the samples is under tension and compression stresses when it rotates. Fatigue testing machine components are illustrated in Figure(4). Value of the load (P) is measured by Newton (N) and applied to a specimen for known value of stress (σ) is measured by (MPa) and can be obtained from applying relation below:

$\sigma = (32 \times L\,P) / (\pi \times d^3)$

where

P = force in Newton.

L = arm of the force which is equal to (125) mm.

d = diameter of the specimen in (mm).

The constant fatigue test was accomplished in the laboratory air (the relative humidity was 25_30%) at the room temperature on (HSM 20) rotary fatigue bending machine as illustrated in figure (4), with the stress ratio of R= -1. Cycle frequency was 50 Hz and rotating speed used is 3000 cycle/min.

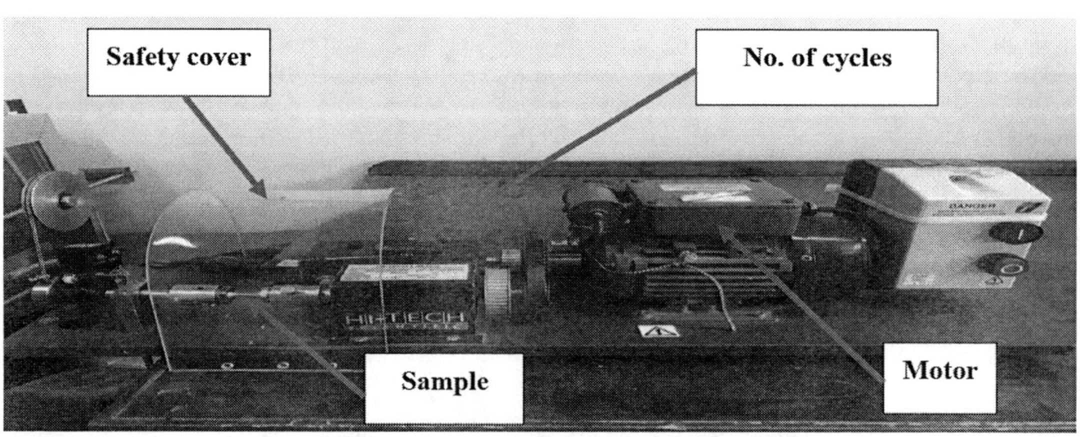

Figure 4: Fatigue testing machine HSM20 (rotary bending)

IOP Conf. Series: Materials Science and Engineering **765** (2020) 012065 doi:10.1088/1757-899X/765/1/012065

3. Result and discussions

3.1. Microscopic examination results

From Figure (5), it can be noticed that the microstructure of low carbon steel AISI (1011) after being subjected to stress relief process at 200 °C for three hours using an eclectic furnace consists of granules dark region (pearlite) and other light region (ferrite).

Figure 5: The microstructure of low carbon steel AISI (1011) magnification (X400) shows two phases (ferrite and pearlite)

3.2. Results of XRD test

XRD analysis was carried out on the tested samples, which are untreated (as received) and treated samples (nickel coating, carburizing), to examine the phases that exist in the material before and after surface treatment. Figure 6 illustrates XRD pattern for all the cases.

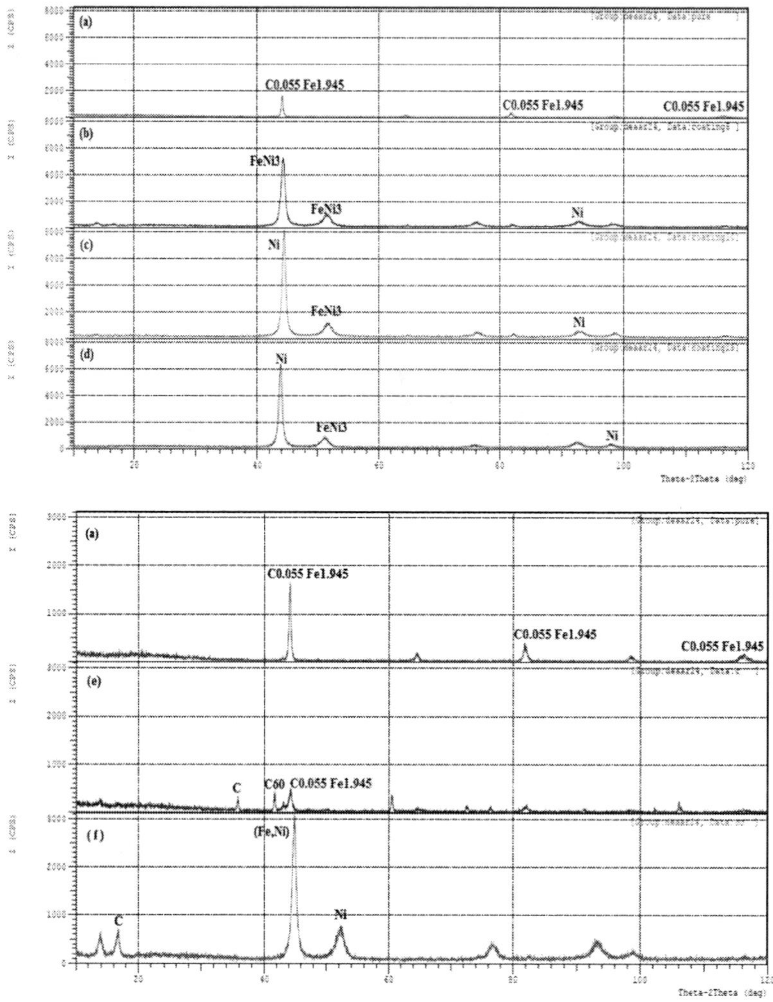

Figure 6: XRD pattern of a) Pure AISI 1011steel; b) Steel coated with Ni at 5 min; c) Steel coated with Ni at 10 min; d) Steel coated with Ni at 15 min; e) Steel carburized; f) Steel carburized and coated with Ni at 15 min

In Table (4), the angles of diffraction (2θ) and the intensities (I) of the X-ray beam for all cases are illustrated.

Cases	2θ (deg)	I
Pure AISI 1011 steel	44.11, 81.84, 116.29	100, 18, 7
Steel coated with Ni at 5 min	44.26, 51.54, 92.59	100, 16, 7
Steel coated with Ni at 10 min	44.40, 51.65, 92.69	100,11, 4
Steel coated with Ni at 15 min	43.83, 51.14, 98	100, 10, 4
Steel carburized	44.22, 41.60, 34.91	100, 83, 5
Steel carburized and coated with Ni at 15 min	44.86, 52.17, 16.75	100, 20, 15

IOP Conf. Series: Materials Science and Engineering **765** (2020) 012065 doi:10.1088/1757-899X/765/1/012065

3.3. SEM test results

The coating thickness and coating solution temperature between (55-60° C) is influenced by many factors like (Temperature, plating time and current density). From table (5) it can be seen that the plating thickness increases with increasing time and this is due to increasing the rate of deposition as current density increased.

Table (5) Plating thickness-time

groups	Plating thickness (μm)	Time (min)
B	9.73	5
C	14.8	10
D	20.42	15

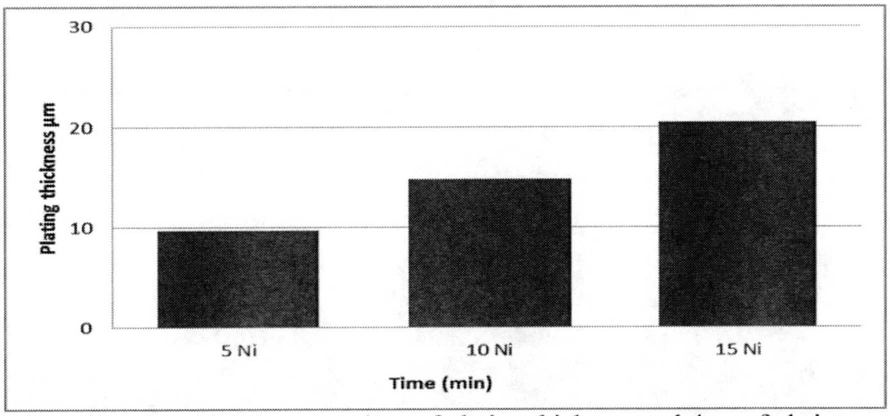

Figure 7: Relationship between values of plating thickness and time of plating

From Figures (8-A,B,C), it can be seen thickness for coating layers of nickel at three times (5,10 and 15) minutes.

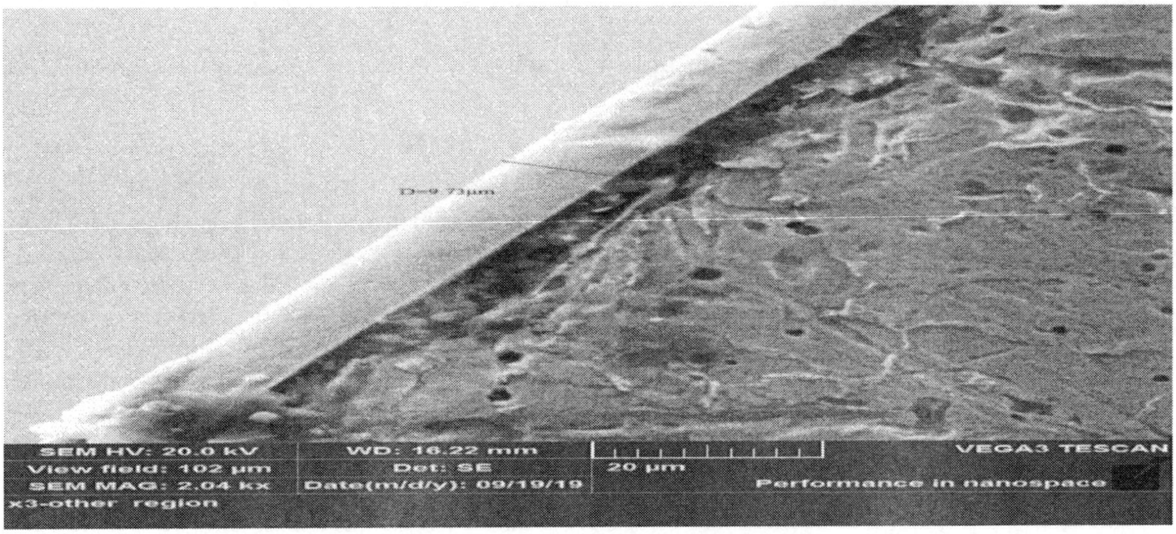

A: specimen of steel AISI (1011) coating with nickel for 5 min

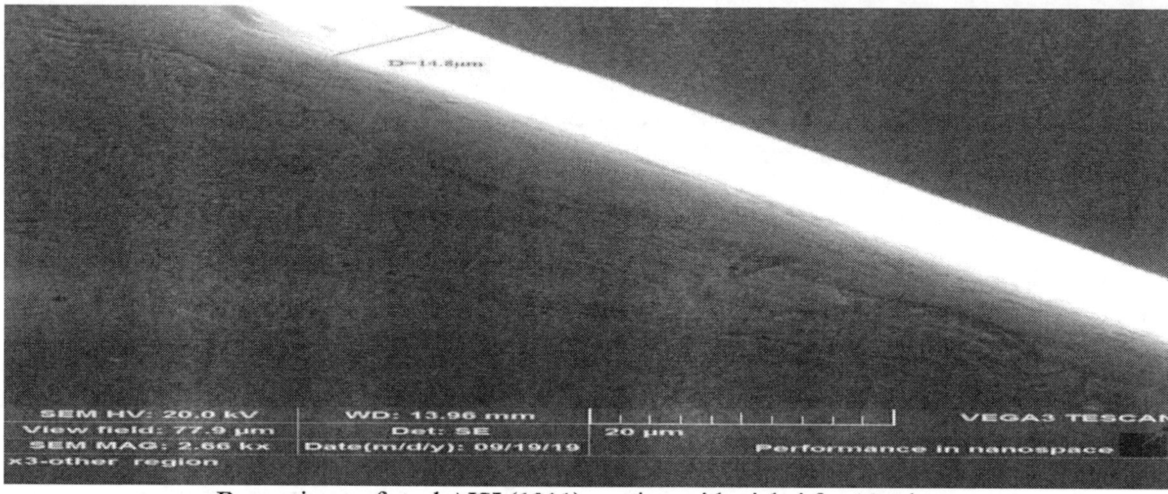

B: specimen of steel AISI (1011) coating with nickel for 10 min.

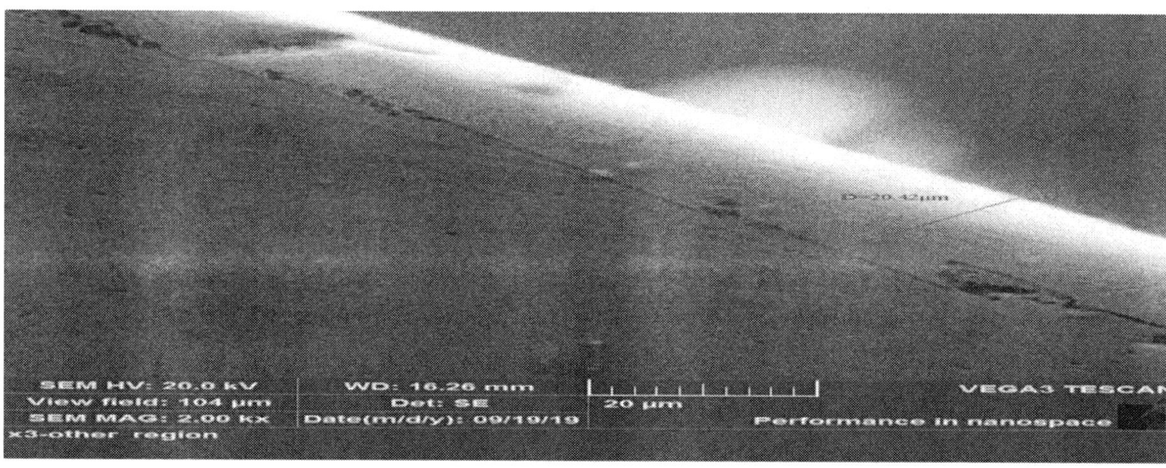

C: specimen of steel AISI (1011) coating with nickel for 15 min

Figures (8-A,B,C). The microstructure of low carbon steel AISI (1011), shows the coating layers thickness

3.4. Microhardness results

Vickers hardness method was performed for specimens , and it has been found that the hardness values increased for all surface treated specimens. The hardness values for specimens treated with nickel plating for (5, 10, 15 min) were higher than those of untreated and this is due to the formation of nickel phosphide precipitates [10]. Furtheremore, hardness of the treated specimens by carburizing process (at 950 °C for 5hr) have higher values than plating specimens by nickel and those of untreated this is due to the formation of carbides, however, carbon diffusion (diffusivity) during the carburizing process increased with time as the relation:

Case depth = $k\sqrt{t}$ where K is constant and t is time So as time increase the diffusivity of carbon increased [11]. The higher carburizing temperature and the higher soaking time will result a large carbide layer and the harder case becomes, while core still tough.

Table (6) The Vickers hardness values for samples of low carbon steel (AISI 1011).

Groups	Treatment	Vickers Hardness (HV)
A	Without treatment	169.65
B	5 min. coating with Ni	183.43
C	10 min. coating with Ni	201
D	15 min. coating with Ni	219.7
E	Carburizing 5 hour	390
F	Carburizing 5 hour + 15 min. coating with Ni	431.65

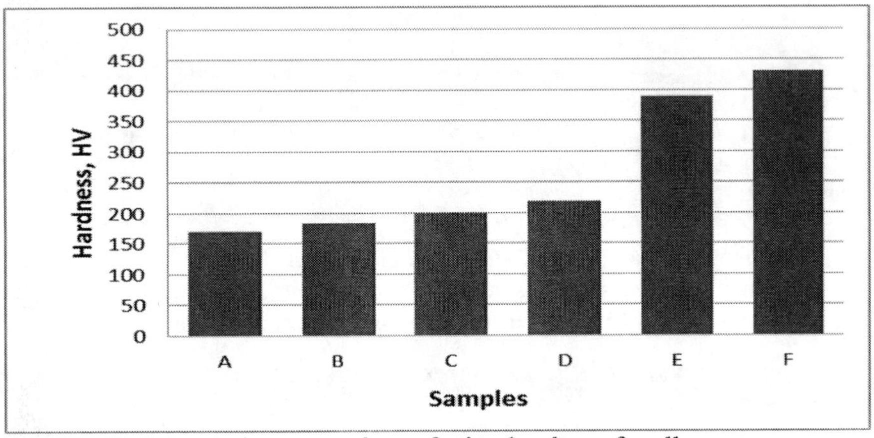

Figure 9: values of microhardness for all groups

3.5. Surface roughness results

The surface roughness decreased for samples that treated with nickel coating, As plating time increased, thickness of nickel layer increased and it will be 20.42 μm within 15 minutes time of plating which leads to surface finer roughness. While, after the carburization process the surface roughness of samples increased. So the surface roughness of carburised specimens became (1.98 μm) at carburising temperature 950°C and time of 5 hours.

Table (7) Roughness measurements for low carbon steel AISI (1011).

Groups	Specimens conditions	Roughness (μm)
A	un treated	0.92
B	Coated with Ni at 5 min	0.61
C	Coated with Ni at 10 min	0.39
D	Coated with Ni at 15 min	0.19
E	Carburizing for (5h, 950°C)	1.98
F	Carburizing for (5h, 950°C) + 15 min nickel coated	1.30

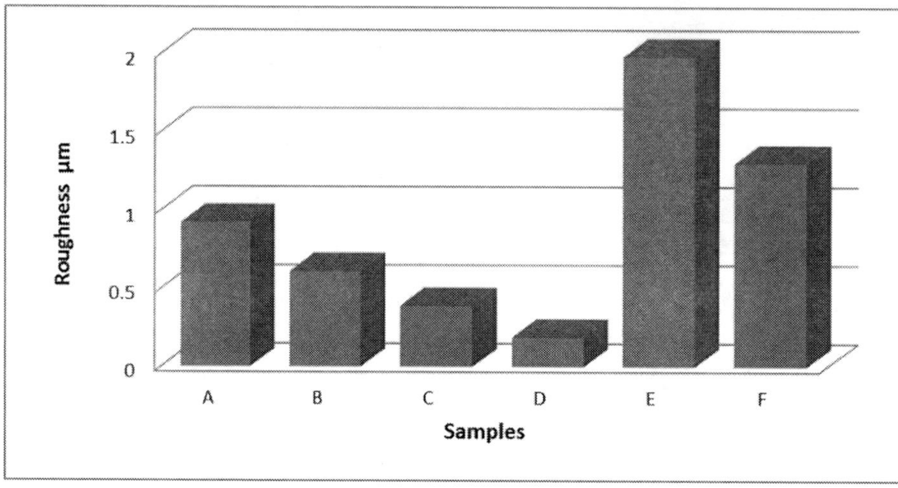

Figure10: values of surface roughness for all groups

3.6. Tensile test results

It can be seen from figure (11) the stress-strain diagram for low carbon steel AISI 1011. This test was carried out for this alloy after has subjected to stress relief process. It can be remarked, that the ultimate tensile strength was (550) MPa and the total percentage elongation became 30 mm.

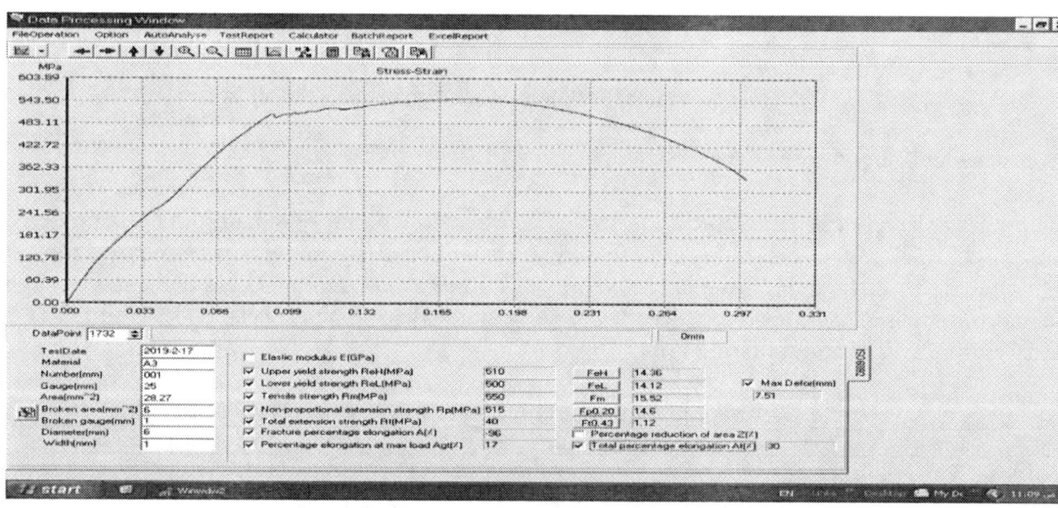

Figure 11: Stress-Strain diagram for low carbon steel AISI 1011.

3.7. Rotating Bending Fatigue Tests

S-N curves for all groups were obtained from the fatigue test as shown in figure (12).
It can be seen that the fatigue strength for low carbon steel AISI (1011) increased for specimens subjected to carburising process as compared with the same metal which don't treated by this process. This improvement of fatigue limit was due to the compressive residual stresses and carbides formation at the surface of steel which may be stopping and blocking the crack. From S-N curves for specimens plating with nickel by electrodeposition process at room temperature, it can be seen that the effects of thickness of coating with nickel after coating process on the fatigue strength and value of fatigue limit

IOP Conf. Series: Materials Science and Engineering **765** (2020) 012065 doi:10.1088/1757-899X/765/1/012065

Furthermore, the characteristics of deposited film itself were examined. As the electroless nickel plating is performed on the metal, their fatigue strength rather increases, as compared with uncoated metal. The fatigue limit for this alloy coated with Ni at 5 minutes became (82.89)Mpa. While the the fatigue limit for this alloy coated with Ni at 15 minutes became (93.21) Mpa. And for untreated alloy was (79.75) Mpa. So It can be seen that fatigue strength improvement against non-plated specimens was between (3.93-16.87) % .

The fatigue limit for carburisied alloy at 950°C for five hours was (98.12) MPa, while the fatigue limit for carburisied alloy at 950°C for five hours and plating with nickel for time of plating 15 minutes became (109.30) MPa.
It can be remarked that the electroless deposited nickel film shows a good adhesion to the substrate. The fatigue damage of plated steels prevented by deposited film. Consequently, the nucleation of fatigue crack is delayed, and there is an improvement in fatigue strength.
In electroplating process, the micro-hardness of deposited film increases, beside that there is an structural changes. This improvement is due to the precipitation hardening of Ni_3P. Furthermore, the tensile residual stress in deposited film is reduced. So the improvement in fatigue limit may be due to compressive residual stresses and largest ductility/ toughness exist in the electroless nickel plating which delayed the crack propagation.

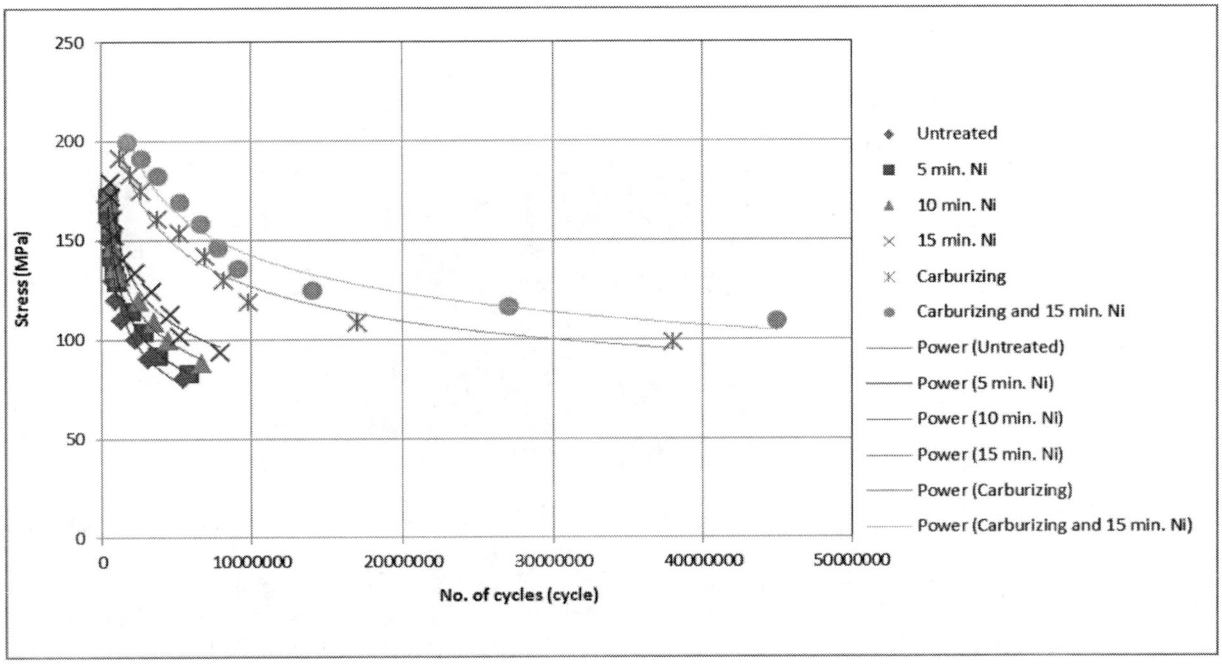

Figure 12: S-N curves for all groups

The surface fractures could be recognize by using scanning electron microscope TESCAN, VEGA 3 LMU device. Figure (13) shows surface fracture. Which occurred in the specimens after sufficient cycles of the bending load.

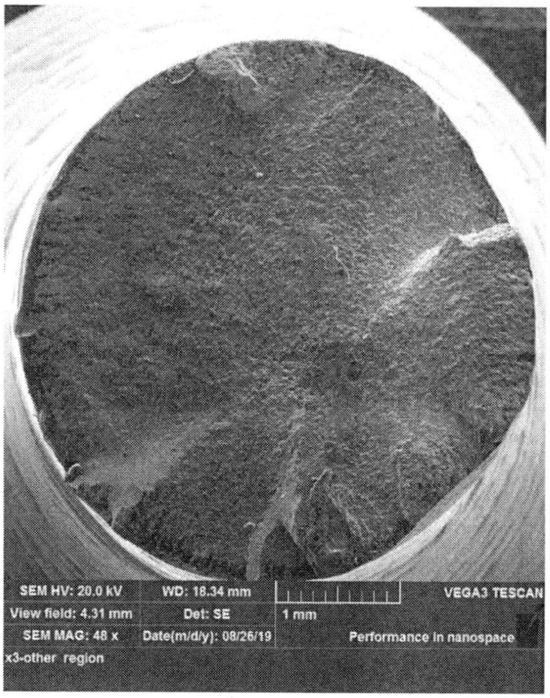

A:

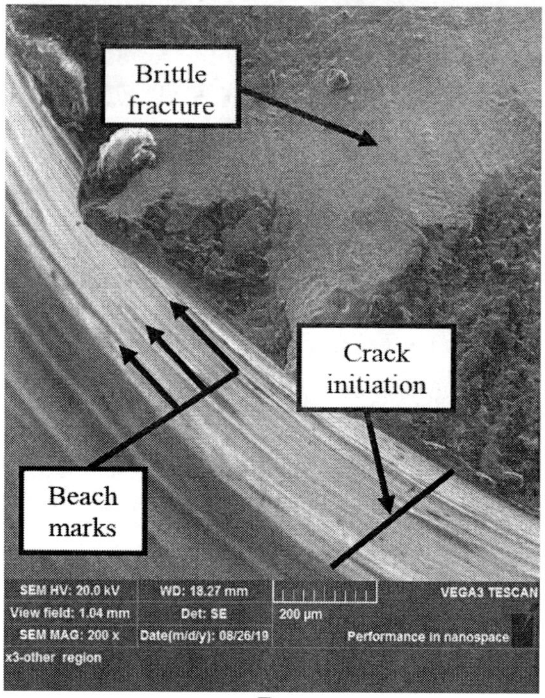

B:

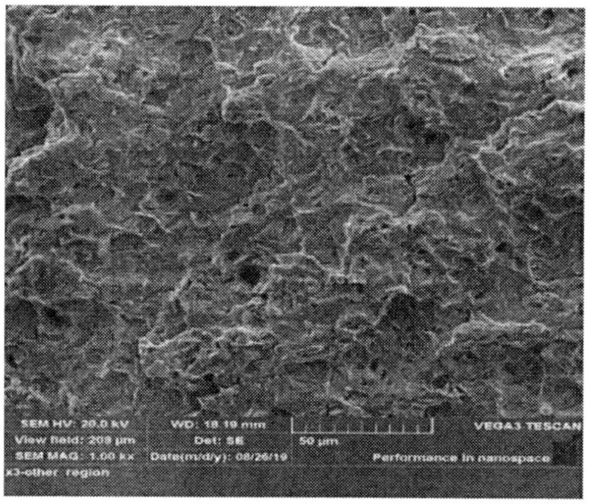

C:

Figure (13): Fatigue fracture surfaces for AISI 1011 steel, A: fatigue fracture, B: crack initiation, C: core of the fracture.

It can be seen the fracture is brittle, beach marks are often formed when the load is alternating during service or when the loading is intermittent. Beach marks always explain a fatigue failure, but unfortunately if there is no beach marks does not rule out fatigue failure [12].

4. CONCLUSIONS

1. Low carbon steel AISI 1011 electroplating by nickel had higher fatigue strength due compressive residual stresses and largest ductility/ toughness exist in the electroless nickel coating that delayed the crack propagation. The maximum fatigue limt became 93.21 Mpa as the alloy plating with nickel at 15 mintes and thickness of coating layer is 20.42 µm. but the best fatigue limit was 109.30 Mpa as the metal subjected to carburizing and plating with nickel at 15 mintes.

2. Carburizing process promote the fatigue strength and the value of fatigue limit was 98.12 MPa as the specimens treated with carburizing temperature is 950°C and time 5 hours.

3. Nickel coating showed increasing in hardness of the metal (steel the AISI 1011) as plating time increased, the maxmuim hardness value was (219.7) Hv at plating time 15 minutes. Furthermore the hardness value became (431.65) Hv at plating time 15 minutes and curburising temperature 950°C with time 5 hours.

4. The thickness of the coating increased with the increasing coating time. The

maximum thickness of coating was 20.42 μm at planting time of 15 min.

5- There is an improvement in surface roughness of samples with plating process , the surface roughness of the metal was 0.19 μm at plating time 15 minutes.

5. References

[1] SA Tukur, MM Usman, Isyaku Muhammad, NA Sulaiman.(2014). Effect of Tempering Temperature on Mechanical Properties of Medium Carbon Steel. International Journal of Engineering Trends and Technology, 9(15), 798-800.

[2] Sanjib kumar jaypuria.(2008) . Heat Treatment of Low Carbon Steel. National Institute of Technology Rourkela.

[3] SA Tukur, MM Usman, Isyaku Muhammad, NA Sulaiman.(2014). Effect of Tempering Temperature on Mechanical Properties of Medium Carbon Steel. International Journal of Engineering Trends and Technology, 9(15), 798-800.

[4] Ralph I. Stephens, Ali Fatemi, Robert R. Stephens, Henry O. Fuchs.(2000). Metal fatigue in ingineering. John Wiley and Sons, second edition .

[5] Mirko Klesnil, P Lukac .(1992). Fatigue of metallic materials. Book, Elsevier, second edition.

[6] Shrikant P. Bhat ,Rahul Patibandla.(2011). Metal Fatigue and Basic Theoretical Models. Alloy Steel - Properties and Use, 204-206.

[7] Vernon John. (1992).Introduction to Engineering Materials. Macmillan International Higher Education, third edition.

[8] Jaykant Gupta.(2009). Mechanical and Wear Properties of Carburized Mild Steel Samples. National Institute of Technology, Rourkela.

[9] George A Di Bari.(2000). Electrodeposition of Nickel. Modern Electroplating 5, 79-114.

[10] Ilangovan soundiah .(2014). Investigation of the mechanical, corrosion properties and wear behaviour of electroless Ni-P plated mild steel. International journal of research in engineering and technology.

[11] J Dossett, GE Totten .(2013). Introduction to Surface Hardening of Steels. ASM Handbook 4, 389-398.

[12] American Society for Metals (ASM Handbook).(1994). surface engineering. ASM International 5 , 690.

IOP Conf. Series: Materials Science and Engineering **765** (2020) 012066 doi:10.1088/1757-899X/765/1/012066

Stability Control Investigation of a Self-Balancing Platform on the Robot Smart Car Using Navigation Parameters

Ghada Adel Aziz

Electromechanical Engineering Department, University of Technology,Baghdad,Iraq

E-mail: 50070@uotechnology.edu.iq

Abstract. A self-balancing platform is constructed on a robot smart car, and its navigation parameters are investigated by running the car outdoors on a straight surface with structured bumps of 3and 5 mm in height. A ball-and-socket joint are supported at the centre of the platform, which is to be freely rotated with the lateral and longitudinal movement of the platform and in each axis, is controlled by three servomotors. Which move according to the recorded data movements from a gyro-accelerometer sensor which controlled by an Arduino UNO. The SimMechanics Matlab simulation helps to give the initial values of all parameters and the additional unexpected signals due to the influence of bumps on the vibration signals, with the proportional-integral-derivative (PID) controller, used to control the relationship of the rapid simulation of the robot smart car and mathematical model between platform and car shock absorber system with respect to the servomotor's rotation angle. These bump is more effective on the pitch and roll angle values when compared to other parameters, with more than 815 degrees and error up to 35 degrees due to vibration signals, and the time required for stabilization increases due to the bump height and the navigation parameters, roll angle from 2 to 6 in mean change value 2 , yaw angle from 43 to 59 in mean change value 8 and pitch angle from 1 to 2 in mean change value 0.5, which indicates that yaw, roll and pitch angle have variation changing in Mission Planner software

Keywords— Stability control, PID controller, Arduino UNO, Self-balancing platform, SimMechanics Matlab .

1. Introduction

Kalman Filter and fitting the mixture of Gaussians is introduced, to decrease the impact of IMU noise are based in one an algorithm, and the results achieve a high accuracy with error is not more 1%, and is immune to electromagnetic interference. The powerful of the low-cost embedded IMU, and GPS could be an Intelligent Vehicles and tracker software which supported by information given from accelerometer and gyroscope [1].

Two-wheel self-balancing robot is constructed using a low-cost components, and controlled with a pair of DC-motor, an Arduino microcontroller board, a single-axis gyroscope, and a 2 axis accelerometer added to determinate the attitude. kinematic parameters is determined experimentally; PID -control is designed to satisfy the requirements of study [2], a new model is built with an addition of two long arms in the center of a four-wheel drive robot, these arms allow it to stand-up for balancing and navigate stairs. The experimental data on the real robot is shown with a single extra degree of freedom increases the field capabilities of self-balancing robots [3].

IOP Conf. Series: Materials Science and Engineering **765** (2020) 012066 doi:10.1088/1757-899X/765/1/012066

This Self balancing car robot unit runs with a PID control loop which improves the stability of the system. On different surfaces, car robot was tested and analyze the stability of the system in a real-time data plot is done in MATLAB and filtered angle versus time by using PID controller, the results show that surfaces like soft rubber is enough and the most suitable surface to be slightly compressed by the weight of the car robot [4] and with F450 Flame Wheel frame kit is assembled mainly with an AurdoPilot type Mega APM 2.6 controller and the Mission Planner as a ground station software. The GPS coordinates accuracy in comparing with the actual measurements has been investigated with Google Satellite coordinates, than the Google earth images with error within the range (0.1 - 5) meters , However, controlling is not so easy to put it into wide scale civilian applications [5].

These model of the robot some time is constructed based on the Lagrangian function, relies upon dynamic balancing systems for balancing and maneuvering for balancing robots , which based on inverted pendulum configuration. DC motor attached to the base frame of the robot which helps for maneuvering , tilt and accelerometer Sensors are used for balancing [6]. Advanced development in the field of robotics is studied with a self-balancing robot which is based on Inverted pendulum theory. An efficient micro controllers and sensors for control system is used to stabilize an unstable system [7] and for local path planning, the Jump Point Search (JPS) algorithm is adopted, and an Information System map is used with consideration of including Global Positioning System position. The optimal safety path is done successfully, and it has a lower time complexity compared with the Vector Field Histogram [8].

The signal processing aspects of position and orientation estimation are studied and used microelectromechanical system with inertial sensors are used and have become widely available, due to their small size and low cost as 3D accelerometers and 3D gyroscopes, inertial sensors are combined with models and additional sensors [9], which very important in design of an autonomous underwater vehicle, which is designed to move quickly without collision and cooperate with the other AUVs, to find the target with the self-organizing map (SOM) neural network is used to build dynamic alliances in real time for getting a good tracking short way cooperated with navigation parameters [10].

The increasing of mass structure weight should be given decreased in eigenvalues and increasing in stresses concentration in the mass positions on the truss structure, which effect on the vibration parameters during the motion of the mass. Therefore the study should not be taken high velocity of mass in the model and if will take a long period of time, there can be a loss of precision solutions [11].

2. Mathematical Modeling Simulation using SimMechanics

The IMU sensors are fixed by the physical orientation of the sensors, and the IMU sensors axes are called a IMU-Fixed frame and all the sensor outputs are referenced to this frame. for the position measurement (x, y, z), the Earth-Fixed frame and utilizing a moving non-inertial frame (u, v, w), IMU Fixed frame with its axes parallel to the IMU sensors axes. A gyroscopes outputs and the accelerometers and magnetometers outputs can combine the advantage of the short-term precision of the gyroscopes and the long-term stability of accelerometers.

The mathematical modeling relationship between platform and servomotor rotation angle is formulated, and the length of the links (L1, L2 and central linkL3) are joining the platform with a motor links (Sr1 and Sr2), which are connected with servomotor shafts are of dimension 12 cm and 3 cm respectively. Servomotor is positioned with an offset length (L3) of 12 cm from center of cart in both the axes [10, 12]. Every time step we are able to know the Euler angles α,φ,θ can keep the order of rotation and follow it every time step as follows:

$$\Delta\theta\,(n+1) = \Delta t \times \left[\frac{\dot{\theta}}{2}\,(n+1) + \frac{\dot{\theta}}{2}\,(n)\right] \ldots\ldots\ldots\; Yaw\ angle \tag{1}$$

$$\Delta\varphi\,(n+1) = \Delta t \times \left[\frac{\dot{\varphi}}{2}\,(n+1) + \frac{\dot{\varphi}}{2}\,(n)\right] \ldots\ldots\ldots\; Pich\ angle \tag{2}$$

$$\Delta\alpha\,(n+1) = \Delta t \times \left[\frac{\dot{\alpha}}{2}\,(n+1) + \frac{\dot{\alpha}}{2}\,(n)\right] \ldots\ldots\ldots\; Roll\ \ angle \tag{3}$$

Where:

IOP Conf. Series: Materials Science and Engineering **765** (2020) 012066 doi:10.1088/1757-899X/765/1/012066

$\Delta\theta$: denotes the incremental angle around the W-axis.

$\Delta\varphi$: denotes the incremental angle around the V-axis.

$\Delta\alpha$: denotes the incremental angle around the U-axis.

Δt: denotes the time step, and n is the time index.
Note that the trapezium rule has been used in equations as numerical integration method. The rotations matrix denoted as M around each particular axis as:

$$M(W,\theta,n+1) = \begin{bmatrix} cos\Delta\theta(n+1) & -sin\Delta\theta(n+1) & 0 \\ sin\Delta\theta(n+1) & cos\Delta\theta(n+1) & 0 \\ 0 & 0 & 1 \end{bmatrix} \quad (4)$$

$$M(V,\varphi,n+1) = \begin{bmatrix} cos\Delta\varphi(n+1) & 0 & sin\Delta\varphi(n+1) \\ 0 & 1 & 0 \\ -sin\Delta\varphi(n+1) & 0 & cos\Delta\varphi(n+1) \end{bmatrix} \quad (5)$$

$$M(U,\alpha,n+1) = \begin{bmatrix} 1 & 0 & 0 \\ 0 & cos\Delta\alpha(n+1) & -sin\Delta\alpha(n+1) \\ 0 & sin\Delta\alpha(n+1) & cos\Delta\alpha(n+1) \end{bmatrix} \quad (6)$$

The general rotation matrix:

$$Rotation(n+1) = M(W,\theta,n+1) \times M(V,\varphi,n+1) \times M(U,\alpha,n+1) \quad (7)$$

The following formulas are used

$$\begin{aligned} pitch &= atan2(-m20, sqrt(m21 * m21 + m22 * m22)) \\ yaw &= atan2(m10, m00) \\ roll &= atan2(m21, m22) \end{aligned} \quad (8)$$

The proportional, integral, and derivative terms are summed to calculate the output of the PID controller. Defining u(t) as the controller output, the final form of the PID algorithm is:

$$u(t) = k_p \times e(t) + k_i \int_0^t e(t)dt + k_d \frac{de}{dt} \quad (9)$$

Where:
K_p: Proportional gain, a tuning parameter.
K_i: Integral gain, a tuning parameter.
K_d: Derivative gain, a tuning parameter.
e_t:Error function corrected with navigation parameters.
t : Time or instantaneous time.
The PID controller is used and tuned by varying these constants K_p ,K_d and K_i and optimizing them. In simulation analysis the SimMechanics is used Figure 1, and the mass and the inertia tensor elements are increased by the same rate when the increment unit mass is up to 15 g, and the inertia tensor is normalized according to this mass unit. so this increment unit and a normalized inertia tensor are to be used in the analysis, and the actuators should be not saturate due to their position and velocity limits, and the input signal of platform motion in terms of accelerations, all of these because to the inverse kinematics function of the system, in order to put the specification verified kinematically, since the motors may saturate due to their position and velocity limits, Variation in angle K_p= 0.85, K_i= 3.2, K_d= 0.1(Balancing Condition).

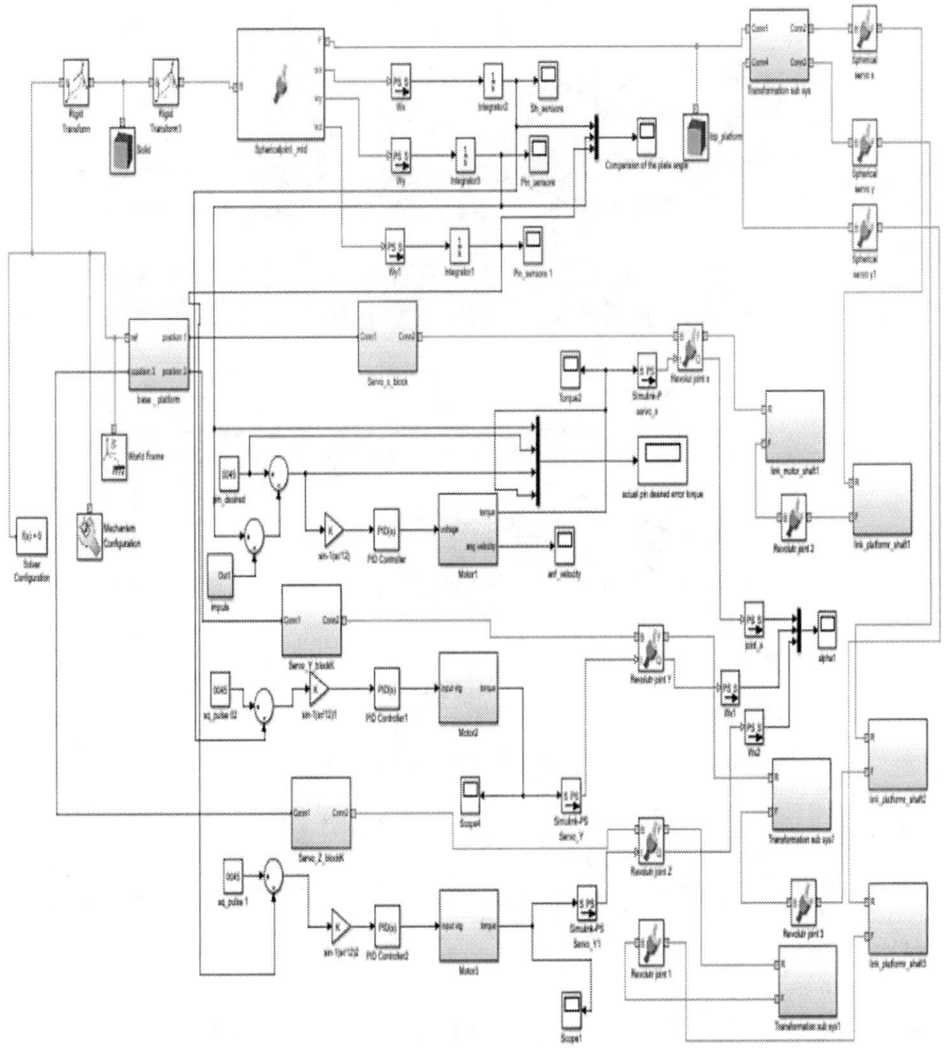

Figure 1. Self-Balancing System Control Model simulation by SimMechanics.

3. Experiment of Physical Modeling of Robot Smart Car

The platform is constructed above the Robot Smart Car and connect the controlling practical parts with it and programming them together to make the system as shown in Figure 2, the platform that works with the movement of the Robot Smart Car which is vibrated because of the unstable movements, these vibration parameters with navigations of location are measured and controlled for a platform self-balancing and comfortable movements. Arduino Uno is a microcontroller board based on the ATmega328P is used Figure 3, the GY-521 MPU6050, 6 DOF 3-Axis Gyroscope and Accelerometer, carrier board based on the MPU-6050 sensor, this sensor contains a MEMS accelerometer and a MEMS gyro in a single chip, at the same time the Micro Electro-Mechanical Systems (MEMS) captures x, y, and z channel. This helps to measure acceleration, velocity, orientation, displacement, and other motion parameters of the system. The schematic of GY 521 connection with Arduino UNO and the electronic circuit of GY 521 as shown in Figure 4.

The sensors are exactly in the same as APM flight controller, for measurements the navigation parameters, however this has an option to be used for built in the compass, with accelerometer, gyroscope sensor which used to detect the attitude of the vehicle (pitch , roll , yaw) by using gyroscope and the

IOP Conf. Series: Materials Science and Engineering **765** (2020) 012066 doi:10.1088/1757-899X/765/1/012066

Figure 2. Robot Smart Car with a Constructed Platform.

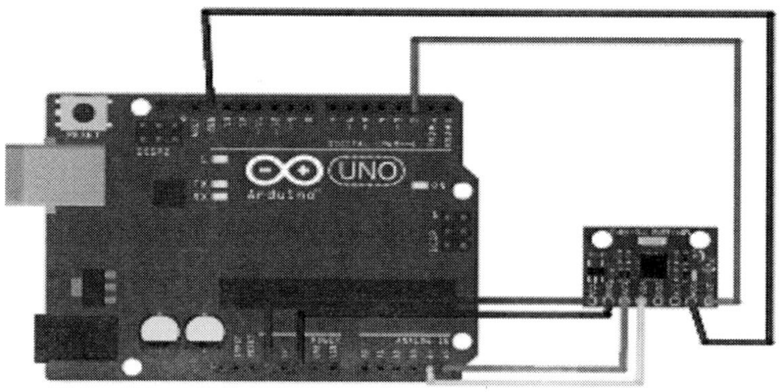

Figure 3. The Connection Gyroscope with Arduino UNO Pins.

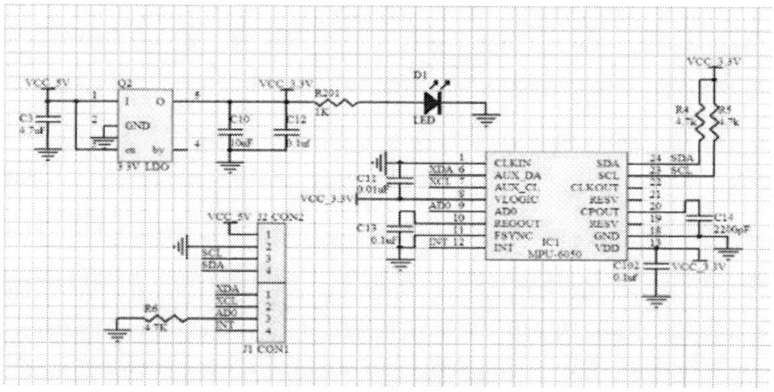

Figure 4. The Electronic Circuit of GY 521.

vibration in x, y and z axis by using accelerometer, which are used to calculate the least vibration happening on the body of car and handle the error and minimize it . An collection navigation system is used in a small car sample designed for self-balancing and self-guidance control under previously prepared programming. The APM 2.8 is a complete open source for autopilot system, and it is connected with others parts in this system as shown in Figure 5 as an unmanned aerial vehicle system competition, which allows the user to turn any fixed, rotary wing or multi rotor cars and boats, into a fully autonomous vehicle and capable of performing programmed missions with waypoints by GPS , the Mission Planner software is used and very powerful measuring the navigation parameters, and more flexibility in positioning location,

IOP Conf. Series: Materials Science and Engineering **765** (2020) 012066 doi:10.1088/1757-899X/765/1/012066

so that GPS/Compass unit can be mounted further from the noise sources than APM itself. All the Measurement is taken and recording from the sensors, creating platform motion demand for self-balancing, inverse kinematics solution, generating and delivering control output signals to the drivers are done by Arduino UNO in real-time connected with servomotor expects a coded signal with every few seconds for required balancing displacements. The servo motor that we have selected has a torque of 15kg/cm, operating voltage of 4.8 V to 6 V, speed of 60 degree/0.20 sec, dimensions: length 49.3 mm, width-25.4mm, height-42.9 mm and weight about 80g.

The study of the navigation parameters under the vibration and control is apply the principles of navigation parameters, vibration and control navigation parameters, which coming from gyro-accelerometers and locations by GPS during the motion between different points, and understanding the control system programming using sensors for a suitable signal according to the direction guidance using navigation parameters by correction with a Mission Planner software, records navigation parameters in order to reach the stability control for x, y, and z axis with roll, yaw, and pitch angles with effect of vibrations. as shown in Figure 6, Which include the calibration of accelerometer, Radio Control, Electronic Speed Controller (ESC), Compass Calibration.

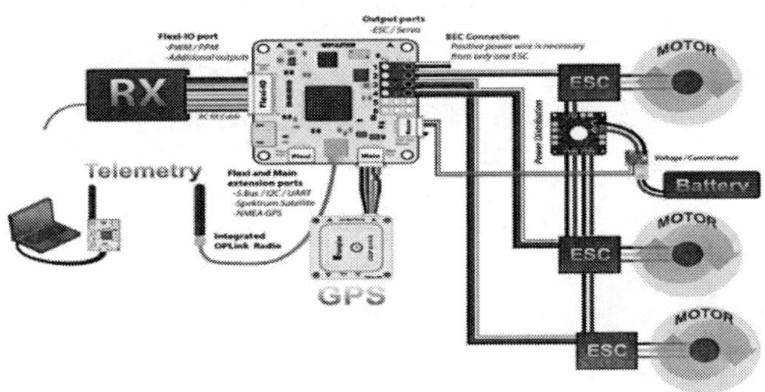

Figure 5. System Parts Connected with APM.

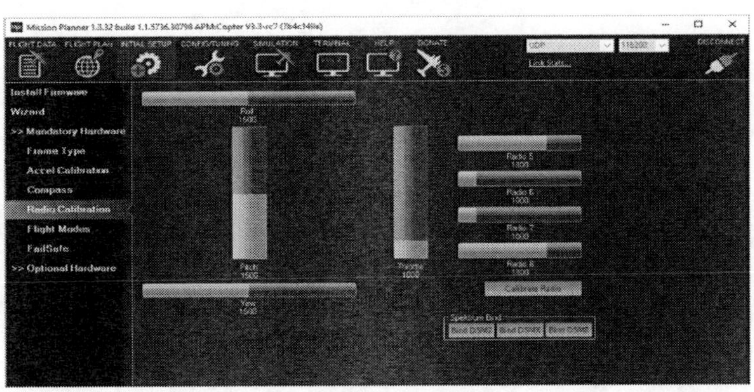

Figure 6. Mission Planner Radio Calibration Screen.

Among these problems, the vibration effect which arises for several reasons, and made imbalance state with defect in the mechanical equipment for guidance of the installation of parts in the not correct axes direction .These mechanical system vibrations cause an error in navigating parameters values during the motion of these vehicles, especially in the gyroscope and accelerometer sensor, these problems and their causes increase the cost of maintenance and repair damage if the accidents has been accrued because of the wrong resulted reading values of these devices. To solve the problem, it must to take into account the

IOP Conf. Series: Materials Science and Engineering **765** (2020) 012066 doi:10.1088/1757-899X/765/1/012066

errors effected coming from the vibration and the around noises on the signals during the motion with self-balancing to re-correct the values of all navigation parameters used.

4. Results and Discussions

Table 1. PID Simulation Values.

	Roll angle	Yaw angle	Pitch angle
P	0.0832	0.0101	0.0902
I	0.0311	0.0011	0.0421
D	0.0021	0.0001	0.0022

Regarding the simulation test of Figure 7 for more than 5 mm, the system needs nearly 1.6 sec to get back to the stability state. Besides, Figure 8 shows that the pitch angle more sensitive and more effective on the system stability during the motion. The motion of robot smart car with speed bump 3 mm and 5mm height is experimented.

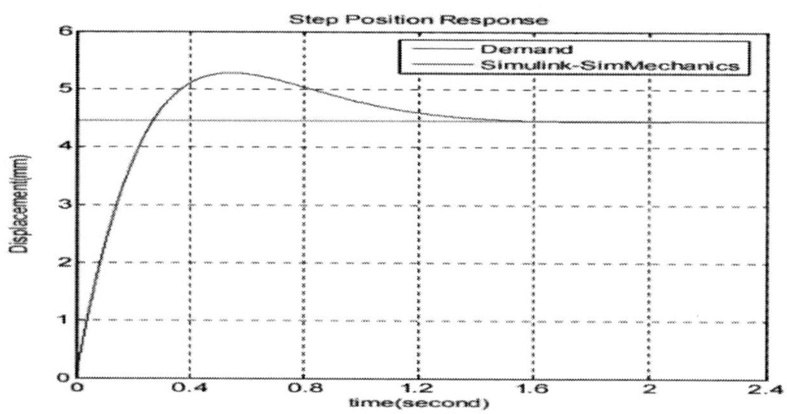

Figure 7. Servomotor Signal Simulation from Simmechanical.

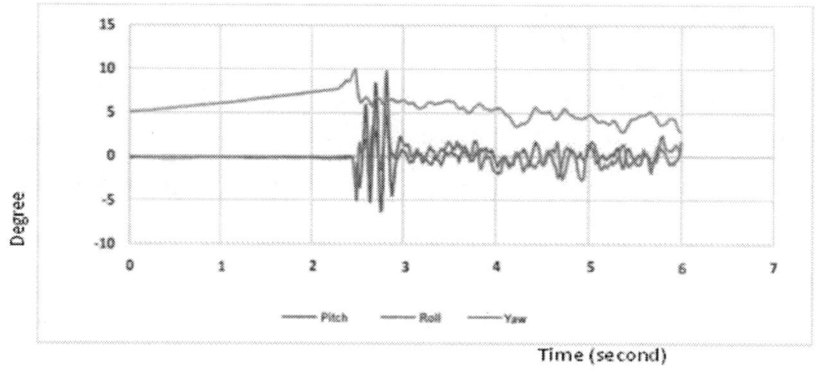

Figure 8. Gyro-Accelerometer Measurements During Random Motion with Vibration Effect.

According to the pitch angle as shown in Figures 9 and 10, almost the angle changes as roll angle but only stable after first point to end point on the bump surface. The unbalance initializing of the platform make the angle needs to get its stabilization within 1.4 seconds. While, with bump 5 mm it takes more time for balancing and stabilizing. Therefore, it has more sensitive for any changing in the stability of the system during the motion.

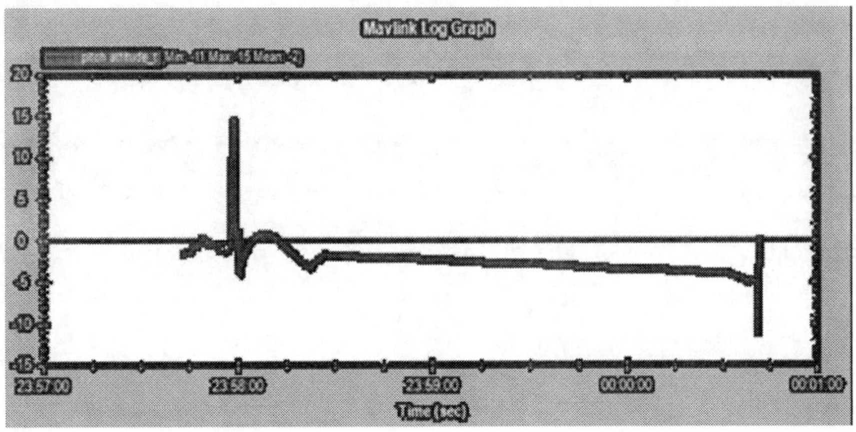

Figure 9. The Pitch Angle, pump 3 mm (Angle between the Longitudinal Axis.)

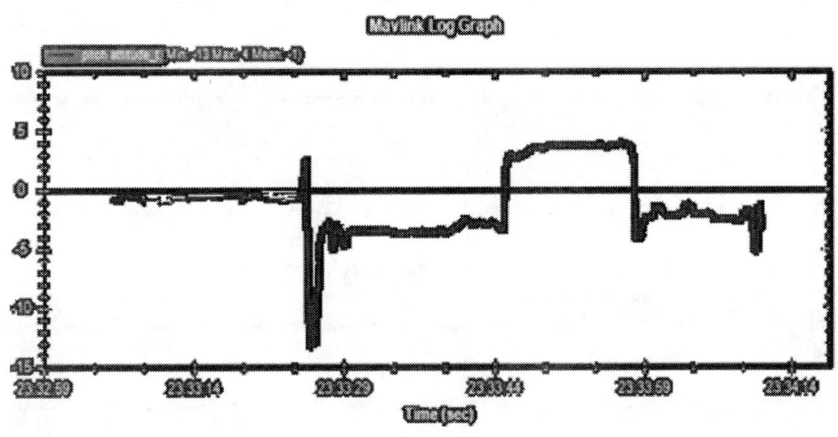

Figure 10. The Pitch Angle, pump 5 mm (Angle between the Longitudinal Axis)

Figures 11 and 12 show that the yaw angle is almost stable. The unbalance initializing of the platform make the angle needs to get its stabilization within 1.2 seconds. While, with bump 5 mm it is more stabilized because of the shape of bump.

According to the simulation results which is specified by the roll angle as shown in Figures 13 and 14, the angle changes between the three angles due to the moving of the vehicle with 3 and 5 mm steps. But, it has not more effect on the stabilization of the system, which starts from initialization state, then start moving, unbalancing of the platform and back to the stability initial value, This process occurs because of the vibration effect had suppled on the surface of motion and the roll angle needs to get its stabilization within 2.1 seconds. Moreover, in case of the bump 5 mm, it has nearly the same time for balancing and stabilizing with bump 3 mm.

The correct values have got from Micro Electro-Mechanical Systems which has been indicated in Figure 15 taking into consideration the vibration and noise signals effect. It has noticed that the sequences pitch,

IOP Conf. Series: Materials Science and Engineering **765** (2020) 012066 doi:10.1088/1757-899X/765/1/012066

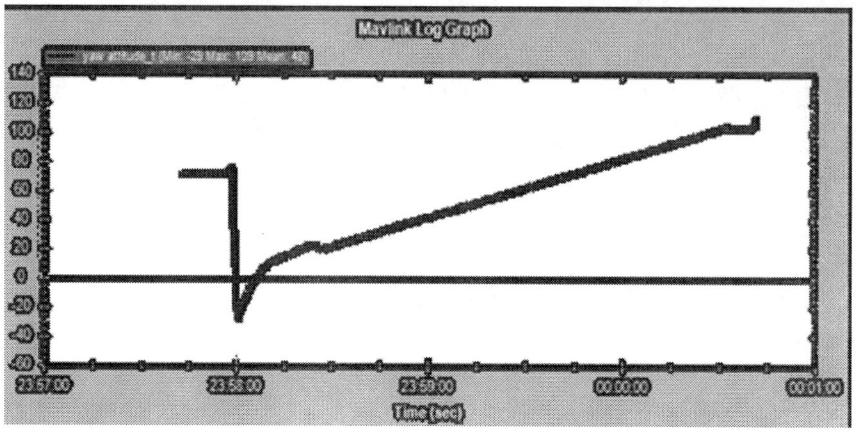

Figure 11. The Yaw Angle, pump 3 mm (Angle between the Vertical Axis).

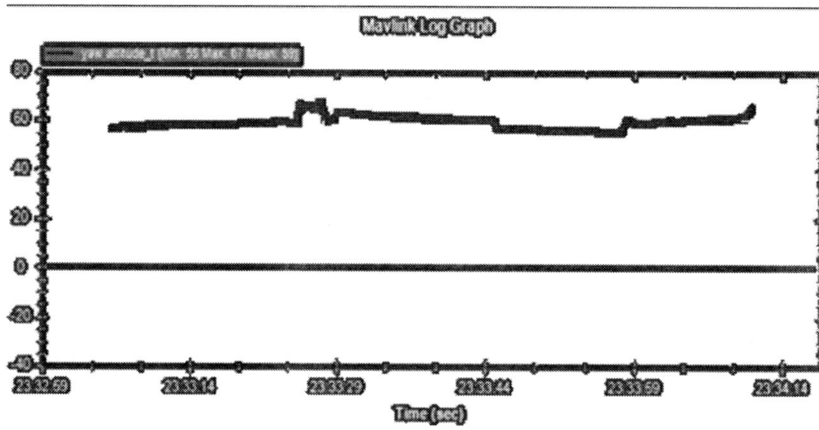

Figure 12. The Yaw Angle, pump 5 mm (Angle between the Vertical Axis).

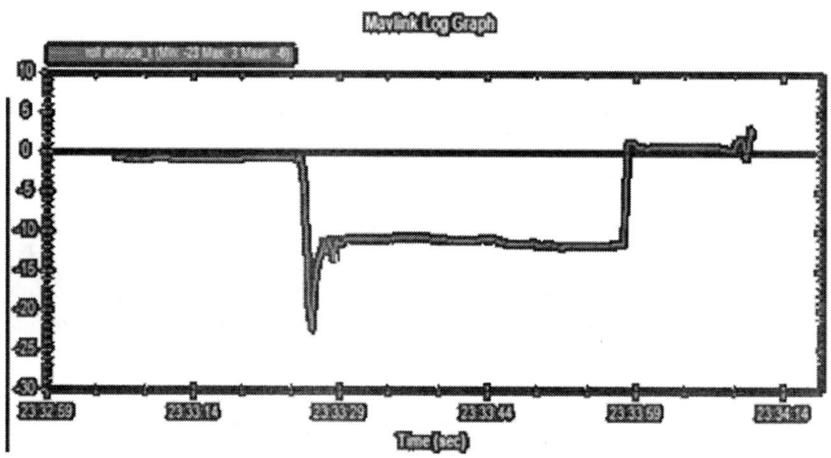

Figure 13. The Roll Angle, pump 3 mm (Angle between the Lateral Axis).

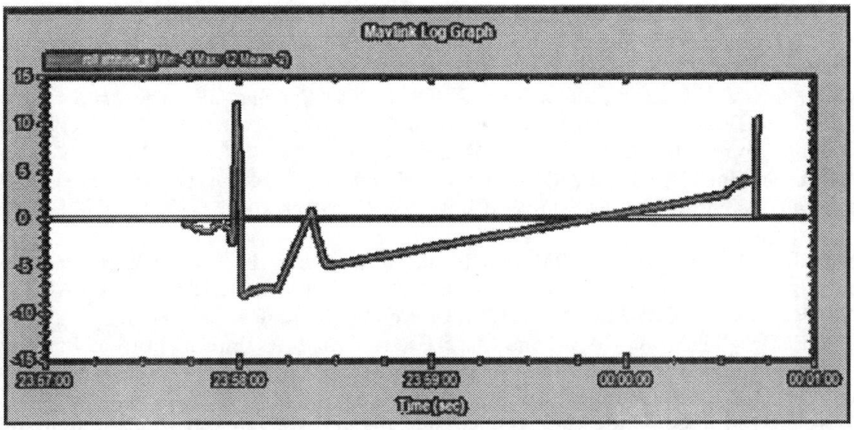

Figure 14. The Roll Angle, pump 5 mm (Angle between the Vertical Axis).

roll and yaw angle have variation change, which should be taken by PID controller for correction stability of the platform.

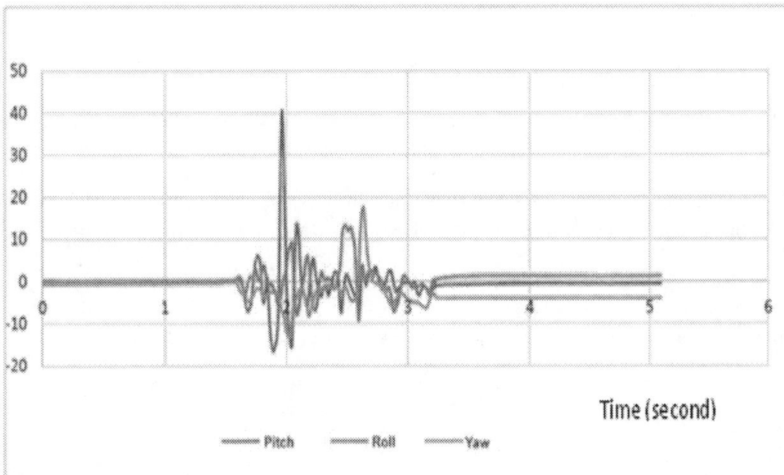

Figure 15. Gyroscope Measurements During Random Motion with Anti-Vibration System Effect.

5. Conclusion

The most critical frequency is the lowest frequency because of the input signal of platform motion in terms of accelerations in that frequency. The PID controller is suitable than PI or PD according to this application which selected before choosing the type of controller based on previous works with an initial acceleration for 3-axis directions as X = 0.1 m/s^2 , Y = 0.1 m/s^2 and Z = 0.1 m/s^2 . The roll and pitch angles are limited to 45 degrees and the yaw angle is limited to 180 degrees. It was able to balance the car smoothly with a maximum roll and pitch angle error of 3-5 degrees. The balancing was also determined with some limitations. Keeping in mind, without a significant movement its unable to get the balancing by car itself after bump step position within a range of -1 mm and +1 mm around the balancing spot in order to balance itself on the flat surfaces. Among these problems, the vibration effect which made imbalance state with defect in the mechanical equipment for guidance of the installation of parts in the not correct axes direction.

References

[1] Almazan J, Bergasa LM, Yebes JJ, Barea R, Arroyo R. Full auto-calibration of a smartphone on board a vehicle using IMU and GPS embedded sensors. 2013 IEEE Intelligent Vehicles Symposium (IV) [Internet]. IEEE; 2013 Jun; Available from: http://dx.doi.org/10.1109/ivs.2013.6629658.

[2] Juang H-S, Lurrr K-Y. Design and control of a two-wheel self-balancing robot using the arduino microcontroller board. 2013 10th IEEE International Conference on Control and Automation (ICCA) [Internet]. IEEE; 2013 Jun; Available from: http://dx.doi.org/10.1109/icca.2013.6565146.

[3] ROBILLARD, Dominic; LANTEIGNE, Eric. Development of Compact Self-Balancing Robot with Self-Standing and Stair-Climbing Capability.2013.CCToMM Mechanisms, Machines, and Mechatronics (M3) Symposium.

[4] Jisha Kuruvilla , Jithin Abraham, Midhun S , Ranjini Kunnath , Rohin Reji Paul.Two Wheeled Self Balancing Robot For Autonomous Navigation.2014. INTERNATIONAL JOURNAL OF ELECTRICAL ENGINEERING and TECHNOLOGY (IJEET), Volume 5, Issue 8, pp. 13-20.

[5] ABDULMAJEED, Wael R., ATHAB, Omar A., SATTAM, Ihab A. Implementation And Evaluation Of Apm 2.6 - Controlled Quadcopter With Aerial Imagery As A Case Study.2016.Journal of Engineering and Applied Sciences, Vol. 11, No. 19, October 2016.

[6] Srinath. G, V. P. Haridasan. Two-wheeled self-balancing autonomous robot. 2016. Journal of Chemical and Pharmaceutical Sciences, Department of Mechanical Engineering, SRM UNIVERSITY Kattankulathur Chennai -603203, JCPS Volume 9 Issue 4, 2016.

[7] SHAON, AKM Asfaque Shakil, et al. Design And Implementation Of A Self-Balancing Robot.2017. Proceedings of the International Conference on Mechanical Engineering and Renewable Energy 2017 (ICMERE2017) 18 20 December, Chittagong, Bangladesh, 2017.

[8] Zhou K, Yu L, Long Z, Mo S. Local Path Planning of Driverless Car Navigation Based on Jump Point Search Method Under Urban Environment. Future Internet [Internet]. MDPI AG; 2017 Sep 12;9(3):51. Available from: http://dx.doi.org/10.3390/fi9030051.

[9] Kok M, Hol JD, Schn TB. Using Inertial Sensors for Position and Orientation Estimation. Foundations and Trends in Signal Processing [Internet]. Now Publishers; 2017;11(1-2):1153. Available from: http://dx.doi.org/10.1561/2000000094.

[10] Ni J, Yang L, Shi P, Luo C. An Improved DSA-Based Approach for Multi-AUV Cooperative Search. Computational Intelligence and Neuroscience [Internet]. Hindawi Limited; 2018 Dec 2;2018:113. Available from: http://dx.doi.org/10.1155/2018/2186574.

[11] ABRAHEM, Hussain Abdulaziz; HASSOUN, Enaam Obeid. A Variation Method In Simulation Of The Slow Moving Concentrated Mass On Truss Structure.2015.Iraqi journal of mechanical and material engineering. 15(1): 1-11.

[12] Skrobek D, Cekus D, Zajc T. Control of the mobile robot using controllers of types P, PI, PID. Journal of Applied Mathematics and Computational Mechanics [Internet]. Czestochowa University of Technology; 2018 Mar;17(1):6978. Available from: http://dx.doi.org/10.17512/jamcm.2018.1.07.

Experimental and Modelling the Combined Measurements Technique for Microwave Absorbing Coatings Testing with Finite Difference Time Domain Method

Abdulkadhim A. Hassan[1] **and Janan H. Saadie**[2]

[1]Electrical Engineering Department,Faculty of Engineering,Kufa University,Al-Najaf,Iraq
[2]Materials Engineering Department,Faculty of Engineering,Kufa University,Al-Najaf,Iraq

E-mail: abdulkadhim.shlash@uokufa.edu.iq

Abstract. Because of the structure property of radar absorbing coatings, using of the existing methods for testing these materials is quite difficult. In this paper, a combined-measurement technique was proposed for broadband characterization of microwave absorbing coatings using flanged rectangular waveguide sensor. The technique combines both frequency sweep reflections measurement and two-layer method to provide two different test conditions of the needed reflection coefficients necessary to extract both complex permittivity and complex permeability over a given frequency range. The first set of reflection coefficients is obtained by directly placing the sensor alongside the coating material and performing broadband measurement using frequency sweep over the selected frequency points of frequency range. The same procedure is applied on a combination of known material followed by the test material to measure the second set of the respective reflection coefficients. Finite Difference Time Domain (FDTD) method was employed to numerically calculate the sensor reflection coefficient and analyse the influence of the sensor geometry and coating material thickness on measurement accuracy. The measurement results of both complex permittivity and permeability for various samples of microwave absorbing coatings were agreed well with reference data. Simple interpolation approximation was introduced in their extraction process from multiple reflection coefficient data since they are frequency dependent parameters . The proposed technique speeds up and simplifies coating materials testing, while it decreases the error due to repeatability of measurement.

Keywords— Layered structure; FDTD; Coatings material; EM-Properties; Microwave measurement.

1. Introduction

Radar absorbers are commonly used in a diverse area of applications to reduce unwanted reflections from metal surfaces or in electromagnetic compatibility and shielding. To solve these problems, especially to reduce unwanted reflections, a layer (or layers) of wideband radar absorbing composite materials is placed or coated on metallic structures. These materials are characterized by both complex permittivity ($\epsilon_r = \epsilon$ (1 jtanδ_ϵ)) and permeability ($\mu_r = \mu$ (1 jtanδ_ϵ)). For the cases where misapplication or damage of radar absorbing coatings, the in situ or in the field electromagnetic (EM) properties measurement over broadband of frequencies is extremely challenging. This is due to that it is quite difficult to test radar absorbing materials coated on metallic structure using the existing methods since these methods need to machine the sample to be tested according to the requirements of measurement process [1, 2, 3, 4, 5]. Moreover, most of these method are destructive, therefore, they are not suitable to be used for coating materials testing.

For nondestructive coating materials testing, it is necessary that the method of measurement used to be efficient enough, fast, accurate and directly applied. The Open-ended rectangular waveguide sensor is one of the promising techniques, which can be used for this purpose for many reasons. The openness in its structure, which diminishes some restrictions to sample preparation, as well as handling high power necessary to perform high loss material testing are the main reasons. This technique has been theoretically and technically studied and analysed and used for simultaneously measuring electromagnetic properties of compound materials among the others [6, 7, 8, 9, 10]. For r and r extraction using reflection only sensors, two-independent reflection coefficients are needed using three developed methods. The first one is called thickness-varying method (TVM) [10]. It makes two reflections measurements with two samples of same material of different thickness. Hence, this method is unsuitable for coating materials testing. The second method is sample-varying or two-layer method (SVM) [10]. In this method, the material is used to be tested first, then a combination of this material and another with known ϵ and μ to test again. For the two mentioned methods, both ϵ_r and μ_r are measured at single frequency point. Thus, using them for broadband measurement requires to repeat the measurement at each frequency point over a given frequency band of interest. The third method is called frequency-varying method (FVM) [11]. In this method, the two reflection coefficients can be measured at two neighbour frequencies of f1and f2 respectively. These frequencies must be separated by small value of frequency interval selected by rule of thumb. Hence. it becomes necessary to develop a technique for coatings material testing, which keeps the above mentioned merits of the sample-varying method as well as gets ride of frequency-varying method demerits.

Numerous measurement techniques have been developed to characterize coating materials as fabricated. These methods can be performed either directly [12, 13] or indirectly [14]. In [14] a loading fraction table, which relates test material thickness to both r and r, is utilized to determine these parameters. Also in [12] and [13] a direct method using horn antenna has been employed for measuring the reflection parameters from the incident plane wave on radar absorbing coatings. These methods can be applied to measure these parameters at single frequency point. In this work, a combined measurements technique was proposed for complex permittivity and complex permeability determination of microwave absorbing coatings. The technique is based on using of frequency sweep measurement technique in conjunction with two-layer method to directly measure the needed reflection coefficients over a given frequency range in only two steps. A set of reflection coefficients data is measured for test material first over the desired frequency points, then a combination of known r, r and thickness material followed by test material is tested again to obtain another set of reflection coefficients data for the respective frequency points. The technique is aimed to simplify and speed up measurement process and improve accuracy by reducing repeatability of measurement if other existing methods are used. FDTD method was employed to calculate input reflection coefficients and extract both r and r of test material by inverse problem since using of analytical procedures, in this case, becomes quite difficult. Simple interpolation approximation was used in extraction process of r and r from the data sets of reflection coefficients since both r and r are frequency dependent parameters . Error analysis was performed to access the influence of sensor flange size, measurement frequency and coating thickness on the measured reflection coefficients. To evaluate the performance of the proposed technique, both ϵ_r and μ_r of selected microwave absorbing coatings were measured over X-band. The obtained results of ϵ_r and μ_r measurements were validated by comparing with reference data published by companies and literatures to demonstrate the feasibility of the proposed technique.

2. Measurement Principles

The problem geometry is shown in Figure 1, where rectangular waveguide sensor with broad and narrow dimension of a and b respectively is placed against a single layer radar absorbing coatings of thickness d as depicted in Figure 1-a. The energy radiated from the sensor penetrates through the test material and then it reflects into the aperture. The reflection coefficient (Γ) measured at the aperture of the sensor carries the desired coating material electromagnetic properties and its thicknesses information described as Γ (f, a,b,$\epsilon(f)$,$\mu(f)$,d) [6]]. Usually, to extract the parameters ϵ_r and μ_r of a test material, it needs at least two independent reflection coefficients measured under different test conditions.

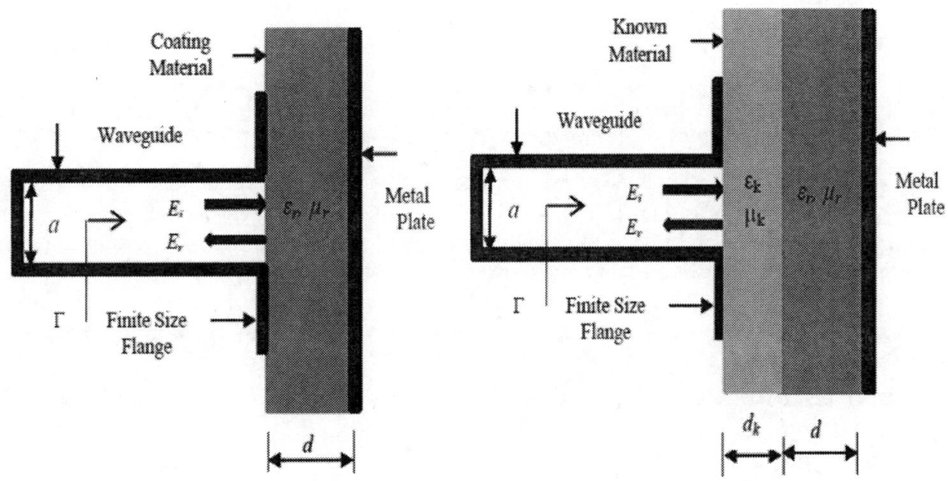

Figure 1. The configuration of the proposed technique (a) single layer Coating material (b) The combination of known material and coating material.

As can be seen from sensor reflection coefficient, the proposed technique is based on two facts: Firstly, for a given sensor dimension, both measurement frequency (f) and thickness of test material (d) are independent variables of the reflection coefficient, thus, by changing both of them, the reflection coefficient can be independently measured. Secondly, the electromagnetic properties of most of the solid materials (ϵ_r and μ_r) are varied slowly alone with frequency [15], hence, both ϵ_r and μ_r) can be accurately determined over frequency interval (Δf) selected between two neighbor frequency points of a given frequency range. The selection of Δf is a critical task since both ϵ_r and μ_r of test material are frequency dependent. Thus, it should be appropriately selected in order for the reflection coefficient is to be changed enough so that it can be distinguished by measuring instrument. Consequently, simple interpolation technique is used in extraction process of these parameters from a set of reflection data measured by frequency sweep using Equation 1.

$$\begin{aligned} \varepsilon_r(f) &\approx \varepsilon_r(f + \Delta f) \\ \mu_r(f) &\approx \mu_r(f + \Delta f) \end{aligned} \tag{1}$$

The polynomial given in Equation 1 needs the reflection coefficients measured at least two frequency points (f_i and f_{i+1} in order for both r and r to be reconstructed and the obtained results are taken as the actual interpreted values averaged over Δf such that:

$$\varepsilon_r(f_i) \approx \varepsilon_r(\frac{f_i + f_{i+1}}{2}) \approx \varepsilon_r(f_{i+1}) \tag{2}$$

$$\mu_r(f_i) \approx \mu_r(\frac{f_i + f_{i+1}}{2}) \approx \mu_r(f_{i+1}) \tag{3}$$

The proposed technique presented in this paper combines both frequency sweep technique of reflection coefficient measurement and two-layer method to provide two different conditions of testing. The first set of reflection coefficients data (both measured and calculated) over a given frequency range is obtained by placing the sensor against the test material as depicted in Figure 1-a. The second set of reflection coefficient data is obtained by placing the sensor against a combination of known material followed by test material as depicted in Figure 1-b. The two processes of measurement are described by simultaneous equations set using Equations 4 and 5.

$$\Gamma^{i}_{meas1}[f_i+\Delta f,\varepsilon_r(f_i+\Delta f),\mu_r(f_i+\Delta f),d]=\Gamma^{i}_{calc1}[f_i+\Delta f,\varepsilon_r(f_i+\Delta f),\mu_r(f_i+\Delta f),d] \quad (4)$$

$$\begin{aligned}\Gamma^{i}_{meas2}[f_i+\Delta f,\varepsilon_k,\varepsilon_r(f_i+\Delta f),\mu_k,\mu_r(f_i+\Delta f),d_k,d]=\\ \Gamma^{i}_{calc2}[f_i+\Delta f,\varepsilon_k,\varepsilon_r(f_i+\Delta f),\mu_k,\mu_r(f_i+\Delta f),d_k,d]\end{aligned} \quad (5)$$

Where, Γ_{meas} and Γ_{calc} are the measured and the calculated reflection coefficients at frequency points of f_i and (f_i+Δf respectively and i= 1, 2, 3....N-1 is the number of frequency points over a given frequency range. The other symbols are pictorially defined as depicted in Figure 1. When implementing the proposed technique, the broadband frequency sweep of reflection coefficients is measured, then from the set reflection coefficients data, both ϵ_r and μ_r can be extracted with values averaged over two neighbour frequency points. The same process is used throughout the entire range of the frequency.

3. Numerical Formulation of the Problem

3.1. Reflection Coefficient Calculation Using FDTD Method

The computational domain of FDTD method of the problem is shown in Figure 2. It consists of finite size flange rectangular waveguide sensor placed against the combination of known EM parameters and thickness material followed by radar absorbing coating to be tested with known thickness and unknown ϵ_r and μ_r.

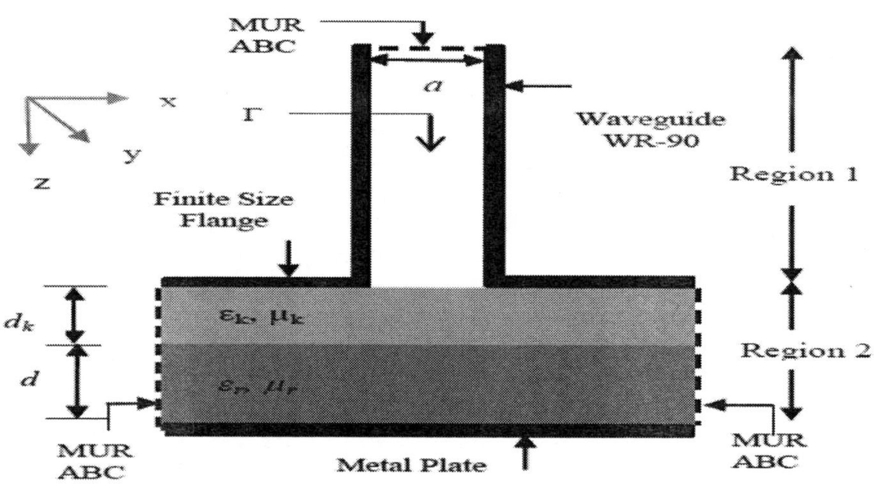

Figure 2. FDTD computational domain of the problem.

Considering only the dominant(TE_{10}) mode, the FDTD method adopts numerically a full-wave direct solution of time-dependent curl Maxwell's equations, which describe the near-field interaction between the electromagnetic field and a medium using Equations 6 and 7.

$$\frac{\partial H}{\partial t}=\frac{-1}{\mu}(\nabla x E) \quad (6)$$

$$\frac{\partial E}{\partial t}=\frac{1}{\varepsilon}(\nabla x H) \quad (7)$$

To solve Equations 6 and 7, accurate central difference approximations of second order are applied for the derivatives of the space and time of both electric (E) and magnetic (H) fields according to Yee procedure [16]. Following Yee procedure, the vector Equations 6 and 7 are translated into a six scalar first-order set of partial differential equations in x, y and z coordinates. For x coordinate, the E and the H field's components, are given using Equations 8 and 9 respectively.

$$\frac{\partial E_x}{\partial t} = \frac{1}{\epsilon_r \sqrt{\epsilon_o \mu_o}} (\frac{\partial H_z}{\partial y} - \frac{\partial H_y}{\partial z}) \tag{8}$$

$$\frac{\partial H_x}{\partial t} = \frac{-1}{\mu_r \sqrt{\epsilon_o \mu_o}} (\frac{\partial E_y}{\partial z} - \frac{\partial E_z}{\partial y}) \tag{9}$$

The same procedure is applied using Equations 6 and 7 for y and z coordinates. The six scalar equations of Equations 6 and 7 represent the FDTD method basis. As shown in Figure 2, the computational domain of the problem is divided into two regions. Region 1 which includes the interior of the probe space surrounded by waveguide walls, which are assumed to be perfect conductors. Region 2 includes the two-layer structure with different EM-properties and thicknesses backed by perfect conductor. The constitutive parameters r and r of test material used for the problem under consideration can be constant, or change with frequency. This allows modeling of material properties for lossless and lossy materials over required broadband analysis and measurement. The total spaces, including the two regions are discretized using Yee procedure into cells. The field propagation through a one cell distance requires time step t of minimum value calculated according to the Courant condition for stability using Equation 10.

$$\Delta t \leq \frac{1}{c[\frac{1}{\Delta x^2} + \frac{1}{\Delta y^2} + \frac{1}{\Delta z^2}]^{1/2}} \tag{10}$$

In this work, a 3D FDTD special grids cell size ofΔ_x, Δ_y, and Δ_z were used to increase the FDTD computation accuracy. Constitutive parameters were used with effective values averaged at different interfaces to calculate field components. Mur absorbing boundary condition (ABC) of the first-order [17] was used for the problem spaces of at radial boundary to limit the domain of the computational. A Gaussian pulse of forward-moving was taken as excitation source lunched at the plane of excitation located a distance away from the aperture and normally incident on the material under test. The distribution of the fields in the test material and the sensor were calculated after the final steady state has been reached. The sensor aperture input admittance was obtained from the calculated components of the fields within test material. The sensor complex reflection coefficient, Γ, was calculated using Equation 11.

$$\Gamma = \frac{Y_a - Y_0}{Y_a + Y_0} \tag{11}$$

where, Y_a is the aperture admittance and Y_0 is the waveguide equivalent characteristic admittance.

3.2. Numerical Validation and Analysis

To validate and assess the accuracy of the FDTD algorithm, the complex reflection coefficient of the sensor (both magnitude and phase) is to be calculated for the sensor terminated with the test material using a 3D code developed for this purpose. A series of numerical simulations and experiments were conducted at (8.2 12.4 GHz) frequency range using rectangular waveguide sensor (WR-90) and network analyzer (HP8510B). The sensor flange dimension is taken to be 50.0 mm [18]. Two different radar absorbing coatings were used for this purpose. The first test material was GEC-9052 radar absorbing coatings with thickness of 2.44 mm while the second one is 1164 radar absorbing coating of 1.16 mm thickness. The FDTD simulations and measurements were carried out at frequency of 10 GHz and the results are shown in Table 1. From the Table 1, it can be seen that the data obtained by the FDTD method is closely followed the measured results. The obtained results validate the computational tool.

An analysis was performed using FDTD simulations to evaluate sensor flange size influence on the accuracy of reflection coefficient measurement for a given operating frequency. Figure 3, shows the variations of the sensor reflection coefficient terminated by coating material sample as the flange size varies from the minimal (30 mm) to largest considered values calculated at 9 GHz and 11 GHz respectively. As shown in the figure, the reflection coefficient variation at 11 GHz is less than that obtained at 9 GHz. This is reasonable since that the radiation losses increases at higher frequencies. Another test was performed to investigate coated material thickness on the measured reflection coefficient.

Table 1. The FDTD results versus experiments data of reflection coefficient for different radar absorbing coatings.

Sample	Method	Γ (Magnitude)	Γ (Phase)
GEC- 9052	FDTD Model	0.4306	-135.37
	Experiment	0.4261	-132.97
1164	FDTD Model	0.8095	-61.19
	Experiment	0.8036	-63.56

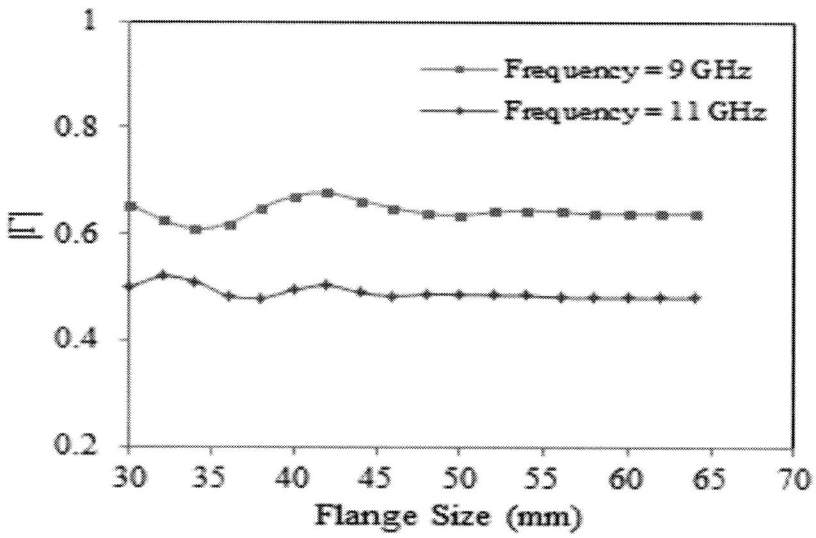

Figure 3. Variation of reflection coefficient versus sensor flange size at different frequencies.

Figure 4 shows the deviation (in %) in reflection coefficient compared with the result calculated for the case of the sensor with infinitely large flange due to the variations in the size of the flange normalized to the test material wavelength (λ_m). Two thicknesses of radar absorbing coating of 0.1 λ_m and 0.2 λ_m were considered in this analysis. As shown in the figure, the effect of size of the flange on reflection coefficient for thick sample case is less than that obtained for the thin sample. This is due to that the measured reflection coefficient for high loss materials (as in the case of radar absorbing coating) decreases for thick coated material.

From Figures 3 and 4, it can be seen that the variation of sensor flange size has negligible effect on sensor reflection coefficient provided that tested material is lossy enough. This a necessary condition to ensure that any reflection of the spurious signal from the flange boundaries has no effect on the measurement. For the cases examined in this analysis, the variations of the reflection coefficient have a dependence of oscillatory damped with the flange size of the sensor.

4. Experimental Verifications

The measurement set up is depicted in Fig. 5, where the sensor is pressed alongside the coating material. and then the reflection coefficient (Γ) is measured using network analyzer. The measured and FDTD calculated reflection coefficients are used to determine both the ϵ_r and μ_r parameters of test material by numerical inversion process. As shown in Fig. 5, the system of measurement consists of two parts. One is the hardware consisting of waveguide and network analyzer used to transfer the desired ϵ_r and

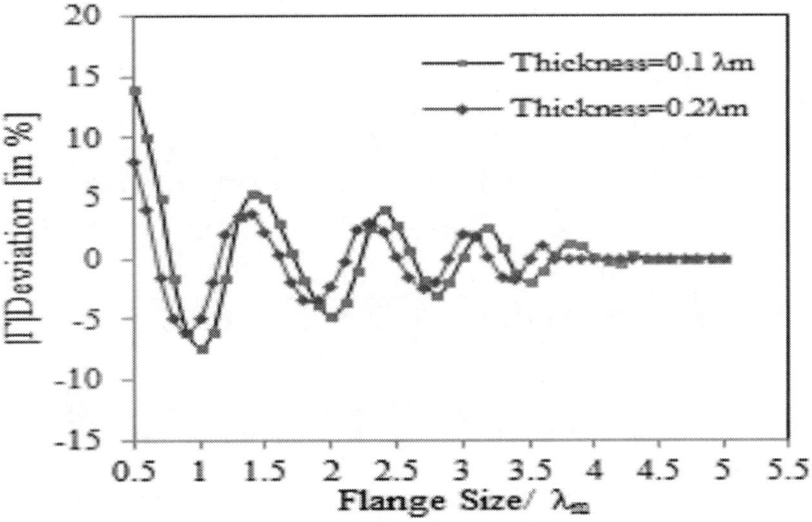

Figure 4. Variation of reflection coefficient versus flange size for lossy material of different thickness.

μ_r information from the measurement results of reflection coefficient (Γ). The other part is the software algorithms developed to calculate sensor reflection coefficients at different conditions of testing and extract the desired ϵ_r and μ_r from measured results of Γ.

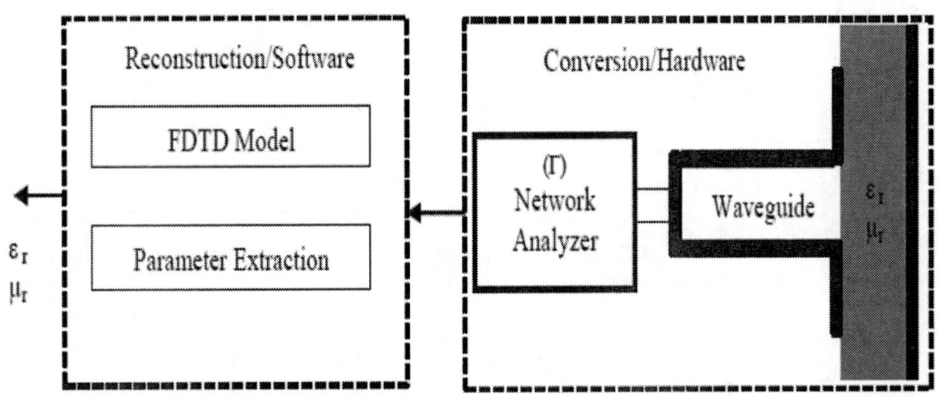

Figure 5. Measurement setup.

A series of measurements were carried out on various microwave absorbing coatings using the proposed technique to determine both ϵ_r and μ_r of several coating materials at X-band. The sensor used was rectangular waveguide (WR-90) and the frequency sweep measurement of reflection coefficients was performed using a HP 8510B network analyzer. The inverse problem of ϵ_r and μ_r extraction can be solved by Newton-Raphson method or Levenberg-Marquardt method [19]. Starting with good initial guess, the given iterative algorithms, in most cases, guarantee convergence. The first test was carried out on 1164 radar absorbing coatings of 1.2 mm thickness to evaluate the influence of flange size on the measured values of ϵ_r and μ_r at 10 GHz. Two flange sizes of 40 mm and 50 mm were selected for this purpose. The known material used in the measurement is Teflon with thickness of 8 mm. This value is experimentally chosen to obtain two reflection coefficients carrying different information for a range of coating material thickness from 1 to 5 mm. The reasons for choosing Teflon as known material are that it is semi-rigid with low loss and low dispersion making it suitable for this purpose. Table 2 shows the obtained results.

Table 2. Comparison of the calculated values of ϵ_r and μ_r for flange with different sizes.

Flange Size (mm)	ϵ_r	**tan**$\delta\epsilon_r$	mu_r	**tan**$\delta\mu_r$
40 mm	14.89	0.023	1.26	1.21
50 mm	14.92	0.0220	1.31	1.22
Reference Data \cite{10}	14.95	0.0222	1.35	1.22

From the Table 2, it is obvious that a 50 mm flange size results agreed well with the reference data [10] as compared to the results of 40 mm flange size. The obtained results clearly show the influence of the sensor flange size on both ϵ_r and μ_r measurement. A broadband frequency sweep measurement [20] was carried out on ECCOSORB MF.116 radar absorbing coatings with 1.22 mm thickness to evaluate the performance of the proposed technique compared to the two-layer method results. Eight frequency points were selected over the given frequency band with frequency interval (Δf) of 0.5 GHz.

The measured results of both ϵ_r and μ_r and reference data are shown in Figure 6. Figure 6-a shows the real parts of both ϵ_r and μ_r while the loss tangents are shown in Figure 6-b. As shown in the figure, although the difference between the results of the two methods is obvious, the whole tendency of the measured results using the proposed technique agrees well with the reference data. It is obvious that two-layer method results exhibit marked diversity in both real and imaginary parts of the measured ϵ_r and μ_r if it is used alone. This is may be due to the repeatability of measurement process, where two mechanical operations are required at each frequency point , while the proposed technique shows results a relatively reasonable and smooth for a given range of frequency.

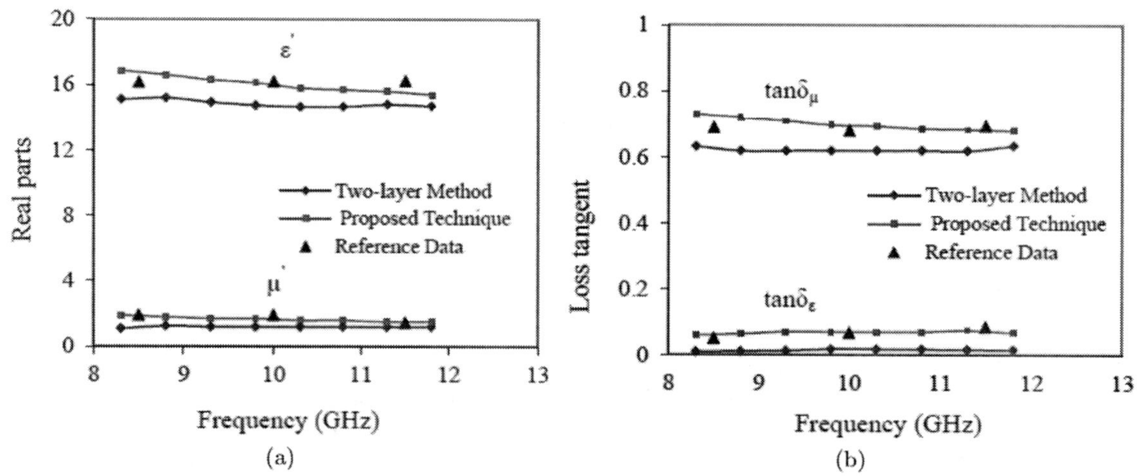

Figure 6. Broadband measurement of complex permittivity and complex permeability using the proposed technique and Two-layer method for MF-116 radar absorbing coating.

Another experiment was performed to evaluate coating material thickness influence on ϵ_r and μ_r measurement accuracy. The ϵ_r and μ_r measured results of a coating material with three different thicknesses using the proposed technique are presented in Figure 7 . Figure 7-a shows the results of real parts of ϵ_r and μ_r and Figure 7-b shows the results of their loss tangents. The thicknesses of the used samples are 2.08 mm, 4.16 mm and 6.24 mm respectively. Eleven frequency points were selected over the given frequency range with frequency interval of 0.4 GHz. It is clear that good agreement is achieved between the results of measurement and the reference values, but the effect of thickness on measurement accuracy and sensitivity is obvious also. The 6.24 mm thickness results of coating showed a deviation relatively larger in the loss tangents compared to 4.16 mm and 2.08 mm sample thicknesses. For this sample thickness, the loss tangent ($\tan\delta\epsilon$) becomes minus. The sample with small loss tangent ($\tan\delta\epsilon \leqslant 0.1$) and too thick are the main reasons for obtaining this result. Furthermore, the waveguide sensor with the open end is suitable to be used for high-loss materials measurement. So, getting a reasonable accuracy for loss tangent ($\tan\delta$) with a value of less than 0.1 is difficult (In this work, the actual value of loss tangent is about 0.06). The obtained results of the experiments indicate that the proposed technique presented in this paper is reliable and effective for in situ determining simultaneously both ϵ_r and μ_r of radar absorbing coatings. Further research work such as analysis of sensitivity and improvement of inverse algorithms should be addressed.

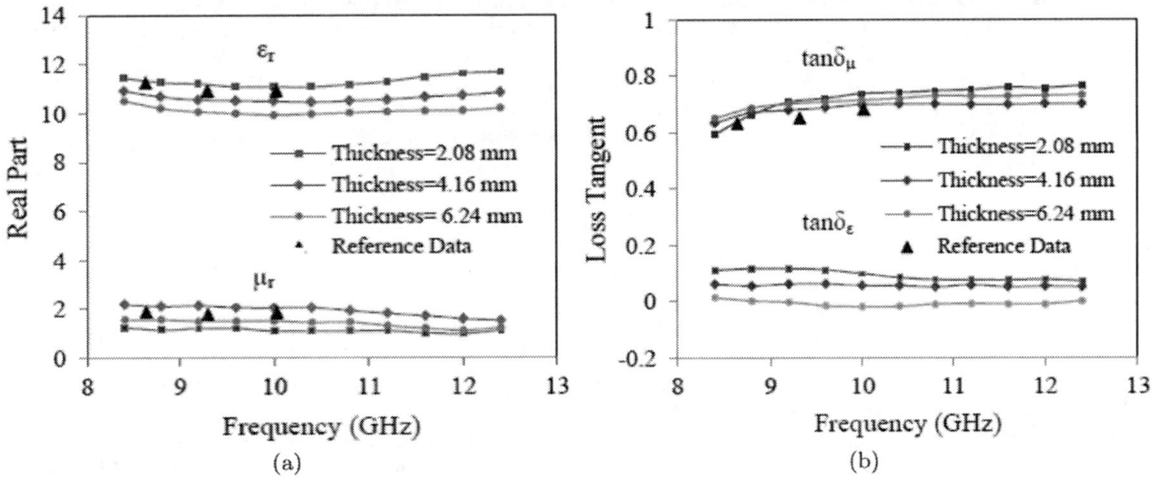

Figure 7. Measured results of a ferrite absorber with three different thicknesses (a) Real parts (b) Loss tangent.

5. Conclusion

A non-destructive measurement technique for broadband determination of complex permittivity and complex permeability of radar absorbing coating materials was presented. The technique was employed based on combining of both frequency sweep of reflection coefficient measurement and two-layer method using flanged rectangular waveguide sensor. FDTD method has been successfully employed to numerically model the problem geometry, where using of the analytical approaches, in this case, is quite difficult. Analysis results of uncertainty of sensor reflection coefficient measurement performed using FDTD simulations and experiments have shown that sensor flange size, measurement frequency and coating material thickness are the dominant factors affecting the accuracy and sensitivity of ϵ_r and μ_r measurements. Results of measured r and r of selected microwave absorbing coatings have demonstrated, in comparison with published data, the feasibility of the proposed technique for testing of high loss coating materials with a thickness of several millimetres. The technique has the advantages that it performs measurement of independent reflection coefficients sets with only two mechanical operations over a given frequency band of interest, thus speeding up and simplifying the measurement process and error factors due to repeatability of measurement can be reduced. More over the technique can be extended to be used for multiparameter measurement of multilayer radar absorbing coatings. .

References

[1] Lin M, Duane MH, Afsar MN. Cavity-Perturbation Measurement of Complex Permittivity and Permeability of Common Ferrimagnetics in Microwave-Frequency Range. IEEE Transactions on Magnetics [Internet]. Institute of Electrical and Electronics Engineers (IEEE); 2006 Oct;42(10):28857. Available from: http://dx.doi.org/10.1109/tmag.2006.879885

[2] Gonalves F, Pinto A, Mesquita R, Silva E, Brancaccio A. Free-Space Materials Characterization by Reflection and Transmission Measurements using Frequency-by-Frequency and Multi-Frequency Algorithms. Electronics [Internet]. MDPI AG; 2018 Oct 18;7(10):260. Available from: http://dx.doi.org/10.3390/electronics7100260.

[3] Wolfson BJ, Wentworth SM. Complex permittivity and permeability measurement at elevated temperatures using rectangular waveguide. Microwave and Optical Technology Letters [Internet]. Wiley; 2003 Jul 31;38(6):44953. Available from: http://dx.doi.org/10.1002/mop.11086.

[4] Salahun E, Queffelec P, Le Floch M, Gelin P. A broadband permeameter for in situ measurements of rectangular samples. IEEE Transactions on Magnetics [Internet]. Institute of Electrical and Electronics Engineers (IEEE); 2001 Jul;37(4):27435. Available from: http://dx.doi.org/10.1109/20.951293.

[5] Pinho MS, Gregori ML, Nunes RCR, Soares BG. Performance of radar absorbing materials by waveguide measurements for X- and Ku-band frequencies. European Polymer Journal [Internet]. Elsevier BV; 2002 Nov;38(11):23217. Available from: http://dx.doi.org/10.1016/s0014-3057(02)00118-0.

[6] Ghasr MT, Simms D, Zoughi R. Multimodal Solution for a Waveguide Radiating Into Multilayered Structures-Dielectric Property and Thickness Evaluation.2009. IEEE Transactions on Instrumentation and Measurement [Internet]. Institute of Electrical and Electronics Engineers (IEEE); 2009 May;58(5):150513. Available from: http://dx.doi.org/10.1109/tim.2008.2009133.

[7] Stewart JW, Havrilla MJ. Electromagnetic Characterization of a Magnetic Material Using an Open-ended Waveguide Probe and a Rigorous Full-wave Multimode Model. Journal of Electromagnetic Waves and Applications [Internet]. Informa UK Limited; 2006 Jan;20(14):203752. Available from: http://dx.doi.org/10.1163/156939306779322693.

[8] Karli R, Ammor H, Terhzaz J, Chabi M, Mediavilla Snchez . Design and construction of miniaturized UWB microstrip antenna with slots for UWB applications. Microwave and Optical Technology Letters [Internet]. Wiley; 2014 Dec 18;57(2):4603. Available from: http://dx.doi.org/10.1002/mop.28869.

[9] Seal MD, Hyde IV MW, Havrilla MJ. NONDESTRUCTIVE COMPLEX PERMITTIVITY AND PERMEABILITY EXTRACTION USING A TWO-LAYER DUAL-WAVEGUIDE PROBE MEASUREMENT GEOMETRY. Progress In Electromagnetics Research [Internet]. EMW Publishing; 2012;123:12342. Available from: http://dx.doi.org/10.2528/pier11111108.

[10] Niu Maode, Su Yong, Yan Jinkui, Fu Chenpeng, Xu Deming. An improved open-ended waveguide measurement technique on parameters ϵ/sub γ/ and μ/sub γ/ of high-loss materials. IEEE Transactions on Instrumentation and Measurement [Internet]. Institute of Electrical and Electronics Engineers (IEEE); 1998 Apr;47(2):47681. Available from: http://dx.doi.org/10.1109/19.744194.

[11] Shoujun Wang, Maode Niu, Deming Xu. A frequency-varying method for simultaneous measurement of complex permittivity and permeability with an open-ended coaxial probe. IEEE Transactions on Microwave Theory and Techniques [Internet]. Institute of Electrical and Electronics Engineers (IEEE); 1998;46(12):21457. Available from: http://dx.doi.org/10.1109/22.739296.

[12] Zhou J, Bie S, Wan D, Xu H, Xu Y, Jiang J. Realization of Thin and Broadband Magnetic Radar Absorption Materials With the Help of Resistor FSS. IEEE Antennas and Wireless Propagation Letters [Internet]. Institute of Electrical and Electronics Engineers (IEEE); 2015;14:247. Available from: http://dx.doi.org/10.1109/lawp.2014.2349533.

[13] Guan S, Wang Y, Jia D. A Field Performance Evaluation Scheme for Microwave-Absorbing Material Coatings. Coatings [Internet]. MDPI AG; 2017 Mar 2;7(3):38. Available from: http://dx.doi.org/10.3390/coatings7030038.

[14] Rothwell EJ, Temme A, Frasch LL. Characterisation of properties of conductor-backed MagRAM layer using reflection measurement. Electronics Letters [Internet]. Institution of Engineering and Technology (IET); 2012 Aug 30;48(18):11313. Available from: http://dx.doi.org/10.1049/el.2012.1184.

[15] Yang RB, Hsu SD, Lin CK. Frequency-dependent complex permittivity and permeability of iron-based powders in 218 GHz. Journal of Applied Physics [Internet]. AIP Publishing; 2009 Apr;105(7):07A527. Available from: http://dx.doi.org/10.1063/1.3068039.

[16] YEE, Kane. Numerical solution of initial boundary value problems involving Maxwell's equations in isotropic media.1966. IEEE Transactions on antennas and propagation.14(3):302-307.

[17] Mur G. Absorbing Boundary Conditions for the Finite-Difference Approximation of the Time-Domain Electromagnetic-Field Equations. IEEE Transactions on Electromagnetic Compatibility [Internet]. Institute of Electrical and Electronics Engineers (IEEE); 1981 Nov;EMC-23(4):37782. Available

IOP Conf. Series: Materials Science and Engineering **765** (2020) 012067 doi:10.1088/1757-899X/765/1/012067

from:http://dx.doi.org/10.1109/temc.1981.303970.

[18] Takanobu Ohno, Kouji Omata, Kouji Wada, Osamu Hashimoto. A tunable bandpass filter using tapped $\lambda/4$ resonators loaded with inductive variable capacitor. 2006 Asia-Pacific Microwave Conference [Internet]. IEEE; 2006 Dec; Available from: http://dx.doi.org/10.1109/apmc.2006.4429639.

[19] PRESS, W. H. Teukolsky SA, Vetterling WT, and Flannery BP. Numerical recipes in C.1992.the art of scientific computing.657-659.

[20] Abdulkadhim Hassan, Al-Modaffer, Janan Saadee. Broadband Thin Sheet Radar Absorbers Characterization Using Frequency-swept Open-ended Waveguide Probe Technique and FDTD Modeling.2019.INTERNATIONAL JOURNAL OF MICROWAVE AND OPTICAL TECHNOLOGY.14(3).175-184.

IOP Conf. Series: Materials Science and Engineering **765** (2020) 012068 doi:10.1088/1757-899X/765/1/012068

Face Recognition using Hybrid Methods of Transformation and Statistical Measures

Hayder Ali Ameen[1] and Asaad Noori Hashim[2]

[1,2]2Computer science department, Faculty of Computer Science and Mathematics, University of Kufa, Iraq

E-mail: [1]haider19992@gmail.com, [2]asaad.alshareefi@uokufa.edu.iq

Abstract. One of the most interesting problems and challenging issues within the pattern recognition computer vision is face recognition. Face recognition has gained special attention in the past few years due to its importance in relation to current applications such as security, forensic analysis, and surveillance systems. The whole system can be explained as the follows, after pre-processing, the first step is edge detection for the input image using the proposed filter, the second step is extraction the matrix of the Local Ternary Pattern (LTP) of the input image, the third step is capturing the features based on The Singular Value Decomposition (SVD). Fourth, extracted features are merged into a single vector. Finally, the matching has been performed between the features stored at the database and the features of the reference (test) image. The Proposed System has been applied to the variety databases. In the proposed system, the recognition ratio is 98% using city block distance and 98.5% using Structural Similarity (SSIM) as a classification method, the ORL (AT&T) database is used.

Keywords— Face recognition, Local Ternary Pattern (LTP), Singular Value Decomposition (SVD), Structural Similarity Measure.

1. Introduction

The human face is a set of features that vary from person to person, and these features can change over time, therefore, the database must be updated from time to time. There are many applications that can use the face recognition application, including access control, banking applications and monitoring systems and many other applications, therefore face recognition is a very useful application.

Face recognition has become one of the most popular applications in the field of image analysis and computer vision because of the easy acquisition of information compared to other biometrics systems where the face recognition system has been integrated into a wide range of safety systems based on biometrics. Human beings have an effective ability to recognize different faces. Face recognition is the recognition of a particular face from a variety of different faces. The face has an important role in communication between humans in that each person has a distinctive face to others. One of the major problems in human interaction is the problem of face recognition between Human and Machines [1].

2. Related Work

Over the last decade, several papers have been published about face recognition.

Zhang et al. [2] proposed two algorithms. In the primary algorithm, the original image is linearly combined with its derived image gotten by perturbing the image matrix's singular values, and then principal component analysis (PCA) is performed on the associated images. In the next algorithm, the derived images are regarded as independent images that could augment the training image set, and then

IOP Conf. Series: Materials Science and Engineering **765** (2020) 012068 doi:10.1088/1757-899X/765/1/012068

PCA is performed on all the training images available, including the original ones and the derived ones. Cevikalp et al. [3] proposed a new face recognition method called the Discriminative Common Vector method based on the diversity of Fisher's analysis of linear discrimination. Sharif et al., [4] proposed the technique for enhancing the SVD algorithm depend on the pre-processing stage for face recognition and this research focused on decreasing the challenge for expression and illumination variations by using fusion between discrete wavelet transforms and SVD perturbation. Shyam and Singh [5] proposed a new method for recognizing faces under non-controlling conditions, which works in a local binary pattern for facial images, and then calculating the contrast between them using Bray Curtis method, and then developed a new method for filtering the fabric using the LBP algorithm, and the new method was named ALBP, a new method to identify texture. Kar et al. [6] proposed the method for face recognition technique by integrated multiple feature sets depending on significant a blocks of a gradient image, debate the utilize of novel morphological, gray-level co-occurrence matrix GLCM and LDP based feature extraction technique to recognize human faces. Jha et al. [7] reported a new method to recognize faces by optimizing the image and then using Fourier Transform, and then using the SVD algorithm results extracted from Fourier Transform, where the SVD algorithm further produces a single numerical value. Hashim and Hussain [8] proposed an effective way to recognize faces in computer vision and electronic images using a new way to extract features from images, where technologies are applied to extract features from face images using Zernike moments and structural similarity measure (SSIM) with local and semi-global blocks. Pancholi and Rana [9] considered an efficient method to recognize faces using the Hidden Markov Model (HMM) and SVD algorithm. Anggo and Arapu [10] proposed a method for facial recognition using the Fisherface technique. In the first step, a PCA algorithm is used and then the Fisherface technique is applied. The PCA algorithm is used to reduce face area, and then Fisherface technology is used to identify faces. Arisandi et al.,[11] Developed a phone application for face recognition for undergraduate students to know their CV, the application depends on the techniques of Fisherface to recognize faces and the face recognition rate is 90%. Rustam and Rika [12] proposed a hybrid method to recognize faces using the Fuzzy Kernel Learning Vector Quantification and Learning Vector Quantification methods, the recognition rate was 89.33%. Li and Xie [13] proposed method are develop Binary gradient pattern (BGP) for face recognition, this algorithm (BGP) was developed based on heuristic information, the performance of the BGP algorithm has been significantly improved to recognize a face. Table 1 shows a comparison between some of the previous work and the proposed system.

Table 1. Comparison between some of the previous work and the proposed system.

Papers	Methods	Database Used	Recognition Rate
Sharif et al (2012) [3]	Enhanced SVD	ORL(AT &T)	96 %
Shyam and Singh (2015) [4]	Augmented LBP	Local Database	84.81 %
Arisandi et al (2018) [13]	Fisherface	Local Database	90 %
Rustam and Rika (2018) [14]	LVQ + FKLVQ	LFW-a	89.3 %
Proposed System	**LTP + SVD and Proposed Filter + SVD**	**ORL(AT &T)**	**98 %**

3. Proposed System

This system introduces another technique to recognize the faces using a hybrid method by combining two different methods. The first method is Local Ternary Pattern (LTP) that it is a useful statistical method in many scientific fields such as image processing applications, biomedical field, industrial defect detection systems, biomedical field, etc. The second method is the singular-value decomposition (SVD) that is a method of decomposing an array into three other matrices: A = UDVT where the matrix D is diagonal with positive real entries and the columns of U and V are orthonormal. The first step of the proposed system, LTP technology is applied to the image produced after preprocessing, the second step, and the proposed filter is applied to the image produced after preprocessing. The SVD algorithm is then applied to both the first and second steps to extract the features from them. After the features are extracted by the SVD algorithm for each of the previous two steps, the features are merged and stored in a single vector. The above steps are applied for both the testing and training phase. The images database used is (ORL). These images are different in form and appearance. The ORL database includes 40 person with 10 images per person, the first five images of the designated person are used for the training phase, while the second five images are used for the testing phase, which contains 400 pictures a total, and these images are captured under certain lighting conditions and different directions for each person. Matlab (2018) has been used as a working environment. Figure 1 shows parts of face image samples.

Figure 1. Parts of face image samples.

3.1. Local Ternary Pattern (LTP)

This is the development of a local binary pattern (LBP) method. Where LTP, its result is three values 1, 0, or -1, unlike LBP, where its result either 0 or 1, The LTB threshold is determined by the user. The result is 0 when the threshold between c-t and c+t where c is the pixel centre intensity in a specific window. For example, when the pixel intensity is 34, the range ranges between (29 and 39) when the threshold is 5, where each value in the window when it is less than 29 becomes -1 and each value greater than 39 becomes 1 . The mathematical expression of the LTP can be described is as shown in Equation 1.

$$f(x) = \begin{cases} 1 & x \geqslant c+t \\ 0 & c-t < x < c+t \\ -1 & x < c-t \end{cases} \quad (1)$$

After completing the identification of the ternary pattern, the resulting figures are divided into lower and upper patterns as follows: Each number (-1), becomes (one) in the lower pattern case and (zero) in the upper pattern case, while each number 1 becomes (zero) in the lower pattern and remains (one) in the upper pattern case [14].

The last step of this pattern is reading the bit pattern beginning from the east neighbourhood with respect to the centre (row 2, column 3), then it goes around counter-clockwise as shown in Figure 2.

An example of the LTP method is shown in Figure 3.

3.2. The Singular Value Decomposition (SVD)

The Singular Value Decomposition (SVD) is considered a promising transformation to the decomposition of the input matrix into three matrices. SVD of array A is the factorization of A into the give of three

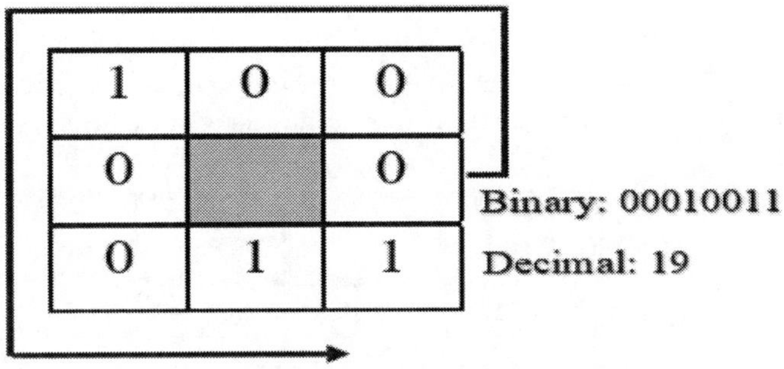

Figure 2. The order of the binary numbers and then converts them to a decimal number [14].

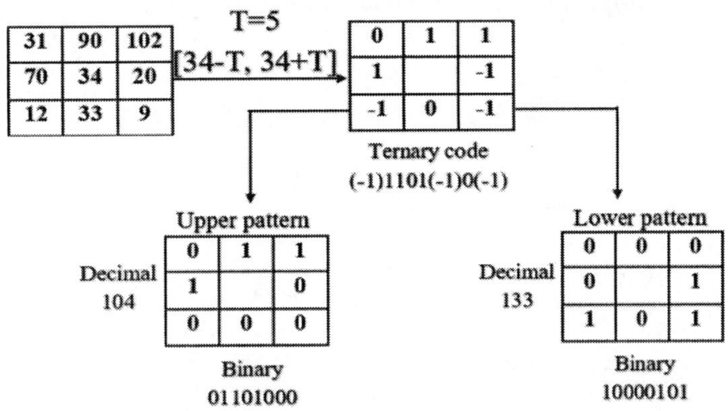

Figure 3. Illustration of the work of LTP operator [14].

matrices $A_{mn} = U_{mm}\ D_{mn}\ V_{nn}^T$, where the columns of U and V are orthonormal and the array D is diagonal with positive real. In many applications the data of array A is near to an array of low rank. It is shown that the singular value decomposition of A can get the matrix B of rank k which best approximates A. Singular value decomposition is defined for all matrices (rectangular or square) not similarity the more commonly used spectral decomposition in linear Algebra. The reader familiar with eigenvectors and eigenvalues required conditions on the array to ensure orthogonality of eigenvectors. The columns of V in the singular value decomposition are called the right singular vectors of A, always form an orthogonal set with no assumptions on A. The columns of U are called the left singular vectors. A simple consequence of the orthogonality is that for a square and invertible array A, the inverse of A is $VD^{-1}\ U^T$ [15].

3.3. The structural similarity (SSIM)

The structural similarity (SSIM) index is a method to recognize the tangible quality of digital and movie images, as well as other types of images and videos. The SSIM is a way to recognize the similarity between two images, and SSIM is a full reference metric; the image is measured based on the original uncompressed image and noise, and used as a reference [16]. SSIM can represent is as shown in Equation 2 :

$$\rho(x.y) = \frac{(2\mu_x\mu_y + C_1)(2\sigma_{xy} + C_2)}{(\mu_x^2 + \mu_y^2 + C_1)(\sigma_x^2 + \sigma_y^2 + C_2)} \tag{2}$$

IOP Conf. Series: Materials Science and Engineering **765** (2020) 012068 doi:10.1088/1757-899X/765/1/012068

3.4. The Proposed Filter for Edge Detection

The proposed filter is a filter that was suggested by assuming variable values by trial and error and better results were obtained for a deal with the SVD algorithm compared to filters currently used to detect the edge. The proposed filter was compared with a Canny filter and indicated better results than the Canny filter as well as with Sobel filter and Prewitt filter by applying the proposed filter with the images within the database used and gave a better distinction than using the Canny filter as shown in Figure 4.

-11	-2	-11
-2	56	-2
-11	-2	-11

Figure 4. The proposed filter for edge detection.

3.5. Algorithm of Proposed Method

This section explains the algorithm of face recognition, which relies on SVD and LTP algorithm to features extraction. Also, City Block is uesd to classification. Figure 5 shows the operation of distinguishing between the training images and the set of test images.

3.5.1. Input and Output Algorithm

- Input: images training, testing image, and number of personal.
- Output: Recognition face.

3.5.2. Pre-processing:

- Step 1: Convert image into grayscale image.
- Step 2: Convert the result of step 1 into double class.
- Step 3: Resize of the result of step 2 into (100 x 100).
- Step 4: Applying edge detection for image using proposed filter and fusion between resulting image of step 3 and resulting image of edge detection using average method.
- Step 5: Applying the LTP algorithm on the original image.

3.5.3. Feature extraction:

- Step 6: Split the image into 10*10 parts for step 4 and 5.
- Step 7: Applying the SVD algorithm on the array of the step4 results after step6 is applied, and store SVD features in a single vector.
- Step 8: Applying the SVD algorithm on the array of the step5 results after step6 is applied, and store SVD features in a single vector.
- Step 9: Merge the resulting vector in step7 with the resulting vector in step8, to create a single vector of features for each image in the database called (Vectorl).

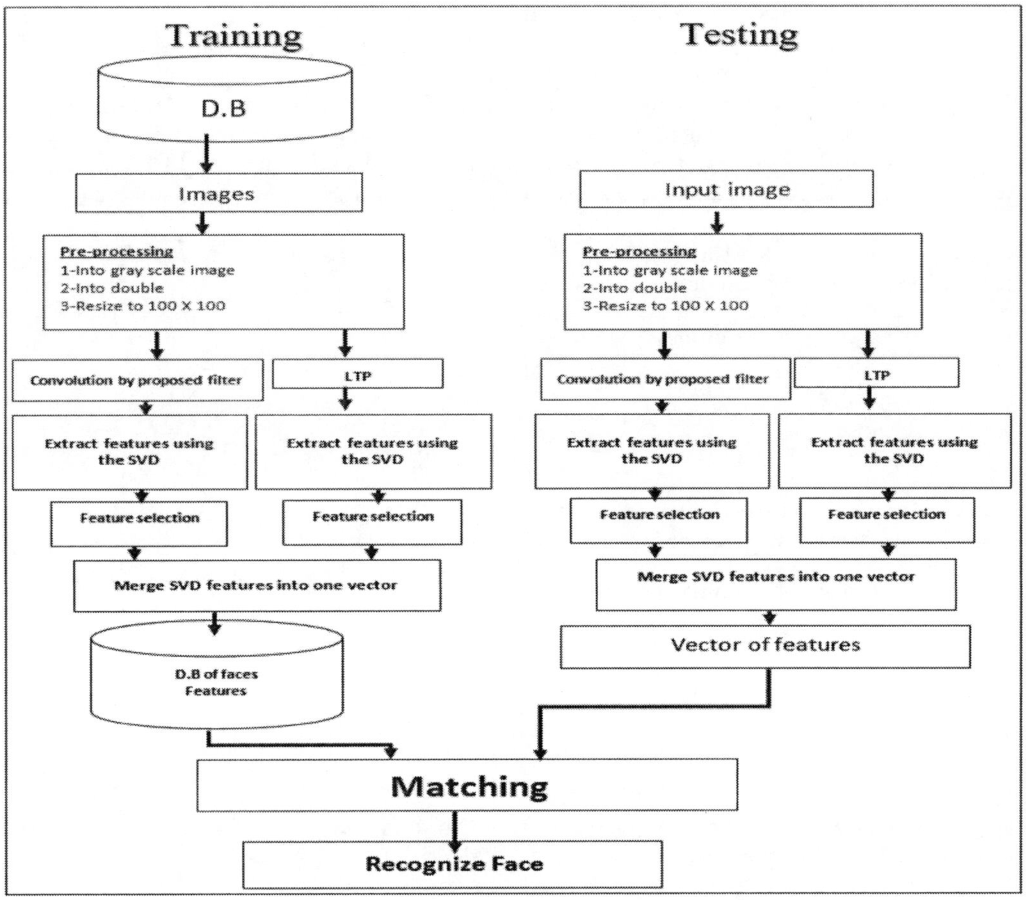

Figure 5. Block diagram of proposed method.

- Step 10: Store vectors in a database to include image features for the training phase.
- Step 11: Read image as testing phase.
- Step 12: Repeat the steps from step 2 to step 5 for pre-processing.
- Step 13: Repeat the steps from step 6 to step 9 for feature extraction to create (Vector2).

3.5.4. Classification:

This step compares two faces of training and testing images, then City Block distance is applied between these two images to measure minimum of distance. The classification section is applied through these steps:

- Extract features from training images.
- Extract features from testing images.
- After extracting the features from the total training images are stored in the vector and then the features of each image are extracted from the test images, then apply City Block distance between all the features of the test images and with all the training pictures, then take minimum distance from the resulting distances and thus be the closest image of the input image.

4. The Results Analysis

At beginning for the proposed system are edge detection for the input image using the proposed filter, and then extract the matrix (LTP) of the input image as well, capturing the features based on SVD for those two steps, after SVD steps to extracted features are merged into a single vector. Applying the same steps above for images in the training and testing phase. The final step performed based on matching between the features stored at the database and the features of the reference (test) image. In the proposed system the recognition ratio was 98% using city block distance and 98.5% using Structural Similarity (SSIM) for ORL database in case five images for testing phase and five images for the training phase.

The following are the cases that can be applied to the proposed system in terms of the number of test images and the number of training images per person:

- Case1 :(1: Testing, 9: Training),
- Case2 :(2: Testing, 8: Training),
- Case3 :(3: Testing, 7: Training),
- Case4 :(4: Testing, 6: Training),
- Case5 :(5: Testing, 5: Training),
- Case6 :(6: Testing, 4: Training),
- Case7 :(7: Testing, 3: Training),
- Case8 :(8: Testing, 2: Training),
- Case9 :(9: Testing, 1: Training).

The recognition rate is in order as: 100%, 100%, 100%, 99.37%, 98%, 95.41%, 94.28%, 85.93%, 77.77%.

Figure 6 shows the cases for the number of test and training images for each person, in order to know the ratio of recognition in the case of using the SVD algorithm only.

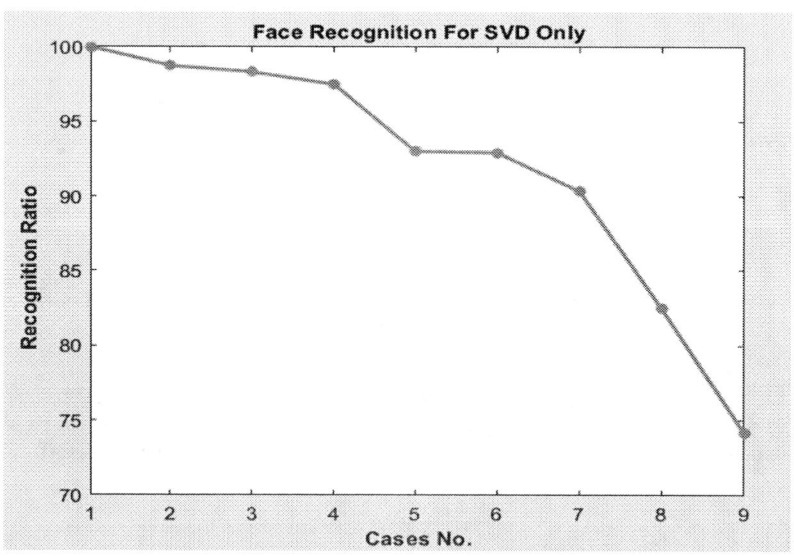

Figure 6. The cases for SVD only.

Figure 7 shows the cases for the number of test and training images for each person, and to know the ratio of recognition in the case of using the SVD and the proposed filter.

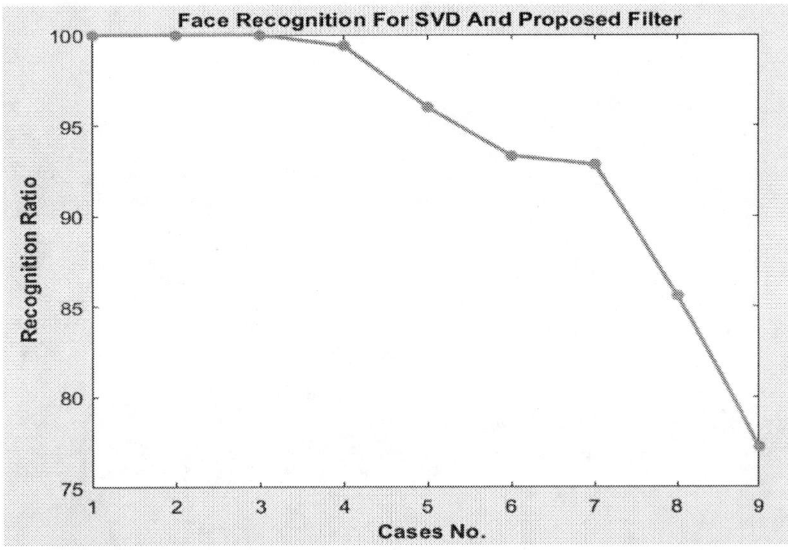

Figure 7. The cases for the SVD and the proposed filter.

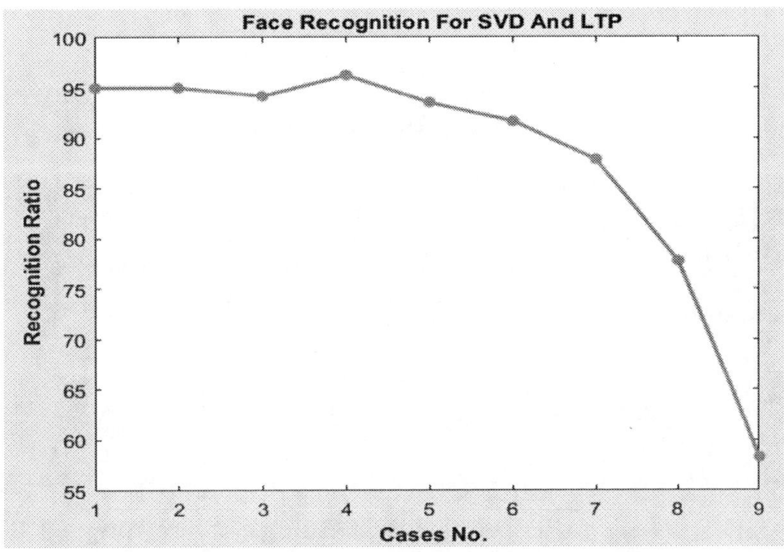

Figure 8. The cases for the SVD and LTP.

Figure 8 shown the cases for the number of test and training images for each person, and to know the ratio of recognition in the case of using the SVD and LTP.

Figure 9 shows the cases for the number of test and training images for each person, and to know the ratio of recognition in the case of using the Proposed System.

In Table 2 and 3, the face recognition ratio for each case is illustrated in terms of the number of images in the training phase and the number of images in the testing phase, it shows only the recognition cases for SVD, then the SVD and the LTP, then the SVD and the Proposed filter, finally, the cases for the Proposed System.

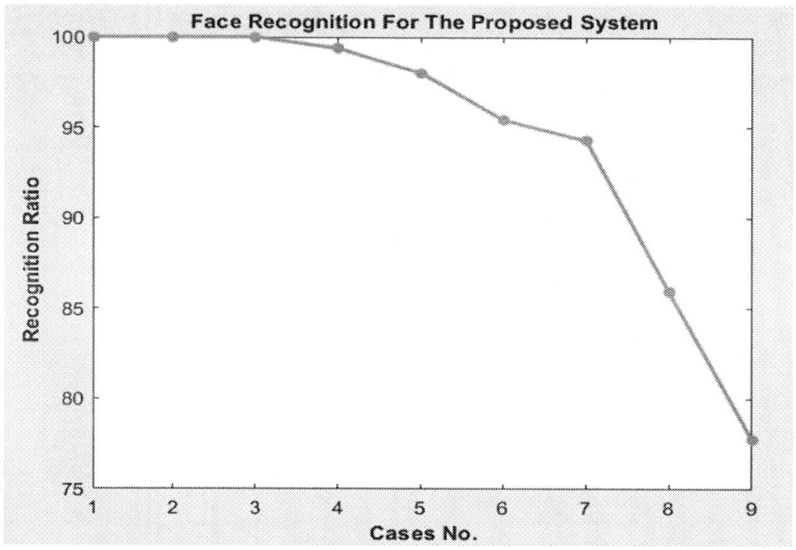

Figure 9. The cases for the Proposed System.

Table 2. Recognition accuracy ratio when change training set and testing set for each method from Case 1 to Case 5.

Methods	1 Test: 9 Training Case 1	2 Test : 8 Training Case 2	3 Test : 7 Training Case 3	4 Test : 6 Training Case 4	5 Test : 5 Training Case 5
SVD	100 %	98.75 %	98.33 %	97.5 %	93 %
SVD + LTP	95 %	95 %	94.16 %	96.25 %	93.5 %
SVD + Proposed Filter	100 %	100 %	100 %	99.37 %	96 %
Proposed System	100 %	100 %	100 %	99.37 %	98 %

Table 3. Recognition accuracy ratio when change training set and testing set for each method from Case 6 to Case 9.

Methods	6 Test: 4 Training Case 6	7 Test : 3 Training Case 7	8 Test : 2 Training Case 8	9 Test : 1 Training Case 9
SVD	92.91 %	90.35 %	82.5 %	74.16 %
SVD + LTP	91.66 %	87.85 %	77.81 %	58.33 %
SVD + Proposed Filter	93.33 %	92.85 %	85.62 %	77.22 %
Proposed System	95.41 %	94.28 %	85.93 %	77.77 %

5. Results Discussion

This work propose hybrid similarity method to recognition face that base on the LTP and the SVD. The proposed system is applied on the database ORL and obtain on 98% as ratio recognition in the case of five test images and five training images per person by using City Block distance, while one can obtain on 98.5% when the use of the SSIM method for matching. While, in the case of three images of the test group and seven images of the training group, the ratio of recognition is 100%.

6. Conclusion

- The proposed system satisfied high recognition rate with fusion of features resulted from the SVD and LTP.
- The system utilized the statistical features of LTP to overcome some problems result for using SVD as alone.
- Choosing singular array (S) of the SVD increase the recognition rate, since, it has the most significant value of the array.
- Locally approach hybrid on applying the proposed system on the sub-image of the enrolled database produces a high recognition rate as a result to discover silent features embedded in the images.

References

[1] Shnain N, Hussain Z, Lu S. A Feature-Based Structural Measure: An Image Similarity Measure for Face Recognition. Applied Sciences [Internet]. MDPI AG; 2017 Aug 3;7(8):786. Available from: http://dx.doi.org/10.3390/app7080786.
[2] Zhang D, Chen S, Zhou Z-H. A new face recognition method based on SVD perturbation for single example image per person. Applied Mathematics and Computation [Internet]. Elsevier BV; 2005 Apr;163(2):895907. Available from: http://dx.doi.org/10.1016/j.amc.2004.04.016.
[3] Cevikalp H, Neamtu M, Wilkes M, Barkana A. Discriminative common vectors for face recognition. IEEE Transactions on Pattern Analysis and Machine Intelligence [Internet]. Institute of Electrical and Electronics Engineers (IEEE); 2005 Jan;27(1):413. Available from: http://dx.doi.org/10.1109/tpami.2005.9.
[4] SHARIF, Muhammad, et al. Enhanced SVD Based Face Recognition. Journal of Applied Computer Science and Mathematics.2012.12.
[5] Shyam R, Singh YN. Face recognition using augmented local binary pattern and Bray Curtis dissimilarity metric. 2015 2nd International Conference on Signal Processing and Integrated Networks (SPIN) [Internet]. IEEE; 2015 Feb; Available from: http://dx.doi.org/10.1109/spin.2015.7095267.
[6] KAR, Arindam, et al. An adaptive block based integrated LDP, GLCM, and Morphological features for Face Recognition.2013. arXiv preprint arXiv:1312.1512.
[7] Jha AK, Gupta R, Saini D. Face Recognition: A Fourier Transform and SVD Based Approach. 2013 5th International Conference on Computational Intelligence and Communication Networks [Internet]. IEEE; 2013 Sep; Available from: http://dx.doi.org/10.1109/cicn.2013.51.
[8] Noori A, M. Z. Local and Semi-Global Feature-Correlative Techniques for FaceRecognition. International Journal of Advanced Computer Science and Applications [Internet]. The Science and Information Organization; 2014;5(8). Available from: http://dx.doi.org/10.14569/ijacsa.2014.050824.
[9] Rana N, Pancholi B. HMM-Based Face Recognition System with SVD Parameter. International Journal of Trend in Scientific Research and Development [Internet]. South Asia Management Association; 2018 Jun 30;Volume-2(Issue-4):36973. Available from: http://dx.doi.org/10.31142/ijtsrd12938.
[10] Anggo M, La Arapu. Face Recognition Using Fisherface Method. Journal of Physics: Conference Series [Internet]. IOP Publishing; 2018 Jun;1028:012119. Available from: http://dx.doi.org/10.1088/1742-6596/1028/1/012119.
[11] Arisandi D, Syahputra MF, Putri IL, Purnamawati S, Rahmat RF, Sari PP. A real time mobile-based face recognition with fisherface methods. Journal of Physics: Conference Series [Internet]. IOP Publishing; 2018 Mar;978:012038. Available from: http://dx.doi.org/10.1088/1742-6596/978/1/012038.
[12] Rustam Z, Rika. Face recognition using fuzzy kernel learning vector quantization. Journal of Physics: Conference Series [Internet]. IOP Publishing; 2018 Nov;1108:012068. Available from: http://dx.doi.org/10.1088/1742-6596/1108/1/012068.
[13] Li L, Xie H. Cascading BGP face recognition method based on heuristic information. Journal of Physics: Conference Series [Internet]. IOP Publishing; 2019 Feb;1168:042006. Available from: http://dx.doi.org/10.1088/1742-6596/1168/4/042006.
[14] Rassem TH, Khoo BE. Completed Local Ternary Pattern for Rotation Invariant Texture Classification. The Scientific World Journal [Internet]. Hindawi Limited; 2014;2014:110. Available from: http://dx.doi.org/10.1155/2014/373254.

[15] XUE, Jian; LI, Jinyu; GONG, Yifan. Restructuring of deep neural network acoustic models with singular value decomposition. In: Interspeech. 2013. p. 2365-2369..

[16] Xu Y, Zhuang Z, Li W, Fanga Y. Fast Implementation of Image Structural Similarity Algorithm. IOP Conference Series: Materials Science and Engineering [Internet]. IOP Publishing; 2019 May 30;533:012057. Available from: http://dx.doi.org/10.1088/1757-899x/533/1/012057.

Performance Evaluation of Grid Wireless Sensor Network with Different Packet Size for Pipeline in Downstream of Oil and Gas Industry

A S A Azman, M Y Lee, S K Subramaniam and F S Feroz

Department of Electronics and Computer Engineering, Universiti Teknikal Malaysia Melaka, Malaysia.

E-mail: m021820013@student.utem.edu.my, m021820002@student.utem.edu.my, siva@utem.edu.my, shahnaz@utem.edu.my

Abstract. In the past few decades, sensed data such as wind, vibration, temperature, and humidity were sent through the costly wire cable. With recent advancement and innovation in the wireless field, wireless sensor network (WSN) has emerged to benefit many industries in different aspects primarily in term of cost. WSN has been widely used in various applications due to its wide coverage area, particularly in oil and gas facility monitoring. Hence, the usage of grid WSN fits the wide coverage area mainly in the pipeline of downstream of oil and gas process. The main objective of this paper is to give a better understanding of the impact of different packet sizes, routing protocols, and the number of nodes towards a grid WSN as well as to provide open research issues in this field. Two types of routing protocols have been presented in this paper; reactive (AODV) and proactive (DSDV) with different packet sizes using a grid node arrangement. The simulated results have shown that with the increasing size of the network, more performance degradation occurred regardless of the packet size used. This issue can be seen through the packet loss and throughput loss in the network prominently after the deployment of 80 nodes and above. Apart from that, DSDV has shown a better delivery ratio, less number of passive nodes existence and less routing overhead produced in the network as compared to the AODV routing protocol.

Keywords— Wireless Sensor Network (WSN), Oil and Gas monitoring, Packet Size, Routing Protocols, Reactive (AODV), Proactive (DSDV)

1. Introduction

Numerous industrial, scientific, and environmental applications demand real-time data of physical events such as humidity, temperature, or pressure. Such events, in the past few decades, were collected and sent to the collecting point through costly, cumbersome cables. In recent years, WSN has emerged with a more promising return with its cheap, simple yet sophisticated solution to most of the industries. WSN has proved its capability in collecting, processing and communicating in various application particularly in the oil and gas industry [1].

WSN has been used in oil and gas facilities in upstream, midstream and downstream sectors to monitor health issues, to monitor pipeline integrity, and to enhance production. As shown in Figure 1, the upstream sector is where the discovery and extraction of raw materials process take place either the sources from the underground or underwater [2]. Midstream is where the extracted material is stored and transported to the downstream process usually via truck, barge, rail or pipelines. Downstream, on the other hand, is where the transported materials are processed, refined, marketed and commercialized.

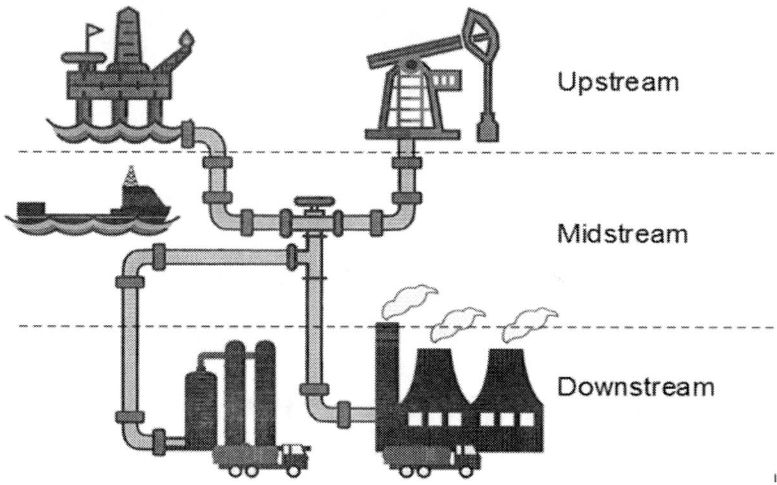

Figure 1. Sectors in oil and gas operation.

There are two types of pipeline distribution which are linear (midstream) and spread out distribution (downstream) as shown in Figure 2. This paper focuses on the WSN application of pipelines in the downstream sectors where the coverage area and the pipeline distribution fit the grid node arrangement. Since most of the assets were placed in remote locations, continuous remote monitoring in these three sectors is highly required to protect the assets and to maintain the production in the oil and gas industry.

Figure 2. Linear (left) and spread-out (right) pipeline distribution.

The nodes deployed in the oilfield environment utilize sensors such as pressure or leakage sensor that gather crucial industrial data allowing new intuitions for plant operations and advanced solutions. It also helps in reducing downtime, operating and production costs, optimizing operations, enhancing platform safety by hindering problems and tolerating any errors. For example, in pipeline transmission, the operations usually dealing with harsh environments, highly flammable material and high pressure of liquid material. Hence, any failure during these operations could cause a tragic incident to the people surrounding and to the environment as well as critical financial loss. This is where WSN comes in handy where it reduces the intervention of humans in oil and gas operation.

In wireless communication, there are 7 layers in the Open Systems Interconnection (OSI) model as shown in Figure 3 and this paper presents the performance evaluation comprising the network layer where the routers operate. In this layer, the data is packaged into IP datagrams consisting of destination and source IP address [3]. IP datagrams are responsible for the data forwarding process across the network. A routing algorithm or routing protocol determines how the data is routed in the network to exchange information. The aim of this paper is to evaluate the performance of the existing WSN routing protocol as well as implementation of different packet sizes towards the routing protocols and to present the lacking factors that affect the performance of the network as the network grows with an arrangement that mimics the spread-out distribution of the pipeline in downstream sector of oil and gas industry which is the grid node arrangement.

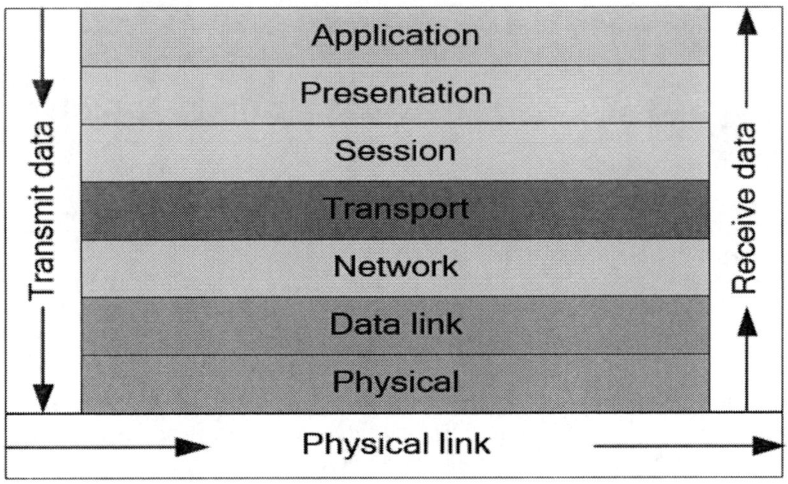

Figure 3. The 7 layers in the OSI model.

1.1. Routing protocols

With the high demand and fast-emerging technologies in WSN, several challenges stem out and created more opportunity window for researchers to solve. Packet loss, throughput loss, and high routing overhead have contributed to the performance degradation primarily in a high-density network. These issues could be caused by the congestion in the network, hence making routing one of the tangible issues that need to be solved. In the network layer or routing layer, different types of routing protocols have been used [4] as shown in Figure 4.

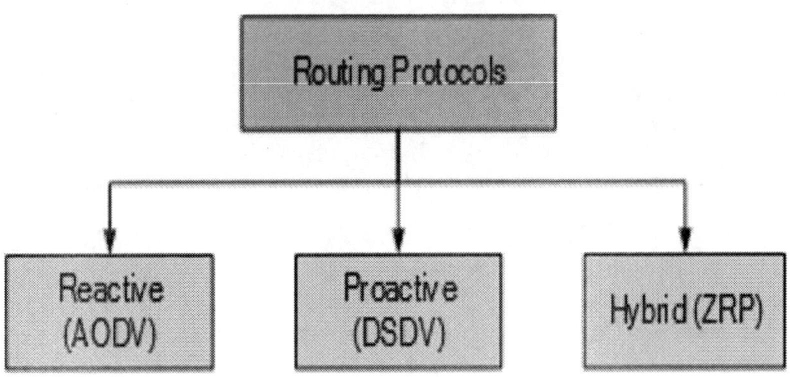

Figure 4. Three routing protocols in the network layer.

The proactive routing protocol is often known as a table-driven approach making the routing in pre-defined mode [5]. With this feature, the delay for packet forwarding is reduced by but the network resources will deplete due to the route update and maintenance for the routing table. One of the examples of proactive routing protocol is Destination-sequenced Distance Vector (DSDV) routing protocol. Reactive routing protocol, on the other hand, uses an ad-hoc method of routing where the path will be discovered only when needed [6]. This has caused the delay in the data forwarding but fewer routing overhead will be drawn. One of the examples of proactive routing protocol is Ad-hoc on Demand Distance Vector (AODV) routing protocol. Hybrid routing protocol combines the best features in the proactive and reactive routing protocol. One of the examples of the hybrid routing protocol is the Zone Routing Protocol (ZRP). There are more examples of routing protocols that have been proposed by past researchers which can be seen in the next section.

2. Related research

The researchers in [7] have presented a routing overhead of the reactive routing protocol study based on Wireless Multihop Networks (WMhNs). The network performance has been improved by implementing routing overhead monitoring methods to Dynamic Source Routing (DSR), AODV, and Dynamic MANET On-demand (DYMO) routing protocol. As a result, the researchers have found that AODV shows the best performance during high-density traffic by considering scalability. On the other hand, the DSR routing protocol shows the lowest end-to-end delay and control packets. The researchers also highlighted that DSR works best in a network with a fewer number of hop.

The researchers in [8] have proposed a Novel Hybrid Distance Vector (NHDV) routing protocol which combines AODV and DSDV routing protocols to reduce route discovery time, reduce congestion, and provide better scalability in the hybrid mesh network. NHDV uses clustering by appointing the gateway as the cluster head by using the Hello packet. As a result, NHDV has improved the network performance from the clustering technique and has enhanced the path quality from the implementation of DSDV.

In [9], the researchers have investigated the disadvantage Multicast Ad-hoc On-demand Distance Vector (MAODV) and with that, the researchers proposed a new routing protocol MAODV-BB (Backup Branches). MAODV-BB is the routing protocol that combines tree and mesh topology. The tree structure uses low network resources, efficient in forwarding but has poor robustness while the mesh structure helps in load handling during high load network. MAODV-BB has shown a significant improvement in the Quality of Services (QoS) during high load networks.

The authors in [10] have proposed Neighbors Perception ZBR (NP-ZBR) routing protocol, a novel ZigBee Routing Protocol (ZBR). As known by many, ZBR has good average hop, delay, and transmission rate. However, ZBR uses a high amount of network resources which then increases the load and energy consumption due to congestion. With NP-ZBR, a better performance resulted in terms of hop average and transmission delay while maintaining the same average end-to-end delay, throughput, and transmission rate as in ZBR.

In conclusion, there are few kinds of research that have been done on grid topology and none of the researches showed an implementation of high number of nodes in the network. Thus, this has led to the study of different numbers of nodes, packet size, and routing protocols towards the performance of grid WSN.

3. Simulation

In the oil and gas industry, there are two types of pipeline distribution which are linear and spread-out. This paper focuses on the spread-out distribution, hence the usage of grid node arrangement. It is important to simulate the network in order to determine the lack in the network using the specified environment which in this case, grid topology with a varied number of nodes. Table 1 shows the parameters as well as the routing protocols used in the simulation using Network Simulator 2 (NS2).

The resulted performance metrics such as delivery ratio, throughput, passive nodes, routing overhead, and fairness index from the simulation have been recorded to show the behaviour of the simulated network. Some statements from a few research papers were included to support the reasoning behind every presented result.

Table 1. Parameters used in the simulation

Parameter	Value
Routing protocol	*AODV, DSDV*
Bandwidth	*2Mbps*
Transport agent	*TCP*
Packet size	*32, 128, 512*
Topology	*Grid with nodes formation of 6x4, 8x6, 10x8, 12x10, 14x12, 16x18 and 18x16*
Number of nodes	*24,48, 80, 120, 168, 224, 288*
Node distance	*50 m*
Propagation model	*Two ray ground*
Simulation time	*500 s*

3.1. Simulation results

As the network size increases, the delivery ratio decreased as illustrated in Figure 5. This indicates more packet loss, as well as the throughput degradation (as in Figure 6) in the network as the number of node increases mainly after 80 number of nodes onwards. This is due to the overloaded traffic which causes congestion in the network as the network density and size increases [11]. Apart from that, the queue length factor plays an important role in this case. In the simulation, the queue length is 50 and the queue type is tail drop queue, hence, without considering the time to live of a packet, any packet after the 50th enqueued packets will drop causing a packet loss. From the results, although the trend in Figure 5 shows a similar pattern, it can be said that DSDV shows a better delivery ratio as compared to AODV. In addition, the usage of 32 bytes of packet size causes fewest packet loss and 512 bytes of packet size causes the most packet loss in the network.

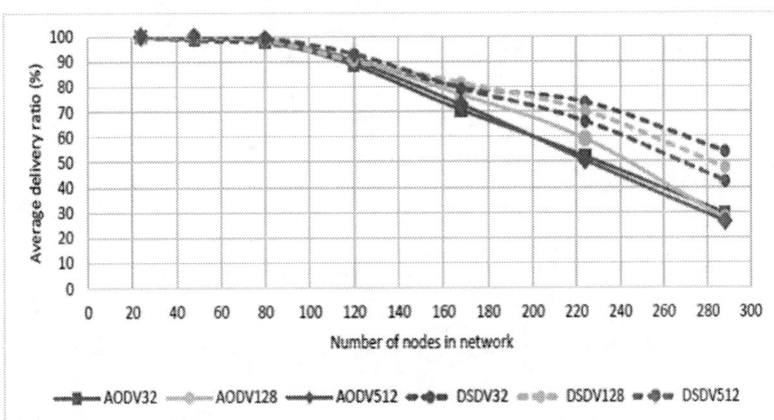

Figure 5. Average delivery ratio over number of nodes.

Packet loss issue that affects delivery ratio as shown in Figure 5 has contributed to the throughput degradation which can be seen in Figure 6. There is a control mechanism known as the TCP congestion control mechanism. In order to reduce the traffic load, the congestion control mechanism reduces the transmission rate or drops the packet [12]. This issue causing packet loss and throughput degradation. As shown in figure 6, a slight throughput reduction shown for packet size 32 and 128 bytes. However, for packet size 512 bytes, the throughput clearly starts to degrade after 80 number of node onwards.

IOP Conf. Series: Materials Science and Engineering **765** (2020) 012069 doi:10.1088/1757-899X/765/1/012069

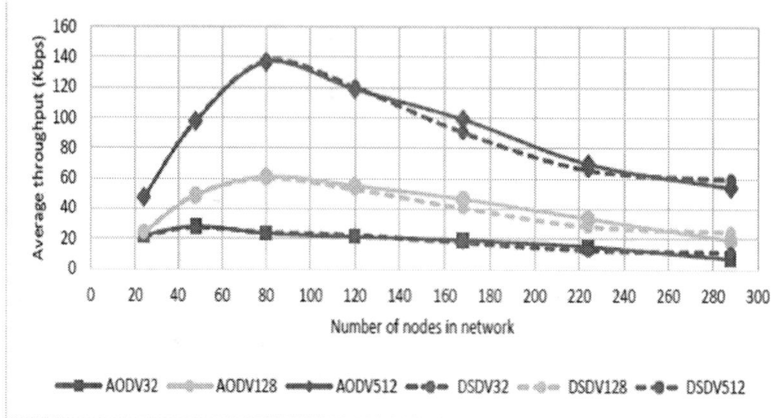

Figure 6. Average throughput over number of nodes.

Theoretically, the smaller the packet size, the more routing overhead the network has and vice versa [13]. This statement can be proved by observing the results shown in Figure 7, 32 bytes of packet size produced the most overhead as compared to 128 and 512 bytes. To be well described, the packages delivery using a truck is more cost saving as compared to the delivery using a standard car due to the fewer fuel consumption and fewer number of trip. For example, to send a big machinery equipment to a place, a truck would only take at most two trips, but a car would take more than that. It can be said that the truck utilize less fuel (routing overhead) consumption to deliver the equipment as compared to the car.

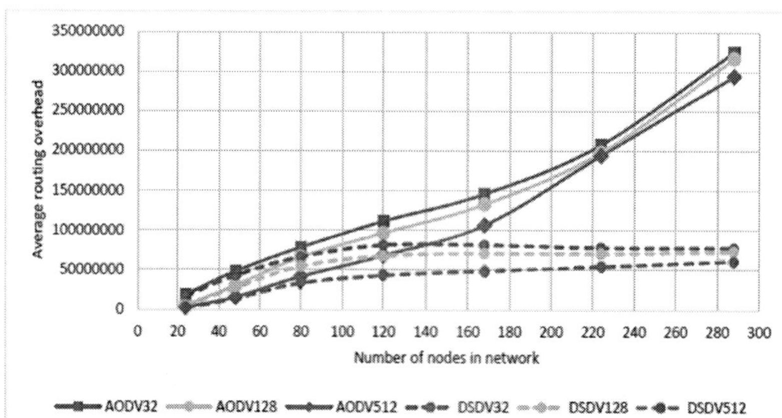

Figure 7. Routing overhead over number of nodes.

Apart from that, it can be clearly seen that the AODV routing protocol produced more routing overhead as compared to the DSDV routing protocol. Studies in [14, 15] have shown that AODV produces more routing overhead as compared to DSDV. Since AODV is on-demand protocol, the routing overhead will increase as the number of nodes increases. Due to high traffic, queue factor and time to live of the packet, ad-hoc routing protocol has to retransmit the dropped packet during congestion at cost of rebroadcasting of unwanted packets. Eventually, routing overhead will increase cannibalizing the network capacity. While for DSDV, during an event of packet drop, all of the required information is already stored in the routing table. Hence, the source does not have to rebroadcast the unnecessary packets to re-establish the path to the destination during high traffic. The packet loss through the delivery ratio of AODV and DSDV as shown in Figure 5 reflects the amount of routing overhead as shown Figure

7, showing the drawback of packet retransmission by both AODV and DSDV routing protocols as the network grows.

With the increasing number of overhead in the network, the network resources will be used up and causing severe unfairness in the network [7, 16]. As shown in Figure 8, there are no significant differences in the fairness index trend between AODV and DSDV routing protocols. However, it can be seen that 32 bytes of packet size produced the worst fairness index and 512 bytes of packet size produced the best fairness index in the network. These situations can be said that by using smaller packet sizes, more network resources will be consumed as discussed on the routing overhead topic earlier.

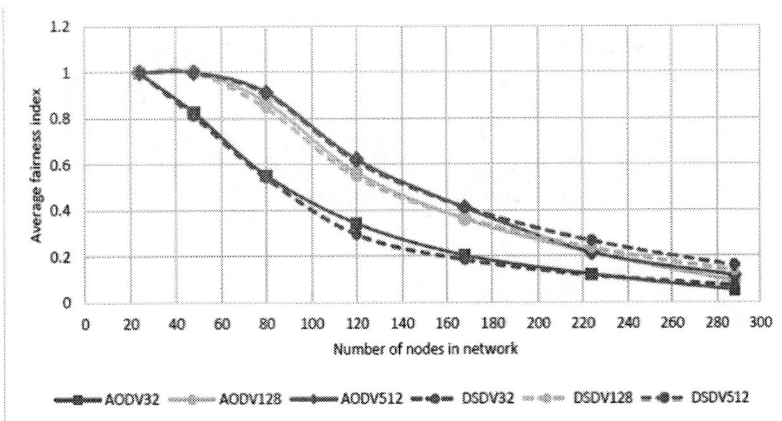

Figure 8. Fairness index over number of nodes.

Apart from the fairness issue, passive nodes' existence also one of the drawbacks of resource wastage. The network resources were not allocated equally, hence, some of the nodes had the chance to transmit or receive the data and some do not have the chance. As illustrated in Figure 9, the number of passive nodes increased as the number of nodes increases. AODV routing protocol produced more passive nodes as compared to DSDV routing protocol. This is due to the consumption of resources by AODV is more as compared to DSDV by taking routing overhead as an indicator to this issue (Figure 7). Passive nodes also existed the highest when the network is using 512 bytes of packet size and the fewest when using 32 bytes of packet size. Due to the unfairness and the presence of passive nodes, the delivery ratio as well has been affected. As can be seen in Figure 5 and Figure 9, the degradation in the delivery ratio and the passive node existence started to occur from 80 number of nodes onward.

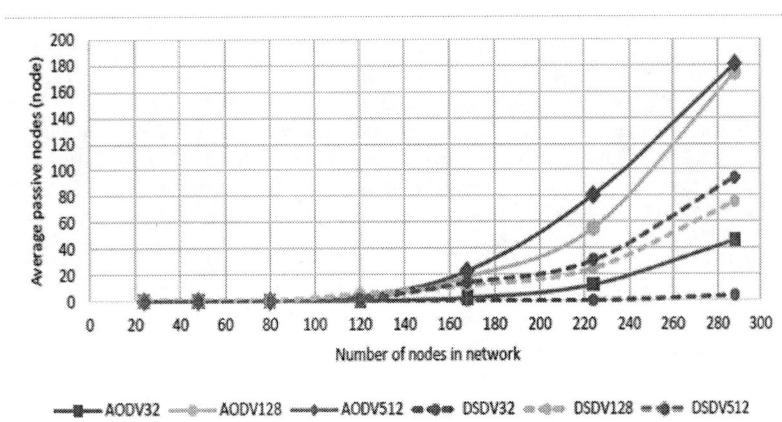

Figure 9. Passive nodes over number of nodes.

IOP Conf. Series: Materials Science and Engineering **765** (2020) 012069 doi:10.1088/1757-899X/765/1/012069

4. Conclusion

WSN has shown its capability in various application mainly in the oil and gas industry. There are three processes involved in the oil and gas industry which are upstream, midstream, and downstream processes. In this paper, the pipeline in the downstream sector is the focused sector since the spread out distribution of the pipeline suits the grid node arrangement. In order to investigate the lack in the performance of such configuration, a simulation has been done using an increasing number of nodes. The results showed that most of the performance metrics start to degrade after 80 number of nodes onwards. The delivery ratio is high if a smaller packet size is used and a better transmission rate will the network has. Hence, fewer packet loss will occur if a smaller packet is used. The throughput of the network totally depends on the packet sizes; higher the packet size, the more throughput. However, as the network size increases, the throughput will degrade prominently after 80 number of nodes onwards. Apart from that, the usage of a smaller packet will cause more routing overhead to be drawn in the network, making the network resources to be wasted and causing the severe fairness issue in the network. In addition to that, more passive nodes will be resulted due to resource wastage. On the other hand, in terms of routing protocols, DSDV shows a better delivery ratio, routing overhead and number of passive nodes as compared to AODV routing protocol. Thus, DSDV outperforms AODV by three performance metrics out of five as the network grows by considering the grid node arrangement.

Acknowledgement

The authors would like to thank the Ministry of Higher Education - Malaysia, Universiti Teknikal Malaysia - Melaka for their support, lab facilities, sincere encouragement, and assistance.

References

[1] Khan WZ, Aalsalem MY, Gharibi W, Arshad Q. Oil and Gas monitoring using Wireless Sensor Networks: Requirements, issues and challenges. 2016 International Conference on Radar, Antenna, Microwave, Electronics, and Telecommunications (ICRAMET) [Internet]. IEEE; 2016 Oct; Available from: http://dx.doi.org/10.1109/icramet.2016.7849577.

[2] Aalsalem MY, Khan WZ, Gharibi W, Khan MK, Arshad Q. Wireless Sensor Networks in oil and gas industry: Recent advances, taxonomy, requirements, and open challenges. Journal of Network and Computer Applications [Internet]. Elsevier BV; 2018 Jul;113:8797. Available from: http://dx.doi.org/10.1016/j.jnca.2018.04.004.

[3] Rayes A, Salam S. The Internet in IoT-OSI, TCP/IP, IPv4, IPv6 and Internet Routing. Internet of Things From Hype to Reality [Internet]. Springer International Publishing; 2016 Oct 23;3556. Available from: http://dx.doi.org/10.1007/978-3-319-44860-2-2.

[4] Rathore S, Naiyar M, Ali A. Comparative study of entity and group mobility models in MANETs based on underlying reactive, proactive and hybrid routing schemes. 2012 15th International Multitopic Conference (INMIC) [Internet]. IEEE; 2012 Dec; Available from: http://dx.doi.org/10.1109/inmic.2012.6511451.

[5] Al-Saadi A, Setchi R, Hicks Y, Allen SM. Routing Protocol for Heterogeneous Wireless Mesh Networks. IEEE Transactions on Vehicular Technology [Internet]. Institute of Electrical and Electronics Engineers (IEEE); 2016 Dec;65(12):977386. Available from: http://dx.doi.org/10.1109/tvt.2016.2518931.

[6] Cana E. Comparative Performance Simulation of DSDV AODV and DSR MANET Protocols in NS2. International Journal of Business and Technology [Internet]. University for Business and Technology; 2013 May;2(1):2431. Available from: http://dx.doi.org/10.33107/ijbte.2013.2.1.4.

[7] Mahmood D, Javaid N, Qasim U, Khan ZA. Routing Load of Route Discovery and Route Maintenance in Wireless Reactive Routing Protocols. 2012 Seventh International Conference on Broadband, Wireless Computing, Communication and Applications [Internet]. IEEE; 2012 Nov; Available from: http://dx.doi.org/10.1109/bwcca.2012.90.

[8] Iqbal M, Shafiq M, Choi J-G, Attaullah H, Akram K, Wang X. Design and Analysis of a Novel Hybrid Wireless Mesh Network Routing Protocol. International Journal of Adaptive, Resilient and Autonomic Systems [Internet]. IGI Global; 2014 Jul;5(3):2039. Available from: http://dx.doi.org/10.4018/ijaras.2014070102.

[9] Kumbarwar P, Padiya P. Result analysis of optimized MAODV-BB tree based protocol in MANETs. 2016 International Conference on Signal Processing, Communication, Power and Embedded System (SCOPES) [Internet]. IEEE; 2016 Oct; Available from: http://dx.doi.org/10.1109/scopes.2016.7955722.

[10] ZHAO, Jianli; JIANG, Xiaofei. NP-ZBR: an improved ZBR routing algorithm based on neighbors perception. Sensors and Transducers, 2014, 172.6: 67..

[11] Rajesh M, Gnanasekar JM. CONGESTION CONTROL USING AODV PROTOCOL SCHEME FOR

IOP Conf. Series: Materials Science and Engineering **765** (2020) 012069 doi:10.1088/1757-899X/765/1/012069

WIRELESS AD-HOC NETWORK. Advances in Computer Science and Engineering [Internet]. Pushpa Publishing House; 2016 Aug 10;16(1-2):1937. Available from: http://dx.doi.org/10.17654/cs016120019.

[12] Thorat MA, Deshpande VS. Handling fairness in wireless sensor network. 2012 World Congress on Information and Communication Technologies [Internet]. IEEE; 2012 Oct; Available from: http://dx.doi.org/10.1109/wict.2012.6409206.

[13] Yigit M, Yildiz HU, Kurt S, Tavli B, Gungor VC. A survey on packet size optimization for terrestrial, underwater, underground, and body area sensor networks. International Journal of Communication Systems [Internet]. Wiley; 2018 May 6;31(11):e3572. Available from: http://dx.doi.org/10.1002/dac.3572.

[14] Rampurkar K, Lavande N, Shilgire S, Mane SN. Study of Routing Overhead and its Protocols. INTERNATIONAL JOURNAL OF ADVANCED ENGINEERING AND MANAGEMENT [Internet]. International Journal of Advanced Engineering and Management (IJOAEM); 2017 Feb 1;2(2):52. Available from: http://dx.doi.org/10.24999/ijoaem/02020015.

[15] KHIAVI, Mina Vajed; JAMALI, Shahram; GUDAKAHRIZ, Sajjad Jahanbakhsh. Performance comparison of AODV, DSDV, DSR and TORA routing protocols in MANETs. International Research Journal of Applied and Basic Sciences, 2012, 3.7: 1429-1436.

[16] Abbas G, Manzoor S, Hussain M. A stateless fairness-driven active queue management scheme for efficient and fair bandwidth allocation in congested Internet routers. Telecommunication Systems [Internet]. Springer Science and Business Media LLC; 2017 Apr 4;67(1):320. Available from: http://dx.doi.org/10.1007/s11235-017-0306-3.

IOP Conf. Series: Materials Science and Engineering **765** (2020) 012070 doi:10.1088/1757-899X/765/1/012070

Performance Analysis of Linear Topology Wireless Sensor Network in Oil and Gas Industry

M.Y. Lee, A.S. Azman, S.K. Subramaniam, and F.S. Feroz

Department of Electronics and Computer Engineering, Universiti Teknikal Malaysia Melaka, Durian Tunggal, Malaysia

E-mail: M021820002@student.utem.edu.my,M021820013@student.utem.edu.my, Siva@utem.edu.my, Shahnaz@utem.edu.my

Abstract. Oil and gas industry is one of the largest corporations in the world. The process and machinery involved for the petroleum product to be extracted, refined and transported are usually critical and complex. Hence, it is crucial to have a stable and reliable system to control, monitor and manage the security the industrial assets. Since Wireless Sensor Network (WSN) can sense, process and communicate, it is one of the best and popular solution to the problem existed in the oil and gas industry. In this paper, a detailed performance analysis of AODV, DSDV and FIXRT with 32 and 128 packet size is presented by using linear topology in accordance with IEEE 802.11 standards by using ns2.35 simulator. The results showed that FIXRT protocol increases the overall performance of the network by almost 30%.

Keywords— Wireless Sensor Network (WSN), Oil and Gas monitoring, Packet Size, Linear Topology , IEEE 802.11 standards, ns2.35 Simulator

1. Introduction

WSN is a group of sensors each has the capability to sense, communicate and process, but form a network which mutually monitors the state of the physical world when deployed in a group. Each of these sensors has the ability to gather and process data while communicate with each other wirelessly [1]. This technology will affect the industry world especially the oil and gas since the procedure and environment involved in this industry are usually complex dangerous. With the capability of WSN is undoubtedly significant, the application of this technology is unlimited form agricultural sector, acoustic sector, environmental management and even oil and gas [2, 3, 4, 5].

Oil and gas industry are known for their dangerous and critical procedure to ensure the assets are in perfect condition. The operation is separated into three main levels, upstream, midstream and downstream. Upstream is where exploration occurred while midstream is the storage and transportation of the crude oil. Downstream, on the other hand, is where the reefing and marketing process took place. The common transportation for the products is pipelines, ship or truck.

As predicted by U.S. Energy Information Administration Office [6], the future demand of energy will increase by 7% per year from 2018 to 2020, which is greater than the 4% average growth rate form the year 2005 to 2015. The report also forecasted that by 2030, natural gas plant liquid (NGPL) production will have a shocking increment by 38% compared to 2018 level. Figure 1 illustrates the subdivisions of oil and gas industrial operation.

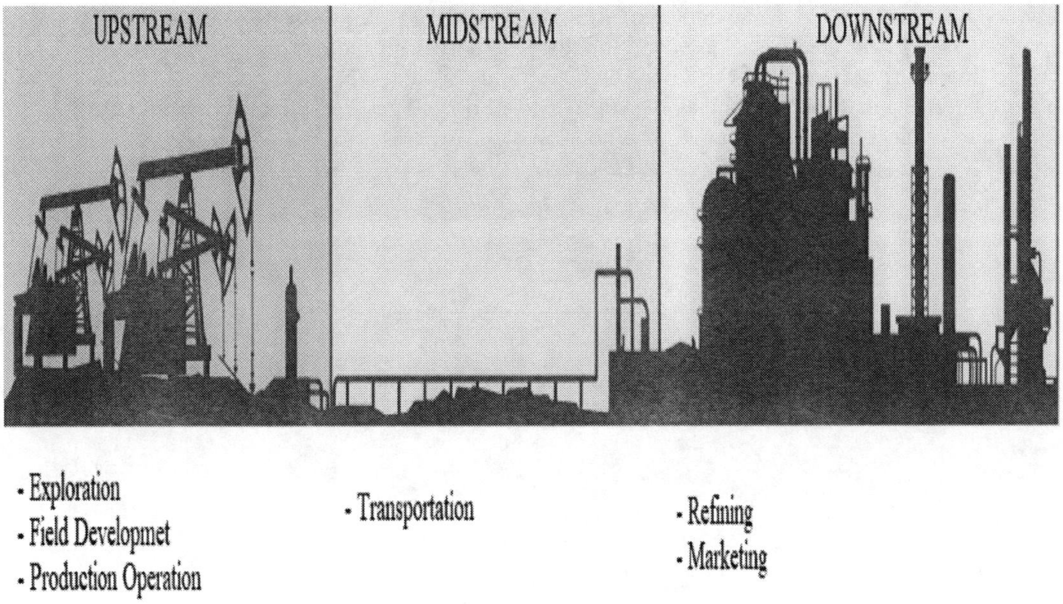

Figure 1. Subdivisions of Oil and Gas Industrial Operation.

1.1. Challenges in Deployment of WSN

Despite having such advantages, there is no perfect solution in this world. There are several challenges and obstacles in deploying WSN. The energy consumption is the most common problem faced by the researches and it is the most critical issues faced by WSN. This is due to the network lifetime is a major criterion, especially to the linear topology deployment as energy, are needed to gather, process and transmit each data packets. The energy consumption increases as the size of the network are increased. Since WSN are deployed to work with limited human intervention, any person can attack the network and the entire system will be disrupted [7]. Other than the mentioned challenges, the reliability and Scalability of the network are also a serious problem faced by the researches. In the oil and gas industry, a consistent communication system is vital. When unpredictable event is detected, the nodes are required to gather and sent the data within a period as failure to do so will lead to calamitous accidents. Figure 2 describes the overall view of challenges faced in the deployment of WSN.

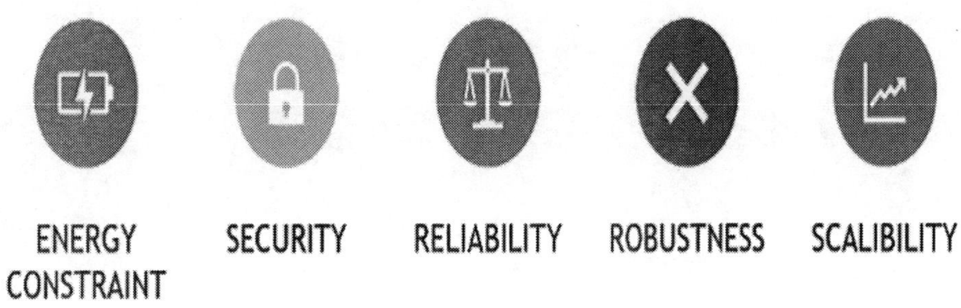

Figure 2. Overall views of challenges in the deployment of WSN.

1.2. Network Characteristic

To reduce the risk involved in the deployment of WSN, several techniques are applied. One of the popular techniques used is the manipulation of the topology. A network topology is referred as the way nodes are linked to one another in a network. Since the main focus in this paper is linear topology, the architecture of a multi-hop linear topology is classified into two major structure which is the flat one-tier topology and the hierarchical multi-tier topology as illustrated in Figures 3 and 3 [8].

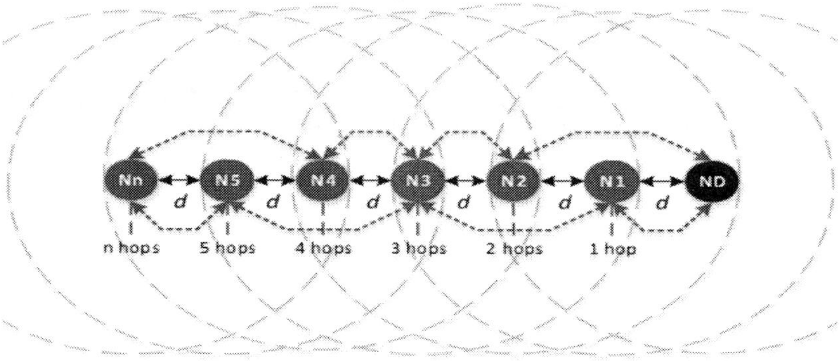

Figure 3. Multi-hop linear WSN.

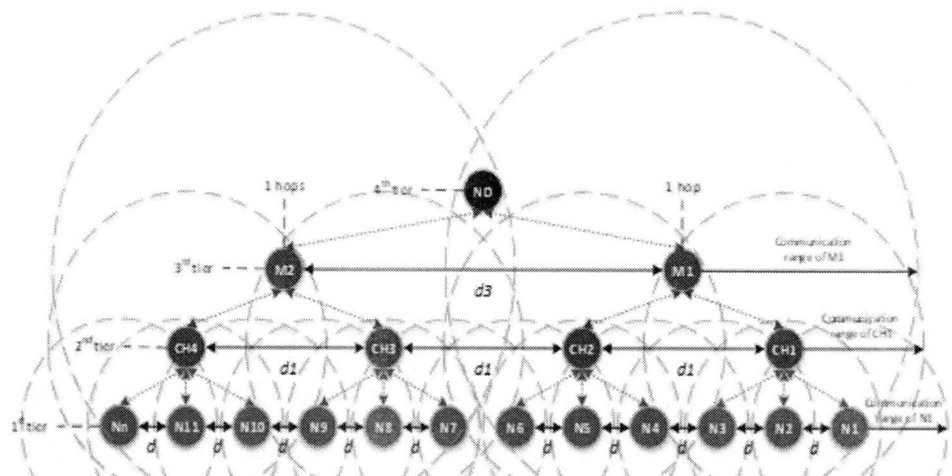

Figure 4. Hierarchical WSN.

The other manipulation that can be considered is the wireless standards. The wireless standards have a significant influence in term of performance and outcome to the nature of topology implemented. The two standard devices that are potential for a real scale implementation on linear topology is IEEE 802.11 and IEEE 802.15.4. The brief technical operation of both standards is as shown in Table 1.

IOP Conf. Series: Materials Science and Engineering **765** (2020) 012070 doi:10.1088/1757-899X/765/1/012070

Table 1. Comparison between IEEE 802.11 and IEEE 802.15 standard.

Parameters	Wifi	ZigBee	WirelessHART	Z-Wave
IEEE standard	*802.11a/b/g/n/ac*	*802.15.04*	*802.15.04*	*802.15.04*
Operational frequency	*5GHz(a,ac)/2.4GHz(b,g)/ 2.4GHz-5GHz(n)*	*2.4 GHz*	*2.4 GHz*	*868.42MHz*
Nodes per maste	*2007*	*> 65000*	*500*	*232*
Range	*100 meter(a,b,g)/ 250 meter(n)*	*1600 meter*	*250 meters*	*100 meters*
Data rate	*11Mbps(b)/54Mbps(a,g)/ 450Mbps(a,c)/600Mbps(n)*	*250 kbps*	*250 kbps*	*100 kbps*
Battery Life/Cost	*Days-weeks High*	*Months-years Low*	*Months-years Moderate*	*Months-years Low*

2. Previous Study and Research

Lately, remote monitoring of pipeline using WSN are becoming a trend due to the speedy improvement of technology on communication protocols, efficiency, capability and implementation cost on the various application [9, 10, 11, 12]. Since the characteristic of the network layout is unique, consequences such as network destabilization, throughput unfairness, high delay and power consumption are formed. The authors in [13] proposed an algorithm which combines the advantages of both clustering and Compressive Sensing-based strategy. The authors also stated that the reason for WSN lifetime to be reduced is mainly because of Hot Spot Problem and the alternation of the role for Cluster Head. To overcome this, the authors suggested that a Backup Cluster Head along with a new mechanism to replace the role of Cluster Head which is called Energy-Efficient Compressive Sensing-based clustering Routing (EECSR). This study showed that the EECSR has better power efficiency and lifespan compared to the existing clustering algorithm.

In [14] the authors proposed a system that cartels JenNet network with IEEE 802.15.4 standard. This proposed system can be used on different network topologies such as linear, tree and star. To increase the robustness of the network, the proposed solution has a self-healing feature. To tackle the power limitation, this system contains sleep mode capability. To make the workers emergency maintenance easier, a wind sensor is added in the system to indicate the wind speed and other information.

The authors in [8] proposed a novel routing method, Odd-Even Linear Static Routing Path (OE-LSRP) to achieve noteworthy improvements in overall network performance. By splitting the node into two different paths, the snowball effect towards the destination nodes are reduced and this gives the performance of the network a significant boost. The authors tested the proposed solution in accordance with the IEEE 802.11 standard and the architecture of the nodes are in linear topology.

Next, the authors [15] presented a method which is called SimpliMote. This system uses IEEE802.15.04 standard to detect leakage in oil and gas pipelines. The author combined a low power custom sensor board with algorithms that detect leakages, localization, actuation and parameter sensing. A testbed of 3-meter-long pipeline that is filled with pressurized fluid is prepared to simulate the oil and gas pipeline. To create a real-life leak, valves are used.

3. Result and Discussion

In this research, certain parameters are being considered. The number of nodes that are being simulated is 200 nodes with 50m distance each and 125 m of communication range. The simulation was run with five cycles where the seeds number varies. The bandwidth used is 2 Mbps and two packet size are considered which are 32 and 128 bytes. The protocols involved are AODV, DSDV and FIXRT. The simulator that is being used is NS2.35 and the topology that is being considered is linear topology.

The results shown in Figure 5 indicates that DSDV with 128 packet size has a better delivery ratio compared to AODV protocol. The network with different sizes suffers from data loss but this problem

IOP Conf. Series: Materials Science and Engineering **765** (2020) 012070 doi:10.1088/1757-899X/765/1/012070

is reduced by DSDV with 128 packet size. However, with FIXRT protocol, we can clearly see that the delivery ratio of the network has significantly increased to almost 30%.

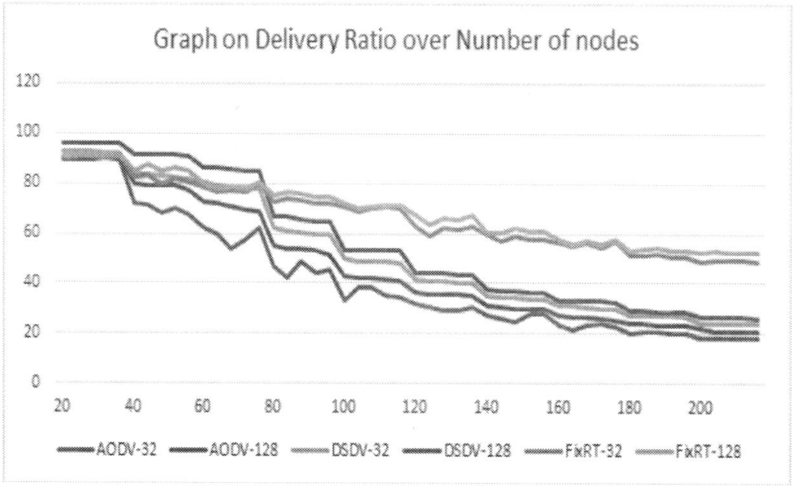

Figure 5. The graph on Delivery Ratio over Number of nodes.

This results can be validated with Figure 6 which indicates that FIXRT with 128 packet size has the highest throughput value. With higher delivery ratio, the network will have a higher throughput. Since the network capacity for packet size 128 is higher, a better delivery ratio is produced.

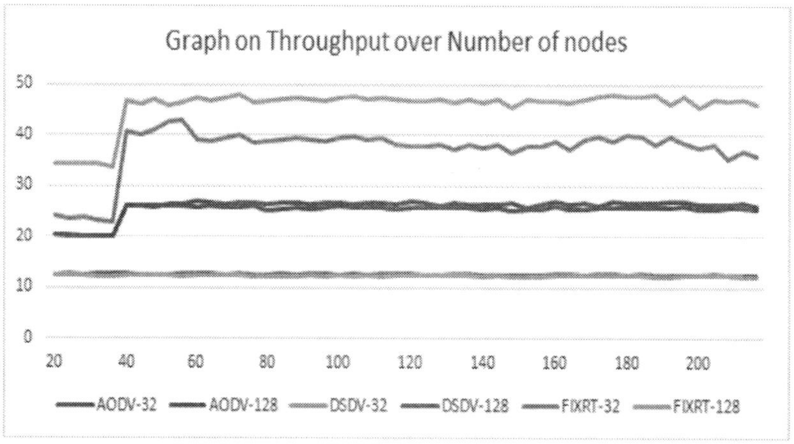

Figure 6. The graph on Throughput over Number of nodes.

In Figure 7, it is clearly shown that the passive nodes for AODV with 128 packet size are higher compared to others. Since AODV has higher passive nodes, the delivery ratio of the network will also be affected. This result has clearly indicated that FIXRT has significantly reduced the number of passive nodes hence increasing the overall network performance of the WSN.

The results in Figure 8 indicate that packet size 128 has a higher routing overhead compared to 32. However, among the routing protocols that has 128 packet size, AODV is higher compared to DSDV. When the routing overhead is high, network recourse wastage will occur. When routing overhead is increasing, the usage of bandwidth is limited, and this will increase the queuing factor of the network.

IOP Conf. Series: Materials Science and Engineering **765** (2020) 012070 doi:10.1088/1757-899X/765/1/012070

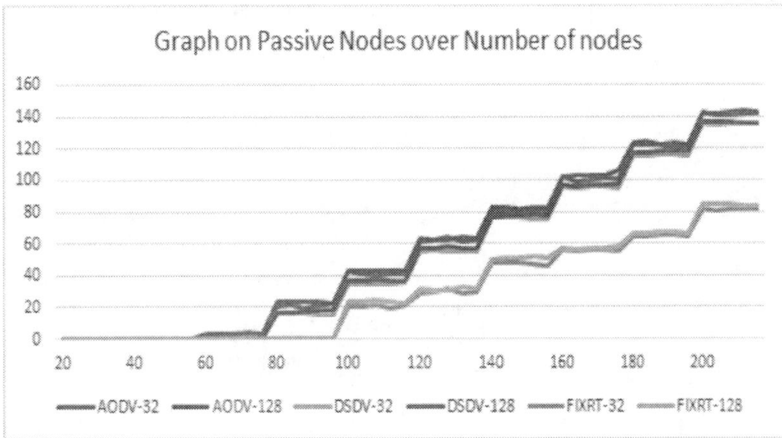

Figure 7. The graph on Passive Nodes over Number of nodes.

Packet or data loss will occur more frequent and this will reflect on the performance of the network. However, among the 3 routing protocols involved in the comparison, FIXRT showed the most promising result as it has the lowest value of routing overhead over number of nodes.

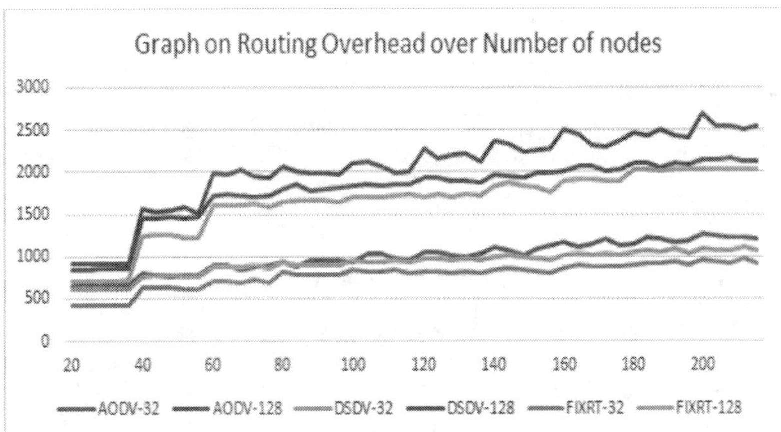

Figure 8. The graph on Routing Overhead over Number of nodes.

This will reflect on Figure 9 which indicates that AODV has lower fairness index compared to DSDV and FIXRT. Among the packet size, 32 packet size has lower fairness index compared to 128 packet size. With lower fairness index, higher passive nodes will be produced, and this will affect the delivery ratio and the overall performance of the network.

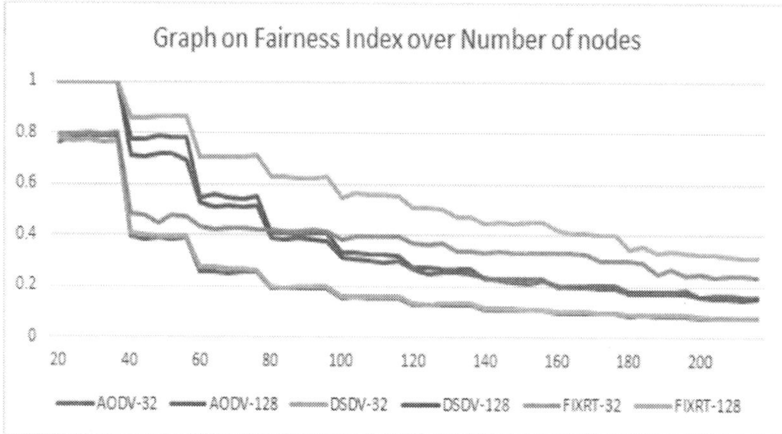

Figure 9. The graph on the Fairness Index over Number of nodes.

4. Conclusion

Wireless Sensor Network (WSN) technology has the potential to provide a proper and efficient monitoring system, especially in the oil and gas industry where all the information is critical to avoid unpredictable disaster. The main challenges faced in the industry are mainly energy consumption, security, reliability, robustness and scalability of the WSN.

As a conclusion, the major attribute that is being focused on this paper to evaluate the overall performance of the network is delivery ratio. This is due to the main function of the WSN is to deliver information. As presented above, the FIXRT has a higher value of delivery ratio compared to other protocols with almost 30% improvement. This result is supported with higher value of throughput and fairness index. The passive nodes and routing overhead for FIXRT is also lower compared to AODV and DSDV protocols.

References

[1] Khan WZ, Aalsalem MY, Gharibi W, Arshad Q. Oil and Gas monitoring using Wireless Sensor Networks: Requirements, issues and challenges. In2016 International Conference on Radar, Antenna, Microwave, Electronics, and Telecommunications (ICRAMET) 2016 Oct 3 (pp. 31-35). IEEE.

[2] Srbinovska M, Gavrovski C, Dimcev V, Krkoleva A, Borozan V. Environmental parameters monitoring in precision agriculture using wireless sensor networks. Journal of cleaner production. 2015 Feb 1;88:297-307.

[3] Parra L, Sendra S, Lloret J, Rodrigues JJ. Low cost wireless sensor network for salinity monitoring in mangrove forests. InSENSORS, 2014 IEEE 2014 Nov 2 (pp. 126-129). IEEE.

[4] Felemban E, Shaikh FK, Qureshi UM, Sheikh AA, Qaisar SB. Underwater sensor network applications: A comprehensive survey. International Journal of Distributed Sensor Networks. 2015 Nov 1;11(11):896832.

[5] Sheng Z, Yang S, Yu Y, Vasilakos AV, McCann JA, Leung KK. A survey on the ietf protocol suite for the internet of things: Standards, challenges, and opportunities. IEEE Wireless Communications. 2013 Dec;20(6):91-8.

[6] EIA U. Annual energy outlook 2015: with projections to 2040.

[7] Khan WZ, Hossain MS, Aalsalem MY, Saad NM, Atiquzzaman M. A cost analysis framework for claimer reporter witness based clone detection schemes in WSNs. Journal of Network and Computer Applications. 2016 Mar 1;63:68-85.

[8] Subramaniam SK, Nilavalan R, Balachandran W, Khan SM. Network performance optimization using Odd and Even routing algorithm for pipeline network. In2016 8th Computer Science and Electronic Engineering (CEEC) 2016 Sep 28 (pp. 118-123). IEEE.

[9] reza Akhondi M, Talevski A, Carlsen S, Petersen S. The role of wireless sensor networks (WSNs) in industrial oil and gas condition monitoring. In4th IEEE International Conference on Digital Ecosystems and Technologies 2010 Apr 13 (pp. 618-623). IEEE.

IOP Conf. Series: Materials Science and Engineering **765** (2020) 012070 doi:10.1088/1757-899X/765/1/012070

[10] Al-Ghamdi AJ, Abdullah A, Khan AM, Daraiseh A. Improving Safety in Oil and Gas Pipelines and Offshore Project Using Wireless Sensors Networks. InProceedings of the Pipeline Technology Conference 2010 (pp. 1-8).
[11] Ali S, Qaisar S, Saeed H, Khan M, Naeem M, Anpalagan A. Network challenges for cyber physical systems with tiny wireless devices: A case study on reliable pipeline condition monitoring. Sensors. 2015 Apr;15(4):7172-205.
[12] Subramaniam SK, Khan SM, Nilavalan R, Balachandran W. Enhancing Pipeline Network Performance Using Dual Interleaving Cluster Head Routing Protocol.
[13] Wang Q, Lin D, Yang P, Zhang Z. An Energy-Efficient Compressive Sensing-Based Clustering Routing Protocol for WSNs. IEEE Sensors Journal. 2019 Jan 21;19(10):3950-60.
[14] Guo X, Wang Y, Wei X. Design of wsn-based environment monitoring system on repair of gas leakage. InThe 27th Chinese Control and Decision Conference (2015 CCDC) 2015 May 23 (pp. 3340-3344). IEEE.
[15] Ali S, Ashraf A, Qaisar SB, Afridi MK, Saeed H, Rashid S, Felemban EA, Sheikh AA. SimpliMote: A wireless sensor network monitoring platform for oil and gas pipelines. IEEE Systems Journal. 2016 Aug 23;12(1):778-89.

Robust Visible Digital Stamp for Instant Documents Authentication and Verification

Hassan Al-Maksousy(1), **Hussein Abdulhussein**(2)
(1) Iraqi Prime Minister Office, Baghdad, Iraq
(2) University of Technology, Baghdad, Iraq
E-mail: (1)halma002@odu.edu, (2)50179@uotechnology.edu.iq

Abstract. Confidential documents, especially the ones issued by educational institutions, large companies, government departments, military organizations, and intelligence agencies are facing great challenges of being used, copied, and redistributed without authorization such as being used for forgery cases. This puts great threats to businesses and any legislature, executive, and judiciary forms in any community. In this paper, a novel model called robust visible digital stamp, RVDS, is introduced to secure and protect important / sensitive documents with the aid of low-power computers that do not require any type of network connections. The proposed model in this paper converts the document's information into a coded form of a customized quick response code, QR code. This code is then bound to the document for the verification process. This model uses a combination of keys for encryption and the authentication check process. The model does not need network connection, which makes it an off-line process that guarding users from the risk of losing their privacy.

Keywords: classified documents, documents authentications, QR code, network-independent authentication, cryptography, information encryption, data transmission

1. Introduction

Nowadays technological progress has made the flexibility to make the counterfeited documents much easier. Documents forgery is a challenging problem for any printed document [1]. Securing the important documents by adding combination of materials-based security features is possible. In many cases, counterfeiting is inexpensive and easy to be done, it only requires time for fraudsters to copy or manipulate these features. People may change information or put false ones into documents to create forged documents for example: government publications, real estate documents, graduation certificates, and identity documents. Documents forgery may help people getting unfair employment opportunities. In addition, some communities may be under a serious security threat if an unauthorized access to important security related documents is granted, and sensitive or life threatening details are leaked [2]. Usually some forms of forgery can be detected or prevented by using a conventional ways like stamps and signatures as from of authentication methods. However, in case of professional forgery where the forger has knowledge about the name of the document issuer and the way the documents are signed, detecting the forgery or any alteration is going to be harder [3, 4]. In addition to that, changing the designs, printing materials, and adding ink based features are used widely to prevent documents manipulation. Other methods like network authentication and verification systems will be affected by many factors such as software vulnerabilities and service availability time. In addition, using network dependent systems may affect users' privacy in addition to the cost of the network infrastructure or services providing. In this paper, a novel model called robust visible digital stamp, RVDS, is developed to make important / sensitive documents secured and protected with the aid of any computer that does not require network connections nor shared databases. The proposed model in this paper encrypts the document's information into a customized QR code which is then attached to the document for the verification process. This model uses a private one-time usage key for the encryption process and a public key for the verification process. The self-authenticated document will include both the document content and the QR Code, which makes it easier for the receiving party to verify the document as the QR Code holds all the document information, but in an encrypted form. What makes this method novel is that the process of converting the document contents into a code does not require databases or third party solutions involved to provide encryption / encoding keys. The model does not use any network-dependent processes, which makes it an off-line and secure process that helps maintaining users' information privacy. The proposed method in this paper provides an efficient approach in the data embedding algorithm to secure the documents content in the QR code form while keeping the QR code valid and readable. Different from the other methods, the proposed algorithm in this paper encrypts the documents with very low error and high capacity QR code which means higher amount of data being transferred and protected.

2. Related Work and History

In 1994, a team at Denso Wave invented and developed a 2D barcode called quick response code, QR Code, to help in the manufacturing process [5]. At that time, it was capable to only hold limited number of alphanumeric characters. It was then developed to pack as much information as possible. In 2000, QR Code became very popular in Japan. It was then approved as an international standard by ISO [6]. QR code is a fast and cost effective way of data transmission. It is commonly used on daily bases. Any QR code reader can easily decode the documents information that is embedded in. A paper-based document authentication is introduced in [7], yet their method did not add encryption to the original text. This leads to privacy / security breaches when delivering the embedded documents in QR code. Therefore, documents content have to be encrypted before embedding into the QR code. In addition to that, optical character recognition, OCR, is used to generate a hashed value, but it ignored right to left languages and other languages that are not supported by OCR. A paper-based documents authentication was proposed as a solution for encrypted authentication [8, 9]. However, their service model requires having a third party to handle verifying documents or credentials to be used by the algorithm. In [10], a system is used to store information in a central database. Meaning a non-offline solution is used while the original text is not encrypted nor compressed. Therefore, it cannot maintain data privacy nor can handle a lot of information. In addition to data / information privacy, the amount of data may affect the size of the used QR code. A color document authentication based on using QR code and embedding digital watermarking was implemented [11] where storing a signature requires multi QR codes. Their approach uses the color map of the document, which makes the size of the document image affects the size of the signature. In many cases, QR code can be tampered in order to change the documents related information. Dual anti-counterfeiting was introduced [12] to verify the legitimacy of users and documents copyrights by encoding encrypted user information in addition to embedding digital watermark, which they will be extracted and decrypted later to be compared together for document verification.

In this paper, the proposed method, RVDS, does not require having third party nor having any intermediate services between the sender and the receiver. In addition, the resulted QR code has almost always constant size. Also, in this paper, the verification process only involves scanning the QR code and extracting the original content as the QR code includes the digital signature and the document contents in compressed and encrypted form.

IOP Conf. Series: Materials Science and Engineering **765** (2020) 012071 doi:10.1088/1757-899X/765/1/012071

3. Method Framework and Motivation

The motivation behind designing the proposed method is to create a digital stamp that is printed within the actual hard copy document which makes authentication and verification are done without the help of any type of network related operations nor having any prior knowledge (also known as zero knowledge) about the document information. The proposed RVDS method staged into two parts, authentication stage, and verification stage. The stages are explained as in the following:

3.1. Authentication Stage

This stage is implemented at the documents issuer side at which the document is created, printed, and then stamped with QR code, such as governments' agencies, universities, and institutions require having their documents protected. This stage includes four main steps as shown in figure (1) and detailed as in the following:

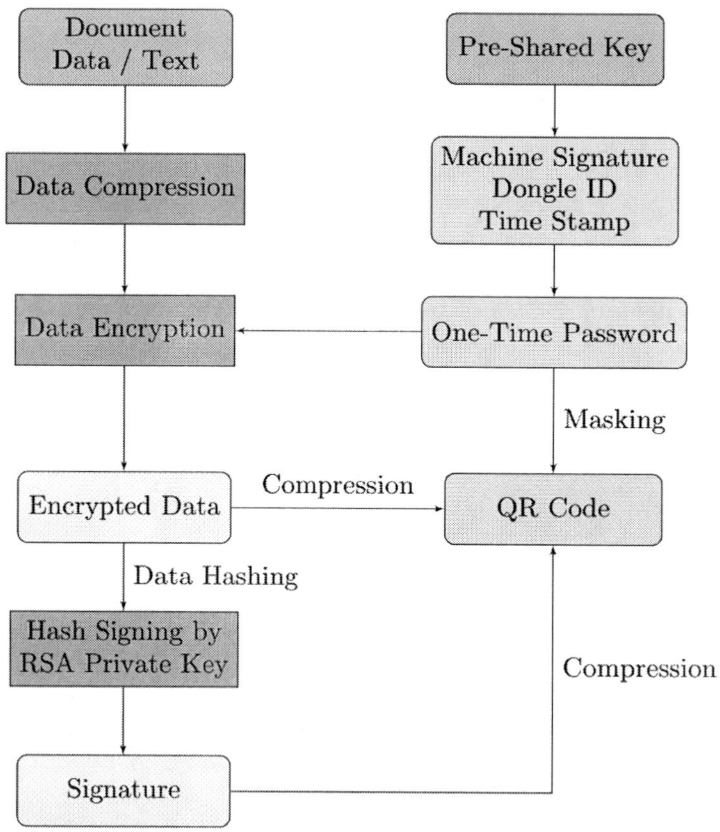

Figure 1: Authentication Stage (red: user entry, orange: machine controlled entry, yellow: intermediate output, green: stage output.)

3.1.1. Data Compression: It involves changing how data is represented so that the compressed results take less space to be stored, which means less time to process and transmit the information. In this step, documents information such as the main text for up to 450 words per page is compressed to reduce the actual data size that efficiently represents the entered document information. Such step can be done using any simple text optimized lossless compression algorithms. For this method, dictionary-based text reduction for limited common or popular

IOP Conf. Series: Materials Science and Engineering **765** (2020) 012071 doi:10.1088/1757-899X/765/1/012071

words is used, where each used text language has its own table. Other than that, Lempel-Ziv-Markov chain algorithm (LZMA) is used for text compression.

3.1.2. Data Encryption: It translates data into another configuration, where it can be only accessed and read by the parties having the password that is used in the encryption (decryption key). Securing important and sensitive data can be done in many methods, of which encryption is the most effective and popular one. In this step, the compressed text data will be encrypted using a unique encryption key. The encryption key is a one-time password because it consists of a unique combination of pre-shared key, machine signature (or identification number of the machine where the encryption process is being executed), and the time stamp (epoch time at which the encryption process is initiated). The result of this process will be an encrypted form of data consists of the compressed text and the machine signature. The encryption process can be mathematically represented by the following equation:

$$Y = E(K, X) \tag{1}$$

Where:

E: is the encryption algorithm

K: is the encryption key

X: is the text representation in compressed form

3.1.3. Data Signature: Data signatures are like fingerprints in electronic forms. In this step, the encrypted data are signed. However, the encrypted data has an arbitrary size. In order to get the data in a fixed size form, data mapping is implemented. This can be done by calculating the hash of the encrypted data. Any available hash function can be used depending on the complexity level; however, for accuracy and simplicity, the hash type that is used in this method is SHA-3 with standard size of 256 bit. Then RSA private key is used to encrypt the hash itself, which means signing the encrypted data. Equation (2) represents the mathematical representation of this step. 2048-bit RSA key size is recommended to be used until 2030 [13].

$$S = E_{RSA}(Hash, Prk) \tag{2}$$

Where:

E_{RSA}: is the signing/encryption algorithm

Hash: is the SHA-3 hash of the encrypted data

Prk: is the RSA private key

3.1.4. QR Code Generation: The QR code in this method has the encrypted data and the data signature embedded in. Meaning, the result of equation (1) is appended to result from equation (2). Together they form one text line. In addition to that, encryption key will be masked into the same text line. The final text line will be compressed using lossless data compression called Lempel-Ziv-Markov chain algorithm (LZMA) as 7-Zip. This makes decrypting the document data much faster at the receiver party and it does not require network access for the encryption key. The compressed information is then converted to QR code form using Zebra Crossing, ZXing, library. This QR code is printed anywhere on the document as required by the issuer. The QR code in this method is considered as a digital stamp that replaces the traditional ink based stamps.

IOP Conf. Series: Materials Science and Engineering **765** (2020) 012071 doi:10.1088/1757-899X/765/1/012071

3.2. Verification and Decryption Stage

This stage is implemented at the documents receiver side at which the document is checked to be verified for authenticity purposes. This stage is usually done at the time where the received documents have QR code printed on them and submitted for verification such as degree certificates submitted for job applications. This stage includes two main steps: verification, and decryption step. These steps are detailed as in the following:

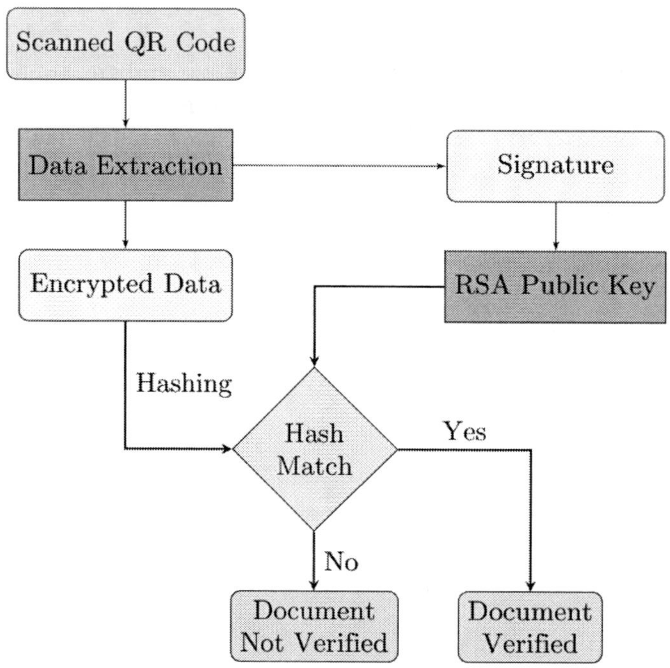

Figure 2: Verification Step (red: user entry, orange: machine controlled entry, yellow: intermediate output, green: stage output.)

3.2.1. Verification: In this step, the verification process is done by extracting encrypted data and the signature from the scanned QR code. Then RSA public key is used to decrypt the hash value from the extracted signature. The resulted hash should represent the original encrypted data. If the resulted hash is the same as the hash for the extracted encrypted data, the document is verified and has not been manipulated in any means. The flow chart of this step is shown in figure (2). To sort this step mathematically, equations (3) and (4) are applied to the extracted QR content (encrypted data and the signature only). If they yield the same results (H_1=H_2), the document is considered verified and valid.

$$H_1 = Hash(EData) \tag{3}$$

Where:

Hash: is the SHA-3 hash function

EData: is the extracted encrypted data

IOP Conf. Series: Materials Science and Engineering **765** (2020) 012071 doi:10.1088/1757-899X/765/1/012071

$$H_2 = D_{RSA}(Sig, Puk) \quad (4)$$

Where:

D_{RSA}: is the decryption algorithm

Sig: is the extracted signature

Puk: is the RSA public key

3.2.2. Data Decryption: After the document is being processed in the verification step, a second process, data decryption, is done. This process requires having the original encryption key used to decrypt the original entered data. In other words, the data was encrypted with one-time password (called encryption combinations) which is a combination of pre-shared key, machine signature, and time stamp. However, these codes are embedded into the QR code in masked forms. So, in addition to the extracted encrypted data and the signature, encryption combinations are extracted and unmasked to be used as key for the decryption process. The decryption process yields the compressed data that should match the original entered data at the issuer party. The flow chart of this step is shown in figure (3).

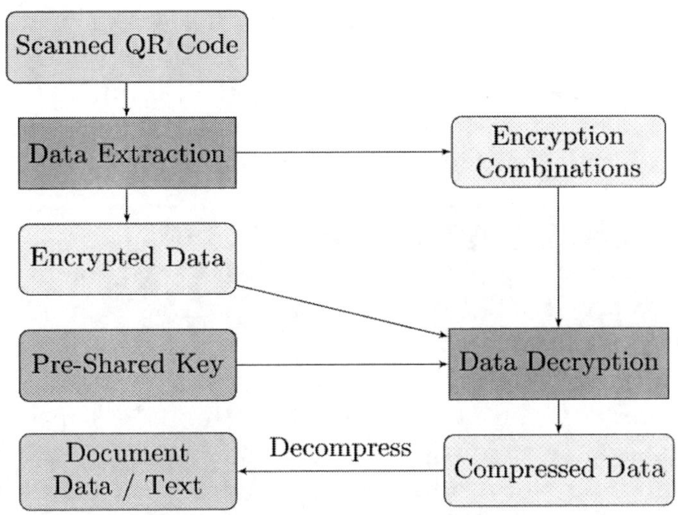

Figure 3: Data Decryption Step (red: user entry, yellow: intermediate output, green: stage output.)

4. Experimental Results

The algorithm is developed using C# programming language and .NET framework. Since the majority of the institutions are using word processor such as Microsoft Word to issue the documents, the prototype is designed to insert the text and the QR code into their Microsoft Word template. Experiments done at the documents issuer side were all conducted by a machine with specifications of: Intel Celeron N3350 2.4 GHz CPU, 2GB RAM, 32GB eMMC flash memory, and 32-bit Windows 10 operating system. On the recipient side, the machine however can be either a low-power computer as in the issuer side or any commercially available off-the-shelf (COTS) devices such as mobile devices with Android operating system (the program can be cross compiled into different platforms). Also, two languages were used during the test, English and Arabic. Figure (4) shows the graphical user interface (GUI) of the proposed method while being used in authentication stage.

IOP Conf. Series: Materials Science and Engineering **765** (2020) 012071 doi:10.1088/1757-899X/765/1/012071

Figure 4: RVDS Program in Authentication Stage: user needs to enter the pre-shared key and the document content only.

Iraq's GDP growth slowed to 1.1% in 2017, a marked decline compared to the previous two years as domestic consumption and investment fell because of civil violence and a sluggish oil market. The Iraqi Government received its third tranche of funding from its 2016 Stand-By Arrangement (SBA) with the IMF in August 2017, which is intended to stabilize its finances by encouraging improved fiscal management, needed economic reform, and expenditure reduction. Additionally, in late 2017 Iraq received more than $1.4 billion in financing from international lenders, part of which was generated by issuing a $1 billion bond for reconstruction and rehabilitation in areas liberated from ISIL. Investment and key sector diversification are crucial components to Iraq's long-term economic development and require a strengthened business climate with enhanced legal and regulatory oversight to bolster private-sector engagement. The overall standard of living depends on global oil prices, the central government passage of major policy reforms, a stable security environment post-ISIS, and the resolution of civil discord with the Kurdish Regional Government (KRG).

Figure 5: Report Authenticated by RVDS with the QR code posted on paper.

In this application, the only required pieces of information are the clerk name (sometimes can be replaced with the pre-shared key) and the document content. The result will be as QR code form of the encrypted document information posted on the document as shown in figure (5). For statistical and future comparison purposes, the QR capacity (number of words per page), language type, and the size of printed QR code are calculated for each document in the experiments. It can be noticed in table (1), where one language is used, that the size of the QR code is almost always the same and it goes up to 4.2x4.2 cm. However, using multi languages at the same time may increase the QR code size up to 5x5 cm, shown in table (2).

Capacity (words/page)	QR Size (cm^2)
100	2.4^2
200	2.5^2
300	3.2^2
400	3.9^2
450	4.2^2
500	N/A

Table 1: One language (Arabic or English)

Capacity (words/page)	QR Size (cm^2)
100	2.5^2
200	2.7^2
300	3.6^2
400	4.1^2
450	4.9^2
500	N/A

Table 2: Two languages (Arabic & English)

5. Protection Evaluation

The proposed method has a very high accuracy in distinguishing between tampered and genuine documents. It is a significant contribution in documents tamper detection since it uses hash functions to represent all the document information. Thus, any minor alteration even if it is done by one letter, the hash followed by the QR code would all be different than the original ones. This paper also helps in forensics document analysis. Figure (6) shows some results of the original and forged sentence detected by RVDS.

The quick brown fox jumps over the lazy dog

(a) Original Sentence

The quick brown foxes ← jumps over the lazy dog

(b) Forged Sentence

Figure 6: Genuine and Tampered Sentences: the difference between them is letters (ex) which are added to word (fox) as an intentional modification.

6. Conclusion

Documents authentication and verification are very important. There are many cases where documents can be altered, leaked, or accessed without proper authorization. Thus, RVDS is proposed as a solution to secure and protect important documents. This method is robust and very fast, it does not require having network access which makes it mobile and offline solution. In addition, one of the main contributions in this paper, the QR code (the stamp) has almost always the same size no matter how the data changes. This method creates digital authentication stamp

IOP Conf. Series: Materials Science and Engineering **765** (2020) 012071 doi:10.1088/1757-899X/765/1/012071

that is printed within the actual document. Verification is done instantly with zero knowledge about the document information. This method does not require having any form of network related operations nor involving any third party solutions to provide encryption / decryption keys transfers.

References

[1] Garain U, Halder B. On automatic authenticity verification of printed security documents. In: Proceedings - 6th Indian Conference on Computer Vision, Graphics and Image Processing, ICVGIP 2008; 2008. p. 706--713.

[2] Abeyratne R. Security Issues. In: Strategic Issues in Air Transport. Berlin, Heidelberg: Springer Berlin Heidelberg; 2012. p. 165--278.

[3] Lowe R, Walske C, Royal Philatelic Society (Great Britain). The work of Jean De Sperati II : including previously unlisted forgeries. Royal Philatelic Society; 2001.

[4] Verma R, Rao DS. Offline Signature Verification and Identification Using Angle Feature and Pixel Density Feature And Both Method Together; 2013. 4.

[5] DENSO CORPORATION TCRII. Two-dimensional Code; 1994. Patent JP2938338B2.

[6] ISO. ISO/IEC 18004:2000 - Information technology - Automatic identification and data capture techniques - Bar code symbology - QR Code. ISO Standards. 2000;2000:122.

[7] Warasart M, Kuacharoen P. Paper-based Document Authentication using Digital Signature and QR Code. 4TH International Conference on Computer Engineering and Technology. 2012;40(January):94--98.

[8] Li CM, Hu P, Lau WC. AuthPaper: Protecting paper-based documents and credentials using Authenticated 2D barcodes. In: IEEE International Conference on Communications. vol. 2015-Septe. Institute of Electrical and Electronics Engineers Inc.; 2015. p. 7400--7406.

[9] Ahmed HA, Jang JW. Document certificate authentication system using digitally signed QR code tag. In: ACM International Conference Proceeding Series. Association for Computing Machinery; 2018. .

[10] Singhal A, S Pavithr R. Degree Certificate Authentication using QR Code and Smartphone. International Journal of Computer Applications. 2015 jun;120(16):38--43.

[11] Mohsin Arkah Z, Alzubaidi L, Ali AA, Abdulameer AT. Digital color documents authentication using QR code based on digital watermarking. In: Advances in Intelligent Systems and Computing. vol. 940. Springer Verlag; 2020. p. 1093--1101.

[12] Xun Y, Li Z, Zhong X, Li S, Su J, Zhang K. Dual Anti-counterfeiting of QR Code Based on Information Encryption and Digital Watermarking. In: Lecture Notes in Electrical Engineering. vol. 543. Springer Verlag; 2019. p. 187--196.

[13] Barker E. Recommendation for Key Management – Part 1: General. NIST Special Publication 800-57. 2016;p. 1--142.

Image Compression Using High Level Wavelet Transformer with Non-Uniform Quantizer and Different Levels Huffman Codes

Rajaa Khalaf Gaber, Ahmed Abdulqader Hussein, Manal Kadhim Oudah and Ahmed Hameed Reja

Electromechanical Engineering Department, University of Technology,Baghdad,Iraq

E-mail: 50049@uotechnology.edu.iq,ahmedabdulqaderhussein@gmail.com, 50030@uotechnology.edu.iq,ahmad8171@yahoo.com

Abstract. Image compression solutions are a pivotal field that gain over a remarkable research interest and called upon to extend widely with respect to the increasing of its applications. The considerable challenge of this solution is related to the storage and bandwidth requirements. A highly compression rate with a high image resolution is considered to be the key factors for implementing a robust image compression technique. In this paper, the proposed method is deemed to be simple and effective due to using different levels of compression process which is starting from dividing the image into sub-images and using high level discrete wavelet transform (DWT). In addition, the Huffman code with non-uniform Quantizer has been used to minimize the compression data rate with high quality reconstructed image. The promising simulation results of the proposed method achieved the most two significant criteria which are denoted by high image quality and a high image compressed ratio. For instance, the proposed strategy simulation results of Lena image achieved PSNR equals to 42.4094 dB with a compression ratio (CR) equals to 47.5435. Moreover, the simulation results of Flower image accomplished PSNR equals to 42.8348 dB with a compression ratio equals to 55.3389 under the same simulation conditions.

Keywords— Color Image Compression, Wavelet Transform, Huffman Code, Non-uniform Quantizer, Compression Ratio.

1. Introduction

Compression alludes to diminish the amount of information used to epitomize an image or video contents without lopsidedly decreasing the nature of the first image. The primary motivation behind image compression is to lessen the repetition and insignificance present in the image, so it very well might be stored and transmitted efficiently. A less number of bits which are matching to the original image are denoted as the compressed image [1].

There are numerous compression procedures, for instance, transform coding quantization, differential encoding, predictive image coding. However, transform coding is the most proficient particularly at low bit rate. The main transformation techniques are utilized discrete cosine transform (DCT) and discrete wavelet transform (DWT) [2].

The compression methods are classified into two types: the reconstructed image with loss which is called a lossy compression technique and reconstructed image without loss specified by a lossless compression technique. In lossy compression pattern, a relative degradation with respect to the original image has been occurred during the reconstruction process of the image. Along these lines, this strategy

IOP Conf. Series: Materials Science and Engineering **765** (2020) 012072 doi:10.1088/1757-899X/765/1/012072

leads to image quality degradation either during the compression or decompression phase. Generally, lossy solutions outfit maximal compression ratios as compared to the lossless strategies [3, 4].

Regarding the JPEG standard image coders, compression process based on Wavelet technique can be surely outperform the traditional DCT solution. The performance of the image coder that relied on wavelet mechanism is depending on the considerable range of the selected wavelet family. In spite of standard wavelets are completed well with photographic images, most of the images do not possess a similar statistical properties such as medical images, fingerprints, scanned documents and satellite images etc. [1, 5]. The objective criteria as peak signal to noise ratio (PSNR) and mean square error (MSE) measures degrade the reconstructed image quality. The Compression Ratio (CR) is used as compress image measures degrade [6].

In this paper the lossy compression techniques have been proposed. The proposed method used high levels DWT (Harr Transform) for sub-image, then threshold and quantized coefficients by variable non-uniform quantizer. After that, the process values coded by variable Huffman codes to reduce data information redundancy.

2. Proposed Compression Method

The paper proposed a new compression of RGB color image. The encoding principle can be explained by the block diagram as shown in Figure 1. The proposed compression method exploits three types of redundancies in order to compress the image. The proposed method has been implemented in a view of accomplishing an optimal point among the CR and PSNR for the desired images.

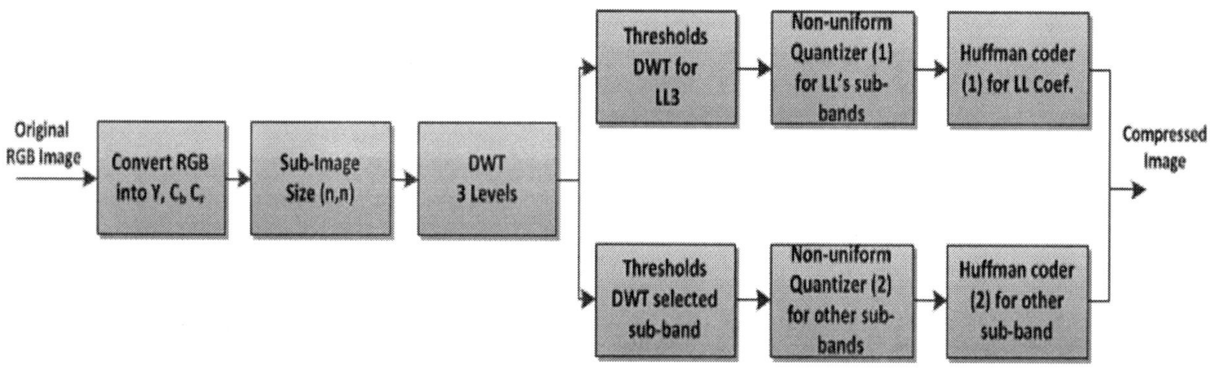

Figure 1. Image Compression Block Diagram of the Proposed Method.

2.1. Convert color image (RGB to $Y\,C_b\,C_r$)

The color image has more than one component Red, Green, and Blue (RGB) planes of a color image each component is individually shifted. The three components dcor related using a nonreversible linear combination of the components. The RGB image convert to Luminesces, Chrominance Blue, and Chrominance Red (Y C_b C_r) color video have been transformed according to the equations 1, 2 and 3 [5].

$$Y(i,j) = 0.299R(i,j) + 0.587G(i,j) + 0.114B(i,j) \tag{1}$$

$$C_b(i,j) = -0.16875R(i,j) - 0.331G(i,j) + 0.5B(i,j) \tag{2}$$

$$C_r(i,j) = 0.5R(i,j) - 0.1869G(i,j) - 0.0813B(i,j) \tag{3}$$

The goal of the transformation is to improve compression efficiency. The transformed components Y, C_b and C_r, are difference images whose histograms are highly peaked around zero [6].

2.2. Wavelet Decomposition

The Discreet Wavelet transform (DWT) can decompose images into various multi-resolution sub bands [6]. Basically, the low frequency coefficients contain the most energy of the image, while the high frequency coefficients reflect the significant visual information.

The vital key role of the decomposition procedure is separating the low frequency coefficients from the high frequency to a small number of coefficients. Meanwhile, the remaining coefficients have been quantized indecently or minimized to zero with insignificant image distortion. Sub bands for 3-levels Haar filter as a basic function of the wavelet decomposition can be shown in Figure 2 [7, 8].

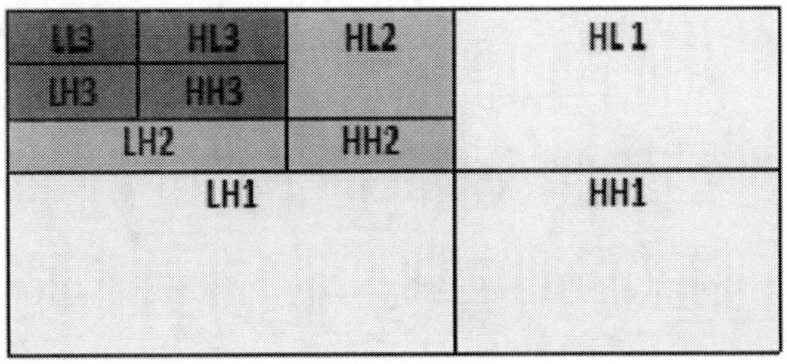

Figure 2. 3-level Wavelet Decomposition Sub-bands.

The energy is mostly concentrated in LL3 sub band and the other sub bands represent the details bands in level-3 are highly correlated with the edge estimated in the guesstimate band. The new propose method decomposes three colors (Y,C_b and C_r) into 3-level Harr transform. The first level of compression has been done by select LL3 sub bands from C_b and C_r and LL3, LH3, HL3, LH2, and HL3 sub bands from Y. when the other sub band has been neglected [9].

2.3. DWT Filtering

The selected sub bands for the three components of the proposed strategy has been achieved by the threshold V_{Th} which is determined by Equation 4 :

$$X(i,j) = \begin{cases} X(i,j) \mid X(i,j) \mid\geqslant V_{Th} \\ 0 \qquad\quad \mid X(i,j) \mid< V_{TH} \end{cases} \tag{4}$$

The threshold step decreases the coefficient variance and correlated the values. Regarding to the new proposed method, it has been selected two different values of thresholds, one for threshold LL3 sub bands while the other one for other sub bands [10].

2.4. Non uniform Quantizer

The non-uniform quantizer design has been implemented in order to select the best quantized levels (s_i) and the reconstructed level (t_i) as well. This approach has been designed for accomplishing a particular optimization criterion and input probability density function P(s) of the non-uniform quantizer as shown in Figure 3. Moreover, the optimization criterion which is specified by a statistical or psych visual measure is defined the minimization of the mean-square quantization error [5, 11].

The non-uniform quantizer Gaussian distribution has been designed to quantized selected coefficients, this step minimize data variables with increased the frequency of data. In this paper two different quantizers has been designed first one for quantized LL3 sub bands coefficients while the second one for other sub bands coefficients.

IOP Conf. Series: Materials Science and Engineering **765** (2020) 012072 doi:10.1088/1757-899X/765/1/012072

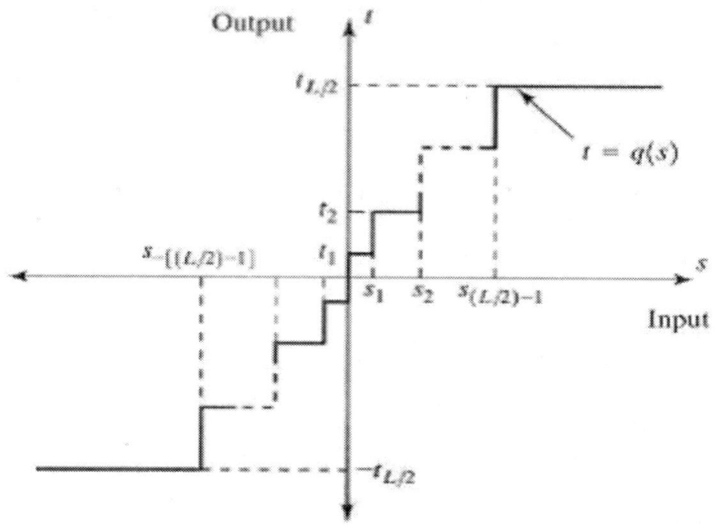

Figure 3. Non-uniform Quantization Function[4].

2.5. Huffman Encoder

This kind of coding strategy is essentially relied on frequency measurement of an information variable. The rule behind this method is based on the data bit occurrence during the encoding process. In particular, high frequent data bits are encoded with a less number of bits. Meanwhile, low frequent data bits are encoded with less number of bits as well. Moreover, pivotal role of the Huffman code dictionary is to implement every data symbol with a codeword. The essential part of this property has been achieved during the encoding procedures in which that each codeword carries a prefix. Keeping in mind that each codeword will carry a different prefix in the Huffman code dictionary [12].

Definitely, Huffman codes should be conducted as an input to encoders. Along these lines, Huffman tables form is epitomized with respect to the indirect specification of the data stream. However, the construction of Huffman tables is carried out by the decoder in priori of the image decompression process. The average codeword LC denoted the average Huffman code per samples which is calculated according to Equation 5 [5, 9].

$$L_c = \sum_{i=1}^{N} l_i P(X_i) \tag{5}$$

Where:

l_i is refers to the samples codeword length.

$P_{(X_i)}$ is refers to the samples probability.

3. Proposed Decompression Method

The implemented decompression block diagram of the proposed method is illustrated in Figure 4. The Huffman dictionary has been implemented in order to reconstruct the coder samples. The reconstructed image received after 3-level inverse Discrete Wavelet Transform (IDWT) by full zeros for every unselected sub bands in three components and regrouped the sub-images. The inverse color module has been implemented (convert Y, C_b, C_r into RGB) to reconstructed original image. The reconstructed image compare with original image by calculate CR and PSNR. The proposed compression and decompression algorithms which have been implemented in this work are illustrated in Algorithms 1 and 2 respectively.

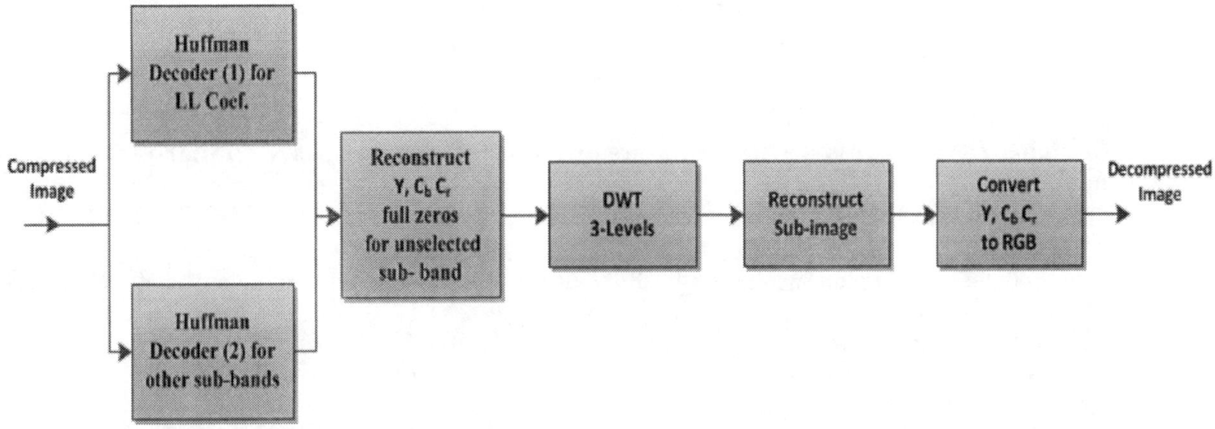

Figure 4. Image Decompression Block Diagram of the Proposed Method.

Algorithm 1 Image Compression Algorithm

Input: Color Image File (RGB Image)
Output: Compressed Image File
Begin
Open Image File
- Convert color module RGB into Y,C_b,C_r
- Divide the three components into sub-block (n,n)
- Apply 3-level wavelet decomposition to the sub-image blocks
- Select wavelet sub band (LL3 for three components Y,C_b,C_r) and (LH3, HL3, LH2,and HL2 for Y component)
- Filter the selected LL3 sub band by V_{Th1} and other sub band by V_{Th2} according to Equation (4)
- Choose threshold VTh (mean of values in wavelet band)
- Threashold values X(x,y), repeat

  if X(i,j) $\geq V_{Th}$ **then**
    $X(i,j) = X(x,y)$
  else
    $X(i,j) = 0)$
  end if

- Quantized the filter selected values by two non-uniform quantizers
- Design two Huffman coder dictionary for every groups, and calculate L_c according to Equation(5)

End
Close Original Image File
Close Compressed Image File
END

IOP Conf. Series: Materials Science and Engineering **765** (2020) 012072 doi:10.1088/1757-899X/765/1/012072

Algorithm 2 Image Decompression Algorithm

Input: Compressed Image File
Output: Color Image File (RGB Image)
Begin
Open Compressed Image File
- Decoding the compressed data using first Huffman coder dictionary to find LL3 for three components Y,C_b,C_r
- Decoding the compress data using second Huffman coder dictionary to find LH3, HL3, LH2, and HL2 for Y component
- Regroup the reconstruct sub-image by full unselected sub band by zeros to reconstruct three components
- Convert color module Y,C_b,C_r to RGB
- Calculate The CR and PSNR
End
Close Compressed Image File
Close Original Image File
END

4. Design Metrics

4.1. Fidelity Criteria

There are two types of image fidelity criteria namely, the objective and subjective criteria [8]. The first one depends on equations that are used to measure the amount of the error in the reconstructed image, PSNR is measured in db according to the Equation 6 [9].

$$PSNR(dB) = 10log_{10}\left(\frac{(L-1)^2}{\frac{1}{M*N}\sum_{i=1}^{R}\sum_{j=1}^{C}[Y(i,j)-X(i,j)]^2}\right) \tag{6}$$

Where:
(L-1) refers to the maximum gray level (in Gray-level image equal to 255).
$X(i,j)$ is the orignal pixel.
$Y(i,j)$ is the reconstructed pixel.
M and N are the image number of rows and columns respectively.

While the second require the definition of qualitative scale to assess image quality and this scale can then be used by human test subjective to decide image fidelity [7].

4.2. Compression Ratio (CR)

The Compression ratio is defined as the ratio between the original image sizes compared with compressed image size as represented in Equation 7 [11].

$$CR = \frac{OriginalSize}{CompressedSize} \tag{7}$$

5. Simulation Results of the Proposed Method

The implementation of the proposed image compression strategy has been accomplished through utilizing the Matlab software. A four different RGB color image datasets of size (512*512*3) which are specified by the images (Lena, Flower and street) has been used in order to achieve the new proposed image compression method.

The new image compression approach steps for image Lena as shown in Figure 5(a to h) have been implemented as (a) denotes the Original RGB Lena image (b) convert color module to (Y C_b C_r), (C) refers to the 3-level wavelet transform for sub-image (256*256), (d) represents the 3-level wavelet transform for sub-image (512*512), (e) denotes the Selected wavelet sub band and filter the selected component by V_{th1}=20 and V_{th2}=20 (for sub-image 256*256), (f) represents the Selected wavelet sub band and filter the selected component by V_{th1}=20 and V_{th2}=20 (for sub-image 512*512). The first non-uniform quantizer of 64 levels when the second of 16 levels, (g) denotes the Reconstructed image and (h) is the image different between original and reconstructed images. The promising results of the proposed algorithm have been evaluated via using the fidelity criteria and the compression ratio metric as well. The obtained simulation results of the proposed technique in case of the threshold value equals to 20 (V_{th}=20) are CR=47.5435 and PSNR= 42.4095dB for sub-image size of (256*256).

In addition, the results of the other dataset images (Flower and street) of size (512*512) have been illustrated in Figure 6. The implemented images group depict the original image, the reconstructed image and the differences between them. Meanwhile, the simulation results of the Flower and street images have been performed under the same parameters and conditions which have been implemented for the Lena image. These simulation parameters can be specified by the threshold V_{th1}=20 , V_{th2}=20 and for the same quantizer level as well .

Figure 7 shows the effect of changing V_{th1} and V_{th2} at PSNR (dB) for images Lena, Flower, and street at fixed Sub-image (256*256). The PSNR has been reduced with the increasing of V_{th}.

In accordance with the results of Figure 7, the simulation results for the Flower image with increasing the V_{th} from 5 to 60, it can be found that the PSNR drops from 43.0636 dB to 42.4742 dB (decreased by 0.589 dB). Moreover, the obtaining results in case of Lena image, the PSNR drops from 42.6155 dB to 41.8512 dB and in case of Street image, the PSNR drops from 40.442 dB to 40.03725 dB. It is clear that the PSNR of the Flower image has better performance as compared to the obtained simulation results for the images of Lena and Street as well.

However, colored images are transferred to various color space and PSNR is notified versus each channel of that color space. Meanwhile, each image related to the implemented dataset contains different color spaces and details as well. Therefore, acquired PSNR values differ for each image and for any proposed algorithm. In addition, the impact of increasing the value of the threshold has been discussed in this work. It has been observed that increasing the threshold values will slightly decrease the obtained PSNR. Keeping in mind that the obtained results are considered to be within the acceptable values for wireless transmission quality loss. Table 1 summarizes the obtained PSNR results with respect to the changing of the threshold values.

Table 1. The Proposed Method Simulation Results Specified by PSNR (dB) with respect of Different Threshold Values.

Threshold Value	**PSNR (dB) Lena Image**	**PSNR (dB) Flower Image**	**PSNR (dB) Street Image**
5	*42.6155*	*43.0636*	*40.442*
10	*42.5669*	*42.9623*	*40.4247*
20	*42.4095*	*42.8348*	*40.371*
30	*42.2191*	*42.7239*	*40.3006*
40	*41.964*	*42.6108*	*40.2237*
50	*41.9577*	*42.5386*	*40.1332*
60	*41.8512*	*42.4742*	*40.03725*

Through following the same strategy, Figure 8 shows the effect of changing V_{th1} and V_{th2} from 5 to 60 at the evaluated performance of the compression ratio (CR) for images Lena, Flower and street at fixed Sub-image (256*256). The compression ratio of the proposed strategy has been improved with the increasing of V_{th}.

Regarding the simulation results, the compression ratio (CR) in case of Flower image has been improved from 46.4079 to 58.7711 (improved by 12.363 times). In addition, the compression ratio in case of Lena image has been enhanced from 38.2517 to 54.0139 and in case of Street image the compression ratio has been improved from 36.1681 to 54.6932. Obviously, the performance evaluation of the proposed approach in terms of compression ratio for the Flower image outperforms the two other images.

In light of view the promising simulation results of the proposed strategy as shown in Figure 8, it is noticed that the compression ratio (CR) has been increased with the increasing of the threshold values. Meanwhile, Flower image achieves higher compression ratio as compared with other dataset which is indicated with respect to Lena and Street images respectively. Table 2 summarizes the obtained compression ratio with respect to different threshold values.

Table 2. The Proposed Method Simulation Results Specified by Compression Ratio (CR) with respect of Different Threshold Values.

Threshold Value	Compression Ratio (CR) Lena Image	Compression Ratio (CR) Flower Image	Compression Ratio (CR) Street Image
5	*38.2517*	*46.4079*	*36.1681*
10	*42.8644*	*51.7057*	*40.5151*
20	*47.5435*	*55.3389*	*45.5103*
30	*50.2768*	*56.8622*	*48.8384*
40	*51.9641*	*57.9009*	*51.2793*
50	*53.166*	*58.4417*	*53.1917*
60	*54.0139*	*58.7711*	*54.6932*

Along these lines, the impact of changing the threshold values on the performance of the proposed strategy has been specified clearly. A very high compression ratio will lead to a lower PSNR. Therefore, the selection of the proper threshold value should be taken into consideration due to its impact on the PSNR and compression ratio as well. Since the compatibility of the obtaining result metrics is considered to be the pivotal role, a harmonious among the compression ratio and the PSNR has been accomplished for implementing the proposed algorithm.

The performance metrics of the proposed method with respect to Lena image has been compared with the state of art works based on PSNR and CR criteria. The comparison of the simulation results have been performed under the same conditions. For instance, the threshold value equals to 20 , image size of (256*256) and sub-image size of (256*256) as well. Table 3 illustrates the simulation results comparison of the proposed mechanism with the other state of art relevant works.

Table 3. Comparison of the Proposed Strategy with Other Image Compression Schemes.

Method	PSNR (dB)	Compression Ratio (CR)
Hashim AT. et al. \cite{13}	38.28	31.33
Sreelekha G. et al. \cite{14}	34.38	27
Zhang C. et al. \cite{15}	33	23.7
Kahu SY. et al. \cite{16}	33.6	81.3
Proposed Method	**42.4095**	**47.5435**

Regarding to the compared results of Table 3, the PSNR values of the proposed approach obtains around 42.41 dB which is outperformed the PSNR of the compared state of art works [13, 14, 15, 16] by about 4.13 dB, 8.03 dB, 9.41 dB and 8.81 dB respectively. It can be observed that the proposed method achieves better performance than the compared works.

On the other hand, the compression ratio (CR) of the proposed strategy is approximately 47.54 which is outperformed the compression ratio of the compared state of art works [13, 14, 15] by about 34%, 43.2% and 50.1% respectively. However, the state of art work [16] has been accomplished better performance specified by the compression ratio (CR) than the proposed method by about 41.5%.

Keeping in mind, the relevant work [16] obtains high compression ratio with a low PSNR value which is affected on the quality of the compressed image. Whilst, the proposed approach achieves a promising results with a compatible between the PSNR and the compression ratio. Hence, the compatible results of the proposed algorithm accomplishes good compression ratio with an acceptable quality of the compressed image

IOP Conf. Series: Materials Science and Engineering **765** (2020) 012072 doi:10.1088/1757-899X/765/1/012072

6. Conclusion

An effective image compression strategy for different color image datasets has been proposed in this work. The implementation of the proposed algorithm has been achieved based on several steps. The first step can be described by dividing the image into sub-images. The second step is utilizing 3-level DWT technique which is filtered by variable threshold values. Finally, the third step can be specified by coding the result through the Huffman code after being quantized by non-uniform quantizer . The proposed method provides satisfactory high compression ratios with no significant degradation of image quality. The efficiency of this approach has been justified through using a set of real images. Regarding the experimental results it is evident that the proposed compression method gives a better performance, with increasing the threshold and sub-image size. In addition, the proposed image compression technique maintains a better image quality with a less complexity. Finally, the simulation results of the proposed algorithm have been compared with the state of art relevant works under the same simulation conditions. It has been concluded that the performance of the proposed strategy outperforms the performance of the other works with a harmonious achievements between the image quality and the compression ratio as well.

References

[1] Joshi MA, Raval MS, Dandawate YH, Joshi KR, Metkar SP. Image and video compression: Fundamentals, Techniques, and Applications. Chapman and Hall/CRC; 2014 Nov. 17.

[2] Gangwar RK, Kumar M, Jaiswal AK, Saxena R. Performance analysis of image compression using fuzzy logic algorithm. Signal and Image Processing: An International Journal (SIPIJ) Vol.(5), 2014.

[3] Taubman D, Marcellin M. JPEG2000 image compression fundamentals, standards and practice: image compression fundamentals, standards and practice. Springer Science and Business Media; 2012 Dec 6.

[4] Vidhya K, Karthikeyan G, Divakar P, Ezhumalai S. A Review of lossless and lossy image compression techniques. Int. Res. J. Eng. Technol.(IRJET). 2016;3(4):616-7.

[5] Umbaugh SE. Digital Image Processing and Analysis: Applications with MATLAB and CVIPtools. CRC Press; 2017 Nov 30.

[6] Kavitha T, Sankar KJ. An efficient compression technique for ITU-T group 3 coded images using variable length codes with reduced average length. In2016 1st India International Conference on Information Processing (IICIP) 2016 Aug 12 (pp. 1-6). IEEE.

[7] Rajan PV, Fred AL. An Efficient Compound Image Compression Using Optimal Discrete Wavelet Transform and Run Length Encoding Techniques. Journal of Intelligent Systems. 2019 Jan 28;28(1):87-101.

[8] Sikka N, Singla S, Singh GP. Lossless image compression technique using Haar wavelet and vector transform. In2016 International Conference on Research Advances in Integrated Navigation Systems (RAINS) 2016 May 6 (pp. 1-5). IEEE.

[9] Ghorpade A, Katkar P. Image Compression Using Haar Transform And Modified Fast Haar Wavelet Transform. International Journal of Scientific and Technology Research. 2014;3(7):302-5.

[10] Alsayyh MA, Mohamad D, Saba T, Rehman A, AlGhamdi JS. A Novel Fused Image Compression Technique Using DFT, DWT, and DCT. Journal of Information Hiding and Multimedia Signal Processing. 2017;8(2):261-71.

[11] Scarmana G. Lossless data compression of grid-based digital elevation models: A PNG image format evaluation. ISPRS Annals of the Photogrammetry, Remote Sensing and Spatial Information Sciences. 2014 Jun 24;2(ISPRS Technical Comm):313-9.

[12] Saroya N, Kaur P. Analysis of image compression algorithm using dct and dwt transforms. International Journal of Advanced Research in Computer Science and Software Engineering. 2014 Feb;4(2).

[13] Hashim AT, Ali SA. Color Image Compression Using DPCM with DCT, DWT and Quadtree Coding Scheme. Engineering and Technology Journal. 2016;34(4 Part (B) Scientific):585-97.

[14] Sreelekha G, Sathidevi PS. An HVS based adaptive quantization scheme for the compression of color images. Digital Signal Processing. 2010 Jul 1;20(4):1129-49.

[15] Zhang C, He X. Image compression by learning to minimize the total error. IEEE Transactions on Circuits and Systems for Video Technology. 2012 Aug 3;23(4):565-76.

[16] Kahu SY, Bhurchandi KM. JPEG-based Variable Block-Size Image Compression using CIE La* b* Color Space. KSII Transactions on Internet and Information Systems. 2018 Oct 1;12(10).

IOP Conf. Series: Materials Science and Engineering **765** (2020) 012072 doi:10.1088/1757-899X/765/1/012072

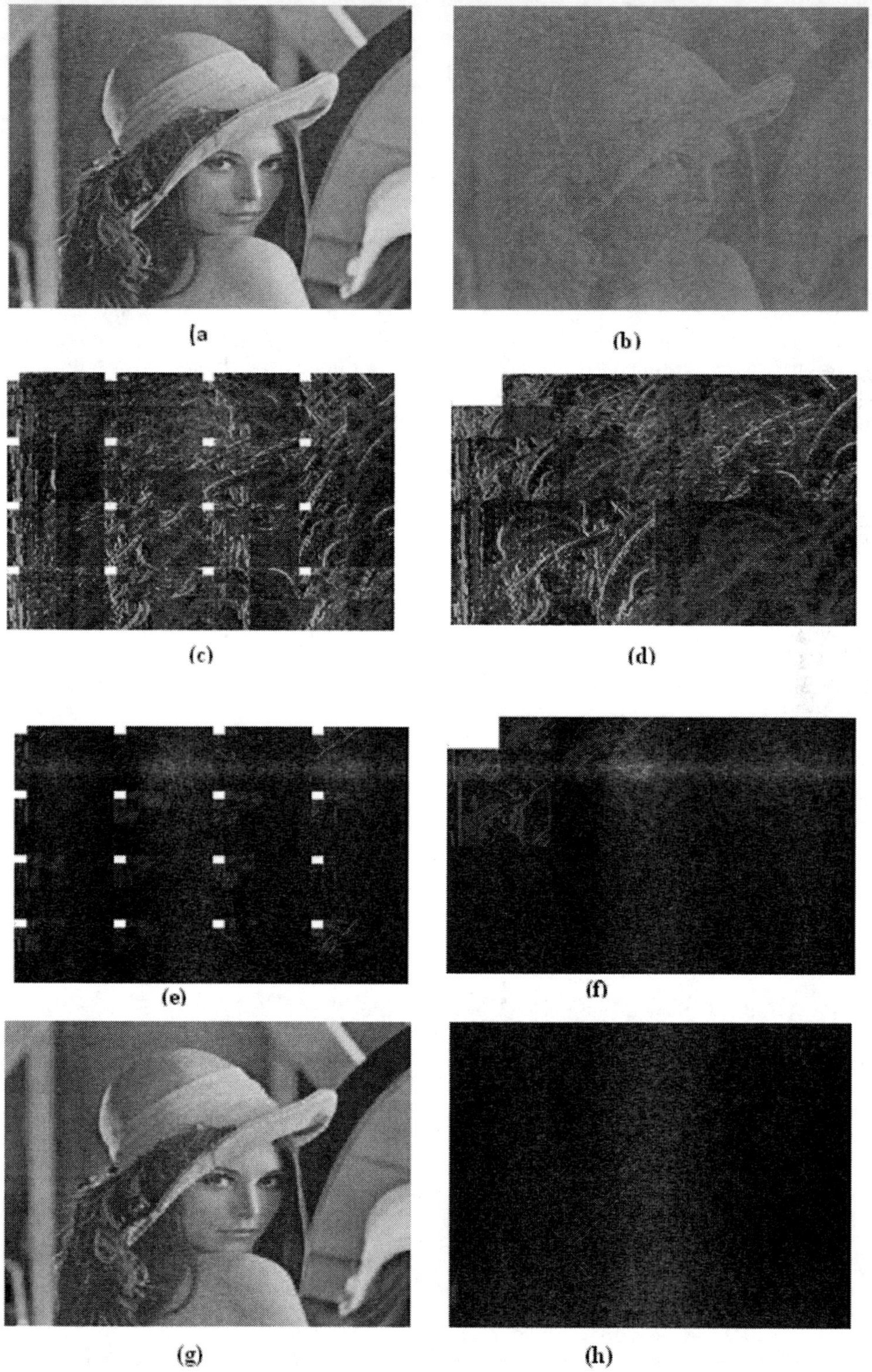

Figure 5. The Propose Method Image Compression Results CR=47.5435 PSNR= 42.4095 dB. a) Original Image Lena of Size (512*512). b) Converting Color Module. c) 3-level DWT (256*256). d) 3-level DWT (512*512). e) DWT after Select and Threshold (256*256). f) DWT after Select and Threshold (512*512). g) Reconstructed Image. h) The Differences between Original Image and the Reconstructed Image.

Figure 6. The Propose Method Image Compression Results for Different Color Images of Size (256*256), the Sub-image Size (256*256), with Threshold Values V_{th1}=20 and V_{th2}=20. a) Flower Image PSNR=42.8348 dB and CR=55.3389. b) Street Image PSNR= 40.371dBand CR=45.5103.

IOP Conf. Series: Materials Science and Engineering **765** (2020) 012072 doi:10.1088/1757-899X/765/1/012072

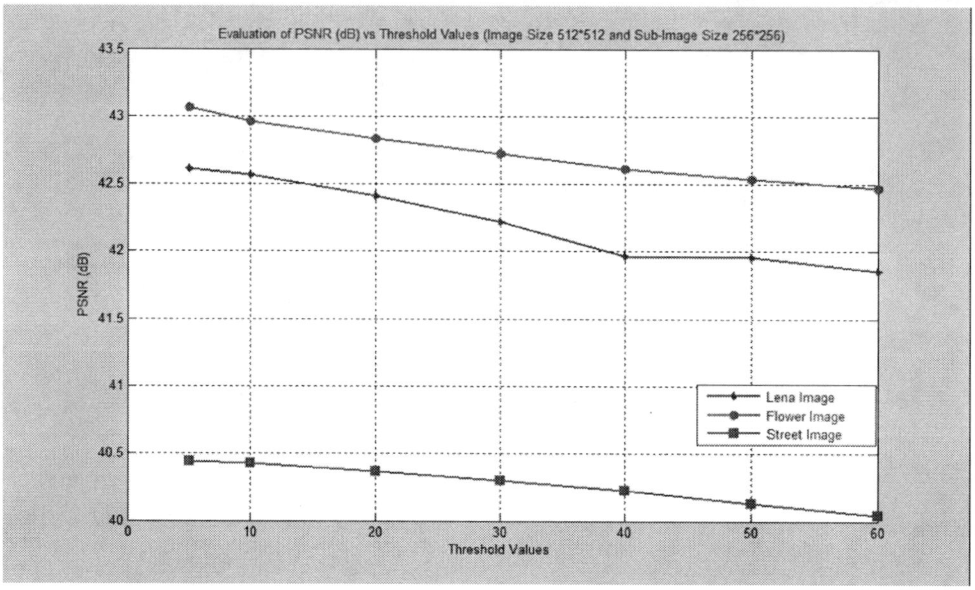

Figure 7. The Effect of V_{th} on the PSNR for Three Images (Lena, Flower, and street) of Size (512*512) and Sub-image Size (256*256).

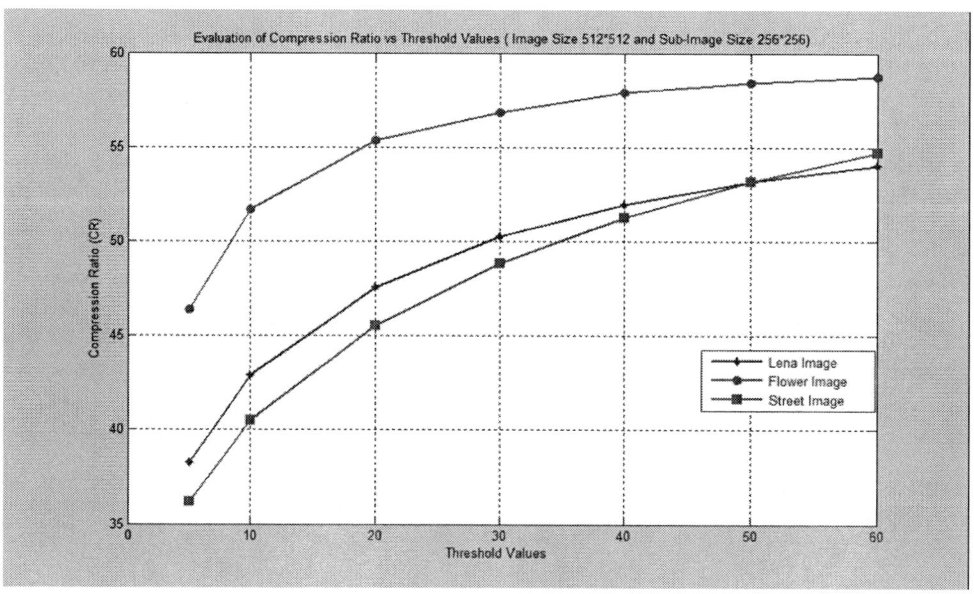

Figure 8. The Effect of V_{th} on the Compression Ratio (CR) for Three Images (Lena, Flower, and street) of Size (512*512) and Sub-image Size (256*256).

IOP Conf. Series: Materials Science and Engineering **765** (2020) 012073 doi:10.1088/1757-899X/765/1/012073

Design and Software Implementation of Analog and Digital Filters For Coherent Receivers under Noisy Channels

Ahmed A. Thabit

Departmentof Communications and computer Engineering, AL-Rafidain University College. Baghdad, Iraq

E-mail: ahmed_a@coalrafidain.edu.iq

Abstract. One of the most important stations in any communications system is the filtering station. Due to the noise power that effects on the signal, the system must include filtering process to eliminate this effect. In this paper, analog and digital filters are designed and implemented using Matlab tools to solve the problems occurred during receiving the signal that effects on the synchronization such as those for coherent receivers. The offsets of Phase and frequency are major problems in the identification of signals that results reduction in the performance; such as, whenever higher order modulations are used. Two systems are designed and simulated. The first one is analog system and the second one is digital system. 3rd order Butterworth filter is selected for system one and digital IIR filter is selected for system two. Filter order was selected according to diagnosis the level of noise power and the transmitted frequency to meet certain frequency domain design specifications. Obtained results showed excellent performance in the receiving and demodulating the received signal. Evaluation of the results is basically depend on comparison between the transmitted and received samples.

Keywords— Butterworth, IIR, PLL, VCO .

1. Introduction

The signal processing is a mathematical operation to modify or derive the response which basically depends on the operation of the filters. The filtering processes are used so as to produce a linearity invariant time system used to perform spectral shaping [1]. This process is largely used in schemes, like to remove the unwanted noisy signals from the signals, shaping of the spectral like the adaptation of channels, signal detection and identification in cognitive radio, communications and for obtaining the analysis of the spectrum for different signals [2]. Signal processing may be programming and reprogrammed with no need to great change in hardware. Signal processing include properties like the accuracy, perfect re-producibility, the flexible and good quality. Generally, DSP signal is derived from the analogy signal that sampled within a certain time to change it into digitally form [2].

The filter is the heart of any communication system at the signal processing stage. As mentioned early, there are two filters are used in this paper: Butterworth and IIR filters. The first one (Butterworth filter) is designed due to its flat frequency response in the passband. It is first discovered at 1930 due to British engineer and the physicist Stephen. It is known as flat magnitude. It has no ripple. Its becoming flat from passband to stopband. As increasing number of filter order (n), the phase response of the Butterworth filter becomes more nonlinear. Several filter orders along with the ideal frequency response which is known as a brick wall. If the filter order nth increase, then brick wall response and filters gets closer as shown in Figure 1 [3, 4, 5].

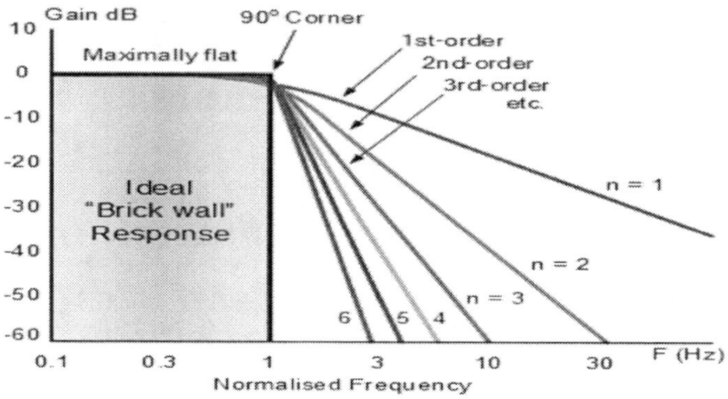

Figure 1. Frequency response of Butterworth filters [3].

Equation 1 is frequency response of low pass Butterworth filter.

$$G(w) = |H(jw)| = \frac{1}{\sqrt{1+w^6}} \tag{1}$$

The filters are very important schemes in DSP design in order to change frequency properties of the signal so as to produce the design properties. Digital filters considered to remove noise, spectrum shaping and interference reduction in communication architectures [6, 7]. Generally, the classes of the filters in the digital form are:

(i) FIR: Finite Impulse response filter
This filter has impulse response is in finite duration, because it approximate to zero in specified time. This filter is designed and implemented usually using numbers of adders, multipliers, delay and adders. FIR filters also called as non-recursive feed forward. It has a structure which may be used to implement most of any sort of freq. response digitally. The unit of impulse response is finite; so FIR filters are stable system [6, 7].

(ii) IIR: Infinite Impulse response filters
IIR filters can be designed and implemented as analog or digital form. In the digital forms, the output feedback is immediately seen on the equations defining the output [8, 9]. Digital IIR filters Design is greatly depend on analogy counterparts because there are plenty of works and straightforward design methods concerning analogy feedback filter design. Unlike FIR filters, in the design it is necessary to be carefully consider the time zero case in which the filter outputs has not yet clearly defined [10, 11, 12].

The designed filters in this paper submitted to coherent receivers to recover the original transmitted signal. As mentioned early, this paper is divided into two parts: the first one related to analog system and the second one is the digital system.

2. Proposed system design

2.1. System 1

In this system, the Butterworth filter is used together with the phase locked loop (PLL) to recover the signal that changed due to the phase and frequency offsets; however, the PLL is a feedback loop has the ability to recover phase and frequency of the received signals. The PLL is combined from: multiplier, voltage controlled oscillator and the loop filter. Figure 2 shows the basic construction of PLL [11, 12].

In Figure 2, the input to the multiplier is the received noisy signal and the output of the vco. This multiplier is simply an XOR gate that compare between two input to produce the output, vco is an oscillator that controlled by input voltage [14, 15, 16, 17]. The loop filter is what we want to design in this paper. The Butterworth is preferred here because it is not has a ripple. In general, two PLL is used

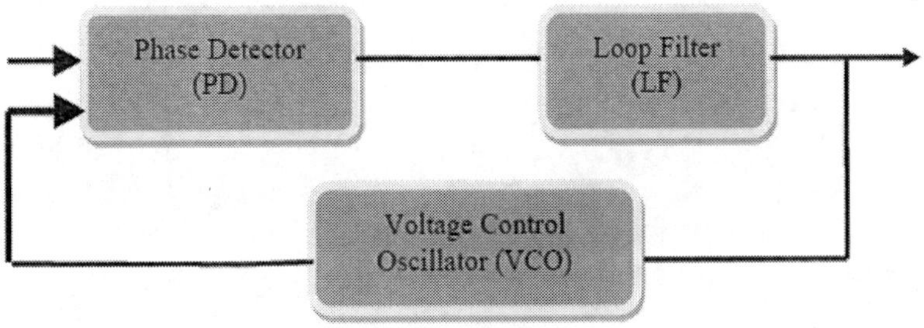

Figure 2. PLL basic construction [12].

here to accommodate the complex received signal; therefore, those two PLL will separated by shifting 90obetween the output of vco as shown in Figure 3.

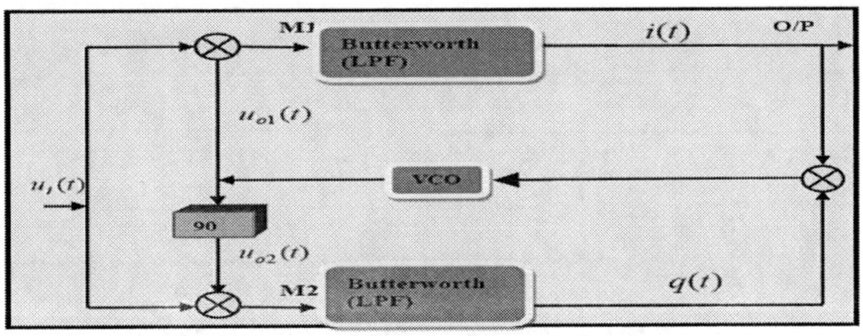

Figure 3. Combinational of two PLL [15].

Where: $i(t)$, $q(t)$ are In phase and Quadrature phase respectively $u_i(t)$ is the received noisy signal. $u_{o1}(t)$ and $u_{o2}(t)$ are the output of VCO for I and Q channels respectively. M1, M2 are the multipliers on I and Q channels respectively.

The block diagram of Figure 3 consists of two PLL shifted by 90o. The two arm filters is selected to be Butterworth filter. Those filters (I and Q arms) are selected carefully to be the exactly the same using filter design and Analysis Tool (FDA) in Matlab program. The design parameters of the proposed system and the characteristics of each filter are shown below in Table 1:

Table 1. Characteristic of the proposed system.

Filter type	**Butterworth filter**
Filter order	Third Order
Cutoff frequency	10 KHz
Modulation type	BPSK
Channel type	AWGN

The structure of the detailed proposed system and the designed filter based on Matlab program (i.e. FDA tool) is shown below on Figure 4.

2.2. Simulation Results of System 1

The simulation results obtained from Figure 4 are shown below:

IOP Conf. Series: Materials Science and Engineering **765** (2020) 012073 doi:10.1088/1757-899X/765/1/012073

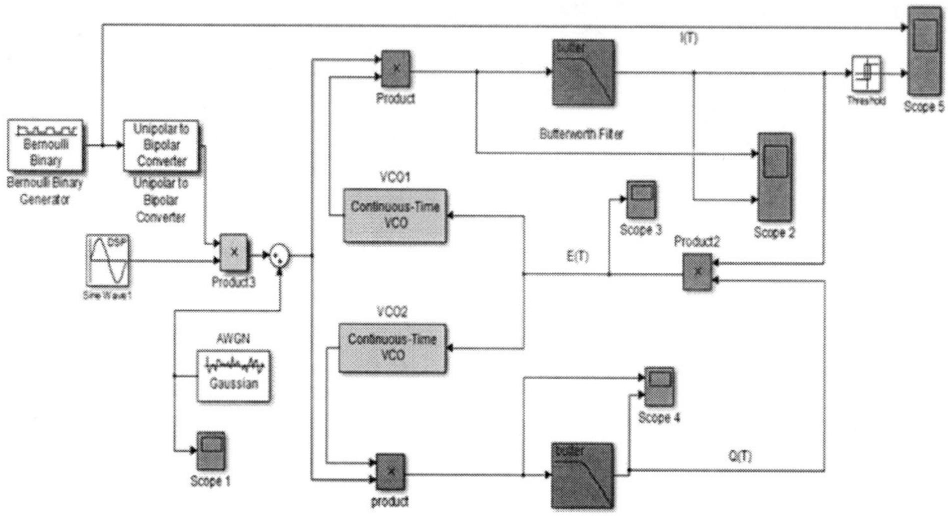

Figure 4. The Block of the proposed system 1.

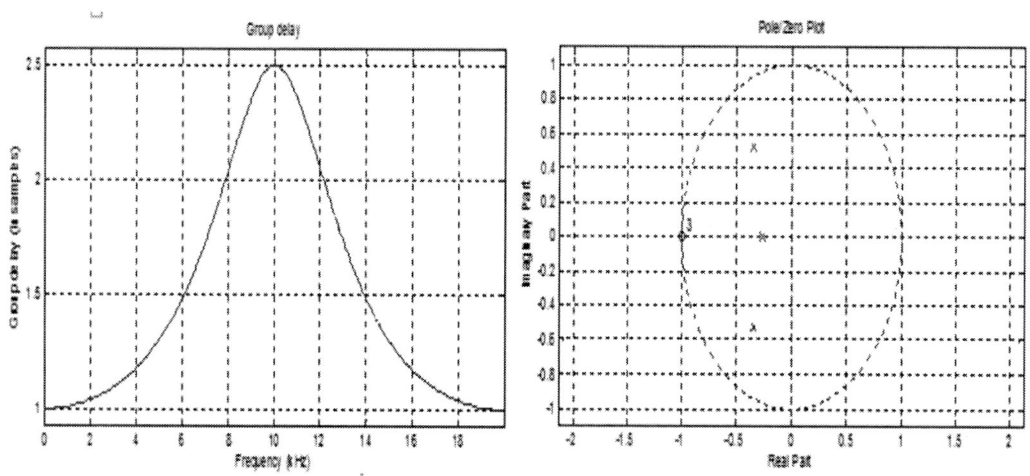

Figure 5. Group delay and pole/zero plot.

First of all, the general characteristics of the designed filter are seen from the Figures 5 and 6. Figure 6 represents the magnitude response of this filter. It is seen that at -3dB the cutoff frequency is 10 KHz as planned before. Figure-6 is the group delay and the pole/zero location plot respectively. The designed filter is stable as shown in the lower plot of Figure 5.

AWGN is additive to the signal as shown in Figure 7 that shows the generated AWGN on the scope 1.

Figure 8 represents input/ output of the Butterworth filter that designed in this paper. As mentioned above, the order of this filter is 3rd order for both sides (I and Q) to meet the requirements for the coherent receiver that proposed in this paper. The order of the filter is selected according to the experimental results to obtain the best output at specified cutoff frequency.

As seen from Figure 8, the upper plot represents the signal that produced from the multiplication between the output of the first VCO and the received signal in the I-side (before filter), whereas the lower plot represents the output signal from the filter that appear on scope 2. The quality of the PLL

IOP Conf. Series: Materials Science and Engineering **765** (2020) 012073 doi:10.1088/1757-899X/765/1/012073

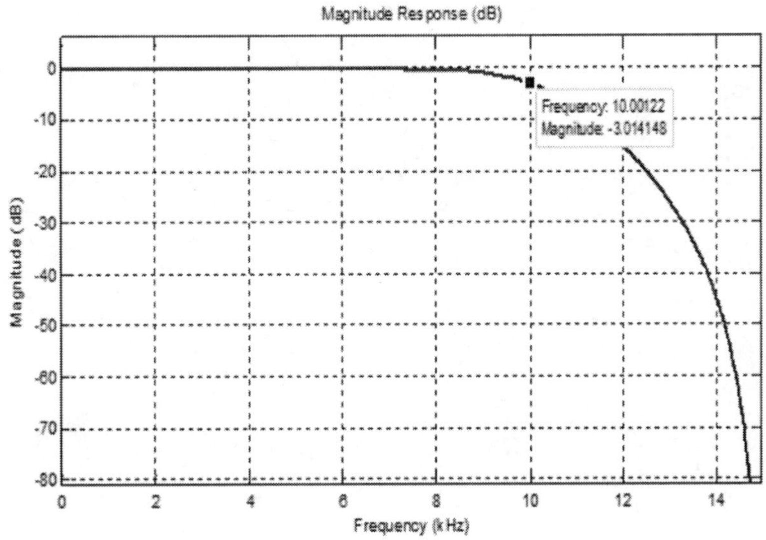

Figure 6. Magnitude response of 3rd Butterworth filter.

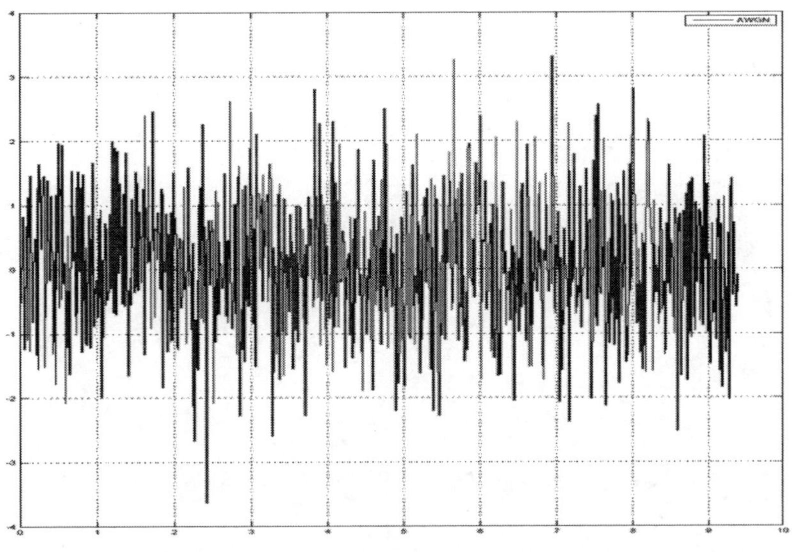

Figure 7. The generated AWGN.

is evaluated according to ability of the proposed block to track the phase offset and freq. offset of the transmitted signal; however, at the time when output is measured on I-side, the Q-side is approximately is zero. This idea can be shown in Figure 9 that plot the output of the Butterworth filter in Q-side which is approximately is zero.

The error that generated on scope (3) of Figure 4 is stand for the DC offset voltage which is the input to the VCO. This error is preferred to be zero at the lock state that determines that the true signal is received. This error is seen on Figure 10.

The designed system shown in Figure 4 represents coherent receiver for the BPSK signal. The basic job of this system is recover the phase and frequency offset occurs due to the synchronization errors. The heart of this system is the filter; however, the real test for the performance evaluation of this system is the comparison between the transmitted and the received (demodulated) symbols of the proposed system.

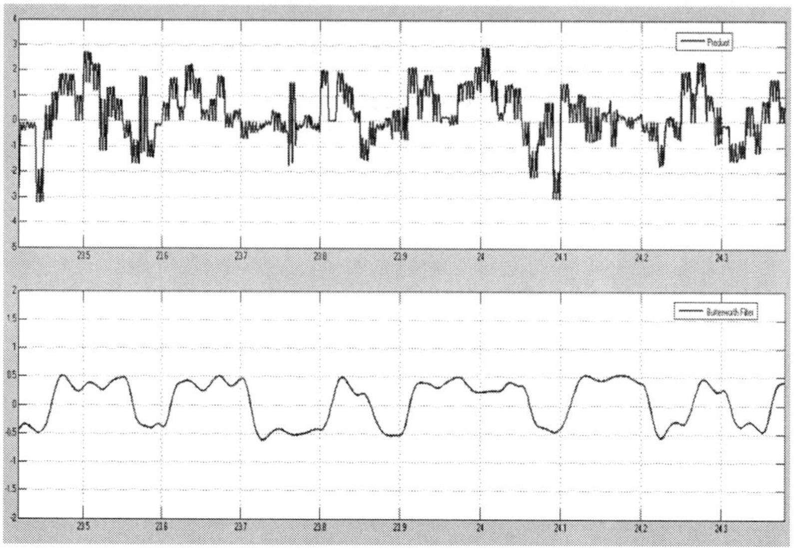

Figure 8. Input and output of the Butterworth filter.

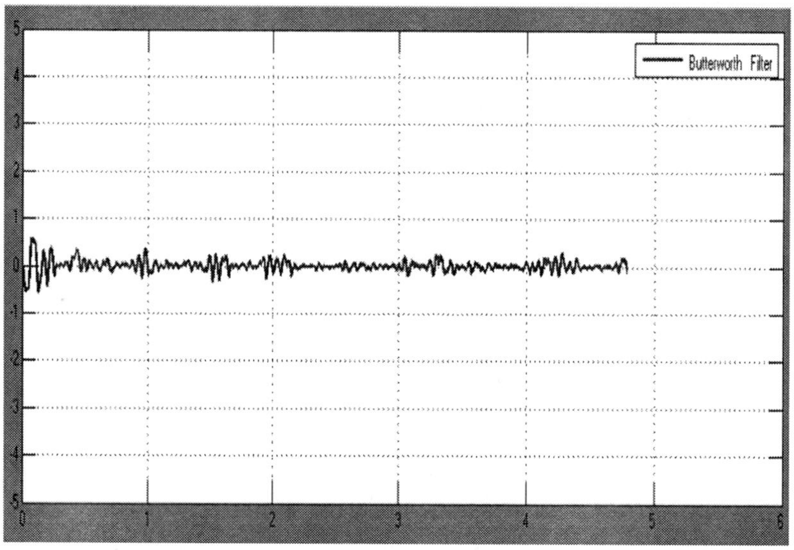

Figure 9. The output of filter on Q-side.

Figure 11 represents the comparison between the transmitted and the output of the system.

As clearly seen in red colour from Figure 11, the system need a time to reach and get the fully synchronization between the receiver and the transmitter. The system will stay in lock state and therefore the error is zero as shown in Figure 12 after rerun the system, the output is fully synchronized with the transmitted signal. This fully compatible was not obtained without the designed filter.

2.3. system 2 (digital form)

The same block diagram that proposed in the system 1 (Figure 4), is designed except replacing the filter by digital IIR filter and discrete VCO instead of continuous VCO. The designed was based on Matlab program tool (FDA tool). IIR filter is designed according to Table 2.

Figure 13 represents the proposed system 2 that consists of digital IIR filter, discrete VCO and digital

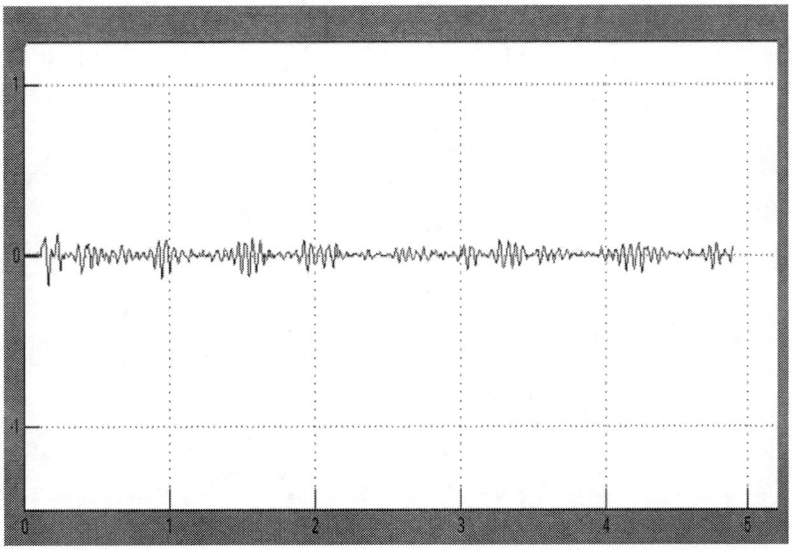

Figure 10. Obtained error between I and Q sides.

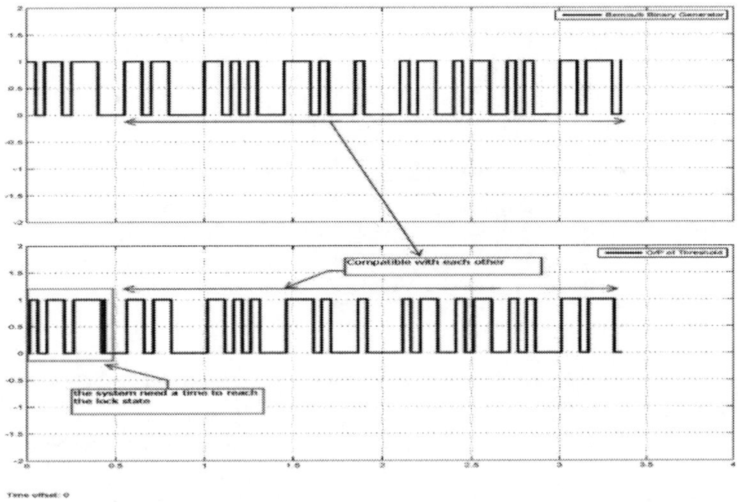

Figure 11. Comparison between input and output.

received signal.

This diagram is implemented by Matlab program and the values of each block is selected to obtain the results as shown in Table 2. However, the simulation results are shown in the next section.

2.4. Simulation results of system 2

The generated sample is modulated using carrier signal. The resultant BPSK is additive by AWGN to reach the coherent receiver that will have the ability to recover the original signal without noise and then finally demodulate the original samples. The core of this diagram is the infinite impulse response (IIR) filter that is low pass filter. The characteristics of designed IIR filter is shown in Table 3.

The structure of this IIR filter is basically based on Z-Transform that consists of numbers of delays as shown in Figure 14.

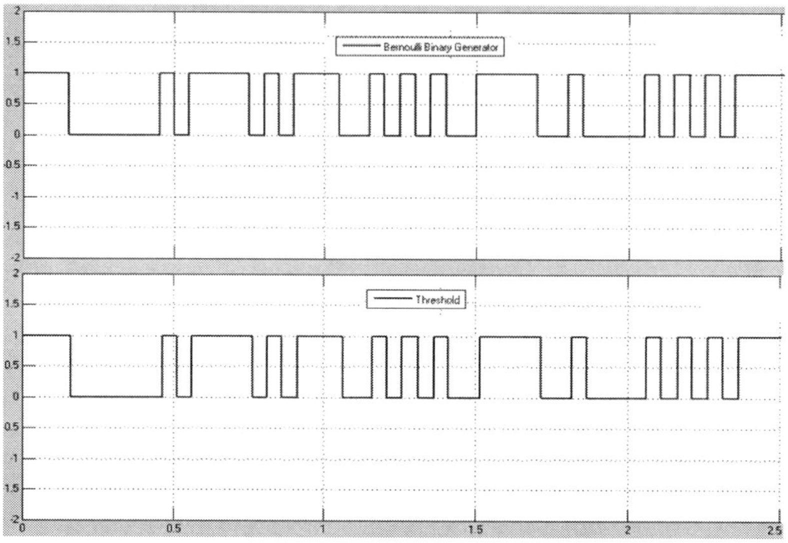

Figure 12. Compatible between input and output.

Table 2. Characteristic of proposed system 2.

Parameters	**Type / value**
Filter type	IIR digital filter
Filter order	3rd
Cutoff frequency	10 KHz
Modulation type	BPSK
Channel type	AWGN
Sampling frequency	30 KHz
VCO	Discrete VCO
Sensitivity of VCO	100 Hz/v

The filter is stable filter consists of two section as shown in Figure 14, is 3rd order at cut off frequency of 10 KHz. The magnitude response is shown in Figure 15.

Figures 16 and 17 represent the group delay and the zero pole plot respectively.

It is very important to check the response of the filter and evaluate the performance. Figure 18 shows this test. The upper represents the received signal before filtering, the lower plot represents the filtered signal, it is seen that the filter is working well.

As mentioned early in this paper, when the lock state is happen, the Q-side will be approximately zero. This is can be seen in Figure 19 that represent the signal after the filtering process of the Q-side. The red rectangle means that the system need a time to get the lock state and the error to be zero.

Final and real test for the performance of the proposed system is to compare between the generated and the recovered samples. This shown in Figure 20.

As expected, the system need a time to reach the lock state. This clearly seen in the start of lower part of Figure 20 which indicated by red rectangle. Finally, it is necessary to see the freq. spectrum of the received signal as shown in Figure 21.

3. Conclusion

As a results from the designed systems of this paper it is easily to see the good response and good filtering process. Designed filters and selection the orders of the filters is very important to obtain the required

IOP Conf. Series: Materials Science and Engineering **765** (2020) 012073 doi:10.1088/1757-899X/765/1/012073

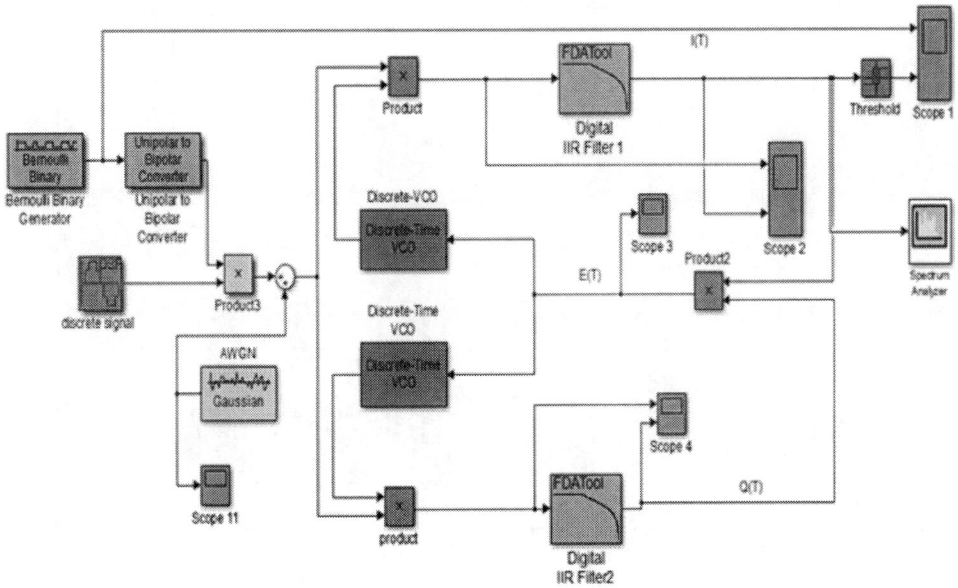

Figure 13. Block diagram of the proposed system 2.

Table 3. Characteristics of designed IIR

Discrete Time IIR Filter (real)	
Filter Structure	Direct Form -II , Second Order Sections
Number of Sections	2
Stable	Yes
Linear Phase	No
Scale Norm	No Scaling
System Object	False
Design Specifications	
Sample Rate	N/A (normalized frequency)
Response	Lowpass
Specification	N, Fp, Ap
Filter Order	3
Passband Edge	0.4
Passband Ripple	1 dB
Measurements	
Sample Rate	N/A (normalized frequency)
Passband Edge	0.4
3-dB Point	0.42779
6-dB Point	0.45816
Stopband Edge	Unknown
Passband Ripple	1 dB

output. In communication system, receiving the same phase and frequency of the transmitted signal is very important. Therefore, the PLL is a device that must be founded in the system to recover the signal.

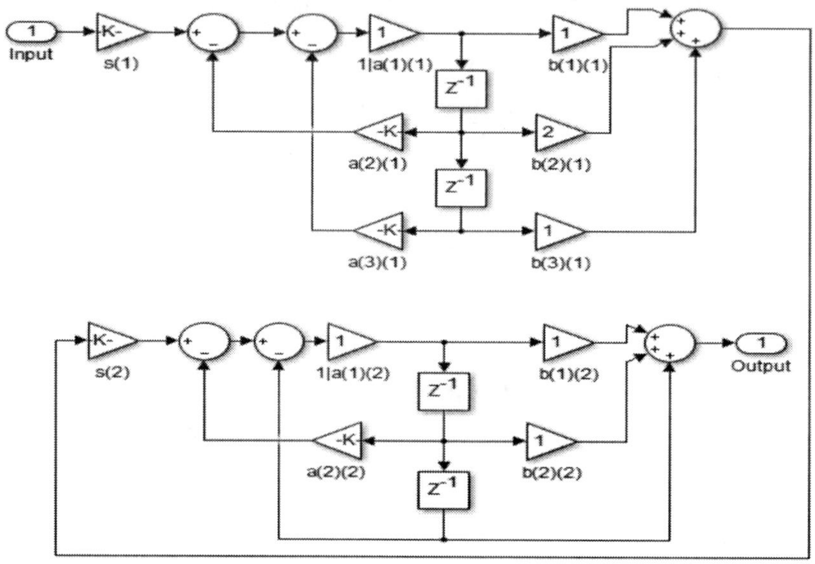

Figure 14. Structure of designed IIR filter.

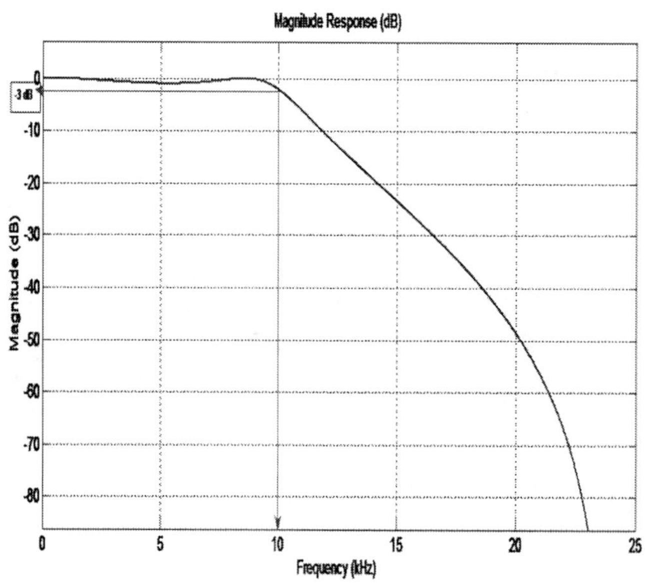

Figure 15. Magnitude response of 3rd order IIR filter.

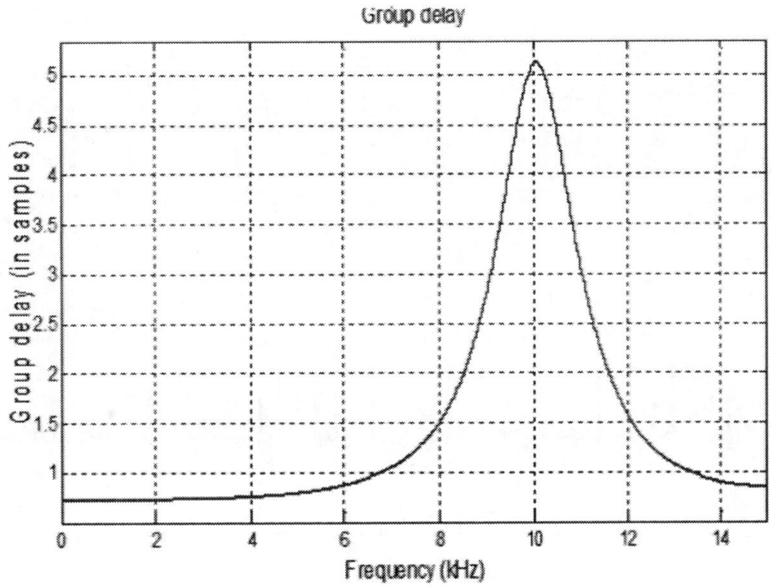

Figure 16. Group delay of IIR filter.

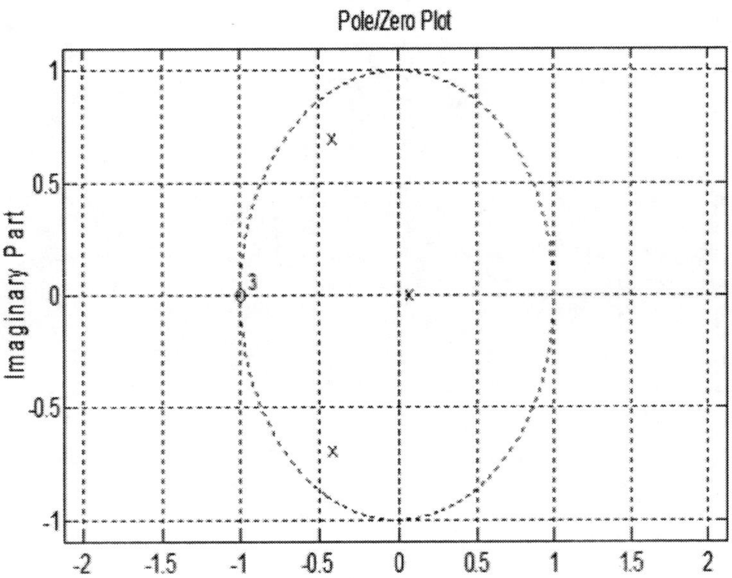

Figure 17. The pole/zero plot.

IOP Conf. Series: Materials Science and Engineering **765** (2020) 012073 doi:10.1088/1757-899X/765/1/012073

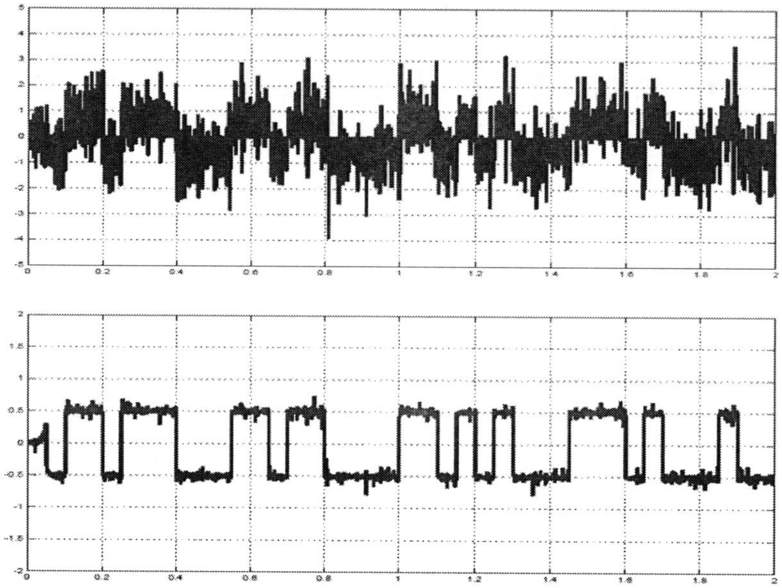

Figure 18. Received noisy signal before and after the filter.

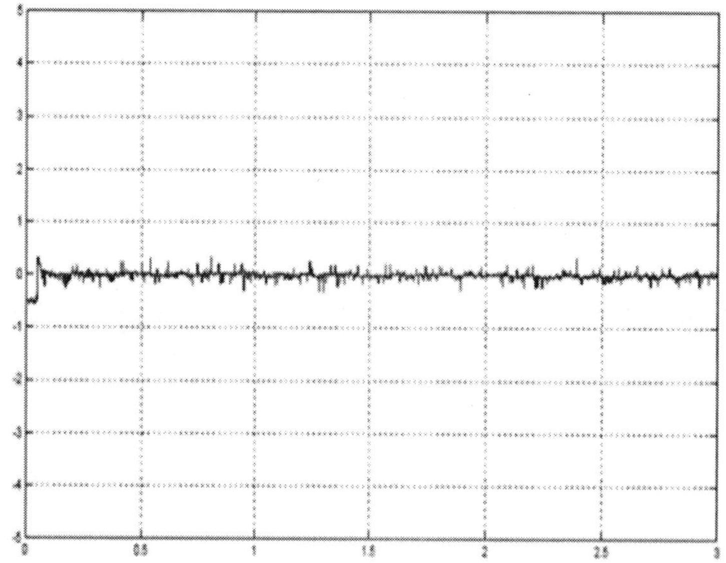

Figure 19. Received signal on the Q-side at lock state.

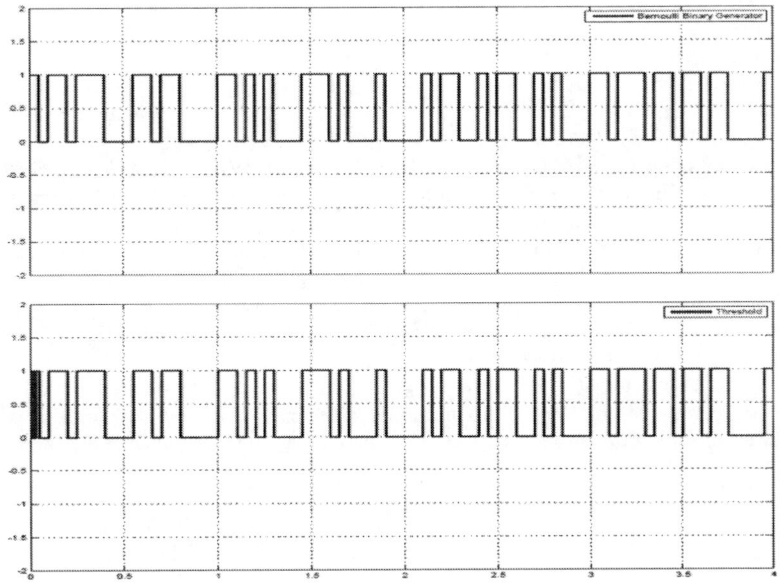

Figure 20. Comparison between the generated and received samples.

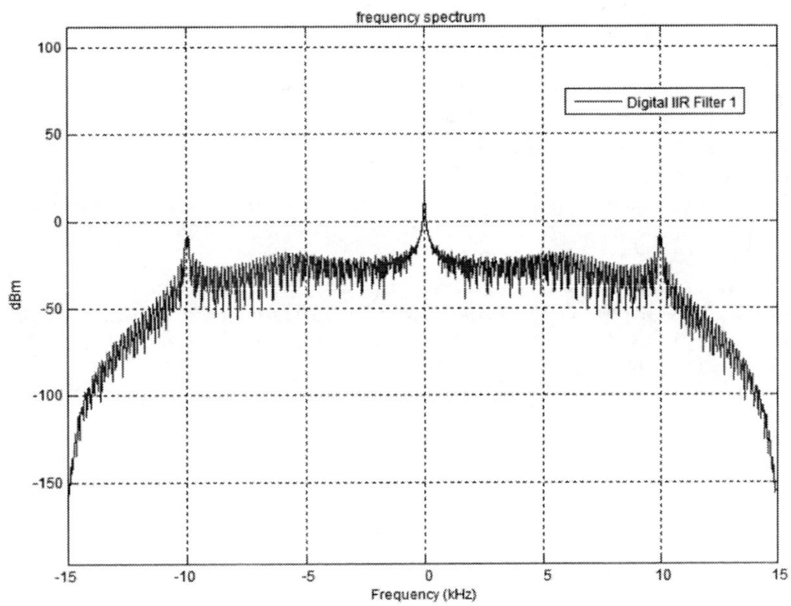

Figure 21. The frequency spectrum of the received signal.

References

[1] Pandey S, Choudhary S, Rahi PK. Design of Low Pass Fir Filter Using Rectangular, Bartlett and Blackman-Harris Window Techniques. IJARCCE [Internet]. Tejass Publisheers; 2017 Mar 30;6(3):43441. Available from: http://dx.doi.org/10.17148/ijarcce.2017.63100.

[2] Parvin KN, Hussain MZ. Multiplication techniques for an efficient FIR filter design for hearing aid applications. 2018 2nd International Conference on Inventive Systems and Control (ICISC) [Internet]. IEEE; 2018 Jan; Available from: http://dx.doi.org/10.1109/icisc.2018.8398944.

[3] IDENTIFICATION OF PHARMACEUTICAL SUBSTANCES WITH RAMAN SPECTROSCOPY. Thesis . March 2016 Department of Laser and Photonics Ruhr Universitt Bochum Germany.

[4] Fadloullah I, Mechaqrane A, Ahaitouf A. Butterworth Low Pass filter design using evolutionary algorithm. 2017 International Conference on Wireless Technologies, Embedded and Intelligent Systems (WITS) [Internet]. IEEE; 2017 Apr; Available from: http://dx.doi.org/10.1109/wits.2017.7934661.

[5] Cheab S, Wong PW, Soeung S. Design of Multi-Band Filters Using Parallel Connected Topology. Radioengineering [Internet]. Brno University of Technology; 2018 Apr 13;27(1):18692. Available from: http://dx.doi.org/10.13164/re.2018.0186.

[6] Smith, Steven W. "The scientist and engineer's guide to digital signal processing.2000. Second Edition.

[7] Alessio SM. IIR Filter Design. Digital Signal Processing and Spectral Analysis for Scientists [Internet]. Springer International Publishing. 2015 Dec 10;263367. Available from:http://dx.doi.org/10.1007/978-3-319-25468-5-8.

[8] Xiaojie Su, Peng Shi, Ligang Wu, Yong-Duan Song. A Novel Approach to Filter Design for TS Fuzzy Discrete-Time Systems With Time-Varying Delay. IEEE Transactions on Fuzzy Systems [Internet]. Institute of Electrical and Electronics Engineers (IEEE); 2012 Dec;20(6):111429. Available from:http://dx.doi.org/10.1109/tfuzz.2012.2196522.

[9] Pandey S, Choudhary S, Rahi PK. Design of Low Pass Fir Filter Using Rectangular, Bartlett and Blackman-Harris Window Techniques. IJARCCE [Internet]. Tejass Publisheers; 2017 Mar 30;6(3):43441. Available from: http://dx.doi.org/10.17148/ijarcce.2017.63100.

[10] Ahmed. A.Thabit, Ziboon H.Design and Simulation of Digital PLL Synchronizer for BPSK and QPSK Based on Software Defined Radio.2009. Engineering and Technology Journal.27(10).

[11] Kumpera A, Malik R, Lorences-Riesgo A, Andrekson PA. Parametric coherent receiver. Optics Express [Internet]. The Optical Society; 2015 May 8;23(10):12952. Available from: http://dx.doi.org/10.1364/oe.23.012952.

[12] Nishant Kumar.Optimal Design of FIR and IIR Filters using some Evolutionary Algorithms.2013. Thesis.

[13] SHI, Kai; SILLEKENS, Eric; THOMSEN, Benn C. 246 GHz digitally stitched coherent receiver.2017.In: Optical Fiber Communication Conference. Optical Society of America.

[14] Basiri M MA, Sk NM. Configurable Folded IIR Filter Design. IEEE Transactions on Circuits and Systems II: Express Briefs [Internet]. Institute of Electrical and Electronics Engineers (IEEE); 2015 Dec;62(12):11448. Available from: http://dx.doi.org/10.1109/tcsii.2015.2468917.

[15] SKLAR, Bernard; HARRIS, Fredric J. Digital communications: fundamentals and applications.1988.Englewood Cliffs, NJ: Prentice-hall.

[16] Cameron R. Advanced Filter Synthesis. IEEE Microwave Magazine [Internet]. Institute of Electrical and Electronics Engineers (IEEE); 2011 Oct;12(6):4261. Available from: http://dx.doi.org/10.1109/mmm.2011.942007.

[17] Tanimura T, Takeshi Hoshida, Shoichiro Oda, Hisao Nakashima, Masahiro Yuki, Zhenning Tao, et al. Digital clock recovery algorithm for optical coherent receivers operating independent of laser frequency offset. 2008 34th European Conference on Optical Communication [Internet]. IEEE; 2008; Available from: http://dx.doi.org/10.1109/ecoc.2008.4729126.

[18] Hisao Nakashima, Takahito Tanimura, Takeshi Hoshida, Shoichiro Oda, Rasmussen JC, Lei Li, et al. Novel wide-range frequency offset compensator demonstrated with real-time digital coherent receiver. 2008 34th European Conference on Optical Communication [Internet]. IEEE; 2008; Available from: http://dx.doi.org/10.1109/ecoc.2008.4729128.

AUTHOR INDEX

Institute of Physics
Dirac House, Temple Back
Bristol BS1 6BE UK

ISSN: 1757-8981
ISBN 978-1-7138-3502-8